AËDES AEGYPTI (L.)

THE YELLOW FEVER MOSQUITO

AËDES AEGYPTI (L.)
THE YELLOW FEVER MOSQUITO

ITS LIFE HISTORY, BIONOMICS AND STRUCTURE

BY

SIR S. RICKARD CHRISTOPHERS

C.I.E., O.B.E., F.R.S., I.M.S. (retd.)

Member of the Malaria Commission of the Royal Society and Colonial Office, 1899–1902. Late Officer in Charge Central Malaria Bureau, India

CAMBRIDGE
AT THE UNIVERSITY PRESS
1960

CAMBRIDGE UNIVERSITY PRESS
Cambridge, New York, Melbourne, Madrid, Cape Town, Singapore, São Paulo, Delhi

Cambridge University Press
The Edinburgh Building, Cambridge CB2 8RU, UK

Published in the United States of America by Cambridge University Press, New York

www.cambridge.org
Information on this title: www.cambridge.org/9780521113021

First published 1960
This digitally printed version 2009

A catalogue record for this publication is available from the British Library

ISBN 978-0-521-04638-1 hardback
ISBN 978-0-521-11302-1 paperback

CONTENTS

CONTENTS

CONTENTS

PREFACE

A few words of explanation seem desirable regarding the ground covered by this book and to make acknowledgement of help received.

As the title indicates, the book gives an account of the mosquito *Aëdes aegypti* (L.). Such an account of an important and much-studied species of mosquito it is hoped may be of interest and useful as a work of reference to those engaged in work on the mosquito in the laboratory or in the many connections in which *A. aegypti* now has a place.

As a type for much that is known about mosquitoes *A. aegypti* has many qualifications. Its history is closely linked with that of early research on mosquitoes and their systematic study. Its distribution and possible source of origin are still subjects of great interest. More than any other single species of mosquito, *A. aegypti* has been cultivated and used in the laboratory for research on mosquito structure and bionomics and on mosquito behaviour and reactions. The species is extensively used as a test insect in the evaluation of insecticides and repellents and in general as a useful type for a variety of forms of research. In public health work it is notorious as the vector of urban yellow fever and is also the common vector of dengue. In both these latter capacities it has been and still is the object of control measures on an immense scale and has largely formed the basis for quarantine regulations in ports and in aeroplane traffic in many parts of the world.

In the present volume the author has attempted to give a full and up-to-date account of what is known of this species. In chapter I is given an account of the history of research on mosquitoes with a section on that of early work on *A. aegypti*. In chapter II is discussed the systematics of the species including its known distribution and the influences responsible for this. Chapter III gives a general account of the life history of *A. aegypti* and of the natural enemies and parasites of mosquitoes. In chapter IV are described the relation of the species to disease transmission and the methods used in its control. A chapter is then devoted to technique, describing how the species can be reared and used in the laboratory with an account of useful apparatus to this end. There follows a chapter on the egg and its structure with changes following oviposition and one on the complicated issues connected with hatching, with a further chapter giving a full and detailed description of the embryology of the mosquito. In succeeding chapters are described the different larval instars, the pupa and the imago with their physical characters, mechanisms, reactions and structure, including much new information such as data regarding weights and measurements of the different stages, critical studies of head sclerites and other systems. Here, for example, is given the first published description ever given of the mosquito brain.

There are chapters on growth, on flight, on viability, on the action of insecticides and other subjects pertaining to mosquitoes and in the last chapter are brief accounts of recent work on such subjects as hormones, reserves and other physiological processes in mosquitoes.

ACKNOWLEDGEMENTS

Much of the author's work here recorded was carried out in the Zoological Laboratories, Cambridge University, and I wish to acknowledge the library and other facilities so kindly made available to me by Professor Sir James Gray, C.B.E., M.C., F.R.S. To Professor V. B. Wigglesworth, C.B.E., F.R.S., in whose Department I worked for some time, I am especially grateful for his kindness, warm interest and help so freely given and the unfailing helpfulness and courtesy of the laboratory staff.

I am also greatly indebted to Professor D. Keilin, F.R.S., for use of the library and other facilities at the Molteno Institute, Cambridge, and for his kindness and help in many ways.

To Dr P. Tate of the Molteno Institute I am indebted for going over the manuscript and for many suggestions and improvements. I have also to thank him for the interest he has taken in my work and kind help in many ways.

To Mr P. F. Mattingly, whose help has been largely responsible for some of the sections, I am deeply indebted for the great trouble he has taken, not only in going over portions of the manuscript and making suggestions, but in advising me regarding recent advances in synonymy, systematics and genetics of *A. aegypti*, thus enabling sections in which these subjects are dealt with to be up-to-date. In this connection too I wish to thank Dr Harold Trapido of the Rockefeller Foundation for his interest, kind help and suggestions.

I also wish to acknowledge help received from Dr A. N. Clements, of the Department of Zoology, Birkbeck College, London University, and from Professor Colvard Jones, of the University of Maryland, in drawing my attention to important publications I might otherwise not have seen. Also my thanks are due to those who have been kind enough to send me reprints or have kept me informed of their researches and whose names, which space prevents me giving here, will be found with their work in due context in the text.

To Dr Ann Bishop of the Medical Research Council and Molteno Institute, Professor R. M. Gordon of the Liverpool School of Tropical Medicine, and Professor P. A. Buxton, F.R.S. and Dr H. S. Leeson of the London School of Tropical Medicine I am indebted for eggs of strains of *A. aegypti* on which the present studies have been based.

To the Press I wish to express my thanks for the careful checking of the text and especially for the very expert and informed checking of the lists of references which this work has demanded.

Lastly I wish to acknowledge the very helpful grant from the Royal Society towards the expenses of my work on *A. aegypti* and the very generous and greatly appreciated grant from the Rockefeller Foundation which has made it possible to publish this book.

S.R.C.

I

HISTORICAL

(*a*) THE NAME 'MOSQUITO'

Mosquito is now the name in common use in English for biting flies of the family Culicidae, suborder Nematocera, order Diptera. Formerly the name for the mosquito in this country was 'gnat' (O.E. *gnaet*). The change took place about 1900, when as a result of Ross's discovery of the mosquito cycle in malaria the importance of these insects to man became realised for the first time and knowledge concerning mosquitoes more generally diffused. Thus Kirby and Spence (1870) in their People's Edition are not at all sure that the foreign 'mosquito' is the same genus as our English *Culex pipiens*. Hurst (1890) uses for his paper the title *The Life Development and History of a Gnat* (*Culex*) and Giles (1900) entitles his book *Handbook of Gnats or Mosquitoes*. But Theobald (1901) and practically all English writers since use simply the name mosquito. It is well to remember therefore when looking up old literature that the reference in the index may be to gnats or Culex. If the name mosquito is used, it probably meant something believed to be fiercer than our English gnats, namely the mosquito of travellers' tales.

The French name is now 'moustique', according to the *Oxford English Dictionary* the metathetic equivalent of the Spanish 'mosquito'. But prior to about 1900, unless described as Culex or Culicidae, mosquitoes were almost invariably referred to by French writers as 'Les cousins' or sometimes 'Les moucherons'. Another French name still given in some dictionaries is 'maringouins'. In German the name 'Stechmücken' is now in general use, the equivalent meaning to our mosquito. But formerly 'Mücke' or 'Schnacke' or sometimes 'Gelse' were not infrequently used. The Italian name 'zanzare' is still used in the literature, though Italian dictionaries mostly give 'muskito'. Other names seen in the older literature and still to be found in dictionaries are: myg (Dano-Norwegian); mygga (Swedish); cinife, cenzalo, zancuda (Spanish); comaro, konops (Russian); konopus (Greek). The word 'empis' in Aristotle has usually been taken by translators as referring to the mosquito (translated as Mücke or gnat according to the language of the translation).

It is interesting to trace the origin and history of this word mosquito and why it of all others should have come to be used. Clearly the word mosquito is of Spanish or Portuguese origin and it is probably correct to say that it must have come originally from Spanish or Portuguese America. It is perhaps more probable that its modern use has come from North America. We first find the name, often with variations in spelling, used in accounts of travellers. Such use dates back to the sixteenth century, for Phillips (1583) in Hakluyt's *Voyages* (1589), p. 568, is quoted in the *Oxford English Dictionary* as saying 'We were almost oftentimes annoyed with a kind of flie, the Spaniards call musketas', and a number of other such early instances of the use of the word by travellers is given under the word mosquito in this dictionary. It is necessary, however, to be cautious in always ascribing such references to mosquitoes as we now use the word. This is made very clear in Humboldt's travels in Equinoxial America (Orinoco). Humboldt (1819) has much to say about mosquitoes, but he is referring to Simulidae, as he definitely says, and the word used by the natives for mosquitoes he states was 'zancudos' and for a smaller kind 'tempraneros'. The word mosquito he gives as meaning 'little fly' and 'zancudos' as 'long-legged'. The same confusion seems to extend to the use of the French 'moustique', for Humboldt says the French call *Simulium* moustiques and the zancudos 'maringouins'. A curious example of this inversion of names will be found in a letter by Combes (1896). Seemingly this letter might be taken as describing mosquitoes attacking another species of fly as this emerged on the water surface from the pupa. But Combes takes care to say that the 'moustique' is not 'le cousin' of France and the larva is not aquatic, and that the name for 'le cousin' is 'maringouin'. So that it was not the mosquito which did the attacking but the *Simulium*. Indeed, one would scarcely expect the word mosquito to have any exact significance as a native word, for anyone familiar with the usual native's reaction to requests for names of insects could well understand that this might be 'a little fly'.

The word mosquito was evidently used from very early times in North America and often in a context that does not obviously suggest direct origin from South America. Thus the *Oxford English Dictionary* quotes Whitbourne (1623) writing of Newfoundland as saying 'There is a little nimble fly which is called musketo', and Wood (1634) referring to a pest in New England as 'a musketo which is not unlike our gnats in England'. Only in America in fact does general use of the word mosquito date back as far as the literature makes mention of these insects. Looking at a bibliography of the older literature on mosquitoes it is possible to tell at a glance, with occasional exceptions, whether the writer was English or American by the respective use of gnat or mosquito in the title. Woodward (1876) goes so far as to write 'On the body scales of the English gnat and American mosquito...'. Manson, as did Ross, always spoke of the 'mosquito'. Manson, however, had worked for many years in China. It is not impossible that the use of the word by these two authors and later by those carrying out investigations on yellow fever in tropical America played a decisive role in bringing about general adoption of the name mosquito in modern writings.

We are on firmer ground in noting that the name *Culex mosquito* was given by Robineau-Desvoidy in 1827 to a mosquito from Cuba which he says 'indigenes vocant mosquito, sicut mihi retulit dominus Poey'. This species is now the *Aëdes aegypti* of present nomenclature and *Culex mosquito* Desv. was the name by which it was known to Finlay and under which he considered on various grounds that it was the cause of yellow fever. Thus there is some excuse for regarding *Aëdes aegypti*, the mosquito to which this volume is dedicated, as to some extent responsible for the name mosquito being that which has come into general use.*

What is the correct plural of mosquito? The plural of the Spanish word would be 'mosquitos' and this form has been used by some English writers as being more correct than 'mosquitoes'. As a thoroughly anglicised word, however, it would seem that the latter form must be considered as sanctioned by general acceptance and usage along with 'potatoes' (Spanish patata) and 'tomatoes' (Spanish tamata), these being along with 'mosquitoes' the plurals as given in English dictionaries.

* Curiously enough the name 'Culex mosquito' occurs on p. 382 of Aldrovando (1602). See next section regarding this author.

(b) EARLY WRITINGS ON THE MOSQUITO

Among the earliest references in the literature to the mosquito are certain passages in Aristotle (384–322 B.C.) relating to *empis*, a name which as previously noted is generally accepted as signifying the mosquito. Thus in Aristotle's *Historia Animalium** it is stated that many animals live at first in water and afterwards change their form as is the case with *empis* (Book I, cap. I as given in parallel column with the Greek text in Aubert and Wimmer, p. 197, lines 8–12). In another passage (Peck, p. 47), this time in *De Generatione Animalibus*, *empis* is given as being among certain creatures which, coming from putrifying liquids, are neither produced out of other animals nor do they copulate, i.e. they were thought to arise by spontaneous generation of life, a belief in regard to the mosquito larva which still held some two thousand years later.†

About 350 years after Aristotle are the writings of the Roman author C. Plinius Secundus (A.D. 23–79). In the thirty-seven books of his great natural history are some dozen or so references to 'culices'. One passage describes the gnat as an example of the wonderfulness of nature in providing on such a small scale the organs required for the five senses and other life requirements, and among other things comments on the fineness and perfection of the small and sharp sucking tube used to suck in and convey the blood. Other references are to swallows feeding on gnats: 'that gnats love sour things, but to sweet things they do not come near...; that gnats keep a foul stir in gardens with some small trees where water runneth through—but can be chased away by burning a little galbanum'. This reference to water is the nearest to any mention by the author that the insects have an aquatic stage.‡

From A.D. 200 to A.D. 1200 has been described as the dark ages for biology, and following on Plinius Secundus it is not until the seventeenth century that naturalists again begin to write about the mosquito. Though from now till Ross's discovery the mosquito was never considered of any importance it still had features which gave it a special interest to the early naturalists. These centred especially about the origin of the fly from the worm-like aquatic stage, its complicated organ of puncture, the proboscis, and a character for which the insect has always been notorious, its hum. It will be seen from what follows that each of these in its turn has been the subject of observation and comment.

The earliest work dealing with the mosquito in what may perhaps be called medieval literature is Aldrovando (1602), a massive work entitled *De Animalibus Insectis* a copy of which is in the University Library, Cambridge. Though the earliest work of any seen in this period, it more than any other of these early works approaches in arrangement a modern text-book of entomology. It deals with insects generally, but is divided into chapters one of which is headed 'De Culicibus'. This again is subdivided into sections such as 'Synonymia', 'Locus', 'Genus', 'Differentia descriptio' and others. It is illustrated with text-figures consisting of coarsely executed but very life-like 'sketches' of insects, in one of

* Those interested in Aristotle should consult Nordenskiold (1929) who gives a very good account of the works of this author. There are ten books of the *Historia Animalium* (three considered spurious); four of *De Partibus Animalium*; five of the *De Generatione Animalibus*; and three on the *Psyche*. Other accounts of Aristotle and other early naturalists are given by Miall (1912) and by Singer (1931). An English translation of the *De Generatione Animalibus* is given by Peck (1943). The standard German translation of the *Historia Animalium* is that of Aubert and Wimmer (1868). For details see references.

† See references later to Bonanni, Joblot, Hooke and others.

‡ An English translation is *The History of the World commonly called the Natural History of C. Plinius Secundus*, Adam Islip, London, 1601. A copy is in the Balfour Library, Cambridge.

which along with easily recognised figures of Panorpids, Ephemerids, etc. is the figure of a mosquito in flight given as *Culex communis*. There seem to be no figures depicting the larva such as are so characteristic of later publications.

Another early work is Moufet, *Insectorum sive minimorum animalium theatrum* (1634), with a chapter of seven pages also headed 'De Culicibus'. He collects references to mosquitoes made by various classical authors including Herodotus and Pausanias. Another author mentioned by Howard, Dyar and Knab (1912)* giving classical references is Cowan (1865), who quotes from Ammianus Marcellinus a statement describing how swarms of mosquitoes in Mesopotamia by attacking the eyes of lions cause these to seek refuge in the rivers and to drown or become mad.

So far such works as have been mentioned are mainly collections of statements about mosquitoes from classical and other authors. Of a different character are the text and plates of Hooke published in 1665. This author, writing at a time when the use of lenses was first revealing new and exciting fields of observation, sets out to describe and figure a great variety of objects as these were seen with the aid of magnifying glasses. Among such he describes and figures the *Culex* larva, seemingly the first author to do so. He describes this as having jaws rather like a crustacean, which can be seen extracting invisible nourishment from the water. The plate of the larva shows this about a foot long, somewhat crude but unmistakeable. The pupa is also depicted and the author observed the sequence of larva, pupa and imago. The adult figured, however, is *Chironomus* which is given as the brush-horned gnat. The copper plates were republished with fresh text anonymously in 1745 under the title *Micrographia restaurata*.†

Other early writers are Wagner (1684); P. Bonanni (1691); San Gallo (1712) and Reviglas (1737). The papers by Wagner, San Gallo and Reviglas are all in the same German journal which in the course of time has several times changed its name, though usually retaining the word *curiosa*. All the papers are short, a few pages only. Wagner, besides giving references to earlier classical authors, describes the stages of metamorphosis, though the eggs described are those of *Chironomus*. San Gallo describes the larva and figures the male and female adult. Reviglas deals with the structure of the proboscis as does Barth (1737). P. Bonanni figures the larva and pupa of *Culex* and discusses the question of spontaneous generation, a belief which died hard. Hooke also refers to spontaneous generation, but thinks it more likely that the mosquito drops its eggs upon the water. Another author named Bonanni (F. Bonanni, 1773) has also written on the mosquito and has described mosquito scales, but almost unrecognisable as such and very different to the clear drawing given by Hogg, referred to later as the first to describe these structures. Another early account is that by Godeheu de Riville (1760), who was the first to describe copulation in the mosquito, an observation that has a special interest in that from the author's description the species was almost certainly *Aëdes aegypti*.

A more modern approach is seen in the work of the great naturalist Swammerdam. Even as early as 1669, in his *Historia insectorum generalis*, he figures the larva, pupa and adult. In his famous work *Biblia naturae* (1737) he deals systematically with the larva and nymph of various insects, showing that the parts of the imago are already present beneath

* These authors give a very full account of the early history of mosquitoes in their first volume which is all the more valuable as they quote *in extenso* translations from many of the authors cited.

† A copy of Hooke's original work is in the library of the Royal Society and one of the *Micrographia restaurata* with the copper plates in the Balfour Library, Cambridge.

the skin and can be shown by dissection.* His description of the mosquito and its life history is remarkably like what might have been written as a short account at the present time. A footnote to the description of the larva is of special interest in relation to later work on insect physiology and is almost prophetic. In the English translation of the work this footnote reads as follows:

> There is not in all the insect world a creature more happily suited to show the several operations of life than this. A moderate microscope discovers to us very clearly what passes within the transparent body.—At this time [i.e., as he explains, when the larva is especially transparent at ecdysis] the beats of the heart and the motion of the stomach and intestines are perfectly seen and the two principal pulmonary tubes may be traced along their whole length.

The most complete account among the early writers is, however, that by Reaumur (1738). Reaumur's *Mémoire XIII. Histoire des cousins* appeared to leave little to be further observed, and for many years remained unchallenged by any other work of a like nature dealing with the mosquito.

(*c*) THE MOSQUITO AS A NATURAL HISTORY OBJECT

During the period comprised in the latter half of the eighteenth and first half of the nineteenth century, roughly 1750–1850, the mosquito was mostly written about as a natural history object, being dealt with sometimes in a few lines, sometimes more extensively, in the many natural histories that characterised this period in zoology. Its systematic position, its life history and stories as to its attacks upon travellers were given with monotonous sameness when its place in the Diptera called for its mention. Among accounts more especially deserving of notice is the section 'Des Cousins' in the work of De Geer (1776) figuring the larva of a species of *Aëdes*. Goldsmith (1779) gives a very original account (seven pages) including reference to mosquitoes retiring into caves and with the curious statement that they are capable of parthenogenetic development. In a supplement to Rosel's *Insecten Belustigungen* are additions to Reaumur's work with plates by Kleeman (1792). In the first two editions of Cuvier's great natural history, *Le règne animal* (1817, 1829) are some few pages by Latreille. In the third edition of 1849 (the so-called Apostle's Edition) is a section containing some observations on the structure of the proboscis by E. Blanchard. A quite interesting account of the 'gnat' is given in an addendum by Griffith and Pidgeon in the English edition of the same work, 1832.

Many other short accounts of gnats, culices, cousins or other names occur during this period in the literature,† many of them little more than a page or two in encyclopaedias or natural histories. The larva is now frequently referred to, often as a grub or worm-like creature, for example, 'Der Schnackenwurm' (Ledermuller, 1761), 'Wurm von der Singschnacke' (Slabber, 1781), 'Ver du cousin' (Joblot, 1754). Joblot figures the *Culex* larva in several stages of growth, saying that he only recognised that they were the aquatic stage of 'le cousin' when he had hatched out 'moucherons'. Joblot in the same paper was

* Swammerdam uses the term nymph for the pre-imaginal stage of insects as do many other early authors. The term pupa now used for this stage in holometabolous insects was given by Linnaeus to the chrysalis of Lepidoptera from its resemblance to a baby that is swathed or bound up as is customary with many peoples (pupa = Latin for a girl or doll). (Comstock, 1920, p. 186.)

† See: Geoffrey (1764), Olivier (1791), Jordens (1801), Latreille (1805–25), Guérin (1835), Packard (1869), Pagenstecher (1874).

the first to describe the larva of *Anopheles* with a large and excellent figure such as might have been a recent representation of this now much studied form. According to Burmeister (1832) the larva of *Anopheles* was first described by Goeze (1775) and later by Lichtenstein (1800), but was only identified as the larva of *Anopheles* by Fischer (1812). There can be no doubt, however, but that Joblot, as given above, figured the *Anopheles* larva as early as 1754. Apart from his figure, he states that the head of the new creature was 'très mobile, tournant à droite & à gauche comme sur un pivot'. It was over 100 years later that Jourdain (1893) described as the larva of a culicine of which he had not determined the genus or species a creature with such a singular movement of rotation of the head. Furthermore, according to Howard (1900), what Fischer described as the larva and pupa of *Culex claviger* Fabr. were those of a species of *Corethra*. Dobell in his life of Leeuwenhoek states that Leeuwenhoek in a letter dated 1700, besides giving a description of the gnat larva, distinguished the attitude of the larva of *Anopheles* from that of *Culex*, but no details are given by the biographer. Brauer (1883) is stated by Nuttall and Shipley (1901), p. 48, to have mistaken the larva of *Anopheles* for that of *Dixa*. Brauer gives an excellent line drawing of an *Anopheles* larva showing the palmate hairs. He gives it, however, as the larva of *Dixa* sp. (*Culex nemorosus* Heeg). The genus *Anopheles* was erected by Meigen, 1818, and the larvae of the two forms clearly described by Mienert, 1886.

Characteristic of the period 1850–1900 are accounts of mosquitoes as encountered in northern latitudes and in the tropics by travellers. Many of the references to polar regions really say very little except that mosquitoes were found extremely tormenting. Regarding the tropics one of the most important travellers' descriptions are passages in Humboldt when describing experiences in the Orinoco. The passages are given *in extenso* by Howard, Dyar and Knab (1912). Another feature of the literature are references to 'swarms' of mosquitoes. Those interested will find many accounts of such in the various editions of Kirby and Spence, especially the People's Edition (1870).

Whilst the above gives an outline of the general trend in the literature on mosquitoes prior to what may be called the modern period, such an account would be incomplete if it failed to take note of letters and brief communications to journals more especially from 1850 to 1900, such as *Nature*, *Science Gossip*, *Entomologist's Monthly Magazine*, and especially *Insect Life*. At first sight these brief and often ephemeral communications may seem of little value. Historically, however, they cannot be neglected, for though a considerable proportion relate to seemingly rather trivial matters, some give the first indication of observations or ideas which have later become important. Among such may be mentioned references by Howard to the first use of oil against larvae, the first use of larvivorous fish, the first reference to natural enemies of mosquitoes and early observations on remedies and preventives against mosquito bites.

These communications further show very strikingly how complete has been the change in character of the literature on mosquitoes from the year 1900 or so onwards. This is very notably so in relation to the subject of our present study, the now notorious vector of yellow fever. For, apart from various names given by systematists, often without any significance attaching to the mosquito they were naming, almost the only author attaching special importance to this species was Finlay, who as early as 1881 had associated it and another mosquito with the transmission of yellow fever and had made observations on its habits. The larva does not seem ever to have been described or figured until it was depicted by James in 1899 as the larva of the 'tiger' mosquito. From 1900 onwards, however, whilst

Anopheles became widely known as 'the malarial mosquito' and has become the subject of an enormous literature, *Stegomyia fasciata* from about the same time became 'the yellow fever mosquito' and under this name and that of *Aëdes aegypti* has since acquired a literature almost as great.

(*d*) EARLY OBSERVATIONS ON STRUCTURE

Reference has already been made to the fact that one of the features relating to the mosquito that had a special interest for the early naturalists was the structure of the proboscis, and in the early literature are many references to the number and nature of its components. Most of these observations are now purely of historical interest, being made under conditions very different to those of the present day with the modern miscroscope.* Swammerdam (1737–8) describes without naming five stylets plus the sheath. All the stylets are shown ending similarly with fusiform swollen ends devoid of any further detail. Authors up to a century or more later gave little more detail than this and in the naming of the parts were not always correct. Thus Blanchard (1849) in his figure of the mouth-parts of *Culex pipiens* shows labrum (upper lip), labium (lower lip), maxillae (labelled as mandibles) and a central bifid structure labelled mâchoires (maxillae). Gerstfeldt (1853) describes a sheath formed of the lower lip only, containing six setae, but like many other writers also gives the maxillae as mandibles. Many other authors have referred to or given some description of the mouth-parts, including Becher (1882) who summarises work to that date in this respect. For all practical purposes, however, the first describer of the mouth-parts in any adequate modern sense was Dimmock (1881) in his classical treatise on the mouth-parts of some Diptera, and later in his paper in *Psyche* on the mouth-parts and suctorial apparatus of *Culex*. After surveying work already done in some detail this author goes on to describe fully and accurately all the parts and gives a plate showing sections through the proboscis at different levels. He describes also the mouth-parts in the male (1884). The parts were briefly described in the same year by Meinert and more fully later by Macloskie (1887, 1888). There are also contributions by Muir (1883); Murphy (1883); Smith (1890, 1896); and Kellog (1899); the last mentioned two authors tracing the parts through their development from the larva. The complete working out of the problems connected with the minute structure and functioning of the parts was not, however, achieved until much later as described in a subsequent chapter.

Beyond drawings of the egg-raft of *Culex pipiens* by Reaumur and others little is to be found in the literature as to any details of egg structure until Howard (1900) and Theobald (1901) figure the eggs of one or two species of Culicines, and Grassi, Nuttall and Shipley and others give descriptions of the eggs of *Anopheles*. That the eggs of *Stegomyia fasciata* were laid singly on the water was noted by Finlay in 1886, by Daniels and by Ross (both in Theobald, 1901), the figure of the egg by Daniels being one of the first, if not the first, to be published. It shows the chorionic bodies on the outline of the egg, but incorrectly as air chambers as many at first thought them to be.

In spite of many representations of the larva in the early literature accurate and detailed figures are not given until relatively late. The siphon appears to have been first described in detail by Haller (1878). In 1886 came the classic by Meinert on the eucephalous larvae of Diptera giving descriptions of mosquito larval structure. A paper in quite modern style

* See: Barth (1737), Sulzer (1761), Roffredi (1766–9), Savigny (1816), Durkee (1855).

is that by Raschke (1887) in which he describes the mouth-parts and other structural features of the larva of *Culex* (now *Aëdes*) *nemoralis*. The plate accompanying the paper is noticeable for the delineation given of the sclerites at the base of the flabella or feeding brushes. At the end of his paper he gives a list of some sixteen references to earlier writers on larval structure. Other authors giving early descriptions of larval characters or structure are: Lacordaire (1838), one of the very early writers on structure of the mosquito and one who has even described the micropilar apparatus on the egg of *Culex*; Pouchet (1847); Kraepelin (1882); Brauer (1883); Wielowiejski (1886); Miall (1895); Howard (1900). Pouchet describes the eight vesicular stomachs (caeca) of the larva. Wielowiejski describes the pericardial cells, oenocytes and fat body. Howard's figures of the larval parts were widely copied and like his figures of the adult will be found repeated in many authors' works. Nuttall and Shipley's study of the *Anopheles* larva, as also Grassi's work, more correctly fall in the modern period and will be referred to later.

An outstanding contribution which will be extensively referred to later is that of Hurst (1890) on the structure of the larva and pupa of *Culex* and the anatomical changes connected with pupal ecdysis and emergence.

Of works relating to the structure of the adult may be mentioned Dufour (1851) on the structure of the digestive tract and generative organs of *Culex* (now *Theobaldia*) *annulata*; Schindler (1878) who first described the mid-gut and Malpighian tubules; Lecaillon (1899) on the filamentous processes in the latter structures and, 1900, on structure of the ovary; and later others whose work will be noted when dealing with the adult structure. Of papers of outstanding character are those by Johnston (1855) and Child (1894) describing the organ at the base of the antenna now known as Johnston's organ, and that by A. M. Mayer (1874) showing that the hairs of the male antenna responded to particular notes and that the antennae acted as sensory organs enabling the male to locate the female. An author, P. Mayer (1879), has also written on certain antennal sense organs in Diptera in Italian, but I have so far been unable to see this work. Mention should also be made of Kowalevsky (1889) who described the pericardial cells, and Hogg (1854, 1871) who described the scales of mosquitoes, followed by Anthony (1871), Newman (1872) and Woodward (1876). In 1899 Ficalbi had given an account of the external structure describing many of the characters used later by systematists, such as those of the palpi, tarsal claws, wing venation and even some indication of genitalic characters. With the discovery of the part played by the mosquito as intermediary host in the life history of the malaria parasite the internal structure of the mosquito assumed an enhanced importance and, at the close of the period we have taken as covering early research, was dealt with for the first time systematically in accounts by Grandpré and Charmoy (1900); Grassi (1900); Christophers (1901); and very fully by Nuttall and Shipley (1901).

(*e*) EARLY SYSTEMATIC WORK ON MOSQUITOES

The main objectives of systematists may be given as: (1) the correct naming of species in accordance with the rules of zoological nomenclature; (2) their identification; and (3) their natural classification. The naming of mosquitoes on the binomial system, as does that for all forms of animal life, dates from 1758, the year of publication of the 10th edition of Linnaeus's *Systema Naturae*. It may be worth while mentioning in this respect that a photostat facsimile of the relevant first volume of this work has been published by the British Museum (Brit. Mus. Publ., 1939), a form in which this famous work may be studied for all practical purposes in the original. Some mosquitoes it is true had been given names before this crucial date, and some of these look very like names on the binomial system, though these must be considered as descriptive names only and invalid. This point has some interest in connection with the synonymy of our species, because in the first edition of Hasselquist *Aëdes aegypti* had already been described as 'Culex (aegypti) articulationibus candidis etc.'. But it is only in the second edition (*Reise nach Palestina*, 1762) edited in respect to nomenclature by Linnaeus that the name given as *Culex aegypti* is valid in form and date.* In 1805 Fabricius in his revision of the Diptera, classified on the basis of their mouth-parts, gives a list of described species of mosquitoes. They number fifteen, including one or two that were possibly not mosquitoes. In 1818 Meigen described thirteen more species and erected two further genera in addition to the original genus *Culex* of Linnaeus, namely *Anopheles* and *Aëdes*, the latter for the European species *Aëdes cinereus* which remained almost the only species in the genus for many years until, following Dyar and Knab (1906), Dyar (1922) and Edwards (1932), the genus *Aëdes* was expanded to cover as subgenera a large number of previously erected genera, so that it now includes probably a quarter or more of the known species of mosquito. By 1889 about fifty-six species of mosquito had been described from Europe, North Africa and Egypt.

In the same period a number of species had been added to the list by Wiedemann (1821) from the East; Robineau-Desvoidy (1827) from South and Central America; Walker (1848–65) from material received at the British Museum from various countries; Skuse (1889) from Australia; Arribalzaga (1891) from South America; which with a certain number described by Loew, Van der Wulp, Macquart, Doleschall, Coquillett and others brought the total number of valid species as given by Theobald prior to publication of his first two volumes (1901) to about 164, that is, something like a tenth of the number now known. To these may be added about eighty-two names placed by Theobald as synonyms, thus giving a total of about 250 namings prior to 1901, though this number might have to be added to somewhat on a close study of the literature.

This summary, however, gives no hint of the confusion and lack of co-ordination regarding the tropical species. The reasons for this are not far to seek. Almost the only systematic work devoted wholly to mosquitoes prior to Theobald was a small volume of reprinted papers by Ficalbi (1899) giving a revision of the European forms. The descriptions given by dipterologists were commonly quite inadequate. Those given by Walker for the considerable number of species he described averaged four to seven lines. Synoptic tables such as are now in such extensive use scarcely existed, and had such existed they would have been useless, for they could have included but a small fraction of the species actually existing. It is not surprising therefore that *Aëdes aegypti* as a world species escaped recognition. It had been described often enough, but the trouble was that it had so many aliases, almost one for every country and systematist, that as a species it is no exaggeration to say that up to 1900 it was still virtually unknown.

On the discovery of the mosquito cycle of malaria by Ross and the work of Grassi and other Italian workers, followed within a year or so by the proof by Reed and his co-workers that *Aëdes aegypti* was the agent in transmission of yellow fever, interest in mosquitoes became general. Collections began to pour in from different tropical countries. Medical

* See, however, remarks later under synonymy of this species.

men and others interested in the new developments not only collected but intensively studied the many species in their natural surroundings. For the first time careful detailed descriptions were given and structural characters studied and made use of in identification and classification.

The first to attempt to correlate the mass of published descriptions in a treatise on world species was Giles, who in 1900 published his *Handbook of the Gnats or Mosquitoes* and himself added some seventeen new species mostly from India. There can be no doubt, however, that it was Theobald who, in his gigantic task of grappling with the Culicidae of the world, in the five volumes of his monograph published over the years 1901–10 opened up the study of mosquitoes to workers all over the world. We have already noted that in his first two volumes this author records 164 previously described valid species with eighty-two synonyms. To these were added in his first two volumes 132 new species under his own name, which with twenty-eight described by Giles and some other contemporary authors brought the total of species described in the two years 1900 and 1901 to about as many as had been described in the previous century and a half. The total of known species at this time was about 320 or with synonyms 400 namings.

Though we are here not so much concerned with systematics as such it is perhaps of interest to note that the final number of species as given by Smart (1940) for Theobald's revision (1910) was 1050, and that in Edwards's *Genera Insectorum* (1932) 1400. Probably the number at the present time is about 2000. This great increase in the number of known species is especially noticeable in the case of certain countries. The number of species described from Africa and its islands prior to 1900 was five. In 1941 the number listed in Edwards's monograph of Ethiopian mosquitoes was 405 (*Anopheles*, 86; *Aëdes*, 132; *Culex*, 99; other genera, 88). The number known from India before 1900 (that is before Giles's 1st edition) seems to have been about three. In Barraud's revision (1934) it was 245 (sixty-eight being new species under that author's name).

Besides this increase in the number of described species much more attention has been given to classification. To some seven genera described prior to 1900, mostly relating to outstandingly distinct forms (*Megarhinus*, *Sabethes* and *Psorophora* by Robineau-Desvoidy; *Janthinsoma*, *Taeniorhynchus* and *Uranotaenia* by Arribalzaga; *Haemagogus* by Williston), Theobald in his first two volumes added some ten further genera (subject, however, to considerable modification later), among which occurs for the first time the now familiar name *Stegomyia*, a name which though later subordinated to the rank of a subgenus is still valid.

It was as the type species of this genus and as *Stegomyia fasciata* Fabr. that *Aëdes aegypti* was first introduced to the world by Theobald in its proper perspective, a name by which it was familiarly known until those changes in nomenclature took place which are discussed in the section on synonymy and which after many vicissitudes finally by general agreement ended in the present designation of *Aëdes aegypti*.

(*f*) RESEARCH ON *AËDES AEGYPTI* AND ITS ROLE IN DISEASE

As already noted there is a large literature relating to *Aëdes aegypti*, its systematics, its distribution, its breeding places, its bionomics, its relation to disease, its control, its use in the laboratory as a test animal and much else. All these subjects will be dealt with in detail later in this volume. We may, however, here follow briefly the general course of research on the species, thus completing our historical survey.

There can be no doubt but that it was the discovery that they were vectors of disease that stimulated and was responsible for the immense amount of research on mosquitoes that dates from the opening of what has been referred to as the modern period of research on these insects. And whilst this applies pre-eminently to the subfamily Anophelini owing to their role as vectors of malaria, it no less applies to the Culicini and especially to the genera *Culex* and *Aëdes*, both of which include important vectors of disease in man and animals. Moreover, this applies especially to the single species *A. aegypti*, for it is almost world-wide within the tropical and subtropical zones, and of the four important human diseases transmitted by mosquitoes, namely malaria, yellow fever, dengue and filariasis, *A. aegypti* is the usual vector species for two, and has been the subject of considerable research before it could be known that it did not play an important role as vector of a third, filariasis. Further, its reputation as the yellow fever mosquito has given it an importance to the sanitarian, especially in the New World, possessed by scarcely any other single species of mosquito.

Knowledge that insects and other arthropods were concerned in the spread of diseases came relatively late. Translated extracts are given by Agramonte (1908) from an article by Beauperthuy (1854) in which this author points to the mosquito 'hypothetically considered' as the agent responsible for yellow fever. Beauperthuy in this and some other writings held that yellow fever (and the intermittents, etc.) were the result of the direct injection of a poison by mosquitoes much on the analogy of snake-bite, though he did not regard this poison as due to the mosquito itself but as a virus derived from swamps. However mistaken he may have been he was correct in the deduction that mosquitoes were responsible for yellow fever, and presumably his theory applied also to malaria. In regard to the former disease it is interesting that he specifically mentions what is evidently *A. aegypti*, namely as the 'zancudo bobo' with legs striped with white which he says may be regarded as more or less the house-haunting kind, one of the earliest references to this species in the tropics. Possibly other hypothetical deductions regarding mosquitoes as conveyers of disease exist even before this. But when King in two rather remarkable papers (1882, 1883) had given reasons for regarding mosquitoes as responsible for malaria, he did not suppose that these were agents in man-to-man infection, as is now understood, but as conveying disease from some outside source. Even as late as the early days following the discovery of the mosquito malaria cycle the true appreciation of the essentially infectious nature of malarial infection was still a new idea.

The first concrete evidence that mosquitoes were concerned in the transmission of disease was the discovery by Manson in 1878* in China, confirmed in the same year by Lewis in India (who first discovered *Filaria* in the blood), that the *Filaria sanguinis hominis* underwent development in a brown mosquito which laid its eggs in masses (*Culex fatigans*). Manson observed the early developmental changes of *Filaria* with casting of the sheath, the change to inert sausage-like bodies and a later and larger active form one-fifteenth of an inch in length. By 1884 Manson, and later (1899) Bancroft, had followed development to all but the last eventual location of the fully developed embryo in the proboscis, a dis-

* Or more correctly 1877, since Paul Russell in his recent book *Man's Mastery of Malaria* gives Manson's discovery of development of *Filaria* in the mosquito as having been first published in the *China Customs Medical Reports* (1877). Manson's paper given in the list of references as published in 1879 in the *Journal of the Linnean Society* was read in March 1878 and reviewed in the *Medical Times and Gazette* and in *Nature* in the same year.

covery later made independently by Low (1900) in *Culex* in England from material sent by Bancroft from Australia, and by James in the same year in India in both *Culex* and *Anopheles*. In 1893 Smith and Kilbourne discovered the hereditary transmission of piroplasmosis of cattle (Texas fever) by the tick. In 1895 the trypanosome of nagana, the tsetse-fly disease, was discovered by Bruce. The discovery of the mosquito cycle of malaria, which revolutionised the outlook on mosquitoes and other biting insects and practically founded modern tropical medicine, was made by Ross in 1898.* The discovery by Reed and his collaborators that yellow fever was transmitted by *Aëdes aegypti* followed in 1901. For a summary of what was known of insect transmission of disease at this time with bibliography see Nuttall's classical paper 'On the role of insects, arachnids and myriapods as carriers in the spread of bacterial and parasitic diseases of man and animals'. See also Stiles who gives a list of workers. A useful bibliography is given by Doane in his book *Insects and Disease* (1910). See also Mackerras (1948). For another recent account see Brumpt, 'Précis de parasitologie'.

Though the discovery that *A. aegypti* conveyed yellow fever by its bite after an interval was solely the work of Reed, Carroll, Agramonte and Lazear, who first conveyed the disease to volunteers by the bite of *A. aegypti* previously fed on cases of the disease, it is impossible to overlook the work of Finlay, who from 1881 onwards was led to investigate the habits and bionomics of the species through his belief that it was this mosquito which conveyed yellow fever by its bite and that many facts in the epidemiology of the disease were to be explained in relation to its life history. He says (1886) that two species of mosquito were common in Havana, *Culex cubensis* La Sagra, laying boat-shaped egg-rafts, and *C. mosquito* Rob.-Desv., laying its eggs singly. He says that he was told that the latter species had lately been described as *C. fasciatus*. He describes the chief structures of the proboscis, the bending back of the sheath (labium) when biting, the usual distance of penetration, the time taken for engorgement, the effect of temperature on the mosquito's activity and other features of its life history, seemingly the first systematic observations ever made on the bionomics of any tropical species of mosquito. Following on the work of the American Commission in Havana was that of the observers in the French Commission, Marchoux, Salimbeni and Simond, and somewhat later of the German observers Otto and Neumann working in Brazil. Much of the main facts as we now know them regarding transmission of the disease was established at this time, further advances being chiefly in the direction of increased knowledge of the properties of the virus and the findings in respect to transmission by other species than *Aëdes aegypti* in jungle yellow fever.

That dengue was conveyed by *A. aegypti* has followed as the result of many observations in different parts of the world, but especially in Australia where Bancroft in 1906 gave strong reasons for regarding *A. aegypti* as the vector and had infected two volunteers, though in an area where infection might have been otherwise contracted. Complete proof was given in the 1916 epidemic in Australia by Clelland, Bradley and McDonald, who infected volunteers in a dengue-free area by the bites of mosquitoes fed on dengue cases elsewhere.

In regard to filariasis the onus of proof has lain rather with showing that *A. aegypti* is not an efficient vector of human filariasis, which is mainly conveyed by *Culex fatigans*.

* Ross published the complete cycle in bird malaria in November 1898, having previously seen pigmented oocysts in human malaria in *Anopheles* in August 1897. A very full and interesting account of this great discovery is given by Russell (1955).

It has, however, been found to be the usual vector of *Dirofilaria immitis* of the dog. For further information regarding diseases transmitted by *Aëdes aegypti* see the chapter on relation to disease.

With the knowledge that mosquitoes were important vectors of disease there came a greatly increased interest in methods by which they might be prevented from breeding or destroyed and in means of protection from their bites. Reference to a writer (Delboeuf) who in 1847 speaks of destroying mosquitoes by pouring oil on water where they breed is made by Howard, Dyar and Knab (1912). The same suggestion is made by Southey (1812). The first recorded experiments, however, in this direction appear to have been those by Howard (1893–4) who observed the effect of such treatment upon a small selected pond. Oil was for many years the chief method used in the treatment of breeding places, but has now been largely replaced by more effective larvicides. It was not long after the discovery of the role of *A. aegypti* in the transmission of yellow fever before control measures were undertaken on a large scale in seaports and towns in the yellow fever zone of America and later elsewhere. Still more recently measures directed especially against the adult mosquito have been greatly developed in connection with aeroplane traffic. These measures have involved organisation and legislation, as also research into the methods of using sprays, lethal gases, aerosols, the effectiveness of different larvicides, use of larvivorous fish and increasingly effective insecticides used against the adult mosquito; also, too, research into methods of personal and communal protection.

Whilst the part played by mosquitoes in disease transmission has been a major reason for such research on methods of control, it has also emphasised the importance of systematic and bionomical research, observations on geographical distribution and much else. Reference to the many observers who have contributed in this respect to knowledge of *A. aegypti*, its life history, systematics, behaviour, structure and physiology, will be found in later chapters of this work.

A. aegypti too has been extensively used as a test animal for research in many fields. It has been used on a large scale in testing insecticides and repellents, in trying out essential food requirements, in work on genetics and in other studies. For such laboratory work *A. aegypti* has many outstanding advantages; its hardihood, its readiness to feed and the ease with which it can be reared, with the great advantage that eggs can be stored for months if necessary without losing vitality, make it almost uniquely useful for such a purpose.

REFERENCES

(*a*) THE NAME MOSQUITO

ALDROVANDO, U. (1602). See under (*b*).

ARISTOTLE. See under (*b*).

COMBES, P. (1896). Les moustiques de l'île d'Anticosta. *Rev. sci., Paris* (4), **6**, 751–3.

GILES, G. M. (1900). See under (*e*).

HUMBOLDT, A. VON (1819). *Relation historique. Voyage aux régions équinoxiales du Nouveau Continent, fait en 1799, etc.* Paris. Vol. II, pp. 333–49. See also below, under MACGILLIVRAY.

HURST, C. M. (1890). See under (*d*).

KIRBY, W. and SPENCE, W. (1815–70). See under (*c*).

MacGillivray, W. (1832). *The Travels and Researches of Alex. von Humboldt.* Oliver and Boyd, Edinburgh, and Simpson and Marshall, London, pp. 247–9.

(*The*) *Oxford English Dictionary.* Oxford Univ. Press.

Robineau-Desvoidy, J. B. (1827). See under (*e*).

Theobald, F. V. (1901–10). See under (*e*).

Woodward, J. J. (1876). See under (*d*).

(*b*) EARLY WRITINGS ON THE MOSQUITO

Aldrovando, U. (1602). *De animalibus insectis libri septem.* Bononiae. Chapter v, pp. 382–402. (Copy in Univ. Library, Cambridge.)

Aristotle. See Aubert, H. and Wimmer, F. (1868); Peck, A. L. (1943).

Aubert, H. and Wimmer, F. (1868). *Aristoteles Thierkunde.* W. Engelmann, Leipzig, p. 197, lines 8–12. (Greek text with German transl.)

Barth, J. M. (1737). *Dissertatio de culice.* Ratisbonae.

Bonanni (or Buonani), F. (1773). *Rerum naturalium historia nempe quadrupedum insectorum etc.* Tempelliano, Rome. Vol. I, XI Culex, pp. 45–51.

Bonanni, P. (1691). *Micrographia curiosa.* Rome. Part II, p. 370.

Comstock, J. H. (1920). *An Introduction to Entomology.* Ed. 2. Comstock Publ. Co. Ithaca, New York.

Cowan, F. (1865). *Curious Facts in the History of Insects, etc.* Philadelphia.

Godeheu de Riville (1760). Mémoire sur l'accouplement des cousins. *Mém. Acad. Sci., Paris,* **3,** 617–22.

Hogg, J. (1854). See under (*d*).

Hooke, R. (1665). *Micrographia, or some Physiological Descriptions of Minute Bodies made by Magnifying Glasses, etc.* London. Folio, pp. 185–95 with plates 27 and 28. The plates republished with anonymous text in 1745 as *Micrographia restaurata.* John Bowles, London.

Howard, L. O., Dyar, H. G. and Knab, F. (1912). *The Mosquitoes of North and Central America and the West Indies.* Carnegie Institute, Washington. Vol. I. A very full historical account with bibliography.

Joblot, L. (1754). See under (*c*).

Miall, L. C. (1912). *The Early Naturalists (1500–1789).* Macmillan and Co. London.

Moufet (or Moffet), T. (1634). *Insectorum sive minimorum animalium theatrum.* London. (Copy in Brit. Mus. Nat. Hist. Library.)

Nordenskiold, E. (1929). *The History of Biology.* Kegan Paul, Trench, Trubner and Co., London.

Peck, A. L. (1943). *Aristotle's Generation of Animals.* Heinemann Ltd. and Harvard Univ. Press, p. 47.

Plinius Secundus, C. (1525). Naturalis historiae. Joannes Camerton Minoritanum. Also English transl. by P. Holland (1601), *The History of the World, commonly called the Natural History of C. Plinius Secundus.* Adam Islip, London. (Copy in the Balfour Library, Cambridge.)

Reaumur, R. A. F. de (1738). *Mémoires pour servir à l'histoire des insectes. Mémoire XIII. Histoire des cousins.* Vol. IV, pp. 573–636.

Reviglas, D. (1737). Observatio de culicum generatione. *Acta Acad. Nat. Curiosa,* **4,** 14–20. (Copy in Royal Society Library, London.)

San Gallo, P. de (1712). Esperienze intorno alle generatione delle zanzare. *Ephem. Acad. Natur. Curiosa,* Cent. 2, pp. 220–32. (Copy in Royal Society Library, London.)

Singer, C. (1931). *A Short History of Biology.* Clarendon Press, Oxford.

Swammerdam, J. (1669). *Historia insectorum generalis.* Utrecht. With later editions 1682 and 1692. Also published at Batavia 1685, with a later edition 1733.

Swammerdam, J. (1737–8). *Biblia naturae; sive historia insectorum, etc.* Leyden. Two volumes with Latin and Dutch in columns; contains 'historia culicis'. Vol. I, pp. 348–62 with plates 31 and 32. An edition published in 1752 at Leipzig contains a section 'Die Mucke', pp. 144–8.

REFERENCES

SWAMMERDAM, J. (1758). *Swammerdam's Book of Nature or the History of Insects*. English transl. by Floyd and Hill. Seffert, London. Contains section on 'Culex or gnat'. Vol. I, pp. 153–9. (Copy in Balfour Library, Cambridge.)

WAGNER, J. J. (1684). De generatione culicum. *Misc. Curiosa sive Ephem. Phys. etc.* Frankfurt and Leipzig. Dec. II, Anno 3, pp. 368–70. (Copy in Royal Society Library, London.)

(*c*) THE MOSQUITO AS A NATURAL HISTORY OBJECT

BLANCHARD, E. (1849). 'Cousin' in Cuvier, *Le règne animal*. Ed. 3, part VI. Insectes. Vol. II, pp. 306–10 and plate 161.

BRAUER, F. (1883). Systematische Studien auf Grundlage der Dipteren-larven, etc. *Denkschr. Akad. Wiss. Wien*, **47**, 1–99.

BURMEISTER, H. (1832). *Handbuch der Entomologie*. G. Reimer, Berlin. Vol. I, p. 182.

CUVIER, BARON G. (1817). *Le règne animal*. Ed. 2, 1829; Ed. 3, 1836–49. Includes sections by Latreille. See also E. BLANCHARD (1849).

CUVIER, BARON G. (1832). *The Animal Kingdom* (English ed.). Whittaker, Treacher and Co. London. Vol. XV. Insecta. Has additions by Griffith and Pidgeon including 'Culices', pp. 726–37.

DE GEER, C. (1776). *Mémoires pour servir à l'histoire des insectes*. P. Hesselberg, Stockholm. Vol. VI, pp. 298–324.

DOBELL, C. (1932). *Antony von Leeuwenhoek and his Little Animals*. Staple Press, London.

FINLAY, C. J. (1881–1904). See under (*f*).

FISCHER, G. (1812). Observations sur quelque Diptéra de la Russie. Notice sur la larve de *Culex claviger* de Fabricius. *Mém. Soc. Imp. Nat., Moscow*, **4**, 167–80, figs. 1–16.

GEOFFREY, E. F. (1764). 'Le cousin commun' in *Histoire abrégée des insectes*, **2**, 573–80.

GOEZE (1775). *Besch. berliner Gesch. Naturf. Freunde*, **1**, 359, pl. 8.

GOLDSMITH, O. (1779). *An History of the Earth and Animated Nature*. J. Nourse, London. Vol. VIII, pp. 151–9.

GRIFFITH and PIDGEON. See under CUVIER (1832).

GUÉRIN, F. E. (1835). *Dictionnaire pittoresque d'histoire naturelle et des phénomènes de la nature*, **2**, 358–60 with plate 127.

HOWARD, L. O. (1900–17). See under (*d*).

HOWARD, L. O., DYAR, H. G. and KNAB, F. (1912). Vol. I. See under (*b*).

JAMES, S. P. (1899). Collection of mosquitoes and their larvae. *Indian Med. Gaz.* **34**, 431–4.

JOBLOT, L. (1754). *Observations d'histoire naturelles faites avec le microscope sur un grand nombre d'insectes*. Vol. I, pp. 121–4. (Copy in the Univ. Library, Cambridge.)

JORDENS, J. H. (1801). *Entomologie und helminthologie des menschlichen Körpers*. G. A. Grau. Vol. I, pp. 160–74. (Figures micropile of *Culex* egg.)

JOURDAIN, J. H. (1893). Note sur un mouvement de rotation singulier de la tête chez une larve de culicide. *C.R. Soc. Biol., Paris*, n.s. **5**, 249–50.

KIRBY, W. and SPENCE, W. (1815–70). *An Introduction to Entomology*. London. Some eight editions, some with 2–4 volumes, all containing references to mosquitoes. Preface to Ed. 5 (1828) gives a short history of entomology mentioning Reaumur, De Geer and others. The People's Edition (1870) refers to breeding in fountains and use of fish. Vol. I, pp. 59–63.

KLEEMAN, C. F. C. (1792). *Beiträge zur Natur- und Insekten-geschichte*. Theil I, *Anhang zu den Roselischen Insekten-Belustigungen*. Nürnberg. Vol. I, pp. 125–48.

LATREILLE, P. A. See under CUVIER (1817). Also short sections in various publications: (1805) *Buffon's Hist. Nat.* **92** (**14** of Insectes), 284; (1807) *Genera Crust. et Insectes*, **4**, 246; (1810) *Consid. générales sur l'ordre des animaux, etc.* (under 'Cousin'), pp. 376–9; (1825) *Fam. nat. du règne animal*, p. 482. Paris.

LEDERMULLER, M. F. (1761). Der Schnackenwurm ein Schlammwasser Insekt. *Micr. Gemüths- und Augen-Ergötzung*, **2**, 154–6.

LEEUWENHOEK, A. See under DOBELL (1932).

LICHTENSTEIN, A. A. H. (1800). Beschreibung eines neu entdeckten Wasserinsekts. *Arch. Zool. und Zoot.* **1**, 168.

MEIGEN, J. W. (1818). See under (*e*).

MEINERT, F. (1886). See under (*d*).

NUTTALL and SHIPLEY (1901). See under (*d*).

OLIVIER, G. A. (1791). 'Cousin Culex' in *Histoire naturelle des insectes*. Paris. Vol. VI. Encyclopédie méthodique, etc.

PACKARD, A. S. JN. (1869). *Guide to the Study of Insects*. Salem, pp. 368–70.

PAGENSTECHER, A. (1874). Ueber d. Schnacke (*Culex pipiens*). *Fühlings Landw. Ztg., Jena*, **23**.

SLABBER (1781). *Culex pipiens* oder Wurm von der Singschnacke. *Physikal Belustigungen oder microscopische Wahrnehmungen, etc.* Nürnberg.

(*d*) EARLY OBSERVATIONS ON STRUCTURE

ANTHONY, J. (1871). Note on paper by Woodward. *Amer. Mon. Micr. J.* **6**, 256–7.

BARTH, J. M. See under (*b*).

BECHER, E. (1882). Zur Kenntnis der Mundteile der Dipteren. *Denkschr. Akad. Wiss. Wien*, **45**, 123–60.

BLANCHARD, E. (1849). See under (*c*).

BRAUER, F. (1883). See under (*c*).

CHILD, C. M. (1894). Ein bisher wenig beachtetes antennales Sinnesorgane der Insekten mit besonderer Berücksichtigung der Culiciden und Chironomiden. *Z. wiss. Zool.* **58**, 475–530.

CHRISTOPHERS, S. R. (1901). Anatomy and histology of the female mosquito. *Repts. Mal. Comm. Roy. Soc.* Ser. 4.

DIMMOCK, G. (1881). *Anatomy of the Mouth-parts and Sucking Apparatus of some Diptera.* Williams and Co., Boston.

DIMMOCK, G. (1883). Anatomy of the mouth parts and suctorial apparatus of *Culex*. *Psyche*, **3**, 231–41.

DIMMOCK, G. (1884). Male *Culex* drinking. *Psyche*, **4**, 147.

DUFOUR, L. (1851). Recherches anatomique et physiologique sur les Diptères. *Mém. Acad. Sci., Paris*, **11**, 205–10.

DURKEE, S. (1855). On the sting of *Culex pipiens*. *Proc. Boston Soc. Nat. Hist.* **5**, 104, 106.

FICALBI, E. (1899). See under (*e*).

FINLAY, C. J. (1886). See under (*f*).

GERSTFELDT, G. (1853). *Ueber die Mundteile der saugenden Insekten.* Dorpat.

GRANDPRÉ, A. D. and CHARMOY, D. DE E. (1900). *Les moustiques: anatomie et biologie.* The Planter's and Commercial Gaz. Press, Port Louis, Mauritius.

GRASSI, B. (1900). *Studi di uno zoologo sulla malaria.* Rome.

HALLER, G. (1878). Kleinere Bruchstücke zur vergleichenden Anatomie der Arthropoden. I. Ueber das Athmungsorgan der Stechmückenlarve. *Arch. Naturgesch.* Jahrg. 44, **1**, 91–5.

HOGG, J. (1854). *The Microscope, its History, Construction and Application.* Plate IX, fig. 7. Also many later editions.

HOGG, J. (1871). On gnat's scales. *Amer. Mon. Micr. J.* **6**, 192–4.

HOOKE, R. (1665). See under (*b*).

HOWARD, L. O. (1900). Notes on the mosquitoes of the United States giving some account of their structure and biology with remarks on remedies. *U.S. Dept. Agric., Div. Ent.* Bull. 25.

HURST, C. M. (1890). *On the life history and development of a gnat* (*Culex*). Guardian Press, Manchester. For other papers see references in later chapters dealing with structure of the larva and pupa.

JOHNSTON, C. (1855). Auditory apparatus of the *Culex* mosquito. *Quart. J. Micr. Sci.* (old series), **3**, 97–102.

KELLOG, V. L. (1899). The mouth-parts of the Nematocerous Diptera. *Psyche*, **8**, 355–9.

KOWALEVSKY, A. (1889). Ein Beitrag zur Kenntnis der Excretions-organe. *Biol. Zbl.* **9**, 42–7.

KRAEPELIN, K. (1882). Ueber die Mundwerkzeuge der saugenden Insekten. *Zool. Anz.* **5**, 547–9.

LACORDAIRE, T. (1838). *Introduction à l'entomologie.* Libr. Encyc. de Roret. Paris. Vol. II. With two plates showing digestive organs, tracheae, heart, etc.

REFERENCES

LECAILLON, A. (1899). Sur les prolongements ciliformes de certaines cellules du cousin adulte, *Culex pipiens* L. (Dipt.). *Bull. Soc. ent. Fr. 1899*, pp. 353–4.

LECAILLON, A. (1900). Recherches sur la structure et le développement post-embryonnaire de l'ovaire des insectes. I, *Culex pipiens*.... *Bull. Soc. ent. Fr. 1900*, pp. 96–100.

LINNAEUS, C. (1758). *Systema Naturae*. Facsimile reprint of first volume. British Museum Publications (1939).

MACLOSKIE, G. (1887). Poison fangs and glands of the mosquito. *Science*, **10**, 106–7.

MACLOSKIE, G. (1888). The poison apparatus of the mosquito. *Amer. Nat.* **22**, 884–8.

MACLOSKIE, G. (1898). Editorial, *Brit. Med. J.* **2**, 901–3. Gives Macloskie's figures.

MAYER, A. M. (1874). Experiments on the supposed auditory apparatus of the mosquito. *Amer. Nat.* **8**, 577–92.

MAYER, P. (1879). *Sopra certi organi di senso nelle antennae dei ditteri*. Rome.

MEINERT, F. (1881). *Fluernes munddele trophi dipterorum*. Copenhagen, pp. 36–40 and plate 1.

MEINERT, F. (1886). De eucephale myggellarven. *K. Danske vidensk. Selsk. Skr.* (6), **3**, 373–493.

MIALL, L. C. (1895). *The Natural History of Aquatic Insects*. Macmillan and Co., London. Reprinted 1903, 1912, 1922, 1934.

MUIR, W. (1883). The head and sucking apparatus of the mosquito. *Canad. Nat. Quart. J. Sci.* (2), **10**, 465–6.

MURPHY, E. (1883). The proboscis and sucking apparatus of the mosquito genus *Culex*. *Canad. Nat. Quart. J. Sci.* (2), **10**, 463–4.

NEWMAN, E. (1872–3). Scales in Diptera. *Entomologist*, **6**, 9–11.

NUTTALL, G. H. F. and SHIPLEY, A. E. (1901–3). *J. Hyg., Camb.*, **1**, 45–77, 269–76, 451–84: **2**, 58–84; **3**, 166–255.

POUCHET, G. (1847). Sur l'apparat digestif du cousin (*Culex pipiens* Linn.). *C.R. Acad. Sci., Paris*, **25**, 589–91.

RASCHKE, E. W. (1887). Die Larve von *Culex nemorosus*. Ein Beitrag zur Kenntnis der Insekten-Anatomie und Histologie. *Arch. Naturgesch.* Jahrg. 53, **1**, 133–63. Also *Zool. Anz.* **10**, 18–19 (abstract only).

REAUMUR, R. A. F. DE (1738). See under (*b*).

ROFFREDI, D. M. (1766–9). Mémoire sur le trompe du cousin et sur celle du taon. *Misc. Taurin. Melang. Phil. Math. Soc. R. Turin*, **4**, 1–46. (Copy in Royal Society Library, London.)

SAVIGNY, J. C. (1816). *Mémoires sur les animaux sans vertèbres*. Gabriel Dufour, Paris. Part I, Premier fasc. Mém. 1, p. 25.

SCHINDLER, E. (1878). Beiträge zur Kenntnis der Malpighi'schen Gefässe der Insekten. *Z. wiss. Zool.* **30**, 587–660.

SMITH, J. B. (1890). A contribution towards a knowledge of the mouth-parts of the Diptera. *Trans. Amer. Ent. Soc.* **17**, 319–39.

SMITH, J. B. (1896). An essay on the development of the mouth parts of certain insects. *Trans. Amer. Phil. Soc.* (n.s.) **19**, 175–98.

SULZER, J. H. (1761). *Die Kennzeichen der Insekten*. Heidegger, Zürich.

SWAMMERDAM, J. See under (*b*).

THEOBALD, F. V. (1901–10). See under (*e*).

WIELOWIEJSKI, H. R. VON (1886). Über das Blutgewebe der Insekten. *Z. wiss. Zool.* **43**, 512–36. (Describes larval oenocytes.)

WOODWARD, J. J. (1876). On the marking of the body scales of the English gnat and the American mosquito. *Amer. Mon. Micr. J.* **15**, 253–5.

(*e*) EARLY SYSTEMATIC WORK ON MOSQUITOES

ARRIBALZAGA, E. L. (1891). Dipterologia Argentina. *Rev. Mus. la Plata* (*Culicidae*), **1**, 345–77; **2**, 131–74.

BARRAUD, P. J. (1934). *Fauna of British India. Diptera*. Taylor and Francis, London. Vol. v.

COQUILLETT, D. W. (1896). New Culicidae from North America. *Canad. Ent.* **28**, 43.

DOLESCHALL, C. L. (1858). Derde Bijdrage tot de kennis der Dipterologische Fauna van Nederlandsch Indie. *Natuurwet. Tijdschr. Ned.-Ind.* **17**, 73–128.

DYAR, H. G. (1922). Mosquitoes of the United States. *Proc. U.S. Nat. Mus.* **62**, no. 2447, 1–119.

DYAR, H. G. and KNAB, F. (1906). Mosquitoes of the United States. *J. N.Y. Ent. Soc.* **14**, 188.

EDWARDS, F. W. (1932). *Genera insectorum.* Brussels. Fasc. 194. Fam. Culicidae.

EDWARDS, F. W. (1941). *Mosquitoes of the Ethiopian Region.* Vol. III. Brit. Mus. (Nat. Hist.), London.

FABRICIUS, J. C. (1805). *Systema antliatorum.* Brunsvegae, pp. 33–6.

FICALBI, E. (1899). *Venti specie de zanzare (Culicidae) Italianae.* Florence. Reprint of papers in *Bull. Soc. Ent. Ital.* **28**, 108–313 (1896) and **31**, 46–234 (1899).

GILES, G. M. (1900). *A Handbook of the Gnats or Mosquitoes.* John Bale Sons and Danielsson, London. Ed. 2, 1902.

GRASSI, B. (1900). See under (*d*).

HASSELQUIST, F. (1757). *Iter Palestinum eller Resa til Heliga Landet,* 1749–52. Lars Salvii, Stockholm.

HASSELQUIST, F. (1762). *Reise nach Palestina.* Later edition of Hasselquist, 1757, edited in respect to nomenclature by Linnaeus. J. Christian Koppe, Rostok.

LINNAEUS, C. (1735). *Systema Naturae.* Lugduni, Batavia. Further editions in 1756 (Ed. 9); 1758 (Ed. 10); 1766–8 (Ed. 12).

LINNAEUS, C. (1746). Fauna Suecica. Batavia. Further editions in 1761 (Ed. 2), Stockholm; 1789 (Entomologia: fauna Suecica descr. aucta.), **3**, 562.

LINNAEUS, C. (1806). *Linnaeus's General System of Nature* (English transl.). Lackington Allen and Co., London.

LOEW, H. (1845). *Dipterologische Beiträge.* Posen, pp. 3–5.

LOEW, H. (1873). *Beschreibungen europäischen Dipteren.* Halle. Vol. III.

MACQUART, J. (1834). *Histoire naturelle des insectes. Diptères.* Vol. I, pp. 28–36.

MACQUART, J. (1838). *Diptères exotiques nouveaux ou peu connus.* Vol. I, part 1, pp. 29–36; part 2, suppl. 176.

MACQUART, J. (1839). In Webb and Berthelot. *Histoire naturelle des Iles Canaries.* Paris, p. 99.

MACQUART, J. (1854). Diptères exotiques (Ins. Dipt.) nouveaux ou peu connus. *Mém. Soc. Imp. Sci. Agric. de Lille.*

MEIGEN, J. W. (1818). *Syst. Beschr. bekannt. europ. zweifl. Insekten.* Aachen. Vol. I, pp. 1–12. Further editions 1830, vol. VI, pp. 241–3; 1838, vol. VII, pp. 1–2.

ROBINEAU-DESVOIDY, J. B. (1827). Essai sur la tribu des Culicides. *Mém. Soc. Hist. Nat., Paris,* **3**, 390–406.

SKUSE, F. A. A. (1889). Diptera of Australia. Part 5, Culicidae. *Proc. Linn. Soc. N.S.W.* (2), **3**, 1717–64.

SMART, J. (1940). Entomological systematics examined as a practical problem. In Huxley's *New Systematics.* Clarendon Press, Oxford. Pp. 475–92.

THEOBALD, F. V. (1901–10). *A Monograph of the Culicidae of the World.* Vols. I–V (Bibl. in vol. II). Brit. Mus. (Nat. Hist.), London.

VAN DER WULP, F. M. (1884). Exotic Diptera. *Notes from the Leyden Museum,* **6**, 248–256. Vol. VI.

WALKER, F. R. (1848). *List of the Species of Dipterous Insects in the Collection of the British Museum.* London, Part 1.

WALKER, F. R. (1851–6). *Insecta Britannica: Diptera.* London. Vol. III (Culicidae), pp. 242–53.

WALKER, F. R. (1857–65). *J. Proc. Linn. Soc., London,* 1857, **1**, 6; 1859, **3**, 77; 1860, **4**, 90; 1861, **5**, 144, 229; 1865, **8**, 102.

WIEDEMANN, C. R. W. (1821). *Diptera exotica.* Kiliae, Part 1.

WILLISTON, S. W. (1896). On the Diptera of St Vincent (West Indies). *Trans. Ent. Soc. Lond. 1896*, pp. 270–2 (Culicidae).

(*f*) RESEARCH ON *AËDES AEGYPTI* AND ITS ROLE IN DISEASE

AGRAMONTE, A. (1908). A pioneer of research in yellow fever. *Brit. Med. J.* **1**, 1306.

AGRAMONTE, A. (1908). An account of Dr Louis David Beauperthuy, a pioneer in yellow fever research. *Boston Med. Surg. J.* **158**, 927–30.

REFERENCES

BANCROFT, T. L. (1899). On the metamorphosis of the young form of *Filaria bancrofti* Cobb. (*Filaria sanguinis hominis* Lewis; *Filaria nocturna* Manson) in the body of *Culex ciliaris* Linn., the house mosquito of Australia. *J. Proc. Roy. Soc. N.S.W.* **33**, 48–62. Reprinted in *J. Trop. Med.* (*Hyg.*), **2**, 91–4.

BANCROFT, T. L. (1906). On the aetiology of dengue fever. *Aust. Med. Gaz.* **25**, 17.

BEAUPERTHUY, L. D. (1854). Transmission of yellow fever and other diseases by the mosquito. *Gaz. Oficial de Cumana.* Anno 4, no. 57. Quoted by AGRAMONTE (1908).

BRUCE, D. (1895). *Preliminary Report on the Tsetse-fly Disease or Nagana in Zululand.* Bennet and Davis, Durban.

BRUMPT, E. (1936). *Précis de parasitologie.* Ed. 5. Masson et Cie, Paris.

CLELLAND, J. B. *et al.* 1918. See references under 'dengue' in chapter IV.

DELBOEUF, J. (1847). *Mag. Pittoresque*, p. 180. Article referred to in 1895, *Rev. Sci., Paris* (4), **4**, 729, use of oil.

DOANE, R. W. (1910). *Insects and Disease.* Constable and Co. Ltd, London.

FINLAY, C. J. (1881). El mosquito hipoteticamente considerado como agente de transmission de la fiebbre amarilla. *R. Acad. Ciencias Habana*, **18**, 147–69 (transl. in *Rev. Med. Trop. y Parasit, Hanaba*, **4**, 163–84, 1938).

FINLAY, C. J. (1886). Yellow fever, its transmission by means of the *Culex mosquito. Amer. J. Med. Sci.* **92**, 395–409.

FINLAY, C. J. (1902). Agreement between the history of yellow fever and its transmission by the *Culex mosquito* (*Stegomyia* of Theobald). *J. Amer. Med. Ass.* **38**, 993–6.

FINLAY, C. J. (1904). New aspects of yellow fever etiology arising from the experimental findings of the last three years. *J. Amer. Med. Ass.* **42**, 430–1.

HOWARD, L. O. (1893). An experiment against mosquitoes. *Insect Life*, **5**, 12–14, 109–10, 199.

HOWARD, L. O. (1894). Another mosquito experiment. *Insect Life*, **6**, 90–1.

HOWARD, L. O., DYAR, H. G. and KNAB, F. (1912–17). *The Mosquitoes of North and Central America and the West Indies.* Vols. I–IV. (1912, vol. I, General, including yellow fever and mosquito bionomics; vol. II, Plates. 1915, vol. III, Systematic. 1917, vol. IV, Systematic, including *Aëdes aegypti.*)

JAMES, S. P. (1900). On the metamorphosis of the *Filaria sanguinis hominis* in mosquitoes. *Brit. Med. J.* **2**, 45, 533–7.

KING, A. F. A. (1882). The prevention of malaria disease, etc. *Trans. Phil. Soc. Wash.* Read 10 February 1882 (printed only in abstract).

KING, A. F. A. (1883). Insects and disease: mosquitoes and malaria. *Pop. Sci. Mon.* **23**, 644–58.

LEWIS, T. R. (1878). Remarks regarding the haematozoa found in the stomach of the *Culex* mosquito. *Proc. Asiat. Soc. Beng.* 1878, pp. 89–93.

LOW, G. C. (1900). A recent observation on *Filaria nocturna* in *Culex*: probable mode of infection in man. *Brit. Med. J.* **1**, 1456–7.

MACKERRAS, I. M. (1948). *Australia's Contribution to our Knowledge of Insect-borne Disease.* The Jackson Lectures. Austral. Med. Publ. Co., Sydney.

MANSON, P. (1879). On the development of *Filaria sanguinis hominis* and on the mosquito considered as a nurse. *J. Linn. Soc.* (*Zool.*), **14**, 304–11. Read on 7 March 1878.

MANSON, P. (1884). The metamorphosis of the *Filaria sanguinis hominis* in the mosquito. *Trans. Linn. Soc.* (2) *Zool.* **2**, part x, 367–88.

MARCHOUX, E., SALIMBENI, A. and SIMOND, P. L. (1903). La fièvre jaune. Rapports de la Mission Française. *Ann. Inst. Pasteur*, **17**, 665–731.

NUTTALL, G. H. F. (1899). On the role of insects, arachnids and myriapods as carriers in the spread of bacterial and parasitic diseases of man and animals. *Johns Hopk. Hosp. Rep.* **8**, 1–154.

OTTO, M. and NEUMANN, R. O. (1905). Studien über Gelbfieber in Brasilien. *Z. Hyg. InfektKr.* **51**, 357–506.

*REED, W. (1901). The propagation of yellow fever: observations based on recent researches. *Med. Rec., N.Y.*, **60**, 201–9.

*REED, W. (1902). Recent researches concerning the etiology, propagation and prevention of yellow fever by the U.S. Army Commission. *J. Hyg., Camb.*, **2**, 101–19.

HISTORICAL

*REED, W. and CARROLL, J. (1901). Experimental yellow fever. *Boston Med. Surg. J.* **144**, 586. Also: The prevention of yellow fever. *Med. Rec., N.Y.*, **60**, 641–9. Also (1902): The etiology of yellow fever, a supplementary note. *Amer. Med.* **3**, 301–5.

*REED, W., CARROLL, J. and AGRAMONTE, A. (1901). Papers on yellow fever in many journals including: *Rev. Med. Trop., Habana*, **2**, 17–34; *Amer. Med.* **2**, 15–33; *Trans. Ass. Amer. Phys.* **16**; *J. Amer. Med. Ass.* **36**, 431–40; *Med. Rec., N.Y.*, **59**, 269–70.

*REED, W., CARROLL, J., AGRAMONTE, A. and LAZEAR, J. W. (1900). The etiology of yellow fever. A preliminary note. *Philad. Med. J.* **6**, 790–6. Also: Etiologia de la fiebre amarilla. Nota preliminar. *Rev. Med. Trop., Habana*, **1**, 49–64.

ROSS, R. (1898). Report on the cultivation of *Proteosoma* Labbe in grey mosquitoes. *Indian Med. Gaz.* **33**, 173–5, 401–8, 448–51.

RUSSELL, P. F. (1955). *Man's Mastery of Malaria.* Oxford Univ. Press, London, New York and Toronto.

SMITH, T. and KILBOURNE, F. L. (1893). Investigations into the nature, causation and prevention of Texas or Southern Cattle Fever. *Bur. Anim. Ind., U.S. Dept. Agric.* Bull. 1, p. 301.

SOUTHEY, R. (1812). *Omniana* or *horae otiosiores.* Vol. I, pp. 55–6, use of oil. (Copy in the Univ. Library, Cambridge.)

STILES, C. W. (1901). Insects as disseminators of disease. *Virginia Med. (Semi-)Mon.* **6**, 53–8.

* Many papers relating to the work of the U.S. Yellow Fever Commission have been reprinted in 1911 in a Government publication (Senate document). Government Printing Office, Washington, D.C.

II
SYSTEMATIC

(*a*) SYNONYMY

Though the name *Aëdes aegypti* (L.) is now in general use and has been for more than two decades, the species has appeared under many other names in the past. Recently the question of nomenclature has again arisen (see Mattingly, 1957), and some account of previous changes in this respect seems called for.

Identified by Theobald (1901) as *Culex fasciatus* Fabr. 1805 (type ♀(?), non-existent, West Indies) and made the type species of his genus *Stegomyia*, the species was for many years known as *Stegomyia fasciata*. As early as 1905, however, Blanchard had contended that the name *fasciata* was pre-occupied by *fasciatus* given to a mosquito by de Villers (1789) and that the name had also been previously used by Meigen (1804) for a European species, *Culex fasciatus* Meigen, 1804 (now thought to be *Aëdes communis* (Deg.)). Therefore the name *fasciata* as applied by Theobald was invalid, the correct specific name being *calopus* Meigen, 1818 (type ♀♂, non-existent, Portugal). By 1913, too, the amended comprehensive genus *Aëdes* had been established including *Stegomyia* as a subgenus and thus the name became *Aëdes calopus* Meigen, 1818, the name *Stegomyia*, if used, being placed in brackets as the subgenus.

The change of name to *calopus* was not, however, to escape criticism, and Theobald (1907) pointed out that De Villers' description was insufficient to enable his species to be identified and that the type was non-existent. Further, Austen (1912) drew attention to the fact that De Villers had not described a new species, but had referred to one described by Müller in 1764. Müller was describing the fauna of Fridrichdal and from his description Edwards (1933) considered Müller's species to be most probably *Culex pipiens*, as have other authors since, for example Natvig (1948) in his excellent work on the Scandinavian mosquito fauna. There remained, however, little doubt but that the name *fasciatus* as applied to the species was invalid. Even if, under some special exempting clause in the rules of nomenclature, *fasciatus* Fabr., removed to the genus *Aëdes*, could be held not to have become permanently invalid as a homonym, it still would be invalid as antedated by *fasciatus* Mg., now also in the genus *Aëdes*. The name *calopus*, indeed, though never in such familiar and universal use as the name it displaced, became increasingly employed from about 1907, indeed generally so by the more systematically-minded observers. The name was, however, eventually by-passed by the finding by Knab (1916) of a name and description which clearly applied to the species and pre-dated both *fasciatus* and *calopus*, namely *Culex argenteus* Poiret, 1787.

Poiret's description leaves no doubt as to the species described. After saying that it is the commonest mosquito in Barbary he continues: 'Il est de la grosseur du nôtre, mais si richement paré, que je lui ai souvent pardonné ses piqûres pour le plaiser de l'admirer. Tout son corps, particulièrement le dos, est couvert d'écailles argentées, placées sur lui comme autant de paillettes orbiculaires et brillantes. Ses pattes sont ornées de bandes alternatives brune & argentées.' The name *Aëdes argenteus* was adopted by Edwards (1921) in his revision of Palaearctic mosquitoes and was in general use by English writers for some ten years.

American writers, however, very largely followed Dyar, who in 1920 put forward a still earlier name, namely *aegypti* given by Linnaeus in 1762, thus leaving only four years to the date 1758, before which any name that might have been given would have no validity. It was this last name that was eventually by general agreement accepted, being used by Edwards in his revision in the *Genera Insectorum* (1932) and from 1934 being indexed in place of *argenteus* in English journals. As noted in chapter I, section (*e*), the name *aegypti* first appears in 1757 in Hassellquist's account of his travels in Palestine as *Culex* (*aegypti*) followed without a break by a short Latin description. In the later edition of this work in 1762, edited as to nomenclature by Linnaeus, the name appears in correct form as *Culex aegypti*. In Opinions 5 and 57 of the International Committee of Zoological Nomenclature the question of the validity of re-edited pre-Linnaean names is considered, and in the last-mentioned Opinion it was given that the German translation by Gadebusch, published in 1762, does not give validity to the names published in the original edition. A later Opinion, Opinion 175, however, states that: 'A work the names of which on account of date were invalid is considered as far as nomenclatural requirements go as having never been published.' Opinion 5, however, still required that reprinted pre-Linnaean names should be reinforced by adoption or acceptance by the author who published the reprint. It has generally been accepted that Linnaeus reinforced by adoption and acceptance the name *aegypti* and that the name *Aëdes aegypti* is by the rules of nomenclature correct.

Nevertheless, the question of the correct nomenclature has again been raised. In the first place, Mattingly (1953) (see also Mattingly and Knight, 1956) has expressed the opinion that it was probably the pale form, var. *queenslandensis* (see section (*c*)), that was described by Linnaeus, since this form occurs in Egypt and would seem to be indicated by the description. If this were so, it would be the pale form that was the type form and that usually accepted as the type form that would be the variety. An even more serious objection to the name *aegypti* has since been put forward by Mattingly (1957). This author points out that at the time when adoption of the name *aegypti* was under consideration and was used by Edwards, the objection was put forward by Patton (1933) that the species described by Linnaeus was more probably another species very common in Egypt, namely *Aëdes* (*Ochlerotatus*) *caspius*. Mattingly further points out that the wording of the Latin description by Linnaeus in his opinion leaves no doubt whatever but that the species described was that now known as *A.* (*Ochlerotatus*) *caspius* Pallas, 1771.

A few words may here be said regarding this latest point in the synonymy and the literature relating to it. One of the earliest authors to give a systematic description of Egyptian mosquitoes was Gough (1914). Gough in his list of species gives *Stegomyia fasciata* Fabr., clearly indicating the species now under consideration. He says it is not rare in Egypt and gives localities from which it has been recorded. He also notes that a specimen of var. *queenslandensis* has been taken at Suez. The name *aegypti* is also given, but having as a synonym *Aëdes* (*Ochlerotatus*) *dorsalis* Meigen, a species very close to *Aëdes caspius*. Later Kirkpatrick (1925) in his monograph on the mosquitoes of Egypt records the species as *Aëdes* (*Stegomyia*) *argenteus* (the name then in general use) and in a list of synonyms beneath the name gives '? *Culex aegypti* Linné'. When later the name *argenteus* was beginning to give place to *aegypti* an interesting discussion took place in

1933 between Patton, who pointed out that Linnaeus's description could not apply to the species but was a description of *Aëdes caspius*, and Edwards, who was unwilling to change the name *aegypti* and taking the various points in the description maintained that this might well refer to the species. The name *aegypti*, indeed, continued to be used and has, without further challenge, remained the accepted name up to the present time. Should the wording of the original description indicate, as Mattingly maintains, not *A. aegypti* as now known, but *A. caspius*, an interesting point in synonymy will arise, more especially as to how far (1) the evidence for a change can be proven, and, should this be so, (2) how far can long and wide use of a name in the literature be held as valid for an Opinion of the International Committee of Zoological Nomenclature to establish retention of the name in use.*

For the verbatim text of early descriptions by authors see volume 4 of Howard, Dyar and Knab (1917). For information regarding types see Edwards (1941). For literature relating to recent questions in synonymy see Seidelin (1912); Gough (1914); Knab (1916); Dyar (1920); Kirkpatrick (1925); Patton (1933); Edwards (1933); Mattingly (1957). A useful list of the International Rules of Zoological Nomenclature is given by Wenyon (1926).

Other names than those referred to above that have been given to the species are now regarded as synonyms. About twenty-four such are listed by Dyar (1928) and are given here in an accompanying list with the country from which the species under the name was described. For varietal names see section (*c*).

List giving localities of synonyms.†

Culex aegypti Linnaeus 1762. Egypt.
Culex argenteus Poiret 1787. North Africa.
Culex fasciatus Fabricius 1805. West Indies.
Culex calopus Meigen 1818. Portugal.
Culex mosquito Rob.-Desvoidy 1827. Cuba.
Culex frater Rob.-Desvoidy 1827. West Indies.
Culex taeniatus Wiedemann 1828. Savannah (Georgia U.S.).
Culex kounoupi Brulle 1836. Morea (Greece).
Culex niveus Eichwald 1837. Caspian area.
Culex annulitarsis Macquart 1839. Canary Islands.
Culex viridifrons Walker 1848. Greece.
Culex excitans Walker 1848. Georgia.
Culex formosus Walker 1848. West Africa.
Culex inexarabilis Walker 1848. West Africa.
Culex exagitans Walker 1856. Para (Brazil).
Culex impatibilis Walker 1860. Makassar (Celebes).
Culex zonatipes Walker 1861. New Guinea.
Culex bancrofti Skuse 1889. Australia.
Culex elegans Ficalbi 1889. Italy.
Culex rossi Giles 1899. India.
Stegomyia nigeria Theobald 1901. Nigeria.
Culex anguste-alatus Becker 1908. Canary Islands.
Culex albopalposus Becker 1908. Canary Islands.
Duttonia alboannulis Ludlow 1911. Philippines.

* Since the above was written Mattingly (*Proc. R. Ent. Soc. Lond.* (c), **22**, 23, 1957) notes that application has been made to the International Commission to fix the name *aegypti* by attaching it to a neotype. As noted, this will involve some further problems in connection with the choice of neotype. The important point in the present author's opinion is that so well established a name for this common and important mosquito should not be changed.

† For references see Dyar (1928).

(*b*) SYSTEMATIC POSITION AND IDENTIFICATION

SYSTEMATIC POSITION

Aëdes aegypti is the type species of the subgenus *Stegomyia* Theobald, 1901, of the genus *Aëdes* Meigen, 1818, as amended by Edwards (1932) to include some fourteen previously erected genera and about 400 species. Though *Aëdes aegypti* is not the type for the genus *Aëdes*, this being *A. cinereus* Mg., it is very representative in many respects of this large section of the Culicini, which contrasts with another large section, namely the genus *Culex* and allied genera, notably in: (*a*) the retracted eighth abdominal segment in the female; (*b*) the possession of toothed claws in the female; (*c*) the absence of pulvilli; (*d*) the laying of eggs singly and not in rafts; (*e*) being composed largely of dark and often highly ornamented species, especially so in the subgenus *Stegomyia*.

Synoptic tables for distinguishing the many subgenera of *Aëdes* are given by Barraud (1934) for the genus as a whole, and by Edwards (1941) for subgenera occurring in the Ethiopian Region, but precise characterisation is difficult. Distinction depends largely on genitalic characters and the form of the male palps (Edwards, 1941, p. 107). The distinctive characters of the subgenus *Stegomyia* as given by Theobald are the flat scaling of the vertex, except for a narrow row of upright scales posteriorly, and the flat scaling of all three lobes of the scutellum. Other characters distinguishing the subgenus *Stegomyia* (including those of the male genitalia) are noted by Edwards (1941) when indicating the subgenus (p. 125).

Some seventy or more species have been described in the subgenus, the classification and grouping of which are discussed by Edwards (1932) and more recently by Mattingly (1952, 1953); Knight and Hurlbut (1949); and by Marks (1954). Synoptic tables for adults and fourth-stage larvae are given by Edwards (1941) and by Mattingly (1952) for Ethiopian species; by Barraud for species from India and Burma; by Bonne-Wepster and Brug (1932) for species of the Malay Archipelago; and by Marks for the *scutellaris* group of the eastern Oriental and Australasian Regions. For larvae of the Ethiopian Region see also Hopkins (1936, 1951).

Recent work has added very greatly to knowledge of many of the less-known species, and this applies to species closely related to or closely resembling *Aëdes aegypti* such as are discussed under Identification, as also to some of the varietal forms dealt with later. But much regarding some of these forms has still to be worked out, making the subject one that must be regarded as yet to some extent still under review (see Mattingly, 1952, 1953).

IDENTIFICATION

Fortunately identification of *A. aegypti* normally offers few difficulties, since it has very characteristic thoracic markings that almost alone serve to distinguish it. In addition the marking of the mid-femur, namely a white line on the anterior surface extending from the base almost to the tip, is peculiar to this species of *Stegomyia*, as is almost equally so the patch of white scales on each side of the clypeus in the female. The clypeus of the male is usually, however, bare or has only a few scales.

The thoracic markings as normally present consist of two crescentic patches of white scales, one on each side of the anterior half of the scutum, between which in the middle line of the thorax and passing back nearly to the scutellum are two narrow parallel white lines (Fig. 46). In addition there are two fine white lines passing from the posterior ends of the crescentic marks to the lateral lobes of the scutellum, thus completing the so-called 'lyre-shaped' ornamentation. In the pale form, var. *queenslandensis* (see next section), markings may be obscured by pale scaling and, though this would appear to be unusual, Edwards refers to the white median lines as being sometimes yellowish.

A number of other species of *Stegomyia* have white shoulder marks and some approximation to the ornamentation in *Aëdes aegypti*. But in most species where such resemblance occurs the shoulder marks are more wedge-shaped or rounded, and, in place of the twin median parallel white lines, these are yellow or a single median white line is present. Some useful figures showing the thoracic ornamentation of a number of species are given by Edwards (1941). A figure of the markings of the common African species, *A. africanus*, with one of *A. pseudo-africanus*, is given by Mattingly and Bruce-Chwatt (1954). The species most nearly resembling *A. aegypti* in their thoracic markings are: *A. subargenteus* Edw., 1925, and its varietal form, or possibly a distinct species, var. *kivuensis* Edw., 1941;* *A. woodi* Edw., 1922; certain species that might be confused with the pale variety of *A. aegypti*, namely *A. mascarensis* and *A. vinsoni* both from Mauritius.†

In *A. subargenteus* the crescentic shoulder marks are described as abbreviated anteriorly and the fine lines passing from these to the scutellum are yellow. The mid-femur, instead of being uniformly striped, has a separate white spot about its middle. Var. *kivuensis* differs from *subargenteus* in the tarsal markings, the fourth hind tarsal segment being all white instead of being pale at the base only. Also the median lines of the thorax are deep yellow (see recent description given by Mattingly, 1952). *A. woodi* has the lateral lobes of the scutellum clothed with dark scales, which at once distinguishes this species from *A. aegypti* where all three lobes are white-scaled. All the above forms were originally described from Nyasaland, the first-mentioned being recorded later from a considerable number of widely dispersed localities on the east of the African continent from Kenya to Cape Province. Var. *kivuensis* has been recorded up to date from only one locality in the Belgian Congo. *A. woodi* has been recorded from Nyasaland, Kenya and Mozambique. (For recent information see Mattingly, 1953.)

It should be noted that some species of subgenus *Finlaya* of the genus *Aëdes* have thoracic ornamentation with some resemblance to that of *A. aegypti*. Little difficulty, however, should arise except in rubbed specimens since the scutellum would fail to show the flat scaling of all three lobes characteristic of *Stegomyia*.‡ Also superficially resembling *Stegomyia* in Central and South America is the important genus *Haemogogus*. In this case the large and approximated pronotal lobes suffice to distinguish from *Stegomyia*. This character also serves for the forest genus *Heizmannia*, in which in addition the postnotum carries a group of small hairs. For accounts of these New World species see Dyar (1928).

White scaling on the clypeus. White scaling on the clypeus is given by Ludlow (1911) in her original description of *Duttonia alboannulis*, and is suggestive of the correctness of the sinking of this species under *Aëdes aegypti* (see list of synonyms in preceding section). White scales on the clypeus are also recorded by Dyar and Knab (1910), for some pale specimens among a number of normal *A. aegypti* received by these authors from Malaga,

* Now given by Mattingly (1953), p. 9, as a distinct species.

† *Aëdes* (*S.*) *simpsoni* should also be given here. See figure of this species given by Muspratt (1956), p. 49.

‡ A few *Aëdimorphus* with shoulder marks somewhat like *Aëdes aegypti* may have flat scaling of the head or scutellum, but the clubbed style of the coxite in the male would distinguish any such with help of synoptic tables for the subgenus.

Spain. Presence of white scales on the clypeus in the female is almost diagnostic of *A. aegypti*, but is described in *A.* (*Stegomyia*) *vittatus* and possibly in *A.* (*Stegomyia*) *trinidad* from Fernando Po (Edwards, 1941).*

Banding of the proboscis. In the majority of species of the subgenus, as in *A. aegypti*, the proboscis is without pale banding. Pale banding, however, is usually present in *A. vittatus*.

Leg markings. The marking of the mid-femur as already described would seem quite peculiar to *A. aegypti*. The tibiae in *Stegomyia* are usually dark. The tarsal banding is used in synoptic tables to differentiate between some species, the fourth hind tarsal segment being white at the base in a number of species including *subargenteus*, but all white in *kivuensis* and all black in *woodi* and *simpsoni*. The extent of banding of the third hind tarsal and in some cases other tarsal segments may also be a differential character. The wings are usually all dark-scaled, but in *A. aegypti* and some other species there is a characteristic small pale spot at the base of the costa.

Genitalic characters. Characters differentiating the male terminalia in *A. aegypti* are: (*a*) ninth tergite with two prominent conical lateral lobes carrying a few short hairs, between which is a deeply hollowed out V-shaped embayment; (*b*) the style of the clasper widened in the middle; (*c*) lateral processes of the paraprocts nearly as long as the terminal processes; (*d*) lateral plates of the phallosome with fine teeth only. The basal lobe in *A. aegypti* and related species is a flattish plaque on the inner membranous aspect of the coxite. In *A. aegypti* it has no projecting process forming a separate claspette or harpago as in some species. Recently Iyengar and Menon (1955) have described three or four characteristic hooked spines arising from the lobe in *A. aegypti* (see also Ross and Roberts, 1943). A full description of the parts in *A. aegypti* is given later when describing the structure of the species in detail (see p. 458).

In *A. subargenteus*, as noted by Edwards (1941), the ninth tergite is convex, with the approximated rather flat lobes carrying a row of large hairs, and the basal arms of the proctiger are short, not long as in *A. aegypti*. The coxite also differs, having a small hairy lobe towards the base on the dorsal surface, and the style is somewhat curved and tapering, with the terminal spine stout, not thin as in *A. aegypti*. The phallosome is as in *A. aegypti*. The male terminalia of *kivuensis* have not up to date of writing been described; those of *woodi* are stated by Mattingly (1952) to be virtually identical with those of *simpsoni*.

Pupal characters. Those distinguishing *A. aegypti* are: (*a*) trumpets triangular, without transverse folds (see Theodor, 1924); (*b*) paddles with a single hair about one-quarter of the length of the paddle; (*c*) no accessory hair; (*d*) chaeta A on abdominal segments II–VI spine-like. Synoptic tables for pupal characters in a number of species of *Stegomyia* are given by Ingram and Macfie (1917); Theodor (1924); Baisas (1938); Edwards (1941).

Larval characters. Larval characters distinguishing *A. aegypti* are: (*a*) antennae smooth, antennal hair simple; (*b*) comb teeth 8–12 in a single row with strong basal lateral denticles; (*c*) pecten teeth on the siphon 15–20, none widely spaced; (*d*) pleural hairs of the meso- and metathorax with conspicuous thorn-like processes. Recently Iyengar and Menon described

* Mattingly (1952), however, states that the type of *Aëdes trinidad* (now sunk as a synonym of *A. dendrophilus* Edw., see Mattingly, *loc. cit.*) has not got scale tufts on the clypeus, frosting of the integument having been mistaken in the original description for these, and that *A. aegypti* is distinguished from all other *Stegomyia* by this character.

a single pecten tooth well separated and situated distal to the siphon hair, a character previously thought to be diagnostic of *A. vittatus*. These authors note that, very rarely, there may be an additional irregular series of pecten spines situated ventral to the main row. Additional teeth on the siphon may, however, as noted by Mattingly (1955), occasionally occur in other species. Very similar larval characters are present in *A. mascarensis* (but pecten spines only 8–12) and in *A. metallicus* (antennae as in *A. aegypti*, but the spines of the pleural hairs much longer). Synoptic tables for larvae of African species of *Stegomyia* are given by Hopkins (1936, 1951) and by Mattingly (1952). The latter author, in 1953, also gave recent information regarding the larvae of a number of the less known species of the subgenus.

(*c*) VARIETIES, STRAINS, HYBRIDS, GYNANDROMORPHS AND MUTANTS

There is a considerable recent literature on the systematics and genetics of *A. aegypti* and its different forms, and space does not permit of more than the main outline of such work being here given. For further information the references given in the text may be consulted. A good and interesting account of recent ideas on the origin and build up of the species will be found in the paper given as in the press by Mattingly (1957).

VARIETIES

Few named varieties have been described that are not now regarded as synonyms. Theobald (1901) refers to var. *mosquito* (*Culex mosquito* Rob.-Desv.) as somewhat distinct (thorax with semilunar lines only) but sinks it under the species. Var. *mosquito* Arrib. he states is typical of the species. Var. *persistans* Banks is also stated by Theobald (1910) not to differ in the description given from the type form. Of the two varieties described by Theobald himself, var. *luciensis* (Theobald, 1901, p. 297) has the apical tip of the last tarsal segment black, whereas it is white in the type form. The variety was originally described from Demerara and St Lucia, but has been noted as occurring occasionally in India (Barraud, 1934) and in Malaya (Stanton, 1920). The dark tarsal tip, however, is not unusual in the species from many areas (see later).

The second of Theobald's varieties, var. *queenslandensis* (Theobald, 1901, p. 297), was described and named from specimens sent by Dr Bancroft from Burpengary, South Queensland as *Culex bancrofti* Skuse. *C. bancrofti* is described by Skuse as having the thorax with lyre-shaped markings, the subjacent cuticle dark brown and covered with brown scales, the description conforming to what is usually regarded as the type form under which the species was sunk by Theobald. Var. *queenslandensis* Theobald describes as having the thoracic scaling golden brown, the mid-lobe of the scutellum covered with a patch of deep purple scales, and a broad band of the same down the dorsum of the abdomen. In one specimen the entire abdomen was covered with creamy scales. Theobald (1903, p. 144) also gives as this variety a form from the Seychelles, the abdomen being as in the Australian form, but the hind tarsi mostly with a distinct black apical band.

A form var. *atritarsis* is described by Edwards (1920). It has the tarsi of the forelegs almost entirely black, the hind tarsi with very narrow white rings at bases of segments 1–3 and 5, segment 4 being almost entirely black. The palpi in the male have the white rings narrower than usual. The terminalia are given by Edwards (1941) as of the type form; he states that the variety has been found only on the Gold Coast and may possibly be a genuine local form rather than a sporadic variation.

Two forms of the species are noted by Hill (1921) to occur in northern Queensland, namely a small dark form occurring away from habitations and a larger pale form frequenting the hospital buildings. The dark form was collected from a tin containing about five inches of water and decaying leaves in dense scrub on an island 600 yards from the nearest seaside dwellings about four miles from Townsville. Several adult *Stegomyia* were captured at the same time attempting to bite. A similar dark form was collected from a tree-hole in the hospital grounds seventy yards from the nearest dwelling. In the hospital building the species was paler and larger. The 'dark' form collected from the tin when bred for four generations produced in each generation after the first a proportion of individuals of both forms as well as intermediate forms. The 'light' form bred true. The males in both cases were dark and of the usual size.

Two forms are described by Legendre (1927) as 'geographical races', namely a small form having thin pale larvae with contrasting dark siphon, Oceano-Indian in distribution and occurring in Hanoi, Madagascar and Syria; and a large form with thick dark larvae occurring on the west coast of Africa.

In Lagos Summers-Connal (1927) observed the species to be subject to considerable variation. Some specimens have the abdomen entirely black except for the lateral spots, some completely covered with white scales or scattered white or yellow scales giving a brindled effect. The hind tarsi may be as in the type form, or the tip, or a quarter to three-quarters, of the last segment may be dark. The material used consisted of 1000 specimens of each sex bred consecutively from larvae collected in Lagos by the sanitary authorities. The author found hereditary transmission of a number of the variations noted.

Describing the mosquitoes of British Somaliland, G. R. C. van Someren (1943) notes that the species varied from jet black to pale brown, the latter being referred to var. *queenslandensis* Theo. It occurred breeding in water tanks on dhows. A similar pale form regarded as var. *queenslandensis* was described by Lewis (1947) at Port Sudan on the Red Sea Coast breeding in water receptacles on sea-going dhows. Inland the species was darker along with intermediate and normal forms. The habits of the pale form were stated by Lewis not to differ from those of the type form and to be capable of harbouring the virus of yellow fever. This characteristic pale form has now been recorded in many coastal areas of the East African continent and elsewhere (see distribution given later in this section).

Recently the work of Haddow and others has brought to light the occurrence in Uganda, Kenya and elsewhere in Ethiopian Africa of another form, namely a 'dark' form found breeding in deep forest far removed from habitations and which differs from the type form in being noticeably very black with clear-cut markings. Regarding this wild form Haddow (1945*a*, *b*) notes that, though larvae of *Aëdes aegypti* are not uncommon in the Bwamba district of Uganda, adults are very rarely taken biting and that the evidence suggests that the species may be largely sylvan and zoophilic in the area. Haddow, Gillett and Highton (1947, p. 319) also note that *A. aegypti* occupies a peculiar position in Bwamba, being present in some uninhabited areas of the Semliki Forest, and that Garnham, Harper and Highton (1946) also found it scattered rather profusely through the Kaimosi Forest in Kenya. Regarding the biting habits of the species under these conditions these authors note that it was not represented in forty catches made in the forest, but that on one occasion thirteen specimens were taken biting in the sunset period at heights over fifty feet.

A similar reference to a 'pale' and a 'dark' form occurring in Kenya is made by van Someren, Teesdale and Furlong (1955). They give the following description of the forms:

In the pale form the ground colour of the scutum varies from dark brown to pale yellow. The vertex is sometimes mainly pale, but usually with a small patch of dark scales on each side, or rarely these patches may be large. The first abdominal segment has usually a large patch of white scales above, the other segments being lightly or heavily sprinkled with white scales, or rarely nearly all scales are white. Occasionally the tergites are dark with narrow apical white bands, or more often with broad apical and basal white or yellow, or rarely indistinct pale, bands. The hind tarsi are usually as in the type form, but sometimes segment 4 has a narrower basal white band and segment 5 a dark band at the tip.

Of the dark form they note that many hundreds of specimens bred out from larvae taken from containers in the bush were consistently dark (black with white markings). The following variations in colour were noted: the white bands on the abdominal segments were usually well marked, rarely narrow, much reduced or absent; the first tergite was usually all black above, but sometimes with a small spot of white scales medianly. The hind tarsal segments 1–3 were white on their basal third or fourth, segment 4 three-quarters white and segment 5 all white.

As with the species in the Uganda forests the two forms in the Mombasa area of Kenya differed also with respect to their association with man, the dark form tending to be a wild form occurring in the bush, whilst the paler form behaved like the type form, being found mainly in habitations. The following are captures made respectively in the bush and in houses in the area by Mrs E. C. C. van Someren as given by Mattingly (1956):

In bush: 125 dark (including 19 males), 5 pale (all females).
In houses: 7 dark (including 2 males), 232 pale (all females).

Summarising the forms of the species that are at present usefully to be recognised, Mattingly (1957) gives the following:

1. *A. aegypti* (L.) s.str., the type form. A brown form, somewhat variable in depth of colour, but always distinctly paler and browner (at least in the female) than the black African subspecies (no. 3). Extension of pale scaling, if any, limited to bleaching of the two dark areas on the back of the head, or the presence of pale scaling on the first abdominal tergite or to both together.

2. *A. aegypti* var. *queenslandensis* Theo. Any form showing any one of the following: bleaching of the dark scales of the mesonotum from mid-brown through shades of buff to almost white; encroachment of the pale basal bands on the abdominal tergites on to the apices of the preceding segments, or extension of these bands in the mid-line to form pale basal triangles or a continuous median pale line on the abdomen; presence of scattered pale scales on the dark areas of the tergites or normally dark areas on the legs.

3. *A. aegypti* s.sp. *formosus* Walker. A form confined to Africa south of the Sahara and the only known form in this area except in coastal districts or one or two inland-trade-frequented areas. This differs from the type form in the markedly black appearance of the dark areas of the thorax and abdomen and entire absence of bleaching or extension of pale scaling on any part of the body. The name *formosus** used by Mattingly is that

* The name has no relation with Formosa, being evidently the Latin word for 'beautiful'.

given by Walker (1848, p. 4) to a specimen from Sierra Leone (type in the British Museum) which shows the characters described above, and is the first description of any form from the area given as the distributional area of the subspecies.

In all the forms the characters given relate to the female. The male in all forms is usually darker.

The distribution of these three forms, so far as known at present, is given when dealing with the distribution of the species in sections (*d*) and (*e*) of this chapter. Very briefly it is the type form (associated in many areas with var. *queenslandensis*) that gives the world-wide character to distribution of the species as described later, ssp. *formosus* being restricted to Africa south of the Sahara and so not affecting the world extension of the species as a whole. The distribution of var. *queenslandensis* is less easy to specify. It is found along with the type form very commonly as a coastal species on the east of the African continent and neighbouring countries, as also in the Mediterranean area, notably in North Africa, and on the west coast of this continent. It appears to be less common in the Indomalayan area where the type form is dominant. It is recorded from Australia (Queensland) along with the type form and appears to be a feature of the species in America (Mattingly: *in lit.*). More details are given later in sections (*d*) and (*e*).

Especially in regard to s.sp. *formosus* more information would be very desirable as to its exact status and especially regarding structural features other than ornamentation. Up to the time of writing there does not appear to be in the literature any careful description of the genitalic or larval characters or whether the characteristic scale tufts on the clypeus or the stripe on the mid-femora are present.

Regarding var. *queenslandensis* there seems little doubt but that this is specifically a form of *A. aegypti* and Dyar and Knab (1910) note the presence of scale tufts on the clypeus in the pale form described by them from Malaga, Spain. Its wide distribution along with the type form is suggestive that it is a sporadic variation of the species or the effect on this of some particular environment. A very characteristic feature of the form is its occurrence as a coastal and desert form and, as noted by the present writer (Christophers, 1933), specimens of *Anopheles* from semi-desert conditions, where the species has bred in shallow, often saline pools in the open, commonly show a reduced pigmentation (flavism or hypomelanism), whilst melanic forms are commonly encountered in shady places with abundant food supply. Of interest in this respect is the occurrence (as noted by Edwards, 1911, p. 248) in Algeria, where the pale variety is recorded as common, of an ochreous form of *Culex pipiens* and a pale form of *Och. nemorosus* (*punctor*).

The pale form of *Aëdes aegypti*, however, has been shown to retain, at least to some extent, its characters when bred for some generations in the laboratory (Summers-Connal, 1927; Shidrawi, 1955). Also Mattingly (1953) notes that not all localities on the east coast of Africa from which the form has been recorded are areas of low rainfall. Thus it would appear that even if the form is the result of some special environment the effect is not one merely of a temporary nature but is associated with genetical change. The same might be said regarding the occurrence of the form in America referred to later.

On the species in general Mattingly and Knight (1956) suggest that the dark form (s.sp. *formosus*) represents the original non-domestic species in its original home and conditions and that the so-called type form is a later development, possibly largely hybrid in character, thus explaining the genetic differences in some geographical forms of the species that have been described.

STRAINS

Besides the forms described in the last section particular stocks of the species maintained in laboratories may be separately indicated or named as strains depending upon the source from which the species was originally obtained. Such strains have been found as a rule to agree in their main behavioural characters. Brug (1928), studying Cuban and Javanese strains, observed no constant differences, except that the head scales in the former were lighter. Mathis (1934) studied strains from Athens, Cuba, Dakar and Java and found their biology similar. Strains differing in certain respects have, however, been described.

Examples of differences in biting capacity in strains from different countries are not infrequently described in the literature; for example, Gillett and Ross (1955) found difficulty in getting Malayan *A. aegypti* to feed on monkeys, whilst African *aegypti* fed readily. Differences have also been frequently referred to in their behaviour when employed for crosses between strains as noted in a later section. Recently a study has been made by Gillett (1955) of differences in behaviour in respect to ovulation as between a strain from Lagos and one from Newala, Tanganyika. In the Lagos strain unmated females usually laid eggs within five days of taking a blood meal (86 per cent ovipositing). In the Newala strain females rarely laid eggs within this period (8 per cent ovipositing). Females of the Newala strain when mated with sexually exhausted males also did not lay infertile eggs to the same extent as the Lagos females.

Other differences relate to capacity to transmit diseases or harbour parasites. Hindle (1929) records a strain from Africa maintained for three years in the Wellcome laboratories which was a typical rather dark race transmitting yellow fever readily and one from India reared by MacGregor that was lighter in colour and a less effective vector. Roubaud, Colas-Belcour, Toumanoff and Treillard (1936) found strains behaved differently in their capacity to transmit the nematode, *Dirofilaria immitis* (see also Kartman, 1953).

The term strain, however, would appear to have a very vague implication. In many cases strains have been maintained in culture for long periods, often many years, under different techniques, and it would not seem safe to assume that the behaviour of a strain that happened to have originally been brought from some particular country necessarily represented the naturally occurring form in that country.

HYBRIDS

Particularly since Toumanoff's observations on the result of mating the two distinct species *Aëdes aegypti* and *A. albopictus* and the early work of Marshall and Staley (1937) and Marshall (1938) on the crossing of nearly related forms in what is now referred to as the *Culex pipiens* complex, a great deal of work has been carried out relating to the genetics of, and hybridisation in, mosquitoes. For a very full account of mosquito genetics the comprehensive paper by Kitzmiller (1953) may be consulted, as also the critical review of findings to date by Mattingly (1956) under the title 'species hybrids in mosquitoes', with which is a very full list of references to recent authors in this connection.

Toumanoff (1937), using female *A. albopictus* from Hanoi and male *A. aegypti* from Calcutta, obtained fertile offspring for a number of generations. In the opposite direction, that is female *aegypti* and male *albopictus*, crossing was unsuccessful. The offspring of the fertile crossing with a single exception up to generation F_4 resembled the female parent. Hoang-Tich-Try (1939), with the same species in Indochina, obtained the same result.

Downs and Baker (1949) and Bonnet (1950), in New York and Hawaii respectively, also obtained fertile offspring from crossing these two species, but in an opposite direction, that is *aegypti* females with *albopictus* males gave fertile offspring, while *albopictus* females with *aegypti* males failed to do so. The same result was obtained by Kartman (1953), *aegypti* females and *albopictus* males giving fertile offspring resembling the female parent, whilst crossing in an opposite direction failed. Not all observers, however, have obtained positive results with crossing of these two species. MacGilchrist (1913) in Calcutta found that, though copulation took place, crossing was unsuccessful. De Buck (1942), though he found spermatozoa in the spermathecae, obtained no embryonic development in eggs laid by *albopictus*, and one larva only that hatched from eggs of *aegypti* died in a few hours.

Later work by Toumanoff (1938–1950) on the crossing of these species and that by a number of authors on the crossing of nearly related forms of the *Culex pipiens* complex has shown that transmission of hereditary characters in hybrid forms of mosquitoes is, as generally considered, the result of a rare, but recognised, condition known as *cytoplasmic inheritance*, a condition in which inheritance is not of the Mendelian type, but is characterised by transmission of the female parent's characters only, *female dominance*, and believed to be the result of transmission through the cytoplasm independent of nuclear influence. Not all authorities agree, however, that this is so in all cases.

The *C. pipiens* complex consists of a number of related species more or less resembling *C. pipiens* and various forms and local strains of these. It includes *C. pipiens*; *C. fatigans*; *C. quinquefasciatus*; *C. pipiens* var. *pallens*; and *C. molestus*. The last-mentioned species occurs in a number of local forms, pairing between which may in some cases be unproductive, in some successful in both directions and in some successful only in one. The first crosses in this group were those by Marshall and Staley and by Marshall with local forms of *molestus* (see also Tate and Vincent, 1936). More extensive observations in this group have since been made by other observers, opening up a wide field of inquiry into the genetics and hybridisation of mosquitoes. A review of these findings has recently been given by Mattingly (1956).*

The work of Laven (1953) on the crossing of various local forms of *C. molestus* appears definitely to prove the existence in this group of cytoplasmic inheritance. Thus, by back-crossing offspring from crosses in both directions, Laven was able to demonstrate the occurrence of a purely maternal type of inheritance, hybrid females being mated with males of the male parent form through twenty successive generations without the offspring showing other than maternal characters and so giving no indication throughout of nuclear effect.

Mention should also be made of recent observations on the crossing of forms in the *scutellaris* group of *Stegomyia*. Here also compatibility in one direction has been found associated with incompatibility in the other, as also has inheritance of female characters only, though sometimes with intergrading.†

The results recorded above obtained from the crossing of *Aëdes aegypti* and *A. albopictus* would appear therefore to be in accordance with other findings of this form of inheritance, which is clearly of common occurrence in mosquitoes. Mattingly (1956), however,

* See also Laven (1951, 1955, 1956*a*), Laven and Kitzmiller (1954), Knight (1953), Knight and Abdel Malek (1951), Kitzmiller (1953), Barr and Kartman (1951), Callot (1947, 1954), and others given in Mattingly's list.

† See Woodhill (1949, 1950, 1954), Smith-White (1950), Perry (1950), Rozeboom (1954*b*).

expresses some doubt whether in the case of the two species mentioned absence of nuclear effect has been proved. His view in this respect is based on the fact that both in Toumanoff's series and in Bonnet's there occurred after some generations a stray form resembling the male parent, making further observations on crossing of these two species very desirable. This is especially so, since in this case there can be no doubt whatever but that the two forms interbred are two species each with very pronounced characters.

As a means of studying more precisely the transmission of characters in crossings Laven (1956*b*) has recently made use of artificially induced mutations, the result of exposure to radiation. In one of these mutant forms the terminal segment of the male palps is changed into the appearance of a sort of hinged claw and in another the male palps are shortened and the terminal joint lost. In other forms changes have been brought about in the wing neuration.

Other means of approach to such problems have been the study of susceptibility of hybrid forms to infection by *Dirofilaria immitis* (Kartman, 1953), the study of transmission of characters in naturally occurring mutant forms (see under mutants) and chromosomal and cytogenetic studies (see section on chromosomal structure in chapter XXXI). A very large field of inquiry is in fact opened up in which research on *Aëdes aegypti* should play a part, not only because it is well adapted to laboratory research, but also on account of its importance as a disease vector, and so a species regarding which knowledge of its possible development of resistance to insecticides is very desirable (see Rozeboom, 1954*a*).

GYNANDROMORPHS

A not uncommon occurrence in mosquitoes, including *A. aegypti*, are forms showing in various degrees mixed sex characters, such forms being termed gynandromorphs. Sex determination in insects is not brought about as in mammals through circulating hormones, but is the result of the chromosomal condition of individual cells in development leading to patches of one sex or another in the sexual mosaic in the same individual (Wigglesworth, 1942). The effects are very haphazard. Sometimes the two lateral halves of the body are of opposite sex, or occasionally partition is transverse, or a body of one sex may have the head or genitalia partly or wholly of the opposite sex, or patches of a particular sex may be scattered indiscriminately. Besides the appearance of the male and female parts in the same individual there may be duplication of the antennae or other parts (Lengerken, 1928). Gynandromorphs have been described in a number of species of mosquitoes. A list is given by Komp and Bates (1948) and by Bates (1949). Kitzmiller (1953) gives thirty-four examples recorded in the literature as given below. For an account of the origin of gynandromorphs see Morgan and Bridges (1919).

Of the five gynandromorphs of *A. aegypti* that described by Martini had the head wholly male and the apical portion of the abdomen wholly female with some other less noticeable parts affected. In the three forms described by Roth and Willis the antennae in all were of male character and the abdomen and genitalia female. Other parts were of varied sex character. In two cases one wing was shorter than the other, indicating male character. In the specimen recorded by Smyly the left side of the head was male and the right female.

A study of the behaviour of the three forms described by them has been made by Roth and Willis. All three forms were indifferent to the human arm and two exposed to females made no male response. They behaved like males, however, in that they remained flying

SYSTEMATIC

Species	Number described	Author
*Aëdes punctor**	5	Edwards, 1917 (3); Shute, 1926 (1); Der Brelje, 1923 (1)
Aëdes detritus	1	Marshall, 1938
Aëdes pullatus	1	Felt, 1905
Aëdes abserratus	1	Felt, 1904
Aëdes canadensis	1	Carpenter, 1948
Aëdes aegypti	5	Martini, 1930 (1); Smyly, 1942 (1); Roth and Willis, 1952 (3)
Culex pipiens	1	Marshall, 1938
Culex molestus	6	Marshall, 1938 (1); Weyer, 1938 (1); Gilchrist and Haldane, 1947 (3); Gratz, 1954 (1)†
Culex fatigans	2	Middlekauf, 1944
Culex coronator	1	Komp and Bates, 1948
Culex salinarius	2	Roth, 1948
Culex nigripalpis	2	Rings, 1946 (1); Warren and Hill, 1947 (1)
Culex theileri	1	Bedford, 1914
Theobaldia annulata	1	Classey, 1942
Orthopodomyia signifera	1	Roth, 1948
Orthopodomyia fascipes	1	Roth, 1948
Haemagogus spegazinnii	2	Bates, 1949
Megarhinus brevipalpis	1	Muspratt, 1951

* Formerly often called *A. nemorosus* or *A. meigenensis*.
† Record by Gratz is additional to Kitzmiller's list.

backwards and forwards in the cage after the females had settled. Normal males made attempts to pair with the gynandromorphs having normal female wings, showing that the sound produced by them resembled that of a female. A few attempts were made to pair with one gynandromorph that had a slightly shorter wing on one side, but none with the other form with a short wing. Tested with tuning forks two were more sensitive than females and took to flight. They failed to give seizing response, that is, clinging to the cloth of the cage near the source of sound with the wings vibrating rapidly. But they gave clasping response, that is, flexing of the abdomen so that the genitalia touched the cloth.

Certain cases where male *A. aegypti* have been recorded as attempting to suck human blood (for example, as described by Howard, Dyar and Knab, 1912, p. 109) may possibly have been gynandromorphs. In such cases it is necessary to distinguish between attempts to feed and males attracted by and sucking sweat, a common happening when feeding mosquitoes in cages.

MUTANTS

Besides gynandromorphs certain changes due to the action of genes lead to forms which differ in some particular from the normal species characters. A short account of such gene action is given by Wigglesworth in his chapter on growth. The condition has been especially studied in certain Lepidoptera and in *Drosophila* in Diptera. Two modes of action are recognised, namely (1) action confined to the cells in which the genes occur as in determining the nature of the wing ornamentation in the moth *Ephestia*; and (2) action in which the cells are caused to liberate chemical substances which exert their action at a distance and determine the character of other tissues. The latter type of action is responsible for the black-eye and red-eye forms of *Ephestia* and for the vermilion-eye form of *Drosophila*.

The only described instance in mosquitoes appears to be the white-eye mutant of *Culex molestus* described by Gilchrist and Haldane. The condition in this case consists in the normally dark eyes of the imago being devoid of pigment so that they appear a dull white. The strain described by the above authors was bred from four females and ten males with white eyes occurring in a cage of several hundred *C. molestus*. The lack of eye pigment is also present in the larva and in the pupa. The white-eye mutants showed normal phototaxis, but did not respond to moving shadows as did the normal individuals. This is attributed by the authors to the fact that, the ommatidia not being isolated by pigment from one another, visual acuity was lost. The affected larvae and pupae, however, gave a well-marked shadow response. They also gave response to changes in light intensity, which was considered the reason they gave a shadow reaction. It is not stated whether the ocelli were without pigment. As these are the functional visual organs in the larva, and possibly even in the pupa, it would be of interest to note their condition.

The authors note that the white eye in *C. molestus* is a recessive character giving normal Mendelian relations and that it is partially linked with sex.

No mutant forms appear to have been described in *Aëdes aegypti*.

(*d*) GEOGRAPHICAL DISTRIBUTION

GENERAL FEATURES OF THE DISTRIBUTION

In giving an account of the distribution of the species note must be taken of the different forms now recognised and described under varieties in section (*c*) of this chapter. It will be most convenient, however, to deal first with certain features in the distribution of the species as a whole and to give later what is known of the distribution of the three forms. It will be evident from what has already been said that in such questions as the limits of the distribution of the species, and a number of other features connected with its occurrence, it will be the type form that in the main is being considered and that to a very large extent the pale form, var. *queenslandensis*, is in varying degree associated with this. Only in Africa south of the Sahara does the account of the distribution need to take note of a distinctive area occupied by one of the forms, the non-human frequenting s.sp. *formosus*.

A large number of early records of localities for the species throughout the world are given by Theobald (1901–10) and a very complete summary of records with 320 references by Kumm (1931). A map of the distribution as known up-to-date is given by Brumpt (1936). Among authors giving the distribution for different countries may be mentioned: Dyar (1928) for the Americas; Edwards (1941) for Ethiopian Africa; Muspratt (1956) for South Africa; Mattingly and Knight (1956) for Arabia; Barraud (1928) for India and Burma; Reid (1954) and Macdonald (1956) for Malaya; Brug (1926) and Bonne-Wepster and Brug (1932) for the Malay Archipelago; Ferguson (1923) for Australia; and Farner *et al.* (1946) for the Pacific. There are, however, very numerous other authors recording the species, some of whom will be found mentioned in the text. Many references to recent records will be found in *Review of Applied Entomology*, ser. B and in reference lists in recent papers by Mattingly (1952, 1953), Marks (1954) and Muspratt (1956).

As with 'useful plants' the distribution of *Aëdes aegypti* (except for the form s.sp. *formosus*) has no very obvious relation to the usual zoogeographical realms. The species is almost the only, if not the only, mosquito that, with human agency, is spread around the whole globe. But in spite of this wide zonal diffusion its distribution is very strictly limited by latitude and as far as present records go it very rarely occurs beyond latitudes of 45° N. and 35° S. Altitude is a further modifying circumstance, as also some other factors, notably distance from the sea, desert conditions and isolation from human intercourse.

THE NORTHERN LIMITS

In the accompanying map (Fig. 1) are given the northern and southern limits of distribution and the relation of these to latitude and the January and July isotherms, respectively for the northern and southern hemispheres, of 50° F. (10° C.). The numbers on the map indicate localities for which furthest north or south records have been made.

In America *A. aegypti* has been recorded far more frequently, and further north, in the east than in the west. There is, however, a record (Good, 1945) stating that *A. aegypti* used to occur in British Columbia, but has not been recorded for thirty years. At the present

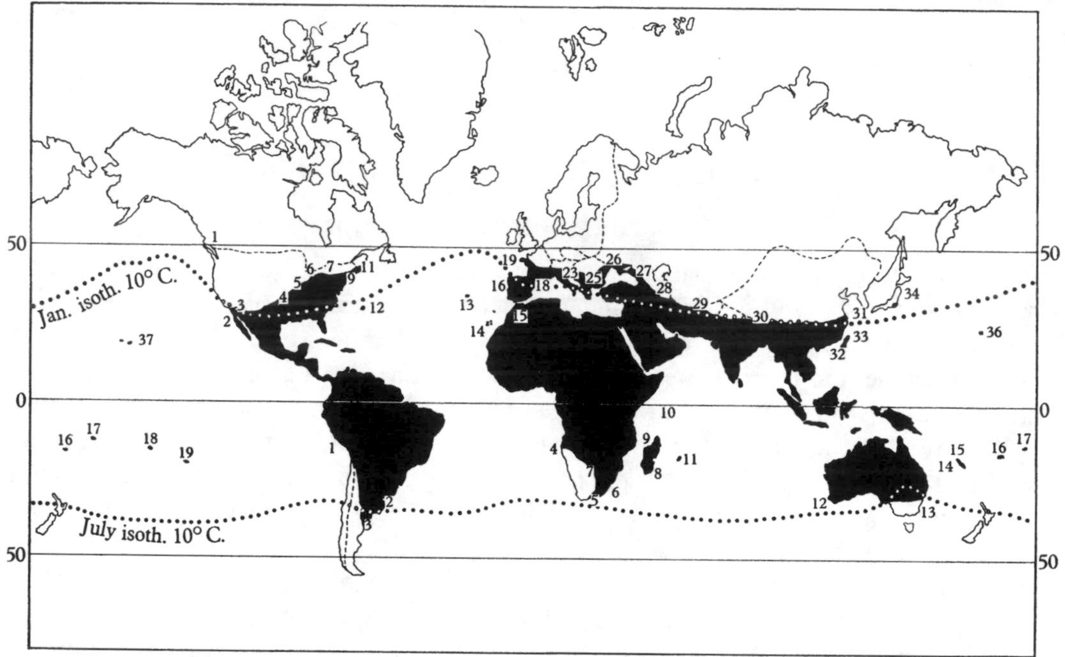

Figure 1. Map showing limits of distribution.

Map of the world showing northern and southern limits of the recorded distribution of *Aëdes aegypti* and isotherms of 10° C.

The numbers are those given to recorded localities as set out in Tables 1 and 2.

Figures in brackets in the tables give the numbers in the list of references to geographical distribution on pp. 50–2.

time the most northerly record on the west appears to be Lower California, at most about 32° N. It is recorded from Tucson, Arizona by Bequaert (1947), who states that it has not been recorded from Colorado or Utah. Passing eastwards *A. aegypti* is recorded from Oklahoma (Rozeboom, 1941) where it is, as noted later, evidently near its temperature limit. Eastwards of this it occurs in most of the States up to about 40° N. and formerly as far north as Boston, 42° 27′ N.

A. aegypti has been recorded from the Bermudas, the Azores and the Canary Islands. In Teneriffe it was common, but was not found by me in Madeira (Christophers, 1929).

In western Europe the species is mainly restricted to the more southern parts and is not normally recorded north of 45° N. There is a record, however, given by Kumm for Brest, 48° 24′ N. The single record for England is probably a laboratory error as otherwise it has

Table 1. *Recorded northern limits of distribution.**

	North America
1	British Columbia (37), 50° N.†
2	Lower California (62), 32° N.
3	Arizona (9), 32° N.
4	Oklahoma (57), 36° N.‡
5	Missouri (40, 62), 38° N.
6	Indiana (40, 62), 40° N.
7	Virginia (40, 62), 38° N.
8	Maryland (40, 62), 39° N.
9	Philadelphia (16), 40° N.†
10	New York (16), 40° 52′ N.†
11	Boston (16), 42° 27′ N.†
	Atlantic Islands
12	Bermuda (6), 32° 20′ N.
13	Azores (62), 38° 35′ N.
14	Canary Islands (18, 47), 28° 40′ N.
15	Morocco (Tangier) (17), 35° 42′ N.
	Europe
16	Portugal (Badajos) (35), 38° N.
17	Spain (north-west point) (13), 43° N.
18	Spain (Ebro) (27), 41° N.
	Europe (*cont.*)
19	France (Brest) (42), 48° 24′ N.
20	France (Dol) (42, 62), 48° 31′ N.
21	France (Bordeaux) (59), 44° 48′ N.
22	Italy (Genoa) (42), 44° 25′ N.
23	Italy (Ravenna) (42), 44° 25′ N.
24	Bosnia (3), 44° N.
25	Macedonia (42), 41° N.
26	Russia (Odessa) (41), 46° 30′ N.‡
27	Russia (Sukkum) (58), 43° N.
28	Russia (Baku) (1), 40° 25′ N.
	Asia
29	India (Peshawar) (8), 34° N.
30	India (Assam) (8), 26° N.
31	China (Shanghai) (29), 31° 10′ N.
32	China (Amoy) (30), 24° 40′ N.
33	Formosa (29, 42), 24° N.
34	Japan (Tokyo) (42, 62), 35° 40′ N.
35	Japan (Okayama) (42), 34° 22′ N.
	North Pacific Islands
36	Wake Island (56), 20° N.
37	Hawaii (11, 39), 20° N.

* The figures in brackets refer to authors as given in the list of references on pp. 50–52. The numbers preceding the name are those entered on the map.

† In past times only.

‡ At biological limit.

never been recorded. In eastern Europe it is again seen at its temperature limit at Odessa (Korovitzkyi and Artemenko, 1933). It is recorded from Transcaucasia up to about 43° N. In Asia its distribution to the north is largely restricted by the Himalayan and other mountain chains. Records north of the great chains in the central regions of Asia are entirely wanting. Its occurrence in Japan is rather vaguely given by Yamada in his revision of the mosquitoes of Japan as limited by the frost line. The main island of Japan lies roughly between the January isotherms of 30° F. and 40° F. so that the limit might well be shown by the record for Tokyo. In the northern Pacific *A. aegypti* has recently been recorded as newly introduced in Wake Island and it is recorded as present in some of the islands of the Hawaii group.

THE SOUTHERN LIMITS

Records for the southern limits are given in Table 2. In general the distribution does not extend to latitudes as high as those recorded for the northern hemisphere. For South America on the west there are relatively few records, the most southerly being for Peru. There appear to be no records for Chili. On the east Brumpt's map is shaded for some distance south of Buenos Ayres, possibly to cover a record for Bahia Blanca. In the south Atlantic there are the islands of Trinidad, 20° 30′ S., Ascension, 8° S., St Helena, 14° 30′ S., Tristan da Cunha, 37° S., and Gough Island, 40° 19′ S., all of which are at latitudes compatible with the possible presence of the species. I have not, however, been able to find any records.

In the south of the African continent Edwards (1941) shows *A. aegypti* as recorded from every region except Cape Province, South West Territory and Bechuanaland, but in his text he notes that specimens are in the British Museum from Grahamstown and Port St John. More recently a number of records as given by De Meillon are noted by Mattingly (1952) for South-west Africa (Tsumeb, 19° 15′ S., Windhoek, 22° 35′ S., Franzfontein, 20° 10′ S., Okimbahe, 21° 25′ S., Karabib, 21° 52′ S., and in the interior to the east in Bechuanaland, Upington and Kimberley). Still more recently Muspratt (1956) has given detailed records with a map for southern Africa showing numerous records along the east coast as far as Grahamstown. Inland are many records for the Transvaal, but only a few south of this and none at all either coastal or inland in the western half of Cape Province where the species would appear not to occur. On the west the most southerly records are those in South-west Africa referred to above, all north of 23° S.

Table 2. *Recorded southern limits of distribution.**

South America

1 Southern Peru (23), 15° S.
2 Monte Video (20), 34° 30′ S.
3 Buenos Aires (13, 42), 34° 30′ S.

Africa

4 South-west Africa (22, 49), 22° 30′ S.
5 Grahamstown (25), 33° 20′ S.
6 Port St John (25), 31° 30′ S.
7 Kimberley (49), 29° S.

Indian Ocean

8 Madagascar (25), 20° S.
9 Comoro Islands (45), 11° 30′ S.
10 Seychelles (38), 4° S.
11 Mauritius (42), 20° 30′ S.

Australia

12 Fremantle (61), 32° S.
13 New South Wales (19, 31, 33, 64), 32° S.

South Pacific Islands

14 New Caledonia (14, 32, 52, 55), 21° S.
15 New Hebrides (14, 52, 53), 17° S.
16 Fiji (14, 32, 43), 17° S.
17 Samoa (14, 15, 32), 14° S.
18 Cook Islands (14, 21, 46), 20° S.
19 Pitcairn Island (14, 42), 25° S.

* The figures in brackets refer to authors as given in the list of references on pp. 50–52. The numbers preceding the name are those entered on the map.

In the Indian Ocean there are records for the Comoro Islands and Madagascar, the Seychelles, Mauritius, the Maldive Islands and the Netherland Indies. I have not found records for the Chagos Islands, Cocoa Islands or Christmas Island, though such may have been given in recent literature to which in some cases access is difficult. In Australia the species has been recorded both on the east and the west almost to the extreme southern limits of the continent. It does not appear to have been recorded from Tasmania. In the South Pacific are records for some of the islands in a number of the larger groups.* There is a record for Pitcairn Island, 25° S. *A. aegypti* does not appear to occur in New Zealand as it has not been found in North Auckland after special search (Miller, 1920).

TEMPERATURE IN RELATION TO LIMITS OF DISTRIBUTION

In the main the northern and southern limits of distribution appear to be related to temperature. The relation, however, is not altogether a simple one. In America the northern limit corresponds fairly well to a January isotherm of 35° F. and a July isotherm of 75° F. In Europe, except in the west, it corresponds to a January isotherm of 40° F. and a July isotherm of 75° F. Such a winter isotherm would, however, include parts of Great Britain

* See Backhouse and Heydon (1950); Bohart and Ingram (1946); Farner *et al.* (1946), who give maps illustrating the distribution of mosquitoes of medical importance in this area.

and Ireland which are well north of the limit and where only one doubtful record of the species has ever been made. For China the figures are 50° F. and 80° F., and for Australia for July and January respectively (the winter and summer periods being reversed) 50° F. and 70° F. In degrees centigrade the above data are as follows:

	° C. January	° C. July		° C. January	° C. July
N. America	1·8	23·9	S. America	15·6	26·7
Europe	4·4	23·9	Africa	15·6	26·7
China	10·0	26·7	Australia	10·0	21·1
Mean	5·4	24·8	Mean	13·7	24·8

Experimental data in the laboratory as given in a later chapter point to a temperature of about 10° C. as that just lethal to the adult and larva. A time of reduction in winter to this temperature would on such a basis presumably limit establishment of the species in so far as it would be lethal to both adult and larva, even if, as Cossio (1931) notes for Uruguay, the species were dependent on passing the winter months in the larval stage. The egg, however, has been shown to be much more resistant to low temperature and several authors have described the species as over-wintering in the egg stage. Ferguson (1928) notes that the species in the southern parts of Queensland and in New South Wales breeds only during the warmer months and passes the winter in the egg. Roukhadzé (1926) similarly says in regard to Georgia (Transcaucasus) that the species oviposits in the autumn and that the winter is passed in the egg stage.

It would not appear, however, that resistance of the egg to cold can be a very important asset to the species in greatly extending its area of distribution. For if the temperature is not sufficiently high for a sufficient length of time in the summer to allow for at least one brood laying eggs the species would still not be established. Korovitzkyi and Artemenko (1933) speaking of the summer on the Black Sea coast note that, at the temperature prevailing, eighteen days elapse between pairing and oviposition and that most females die before their eggs mature. Rozeboom (1941) speaking of the over-wintering of *A. aegypti* in Oklahoma considers that there are, towards the limit of distribution, three zones, namely (1) a zone in which the temperature allows continuous breeding, (2) a border-line zone where the species hibernates as the egg, and (3) a temporary summer zone in which the species breeds and spreads in the warm weather but dies out. Viosca (1924) notes that in New Orleans no adults were found between January and April inclusive.

Whether restriction of the species is due to a low winter temperature or to insufficiently high and long summer temperature is therefore uncertain. It would appear rather more probable that it was insufficient summer temperature that was mainly responsible. That the limit should appear to be related to a critical low winter temperature would largely follow since the two temperatures are clearly to a considerable extent correlated, a certain degree of low winter temperature being probably accompanied by a corresponding low and short summer temperature.

THE EFFECT OF ALTITUDE

An important factor influencing the prevalence of the species in particular localities and even in the major features of the world distribution is altitude. Altitude is clearly responsible for the southerly position of the northern limit in South Asia, there being no records for the species north of the line of the Himalayas and country to the east of this, though extensive tracts lie well within the latitude of 40°. The highest altitude recorded would appear to be that of Mara (1946), for presence of the species on the summit of Mt Bizen, Eritrea, at an altitude of 7800 feet. Garnham, Harper and Highton (1946) also found the species in the Kaimosi Forest, Kenya, at 6500 feet altitude. Both these records probably relate to the wild form, though very similar altitudes could probably be found for the normal human associated form, for example Dalhousie, given as a record by Barraud, is over 7000 feet. In general, however, an elevation of 6000–8000 feet is likely to limit distribution of the species, or even lower levels in temperate latitudes. On Mt Bizen the temperature is recorded as ranging from 15·3° C. to 19·7° C., the lower limit being not very far from the temperature limiting the species by latitude. Much may depend upon the situation not only as regards latitude, but also upon the physical features of an area. Thus much of the Transvaal, where Muspratt (1956) gives many altitudes in his records of over 3000 feet and one of 5900 feet (Johannesburg), though elevated, is a plateau with corresponding high January temperatures, whilst in steep mountainous country not only climatic conditions but those associated with man may be more hostile.

SOME OTHER FACTORS AFFECTING DISTRIBUTION

Whilst the extreme limits of distribution of the species have been indicated, it by no means follows that the species is necessarily present everywhere within such limits, and other conditions than temperature may decide its presence. Thus it is absent or rare in many desert areas. Lewis (1947) notes that *A. aegypti* is rare in the desert-like country north of Khartoum where it is scarcely ever found despite careful search. Baque and Kieffer (1923) refer to its scarcity in the Sahara; and the absence of records over extensive tracts of desert country in the western regions of America already alluded to, as also in the Kalahari Desert in Southern Africa, are noticeable features in its distribution. There are extensive areas of forest where the species does not occur. Aragao (1935), speaking of jungle yellow fever in Brazil, states that *A. aegypti* is not found in the forest unless brought by the inhabitants from the towns, and other authors refer to general acceptance that in the great forests of America the species is not found unless introduced. In Africa south of the Sahara the species is recorded from primeval forest, but it appears that here it is not the usual form, but the s.sp. *formosus* that is encountered.

On the other hand are conditions favouring the species. Thus, marked features in its distribution are its relative abundance in moist warm climates, its close association with coastal conditions and its special prevalence in towns and especially the great sea ports. Its prevalence in moist warm climates is in keeping with experience in the laboratory, a high humidity being especially favourable to the life and activity of the adult, besides providing in the form of rainfall abundant facilities for breeding. In the last respect, however, a dry climate and domestic habits that lead to storage of water may also be favourable. High humidity is obviously a factor favouring coastal distribution, and in the conditions formerly largely present in ships and dhows is a further factor that has favoured dispersion. Chandler (1945) notes that when ships offered good breeding places, as they did in the past, *A. aegypti* was found established at places like Philadelphia, New York and Boston at present beyond the species area of distribution.

SPREAD AND ASSOCIATION WITH MAN

Whilst the influences mentioned have in various degrees played their part in modifying distribution and prevalence within the limits imposed by temperature, the one all-important cause of the present extended distribution of the species is very clearly its spread by man. As a corollary, where man has not introduced the species it may often be absent though conditions are favourable. Such areas may cover considerable tracts or may be quite small. The species is common in Ceylon, but Senior White (1920) in his mosquito survey of a rubber estate in the less frequented eastern part of the island found *A. aegypti* entirely absent, though altitude and temperature conditions were not unfavourable. In an *Aëdes* survey of Hog Island, a low-lying island within sight of the city of Bombay, but private and secluded and largely covered with dense forest, Afridi (1939) found *A. aegypti* absent, though other species of *Aëdes* were abundant.

In the Pacific the part played by man in spreading the species is particularly evident. Buxton and Hopkins (1927), referring to Samoa, note that the species is abundant in Apia, a good harbour with a considerable European community, but has never been found far from houses of Europeans and has failed to colonise the forests in spite of suitable breeding places. It has been recorded from the Society Islands, Tonga, Hawaii and other island groups, but usually only in the more frequented islands in such groups. Recently many interesting instances have been recorded of occurrence of the species only after Japanese occupation during the last war. Thus Perry (1949) notes that in Treasury Island (British Solomons) *A. aegypti* had not been known up to 1945, but was present at the time of his investigation. Rosen, Reeves and Aarons (1948) record the presence of the species in Wake Island in 1947, after Japanese occupation, though it had not been present up to 1941.

The species would appear also to have become more diffused recently throughout Malaya. Macdonald (1956), as a result of a recent survey, notes that the original largely coastal distribution of the species has been rapidly vanishing and that, though there are still some rural areas, especially inland, where *A. aegypti* is absent, the species is now widely spread throughout the country.

THE ORIGINAL HOME OF THE SPECIES

It has been suggested that the original home of *A. aegypti* was the New World. Dyar (1928), however, notes that there are no nearly related species in the American continent, but many such in the Old World, especially in Africa, and he considered that it was probably the African continent from which the species originated. Recent findings have strengthened this view. Not only have numerous species of the same subgenus been recorded from the Ethiopian Region (some thirty-seven as against thirty from the Oriental Region, the next richest in number, with but few elsewhere), but there are in this region some very closely related forms. Furthermore, the species has been found in Africa breeding in forest areas independent of man, a very unusual feature elsewhere. If the species originally spread from Africa there might well be the general picture of the present distribution.

This would seem to be an appropriate place to refer to the origin of the distribution of *A. aegypti* in the New World, about which, beyond giving its limits, nothing has so far been said. That *A. aegypti* had a wide distribution in the Americas was evident from the large amount of work carried out in the early part of the century in the control and elimination of the species as the vector of urban yellow fever. Thus its presence throughout the southern United States, Central America, the West Indies and parts of Brazil was sufficient at one time, as we have seen, to suggest that this was its original home. Yet it seems unlikely that the species could have been introduced before the discovery of America and it is remarkable that so extensive a spread could have taken place in what must have been relatively late historical times. Yet the reasons for thinking so are very strong. It is possible that, once introduced, such active dissemination may have been partly at least the result of a relatively high level of civilisation with corresponding trade and other activities. There might here be a certain irony in such a cause for the species in the fact that this carried with it its own Nemesis. For in no part of the world have there been such active measures carried out against the species. So much so that this has led to its elimination from much of the areas once occupied.* Referring to the present distribution of *A. aegypti* in America Severo (1956) gives a map showing areas from which the species has now been eradicated and where it is still present. A striking feature of the map are the narrow strips of coastal distribution in Central America and some other parts where the species is still present. Absence of the species from the vast tracts of forest in Brazil is in remarkable contrast to the condition in Ethiopian Africa.†

DISTRIBUTION OF DIFFERENT FORMS OF THE SPECIES

A good deal is now known of the distribution of the forms of the species as already described. But more information is very desirable, especially with precise identification of these forms as encountered in different areas and conditions. Briefly the distribution as at present shown by records is as follows:

For the type form the distribution is as has been given above for the species as a whole, except that in Africa south of the Sahara the only form present in the interior, except in some special areas, is considered by Mattingly (*in lit.*) to be the subspecies *formosus*. Otherwise it is the type form, associated to a varying degree with the pale form, var. *queenslandensis*, that represents the species throughout the distributional area. The type form appears to be the dominant form in the Indo-Malayan region and in the Pacific, though some examples of pale forms have been recorded as given below when dealing with this form. In the Mediterranean and on the East African coast the type form occurs with a considerable number of records for the pale form. The type form is also recorded from the west coast of Africa where in some areas it may show marked variation and presence of pale forms as in the species studied at Lagos by Summers Connal. In southern Africa it is noted by Muspratt (1956) as present to the limits already given for the species, neither var. *queenslandensis* nor the dark form var. *atritarsis* Edw. being known except from Durban and Zululand where the pale form has been recorded by Mattingly (1953). In Australia the type form appears to be the usual form (*A. bancrofti* of Skuse) with var. *queenslandensis* present in certain localities in Queensland. In America the species exists as a mixture of type form and var. *queenslandensis*, the latter being apparently very abun-

* See Soper, Wilson, Lima and Antunes (1943) on measures undertaken in Brazil, and many other authors in the literature. See also under 'control' in chapter IV, section (*b*).

† A map of North and South America showing the distribution in relation to control measures is also given in the *Annual Report of the Director PanAmerican Sanitary Bureau, WHO for 1956* (see in references under Soper, 1956).

dant in the New World, especially in the southern United States and the West Indies (Mattingly, *in lit.*).

Besides the original locality in south Queensland, var. *queenslandensis* has been recorded from many localities on the east coast of Africa and neighbouring countries. Mattingly (1953, p. 60) gives the following list of countries in this area from which the form has been recorded: the Sudan, Abyssinia, Eritrea, British Somaliland, Kenya, Tanganyika, Zanzibar, Zululand, Natal. The same author records this form from the Seychelles, Mauritius, the Aldebara Islands. It has also been recorded from Egypt (see section (*a*)). Lewis (1953) marks a limited coastal area in the Sudan, inland to which the species is darker. In Arabia it has been recorded from localities in the Eastern Aden Protectorate and in Yemen, Aden, Kedda, Bahrein Island in the Persian Gulf (Mattingly and Knight, 1956). Mattingly (1953) also records the form from Kameran Island, Amd and Tarim, the last-mentioned locality being in the region of the south Arabian desert about 200 miles inland.

Var. *queenslandensis* is apparently rare in the Indo-Malayan region, but has been recorded from Bangalore, South India (Barraud, 1934); from Java (Brug); and from Singapore (Stanton). It is a recognised Mediterranean form and is recorded from Algeria (Edwards, 1911); and from Malaga, Spain (Dyar and Knab, 1910). It is recorded from the West Coast of Africa: Lagos (Summers-Connal, 1927); Principe Island (Gulf of Guinea), Lokoja and Degemma (Mattingly, 1953); and from Bonny (Theobald, 1910). It is stated by Muspratt (1956) not to be known from southern Africa except from Durban and Zululand. It is apparently very abundant in the New World as already noted.

The area of distribution of s.sp. *formosus* is given by Mattingly as confined to Africa south of the Sahara. It has been recorded from the forest or bush apart from man and breeding in natural breeding places from Kenya (Garnham, Harper and Highton, 1946; Van Someren, Teesdale and Furlong, 1955) and from Uganda (Haddow, 1945; Haddow, Gillett and Highton, 1946; Haddow and Mahaffy, 1956). Muspratt (1956) records *A. aegypti* as occurring in the Transvaal away from man in the bush and as commonly breeding in tree-holes, but does not specify any particular form. *A. aegypti* appears to be unrecorded as a wild or forest form in the New World.

(*e*) PALAEONTOLOGICAL RECORD

Though *A. aegypti* has not been recorded among fossil mosquitoes it may be of interest, as possibly throwing some light on the past history of this species, to note what has been recorded regarding fossil Culicidae. A study of what is known of these forms has been given by Edwards (1923). Short accounts are also given by Martini (1929) and by Marshall (1938). One of the most productive sources of fossils of Nematocera including mosquitoes is the famous 'insect bed' of Oligocene age at Gurnet Bay, Isle of Wight, from which a large number of the specimens in the British Museum have been obtained.

True Culicidae are known from the Eocene, one possibly a *Culex*. Among genera recognised by Edwards from the Oligocene are: *Aëdes*, *Culex*, *Taeniorrhynchus*; possibly *Theobaldia*, *Chaoborus* and *Dixa*. Commenting on his studies Edwards states that probably all the main divisions of the Culicinae existed in Mid-Tertiary times as they do today with almost identical characters.

SYSTEMATIC

The following are fossil species recorded as mosquitoes in the literature with remarks by Edwards as to their probable nature:

JURASSIC (Purbeck of Dorset or Wiltshire). No certain Culicidae have been recorded. *Culex fossilis* Brody, 1845, is certainly a Tipulid. *Rhyphus priscus* Brody, 1845, is probably not a Rhyphus and may be a Culicid.

EOCENE (Green River, U.S.)
Culex damnatorum Scudder, 1890. A true Culicid and may be a *Culex*.
Culex winchesteri Cockerell, 1920. Possibly a *Culex*.
Culex proavitus Scudder, 1890. A *Phlebotomus* or other small Psychodid.

LOWER OLIGOCENE (Baltic amber)
Mochlonyx sepultus Meunier, 1902. Correctly given as *Mochlonyx*.
Corethra ciliata Meunier, 1904. A true *Chaoborus*.
Dixa succinia Meunier, 1906. Differs only in small details from recent species.

MIDDLE OLIGOCENE (Gurnet Bay, Isle of Wight)
Aëdes protolepis Cockerell. Forty specimens in the British Museum.
Aëdes petrifactellus Cockerell. Male of the above.
Aëdes sp. Cockerell.
Culex protorhinus Cockerell. Doubtful if a *Culex*. Might be *Theobaldia*.
Culex vectensis Edwards, 1923. Probably *Culex* in the modern sense.
Taeniorrhynchus cockerelli Edwards, 1923. *Taeniorrhynchus*.

UPPER OLIGOCENE (Aix-en-Provence)
Culicites depereti Meunier, 1916. Typical culicine mosquito.
Culicites tertiarius Heyden, 1862. A *Chaoborus* (*Corethra*).
Culex ceyx Heyden, 1870. Possibly an *Aëdes*.
Corethra sp. Hope, 1847. (From West Germany.)
Eriopterites tertiaria Meunier, 1916. A species of *Dixa*.

MIOCENE. No culicines recorded.

QUARTERNARY
Culex ciliaris Linn. Bloch. 1776. No locality given.
Culex flavus Gistl, 1931. Brazil.
Culex loewii Giebel, 1856. No locality given.

That mosquitoes have retained generic characters since Oligocene times would seem to make any attempt to trace the history of our species with the data available purely a matter of conjecture. The only hope would seem to lie in tracing the history of a fauna to which it appeared to belong. If in truth the original home of the species was the Ethiopian Region, then assuming that *A. aegypti* or its predecessor species originated as a member of the Pliocene fauna of which the Ethiopian fauna is the modern representative it might be possible that long before man travelled about the earth this species was present in Europe and South-west Asia as part of the Pliocene fauna. This originally extended over Europe and parts of Asia, but eventually became the present-day more restricted fauna of the Ethiopian Region, though relics may have been left in some areas.

For a general account of fossil insects see Handlirsh (1908). For an account of tertiary land and faunal changes see Suess (1924); also Zittel (1925); Reinig (1937); Zeuner (1945, 1946). For information on zoogeographical realms see Beddard (1895); Lydekker (1896); Heilprin (1907); Newbigin (1913); Lindtrop (1929); De Beauford (1951). Some observations relating to the distributional history of subfamily Anophelini are given by the present author (Christophers, 1920, 1921, 1933), and some remarks on the evolution of mosquitoes in relation to mammals are given by Ross (1951). A discussion on problems of distribution of plants and animals in Africa by a number of authors including a paper relating to mosquitoes by Mattingly is here given in the references under the name of Milne Redhead *et al.* (1954).

REFERENCES

(*a–b*) SYNONYMY, SYSTEMATIC POSITION AND IDENTIFICATION

AUSTEN, E. E. (1912). Nomenclature of *S. fasciata*. *Yell. Fev. Bur. Bull.* **2**, 3.

BAISAS, F. E. (1938). Notes on Philippine mosquitoes. *Mon. Bull. Bur. Hlth P. I.* **18**, 175–232.

BARRAUD, P. J. (1934). *Fauna of British India. Diptera.* Vol. V, *Culicidae*. Taylor and Francis, London.

BLANCHARD, R. (1905). *Les moustiques.* Rudeval, Paris.

BONNE-WEPSTER, J. and BRUG, S. L. (1932). The subgenus *Stegomyia* in Netherland India. *Geneesk. Tijdschr. Ned.-Ind.* Bijblad **2**, 35–119.

DE VILLERS, C. (1789). *C. Linnaei entomologia faunae Suecicae descriptionibus aucta.* Lugduni. Vol. III, pp. 562–6.

DYAR, H. G. (1920). The earliest name for the yellow fever mosquito (Diptera: Culicidae). *Insec. Inscit. Menst.* **8**, 204.

DYAR, H. G. (1928). *The Mosquitoes of the Americas.* Carnegie Institute, Washington.

DYAR, H. G. and KNAB, F. (1910). On the identity of *Culex pallidohirta* (Diptera: Culicidae). *Proc. Ent. Soc. Wash.* **12**, 81.

EDWARDS, F. W. (1921). A revision of the mosquitoes of the Palaearctic Region. *Bull. Ent. Res.* **12**, 325.

EDWARDS, F. W. (1922). Mosquito notes III. *Bull. Ent. Res.* **13**, 82 (*Aëdes woodi*).

EDWARDS, F. W. (1925). Mosquito notes V. *Bull. Ent. Res.* **15**, 262 (*Aëdes subargenteus*).

EDWARDS, F. W. (1932). *Genera insectorum.* Brussels. Fasc. 194. Fam. Culicidae.

EDWARDS, F. W. (1933). Identity of *Culex aegypti*. *Ann. Trop. Med. Parasit.* **27**, 357–60.

EDWARDS, F. W. (1941). *Mosquitoes of the Ethiopian Region.* III, *Culicine Adults and Pupae.* Brit. Mus. (Nat. Hist.), London.

FABRICIUS, J. C. See under references, chapter I, section (*e*).

GOUGH, L. H. (1914). Preliminary notes on Egyptian mosquitoes. *Bull. Ent. Res.* **5**, 133–5.

HASSELQUIST, F. See under references, chapter I, section (*e*).

HOPKINS, G. H. E. (1936). *Mosquitoes of the Ethiopian Region.* Vol. I, *Larval Bionomics of Mosquitoes and Taxonomy of Culicine Larvae.* Brit. Mus. (Nat. Hist.), Ed. 2. London. 1951. With notes by Mattingly.

HOWARD, L. O., DYAR, H. G. and KNAB, F. (1912–17). *Mosquitoes of North and Central America and the West Indies.* Vols. I–IV. Carnegie Institute, Washington.

INGRAM, A. and MACFIE, J. W. S. (1917). The early stages of certain West African mosquitoes. *Bull. Ent. Res.* **8**, 135–54.

INTERNATIONAL RULES OF ZOOLOGICAL NOMENCLATURE (1913). *Proc. 9th Internat. Congr. Zool. held at Monaco in 1913.* The rules are given by Wenyon, 1926 (*q.v.*), and (slightly revised) in *Proc. 10th Internat. Cong. Zool. 1927.* 'Opinions and Declarations' are published by, and on sale at, Office of the Committee, 41 Queen's Gate, London, S.W. 7.

IYENGAR, M. O. T. and MENON, M. A. U. (1955). Mosquitoes of the Maldive Islands. *Bull. Ent. Res.* **46**, 1–9. With appendix by P. F. Mattingly, pp. 9–10.

KIRKPATRICK, T. W. (1925). *The Mosquitoes of Egypt.* Government Press, Cairo.

KNAB, F. (1916). The earliest name of the yellow fever mosquito. *Insec. Inscit. Menst.* **4**, 59–60.

KNIGHT, K. L. and HURLBUT, H. S. (1949). The mosquitoes of Ponape Island, Eastern Carolines. *J. Wash. Acad. Sci.* **39**, 20–34.

LINNAEUS, C. (1762). In Hasselquist's *Reise nach Palestina*, etc., p. 470. The German transl. 1757, edited as regards nomenclature by Linnaeus. J. Christian Koppe, Rostok. See also under references, chapter I, section (*e*).

LUDLOW, C. S. (1911). Philippine mosquitoes. *Psyche*, **18**, 132.

SYSTEMATIC

Marks, E. N. (1954). A review of the *Aëdes scutellaris* subgroup with a study of variation in *Aëdes pseudoscutellaris* (Theobald). *Bull. Brit. Mus.* (*Nat. Hist.*) *Ent.* **3**, no. 10.

Mattingly, P. F. (1952). The sub-genus *Stegomyia* (Diptera: Culicidae) in the Ethiopian Region. Part i. *Bull. Brit. Mus.* (*Nat. Hist.*) *Ent.* **2**, no. 5.

Mattingly, P. F. (1953). The subgenus *Stegomyia* (Diptera: Culicidae) in the Ethiopian Region. Part ii. *Bull. Brit. Mus.* (*Nat. Hist.*) *Ent.* **3**, no. 1.

Mattingly, P. F. (1955). Appendix to Iyengar and Menon, 1955, *q.v.*

Mattingly, P. F. (1957). Genetical aspects of the *Aëdes aegypti* problem. *Ann. Trop. Med. Parasit.* **51**, 392–98 and **52**, 5–17.

Mattingly, P. F. and Bruce-Chwatt, L. J. (1954). Morphology and bionomics of *Aëdes* (*Stegomyia*) *pseudoafricanus* Chwatt (Diptera: Culicidae) with some notes on the distribution of the subgenus *Stegomyia* in Africa. *Ann. Trop. Med. Parasit.* **48**, 183–93.

Mattingly, P. F. and Knight, K. L. (1956). The mosquitoes of Arabia. *Bull. Brit. Mus.* (*Nat. Hist.*) *Ent.* **4**, no. 3.

Meigen, J. W. (1804). *Klassifikation und Beschreibung der Europäischen zweiflügeligen Insekten.* Braunschweig. Vol. i, p. 4.

Meigen, J. W. (1818). *Systematische Beschreibung der bekannten Europäischen zweiflügeligen Insekten.* Aachen. Vol. i, pp. 1–12.

Müller, O. F. (1764). *Fauna insectorum Fridrichdalina.* Hafn et Lipsiae. p. 87.

Muspratt, J. (1956). See under (*c*).

Natvig, L. R. (1948). Contributions to the knowledge of the Danish and Fennoscandian mosquitoes. *Norsk ent. Tidsskr.* Suppl. 1, p. 436.

Pallas, P. S. (1771). *Reisen durch Veisch der Russisch. Reiche.* Petersburg.

Patton, W. S. (1933). Identity of *Culex aegypti* L. *Ann. Trop. Med. Parasit.* **27**, 182–4.

Poiret, l'Abbé (1787). Mémoire sur quelques insectes de Barbarie. *J. Physique*, **30**, 245.

Ross, E. S. and Roberts, H. R. (1943). Mosquito Atlas. Part i. *Amer. Ent. Soc.* Quoted by Iyengar and Menon (1955).

Seidelin, H. (1912). A note on the nomenclature of *Stegomyia fasciata*. *Yell. Fev. Bur. Bull.* no. 11, 365–6.

Theobald, F. V. (1901). *Monograph of the Culicidae or Mosquitoes.* Vol. i, pp. 283, 289, 294.

Theobald, F. V. (1907). *Monograph of the Culicidae or Mosquitoes.* Vol. iv, p. 177.

Theodor, O. (1924). Pupae of some Palestinian Culicides. *Bull. Ent. Res.* **14**, 341–5.

Wenyon, C. M. (1926). *Protozoology.* Baillière, Tindall and Cox, London. Vol. ii, pp. 1336–49.

(*c*) VARIETIES, STRAINS, HYBRIDS, GYNANDROMORPHS AND MUTANTS

Varieties and strains

Arribalzaga, E. L. (1891). Dipterologia Argentina. *Rev. Mus. La Plata* (Culicidae).

Banks, C. S. (1906). A list of Philippine Culicidae with descriptions of some new species. *Philipp. J. Sci.* **1**, 977–1005.

Barraud, P. J. (1934). See under (*a–b*).

Brug, S. L. (1928). Remarks on the paper by Prof. W. H. Hoffmann. *Meded. Volks. Ned.-Ind.* Foreign ed. **17**, 184–5.

Christophers, S. R. (1933). *Fauna of British India. Diptera.* Vol. iv, *Tribe Anophelini.* Taylor and Francis, London.

Dyar, H. G. and Knab, F. (1910). On the identity of *Culex pallidohirta* (Diptera: Culicidae). *Proc. Ent. Soc. Wash.* **12**, 81.

Edwards, F. W. (1911). The African species of *Culex*. *Bull. Ent. Res.* **2**, 241–68.

Edwards, F. W. (1920). Mosquito notes. *Bull. Ent. Res.* **10**, 129–37.

Edwards, F. W. (1941). See under (*a–b*).

Garnham, P. C. C., Harper, J. O. and Highton, R. B. (1946). The mosquitoes of the Kaimosi Forest, Kenya Colony, with special reference to yellow fever. *Bull. Ent. Res.* **36**, 473–96.

Gillett, J. D. (1955). Behaviour differences in two strains of *Aëdes aegypti*. *Nature, Lond.*, **176**, 124–6.

REFERENCES

GILLETT, J. D. and ROSS, R. W. (1955). The laboratory transmission of yellow fever by *Aëdes (Stegomyia) aegypti* (Linnaeus) from Malaya. *Ann. Trop. Med. Parasit.* **49**, 63–5.

HADDOW, A. J. (1945*a*). On the mosquitoes of Bwamba County, Uganda. I. Description of Bwamba with special reference to the influence of mosquito ecology. *Proc. Zool. Soc. Lond.* **115**, 1–13.

HADDOW, A. J. (1945*b*). On the mosquitoes of Bwamba County, Uganda. II. Biting activity with special reference to the influence of microclimate. *Bull. Ent. Res.* **36**, 33–73.

HADDOW, A. J., GILLETT, J. D. and HIGHTON, R. B. (1947). On the mosquitoes of Bwamba County, Uganda. V. Vertical distribution and biting cycle of mosquitoes in rain forest. *Bull. Ent. Res.* **37**, 301–30.

HADDOW, A. J. and MAHAFFY, A. F. (1949). On the mosquitoes of Bwamba County, Uganda. VII. Intensive catching on tree platforms with further observations on *Aëdes (Stegomyia) africanus* Theobald. *Bull. Ent. Res.* **40**, 169–78.

HILL, G. F. (1921). Notes on some unusual breeding places of *Stegomyia fasciata* in Australia. *Ann. Trop. Med. Parasit.* **15**, 91–2.

HINDLE, E. (1929). The experimental study of yellow fever. *Trans. R. Soc. Trop. Med. Hyg.* **22**, 405–30.

KARTMAN, L. (1953). Factors influencing infection of the mosquito with *Dirofilaria immitis* (Leidy, 1956). *Exp. Parasit.* **2**, 27–78.

LEGENDRE, J. (1927). Races de *Stegomyia fasciata* et fièvre jaune. *C.R. Acad. Sci., Paris,* **185**, 1224–6.

LEWIS, D. J. (1947). Mosquitoes in relation to yellow fever in the Sudan. *Bull. Ent. Res.* **37**, 543–66.

MACDONALD, W. W. (1956). *Aëdes aegypti* in Malaya. I. Distribution and dispersal. *Ann. Trop. Med. Parasit.* **50**, 385–98.

MATHIS, M. (1934). Biologie comparée en conditions expérimentales de quatre souches du moustique de la fièvre jaune. *C.R. Soc. Biol.* **117**, 878–80.

MATTINGLY, P. F. (1952, 1953). See under (*a–b*).

MATTINGLY, P. F. (1956). Species hybrids in mosquitoes. *Trans. R. Ent. Soc. Lond.* **108**, 21–36.

MATTINGLY, P. F. (1957). See under (*a–b*).

MATTINGLY, P. F. and KNIGHT, K. L. (1956). See under (*a–b*).

MUSPRATT, J. (1956). The *Stegomyia* mosquitoes of South Africa and some neighbouring territories. *Mem. Ent. Soc. S. Afr.* no. 4.

REID, J. A. (1954). A preliminary *Aëdes aegypti* survey. *Med. J. Malaya,* **9**, 161–8.

ROBINEAU-DESVOIDY, J. B. (1827). Essai sur la tribu des culicides. *Mém. Soc. Hist. Nat. Paris,* **3**, 390–416.

ROUBAUD, E., COLAS-BELCOUR, J., TOUMANOFF, C. and TREILLARD, M. (1936). Recherches sur la transmission de *Dirofilaria immitis* Leidy. *Bull. Soc. Path. Exot.* **29**, 1111–20.

SHIDRAWI, G. R. (1955). Hybridization of *Aëdes aegypti* and *Aëdes aegypti* var. *queenslandensis*. *Trans. R. Soc. Trop. Med. Hyg.* **49**, 196.

SKUSE, F. A. A. (1888). Diptera of Australia. *Proc. Linn. Soc. N.S.W.* (2), **3**, 1740.

STANTON, A. T. (1920). Mosquitoes of the Far Eastern Ports. *Bull. Ent. Res.* **10**, 340.

SUMMERS CONNAL, S. L. M. (1927). On the variations occurring in *Aëdes argenteus* Poiret in Lagos, Nigeria. *Bull. Ent. Res.* **18**, 5–11.

TEESDALE, C. (1955). Studies on the bionomics of *Aëdes aegypti* L. in its natural habitats in a coastal region of Kenya. *Bull. Ent. Res.* **40**, 711–42.

THEOBALD, F. V. (1901–10). *Monograph of the Culicidae*: 1901, **1**, 297 (var. *luciensis*; var. *queenslandensis*); 1903, **3**, 144; 1910, **5**, 159.

VAN SOMEREN, E. C. C., TEESDALE, C. and FURLONG, M. (1955). The mosquitoes of the Kenya coast: records of occurrence, behaviour and habitat. *Bull. Ent. Res.* **46**, 463–93.

VAN SOMEREN, G. R. C. (1943). Notes on the mosquitoes of British Somaliland. *Bull. Ent. Res.* **34**, 323–8.

WALKER, F. (1848). *List of Species in the British Museum, London (formosus).*

SYSTEMATIC

Hybrids

BARR, A. R. and KARTMAN, L. (1951). Biometrical notes on the hybridization of *Culex pipiens* L. and *C. quinquefasciatus* Say. *J. Parasit.* **37**, 419–20.

BONNET, D. D. (1950). The hybridization of *Aëdes aegypti* and *Aëdes albopictus* in Hawaii. *Proc. Hawaii Ent. Soc.* **14**, 35–9.

CALLOT, J. (1947). Études sur quelques souches de *Culex pipiens* (sensu lato) et sur leurs hybrides. *Ann. Parasit. Hum. Comp.* **22**, 380–93.

CALLOT, J. (1954). Le rapport trompe/palpes des biotypes du complexe *Culex pipiens* et leurs hybrides. *Ann. Parasit. Hum. Comp.* **29**, 131–4.

DE BUCK, A. (1942). Kreuzungsversuche mit *Stegomyia fasciata* Fabricius und *S. albopictus* Skuse. *Z. angew. Ent.* **29**, 309–12.

DOWNS, W. G. and BAKER, H. H. (1949). Experiments in crossing *Aëdes* (*Stegomyia*) *aegypti* L. and *Aëdes* (*Stegomyia*) *albopictus* Skuse. *Science*, **109**, 200–1.

HOANG-TICH-TRY (1939). Essai de croisement de *Stegomyia albopicta* et *S. fasciata* en espèce restreint. *Bull. Soc. Path. Exot.* **32**, 511–13.

KARTMAN, L. (1953). Factors influencing infection of the mosquito with *Dirofilaria immitis* (Leidy, 1856). *Exp. Parasitol.* **2**, 27–78.

KITZMILLER, J. B. (1953). Mosquito genetics and cytogenetics. *Rev. Bras. Malariol.* **5**, 285–359.

KNIGHT, K. L. (1953). Hybridization experiments with *Culex pipiens* and *C. quinquefasciatus* (Diptera: Culicidae). *Mosquito News*, **13**, 110–15.

KNIGHT, K. L. and ABDEL MALEK, A. A. (1951). A morphological and biological study of *Culex pipiens* in the Cairo area of Egypt. *Bull. Soc. Fouad Ier. Ent.* **35**, 175–85.

LAVEN, H. (1951). Crossing experiments with *Culex* strains. *Evolution*, **5**, 370–5.

LAVEN, H. (1953). Reziprok unterschiedliche Kreuzbarkeit von Stechmücken (Culicidae) und ihre Deutung als plasmatische Vererbung. *Z. indukt. Abstamm.- u. VererbLehre*, **85**, 118–36.

LAVEN, H. (1955). Erbliche intersexualität bei *Culex pipiens*. *Naturwissenschaften*, **42**, 517.

LAVEN, H. (1956*a*). Cytoplasmic inheritance in *Culex*. *Nature, Lond.*, **177**, 141–2.

LAVEN, H. (1956*b*). X-ray induced mutations in mosquitoes. *Proc. R. Ent. Soc. Lond.* (A), **31**, 17–19.

LAVEN, H. and KITZMILLER, J. B. (1954). Kreuzungsversuche zwischen europäischen und amerikanischen Formen des *Culex-pipiens*-complexes. *Z. Tropenmed. u. Parasit.* **5**, 317–23.

MACGILCHRIST, A. C. (1913). *Stegomyia* survey, Port of Calcutta. *Proc. 3rd Meet. Gen. Mal. Comm., Madras, 1912*, pp. 193–6.

MARSHALL, J. F. (1938). *The British Mosquitoes*. London.

MARSHALL, J. F. and STALEY, I. (1937). Some notes regarding the morphological and biological differentiation of *Culex pipiens* Linnaeus and *Culex molestus* Forskal (Diptera: Culicidae). *Proc. R. Ent. Soc. Lond.* (A), **12**, 17–26.

MATTINGLY, P. F. (1956). Species hybrids in mosquitoes. *Trans. R. Ent. Soc. Lond.* **108**, part 2, 21–36.

PERRY, W. J. (1950). Biological and cross-breeding studies on *Aëdes hebrideus* and *Aëdes pernotatus*. *Ann. Ent. Soc. Amer.* **43**, 123–36.

ROZEBOOM, L. E. (1954*a*). The genetic relationships of *Aëdes pseudoscutellaris* Theobald and *A. polynesiensis* Marks (Diptera: Culicidae). *Amer. J. Hyg.* **60**, 117–34.

ROZEBOOM, L. E. (1954*b*). Hybridization among mosquitoes and its possible relation to the problem of insecticide resistance. *J. Econ. Ent.* **47**, 383–7.

SMITH-WHITE, S. (1950). A note on non-reciprocal fertility in matings between subspecies of mosquitoes. *Proc. Linn. Soc. N.S.W.* **75**, 279–81.

TATE, P. and VINCENT, M. (1936). The biology of autogenous and anautogenous races of *Culex pipiens* L. (Diptera: Culicidae). *Parasitology*, **28**, 115–45.

TOUMANOFF, C. (1937). Essais préliminaires d'intercroisement de *St. albopicta* Skuse avec *St. argenteus* Poiret (*S. fasciata* Theobald). *Bull. Soc. Méd. Chir. Indochine*, **15**, 964–70.

TOUMANOFF, C. (1938). Nouveaux faits au sujet de l'intercroisement de *St. albopicta* Skuse avec *St. argentea s. fasciata* Theob. *Rev. Méd. Franc. Extr. Orient.* **17**, 365–8.

REFERENCES

TOUMANOFF, C. (1939). Les races géographiques de *St. fasciata* et *St. albopicta* et leur intercroisement. *Bull. Soc. Path. Exot.* **32**, 505–9.

TOUMANOFF, C. (1950). L'intercroisement de l'*Aëdes* (*Stegomyia*) *aegypti* L. et *Aëdes* (*Stegomyia*) *albopictus* Skuse. Observations sur la mortalité des générations hybrides F_1 et F_2 de ces insectes. *Bull. Soc. Path. Exot.* **43**, 234–40.

WOODHILL, A. R. (1949). A note on experimental crossing of *Aëdes* (*Stegomyia*) *scutellaris scutellaris* Walker and *Aëdes* (*Stegomyia*) *scutellaris katherinensis* Woodhill (Diptera: Culicidae). *Proc. Linn. Soc. N.S.W.* **74**, 224–6.

WOODHILL, A. R. (1950). Further notes on experimental crossing within the *Aëdes scutellaris* group of species (Diptera: Culicidae). *Proc. Linn. Soc. N.S.W.* **75**, 251–3.

WOODHILL, A. R. (1954). Experimental crossing of *Aëdes* (*Stegomyia*) *pseudoscutellaris* Theobald and *Aëdes* (*Stegomyia*) *polynesiensis* Marks (Diptera: Culicidae). *Proc. Linn. Soc. N.S.W.* **79**, 19–20.

Gynandromorphs and mutants

BATES, M. (1949). *The Natural History of Mosquitoes.* Macmillan and Co., New York.

BEDFORD, G. A. H. (1914). A curious mosquito. *Trans. R. Soc. S. Afr.* **4**, 143.

CARPENTER, S. J. (1948). Gynandromorphism in *Aëdes canadensis*. *J. Econ. Ent.* **41**, 522–3.

CLASSEY, E. W. (1942). Gynandromorphism in *Theobaldia annulata* Schrank (Diptera: Culicidae). *Entomologist*, **75**, 181.

DER BRELJE, R. VON (1923). Ein Fall von Zwitterbildung bei *Aëdes meigenensis*. *Arch. mikr. Anat.* **100**, 317–43.

EDWARDS, F. W. (1917). Notes on Culicidae with descriptions of new species. *Bull. Ent. Res.* **7**, 201–29.

FELT, E. P. (1904). Mosquitoes or Culicidae of New York State. *Bull. N.Y. State Mus.* **79**, 241–400.

FELT, E. P. (1905). Studies in Culicidae. *Bull. N.Y. State Mus.* **97**, 442–97.

GILCHRIST, E. M. and HALDANE, J. B. S. (1947). Sex linkage and sex determination in a mosquito, *Culex molestus*. *Hereditas*, **33**, 175–90.

GRATZ, M. G. (1954). A gynandromorph of *Culex pipiens molestus* (Forsk.). *Mosquito News*, **14**, 22–3.

HOWARD, L. O., DYAR, H. G. and KNAB, F. (1912). See under references, chapter I, section (*f*), p. 109.

KITZMILLER, J. B. (1953). See under *Hybrids*.

KOMP, W. H. W. and BATES, M. (1948). Notes on two mosquito gynandromorphs from Colombia (Diptera: Culicidae). *Proc. Ent. Soc. Wash.* **50**, 204–6.

LENGERKEN, J. (1928). Über die Entstehung bilateral-symmetrischer Insekten-gynander aus verschmolzenen Eiern. *Biol. Zbl.* **48**, 475–509.

MARSHALL, J. F. (1938). See under *Hybrids*.

MARTINI, E. (1930). Culicidae. In Lindner's *Die Fliegen der Palaearktischen Region.* Stuttgart. Parts 11 and 12, p. 248.

MIDDLEKAUFF, W. W. (1944). Gynandromorphism in recently collected mosquitoes. *J. Econ. Ent.* **37**, 297.

MORGAN, T. H. and BRIDGES, C. B. (1919). *The Origin of Gynandromorphs.* Publications Carnegie Institute, no. 278. Washington.

MUSPRATT, J. (1951). A gynandromorph of a predatory mosquito. *J. Ent. Soc. S. Afr.* **14**, 24–5.

RINGS, R. W. (1946). Gynandromorphism in *Culex nigripalpis*. *J. Econ. Ent.* **39**, 415.

ROTH, L. W. (1948). Mosquito gynandromorphs. *Mosquito News*, **8**, 168–74.

ROTH, L. W. and WILLIS, E. R. (1952). Notes on three gynandromorphs of *Aëdes aegypti*. *Proc. Ent. Soc. Wash.* **54**, 189–93.

SHUTE, P. G. (1926). Intersexual forms of *Ochlerotatus punctor* Kirby var. *meigenensis*. *Entomologist*, **59**, 12–13.

SMYLY, W. J. P. (1942). A gynandromorph of *Aëdes aegypti* L. (*Stegomyia fasciata*). Diptera. *Proc. R. Ent. Soc. Lond.* **92**, 111–12.

WARREN, M. and HILL, S. O. (1947). Gynandromorphism in mosquitoes. *J. Econ. Ent.* **40**, 139.
WEYER, F. (1938). Ein Zwitter von *Culex pipiens*. *Zool. Anz.* **123**, 184–92.
WIGGLESWORTH, V. B. (1942). Principles of insect physiology. Ed. 2. Methuen and Co., London. (See also subsequent editions.)

(*d*) GEOGRAPHICAL DISTRIBUTION

(1) ACHUNDOW, J. (1932). Zur Frage der Biologie und das Verkommen von *Stegomyia fasciata* in Baku (Aserbeidschan U.S.S.R.). *Arch. Schiffs.- u. Tropen-Hyg.* **36**, 31–3.
(2) AFRIDI, M. K. (1939). *Aëdes* survey of Hog Island. *J. Malar. Inst. India*, **2**, 113–14.
(3) APFELBECK, V. (1925). Recherches et observations sur les arthropodes pathogènes de l'homme et des animaux. *Edit. Inspect. Minist. Santé Publ. Sarajevo*, no. 17: abstract in *Rev. Appl. Ent.* **16**, 180.
(4) ARAGAO, DE B. H. (1935). Mosquitoes and yellow fever virus. *Mem. Inst. Osw. Cruz*, **34**, 565–81.
(5) BACKHOUSE, T. C. and HEYDON, G. A. M. (1950). Filariasis in Melanesia: observations at Rabaul relating to incidence and vectors. *Trans. R. Soc. Trop. Med. Hyg.* **44**, 291–300.
(6) BALFOUR, A. (1928). Health lessons from Bermuda. *Brit. Med. J.* **1**, 447–8.
(7) BAQUE, B. and KIEFFER, J. J. (1923). Existence de *Stegomyia fasciata* au Sahara à 500 kilomètres de la côte Méditerranéenne. *Arch. Inst. Pasteur Afr. Nord*, **3**, 169–71.
(8) BARRAUD, P. J. (1928). The distribution of *Stegomyia fasciata* in India with remarks on dengue and yellow fever. *Ind. J. Med. Res.* **16**, 377–88.
(9) BEQUAERT, J. (1947). *Aëdes aegypti* (Linnaeus), the yellow fever mosquito in Arizona. *Bull. Brooklyn Ent. Soc.* **41**, 157.
(10) BOHART, R. M. and INGRAM, R. L. (1946). Mosquitoes of Okinawa and islands in the Central Pacific. *Nav. Med. Bull. Wash.* p. 1055. Wash., D.C.
(11) BONNET, D. D. (1947). The distribution of mosquitoes breeding by type of container in Honolulu, T.H. *Proc. Hawaii Ent. Soc.* **13**, 43–9.
(12) BRUG, S. L. (1926). The geographical distribution of mosquitoes in the Malayan Archipelago. *Meded. Volks. Ned.-Ind.* (foreign ed.), no. 4, pp. 471–86.
(13) BRUMPT, E. (1936). *Précis de Parasitologie*. Masson et Cie, Paris.
(14) BUXTON, P. A. and HOPKINS, G. H. E. (1927). Researches in Polynesia and Melanesia. Parts I–IV. *Mem. Lond. Sch. Hyg. Trop. Med.* Mem. no. 1.
(15) BYRD, E. E., ST AMANT and BROMBERG, L. (1945). Studies in filariasis in the Samoan area. *Nav. Med. Bull., Wash.*, **44**, 1–20.
(16) CHANDLER, A. C. (1945). Factors influencing the uneven distribution of *Aëdes aegypti* in Texas cities. *Amer. J. Trop. Med.* **25**, 145–9.
(17) CHARRIER, H. (1924). Le *Stegomyia fasciata* dans la région de Tanger (Maroc). *Bull. Soc. Path. Exot.* **17**, 137–42.
(18) CHRISTOPHERS, S. R. (1929). Note on a collection of Anopheline and Culicine mosquitoes from Madeira and the Canary Islands. *Ind. J. Med. Res.* **17**, 518–30.
(19) CLELAND, J. B. (1928). Dengue fever in Australia. *Rev. Prat. Mal. Pays Chauds*, **8**, 509–14.
(20) COSSIO, V. (1931). Observaciones sobre al *Aëdes aegypti* (*Stegomyia*) mosquito de la febbre amarilla en Montevideo. *Bol. Cons. Nac. Hyg. Publ.* **23**, 1664. Abstract in *Bull. Inst. Pasteur*, **29**, 895.
(21) DAVIS, T. R. A. (1949). Filariasis control in the Cook Islands. *N.Z. Med. J.* **48**, 262–70.
(22) DE MEILLON, B. (1940). Report on the Department of Entomology. *Rep. S. Afr. Inst. Med. Res. for 1939*, pp. 30–7. Abstract in *Rev. App. Ent.* **28**, 211.
(23) DUNN, L. H. (1923). Prevalence of the yellow fever mosquito *Aëdes calopus* in the Southern parts of Peru. *Amer. J. Trop. Med.* **3**, 1–8.
(24) DYAR, H. G. (1928). *The Mosquitoes of the Americas*. Carnegie Institute, Washington.
(25) EDWARDS, F. W. (1941). *Mosquitoes of the Ethiopian Region*. Vol. III, *Culicine Adults and Pupae*. Brit. Mus. (Nat. Hist.), London.
(26) EKBLOM, T. (1929). Some observations on *Stegomyia fasciata* during a visit to Greece in the autumn of 1928. *Acta Med. Scand.* **70**, 505–18.

REFERENCES

(27) Elvira, J. (1930). Nota a cerca de los Culicidos encontrados en la cuenca del Ebro. *Med. Paisos Calidos*, **3**, 63.

(28) Farner, D. S., Dicks, R. J., Sweet, G., Isenhour, L. and Hsias, T. Y. (1946). Distribution of mosquitoes of medical importance in the Pacific area. *Nav. Med. Bull., Wash.*, p. 983. Wash., D.C.

(29) Faust, E. C. (1926). A preliminary check list of the mosquitoes of the Sini-Japanese areas. *China Med. J.* **40**, 142–3.

(30) Feng, Lan-Chou (1933). Household mosquitoes and human filariasis in Amoy. *China Med. J.* **47**, 168–78.

(31) Ferguson, E. W. (1923). On the distribution of insects capable of carrying disease in Eastern Australia. *Med. J. Aust.* **2**, 339.

(32) Ferguson, E. W. (1926). On the distribution of insects capable of carrying disease in Eastern Australia. *Proc. Pan-Pacific Sci. Congr. Australia 1923*, **2**, 1477–86.

(33) Ferguson, E. W. (1928). Dengue fever: the 1925–6 outbreak in New South Wales. *Rep. Direct. Publ. Hlth N.S.W. 1926*, pp. 154–64.

(34) Garnham, P. C. C., Harper, G. A. and Highton, R. B. (1946). The mosquitoes of the Kaimosi forest, Kenya Colony, with special reference to yellow fever. *Bull. Ent. Res.* **36**, 473–96.

(35) Gil Collado, J. (1930). Datos actuales sobre la distribucion geographica de los Culicidos Espanioles. *Eos*, **6**, 329–47.

(36) Goldberger, J. (1907). Yellow fever. Etiology, symptoms and diagnosis. *Yell. Fev. Inst. Bull.* no. 16.

(37) Good, N. E. (1945). A list of mosquitoes of the district of Columbia. *Proc. Ent. Soc. Wash.* **47**, 168–79.

(38) Harper, J. O. (1947). A mosquito survey of Mahé, Seychelles. *E. Afr. Med. J.* **24**, 25–9.

(39) Hoffman, F. L. (1916). *The Sanitary Progress and Sanitary Statistics of Hawaii.* Prudential Press, Newark, N.Y.

(40) Howard, L. O. (1905). Concerning the geographic distribution of the yellow fever mosquito. *Suppl. to Publ. Hlth Rep.* **18**, no. 46, ed. 2.

(41) Korovitzkyi, L. K. and Artemenko, V. D. (1933). Zur Biologie des *Aëdes aegypti*. *Mag. Parasit., Lening.* **2**, 400–6. Abstract in *Rev. Appl. Ent.* **22**, 78.

(42) Kumm, H. W. (1931). The geographical distribution of the yellow fever vectors. *Amer. J. Hyg. Monogr. Ser.* no. 12, pp. 5–45.

(43) Lever, R. J. A. W. (1946). Entomological notes. *Agric. J. Fiji*, **17**, 9–15.

(44) Lewis, D. J. (1947). Mosquitoes in relation to yellow fever in the Sudan. *Bull. Ent. Res.* **37**, 543–66.

(45) McCarthy, D. D. and Wilson, D. B. (1948). Dengue in East African Command. Incidence in relation to *Aëdes* prevalence and some clinical features. *Trans. R. Soc. Trop. Med. Hyg.* **42**, 83–8.

(46) McKenzie, A. (1925). Observations on filariasis, yaws and intestinal helminthic infections in the Cook Islands with notes on the breeding habits of *Stegomyia scutellaris*. *Trans. R. Soc. Trop. Med. Hyg.* **19**, 138–49.

(47) Macquart, J. (1839). In Webb and Berthelot: *Histoire naturelle des Canaries*, p. 99.

(48) Mara, L. (1946). Considerations sul rinvenimento dell' *Aëdes aegypti* L. (Diptera: Aëdinae) ad altitudini d'accazione e brevi note sulla fauna del M. Bizen (Eritrea A.O.). *Boll. Soc. Ital. Med.* (*Sez. Eritrea*), **5**, 189–98. Abstract in *Rev. Appl. Ent.* **30**, 40.

(49) Mattingly, P. F. (1952, 1953, 1957). See under (*a–b*).

(50) Miller, D. (1920). *Report on the Mosquito Investigations Carried out in the North Auckland Peninsula of New Zealand during the Summer of 1918–19.* New Zealand Dept. Hlth. Publ. 3.

(51) Newton, W. L., Wright, W. H. and Pratt, I. (1945). Experiments to determine potential mosquito vectors of *Wucheria bancrofti* in the continental United States. *Amer. J. Trop. Med.* **25**, 253–61.

(52) Oman, P. W. and Christensen, L. D. (1947). Malaria and other insect-borne diseases in the South Pacific Campaign 1942–1945. *Amer. J. Trop. Med.* **27**, 91–117.
(53) Perry, W. J. (1948). The dengue vector in New Caledonia, the New Hebrides and the Solomon Islands. *Amer. J. Trop. Med.* **28**, 253–9.
(54) Perry, W. J. (1949). The mosquitoes and mosquito-borne diseases of the Treasury Islands (British Solomon Islands). *Amer. J. Trop. Med.* **29**, 747–58.
(55) Perry, W. J. (1950). The mosquitoes and mosquito-borne diseases in New Caledonia, an historic account, 1888–1946. *Amer. J. Trop. Med.* **30**, 103–14.
(56) Rosen, L., Reeves, W. C. and Aarons, T. (1948). *Aëdes aegypti* in Wake Island. *Proc. Hawaii Ent. Soc.* **13**, 255–6.
(57) Rozeboom, L. E. (1941). The over-wintering of *Aëdes aegypti* L. in Stillwater, Oklahoma. *Proc. Okla. Acad. Sci.* **19**, 81–2.
(58) Roukhadzé, N. (1926). L'hibernation des *Anopheles* et des *Stegomyias* en Abasie (Georgie Maritime). *Bull. Soc. Path. Exot.* **19**, 480–7.
(59) Seguy, E. (1923). *Histoire naturelle des moustiques de France.* Paul le Chevalier, Paris.
(60) Senior White, R. (1920). A survey of the Culicidae of a rubber estate. *Ind. J. Med. Res.* **8**, 304–25.
(61) Taylor, F. H. (1942). Contributions to a knowledge of Australian Culicidae, no. v. *Proc. Linn. Soc. N.S.W.* **67**, 277–8.
(62) Theobald, F. V. (1901–10). *Monograph of the Culicidae of the World.* Vols. i–v. Brit. Mus. (Nat. Hist.), London.
(63) Viosca, P. (1924). A bionomical study of the mosquitoes of New Orleans and S.E. Louisiana. *Dep. Bd Hlth Orleans Parish, 1924*, 35–52.
(64) Woodhill, A. R. (1949). Observations on the comparative survival of various stages of *Aëdes* (*Stegomyia*) *aegypti* (Linnaeus) at varying temperatures and humidities. *Proc. Linn. Soc. N.S.W.* **73**, 413–18.
(65) Yamada, S. (1925). A revision of the adult anopheline mosquitoes of Japan. *Sci. Rep. Gov. Inst. Infect. Dis.* **3**, 215–41.

Additional references to section (d)

Barraud, P. J. (1934). See under (*a–b*).
Bonne-Wepster, J. and Brug, S. L. (1932). See under (*a–b*).
Dyar, H. G. and Knab, F. (1910). See under (*c*), *Varieties and strains.*
Edwards, F. W. (1911). See under (*c*), *Varieties and strains.*
Haddow, A. J. (1945). See under (*c*), *Varieties and strains.*
Haddow, A. J., Gillett, J. D. and Highton, R. B. (1947). See under (*c*), *Varieties and strains.*
Haddow, A. J. and Mahaffy, A. F. (1949). See under (*c*), *Varieties and strains.*
Lewis, D. J. (1953). The *Stegomyia* mosquitoes of the Anglo-Egyptian Sudan. *Ann. Trop. Med. Parasit.* **47**, 51–61.
Macdonald, W. W. (1956). *Aëdes aegypti* in Malaya. *Ann. Trop. Med. Parasit.* **50**, 385–98.
Marks, E. N. (1954). See under (*a–b*).
Mattingly, P. F. and Knight, K. L. (1956). See under (*a–b*).
Muspratt, J. (1956). See under (*c*), *Varieties and strains.*
Reid, J. A. (1954). See under (*c*), *Varieties and strains.*
Severo, O. P. (1956). Eradication of the *Aëdes aegypti* mosquito from the Americas. *Mosquito News*, **16**, 115–21.
Soper, F. L. (1956). *Annual Report of the Director Pan-American Sanitary Bureau, WHO* Washington, D.C. Map, p. 42.
Soper, F. L., Wilson, D. B., Lima, S. and Antunes, W. Sá. (1943). *The Organisation of Permanent Nation-Wide Anti-*Aëdes aegypti *Measures in Brazil.* Rockefeller Foundation, New York.
Stanton, A. T. (1920). See under (*c*), *Varieties and strains.*
Summers Connal, S. L. M. (1927). See under (*c*), *Varieties and strains.*
van Someren, E. C. C. *et al.* (1955). See under (*c*), *Varieties and strains.*

REFERENCES

(e) PALAEONTOLOGICAL RECORD

BEDDARD, F. E. (1895). *A Text-book of Zoogeography*. Cambridge Univ. Press.

BRODIE, P. B. (1845). *A History of the Fossil Insects in the Secondary Rocks of England*. Van Voorst, London, p. 34.

CHRISTOPHERS, S. R. (1920). A summary of recent observations upon the Anophelini of the Middle East. *Ind. J. Med. Res.* **7**, 710–16.

CHRISTOPHERS, S. R. (1921). The geographical distribution of the Anophelini. *Trans. 4th Congr. F.E.A.T.M.* **1**, 421–30.

CHRISTOPHERS, S. R. (1933). Pp. 357–59. See under (*c*), *Varieties and strains*.

DE BEAUFORD, L. F. (1951). *Zoogeography of the Land and Inland Waters*. London.

EDWARDS, F. W. (1923). Oligocene mosquitoes in the British Museum, with a summary of our present knowledge concerning fossil Culicidae. *Quart. J. Geol. Soc. Lond.* **79**, 139–55.

HANDLIRSCH, A. (1908). *Die fossilen Insekten*. W. Engelmann, Leipzig.

HEILPRIN, A. (1907). *The Geographical and Geological Distribution of Animals*. London.

LINDTROP, G. T. (1929). *Russ. J. Trop. Med.* **7**, 153–191 (in Russian).

LYDEKKER, R. (1896). *A Geographical History of Mammals*. Cambridge Univ. Press.

MARSHALL, J. F. (1938). *The British Mosquitoes*. London.

MARTINI, E. (1929). *Lehrbuch der medizinische Entomologie*. Gustav Fischer, Jena.

MATTINGLY, P. F. (1957). See under (*a–b*).

MILNE REDHEAD, E. *et al.* (1954). Discussion on the problem of distribution of plants and animals in Africa. (Different authors.) *Proc. Linn. Soc. Lond.* **165**, 24–74.

NEWBIGIN, M. L. (1913). *Animal Geography*. Clarendon Press, Oxford.

REINIG, W. F. (1937). *Die Holarktis*. Jena.

ROSS, H. H. (1951). Conflict with *Culex*. *Mosquito News*, **11**, 128–32.

SUESS, E. (1924). *The Face of the Earth*. Oxford Univ. Press. Vol. IV, pp. 637–73.

ZEUNER, F. E. (1945). *The Pleistocene Period, its Climate, Chronology and Faunal Succession*. Ray Society, London.

ZEUNER, F. E. (1946). *Dating the Past*. London.

ZITTEL, A. A. VON (1925). *A Text-book of Palaeontology*. Macmillan and Co., London. Vol. III, pp. 290–310.

III
BIONOMICAL

(*a*) LIFE HISTORY

The bionomics of *Aëdes aegypti* will be studied in later chapters in detail in relation to the different stages of development. It seems desirable, however, at this stage to say something regarding its life cycle as a whole and on the conditions relative to this in nature.

From the *egg*, hatching or *eclosion* gives rise to the *first instar larva*. This is followed by three successive moults or *ecdyses*, leading to the respective stages of *second*, *third* and *fourth instar larva*. A fourth ecdysis, or *pupation*, gives rise to the *pupa*, and a final casting of the pupal skin, *emergence*, results in the appearance of the *imago*, in mosquitoes commonly spoken of as the adult, male or female. To complete the life cycle there is pairing of the sexes, *copulation*, leading to *fertilisation* of the female, followed in due course, if a sufficient blood meal has been obtained, by *oviposition*.

Continued propagation of the species is dependent upon: (1) the obtaining of a blood meal by the female, without which eggs are not formed; (2) the presence of water suitable for larval life; (3) a temperature that will permit the species to exhibit the activities necessary to propagate. For the adult, the degree of activity and duration of life is largely dependent on a suitable humidity, as well as upon shelter and, to judge from experience in the laboratory, upon access to water to drink, for the species soon dies in captivity, even in a relatively moist atmosphere, if it is unable to drink. For the larva there must exist breeding places where the larva can obtain food under conditions for which it is adapted, that is, neither very foul water, in which the species is rarely found, nor open natural water, for which some mosquitoes are adapted,* but not *A. aegypti*. The larva must also find safety from natural enemies to which, with its soft and conspicuous body, it has been shown to be very susceptible, so much so that as noted later a measure in its control has been the use of small larvivorous fish and other enemies including a voracious species of cannibal mosquito larva.

Conditions favourable in all these respects are normally found in the neighbourhood of man and especially in man's domestic arrangements. Only in the forest form as described in a previous chapter is the species usually found living apart from such conditions. And whilst *Anopheles* is very commonly the form found associated with man in his most primitive state living near jungle and swamp, *Aëdes aegypti* is seen as a rule in greatest profusion in the dwellings of man at a higher level of civilisation and very conspicuously of man as a town dweller. Here, with the domestic water pot, the cistern, the anti-formica, the flower vase and shaded rooms with cupboards and hanging equipment and with man to feed upon, *Aëdes aegypti* as usually encountered finds a very suitable *locus*. Also at the dock side and water front of teeming harbours in moist and rainy tropical climates, with miscellaneous collections of water in machinery, country boats and even in the old days

* Such mosquitoes commonly have an especially large pharynx with corresponding large head. See chapter XIII, p. 289.

in ships, whereby it is spread to other ports and harbours, the species obtains conditions optimal for its proliferation. It does not find conditions so suitable where communities are well housed with permanent water supply and good drainage system and tidier house habits.

Spread of the species is ensured by a variety of means: by transport of the adult, or of the larva, or by a method that is not often considered, but may well happen, the transport of eggs characteristically cemented as described later on the sides of pots or other receptacles.

Establishment of the species in a new area would not seem to need any very special requirements other than introduction. But some efforts to introduce the species experimentally in Malaya described by Macdonald (1956) were not so successful as expected, and it is possible that under such conditions dispersion by too great diffusion of individuals may be an adverse influence.

Of climatic conditions it is unlikely that any natural climate would through high temperature alone limit existence of the species. Creac'h (1947) records the species in the Lake Chad area where the temperature from March to May regularly exceeds 40° C. In the hot dry season in the Punjab and Rajputana the air temperature for many hours in the day may be 115° F. (46° C.), but the species is not eliminated. Combined, as it usually is with aridity, such a climate is not favourable, as are moist humid conditions, but within dwellings the species by choice of resting-place for the adult may escape the full effect, for example, by its common habit of resting on porous water chatties or damp places. The larva, too, living in receptacles (often porous) is not exposed to the full temperature effect.

Low temperature may, however, severely limit the activity and even existence of the species as has been shown in regard to its geographical distribution. The actual limiting temperature for activity of the adult appears to be round about 15° C. Charrier (1924), in Morocco, notes that at 18° C. man is freely attacked and mating observed morning and evening. But when the mean daily temperature is about 11–12° C. the species disappears. Cossio (1931) records the species biting at 15° C. at Monte Video. Creac'h, in the Lake Chad area, found numerous adults in the huts when the temperature was 15° C. But again, whilst temperature in the open may be low, conditions in buildings may afford protection. Adrien (1918) notes that in Syria the species still breeds in cold weather in cisterns indoors, attacking especially at night.

It seems doubtful if either the adult or the larva undergoes true hibernation such as is exhibited by adults of *Culex pipiens* or larvae of *Anopheles plumbeus*. Over-wintering through the egg has been described and was discussed when dealing with the geographical distribution. Korovitzkii and Artemenko (1933) note that eggs were viable for six months in a room with a temperature sometimes as low as 16° C., but none survived the winter in the open with temperatures between −4·5 and 17·2° C. Effects of temperature as studied experimentally in the laboratory will be described in due course later.

One important effect of temperature indirectly affecting the species adversely is the slowing up of development to such a degree that time is not given in the warm season for completion of the life cycle, so that the species is prevented from permanently establishing itself. Normally in the tropics the period from hatching to emergence is a week or so. Marchoux, Salimbeni and Simond (1903), speaking of Cuba, say that at 26–27° C. night temperature and 28–31° C. day temperature the period from hatching to adult took nine days at the earliest, but for the most part ten days. In the French Sudan Boulford (1908) for the same period of development gives twelve to fifteen days in the rains and eighteen to twenty in the cold season. If three days be added for maturation of the eggs and three for time of emergence to the first blood meal and three for hatching time from oviposition, this gives a rough life cycle time of about sixteen to thirty days. In contrast to this are the observations of Gutzevich (1931) on *Aëdes aegypti* in nature on the south Black Sea coast (temperature from May to October 21° C. average). The species here undergoes in the summer only three or four generations, the author giving forty to forty-five days for the life cycle. Korovitzkii and Artemenko record for Jugoslavia between May and the end of August (107 days) three generations out of doors and four in buildings, or about thirty-five days for the cycle out of doors. Further lengthening of the period it is evident might prevent the completion of even one life cycle. Aragao (1939), for a plateau region in Brazil, records that eggs take twenty days to hatch and oviposition sixteen to twenty days from time of the first blood meal, giving a life cycle period of thirty-nine to forty-three days or about the same length as at the limit of the species on the Black Sea coast.

Starvation by prolonging the period of larval development may in some cases help to tide the species over periods of adversity, all the more so since the larva of *A. aegypti* displays, as will be described later, a rather remarkable resistance to starvation, continuing to exist and drawing on its food reserves for considerable periods in almost clean water. Nevertheless, undue prolongation of life from such a cause carries its own danger, for the result is apt to give rise to adults too small and weak to propagate the species. The effect of starvation on size raises the question of the normal size of the species in nature. In the laboratory considerable variation in size occurs when the species is bred artificially unless conditions are optimal. In nature in the author's experience the species is usually of about the size as cultured under fairly good conditions in the laboratory. Any larger size than well-bred specimens in the laboratory is unusual. Mathis (1938) notes that in the warm rainy season size and fertility in nature equal, but are not greater than, those in the laboratory.

In nature *A. aegypti*, in marked contrast to other house-frequenting species, is usually most active by day, the reason why in many tropical countries it is so annoying a pest. That it is active by day and less so, if noticeable at all, by night is such common experience that it is commonly described as the day-biting mosquito. Nevertheless, a number of authors have described the species as biting by night and the question how far it is active by night has been a matter of some controversy. It is also important where yellow fever occurs, since night activity may be necessary to consider in prevention. Marchoux, Salimbeni and Simond (1903), in contrast to usual ideas at the time, stated that the female is at first diurnal in biting, but later attacks by night, and Aragao notes that whilst the primiparous female attacks in the day, nocturnal feeding supervenes in females that have laid one batch of eggs. In the laboratory adults that have escaped behave with all the characteristic behaviour shown in the tropics as day-feeding mosquitoes. In the experimental cage, however, they feed apparently much the same, whether by day or night. The point has received additional importance recently since many records have been given of *A. aegypti* being captured feeding at night in dark native huts in Africa. Various points in connection with feeding will be considered later. Meantime, the apparently conflicting evidence as to the feeding habits of the species is probably due to the fact that feeding is not the result of a simple taxis, but involves more than one form of stimulus. Thus in a cage it is warmth of the attractive object that is mainly responsible. But, as described in a later chapter, localisation of a victim at a distance is largely through visual stimuli of a dark, moving object. It is a commonly held belief that mosquitoes find their prey through the sense of smell and evidence regarding this is considered later. It would seem, however, that with *A. aegypti* such a method of locating their victim is at most secondary to other stimuli. Behaviour is by no means necessarily the same, however, in all mosquitoes and in many respects *A. aegypti* differs markedly from some other forms that have been investigated.

A final word may be said regarding length of life in nature. What this may be is very uncertain. In captivity, as shown later, females of the species may be kept alive for a year or more. In nature it is more likely to be a matter of two or three ovipositions, but methods of determining the age of mosquitoes caught in nature are as yet incapable of eliciting more than a very limited amount of information.

(*b*) BREEDING PLACES

In describing the breeding places of *A. aegypti* it will be most convenient to describe first the type of breeding place which the species almost invariably adopts as it is ordinarily encountered in association with man and where natural breeding places play a quite subsidiary part, and to follow this by some reference to occurrence of the species recently described as a wild form remote from man in forest or other uninhabited wild areas.

As seen in association with man the breeding places of *A. aegypti* are almost entirely confined to artificial collections of water: small artificial collections of water (Boyce, 1910); water receptacles of all kinds (Barraud, 1934); artificial containers or things resembling these (Carter, 1924); any receptacle holding fresh water (Hamlyn-Harris, 1927); 92·5 per cent of domestic receptacles at Lagos examined by Graham in 1911; 84·4 per cent of those collected by Macfie and Ingram (1916) in Accra. See also Soper, Wilson, Lima and Antunes (1943) on breeding places as encountered during the campaign against *A. aegypti* in Brazil.

Such breeding places may be in the dwelling itself, for example the earthenware pots (chatties) or calabashes almost universal in native huts for storing water; anti-formicas, flower vases, neglected cups or jugs or other household collections of water in better-class houses. Or another type of breeding place about the house arises from defects or neglected features of the building itself, for example uncovered cisterns (Connor, 1922; Rigollot, 1927), roof gutters (Jones, 1925), cracks in the masonry, traps of drains, or flush tanks or pans of water-closets when temporarily out of use.

Or, especially during rainy periods, breeding places may be receptacles of various kinds about the vicinity of habitations, for example barrels for rain-water, rain-filled empty cans or food tins, flower pots, broken bottles or the hollows of upturned bottles used for decorative purposes, garden tanks or dumps (Thomas, 1910; Horne, 1913; Mhaskar, 1913). In towns breeding may occur in stagnant ornamental fountains, horse troughs, etc. Especially prolific in breeding places are engineering and other dumps, due to rain-water collecting in hollows in iron girders, disused boilers, iron drums, motor-car parts, etc.

The species is pre-eminently associated with shipping and harbours. Here it breeds in water-holding pockets in cranes, in bilge water in boats, unsealed pontoons and in drain traps. In ships, especially those long in tropical waters, larvae used frequently to be found in the containers for wash basins in cabins and in various places in the hold and elsewhere (Lutz and Machado, 1915; Chabrillat, 1934). Breeding is particularly common in stored water in barrels or tanks on country boats, dhows, etc. (Bana, 1936; Lewis, 1947). Water left in unused canoes and boats may give rise to heavy breeding (Dutton, 1903; Macfie and Ingram, 1916).

Wells have been recorded as breeding places: in Bombay (Liston and Akula, 1913); 51 per cent of the wells in Lagos (Dalziel, 1920); the only breeding places in the Sahara (Baque and Kieffer, 1923). Where habitations in areas of heavy rainfall are closely surrounded by vegetation the species may breed in water collected in leaf axils in such situations as *Colocasia* leaves (Carter, 1924; Haddow, 1948); in pineapple leaves or in the axils of banana leaves (Teesdale, 1941; Haddow, 1948). It is recorded breeding in stumps of recently felled trees (Garnham, Harper and Highton, 1946); in bamboo stumps (Clare, 1916; Shannon, 1931); in slots cut in coconut trees. Haworth (1922) at Dar-es-Salaam recorded breeding in leaf axils at the top of coconut trees, but after careful search in the same locality Lester (1927) failed to find breeding in such a situation and thought it probable there had been some mistake. The species has been found breeding at certain water levels in crab-holes: at Lagos (Dalziel, 1920); at Dakar (Riqueau, 1929); at Togo (Cheneveau, 1934). On Banana Island at the mouth of the Congo Wanson (1935) found only resting adults of *Aëdes aegypti*, though some other species were breeding. Tree-holes may be an important source of breeding for the species in some areas. Dalziel records 28 per cent of tree-holes at Lagos as showing larvae, and Allan (1935) in Sierra Leone found 749 out of 1289 samples from tree-holes positive. However, Barraud states that in India it is unusual to find larvae of the species in tree-holes.

A noticeable feature is that the species is rarely found breeding in pools or collections of water on the ground such as borrow-pits or earth drains in which all the edges are of earth or mud. Absence of breeding in such situations is ascribed by Gordon (1922) to the vulnerability of larvae to natural enemies. Also such waters are clearly not very attractive to the species, which prefers conditions where it can lay its eggs in some sheltered or confined situation and especially where these can be laid at the edge of the water cemented on some solid surface (see chapter XXII under oviposition). Breeding in rock pools, however, is a feature of the wild form as noted later.

The above records relate mainly to the species as it occurs in association with man, natural breeding places being utilised only when close to or at a moderate distance from habitations. In the case of the wild form breeding in forests and other uninhabited areas in Africa the breeding places described are chiefly tree-holes, rock pools, water courses with leaves or in receptacles placed as traps (Dunn, 1927; Haddow, 1945, 1948; Haddow *et al.* 1951). Tree-holes, however, are not necessarily the most frequent form of breeding place in such circumstances. Garnham, Harper and Highton (1946), in the Kaimosi Forest, Kenya (5500 feet altitude), give the following numbers for the different breeding places in which the wild form was found: tree-holes 2, rock pools in the shade 5, rock pools in sunlight 8, tree stumps 11, pools in river bed 34. Bamboo pots were negative. In the use of traps Haddow gives a warning that observations may be vitiated, unless care be taken, by eggs being present in the pots when set out.

The water of breeding places is most commonly clean or with a moderate amount of organic matter present: clean (Ingram, 1919); not very dirty (Howlett, 1913); dirty and clean water equally (Graham, 1911); all but very foul water (Stephanides, 1937); never found in cess-pits (Mhaskar, 1913); or in water polluted with urine or faeces (James, Da Silva and Arndt, 1914). The larvae can exist in water with a very poor food supply (Aders, 1917), but very frequently, though the water is clean, there are fragments of food or other organic debris at the bottom. The pH of the water within wide limits is unimportant (MacGregor, 1921; Rudolfs and Lackey, 1929). Though the species may be found in brackish water (Darling, 1910; Drake-Brockman, 1913), sea water in any concentration over 50 per cent (1·6 per cent salt) is lethal (Macfie, 1922). Much information on breeding in saline waters is given by Balfour (1921). Cossio (1931) notes that in Uruguay larvae are found especially in water of alkaline reaction.

(*c*) NATURAL ENEMIES

A very full account of natural enemies of mosquitoes recorded up to date of that publication has been given in the first volume of Howard, Dyar and Knab (1912).* Though a great many natural enemies of mosquitoes have been recorded in the literature not many of these relate to *Aëdes aegypti*, since breeding in this case is to a large extent restricted to domestic utensils or other situations that are relatively free from predaceous forms, whilst the adult is largely protected by its indoor habits. It may be useful, however, to give an account of natural enemies of mosquitoes in general and when doing so note such as have actually been recorded attacking *A. aegypti* under experimental conditions or in nature. It will be convenient to consider natural enemies as divided into those active against the aquatic stages (larva and pupa), those destroying the adult mosquito and those attacking the egg.

NATURAL ENEMIES OF THE AQUATIC STAGES

Plants and lower forms of animal life. Among forms destructive to larvae are certain aquatic carnivorous plants, notably the bladder-worts (*Utricularia*) which have small trap-like receptacles into which insects, including mosquito larvae, can enter but not return (Darwin, 1875; Galli-Valerio and Rochaz de Jongh, 1909; Eysell, 1924). A species has been described as destructive to larvae in pools in Trinidad (Hart, 1906) and one in bromeliaceous plants in Brazil (Krause, 1887). The plant *Genlisea ornata* in Brazil acts in the same way and a species of submerged sundew (*Aldrovanda vesciculosa*) traps larvae in Europe (Eysell). A good account of the capture of larvae by plants of the genus *Utricularia* is given by Brumpt (1925).

Certain plants that are not actively destructive may still act as enemies by covering the surface of the water or through their products. Among the first-mentioned are the duckweeds, notably the rootless duckweeds, *Woolfia arhiza* and *Azolla* (J. B. Smith, 1910; Bentley, 1910). Water covered by these forms may be rendered quite unsuitable for breeding. Extensive sheets of water may be rendered unsuitable (except for certain species which it may favour) by the water lettuce, *Pistia*, or the water hyacinth. Certain species of *Chara* are credited with a deterrent effect, apparently due to liberation of oxygen (Matheson and Hinman, 1928, 1931). Certain plants are said to be favourable, for example *Nasturtium officinalis* and *Spirogyra* (Waterston, 1918).

Among low forms of animal life *Hydra* may in certain circumstances be destructive to

* See also accounts by Wilson (1914), Waterston (1918), Purdy (1920), MacGregor (1920), Young (1921), Eysell (1924), Hayes (1930), and the summaries of recorded forms with bibliography by Hinman (1934). A very interesting and full description of natural enemies encountered in New Britain is that by Laird (1947). Other authors recording natural enemies are given in the text. Some of the forms here treated as parasites (see (*d*) below) may also be considered as natural enemies.

larvae (Dyar in Howard, Dyar and Knab, 1912). *Vorticella*, by excessive growth on the bodies of larvae, may cause death (Seguy, 1923). Larvae so affected appear white and gelatinous or fluffy. A planarian has been recorded as destroying larvae of *Aëdes aegypti* in a glass tank (Lischetti, 1919). Water snails have been found to have an adverse effect and may feed on eggs or egg-rafts (Hamlyn-Harris, 1929; Gasanov, 1938). Water fleas (*Daphne*) may destroy larvae, biting or pulling off their long lateral hairs so that the larvae are disabled and drown (Wilson, 1914).

None of the above would appear to have any importance as enemies of *Aëdes aegypti*.

Predaceous arthropods. A great many predaceous aquatic insects, as also some hydrachnids, have been noted as attacking larvae, though the fact that they do so under artificial conditions in the laboratory does not always provide a measure of their effectiveness in nature.

A predaceous mite, *Limnesia jamurensis*, is described by Laird as actively attacking young larvae and eggs experimentally and as appearing to play a useful role in keeping down the number of larvae in pools where it was abundant in New Britain. The same author notes that Hearle (1926) refers to a predaceous mite (hydrachnid) in swamps in British Columbia making mosquito larvae its chief food.

Dragonfly larvae have been considered important natural enemies by Lamborn and his collaborators (1890). But the fact that they are bottom feeders has been thought to make this doubtful on any scale in nature. Ephemerid larvae have also been observed attacking mosquito larvae (Foley and Yvernault, 1908).

In the order Coleoptera both adults and larvae of Dytiscidae and the larvae of Hydrophilidae, as also Gyrinidae, are actively predaceous (J. B. Smith, 1910; Viereck, 1904; Britton and Viereck, 1905; Laird, 1947). Laird, after introducing 100 gyrinid beetles to a temporary rain pool about 12 by 4 feet containing about 500 larvae of *Anopheles punctulatus*, found only forty-six larvae after 24 hours and none after a further 24 hours. In Alaska Sailer and Lienk (1954) describe the beetle *Agapus* as destroying in twenty-one days sixty mosquito larvae and ten pupae. Not all Dytiscidae are, however, effective in this respect (D'Emmerez de Charmoy, 1902; Derivaux, 1916). Macfie (1923) records the tiger beetle, *Cicindela octoguttata*, as preying on larvae and pupae at the edges of pools in the Gold Coast, gripping the larvae by the posterior extremity and the pupae by the cephalothorax. A natural enemy acting indirectly are the small beetles (Halticinae) which Hamlyn-Harris (1930) describes in Australia as poisoning not only the vegetation, but also water in the neighbourhood making this fatal to larvae.

Aquatic bugs, *Notanecta*, *Corixa* and other genera, have been noted as actively destroying larvae (Dempwolff, 1904, and other authors). The bug *Laccotrophes kohlii* (Nepidae) has been experimented with as a control measure in China (Hoffmann, 1927); a belostomid (*Sphaerodema*) also gave some results. Laird found that the notonectid *Enithares* sharply reduced larvae of *Aëdes scutellaris* when introduced into a water drum and mentions it as one of the natural enemies that might be of practical use.

Anthomyid flies of the genus *Lispa* have been observed attacking larvae, pupae and emerging adults at the water surface: *L. sinensis* at Hong Kong (Atkinson, 1909); *L. tuberculata* acting similarly in Europe (Dufour, 1864); *L. tentaxulata* var. *sakhalensis* in Japan (Yamada, 1927); as also certain Dolichopodidae observed behaving in a similar manner at Panama (Osterhout). *Simulium* has been recorded attacking newly emerged mosquitoes (Osterhout; Combes, 1896).

That the larvae of *Aëdes aegypti* are extremely vulnerable to the attacks of such natural enemies as many of the above under experimental conditions has been shown by Gordon (1922). Nevertheless, few of the forms mentioned are likely to be of importance against *A. aegypti* under natural conditions owing to the type of breeding place in which the species is found. A certain number of predatory forms attacking *A. aegypti* in nature have, however, been described. Shannon (1929) observed the larva of a tipulid, *Sigmatomera*, preying upon *Aëdes aegypti* larvae in tree-holes in Brazil. Carter (1919) records larvae of *A. aegypti* as being preyed upon in west Africa by a species of *Forcipomyia* (Ceratopogoninae). Much the most effective and important of all insect enemies of the species are the cannibal larvae of certain species of mosquito which frequent many of the breeding places utilised by *A. aegypti* and when present may decimate this species.

Cannibal mosquito larvae. Where *A. aegypti* is breeding in bamboo stumps, a common breeding place near dwellings in many areas, or breeding places of a similar type, an active and common enemy are cannibal larvae of various species of *Megarhinus*. Such larvae are extremely voracious (Green, 1905; Morgan and Cotton, 1908; Senior White, 1920; Pruthi, 1928). A single megarhine larva was observed by Garnham *et al.* to devour twenty-one *Aëdes aegypti* larvae in a night. Laird in New Britain notes that two larvae of *Megarhinus inornatus* in five days destroyed sixty-eight third and fourth instar larvae of *Aëdes albolineatus*, fifty-nine of *Aëdes scutellaris* and thirty of the larger larvae of *Armigeres lacuum*, but only seven pupae under similar conditions. He notes that the species breeds in coconut husks and other small containers in heavy shade just within the edge of the jungle. Paiva (1910) describes the larva of *Toxorhynchites immisericors* (*Megarhinus splendens*) as commonly preying on larvae of *Aëdes aegypti* in earthenware water storage pots in Calcutta. *Megarhinus* has been introduced as a means of controlling *Aëdes* spp. into areas outside its normal range. Attempts to introduce it into Hawaii do not appear to have been successful (Swezey, 1931), but *M. splendens*, not previously recorded east of New Britain, was successfully established in Fiji as a measure directed against *Aëdes scutellaris*, a vector of dengue (Paine, 1934).

Another effective cannibal genus is *Lutzia*. Jackson (1953) records that one larva of the African species *Lutzia tigripes* ate eighty-eight larvae of *Aëdes aegypti*, preferring this to other species. In India the larvae of *Lutzia* are most commonly seen in association with *Culex* larvae in garden sumps and when present leave very few of the victim species to emerge. But they may also be found with the larvae of *Aëdes aegypti* in certain types of breeding place. Haddow (1942) found *Culex tigripes* reducing the larvae of *Anopheles* in borrow pits in Kenya. Howard (1910) notes that *Lutzia bigotii* effectively destroyed larvae of *Aëdes aegypti* breeding in artificial containers in Rio de Janeiro.

Still another genus of mosquito with cannibal larvae is *Mucidus*. The larvae of *M. scataphagoides* are found in India in natural pools (Christophers, 1906; Bedford, 1919; Barraud, 1934). *Psorophora* and *Lestiocampa* (*Goeldia*) are New World forms with similar cannibalistic habits, the latter a sabethine mosquito occurring in collections of water in bromeliaceous plants (Howard *et al.* 1912). The larvae of a number of non-biting Culicidae are actively predaceous: *Corethra* and *Eucorethra* (Underwood, 1903); *Mochlonyx* and *Chaoborus* (Sailer and Lienk, 1954); as also are the larvae of some Chironomidae, for example *Tanypus* (Knab, 1912).

Vertebrate enemies of the aquatic stages. Probably the most effective of all natural enemies of mosquitoes in nature are various small larvivorous fish. Small fish of this kind are indeed the only natural enemy that has been extensively used with success in the control of mosquito breeding as a sanitary measure. This applies to the control of both *Anopheles* and, to an even greater extent, of *Aëdes aegypti.* The use of fish in control of *A. aegypti* breeding is too important a measure to be adequately dealt with here and will be gone into more fully in the section on control.

Certain batrachians have been listed as natural enemies of the larva: the frog, *Discoglossus pictus*, in North Africa (Galli-Valerio and Rochaz de Jongh, 1908); tadpoles of the Spadefoot Toad, *Scaphiosus hammondi* (Barber and King, 1927, 1928); the Western Salamander, *Diemyctylus tortosus* (Howard *et al.* 1912); the Vermilion Spotted Newt, *Diemyctylus viridescens* (Matheson and Hinman, 1929). Frogs and toads and their tadpoles are not, however, as a rule larvivorous (Galli-Valerio and Rochaz de Jongh).

Lewis (1942) in the Sudan found small water tortoises, *Pelomedusa galeata*, commonly present in stored water pots. These fed voraciously on larvae and pots containing them were free from larvae.

Ducks and other aquatic birds have been shown to destroy larvae under certain circumstances and ducks have been suggested and experimentally proved effective as a control measure (editorial, *J. Trop. Med.* (*Hyg.*), 1916).

NATURAL ENEMIES OF THE IMAGO

Arthropod enemies. Among arthropods that are natural enemies are spiders and other arachnids and various predaceous insects.

Spiders have been considered very important natural enemies by a number of observers (J. B. Smith, 1910; McCook, 1890, and others in America; Leon, 1910, in Roumania; Nichols, 1912, in the West Indies; O'Connell, 1912). The jumping spiders, *Salticus*, are especially referred to by Marchoux *et al.* (1903) and by Nichols. Bacot (1916) notes a spider, *Uloborus feniculatus*, the arachnids, *Monomorium pharaonis* L. and *Solenopsis geminata* F., also a scorpion, *Isometrus maculatus*, as attacking *Aëdes aegypti.* Mathis and Berland (1933) describe an African domestic spider, *Plexippus paykulli*, as capturing most of the mosquitoes bred out in a cage, and the same spider is also referred to by Mathis (1938). Spiders of the genera *Epeira* and *Meta*, which spin webs over pools in New Britain, are given as of some importance as predators by Laird. Among acarids may be mentioned the larvae of water mites (Hydrachnidae), commonly found attached by their proboscis to the adults of mosquitoes, which breed in some natural waters and which are described later among parasites affecting mosquitoes.

Among insects are: dragonflies (Lamborn and collaborators); Panorpidae (Eysell); the reduviid bug, *Emesia longipes*, which infests houses in certain parts of the United States (Howard *et al.*); predaceous Hymenoptera (Bates, 1863; Ferton, 1901); predaceous Diptera such as Empidae (Wahlberg, 1847); Asilidae (Eysell); and Scatophagidae (Wahlberg). In Cuba a wasp, *Monedula*, has been observed capturing adult *Aëdes aegypti*, one specimen capturing on the wing twenty in five minutes (Howard *et al.*). A wasp in Corsica which stocks its nest with mosquitoes is described by Ferton. Drake-Brockman (1913) in Somaliland observed ants attacking newly emerged *A. aegypti.* Here also should be mentioned the small midges, *Culicoides*, which attack and suck the blood from gorged mosquitoes and which, since they are usually found attached by their proboscis to their victim, are dealt with in the section on parasites.

Vertebrate natural enemies. Liston (1901) sometimes found mosquitoes in the stomach of frogs in Bombay. Geckoes and insectivorous lizards have been observed feeding on mosquitoes (Giles, 1902; Shannon, Burke and Davis, 1930; Hayes, 1930). Other important vertebrate enemies are insectivorous birds and bats. Waterston in Macedonia considered swallows to be very important natural enemies and Theobald (1901) observed house martins catching *Anopheles* in flight in England. Goat-suckers (Harvey, 1880) and the Cuban night-hawk, *Chordeiles* (Jennings, 1908), which feed at dusk, have also been recorded as feeding on mosquitoes.

Though insectivorous bats are credited with great destructive power against mosquitoes and have been recommended for control purposes, few careful observations seem to have been made and, whether effective or not in nature, their use in control seems generally to have been found impracticable (Herms and Gray, 1940, p. 238). For references to the literature dealing with the use of bats in control see Covell (1941).

NATURAL ENEMIES ATTACKING THE EGGS

In the laboratory the eggs of *Aëdes aegypti* are open to serious attack when left unprotected in a stored condition. Bacot (1916) found the chief enemy of stored eggs to be a psocid. In the present author's experience precautions are necessary, not only against psocids, but even more so against cockroaches if such be present. The latter, if they gain access, browse on the deposit of eggs, completely removing these and leaving only the characteristic cockroach faeces behind. If eggs are kept moist, mites are apt to occur in large numbers, but whether they attack the eggs is not known. An adult hydrachnid mite, *Limnesia jamurensis*, was found by Laird in New Britain to consume eggs readily experimentally. Buxton and Hopkins (1927) found eggs of *Aëdes variegatus* being carried off by ants.

(*d*) PARASITES

Except for some early references to moulds attacking mosquitoes and a possible reference to nematodes by Leuchert (see Howard *et al.* p. 162) the first to record parasites in mosquitoes was Ross, who in 1895 noted the presence of gregarines in a species of *Aëdes* and subsequently published a list of other parasites encountered by him from 1895 to 1898 (Ross, 1898, 1906). Since then many authors have recorded and described parasites in the larva and adult form of mosquitoes so that there is a very considerable literature in this respect. Of authors giving lists of recorded parasites may be mentioned especially Speer (1927) who gives a complete documented list of the parasites of mosquitoes up to the date of his publication.*

As with natural enemies, comparatively few of the long list of parasites have been recorded from *A. aegypti*, which from the nature of its usual breeding places is less exposed to many forms of parasite than are anophelines and many culicines that habitually breed in natural waters. It will probably, however, be most useful to note what parasites of

* See also Dye (1905), Howard *et al.*, Patton and Cragg (1913), Bresslau and Buschkiel (1919), Martini (1920), Séguy (1923), Missiroli (1928, 1929, Italy), Feng (1933*a*, *b*, China), Brumpt (1936), Iyengar (1938, India), Steinhaus (1946), Bates (1949), Christophers (1952). A list of recorded fungal and protozoan parasites is given by Keilin (1921*a*, *b*). Other authors describing particular forms are mentioned below in the text.

mosquitoes have been described and to point out and describe somewhat more fully those recorded from *A. aegypti*. Both vegetable and animal forms have been recorded as parasites of mosquitoes, notably in the first case moulds, and in the case of animal parasites the protozoa. But worms and mites are important endo- and ecto-parasites respectively and there is one semi-parasitic insect, the tiny voracious midge, *Culicoides*, more correctly perhaps considered a predator, which sucks the blood from the gut of the successful predator mosquito. Taken *seriatim* the parasites of mosquitoes can be enumerated as follows.

Bacteria, rickettsia and yeasts. The only bacterial parasite recorded, apart from disease organisms, is a bacterium resembling a *Leptothrix* described by Perroncito (1900) as parasitic and pathogenic in the larva of *Anopheles maculipennis*.

Rickettsial forms are described by Sellards and Siler (1928) in *Aëdes aegypti* infected with the virus of dengue (see also Montoussis, 1929). The authors note that certain stages in the gregarine, *Lankesteria*, may simulate these forms. An intracellular *Rickettsia* is described by Brumpt (1938) in the stomach of *Culex fatigans* (*Rickettsia culicis* Brumpt).

Yeasts may be present in the gut and air diverticula, especially in mosquitoes fed on fruit or sugar. Yeasts are recorded as present in the coelomic cavity in *Anopheles maculipennis* by Laveran (1902) and by Marchoux *et al.* in *Aëdes aegypti*.

Moulds and fungi. Many records of mycelial fungi of various kinds in the larva and adult of mosquitoes are given in the literature. Adult mosquitoes under certain conditions are liable to be attacked and destroyed by fungi of the *Empusa* type (*E. culicis* Braun; *E. papillata*) and by *Entomophora* (Braun, 1855; Thaxter, 1888; Pettit, 1903); see account by Howard *et al.* Leon (1924) describes tumours with mycelium in *Anopheles maculipennis* following bites by *Culicoides*. A number of fungal forms are also described parasitising the larva. Liston (1901) records a form in Bombay resembling *Trichophyton* on the surface of the larva. Macfie (1917) describes the larva of *Aëdes* sp. in the laboratory as covered with fungal hyphae (*Nocardia* and an unidentified fungus), also as infected in nature with a brown fungal mass in the thorax (*Fusarium*). There are also forms described by Leger and Duboscq (1903), Langeron (1929), Chorine and Baranoff (1929). A fungus, *Polyscytalum*, has been found covering the eggs of *Psorophora* (Howard *et al.*). Galli-Valerio and Rochaz de Jongh (1905) infected the larvae of both *Culex* and *Anopheles* with *Aspergillus niger* and *A. glaucus*, such infection causing the peritrophic membrane and faeces to pass out in long lengths in place of breaking off. Dye (1905) describes a mould in the digestive tube and coelome of *Aëdes aegypti*. According to Dye, Vaney and Conte found 'altises' (*Culex pipiens*?) adults and larvae destroyed by the fungus *Botrytis bassianae*, but he gives no date or reference.

Keilin (1921 *b*) described as *Coelomomyces stegomyiae* a fungus in *Aëdes scutellaris* larvae from Federated Malay States sent to him by Lamborn. This occurs as a septate mycelium ramified over the viscera with terminal thickenings which become detached and form oval multi-nucleated sporangia with very thick walls. The sporangia, which resemble nematode eggs or *Coccidia*, measure some 37–57μ in length by 20–30 μ in breadth. They occur in large numbers packed in the anal papillae and other parts Similar forms have been described by Iyengar (1935) in *Anopheles* larvae in India (*C. indiana* and *C. anophelesica*); by Walker (1938) in *A. gambiae*; and in the same species causing heavy mortality by Muspratt (1946). The forms described by Manalang (1930) as probably *Coccidia* may also be this fungus, *Coelomomyces* being clearly a frequent and important parasite of *Anopheles* larvae. Van Thiel (1954) has described sporangia of a species of the same fungus in the ovary of an adult *Anopheles tessellatus*. The *Coelomomyces* parasitic in mosquito larvae have been reviewed by Couch (1945) who describes a species, *C. psorophorae*, in *Psorophora ciliata* larva; see also Couch and Dodge (1947). A form distinct from but close to *Coelomomyces psorophorae* is described by Laird (1956) in larvae of *Aëdes* (*pseudoskusea*) *australis* in Tasmania (*Coelomomyces tasmaniensis*). The larvae in this case are a brackish-water species. As noted above, the oval sporangia are liable to be taken for nematode eggs.

Spirochaetes. Spirochaetes have been recorded in the adult and larva of various species of mosquito: Schaudinn (1904); Ed. and Et. Sergent (1906) in the intestine of both *Culex* and *Anopheles*; Patton (1907) in *Culex fatigans* in India; Jaffe (1907) as *Sp. culicis* in the digestive tract of the larva and in the Malpighian tubules of the adult of *Culex* in Germany. A large spirochaete has been described by Wenyon (1911) in the gut of the larva and Malpighian tubules of the adult of *Aëdes aegypti*. Spirochaetes have also been recorded in the intestine of this species by Noc and Stevenel (1913) and in the Malpighian tubules by Noc (1920) in Martinique. The first-mentioned species resembled *Sp. refringens*, those in the Malpighian tubules were very fine and small forms. Patton and Cragg note spirochaetes as common in the Malpighian tubules of *Aëdes aegypti* in Madras. Spirochaetes are described in the salivary glands of *Anopheles funestus* in French West Africa by Masseguin and Palinacci (1954) and in the salivary glands of *A. gambiae* in the same area by Masseguin, Palinacci and Brumpt (1954). An account of spirochaetal infections of mosquitoes is given by Sinton and Shute (1939) and a list of recorded findings by Colas-Belcour (1954).

Gregarines. A gregarine, *Lankesteria culicis* (Ross), was first described from India in a species of *Aëdes* (Ross, 1898, 1906), and later by Marchoux *et al.* in Brazil. The same species was noted by Wenyon (1911) in Mesopotamia. Its life history has been described by Wenyon (1926). See also Ray (1933); Ganapati and Tate (1949).

Wenyon describes mobile pear-shaped or elongate forms in the gut of the larva, which, when the larva pupates, pass into the Malpighian tubules where they associate in pairs and encyst. Cysts with numerous oocysts, each with eight sporozoites, are present in the Malpighian tubules of the adult. Spores brought with eggs from West Africa developed in England (Stevenson and Wenyon, 1915). The same gregarine is recorded in *Aëdes aegypti* by Bacot (1916) and by Macfie (1917) in West Africa; by Feng (1930) in various species of *Aëdes* in China; by Martirano (1901) in Italy; by Mathis and Baffet (1934) in France; as also still more recently by Ganipati and Tate in *A. geniculatus* in England. In *A.* (*Finlaya*) *geniculatus*, as described by the last-mentioned authors, the earlier stages are intra-epithelial in the anterior portion of the gut in the larva. After a period of growth the trophozoites are liberated into the gut and attach themselves to the epithelium by a well-developed epimerite, which functions as a sucker. Cyst formation and further stages in sporogony take place in the Malpighian tubules when the larva becomes the pupa. In the adult mosquito only ripe sporocysts are found, packed in the Malpighian tubules and scattered in the hind gut. The spores escape through the alimentary canal.

A gregarine, *Diplocystis*, from the body cavity of a larva of a species of *Culex* is described by Leger and Duboscq, and a species of gregarine on the outer wall of the mid-gut in *Anopheles maculipennis* adult resembling oocysts of the malaria parasite is described by Johnson (1903). A species, *Caulleryella anophelis*, has been described in the intestine of the larva of *Anopheles bifurcatus* (*A. claviger*) by Hesse (1918), and still another species, *Caulleryella pipientis*, in the larva of *Culex pipiens*, and other species of mosquito by Bresslau and Buschkiel, as also *Caulleryella maligna* in an *Anopheles* in Brazil by Godoy and Pinto (1923). Mathis and Baffet describe a method by which an infected strain of *Aëdes aegypti* can be obtained free from such parasites. Isolated females were fed on a rabbit and after passing faeces were transferred to a fresh cage and allowed to oviposit on wet wool.

Flagellates. Flagellates commonly occur in the gut and rectum of adult mosquitoes and in the aflagellate stage in the larva. There are two types, *Leptomonas* (*Herpetomonas*) without an undulating membrane, and *Crithidia* with short undulating membrane. Both forms have been recorded in many culicine and anopheline mosquitoes in many countries: Chatterjee (1901, Crithidial forms); Leger (1902, *C. fasciculatus* in *Anopheles maculipennis*); Johnson (1903); Novy, MacNeal and Torrey (1907, *H. culicis*); Patton (1907, 1912); Missiroli (1928). Besides the usual record of forms in the gut, Mathis (1914) describes a trypanosome-like form in the salivary glands of a species of *Culex*. Trypanosomes have also been found in the salivary glands of *C. fatigans* in India by Viswanathan and Bhatt (1948) and by Jaswant Singh, Ramakrishnan and David (1950). In the case of the last-mentioned authors the trypanosomes occurred along with sporozoites in a laboratory-bred *C. fatigans* fed on a sparrow with cryptic infection (see also Ramakrishnan, David and Nair (1956) who describe the trypanosome stages).

In *Aëdes aegypti* leptomonad forms have been recorded by Durham and Myers (1902) in Brazil and by Ed. and Et. Sergent (1906) in Algeria (*L. algeriensis*). Leptomonad forms were found in the larva of *Aëdes aegypti*, but not in the adult, by Wenyon (1911) in Mesopotamia. Noc and Stevenel record *Herpetomonas* in the digestive tract of *Aëdes aegypti* in Martinique. Patton (1912) and Patton and Cragg record *Herpetomonas culicis* in *Aëdes* in Madras.

The growth characters on media of different flagellates have been described by Noller (1917). For summaries dealing with the flagellates of mosquitoes see Thompson and Robertson (1925) and Wallace (1943).

Thelohania and other microsporidial forms. Various microsporidia have been recorded in the adult and larva of mosquitoes. A very full study has been made by Kudo (1921–5), see also Iyengar (1930). Under this head are included *Nosema culicis* in the body cavity of *Culex pipiens* larva (Bresslau and Buschkiel); *Nosema stegomyiae* in the intestine of the imago (Lutz and Splendore, 1908); *Nosema* sp. in *Aëdes nemorosus* adult and larva (Noller, 1920); *Nosema anophelis* in adult *Anopheles quadrimaculatus* in North America (Kudo) and in *A. maculipennis* in Italy (Missiroli, 1929).

A characteristic type is *Thelohania* with pansporoblasts containing eight spores occurring in various tissues including the developing ova in the ovary (Christophers, 1901). Kudo recognises a number of species, namely *T. legeri* Hesse, *T. obesa*, *T. pyriformis*, *T. anophelis*, all in larvae of *Anopheles* sp.; also *T. rotunda*, *T. opacita*, *T. minuta* in culicine species. All but the first have been named by the author. Missiroli refers to these as also to another species which he does not name. A study of the development and extrusion of the filament from the spores is given by Iyengar. According to this author the spores are small and not distinguishable under a 2/3 objective. Though *Thelohania* forms are extremely common in various species of *Anopheles* the absence of records for *Aëdes aegypti* is noticeable. Ross (1906) has recorded a form with eight spores in the adult of a small species of *Stegomyia* in India as well as in *Culex fatigans*.

Under the name of *Glugea* are some ambiguous forms including *Myxosporidium stegomyiae* of Parker, Beyer and Pothier (1903) and *Glugea* (*Nosema* or *Pleistophora*) *stegomyiae* of Marchoux *et al.* found in the body cavity and tissues of *Aëdes aegypti* in Cuba and Brazil respectively. Another form of doubtful nature is *Serumsporidium* (*Nosema*) sp. in *Aëdes serratus* in Brazil (Lutz and Splendore). Besides *Nosema anophelis*, Missiroli (1928, 1929) describes a sarcosporidial form, *Sarcocystis anophelis*. Recently Canning (1957) has given a description of a species *Plistophora culicis* Weiser parasitic in the Malpighian tubules of *Anopheles gambiae*. The parasite forms oval sporonts 25–26·5 μ in diameter with closely packed nuclei, which may be mistaken for oocysts. It also occurs in the tissues and fat-body. It was originally described by Weiser in *Culex pipiens*.

Ciliates. A ciliate, *Lambornella stegomyiae* Keilin, is recorded by Lamborn (1921) in the larva of *Aëdes scutellaris* (*albopictus*) from Malaya and its characters and life history have been described by Keilin (1921*a*). It consists of oval, actively motile forms 50–70 μ in diameter which are present in large numbers, especially in the anal papillae, but also in other parts. It passes out of the body through ruptures and encysts on the larval cuticle and probably elsewhere. A ciliate, *Glaucoma* sp. (probably *G. pyriformis*), is described by MacArthur (1922) in *Theobaldia annulata* in England (see also Wenyon, 1926). The parasites are mostly 25–40 μ in length with large cytostome. They are commonly found in the head and a feature of the infection is destruction of the eyes. As many as a thousand may be present in a single larva.

An interesting account of a parasitic ciliate, probably a form of *Glaucoma pyriformis* (*G. pyriformis* Ehrenberg), found in the body cavity of several species of tree-hole-breeding culicines in Northern Rhodesia is given by Muspratt (1945). It occurs in two forms, *A* and *B*. *A* multiplies rapidly by binary fission in the haemocoele of the larva finally causing death. *B* seldom divides, if at all, does not kill the larva, and has been recovered from the adult. The parasites escape from the body of the larva by rupture of the anal papillae. In some species of mosquito, including *Aëdes aegypti*, the anal papillae are tough and do not rupture. On this account, though *A. aegypti* can be infected, it is not a suitable host and so seems to be biologically distinct from *Lambornella stegomyiae* Keilin. Also it does not attack the eyes of the larva but tends to congregate round the heart when it first enters, thus differing from *Glaucoma pyriformis* MacArthur, 1922. Infection is by free-living ciliates which encyst on the larval cuticle just before ecdysis and penetrate to the haemocoele through the new cuticle when this is soft.

Trematodes. Of helminths, encysted trematodes (*Agamodistomum*) are not uncommon in the larva and adult of *Anopheles* (Chatterjee, 1901; Martirano, 1901; Schoo, 1902; Ruge, 1903; Alessandrini, 1909; Sinton, 1917; Soparkar, 1918; Joyeux, 1918; van Thiel, 1921–54; Eckstein, 1922; Iyengar, 1930; Corradetti, 1937; and probably others). The species recorded include *A. sintoni* in various species of *Anopheles* in India and *A. anophelis* in *Anopheles maculipennis*. The life history has been worked out very completely by van Thiel.

The cysts are large, about 150–200 μ in diameter, and usually show a trematode bent sharply upon itself. They are found in the body cavity and in the tissues of various parts. The corresponding Cercaria of *A. anophelis* discovered by van Thiel (1922, 1925) develops into a sporocyst in the liver of the mollusc *Planorbis vertex* and later (van

Thiel, 1930*a*, *b*) into the fluke *Pneumonoeces variegatus* parasitic in the lung of the frog, *Rana esculenta* (van Thiel, 1954).

Nematodes. Immature nematode parasites (*Agamomermis*) have been described in the larva and adult of various species of mosquito (Johnson, 1903; J. B. Smith, 1903; Stiles, 1903; Gendre, 1909; Iyengar, 1930, 1935). The species referred to by Johnson and J. B. Smith is *Agamomermis culicis* Stiles, parasitic in *Culex sollicitans* in America. The species described by Gendre is from *Aëdes aegypti*. It occurs in the body cavity in pairs, a large and a small form, and escapes through the perianal membrane when the larva is about to pupate. The species described by Iyengar in *Anopheles* occurs as a single large adult worm, smaller worms in the same larva dying off. The fully grown worm escapes into the water by rupturing the body wall of the larva, which dies immediately. In the water the worms become sexually mature and give birth to numerous young larvae which swim about actively. These move about on the surface of the water and hold on to young larvae whilst piercing the cuticle to gain entrance to the haemocoele. Iyengar also notes that certain other large species of *Mermis* are found in adult mosquitoes. They are sexually mature when they emerge. A mermethid nematode is described in *Aëdes communis* in Canada by Jenkins and West (1954). The larval stages of the nematodes *Foleyella ranae* and *F. dolichoptera* infesting certain frogs in North America have been obtained by Causey (1939) in *Aëdes aegypti*, *Culex fatigans* and *C. pipiens*. Developing and mature infective larvae were present in the head, abdomen and thoracic muscles of mosquitoes fed on infected frogs. A parasitic nematode (*Agamomermis*) is described by Muspratt (1947) in larvae of tree-hole breeding species of mosquito in Northern Rhodesia and has been successfully cultured in the laboratory and *Aëdes aegypti* infected.

Among nematodes parasitic in mosquitoes are also those giving rise to filariasis in man and animals, including various species of *Filaria* in man and *Dirofilaria* of the dog (see Ch. IV (*a*)).

Mites. Acarids (Hydrachnidae larvae) are common on adult *Anopheles* and various culicines that breed in weedy waters. They occur, attached by their rostrum, clustered about the under side of the body, especially about the neck. For information about such forms see Edwards (1922); Balfour (1923); W. H. Dye (1924). Records of such mites have been given by: Hodges (1902); Gros (1904); Ed. and Et. Sergent (1904); Galli-Valerio and Rochaz de Jongh (1907); Macfie (1916); Boyd (1922). The acarids are larval forms with six legs only which attach themselves to the adult mosquito in the act of emergence. The life history of a Japanese water-mite, the larvae of which attach themselves to the pupa and transfer to the adult mosquito at emergence, is described by Uchida and Miyazaki (1935). A larval mite, *Limnesia*, described by Laird (1947) from New Britain which attaches itself in this way is in the adult stage a predator of young larvae and eggs (see (*c*)).

Where such mites have been attached, peculiar tubular tunnels lined with chitin develop (Marshall and Staley, 1929; Feng, 1930; Feng, 1933*a*, *b*; Feng and Hoeppli, 1933).

Diptera. Since the minute blood-sucking midges, *Culicoides*, are found attached more or less permanently by their proboscis to their victim they may be considered as semi-parasitic. An account of these forms is given by Edwards (1922). In all cases seen by this author the species was the same, namely *C. anophelis* Edw., and records by others are, when named, given as this species. Usually records are from oriental tropical countries (India, Burma, Malaya, Indochina). Leon, however, records such mites from Roumania.

In almost all cases too the mosquito attacked is an *Anopheles*, but a specimen of *Armigeres lacuum* so infested is described by Laird (1946) who gives an excellent review of the literature with a figure of the midge. The midges are usually found attached on the abdomen of the gorged female with their mouth-parts deeply embedded. They are not found on males and there seems no reason to doubt but that their objective is the blood in the gut of the gorged mosquito.

Though not strictly a case of parasitism the oestrid fly *Dermatobia hominis*, the human warble fly, utilises mosquitoes, especially *Psorophora*, by attaching its eggs to the mosquitoe's abdomen so that these may be carried to their host to hatch out when the mosquito feeds, see Sambon (1922).

The black spores of Ross. A word may be said about the black spores of Ross and other authors (see description by Brumpt, 1938) once thought possibly to be some stage in the malaria parasite cycle. The appearances are now known to be formed by deposition of chitin on dead structures in the body of the mosquito, including dead oocysts or their contents. They have been noted by Brumpt as occurring in *Aëdes aegypti* infected with *Plasmodium gallinaceum*. A similar 'chitinisation' causes the tubes left by the mites described above.

Mosquitoes as parasites and vectors of parasitic diseases. Larvae of *Culex pipiens* have been recorded as being passed with human faeces and so considered as parasites (E. Blanchard, 1890; Tosatto, 1891). See also Ficalbi (1891) and R. Blanchard (1901), who dispose of such a view.

A case almost amounting to mild parasitism is that of mosquitoes of the genus *Harpagomyia* which obtain food from ants which they stroke to cause them to make the necessary disgorgement and for which purpose they have a markedly specialised proboscis (Jacobson, 1911; Edwards, 1932).

In addition to being parasitised in the strict sense mosquitoes are pre-eminent as intermediate hosts of parasites causing disease in man and animals; see section on role in disease, pp. 77–84.

REFERENCES

(*a*) LIFE HISTORY

Adrien, C. (1918). Dengue Méditerranéenne observée à l'île Rouad (Syrie). *Arch. Méd. Pharm. Nav.* **105**, 275–307.

Aragao, de B. H. (1939). Mosquitoes and yellow fever virus. *Mem. Inst. Osw. Cruz*, **34**, 565–81.

Boulford, G. (1908). Le *Stegomyia fasciata* au Soudan français. *Bull. Soc. Path. Exot.* **1**, 454–9.

Charrier, H. (1924). La *Stegomyia fasciata* dans la région de Tanger (Maroc). *Bull. Soc. Path. Exot.* **17**, 137–42.

Cossio, V. (1931). Observations sobre al *Aëdes aegypti* (*Stegomyia*) mosquito de la febbre amarilla en Montevideo. *Bol. Cons. Nat. Hig. Uruguay*, **23**, 1664. Abstract in *Rev. Appl. Ent.* **19**, 230.

Creac'h, P. (1947). Notes succinctes sur *Stegomyia fasciata* en saison fraîche 1938–9 à Fort-Lamy (Tchad). *Encyc. Ent.* B 11, Dipt. x, 152–4.

Gutzevich, A. V. (1931). The reproduction and development of the yellow fever mosquito under experimental conditions (in Russian). *Mag. Parasit. Leningrad*, **2**, 35–54. Abstract in *Rev. Appl. Ent.* **21**, 2.

KOROVITZKII, L. K. and ARTEMENKO, V. D. (1933). Zur Biologie des *Aëdes aegypti*. *Mag. Parasit. Leningrad*, **2**, 400–6 (in Russian). Abstract in *Rev. Appl. Ent.* **22**, 78.

MACDONALD, W. W. (1956). *Aëdes aegypti* in Malaya. *Ann. Trop. Med. Parasit.* **50**, 385–98.

MARCHOUX, E., SALIMBENI, A. and SIMOND, P. L. (1903). La fièvre jaune. Rapports de la Mission française. *Ann. Inst. Pasteur*, **17**, 665–731.

MATHIS, M. (1938). Influence de la nutrition larvaire sur la fécondité du *Stegomyia* (*Aëdes aegypti*). *Bull. Soc. Path. Exot.* **31**, 640–6.

(*b*) BREEDING PLACES

ADERS, W. M. (1917). Insects injurious to man and stock in Zanzibar. *Bull. Ent. Res.* **7**, 391–401.

ALLAN, W. (1935). *Rep. Med. San. Dept. Sierra Leone for 1934*, pp. 28–32.

BALFOUR, A. (1921). Mosquitoes breeding in saline waters. *Bull. Ent. Res.* **12**, 29–34.

BANA, F. D. (1936). A practical way of dealing with *Aëdes aegypti* (*Stegomyia fasciata*) mosquito breeding in country craft. *Indian Med. Gaz.* **71**, 79–80.

BAQUE, B. and KIEFFER, J. J. (1923). Existence de *Stegomyia fasciata* au Sahara à 500 kilomètres de la côte Méditerranéenne. *Arch. Inst. Pasteur, Afr. Nord*, **3**, 169–71.

BARRAUD, P. J. (1934). See references in chapter II (*a–b*), p. 224.

BOYCE, R. (1910). The distribution and prevalence of yellow fever in West Africa. *J. Trop. Med.* (*Hyg.*) **13**, 357–62.

CARTER, H. R. (1924). Preferential and compulsory breeding places of *Aëdes* (*Stegomyia*) *aegypti* and their limits. *Ann. Trop. Med. Parasit.* **18**, 493–503.

CHABRILLAT, M. (1934). Note sur la fièvre de trois jours. *Bull. Soc. Path. Exot.* **27**, 762–6.

CHENEVEAU, R. (1934). Note sur les trous de crabes à Anecho (Togo). *Bull. Soc. Path. Exot.* **27**, 590–3.

CLARE, H. L. (1916). Report to Surgeon-General 1914–15. *Trinidad and Tobago Counc. Pap.* no. 154. Abstract in *Rev. Appl. Ent.* **4**, 69.

CONNOR, M. E. (1922). Final report on the control of yellow fever in Merida, Yucatan, Mexico. *Amer. J. Trop. Med.* **2**, 487–96.

COSSIO, V. (1931). See under (*a*).

DALZIEL, J. M. (1920). Crab-holes, trees and other mosquito sources in Lagos. *Bull. Ent. Res.* **11**, 247–70.

DARLING, S. T. (1910). Factors in the transmission and prevention of malaria in the Panama Canal Zone. *Ann. Trop. Med. Parasit.* **4**, 179–223.

DRAKE-BROCKMAN, R. E. (1913). Some notes on *Stegomyia fasciata* in the coast towns of British Somaliland. *J. Lond. Sch. Trop. Med.* **2**, 166–9.

DUNN, L. H. (1927). Tree-holes and mosquito breeding in West Africa. *Bull. Ent. Res.* **18**, 139–44.

DUTTON, J. E. (1903). Report of the malaria expedition to the Gambia. *Thomp. Yates Labs. Rep.* **5**. And as *Lpool Sch. Trop. Med.* Mem. X.

GARNHAM, P. C. C., HARPER, J. A. and HIGHTON, R. B. (1946). The mosquitoes of the Kaimosi Forest, Kenya Colony, with special reference to yellow fever. *Bull. Ent. Res.* **36**, 473–96.

GORDON, R. M. (1922). Note on the bionomics of *Stegomyia calopus* Meigen in Brazil. *Ann. Trop. Med. Parasit.* **16**, 425–39.

GRAHAM, W. M. (1911). Results obtained from a monthly examination of the native domestic receptacles at Lagos, Southern Nigeria in 1910–11. *Bull. Ent. Res.* **2**, 127–39.

HADDOW, A. J. (1945). On the mosquitoes of Bwamba County, Uganda. I. Description of Bwamba with special reference to mosquito ecology. *Proc. Zool. Soc. Lond.* **115**, 1–13.

HADDOW, A. J. (1948). The mosquitoes of Bwamba County, Uganda. VI. Mosquito breeding in plant axils. *Bull. Ent. Res.* **39**, 185–202.

HADDOW, A. J., VAN SOMEREN, E. C. C., LUMSDEN, W. H. R., HARPER, J. O. and GILLETT, J. D. (1951). The mosquitoes of Bwamba County, Uganda. VIII. Records of occurrence behaviour and habitat. *Bull. Ent. Res.* **42**, 207–38.

HAMLYN-HARRIS, R. (1927). Notes on the breeding places of two mosquitoes in Queensland. *Bull. Ent. Res.* **17**, 411–14.

REFERENCES

HAWORTH, W. E. (1922). A new breeding place for mosquitoes. *Trans. R. Soc. Trop. Med. Hyg.* **16**, 201.

HORNE, J. H. (1913). Notes on distribution and habits of *Stegomyia* mosquitoes in Madras. *Proc. 3rd Meet. Gen. Mal. Comm. Madras 1912*, pp. 197–9.

HOWLETT, F. M. (1913). *Stegomyia fasciata. Proc. 3rd Meet. Gen. Mal. Comm. Madras 1912*, p. 205.

INGRAM, A. (1919). The domestic breeding mosquitoes of the Northern Territories of the Gold Coast. *Bull. Ent. Res.* **10**, 47–58.

JAMES, S. P., DA SILVA, W. T. and ARNDT, E. W. (1914). Report on a mosquito survey of Colombo and the practicability of reducing *Stegomyia* and some other mosquitoes in the sea-port. Government Record Office, Colombo.

JONES, H. L. (1925). Report of the Health Depart for year ending 30 June 1922. *Northern Territory Australia*: *Rep. Adminstr. 1921–2*, pp. 18–20. Abstract in *Rev. Appl. Ent.* **13**, 145.

LESTER, A. R. (1927). The coconut palm: its potentialities in providing breeding places for mosquitoes. *J. Trop. Med.* (*Hyg.*) **30**, 137–45.

LEWIS, D. J. (1947). General observations on mosquitoes in relation to yellow fever in the Anglo-Egyptian Sudan. *Bull. Ent. Res.* **37**, 543–66.

LISTON, W. G. and AKULA, T. G. (1913). A *Stegomyia* survey of the City and Island of Bombay. *Proc. 3rd Meet. Gen. Mal. Comm. Madras 1912*, pp. 187–8.

LUTZ, A. and MACHADO, A. (1915). Viajem pelo rio S. Francisco etc. *Mem. Inst. Osw. Cruz*, **7**, 5–50.

MACFIE, J. W. S. (1922). The effect of saline solutions and sea water on *Stegomyia fasciata*. *Ann. Trop. Med. Parasit.* **15**, 377–80.

MACFIE, J. W. S. and INGRAM, A. (1916). The domestic mosquitoes of Accra. *Bull. Ent. Res.* **7**, 162.

MACGREGOR, M. E. (1921). The influence of the hydrogen-ion concentration in the development of mosquito larvae. *Parasitology*, **13**, 348–51.

MHASKAR, K. S. (1913). *Stegomyia* survey of Karachi. *Proc. 3rd Meet. Gen. Mal. Comm. Madras 1912*, pp. 189–92.

RIGOLLOT, S. (1927). A propos de la prophylaxie de la fièvre jaune à la Côte Occidentale d'Afrique. *Bull. Soc. Path. Exot.* **20**, 859–65.

RIQUEAU (1929). Les trous de crabes, gîtes à larves. *Bull. Soc. Path. Exot.* **22**, 175–9.

RUDOLFS, W. and LACKEY, J. B. (1929). The composition of water and mosquito breeding. *Amer. J. Hyg.* **9**, 160–80.

SHANNON, R. C. (1931). The environment and behaviour of some Brazilian mosquitoes. *Proc. Ent. Soc. Wash.* **33**, 1–27.

SOPER, F. L., WILSON, D. B., LIMA, S. and ANTUNES, W. S. (1943). *The Organisation of Permanent Nation-wide anti-*Aëdes aegypti *Measures in Brazil*. Rockefeller Foundation, New York.

STEPHANIDES, T. (1937). The mosquitoes of the island of Corfu, Greece. *Bull. Ent. Res.* **28**, 405–7.

TEESDALE, C. (1941). Pineapple and banana plants as sources of *Aëdes* mosquitoes. *E. Afr. Med. J.* **18**, 260–7.

THOMAS, H. W. (1910). The sanitary conditions and diseases prevailing in Manaos, North Brazil, 1905–1909. *Ann. Trop. Med. Parasit.* **4**, 7–55.

WANSON, M. (1935). Note sur les trous de crabes, gîtes larvaires. *Ann. Soc. Belge Méd. Trop.* **5**, 575–85.

(*c*) NATURAL ENEMIES

ATKINSON, M. (1909). *J. Trop. Med.* (*Hyg.*) **12**, 17.

BACOT, A. W. (1916). *Rep. Yell. Fev. Comm., London*, **3**, 1–191.

BARBER, M. A. and KING, C. E. (1927). *Publ. Hlth Rep., Wash.*, **42**, 3189–93.

BARBER, M. A. and KING, C. E. (1928). *Trans. R. Soc. Trop. Med. Hyg.* **21**, 429.

BARRAUD, P. J. (1934). See reference in chapter II (*a–b*), p. 146.

BATES, H. W. (1863). *A Naturalist on the River Amazon*. London. (6th ed. 1891.)

BEDFORD, G. A. H. (1919). *U.S. Afr. Dept. Agric. 5th and 6th Ann. Rep. Vet. Res.* pp. 739–43. Abstract in *Rev. Appl. Ent.* **8**, 9.

BIONOMICAL

BENTLEY, C. A. (1910). *J. Bombay Nat. Hist. Soc.* **20**, 392–422.

BRITTON, W. E. and VIERECK, H. L. (1905). *4th Rep. State Ent. Connecticut Agric. Exp. Sta. for 1904*, pp. 253–310.

BRUMPT, E. (1925). *Ann. Parasit. Hum. Comp.* **3**, 403–11.

BUXTON, P. A. and HOPKINS, G. H. E. (1927). *Mem. Lond. Sch. Hyg. Trop. Med.* no. 1.

CARTER, H. F. (1919). *Ann. Trop. Med. Parasit.* **12**, 289–302.

CHRISTOPHERS, S. R. (1906). *Sci. Mem. Off. San. Med. Dept., Govt. India*, no. 25.

COMBES, P. (1896). *Rev. Scient.* (4), **6**, 751–3.

COVELL, G. (1941). *Malaria Control by Anti-mosquito Measures.* Ed. 2. Thacker and Co., London.

DARWIN, C. (1875). *Insectivorous Plants*, pp. 408–9, 430.

D'EMMEREZ DE CHARMOY, D. (1902). *Rep. Mal. Enq. Comm., Mus. Desjardins.* Port Louis, Mauritius.

DEMPWOLFF (1904). *Z. Hyg. InfektKr.* **47**, 81–132.

DERIVAUX, R. C. (1916). *U.S. Publ. Hlth Rep.* **31**, 1228–30.

DRAKE-BROCKMAN, R. E. (1913). See under (*b*).

DUFOUR, L. (1864). *Ann. Soc. Ent. France 1864*, p. 633.

EDITORIAL (1916). *J. Trop. Med.* (*Hyg.*) **19**, 34.

EYSELL, E. (1924). *Die Stechmucken.* In Mense's *Handb. TropenKr.* ed. 3, **1**, 176–303.

FERTON, C. (1901). *Ann. Soc. Ent. France 1901*, pp. 83–145.

FOLEY, F. H. and YVERNAULT, A. (1908). *Bull. Soc. Path. Exot.* **1**, 172–3.

GALLI-VALERIO, B. and ROCHAZ DE JONGH, J. (1908). *Zbl. Bakt.* (*Abt. 1, Orig.*), **46**, 134.

GALLI-VALERIO, B. and ROCHAZ DE JONGH, J. (1909). *Zbl. Bakt.* (*Abt. 1, Orig.*), **49**, 557.

GARNHAM, P. C. C., HARPER, J. O. and HIGHTON, J. (1946). See under (*b*).

GASANOV, A. P. (1938). *Med. Parasit., Lening.*, **7**, 617. Abstract in *Rev. Appl. Ent.* **27**, 76.

GILES, G. M. (1902). *A Handbook of Gnats or Mosquitoes.* Ed. 2. John Bale Sons and Danielsson, London, p. 149.

GORDON, R. E. (1922). *Ann. Trop. Med. Parasit.* **16**, 425–39.

GREEN, E. E. (1905). *Spol. Zeyl.* **2**, 159–64.

HADDOW, A. J. (1942). *Proc. R. Ent. Soc. Lond.* **17**, 71–98.

HAMLYN-HARRIS, R. (1929). *Proc. R. Soc. Queensl.* **41**, 23–38.

HAMLYN-HARRIS, R. (1930). *Bull. Ent. Res.* **24**, 159.

HART, J. H. (1906). *Bull. Misc. Inform., Trinidad Bot. Dept.* no. 50, p. 37.

HARVEY, F. L. (1880). *Amer. Nat.* **14**, 896.

HAYES, T. H. (1930). *U.S. Nav. Med. Bull.* **28**, 194–222.

HEARLE, E. (1926). *Canad. Nat. Res. Council Rep.* **17**, 1–94.

HERMS, W. B. and GRAY, H. F. (1940). *Mosquito Control.* Commonwealth Fund, New York and Oxford Univ. Press, London.

HINMAN, E. H. (1934). *J. Trop. Med.* (*Hyg.*) **37**, 129–34, 145–50. With bibliography.

HOFFMANN, W. E. (1927). *Lingnaan Agric. Rev.* **4**, 77–93. Abstract in *Rev. Appl. Ent.* **15**, 233.

HOWARD, L. O. (1910). *U.S. Dept. Agric., Bur. Ent. Bull.* **88**, 1–126.

HOWARD, L. O., DYAR, H. G. and KNAB, F. (1912). *The Mosquitoes of N. and C. America and the West Indies.* Carnegie Institute, Washington. Vol. I, pp. 156–79.

JACKSON, N. (1953). *Proc. R. Ent. Soc. Lond.* (A), **28**, 10–12.

JENNINGS, A. H. (1908). *Proc. Ent. Soc. Wash.* **10**, 61–2.

KNAB, F. In HOWARD, DYAR and KNAB (1912), p. 170.

KRAUSE, E. (1887). *Kosmos, Lwów*, **1**, 80.

LAIRD, M. (1947). *Trans. R. Soc. N.Z.* **76**, 453–76.

LAMBORN, R. H. *et al.* (1890). *Lamborn Prize Essays.* New York.

LEON, N. (1910). *Zbl. Bakt. Abt. 1, Orig.* **57**, 148–54.

LEWIS, D. J. (1942). *Sudan Notes*, **25**, 141. Abstract in *Rev. Appl. Ent.* **31**, 214.

LISCHETTI, A. B. (1919). *Physis, B. Aires*, **4**, 591–5.

LISTON, W. G. (1901). *Indian Med. Gaz.* **36**, 361–6.

McCOOK, H. C. In LAMBORN, R. H. (1890).

MACFIE, J. W. S. (1923). *Bull. Ent. Res.* **13**, 403.

REFERENCES

MacGregor, M. E. (1920). *J. R. Army Med. Cps*, **34**, 248–50.
Marchoux, Salimbeni and Simond. See under (*a*).
Matheson, R. and Hinman, E. H. (1928). *Amer. J. Hyg.* **8**, 279–92.
Matheson, R. and Hinman, E. H. (1929). *Amer. J. Hyg.* **9**, 188–91.
Matheson, R. and Hinman, E. H. (1931). *Amer. J. Hyg.* **14**, 99–108.
Mathis, C. and Berland, L. (1933). *C.R. Acad. Sci., Paris*, **197**, 271–2.
Mathis, M. (1938). *Bull. Soc. Path. Exot.* **33**, 301–5.
Morgan, H. C. and Cotton, E. C. (1908). *Science*, n.s. **27**, 28–30.
Nichols, L. (1912). *Bull. Ent. Res.* **3**, 257.
O'Connell, M. D. (1912). *J. R. Army Med. Cps*, **19**, 491–3.
Osterhout. Cited in Howard, Dyar and Knab (1912).
Paine, R. W. (1934). *Bull. Ent. Res.* **25**, 1–32.
Paiva, C. A. (1910). *Rec. Indian Mus.* **5**, 187–90.
Pruthi, S. (1928). *Ind. J. Med. Res.* **16**, 153–7.
Purdy, M. S. (1920). *Publ. Hlth Rep. Wash.* **35**, 44, 2556–70.
Sailer, R. J. and Lienk, S. E. (1954). *Mosquito News*, **14**, 14–16.
Seguy, E. (1923). *Histoire naturelle des moustiques de France*. Paul Lechevalie, Paris.
Senior White, R. (1920). *Ind. J. Med. Res.* **8**, 319.
Shannon, R. C. (1929). *Encyc. Ent.* B 11, Dipt. **5**, 155–62 (in article by C. P. Alexander).
Shannon, R. C., Burke, A. W. and Davis, N. C. (1930). *Amer. J. Trop. Med.* **10**, 145–50.
Smith, J. B. (1910). *Ent. News*, **21**, 437–41.
Swezey, O. H. (1931). *Rept. Comm. Expt. Sta., Hawaiian Sugar Pl. Assoc. Entom. Sect. for 1929–30*, pp. 23–30.
Theobald, F. V. (1901). *Mono. Culicidae*, **1**, 73.
Underwood, W. L. (1903). *Pop. Sci. Mon. 1903*, p. 466.
Viereck, H. L. (1904). (In Smith, J. B. *Rep. N.J. State Agric. Exp. Sta.* pp. 399–404.)
Wahlberg, H. (1847). *Öfvers. Vetensk.Accad. Förh., Stockh.*, **4**, 257–9.
Waterston, J. (1918). *Bull. Ent. Res.* **9**, 1–12.
Wilson, H. C. (1914). *Ind. J. Med. Res.* **1**, 691–701.
Yamada, S. (1927). *Kontyû*, **2**, 143–54. Abstract in *Rev. Appl. Ent.* **16**, 94.
Young, C. J. (1921). *Ann. Trop. Med. Parasit.* **15**, 301–8.

(*d*) PARASITES

Alessandrini, G. (1909). *Malaria, Lpz.*, **1**, 133–7.
Bacot, A. W. (1916). *Rep. Yell. Fev. Comm., London*, **3**, 1–191.
Balfour, A. (1923). *J. R. Army Med. Cps*, **40**, 122–7.
Bates, M. (1949). *The Natural History of Mosquitoes*. Macmillan and Co., New York.
Blanchard, E. (1890). *Traité de zoologie médicale*, **2**, 497–8.
Blanchard, R. (1901). *Bull. Soc. Zool. France*, **16**, 72–3.
Boyd, J. E. M. (1922). *J. R. Army Med. Cps*, **38**, 459–60.
Braun, A. (1855). *Algarum unicelliularium genera nova et minus cognita*. Leipzig.
Bresslau, E. von and Buschkiel, E. (1919). *Biol. Zbl.* **19**, 325–35.
Brumpt, E. (1936). *Précis de parasitologie*. Ed. 5. Masson et Cie, Paris.
Brumpt, E. (1938). *Ann. Parasit. Hum. Comp.* **16**, 153–8.
Canning, E. V. (1957). *Trans. R. Soc. Trop. Med. Hyg.* **51**, 8.
Causey, O. R. (1939). *Amer. J. Hyg.* **29**, Sect. C, 79–81; Sect. D, 131–2; **30**, Sect. D, 69–71.
Chatterjee, G. C. (1901). *Indian Med. Gaz.* **16**, 371–2.
Chorine, V. and Baranoff, N. (1929). *C.R. Soc. Biol., Paris*, **101**, 1025–6.
Christophers, S. R. (1901). *Rep. Mal. Comm. R. Soc.* (4), London.
Christophers, S. R. (1952). *Riv. Parassit.* **13**, 21–8.
Colas-Belcour, J. (1954). *Bull. Soc. Path. Exot.* **4**, 236–7.
Corradetti, A. (1937). *Riv. Parassit.* **1**, 39–50.
Couch, J. N. (1945). *J. Elisha Mitchell Sci. Soc.* **61**, 123–36.
Couch, J. N. and Dodge, H. R. (1947). *J. Elisha Mitchell Sci. Soc.* **63**, 69–79.

BIONOMICAL

DURHAM, M. E. and MYERS, W. (1902). *Lpool Sch. Trop. Med.* Mem. VII.
DYE, L. (1905). *Arch. Parasit.* **9**, 5–77.
DYE, W. H. (1924). *J. R. Army Med. Cps*, **42**, 87–102.
ECKSTEIN, F. (1922). *Zbl. Bakt.* (*Orig.*) **88**, 128–35.
EDWARDS, F. W. (1922). *Bull. Ent. Res.* **13**, 161–7.
EDWARDS, F. W. (1932). *Genera Insectorum.* Brussels. Fasc. 194. Fam. Culicidae, p. 92.
FENG, LAN CHOU (1930). *Ann. Trop. Med. Parasit.* **24**, 361–2.
FENG, LAN CHOU (1933*a*). *Lingnan Sci. J.* **12**, Suppl. 28.
FENG, LAN CHOU (1933*b*). *Chinese Med. J.* **47**, 168–78.
FENG, LAN CHOU and HOEPPLI, R. (1933). *Chinese Med. J.* **47**, 1191–9.
FICALBI, E. (1891). Quoted by Blanchard, E. (1890).
GALLI-VALERIO, B. and ROCHAZ DE JONGH, J. (1905). *Zbl. Bakt.* (*Abt. 1, Orig.*), **38**, 174–7.
GALLI-VALERIO, B. and ROCHAZ DE JONGH, J. (1905). *Zbl. Bakt.* (*Abt. 1, Orig.*), **40**, 630–3.
GALLI-VALERIO, B. and ROCHAZ DE JONGH, J. (1907). *Zbl. Bakt.* (*Abt. 1, Orig.*), **43**, 470.
GANAPATI, P. N. and TATE, P. (1949). *Parasitology*, **39**, 291–4.
GENDRE, E. (1909). *Bull. Soc. Path. Exot.* **2**, 106–8.
GODOY, A. and PINTO, C. (1923). *Brazil-med.* **37**, 29–33.
GROS, H. (1904). *C.R. Soc. Biol., Paris*, **56**, 56–7.
HESSE, E. (1918). *C.R. Acad. Sci., Paris*, **166**, 569–72.
HODGES, A. (1902). *J. Trop. Med.* (*Hyg.*), **5**, 293–300.
HOWARD, DYAR and KNAB. See under (*c*).
IYENGAR, M. O. T. (1930). *Trans. F.E.A.T.M. 7th Congr. Calcutta*, **3**, 128–42.
IYENGAR, M. O. T. (1935). *Parasitology*, **27**, 440–9.
IYENGAR, M. O. T. (1938). *Proc. Nat. Inst. Sci. India*, **4**, 237–9.
JACOBSON, E. (1911). *Tijdschr. Ent.* **54**, 157–61.
JAFFE, J. (1907). *Arch. Protistenk.* **9**, 100–7.
JASWANT SINGH, RAMAKRISHNAN, S. P. and DAVID, A. (1950). *Indian J. Malar.* **4**, 189–92.
JENKINS, D. W. and WEST, A. S. (1954). *Mosquito News*, **14**, 138–43.
JOHNSON, H. P. (1903). *Rep. Ent. Dep. N.J. Agric. Exp. Sta. for 1902*, Appendix, pp. 559–93.
JOYEUX, C. (1918). *Bull. Soc. Path. Exot.* **11**, 530–47.
KEILIN, D. (1921*a*). *Parasitology*, **13**, 216–24.
KEILIN, D. (1921*b*). *Parasitology*, **13**, 225–34.
KUDO, R. (1921). *J. Morph.* **35**, 153–93.
KUDO, R. (1922). *J. Parasit.* **8**, 70–9.
KUDO, R. (1924*a*). *Illinois Biol. Monogr.* **9**, nos. 2, 3.
KUDO, R. (1924*b*). *Arch. Protistenk.* **49**, 147–62.
KUDO, R. (1924*c*). *J. Parasit.* **11**, 84–9.
KUDO, R. (1925). *Zbl. Bakt.* (*Abt. 1, Orig.*), **96**, 428–40.
LAIRD, M. (1946). *Trans. R. Soc. N.Z.* **76**, 158–61.
LAIRD, M. (1947). *Trans. R. Soc. N.Z.* **76**, 471.
LAIRD, M. (1956). *J. Parasit.* **42**, 53–5.
LAMBORN, W. A. (1921). *Parasitology*, **13**, 213–15.
LANGERON, M. (1929). *Ann. Parasit. Hum. Comp.* **7**, 107–11.
LAVERAN, A. (1902). *C.R. Soc. Biol., Paris*, **54**, 233–5.
LEGER, L. (1902). *C.R. Soc. Biol., Paris*, **54**, 354–6.
LEGER, L. and DUBOSCQ, O. (1903). *C.R. Assoc. Franc. Av. Sci.* Sess. 31, 1902, pp. 703–4.
LEON, N. (1924). *Ann. Parasit. Hum. Comp.* **2**, 211–13.
LISTON, G. L. (1901). *Indian Med. Gaz.* **36**, 364.
LUTZ, A. and SPLENDORE, A. (1908). *Zbl. Bakt.* (*Abt. 1, Orig.*), **46**, 311–15, 652.
MACARTHUR, W. P. (1922). *J. R. Army Med. Cps*, **38**, 83–92.
MACFIE, J. W. S. (1916). *Rep. Accra Lab. for year 1915*, pp. 76–9. Abstract in *Rev. Appl. Ent.* **5**, 47.
MACFIE, J. W. S. (1917). *Rep. Accra Lab. for year 1916*, pp. 67–75. Abstract in *Rev. Appl. Ent.* **6**, 16.

REFERENCES

MANALANG, C. (1930). *Philipp. J. Sci.* **42**, 279–80.
MARCHOUX, SALIMBENI and SIMOND. See under (*a*).
MARSHALL, J. F. and STALEY, J. (1929). *Parasitology*, **21**, 158–60.
MARTINI, E. (1920). *Arch. Schiffs.- u. Tropenhyg.* **24**, Beih. 1.
MARTIRANO, F. (1901). *Zbl. Bakt.* (*Abt. 1, Orig.*), **30**, 849–52.
MASSEGUIN, A. and PALINACCI, A. (1954). *Bull. Soc. Path. Exot.* **47**, 391–2.
MASSEGUIN, A., PALINACCI, A. and BRUMPT, V. (1954). *Bull. Soc. Path. Exot.* **47**, 234–6.
MATHIS, C. (1914). *C.R. Soc. Biol., Paris*, **77**, 297–300.
MATHIS, M. and BAFFET, O. (1934). *Bull. Soc. Path. Exot.* **27**, 435–7.
MISSIROLI, A. (1928). *Riv. di Malariol.* **7**, 1–3.
MISSIROLI, A. (1929). *Riv. di Malariol.* **8**, 393–400.
MONTOUSSIS, K. (1929). *Arch. Schiffs.- u. Tropenhyg.* **33**, 330–3.
MUSPRATT, J. (1945). *J. Ent. Soc. S. Afr.* **8**, 13–20.
MUSPRATT, J. (1946). *Ann. Trop. Med. Parasit.* **40**, 10–17.
MUSPRATT, J. (1947). *J. Ent. Soc. S. Afr.* **10**, 131.
MUSPRATT, J. (1947). *Parasitology*, **38**, 107–10.
NOC, F. (1920). *Bull. Soc. Path. Exot.* **13**, 672–9.
NOC, F. and STEVENEL, L. (1913). *Bull. Soc. Path. Exot.* **6**, 708–10.
NOLLER, W. (1917). *Arch. Schiffs.- u. Tropenhyg.* **21**, 53–94.
NOLLER, W. (1920). *Arch. Protistenk.* **41**, 169–88.
NOVY, F. G., MACNEAL, W. J. and TORREY, H. N. (1907). *J. Infect. Dis.* **4**, 223–76.
PARKER, H. B., BEYER, G. E. and POTHIER, O. L. (1903). *Yell. Fev. Inst. Bull.* no. 13.
PATTON, W. S. (1907). *Brit. Med. J.* **2**, 78–80.
PATTON, W. S. (1912). *Sci. Mem. Off. Med. San. Dept., Govt. India* (n.s.), no. 57.
PATTON, W. S. and CRAGG, F. W. (1913). *Textbook of Medical Entomology.* Christian Lit. Soc. for India, London and Madras.
PERRONCITO, E. (1900). *G. Accad. Med. Torino*, (4), **6**, 387–8.
PETTIT, E. H. (1903). *Michigan Agric. Exp. Sta.* Special Bull. no. 17.
RAMAKRISHNAN, S. P., DAVID, A. and NAIR, C. P. (1956). *Indian J. Malar.* **10**, 313–15.
RAY, H. (1933). *Parasitology*, **25**, 392–5.
ROSS, R. (1895). *Trans. S. Indian Branch Brit. Med. Ass.* **6**, 334–50.
ROSS, R. (1898). *Trans. S. Indian Branch Brit. Med. Ass.* **9** (reprinted *Ind. Med. Gaz. 1898*).
ROSS, R. (1906). *J. Hyg., Camb.*, **6**, 101–9.
RUGE, R. (1903). *Festschr. R. Koch*, p. 174.
SAMBON, L. W. (1922). *J. Trop. Med.* (*Hyg.*), **25**, 170–85.
SCHAUDINN, F. (1904). *Arb. GesundhAmt., Berlin*, **20**, 387–439.
SCHOO, H. J. M. (1902). *Ned. Tijdschr. Geneesk.* **1**, 283.
SÉGUY, E. (1923). *Histoire naturelle des moustiques de France.* Paris, pp. 45–51.
SELLARDS, A. W. and SILER, J. F. (1928). *Amer. J. Trop. Med.* **8**, 299–304.
SERGENT, ED. and ET. (1904). *C.R. Soc. Biol., Paris*, **56**, 100–2.
SERGENT, ED. and ET. (1906). *C.R. Soc. Biol., Paris*, **60**, 291–3.
SINTON, J. A. (1917). *Ind. J. Med. Res.* **5**, 192–4.
SINTON, J. A. and SHUTE, P. G. (1939). *J. Trop. Med.* (*Hyg.*), **42**, 125–6.
SMITH, J. B. (1903). *Rep. Ent. Dep. N.J. Agric. Exp. Sta. for 1902*, pp. 509–93.
SOPARKAR, M. B. (1918). *Ind. J. Med. Res.* **5**, 512–15.
SPEER, A. J. (1927). *Hyg. Lab. Bull. Wash.* no. 146.
STEINHAUS, E. A. (1946). *Insect Microbiology.* Comstock Publ. Co., Ithaca, New York.
STEVENSON, A. C. and WENYON, C. M. (1915). *J. Trop. Med.* (*Hyg.*), **18**, 196.
STILES, C. W. (1903). *Hyg. La. Publ. Hlth, Marine Hosp. Serv. Bull.* 13, pp. 15–17.
THAXTER, R. (1888). *Mem. Boston Soc. Nat. Hist.* **4**, 133–201.
THOMPSON, J. G. and ROBERTSON, A. (1925). *J. Trop. Med.* (*Hyg.*), **28**, 419–24.
TOSATTO, E. (1891). *Insect Life*, **4**, 285.
UCHIDA and MIYAZAKI (1935). *Proc. Imp. Acad. Japan*, **11**, 73–76.
VAN THIEL, P. H. (1921). *Tijdschr. verglijk. Geneesk.* **6**, no. 4.

VAN THIEL, P. H. (1922). *Tijdschr. verglijk. Geneesk.* **7**, no. 25.
VAN THIEL, P. H. (1925). *Arch. Schiffs.- u. Tropenhyg.* **29**, Beih. 1, 396–400.
VAN THIEL, P. H. (1930*a*). *Zbl. Bakt.* (*Abt. 1, Orig.*), **117**, 103–12.
VAN THIEL, P. H. (1930*b*). *Acta leidensia*, **5**, 238–53.
VAN THIEL, P. H. (1954). *J. Parasit.* **40**, 271–9.
VANEY, C. and CONTE. Referred to by DYE (1905).
VISWANATHAN, D. E. and BHATT, H. R. (1948). *J. Nat. Mal. Soc.* (*India,*) **7**, 207–11.
WALKER, A. J. (1938). *Ann. Trop. Med. Parasit.* **32**, 231–41.
WALLACE, F. G. (1943). *J. Parasit.* **29**, 196–205.
WENYON, C. M. (1911). *Parasitology*, **4**, 273–344.
WENYON, C. M. (1926). *Protozoology*. Baillière, Tindall and Cox, London.

IV
MEDICAL

(*a*) RELATION TO DISEASE

As the chief vector of yellow fever *Aëdes aegypti* has long been notorious as the 'yellow fever mosquito'. Nor is this its only role as a disease transmitter, for it is also the chief vector of dengue; besides being known or suspected of conveying some other diseases of man, it is the chief agent in several important virus diseases of animals and the vector of *Dirofilaria* in dogs.

YELLOW FEVER

Following experimental proof by Reed and his collaborators *A. aegypti* has now for some fifty years been recognised as the chief vector of yellow fever, at least in its urban form. A short account of researches leading to this discovery has already been given in chapter I. A large literature has since grown up giving the results of countless researches on its epidemiology, the behaviour of the virus, the bionomics of its vectors and other aspects of the disease. It is not possible to give here a detailed account of yellow fever in all these relations, but the main facts regarding its epidemiology and the part played in this by *A. aegypti* and some other mosquitoes may be noted.

For an account of early work on the epidemiology and control of yellow fever see Boyce (1909) and Howard, Dyar and Knab (1912). A good later account of the disease is that by Brumpt (1936) and a review of jungle yellow fever is given by Bates (1949). A very complete account dealing with cultivation of the virus and other aspects is that by Strode (1951) incorporating articles by various authorities describing the work carried out during the previous thirty years by the International Health Division of the Rockefeller Foundation (see also review and abstract of this work by Findlay, 1952).

Yellow fever in the past was chiefly known from outbreaks and epidemics in seaports and towns in tropical America, sometimes occurring as far north as Philadelphia, and as cases on ships or in harbours frequented by shipping from this source.

It is now known that it also occurs as an endemic disease not only in America but throughout a large part of Africa, two forms of the disease being recognised, namely the *urban* and the *sylvan* form. A third form, *rural* yellow fever, has been given by Kirk (1943) for certain types of its occurrence.

Urban yellow fever occurs as the typical form of the disease in epidemics in towns or similar situations, the vector being *A. aegypti* and infection being from man to man. It was this form in which yellow fever occurred as epidemics in America and against which control measures were mainly directed. Many references to such outbreaks and the measures carried out in their control and prevention are given by Carter, Connor and others in the literature. Outbreaks of the urban type with a similar epidemiology are also known to occur in Africa. Findlay and Davey (1936) refer to an epidemic in the Gambia in 1934, and that there have been others in other parts of Africa since, and state that at least one of these was similar to the American outbreaks with *A. aegypti* as the principal vector.

Sylvan or forest yellow fever is largely endemic in character and evidenced chiefly by presence of immunity in the indigenous population, which may sometimes be as high as 60 per cent or more. In this form *A. aegypti* is not the vector, or takes but little part, and infection is not primarily derived from man. Sylvan yellow fever is present throughout large parts of tropical America in the forest regions and throughout a large part of Africa south of the Sahara. The endemic area in Africa as given by the World Health Organisation (1952) is that lying between the parallels of 15° N. and 10° S. with an outlying area in Northern Rhodesia (Barotse Province). The disease is not known in any form east of the African continent.

The rural form is a term usefully applied where outbreaks of a mixed character occur, *A. aegypti* often taking some part, such as are met with in some African areas. As noted later much has still to be worked out regarding the epidemiology in the sylvan and rural forms.

In America it was at first the urban form that constituted 'yellow fever' and the forest form largely became recognised only as control measures eliminated or greatly controlled the urban epidemics. It then became evident that there existed the forest form and that urban epidemics commonly started from infection introduced by cases that had contracted the disease in the forest, the epidemic then being carried on as a typical urban outbreak by *A. aegypti* (see Bauer, 1928; Soper *et al.* 1933; Soper, 1938; and Aragao, 1939, who gives an interesting description of such conditions in Brazil. See also description of the situation by Sawyer, 1942).

Recognition of sylvan yellow fever in Africa had a somewhat different history. Though yellow fever had been recorded from West Africa, there was in the early stages of research upon this disease no realisation of its extent and importance in Africa, or even certainty of its existence. Boyce (1910), however, clearly showed that yellow fever had long been, and still was, present in West Africa. Later observations not only confirmed Boyce's findings, but showed that throughout a large part of tropical Africa and extending far to the east in that continent there existed in the indigenous population a considerable percentage of those showing immunity to the disease, indicating that yellow fever must quite recently have been present, if not still so, over large areas where its existence had never before been suspected. Further it became clear that the epidemiology of the disease under these conditions was quite different from that of the urban epidemics. That the presence of immunity over so great an area did in fact indicate existence of the disease was very clearly shown shortly after this discovery by the occurrence in the Anglo-Egyptian Sudan, far to the east, of an epidemic of yellow fever in which *A. aegypti* was not an important vector (Kirk). Soon observations from Uganda, Kenya and other areas established with certainty the wide occurrence of sylvan yellow fever. (See Beeuwkes, Bauer and Mahaffy (1930).)

The above findings in America and Africa greatly stimulated research, which has continued with increasing intensity up to the present time. Such research has included the determination of areas where immunity exists in indigenous populations, the species of mosquito in which the virus can be demonstrated in wild-caught specimens, or which can be shown to transmit the disease experimentally, the forest animals that provide the source of infection and how infection of man is brought about, as also the whole question of the nature and properties of the virus and how these may be made use of in man's protection.

Essential to all such work has been the fact that monkeys, as first noted by Stokes, Bauer and Hudson (1928), were susceptible to the disease and could be used experimentally in research on the virus and in other ways. Another important finding, upon which much of modern research has depended, has been the discovery that mice were susceptible and could be used, not only as a means of ascertaining the presence of the virus, but also as a test for the existence of immunity in man or animals, the blood of such humans or animals as possessed immunity being capable of preventing the disease in mice inoculated with the virus. It is this last test, known as the 'mouse protection test', that has been so extensively used in mapping the endemic area of yellow fever both in America and in Africa. It also enables the presence of immunity in animals suspected to be sources of the virus to be determined. The susceptibility of mice to infection with the virus, as will be seen later, also enters largely into the preparation of suitable strains of the virus for use in protection.

Various techniques are given in the literature for determining the presence of the virus in mosquitoes. Briefly the basis of such procedure is that the mosquitoes are ground up in a diluent and the product after filtration injected intra-cerebrally or otherwise into mice (see Waddell, 1945; Smithburn and Haddow, 1946; Smithburn, Haddow and Lumsden, 1949; Ross and Gillett, 1950). The following records of recovery of the virus from wild-caught forest mosquitoes in America are given by Bates:

Haemagogus spegazinnii	17 occasions
Aëdes leucocelaenus	3 occasions
Haemagogus lucifer and/or *equinus*	1 occasion
88 various Sabethine mosquitoes	1 occasion

In Africa Lumsden (1955), summarising results to date, notes that, though many species of mosquito have been shown capable of transmitting yellow fever virus experimentally, only three have been incriminated in the field in Africa. These are:

Aëdes aegypti	Beeuwkes and Hayne (1931), Taylor (1951).
A. simpsoni	Mahaffy, Smithburn, Jacobs and Gillett (1942).
A. africanus	Smithburn, Haddow and Lumsden (1949).

Besides these proved vectors, a number of species in American and African forests have been shown capable of transmitting yellow fever virus experimentally and others, though incapable of conveying the virus, have been shown to harbour the virus when fed on infected material. Lists of such species will be found given under 'yellow fever' by Manson-Bahr (1954). A list of African species classified as above is also given by Muspratt (1956). Most of the species shown able to transmit experimentally are in the genus *Aëdes* or (in America) *Haemagogus*. But *Eretmapodites chrysogaster* and *Culex thalassius* in Africa and *Trichoprosopon frontosus* in America also have been shown to transmit experimentally and among those harbouring the virus are species of *Psorophora*, *Taeniorhynchus* and other genera. Even *Culex fatigans* has been shown to be capable in some degree of transmission (Davis, 1933).

Of forest animals that may be sources of the virus are especially species of monkey. In tropical America it is now known that epidemics occur periodically among the monkeys in forest areas, the human disease being a secondary phenomenon to this. In Africa baboons have been found sometimes showing immunity and among other primates the small African 'bush-babies' (*Galago*) have been suspected of possibly playing a part in transmission (Bailey, 1947; Smithburn and Haddow, 1949; Haddow, Dick, Lumsden and Smithburn, 1951). Animals other than primates have also been noted as showing immunity

(Kirk and Haseeb, 1953). Of such animals in Africa Muspratt gives: *Hyrax*, the red river hog, mongoose, genets, and the pottos. Birds (heron, kingfisher) may also show immunity (Findlay and Cockburn, 1943).

A curious feature of the epidemiology of forest yellow fever is that the vertical distribution of species of mosquito and of the mammalian fauna has been found to be important. Thus in the American forests it is especially when trees are felled, bringing down their high-level mosquito fauna, that the chief danger of infection arises. In Africa *Aëdes africanus* plays the chief role in maintaining, receiving and transmitting infection in the high-level monkey community, though it is *A. simpsoni*, a frequenter of banana plantations where it breeds, that in many areas transmits infection to man. Thus the process envisaged is that infected monkeys raiding plantations may be bitten by *A. simpsoni*, or *A. africanus* may bite man in or near the forest and he in turn give infection to *A. simpsoni*. Or there are other possibilities, much still requiring to be worked out in different areas (see Gillett, 1951, 1955).

Studies on the virus have shown that in the mammalian host it is filterable. In the crushed vector it has been said not to be (Brumpt, 1936), though presumably it is so in the fluid commonly used in testing. According to Davis and Shannon (1930) the virus is generally distributed in the body of the mosquito, being obtained from the legs, ovary, salivary glands, mid-gut and hind-gut and occasionally in the dejecta. Aragao and Costa-Lima (1929) also state that the virus may be present in the faeces and that it may be transmitted to monkeys through the broken skin. Attempts to transmit by interrupted feeding, as also by hereditary transmission through the egg, have given negative results (Philip, 1929, 1930; Davis and Shannon, 1930; Stokes *et al.* 1928; Gillett *et al.* 1950).

The time of infectivity in cases of yellow fever in man is during some three to four days in the height of the fever, but the virus has been isolated up to the eighth day (Macnamara, 1954). The time required for the mosquito to become infective following feed on an infective case depends largely upon the temperature. In *A. aegypti* at 24–26° C. Lewis, Hughes and Mahaffy (1942) found it to be eight days, but it may be longer.

An important aspect of virus research has been the development of a safe and effective vaccine which can be used in protection. A vaccine 17 D, prepared from virus cultivated in chick and mouse medium and which has lost pathogenicity to man, has now been widely used (see Theiler and Smith, 1937; Smith and Theiler, 1937; Strode; Rivers, 1952). A vaccine from virus grown on mouse brain is also in use in some countries (Rivers, 1952; Cannon and Dewhurst, 1955).

An outstanding and curious feature of yellow fever epidemiology is absence of the disease from countries to the east of the African continent, in spite of the fact that *A. aegypti* occurs widely and may locally be extremely abundant. It has been considered possible that *A. aegypti* in these countries may not be an efficient vector. That the virus is capable of being transmitted by strains from Java has, however, been shown by Dinger, Schuffner, Snyders and Swellengnebel (1929), and by Hindle (1930). Hindle (1929) has also shown that a strain from India was capable of transmitting experimentally, and recently Gillett and Ross (1955) have shown that a strain from Malaya was capable of transmission.

Whilst *A. aegypti* plays little part in forest or sylvan yellow fever, it still remains the sole vector and cause of urban epidemics and it is the chief agent against which control measures are directed, whether against the insect locally or in connection with its possible spread through aeroplanes and other means.

RELATION TO DISEASE

DENGUE

Good accounts of the epidemiology of dengue are given by Clelland, Bradley and MacDonald (1918) and by Siler, Hall and Hichens (1926)* A detailed survey of the literature to the date of his paper is given by Lumley (1943).

The disease is widespread and commonly occurs as small or larger epidemics, but is stated to be endemic in some areas. In Europe a serious epidemic in Greece in 1928 has been very fully described by Cardamatis (1929), and by Blanc and Caminopetros (1930). Ekblom (1929) gives a map showing the known distribution of dengue at this time. In the Mediterranean, Adrien (1918) describes an epidemic in the island of Rouad (Syria). Dengue is stated to be endemic in West Africa by Muspratt and in Eritrea (Spadero, 1952). Lewis (1955) notes that dengue was formerly present in Port Sudan and McCarthy and Wilson (1948) refer to it as occurring in British East Africa (Kenya), though there have been no known epidemics. An outbreak of a dengue-like fever occurred in 1952 at Newala (Tanganyika Territory) (Lumsden, 1955) and serious epidemics have occurred in and about Durban in 1892, 1926 and 1927, there being in the last outbreak 50,000 cases and sixty deaths (Muspratt, 1956). Further to the east, south Asia through the south Pacific and Australasia would appear particularly to be an area where dengue is widespread and common. It occurs in Madagascar (Legendre, 1918); in parts of India (Barraud, 1928), and especially in such places as Calcutta and Madras where *A. aegypti* is abundant; in the Philippine Islands (Siler *et al.*); in New Guinea (Mackerras, 1946); in New Caledonia (Oman and Christiensen, 1947; Perry, 1950); in the New Hebrides and in a number of islands in the South Pacific (Perry, 1948). In Australia it is recorded by Cleland (1928) as occurring in the east to about 100 miles north of Sydney. In western Australia it occurs only in the tropical parts such as Broome and Kimberley (Jenkins, 1945). Dengue is present also in the New World. Severe epidemics in the past have occurred (Bishopp, 1923; Chandler and Rice, 1923). It was formerly especially prevalent on the Gulf Coast and in former times outbreaks occurred as far north as Boston (Chandler, 1925). On the whole there is a very definite suggestion in its distribution of a relation to that of *A. aegypti*.

A. aegypti is indeed now generally recognised as the usual vector of dengue. Species of *Culex* were at one time thought to be possible vectors (Graham, 1902; Ashburn and Craig, 1907), but the result of experimental work by many authors has not confirmed this. Positive results with *Aëdes aegypti* have been obtained by Cleland, Bradley and MacDonald (1918), by Chandler and Rice (1923), and by Schule (1928).

Other species of *Stegomyia* have, however, been incriminated and may act as efficient vectors, notably *Aëdes* (*Stegomyia*) *scutellaris* which is the vector in New Guinea in areas remote from occurrence of *A. aegypti* and specimens of which transported to Australia produced infection in volunteers, results with some other species being negative (Mackerras, 1946). *A. albopictus* (Simmons, St John and Reynolds, 1930) and *A. hebrideus* (Farner and Bohart, 1948) have been thought to be vectors.

Incubation of the disease may be as short as eight days, but is usually eleven to fourteen days (Schule). It is infective six to eighteen hours before onset and for three or more days after the attack. Mosquitoes become infective eight to ten days from feed and maintain the virus for a long period (Schule). As in yellow fever the virus is generally distributed in

* See also Chandler and Rice (1923), Armstrong (1923, with bibliography), Simmons (1931), Findlay (1951), Muspratt (1956).

the tissues of the mosquito (Holt and Kintner, 1931). Mosquitoes can become infective if fed on infective blood or macerated mosquitoes (St John, Simmons and Reynolds, 1930). Infection is not conveyed, or only with great difficulty, by interrupted feeding or from mosquito to mosquito in copulation (Simmons, St John, Holt and Reynolds, 1931). Small animals are susceptible by bite or from injection of crushed mosquito tissue.

OTHER DISEASES OF MAN AND ANIMALS

Virus diseases of man. A number of virus diseases in man other than yellow fever and dengue have been shown or suspected to be conveyed by mosquitoes, in some of which *A. aegypti* has been shown or suspected to act as vector. Coggleshall (1939) and Milzer (1942) record conveyance of the virus of lymphocytic choriomeningitis, a sporadic disease of man, by mosquitoes, *A. aegypti* being infective seven to thirty-eight days after the infective feed. Murray Valley encephalitis, which in 1951 caused a serious epidemic in Australia, was thought to be mosquito-borne from birds as natural reservoirs. Japanese B.- and St Louis encephalitis have also been considered possibly mosquito-borne (Marks, 1954).

A virus, West Nile virus, causing disease in children near Cairo has been isolated from *Culex* by Taylor and Hurlbutt (1953) and has been shown by Goldwasser and Davies (1953) and by Davies and Yoshpe-Purer (1953, 1954*a*) to be capable of transmission by *A. aegypti*, wild birds being found to be susceptible and thought to be possible reservoirs of the virus in nature (Work, Hurlbutt and Taylor, 1955).

A human epidemic caused by Venezuelan equine encephalitis in Colombia is recorded by Sanmartin-Barberi, Groot and Osarno-Mesa (1955), *A. aegypti* being found in 687 out of 2295 houses, and found infected.

Altogether no less than seven new viruses isolated from man, animals or mosquitoes are listed by Dick (1953) as recorded in 1937–48 from Uganda, namely Bwamba fever, West Nile virus, Semliki Forest virus, Bunyamwera, Ntaya, Uganda S. and Zika. Semliki Forest virus did not develop in *A. aegypti* which was not thought to be significant in its transmission (Davies and Yoshpe-Purer, 1954*b*).

Forms of encephalitis affecting human beings and horses in Ecuador acquired in jungle areas are noted by Levi-Castillo (1952) as having been shown to be transmissible by several species of mosquito, *A. aegypti* not being among these.

Virus diseases of animals. An important disease of animals shown to be transmitted by *A. aegypti* is equine encephalitis which exists in Africa in two forms, a western and an eastern form, the former being more readily transmitted. The virus is thought to be normally maintained as an infection of birds (Kelser, 1933; Merrill and Ten Broek, 1940; and others). The virus has been shown to multiply a hundred times or more in the mosquito.

A. aegypti has been shown to transmit haemorrhagic septicaemia of buffaloes (Nieschulz and Kraneveld, 1929). It also transmits the virus of fowl-pox. The virus at first is generally distributed throughout the body of the mosquito, but later collects on and about the proboscis (Rivers, 1929; Stuppy, 1931; Brody, 1936).

An important disease of sheep and to a less extent of cattle in Africa is Rift Valley fever (enzootic hepatitis). It causes abortion in sheep and cattle and human cases are contracted by contact with diseased animals or their organs. The virus is present in wild mosquitoes including *A. africanus* (Smithburn and Gillett, 1948; Smithburn, Haddow and Lumsden,

1949). A neurotropic strain from mice has been used as a vaccine for protection of sheep (see Smithburn, 1949).

A. aegypti has been suspected of transmitting horse sickness (Balfour, 1912), but many attempts to transmit this disease experimentally by the species have given negative results (Nieschulz, Bedford and Du Toit, 1934; Du Toit, 1944).

Filariasis. It has been shown by numerous observations that *A. aegypti* is not an efficient vector of *Wucheria bancrofti*. Though some development takes place it is very rarely that any embryo is found in the proboscis (Aders, 1917; Francis, 1919; Edwards, 1922; Connal, 1931; Hicks, 1932; Newton, Wright and Pratt, 1945; and others).

Nor does *A. aegypti* transmit *Filaria ozzardi*, a single embryo only being found by N. C. Davis (1928) in his investigation of filarial diseases in the Argentine.

A. aegypti is, however, an active vector of *Dirofilaria immitis* in the dog (Bernaud and Bauche, 1913 (*D. repens*); Feng, 1930; Hinman, 1935; del Rosario, 1936; Roubaud, 1937; Galliard, 1937, 1942; Summers, 1943; Kartman, 1953). Kershaw, Lavoipierre and Chalmers (1953), investigating the intake of embryos by *A. aegypti*, found a considerable reduction in survival rate of the mosquitoes following feeding on an infected, as against an uninfected, dog. Those mosquitoes that took in more than a certain number failed to live long enough to become infective.

Other miscellaneous diseases. Philip, Davis and Parker (1932) obtained transmission of tularaemia by interrupted feed in one positive result with *A. aegypti*. Under certain conditions Roubaud and Lafont (1914) thought transmission of human trypanosomiasis might be possible. It was only transmitted, however, with very heavy infections and a yard's distance sufficed to protect. *A. aegypti* has been suspected of transmitting leprosy (Blanchard, 1905; Noc, 1912; Gomes, 1923). Archibald (1923) found undulant fever could be conveyed to monkeys by the bite of *A. aegypti*.

A. aegypti failed to transmit avian plague (Nieschulz, Bos and Tarip, 1931) or surra to rats by interrupted feeding (Kelser, 1927), or ephemeral fever in Australian cattle, a disease related to dengue (Mackerras, Mackerras and Burnet, 1940). *A. aegypti* has been shown to convey myxomatosis in wild rabbits (Aragao, 1943; Bull and Mules, 1944), though not apparently specifically.

Blood parasites of birds and mammals. *A. aegypti* is the normal vector of *Plasmodium gallinaceum* Brumpt of fowls (Brumpt, 1936; James and Tate, 1938); Lumsden and Bertram, 1940; Huff and Coulston, 1944; and others). It is also capable of acting as host to some other plasmodia of birds, being susceptible to *P. cathemerium* and *P. inconstans*, but negative to *P. praecox* (Huff, 1927). Et. Sergent (1942) also notes that compared with *Culex pipiens* it is a poor host to *Plasmodium relictum*. A very small percentage produced oocysts (but not sporozoites) with *P. lophurae* (Coggleshall, 1941; Laird, 1941). Trager (1942), however, produced a strain by selective breeding that was susceptible to this parasite.

Sepsis. Apart from transmission of disease in the usually accepted sense the bites of *Aëdes aegypti* may cause local reaction which may in certain circumstances be severe or be followed by septic infection. The effect of mosquito bites on the human subject is dealt with in a later chapter.

Not to be minimised also is the irritation of the bites of this species, especially to newcomers in the tropics, and the effect of the perpetual attacks in day-time which make *A. aegypti* when at all numerous one of the most troublesome pests in the tropics.

(*b*) CONTROL AND PROTECTION

A distinction may be made between methods which have the object of destroying or preventing the breeding of mosquitoes, which may be termed measures of control, and those concerned with protection against their bites, or measures of protection. Control measures are by their nature usually such as are carried out by public health or other organised bodies, though the individual may by his co-operation greatly assist. Protective measures more usually are the concern of the individual, though one such measure, namely screening, may in certain cases be carried out on a large scale as a public health measure.

CONTROL

Anti-*Aëdes aegypti* campaigns. Control measures of the nature of anti-*A. aegypti* campaigns have been extensively carried out, more especially in those parts of the New World where yellow fever was liable to occur in epidemic form. Some of these have been on a very large scale, aiming at eradication of the species from whole towns or even whole countries.

Many accounts of such campaigns are to be found in the literature. One of the most detailed is that by Soper, Wilson, Lima and Antunes (1943) describing the operations and technique employed in the nation-wide anti-*A. aegypti* campaign undertaken in Brazil. Another very full account of control measures against *A. aegypti* is given by Herms and Gray (1940) in their chapter on mosquito control in urban areas.*

All the above pre-date the discovery and use of DDT and other synthetic insecticides in residual spraying. The methods and technique used have been those now described as 'classical methods of control' and the question arises how far such methods are now to be considered outmoded. De Caires (1947), describing *A. aegypti* control in Georgetown, British Guiana, under the adverse condition, as he notes, of a lack of a piped water supply, observes that classical methods had been in use up to 1945, but *A. aegypti* was not eradicated and the numbers of the species continued to fluctuate with rainfall. At the beginning of 1946 residual spraying with 5 per cent DDT in kerosene giving 100 mg. DDT residual film per sq. ft. was substituted, and the *A. aegypti* index was reduced to zero by the end of the year. Giglioli (1948), speaking of measures taken in British Guiana, notes that malaria, filariasis and yellow fever have all been controlled by residual spraying with DDT and that *Anopheles darlingi* and *Aëdes aegypti* have been eradicated from 200 miles of coast land and estuary banks by the same methods. Nor are these unique as claims for the remarkable effectiveness of this new approach to the problem and there can be no doubt that in *A. aegypti* control as in malaria control the use of the new compounds, especially the

* For accounts of earlier examples see Boyce (1906, prophylaxis in New Orleans in 1905), James (1914) and James, Da Silva and Arndt (1914, survey of the Port of Colombo), Howard, Dyar and Knab (1912, pp. 252–5 and 429–36), Carter (1922, epidemic of 1919–20 in Peru), Connor and Monroe (1923, *Stegomyia* indices), Connor (1924, suggestions for developing a campaign of control), Dunn (1923, Peru), Dunn and Hanson (1925, Colombia), Hanson (1925, Colombia), Rigollot (1927, Dakar), Soper (1938, South America), Sneath (1939, British Guiana).

method of residual spraying, will both supplement and to a large degree modify or even render obsolete much of the previous technique as used against all mosquito-borne disease.

Whilst, however, residual spraying is a very effective method, it does not necessarily rule out all other action. An effective piped water supply is still an important measure in control, the education of the inhabitants of a city regarding use of certain simple precautions is still desirable, water tanks and certain other collections should by legislation be suitably protected and, above all, adequate inspection is essential. Certain circumstances too may still make some of the classical methods applicable. Thus the type of house is not always the native hut or simple dwelling so often to be dealt with in malaria work, and heavy infestation with *A. aegypti* may be present in large cities in relatively better-class houses where the older methods still have a place. Whatever methods be used, success, as in other public health work, depends upon good administration, aided by all the many methods which make public health work effective, including trained personnel, legislation, propaganda and education of the public where this can be brought to bear.

That such action may result in virtual disappearance of the species over large areas has been shown by the result of operations in Brazil (Soper *et al.* 1943) and Texas (Chandler, 1956). See also Severo (1956*a*), who, as previously noted, gives a map showing the present greatly restricted distribution of *A. aegypti* in America.

Control of breeding places. The nature of the breeding places of *A. aegypti* has already been indicated. Many of these can be disposed of by emptying or destroying receptacles. When water pots cannot be emptied, as the water is required for human consumption, some form of covering is a simple precaution. Cisterns and other water storage collections should be made impervious to mosquitoes. Where water is not required for drinking or cooking and cannot be otherwise dealt with, some form of larvicide is called for. A number of insecticidal substances, and even some natural enemies, have been suggested for such conditions, but for most purposes in or about the house the commonly sold preparations of cresol or kerosine are simple to use and effective. In better-class houses attention should be given especially to anti-formicas, a very common form of breeding place, and flower vases. Disused water-closets may be a potent source of breeding. In rural areas, especially in the rainy season, disused pots and various other receptacles holding rain-water can be destroyed or otherwise dealt with. Where water is precious Lewis (1955) refers to the use of benzene hexachloride and cement tablets as devised by Bruce-Chwatt (1953). Breeding places about docks, lighters, river craft, machinery dumps, waste ground, cemeteries, etc. are usually most simply treated with some form of crude petroleum. A very complete account of the properties of different oils used as larvicides is given by Moore and Graham (1918) and by Murray (1936); see also the very full account given by Herms and Gray (1940) on oils and larvicides. However, as noted later, the newer forms of insecticide may be used. Breeding in ships is an important special case (Lutz and Machado, 1915; Tanner, 1931; Chabrillat, 1934) and a very full account, with apparatus likely to be useful, is given by Blacklock and Wilson (1942). Breeding in country craft is dealt with by Bana (1936).

For some purposes poisonous gases and vapours have been used. Macfie (1916) found chlorine effective. Williamson (1924), testing gases and vapours against breeding in wells, found H_2S and chloropicrin (1 in 200,000) effective against some species, but first-stage *A. aegypti* larvae were exceptions. Barber (1944) found carbon bisulphide, toluene, benzene and carbon tetrachloride the most rapidly fatal of vapours.

A measure much used, especially in the earlier campaigns against the species, has been the employment of larvivorous fish. Seal (1908) is stated by Molloy (1924) probably to be the first to recommend use of such fish on a large scale. The first suggestion to use fish in control of the breeding of mosquitoes was, however, made as early as 1892 by Russell. A digest of literature on the use of fish for mosquito control with a bibliography of 217 titles is published by the International Health Board, Rockefeller Foundation, 1924. An account of Indian larvivorous fish is given by Prashad and Hora (1936). Species of *Gambusia*, or 'millions', have been successfully used in wells, cisterns and tanks (Myers, 1926; Le Van, 1940), but other small fish have been preferred for particular purposes, for example the native fish 'chalaco' (*Dermitator latifrons*) found hardier by Connor (1921) in Ecuador. Silver bait (*Tetragonopterus*, *Charax* and *Hemigrammus*) which can live in the absence of larvae were found of great value by Haslam (1925) in British Guiana. This author also found *Haplosternum littoralis* and *Chicosoma bimaculatum* useful for small collections of water. *Astranax bimaculatum* was found effective for large, and *Hemigrammus unilineatus* for smaller collections of water by Ihering (1933). The poeciloid fish *Rivulus* common in American jungles is suggested by Myers.

How far many of the above means of controlling breeding are still likely to be found of use is doubtful. Not only is it whether DDT or some other synthetic chlorinated compound is that to be used, but the whole question of the use of larvicides for the purpose of control has been brought into question, or at least modified, by the results obtained from residual spraying. DDT as a larvicide is used as a wettable powder, as an emulsion or dissolved in diesel oil or other solvent. Floch and Layudie (1946) found DDT powder highly toxic to larvae of *Culex* spp. and *Aëdes aegypti* and more effective than the plant poisons previously in use. In a U.S. Army medical bulletin, 1945, entitled *The Prevention of Dengue Fever*, a powder containing 10 per cent DDT or a 5 per cent solution or emulsion is recommended for artificial containers not holding water for drinking or cooking. In the same bulletin tests with different waters showed that DDT killed 70–100 per cent of larvae in such low dilution as 0·03 per million. Lever (1946) found sawdust impregnated with DDT in diesel oil suspended in a disused swimming bath prevented breeding of *Culex sitiens*. O'Kane (1947) notes that a dust containing 0·5 per cent gamma isomer benzene hexachloride on the surface of water at 10 lb. per acre gave complete control of breeding of *Aëdes* species including *A. aegypti*.

Measures against the adult mosquito. Of measures against the adult mosquito one of the most important and usual is the use of sprays or aerosols. These have been greatly developed recently and there is a large literature dealing with the physical factors involved and the action of the droplets. The use of very finely divided droplet sprays (aerosols) has been specially developed.*

An important point in respect to aerosols is that the dose taken up by the insect is largely proportional to the extent to which the wings are used in flight. To give rapid knock-down pyrethrum, usually in the form of the pyrethrins, holds a high position. Synthetic compounds of the same type as the pyrethrins are also effective (see under insecticides in ch. xxv).

* See Sinton and Wats (1935), David (1946*a*, *b*), Bishopp (1946), Goodhue (1946), David and Bracey (1946), Muirhead Thomson (1947). A very full and technical description of the apparatus and parts used is given in the 6th Report of the Expert Committee on Insecticides of the World Health Organisation, 1956.

Fumigation is a method formerly much employed, though now largely replaced by spraying. A number of fumigants are given by Howard (1917); see also Covell (1941).

The use of traps against *A. aegypti* is also a measure that has not been much employed. For an account of mosquito traps see Herms and Gray (1940), p. 92. Other methods such as the use of natural enemies (see ch. III), attraction of the male by sound (see Kahn and Offenhauser, 1949), though they should be mentioned, are now but little used, or do not apply to control of *A. aegypti*.

Almost all direct attack on the adult is now by sprays or aerosols containing DDT or one of its analogues combined with a suitable knock-down component (see section on insecticides in ch. XXV). Or residual spraying is employed.

Residual spraying. Residual spraying consists in spraying or otherwise coating surfaces such as walls of habitations with fluid that will leave a layer of the non-volatile insecticide employed, the constitution of the sprayed material being commonly referred to as a 'formulation'. The compounds most usually employed are DDT or gamma BHC (gammexane) usually referred to as BHC. An analogue of DDT, namely dieldrin, has also been found very effective and has increasingly come into use. Another compound of the same type, chordane, much used against the housefly has also sometimes been used. For an account of the nature of these substances, usually described as synthetic chlorinated compounds, see Busvine (1952), who gives a brief, but adequate, account of compounds later than DDT and BHC. Other references dealing with these new insecticides, including several helpful books, are given in the section on insecticides referred to above. Regarding various points connected with the use of the compounds in residual spraying a good recent factual summary is given by Jaswant Singh and Rajindar Pal (1952). Much practical information on details in their employment is given by Davidson (1952).* The literature is so large dealing with various aspects of the use of the compounds that it is difficult to point out those most likely to be helpful. Many of those dealing with practical problems will be found in the *Indian Journal of Malariology*, 1952–4; recent numbers of the *American Journal of Tropical Medicine*; *Bulletin of Entomological Research*; *Transactions of the Royal Society of Tropical Medicine and Hygiene*; and abstracted in the *Review of Applied Entomology* B.

Certain problems in connection with spraying require brief mention here, sinee it would be a mistake not to take note that these exist.

The dosage and time that a residual film may be expected to retain efficiency are important. After an appropriate dosage (say 200 mg. of DDT per sq. ft.) this may be effective up to six months. A dosage of 100 mg. per sq. ft is more usual with corresponding reduction in the duration. Turner (1946) gives some experimental data in regard to *Anopheles* on the West Coast and still smaller dosage. One per cent DDT used to give a calculated deposit of 38·5 mg. per sq. ft. gave a relative absence of mosquitoes (1 to 4 as against the control) for eighteen weeks. When the dosage was reduced to 11·9 mg. per sq. ft. the same degree of reduction lasted four weeks. For reasons given later it is undesirable to employ sublethal dosage and a larger dosage than that just given might have been better. In India a dosage of 50–100 mg. per sq. ft. and repeating the application twice or even three times in the malaria season is practised (Jaswant Singh and Rajindar Pal, 1952).

* See also Parker and Green (1947), Muirhead Thomson (1947), Andrews and Simmons (1948) (bibliography of 134 references), Bishopp (1951), Hadaway and Barlow (1952).

Another point of importance is the effect of repellency by the film, which may result in mosquitoes not remaining sufficiently long in contact to receive a lethal dose and even possibly becoming excited so that, after feeding or not, they leave the house without being killed (Kennedy, 1947; Muirhead Thomson, 1947).

Related to the above is the question of choice of insecticide as between DDT and BHC or other analogue. Results by workers are not always identical, but the general view is that BHC has a stronger initial effect so that dosage on contact is almost certainly lethal. At the same time repellent effect is greater and being more volatile the period of efficiency of the film is less. Many observers for these reasons use both substances in combination (Jaswant Singh and Rajindar Pal).

Another important point is the nature of the deposit as laid down by different formulations, since the effectiveness of the film depends to some extent on whether DDT is used as a wettable powder or in solution and upon the size and character of the crystals as laid down by different solvents and procedures (Hadaway and Barlow, 1952; Rajindar Pal, 1954*b*). In particular loss of effectiveness may be very great if spraying is carried out on mud walls (Rajindar Pal, 1954*a*; Rajindar Pal and Sharma, 1952) or on a whitewashed surface (Hadaway and Barlow, 1947).

Also residual spraying, if long continued, may lead to development of resistant strains of mosquito. This condition was first reported in mosquitoes from Italy by Missiroli (1947) and by Mosna (1947), a resistant strain of *Culex pipiens* being developed following five years of spraying with DDT. A good account of the history of the condition with bibliography is given by Rajindar Pal, Sharma and Krishnamurthy (1952), who also record a resistant strain of *C. fatigans* in an area that had been sprayed for six years. A considerable number of other instances have now been recorded both in Culicines and in species of *Anopheles*. A troublesome feature is that resistance acquired to one chlorinated insecticide may be displayed in some degree to other compounds of this nature. A very full and complete account of such resistance with details of recorded instances has recently been given by Busvine (1957). Resistance in *Aëdes aegypti* has recently been noted in Trinidad where some cases of yellow fever occurred in 1954 (H. P. S. Gillette, 1956; Busvine, 1957). See also Severo (1956*b*), who refers to acquired resistance of *A. aegypti* to DDT in the Dominican Republic.

Some further information on the nature of insecticides, the way in which they act and some remarks on the nature of acquired resistance are given in chapter XXV when dealing with viability of the adult under different conditions.

Disinfection of aircraft. A special importance attaches to measures against mosquitoes in aircraft. This applies in particular to *A. aegypti* as a possible vector of yellow fever and recently as a possible vector of dengue in the South Pacific where there is a danger of introducing this species into areas where it has as yet not been established. Here also the use of residual spraying is replacing other methods.

Disinfestation (or disinsectisation, a term often used) has been commonly carried out by aerosol spraying with pyrethins, lethane and other insecticides put up in suitable formulations for use with sparklet or pressure aerosol equipment. A description of methods is given by Duguet (1949), and a discussion of disinfestation under tropical conditions is given by Bruce-Chwatt and Coi (1950). A recent account of methods of aircraft disinfection is that by Laird (1951), who notes that this may be carried out by fumigation, by aerosol mists, or by residual insecticides. Fumigation, though very effective, is too time consuming for wide use. For spraying with aerosol mists the medium used must be both free from inflammable material and non-damaging to fabrics, paper, etc., and also to perspex windows in pressurised aircraft. He recommends aerosol spraying before flight and monthly treatment with residual DDT. Aeroplanes were treated with DDT in 1945 by Madden, Linquist and Knipling.

PROTECTIVE MEASURES

Screening. Screening of houses and buildings is a method that has been widely employed as a protective measure. It requires to be carried out with due regard to a number of considerations. An effective mesh and non-corrosible, sufficiently strong material with complete closure of entry are essential, as well as attention to the amenities and comfort of life. A good account of specifications for wire gauze and precautions to be taken is given by Covell (1941), and by Herms and Gray (1940). Mesh should be counted along two sides of a square inch counting the corner square twice, but is often given as meshes to the linear inch. Thickness of wire must be taken into account in considering size of aperture and to provide suitable strength to the screen. A mesh of 14, or better 16, to the linear inch and 30 I.S.W.G. (Imperial Standard Wire Gauge), that is a thickness of 0·0124 in. (0·315 mm.), is generally considered suitable. Or a wire thickness of 0·015 in., giving an aperture of 0·056 in., is better. A non-corrosive metal is also very important; see Blacklock (1937); Galli-Valerio (1925); MacArthur (1923); Brown (1934); Mooij (1940); Covell; Herms and Gray. The screening of ships is dealt with by Melville-Davison (1912).

Personal protection. At least as important as any other form of protection is the use by the individual of the mosquito net and other precautions against being bitten. Most of these precautions are matters of common sense, yet are curiously neglected. Especially important is the use of suitable protective clothing, especially protection of the ankles (double socks) and seat and knees (under garment) where the insect is prevalent. A small hand net convenient to handle is useful to despatch a stray annoying insect (see also Legendre, 1913).

Where all else fails some form of repellent preparation may be used. The preparations most likely to be useful against *A. aegypti* under ordinary circumstances are not so much the long-lasting non-volatile repellents such as dimethyl phthalate (DMP) used where mosquitoes are attacking voraciously on active service and the like, but certain more pleasant temporarily effective substances such as a good Java citronella oil, or citronellol, the active principle. A good citronella oil is very useful for protection by day in the house or office, as it need not be spread over all exposed skin surfaces as with DMP preparations, but can be applied here and there on clothing, etc. Where mosquitoes are numerous, and more effective methods of protection are necessary, DMP should be used. This is most effective neat, covering all exposed skin surfaces after anointing the palms, but may be used for tender skins or greater comfort as a cream. DMP does not readily form a good stable cream, but one recommended as stable and pleasant is that given by the present author in a report on mosquito repellents (Christophers, 1947) which is as follows:

Dimethyl phthalate	12·5 c.c.
White wax (Cera alba B.P.)	9 gm.
Arachis oil	27·5 c.c.

Melt the wax on a water bath with the arachis oil and stir in the DMP while hot. Filter hot through wool if necessary and put up in suitable wide mouth bottle as a solid cream.

The preparation can be modified for different climates by slightly increasing or decreasing the proportion of white wax. It has no waxy feel when applied, which its composition might suggest. It is permanent and spreads thinly and readily.

The use of zinc oxide, included as an adsorbent of the DMP, as employed in some preparations on the market, is not recommended as such are cosmetically unpleasant. So far no substance taken internally has been found effective. Thiamin chloride was stated to be so by W. R. Shannon (1943), but was not found so by Wilson, Matheson and Jackowski (1944). For an account of research on repellents see Travis, Morton and Cochran (1946).

A measure of very doubtful efficacy is the use of certain plants that have been popularly supposed to have a repellent effect on mosquitoes, for example *Ocimum viride*, a plant with a strong smell of lemon thyme if rubbed or chewed (Holmes, 1878, 1879); the castor oil plant, claims for which seem to be without any real foundation (see *Insect Life*, **5**, 359, 1893). Various forms of mosquito trap are described, but they are probably more suitable for *Culex* than *Aëdes aegypti*, though special types might perhaps be designed.

Whilst all the above are available to the individual for his self-protection, the prime procedure under normal conditions in the tropics, when protection is called for against *A. aegypti* as a troublesome day-biting mosquito in a bungalow or quarters, is to make a thorough search for any breeding places of the pest there may be about the premises and take appropriate action against these. Often days or weeks of annoyance may be done away with by such a search whenever presence of the insect begins to be noticeable. If the source of supply is from outside and irremovable, then screening of the house or quarters is indicated.

REFERENCES

(*a*) RELATION TO DISEASE

Yellow fever

ARAGAO, H. DE B. (1939). Mosquitoes and yellow fever virus. *Mem. Inst. Osw. Cruz*, **34**, 565–81.

ARAGAO, H. DE B. and COSTA-LIMA, A. (1929). Sur la transmission du virus de la fièvre jaune par les déjections de moustiques infectés. *C.R. Soc. Biol., Paris*, **102**, 53–4, 477–8.

BAILEY, K. P. (1947). Preliminary note on the sylvan mosquitoes of Gede. *E. Afr. Med. J.* **24**, 38–41.

BATES, M. (1949). *The Natural History of Mosquitoes.* Macmillan and Co., New York.

BAUER, J. (1928). The transmission of yellow fever by mosquitoes other than *Aëdes aegypti. Amer. J. Trop. Med.* **8**, 261–82.

BEEUWKES, H., BAUER, J. H. and MAHAFFY, A. F. (1930). Yellow fever endemicity in West Africa with special reference to protection tests. *Amer. J. Trop. Med.* **10**, 305–33.

BEEUWKES, H. and HAYNE, T. B. (1931). An experimental demonstration of the infectivity with yellow fever virus of *Aëdes aegypti* captured in an African town. *Trans. R. Soc. Trop. Med. Hyg.* **25**, 107–10.

BOYCE, R. (1909). *Mosquito or Man.* John Murray, London.

BOYCE, R. (1910). The distribution and prevalence of yellow fever in West Africa. *J. Trop. Med.* (*Hyg.*), **13**, 357–62.

BRUMPT, E. (1936). *Précis de parasitologie.* Ed. 5. Masson et Cie, Paris.

CANNON, D. A. and DEWHURST, F. (1955). The preparation of 17D virus yellow fever vaccine in mouse brain. *Ann. Trop. Med. Parasit.* **49**, 174–82.

CARTER, H. R. (1922). Yellow fever in Peru. Epidemic of 1919 and 1920. *Amer. J. Trop. Med.* **2**, 87–106.

REFERENCES

CONNOR, M. E. (1923). Notes on yellow fever in Mexico. *Amer. J. Trop. Med.* **3**, 105–16.

DAVIS, N. C. (1933). Transmission of yellow fever virus by *Culex fatigans* Wied. *Ann. Ent. Soc. Amer.* **26**, 491–5.

DAVIS, N. C. and SHANNON, R. C. (1930). The location of yellow fever virus in infected mosquitoes and the possibility of hereditary transmission. *Amer. J. Hyg.* **11**, 335–44.

DINGER, J. E., SCHUFFNER, W. A. P., SNYDERS, E. P. and SWELLENGREBEL, N. H. (1929). Onderzoek over gele koorts in Nederland (derde mededeling). *Ned. Tijdschr. Geneesk.* **73** (quoted by Strode, 1951).

FINDLAY, G. M. (1952). Reviewer's abstract to Strode (1951). *Trop. Dis. Bull.* **49**, 337–9.

FINDLAY, G. M. and COCKBURN, T. A. (1943). The possible role of birds in maintenance of yellow fever in West Africa. *Nature, Lond.*, **152**, 245.

FINDLAY, G. M. and DAVEY, T. H. (1936). Yellow fever in the Gambia. I. Historical. *Trans. R. Soc. Trop. Med. Hyg.* **29**, 667–78.

GILLETT, J. D. (1951). The habits of the mosquito *Aëdes* (*Stegomyia*) *simpsoni* Theobald in relation to the epidemiology of yellow fever in Uganda. *Ann. Trop. Med. Parasit.* **45**, 110–21.

GILLETT, J. D. (1955). Further studies on the biting behaviour of *Aëdes* (*Stegomyia*) *simpsoni* Theobald in Uganda. *Ann. Trop. Med. Parasit.* **49**, 154–7.

GILLETT, J. D. and ROSS, R. W. (1955). The laboratory transmission of yellow fever by Malayan *Aëdes aegypti*. *Ann. Trop. Med. Parasit.* **49**, 63–5.

GILLETT, J. D., ROSS, R. W., DICK, G. W. A., HADDOW, A. J. and HEWITT, L. E. (1950). *Ann. Trop. Med. Parasit.* **44**, 342–50.

HADDOW, A. J., DICK, G. W. A., LUMSDEN, W. H. R. and SMITHBURN, K. C. (1951). Mosquitoes in relation to the epidemiology of yellow fever in Uganda. *Trans. R. Soc. Trop. Med. Hyg.* **45**, 189–224.

HINDLE, E. (1929). An experimental study of yellow fever. *Trans. R. Soc. Trop. Med. Hyg.* **22**, 405–30.

HINDLE, E. (1930). The transmission of yellow fever. *Lancet*, **2**, 835.

HOWARD, L. O., DYAR, H. G. and KNAB, F. (1912). See chapter I (*b*), p. 4.

KIRK, R. (1943). Some observations on the study and control of yellow fever in Africa with particular reference to the Anglo-Egyptian Sudan. *Trans. R. Soc. Trop. Med. Hyg.* **37**, 125–50.

KIRK, R. and HASEEB, M. A. (1953). Animals and yellow fever infection in the Anglo-Egyptian Sudan. *Ann. Trop. Med. Parasit.* **47**, 225–31.

LEWIS, D. J., HUGHES, T. P. and MAHAFFY, A. F. (1942). Experimental transmission of yellow fever by three common species of mosquitoes from the Anglo-Egyptian Sudan. *Ann. Trop. Med. Parasit.* **36**, 34–8.

LUMSDEN, W. H. R. (1955). Entomological studies relating to yellow fever epidemiology at Gede and Taveta, Kenya. *Bull. Ent. Res.* **46**, 149–83.

MACNAMARA, F. N. (1954). Isolation of the virus as a diagnostic procedure for yellow fever in West Africa. *Bull. World Hlth Org.* **11**, 391–401.

MAHAFFY, A. F. (1949). The epidemiology of yellow fever in Central Africa. *Trans. R. Soc. Trop. Med. Hyg.* **42**, 511–24.

MAHAFFY, A. F., SMITHBURN, K. C., JACOBS, H. R. and GILLETT, J. D. (1942). Yellow fever in Western Uganda. *Trans. R. Soc. Med. Trop. Hyg.* **36**, 9–20.

MANSON-BAHR, P. H. (1954). *Manson's Tropical Diseases*. Ed. 14. Cassell and Co., London.

MATTINGLY, P. F. (1949). Studies on West African forest mosquitoes. Part 1. The seasonal distribution and biting cycle and vertical distribution of four of the principal species. *Bull. Ent. Res.* **40**, 149–68.

MUSPRATT, J. (1956). The *Stegomyia* mosquitoes of South Africa and some neighbouring territories. *Mem. Ent. Soc. Sthn. Afr.* no. 4.

PHILIP, C. B. (1929). Possibility of hereditary transmission of yellow fever virus by *Aëdes aegypti* (Linn.). *J. Exp. Med.* **50**, 703–8.

PHILIP, C. B. (1930). Possibility of mechanical transmission by insects in experimental yellow fever. *Ann. Trop. Med. Parasit.* **24**, 493–501.

RIVERS, T. M. (1952). *Viral and Rickettsial Infections of Man.* Ed. 2. J. B. Lippincott Co., Philadelphia.

ROSS, R. W. and GILLETT, J. D. (1950). The cyclical transmission of yellow fever virus through the Grivet monkey *Cercopithecus aethiops centralis* Neumann and the mosquito *Aëdes (Stegomyia) africanus* Theobald. *Ann. Trop. Med. Parasit.* **44**, 351–6.

SAWYER, W. A. (1942). The yellow fever situation in the Americas. *Bol. Ofic. Sanit. Pan-amer.* no. 173, pp. 1–15.

SMITH, H. H. and THEILER, M. (1937). The adaptation of unmodified strains of yellow fever virus to cultivation *in vitro*. *J. Exp. Med.* **65**, 801–8.

SMITHBURN, K. C. and HADDOW, A. J. (1946). Isolation of yellow fever virus from African mosquitoes. *Amer. J. Trop. Med.* **26**, 261–71.

SMITHBURN, K. C. and HADDOW, A. J. (1949). The susceptibility of African wild animals to yellow fever. *Amer. J. Trop. Med.* **29**, 389–405.

SMITHBURN, K. C., HADDOW, A. J. and LUMSDEN, W. H. R. (1949). Outbreak of sylvan yellow fever in Uganda with *Aëdes (Stegomyia) africanus* Theobald as principal vector and insect host of virus. *Ann. Trop. Med. Parasit.* **43**, 74–89.

SOPER, F. L. (1938). Yellow fever: the present situation (October 1938) with special reference to South America. *Trans. R. Soc. Trop. Med. Hyg.* **32**, 297–332.

SOPER, F. L., PENNA, H., CARDOSO, E., SERAFIM, JR. J., FROBISHER, JR. M. and PINHEIRO, J. (1933). Yellow fever without *Aëdes aegypti*. *Amer. J. Hyg.* **18**, 555–87.

STOKES, A., BAUER, J. H. and HUDSON, N. P. (1928). Experimental transmission of yellow fever to *Macacus rhesus*. Preliminary note. *J. Amer. Med. Assoc.* **90**, 253.

STRODE, G. K. (1951). *Yellow Fever.* McGraw-Hill Publ. Co., London.

TAYLOR, R. M. (1951). *Epidemiology.* In STRODE (1951).

THEILER, M. and SMITH, H. H. (1937). The use of yellow fever virus modified by *in vitro* cultivation for human immunisation. *J. Exp. Med.* **65**, 789–800.

WADDELL, M. B. (1945). Persistence of yellow fever virus in mosquitoes after death of the insect. *Amer. J. Trop. Med.* **25**, 329–32.

WORLD HEALTH ORGANISATION (1952). *Chron. World Hlth Org.* **6** (11), p. 338.

Dengue

ADRIEN, C. (1918). Dengue Méditerranéenne observée à l'île Rouad (Syrie). *Arch. Méd. Pharm. Nav., Paris,* **105**, 275–307.

ARMSTRONG, C. (1923). Dengue fever. *Publ. Hlth Rep., Wash.,* **38**, 1750–84.

ASHBURN, P. M. and CRAIG, C. F. (1907). Experimental investigation regarding the aetiology of dengue fever. *Philipp. J. Sci.* Sect. B, **2**, 93–147.

BARRAUD, P. J. (1928). The distribution of *Stegomyia fasciata* in India with remarks on dengue and yellow fever. *Ind. J. Med. Res.* **16**, 377–88.

BISHOPP, F. C. (1923). Dengue fever and mosquitoes in the south. *J. Econ. Ent.* **16**, 97.

BLANC, G. and CAMINOPETROS, J. (1930). Recherches expérimentales sur la dengue. *Ann. Inst. Pasteur,* **44**, 367–436.

CARDAMATIS, J. P. (1929). La dengue en Grèce. *Bull. Soc. Path. Exot.* **22**, 272–92.

CHANDLER, A. C. (1925). The transmission and etiology of dengue. A critical review. *Indian Med. Gaz.* **60**, 460–2.

CHANDLER, A. C. and RICE, L. (1923). Observations on the etiology of dengue fever. *Amer. J. Trop. Med.* **3**, 233–62.

CLELAND, J. B. (1928). Dengue fever in Australia. *Rev. Prat. Mal. Pays Chauds,* **8**, 509–14.

CLELAND, J. B., BRADLEY, B. and MACDONALD, M. (1918). Dengue fever in Australia. Its history and clinical course: its experimental transmission by *Stegomyia fasciata* and the results of inoculation and other experiments. *J. Hyg., Camb.,* **16**, 317–418.

EKBLOM, T. (1929). Some observations on *Stegomyia fasciata* during a visit to Greece in the autumn of 1928. *1929 Acta Med. Scand.* **70**, 505–18.

FARNER, D. S. and BOHART, R. M. (1948). A preliminary revision of the *scutellaris* group of the genus *Aëdes*. *U.S. Nav. Med. Bull.* **44**, 37–53.

REFERENCES

FINDLAY, G. M. (1951). In Bank's *Modern Practice in Infectious Fevers*. Vol. II. Butterworth and Co., London.

GRAHAM, H. (1902). Dengue: a study of its mode of propagation and pathology. *Med. Rec., N.Y.*, **61**, 204–7.

HOLT, R. L. and KINTNER, J. H. (1931). Location of dengue virus in the body of mosquitoes. *Amer. J. Trop. Med.* **1**, 103–11.

JENKINS, C. F. H. (1945). Entomological problems of the Ord River Irrigation Area. *J. Dep. Agric. W. Aust.* (2), **22**, 131–45.

LEGENDRE, J. (1918). Note sur les Stegomyias de Tamatave. *C.R. Soc. Biol., Paris*, **81**, 832–3.

LEWIS, D. J. (1955). The *Aëdes* mosquitoes of the Sudan. *Ann. Trop. Med. Parasit.* **49**, 164–73.

LUMLEY, G. F. (1943). Dengue. Part I. Medical. *Serv. Publ., Dep. Hlth Australia* (*Sch. Publ. Hlth and Trop. Med.*), no. 3, pp. 1–141.

LUMSDEN, W. H. R. (1955). An epidemic of virus disease in Southern Province, Tanganyika Territory in 1952–3. *Trans. R. Soc. Trop. Med. Hyg.* **49**, 28–32.

MCCARTHY, D. D. and WILSON, D. B. (1948). Dengue in the East African Command: incidence in relation to *Aëdes* prevalence and some clinical features. *Trans. R. Soc. Trop. Med. Hyg.* **42**, 83–8.

MACKERRAS, I. M. (1946). Transmission of dengue fever by *Aëdes* (*Stegomyia*) *scutellaris* Walk. in New Guinea. *Trans. R. Soc. Trop. Med. Hyg.* **40**, 295–312.

MUSPRATT, J. (1956). See under (*a*) 'Yellow fever'.

OMAN, P. W. and CHRISTIENSEN, L. D. (1947). Malaria and other insect-borne diseases in the South Pacific campaign 1942–45. Entomology. *Amer. J. Trop. Med.* **27**, 91–117.

PERRY, W. J. (1948). The dengue vector in New Caledonia, the New Hebrides and the Solomon Islands. *Amer. J. Trop. Med.* **28**, 253–9.

PERRY, W. J. (1950). The mosquitoes and mosquito-borne diseases in New Caledonia, an historical account 1888–1946. *Amer. J. Trop. Med.* **30**, 103–14.

SCHULE, P. A. (1928). Dengue fever: transmission by *Aëdes aegypti*. *Amer. J. Trop. Med.* **8**, 203–13.

SILER, J. F., HALL, M. W. and HICHENS, A. D. (1926). Dengue: its history, epidemiology, mechanism of transmission, etiology, clinical manifestations, immunity and prevention. *Philipp. J. Sci.* **29**, 1–304.

SIMMONS, J. S. (1931). Dengue fever. *Amer. J. Trop. Med.* **11**, 77–101.

SIMMONS, J. S., ST JOHN, J. H., HOLT, R. L. and REYNOLDS, F. H. K. (1931). Possible transfer of dengue virus from infected to normal mosquitoes during copulation. *Amer. J. Trop. Med.* **11**, 199–216.

SIMMONS, J. S., ST JOHN, J. H. and REYNOLDS, F. H. K. (1930). Dengue fever transmitted by *Aëdes albopictus* Skuse. *Amer. J. Trop. Med.* **10**, 17–21.

ST JOHN, J. H., SIMMONS, J. S. and REYNOLDS, F. H. K. (1930). Transmission of dengue virus from infected to normal *Aëdes aegypti*. *Amer. J. Trop. Med.* **10**, 23–4.

SPADARO, O. (1952). Osservazioni e considerazioni sulle principali malattie infettive dell Erytrea. *Arch. Ital. Sci. Med. Trop. Parasit.* **33**, 666.

Other diseases of man and animals

ADERS, W. M. (1917). Economic biology. *Zanzibar Protect., Ann. Rep. Publ. Hlth Dep. for 1916*. Abstract in *Rev. Appl. Ent.* **6**, 123.

ARAGAO, H. DE B. (1943). O virus do mixoma no coelho de mato (*Sybvilagus minensis*) sua transmissao pelos *Aëdes scapularis* e *aegypti*. *Mem. Inst. Osw. Cruz*, **38**, 93–9.

ARCHIBALD, R. G. (1923). An unusual and fatal case of undulant fever contracted in Khartoum. *J. Trop. Med.* (*Hyg.*), **26**, 55–7.

BALFOUR, A. (1912). Current notes. *Bull. Ent. Res.* **2**, 179.

BERNARD, P. N. and BAUCHE, J. (1913). *Stegomyia fasciata* hôte intermédiaire de *Dirofilaria repens*. *Bull. Soc. Path. Exot.* **6**, 89–99.

MEDICAL

BLANCHARD, R. (1905). *Les moustiques; histoire naturelle et médicale*. Rudeval, Paris.

BRODY, A. L. (1936). The transmission of fowl-pox. *Mem. Cornell Agric. Exp. Sta.* no. 195. Abstract in *Rev. Appl. Ent.* **25**, 154.

BRUMPT, E. (1936). Réceptivitié de divers oiseaux domestiques et sauvages au parasite (*Plasmodium gallinaceum*) du paludisme de la poule domestique etc. *C.R.Acad. Sci., Paris*, **203**, 750–2.

BULL, L. B. and MULES, M. W. (1944). An investigation of myxomatosis cuniculi etc. *J. Coun. Sci. Industr. Res. Aust.* **17**, 79–93.

COGGLESHALL, L. T. (1939). The transmission of lymphocytic choriomeningitis by mosquitoes. *Science*, **89**, 515–16.

COGGLESHALL, L. T. (1941). Infection of *Anopheles quadrimaculatus* with *Plasmodium cynomolgi*, a monkey malaria parasite, and with *P. lophurae*, an avian malaria parasite. *Amer. J. Trop. Med.* **21**, 525–30.

CONNAL, A. (1931). *Ann. Med. San. Rep. Nigeria for 1930*, pp. 71–97. Abstract in *Rev. Appl. Ent.* **20**, 66.

DAVIES, A. M. and YOSHPE-PURER, Y. (1953). *Aëdes aegypti* as a vector of West Nile virus. *Bull. Res. Coun. Israel*, **3**, 127–8.

DAVIES, A. M. and YOSHPE-PURER, Y. (1954*a*). Observations on the biology of West Nile virus with special reference to its behaviour in the mosquito *Aëdes aegypti*. *Ann. Trop. Med. Parasit.* **48**, 46–54.

DAVIES, A. M. and YOSHPE-PURER, Y. (1954*b*). The transmission of Semliki Forest virus by *Aëdes aegypti*. *J. Trop. Med.* (*Hyg.*), **57**, 273–5.

DAVIS, N. C. (1928). A study of the transmission of *Filaria* in northern Argentine. *Amer. J. Hyg.* **8**, 457–66.

DEL ROSARIO, F. (1936). *Dirofilaria immitis* Leidy and its culicine intermediate hosts in Manila. *Philipp. J. Sci.* **60**, 45–57.

DICK, G. W. A. (1953). Epidemiological notes on some viruses isolated in Uganda. *Trans. R. Soc. Trop. Med. Hyg.* **47**, 13–43.

DU TOIT, R. M. (1944). The transmission of blue tongue and horse sickness by *Culicoides*. *Onderstepoort J. Vet. Sci.* **19**, 7–16.

EDWARDS, F. W. (1922). The carriers of *Filaria bancrofti*. *J. Trop. Med.* (*Hyg.*), **25**, 168–70.

FENG, LAN-CHOU (1930). Experiments with *Dirofilaria immitis* and local species of mosquitoes in Peiping, North China etc. *Ann. Trop. Med. Parasit.* **24**, 347–62.

FRANCIS, E. (1919). Filariasis in Southern United States. *U.S. Publ. Hlth Serv., Hyg. Lab. Bull.* **117**, 36 pp.

GALLIARD, H. (1937). L'évolution de *Dirofilaria immitis* Leidy chez *Aëdes* (*Stegomyia*) *aegypti* et *A. albopictus* au Tonkin. *C.R. Soc. Biol., Paris*, **125**, 130–2.

GALLIARD, H. (1942). Recherches sur le mécanisme de la transmission filaires par les Culicides. *Ann. Parasit. Hum. Comp.* **18**, 209–14.

GOLDWASSER, R. A. and DAVIES, A. M. (1953). Transmission of a West Nile-like virus by *Aëdes aegypti*. *Trans. R. Soc. Trop. Med. Hyg.* **47**, 326–7.

GOMES, E. (1923). Sobre a transmissao da lepra pelos mosquitos. *Brazil-med.* **37**, 379–91.

HICKS, E. P. (1932). The transmission of *Wucheria bancrofti* in Sierra Leone. *Ann. Trop. Med. Parasit.* **26**, 407–22.

HINMAN, E. H. (1935). Studies on the dog heart-worm, *Dirofilaria immitis*, etc. *Amer. J. Trop. Med.* **15**, 371–83.

HUFF, C. G. (1927). Studies on the infectivity of *Plasmodia* of birds for mosquitoes, etc. *Amer. J. Hyg.* **7**, 707–34.

HUFF, C. G. and COULSTON, F. (1944). The development of *Plasmodium gallinaceum* from sporozoites to erythrocytic trophozoites. *J. Infect. Dis.* **75**, 231–49.

JAMES, S. P. and Tate, P. (1938). Exoerythrocytic schizogony in *Plasmodium gallinaceum* Brumpt 1935. *Parasitology*, **30**, 128–39.

KARTMAN, L. (1953). Factors influencing infection of the mosquito with *Dirofilaria immitis* (Leidy). *Exp. Parasit.* **2**, 27–78.

REFERENCES

KELSER, R. A. (1927). Transmission of surra among animals of the equine species. *Philipp. J. Sci.* **34**, 115–41.

KELSER, R. A. (1933). Mosquitoes as vectors of the virus of equine encephalomyelitis. *J. Amer. Vet. Med. Ass.* **82** (new ser. 35), 767–71.

KERSHAW, W. E., LAVOIPIERRE, M. M. J. and CHALMERS, T. A. (1953). Studies on the intake of microfilariae by their insect vectors, etc. I. *Dirofilaria immitis* and *Aëdes aegypti*. *Ann. Trop. Med. Parasit.* **47**, 207–24.

LAIRD, R. L. (1941). Observations on mosquito transmission of *Plasmodium lophurae*. *Amer. J. Hyg.* **34**, 163–7.

LEVI-CASTILLO, R. (1952). The problem of human and equine encephalomyelitis. *Acta Tropica*, **9**, 77–80.

LUMSDEN, W. H. R. and BERTRAM, D. S. (1940). Observations on the biology of *Plasmodium gallinaceum* Brumpt 1935 in the domestic fowl with special reference to the production of gametocytes and the development in *Aëdes aegypti*. *Ann. Trop. Med. Parasit.* **34**, 135–60.

MACKERRAS, I. M., MACKERRAS, M. J. and BURNET, F. M. (1940). Experimental studies of ephemeral fever in Australian cattle. *Bull. Coun. Sci. Industr. Res. Aust.* no. 136, 116 pp.

MARKS, E. N. (1954). Research on Australian mosquitoes. Presidential Address. *Ent. Soc. Queensland.*

MERRILL, M. H. and TEN BROECK, C. (1940). The transmission of equine encephalomyelitis virus by *Aëdes aegypti*. *J. Exp. Med.* **62**, 687–95.

MILZER, A. (1942). Studies on the transmission of lymphocytic choriomeningitis virus by Arthropods. *J. Infect. Dis.* **70**, 152–72.

MUSPRATT, J. (1956). See under (*a*) Yellow fever.

NEWTON, W. L., WRIGHT, W. H. and PRATT, I. (1945). Experiments to determine potential vectors of *Wucheria bancrofti* in the continental United States. *Amer. J. Trop. Med.* **25**, 253–61.

NIESCHULZ, O. and KRANEVELD, F. C. (1929). Experimentelle Untersuchungen über die Übertragung der Büffelseuche durch Insekten. *Zbl. Bakt.* (*Orig.*), **113**, 403–17.

NIESCHULZ, O., BEDFORD, G. A. H. and DU TOIT, R. M. (1934). Investigation of the transmission of horse sickness at Onderstepoort during the season 1931–2. *Onderstepoort J. Vet. Sci.* **3**, 275–334.

NIESCHULZ, O., BOS, A. and TARIP, R. M. (1931). Uebertragungsversuche mit Geflügelpest und *Stegomyia aegypti*. *Zbl. Bakt.* (*Orig.*), **121**, 413–20.

NOC, F. (1912). Remarques et observations sur le rôle des moustiques dans la propagation de la lèpre. *Bull. Soc. Path. Exot.* **5**, 787–89.

PHILIP, C. B., DAVIS, G. E. and PARKER, R. R. (1932). Experimental transmission of tularaemia. *Publ. Hlth Rep., Wash.*, **47**, 2077–88.

RIVERS, T. M. (1929). *Filterable Viruses.* Williams and Wilkins Co., Baltimore.

ROUBAUD, E. (1937). Nouvelle recherches sur l'infection du moustique de la fièvre jaune, *Dirofilaria immitis* Leidy. Les races biologiques d'*Aëdes aegypti* et l'infection filarienne. *Bull. Soc. Path. Exot.* **30**, 511–19.

ROUBAUD, E. and LAFONT, A. (1914). Expérience de transmission de trypanosomes humains d'Afrique par les moustiques des habitations (*Stegomyia fasciata*). *Bull. Soc. Path. Exot.* **7**, 49–52.

SANMARTIN-BARBERI, G., GROOT, H. and OSARNO-MESA, E. (1955). Human epidemics in Colombia caused by the Venezuelan equine encephalomyelitis. *Amer. J. Trop. Med. Hyg.* **3**, 283–93.

SERGENT, ET. (1942). Paludisme des oiseaux. Réceptivité des moustiques à l'infection du *Plasmodium relictum*. *Arch. Inst. Pasteur Algér.* **18**, 214–15.

SMITHBURN, K. C. (1949). Rift Valley fever: neurotropic adaptation of the virus and experimental use of this modified virus as a vaccine. *Brit. J. Exp. Path.* **30**, 1–16.

SMITHBURN, K. C. and GILLETT, J. D. (1948). Rift Valley fever: isolation of the virus from wild mosquitoes. *Brit. J. Exp. Path.* **29**, 107–21.

SMITHBURN, K. C., HADDOW, E. J. and LUMSDEN, W. H. R. (1949). Rift Valley fever; transmission of the virus by mosquitoes. *Brit. J. Exp. Path.* **30**, 35–47.

STUPPY, C. (1931). Uebertragung von Hühnerpocken durch Mücken. *Zbl. Bakt.* (*Orig.*), **123**, 172–8.

SUMMERS, W. A. (1943). Experimental studies on larval development of *Dirofilaria immitis* in certain insects. *Amer. J. Hyg.* **37**, 173–8.

TAYLOR, R. M. and HURLBUTT, H. S. (1953). Isolation of West Nile virus from *Culex* mosquitoes. *J. R. Egypt. Med. Ass.* **36**, 199–208.

TRAGER, W. (1942). A strain of the mosquito *Aëdes aegypti* selected for susceptibility to the avian parasite *Plasmodium lophurae*. *J. Parasit.* **28**, 457–65.

WORK, T. H., HURLBUTT, H. S. and TAYLOR, R. M. (1955). Indigenous wild birds of the Nile Delta as potential West Nile virus circulating reservoirs. *Amer. J. Trop. Med.* **4**, 872–8.

(*b*) CONTROL AND PROTECTION

ANDREWS, J. M. and SIMMONS, S. W. (1948). Developments in the use of the newer organic insecticides of public health importance. *Amer. J. Publ. Hlth*, **38**, 613–31.

ARMY MEDICAL DEPT. U.S.A. (1945). The prevention of dengue fever. *Bull. U.S. Army Med. Dep.* **4**, 535–9.

BANA, F. D. (1936). A practical way of dealing with *Aëdes aegypti* (*Stegomyia fasciata*) mosquito breeding in country craft. *Indian Med. Gaz.* **71**, 79–80.

BARBER, M. A. (1944). A measurement of the toxicity to mosquito larvae of the vapours of certain larvicides. *Publ. Hlth Rep.*, *Wash.*, **59**, 1275–8.

BISHOPP, F. C. (1946). The insecticidal situation. *J. Econ. Ent.* **39**, 449–59.

BISHOPP, F. C. (1951). Insecticides and methods of control. *Rep. 9th Internat. Congr. Entomology*. WHO/Mal./73.

BLACKLOCK, D. B. (1937). Screencloth for houses in the tropics. *Ann. Trop. Med. Parasit.* **31**, 447.

BLACKLOCK, D. B. and WILSON, C. (1942). Apparatus for the collection of mosquitoes in ships with notes on methods of salivary gland dissection. *Ann. Trop. Med. Parasit.* **36**, 53–62.

BOYCE, R. (1906). Yellow fever prophylaxis in New Orleans in 1905. *Lpool Sch. Trop. Med.* Mem. 19.

BROWN, J. Y. (1934). Safe mosquito nets for use in the tropics. *W. Afr. Med. J.* **7**, 147–8.

BRUCE-CHWATT, L. J. (1953). *Colonial Med. Res. Comm. 1953. Eighth Ann. Rep. 1952–3*, p. 126.

BRUCE-CHWATT, L. J. and COI, K. H. (1950). Report on trials on aircraft disinsectization in West Africa. *WHO Expert Comm. on Insecticides*/6, dated 30 May 1950.

BUSVINE, J. R. (1952). The newer insecticides in relation to pests of medical importance. *Trans. R. Soc. Trop. Med. Hyg.* **46**, 245–52.

BUSVINE, J. R. (1957). Insecticide resistance. Strains of insects of public health importance. *Trans. R. Soc. Trop. Med. Hyg.* **51**, 11–31.

CARTER, H. R. (1922). Yellow fever in Peru. Epidemic of 1919 and 1920. *Amer. J. Trop. Med.* **2**, 87–106.

CHABRILLAT, M. (1934). Note sur la fièvre de trois jours. *Bull. Soc. Path. Exot.* **27**, 761–6.

CHANDLER, A. C. (1956). History of *Aëdes aegypti* control work in Texas. *Mosquito News*, **16**, 58–63.

CHRISTOPHERS, S. R. (1947). Mosquito repellents. *J. Hyg.*, *Camb.*, **45**, 176–231.

CONNOR, M. E. (1921). Fish as mosquito destroyers. An account of the part they played in control of yellow fever at Guayaquil, Ecuador. *Nat. Hist.* **21**, 279–81. New York.

CONNOR, M. E. (1924). Suggestions for developing a campaign to control yellow fever. *Amer. J. Trop. Med.* **4**, 277–307.

CONNOR, M. E. and MONROE, W. M. (1923). *Stegomyia* indices and their value in yellow fever control. *Amer. J. Trop. Med.* **3**, 9–19.

COVELL, G. (1941). *Malaria Control by Antimosquito Measures.* Thacker and Co., London.

DAVID, W. A. L. (1946*a*). Factors influencing the interaction of insecticidal mists and flying insects. Part I. *Bull. Ent. Res.* **36**, 373–93.

REFERENCES

David, W. A. L. (1946*b*). Factors influencing the interaction of insecticidal mists and flying insects. Part II. *Bull. Ent. Res.* **37**, 1–28.

David, W. A. L. and Bracey, P. (1946). Factors influencing the interaction of insecticidal mists and flying insects. Part III. *Bull. Ent. Res.* **37**, 177–90.

Davidson, G. (1952). Experiments on the use of DDT, BHC and dieldrin against adult mosquitoes at Taveta. *Nature, Lond.*, **170**, 702.

De Caires, P. F. (1947). *Aëdes aegypti* control in absence of a piped water supply. *Amer. J. Trop. Med.* **27**, 733–43.

Duguet, J. (1949). Disinsectization of aircraft. *Bull. World Hlth Org.* **2**, 155–91.

Dunn, L. H. (1923). Prevalence of yellow fever mosquito *Aëdes calopus* in the southern parts of Peru. *Amer. J. Trop. Med.* **3**, 1–8.

Dunn, L. H. and Hanson, H. (1925). Prevalence of the yellow fever mosquito *Aëdes aegypti* in Colombia. *Amer. J. Trop. Med.* **5**, 401–18.

Floch, H. and De Layudi, P. (1946). Études des propriétés insecticides de plantes guyanaises. *Publ. Inst. Pasteur Guyane*, no. 127, 7 pp. Cayenne.

Galli-Valerio, B. (1925). Beobachtungen über Culiciden etc. *Zbl. Bakt.* (*Orig.*), **94**, 309–13.

Giglioli, G. (1948). Malaria, filariasis and yellow fever in British Guiana. Control by residual DDT method, etc. Georgetown. Abstract in *Rev. Appl. Ent.* **37**, 40.

Gillette, H. P. S. (1956). Yellow fever in Trinidad: a brief review. *Mosquito News*, **16**, 121–5.

Goodhue, L. D. (1946). Aerosols and their application. *J. Econ. Ent.* **39**, 506–9.

Hadaway, A. B. and Barlow, F. (1952). Some physical factors affecting the efficiency of insecticides. *Trans. R. Soc. Trop. Med. Hyg.* **46**, 236–44.

Hanson, H. (1925). General report on the yellow fever campaign in Colombia. *Amer. J. Trop. Med.* **5**, 393–400.

Haslam, J. F. C. (1925). Observations on the experimental use of fish indigenous to British Guiana for control of mosquitoes breeding in vats, tanks, barrels and other water containers. *J. Trop. Med.* (*Hyg.*), **28**, 284–8.

Herms, W. B. and Gray, H. F. (1940). *Mosquito Control.* The Commonwealth Fund, New York, and Oxford Univ. Press, London.

Holmes, E. M. (1878, 1879). Notes on the medicinal plants of Nigeria. *Pharm. J.* **8**, 563–4; **9**, 853.

Howard, L. O. (1917). Remedies and preventions against mosquitoes. *U.S. Dep. Agric. Farmer's Bull.* no. 444, pp. 1–15.

Howard, Dyar and Knab. See chapter I (*b*).

Ihering, R. von (1933). Os heixes larbophagos utilizado no combate a febre amarilla et a malaria. *Rev. Med.-cirurg. Brasil*, **41**, 221–34.

International Health Board. Rockefeller Foundation (1924). *The Use of Fish for Mosquito Control.* Bibliography of 217 titles.

James, S. P. (1914). Summary of a year's mosquito work in Colombo. *Ind. J. Med. Res.* **2**, 227–67.

James, S. P., Da Silva, W. T. and Arndt, E. W. (1914). Report on a mosquito survey of Colombo and the practicability of reducing *Stegomyia* and some other mosquitoes in that seaport. Govt. Record Off. Colombo.

Jaswant Singh and Rajindar Pal (1952). Organic synthetic insecticides employed in malaria control. *Ind. J. Malar.* **6**, 219–23.

Kahn, M. C. and Offenhauser, W. (1949). The first field tests of recorded mosquito sounds used for mosquito destruction. *Amer. J. Trop. Med.* **29**, 811–25.

Kennedy, J. S. (1947). The excitant and repellent effect on mosquitoes of sublethal contact with DDT. *Bull Ent. Res.* **37**, 593–607.

Laird, M. (1951). Insects collected from aircraft arriving in New Zealand from abroad. *Zool. Publ. Vict. Univ. Coll.* no. 11, pp. 1–30.

Le Van, J. H. (1940). Measures instituted for the control of *Aëdes aegypti*. *Amer. J. Publ. Hlth*, **30**, 595–9.

Legendre, J. (1913). Destruction des Culicides à l'aide du filet. *Bull. Soc. Path. Exot.* **6**, 43–7.

LEVER, R. J. A. W. (1946). Entomological notes. *Agric. J. Fiji*, **17**, 9–15.
LEWIS, D. J. (1955). The *Aëdes* mosquitoes of the Sudan. *Ann. Trop. Med. Parasit.* **49**, 164–73.
LUTZ, A. and MACHADO, A. (1915). Viajem pelo rio S. Francisco, etc. *Mem. Inst. Osw. Cruz*, **7**, 5–50.
MACARTHUR, W. P. (1923). Mosquito netting. *J. R. Army Med. Cps*, **40**, 1–11.
MACFIE, J. W. S. (1916). Chlorine as a larvicide. *Rep. Accra Laboratory for the year 1915*, p. 71. Abstract in *Rev. Appl. Ent.* **5**, 47.
MADDEN, A. H., LINDQUIST, A. W. and KNIPLING, A. F. (1945). DDT treatment of aeroplanes to prevent introduction of noxious insects. *J. Econ. Ent.* **38**, 252–4.
MELVILLE-DAVISON, W. (1912). Mosquito screening of ships. *Yell. Fev. Bur. Bull.* **2**, 200–23.
MISSIROLI, A. (1947). Riduzione o eradicazione degli Anofeli. *Riv. Parasit.* **8**, 141–69.
MOLLOY, D. M. (1924). Some personal experiences with fish as antimosquito agencies in the tropics. *Amer. J. Trop. Med.* **4**, 175–94.
MOOIJ, W. (1940). Malaria-prophylaxis in het Koninklijk Nederlandsch-Indische Leger. *Geneesk. Tijdschr. Ned.-Ind.* **80**, 223–4.
MOORE, W. and GRAHAM, S. A. (1918). A study of the toxicity of kerosene. *J. Econ. Ent.* **11**, 70–5.
MOSNA, E. (1947). Su una caratteristica biologica del *Culex pipiens autogeneticus* di Latina. *Riv. Parasit.* **8**, 125–6.
MUIRHEAD THOMSON, R. C. (1947). The effect of house spraying with pyrethrum and with DDT on *Anopheles gambiae* and *A. melas* in West Africa. *Bull. Ent. Res.* **38**, 449–64.
MURRAY, D. R. P. (1936). Mineral oil in mosquito larvicides. *Bull. Ent. Res.* **27**, 289–305.
MYERS, G. S. (1926). Fishes and human disease. *Fish Culturist*, **5**, 27–9.
O'KANE, W. C. (1947). Results with benzene hexachloride. *J. Econ. Ent.* **40**, 133–4.
PAL, RAJINDAR. See RAJINDAR PAL.
PARKER, E. A. and GREEN, A. A. (1947). DDT residual films. I. The persistence and toxicity of deposits from kerosene solutions on wall-board. *Bull. Ent. Res.* **38**, 311–25.
PRASHAD, B. and HORA, S. L. (1936). General review of the problem of larvivorous fishes of India. *Rec. Mal. Surv. India*, **6**, 631–48.
RAJINDAR PAL (1954*a*). Sorption of insecticides on porous surfaces. *Bull. Nat. Soc. Ind. Mal. Mosq. Dis.* **2**, 177–9.
RAJINDAR PAL (1954*b*). Particle size of insecticides in relation to biological efficiency. *Bull. Nat. Soc. Ind. Mal. Mosq. Dis.* **2**, 180–1.
RAJINDAR PAL and SHARMA, M. I. D. (1952). Rapid loss of biological efficiency of DDT applied to mud surfaces. *Ind. J. Mal.* **6**, 251–63.
RAJINDAR PAL, SHARMA, M. I. D. and KRISHNAMURTHY, B. S. (1932). Studies on the development of resistant strains of houseflies and mosquitoes. *Ind. J. Mal.* **6**, 303–15.
RIGOLLOT, S. (1927). A propos de la prophylaxie de la fièvre jaune à la Côte Occidentale d'Afrique. *Bull. Soc. Path. Exot.* **20**, 859–65.
RUSSELL, C. H. (1892). The best mosquito remedy. *Insect Life*, **4**, 223–4.
SEAL, W. P. (1908). Fishes and the mosquito problem. *Sci. Amer.* Suppl. **65**, 351–2.
SEVERO, O. P. (1956*a*). Eradication of the *Aëdes aegypti* mosquito from the Americas. *Mosquito News*, **16**, 115–21.
SEVERO, O. P. (1956*b*). Acquired resistance to DDT in the Dominican Republic. *Chron. World Hlth Org.* **10**, 347–54.
SHANNON, W. R. (1943). An aid in the solution of the mosquito problem. *Minn. Med.* **26**, 799.
SINTON, J. A. and WATS, R. C. (1935). The efficacy of various insecticidal sprays in the destruction of adult mosquitoes. *Rec. Mal. Surv. India*, **5**, 275–306.
SNEATH, P. A. T. (1939). Yellow fever in British Guiana. *Trans. R. Soc. Trop. Med. Hyg.* **33**, 241–52.
SOPER, F. L. (1938). See under (*a*) Yellow fever.
SOPER, F. L., WILSON, D. B., LIMA, S. and ANTUNES, W. SÁ (1943). *The organisation of permanent Nation-wide anti-*Aedes aegypti *measures in Brazil.* Rockefeller Foundation, New York.

REFERENCES

TANNER, W. F. (1931). Inspection of ships for determination of mosquito infestation. *Publ. Hlth Rep., Wash.* **46**, 2306–20.

THOMSON, M. R. C. See MUIRHEAD THOMSON.

TRAVIS, B. V., MORTON, F. A. and COCHRAN, J. H. (1946). *J. Econ. Ent.* **39**, 627–30.

TURNER, J. G. S. (1946). *Rep. Med. Dep. Gold Coast*, 1945. Abstract in *Rev. Appl. Ent.* **37**, p. 52.

WILLIAMSON, K. B. (1924). The use of gases and vapours for killing mosquitoes breeding in wells. *Trans. R. Soc. Prop. Med. Hyg.* **17**, 485–519.

WILSON, C. S., MATHESON, D. R. and JACKOWSKIE, L. A. (1944). Ingested thiamin chloride as a mosquito repellent. *Science*, **100**, 147.

WORLD HEALTH ORGANISATION. TECHNICAL REPORT SERIES, no. 110. Expert Committee on Insecticides. 6th Report. Technique of Spraying and Spray Apparatus.

V
TECHNIQUE

The particular technique employed by an observer naturally largely depends upon the facilities such observer has at his disposal, upon the nature of the work to be carried out and upon the personal preferences of the individual. In what follows various points in the technique of rearing and utilising *Aëdes aegypti* are given as they have been found suitable by the author using the species on a considerable scale over a number of years on work in determining the effectiveness of different repellents and of their preparations and in studying the bionomics of the species (Christophers, 1947). That the apparatus and methods described will necessarily be those best suited for some particular purpose of the reader can scarcely be expected. Thus it may at times be required merely to breed out small numbers of the species using little more than a suitable cage and a few ordinary laboratory articles and much that is said here will be irrelevant. On the other hand, much that is ambiguous or even contradictory in the literature may be largely avoided if care be taken to see that the material used is properly reared, of good quality and the conditions under which it is used, such as age from emergence and other relevant facts, given. Even if the technique here given is not followed there may be points of interest or usefulness to those commencing study of the species. The reader must therefore make his own selection in what follows of those items that may serve or assist in his technique. Besides the author's own methods a number of references to the work of others will be found which may be useful to consult. Special techniques relating to particular methods of experimentation or preparation of material not dealt with here are given in later chapters. If more attention were given than is often the case to the use of standard material, well bred, with wing length or weight recorded, with the day from emergence or the blood feed noted (not a batch of unknown age fed on sugar!) much of the troublesome ambiguities so often recorded in the literature with this species would be avoided.

(*a*) APPARATUS

Breeding jars. For rearing larvae the usual laboratory glass basins are very suitable (Fig. 2 (3 A)). A size of about 15–20 cm. in diameter and about 10 cm. in height are convenient for early stages of the culture or for a total of larvae up to about 500. Larger basins up to 30 cm. or more if they will go conveniently into the incubator may be used for larger numbers. A graduated wooden support in steps, as shown in the figure, is useful in collecting the pupae with the dish tilted.

Meat dishes. Small glass dishes (potted meat dishes)* about 12 cm. in diameter and some 5–6 cm. deep, or some suitable substitute, are almost essential for collecting the pupae for hatching out in the cages, for taking samples of larvae and many other purposes. They may usually be obtained in a Woolworth's stores (Fig. 3 (7)).

* Also used as butter dishes.

Collecting pipette. For removing larvae, collecting pupae, etc. a wide-mouthed pipette is useful. This consists of 6 in. of glass tubing of not more than 6 mm. bore provided with a rubber ball-teat preferably of 4 cm. diameter (Fig. 2 (3C)). The distal opening may be very slightly narrowed and the cut edges smoothed in the flame.

Collecting sieve. This is used to collect larvae or pupae in bulk. It consists of a 2 in. length of 1 in. diameter glass tubing with mosquito netting over one end held in place by a rubber band (Fig. 2 (3B)). When in use the end with the netting should be held beneath the surface to avoid damaging the larvae or pupae as these are pipetted in.

Earthenware pots. These are of unglazed earthenware (known as rabbit feeding dishes). They are about 12 cm. in diameter and 7 cm. high with turned-in rim (Fig. 3 (5)). They are invaluable, not only for egg deposition and storage, but as humidifiers during emergence. Having a turned-in edge they give, when partly filled with water, conditions closely resembling those of the native water pot, a favourite breeding place in nature. Being rough they give good foothold and being porous they retain moisture after egg laying for some time after fluid has been removed, and, filled with water, they can be placed over other pots with eggs to keep these moist for the desired time. Eggs are laid in a heavy ring round the edges of the water and after conditioning and drying the pots can be broken and suitable sized fragments with their attached eggs used for starting cultures. As humidifiers when filled with water they are much more effective than dishes of water, even if these are provided with wet wool. They can be obtained from time to time at Woolworth's stores.

Air bubbling apparatus. Owing to the danger of scum formation a great deal of trouble will be saved if bubbling air can be used when culturing larvae. If available in the laboratory current air may be brought into the incubator through the thermometer hole and linked up by T-pieces with as many distributing tubes as required. The rate of bubbling should be controlled by screw clips, and it is a great convenience to have these set firmly in a vertical position on the rubber tube between two staples fixed on a largish cork, which has been cut to ride on the edge of the breeding dish (Fig. 2 (2)). It is important to insert an air gauge in the course of the main air supply tube (Fig. 2 (1)). This ensures uniform pressure. If this is not done, air will escape only in such a distributing tube as lies highest in water, making regulation difficult.

If current air is not provided, it is possible to arrange for this by a small motor. Small motors for the purpose are to be obtained from firms catering for amateur aquaria (for example, Gamage's). If air bubbling is not available, some other device must be substituted (see section on the culture fluid), but it is a great and almost necessary convenience if work on any scale is being carried out.

Cages. The simplest form of cage consists of a light wooden frame with sides of mosquito netting, having a wooden or metal base and with one side provided with a sleeve (see regarding construction of such below). The netting should be a rather fine mosquito net. That in ordinary use in the tropics may be used (about 14 by 16 meshes to the inch). But a finer netting as used in dress-making can be obtained and is probably safer (mesh about 18 by 20 to the inch). The actual mesh aperture varies with the thickness of the cotton enclosing the mesh. A black netting of this kind is also available and is preferred by some as giving better visibility. The sleeve should be at least 18 in. long and rather voluminous. A glass side is also useful.

Recently cages have come into use with the sides, or the side with the sleeve, of plastic sheet. In this case the sleeve is usually attached to the edges of a circular opening cut in the sheet by being stitched to a length of slit rubber tubing. When the slit rubber is adjusted to the edge of the opening the rubber firmly grips the plastic. The sleeve attached to the frame, however, is more convenient for introducing and taking out dishes and is to be preferred.

Cages need not be large, except for special purposes. An all-purpose cage, for breeding and use of a moderate number of mosquitoes, about 12 by 2 by 12 in. is very convenient as this will go easily in the incubator and allows of dishes being taken in and out. It is also not too large to make capturing mosquitoes by suction difficult.

Whilst the above will serve for small-scale work any more extended research will be greatly facilitated by use of several sizes of cage designed for special purposes. Thus, for emergence, cages small enough for several to go at the same time into an incubator are very convenient. But these might not be suitable for later feeding or other purposes, for which larger and perhaps quite different forms of cages might be desirable. Hence for this and other reasons some method of being able readily to transfer the mosquitoes in bulk is a great convenience and only requires some thought given to construction of the cages to be used (see Fig. 3, which shows cages as used by the author). Another method which serves a slightly different purpose, namely removing mosquitoes as required from a stock cage, is that described by David, Bracey and Harvey (1944), who use a cage with a wooden roof in which is an opening closed by a sliding door.

Large cages used for stock, except for some special reason, are not very desirable for ordinary work, but may serve a purpose. A cage designed to give a continuous supply of eggs for a large laboratory is described by Leeson (1932) and has proved very useful. It is

---

Figure 2. Technique.

1 Air pressure regulating valve. The arrow indicates direction of air flow.

2 Arrangement for pinchcock fixed on cork by staples for regulating rate of air bubbling on each air distributing tube.

3 Apparatus used in collection of larvae and pupae. A, glass culture basin sloped for collection. The arrow shows direction of light from window; B, sieve; C, collecting pipette.

4 Suction pipette for collection of adults. Simple form. *a*, collecting tube; *b*, glass mouthpiece at end of rubber suction tube.

5 A more elaborate pipette for collecting and accurately chloroforming adults. Lettering as in 4.

6 Specimen tube with filter paper strip resting in water as used for observing time of oviposition, number of eggs laid, etc. The double rubber band shown is a precaution-measure against spontaneous breaking of the small rubber bands used which sometimes occurs with loss of specimen.

7 Mosquito hotel. Jar with cardboard shown lifted from hollow lid furnished with water, and paper-slip on cork.

8 Method of mounting mosquitoes for transport. A, single mosquito mounted on cork strip; B, same put in tube with opening in cork closed by light wool pledget; C, tube put up with a number of mounted specimens; D, showing how cork strip is pinned to cork.

9 Apparatus for rearing and feeding mosquitoes as used at the Molteno Institute (Tate). *a*, hole in mosquito netting cover plugged with wool for removal of mosquitoes by suction pipette; *b*, rivet holding bent metal sheet in place and giving some latitude to movements of lower part; *c*, one of three supporting nuts on inner side of metal to engage over milk dish (unshaded portion below).

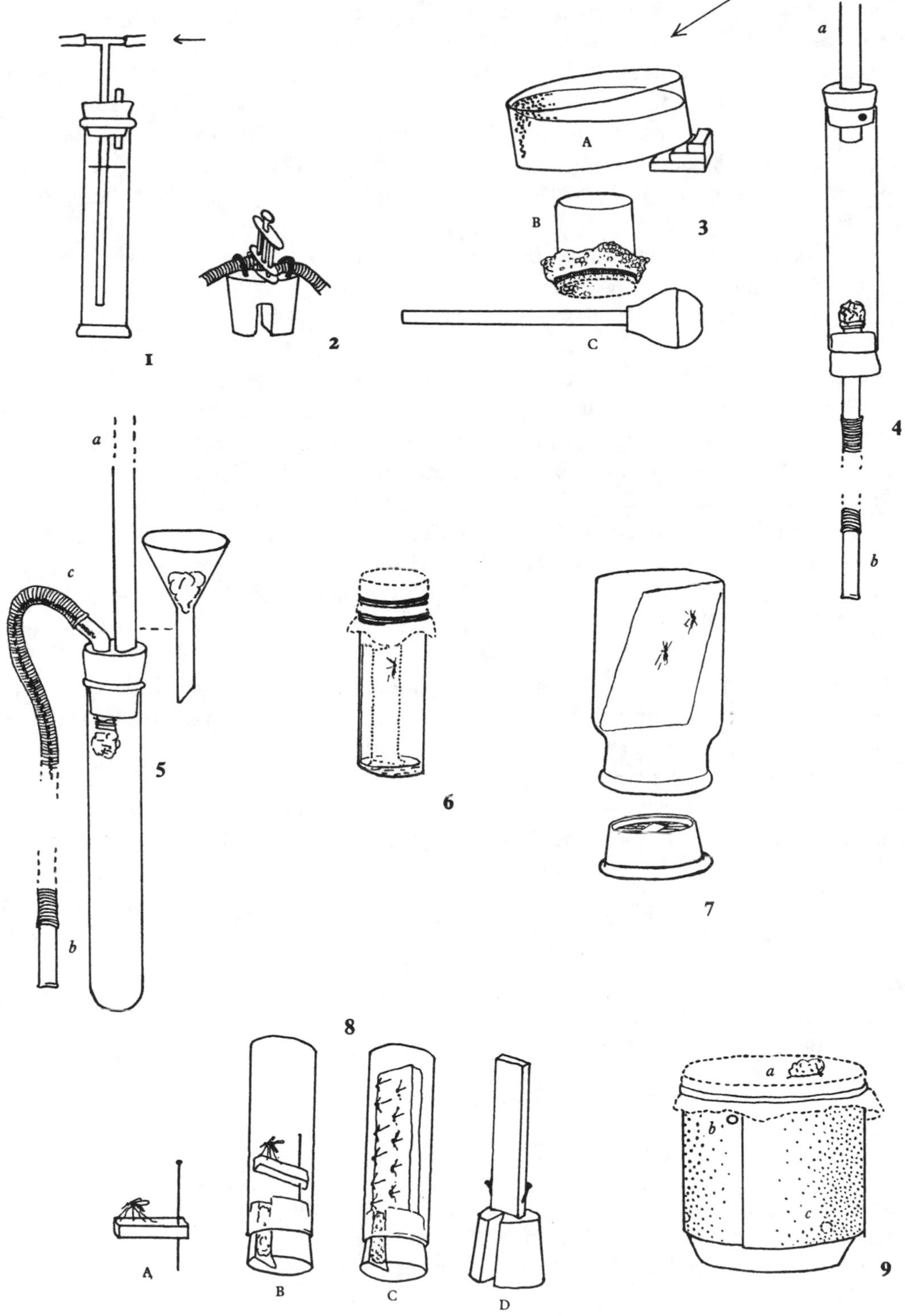

Figure 2

a large cage with permanent accommodation for a guinea-pig. The roof of the animal's quarters forms a shelf on which is a breeding tank kept automatically at a desired temperature by means of a thermostatically controlled electric bulb. Continuity of supply may also be arranged for through ability of *Aëdes* eggs to retain vitality for several weeks or longer (see later on manipulation of eggs). For description of equipment for continuous rearing of *A. aegypti*, see Bacot (1916); Buxton and Hopkins (1927); Putnam and Shannon (1934); H. A. Johnson (1937); David *et al.* (1944); Christophers (1947); Burchfield, Redder, Storrs and Hilchey (1954).

Special cages and equipment may be required for special purposes. Stokes, Bauer and Hudson (1928) describe a cage used for feeding monkeys in yellow fever work; see also Sellards (1932); Shannon (1939); Barnett (1955); Ross (1956). Snijders, Dinger and Schuffner (1931) describes cages used in work on dengue transmission. Nieschulz and Du Toit (1934) describe cages used in veterinary studies using *A. aegypti*.

Not all authors, however, in their techniques use cages of the ordinary kind. For some purposes the simple apparatus and procedures given by Stephens and Christophers (1908) in handling a few mosquitoes still hold good and are referred to later. Yolles and Knigin (1943) use a celluloid cylinder with a rubber diaphragm and a slit for inserting a tube. Young and Burgess (1946) use a plastic sheet 22 by 10 in. bent into a semi-cylinder and tacked on a wooden base 10 in. square, the back being of transparent celluloid and the front of netting. Trembley (1944*a*) uses for oviposition lamp glasses kept over petri dishes with wet wool. Fifty fed females can be added or mosquitoes can be fed through the netting covering the top.

A system which is in use at the Molteno Institute, Cambridge, for work on bird malaria and other studies and found very satisfactory employs for emergence, feeding and oviposition cylinders of sheet metal 6 in. in diameter and 6 in. high resting on flat enamel pans (milk dishes) (see Tate and Vincent, 1936). The cylinder is made of a metal sheet bent into a cylinder held by a rivet only at the top. The bottom of the cylinder can therefore be readily adjusted to grip the dish. The grip if necessary can be further strengthened by a rubber band or clip. Three small projections (rivets) near the bottom prevent the edge of the cylinder sinking more than a certain distance beyond the rim of the dish (Fig. 2 (9)).

Figure 3. Technique.

1 Type *B* cage. *a*, sleeve; *b*, sliding zinc door. Note handle formed by outer edge rolled *backwards*; *c*, edge of cardboard strip as used to give socket room for door and glass planes; *d*, nut and bolt to prevent door being accidentally pulled from its socket; *e*, screw to retain glass pane in position.
The roof is shown reinforced by wire netting for use with rabbit feeding attachment.

2 Type *A* cage. Lettering as for 1.

3 To show details of construction of cages. A, screw for glass panes; B, showing sliding socket for door made from strips of bent zinc sheet; or socket may be merely gap left when outer slat is separated from the cage frame by a thick strip of cardboard; C, socket for glass pane formed by strip of cardboard (*c*); D, bolt near edge of door to prevent total withdrawal.

4 Rabbit feeding attachment. The dotted areas are sorbo rubber resting on wooden floor. Twice scale of 1.

5 Porous pot (rabbit feeding pot) referred to as 'egg-pot'.

6 Fragment of same with portion of egg rim as used for starting a culture.

7 Glass dish (meat dish; butter cooler dish).

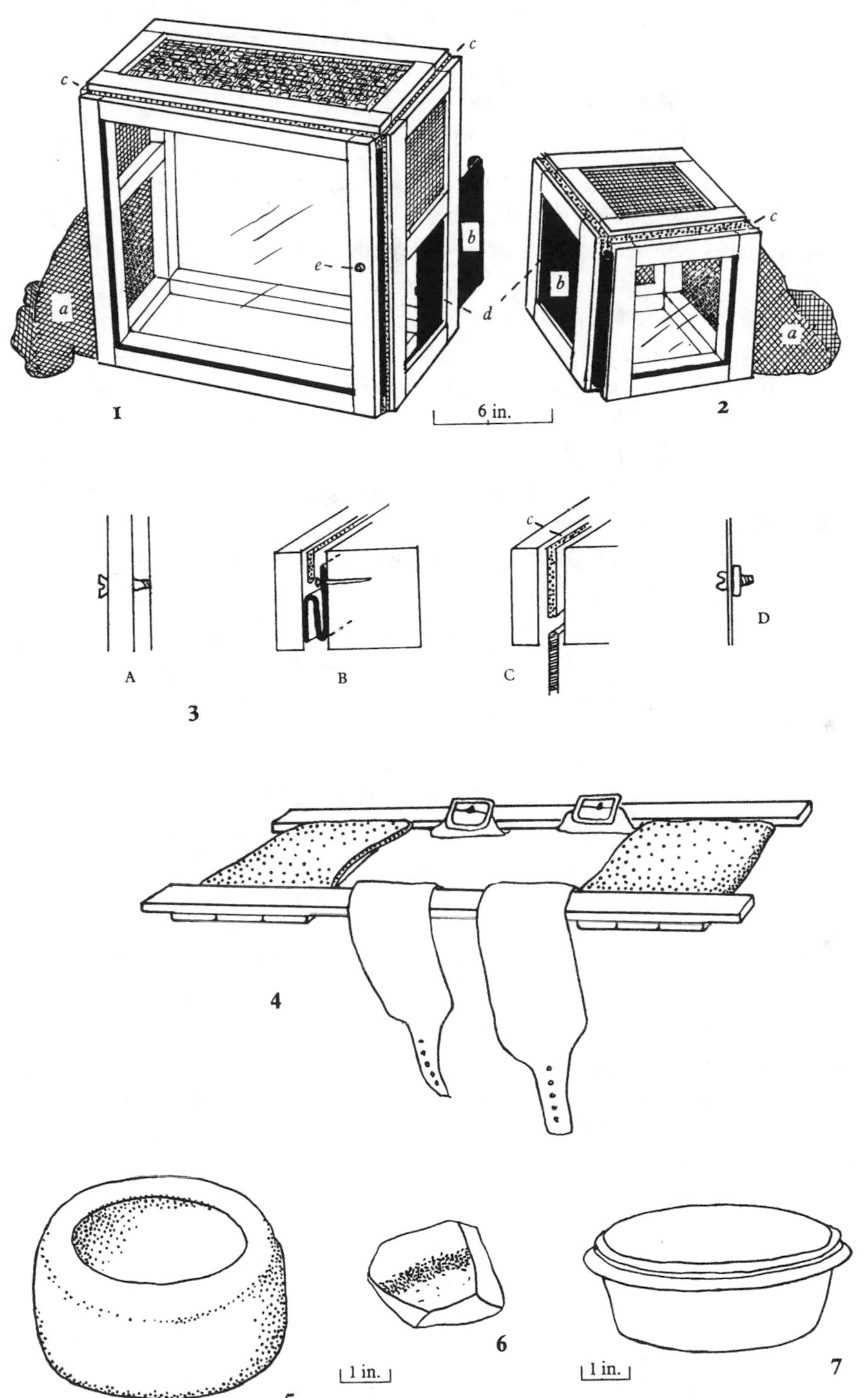

Figure 3

Pupae are placed in the dish to hatch out in the incubator. Feeding is done by the bird or small animal being laid on the netting at the top of the cylinder with some portion of bare skin touching this. The mosquitoes oviposit freely around the dish giving an abundant supply for immediate breeding or holding in stock.

Cages as used by the author. The cages are of two types, namely, small cages of which a number can be used if necessary in the incubator at the same time for emergence and for holding adults until ready for testing or experiment (type *A* cage); and larger cages of a size found suitable for testing, feeding, oviposition or such other purposes as may be carried out (type *B* cages). Manipulation is avoided by having both types of cage with a sliding sheet metal door and a frame which will allow the doors of any two cages to be brought flush. Both type *A* and type *B* cages are made with a matchboard base and a wooden frame of 1 in. scantling. Type *A* cages are small cubical cages measuring 8½ by 8½ by 8½ in. One side is furnished with a sleeve 18 in. long and of ample width. Opposite to this is the sliding door. The remaining sides are covered with mosquito netting or a substitute and, especially if muslin is used, one side is fitted with a glass sheet. All fittings are by means of light wooden slats lightly nailed in place. Type *B* cages are of similar construction but measure 14 by 8½ by 12 in. high. One side or better both are of glass. The cages can be made by any amateur carpenter and stand up well against rough wear. The floor should be of smooth wood. The door may have its inner end bent into a roll to give convenient grip and should have a small bolt and nut set near its outer edge to prevent accidental withdrawal of the door from its bed when in use. The construction of the cages will be clear from the figure (Fig. 3 (1–3)).

With type *A* cages the hatched-out adults from the pupae placed in the cage can be retained under suitable conditions until such time as they are ready for experiment. Up to 200 adults are readily accommodated without overcrowding. As many as eight such cages could, if required, be accommodated in a large incubator. At time of experiment a few moments suffices to transfer all the mosquitoes in the small cage to some form of type *B* cage. All that is necessary is to place the larger cage against some support and, holding the smaller cage firmly in place with the door frames in apposition, to open both doors and agitate the sleeve until all mosquitoes have been transferred. That the door of the large cage occupies only the lower half of the end favours the non-return of any mosquito that has once entered the large cage. To facilitate opening the doors those on type *A* cages can be set in an opposite direction to those on type *B* cages. A clamp to hold both cages firmly together can be used (Christophers, 1947), but is not necessary. By a very simple adaptation, as noted later, a cage of type *B* can be fitted for feeding mosquitoes on a rabbit for oviposition. By the adoption of such a system as that described and with the aid of one or two minor accessory pieces of apparatus techniques of almost any kind can be greatly simplified.

Mosquito aspirator pipette. Apart from transfer in bulk it is often required to remove a few specimens for observation or certain forms of experiment. For this purpose Buxton (1928) uses a length of 1 in. diameter glass tube with a rubber cork at each end into which are inserted respectively a piece of glass tube of suitable length and bore for aspirating adults from a cage and a short piece to which can be attached a length of rubber tubing for suction. This last-mentioned piece of tubing has its inner end protected by a piece of mosquito netting held in place by a small rubber band (Fig. 2 (4)).

A somewhat more elaborate, but very useful arrangement of this kind has been found suitable for many purposes and is illustrated in Fig. 2 (5). This consists of a 6 in. boiling tube with a rubber cork in which is set both a long collecting and a short sucking tube which is either bent sharply to face back to the operator or a piece of wire is inserted in the rubber to prevent it buckling. The collecting tube should fit in the cork hole easily so that it can be readily replaced by a small funnel by means of which mosquitoes in the tube can be accurately anaesthetised by placing a small piece of wool in the funnel and adding a drop or so of chloroform. By sucking gently and watching the mosquitoes they can be anaesthetised to any desired degree and turned out. Or the funnel can be replaced by a bent tube and the mosquitoes without being anaesthetised blown gently into any desired receptacle.

Attachment for feeding on rabbit. For feeding mosquitoes in cages it is now a common practice to immobilise some small animal (guinea-pig or rabbit) with nembutal and place in the mosquito cage. In this case no special apparatus is required other than a small cradle to facilitate the operation (see later under feeding for stock). It has been found, however, that better feeds are given using a rabbit and that with so large an animal it is convenient, whether the animal is immobilised or not, to use a suitable attachment that can be fixed when required on a cage *B* type, the mosquitoes feeding through the gauze roof. Or one such cage can be kept permanently fitted with the device. The attachment consists of a padded board made to fit and to some extent to grip the top of a type *B* cage which has been previously fitted with wire netting under the muslin of the roof. A suitable space is left in the middle to the sides of which are attached canvas straps, each about 4½ in. wide with the ends cut narrower to enter the buckles (Fig. 3 (1)). A rabbit is placed on the board and the buckles adjusted to hold the animal firmly but comfortably in position. To ensure this the straps should be brought up from the inner sides of the frame as shown in the figure. The hair over a portion of the abdomen is previously clipped or shaved, or preferably removed by one of the patent hair removers. A period of fifteen minutes is allowed for feeding. Very heavy feeds may be obtained in this way.

Killing chamber. It is not easy to kill off mosquitoes in a cage without some adequate form of killing chamber or having cages out of action for sufficiently long to ensure death of all the mosquitoes which may take several days. Such a chamber can be made of ply-wood on a strong wooden frame just large enough to take type *B* cages. It is convenient to have a down-closing sliding door. Under the roof there should be sufficient space to fix a wire netting partition on the top of which can be placed a thickish pad of filter paper or other absorbent. Two holes are cut in the roof furnished with corks for adding the lethal agent. The amount of this substance (chloroform or, cheaper, carbon tetrachloride) necessary to ensure complete lethal effect in, say, 15 minues can be ascertained and this amount used as a routine. For a killing chamber just large enough to take cages of the size noted 30 ml. of carbon tetrachloride was found necessary.

Forms. In any full-scaled research using considerable numbers of the species suitable printed or duplicated forms with spaces for relevant information, such as date, specification of the batch from which mosquitoes were taken and other information, will not only greatly reduce the labour of keeping notes but will avoid hiatuses in recordings.

(*b*) THE CULTURE MEDIUM

In early observations on the species bred in the laboratory eggs were placed to hatch and develop in conditions supposed to resemble those of their natural breeding places. MacGregor (1915), commencing his study of the most suitable form of culture fluid, placed a new strain he had received in tap water to which had been added 'fallen leaves and dry twigs'. The unsuitability of such a fluid is very apparent from the statement made by the author that some larvae remained in 'suspended development' from early June to late August, that is the conditions were those for rearing 'starvation forms' as described in a later chapter. Later MacGregor (1929) experimented with a number of other media. Those media consisting exclusively of sugar or starches were useless. A medium of lacto-peptone was fairly satisfactory. Media containing egg albumen or egg yolk gave variable results depending upon the degree of contamination, 0·1 per cent egg yolk being lethal if on decomposition it yielded H_2S. Boiled oatmeal was satisfactory but growth slow. Powdered bean was fairly good. Of all the materials tried the most successful were bemax (excellent at 0·1 to 3·0 per cent if the medium did not become acid) and different forms of bread, namely standard, brown and white (the last not quite so good). Of standard and brown bread the author remarks that the medium apparently contained all that is necessary for vigorous larval growth.

Bacot (1916) early notes that larvae develop readily in water containing a plentiful supply of organic matter.

Buxton and Hopkins (1927) tried out infusion of hay, rice or bran, diluted and allowed to rot. They found the most suitable medium to be 'grass infusion'. It was recognised that bacterial growth took place (Buxton and Hopkins, 1925). These authors made some pertinent comments on the advantages and disadvantages of different 'organic fluids' The advantage of bran, meal and starches is that these substances are easily weighed and have a fairly definite composition. It is almost useless to weigh grass as a large and varying proportion of it is water. Hay also may consist of different grasses. The present author also found that the value of hay for this purpose varied greatly with its age and conditions under which it had been kept. In regard to bran the above authors noted that it has the disadvantage of growing a thick scum. Their studies indicated that on the whole the best medium is dilute grass infusion. They did not claim that the medium is other than a source of bacterial food.

The work of Trager (1935) and others working on essential food requirements of *Aëdes aegypti* larvae (see ch. XI, section (*g*)) later showed that, whilst it is possible to rear larvae up to emergence under sterile conditions there are considerable difficulties such as are entirely absent using fluids with an abundant bacterial flora and that normally the presence of living bacteria and yeasts is necessary to provide certain essential food elements. That solutes or colloidal material in dispersion are to any extent available for food and that larvae can grow to any extent in such substances as sterile nutrient broth has also been negatived by the experience of several authors (for example, Hinman, 1930, 1932).

Briefly, then, the requirements of a good culture medium would appear to be that it should be a fluid providing continued bacterial growth, but without such excessive fermentation and formation of scum and harmful products as will be prejudicial to the larvae.

Bread, usually in the form of dried bread crumbs sprinkled on the surface, has been the food most usually employed. Shannon and Putnam (1934) used tap water with dried bread, either exclusively or with dried blood serum. Trembley (1944*a*) used oat grains cut in half (12–30 grains per tray of 3 litres) alone or with dog biscuit put in 12–24 hours in advance. Or 150 c.c. from an old pan was added daily. David, Bracey and Harvey (1944), whilst they found powdered bread satisfactory, obtained the best results with a breadcrumb infusion. This was allowed to ferment for five days at 20° C., shaking daily and before being used was strained through voile, warmed and aerated. Other forms of food used in the rearing of mosquito larvae, including *A. aegypti*, have been heated yeast (Tate and Vincent, 1936); finely ground lentils (De Boissezon, 1933); piscidin, a proprietary fish food consisting of powdered flies (Martini, 1921; Weyer, 1934; James, private communication); a powdered breakfast food found more convenient than yeast (H. A. Johnson, 1937); and probably others, though the comparative value of these mostly rests on the fact that they did serve for culture, it being rare in the literature for measurement of size or weight of the mosquitoes reared to be given. Weyer (1934), however, notes that a larval diet of blood and liver produced vigorous males and females.

With the present author no food has given such consistently good results as a modification of that used in Professor Buxton's laboratory, namely a finely ground proprietary dog biscuit (see Lumsden, 1947). The modification used has been the addition of bemax and a standardised method of preparation and usage. The biscuit used has been Entwistle's Puppy Meal. This consists as purchased in the market of a mixture of broken fragments of an ordinary yellow hard biscuit and of a brown biscuit which contains a little bone meal and probably other ingredients considered suitable for dogs. The makers, Entwistle Ltd, Vulcan Mills, Liverpool, have, however, on request very kindly supplied these two constituents separately in seven-pound bags. Otherwise if bought on the market the yellow biscuit is in excess. Preparation of the food is as follows.

Equal parts of yellow biscuit, brown biscuit and bemax are separately finely ground and well mixed, fineness being assured by sifting through fine mosquito netting. For large quantities a mincing machine is helpful, though the final grinding is best done by rubbing up in a mortar. This finely ground powder is kept as stock, being used as required. When preparing culture medium a small quantity, say two piled teaspoonfuls, are rubbed up with a little water into a paste and then diluted to form a suspension of which desired quantities are added to tap water. Enough should be added in the first place to give, eventually, a good turbidity, that is, one in which, seen through the glass, larvae are only visible when they approach close to this. The culture fluid may be clear when freshly made, due to settling, but later becomes increasingly turbid. If a heavy scum forms in spite of gently bubbling air, too much of the food has been added for the number of larvae present. If overnight the culture becomes dark and transparent, too little food is being given or there is overcrowding. An appropriate quantity is added daily or from time to time to maintain conditions as they should be. Changing the water daily as is sometimes recommended is to be avoided. But as the larvae increase in size larger receptacles may conveniently be brought into use and so a certain amount of fresh water comes to be added during progress of the culture. Where bubbling air is not used, an attempt may be made to add just sufficient food to keep scum from forming. Close attention is required, however, as one night's scum may kill off the larvae, especially when still in the young stages or as pupae. Another way is to rear at a lower temperature.

(*c*) REARING AND MAINTENANCE

The following is the procedure followed by the author in rearing *Aëdes aegypti.*

A fragment of egg-pot with what is judged to be an appropriate supply of eggs (see section (*d*)) is sunk in a suitably sized deepish glass basin (say a meat dish or suitable vessel to take the piece of pot) filled with tap water to which about one-third of its volume of old culture fluid from a recent brew and a trace of food has been added. The preparation is then left overnight at 28° C. with gently bubbling air. By the morning the great majority of eggs that have been conditioned as described later will have hatched. Older eggs may take somewhat longer, but are best not used if more recent eggs are available. The old culture fluid is not essential, but it helps to ensure rapid and uniform hatching.

Suitable hatching having occurred, the larvae with the fluid in the dish are turned into a larger dish of tap water at 28° C. and a suitable amount of food suspension added. As growth takes place the larvae and contents are transferred *en bloc* if necessary to a larger dish. Dishes up to 5 litres or more may be used, but it is thought that those of about 3 litres are optimum. About up to 250 larvae per litre can be safely reared without any fear of overcrowding. A certain concentration seems desirable. If the number hatched is in excess, the excess should be removed with the pipette where they lie accumulated at the side of the dish away from the light. This may require to be done from time to time as the excess becomes apparent. After a little experience a rough estimate can be made about the second day as to what thinning out is required.

When the majority of larvae have pupated, usually about the sixth day, pupae are pipetted out and placed in clean water at 28° C. in meat dishes in cages of type *A* for emergence in the incubator. To ensure humidity so that there shall be little or no mortality a few porous egg-pots filled with water are placed in the incubator along with the cages.

This is the time when counting for future experimentation is done. As a routine it is convenient to have each small cage put up with, say, 100 pupae of a given sex. It is then always possible later when transferring the mosquitoes to know what numbers are being used. Even if this is not required it is convenient to have small unit cages to work with. This is the time too for putting aside a sufficient stock of the old culture fluid for use later.

Emergence is complete in about 48 hours from collection of the pupae and when so completed cages can be removed to a lower temperature, each cage being provided with a filter paper fan slipped into its dish. This ensures humidity and gives ready access to water for drinking. Later when the dishes are removed counts of pupal skins can be made if desired to check the cage contents. If clean water has been used for the pupae and high humidity ensured during emergence, there should be at this time practically no mortality.

Cages can now be left, say at 25° C. and moderate humidity (70 per cent relative humidity), until the fifth day from emergence when the mosquitoes are suitable for testing or feeding. The use of sugar, fruit, etc. in the cages with the idea of maintaining the mosquitoes alive is objectionable from several points of view. In the first case it is not necessary since the mosquitoes, if treated as above and given water to drink, do not begin to show mortality, except for some males, until about the seventh day and to use mosquitoes beyond the seventh day from emergence, except for a special purpose, is again undesirable. Further to give food other than water is to introduce a distinctly disturbing factor, as it may seriously reduce biting urge and possibly have other effects. Shannon (1939) found that 22, 35 and 56 per cent respectively fed in three series treated as follows: fed on sugar up to time of feed; sugar removed three days before feed; mosquitoes 2–3 days old with only water. Water to drink (and preferably with a filter paper fan), however, is essential and adults soon die if this is not provided.

In a few days the cages are ready for experiment. The fifth day is the most suitable time for this as the mosquitoes are now uniformly hungry and excitable and there should be as yet no appreciable mortality except among the males. Offered food, for example the human arm, the percentage feeding should approach 100 per cent. Normally if being tested the mosquitoes from a small cage of females and one of males are transferred to a cage type *B* and before use are exposed without water for one hour to bright electric light.

A word may be said here as to the use of the expression 'first, second or third day' when referring to age of mosquitoes. By 'first day' is here meant some time on the day following any operation, such as putting eggs to hatch, the day of the operation being day 0. Buxton and Hopkins use the same convention. Most usually the condition described as that on any particular day relates to the time of morning inspection, the operation having been done some time during the previous day. Whilst very rough, it is useful where exactness is not important. Otherwise time is best given in hours.

(*d*) FEEDING FOR EGG STOCK

When arranging for egg supply it is convenient to carry out emergence with a few hundred male and female pupae sieved out in bulk and placed in a meat dish with clean water in a *B*-type cage (that with the rabbit attachment if this has been set aside for the purpose). When emergence is completed filter paper fans are inserted into the dishes and the cage left preferably to the fifth day for feeding. Feeding may be done on the human arm, but use of a rabbit is preferable. The guinea-pig has not been found to give such satisfactory feeds. Fielding (1919) notes the same experience. An hour before the feed, water should be removed and the cage exposed to the light of a lamp.

The animal used for feeding may be immobilised with nembutal or not. In either case it will be found convenient to use the feeding attachment. Hair over the abdomen should be cropped or shaved. The animal is then laid comfortably on the attachment with the abdomen against the gauze roof of the cage. Usually if performed quietly the animal lies still. Fifteen minutes should be given for the feed.

I am indebted to Mr Shute of the Ministry of Health Malaria Laboratory for the following notes on the use of nembutal.

The drug used for immobilising guinea-pigs, rabbits and monkeys is veterinary NEMBUTAL (Pentobarbital sodium). Each c.c. contains one grain of Nembutal alcohol 10 per cent. It can be used intraperitoneally and intravenously. For complete sedation use 1·0 c.c. (one grain) for each 5 lb. body weight. It is a product of Abbot Laboratories. For fairly large animals, such as rabbits or monkeys, it is best to give it intravenously, very slowly, and stop as soon as the animal becomes immobilised, a matter of seconds. For guinea-pigs the drug may be given intraperitoneally and by this route it takes 5 to 15 minutes to work. With monkeys it may take even longer. Immobilisation lasts from 2 to 6 hours. Small cradles may be used for guinea-pigs in which they lie while resting in the mosquito cages. The same animal can be used over and over again if care be exercised to give the right dose according to body weight. A patch over the abdomen of guinea-pigs should be closely cropped, but for monkeys no clipping is necessary.

After the feed, fresh glass meat dishes with water and filter paper fans are left in the cage for the first two or three days and then replaced by earthenware egg pots about one-third filled with water. These are not placed directly on the floor of the cage, but raised on empty meat dishes. Otherwise a great many eggs may be laid on the floor where this has been wetted by leakage from the pots. By not placing the earthenware egg pots in position at the beginning, much contamination with post-feed faeces is avoided. In nature there may be some selection by the female in depositing her eggs on water of different characters. But in the laboratory there does not appear to be any necessity to make provision for this, since females oviposit freely in captivity on clean tap-water. More important is the character of the vessel, a point referred to later when dealing with manipulation of the eggs.

Oviposition is practically completed by the seventh day from feed, when pots are removed, the water pipetted off and the pots, after labelling, put aside for the eggs to be conditioned. Being porous the pots take a day or two to become dry, which allows time for the conditioning of most of the eggs, these having probably been laid on the fourth or fifth day after the feed. To make quite sure, however, the precaution can be taken to place a pot filled with water above one or more of the pots with eggs, removing this in a day or two and allowing the pots with eggs to become quite dry.

The pots are emptied on the seventh day because even the earliest laid eggs (those laid on the third day) will not by then have had time to hatch. Beyond the eighth day there will be heavy losses from hatching.

Eggs being conditioned should not be kept too moist, that is wet, or there may be serious loss from larvae that have cut the egg cap, but have not emerged, and so will eventually die from desiccation when the eggs are allowed to dry.

If filter paper is used for oviposition, special care is required to ensure that the eggs have been kept moist sufficiently long to ensure conditioning. This is probably most readily done by keeping the filter paper on wet wool as is done by the American observers.

Egg pots when stored should be protected from psoci, cockroaches, ants and other possible marauding insects by being placed one above the other standing upon an upturned flower pot standing in a vessel of water. Oil, as less likely to evaporate, can be used instead of water, but is apt to be both messy and dangerous. The time the eggs remain viable depends a good deal on the humidity. The time is much shorter in ordinary room conditions than when kept in a constant temperature room with a humidity of 70. The extent to which stored eggs are still viable can be judged by examining them under a low power. If still plump and rounded, they will certainly be viable. Eggs properly conditioned and kept at reasonable humidity should retain their viability for some weeks at least without appreciable deterioration and give good results in hatching up to three or four months. If kept too moist, they are apt to become infected with psoci or mites.

The resistance of conditioned eggs to desiccation makes this much the best form in which to transmit material for culture. Eggs may be sent by post on dry filter paper or on small pieces of pot from a broken-up egg pot.*

(*e*) MANIPULATION OF EGGS

Shannon and Putnam (1934) have specially investigated the results of different ways of conditioning eggs. They keep those laid on filter paper on wet wool for several days in the open so that they dry very slowly. Trembley (1944*b*) allows eggs to be laid on filter paper

* For egg production Kneirim, Lea, Dimond and De Long (1955) feed mosquitoes on preserved blood, as also do Lang and Wallis (1956). Whether such procedure is useful or not may depend on circumstances.

placed on moist cullocotton pads in the dishes that form the base of her lamp glass cages. They are stored after being kept moist for 2–3 days at room temperature in covered containers and then dried in the open air. H. A. Johnson (1937) in place of filter paper uses pieces of sponge that can be dried and put away. At the Molteno Institute, Cambridge, in the method of employing mosquitoes described by Tate and Vincent (1936), eggs, after feeding of the mosquitoes, are laid on and around the edges of the water in shallow enamel pans used as the bottom of the feeding cage. As these dishes dry up eggs remain moist sufficiently long to become for the most part conditioned. Blanc and Caminopetros (1929) place in the cage of fed mosquitoes porous earthenware vessels containing water. The eggs are deposited on the outside of the pot. Earthenware pots as used by the author have already been noted.

When not properly conditioned a certain number of eggs become dry and shrunken. But during the operation of being put to hatch these eggs swell up and appear as unhatched eggs (see later under 'swollen eggs', p. 145), and may give a false impression that these were resistant eggs, that is still viable eggs that did not hatch.

MICROSCOPIC STUDY OF EGGS

The above remarks relate to a supply of eggs for culture. There are, however, a number of other ways of manipulating eggs when the object is to prepare these for various research purposes. One useful method for many purposes is to collect the film of floating eggs left after oviposition in the meat dish used. If this is to be done, a glass pot (meat dish) should be placed in the cage for oviposition instead of an earthenware pot. The glass sides of the dish being not very suitable for the insect to oviposit from, the great majority of eggs will now be laid on the surface of the water. To collect these a grease-free cover-glass, preferably a long one such as is used for mounting serial sections, is used. This is slipped under the film of eggs and a portion of this lifted out. The eggs being uniformly oriented with the ventral surface uppermost, such preparations are very suitable for making measurements, studying egg structure or for demonstration purposes. This is also a way of obtaining a number of undamaged eggs for research purposes. The cover-glass is examined under the microscope and any damaged or abnormal eggs together with debris removed. The eggs which have passed inspection are then swept off with a fine brush into a pot of water. Never having dried, such eggs will be intact as regards the chorion, which is not the case with eggs detached after being allowed to dry on any surface.

The thick band of eggs deposited round the water's edge in an earthenware egg-pot is also another useful source of eggs whilst these are still moist, for the eggs, not yet being cemented, can be removed with a fine brush without damage and used for many purposes, such as observing hatching under the microscope, weighing eggs, etc. (see pp. 139–40). Usually the eggs from this band, especially after a heavy oviposition, in the still moist condition give very clean preparations of eggs free from debris. Even when dry, however, this band is a very useful source of eggs for some purposes, it being simply necessary to brush them off with a camel hair brush. Though the chorion may be damaged, this does not affect their viability.

Eggs on filter paper are not so suitable for the above purposes as they tend to be adherent to the fibres of the paper. Eggs laid by isolated females on strips of filter paper are, however, very useful for many purposes, such as ascertaining the number of eggs laid, viability of these, etc.

A common trouble with eggs, even those only momentarily dry, is that when placed in water they then tend to float and are difficult to submerge. The following is applicable to either moist or dry eggs in overcoming this difficulty. Brush off the eggs into a little water in a watch-glass or small glass pot. Using a pipette and teat, swish the water in and out until it is seen that the majority of the eggs have sunk. Now give the watch-glass a rotary movement to bring these sunk eggs in one spot. They can now be easily taken up in a pipette and transferred to where required.

To obtain undamaged eggs free from fluid, pipette out eggs on to a fine-grained porous tile. By passing them out of the pipette mouth at one spot a little heap of moist eggs free from fluid can be obtained.

Pipettes used for manipulating eggs should be kept thoroughly cleaned with acid, as otherwise eggs are very liable to adhere to the inside of the pipette and cause difficulty.

To bleach eggs, treat as above and, bringing them together by rotatory movement of the pot or dish, transfer to 1 in 5 diaphanol (or other bleaching fluid), leaving them in this until the desired effect is produced. For bleaching still viable eggs to observe the act of hatching, see p. 158. For sterilisation of viable eggs, see p. 151.

For all observations dealing with the black and opaque eggs of *A. aegypti* some form of direct illumination by a strong beam of light, used from as high an angle as the objective will permit, is essential. The use of a Lieberkuhn's parabolic reflector is recommended by James (1923) for this purpose. The author has used a Baker's 'high intensity low voltage' microscope lamp which serves well. The background should not be white or the glare will give poor definition. A black matt surface gives, when strongly illuminated, a nice grey contrast background. An upright piece of white card to reflect diffused light on the part of the egg in shadow improves the effect.

MOUNTING EGGS

Eggs may be mounted for the purpose of showing the egg characters or to show the later embryological changes. For the first purpose, however, mosquito eggs offer almost insurmountable difficulty as when put up in a medium some of the most striking characters of the exochorion no longer show up. Eggs of *A. aegypti* mounted in gum or other medium show little of the appearances of the unmounted egg suitably illuminated. Suitably bleached and treated egg-shells are most suitable to show the chorionic structure (see under description of egg characters given later). For methods of mounting to show embryological characters, see chapter VII.

Pick (1950), after he had shown that eggs of *A. aegypti* laid on silico-gel medium remain viable up to seven months, has found that eggs may even be mounted in a viable condition in such medium, 50 per cent hatching out after thirty days. The method of preparing the medium is given by the author.

(*f*) MANIPULATION OF LARVAE AND PUPAE

Examination of living larvae. Larvae as required are pipetted out from the culture using the pipette already described. Where a mass sample is being taken at a given time or stage of culture for weighing, larvae may be pipetted out where they have congregated away from the light, and decanted from the pipette into clean water in a dish facing the light (for example, a window), the pipette being emptied on the side towards the light. On entering the water the larvae swim rapidly away from the light, whilst any debris or sediment falls to the bottom where the pipette is emptied. The larvae are then pipetted out from where they have collected free from debris and filtered off in a small funnel on to filter paper. The mass so obtained is then tipped on to a smooth porous tile, one portion being then weighed, whilst another portion is immediately killed by boiling water and is used for determining the percentages of the different instars, etc. Pressure at any stage should be carefully avoided since this may detach anal papillae and so lead to loss of body fluid. By such means, and measurements made as described later, growth in cultures can be followed.

To obtain larvae at the stages immediately before and immediately following upon ecdysis it was found very helpful to pipette out a mass of, say, fifty larvae from a culture in which the desired ecdysis was in active progress and to examine this rapidly on a glass plate under a low power. The pre-ecdysis forms are at once detected by the dark hairs of the oncoming instar wrapped round the thorax under the cuticle as described later. For the post-ecdysis stages pre-ecdysis forms were collected and kept under observation, larvae at the desired stage being removed as they were observed to undergo ecdysis. The larvae at this stage are very strikingly distinguished by the almost completely transparent head. Before measuring it is desirable to give a few minutes for the parts to adjust themselves as they may be deformed somewhat immediately following ecdysis. By pre- and post-ecdysis is always meant forms just about to ecdyse, or that have just completed ecdysis respectively, these being the only times at which measurements of soft parts can be made corresponding to a particular stage in development.

For measuring larvae for body length some method of ensuring immobilisation is necessary. For killing without change or distortion the method described later with formalin vapour is that giving the most perfect result. But though excellent for preparing larvae for mounting, it is too lengthy a process for measuring numbers of larvae at different stages of growth. Living larvae can be immobilised by flooding them on to filter paper, but they are usually not well set out for measurement. Gerberich (1945) describes a method using 10 per cent solution of methyl cellulose in the form known as *methocel* (manufactured by Dow Chemical Co., Midland, Michigan. Viscosity type XX low; see also *Science*, 5 Nov. 1943). One drop of the solution is placed in the hollow of a hollow slide. The larva, with not more than one drop of water, is then added and the water and methocel mixed. The slide is left for 20 minutes for the water to evaporate. The larva at first moves about but later becomes quiescent. Larvae can be kept alive in high humidity in such a condition for 24 hours without injury.

Bates (1949) used iced water for immobilising larvae. The author has found the most generally satisfactory method for use in measuring larvae for growth studies as given later to be very careful chloroforming of the larvae on a glass plate until they just cease moving. For body length those are selected which lie with the body extended in a straight line.

Killing larvae for mounting. For mounting without distortion so that preparations can be made suitable for drawing or for reference, the most satisfactory method of killing is as follows. Larvae are placed in a shallow dish with a suitable depth of 2 per cent formalin to allow them to rest normally at the surface and covered with a glass sheet on which is a filter paper wetted with pure formalin. It may take several hours to kill the larvae, which, however, mainly die still floating at the surface and with the respiratory parts at the end of the siphon fully expanded. After 24 hours the fluid is changed to 4 per cent, care being taken not to sink the larvae. After another 24 hours the larvae are brought through graded alcohols to xylol, using the xylol, and eventually the canada balsam, also in graded proportions. A small opening in an unimportant part of the body wall will greatly facilitate these later operations.

Mounting larvae. Larvae can be mounted in balsam, gum, Bhatia's fluid or other medium. They can also be brought without shrinkage into pure glycerine by the method described by Buxton and Hopkins (1927), and thence mounted in glycerine jelly. Buxton and Hopkins' method consists in placing four volumes of concentrated glycerine in a tube, covering this with a layer of one volume of a mixture of 90 per cent alcohol (nine parts) and ether (one part), transferring the larvae from 90 per cent alcohol to the tube and allowing them to remain until the alcohol and ether have evaporated. However, by making a small opening in the body wall larvae can be mounted in glycerine jelly without much difficulty.

A number of methods for mounting larvae have been described. A simple and rapid method is given by Gater (1929). For mounting fluid Gater uses a medium consisting of water 10; picked gum arabic 8; chloral hydrate 74; glucose syrup 5; and glacial acetic 3, dissolved in the order given, preferably on a water bath at about 50° C. and afterwards filtered through Whatman's paper No. 5 by Buchner's funnel and suction pump. Small quantities can be cleared by centrifuging. The glucose syrup is made by dissolving 98 g. of bacteriological glucose in 100 c.c. water. Larvae are freed from water and mounted direct. The medium is also very useful for mounting small objects generally from water.

Puri (1931) found Gater's medium the most satisfactory for mounting whole larvae. De Faure's chloral gum (Imms, 1929) was not satisfactory for mounting whole larvae, but gave good results with skins. King (1913–14) found a medium made by dissolving small lumps of pale colophonium in rectified oil of turpentine better than xylol balsam. For killing and clearing he used equal parts pure carbolic crystals and absolute alcohol. After 15–30 minutes larvae were ready for mounting in the colophonium medium after removing excess of fluid. Buckner (1934) places larvae for 15–20 minutes in 1·5 per cent $MgSO_4$ solution to clear them of foreign matter and kills with cocaine hydrochloride (2 per cent for 4 hours). Wanamaker (1944) found creosote substituted for xylene a better canada balsam medium. Canada balsam was heated slowly to drive off all xylene and thinned with creosote.

A medium for mounting larvae that has given good results with the author is described by Bhatia (1948). 50 g. crystalline pure resin is dissolved in 75 c.c. oleum eucalypti at room temperature. Allow to stand and decant. Kill larvae in hot water and bring to 90–95 per cent alcohol or kill in alcohol. Transfer to 95 per cent or absolute alcohol, puncture the thorax, remove excess of alcohol and mount in the medium. The larvae become transparent in about half an hour. In 2–3 days the medium sets hard and does not require ringing. Its refractive index (1·497) is less than that of balsam (1·524) and so shows up hairs better.

For the final mounting some form of cell or support for the cover-glass is necessary. For the small first and second instar larvae strips of broken cover-glass are suitable. For larger larvae strips of bristol board on each side of the object, or if thicker material is required, strips of glass slide cut by glazier's diamond. It is usually necessary with the thicker mounts to add more mounting medium until fluid under the cover-glass ceases to shrink.

One difficulty in mounting is due to the siphon not being in line with the body. For many purposes it is sufficient to cut the larva into two at about the middle of the abdomen, mounting the head half on the flat and the tail half on its side.

Mounting larval skins. Many external structural features are specially well displayed in the cast skin. These, when first cast, are much wrinkled and after a time become unduly loose and elongated. By choosing a skin at the right amount of relaxation it can be floated on a slide with the parts in fair position. Such skins can be mounted in chloral gum, Gater or Bhatia medium and are especially valuable for the clear display given of the larval hairs, but also of some other parts, for example the pharynx, and, with a little dissection, mouth-parts, siphon structures, etc. For systematic purposes Puri (1931), for *Anopheles* larvae, makes preparations of specially displayed skins as follows. Skins are treated in 10 per cent KOH for an hour, washed in three changes of water and under the binocular on a slide are cut along the mid ventral line by niggling with needles. They are afterwards passed through alcohols to 70 per cent and spread on a slide. When satisfactorily spread they are flooded with 90 per cent alcohol, which causes them to become flattened and stiff and adherent to the slide. They can then be mounted as they are, or detached and treated as sections.

Fixing larvae and pupae. For general purposes the most satisfactory fixative for larvae and pupae is Bles's fluid (formalin 7; glacial acetic 3; 70 per cent alcohol 90). This was also found by Puri very suitable for *Anopheles* larvae. The fixative penetrates readily and except for some special purpose it is unnecessary to puncture larvae.

For fixing at different hours of culture and as a general fixative the rapid method of killing and fixing used by Tower (1902), in studying ecdysial fluid, is very satisfactory. Larvae at the desired stage are killed by momentary immersion in sat. solution of $HgCl_2$ in 35 per cent alcohol 70 parts, acetic acid 30 parts heated to 60–65° C. They are then transferred to sat. solution $HgCl_2$ in 35 per cent alcohol 99 parts, acetic acid 1 part, for from 1 to 6 hours according to size. Later they are dealt with in the usual way for sublimate fixed material.

In dealing with fixed material it is often convenient to turn out a number of larvae to select material for sectioning or other purposes. A useful piece of apparatus in returning such material to its tube is a perforated lifter made from a piece of perforated zinc or other metal about 2½ by 1 cm. set in a slit in a wooden handle. This cannot be used with Bles or other acid fluids, but a strip of bristol board may be helpful.

Slicing technique in study of structure. Whilst for the study of internal structure serial sectioning is essential (see later subsection in this chapter), very useful subsidiary help may be given by suitable bisecting or slicing of fixed larvae or pupae. For this a suitably fixed larva or pupa should be placed in a suitable position on a slab of paraffin under a low power microscope. When in correct position, this is then cut down upon by a sharp razor. The larva or pupa may be bisected in a vertical or horizontal plane or in any way which it is thought will enable particular structures to be displayed. To prevent the object shifting during the process it may be fixed in position by a fine entomological pin driven through it into the paraffin in such a way as not to interfere with the cutting.

Instead of a razor, Pantin (1946) uses a fragment of blue Gillette razor blade so broken off with pliers as to give a length of cutting edge and a projecting portion by which the fragment can be set in a slit wooden handle. It is essential that it should be firmly held, which can be achieved by sealing the knife in position with sealing wax. In use it can often be steadied by contacting the free end of the blade with a finger of the left hand.

In the two halves of a bisected larva, including the bisected head, the various structures are often marvellously shown. It is usually desirable to lift out with a touch or two of the needle the loosely held alimentary canal in order to display better other structures. Larvae, fixed and bisected or sliced, form excellent material for the study of larval musculature viewed under polarised light (see under muscular system).

A useful technique for studying certain structures is given by Wigglesworth (1942). The larva is slit along the side with scissors and the gut removed through the slit. The body may then be unrolled and mounted flat, all tissues, germinal buds, heart, etc. being displayed.

Another useful method for certain purposes is to select a larva that has just ecdysed and has a completely transparent head. By staining in picro-carmine or other suitable stain and clearing, the brain and other nervous structures are especially clearly shown, the nerves going to the antennae, optic lobes and other cephalic structures especially being shown up in their proper relations with great clearness. For dissection of salivary glands in larvae and pupae, see Jenson (1955).

MANIPULATION OF PUPAE

When collecting pupae from a culture with the intention of separating the sexes it is convenient to place the culture dish facing a window and to tilt this somewhat towards the observer. If rearing is being done on any scale, it is useful to have a graduated wooden support for this purpose (Fig. 2 (3A)). The pupae then arise free from the glass sides of the dish and can more conveniently be pipetted out as they do so, diagnosis of sex being made at the same time. If the culture is optimal, the difference in size and appearance of the two sexes is so great that any mistake can scarcely be made. With poor cultivation the difference is less marked and with starved forms may almost, but never quite, disappear.

If collecting pupae in mass for bulk breeding out or other purpose, the small sieve described in the subsection on apparatus will be found very useful. By holding this in one hand *with the netting submerged* and pipetting in the pupae, even a large culture can be cleared in a few minutes without damage to the insects and without including still unchanged larvae with the pupae. By including larvae when pupae are put up for emergence the nice clean-cut technique where a dish is left with just the number of skins of mosquitoes in the emergence cage no longer holds, nor can freedom from the other sex be assured.

Use of the sieve also simplifies the final putting up of the collected pupae in clean water, since no material amount of culture medium with suspended matter is carried over. If much culture fluid is present in the dishes there is considerable danger that in the 48 hours or so at 28° C. scum may form with mortality.

Pupae fix well in Bles's fluid, but in this as in other lethal fluids the time taken for killing may be considerable. As noted later also pupae do not readily bleach. This is due to an outer oily layer which is removed after momentary dipping in chloroform. Pupae will then allow ready penetration and will bleach readily.

(*g*) MANIPULATION OF ADULTS

Collecting and transferring. Methods of manipulating small numbers of mosquitoes are given by Stephens and Christophers (1908).

Single mosquitoes can be collected in nature or from cages in test tubes or 3 by 1 in. specimen tubes closed by wool plugs or collected by a suction pipette and transferred to tubes by manipulating the wool plugs. They can be transferred from such tubes to a prepared jar by covering the mouth of the jar with a piece of cardboard in which a hole has been cut plugged with wool and inserting the mouth of the tube into this. The mosquito usually flies in without difficulty. If not, it can be propelled in by moving the tube up and down. Yolles and Knigin (1943) use a slit in a rubber diaphragm through which the tube is thrust.

Mosquitoes bred out in a jar may be collected as follows: place over the mouth of the jar an empty jar and, drawing out the netting covering the breeding jar, allow as many mosquitoes as will do so to fly up. Now insert two pieces of cardboard between the jars and separate these each with a piece of cardboard closing it. Replace the pieces of cardboard with netting.

If the mosquitoes are to be killed at once for mounting, etc., replace the cardboard or netting with a piece of cardboard with a hole and wool plug. The mosquitoes can now be killed by dropping a suitable amount of chloroform on the wool plug.

If the mosquitoes so collected are to be kept alive, the jar into which they have been taken should be previously prepared by inserting a piece of cardboard to give foothold (Fig. 2 (7)). Or a 'mosquito hotel' can be made ready for them. For this a wide-mouth bottle with a flat hollow stopper, as is usually available in the laboratory (Fig. 2 (7)) is very convenient. The stopper is filled with water on which is floated a small piece of cork shaving and on this a small piece of paper. Mosquitoes can be fed through netting onto which the bottle can be lifted without danger of any escapes. Such an arrangement is a very useful one on occasions, for example to obtain eggs from a captured female of a rare species. In place of the apparatus as described the jar covered with netting can be stood so that it rests with part of the opening immersed in water in a dish: or a small globe-shaped lampshade can be used. Or alternatively a small cage can be used with a meat dish of water and a filter paper fan.

Ordinarily single specimens of *Aëdes aegypti* are put up for observation, for example on time of digestion, oviposition, number of eggs laid, etc., in 3 by 1 in. specimen tubes with a little water in the bottom and a strip of filter paper, which should touch the water and lie against the side of the tube. A piece of netting fastened over the opening by a small rubber band is much to be preferred to a wool plug for such tubes, as wool plugs in the course of days are apt to get displaced, with loss of the specimen, and also give less ventilation. Usually the mosquito, if netting is used, rests on the netting (see also Gillett, 1955).

When dealing with mosquitoes in bulk as referred to in the subsection on apparatus these are transferred as described by having cages equipped with sliding door, sleeve, etc., and adopting the procedure as described.

Mounting mosquitoes for systematic purposes. For mounting, mosquitoes are best killed by chloroform. Only just sufficient should be used to ensure death as excess stiffens and distorts the insects. When turned out on to a sheet of paper they should lie flaccid with legs and wings extended. The best pins to use are the small stainless steel pins now available for the purpose. The insect should be turned on its back and the pin introduced in the mid-point between the origin of the legs. It should be pushed in only enough just to touch the dorsum, but not to project. Mosquitoes of the subgenus *Stegomyia* are of an awkward shape to pin in this way. Pinning from the side, however, damages important structures of the pleura. In default, mosquitoes can be sent unpinned. In this case a little crumpled tissue paper is to be preferred to wool.

After transfixing with a pin some method of mounting is necessary. Some of the older methods are unnecessarily elaborate. For a single specimen a narrow strip of cork sheet, say $\frac{3}{8}$ by $\frac{1}{4}$ in. or less through one end of which an ordinary pin, or better a no. 16 entomological pin has been passed is suitable as it will take the blunt end of the stainless steel pin satisfactorily and gives a firm support. For museum specimens a small piece of celluloid sheet may be substituted for the cork. It is usually necessary to puncture this first with a fine needle to take the pin (Fig. 2 (8A)).

For putting up specimens in the tropics for transmission by post, etc., see Christophers, Sinton and Covell (1941). Single mounted mosquitoes put up in numbers in a box are extremely liable to be damaged in transit. Another danger is mould which is liable to attack any specimen hermetically sealed up. A method which has served well is the use of 3 by 1 in. specimen tubes with certain precautions. A wedge should be cut out of the cork and the opening so made plugged with wool. As an extra precaution, specimens put up in the tubes after such procedure can be well dried in a desiccator. For a single mosquito all that is necessary is to stick this mounted as above firmly in the cork of the tube. The tube is then wrapped in a small square of wool, a number of such tubes being again enclosed in a suitable box when despatched (Fig. 2 (8B)).

If numbers of mosquitoes have to be dealt with they can be mounted on a cork strip set in a 3 by 1 in. specimen tube. A strip of cork sheet, say $2\frac{1}{4}$ by $\frac{1}{2}$ in. is firmly pinned towards one side of the inner face of the cork. Mosquitoes impaled on the small steel pins are now set along the cork strip in a single or double row. Up to twenty specimens can usually be put in a tube in this way (Fig. 2 (8C, D)).

Observations on adults. For mass observations, such as percentage fed, the average weight for a culture and other bionomical data, the adults in the experimental cage, after being killed in the killing-chamber, are turned out, counted and classified. Where weight is required a sufficient number of the required specification, that is males or females, gorged females, etc., are collected in small stoppered weighing bottles and weighed on the chemical balance. Or single mosquitoes can be weighed on a torsion balance.

For the latter purpose a fine stainless steel pin, as used for mounting, is bent into a hook to hang on the torsion arm. With this the mosquito is impaled through the tip of the abdomen or a wing, allowance being made for the weight of the pin. Considerable numbers can be weighed in this way without undue trouble and the mean taken.

For measurement of wing length the insects should be allowed to dry, as the wing is then readily detached, whilst in the undried insect to detach the wing is often difficult.

For mounting, wings should be dropped first into a watch-glass with xylene, then pipetted out on to a slide and a drop of thin balsam added whilst the xylene is still in excess. This prevents the formation of a very troublesome air bubble which is very apt to form in the hollow of the wing. There is also less danger of scales being detached than if wings are mounted direct into balsam.

Feeding mosquitoes. Mosquitoes in bulk can be fed in experiments or for oviposition on the human arm or on an animal. If observations on the number biting are important some device is desirable to restrict the accessible area to what can be clearly seen. Use of a glove and some other precautions are also useful; see Christophers (1947). The most suitable animal would appear to be the rabbit which gives good feeds and a good egg supply; see under 'Apparatus', p. 107.

For feeding in connection with bird malaria and feeding through a membrane, see Bishop and Gilchrist (1946). For artificial feeding by capillary pipette, see Hertig and Hertig (1927); Patton and Hindle (1927); Kadletz and Kusmina (1929); MacGregor (1930). Artificial feeding is, however, less satisfactory than normal feeding (see p. 490). For technique in virus diseases see Dunn (1932), Davis (1940), Merril and Ten Broek (1940) and Waddell (1945).

A simple method of experimental feeding of single mosquitoes with any desired fluid is described by Mattingly (1946). The mosquitoes to be fed are allowed to emerge from the pupa in ordinary agglutination tubes (which are just about the width to prevent free flight) closed by a plug of cotton wool or gauze through which the water is poured off after emergence. The food is administered on the end of a half-inch long feeding tube cut from glass tubing that will easily slide into the agglutination tube. A small circle of linen gauze is pushed down the inside of the feeding tube to project at the end. Food is dropped on the projecting gauze from a pipette and the tube pushed down towards the mosquito. Kligler (1928) makes use of light to cause the mosquito to fly to the netting. When on the netting the desired fluid is brought into contact with this. Russell (1931) uses a netting cylinder smeared with mango juice to attract the mosquitoes, but not enough for them to feed. This is followed by a mixture of the blood and normal saline, 10 drops to the c.c.

Dissection. For dissection of the mid-gut and salivary glands needles should be fairly stout; or, if fine, mounted short to ensure rigidity. For finer dissections the fine steel pins used for mounting fixed in match sticks and sharpened on a hone work very well. Russell (1931) finds dental instruments, for example nerve canal probes, useful.

For dissection of the gut, the mosquito, after removal of the legs and wings, is placed on its left side with the abdomen pointing to the observer. With a needle held in the left hand transfix the thorax and, holding the mosquito ventral side uppermost, bring the apex of the abdomen to a small drop of saline. Nick the chitin of the abdomen on each side to separate partially the last two segments, and placing the needle point held with the right hand flatly on the partially separated segments, exert gentle traction drawing the separated segments away with the viscera attached. If, as the viscera are drawn out, the gut threatens to break, prod the thorax once or twice with the needle and try again. In a successful dissection the alimentary canal from the proventriculus backwards should be drawn out together with the diverticula (Fig. 63 (1)).

To examine the mid-gut for oocysts, cut through the intestine and Malpighian tubules just beyond the pylorus and remove these, as also any fat-body, etc. Cover, using an excess of fluid. Now remove this with filter paper, allowing the mid-gut to become flattened.

When satisfactorily flattened the pattern of muscles fibres should show up. Examine under the oil immersion. Dissection and staining of the mid-gut is dealt with by Seidelin and Summers-Connal (1914).

When a dissection is made in this way the ovaries, spermathecae and rectum remain attached to the terminal segments and if required can be dissected out. If the ovaries are undeveloped or only partially enlarged, they offer no difficulty and can be readily seen and isolated. If the female is fully gravid, however, only careful dissection enables them to be withdrawn intact. For a physiological saline for *A. aegypti*, see Hayes (1953).

For dissection of the salivary glands the neatest and simplest method depends upon the fact that under slight pressure on the thorax these can be made to bulge at the neck. After removal of the legs and wings, place the mosquito in a drop of saline on its right side with the proboscis towards the observer. Now place the shaft of the left-hand needle towards its point across the thorax and exert gentle pressure, so that the soft parts at the base of the neck are seen to bulge. Now place the shaft of the other needle towards the point behind the head and drag this steadily away from the body, keeping the needle in position when the head leaves the body and simultaneously cutting down to sever the small tag of tissue that has come away attached to the head. In this tag should be the glands, all six of which may be seen as refractive bodies. If not to be seen in the tag, press a little more with the left-hand needle to squeeze out a little more tissue and detach this. For further details, see Christophers *et al.* Or in a method used by Shute (1940) the head is cut off to start with and the tissues at the base of the neck expressed (see Barber and Rice, 1936; Hunter *et al.* 1946).

For detection of sporozoites the glands should be crushed under the cover-glass, as otherwise these bodies, even if present, may not be seen still lying in the gland. The glands in *A. aegypti* are very small and quite different to the voluminous structures in *Anopheles*.

Methods of preparing the material for dissection and study of the mouth-parts, hypopygium and various organs and tissues will be found in the sections dealing with external and internal structure.

(*h*) SECTIONING

Histological technique might not be considered a subject which came within the scope of the present work. But the cutting of good sections is so important in connection with the study of mosquitoes and especially of one like *Aëdes aegypti* so much concerned with disease that some latitude may be allowed. This is especially so since sectioning the mosquito is a rather special technique of which little will be found in the usual text-books, so that some brief description of embedding in this case may be useful.

Mosquitoes are most satisfactorily sectioned by double embedding in celloidin and paraffin, using the Peterfi methyl benzoate method (see Pantin, 1946). This method is quicker and simpler than the original oil of cloves and celloidin method. In only one respect is it less satisfactory, namely, as usually described, for orientation as described later. However, by transferring the specimen from the thin celloidin to a final thick syrup, which is quite easy to make, this disadvantage is done away with. The most satisfactory celloidin is celloidin wool obtained from Gurr. The thick syrup referred to should be about the consistency of golden syrup. The method is as follows:

The Peterfi methyl benzoate method. The material after dehydration (preferably with a final change to copper-sulphated absolute alcohol) is transferred to 1·0 per cent solution of celloidin in methyl benzoate. Or a useful way is to place a layer of alcohol above methyl benzoate in a small tube and allow the specimen to remain until it has sunk into the latter and has become transparent. When transparent the specimen is transferred for 30 minutes to a thick syrup of celloidin in methyl benzoate and then placed in a drop of the syrup on a paraffined cover-slip placed in the bottom of a small glass pot or watch-glass under the dissecting microscope for orientation as described in the next subsection. When oriented, benzene is pipetted carefully into the pot so that it rises gradually over the cover-glass and specimen solidifying this as it does so. By proceeding in this way there is no displacement of the orientation as there is apt to be using the inverted cover-glass dropped on the fluid. When the tablet finally detaches itself it is transferred to benzene, with at least one change to remove all trace of the methyl benzoate which is not very soluble in paraffin. The tablet is now ready for embedding.

Orienting for sectioning. Much of the success of satisfactory sectioning depends upon perfect orientation. To the drop of celloidin syrup in which the object is now lying under the microscope a suitable length of a fairly thick hair should be added and arranged parallel to the specimen. The hair should preferably be thick and dark so as to be easily seen by transmitted light in the final paraffin block. The hairs from a black varnish brush serve very well. Further to assist in arranging the block for cutting, the anterior end of the specimen should be indicated. This can be done when the tablet is being finally blocked as described below and when the object is still visible in the solidifying wax. One way is to draw out a point on the edge of the wax as it solidifies. Another is to insert a small paper wedge and still another is to touch the edge of the wax with a grease pencil. Where accurate orientation is desired it is important that the hair should not only give the line of direction of the object but be as far as possible lying in the same plane, or, what gives a very good indication, two lengths of hair, one on each side of the object and in the same plane as this, can be used.

To embed. In the method described above the specimen, duly oriented and passed through two changes of benzene, will be enclosed in a more or less tablet-like portion of transparent celloidin jelly This is now transferred to the top of the paraffin bath with enough benzene to cover it and shavings of paraffin of the selected melting point added until an approach to pure paraffin is reached when it is transferred to paraffin, after which it can be blocked.

Smear a watch-glass with glycerine (or better melted glycerine jelly). Place the watch-glass on wet filter paper on a slab. Pipette in sufficient melted paraffin to form when cold a suitably sized button. Whilst this is still fluid add the specimen, removing it from the embedding bath with a heated small doll's spoon or wide-mouthed pipette. Whilst the top of the button is still fluid pipette in some more paraffin to make a second button over the first, stopping short before it overflows. Now lower the watch-glass into a basin of cold water, tilting it as it is lowered and blowing upon it at the same time, and at such a rate and tilt that the fluid paraffin does not break through the congealed surface. When fully submerged let it fall to the bottom. When the paraffin floats free scratch on it some identifying number. Later, when putting aside for sectioning, a printed serially numbered slip can be melted on to the block with a hot metal holder.

To cut. It is usually possible to locate by transmitted light the position of the object and guiding hair. Roughly trim the block as desired. Now fix it on the holder of the microtome and cut the upper and lower surfaces so that when the block is being cut the top and bottom edge of the sections will be strictly parallel and the ribbon be straight.* Dip the block momentarily into overheated soft paraffin (about 49° C. melting point is usually satisfactory for making sections adhere), or, using soft paraffin very hot (75–80° C.), smear both upper and lower surfaces, using a small brush. Trim the block at the sides to an appropriate width, making one side oblique so that this will give a useful indication later as to which side of the ribbon may be uppermost. This will prevent some sections being mounted upside down. If the ribbon tends to break, the soft paraffin may be too soft or too hard, or not applied hot enough, so that it is not incorporated with the underlying hard paraffin. Or there may be a crack (if the celloidin syrup was too thick) where the paraffin meets the celloidin. If the ribbon comes off to one side, the faces of the block may not have been parallel, or not set parallel to the knife's edge, or the whole block set at a slant so that the surfaces, though parallel, do not cut so. When trimming, the guide hairs can usually be cut off.

To set out sections. Clean slides in sulphuric acid and potassium bichromate. Wash without handling, using old forceps dipped in the acid. Keep the slides in absolute alcohol until used. Wipe clean immediately before use with a single sweep of a freshly laundered handkerchief. Success largely depends upon the slides being truly grease-free. Mix a camel-hair brushfull of Mayer's glycerine-albumen with about 5 ml. distilled water in a small glass pot. Mark off on the slide with a grease pencil about the limits of the cover-glass to be used. This may help later in case any doubt arises as to which side of the slide the sections are on. With sweeps of the camel-hair brush cover between the marks with an even layer of the diluted fixative. If properly grease-free this should lie as it is placed. Transfer suitable lengths of ribbon cut off with a sharp scalpel to the slide, allowing for some lengthening when these are flattened. Place the slide on a plate just hot enough not to melt the paraffin and allow the sections to flatten out. A very convenient hot-plate for this purpose, and some other uses, is a thickish sheet of copper about 8 by 5 in. laid on the floor of the recess near the bottom of the usual type of embedding bath. It should project about 2 in. When fully flattened, drain off excess of fluid and, after allowing the slide to cool, blot firmly with smooth white filter paper. A trace of diacetin in the diluted fixative may be used to give increased flattening and adhesion (Abul-Nasr, 1950).

Label on the opposite side to the sections at one end of the slide using a writing diamond. If this is done on the same side it is apt to become covered with balsam and invisible. When completed, place the slides to dry, preferably in a slotted trough in a desiccator in the incubator.

To stain and mount. A good stain for general purposes is Ehrlich's haematoxylin-picric-eosin. This gives many distinguishing gradations in colour, depending on the tissue and strongly stained nuclei. When slides are thoroughly dry, remove one at a time from the desiccator to the hot-plate, and when the paraffin is just not melted run over the sections in succession from drop bottles: xylol (xylene), carbol-xylol (melted phenol crystals 1, xylol 4) and copper-sulphated absolute alcohol. Transfer to a glass pot of 90 per cent alcohol. Apart from ensuring certain clearing, the carbol-xylol is largely responsible for complete removal of the celloidin which otherwise stains. Pass through alcohols to 30 per

* Some microtomes have an attachment for doing this, or it is worth while making some small device oneself to do this automatically. The narrower the block is made from above down the more sections per row can be mounted.

cent and flood with Ehrlich's acid haematoxylin. Stain 30–60 minutes. Sections should at this stage be strongly overstained, as they are later decolorised by the picric acid. Wash off the stain with 30 per cent alcohol and pass through alcohols to 90 per cent. Flood for some seconds with saturated solution of picric acid in 90 per cent alcohol. Wash out excess of picric with 90 per cent alcohol and flood with eosin (watery) 1–2000 in 90 per cent alcohol. Wash quickly in 90 per cent alcohol and treat briefly in succession with absolute alcohol, carbol-xylol and xylol, finally mounting in balsam. Success depends on getting the best effects with the three reagents. (See Addenda, p. 718.)

For further information on histological procedure see Bolles Lee (1950); Carleton and Leach (1938); Baker (1950); Pantin (1946). The last-mentioned author gives a particularly clear account of methods suitable for the zoologist including orientation. For methods of orienting small objects see also Patten (1894); Drew (1900). For an alternative method of fixing sections on the slide see Haupt (1930), and for the diacetin method, Abul-Nasr (1950).

Fixing and sectioning of eggs. For fixing and sectioning eggs of insects see Nelson (1915); Gambrell (1933); Butt (1934). Fixation usually requires puncture of the egg and use of a strong fixative, for example Carnoy's fluid. For fixing mosquito eggs best results have been obtained by slicing off the posterior extremity or anterior end, depending on the part to be studied. Later stages are reasonably penetrable. When fixed, cutting is not difficult if care be given to good orientation.

For methods of cutting highly refractory material see Slifer and King (1933); also Tahmisian and Slifer (1942). Slifer and King fix in Carnoy-Lebrun's fluid, wash with iodised alcohol, cut the egg in half (grasshopper egg), soak 24 hours in 4 per cent phenol in 80 per cent alcohol, dehydrate and pass through carbon-xylol to paraffin. Before cutting, the block is trimmed just to expose the yolk and soaked for 24 hours in water. Tahmisian and Slifer fix in Bouin's fluid, transfer without washing to 100 per cent dioxan for 8 hours with one change, then to paraffin with 0·5 beeswax for 8 hours with two changes. After embedding, the block is cut to expose the yolk and soaked 24 hours in water.

Eggs and small first instar larvae, etc., can be oriented in rows on small slabs of hardened liver by dropping glycerine-albumen on to them when lying on the slab in 90 per cent alcohol. Without some such method the cutting of small objects such as mosquito eggs is not very satisfactory.

Help may be obtained by softening the brittle egg-shell by use of diaphanol.

Reconstruction. For methods of making reconstructional drawings or models see Norman (1923); Wilson (1933); Pusey (1939); Pantin (1946).

Drawings. For suggestions for preparing drawings for reproduction see Cannon (1936).

Histochemical studies. For technique in histochemical studies see Lison (1936).

(*i*) OTHER SPECIAL TECHNIQUES

For methods of ensuring desired degrees of humidity see Buxton (1931); Lewis (1933); Buxton and Mellanby (1934); C. G. Johnson (1940); Seaton and Lumsden (1941); Lumsden (1947); Bertram and Gordon (1939). For psychromatic tables of vapour pressure, relative humidity and the dew-point see Marvin (1941). For specific gravity determination by the falling-drop method see Barbour and Hamilton (1926). For methods of determining dissolved oxygen see Senior-White (1926); Iyengar (1930); Theriault (1925); U.S.A. Publ. Hlth Ass. (1920). For method of measuring angle of contact see Beament (1945). For measurement of light intensity see Muirhead-Thomson (1940). For measurement of pH of microscopic objects see Pantin (1923). For use of radio-isotopes see Bugher and Taylor (1949); Brues and Sacher (1952); Terzian (1953); Jenkins (1954); Scott and Peng (1957). For use of chromatography with mosquitoes see Micks (1954). For other special methods connected with the study of mosquito biology see under respective subsections. (See also Addenda, p. 718.)

REFERENCES

(*a–g*) REARING AND MANIPULATION

BACOT, A. W. (1916). See references to chapter VI, p. 155.

BARBER, M. A. and RICE, J. B. (1936). Methods of dissecting and making permanent preparations of salivary glands and stomachs of *Anopheles*. *Amer. J. Hyg.* **24**, 32–40.

BARNETT, H. C. (1955). Cage equipment for the study of mosquitoes infected with pathogenic agents. *Mosquito News*, **15**, 43–4.

BATES, M. (1949). The natural history of mosquitoes. Macmillan and Co., New York.

BHATIA, M. L. (1948). A simple medium for mounting mosquito larvae. *Indian J. Malar.* **2**, 283–4.

BISHOP, A. and GILCHRIST, B. M. (1946). Experiments upon the feeding of *Aëdes aegypti* through animal membranes, etc. *Parasitology*, **37**, 85–100.

BLANC, G. and CAMINOPETROS, J. (1929). Quelques mots sur le mode de conservation des Stegomyies en cage. *Bull. Soc. Path. Exot.* **22**, 440–4.

BOISSEZON, P. DE. See DE BOISSEZON, P.

BUCKNER, J. F. (1934). An improved method for mounting mosquito larvae. *Amer. J. Trop. Med.* **14**, 489–91.

BURCHFIELD, H. P., REDDER, A. M., STORRS, E. E. and HILCHEY, J. D. (1954). Improved methods for rearing larvae of *Aëdes aegypti* (L.) for use in insecticide bio-assay. *Contr. Boyce Thompson Inst.* **17**, 317–31.

BUXTON, P. A. (1928). An aspirator for catching midges. *Trans. R. Soc. Trop. Med. Hyg.* **22**, 179–80.

BUXTON, P. A. and HOPKINS, G. H. E. (1925). Race suicide in *Stegomyia*. *Bull. Ent. Res.* **16**, 151–3.

BUXTON, P. A. and HOPKINS, G. H. E. (1927). Researches in Polynesia and Melanesia. Parts I–IV. *Mem. London Sch. of Hyg. and Trop. Med.* no. 1.

CHRISTOPHERS, S. R. (1947). Mosquito repellents. *J. Hyg. Camb.* **45**, 176–231.

CHRISTOPHERS, S. R., SINTON, J. A. and COVELL, G. (1941). How to do a malaria survey. *Govt. India Hlth Bull.* no. 14. Ed. 4. Delhi.

DAVID, W. A. L., BRACEY, P. and HARVEY, A. (1944). Equipment and method employed in breeding *Aëdes aegypti* for the biological assay of insecticides. *Bull. Ent. Res.* **35**, 227–30.

DAVIS, W. A. (1940). A study of birds and mosquitoes as hosts for the virus of eastern equine encephalomyelytis. *Amer. J. Hyg.* **32**, 45–59.

DE BOISSEZON, P. (1933). De l'utilisation des proteins et du fer d'origin végétale dans la maturation des œufs, chez *Culex pipiens*. *C.R. Soc. Biol., Paris*, **114**, 487–9.

DUNN, L. H. (1932). A simple method of immobilising animals for laboratory purposes. *Amer. J. Trop. Med.* **12**, 173–8.

FIELDING, J. W. (1919). Notes on the bionomics of *Stegomyia fasciata* Fabr. *Ann. Trop. Med. Parasit.* **13**, 259–96.

GATER, B. A. R. (1929). An improved method of mounting mosquito larvae. *Bull. Ent. Res.* **19**, 367–8.

REFERENCES

GERBERICH, J. B. (1945). Quieting mosquito larvae. *J. Econ. Ent.* **38**, 393–4.

GILLETT, J. D. (1955). Mosquito handling—a recent development in technique for handling pupae of *A. aegypti. Rep. Virus Res. Inst. E. Afr. High Comm.* 1954–5, p. 24.

HAYES, R. O. (1953). Determination of a physiological saline solution for *Aëdes aegypti. J. Econ. Ent.* **46**, 624–7.

HERTIG, A. T. and HERTIG, M. (1927). A technique for artificial feeding of sandflies and mosquitoes. *Science*, **65**, 328–9.

HINDLE, E. (1929). The experimental study of yellow fever. *Trans. R. Soc. Trop. Med. Hyg.* **22**, 405–30.

HINMAN, E. H. (1930). A study of the food of mosquito larvae (Culicidae). *Amer. J. Hyg.* **12**, 238–70.

HINMAN, E. H. (1932). The role of solutes and colloids in the nutrition of mosquito larvae. *Amer. J. Trop. Med.* **12**, 263–71.

HUNTER, W. III, WELLER, T. H. and JAHNES, W. G. (1946). An outline for teaching mosquito stomach and salivary gland dissection. *Amer. J. Trop. Med.* **26**, 221–7.

IMMS, A. D. (1929). Some methods of technique applicable to entomology. *Bull. Ent. Res.* **20**, 165.

JAMES, S. P. (1923). Eggs of *Finlaya geniculata* and of other English mosquitoes illuminated with the aid of Lieberkuhn reflector. *Trans. R. Soc. Trop. Med. Hyg.* **17**, 8–9.

JENSEN, D. V. (1955). Method for dissecting salivary glands in mosquito larvae and pupae. *Mosquito News*, **15**, 215–16.

JOHNSON, H. A. (1937). Note on the continuous rearing of *Aëdes aegypti* in the laboratory. *Publ. Hlth Rep. Wash.* **52**, 1177–9.

KADLETZ, N. A. and KUSMINA, L. A. (1929). Experimentelle Studien ueber den Saugprozess bei *Anopheles* mittels einer zwangsweisen Methode. *Arch. Schiffs.- u. Tropenhyg.* **33**, 335–50.

KING, W. V. (1913–14). Note on the mounting of mosquito larvae. *Amer. J. Trop. Dis.* **1**, 403.

KLIGLER, I. J. (1928). Simple method of feeding *Stegomyia* on blood or mixtures containing culture. *Trans. R. Soc. Trop. Med. Hyg.* **21**, 329–31.

KNEIRIM, J. A., LEA, A. O., DIMOND, J. B. and DELONG, D. M. (1955). Feeding adult mosquitoes on preserved blood to maintain egg production. *Mosquito News*, **15**, 176–9.

LANG, C. A. and WALLIS, R. C. (1956). An artificial feeding procedure for *Aëdes aegypti* L. using sucrose as a stimulant. *Amer. J. Trop. Med. Hyg.* **5**, 915–20.

LEESON, H. S. (1932). Method of rearing and maintaining large stocks of fleas and mosquitoes for experimental purposes. *Bull. Ent. Res.* **23**, 25–31.

LUMSDEN, W. H. R. (1947). Observations on the effect of microclimate on the biting of *Aëdes aegypti* (L.) (Diptera: Culicidae). *J. Exp. Biol.* **24**, 361–73.

MACGREGOR, M. E. (1915). Notes on the rearing of *Stegomyia fasciata* in London. *J. Trop. Med. (Hyg.)*, **18**, 193–6.

MACGREGOR, M. E. (1929). The significance of pH in the development of mosquito larvae. *Parasitology*, **21**, 132–57.

MACGREGOR, M. E. (1930). The artificial feeding of mosquitoes by a new method which demonstrates certain functions of the diverticula. *Trans. R. Soc. Trop. Med. Hyg.* **23**, 329–31.

MARTINI, E. (1921). Ueber Stechmücken und Kriebelmückenzucht. *Arch. Schiffs.- u. Tropenhyg.* **25**, 120–1.

MATTINGLY, P. F. (1946). A technique for feeding adult mosquitoes. *Nature, Lond.*, **158**, 751.

MERRIL, M. H. and TEN BROECK, C. (1940). The transmission of equine encephalitis virus by *Aëdes aegypti. J. Exp. Med.* **62**, 687–95.

NIESCHULZ, O. and DU TOIT, R. M. (1934). Handling mosquitoes for experimental purposes under South African conditions. *Onderstepoort J. Vet. Sci.* **3**, 79–95.

PANTIN, C. F. A. (1946). *Notes on Microscopical Technique for Zoologists.* Cambridge Univ. Press.

PATTON, W. S. and HINDLE, E. (1927). Artificial feeding of sandflies. *Proc. R. Soc.* B, **101**, 369.

PICK, F. (1950). L'inclusion temporaire des œufs d'*Aëdes aegypti* à l'aide de technique de silicogel sur larve. *Bull. Soc. Path. Exot.* **43**, 364–72.

PURI, I. M. (1931). Larvae of Anopheline mosquitoes with full description of those of the Indian species. *Indian Med. Res. Mem.* no. 21.

PUTNAM, P. and SHANNON, R. A. (1934). The biology of *Stegomyia* under laboratory conditions. 2. Egg-laying capacity and longevity of adults. *Proc. Ent. Soc. Wash.* **36**, 217–42.

ROSS, R. W. (1956). A laboratory technique for studying the insect transmission of animal viruses employing a bat-wing membrane, demonstrated with two African viruses. *J. Hyg., Camb.*, **54**, 192–200.

RUSSELL, P. F. (1931). A method of feeding blood meals to mosquitoes—male and female. *Amer. J. Trop. Med.* **11**, 355–8.

SEIDELIN, H. and SUMMERS-CONNAL, S. (1914). A simple technique for the dissection and staining of mosquitoes. *Yell. Fev. Bur. Bull.* **3**, 193–7.

SELLARDS, A. W. (1932). Technical precautions employed in maintaining the virus of yellow fever in monkeys and mosquitoes. *Amer. J. Trop. Med.* **12**, 79–92.

SHANNON, R. C. (1939). Methods of collecting and feeding mosquitoes in jungle yellow fever studies. *Amer. J. Trop. Med.* **19**, 131–8.

SHANNON, R. C. and PUTNAM, P. (1934). The biology of *Stegomyia* under laboratory conditions. I. The analysis of factors which influence larval development. *Proc. Ent. Soc. Wash.* **36**, 185–216.

SHUTE, P. G. (1940). Supernumerary and bifurcated acini of the salivary glands of *Anopheles maculipennis*. *Riv. di Malariol.* **19**, Sez. 1, 16–19.

SNIJDERS, E. P., DINGER, E. J. and SCHUFFNER, W. A. P. (1931). On the transmission of dengue in Sumatra. *Amer. J. Trop. Med.* **11**, 171–97.

STEPHENS, J. W. S. and CHRISTOPHERS, S. R. (1908). *The Practical Study of Malaria.* Ed. 3. Williams and Norgate, London.

STOKES, A., BAUER, J. H. and HUDSON, N. P. (1928). Experimental transmission of yellow fever to *Macacus rhesus*. *J. Amer. Med. Ass.* **90**, 253.

TATE, P. and VINCENT, M. (1936). The biology of autogenous and anautogenous races of *Culex pipiens* (Diptera: Culicidae). *Parasitology*, **28**, 115–45.

TOWER, W. L. (1902). Observations on the structure of the exuvial glands and the formation of the exuvial fluid in insects. *Zool. Anz.* **25**, 466–72.

TRAGER, W. (1935). On the nutritional requirements of mosquito larvae (*Aëdes aegypti*). *Amer. J. Hyg.* **22**, 475–93.

TREMBLEY, H. L. (1944*a*). Mosquito culture technique. *Mosquito News*, **4**, 103–19.

TREMBLEY, H. L. (1944*b*). Some practical suggestions for the rearing of *Aëdes aegypti*. *Proc. N.J. Mosq. Ext. Ass.* **31**, 168–72.

WADDELL, M. B. (1945). Persistence of yellow fever virus in mosquitoes after death of the insects. *Amer. J. Trop. Med.* **25**, 329–32.

WANAMAKER, J. F. (1944). An improved method for mounting mosquito larvae. *Amer. J. Trop. Med.* **24**, 385–6.

WEYER, F. (1934). Der Einfluss der Larvalernährung auf die Fortpflanzungsphysiologie verschiedenen Stechmücken. *Arch. Schiffs.- u. Tropenhyg.* **38**, 394–8.

WIGGLESWORTH, V. B. (1942). The storage of protein, fat, glycogen and uric acid in the fat body and other tissues of mosquito larvae. *J. Exp. Biol.* **19**, 56–77.

YOLLES, S. F. and KNIGIN, T. D. (1943). Note on a new transparent cage for collecting and feeding mosquitoes. *Amer. J. Trop. Med.* **23**, 465–9.

YOUNG, M. D. and BURGESS, R. W. (1946). Plastic cages for insects. *Science*, **104**, 375.

(*h*) SECTIONING

ABUL-NASR, S. E. See references to chapter XIV (*i*), p. 354.

BAKER, J. R. (1950). *Cytological Technique.* Ed. 3. Methuen and Co., London.

BOLLES LEE, A. (1950). *The Microtomist's Vade-mecum.* Ed. 11. Churchill, London.

BUTT, F. M. (1934). Embryology of *Sciara* (Sciaridae: Diptera). *Ann. Ent. Soc. Amer.* **27**, 565–79.

CANNON, H. G. (1936). *A Method of Illustration for Zoological Papers.* The Ass. of British Zoologists.

REFERENCES

CARLETON, H. M. and LEACH, E. H. (1938). *Histological Technique.* Ed. 2. Oxford Univ. Press.

DREW, G. (1900). A modification of Patten's method of embedding small objects. *Zool. Anz.* **23**, 170.

GAMBRELL, F. L. (1933). The embryology of the blackfly *Simulium pictipes* Hagen. *Ann. Ent. Soc. Amer.* **26**, 641–71.

HAUPT, A. W. (1930). *Stain Tech.* **5**, 97. Given by BOLLES LEE, p. 634. Ed. 10.

LISON, L. (1936). *Histochimie animale.* Gauthier-Villars, Paris.

NELSON, J. A. (1915). *The Embryology of the Honey Bee.* Oxford Univ. Press. Pp. 251–3.

NORMAN, J. R. (1923). Methods and technique of reconstruction. *J. Roy. Micr. Soc.* **1923**, 37–56.

PANTIN, C. F. A. (1923). The determination of pH of microscopic bodies. *Nature, Lond.*, **111**, 81.

PANTIN, C. F. A. (1946). See under (*a–g*).

PATTEN (1894). *Z. wiss. Mikr.* **11**, 13. Abstract in BOLLES LEE, p. 84. Ed. 10.

PUSEY, H. K. (1939). Methods of reconstruction from microscopic sections. *J. Roy. Micr. Soc.* **59**, 232.

SLIFER, E. H. and KING, R. L. (1933). Grasshopper eggs and the paraffin method. *Science*, **78**, 366.

TAHMISIAN, T. N. and SLIFER, E. H. (1942). Sectioning and staining refractory materials in paraffin. *Science*, **95**, 284.

WILSON, D. P. (1933). A simple block trimmer for the Cambridge rocking microtome. *J. Roy. Micr. Soc.* **53**, (3), 25–7.

(*i*) OTHER SPECIAL TECHNIQUES

BARBOUR, H. G. and HAMILTON, W. F. (1926). The falling drop method of determining specific gravity. *J. Biol. Chem.* **69**, 625–40.

BEAMENT, J. W. L. (1945). An apparatus to measure contact angles. *Trans. Faraday Soc.* **41**, 45–7.

BERTRAM, D. S. and GORDON, R. M. (1939). An insectarium with constant temperature and humidity control, etc. *Ann. Trop. Med. Parasit.* **33**, 279–88.

BRUES, A. M. and SACHER, G. A. (1952). In Nickson's *Symposium on Radiobiology.* Chapman and Hall, London.

BUGHER, J. C. and TAYLOR, M. (1949). Radiophosphorus and radiostrontium in mosquitoes. Preliminary report. *Science*, **94**, 146–7.

BUXTON, P. A. (1931). The measurement and control of atmospheric humidity in relation to entomological problems. *Bull. Ent. Res.* **22**, 431–47.

BUXTON, P. A. and MELLANBY, K. (1934). The measurement and control of humidity. *Bull. Ent. Res.* **25**, 171–5.

IYENGAR, M. O. T. (1930). Dissolved oxygen in relation to *Anopheles* breeding. *Ind. J. Med. Res.* **17**, 117–188.

JENKINS, D. W. (1954). Advances in medical entomology using radio-isotopes. *Exp. Parasit.* **3**, 474.

JOHNSON, C. G. (1940). The maintenance of high atmospheric humidities for entomological work with glycerol-water mixtures. *Ann. Appl. Biol.* **27**, 295.

LEWIS, D. J. (1933). Observations on *Aëdes aegypti* L. (Diptera: Culicidae) under controlled atmospheric conditions. *Bull. Ent. Res.* **24**, 363–72.

LUMSDEN, W. H. R. (1947). See under (*a–g*).

MARVIN, C. F. (1941). *Psychrometric Tables.* U.S. Dept. Commerce. Weather Bureau. (Obtainable from C. F. Casella and Co. Ltd, Regent House, Fitzroy Square, London, W. 1).

MICKS, D. W. (1954). Paper chromatography as a tool for mosquito taxonomy: the *Culex pipiens* group. *Nature, Lond.*, **174**, 217–18.

MUIRHEAD-THOMSON, R. C. (1940). Studies on the behaviour of Anopheles, etc. *J. Malar. Inst. India*, **3**, 265–94.

PANTIN, C. F. A. (1923). See under (*h*).

SCOTT, K. G. and PENG, C. T. (1957). Cerium 144 as a tag for arthropods of medical importance. *Trans. R. Soc. Trop. Med. Hyg.* **51**, 87–8.

SEATON, D. R. and LUMSDEN, W. H. R. (1941). Observations on the effect of age, fertilization and light on biting by *Aëdes aegypti* (L.) in a controlled microclimate. *Ann. Trop. Med. Parasit.* **35**, 23–36.

SENIOR WHITE, R. (1926). Physical factors in mosquito oecology. *Bull. Ent. Res.* **16**, 187–248.

TERZION, L. A. (1953). The effect of X-irradiation in the immunity of mosquitoes to malarial infection. *J. Immunol.* **71**, 202–6.

THERIAULT, E. J. (1925). The determination of dissolved oxygen by the Winkler method. *U.S. Publ. Hlth Serv. Bull.* 151, 43 pp.

U.S.A. PUBL. HLTH ASS. (1920). Standard methods for the examination of water and sewage.

VI
THE EGG

(*a*) EXTERNAL CHARACTERS

EARLY DESCRIPTIONS OF THE EGG

References in the literature to the characters of the egg of *Aëdes aegypti* with any claim to detail are few. Parker, Beyer and Pothier (1903) and Otto and Neumann (1905) describe and figure the white polygonal markings. Mitchell (1907) gives a very good outline drawing of the egg in respect to its shape and figures the polygonal markings. Goeldi (1905) figures the chorionic bodies. These were first described and figured by Theobald (1901) from observations by Daniels, as air chambers, which, seen in profile around the borders of the egg, they much resemble in appearance. Boyce (1911) describes the egg as elongate, blackish in colour and rather sparsely studded with minute hemispherical bodies of whitish secretionary matter. Howard, Dyar and Knab (1912) describe the egg as fusiform, black, very slightly flattened on one side, slightly more tapered towards the micropilar end and with somewhat irregular callosities forming spiral rows. In the plates accompanying their monograph this hummocky appearance is shown in Figure 668, whilst Figure 684 shows an egg covered only with white polygonal markings. No explanation of the relation between the two figures is given. As explained later the seeming lack of agreement in the different descriptions is an indication of the very different appearance given by the eggs depending upon the conditions under which they are viewed. Some later descriptions give more detail, but are still far from describing the essential characters of the egg. It is only in the description by De Buck (1938) that any indication is given that the egg has a ventral and a dorsal surface with entirely different characters, or that the polygonal markings are air channels, or in which the nature of the chorionic bodies is indicated. De Buck's paper describing in detail the structure of the exochorion indeed appears to have been somewhat overlooked and was so by the present writer until very recently.

Of studies of related eggs may be mentioned the very detailed description of the structure of the *Anopheles* egg by Nicholson (1921), of the *Culex* egg by Vishna Nath (1924), and by Christophers (1945), as also the earlier work of Bresslau (1920). The eggs of a number of English species of *Aëdes* are described and figured by James (1922, 1923), and also by Marshall (1938), giving measurements and the points in which they differ, as well as very clearly indicating the existence in many of these species of the peculiar clear refractile chorionic bodies described later, which are such a marked feature of the egg of *Aëdes aegypti*, but which do not figure at all in the anopheline or culicine egg.

DESCRIPTION OF THE EGG

To the naked eye the eggs are small, intensely black, elongate oval, seed-like bodies under a millimetre in length. Under the microscope they appear somewhat torpedo-shaped with one end, the anterior, rather thicker and more abruptly tapered than the other. At the anterior pole is the micropile, indicated by a ring of transparent globular bodies similar to those to be described as characteristic of the upper (ventral) surface. The posterior pole has no special modification of structure such as has been described by me in the *Culex* egg (Christophers, 1945).

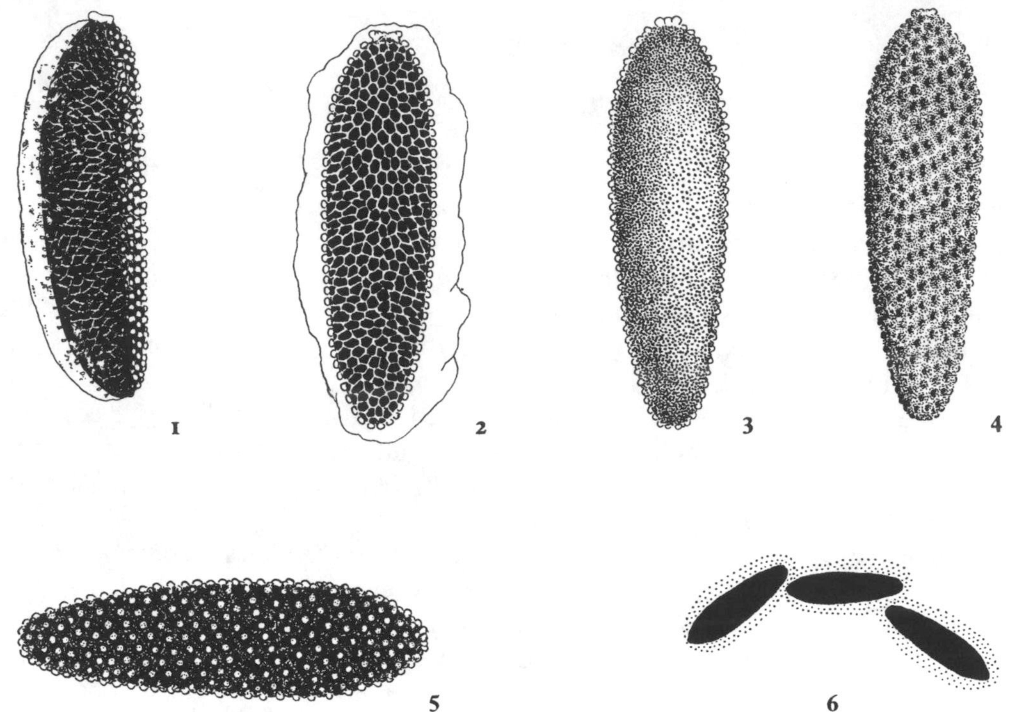

Figure 4. Showing appearance of the egg under different conditions of wetness and moisture.

1 A moist egg seen in lateral view (on edge of filter paper) showing chorionic pad, chorionic bodies and air channel reticulum.

2 A submerged egg. Appearance similar whether dorsal or ventral surface shown. The extent of the chorionic pad when fully swollen shown in outline.

3 Egg as seen in dry state. Lower surface (that is, dorsal in morphological sense). Genera. effect smooth or very finely granular.

4 The same. Upper surface. Appearance as of hummocks with faint indications of polygonal markings.

5 An egg (upper surface) still moist (but not wet) showing the chorionic bodies as brightly refractile bodies in a dark setting.

6 Three eggs floating on the surface of water. The extent of the swollen transparent chorionic pad indicated by lines.

Although the egg has been described above as torpedo-shaped, it is really rather boat-shaped with a flatter upper surface (morphologically the ventral in respect to the developing larva) and a more convex lower surface (morphologically the dorsal). These two surfaces further differ not only in shape but in other respects. Both are covered with polygonal markings which under certain circumstances stand out conspicuously as a network of milky white lines. These are in reality, as first pointed out by De Buck, fine channels at the base of the exochorion filled with air, thus accounting for their striking milk-white appearance when submerged. Similar lines are present on the under surface of some Anopheline eggs.

The upper surface carries the peculiar transparent globular bodies described later as *chorionic bodies*. The appearances shown by this surface depend to a remarkable extent on whether the egg is being examined completely submerged or in various degrees of dryness. Submerged, no matter which surface is being viewed, the egg shows only the network of milky white lines on a dark background (Fig. 4 (2)). Out of the water, and especially as the egg surface dries, the white lines marking the polygonal areas are much less conspicuous and what now become the predominant feature of the upper or ventral surface are the low tuberosities formed by the chorionic bodies occupying the centres of the polygonal areas. In the still wet egg with a film of water over its surface these bosses may catch the light and give the effect of a surface covered with bright points of light, which are merely reflections from the wet tuberosities (clearly what Peryassu (1908) saw, who figures the egg with white points joined by pale lines). On a somewhat drier egg the tuberosities take on the appearance of bright refractile bodies each lying in the centre of a polygonal area. At a suitable stage of dryness and well illuminated these have the appearance of dull diamonds set in dark metal (Fig. 4 (5)). With the completely dry egg the chorionic bodies show up only as dark rounded hummocks. This is the appearance, rather like the surface of a pine-apple, described and figured by Howard and others (Fig. 4 (4)). Seen in profile on the margins of the egg as clear blebs the chorionic bodies may well be taken as a series of tiny air blisters or floats, as for many years they were supposed to be.

The lower surface shows the network of air channels, but is devoid of chorionic bodies and in the dry egg appears uniformly smooth or finely granular (Fig. 4 (3)). The peculiarity of the lower surface is that its epichorion swells in water and forms a gelatinous pad, the *chorionic pad*. Depending upon the time the eggs have been in water and probably on the pH and possibly other conditions the chorionic pad forms a more or less thick pad of jelly-like material so transparent that when its surface has not been made apparent by attached particles of foreign matter it may easily be overlooked. When fully swollen it forms a kind of aureole surrounding the egg, sometimes up to nearly half the width of the egg in thickness (Fig. 4 (2, 6)). For some reason not ascertained it is sometimes not so noticeably swollen, even when eggs have been some time in water, but its presence is always evidenced by the way eggs stick to any surface against which they come to rest. The chorionic pad is in no way an adventitious secretion but represents a definite chorionic adaptation which as shown later has important functions.

SITE OF OVIPOSITION

The eggs of *Aëdes aegypti* appear never to be laid on a dry surface, though a wet surface appears to be preferred to that of water itself. Thus they are readily laid on wet filter paper, or wet sponge; or, by preference to actual water, in such situations as the floor of the cage wetted from a leaky pot. That eggs are laid *in situ* whether on a surface or on water, not merely dropped indiscriminately, is clear from the fact that almost invariably they are laid with the flatter ventral surface uppermost. On this account and for another reason described below it is nearly always the upper surface of the egg that is seen when viewing eggs under the microscope. On filter paper or other uniform wet surface the eggs are usually freely scattered, though often laid in lines or small groups and especially along any irregularity such as where the edge of a filter paper strip rests on glass and often on lines formed by scarcely visible streaks in the glass of a specimen tube. In a vessel containing water and with the water surface freely available eggs are usually laid, not on the water, but on the sides of the vessel just above the water's edge. A certain proportion are, however, laid on the water, especially if the receptacle is of glass which does not give the female a good foothold. The eggs have sometimes been described as drawn to the water's edge by capillarity. Eggs may be so drawn, but the great majority after an oviposition are laid *in situ* in this position and, when dry, are so cemented. When large numbers of females are ovipositing in earthenware pots the massed eggs so laid may form a line around the vessel just above the water's edge a centimetre in width and several eggs deep.

When it is difficult for the female to oviposit on a wet surface, as by use of a glass vessel, eggs may be laid freely on the water where they float ventral surface upwards, often forming patterns as do Anopheline eggs. Though they have no floats or other obvious aid to flotation they are not so easily sunk as might be supposed. Nevertheless, when eggs are laid on water there are usually a certain number which have sunk to the bottom as well as those floating on the surface. Eggs so found after an oviposition are referred to later as 'sunken eggs'. It is thought that eggs, when freshly laid and before changes referred to later occur, may be rather readily sunk and so liable to certain effects which may alter their future behaviour in hatching. This is one reason why, in determining the proportion of eggs hatching, etc., the use of porous earthenware as a surface for oviposition has a certain desirability. As pointed out by Fielding (1919), eggs may also be sunk through larvae nibbling around them. Such eggs would not be necessarily 'sunken eggs' in the sense used above.

Functions of the chorionic pad. The chorionic pad is a most important structure functionally. With eggs laid *in situ* on a wet surface the sticky chorionic pad anchors the egg ventral side upwards to some extent as laid. Further, when the surface dries, the pad becomes a hard cement fixing the egg quite firmly in this position. Should such an egg be detached the exochorion, that is the dried pad, is left behind as a whitish flake and the lower surface of the egg, now denuded of its exochorion, shows the smooth shiny surface of endochorion.

Even if eggs are not deposited originally *in situ* on a surface the sticky chorionic pad may still function. Its effect will be evident if some eggs sunk in water are left for a short time in contact with the glass at the bottom. On pouring off the fluid most of the eggs will be left behind adhering to the glass, and stranded eggs will be brought to rest by the sticky pad oriented with the ventral surface directed upwards.

(*b*) STRUCTURE

Morphologically the egg of *A. aegypti* has the general characters common to many insect eggs and consists, before fertilisation, of the *ovum* and remains of the *nurse cells* surrounded by a covering, the *chorion*, formed from the follicular epithelium and consisting of an outer layer, exochorion, and an inner layer, endochorion, which together form the *egg-shell*.

In the freshly-laid egg the ovum, fertilised in the act of oviposition, consists of cytoplasm packed with yolk granules, whilst towards the anterior end lie the female and male pronuclei, the whole surrounded inside the chorion by the vitelline membrane corresponding to the cell wall of the ovum. In the more mature egg are the various stages of development of the embryo, leading eventually to formation of the *primary larva* and its membranes as described in the chapter dealing with the embryology.

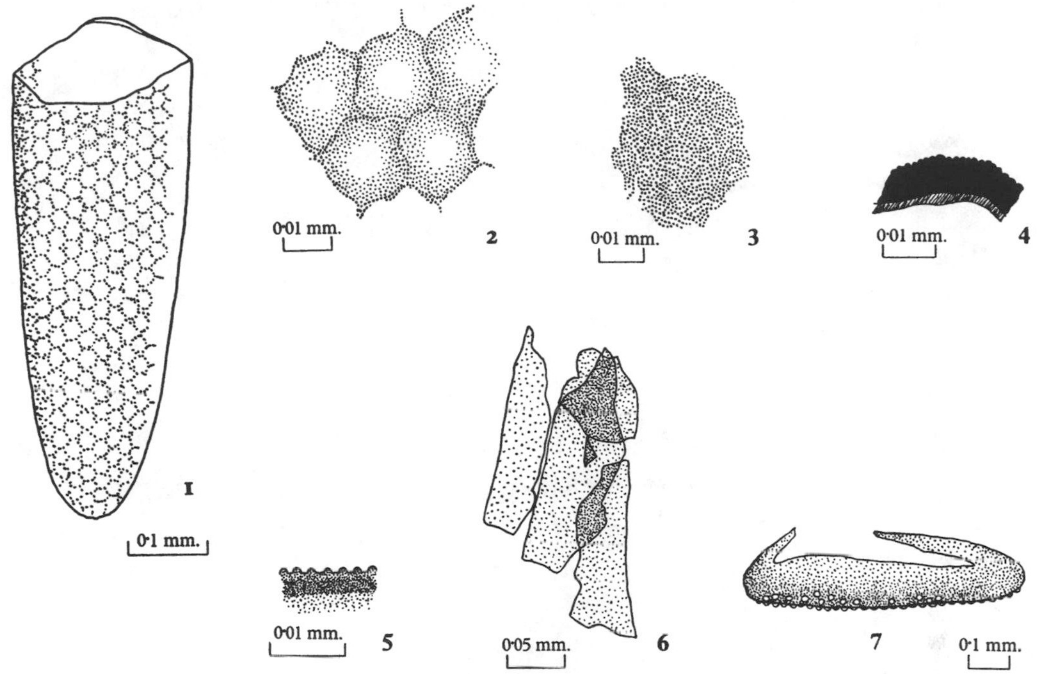

Figure 5. Showing structure of the chorion.

1 Egg-shell bleached by diaphanol showing the pattern formed by minute projections on the surface of the endochorion present only on the upper surface, the lower surface appearing blank though actually uniformly covered with small projections. The pattern is probably visible owing to the larger projections shown in Fig. 2 bounding the polygonal areas.

2 Portion of above highly magnified showing central space where papillae are smaller or absent on which the chorionic bodies of the exochorion rest.

3 Portion of the under surface.

4 Oblique section of endochorion showing small surface projections and an inner cleared zone.

5 Endochorion seen in optical section showing surface projections.

6 Portion of endochorion fractured by crushing. Note tendency to longitudinal and transverse fracture, the longitudinal being in the direction of the length of the egg.

7 'Exploded' egg. All such eggs show the lower (morphologically dorsal surface) burst.

THE EXOCHORION

The exochorion is a delicate, colourless and somewhat transparent membrane, which, especially when the egg is recently laid, is easily damaged or detached. It is also relatively much more easily affected by solvents and chemicals than is the harder more resistant endochorion. To the exochorion is due most of the ornamentation of the egg, so that eggs denuded of their exochorion show a uniform and smooth black surface.

The structure of the exochorion can be demonstrated in detached flakes and in sections, or in the impressions left by eggs which have dried on the slide and been detached.* In

* A technique for study of the exochorion of *Aëdes* eggs is given by Craig (1955).

the centre of each polygonal area, embedded in the substance of the exochorion and occupying its whole thickness, is one of the characteristic refractive bodies already referred to. These are almost globular or slightly flattened bodies, 9–11 μ in diameter. In flat

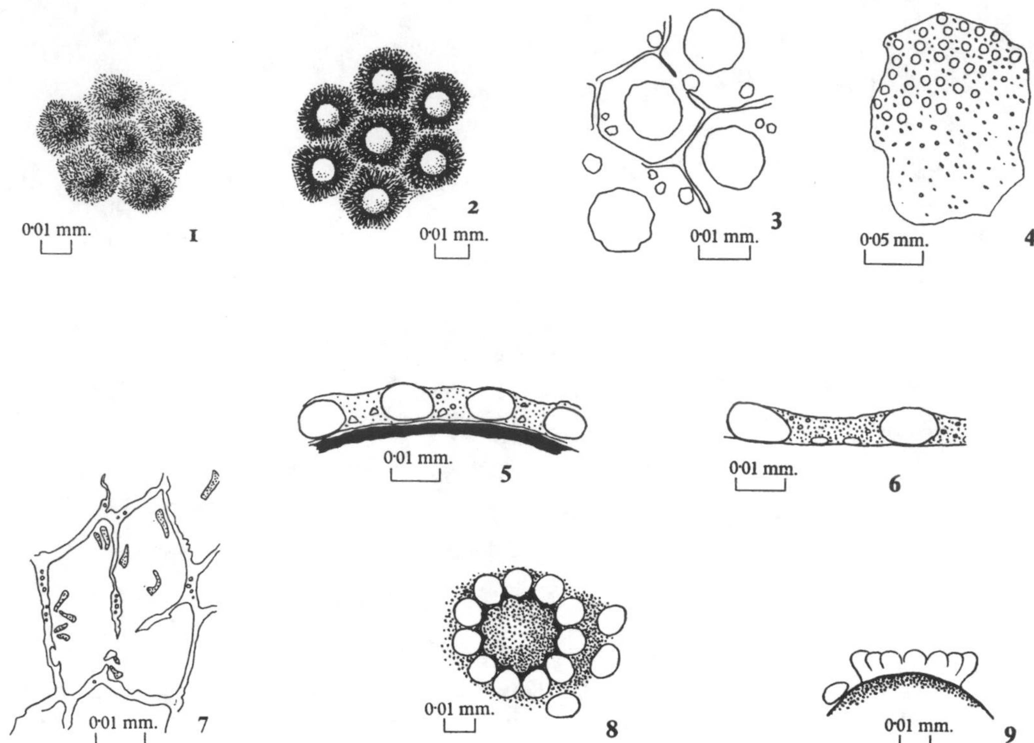

Figure 6. Showing structure of the chorion and the micropilar area.

1 Portion of upper surface as seen in dry egg.
2 The same of moist egg showing clear refractile bodies.
3 Outline drawing showing chorionic bodies and small accessory masses of upper surface with air channels and an air-lock.
4 Portion of exochorion at junction of upper and lower surface showing chorionic bodies and small accessory bodies, the latter alone present on under surface.
5 Section through egg-shell showing chorionic bodies of exochorion and the underlying endochorion. Some small accessory bodies are present in the exochorionic matrix.
6 Section through exochorion showing air channel cut near bifurcation.
7 Portion of exochorion from under surface left as impression on slide from a dried-on egg. Small rod-shaped bodies and air-locks shown. Some of the fine projections on the surface of the endochorion are seen here and there in the line of air-channels which are shown by this preparation to be in the substance of the exochorion.
8 View of micropilar area seen on flat. The outlying bodies are some of those in the first row of the general egg surface.
9 Lateral view of same.

preparations of the chorion they appear as clear, approximately circular bodies with a slightly wavy contour. They are transparent and at the margin of the egg they look very much like small air blebs. They are, however, entirely solid and appear to be composed of a dense transparent almost crystalline material of a protein nature, since they stain pink with Millon's reagent. They are the 'hemispherical bodies of whitish secretionary matter' of Boyce. They are easily shed from the exochorion matrix retaining their shape and appearance. In the fresh state they stain with watery methylene blue or eosin and are deeply and apparently selectively stained by watery thionin. Seen in section they occupy the middle of a slight depression in the endochorion corresponding to the polygonal area in which they lie. Though they appear to occupy the position of the original follicle cell nucleus they stain in section as eosinophil matter and do not take up haematoxylin staining. Under very high magnification they show indications of consisting of agglomerated sago-like masses, an appearance which is accentuated by treatment with concentrated H_2SO_4 or HCl.

Besides these bodies there are present smaller irregular masses which appear to be of the same material. They lie more or less scattered in the polygonal area and on the under surface of the egg especially. They are rod-like or wedge-like and are arranged at right angles to the surface. In fresh preparations of detached exochorion they give the appearance of granules, but in dried flakes most of them are seen lying flat when their rod shape is apparent (Fig. 6 (5, 6, 7)).

The boundaries of the polygonal areas are formed by air channels. These lie in spaces at the base of the exochorion between this and the endochorion. The appearance of air-locks and other characteristic appearances of air leave no doubt as to their nature. They do not appear to have any definite walls, being merely spaces in the matrix of the exochorion with support or anchoring from a line of projection in the endochorion, as described later, surrounding the polygonal area.

The exochorion of the under surface of the egg is very similar in essential structure to that on the upper surface, except that the centrally situated refractive bodies are not present in the polygonal areas. The smaller peg-like bodies are, however, present and rather more developed than on the upper surface. Seen in profile in the swollen chorionic pad these appear as erect wedges presumably giving some support to the jelly-like mass. Here the air channels do not appear to receive any special support from the endochorion, though equally developed with those on the upper surface.

THE ENDOCHORION

The endochorion is a dense, tough and hard membrane some 3–5 μ in thickness, which gives the egg its shape and rigidity. By reflected light it is an intense almost jet black and it is almost black even by transmitted light. It is brittle and fractures in much the same sort of fragments as glass or china, though it is obviously not so rigid as these substances. In sections it may give the appearance of an inner thin layer of lighter colour continued without break into the main thickness of the membrane. Such a distinction into two layers has not, however, been otherwise demonstrable even after prolonged soaking in various mixtures of glycerine and HCl and subsequent soaking in water. The membrane is not dissolved or noticeably affected by heating to 160° C. in saturated KOH for 15 minutes and is generally resistant to chemicals, except chlorine which decolorises and somewhat softens it. When ruptured by the hatching larva the fracture forming the cap is obliquely transverse, but if ruptured artificially or burst by internal pressure the endochorion splits for the most part into more or less longitudinal rectangular splinters (Fig. 5 (6)).

The outer surface of the endochorion is covered by closely aggregated minute papillae. Seen on the flat under a high magnification these appear as a speckling of dots. When seen in profile they seem to be hemispherical in shape projecting about half their diameter. They measure less than 1·0 μ in diameter and are not very apparent in sections except occasionally where the membrane is cut obliquely. On the lower surface of the egg they form intercurving rows so that the whole surface is fairly evenly and closely covered. On the upper surface the rows are more aggregated and the individual papillae are darker towards the boundaries of the polygonal areas where too in the actual line of the boundary the endochorion is sharply ridged. In the centre of the area, corresponding to the position of the refractive body in the overlying exochorion, the papillae are increasingly lighter in tint viewed from above and less numerous, being altogether absent in the middle of the area so that they form a kind of bed on which the refractive body lies (Fig. 5 (2)). As a result of this arrangement, especially when the endochorion has been rendered less dense by bleaching, the upper and lower surfaces of the egg are still indicated on the membrane, the upper surface being marked out in darkly bordered polygonal areas, whereas the lower surface is uniformly and very finely granular with no indication of such areas (Fig. 5 (3)).

When eggs are ruptured from internal pressure ('exploded eggs') it is always the lower or dorsal aspect which is forced out, suggesting that the endochorion is here weaker than elsewhere. In sections, too, the membrane seems slightly thicker on the upper surface where it is slightly hollowed also opposite the polygonal areas.

THE SEROSA

In eggs where development has proceeded up to a certain stage there can be seen on bleaching the egg that there lies within the endochorion and surrounding the embryo a thin membrane, the *serosa*. This is very conspicuous in cleared egg-shells after hatching of the larva. It consists of a very thin hyaline membrane lying loosely within the empty egg-shell. When stained it shows sparsely-situated flattened circular nuclei.

For formation of the egg membranes in the ovary see section dealing with ovarian changes. Some points regarding the chemical nature of the membranes will be found under physiology of the egg. Developmental changes are dealt with in the chapter on embryology.

(*c*) PHYSICAL CHARACTERS

LINEAR DIMENSIONS

Various rough measurements of the egg of *Aëdes aegypti* have been given by different authors which are perhaps unnecessary to give in detail. The most accurate measurements recorded in the literature are those by Buxton and Hopkins (1927), who give the following, based on messurement of 105 eggs.

	Max.	Mean	Min.
Length (mm.)	0·704	0·664	0·624
Breadth (mm.)	0·192	0·170	0·144

These authors also give a frequency curve for length measurements. The curve is very nearly symmetrical with a peak at 0·672 mm. There is, however, some extra loading in the region of small eggs, a point that will be referred to later.

PHYSICAL CHARACTERS

Measurements of length in millimetres for 295 eggs taken at random are given in Table 3. The measurements have been made to the nearest division of the eyepiece scale used (1 div. = 0·01 mm. at mag. used) and include the micropilar ring. Measurements of greatest breadth varied from 0·180 to 0·210 mm. including the refractive bodies. If these are not included, about 0·01 mm. should be deducted. The greatest depth varied from 0·180 to 0·190 mm.

Table 3. *Measurements of the length of the egg*

Date of examination made	0·569	0·588	0·608	0·627	0·647	0·666	0·686	0·706	0·725	0·745	0·765	Total
5. v. 47 (*a*)	—	—	—	1	6	27	19	5	1	—	—	59
27. v. 47 (*b*)	—	4	7	17	17	9	9	1	—	—	—	64
28. v. 47 (*b*)	1	2	5	9	13	7	3	—	—	—	—	40
16. vi. 47 (*c*)	—	—	—	—	2	5	3	6	—	—	—	14
11. viii. 47 (*d*)	—	1	1	2	4	19	20	18	6	1	1	73
12. vii. 47 (*e*)	—	—	2	4	9	4	2	1	—	—	—	22
Totals	1	7	15	33	51	71	56	29	7	1	1	295

(*a*) From isolated females in tubes. (*b*) Dry on cover-glass. (*c*) Just dry on cover-glass. (*d*) Measured on water. (*e*) Kept moist 4 days.

It will be seen that the frequency curve is largely symmetrical with some loading in the region of the smaller eggs as in the Buxton and Hopkins measurements. This may be due to the fact that a small proportion of eggs may still not have been fully swollen even though black and apparently mature (see changes following oviposition). These measurements are practically identical with those of Buxton and Hopkins for the species in nature in Samoa, the mean length in the two cases being 0·662 and 0·664 mm. Breadth is less easy to measure accurately, but there is nothing in the figures to indicate that the results are significantly different.

A few other measurements may be given. The distance of the widest part of the egg from the anterior pole is about one-third of the length (0·22 mm.). The greatest depth is more near the middle of the egg. The egg-cap tear is about 0·25 of the egg length on the upper surface and about 0·19 of this on the lower surface from the anterior pole.

WEIGHT

I have not been able to find in the literature any reference to the weight of any mosquito egg. Those of *A. aegypti* are too small to weigh singly, even on the torsion balance, since one hundred go to the milligramme. They have been weighed in mass on a chemical balance and in numbers from twenty upwards on a 5 mg. torsion balance.

As a result of experience in earlier weighings certain precautions were found to be necessary. Thus eggs weighed in mass though approximately dry may have a certain amount of free moisture included, and if dried further there was the possibility of some eggs becoming collapsed and shrunken. Further, with 'sunken' eggs especially, a certain amount of debris was liable to be attached to the sticky chorionic pad. Eggs collected from the surface of water floated on the cover-glasses could be allowed to dry and any shrunken eggs removed, but there were always scales from the surface of the water present. A number of earlier weighings, however, showed that the mean weight of the egg could not be over at most 0·0117 mg. or under 0·0109 mg. The mean of all weighings up to a certain date was 0·0113 mg. Finally to remove as far as possible all sources of error the procedure as described below was adopted.

THE EGG

Eggs taken still moist from the massive deposit at the water's edge after a single night's oviposition were kept moist for 4 days to ensure thorough conditioning. These were then washed quickly as described in the chapter on technique, transferred to a smooth porous tile and as soon as the fluid was absorbed and the eggs still moist they were spread about with a fine brush to allow them to dry without becoming attached to the tile. After examination and removal of any doubtfully damaged eggs a suitable number were weighed on the torsion balance. The results were very consistent as shown below and gave a mean weight for the conditioned egg in the dry state four days from oviposition of 0·0111 ± 0·0002 mg.

Date	Number of eggs	Weight (mg.)	Mean weight (mg.)
19. 1. 48	79	0·87	0·0110
19. 1. 48	92	1·02	0·0111
26. 2. 48	104	1·15	0·0110
26. 2. 48	213	2·39	0·0112
27. 2. 48	113	1·24	0·0110

Eggs that had been kept some time gave the following results.

Date	Description	Number (mg.)	Weight (mg.)	Mean weight (mg.)
30. xii. 47	15 days dry (3 sets)	488	5·06	0·0104
9. i. 48	25 days dry	186	1·94	0·0104
13. ix. 47	Slightly shrunken	20	0·15	0·0075
13. ix. 47	Shrunken to prismatic form	41	0·24	0·0059
13. ix. 47	Flattened, saucer-like	20	0·08	0·0040

From this it would appear that the loss of water in eggs still possibly viable is about 0·003 mg. or 28 per cent of the original weight.

SPECIFIC GRAVITY

Though eggs are commonly found floating on the surface they are in this position clearly only because some portion of their surface is not wetted. Viewed from above, floating eggs usually show the water line encroached on the upper surface to an extent all round the egg of about one-quarter of its width, that is, very little extension of the wetted area is required to sink the egg. Once the egg is completely wetted by forcible submersion or addition of a surface tension lowering substance to the water it sinks at once.

The specific gravity was determined by ascertaining that strength of solution of sodium sulphate in which selected conditioned eggs neither rose nor sank when introduced into the middle of the fluid by means of a pipette, the eggs being previously sunk in water to avoid any surface action. There was no reason to suppose that the small amount of water introduced with the egg affected the issue. A first observation gave the point as between 1·074 and 1·076 and this was confirmed some months later by a second estimation as given below.

Readings at 20° C.	Corrected to 15° C.	Readings taken as	Result
1·066	1·0668	1·067	All eggs sank
1·070	1·0708	1·071	All eggs sank
1·072	1·0728	1·073	All eggs sank
1·073	1·0738	1·074	Eggs showed little tendency to sink or rise
1·074	1·0748	1·075	Eggs suspended without movement
1·075	1·0758	1·076	Eggs suspended without movement
1·076	1·0768	1·077	All eggs rose

PHYSICAL CHARACTERS

The eggs were used 3 days from collection. The weight of 1·10 c.c. of water at 20° C. is given in tables as 0·998203 g. and at 15° C. as 0·998956 g., giving the correction factor used of 1·00075. The findings give the specific gravity at 15° C. of conditioned eggs 3 days old as 1·075–6. Though such a density may seem high, it is not so great as the density given by Bodine and Robbie (1943), for some stages of the grasshopper egg (see under changes following oviposition). Some larvae which hatched in the fluid seemed slightly lighter than the eggs.

VOLUME

Volume of the egg should be given by the relation weight/density = volume, which taking weight and density as determined would be 0·0103 mm.³. However, it was thought desirable to determine the volume directly if possible and after some tentative trials a modification of the method used by C. G. Johnson (1937) in measuring the volume of the egg of *Notostirus* was employed. The method consists in making use of a U-shaped tube with a bore somewhat under 1·0 mm., one end of which is drawn out to a fine capillary point. This is linked to a thermometer by a fine rubber band, the whole being included in an outer tube kept at a uniform temperature by circulating tap water. The tube is emptied or filled by suction and the application of filter paper to the fine capillary tip by which the meniscus in the opposite limb can be brought to any desired level. For a full description the author's paper should be consulted. In place, however, of determining volume by differences in weight this was done by direct observation of change in position of the meniscus as observed through a binocular microscope in the horizontal position with an eyepiece scale and using a marked portion of the tube and as small a displacement as possible to avoid parallax. Though the apparatus was much the most promising of any used, it required considerable care in manipulation and especially in calibration. It was calibrated by dropping in small measured amounts (1–3 mm.³) of mercury from a blood-measuring pipette after testing this. One difficulty was to get the dry eggs to sink without manipulation. This was finally achieved by the use of butyl alcohol in the tube and using the point of a fine pipette cut off to make a tiny funnel, so preventing eggs when tipped in from sticking on the sides of the tube above the fluid. With butyl alcohol the eggs sank at once and appeared not to be affected when examined later. Kerosene was also effective in allowing eggs to sink at once, but the exochorion was swollen and tended to become detached. A 4 per cent solution of tauroglycocholate was also tried, but was less satisfactory than butyl alcohol. The following are the results obtained.

Date of observation	Fluid used	Displacement of meniscus (divisions of scale)	Total volume (1 div. 0·193 mm.³)	Number of eggs	Mean volume (mm.³)
21. i. 48	4 per cent sodium tauroglycocholate	22·0	4·246	424	0·0100
26. ii. 48	Kerosene	3·0	0·580	64	0·0091
27. ii. 48	Butyl alcohol	15·0	2·895	279	0·0104
27. ii. 48	Butyl alcohol	16·0	3·100	304	0·0102

In each of the last three observations the mean weight of the eggs weighed on the torsion balance was 0·0110 mg.

Though the later observations give figures in keeping with the theoretical value, it was not felt that volume determination direct was sufficiently critical to do more than support values obtained for weight and density. The method used as above appeared, however, one very suitable for volume determination of small objects and probably could have been made very accurate provided error due to varying angle in reading the meniscus was kept small.

In the next subsection reference is made to a model used to supplement information regarding surface area of the egg, the model being of the correct weight in proportion to length and, as far as was possible, the accurate shape of the egg. Displacement of the model was 13·25 ml., and since the volume of the model (see next section) was calculated to be 1,286,000 times that of the egg this gave a value of 0·0103 mm.^{3} for the mean volume of the egg.

SURFACE AREA

An estimation of surface area was made from a camera lucida drawing of a flattened and completely disrupted egg-shell and its cap at a magnification of 170. Including areas of overlapping which were clearly displayed, this gave 0·052 mm.^{2} as the area of the cap and 0·266 mm.^{2} as that of the rest of the shell. The total area on this computation was therefore 0·318 mm.^{2}. The area of the cap in this case was about one-sixth of the total area of the egg.

As a further way of arriving at such an estimate a model was made in plasticine as nearly as possible the shape of the egg. Its length measured by gauge was 72 mm. In this respect it therefore represented an egg (mean length 0·662 mm.) multiplied 108·76 times linear dimensions. The volume should therefore be 108·76^3 or 1,286,000 times that of the actual egg and if the model were of the same density as the egg its weight should have been this number of times the weight of the egg or 14·275 g. The specific gravity of the plasticine was, however, 1·675 whilst that of the egg as previously given was 1·075 giving the weight for the model as 22·242 g. This with very slight adjustment was made the weight.* The circumference of the model was now taken at intervals of 0·5 cm. projection, which plotted on squared paper gave the area enclosed as 35·2 cm.^{2} and this divided by the square of the linear magnification (11828) gave as the area of the egg 0·297 mm.^{2}.

A value of about 0·3 mm.^{2} may, therefore, be given for the surface area of the egg.

WEIGHT AND SPECIFIC GRAVITY OF THE EGG-SHELL

The following observations were made on the weight of the egg-shell and cap.

	Weight (mg.)	Number	Mean weight (mg.)
Egg-shells without cap, dry overnight	0·075	91	0·00082
Egg-shells without cap, dry overnight	0·35	465	0·00079
Egg-shells without cap, dry overnight	0·30	380	0·00079
Egg caps, dry overnight	0·06	377	0·00016
Egg-shells with attached caps, dry	0·02	20	0·001
Egg-shells with attached cap, dry	0·14	140	0·001

The mean weight of the egg-shell complete is, therefore, approximately 0·001 mg. and that of the egg-cap 0·00016 mg. or again about one-sixth of the total egg-shell.

It was thought possible that the shell when forming part of the egg might not be so dry as when treated as above. The following results are, however, in favour of the view that the weight as given above is at least approximately that of the shell as it exists in the egg,

* The weight of the model before this alteration was 22·107 g., so that it required only less than 1 per cent adjustment.

there being no further loss of water after the wet shells have been exposed to the laboratory for 30 minutes.

Direct from porous tile	0·98 mg.
After 15 minutes	0·54 mg.
After 30 minutes	0·29 mg.
After drying overnight	0·30 mg.

Determination of the specific gravity of the egg-shell after hatching was made by transferring egg-shells from water into solutions of sodium sulphate of known density, and allowing a short time for diffusion of contained water into the surrounding medium. The shells sank in all the concentrations used up to a density of 1·200. Placed in a solution of this density egg-shells sank to the bottom in from 5 to 10 minutes. Whole eggs remained floating even overnight.

The specific gravity of keratin in the dry state is given by Robertson (1918) as 1·318 and as the egg-shell is evidently of protein nature and very hard and tough it may well have a density approaching that of keratin.

(*d*) CHANGES FOLLOWING OVIPOSITION

COLOUR CHANGES

When first laid the eggs of *Aëdes aegypti* are of a translucent pure white and seen against a white background such as filter paper are scarcely visible at a casual glance. Whether exposed to light or not they soon darken. O'Connor (1923) gives 15 minutes for becoming grey and 40 for black. Buxton and Hopkins (1927) consider this too short a time, as in the case of a female seen laying eggs at 4.30 p.m. the eggs were grey at 6 and black at 7 p.m. Actually the first change occurs in about an hour (see Table 4), the eggs gradually darkening, rather more rapidly as they are becoming black.

Table 4. *Change in colour and size of the egg following upon oviposition*

Time from oviposition (minutes)	Obs. 1 Colour	Obs. 1 Mean length (mm.)	Obs. 2 Colour	Obs. 2 Mean length (mm.)
In oviduct	White	0·467	White	—
Immediate	White	0·500	White	—
30	White	—	White	—
60	Faint tint of blue	0·522	White	—
75	—	—	Faint tint of blue	—
90	Pale blue	—	Pale blue	—
120	Darkish blue	0·550	Darkish blue	0·560
150	Blue black	—	Dark blue	—
180	Blue black	—	Blue black	0·580
210	Black	—	One black, one blue-black	—
240	Black	0·600	Black	0·620
20 hours	Black	0·684	Black	—

What does not seem to have been recorded is the colour of the darkening eggs which is a distinct and at times a vivid blue. The first tint is a pale watery blue which begins to be apparent in about an hour from oviposition. This colour darkens and in about 3 hours the eggs are a deep blue-black. Complete darkening to black takes about 4 hours. The egg develops colour and darkens uniformly and not as in the *Culex* egg first in a zone around the posterior pole (Christophers, 1945) (see also under 'Physiology of the egg', p. 154).

CHANGES IN VOLUME, WEIGHT AND SPECIFIC GRAVITY

Eggs when first laid are not only white in colour but are much smaller than the mature black egg. In Table 4 are given measurements (together with colour changes) as observed in eggs seen to have been oviposited and kept under observation for some hours. Those in the first series relate to six eggs and those in the second series to two eggs. It will be evident that there is a steady increase in size which continues even after the eggs in about 4 hours have become black. As shown by measurements given in the table there is an increase in the length of the egg when oviposited amounting to 0·66 of its length in the oviduct. A similar increase amounting to 0·1 is noted by Roy and Majundar (1939) for the egg of *Culex fatigans* and a similar lengthening of the egg of *Anopheles subpictus* and *A. annularis* is recorded by Roy (1940). There is then a further increase from the length when laid to that at 20 hours of 1·37. Such an increase, provided there was no change in shape, would correspond to the cube of this amount, that is to 2·57 times the volume when laid. Actually the egg does change slightly in shape, being somewhat more oval and less torpedo-shaped when first laid, the diameter being about the same as that of the mature egg, but the general order of the estimate should hold. This increase can only be brought about by imbibition of water, so that the egg must imbibe about its own volume of water during this period.

The imbibition in the first few hours at least would appear to be fairly uniform, the rate of increase in length being about 0·02 mm. per hour.

Since the volume of the egg increases, which can only be by taking in water, the specific gravity of the newly laid egg should be greater than that of the mature egg. This is actually so. Determination of the specific gravity of white eggs some little time after oviposition gave a density of 1·098. On another occasion white eggs sank in sodium sulphate solution of 1·130 sp.gr. Some black eggs previously laid rose at the same time that the white eggs sank. Fully developed eggs taken from gravid females slowly sink even at 1·130 and even the whole ovary sinks without hesitation when introduced into fluid of this or slightly less density. The condition in the egg of *Aëdes aegypti* would seem, therefore, very similar to that described by Bodine and Robbie (1943) in the grasshopper egg. These authors give the density of the grasshopper egg as varying from about 1·13 to about 1·06, whilst the weight due to absorption of water increases from 3·2 to 6·0 mg. Development in this case takes 18 days to reach diapause and a further 18 days following upon diapause to reach full development. The egg of *A. aegypti* absorbs water and swells to about double its size in a few hours. Before such intake it has a high density, again very similar to that described in the grasshopper egg.

DESICCATION

Provided eggs are kept moist they retain their shape and if first kept moist for a period of 24–72 hours they do so even if exposed to prolonged desiccation. If not kept moist for this initial period they dry and collapse even in a matter of minutes. As noted by MacGregor (1916) two chief forms of collapsed eggs are seen. With a moderate degree of collapse the sides of the egg flatten or fall in leaving a central ridge so that the egg is prismatic or triradiate in section. In a more completely collapsed form the egg is completely flattened in a saucer-shaped fashion. These two forms of collapsed egg are quite characteristic, but all degrees of collapse are seen from a slight dimpling to completely flattened saucer-like eggs as described by MacGregor. The former in profile appear narrower than normal, the latter considerably broader when seen on the flat, but narrowly crescentic seen on edge.

When eggs are removed from a moist atmosphere shortly after being laid all may collapse completely within some minutes. As resistance to desiccation increases, some eggs collapse while others remain unshrunken though these latter may in turn, if the eggs are not fully conditioned, also eventually collapse within a variable time. After full resistance has been achieved, which may require keeping in a moist atmosphere for several days, eggs remain rounded and normal for periods of weeks or months. Eventually, however, perhaps after some months to a year, depending on conditions, these too become collapsed, the number so collapsing increasing with time. The process of keeping eggs in a moist condition until they become resistant to desiccation has been termed by Shannon and Putnam (1934) 'conditioning', and eggs in such a state may conveniently be termed 'conditioned eggs' (see ch. v).

SWOLLEN AND EXPLODED EGGS

When eggs collapsed from desiccation are placed in water they swell up and are scarcely distinguishable from normal unshrunken viable eggs. Such eggs are, however, at once identified when treated with diaphanol or other bleaching agent, since they show only disorganised contents or macerated embryos. They often form the bulk of unhatched eggs left behind with empty egg-shells from which larvae have hatched, and unless the condition is recognised may lead to error from being mistaken for viable eggs that have not yet hatched. Sometimes such swollen-up eggs may appear as specially large eggs and may measure up to 0·8 mm. in length.

What is probably the final stage of such swollen-up eggs are ruptured or 'exploded' eggs. In this case one surface of the egg, apparently always the dorsal, that is the under surface and the one nearest to the yolk remnant, is burst outwards in a long tongue giving with the rest of the egg an appearance much like a safety pin. Or there may be such a tongue at each end of the egg, which then resembles a penknife with two blades half open (Fig. 5 (7)).

Such ruptured eggs never show any indication of the forcing off of an egg-cap, showing that the latter structure is not the result of any structural condition of the chorion (see under 'Mechanism of hatching', p. 162).

Ruptured eggs as described above are evidently what Howard, Dyar and Knab refer to when they say that eggs in prolonged immersion do not remain intact but split open longitudinally. Young (1922) also refers to split and collapsed eggs.

A certain number of eggs, depending upon circumstances, may be unfertilised. These do not differ from normal viable eggs until bleached, when the absence of embryo formation is apparent. A common appearance when bleached is the rounding up of the yolk into one or more globular masses whilst other eggs are showing various stages of embryo formation.

Besides unfertilised eggs, obviously abnormal eggs are sometimes seen. The commonest form of such are small oval eggs. On one or two occasions a very large egg has been seen, half as long again as a normal egg. Among the usual black eggs on the surface of the water newly laid white eggs may be seen and may be mistaken on account of their small size for prematurely laid eggs. Or some of such may be prematurely laid eggs which are not darkening. Prematurely laid eggs may sometimes be seen which have swollen up to resemble a coiled worm, a shape seen when follicles in the ovary at certain stages are treated with potash, that is the wall of the follicle does not yield to swelling evenly but causes the swollen follicle to elongate.

CHANGES DUE TO DEVELOPMENT

Since eggs in their natural condition are intensely black and opaque, nothing of their contents is visible by direct observation without some form of bleaching. After diaphanol or chlorine treatment, however, various stages in the formation of the embryo up to the fully developed primary larva can be seen. A detailed account of the embryology would here be out of place, but a brief description of the embryology of the mosquito as worked out on *Culex molestus* is given later in chapter VII. The following are the most obvious appearances which may be seen in ordinary manipulation of eggs.

Eggs which contain fully developed larvae ready to hatch show when bleached towards the anterior pole three conspicuous dark spots, two lateral, the pigmented ocelli of the larva, and one median dorsal, the egg-breaker. At this stage the large head occupies about one-third of the egg and there is a smaller mass at the posterior pole consisting of the siphon and other terminal parts. Between these two masses are clearly shown the three thoracic and eight abdominal segments, all very compacted like a rouleau of coins (Fig. 9 (1)). Many larvae at this stage will, however, almost certainly be extruded from the egg due to the effect of the bleaching agent as described later. The segments in such cases are no longer compacted but widely expanded and separated by deep notches. It may be noted that Roubaud (1927) terms the larva whilst still contained in the egg, the *primary larva*, a name which may sometimes be usefully employed.

In larvae not quite fully developed eye-spots may be present, but not the spot indicating the egg-breaker, or still earlier even the eye-spots are not to be made out, or as is often the case, these are visible but not conspicuous.

Still earlier stages show various appearances of the developing embryo. A very characteristic stage is one in which the anterior end of the embryo is bifid, due to the presence of the labial groove and another one in which, the groove having everted, there is a bulbous prominence anteriorly. At such stages the body of the embryo is elongate and curved on itself with the caudal segments lying along the dorsal aspect to within a varying distance from the head. The appearances seen depend upon the position in which the egg is viewed. If seen from the dorsal or ventral aspect there is a somewhat sausage-shaped body, bifid or with a protrusion in front as described. If seen lying upon its side the embryo appears as a bulbous head with a narrow bent hairpin-like body. Earlier stages are usually not recognisable except in stained specimens where the blastoderm stages may be characteristically shown.

Segmentation becomes apparent at about 24 hours from oviposition. Eye-spots become visible at about 70 hours at 25° C. Conspicuous eye-spots with egg-breaker are first seen at about 90 hours at this temperature. Eggs between 24 and 48 hours usually show the bifid appearance or globular protrusion as described above. Eggs younger than 24 hours show what appears as a more or less uniform yolk mass, though in sections various stages in cleavage and blastoderm formation are strikingly shown. The times given are for *Culex molestus* for which the full development takes about 56 hours. For *Aëdes aegypti* somewhat longer may be allowed.

(e) VIABILITY OF EGGS UNDER DIFFERENT CONDITIONS

RESISTANCE TO DESICCATION

That the eggs of *A. aegypti* are capable of retaining their viability for long periods of desiccation was first brought to the notice of workers in this country by Theobald (1901), who hatched out in England eggs sent from Cuba which had been dry in a tube for 2 months. But that eggs may remain a long time in the dry state without losing their vitality had been noted as early as 1886 by Finlay, who sent the eggs referred to above to Theobald.

Numerous later observations on the length of time that eggs of *A. aegypti* kept in the dry state can retain their viability have been given in the literature. Francis (1907) found eggs kept dry for 4 months still viable. Newstead and Thomas (1910) obtained larvae from eggs sent from Manaos after these had been dried for 24 hours over $CaCl_2$ and kept 45–47 days in a tightly corked tube. Bacot (1916) found eggs still viable that had been stored under dry conditions for 262 days and the same author (1918) found eggs kept dry at a temperature varying between 6·6° and 17·8° C. on filter paper still viable after 15 months. Fielding (1919) found a percentage of eggs kept at room temperature still living after 257 days (8½ months). Eggs stored 391–579 days did not hatch. Shannon and Putnam (1934) state that, if well conditioned, eggs can be kept stored for six months without high mortality and that about 5 per cent will survive a year or more.

Such statements do not, however, give the whole picture, for as noted by Buxton and Hopkins (1927) the longer eggs are stored the greater the mortality. In Fielding's series the percentage of different batches which hatched at times between 114 and 216 days varied from 21 to 58 per cent with one exception. At 257 days only 1 per cent hatched. Bonne-Wepster and Brug (1932) give the following data obtained by their assistant R. Soebekti for imagos bred out from eggs kept various times in the dry state and placed to hatch in grass infusion.

Period	Percentage of hatchings	Means
1 month	60, 60, 23, 14, 100, 52, 40	46·5
2 months	4, 12, 13, 10, 36, 66, 18, 44, 26, 28, 60, 3, 60	34·0
3 months	65, 0, 0, 8, 12	17·0
4 months	4, 8, 6, 0, 0, 0, 0, 0, 0, 0, 0	1·6
5 and 6 months	0, 0, 0, 0	0·0

In nature Cooling (1924) found that eggs collected from a tree-hole hatched, though they must have been dry for from 12 to 14 weeks.

The longest period in the experience of the present author has been 233 days (8 months) when an egg-pot kept at 25° C. and 70 per cent relative humidity and containing some thousands of eggs yielded only ten larvae on the second day of submergence. Other examples of pots under similar conditions kept for periods over two months are:

Batch number	Days kept dry	Result of placing to hatch
960 (2)	233	Ten larvae on second day
977 (1)	209	One or two larvae only
981 A	166	About 200 larvae from fragment overnight
969 (2)	108	Many in 90 minutes
977 (2)	107	Many
970 (2)	102	Many hundreds from small fragment
968 (3)	71	Normal hatch
968 (2)	66	Normal hatch. A small fragment with sixty-one eggs gave 100 per cent hatch overnight

It is evident that under such conditions and protected from psoci, cockroaches, etc., eggs up to 100 days had largely retained viability. The general experience has been that up to one month eggs have not appreciably lost viability or exhibited delay in hatching, and that no serious loss in viability took place up to 4 months. After this results were uncertain and after six months rarely more than a small percentage hatched. Shannon and Putnam also state that well-conditioned eggs, if used within a month, should serve as well as freshly conditioned eggs for routine rearing.

The time eggs remain viable, however, is largely dependent on the degree of desiccation to which they are exposed and where a relatively high humidity is not present the period for which they remain viable is usually much reduced. In the laboratory keeping eggs in a closed container under damp conditions favours long viability, but care is required since if left without occasional airing mould may cause trouble and if too moist premature and imperfect hatching may occur as described later.

Artificial desiccation greatly reduced the viability period. Fielding found some eggs kept over $CaCl_2$ still living after 19 days, but all were dead after 26 days. Buxton and Hopkins found that eggs kept over H_2SO_4 were all killed in from 30 to 40 days. Eggs kept in the laboratory without precautions showed collapsed eggs very much earlier than when kept in the constant temperature room at a raised humidity.

Development of resistant properties. The dormancy of the egg of *A. aegypti* is dependent on the formation of the primary larva, a term applied by Roubaud (1927) to the larva whilst still in the egg, and is not due to arrest of embryonic development as is more usually the case in diapause of insect eggs. Bacot (1917) seems to have been the first to note that moist conditions were essential for some time after laying if the eggs were to resist desiccation. He gives for this a period of about 30–40 hours by which time he says the larva was fully developed in the egg. Buxton and Hopkins emphasise the importance of this period which they term 'maturation', pointing out that observations relating to viability of eggs undertaken without regard to the period of maturation are valueless. H. A. Johnson (1937) found in dissection of freshly laid eggs that the embryo was not developed. As eggs aged the embryos progressed to the point of hatching, provided eggs were kept moist. If eggs dried out, the young embryos died before maturing.

The time necessary to keep eggs moist to ensure their viability when subjected to desiccation has very generally been given as about 24 hours. MacGregor (1916) gives the following results of drying eggs at different times after oviposition:

Maximum time left on water (hours)	15	20	25	37	43	48	60	65	70
Percentage collapsed after drying 12 hours	100	100	100	62	28	2	0	0	0

At 15 hours eggs collapsed in 15 minutes.

Buxton and Hopkins found that of 295 eggs which had been dried when they were not more than 24 hours old, 197 (67 per cent) hatched. Shannon and Putnam, however, for optimum conditioning give several days on damp filter paper which is placed in the open air to dry slowly.

The following are results obtained with eggs laid by isolated females on filter paper strips in tubes kept moist at the temperature noted and allowed to dry after various periods. The eggs referred to in the table were those laid on the paper slips and not the whole oviposition.

Female	Temperature at which eggs kept (° C.)	Allowed to dry after hours	Collapsed	Not collapsed	Hatched	Per cent hatched
12	25	4	25	0	0	0
7	28	20	5	9	8	
3	28	20	4	23	23	88
8	28	20	3	17	17	
5	28	20	2	63	—	
18	25	24	2	11	9	
13	25	44	1	10	10	91
10	28	44	2	8	8	
21	28	48	—	42	42	
14	25	69	1	12	12	
16	25	69	0	5	5	96
22	28	72	0	6	6	

Though the majority of eggs kept moist for 24 hours after oviposition had become capable of resisting collapse on drying overnight, some eggs were still collapsed which had been kept moist for 48 hours or more. Obviously, however, 48 hours had sufficed to allow protection to be achieved by the great majority of eggs.

In the main, loss of viability runs *pari passu* with collapse, so that long-stored eggs which are seen to be collapsed may be expected to give at the best a small yield of larvae. To what extent collapse is certain evidence of loss of viability was tried out in the following series, which was a sample of fifty-five eggs taken from a pot some 6 months old in which the eggs appeared all more or less collapsed.

One egg apparently unshrunken	Larva hatched in 3 hours
Thirty-one eggs showing various degrees of collapse of prismatic form	Five larvae (four in 3 hours), two dead larvae, two half emerged
Twenty-three eggs flattened or grossly shrunken	No larvae overnight

As a rule therefore eggs showing a marked degree of collapse have ceased to be viable. A small percentage of moderately collapsed eggs may, however, still be viable.

Briefly, everything points to the fact that if eggs are prevented from drying up until the larva is fully formed they have the power of resisting desiccation and retaining their viability for long periods in the dry state. It does not necessarily follow that it is the larva itself which has this power. Indeed, it is almost certain that such resistance is due to waterproofing of the egg membranes at some stage, which one would expect to begin some time before 24 hours from oviposition. The onset of waterproofing must indeed take some hours, for up to some 6 hours or possibly longer the eggs are actively absorbing water (see under 'Changes following oviposition', p. 144).

Duly protected from desiccation the embryo develops to the larva in which state it has the remarkable power to lie dormant until the egg is again submerged, or until, as observed by Buxton and Hopkins, a slow desiccation ultimately destroys the egg or the reserves held by the larva are exhausted. This state of dormancy has been termed by Roubaud (1929) one of secondary diapause, diapause in most insect eggs occurring during, not at the end of, embryonic development. So far there is no evidence that any of the eggs of *A. aegypti* undergo any period of suspended development before the embryo becomes the primary larva.

THE EFFECT OF HEAT AND COLD ON VIABILITY

Bacot (1916) notes the following effects on viability of eggs exposed for 24 hours to different temperatures.

29° F. (−1·7° C.)	81 per cent hatched
75° F. (24·0° C.)	80 per cent hatched
95° F. (35·0° C.)	28 per cent hatched
102° F. (39·0° C.)	12 per cent hatched
107·6° F. (42·0° C.)	0 per cent hatched

Marchoux, Salimbeni and Simond (1903) exposed eggs taken from water the day after they were laid for 5 minutes to temperatures varying from 37° to 49° C. and found they afterwards hatched. Above 48° C. they failed for the most part to hatch and above 49° C. they did not develop. Macfie (1920) found that eggs withstood 5 minutes exposure to temperatures as high as 46° C. They were generally killed at temperatures above this and all were killed at 49° C. The same result was obtained with undried eggs about 24 hours old and eggs on filter paper kept dry for two months. 45° C. had no appreciable effect. Larvae, it may be noted, are rapidly killed at this temperature. Davis (1932) found that fresh eggs were killed by 48 hours exposure to 40° C., mature eggs in less than 7 days. In general it may be taken that the lethal temperature for short exposure to high temperatures is about 49° C., but that temperatures as low as 40° C. are fatal for exposures lasting several days, and that 45° C. was fatal to all eggs, fresh and mature, within 24 hours, mature eggs being somewhat more resistant than those in an early stage of development.

In regard to the effect of cold, Howard, Dyar and Knab state that freezing does not kill *A. aegypti* eggs. Davis (1932), however, found that freezing kills both fresh and mature eggs after 48 hours, fresh eggs being slightly more resistant.

These results are in keeping with observations now made. Whilst eggs are not very susceptible to lethal effect from low temperature, such effect depends largely upon time of exposure. Two distinct issues are concerned, namely the lethal effect of low temperature and the limiting temperature at which development can proceed. In regard to lethal effect eggs exposed 24 hours at 1·0° C. hatched when placed at 25° C. Eggs after eleven days exposure at 7° C. on being placed at room temperature and then at 27° C. gave a 25 per cent hatch. Those exposed at 1·0° C. for nine days gave only a single larva out of some hundreds on being similarly treated. It was thought from these and other observations that the point at which prolonged exposure to low temperature ceased to be lethal was about 10° C. This is in keeping with Bacot's observation that some eggs were viable after being kept some months at a variable cool temperature of from 6·6° to 17·8° C.

The above remarks relate to conditioned eggs, that is eggs in which development has proceeded to formation of the primary larva before exposure. Where eggs are exposed from shortly after oviposition the question of delayed or inhibited development has to be taken into account (see ch. VII). Eggs shortly after oviposition exposed to 18° C. had not hatched after 21 days, but did so almost at once when placed at 28° C. In the case of those eggs placed at 7° C. not only was development entirely inhibited so that after 21 days no developmental changes beyond those seen when the eggs were first taken, but the eggs failed to hatch after 5 days at 28° C., that is exposure for 21 days had not only stopped development but had killed the eggs. This is in keeping with the observations of Davis that 10° C. has a very deleterious effect on fresh eggs. It was thought that about 15° C.

might provisionally be taken below which no development takes place in the egg. These results are of interest in relation to the limits of distribution of the species; see chapter II, section (*d*) on temperature in relation to limits of distribution.

EFFECT OF SALT AND CHEMICALS

Marchoux, Salimbeni and Simond found that eggs of *A. aegypti* on one-sixth seawater hatched in four days, as also on one-fifth seawater (about 0·6 per cent salt). With one-third seawater (about 1·0 per cent salt) eggs were killed. Macfie (1915) notes that on 2 per cent salt eggs of *A. aegypti* do not blacken but remain soft and white. Eggs not only failed to hatch out on 2·3 per cent salt but failed to do so later, that is they were killed.

Owing to their impervious endochorion, once this has become hard and waterproofed eggs of *A. aegypti* are resistant up to a point in such substances as alcohol (Barber, 1927); lysol (Atkin and Bacot, 1917, and others); hexyl resorcinol (Hinman, 1932); mercuric chloride (Trager, 1937); formalin and indeed most other chemicals for a short exposure, a fact which enables the substances noted above to be used for the production of sterile larvae (see next section). Eggs bleached with diaphanol sufficiently to show the larval ocelli and egg-breaker through the membranes are not by such treatment prevented from hatching (see p. 158).

Powers and Headlee (1939) found various mineral oils lethal to eggs of *A. aegypti*. The effect was most pronounced at a certain viscosity, being slight with very low viscosity oils (under 40), reaching a peak at about 108, and falling gradually to viscosities of 300 or more. No penetration of the chorion, as shown by use of oils with oil-soluble dyes, occurred, nor was any chemical change in the endochorion to be detected. As a result of observations on effect of exposing eggs to nitrogen, hydrogen and oxygen gas the authors conclude that the lethal effect of oils was due to oxygen deprivation, those oils whose physical characters best allowed of forming a closely investing layer on the egg being the most effective. Eggs subjected to pure nitrogen for 24 hours were largely killed (81–96 per cent). Hydrogen was fatal in 6 days. CO_2 killed 50 per cent of eggs in 24 hours.

STERILISATION OF VIABLE EGGS

Sterilisation of the eggs of *A. aegypti* as a means of obtaining sterile larvae that can be cultured in sterile media has become an important technique in nutritional and other studies. The first authors to carry out observations in this connection were Atkin and Bacot (1917). Eggs after being washed with a fine jet of tap water in a deep pan were pipetted into 0·5 per cent lysol, vigorously washed with the jet, transferred after a few minutes to 2 per cent lysol for 5–10 minutes and again washed in boiled water.

Barber (1927) dripped 80 per cent alcohol for 2–3 minutes over eggs distributed on a piece of sterile cloth laid on a perforated spoon. The cloth was dried as quickly as possible by draining alcohol away with a wad of sterile asbestos wool placed under the spoon for a few seconds. The cloth was then lifted off with sterile forceps, the alcohol burned away from the spoon and the warmth of this used to assist in drying the cloth held at a safe distance above it. Eggs were transferred to the tube with a fine sterile spatula moistened with sterile fluid. The same method was successful with *Anopheles* eggs. MacGregor (1929) reared larvae on sterile standard bread using soap to sterilise the eggs (see under 'Trager', p. 152).

Roubaud and Colas-Belcour (1929) sterilised eggs using hydrogen peroxide and also mercuric chloride. Hinman (1930) used chlorinated lime, but later, 1932, thoroughly disinfected in hexyl resorcol. Trager (1935), in cultivating mosquito larvae free from living organisms, used a modification of MacGregor's method. Boats made of cover-slips by heating the edges in the flame and sterilising by dry heat were placed in small sterile petri dishes containing 5 per cent solution of Castile soap. Ten to forty eggs were placed in each boat and left in soap solution 5–7 minutes. After being drained of excess fluid they were then transferred to 80 per cent alcohol for 15–17 minutes and thence to a sterile petri dish of distilled water. Finally boats were dropped into a tube of culture medium. One hundred per cent were successful so long as eggs had been laid recently (within 1–3 days) on filter paper half immersed in distilled water. Later the same author (1937) used 15 minutes immersion of eggs in White's solution (see below).

The sterilising fluid most frequently used is that recommended by White (1931) for sterilisation of blowfly eggs (Trager, 1937; De Meillon, Golberg and Lavoipierre, 1945. and others). The following is the composition of the fluid:

Mercuric chloride	0·25 g.
Sodium chloride	6·5 g.
Hydrochloric acid	1·25 c.c.
Ethyl alcohol	250·0 c.c.
Distilled water	750·0 c.c.

The technique used by De Meillon, Golberg and Lavoipierre is as follows. Gravid females were confined singly in sterile test tubes with 2–3 c.c. sterile distilled water in an incubator at 28° C. When oviposition had taken place the water was removed with a sterile pipette and the eggs left stranded. The tubes were then half filled with White's solution and left 15 minutes. The fluid was then removed, the eggs washed several times in distilled water using a sterile pipette to remove the water. In the final wash eggs were dislodged with a platinum loop, poured into a sterile petri dish and covered. Five or more eggs were removed at a time to tubes containing agglutinating broth and stored at 37° C. On about the third day hatching begins. Larvae do not grow in agglutinating broth but remain alive for several days. There is no difference in the ultimate growth whether such larvae when used are 12, 24 or 96 hours old.

An account of various methods that have been used in preparing sterile blowfly eggs for surgical use are discussed by Simmons (1934) with a bibliography. This author found that mercuric chloride with phenol, as also formalin alone, caused agglutination of the eggs; 5 per cent formalin with 1 per cent NaOH for 5 minutes was, however, an excellent sterilising fluid. White found sodium hydroxide followed by 5 per cent formalin for 15 minutes a satisfactory disinfectant for blowfly eggs. A device described for storage of blowfly eggs by White (Simmons) is a jar with a cap (like a canada balsam bottle) with a Gooch crucible fixed in the neck on which sterile gauze with eggs can be placed after disinfection. About 50 c.c. of sterile water is poured slowly over the eggs to remove disinfectant.

It is obvious that a tube into which a mosquito, bred in the ordinary way, has been introduced is no longer sterile. It would be quite possible, however, if thought necessary, without much trouble, to perform the whole process under sterile conditions. The pupa of *Aëdes aegypti* is extremely resistant to chemicals and will live for a considerable time in say 80 per cent alcohol and especially so in more watery solutions of such a substance as mercuric chloride. A disinfected pupa could be washed and placed in a sterile tube with a gauze top to hatch out and could be fed on a sterile skin through this and allowed to oviposit all under sterile conditions.

(*f*) PHYSIOLOGY OF THE EGG

Very little is recorded specifically in the literature regarding physiology of the egg of *A. aegypti* other than in respect to influences affecting hatching which will be dealt with in the next chapter. Much of what is known of the physiology of insect eggs in general, of which a short account is given by Wigglesworth (1942) in the first chapter of his *Principles of Insect Physiology* and later communications by Pryor, Beament and others, will no doubt relate also to the eggs of *A. aegypti*. There are certain points, however, which are of special interest in the present case. These are: (1) the chemical nature of the egg membranes, (2) the nature of the waterproofing of the egg to enable it to resist desiccation to such a marked extent as it does in respect to some other mosquito eggs, and (3) its oxygen requirements.

CHEMICAL NATURE OF THE EGG MEMBRANE

The chorion of the insect egg is often spoken of as though it were a homogenous structure. Actually, as shown by Beament (1946), it may be extremely complex. In the case of *A. aegypti* the two portions described as exo- and endo-chorion are clearly of very different nature. Whilst the former, especially in the fresh moist egg, is very delicate, soft and readily detached, acted upon by hot KOH and by various solvents, the latter, besides being remarkably tough, is resistant alike to strong acids, hot KOH and solvents.

The *exochorion* consists of two distinct elements, the matrix forming the membrane and the chorionic bodies and pegs, seemingly of similar nature, which are embedded in it. Though somewhat resistant to action of KOH in the cold, the matrix completely disappears after 15 minutes' exposure to sat. KOH at 160° C. and is therefore not chitin. Detached fragments and egg-shells with the exochorion *in situ* treated overnight with conc. H_2SO_4 or HCl show the matrix still intact with the chorionic bodies embedded in it. It is eventually, however, destroyed by conc. H_2SO_4 in the cold, but requires one or two weeks for this. In conc. HNO_3 the exochorion is quickly destroyed.

In solvents the matrix commonly swells and becomes more or less disorganised. Treated with ether it becomes after a time fragmented and leaves the egg largely bare. Chloroform causes swelling of the matrix and its detachment. The most marked and curious effects by a solvent is given by acetone, which causes the exochorion to become markedly swollen and opaque with the chorionic bodies projecting beyond its surface. It has already been noted that water causes the exochorion of the under surface of the egg to swell and become gelatinous. The upper surface is not so affected.

When dried, as when left as a flake after a dry egg has been detached, the substance of the exochorion (chorionic pad) appears to become altered as it no longer swells up in water and is more resistant to solvents, but it is dissolved in strong KOH overnight and eggs cemented by it become detached.

The chorionic bodies are noticeable for their transparency and absence of colouring matter. They are resistant to most reagents including conc. H_2SO_4 and HCl. In such acids they become very visible in the clear matrix and seem to become somewhat swollen showing a sago or cauliflower effect. It is possible that they swell somewhat in water appearing 9–11 μ in a dry state and 12–13 μ after 30 minutes in water. They stain with methylene blue and more readily with eosin especially in sections. They stain deeply and selectively with watery thionin. In general the reactions of the chorionic bodies indicate that they are of untanned protein nature and they stain pink with Millon's reaction. Unfortunately, except that the whole exochorion disappeared after heating for 15 minutes in sat. KOH at 160° C. the effect on the chorionic bodies was not further investigated. The exochorion appears to play no part in waterproofing the egg, as eggs that have been largely deprived of the exochorion are still resistant to desiccation. What function, if any, the chorionic bodies subserve is unknown, but they seem to be associated with eggs that are resistant to desiccation and they may play some role in this respect or in subsequent stimulus to hatching when eventually submerged.

The *endochorion* microscopically consists, in the greater part of its thickness at least, of a uniform very dark and tough layer. That it is not chitin is shown by the fact that though it resists heating for 15 minutes to 160° C. in sat. KOH, it does not after such treatment show any indication of dissolving in 3 per cent acetic. It seems certain that it is a tanned protein as shown by Pryor (1940), for the hardened larval skin of the *Calliphora* pupa and by Beament (1946) for the resistant endochorion layer of the *Rhodnius* egg. In support of such a view is the fact that if the newly laid (white) egg be placed in water with some crystals of pyrocatechol the endochorion becomes a beautiful pink indicating the presence of polyphenolase, the ferment concerned in the tanning of protein in insect cuticle. In bleached egg-shells treated with Millon's reagent there is some darkening and a distinct red tinge where the egg-shell is broken and the edge turned up. At the suggestion of Dr Pryor a quantity of egg-shells were submitted for determination of the sulphur content. Only a very small sulphur content was found, indicating that sulphur was not an important constituent of the endochorion.

Some interest attaches to the colour exhibited during the process of darkening (tanning) of the endochorion. As previously noted in *Aëdes aegypti* egg, this is a distinct blue. In *Culex molestus*, as I have previously noted (Christophers, 1945), it is green. Herms and Freeborn (1921) describe the egg of *Anopheles punctipennis* as at first pearly white becoming progressively yellowish and then darker. At 35 minutes it is leaden, at 45 dull black and under the microscope a rich chitinous brown. Blue and green I am informed by Dr Pryor are unusual, the most common colour developing in insect cuticle being yellow or brown.

WATERPROOFING OF THE EGG

Though the endochorion consists of a very hard and tough material this does not suffice to waterproof the egg, since as already noted the endochorion has become fully darkened in 24 hours without the egg being prevented from collapsing from desiccation in a short time, whereas after some three days the egg resists desiccation for months. Wigglesworth (1945) has shown that the waterproofing of insect cuticle is commonly brought about by a thin layer of wax, in which case the insect at a critical temperature, depending upon the point at which the wax layer is disorganised, suddenly ceases to resist desiccation and, as shown by loss of weight, rapidly loses water. Similarly in the egg of *Rhodnius* waterproofing is by waxy layers, the critical temperature at which desiccation becomes rapid being 42·5° C. (Beament, 1946). In the case of different species of tick Lees (1946) found this critical point ranged from 32° to 75° C., those species which are normally more resistant to desiccation having the higher critical temperatures. Another way of determining the critical temperature is by heating in a strong solution of salt. On disorganisation of the waxy layer osmosis, now possible, causes collapse of the egg. *Aëdes aegypti* eggs heated to 50° C. for 24 hours in sat. NaCl showed no change, whereas heated to 55·5° C. collapse in every case took place within 2 hours. The critical temperature lay therefore in this case between 50 and 55·5° C., and it is probable that the waterproofing of the *A. aegypti* egg is by a waxy layer, the position of which, however, has not been determined.

OXYGEN REQUIREMENTS OF THE EGG

It is probable from what is known of respiration of insects in general that, expecially during the period of development of the primary larva, a certain oxygen requirement is necessary. Powers and Headlee (1939) have shown that eggs coated with certain oils are killed and ascribe this effect to deprivation of O_2, the maximum effect being produced by those oils which give the most complete coating. It is of interest in this connection to consider the function of the air channels in the chorion. In the dry condition of the egg it would seem that these channels, lying outside the dense endochorion, could have little usefulness in supplying oxygen. But during the early stages of development and when the egg is wet and the parts soft, such channels may well have a functional role to play, either in respect to hardening of the endochorion or development of the embryo or both these processes. In regard to the latter this may explain some of the apparent anomalies connected with such questions as rate of development and hatching under different conditions. Thus Shannon and Putnam (1934) note that eggs freshly deposited less than 12 hours old if placed immediately in water containing food do not undergo as rapid or as uniform development as those kept in open air on moist filter paper. Howard, Dyar and Knab (1912) refer to eggs perishing which have been submerged shortly after laying. Mature eggs from the ovary and white eggs sunk in fluid may or may not darken. In the latter case they swell only to a moderate extent, fail to hatch, and after remaining in the fluid some days collapse from desiccation on removal.

When the primary larva is in a state of diapause the need for O_2 is almost certainly less marked. In testing for lethal effect by oils, etc., it may therefore be important to state at what stage the eggs used have been taken.

If the views of certain authors as to the effect of reduced oxygen in bringing about hatching be correct (see under next chapter) such reduction must in some way be capable of making itself felt through the extremely impervious egg membranes. It is not known how this could be brought about.

REFERENCES

Atkin, E. A. and Bacot, A. W. (1917). The relation between the hatching of the egg and the development of the larva of *Stegomyia fasciata* and the presence of bacteria and yeasts. *Parasitology*, **9**, 482–536.

Bacot, A. W. (1916). Report of the entomological investigation undertaken for the Commission for the year 1914–15. *Rep. Yell. Fev. Comm.* **3**, 1–191.

Bacot, A. W. (1917). The effect of the presence of bacteria or yeasts on the hatching of eggs of *Stegomyia fasciata* (the yellow fever mosquito). Summary. *J. R. Micr. Soc.* **3**, 173–4.

THE EGG

BACOT, A. W. (1918). A note on the period during which the eggs of *Stegomyia fasciata* (*Aëdes calopus*) from Sierra Leone stock retain their vitality in a humid temperature. *Parasitology*, **10**, 280–3.

BARBER, M. A. (1927). The food of anopheline larvae. Food organisms in pure culture. *Publ. Hlth Rep. Wash.* **42**, 1494–510.

BEAMENT, J. W. L. (1946). The waterproofing process in eggs of *Rhodnius prolixus* Stahe. *Proc. Roy. Soc.* B, **133**, 407–18.

BODINE, J. H. and ROBBIE, W. A. (1943). Physiological characteristics of the diapause grasshopper egg. 2. Changes in density and weight during development. *Physiol. Zool.* **16**, 279.

BONNE-WEPSTER, J. and BRUG, S. L. (1932). See references in ch. II (*a–b*), p. 45.

BOYCE, R. (1911). The prevalence, distribution and significance of *Stegomyia fasciata* (*calopus* Mg.) in West Africa. *Bull. Ent. Res.* **1**, 233–63.

BRESSLAU, E. (1920). Eier und Eizahn der einheimischen Stechmücken. *Biol. Zbl.* **40**, 337–55.

BUXTON, P. A. and HOPKINS, G. H. E. (1927). Researches in Polynesia and Melanesia. Parts I–IV. *Mem. Lond. Sch. Hyg. Trop. Med.* Mem. 1.

CHRISTOPHERS, S. R. (1945). Structure of the *Culex* egg and egg-raft in relation to function (Diptera). *Trans. R. Ent. Soc. Lond.* **95**, 25–34.

COOLING, L. E. (1924). On the protracted viability of eggs of *Aëdes aegypti* and *A. motoscriptus* in a desiccated condition in a state of nature. *Health*, **2**, 51–2.

CRAIG, G. B. JR. (1955). Preparation of the chorion of eggs of Aëdine mosquitoes for microscopy. *Mosquito News*, **15**, 228–31.

DAVIS, N. C. (1932). The effects of heat and cold upon *Aëdes* (*Stegomyia*) *aegypti*. *Amer. J. Hyg.* **16**, 177–91.

DE BUCK, A. (1938). Das Exochorion der *Stegomyia*eier. *Proc. K. Akad. Wet.* **41**, 677–83.

DE MEILLON, B., GOLBERG, L. and LAVOIPIERRE, M. (1945). The nutrition of the larva of *Aëdes aegypti* L. *J. Exp. Biol.* **21**, 84–9.

FIELDING, J. W. (1919). Notes on the bionomics of *Stegomyia fasciata*. Part I. *Ann. Trop. Med. Parasit.* **13**, 259–96.

FINLAY, C. (1886). Yellow fever transmission by means of the *Culex* mosquito. *Amer. J. Med. Sci.* **92**, 395–409.

FRANCIS, E. (1907). Observations on the life cycle of *Stegomyia calopus*. *Publ. Hlth Rep. Wash.* **22**, 381–3.

GOELDI, E. A. (1905). *Os mosquitos no Para*. Weigandt, Paris.

HERMS, W. B. and FREEBORN, S. B. (1921). The egg laying habits of Californian anophelines. *J. Parasit.* **7**, 69–79.

HINMAN, E. H. (1930). A study of the food of mosquito larvae (Culicidae). *Amer. J. Hyg.* **12**, 238–70.

HINMAN, E. H. (1932). The presence of bacteria within the eggs of mosquitoes. *Science*, **76**, 106–7.

HOWARD, L. O., DYAR, H. G. and KNAB, F. (1912). See references in ch. I (*f*), p. 19.

JAMES, S. P. (1922). Exhibit of eggs of culicine mosquitoes found in England. *Trans. R. Soc. Trop. Med. Hyg.* **16**, 267–9.

JAMES, S. P. (1923). Eggs of *Finlaya geniculata* and other English mosquitoes illuminated with the aid of Lieberkuhn reflectors. *Trans. R. Soc. Trop. Med. Hyg.* **17**, 8–9.

JOHNSON, C. G. (1937). The absorption of water and the associated volume changes occurring in the eggs of *Notostira erratica* L. etc. *J. Exp. Biol.* **14**, 413–21.

JOHNSON, H. A. (1937). Note on the continuous rearing of *Aëdes aegypti* in the laboratory. *Publ. Hlth Rep. Wash.* **52**, 1177–9.

LEES, A. D. (1946). Transpiration and the structure of the epicuticle in ticks. *J. Exp. Biol.* **23**, 379–410.

MACFIE, J. W. S. (1915). Observations on the bionomics of *Stegomyia fasciata*. *Bull. Ent. Res.* **6**, 205–29.

MACFIE, J. W. S. (1920). Heat and *Stegomyia fasciata*, short exposures to raised temperatures. *Ann. Trop. Med. Parasit.* **14**, 73–82.

REFERENCES

MacGregor, M. E. (1916). Resistance of the eggs of *Stegomyia fasciata* (*Aëdes calopus*) to conditions adverse to development. *Bull. Ent. Res.* **7**, 81–5.

MacGregor, M. E. (1929). The significance of pH in the development of mosquito larvae. *Parasitology*, **21**, 132–57.

Marchoux, E., Salimbeni, A. and Simond, P. L. (1903). See references in ch. I (*f*), p. 19.

Marshall, J. F. (1938). *The British Mosquitoes*. Brit. Mus. (Nat. Hist.). London.

Mitchell, E. G. (1907). *Mosquito Life*. G. P. Putnam's Sons, New York and London.

Newstead, H. and Thomas, H. W. (1910). The mosquitoes of the Amazon Region. *Ann. Trop. Med. Parasit.* **4**, 141–50.

Nicholson, A. J. (1921). The development of the ovary and ovarian egg of a mosquito, *Anopheles maculipennis* Meigen. *Quart. J. Micr. Sci.* **65**, 395–448.

O'Connor, F. W. (1923). Researches in the Western Pacific. *Res. Mem. Lond. Sch. Trop. Med.* **4**.

Otto, M. and Neumann, R. O. (1905). See references in ch. I (*f*), p. 19.

Parker, H. B., Beyer, G. E. and Pothier, O. L. (1903). A study of aetiology of yellow fever. *Yell. Fev. Inst. Bull.* **13** (*Publ. Hlth Marine Hosp. Serv.*), 21–7.

Peryassu, A. G. (1908). *Os culicideos do Brazil.* Typog. Leuzinger, Rio de Janeiro.

Powers, G. E. and Headlee, T. J. (1939). How petroleum oil kills certain mosquito eggs. *J. Econ. Ent.* **32**, 219–22.

Pryor, M. G. M. (1940). On the hardening of the cuticle of insects. *Proc. Roy. Soc.* B, **128**, 393–407.

Robertson, T. B. (1918). *The Physical Chemistry of the Proteins.* Longmans Green and Co., London and New York.

Roubaud, E. (1927). L'éclosion de l'œuf et les stimulants d'éclosion chez le *Stegomyia* de la fièvre jaune etc. *C.R. Acad. Sci., Paris*, **184**, 1491–2.

Roubaud, E. (1929). Recherches biologiques sur le moustique de la fièvre jaune *Aëdes argenteus* Poiret. *Ann. Inst. Pasteur*, **43**, 1093–1209.

Roubaud, E. and Colas-Belcour, J. (1929). Action des diastases et des facteurs microbiens sur l'éclosion des œufs durable du moustique de la fièvre jaune. *Ann. Inst. Pasteur*, **43**, 644–55.

Roy, D. N. (1940). A study of the bionomics of *Anopheles subpictus* and *A. annularis*. *J. Mal. Inst. India*, **3**, 499–507.

Roy, D. N. and Majundar, S. P. (1939). On mating and egg formation in *Culex fatigans* Wied. *J. Mal. Inst. India*, **2**, 245–51.

Shannon, R. C. and Putnam, P. (1934). See references in ch. V (*a–g*), p. 128.

Simmons, S. W. (1934). Sterilization of blowfly eggs, etc. *Amer. J. Surg.* **25**, 140–7.

Theobald, F. V. (1901). *Monog. of the Culicidae*, **1**, 21.

Trager, W. (1935). The culture of mosquito larvae free from living micro-organisms. *Amer. J. Hyg.* **22**, 18–25.

Trager, W. (1937). A growth factor required by mosquito larvae. *J. Exp. Biol.* **14**, 240–51.

Vishna Nath (1924). The egg follicle of *Culex*. *Quart. J. Micr. Sci.* **69**, 151–75.

White, G. F. (1931). Production of sterile maggots for surgical use. *J. Parasit.* **18**, 133.

Wigglesworth, V. B. (1942). *The Principles of Insect Physiology.* Ed. 2. Methuen and Co. (See also many subsequent editions.)

Wigglesworth, V. B. (1945). Transpiration through the cuticle of insects. *J. Exp. Biol.* **21**, 97–114.

Young, C. J. (1922). Notes on the bionomics of *Stegomyia calopus* Meig. in Brazil. *Ann. Trop. Med. Parasit.* **16**, 389–406.

VII
ECLOSION

(*a*) THE ACT OF HATCHING

No detailed account of the hatching of the egg of *Aëdes aegypti* appears to have been given in the literature. Whilst, therefore, there is still much to be investigated as to the physiological processes involved, a brief account of the act as seen under the microscope will not here be out of place.

As already noted, eggs are normally oviposited and cemented *in situ* with the ventral or flatter surface uppermost in respect to the object on which they are laid. The larva in the egg occupies a position in accordance with the naming of the surfaces here given, that is with its ventral aspect upwards. Since the line of fracture which forms the egg-cap passes further caudal on the ventral than on the dorsal aspect of the egg, it follows that the opening made by detachment of the cap faces obliquely upwards so that looking at eggs that have hatched still cemented *in situ* all the openings to some extent face the observer. Actually there seems to be no difficulty in hatching should the egg by any chance not be in this position or detached and lying loose in the fluid. It is useful to remember, too, that since the crack forming the egg-cap starts on the dorsal aspect, this is the aspect that should be in view if early stages in hatching are to be observed.

The whole process is best watched using the following technique. Eggs from a fragment of pot with massive deposit of non-shrunken mature eggs that are known to hatch shortly after submersion are brushed off into a watch-glass with a fine camel-hair brush, a little hatching fluid added and the eggs rapidly squished in and out with a pipette and teat to ensure that a sufficient number will be sunk in the fluid. Seeing that there are such a number, the watch-glass is moved with a rotatory movement which brings the scattered sunken eggs together in a little heap in the middle. Pipette a suitable number of these on to a hollow-ground slide and add hatching fluid until on sliding a $1\frac{1}{2}$ by $\frac{7}{8}$ in. cover-glass along the slide and over the cavity no air bubble is left. Any moderate excess pushed before the cover-glass will serve as an extra reserve of fluid. If this be done as described the eggs will not be disturbed or moved. Observe under a strong illumination by reflected light, keeping under observation such eggs as have their dorsal surface upwards and the anterior portion of this surface well illuminated. A lamp giving a bright beam of light is very suitable and a warm stage helps, though it is not essential if the room temperature is reasonably warm. It is advisable to use as hatching fluid some undiluted old culture fluid as the evolution of bubbles of oxygen when tap water is warmed is otherwise troublesome. If it is desired to bleach the eggs, they should be covered in the first place with 1 in 5 diaphanol in distilled water for 15 minutes or until at least some eggs show eye-spots and then transferred for a few moments to 1 in 1000 NaOH and thence to tap water. They should be sunk in the diaphanol with a pipette and teat as described above and any egg that is not easily wetted and sinking may be neglected. Select eggs for watching with the egg-breaker well displayed (Fig. 7 (1)).

THE ACT OF HATCHING

The first indication of hatching is the appearance of a slowly widening transverse crack in the egg-shell towards the anterior end of the dorsal aspect of the egg. In the bleached egg on one or two occasions an appearance was seen as if the sharp point of the egg-breaker had previously penetrated the shell, but this requires confirmation. The slit widens slowly and evenly showing in sharp contrast the pale larval tissue. It extends fairly quickly round the egg bringing into view the egg-breaker and, laterally, the two conspicuous dark eye-spots or larval ocelli (Fig. 7 (2)). Active pulsatile movements are now seen within the head as this begins to be exposed, due most probably to muscular movements of the pharynx. Soon it is evident that a cap-like portion of the egg-shell is being pushed off. In a few seconds the bulging head, advancing steadily, pushes the cap before it still attached ventrally so that it is opening like the lid of a box. During this process the cap is firmly impacted like a muzzle over the front of the head up to about the level of the egg-breaker and ocelli. Finally, as the head bulges more and more, the egg-cap, now forced ventrally almost at right angles to the egg axis, is suddenly freed by breaking of the ventral connection and the head with the cap perched on it is rapidly extruded followed by the thorax in a tumultuous-like movement of the soft parts and later by the foremost segments of the abdomen. The appearance in fact is as if some force from behind had pushed the larval parts out from the egg-shell. After a short pause a rapid movement of the now largely free body liberates the larva from the shell and usually detaches the egg-cap from the head. The larva, now with all its hairs erect, usually remains stationary for a moment or two and then swims off. The egg-shell is left as an empty case. Within it when decolorised can be seen a thin loosely attached lining membrane, broken across like the egg-shell itself, the amnion-serosa.

The larva, however, is not yet in a position to take on its normal functions, for it is not until an hour or more from hatching that the head swells from its triangular to a globular shape and the feeding brushes are liberated.

The above is a description of hatching as may be observed in the case of mature eggs submerged in water or other medium. Mention should, however, be made of the appearance seen when hatching occurs under the influence of strong chemicals such as sodium hypochlorite or diaphanol. In this case the larva is usually dead, extruded from the egg with the caudal extremity still lying within the egg-shell. The abdominal segments, closely packed together in the egg, are now widely separated with intersegmental notches. The brushes are expanded and free and the egg-cap is almost always still attached to the egg-shell ventrally (Fig. 7 (6)). It is generally assumed that the chemical has led to the hatching of the live larva, but that this has been killed on emerging. The nature of such hatching is dealt with later.

(*b*) MECHANISM OF HATCHING

In the egg ripe for hatching the fully developed primary larva lies ventral aspect upwards. The head occupies about the anterior third of the egg and almost another third at the posterior end is taken up by the siphon and anal segment with its papillae. In the intermediate portion are the three thoracic and the remaining seven abdominal segments flattened like coins in a rouleau and showing no indication of constrictions between the segments. The long lateral hairs lie directed forwards along the length of the egg, except the prothoracic hairs which are directed backwards. In the mid-line dorsally towards the anterior end of the egg is the egg-breaker and laterally and a little behind its level are the black pigmented ocelli (Fig. 7 (1)).

The egg-breaker in shape and appearance is not unlike a thorn in a young rose shoot, except that it is not curved. It consists of a heavily sclerotised circular plaque which rises conically in the middle to an extremely sharp point (Fig. 7 (4)). Posteriorly the plaque is continued backwards as a rod-shaped extension giving to the whole rather the appearance of a tennis racquet. The hard and thick plaque lies in a circular depression on the dorsum of the head and is surrounded by soft membrane which forms the floor of the fossa (Fig. 7 (5)). It seems clear from the structure that the plaque carrying the conical spine, whilst normally it lies at rest depressed in the hollow and so inoperative, if forced upwards would bring the very sharp point of the spine against the endochorion.

The most obvious way by which the spine of the egg-breaker could be so protruded would seem to be by intracranial pressure and though, as noted by Bresslau (1920) there is a muscle inserted into the plaque (the inner retractors of the *flabella* or feeding brushes) it is difficult to see how this could cause protrusion of the spine unless, acting along with intracranial pressure, it served to draw the spine forwards at the same time as it was protruded. There is just another possibility, namely that these muscles are inserted into the plaque as the only rigid place of origin, the cuticle of the head otherwise being noticeably very soft. It is important that the feeding brushes which at hatching are still folded in the head should be everted and the muscles in question may be more related to this function than to forcing out the spine of the egg-breaker.

Further points regarding the mechanism of hatching may most readily be given as answers to the following questions, namely, (1) How does the larva use the egg-breaker to cut the cap? (2) Does the larva swallow water to increase its bulk when forcing off the egg-cap? and (3) How is final escape from the egg brought about?

HOW DOES THE EGG-BREAKER WORK?

It had previously been assumed by the author that the egg-cap in mosquitoes was cut by the egg-breaker and that the only way that this could be done was by rotation of the head of the larva. In favour of this is the fact that there is no indication in the unhatched egg of

---

Figure 7. Eclosion.

1 Dorsal view of bleached egg showing fully developed primary larva.

2 Lateral view of egg showing first indication of hatching. Note the egg-breaker and larval ocellus showing in the fissure.

3 Larva killed in the egg and extracted. Diaphanol specimen.

4 Lateral view of egg-breaker. Anterior direction is on the left. The line across the drawing shows the level of the general surface of the dorsum of the head.

5 Dorsal view of head of newly emerged larva showing the egg-breaker set in a membranous depression and the bent antennae.

6 Larva 'hatched' in diaphanol. Note extrusion of brushes though the head is as yet not swollen. Only the upper portions of the egg-shell have been shaded to enable position and appearance of the larval abdomen to be seen.

7 Larva immediately after hatching from the egg. Note expansion of the abdomen and general increase in size and length of the larva. An egg drawn at the same time and scale is shown at its side.

8 Head of larva 2 hours after hatching. Same scale as no. 7.

9 Ventral view of head of larva hatched in suspension of Indian ink. Note blackening of brushes, palatum and mandibles.

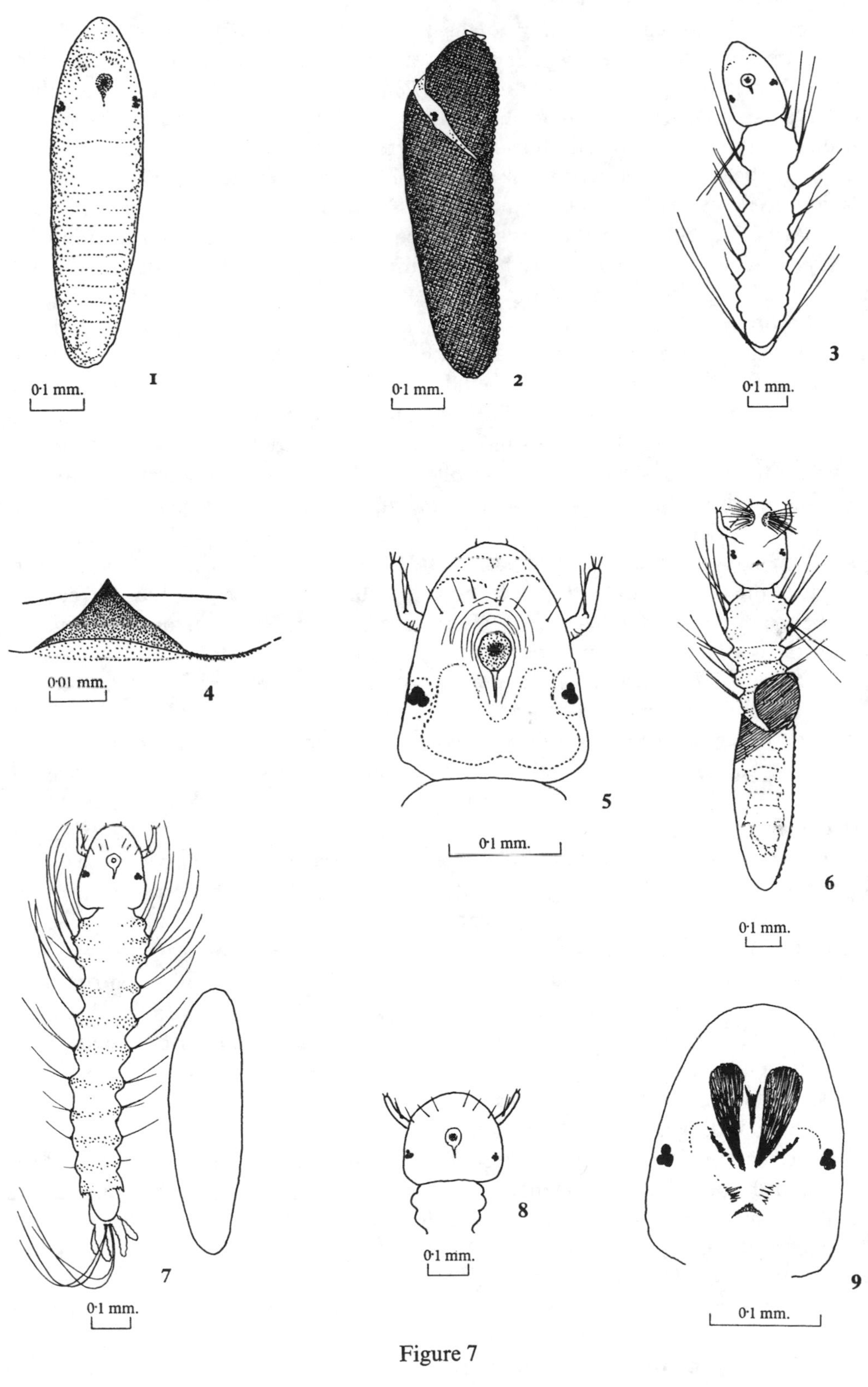

Figure 7

any line of weakness corresponding to the line of detachment of the cap and that in *Anopheles* we know that the larva has the power of rotating the head through as much as 180°. Further, in the case of a mycetophylid (*Bolitophila*) Edwards (1919) notes that the young larva can be observed moving its head up and down and cutting or scratching a slit in the shell. In order to ascertain how the egg-breaker in *Aëdes aegypti* worked, hatching has been observed in eggs bleached with diaphanol to the point where the shell is sufficiently decolorised to allow the larval ocelli and egg-breaker to be seen, a process which evidently had no harmful effect since the larvae escaped from the egg in living condition and seemingly in no way affected. In such preparations movement of the head would be at once evident, but no movement of the egg-breaker preparatory to or during formation of the egg-cap break has ever been seen. All that was seen was that just before the crack started there was an appearance resembling a thin layer of air passing up to the point of the egg-breaker spine. This, however, was found to be due to reflection of light from the very smooth enamel or china-like surface of the egg-breaker at this time. On some occasions it was thought that the point of the spine had been observed to pass through the endochorion, but this requires confirmation. Following the appearance described above a slit appeared at the site of the spine point and passed rapidly round the egg. The spine made no movement.

Since the egg-breaker does not appear to be rotated from its median position during formation of the egg-cap the only conclusions possible seem to be, (1) that the egg-cap is cut some time previous to hatching, (2) that the egg-breaker besides merely perforating the endochorion starts a crack in a particular direction which is followed up by further splitting caused by pressure, or (3) the endochorion under pressure breaks in the way it does once it has been punctured.

Bringing pressure to bear upon eggs in various ways showed that a cap was only to be detached in eggs that were fully ready to hatch, that is those with a conspicuous and presumably fully developed egg-breaker. Eggs that were in this condition when pressed with a needle could usually be made to release an egg-cap in a manner that could almost be described as a 'pop'. Further, examination of the endochorion when larvae were extracted through the posterior end of the egg showed no indication of any previously cut or marked line. It was therefore concluded that mechanical pressure alone could force off an egg-cap once the egg-breaker was fully developed and that natural hatching was possible without any cutting action. It has previously been noted that the endochorion when ruptured by internal pressure (exploded eggs) breaks into longitudinal strips. Nevertheless, it was noted that by crushing eggs some transverse breaking of the endochorion also tended to occur, fragments to a large extent being rectangular in shape showing that this structure had some tendency to split transversely as well as longitudinally.

Seeing that splitting of the endochorion appeared to be started in a fixed direction as a result of perforation by the egg-breaker spine, it was thought that the spine might have a chisel-shaped point, but no evidence of this could be found. Later it was interesting to find that Marshall (1938) in his book *The British Mosquitoes* states that the point of the egg-breaker in the British species other than those of the genus *Aëdes* is chisel-shaped, but that in all species of *Aëdes* it is round.

It therefore seems most probable that the egg-breaker in *Aëdes aegypti* acts merely by puncturing the endochorion and that pressure then is able to start a crack which under continued pressure of the swollen head front is led round the egg in a relatively fixed course. In other words, the structure and forces being the same, what is likely to happen in one egg is likely to happen in all. In favour of this view is the fact that though the cap is roughly equivalent in different eggs its cutting is far from being meticulously accurate as it might be expected to be if prepared and carried out by some more precise method of detachment.

It may be mentioned here in relation to hatching that where eggs have been kept rather wet a certain number may be found with the egg-cap already cut, though the larva has not hatched and often the crack is so narrow as to be easily overlooked. Should the surface on which the eggs are lying then become dry the larva dries and perishes, a fact later sometimes made apparent by a little tuft of mould growing at the crack. Where, however, eggs have been kept in a just critical state of dampness it is possible such larvae may survive, a fact which it is thought may account for the almost immediate hatching of larvae in some cases when eggs are submerged. This, however, requires confirmation.

DOES THE LARVA SWALLOW WATER DURING ECLOSION?

That there is some active movement within the head when this is first exposed in the widening crack in the egg-shell has already been referred to and ascribed to the movements of the pharynx in the swallowing of fluid. That insects in hatching swallow amniotic fluid and later air to cause increase in volume is well established (Sikes and Wigglesworth, 1931). Seemingly the *Aëdes* larval head occupies the whole available space in the anterior portion of the egg, the clear transparent area in front of the darker area formed by the mouth-parts and other structures as seen in the bleached egg being occupied, not by amniotic fluid, but by the swollen clypeal region of the larval head, that part in fact which later is seen covered by the dislodged cap. Further, the swallowing movements appear most active when the head is slowly bulging forward against the separating egg-cap. It seemed probable therefore that the insect at this time is swallowing water from the outside medium.

To ascertain whether this was actually so eggs were allowed to hatch in weak suspensions of Indian ink. Under such conditions, whether as a result of toxicity of the suspension, or by the blocking effect to be described, practically all larvae died in various stages of escape from the egg and most usually with the cap still firmly impacted over the front swollen portion of the head as previously described, that is, covering the parts just up to the level of the egg-breaker and ocelli (Fig. 7 (2)). At this stage the larval head is unexpanded and triangular in shape, the brushes and the bent down hairy labral area (epipharynx) being tucked into the pre-oral cavity as when in the later stages the larva 'swallows' its brushes. Only when the head within an hour or two has swollen into a globular state are these organs free and everted. In larvae hatched in Indian ink suspension the compacted brushes with the hairy epipharynx and the bases of the mandibular hairs appear quite black from accumulation of the black suspended particles. No dark material at this stage could be detected in any part of the alimentary canal. It would seem, therefore, that fluid has been taken in but filtered through the compacted brushes. In some cases where larvae were still alive and removed to clean water they were observed attempting to comb out the brushes with the mandibular combs which at this stage are relatively very large and conspicuous. In time some succeeded in clearing the obstruction with the brushes still impacted and particles of the ink were then seen in the proventriculus or lower down the alimentary canal. It was clear that active imbibition of fluid had taken place before the larva had left the egg, but not clear whether such imbibition had occurred in the earlier stages when the cap was first being pushed off.

Since particulate matter was thus prevented from entering, the effect of hatching in a strong solution of cochineal was tried. This stain was used by Bishop and Gilchrist (1946) for testing food choice by the adult and was found very suitable for the present purpose. The solution, made up by a dispensing chemist, was presumably an alcoholic tincture, but it did not prevent hatching in a 1 to 4 dilution. Larvae pipetted out as rapidly as possible as they hatched into water heated to 80° C. invariably showed the alimentary canal from behind the thorax to the seventh abdominal segment distended with pink fluid. The same condition was found in larvae which had not emerged beyond the first stage of hatching, that is with the cap in process of being pushed off as in Fig. 7 (2). It is certain, therefore, that external fluid is being swallowed during the pushing forward of the cap by the head, further strengthening the conclusion that the pulsatile movements seen in the head at this time are due to muscular movements by the pharynx.

In a 1 to 2 dilution of the stain no larvae hatched, but a number of eggs were found in the stage referred to above, with the egg-cap just beginning to be displaced. These larvae were all dead, but on being extracted from the egg-shell they showed the broad band of pink colour down the abdomen indicating distension of the gut with external fluid. Presumably the larvae began hatching, but having reached a certain stage had swallowed sufficient alcohol to cause their death. This was found to be the readiest way to demonstrate imbibition as it is difficult to catch specimens early enough once they are seen to be hatching to exclude the possibility of their having taken in fluid at some later stage. By the above procedure eggs can be put to hatch in the stain and after a suitable interval examined for the above described stages.

WHAT IS THE MECHANISM OF PROPULSION IN THE LATER STAGES OF HATCHING?

When, with extension of the fissure in the egg-shell, the ventral connection of the egg-cap with the rest of the endochorion finally gives way, there occurs a rapid completion of the process of hatching almost as if the larva were being pushed out by some propelling force behind. This effect continues until the thorax and a part of the abdomen have left the egg-shell, a very similar appearance to that seen in the egg artificially hatched by diaphanol, fixatives, etc. From then on the act of leaving the egg is clearly only a matter of ordinary locomotion. Interest lies in what caused the initial extrusion. The larva whilst in the egg measures in length about the same as the egg, that is about 0·7 mm. of which the head occupies a third. The abdomen and thorax occupy about one-half of the egg length and the abdomen alone about 0·26 mm. If the larva be killed by heat, or other means which coagulate it, and then extracted from the egg-shell, the abdomen is cylindrical in shape with smooth sides, the segments being compressed and resembling a pile of thin coins. This is the condition when the egg-cap has been partially pushed off. Then a few seconds later, when the final connection of the cap with the rest of the shell gives way, the whole appearance of the abdomen changes so that it is now elongate with deep intersegmental notches. In other words it calls to mind the toy jack-in-the-box in which a figure with compressed spring leaps upwards when the cover of the box is opened. From measuring about 0·26 mm. the abdomen now measures quite double that length and its hinder half still occupies most of the length of the egg-shell.

When first hatched the larva is unable to move or use its feeding brushes. The head is bluntly triangular in shape and the antennae elbowed (Fig. 7 (5)). The cuticle appears to be very soft and that on the dorsum of the head is seen in sections to be thrown into wrinkles. The only sclerotised parts are the egg-breaker and the triangular mental sclerite. The long lateral hairs soon take up their normal position. During the first four hours the head, and especially the anterior portion that will become the fronto-clypeus of the later larval stages, swells greatly until it becomes globular (Fig. 7 (8)). During this process the feeding brushes are everted and begin to function. As shown by Wigglesworth (1938) no air is at first present in the tracheal trunks, nor has air appeared to take any part in the hatching process. Very shortly after hatching the young larvae seek the surface and make contact with the surface film. Of a number of newly hatched larvae observed at 27° C. in water of depth 1, 3 and 6 cm. the majority had attached themselves to the surface film in 10 minutes in depth 1 cm. and smaller proportions at 3 and 6 cm. Further information regarding the first instar larva and the changes undergone after hatching will be found later when the different larval instars are described.

(*c*) CONDITIONS AFFECTING HATCHING

SUBMERGENCE

A vital essential to hatching is that eggs should be submerged. It is necessary in view of such a statement, however, to guard against what may be a serious misunderstanding. For though eggs require submergence for the act of hatching, submergence before the eggs are ready for hatching may be fatal to them. Eggs sunk immediately after oviposition may not even undergo the normal processes of swelling and darkening. Even at a later stage 'sunk eggs' under some conditions may not develop and unless eggs, before being placed to hatch, are known to be conditioned (see p. 45) serious error may arise in interpreting their subsequent behaviour when placed to hatch.

Whatever the conditions, hatching never seems to occur unless the eggs are actually bathed in fluid. Though eggs do not hatch in the absence of fluid, in a damp atmosphere they may show incomplete hatching should there be condensation. Under such circumstances eggs are often found with the egg-cap wholly or partially separated. In such circumstances, even though the egg is wet, larvae have never been seen emerged from the egg-shell. If desiccation now occurs, the larvae soon die.

Even incomplete submergence appears to be, in some degree at least, inadequate. Buxton and Hopkins (1927) refer to the fact that they found it advisable, to ensure hatching of eggs on filter paper, to sink the paper beneath the surface. It is a common experience to see eggs on filter paper hatch only where they have become submerged and remaining unhatched on the paper where it is merely wet.

A possible instance of lack of complete submergence is the frequent failure of floating eggs to hatch if undisturbed. Young (1922) notes that in Manaos the majority of floating eggs did not hatch unless submerged. Of eggs floating for 20 days without hatching 96 per cent hatched after being submerged. Fielding (1919), however, found no difference, eighteen of twenty-one submerged eggs and twenty-one of twenty-three floating eggs hatching on the same day.

ECLOSION

In my experience I have found that floating eggs left undisturbed may remain largely unhatched and that their submergence leads to hatching. But as with many other features in the hatching of the eggs of this species no hard and fast line seems to hold good.

With properly conditioned eggs laid by well-reared broods I have found, as have Shannon and Putnam (1934), that the number failing to hatch on submersion under appropriate conditions is very small. As noted by these authors a certain number of the small proportion not hatching on the first submersion may do so if re-dried and again submerged.

Such experiences emphasise the advantage of arranging for oviposition, as is the natural method of choice, to be carried out in porous earthenware receptacles where the eggs are laid *in situ* under optimum conditions. If other methods are used, for example on filter paper, etc., allowance must be made for possible adverse effects.

AGITATION

Many authors have referred to the effect of agitation in bringing about hatching. Dupree and Morgan (1902), working with the eggs of several species of *Aëdes*, noted that agitation seems in some way associated with hatching. Eggs of many species, after remaining upon the surface of the water or upon the bottom of breeding vessels for days, hatch if removed to a phial and shaken, but if left undisturbed will remain unhatched for months. Young (1922) found in Manaos that when eggs of *A. aegypti* were submerged before they were ready to hatch and left undisturbed, they remained unhatched provided no food was added. Buxton and Hopkins also emphasise the fact that agitation is an important stimulus to hatching, shaking being sufficient when eggs were ready for hatching to cause them to hatch immediately.

The following count made of the hatching of floating eggs when subjected to agitation will serve to illustrate such an effect. The hatching in the shaken eggs occurred within less than half an hour from the shaking.

1. Still floating in original vessel	20 per cent
2. Submerged by sinking with a needle	44 per cent
3. Transferred to test tube and shaken	93 per cent

The fact that Young found that not only floating, but also sunken eggs left undisturbed, remained unhatched for 5 months indicates that submergence alone may not ensure hatching under certain circumstances and that absence of agitation may be unfavourable to hatching.

What may be considered as a form of agitation is the nibbling of eggs by larvae already present in the medium as described by Fielding (1919) and others. Apart from nibbling, the presence of larvae in active movement would also act as a form of agitation. This may be the explanation of the observation made by Thomas (1943) that when hatching eggs under sterile conditions, the degree of hatching was increased as the number of eggs was greater. Thus with one egg per tube 6 per cent hatched, with two eggs per tube 21 per cent and with fifty-three to sixty eggs per tube 80–84 per cent hatched.

NATURE OF MEDIUM

There is very general agreement that hatching occurs less readily in clean tap water or distilled water than in media rich in organic matter, such as infusions or fluids containing food material (Bacot, 1917; Young, 1922). Bacot says that in water of low organic content the hatching of dry eggs was often delayed for weeks or months. In two instances eggs hatched after 5 months' submersion. Addition of foul contaminated water, on the other hand, caused a high percentage of hatching in a brief time. To explain such effects Bacot considers the stimulus might be smell or some closely analogous stimulus. Searching for favourable media to produce maximum hatching Buxton and Hopkins tested the effect of dilute solutions of tannin (0·1 per cent), formaldehyde (0·01 per cent), maltose, mannite, saccharose, glucose, glycerine, methyl and ethyl alcohol, citric acid and lactose. With the exception of the last two, the effect of which was uncertain, they found that dilute solutions

Table 5. *Experiments with dried eggs*

	Hatched by		
De cription of experiment	15 minutes	30 minutes	Later
Eggs 4 days dry			
Old culture fluid	Numerous	Numerous	Numerous
Clean tap water	Some	Less numerous	Numerous
Eggs 20 days dry			
Old culture fluid	*c.* 200	Numerous	Numerous
Clean tap water	0	*c.* 100	Numerous
Distilled water	0	A few	Numerous
Eggs 136 days dry			
Old culture fluid	2	10	—
Clean tap water	0	8	—
Distilled water	0	8	—

of all these substances were more effective than distilled water, but none so effective as grass infusion. They found that if the fluid is favourable (for example, grass infusion) there is no difference in the hatching of eggs which have been matured and then dried and those which have not been dried. But if the medium is unfavourable there may be a great difference in the two cases, eggs that have never been dried hatching readily in distilled water, whilst dried eggs which had hatched readily in grass infusion failed to do so. Shannon and Putnam (1934), however, found that a large proportion of eggs hatched even in distilled water.

Our own observations have been much like those of these authors. In general the results in old culture fluid are better than in tap or distilled water. There is also a noticeable difference in the rapidity and uniformity with which hatching takes place in old culture fluid. That properly conditioned eggs hatch to a considerable extent in distilled water is, however, clear (see Table 5). In general it has been the practice to use some diluted old culture fluid when placing eggs to hatch.

PRESENCE OF MICRO-ORGANISMS

Bacot (1916) found that water charged with organic matter and swarming with bacteria and yeasts exerted a powerful stimulus to the hatching of eggs that had not responded to immersion in clean water. Bacot (1917) and Aitken and Bacot (1917) sterilized eggs by exposing them for 5–10 minutes to 2 per cent lysol followed by washing in boiled water. Eggs thus sterilised and kept in sterile water failed to hatch over long periods, but did so when exposed to the action of yeasts or bacterial cultures. Roubaud and Colas-Belcour (1927) found that in sterile or little contaminated water eggs may remain months without hatching. Addition of water rich in organic matter and containing bacteria and yeasts brought about rapid hatching, as also did pure bacterial cultures. On the other hand,

Barber (1927) sterilised eggs by momentary immersion in alcohol (2–3 minutes in 80 per cent) and found no indication that addition of bacteria promoted hatching. Rozeboom (1934) found that of fresh moist eggs 35 per cent hatched in distilled water, 51 per cent in sterile filtered medium and 82 per cent in contaminated medium. Hinman (1932) cultured eggs that had been superficially sterilised in hexyl resorcinol and found only a few of several hundred eggs showed on hatching a growth of organisms. This author appeared to have no difficulty in hatching larvae from such eggs. Trager (1935), using eggs sterilised by the method of MacGregor and later by exposure for 15 minutes to White's fluid (see under sterilisation of viable eggs), found such eggs hatched readily in the sterile medium used (killed yeast plus liver extract). De Meillon, Golberg and Lavoipierre (1945) apparently found no difficulty in rearing larvae from sterilised eggs. It would seem, therefore, that absence of micro-organisms does not necessarily prevent eggs from hatching.

EFFECT OF ENZYMES

Roubaud and Colas-Belcour found that animal and vegetable digestive diastases (pepsin, trypsin, papaine) acted in the same way as cultures of organisms or their filtrates in bringing about hatching. Hatching with these substances took place in a few minutes when in concentration capable of killing the larvae after hatching. After autoclaving they were ineffective.

COOLING

Fielding notes that of 228 eggs which failed to hatch on immersion varying from 5 to 19 days, 128 were induced to hatch after periods of cooling (placed in the ice chest). One hundred and five of these hatched within 2 hours. Bacot (1917) had also noted that a fall of 6–10° F. usually causes a percentage of eggs to hatch, though as a rule the majority do not respond. Other authors have also drawn attention to the effect of cooling in bringing about hatching. Atkin and Bacot (1917), however, say that change in temperature of the water seldom caused more than a few dormant eggs to hatch and sometimes had no effect.

EFFECT OF CHEMICALS

Dupree and Morgan (1902) noted that *Aëdes* eggs which normally remain unhatched will hatch if placed in 1 or 2 per cent formalin. Howard, Dyar and Knab (1912) refer to Agramonte as noting that wood ashes hasten hatching. Fielding found that after short exposure to lysol resistant eggs hatched on being submerged in water. Soft soap 1 in 8000 had the same effect. Roubaud (1927) notes that sulphuric ether, potassium permanganate and hydrogen peroxide can cause hatching, but the effects are inconstant. On the other hand, sodium hypochlorite has an effect comparable with that of the diastases. Eggs kept three months in water without hatching, hatch in less than 24 hours in eau-de-Javelle 1 in 1000 and in 1–6 days in 1 in 10,000. The newly emerged larvae die rapidly in the solutions.

In the above connection it is important to recognise a distinction between what might be called normal hatching and that which occurs with chemicals killing the larvae in the act. Diaphanol causes hatching if extrusion of the larva be considered as such. Normal eggs exposed for a short time to diaphanol and washed may still hatch normally with the living larva, though the egg-shell has been bleached. Eggs left in diaphanol (1 in 5) for half an hour are often not hatched at the time, but are found next morning protruded from the egg-shell nearly always with the hinder part of the abdomen still lying within the egg-shell. The brushes are commonly expanded, even though the larval head has not swollen, and nearly always the egg-cap, instead of being completely detached is still adherent ventrally (Fig. 7 (6)). It would appear doubtful whether such 'hatching' is a vital phenomenon. The fact that the egg-cap has been cut need not indicate living action by the larva, as we know from what has been said above that the cap in an egg ready for hatching can be detached by pressure on the egg by a needle.

REDUCED DISSOLVED OXYGEN

Gjullin, Hegarty and Bollen (1941) found that a reduction in dissolved oxygen caused hatching of eggs of *A. vexans* and *A. lateralis*. King and Bushland (1940) similarly found that a large percentage of eggs of *A. aegypti* will hatch with a normal amount of dissolved oxygen, but only a small percentage of eggs 20 days or more old will do so unless the dissolved oxygen is lowered. Barber (1928), on the contrary, found that eggs embedded in agar or placed in broth recently boiled under vaselin failed to hatch.

Table 6

	Number of eggs hatched at		
	45 minutes	65 minutes	Overnight
Aerobic			
Tap water	4	*c.* 50	Nearly all
Distilled water	10	*c.* 50	Nearly all
One-third old culture in tap water	*c.* 400	—	Nearly all
Anaerobic			
Boiled tap water	*c.* 100	Some hundreds	Nearly all
Boiled distilled water	*c.* 100	Some hundreds	Nearly all

It was thought in view of these findings that a strictly anaerobic test as used for cultures of anaerobic organisms would be of interest. Eggs floated on cover-glasses and kept some days in a moist chamber were so arranged that they could, when anaerobic conditions were fully established in a Bulloch's apparatus, be tipped into boiled water and boiled distilled water and the results compared with eggs of the same batch treated aerobically. The results are given in Table 6.

The larvae in the anaerobic experiment examined next morning were all dead, many half emerged from the egg. There were, however, many seen active and swimming about on the previous evening. The percentage of eggs hatched was 100 per cent in the boiled tap water and 98 per cent in the boiled distilled water. It was clear that anaerobic conditions did not militate against hatching, even if they were not the direct cause.

HEREDITARY INFLUENCE

Recently Gillett (1955*a*, *b*) has studied variation hatching response of *A. aegypti* eggs from the genetical point of view. Using removal from water for one minute as a moderate stimulus, this author found differences in the response percentage to hatching not only in the same batch from a single female, but in batches from the result of pairing different females and males. Especially marked was the difference in eggs laid by two strains of the species, a strain from Nigeria maintained from 1947 (strain W) which gave low response, and one from Tanganyika started in March 1953 (strain E) which gave a high response often approaching 100 per cent, whether used alone, or as females or males with the opposite sex of strain W. The author considered that response of the eggs to hatching was mainly controlled by an inherited property, though the fact that the depth of diapause may vary within very wide limits in a single batch led him to suspect that a multiple factor mechanism is involved.

(*d*) INCUBATION PERIOD AND HATCHING TIME

INCUBATION PERIOD

In many of the earlier observations in the literature the time taken for eggs to hatch is commonly given as from 2 to 4 days. It is evident that most of such statements relate to the time when eggs have been left to hatch from time of oviposition under conditions suitable for hatching. Actually there can be distinguished no less than four periods connected with hatching. These are: (*a*) The period it takes the larva to be formed and before which the egg cannot hatch under any circumstances. This has been termed by Buxton and Hopkins maturation time, but may perhaps be most simply termed *incubation period*. (*b*) The time during which the egg, even after full development, remains unhatched because the conditions for hatching have not been available and for which the term *diapause period* may be used, covering such period as the egg has been left dry or wet as the case may be. (*c*) The time taken for the egg to hatch when placed under such conditions as cause it to hatch. (*d*) The time taken for the act of eclosion. As this last is at most some minutes it need not be further discussed here.

For the incubation period Marchoux, Salimbeni and Simond give 2–3 days from oviposition at 27–29° C. and 4–5 days at 25–26° C. H. A. Johnson (1937) gives 100 hours as the time at 70–75° F. (21–24° C.) from oviposition until the time the primary larva has been formed and is ready for eclosion. Shannon and Putnam give as time of development to the pre-hatching state: 2–3 days at 25° C., longer under 25° C. and 4–5 days at 23·5° C. These authors have also made a very detailed and special investigation as to the time at which eggs kept moist and exposed to air (that is their time of conditioning) hatch on being submerged. At 24° C. larvae conditioned for 40–60 hours still took 18 hours. In those conditioned for 60–81 hours the first larvae hatched in 8 minutes on being submerged, but for all to hatch took another 24 hours. When conditioned for 87–111 hours all hatched in 10 minutes.

The following are the times for incubation period as determined for a number of temperatures:

Temperature (° C.)	Time (hours)	Round figures (days)
28	74	3
25	96	4
23	114	5
18	—	12
7	Development inhibited	
1	Development inhibited	

There is thus no very marked drop in the rate of development as from 28 to 25° C. and only a gradual lengthening of the period to 23° C. But between the last mentioned temperature and 18° C. there is a rapid and disproportionate lengthening and at 7° and lower there is complete inhibition.

The above relates to eggs as laid by the insect on porous earthenware. As previously noted floating eggs may not hatch in the normal time. This, however, is not due to delay in development, for such eggs when submerged and shaken hatch in a few minutes. Eggs submerged from oviposition, according to Shannon and Putnam show retarded development, and as noted by Howard, Dyar and Knab eggs sunk at oviposition are liable to be rendered non-viable.

HATCHING TIME

Once hatching has started as shown by appearance of the fissure in the egg-shell completion of the process quickly follows, usually within a few seconds or minutes. The time from submergence to appearance of the larva may, however, vary from a few minutes to 24 hours or more. With relatively fresh eggs, including those kept up to a month or so in the dry condition, hatching time is usually a matter of minutes or at most an hour or two.

Table 7. *Number of larvae hatching at intervals of* 20 *minutes from eggs kept various times in the dry state and then submerged*

Minutes from submersion	Number of days eggs kept in the dry state						
	7	8	20	66	69	75	77
15	Very many	1178	*c.* 200	—	—	—	—
20	—	—	—	4	0	—	0
30	Less	1604	Larger number	—	—	—	—
40	—	—	—	5	—	—	7
60	Still fewer	841	Still larger	—	*c.* 100	Many	72
80	—	—	—	—	—	—	70
100	—	—	—	—	—	—	39
120	—	1780	—	23	Fair number	Many	25
140	—	—	—	—	—	—	22
160	—	—	—	—	—	—	15
180	—	—	—	—	—	—	7
220	—	—	—	—	—	Much fewer	1
Over 240	—	3	—	38	Many	—	3
Over 480	—	—	—	—	—	—	—

Bacot (1916), after storing eggs 50 hours under moist conditions, found hatching took place in the majority of cases within 30 minutes and that eggs dried 1–7 days hatched out to the extent of 58–84 per cent in from 1 to 4 days. Shannon and Putnam state that well-conditioned eggs on submergence hatch in from 10 seconds to 10 minutes. Conditioned eggs dried 1–4 weeks began to hatch in 30 minutes, but in water in which food had lain for 12 hours they hatched in 7 minutes. With floating eggs eclosion was irregular, requiring 1–5 days or more. Eggs laid round the water's edge were more favourably situated.

My own experience has been very similar to that of Shannon and Putnam. Eggs kept moist for a few days after the period necessary for maturing hatch almost at once. Usually with a pot under such circumstances swarms of larvae are liberated within 5 minutes. Older pots commonly take about 20 minutes for the first larvae to appear and pots kept many months do not start for some hours and may take 24 hours or more for hatching to be complete. In Table 7 counts are given of the number of larvae hatching at intervals of 20 minutes from eggs kept dry for various times. In all cases the fragments of pot with the eggs were placed to hatch in one-third old culture fluid in tap water at 27–28° C.

The effect of temperature on time of hatching does not greatly differ down to about 18° C. Some interest attaches to temperatures lower than this and especially to that at which hatching is inhibited. Marchoux, Salimbeni and Simond give 20° C. as the minimum temperature at which hatching takes place. This, however, probably relates to eggs placed to hatch shortly after oviposition, that is it might be the effect on development.

Conditioned eggs were found to hatch freely at 17° C. (water temperature). A fragment of pot placed overnight to hatch in water cooled to this temperature, and later maintained at it, gave a count of 239 shells and four intact eggs. Massive hatching has also been observed at as low as 13° C., at which temperature the young larvae are still able to live and are motile. Mature eggs placed to hatch at 12° C. showed, however, no normal hatchings up to 48 hours, though at this and even much lower temperatures an occasional dead larva is seen extruded from the egg-shell as previously described. At 7° C. in one experiment a single apparently dead larva was seen on the fifth day and a few macerated larvae on the ninth day. In another experiment one motionless larva was seen on the fifth day and on the tenth day 250 unhatched eggs and four dead half emerged larvae were counted; in still another trial two dead larvae were seen at 48 hours and a few more on the fourth day. These eggs were all conditioned and mature, so that no question of interference with development was at issue and the results relate only to the act of hatching.

At 1·0° C. there were no hatchings of living larvae from two samples of mature eggs up to 9 and 18 days respectively. The fact that a few characteristic half-emerged larvae were still present strongly suggests that such were not a living phenomenon, but like the result of chemicals previously described were artifacts. The act of hatching as shown by these results can take place at a temperature as low as 13° C., below which temperature hatching is inhibited.

(*e*) RESIDUAL EGGS

No account of conditions affecting hatching of the eggs of *A. aegypti* would be complete without a consideration of eggs which do not hatch, or which do not hatch normally, but do so only after abnormally long intervals of time or by some special stimulus. Eggs behaving in this way have been referred to as 'resistant eggs' as a result largely of Roubaud's work. Roubaud (1929), however, terms them 'durable' or sometimes 'inactive' eggs. Buxton and Hopkins (1927) use the term 'residual eggs' and as this designation seems more precisely to describe such eggs, their term has been here adopted. By 'residual eggs' is here meant eggs which are still found unhatched, though viable, when other eggs with which they have been associated, usually the great majority, have hatched normally.

Since observations on such eggs have bulked very largely in the literature, it is very desirable to start with a clear appreciation of the facts. In the first place, most observers' experience has been that under normal conditions the great majority of mature eggs will hatch within minutes or hours if submerged. Buxton and Hopkins, speaking of their massed results, state that 80 per cent of eggs hatched by the first and 95 by the second day and they also make it clear that the great majority of eggs will hatch in any medium. The same would appear to have been the experience of Shannon and Putnam who lay little stress on the small percentage of eggs that do not hatch when submerged under suitable conditions. The chief author with a contrary opinion is Roubaud who has devoted much study to the phenomenon and whose conclusions succinctly stated may be briefly given almost in his own words:

(1) Certain eggs contain 'active' larvae ready for spontaneous hatching in from 3 to 4 days. Even in sterile water these 'active' eggs hatch. This takes place at the least excitation (mechanical agitation, cold, changes in the fluid, etc.). These eggs are unsuitable for long preservation in the dry state.

(2) 'Les œufs durable', more numerous than the above, are unsuitable for spontaneous hatching. They contain larvae which preserve a state of latent life in the dry egg, floating egg or eggs in pure water. Diastases and excitants, physical or chemical, start the process of hatching of such 'inactive' larvae and call forth resumption of active metabolism.

Roubaud further states that if development of the egg in the mother is effected rapidly and if the batch is reared normally and rapidly the eggs produced are in the majority or totality 'active' eggs. On the other hand, if ovular development is tardy in the mother the majority of eggs are 'inactive'. Even in the same female some eggs may be 'active' and others 'inactive'.

Residual eggs have been seen by us under two conditions, namely, as unhatched eggs left in egg-pots after these have been submerged overnight or longer or as eggs left by isolated females in tubes which have failed to hatch. In both cases, whilst resistant eggs in the true sense may well occur, there are two serious causes of possible error. These are (1) that dried collapsed and dead eggs may swell up and appear to be still viable; and (2) the assumption that an egg hatched by some chemical or other exciting agent, but where the larva has been seemingly killed on emerging, is necessarily a vital act. The following experiences may be given.

When rearing *A. aegypti* on a large scale the earthenware pots used for egg-laying were at an early stage in the work reset after being placed for hatching with the object of ascertaining what proportion of eggs were still left unhatched. On examining the residue in such pots, sometimes after 5 or 6 days' resetting and when no larvae were appearing, apparently intact and still unhatched eggs were usually present in considerable numbers. Examined, however, after bleaching with diaphanol most of such eggs were found to contain only debris or macerated embryos of different ages, being in fact collapsed eggs reswollen as described in a previous section. When care was taken to see that the pots after oviposition were kept moist sufficiently long to ensure conditioning, the appearance of large numbers of such apparently unhatched eggs ceased. A particularly interesting condition was seen in pot 981/21. 8. 46(2), that is, one of two pots removed from the cage after oviposition on the date noted. This was put to hatch on 26. 3. 47 (7 months later) but failed to show any larvae. Yet large numbers of seemingly intact eggs were present. They contained the usual macerated embryos and some with eye-spots. Left overnight in diaphanol it showed many 'hatched' larvae with the dead larva projected from the egg-shell, but with the caudal extremity in position and the egg-cap still attached ventrally. The pot was found to have been removed with third-day eggs only, for example, it must have contained only one night's eggs. The pot would have kept moist for some time, but just not long enough to allow the eggs to reach full development and resistance before being desiccated. Yet after seven months they were hatched by what is termed by Roubaud 'stimulants irrésistibles'. Seeing that these eggs must have only just become conditioned and not very resistant it is almost certain that the action of diaphanol was exerted on larvae already dead and desiccated.

What seems to be proof that such hatchings are artifacts is that eggs have frequently been seen after a short exposure to diaphanol with a crack in the egg-shell indicating the first stage of hatching but with the larva dead, evidently killed by the chemical. Observation showed that the larvae were dead, but next morning it was quite common to find the same specimen with the larva fully extruded. If what has been said as to the mechanism of cutting the egg-cap be correct, there is nothing mysterious in the forcing off of the cap in an egg with a mature dead larva provided the egg cutter is fully developed and sharp and the necessary pressure is brought to bear. It is quite conceivable, therefore, that a chemical could cause sufficient swelling of the body of a dead larva, or for that matter of a living larva that might otherwise have been unable to hatch, to bring about the appearances shown. It has already been noted that there are differences to be seen as between the natural hatched larva and the artificial condition referred to.

(*f*) HATCHING RESPONSE OF NORMAL EGGS

That eggs kept under normal conditions hatch on reaching full development if submerged is a common phenomenon and requires no comment. That in eggs subjected to desiccation the primary larva passes into a state of suspended animation which may last 6 months or more, though a striking fact, is also a condition commonly seen in insect life and recognised under the term diapause. A discussion on the large question of diapause would here be out of place. It occurs very commonly as a result of reduced temperature, but also of restricted food, drought, season or other condition. Such a state may occur in the adult insect, but also in the egg. What is unusual in the *A. aegypti* egg is that the state of diapause does not occur in some stage in embryonic development, but as a condition affecting the fully developed larva. Of interest in this connection is the nature of the stimuli which are known in other insects to break diapause. Some of these strongly recall the stimuli already mentioned as bringing about hatching. The one essential stimulus and the one which in the great majority of eggs acts apparently without need of any further assistance is submergence. In most cases other influences appear to be largely adjuvant, though they are nevertheless clearly effective and may much modify the normal response. Some of these influences, such as cold, agitation, action of micro-organisms, presence of larvae, suggest the kind of mechanical or physical stimuli already referred to as effective in diapause in other insects. Reduction in dissolved oxygen could be a cause underlying many of the other observed stimuli. Thus cooling enables water to take up more oxygen and thus reduces the degree of oxygen saturation. Shaking might seem at first a means of increasing dissolved oxygen. But if tap water be shaken it might, at a warmer room temperature, give up rather than take up oxygen. Obviously fluids containing organic matter and bacteria would be ideal for providing a medium with a low dissolved oxygen content. Reduction of oxygen appears to be an important stimulus, but how far it is the essential stimulus, as has been claimed, it is not possible at present to say.

(*g*) HATCHING RESPONSE OF RESIDUAL EGGS

How little importance has been attached by many observers to resistant eggs has been exemplified in the results already given. Shannon and Putnam suggest that Roubaud's material was a different strain from the Brazilian strain which they themselves were using.

On the other hand, the experience of too many authors (Bacot, Fielding and others) has recorded results not unlike Raubaud's and it would seem more probable that the solution lies in the methods of feeding and culture used. This would be in keeping with Roubaud's hypothesis of asthenobiosis. On such a hypothesis resistant eggs might be regarded as eggs in which the primary larva is in some way deficient in power to puncture the egg-shell or in responsiveness to stimuli so that submersion alone does not cause hatching, or not with the same facility as with normal eggs. Such an hypothesis also receives some support from observations by Gillett (1955*a*, *b*) that there may be differences in hatching response as between local strains of *A. aegypti*, the less effective eggs of two strains experimented with being, as judged from the author's description, those of an otherwise less vigorous strain. Differences in feeding activity and other life processes have been recorded as characterising strains from different geographical areas and less vigorous strains may conceivably be more prone to lay asthenobiotic eggs.

Whether resistant eggs are a provision for enabling the species to tide over unusually long periods of drought seems very doubtful since normal eggs appear already well adapted for this and procedures taken in the laboratory to cause hatching may merely cause eggs to hatch that would in nature be casualties.

It is perhaps desirable briefly to sum up the general position, so far as the author's researches have taken him, regarding resistant eggs of *A. aegypti*.

(1) The conditioned eggs of well-bred females when submerged rarely fail to hatch.

(2) The condition of dormancy of the primary larva in the egg is very similar to the diapause state in various insect species and as such its termination is dependent on some form of stimulus. Normally in *A. aegypti* this appears to be submersion, but in certain cases there are evidently accessory stimuli which may hasten, or even be necessary, to make cessation of the diapause state effective.

(3) All deductions based on hatching by 'stimulants irrésistibles' (hypochlorite, etc.) require revision in view of the fact that such hatching does not necessarily indicate that the eggs contained a living larva, or one that would ever have hatched in nature.

(4) Roubaud's conclusion that the cause of 'durable' eggs is asthenobiosis and due to inadequate vitality in some form seems very probably correct. A too hard egg-shell, a too poorly developed hatching spine or a too feeble larva due to asthenobiosis could each be a cause of deficient hatching.

(5) It seems doubtful whether resistant eggs are a provision for tiding the species over periods of drought, and it seems more probable that such eggs under natural conditions might be fatalities and that their hatching in the laboratory might be an artifact.

REFERENCES*

BARBER, M. A. (1928). The food of culicine larvae. *Publ. Hlth Rep. Wash.* **43**, 11–17.

BISHOP, A. and GILCHRIST, B. M. (1946). Experiments upon the feeding of *Aëdes aegypti* through animal membranes, etc. *Parasitology*, **37**, 85–100.

DUPREE, J. W. and MORGAN, H. A. (1902). Mosquito development and hibernation. *Science*, (n.s.) **16**, 1036–8.

EDWARDS, F. W. (1919). A note on the egg-burster of eucephalous fly-larvae. *Ann. Mag. Nat. Hist.* (9), **3**, 372–6.

* For authors mentioned in the text, but not given here, see references to chapter VI.

GILLETT, J. D. (1955*a*). Variation in the hatching response of *Aëdes* eggs (Diptera: Culicidae). *Bull. Ent. Res.* **46**, 241–54.

GILLETT, J. D. (1955*b*). The inherited basis of variation in hatching response of *Aëdes* eggs (Diptera: Culicidae). *Bull. Ent. Res.* **46**, 255–65.

GJULLIN, C. M., HEGARTY, C. P. and BOLLEN, W. B. (1941). The necessity of a low oxygen content for the hatching of *Aëdes* mosquito eggs. *J. Cell. Comp. Physiol.* **17**, 193–202.

KING, W. V. and BUSHLAND, R. C. (1940). Quoted by Gjullin, Hegarty and Bollen (1941).

ROUBAUD, E. (1929). Recherches biologiques sur le moustique de la fièvre jaune, *Aëdes argenteus* Poiret, etc. *Ann. Inst. Pasteur*, **43**, 1093–209.

ROZEBOOM, L. E. (1934). The effect of bacteria on the hatching of mosquito eggs. *Amer. J. Hyg.* **20**, 496–501.

SIKES, E. K. and WIGGLESWORTH, V. B. (1931). The hatching of insects from eggs and the appearance of air in the tracheal system. *Quart. J. Micr. Sci.* **74**, 165–92.

THOMAS, H. D. (1943). Preliminary studies on the physiology of *Aëdes aegypti* (Diptera: Culicidae). 1. The hatching of eggs under sterile conditions. *J. Parasit.* **29**, 324–8.

WIGGLESWORTH, V. B. (1938). The absorption of fluid from the tracheal system of mosquito larvae at hatching and moulting. *J. Exp. Biol.* **15**, 248–57.

VIII
EMBRYOLOGY

(*a*) GENERAL OBSERVATIONS ON DEVELOPMENT

LITERATURE

In the present connection embryology may be defined as the developmental changes taking place in the egg between fertilisation and hatching. Such development has been described for a number of families of Diptera Nematocera. Excluding those chiefly concerned with chromosome characters, paedogenesis, or other special aspects may be mentioned accounts of the embryology: of *Sciara* (Mycetophilidae) by Schmuck and Metz (1932), Du Bois (1933), Butt (1934); of *Simulium* (Simulidae), by Gambrell (1933); of *Chironomus* (Chironomidae), by Kolliker (1843), Weismann (1863), Kuppfer (1866), Ritter (1890), Miall and Hammond (1900), Craven (1909), Hasper (1911), Sachtleben (1918); of *Miastor* (Cecidimyidae), by Metschnikoff (1865). No observations at all appear to have been made up to date on the Culicidae. (See however Addenda, p. 718.)

CHOICE OF MATERIAL

For various reasons the eggs of *Aëdes aegypti* are not the most suitable form on which to study mosquito embryology and the following account is of observations upon *Culex molestus*. In *Aëdes aegypti* the egg coats are quite opaque, the eggs are laid singly and intermittently and hatching is apt, as has been seen, to be rather uncertain and variable. In the *Culex* egg-raft the eggs may be considered as laid almost simultaneously, the egg coats are somewhat transparent and show the eyes and some other details of the stage of development, and hatching in the case of *C. molestus* occurs with great regularity at 18° C. at 54 hours. Further the egg coats in the *Culex* egg are more penetrable to fixatives and less brittle.

TECHNIQUE

A description of the external egg structure of *C. molestus* has been given by me in another publication (Christophers, 1945). For the method of procedure using this species rafts are picked up as soon as possible following oviposition as judged by their colour and kept on water at 18° C. for the required periods. When first laid the rafts are creamy white without any trace of colour or darkening. An hour after being laid the upper surface of the raft (posterior poles of the eggs) shows a faint greenish tinge which becomes greyish green and by 2 hours brownish grey. By 3 hours the raft is definitely dark. These changes are independent of exposure to light. Darkening does not occur uniformly over the egg. The first colour change is round the posterior pole. Later the whole egg becomes slightly darkened with a more intense ring of darkening round the posterior pole. At about 36 hours the eye-spots become visible through the chorion and at about the same time segmentation of the contained embryo can be made out, especially in fixed material. Towards the time of hatching the egg-breaker is visible.

EMBRYOLOGY

Whole eggs containing well-developed embryos are readily fixed without special manipulation. A fixative found very useful for routine use was Bles's fluid (formalin 7, glacial acetic 3, 70 per cent alcohol 90). Good results were also obtained with Carnoy-Le Brun fixative (absolute alcohol, glacial acetic acid and chloroform equal parts, mercuric chloride to saturation), which should be freshly made.

In the early stages no fixative tried was satisfactory in preventing collapse of the egg unless this was made penetrable by puncture or cutting. The most satisfactory procedure was found to be that of placing the eggs in the fixative, removing to a paraffin block under the dissecting microscope and cutting down with a sharp razor to slice off one or other end of the egg according to requirements. The egg is then removed to the fixative for the necessary period. This gave less disturbance than puncturing.

Eggs were sectioned serially by double embedding in celloidin and paraffin (see ch. v). The stain found most generally useful was Ehrlich's haematoxylin-picric acid-eosin method. For the greater part, reconstruction is necessary Useful preparations can, however, be obtained by staining the embryo *in situ* and examining after clearing whilst still within the egg-shell or after dissecting it out.

BRIEF SUMMARY OF COURSE OF DEVELOPMENT

Fertilisation probably takes place when the egg passes the opening of the spermathecal ducts during oviposition. In the early stages development follows closely that described for other insect forms. Following upon fertilisation there is *cleavage*, which results in the formation of the *blastoderm*. Following upon the formation of a uniform layer of blastoderm covering the whole egg surface beneath the egg membranes there appears a thickening of the blastoderm giving rise to the *germ band*. From the edges of the germ band there develop folds which grow until they entirely cover the germ band, eventually forming a double layer, the outer layer being the *serosa*, the inner the *amnion*. Over the anterior end of the embryo the amnion becomes especially thickened, forming a hood-like covering to this portion of the embryo, *cephalic amnion*. A similar hood develops over the posterior end of the embryo, *caudal amnion*. Meanwhile the germ band becomes differentiated and thickened anteriorly, forming a projecting lobe, the *precephalic element*, which will form the future labral structures, and behind this are massive thickenings, the *procephalic lobes*, from which are developed the brain and overlying head capsule. Behind these parts are indications of three thoracic and nine abdominal segments. At about 12 hours there appears on the ventral surface, just caudal of the precephalic element, the medianly situated thickening which is the first indication of the rapidly developing invagination, the *anterior mesenteron rudiment* that forms the *stomodaeum* and eventually the alimentary canal as far back as the termination of the mid-gut. About the same time posteriorly is the invagination forming the *proctodaeum* from which is developed the posterior portion of the alimentary canal including, in the mosquito, the Malpighian tubules. Other changes complete the larval structure as it appears at hatching. In the present case it is these later changes, or *organogeny*, that offer most opportunity for obtaining information regarding certain parts of the larva of which the homologies are uncertain.

So far as observations have been made on development in the egg of *Aëdes aegypti* the changes closely follow those seen in *Culex molestus*. In Table 8 has been set out the time in hours of the more important changes in the last-mentioned species.

Table 8. *Giving hours and stages in development of the embryo in the egg of* Culex molestus

Hours	Stages in development
1	Cleavage up to about eight nuclei
2	Basket arrangement of nuclei formed in yolk
3	Cleavage largely completed. Nuclei appearing at surface
6	Blastoderm forms complete layer over egg
8	Formation of the germ-band
10	Cephalic and caudal amnion folds begin
	Germ-band extends dorsally beyond middle of egg
	Precephalic element and procephalic lobes indicated
	(Pre-stomodaeal stage)
12	Labial groove gives bifid appearance to embryo
	Thickening of stomodaeal area with commencing invagination of anterior mesenteron rudiment
	(Early stomodaeal stage)
24	Labial groove everted giving head appearance to embryo
	Antennal and mouth-part rudiments present
	Anterior mesenteron and proctodaeal invagination well advanced
	Caudal extremity at posterior pole of egg
	(Late stomodaeal stage)
36	Brain mapped out and cerebral commissure formed
	Ventral nerve cord differentiated into future ganglia
	Anterior mesenteron rudiment and proctodaeal invaginations well advanced and Malpighian tubules forming
	Delamination of the head capsule
	Myoblasts evident in segments
	(Organogeny)
48	Egg-breaker begins to be visible
	Setae present on epipharynx and mouth-parts
	Dorsal closure completed
	Muscles of feeding brushes conspicuous
	Salivary glands present
	(Prehatching stage)
54	Larva hatches

(*b*) CLEAVAGE AND FORMATION OF THE BLASTODERM

CLEAVAGE (Fig. 8 (1), (2))

The earliest stage seen was that in which the first cleavage nucleus had been formed. This is situated near the anterior end of the egg. Cleavage proceeds with great rapidity, the nuclei dividing with typical mitotic stages followed by separation of the newly formed nuclei to a distance commonly almost equal to half the diameter of the egg (Fig. 8 (2)). By the time the egg-raft has developed a greenish colour (about an hour from oviposition) there are already present eight cleavage nuclei. At the end of 2 hours fifty or more nuclei were counted and at the end of 3 hours, in an egg where an attempt to count the nuclei was made, the number was approximately 720. At the end of 4 hours, when nuclei are present both in the surface layer and in the yolk, an estimate of their number based on sample counts of portions of the egg gave approximately 2000. By 6 hours a uniform layer of nuclei has been formed over the whole egg surface, the estimated number of nuclei being 5000–6000. Plotting the logs of the figures arrived at for the first 3 hours against time gave practically a straight line and a rate of division equivalent to three powers of 2 (2^3, 2^6, 2^9) per hour, or a nuclear division every 20 minutes.

This rate of division is much greater than that given by Roonwall (1936) for cleavage in the grasshopper egg, where the time for division arrived at in the same way would work out as over one hour. The time taken to form the blastoderm in the mosquito is, however, 6 hours, as against 28 in the grasshopper, and the time to hatching in the two cases 54 hours and 13 days respectively.

After 3 hours the rate of division has fallen off and the logarithmic curve falls short. New conditions are, however, now being introduced, for by the fourth hour the nuclei no longer have the appearance of actively dividing nuclei, but resemble more nuclei of formed blastoderm with resting chromatin and large nucleolus.

At least in the early part of the period of active division all the nuclei divide simultaneously or nearly so (Fig. 8 (2)), the appearance shown in any particular egg depending upon the stage that division happens to be in at the time of fixation. This stage may be that of the resting nucleus or some stage in mitosis. Those nuclei having characters of the resting nucleus are large, vesicular, globular or oval with clearly defined nuclear membrane and chromatin in small granules or beaded threads peripherally disposed. Probably the condition is more correctly described as late telophase. That they have recently divided is shown by the presence of the flare-like appearance in the cytoplasm as described below. Nuclei in process of division have no clearly defined outline or nuclear membrane. They appear as a cluster or clusters of deeply stained chromosomes, around which can be made out an area representing nuclear substance but quite diffuse in its limits. The chromosomes exhibit striking appearances indicating prophase, metaphase and anaphase. In all cases

---

Figure 8. Formation of blastoderm and amnion.

1 Showing three of four pairs of dividing nuclei in yolk towards anterior end of egg. About 1 hour. Reconstruction.

2 Showing 52 nuclei in late telophase in process of forming basket-like network in yolk. 2 hours. Reconstruction. The darkened nuclei are those lying on opposite side of yolk from the observer.

3 Transverse section of egg showing complete layer of blastoderm nuclei. 6 hours. *a*, nuclei and dark staining cytoplasm of blastoderm; *b*, layer of light granular cytoplasm. Periplasm of egg; *c*, nuclei still in yolk showing degeneration.

4 Portion of 3 showing blastoderm forming nuclei and cytoplasm. *a*, *b*, as in 3; *c*, small oval bodies in cytoplasm resembling bacilli.

5 Section through posterior pole of egg showing forming blastoderm and germ cells. About 6 hours. *a*, as in 4.

6 Transverse section of egg showing formation of amnion (cephalic amnion) and serosa. 10 hours. *a*, blastoderm; *b*, amnion; *c*, serosa.

7 Transverse section of egg showing formation of germ band and thickening of cephalic area. 12 hours. *a*, germ band; *b*, amnion; *c*, inward growth of cells forming cephalic lobes.

8 Lateral view of embryo at 10 hours (right side is ventral). Showing formation of cephalic and caudal amnion. *a*, cephalic amnion; *a'*, line of edge of amnion fold; *b*, caudal amnion; *c*, as yet uncovered portion of embryo.

Lettering: *gcc*, germ cells; *y*, yolk.

Note. The dark line surrounding the egg represents endochorion. The exochorion is not indicated.

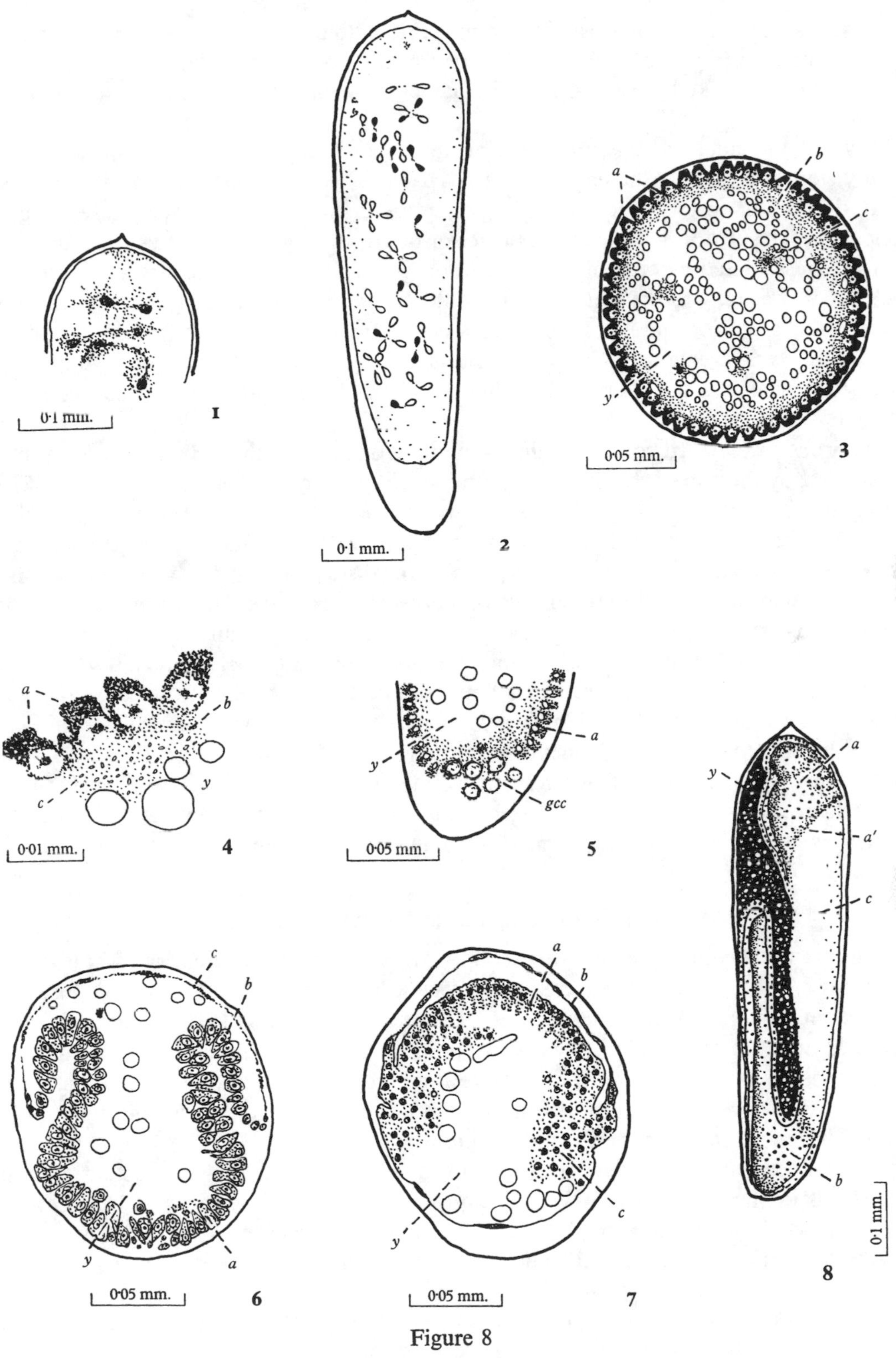
0·1 mm.
1
0·1 mm.
2
a
b
c
y
0·05 mm.
3
a
b
c
y
0·01 mm.
4
y
a
gcc
0·05 mm.
5
c
b
y
a
0·05 mm.
6
a
b
y
c
0·05 mm.
7
y
a
a′
c
b
0·1 mm.
8
Figure 8

there is present in the cytoplasm a pink staining area with radial striations, *attraction sphere*, forming either a spindle or two flare-like areas at the poles of the dividing system. Small pink staining disk-like bodies about 2 μ in diameter, presumably centrioles, are commonly present.

Movement of nuclei at this time certainly appears to be associated with, or a result of, the division process. In the early stages of cleavage (two to four nuclei) the nuclei are situated near the anterior pole of the egg. In the eight-nuclei-stage telophase daughter nuclei are seen spreading towards the posterior pole of the egg. About this time the nuclei with their associated cytoplasm tend to form a globular or oval basket-work of cleavage nuclei. Most usually this is situated in the anterior portion of the egg gradually extending towards the posterior end. In forming this closed hollow cylinder the nuclei divide in a longitudinal, transverse and oblique direction. Those dividing in a transverse direction usually pass round in the plane of the cylinder wall until they reach the opposite side of the cylinder.

The process described continues until a hundred or more nuclei have so arranged themselves. But by the end of the third hour numbers have reached the surface of the yolk. In an egg in this stage a count gave 756 peripherally situated and 549 still in the depths of the yolk. The centrally situated nuclei remain much as before, namely large nuclei with elaborate expanded chromosomes. The peripherally distributed nuclei, which are no longer dividing synchronously with those in the basket-work, now have the appearance of tightly compressed chromosomal knots. These later take on the appearance more characteristic of the blastoderm nuclei. They still do not show any limiting nuclear membrane, but the chromatin is in the form of a reticular network and there is a large nucleolus. They are embedded at intervals in a thin, defined layer of dark cytoplasm, which appears to have been formed by the cytoplasm of the cleavage nuclei passing with the nucleus to the surface of the egg and displacing the light-staining finely reticular cytoplasm which previously formed the cortical layer of the egg (*periplasm*). This latter now forms a considerable and well-defined layer, the *inner cortical layer*, situated internal to the darker outer layer containing the nuclei (Fig. 8 (3, *b*)).

FORMATION OF THE BLASTODERM (Fig. 8 (3), (4))

At 6 hours the characteristic appearance of (early) blastoderm is seen, the whole surface of the egg being uniformly covered with regularly arranged, almost contiguous, large globular nuclei, each with a conspicuous large nucleolus (Fig. 8 (4)). The dark superficial layer has now become divided by furrows apportioning it to the nuclei so that it gives the appearance of epithelium, except that the cells are not yet separated at their bases. In the yolk are still seen some nuclei which have not migrated, but they are indefinite in outline and for the most part apparently undergoing degeneration.

Whether seen in surface view or in section the blastoderm stage is a most striking and characteristic phenomenon entirely unlike any other stage in development. When eventually fully developed it consists of large columnar cells measuring up to 25 μ or more in length with dark-staining coarsely reticular cytoplasm and a large oval nucleus containing a diffuse central mass (karyosome?). A peculiar appearance seen at an early stage is the presence of clear vacuole-like spaces lying between the nucleated layer and the underlying periplasm. In tangential sections these are seen as branching channels or lacunae, possibly collections of oily material.

Differentiation of the blastoderm cells clearly accounts for the appearance of many of the early tissues in the embryo. Thus the cells first forming the embryo are not very dissimilar to those of the blastoderm and those forming the pro-cephalic lobes in their early stage, though largely 'neuroblasts' are not unlike blastoderm cells.

The blastoderm layer, except for a small area at the posterior pole of the egg, shows at one stage little or no difference in thickness. At the posterior pole, however, there are some large cells lying more or less extruded, the *germ cells*. In this area the regular covering of blastoderm cells is deficient, its place being taken by large globular cells lying loosely arranged upon the layer of periplasm. The fate of these cells will be described later.

(*c*) THE GERM-BAND AND EMBRYONIC MEMBRANES

THE GERM-BAND (Fig. 8 (7), (8))

The uniform layer of blastoderm thickens ventrally to form the *germ-band*. The thickening extends also beyond the posterior pole of the egg along the dorsal surface. The result is a thickened band of blastoderm almost completely encircling the egg longitudinally from a point a little dorsal of the cephalic pole along the whole ventral surface, past the posterior pole and along the dorsal aspect almost to the starting-point (Fig. 8 (8)). Elsewhere the blastoderm becomes thinned. But since the area of thickening is wide, the thinned area is relatively small, namely narrow bands laterally on each side linked by an area on the dorsal surface near the head where the posterior end of the germ-band fails to reach to the head end. Over these areas the blastoderm has become so thinned as to be merely a delicate nucleated membrane such as later always limits the yolk where this is exposed at the surface. Over the area occupied by the germ-band, however, the blastoderm now forms a deep layer of long columnar cells (Fig. 8 (7)).

THE AMNION-SEROSA (Fig. 8 (6))

As in other insects the amnion and serosa originate in a fold, the *amnion-serosal fold*. This arises around the edges of the germ-band and, extending over the embryo, forms by its inner and outer layers respectively the *amnion* and the *serosa*. The outer layer of the fold is an extremely thin membrane formed of flattened cells. The inner layer when it is first formed consists of more or less cubical cells. This layer forming the amnion is continuous with the edges of the embryo, that is the germ-band.

As the amnion-serosal folds become more and more extended over the embryo they are much thinned out and the two layers closely approximated. Early in development, however, it is usually possible to distinguish an inner thicker element with nuclei elongate oval in section, the *amnion*, and an outer exceedingly thin membrane, not above 1·0 μ in thickness, with smaller and still more flattened nuclei, the *serosa*. Since the amnion-serosal folds arise around the edges of the embryo and are directed over this, such portions of the egg surface as are outside the germ-band are at first not covered by amnion-serosa, but only by thinned blastoderm, which, however, closely resembles the serosa with which it is continuous. As the area of exposed yolk, however, becomes covered eventually by extension over it of the embryonic margins carrying with them the amnion-serosal folds, the whole egg surface beneath the egg-shell becomes covered with amnion-serosa. Eventually this becomes detached from all underlying connection to become the thin membrane which is left lining the egg-shell after the larva has hatched.

The final stages of the covering-in process, referred to as *dorsal closure*, are not so important in the mosquito as in some insects where the yolk is relatively large in amount compared to the embryo. They will be described later.

THE CEPHALIC AMNION

Though the amnion-serosa takes origin all round the edge of the germ-band it is specially conspicuous and thickened in certain areas, namely at the cephalic and caudal ends of the germ-band where it forms what appear as hood-like folds. The two portions may be distinguished respectively as the *cephalic* and *caudal amnion-serosa* (Fig. 8 (8)). The cephalic amnion-serosa, as it develops, progresses in a ventral direction overriding the cephalic end of the embryo and eventually, with the later developed lateral fold, covers most of the ventral surface of the egg. The cephalic amnion is particularly thick and massive near its origin where it is continuous with the edge of the thickened embryo in its anterior portion. The appearance is here much as if two deeply-thrust-in pockets had been formed, and in transverse sections these two pouches appear as deep fissures around which, radially arranged, are large cells such as later are characteristic of neuroblasts.

Each pouch eventually becomes divided into two, due to a fraenum-like extension of the amnion-serosa, namely an anterior or *labral* pouch which covers the rudiment of the feeding brush of its side, and a posterior larger pouch, *cephalic* pouch, which covers the procephalic lobe of its side. The labral pouches lie on each side of the anterior end of the embryo, a reflection of the amnion forming a tongue dorsally between them. Their thickened floor becomes increasingly conspicuous as the large button-like rudiments of the feeding brushes develop. The cephalic pouches extend backwards nearly to the mid-dorsal line, being separated only by a narrow portion of uncovered yolk. Their thickened floor later forms the brain and head capsule overlying this.

THE CAUDAL AMNION

About the time that the cephalic amnion is forming a fold over the anterior end of the embryo the caudal amnion is similarly forming a thickened hood over the posterior extremity. At this stage the posterior end of the germ-band is, as already noted, situated dorsally not very far behind the dorsal extension of the cephalic amnion (Fig. 9 (1)). Here the germ-band dips inwards forming a deep pocket in a somewhat globular mass of cellular tissue, in the deeper portions of which are now situated the germ cells, apparently moved forwards by extension of the posterior end of the embryo from their former position at the pole of the egg. In this position too there forms a dimple which is the first indication of the proctodaeal invagination. As with the cephalic amnion, the caudal amnion, where it is continuous with the embryo in the region of the pouch, is thick and massive. As it extends backwards to meet the lateral amnion it becomes thin and membranous.

THE LATERAL AMNION

Somewhat later in starting than the cephalic and caudal folds a less massive fold can be traced along the edge of the embryo. Where this joins the cephalic fold, that is just posterior to the cephalic lobe, there is an inconspicuous fraenum serving as a landmark between the parts in the cephalic pocket and those lying posterior to this.

(*d*) FORMATION OF THE EMBRYO

The precephalic element (Fig. 9). Underlying the most anterior portions of the cephalic amnion on each side and forming the floor of the labral pouches are convex areas of blastoderm which soon begin to develop as the future feeding brushes. These form button-like masses composed of long closely packed columnar cells early showing indications of forming the filaments which characterise the mature organs. Throughout later development these large button-like bodies form a striking feature of the embryo, giving at times rather the appearance of the suckers on the head of a tapeworm.

At first the rudiments are separated in the median line ventrally by a deep groove, *labral groove*, indenting the cephalic end of the embryo and extending back on the ventral surface as far as the position of the future stomodaeum (Fig. 9 (5)). The presence of this deep groove gives rise to a very characteristic appearance indicating a stage to which development has progressed, namely an embryo with a bifid anterior end. Later the deep sulcus becomes everted so that a prominent projecting lobe is formed. With the increase in size of the feeding brush rudiments, eversion of the labral groove and further delamination of the amnion dorsally the parts anterior to the procephalic lobes become marked off by a neck-like constriction to form an almost head-like projection. This pseudo-head may be termed the *precephalic element*. With its appearance the embryo takes on a further easily recognised and very characteristic stage. The parts forming the precephalic element are preoral in position and anterior to the procephalic lobes with their extensions forming the antennal rudiments. They give rise to the feeding brushes with their supporting structures and that part of the larval head usually described as the fronto-clypeus. The smooth convex ventral surface of the precephalic element formed by eversion of the labral groove gives rise to the epipharyngeal structures as far back as the larval pharyngeal opening. There can be little doubt that all the structures derived from it represent the labral segment, preoral in position and anterior to the antennae and eyes.

The procephalic lobes. Lying under that portion of the amnion forming the cephalic pouch is an area of blastoderm that early undergoes thickening indicative of the formation of the brain. This area is sharply demarcated dorsally where it comes up against the yolk and the thickened base of the amniotic fold. Anteriorly it abuts upon the head-like precephalic element and at each side of this border it shows swellings which later become prominences, the *antennal rudiments* (Fig. 9 (2), (3)). The procephalic lobes are characterised by the presence of large cells with nuclei containing a central mass surrounded by a clear zone. Cells of this type are seen where brain or ganglionic tissue develops and are precursors of neuroblasts. Originally a single layer of blastoderm cells, the procephalic lobe area, soon becomes greatly thickened to form on each side a cellular mass extending deeply into the yolk. Eventually the masses meet in the middle line to form the rudiment of the *protocerebral commissural tract*. It can be seen too that, whilst in other directions the developing nervous tissue is rather sharply delimited, there is ventrally a passage of such tissue into that developing about the neural groove. These areas eventually form the *cerebral connectives*.

The stomodaeal area (Fig. 9 (5)–(8)). Between the neurogenic areas just described there is situated ventrally what is at first a relatively thin area of germ-band, the *stomodaeal area*. At an early stage this consists of a single layer of columnar cells with a certain number of polygonal cells lying beneath it. Anteriorly this area extends as far forwards as the ventral termination of the labral groove, or, when this becomes everted, to the bulged ventral surface of the precephalic element. Laterally the limits are well-defined, especially where they come up against that part of the cephalic lobes which will eventually develop into the antennal rudiment. Posteriorly the area occupies the notch where the neurogenic tissue of the future ventral nerve cord extends forwards in a forked fashion to link up with the cephalic lobes.

Approximately in the centre of the area fairly early in development and before eversion of the labral groove some of the blastoderm cells show some elongation, thereby indicating where the *stomodaeum* will form. A little later this point is backed by a button-like mass of proliferating polygonal cells with reticular nuclei constituting the beginning of the *anterior mesenteron rudiment* (Fig. 9 (5)). A dimple followed by a deepening depression now starts the invagination that forms the stomodaeum and eventually the anterior portion

---

Figure 9. Formation of the embryo.

1 Lateral view of embryo at 15 hours. Ventral to right. Showing early segmentation and appearance of the anterior mesenteron rudiment. Reconstruction. *a*, anterior pouch of cephalic amnion and indication of feeding brush rudiment; *b*, posterior pouch of cephalic amnion and procephalic lobe; *c*, first indication of stomodaeal invagination.

2 Dorsal view of embryo prior to dorsal closure, 24 hours. *a*, precephalic element with indication of feeding brush rudiments; *b*, procephalic lobe; *c*, thoracic segments.

3 Lateral view of embryo, same stage as 2. Reconstruction. *a*, *b*, *c*, as in 2.

4 Dorsal view of embryo showing stomodaeal and proctodaeal invaginations. Reconstruction. *a*, *b*, as in 2.

5 Ventral view of anterior end of embryo at 12 hours showing labral groove and early anterior mesenteron rudiment with beginning of stomodaeal invagination. *a*, labral groove; *b*, rudiment of feeding brush; *c*, anterior mesenteron rudiment with beginning of stomodaeal invagination.

6 Ventral view of later stage. 24 hours. Showing eversion of labral groove and appearance of antennal rudiments and common rudiment of mandible and maxilla. The same stage is shown in lateral view in 3. Magn. as for 5.

7 Precephalic element. Ventral view. 28 hours. *a*, indication of feeding brush rudiment; *b*, beginning of epipharynx; *c*, slit-like entrance into stomodaeal funnel; *d*, lip-like fold anterior to stomodaeal opening. Rudiment of suspensorium.

On the right the mandibular and maxillary rudiment, as also the antennal rudiment, are shown as cut away.

8 The same dorsal view. *a*, area becoming pre-clypeus; *b*, area becoming clypeus; *c*, area between cephalic lobes dorsally.

9 Inner view of developing brain. About 30 hours. Reconstruction. I, *protocerebrum* with cerebral commissure; II, *deutocerebrum* pointing towards antennal rudiment and later giving rise to antenna nerve; III, *tritocerebrum* with commissure; *g*, sub-oesophageal ganglion with cut-away circum-oesophageal connective.

The arrow shows position of stomodaeal invagination.

10 Tangential section through ventral ganglionic area. *a*, conspicuous twin nuclei. Probably connected with forming inter-commissural tracts; *b*, cellular masses of future ganglion tissue; *c*, mesodermal tissue with forming myoblastic tissue.

Lettering: *A*, anal lobe; *amr*, anterior mesenteron rudiment; *an.r*, antennal rudiment; *anp*, anal papillae; *g*, ganglion; *gcc*, germ cells; *md*, rudiment of mandible; *md'*, cut base of same; *mdx*, common rudiment of mandible and maxilla; *Mt*, Malpighian tubules; *mx*, rudiment of maxilla; *mx'*, cut base of same; *pcd*, proctodaeal invagination; *pce*, precephalic element; *Sd*, stomodaeum; *Sd'*, position of stomodaeal invagination; *y*, yolk.

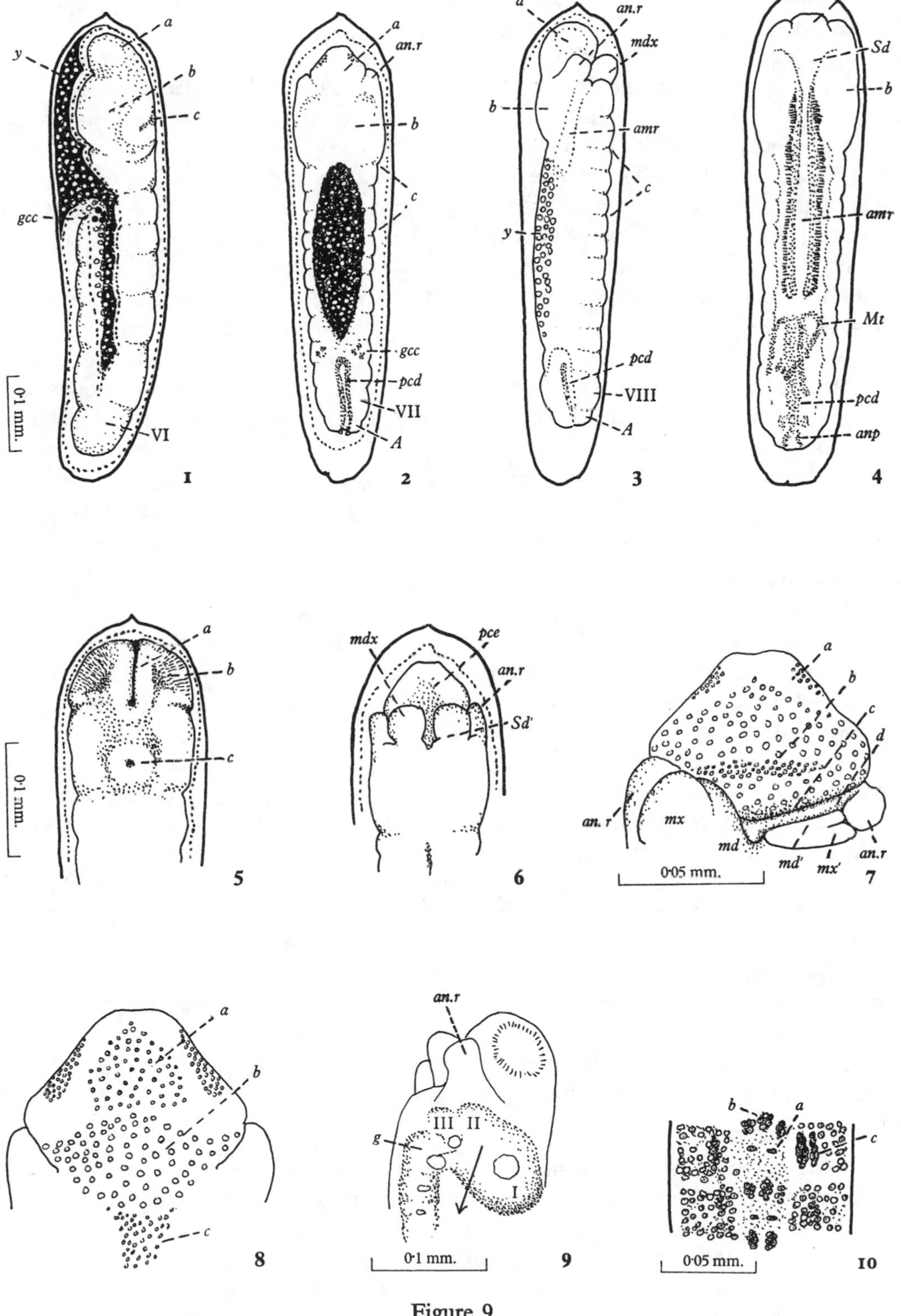

y
a
b
c
gcc
0·1 mm.
VI
1
an.r
gcc
pcd
VII
A
2
mdx
amr
VIII
3
Sd
Mt
pcd
anp
4
0·1 mm.
5
pce
Sd'
6
d
an. r
mx
md
md'
mx'
0·05 mm.
7
8
g
III
II
I
0·1 mm.
9
0·05 mm.
10

Figure 9

of the alimentary canal as far as the termination of the mid-gut, but in the mosquito not the Malpighian tubules.

As the anterior mesenteron rudiment increases in size its cells replace the original blastoderm of the notch between the cephalic lobes, and there is formed an area of some extent in which the cells have wholly the character of those of the rudiment. The stomadaeal opening deepens and widens, forming a flattened funnel. The dorsal wall of this funnel, which will become the roof of the larval pharynx, merges without break into the ventral surface of the precephalic element, now largely formed by the everted labral groove. In each lateral angle of the funnel, close up against the now bulging antennal rudiment eventually to become the antennal prominence, there develops a small notch where the fishhook-like *lateral oral apodeme* of its side will appear. Ventrally the edge of the opening forms a low lip-like ridge, later to become the *suspensorium* of the larval head.

Just ventral to this lip on each side there arise the rudiments of the mandible and maxilla. These commonly commence as a single globular projection on which there later appears a groove or fissure dividing off a dorsal portion, the *mandibular rudiment*, from a more ventral portion, the *maxillary rudiment*. Even up to the end of development, however, these two structures largely have a common base (Fig. 9 (7)). In the median line a little posterior to the ventral lip of the stomodaeum is an area which much later develops into the hypostomal area and mental sclerite of the larva.

Body segmentation. In the early stomodaeal stage, when the germ-band still extends far forwards on the dorsal surface, segmentation becomes indicated by the appearance of slight annular constrictions (Fig. 9 (1)). Starting from the sulcus that indicates the posterior limit of the procephalic lobes eleven more or less normally developed segments can be counted. Of these the first three are the three thoracic segments, at this time no larger or noticeably different from the following eight, which are abdominal segments I–VIII. In addition there is a terminal mass that becomes evident with the retraction of the caudal extremity to the posterior extremity of the egg (Fig. 9 (2), (3)). The eighth abdominal segment later becomes considerably enlarged, especially in its dorsal parts, which are early freed from amnion and eventually form the soft somewhat globular respiratory siphon of the newly hatched larva. The small terminal segment becomes the *anal segment*. With the invagination of the proctodaeum it comes to form a ring around the proctodaeal opening from which there project the *anal papillae*. These are first formed within the proctodaeal opening and only later come to project externally.

Nothing resembling clearly defined mandibular, first maxillary or second maxillary segments, as are present in many insect embryos, has been made out in the embryo of the mosquito. All that can be said is that there is an area posterior to the mandibular and maxillary rudiments and lying between these and the first thoracic segment which might represent such segments.

Dorsal closure. Eventually the segments of the embryonic abdomen become clearly delimited on the ventral aspect by surface constrictions and by segmentation of the mesoderm, followed by the arrangement of developing myoblasts derived from this, as also by the differentiation of the ventral ganglia. Dorsally also there is an indication of the first thoracic segment being closed in just caudal to the procephalic lobes. Abdominal segments VIII, VII and VI, in this order, also early become completed dorsally, followed by partial closure of segment V and to some extent of segment IV. Over the remaining segments, occupying about the middle third of the egg, there remains until a comparatively late stage an oval exposure of the now much diminished yolk (Fig. 9 (2), (3)).

This area of uncovered yolk is small and its final covering in (dorsal closure) is less important in the mosquito than in the case of many insects where the yolk surface may be relatively large. Nevertheless, something requires to be said regarding it. In the very early stages the yolk is left exposed at the surface in three positions, (1) on the dorsal aspect of the cephalic end of the embryo, with an extension backwards between the procephalic lobes; (2) covering the dorsal surface between the procephalic lobes and the caudal amnion dorsally; and (3) in two narrow strips down the sides of the egg between the germ-band on the ventral surface and its continuation on the dorsal surface of the egg.

The first-mentioned area, that part of the yolk exposed near the cephalic pole, is encroached upon and finally covered by backward extension of the cephalic amnion-serosal fold, so giving rise to important structures as described later under organogeny. Areas 2 and 3, after retraction of the caudal extremity of the embryo to the posterior pole of the egg, become merged into a single area. As development proceeds this is progressively closed in as already described by growth from the sides of the last few segments, the general effect being that there is left an oval area of exposed yolk occupying about the middle third of the egg on the dorsal surface. This area gradually becomes covered by growth from the sides and surrounding areas. By this time the anterior mesenteron rudiment has progressed backwards as far as the hinder border of the procephalic lobes and here comes into close connection with the anterior end of the exposed oval area of yolk. It is possible that cells derived from it play some part in the final closure.

The anterior mesenteron rudiment. The anterior mesenteron rudiment, at first a somewhat rounded mass of cells with the stomodaeal depression at its centre, rapidly grows backwards as a hollow finger-like process until it reaches the anterior end of the now much reduced yolk mass. This it does just caudal of the procephalic lobes dorsally. It is at this point that it comes into close connection with, and may help to complete, dorsal closure. Having reached so far, the walls of the rudiment split and continue growing caudal in the form of twin, rapidly growing ribbons. These consist of a layer facing inwards of columnar cells resembling those of the rudiment more anteriorly and a layer facing outwards of what seems to be mesoblast. At about one-third of the embryo's length from the posterior pole the ribbons meet and eventually fuse with the blind extremity of the proctodaeal invagination (Fig. 9 (4)).

The general outline of the alimentary canal can now be clearly seen. Anteriorly is the wide *mouth*, the anterior opening of the stomodaeal funnel, the site of the stomodaeum. Behind this lies the funnel itself which becomes the *pharynx* and its posterior narrow part the *oesophagus*. Still further back is a swollen portion indicating the future *proventriculus*. From the ribbons which grow round and largely enclose the yolk is formed the larval *mid-gut*. The remaining portions of the canal, including the Malpighian tubules, are formed from the proctodaeal invagination.

The proctodaeal invagination. In the prestomodaeal and early stomodaeal stages the caudal extremity of the germ-band lies somewhat curled inwards and limited in this direction by the fornix of the caudal amnion-serosa. Extending inwards from the fornix in the middle line is a button-like mass of cells, the *posterior mesenteron rudiment*. Into this there extends from the amniotic cavity a finger-like hollow which indicates the position of the future *proctodaeum*. In the cell mass are a number (about fifteen) of large cells, *germ cells*, which in the blastoderm stage occupied the posterior pole of the egg and which will end up eventually in the fifth and sixth abdominal segments where they give rise to the gonads. As development proceeds the caudal extremity of the embryo is dragged backwards so that it comes to lie at the posterior pole. By this time the proctodaeal invagination has passed some distance forwards. Later the invagination deepens rapidly and at about one-quarter of the egg length from the posterior end it meets and fuses with the anterior mesenteron rudiment. Before it does so there appear as outgrowths from its inner end five tubules which eventually become the *Malpighian tubules*. At an early stage four tubercles appear just inside the proctodaeal opening. They elongate and project through the anus as sharp pointed outgrowths eventually to become the *anal papillae*.

The neural groove and ventral nerve cord. In the early stomodaeal stage indication of the formation of the ventral nerve cord is given by the appearance in the mid ventral line of a groove, *neural groove*. This is a narrow sharply incised depression made more conspicuous in sections by the occurrence at intervals of characteristic large elongate cells with the rounded body of the cell lying inwards. As previously noted tissue containing such neuroblasts connects up the procephalic lobes with an area lying behind the stomodaeal area. Similar tissue occupies a broad band on the ventral aspect of segments as far back as the caudal extremity and gives rise in due course to the ventral nerve cord.

Formation of the mesoderm. No appearance suggesting formation of a median ventral invagination giving origin to the mesoderm has been seen, though such a stage, if of short duration or modified in character, may have been overlooked.

The mesoderm is first evident in the early stomodaeal stage as a mass of cells on each side of the anterior mesenteron rudiment. A bridge of similar cells links up the two masses anterior to the rudiment. Extensions of loosely arranged cells also pass forwards from these masses towards the precephalic element and posteriorly soon become linked up with mesoblastic tissue along the rapidly developing anterior mesenteron rudiment. Cell masses having a similar appearance also appear about this time on each side of the ventral nerve cord (Fig. 9 (10)).

(*e*) ORGANOGENY

THE LABRAL STRUCTURES (Fig. 9 (7), (8))

As already noted, the precephalic element, owing to its anterior projection, its large size, the presence of the conspicuous feeding brush rudiments and its pinched-in neck, is a very striking feature of the embryo. Before eversion it is responsible for the bifid anterior extremity of the embryo and, after eversion, it might easily be mistaken for the head of the embryo. Following eversion, its ventral surface is much expanded forming a smooth convex area. Across the middle of this area there develops a transverse thickening in the middle of which is a small differentiated area. The latter becomes the so-called *epipharynx* (that is the peculiar median process with a crown of setae so conspicuous in the larva). The transverse thickening becomes the *epipharyngeal bar*. A large part of the ventral surface remains undifferentiated and eventually constitutes the large *post-epipharyngeal lobe* of the larva. Where the ventral surface is continued into the dorsal wall of the stomodaeal funnel a dimple forms which will ultimately develop into the *frontal ganglion*. The position of this ganglion therefore marks very definitely in the larval head where the original embryo surface ends and the stomodaeal invagination begins.

The portion of the element situated apically between the two brush rudiments becomes the hairy lobe that has been termed the *palatum*. The subsequent history of the palatum shows that this is the labrum, *sensu stricto*, its hypodermis eventually forming the trunk-like projection in the pupa which becomes the labrum of the imago.

On the dorsal surface of the precephalic element and situated between the feeding brush rudiments is an area shaped like an inverted heart formed of rather small cubical cells. This can be identified clearly as the rudiment of the future *preclypeus* of the larva. The area caudal of this forming the broad basal bulge of the element is the area previously described as formed by the backward progression of the cephalic amniotic fold. It is formed of cells having the appearance of blastoderm. It eventually forms the *clypeus* of the larva, that is the area lying between the diverging frontal sutures and forming the greater part of the dorsal surface of the head of the larva. It is this area which mainly forms the bulged front portion of the head that is so striking a feature of the young first instar larva and ends as the relatively insignificant clypeus of the imago. Beneath this blastoderm-like area myoblasts early appear which eventually form the large muscular masses working the feeding brushes, and a little posterior to these a specialised area develops that gives rise to the *egg-breaker*.

The rudiments of the feeding brushes occupy the sides of the precephalic element. The brushes develop from button-like masses on the floor of the pouches in which they are first formed and which it may be recalled are the anterior pouches formed by the cephalic amnion. As the brush filaments increase in length they pass inwards and backwards out of these pouches to the position they eventually occupy in the hatching larva.

It seems clear that all the structures mentioned are parts of a preoral segment or lobe corresponding to the primitive labrum of the generalised insect head.

STRUCTURES DERIVED FROM THE CEPHALIC LOBES (Fig. 9 (9))

The area which the cephalic lobes occupy at an early stage in development may be described as rather like a butterfly's wings, expanding from the narrow stem at the bifurcation of the neurogenic area of the germ-band ventrally to the broad posterior borders which approach, but do not meet, each other on the dorsal surface of the embryo. Each lobe forms a massive plaque of thickened epidermal layer containing large modified cells developed around a deep fissure (the cephalic amnion pouch), the outer wall of the fissure being greatly thickened amnion and the inner wall original germ-band. Later the amniotic portion becomes progressively shrunken and to a large extent eventually disappears by delamination. By such delamination the deeper parts are left to form the *brain*. The delaminated layer includes the antennal rudiments, later to become the *antennal prominences* and that portion of the head capsule forming the *genae*. Towards the ventral aspect the procephalic lobe areas abut upon the post-oral area and at this point on each side there develops a deep sulcus. In the larva this becomes the hypostomal suture separating the genal from the labial area.

The brain (Fig. 9 (9)). With the thickening of the procephalic lobe areas and ingrowth of the characteristic neurogenic tissue containing large neuroblasts there come to be formed two large lateral masses. By continued inward growth these eventually meet in the middle line across remnants of the yolk, forming what will later appear as the *cerebral commissure*.

Where the procephalic lobes lie lateral to the stomodaeal area two further neurogenic tracts develop forming the *cerebral connectives*, linking the procerebral lobes with the ventral nerve chain. The main mass of the delaminated lobes becomes the *protocerebrum* (I). Only much later in the fourth instar larva do the large optic lobes develop and become linked with the more central portion of the brain. Forming the most anterior portion of the protocerebral lobes are two prominent portions which become the *deutocerebrum* (II), from which are eventually developed the antennal nerves. Lying somewhat ventral to the last-mentioned lobes there is on each side a small, but well-demarcated, lobe with its own core of neuropile, the *tritocerebrum* (III). From the tritocerebral lobes a small commissural tract crosses the median line close to, but distinct from, the larger connectives passing from the protocerebral lobes to the sub-oesophageal ganglion.

THE VENTRAL NERVE CORD

At an early stage the neurogenic area posterior to the stomodaeum shows three neuropile areas which later fuse to form the heart-shaped sub-oesophageal ganglion. Along the ventral aspect of the embryo the broad band of neurogenic tissue similarly becomes differentiated, indicating the position of the future thoracic and abdominal ganglia. These first become clearly evident as areas of pinkish stained neuropile. In coronal section at about 36 hours these show two conspicuous nuclei, one on each side of the median line at about the middle of the segment (commissural ?), and anterior and posterior to this from two to three compact groups of nuclei which appear to be the forerunners of the cortical areas of the ganglia. The ganglia so indicated are three thoracic and eight abdominal, the eighth abdominal ganglion lying in the eighth abdominal segment. No appearances have been seen of neuropile areas posterior to the eighth segment or indicating that the eighth abdominal ganglion includes more than one such area. When fully formed the ventral neurogenic cord is relatively broad, occupying a large part of the ventral aspect of the segments, as is the case also in the newly hatched larva.

THE MUSCULAR SYSTEM

In the late stomodaeal stage collections of mesoderm cells in various parts of the body take on the appearance of myoblasts and the development of muscular bundles rapidly follows. Among the most conspicuous of such are the large internal and external retractor muscles of the feeding brushes. These develop beneath that portion of the precephalic element which has previously been noted as becoming the clypeus of the larva. Since in the larva the muscles in question arise well back on the dorsum of the larval head, their origin in the precephalic element is confirmatory that the dorsum of the larval head is largely formed from this element, which there seems reason to believe is essentially labral in nature and pre-oral. Other conspicuous bundles form in the thoracic and abdominal segments lateral to the ventral nerve cord (Fig. 9 (10)).

OTHER LATER DEVELOPMENTS

The development of the alimentary canal, and nervous and muscular systems has already been indicated. One or two further details may be mentioned. A conspicuous feature of the embryo in the later stages are the *ocelli*. These first appear, indicated by their pigment, at about 36 hours and later become fully developed as the ocelli of the first instar larva. Another feature conspicuous in the later stages is the *egg-breaker*. This becomes visible through the egg-shell somewhat later than the ocelli (at about 45 hours). It is situated on the precephalic element immediately posterior to the origin of the internal retractor muscles of the feeding brushes. These muscles take origin largely on the anterior border of the egg-breaker and as they pass between this structure and the feeding brushes their contraction, if it fails to move the feeding brushes, would seemingly drag forwards the spine of the egg-breaker. Other features of late development are the structural details and hairs associated with the mouth-parts and body generally. These appear only in the latest stages and have not been followed in detail. For the conditions in this respect at full development see under the section dealing with eclosion.

REFERENCES

Butt, F. M. (1934). Embryology of *Sciara* (Sciaridae: Diptera). *Ann. Ent. Soc. Amer.* **27**, 565–79.

Christophers, S. R. (1945). Structure of the *Culex* egg and egg-raft in relation to function. *Trans. R. Ent. Soc. Lond.* **95**, 25–34.

Craven, W. N. (1909). Some observations on the embryology of *Chironomus*. *Proc. Iowa Acad. Sci.* **16**, 221–8.

Du Bois, A. M. (1933). A contribution to the embryology of *Sciara* (Diptera). *J. Morph.* **54**, 161–92.

Gambrell, F. L. (1933). The embryology of the blackfly *Simulium pictipes*. *Ann. Ent. Soc. Amer.* **26**, 641–71.

Hasper, M. (1911). Zur Entwicklung der Geschlechtsorgane bei *Chironomus*. *Zool. Jb.* **31**, 543–612.

Kolliker, A. (1843). Observationes de prima insectorum genesi, etc. (Mainly *Chironomus*.) *Ann. Sci. Nat.* (2), **20**, 253–84.

Kuppfer, C. (1866). Ueber des Faltenblatt an den Embryonen der Gattung *Chironomus*. *Arch. Mikr. Anat.* **2**, 385–98.

Metschnikoff, E. (1865). Ueber die Entwicklung der Cecidomyienlarven aus dem Pseudovum. Vorläufige Mittheilung. *Arch. Naturgesch.* **31**, 304–10.

Miall, L. C. and Hammond, A. P. (1900). *The Structure and Life History of the Harlequin Fly* (*Chironomus*). Clarendon Press, London.

Ritter, R. (1890). Die Entwicklung der Geschlechtsorgane und Darmes bei *Chironomus*. *Z. wiss. Zool.* **50**, 408–27.

Roonwall, M. L. (1936). Studies on the embryology of the African migratory locust. Part I. *Phil. Trans.* B, **226**, 391–421.

Sachtleben, H. (1918). Über die Entwicklung der Geschlechtsorgane von *Chironomus*, etc. Dissertation, pp. 1–59. München.

Schmuck, M. L. and Metz, C. W. (1932). The maturation divisions and fertilization in eggs of *Sciara coprophila*. *Proc. Nat. Acad. Sci. Wash.* **18**, 349–52.

Weismann, A. (1863). Die Entwicklung der Dipteren im Ei, nach Beobachtungen an *Chironomus*, etc. *Z. wiss. Zool.* **13**, 107–220.

IX
THE LARVA

(a) GENERAL DESCRIPTION

The first author to describe the larva of *Aëdes aegypti* appears to have been James (1899), who figures it as the larva of the *tiger mosquito*; see also Theobald (1901), vol. I, p. 28, where James's description is given as that of the larva of *Culex fasciatus*. Many other authors have since described the larva in various degrees of detail, usually from the point of view of characters of systematic importance. One of the most complete descriptions of its external characters is that by Macfie (1917), who has described the chaetotaxy and other cuticular structures of the different instars.*

To the naked eye the chief distinguishing characteristics of the larva of *Aëdes aegypti* (Fig. 10 (1)) are its rather cylindrical and elongate form and the dead white colour of the body due to the almost entire absence of sclerotised parts other than the head and siphon, which latter being almost black offers a marked contrast to the rest of the insect.

The *head* is rather globular and, at least in the later stages, small in relation to the thorax. Anteriorly are the clypeal structures carrying the feeding brushes (*fb*). The antennae (*an*) are much reduced in size compared with many culicine larvae, small, cylindrical and smooth. They carry a little distal to the middle a small simple antennal hair (Fig. 10 (2) *anh*), which is double in the first instar, and a small simple terminal hair (*ant*). The head hairs are small and inconspicuous. Laterally are the developing imaginal compound eyes (*om*) and the larval ocelli (*oc*).

The *thorax* is roughly globular in shape consisting of the fused three thoracic segments,

* Among other authors who have given descriptions of the larva are Blanchard (1905), Goeldi (1905), Wesche (1910), Doane (1910), Howard, Dyar and Knab (1912), Cooling (1924), Kumm (1931), Barraud (1934), Bonne-Wepster and Brug (1932), Walkiers (1935), Hopkins (1936). For authors describing special structures or systems see the appropriate sections.

Figure 10. Showing external characters of fourth instar larva.

1 Mature fourth instar larva. Dorsal view.
2 Antenna of same. Dorsal view and terminal portion.
3 Terminal segments of same. Lateral view.
4 Comb spines.
5 Pecten spines.
6 Showing dorsal view of base of dorsal (caudal) hairs.
7 Showing basal thickenings of hairs of ventral fan.

Lettering: *A*, anal segment; *Adh*, dorsal or caudal hairs: 1, dorso-internal hairs. 2, ventro-internal hairs; *Adp*, dorsal plate; *Ap*, anal papillae; *Ash*, saddle hair; *an*, antenna; *anh*, antennal hair; *ant*, terminal hair of antenna; *cm*, comb; *cr*, cratal tufts of ventral fan; *cr′*, simple hairs of same; *fb*, flabella (feeding brushes); *lh*, lateral hairs; *MS*, mesothorax; *MT*, metathorax; *oc*, ocellus; *om*, ommatidia of adult eye; *PR*, prothorax; *pct*, pecten; *ph*, pilot hair; *rt.r*, pupal trumpet rudiment; *s*, respiratory siphon; *sh*, siphon hair; *spl*, perispiracular lobe; *st*, spiracular puncta; *tt*, tracheal trunks showing at intersegments.

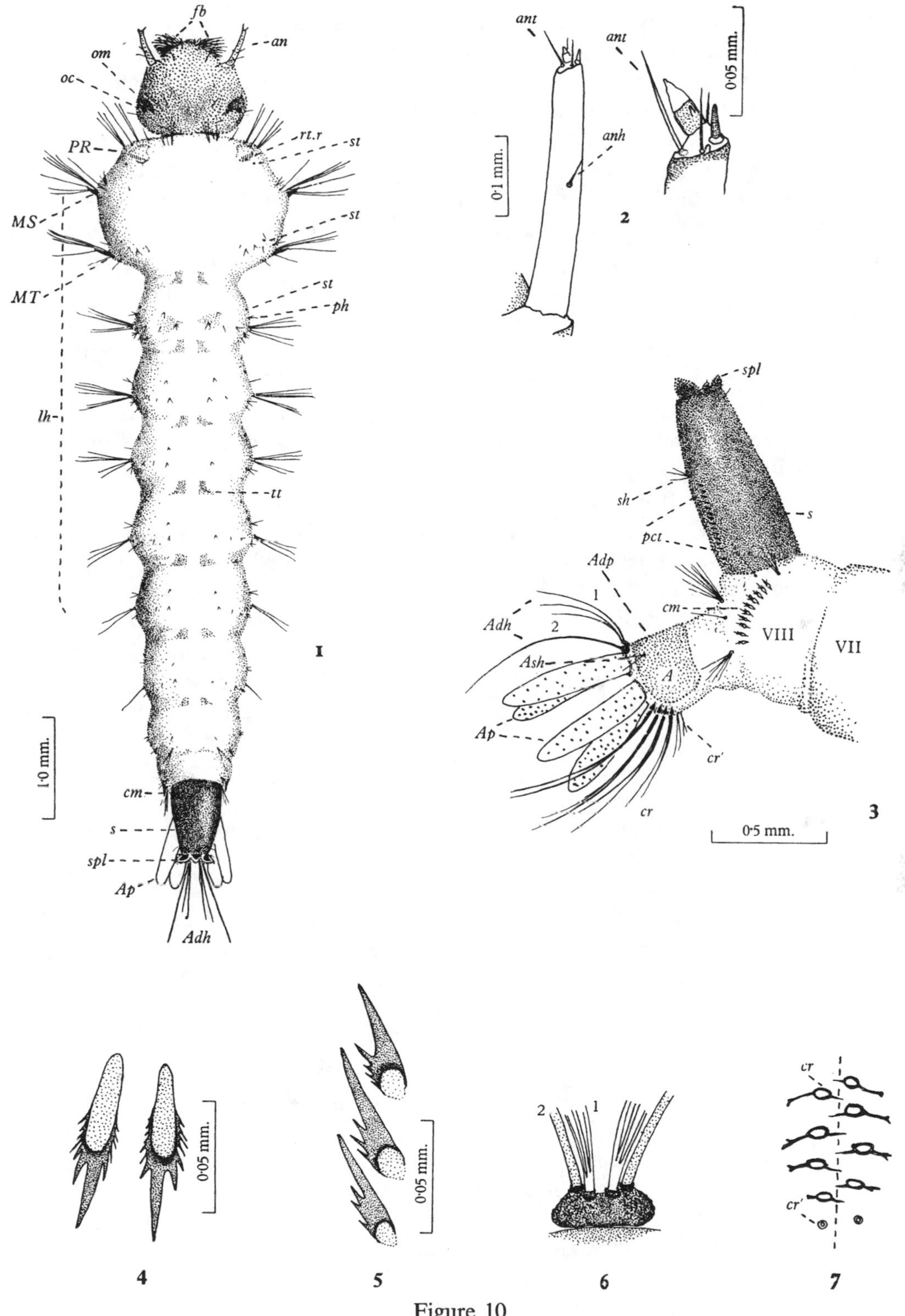

Figure 10

13-2

except just after hatching when these are distinct. The segments are, however, indicated by their hair series. Each of the segments, besides many small hairs, carries long *lateral hairs* arising from small chitinised plates, those of the meso- and metathorax being produced into sharp, thorn-like points very characteristic of the species. In the mounted larva the lateral hairs give the appearance of being closely approximated in tufts, but in the living larva free in water the hairs of each segmental series stand out more or less evenly spaced in a plane transverse to the body length. Though they are movable, the lateral hairs are not supplied with any muscles and if used in swimming must take their movements secondarily from those of the body. Most probably their chief function is sensory and protective. Among the lateral hairs are included the *pleural hairs* of Keilin, representing respectively the pro-, meso- and metathoracic legs. In the fourth instar can be seen, in the shoulders of the thorax, the rudiments of the future pupal respiratory trumpets (Fig. 10 (1) *rt.r*) which as pupation approaches become more and more conspicuous, until, when they finally become yellow, they give an indication that the larva will shortly pupate. Also conspicuous in the fourth instar larva, seen through the cuticle, are the *imaginal discs* or *buds* of the legs, wings and halteres as described in the section on the prepupa (ch. XIII).

The *abdomen*, apart from the small terminal anal segment, consists of eight roughly equal segments increasing in size somewhat to the third and then decreasing to the last or eighth. Each of these segments carries a number of hairs, including a decreasing series of long lateral hairs comparable with those of the thorax. At the posterior lateral aspect of the eighth segment on each side is a conspicuous row of spiny scales, *comb teeth*, forming the *comb* (Fig. 10 (3) *cm*), the number and arrangement of which and the absence or presence of a plate from which they arise being used in the identification of larvae of the genus. In the *Aëdes aegypti* fourth instar larva there is no plate, the comb teeth form a single row of 8–12 teeth and have well-developed basal denticles (Fig. 10 (4)). Just behind the comb at the posterior end of the segment are five hairs, three large and branched and two small and simple, named by Marshall the *pentad hairs*, useful as landmarks and referred to in detail when dealing with the chaetotaxy (Fig. 10 (3)). From above down in the figure the five small hairs are respectively those designated α, β, γ, δ, ϵ by Marshall.

Projecting from the dorsal and posterior portion of abdominal segment VIII is the *respiratory siphon*. At its apex are the *terminal spiracles* surrounded by certain structures, *perispiracular lobes* (*spl*), that play an important role in respiration and are described in detail later. On the posterior aspect of the respiratory siphon in its basal half are two slightly diverging lines of spines, *pecten teeth* (Fig. 10 (5)), constituting the pecten (*pct*). A little apical and internal to each line of pecten teeth is the *siphon hair* (*sh*). In *A. aegypti* above each line there is one hair, characteristic of certain genera, the genus *Culex* having several such hairs. Certain characteristics of the siphon are of importance in identification: the proportion of length to diameter at the base, *siphonic index*, the number and character of the pecten teeth and the situation and number of branches of the hair tuft or tufts. In *A. aegypti* larva in the fourth instar the siphonic index is slightly above 2, the pecten teeth number 12–20 and the siphon hair is single with from 3 to 5 branches.

Ventral to the siphon and at an angle to the rest of the abdomen is the small *anal segment* (*A*) with the opening of the anus terminally, around which are the four transparent finger-like *anal papillae* (*Ap*), two dorsal and two ventral. The anal segment carries dorsally a saddle-shaped plate, dorsal plate (*Adp*), which increases in proportionate size with each successive instar, until in the fourth instar it forms a cuff-like plate nearly surrounding the segment. Towards the posterior end on each side of the dorsal plate, or on the membrane beyond this in early instars, is the *saddle hair* (*Ash*), in *A. aegypti* small and usually bifid. Just posterior to the dorsal plate in the middle line dorsally and at the apex of the segment is a heavily chitinised plate formed from the fused basal plates of two pairs of the long *dorsal* or *caudal hairs* (Fig. 10 (3) *Adh*; (6)). The outer hairs are longer and unbranched whilst the inner and somewhat more dorsal hairs are branched. On the ventral aspect of the segment towards its posterior end is the *ventral fan* (*cr*). This is not present in the first instar, but becomes more and more developed at each successive instar. The hairs forming the fan, *cratal tufts* of Marshall, are mainly two-branched and arise from chitinised bases with thickenings passing from these, the hairs alternating to form a row on either side of the middle line (Fig. 10 (7)). In front of the fan are some hairs (usually two) without thickened bases (*cr'*).* In *A. aegypti* the number of cratal tufts is usually four in each row. In life the larva at rest at the surface carries the anal segment projecting backwards at an angle slightly above the horizontal. It is possible that with the dorsal hairs and ventral fan it plays some part in stabilising the insect. As long ago figured by Howard (1901), the outer dorsal hairs in the living larva diverge laterally at an angle of about 60°.

(*b*) THE HEAD CAPSULE AND RELATED STRUCTURES

THE HEAD CAPSULE (Fig. 11)

Of the many authors who have described or referred to the head capsule of the mosquito larva it will suffice to mention: Thompson (1905), whose paper entitled 'The alimentary system of the mosquito' contains much detailed information on the head capsule; Martini (1929), who devotes sixteen pages of his article on Culicidae in Lindner to the head of the larva; Puri (1931), who gives a very complete and detailed account in his monograph of the head of the *Anopheles* larva and who discusses the segmental relations of the parts; Salem (1931), whose important paper dealing especially with labral structures in the larva of *Aëdes aegypti* is one of the most detailed morphological studies of the head parts in the literature; and Crawford (1933), who writes specifically on the head of some anopheline larvae. In addition are authors already mentioned as describing the larva. Whilst the general characters of the head capsule of the mosquito larva are well known and parts are named with more or less uniformity by authors, there is still a certain amount of uncertainty about some of the parts and room for more detailed description of others.

The structures composing the head capsule are shown in the accompanying figures. In its posterior portion the capsule forms an almost complete box, its cavity opening into the neck through the relatively narrow and almost circular *occipital foramen* (*of*), surrounding which is a thickened collar-like rim, *cervical collar* (*cc*). Anteriorly and ventrally the head capsule is largely closed in by soft structures and is much fenestrated for accommodation of the labral and gnathal parts. Starting from a narrow gap in the mid-line of the collar dorsally are two diverging lines of suture, which have the appearance of being the Y-shaped arms of the epicranial suture of the generalised insect head and which are lines of weakness at ecdysis along which the head capsule splits. These lines, which may provisionally be termed the frontal sutures (*fs*), pass forward on each side of the dorsum of the head

* According to Mattingly (in lit.) absence of precratal tufts is diagnostic of the subgenus *Stegomyia* except in some aberrant forms. The extra hairs referred to have not the characters of true tuft hairs.

internal to the larval eyes to the inner side of each *antennal prominence* (*AP*), that is the lateral portion of the head capsule where this projects on either side to form a prominence from which the antennae arise. These two lines of suture demarcate between them a broad convex median area of the dorsum of the head which at ecdysis is pushed up in a flap-like fashion to allow passage of the head of the succeeding instar and which for reasons given later will be considered to be *frons* (*F*). In the posterior portions of their course the frontal sutures appear only as faint lines of suture, but anteriorly the outer lamella of the suture becomes increasingly developed as an internally directed chitinous crescentic flange which has been named by Thompson from its appearance the scythe-shaped thickening and may be termed the *falciform apodeme* (*fa*). This has the appearance of being the result of the antennal prominence, as it nears the base of the antenna, being pressed against the more median portion of the head capsule, so that the two walls in the area of contact have become an apodeme (Fig. 11 (1)).

The smooth median dorsal surface of the head described as *frons* is continued forwards without break or interruption beyond the level of the bases of the antennae to the front of the head, such portion being considered to be *clypeus* (*CL*). The whole area from the collar forwards forms a rather expanded and ballooned *fronto-clypeus*. The clypeal portion terminates anteriorly at a narrow transverse bar, the *preclypeus* (*PC*), lying at the extreme front of the head and carrying, in practically all culicine larvae, the small curved *preclypeal hairs*. Though often regarded merely as part of the clypeus, the preclypeus is a distinct morphological structure as will be shown later. It is separated from the clypeus by a line of thickening, the *clypeo-labral suture* of Salem.

External to the frontal suture on each side is a broad expanse of surface extending over the lateral aspects of the head and well on to the ventral surface. Assuming that the frontal sutures are branches of the epicranial suture, these areas would be the epicranial plates of the generalised insect head. They are most conveniently on each side termed the *gena* (*G*). Anteriorly each gena is continued into the antennal prominence. On the most prominent part of each gena and just on the mid-dorso-ventral line is the larval *ocellus* (*oc*), formed of a cluster of five pigmented ommatidia. In the fourth instar, in addition, are the

---

Figure 11. The head capsule.

1 Head capsule of fourth instar larva. Dorsal view. The dotted outlines and figures indicate insertion areas of head muscles numbered as given later under the muscular system.

2 Ventral view of same.

3 Posterior view.

4 Lateral view.

Lettering: *AP*, antennal prominence; *an*, antenna; *ap*, flabellar apodeme; *ap′*, anterior arm of same; *bs*, black spot area; *CL*, clypeus; *cc*, cervical collar; *ep*, epipharynx; *ep′*, epipharyngeal spines; *ep″*, epipharyngeal hairs; *epl*, post-epipharyngeal lobe; *F*, frons; *fa*, falciform apodeme; *fb*, flabellum; *fb′*, flabellar plate; *fb″*, scallop; *fs*, frontal suture; *G*, gena; *H*, hypostomal area; *hs*, hypostomal suture; *L′*, labial area; *lp*, lateral plate of clypeus; *Mo*, mouth; *md*, mandible; *mdo*, fenestra for articulation of mandible and maxilla; *ms*, mental sclerite; *mx*, maxilla; *mxp*, maxillary palp; *oc*, ocellus; *of*, occipital foramen; *om*, compound eye; *P*, pharynx; *PC*, preclypeus; *p*, palatum; *pe*, epipharyngeal bar; *pp*, postpalatal bar; *ppl*, postpalatal lobe; *sma*, submaxillary apodeme; *sr*, stirrup apodeme; *sr′*, point of involution of same; *ss*, suspensorium; *ss′*, lateral oral apodeme; *T*, tentorium; *T′*, anterior portion of same; *tm*, tessellated membrane; *tp*, tentorial pit.

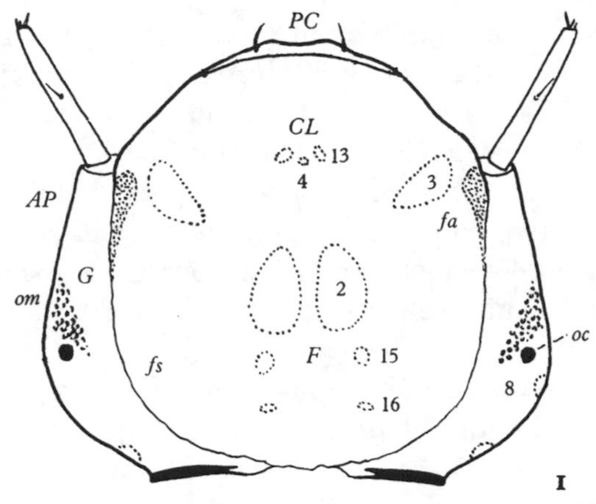

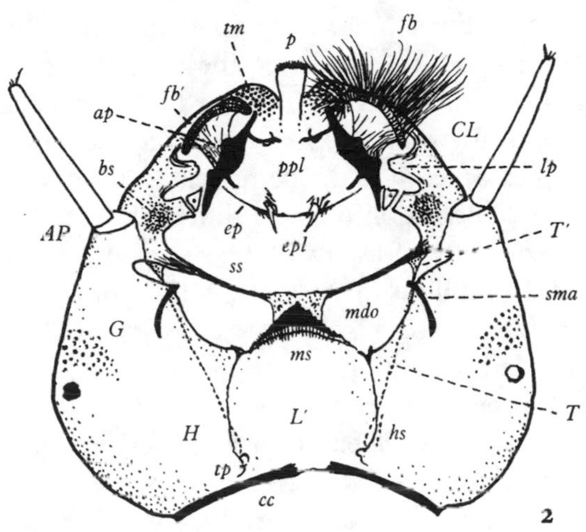

0·5 mm.

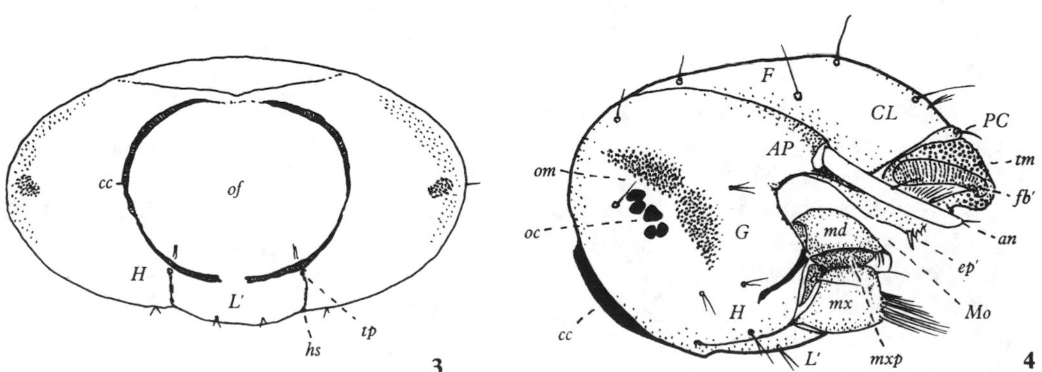

Figure 11

developing *compound eyes* of the future adult (*om*) showing through the cuticle. These commence as narrow eyebrow-like crescentic areas anterior to the ocelli and gradually increase in size by the addition of new ommatidia on their anterior periphery until they form large semilunar areas completely subordinating the ocelli which lie behind them (Fig. 11 (4) *om*). The compound eyes are not, however, organs of the larva at all, but of the future imago, and leave no mark on the cast skin of the head after ecdysis. Posteriorly the genae form rounded shoulder-like angles very characteristic of the mosquito larva. These are the areas which give origin internally to most of the muscles moving the mandibles and maxillae.

Passing to the ventral aspect of the head we find this sharply divided into distinct portions, namely an anterior or *labral area* with complicated apodemes and other structures including the feeding brushes; an intermediate area, which may be termed the *gnathal area*, giving insertion for the mandibles and maxillae; and an extensive smooth posterior area. This smooth posterior portion is formed laterally by the ventral extensions of the genae, which internal to the submaxillary apodeme form the *hypostomal areas* (*H*). Between the hypostomal areas is a median area, the *labial area* (*L'*), marked off by well-defined sutures, the *hypostomal sutures* (*hs*). This median area is bounded anteriorly by a conspicuous triangular toothed structure, the *mental sclerite* (*ms*). In the later stages of larval development there lies beneath it the complicated developing *labial rudiment*. The following is a more detailed description of these different areas.

The *labral area* consists dorsally of the *preclypeus* (Fig. 12 (1) *PC*). The extent laterally of this structure is difficult to define with certainty, but it appears to extend sufficiently far to encircle on their outer side the areas from which the feeding brushes take origin and to end in a conspicuous wrinkled projection on the inner aspect of the head (Fig. 11 (2) *lp*). These projections appear to be the upper processes of the lateral plates of Salem. Each carries a few hairs like those of the feeding brushes. More or less surrounded by the lateral extensions of the preclypeus are the structures at the base of the feeding brushes. Anteriorly extending from the border of the preclypeus is a peculiar bossed membrane, *tesellated membrane* (*tm*). Between the two tessellated membranes in the middle line is the projecting hairy lobe termed *palatum* by Thompson (*p*). Passing more ventrally in continuation of the tessellated membranes are the oval *basal plates of the flabella* from which rise the hairs forming the feeding brushes or *flabella*, and between these and the *flabellar apodemes* (*ap*) is the peculiar corrugated structure which from its resemblance to a scallop

---

Figure 12. Clypeal and labral structure of larva.

1 Showing dorsal view of preclypeal and clypeal hairs.*

2 Flabellar plate and apodemes associated with same. In position with feeding brushes extended. Hairs of plate are not shown.

3 Right flabellum and apodeme. The numbers 1, 2 and 3 indicate hairs of the inner, middle and outer portion respectively.

4 Showing relation of parts.

5 Showing position of parts with the feeding brushes retracted.

6 Showing parts of epipharynx.

7 Terminal portion of hair from inner series of flabellar hairs.

For lettering see under Fig. 11.

* For names of head hairs, see Table 9, p. 211.

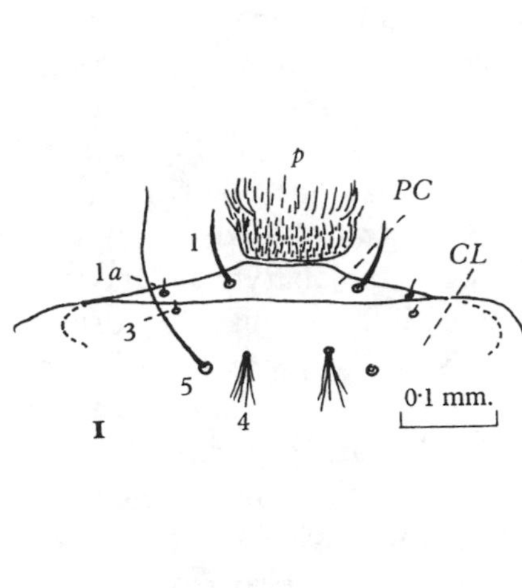
p
PC
1
1a
CL
3
5
4
0·1 mm.
1

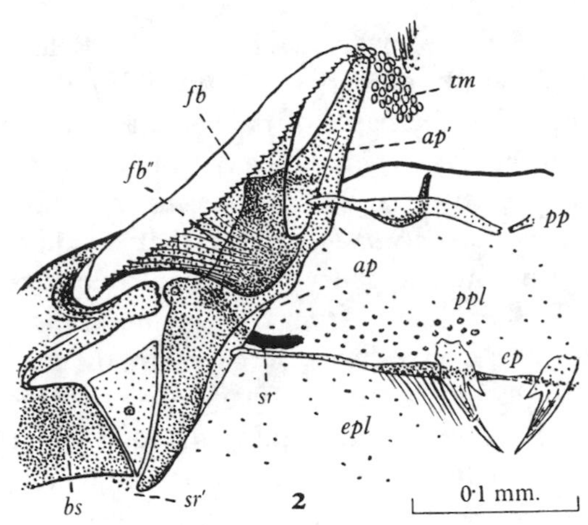
tm
fb
ap'
fb''
pp
ap
ppl
cp
sr
epl
bs
sr'
2
0·1 mm.

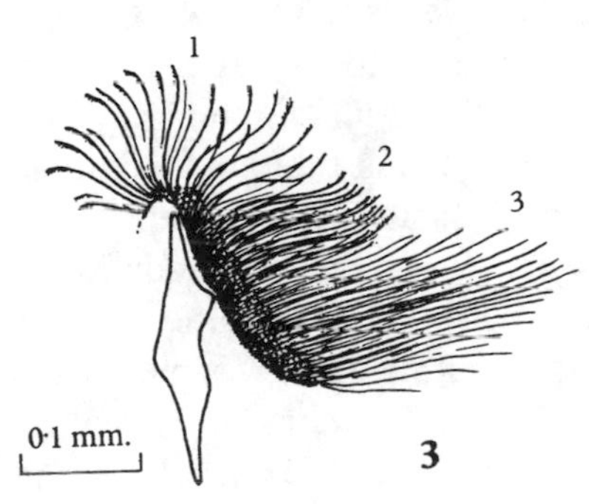
1
2
3
0·1 mm.
3

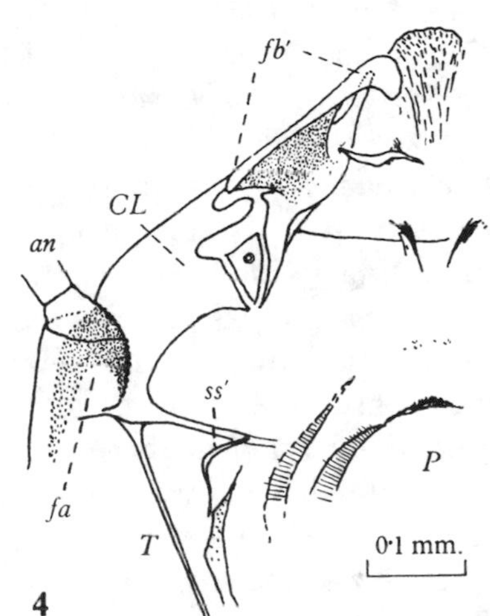
fb'
CL
an
ss'
P
fa
T
0·1 mm.
4

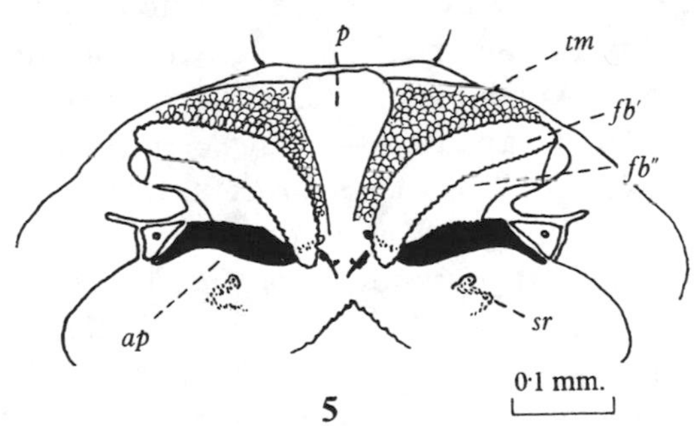
p
tm
fb'
fb''
ap
sr
0·1 mm.
5

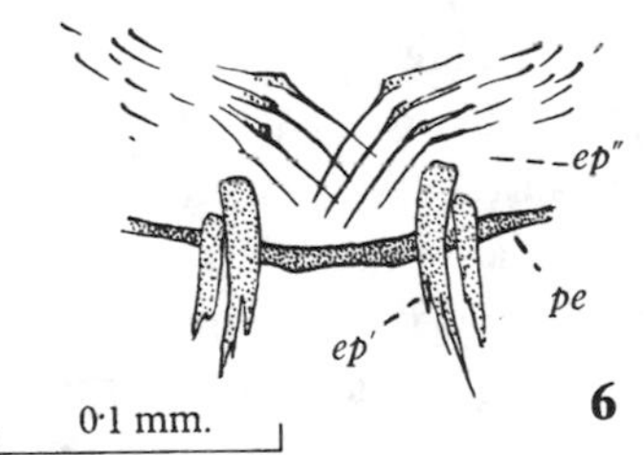
ep''
pe
ep'
0·1 mm.
6

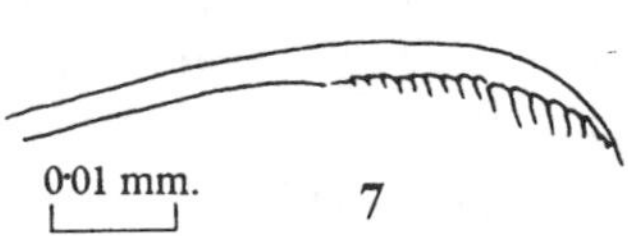
0·01 mm.
7

Figure 12

shell I have termed the *scallop* (*fb''*). Behind the palatum is an area of membrane forming a cushion-like lobe, *post-palatal lobe* (*ppl*), crossed by a sclerotised bar, *post-palatal bar* (*pp*), connecting the bases of the flabella. Behind the post-palatal lobe is another bar, the *epipharyngeal bar* (*pe*), which with the *epipharyngeal spines* (*ep'*) and the *epipharyngeal hairs* (*ep''*) constitutes the *epipharynx* (*ep*). Still further back behind the epipharynx is the *post-epipharyngeal lobe* (*epl*), that overhangs like an upper lip the broad opening (true mouth) into the pharynx. It is capable of being retracted when the pharynx is functioning by muscles arising from the dorsum of the head. The lobe, with the epipharynx, forms the roof of a cavity closed in below by the mandibles and maxillae, the *preoral cavity*, which opens behind into the pharynx by a broad slit-like opening, the *true mouth*, that is the original stomodaeal opening. Though this at rest is slit-like it is widely expanded during the act of feeding, the preoral cavity, true mouth and pharynx forming an almost continuous cave-like passage into the head. Views from the ventral or dorsal aspect give a very imperfect idea of the nature of the larval mouth, which as it would appear from the front and sides is shown in Figs. 17 (2) and 11 (4) (*Mo*).

Behind the turned-in edges of the preclypeus there pass inwards the heavily sclerotised smooth edges of the *clypeus*, forming on each side of the head a buttress-like projection on which is a peculiar black mark present apparently in all culicine larvae and very appropriately named by Thompson the *black spot area* (*bs*). From this there extends a stout process passing to and giving support to the complicated system of apodemes of the feeding brushes described later. From the same black spot area there extends a strong *chitinous bar* (*ss*) passing backwards and inwards to the structures lying behind the mental sclerite. This forms a lip-like lower edge to the pharyngeal opening and gives membranous attachment to the dorsal basal edge of the mandible of its side. The bar clearly represents the *suspensorium* of the generalised insect head. At the outer end of the bar where it joins the turned-in edge of the clypeus forming the black spot area is the funnel-shaped opening of the anterior arm of the tentorium described later.

Posterior to the suspensoria and forming what has been termed the gnathal area is on either side a large fenestrated opening (*mdo*). This opening by its edges gives attachment to both mandible and maxilla. The anterior edge is formed by the suspensorium and gives attachment only to the dorsal edge of the base of the mandible. Posteriorly the openings are limited by the sharp sinuous edges of the genae. About the middle of this edge on each side is a strong spur-like process continued as a partially apodematous thickening on to the ventral surface of the gena, *submaxillary apodeme* (*sma*). This process forms the main articulation both for mandible and maxilla, which are otherwise almost entirely connected with the head capsule by membrane only. At this point there arises the *maxillary palp*, with its small basal crescentic sclerite bearing a hair as described later. The fact that the palp arises in this position and that the hypoderm lying between the submaxillary apodeme and hypostomal suture as far back as the tentorial pit near the cervical collar is included in the area of the developing maxillary rudiment shows that the structure designated the maxilla in the larva represents only such part of this appendage as is apical to the maxillary palp and that the more basal parts (*cardo* and *stipes*) are unrepresented except in the hypoderm. In other words the bases of the maxillae in the larva really extend, as in the adult, practically to the occipital foramen forming the *hypostomal area* (*H*).

The smooth convex median plate between the hypostomal sutures we have termed the *labial area*. Anteriorly this terminates in the mental sclerite. Posteriorly on either side at the ends of the sutures are the *posterior tentorial pits* (*tp*), indicating the points near the cervical collar where the posterior arms of the tentorium join the head capsule. The *mental sclerite* in *Aëdes aegypti* is of the usual triangular shape with, in the fourth instar, 12–13 teeth on either side of the median tooth, the outer three or four teeth being larger and more widely spaced than the inner ones. Though a striking structure in the larva, the mental sclerite appears to have no great morphological significance, being cast with the larval skin and leaving no particular effect upon the hypoderm. Lying ventrally at the base of the mental sclerite is a less conspicuous fold of the cuticle carrying a fringe of peculiar flattened fimbriated hairs.

Lying within the head is the phragmatic structure, the *tentorium* (Fig. 11 (2) *T′*, *T*). This consists of a spur-like process projecting inwards from each of the posterior tentorial pits, *posterior arm*, from the termination of which there passes a narrow tubular strand opening at the anterior tentorial pit of its side, *anterior arm* (*T′*). From the point of the spur there arise certain small muscles of the head among which in *Culex pipiens* is the long thin antennal muscle. In *Aëdes aegypti*, where the antenna is very poorly developed, this muscle is very small and appears to arise only from the shaft of the anterior arm.

Before leaving the description of the head capsule proper a word may be said regarding the structures *frons*, *clypeus* and *preclypeus*, as also a further brief reference to the *tentorium*.

Embryologically the parts in front of the antennal bases are all formed from the everted precephalic groove, that is, the deep median longitudinal groove which at a certain stage of development is everted at the anterior end of the embryo to form a conspicuous bulbous mass (see ch. VII). The ventral surface of this mass becomes the epipharyngeal area. On it laterally are the brush rudiments looking rather like the disks on the head of a tapeworm, and connecting these on the dorsal surface is an inverted heart-shaped area, picked out conspicuously by its special cellular structure, which becomes the *preclypeus*. The surface of the embryo posterior to this area as far back as the antennal lobes becomes what has here been given as *clypeus*. There is also the anatomical arrangement of parts to support the above conclusion. Snodgrass (1935, p. 110) notes that a facial suture that contains the anterior tentorial pits is usually to be identified as the *epistomal suture*, that is the suture between the clypeus and the frons. As already noted the anterior arm of the tentorium opens just internal to the falciform apodeme at the point where the dorsal surface of the head is continued on to the black spot area. If at this level there had been an epistomal suture crossing the dorsal surface, it would be reasonable to term that part of the dorsum behind it frons, and that in front clypeus. Further, the whole ballooned area of the larval dorsal head surface, ballooned to accommodate the vitally important and large pharynx (on which as we shall see the whole life of the larva depends), shrinks in the adult to a small insignificant area more in keeping with the usual appearance of the frons in the insect head. The preclypeus, both from its anatomical relations and its development, would therefore appear to be part of the labrum.

THE FEEDING BRUSHES AND THEIR APODEMES (Fig. 12)

These have been described by Salem, though knowledge of their essential structure and mode of action is still incomplete. Each brush or *flabellum* consists of long specialised hairs arising from an elongate oval plaque, *flabellar plaque* (*fb′*). Each flabellar plaque somewhat resembles in appearance an oval mat of chain mail, due to the regular arrangement of the complicated ridge thickenings giving attachment to the swivel-like basal attachments of the brush hairs. Of these hairs there are at least some 50–60 transverse rows, each row carrying at least 25 hairs, thus making a total of at least 1000 hairs per flabellum. The hairs are about 3–4 μ in diameter and are therefore far coarser than the fine pharyngeal fringe hairs which are the real filtering mechanism. It seems certain that it is not the function of the flabella to entangle food, but to direct a current into the pharynx where the filtering is done. The hairs in *Aëdes aegypti* larva in the outer region of the brush are simple and filamentous, but those towards the inner end are much thicker and, as noted by Salem, are toothed like a comb at the ends, much as in carnivorous larvae. The minute structure of the brushes has been studied by me chiefly in *Culex pipiens*, but in all essentials the brushes of *Aëdes aegypti* appear similar, except that in *Culex pipiens* all the hairs are equal and simple. Each hair at its base has a flattened spear-like head articulated by its point with the plaque ridges (Fig. 12 (3)). The arrangement is extremely regular and each hair base is connected with its neighbours by means of small cross-bars in the basal plaque, apparently to ensure co-ordinate movement. The form taken by the brush is entirely a matter of the state of the plaque. If this is concave or cupped, the hairs tend to come together in a point. Conversely convexity of the plaque causes the hairs to radiate widely. Movement of the hairs in feeding is further strictly co-ordinated, each row in succession rising, becoming erect and falling in regular order and sequence.

The plaques are not attached directly, except at their outer ends, to the preclypeus, but are for the greater part of their external border joined to this structure by the *tessellated membrane* (*tm*). This membrane is covered with block-like thickenings much as might be a fabric covered with bead work. It is not very apparent when the brushes are erected, but shows an extensive area between these and the preclypeus, widening medianly and merging with the palatum when the brushes are flexed. As regards the plaques, each is attached externally by its end to the encircling lateral extension of the preclypeus and at this attachment is relatively fixed. Its inner and anterior end is attached by a ligament to the point of the finger-like apodeme to be described shortly, which apodeme is freely moveable and capable of being dragged down and inwards by the flabellar muscles. Hence the flabellar plate relatively fixed at its outer end, though allowed some hinge-like movement, can be dragged down through an angle approaching 90° by its apodeme, dragging with it the tessellated membrane (Fig. 12 (5)).

Extending inwards from the inner border of the plaque is a smooth dark and wrinkled sheet of thickened membrane which lies over, and by its inner margin is joined to, the apodeme. This sheet is described by Salem as part of the plaque. It seems desirable, however, to consider the plaque with its peculiar structure as distinct. The dark wrinkled membrane might well from its appearance be termed the *scallop* (*fb''*). It seems to be of an elastic nature. It is all the more desirable to give at least descriptive names to these various structures, since they occur with little variation seemingly through most if not all the Culicini, the parts here described being almost identical with those in *C. pipiens*.

The apodeme spoken of above is a conspicuous heavily sclerotised structure lying along the inner side of the turned-in edge of the clypeus and flabellar plaque. Its appearance and connections can best be seen from the figures. Anteriorly it is continued forwards as a partly superficial and partly deep-lying extension to form the finger-like apodeme already referred to. Its thickened middle portion is partly overlaid by the scallop. Its posterior end is continued back to link up, through an intermediate small triangular plate bearing a hair, with the powerful spur coming from the black spot area. Just internal to this junction on the adjoining membrane is a little spicular area which marks the position of a deep invagination leading to another and quite independent apodeme, which from its characteristic appearance may be termed the *stirrup apodeme* (Fig. 12 (2) *sr*).

The main apodeme consists essentially of several involuted folds of cuticle turned in along the margin of the flabellar complex and flanking the median membranous area. The stirrup lies deeper. It has one termination, the more dorsal, thickened to form a knob-like head. The other end is pointed and roughened. Though no doubt connected indirectly there is no obvious linking of the stirrup apodeme with the flabellar apodeme. Curiously enough none of the structures described, except the stirrup apodeme, receive the insertion of any muscles concerned in the movement of the brushes. The muscles moving these structures are, as will be seen, the inner and outer retractors of the flabella, by far the largest and most massive muscles of the head. These two muscles are inserted into the ends of the stirrup apodeme, the large median (inner) retractor into the head and the outer retractor into the more ventrally lying pointed end. The action of these muscles in retracting the brushes is not difficult to understand. By their contraction the whole flabellar complex is dragged backwards and inwards and so into the shelter of the preoral cavity, the flabellar plaques being made concave and the brushes brought to points. Whilst this, however, accounts for the dragging down of the brushes in this way it does not explain, as is noted by Thompson, how the characteristic movement of these structures in feeding is brought about.

Between the two flabellar apodemes, crossing the membrane behind the palatum is a chitinous bar, in part apodematous, *post-palatal bar* (*pp*), which links the two apodemes and indirectly the two flabellar plaques together. Posterior to this is the area of membrane carrying the epipharyngeal structures and posterior again to the epipharyngeal spines is a long narrow chitinous band passing from near the posterior end of one flabellar apodeme to the other. Into the middle of this is inserted the small muscle (no. 4) which takes origin from the dorsum of the head. The band may be termed the *post-epipharyngeal bar* (*pe*). Still further back is the large expanse of membrane forming the *post-epipharyngeal lobe*. All these structures constitute the roof of the preoral cavity.

THE LARVAL MOUTHPARTS (Fig. 13)

A large part of the under-surface of the anterior region of the head, that is especially the extensive area of membrane forming the post-epipharyngeal lobe, forms the roof of what becomes a cavity when closed in below by the mandibles, maxillae and intervening structures. This, however, is not correctly the oral cavity, the true mouth being the opening of the original stomodaeum which in this case is the entrance to the pharynx, the closed-in space referred to being the *preoral cavity*. This receives posteriorly the wide slit-like opening of the pharynx, the true mouth. The floor is mainly formed by the dorsal surfaces of the mandibles and to a limited extent by the maxillae and in the middle line by the dorsal surface of the mental sclerite and the parts lying behind this (Fig. 25 (2), p .228). Excluding the labrum, the mouth parts consist of the *mandibles*, the *maxillae* and a small rudiment behind the mental sclerite representing the undeveloped *labium* and *hypopharynx*.

The characters of the mandibles and maxillae in a number of species of mosquito have been figured by Howard, Dyar and Knab. Those of *Aëdes aegypti* have been figured by Salem and dorsal and ventral views of these organs given in Fig. 13 sufficiently indicate their structure. As already noted the bases of both mandibles and maxillae are accommodated in the same openings in the head capsule and except for membranous connections they are articulated with the capsule by a single common point of attachment, namely with the spur-like process about the middle of the genal edge which is a continuation of the line of submaxillary apodeme (*sma*). Here mandible, maxillary palp and maxilla are all articulated (Fig. 13 (1), (3) *a*). In the case of the maxilla there is a further connection, namely a narrow, only slightly sclerotised process on the inner ventral surface which connects with the upper end of the hypostomal suture near the angles of the mental sclerite (Fig. 13 (3) *b*). There is also an inconspicuous process on the dorsal surface ending in membrane (Fig. 13 (4) *c*).

Of the various structures characterising the mandible may be noted the conspicuous comb-like fringes, the function of which would appear to be combing the brushes. Internal to these are the strongly sclerotised mandibular teeth and behind these the peculiar molar lobe (*mp*), which when the organ is adducted lies in the hollow pouch on either side of the labial plate. Unlike the mandibles of many insects those of the mosquito as shown in *A. aegypti* are rather thin plate-like structures. Apart from the retractors of the flabella the largest muscles of the head are inserted by a common tendon into the base of the mandible towards its inner aspect (Fig. 13 (1) *adt*).

The maxillae are also somewhat plate-like, though not so markedly as the mandibles. The most conspicuous feature of the maxilla of *A. aegypti* is the large brush of hairs at its apex. The ventral surface is relatively smooth, but the dorsal surface has a spinose lobe and internal to this a hairy area. There is a thickened ridge on the dorsal surface passing from the articulation with the hypostomal process to the apex of the organ. This passes between the spinose lobe and the rest of the maxilla. At the apex is a single or double conspicuous hair, but in *A. aegypti* there is no hair on the ventral face which is apparently the maxillary hair of authors. Close to the outer aspect of the base arises the maxillary palp carrying some small apical soft papillae. Below the origin of the maxillary palp and proximal to the maxilla is a small triangular sclerite bearing a hair, possibly the palpiger. As already noted what is usually described as the maxilla is only the apical portion of this structure; the more proximal parts are here considered to form the hypostomal area as previously described.

The *labium* in the larva is completely undeveloped and exists only as a small, but extremely complex-looking strongly chitinised small oval plate situated behind the mental sclerite and comprised in a chitinised sheet formed by the expanded ends of the suspensoria (Fig. 11 (2)). The plate has been figured for several species of mosquito by Marshall (1938). Its

---

Figure 13. Mandible, maxilla and labial rudiment.

1 Right mandible. Ventral view. *a*, point of articulation. External to this is an extensive membranous area.

2 The same, dorsal view. This surface forms the floor of the preoral cavity.

3 Right maxilla and maxillary palp. Ventral view. *a*, main point of articulation as for mandible at anterior end of submaxillary apodeme; *b*, inner subsidiary attachment at hypostomal suture.

4 Right maxilla and maxillary palp. Dorsal view. *c*, roughened thickening forming accessory attachment on dorsal aspect of base.

5 Mental sclerite. Fourth instar.

6 Labial and hypopharyngeal rudiment.

Lettering: *adt*, adductor tendon of mandible; *an*, base of antenna; *bs*, black spot; *hs*, hypostomal suture; *mp*, molar process of mandible; *sd'*, opening of salivary duct; *sma*, submaxillary apodeme; *T*, tentorium.

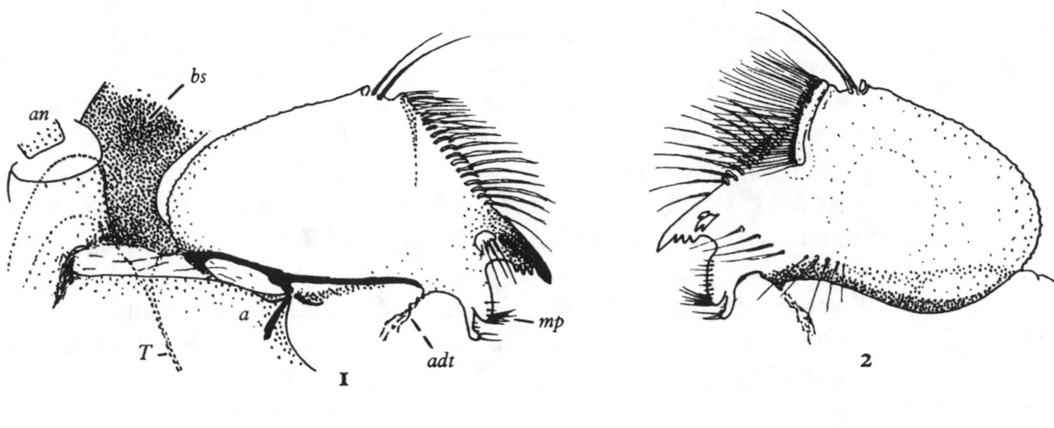

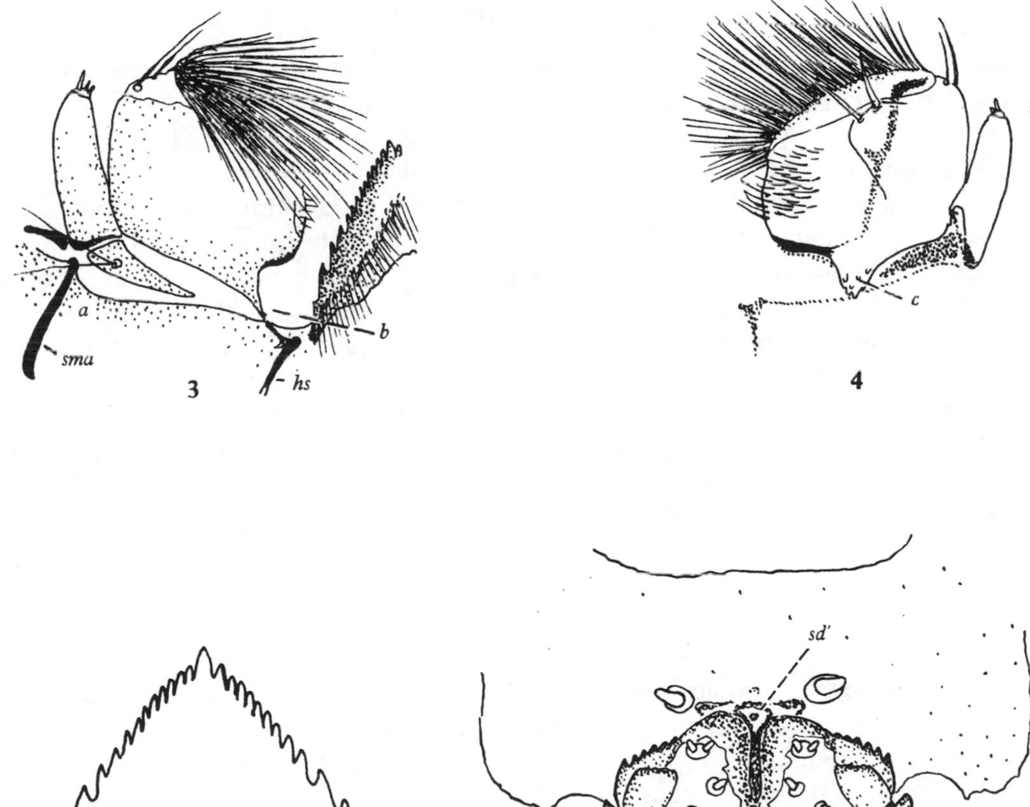

Figure 13

appearance in *A. aegypti* has been figured by Salem and is shown in Fig. 13 (6). It appears very complicated, but actually it seems to consist mainly of two closely approximated conical prominences carrying on their apical membranous apex three small sensory papillae, one double, very like those carried by the maxillary palpi. There seems little doubt that these represent the labial palpi. At the base of the prominences laterally are some peculiar toothed ridges. Beneath the plate are the actively developing labial parts of the pupa, which quickly become very large, occupying a considerable space behind the submaxillary area.

Dorsal to the labial plate (Fig. 13 (6)) is a quadrangular area of chitinisation lying between the ends of the suspensoria with, on each side, near the labial plate, a large sensory papilla between which in the middle line is the opening of the common salivary duct (*sd'*). The plate represents the hypopharynx and lies just before the entrance to the pharynx. On each side it forms the inner wall of a pocket lying behind the corners of the mental sclerite in which in adduction the molar process of the mandible lies.

(*c*) THORACIC AND ABDOMINAL SEGMENTS

THORACIC SEGMENTATION

In the first instar following hatching the three thoracic segments, pro-, meso- and metathorax, are clearly marked off by deep indentations (Fig. 9 (7)). In later stages these segments are compacted to form a single more or less globular mass. The limits of the segments, however, are still indicated by the thoracic hairs and in the case of the last two segments by the position of the puncta or dark spots indicating the position of the repressed spiracles, as also by marks on the surface made by the insertion of muscles, even by slight folds in the cuticle and appearances given by lobes of the parietal layer of fat-body shining through the transparent cuticle. It is characteristic of the larva of *Aëdes aegypti* that, apart from the relatively small basal plates of some hairs, there are no cuticular sclerites indicating the segments.

The surface indications given by the above features show clearly that the greater part of the larval thorax is mesothorax (Fig. 15 (1), (2)), there being only a relatively narrow strip of prothorax anterior to this and a somewhat more extensive but tapering portion ascribable to the metathorax. This is of some importance in judging of the relation to segmentation of internal organs, tracheae, etc.

THE ABDOMINAL SEGMENTS

The abdominal segments, apart from those forming the terminal portion of the abdomen, are, like those of the thorax, entirely devoid of sclerotised plates other than hair bases. The limits of the segments are, however, well defined by folds of the integument. Usually at each intersegment are two dominant folds or creases, an anterior and a posterior, the anterior fold marking the limit of the more anterior segment being usually the more strongly marked and even reinforced by a narrow line of thickening. The two folds mark off between them a fairly broad intersegmental area. They correspond to some extent to the lines at which in many insects the segments are telescoped, the line in front corresponding to the posterior edge of the foremost segment and the second line to that at which the intersegmental membrane is reflected backwards. Usually in the larva there is no telescoping and the intersegment stands out as a slightly bulging annular ring, lying between the broader bulges of the segments. It is into the lines and especially the posterior line that the intersegmental muscles take their origin and are inserted.

Segments 1–7 are all approximately similar in shape being cylindrical or somewhat flattened dorso-ventrally and bulging laterally a little posterior to their middle where the lateral hairs take origin. In size they increase from 1 to 3 and then gradually diminish in size to segment 7, which is the smallest segment apart from the still smaller eighth segment to be described. Each segment towards its anterior end carries laterally the small dark spot indicating the position of the repressed spiracle (Fig. 10 (1); Fig. 15 *st''*) and such representatives of the lateral and other hairs as are denoted in the section on chaetotaxy.

The eighth segment in general resembles the preceding segments except that, besides differences in chaetotaxy, it carries on either side towards its posterior border the row of spinose scales forming the *comb*. As in almost all species of *Aëdes* the comb in *A. aegypti* is devoid of any chitinous plate from which the spines arise. The comb teeth have been figured by Macfie and are shown in Fig. 10 (4). On each side of the median spine is a spine of almost equal thickness behind which are a variable number of much finer spines. The comb scale is symmetrical whereas the somewhat similar pecten scale has asymmetrical denticles, these being present in the main only on the proximal side of the main spine (Fig. 10 (5)). The pentad hairs are a conspicuous feature of the segment. Of these branched hairs 1 and 5 are situated a little beyond either end of the comb opposite the siphon and anal segment respectively, whilst 3 is behind the comb about its middle and half way between the siphon and anal segment. The two small simple hairs 2 and 4 are situated a little ventral of 1 and 3 respectively. Between the eighth and the succeeding anal segment is a fairly broad intersegmental area.

The anal segment is largely composite since at its base in the hypoderm there develops the ninth sternite and from its more distal parts are eventually derived the tenth segment and cerci of the adult. The caudal hairs, four in number, are arranged in two pairs, an inner and an outer, the hairs of each side arising close together from a convex rounded half of the fused basal plates. The outer hair is unbranched (or sometimes double) and longer than the two to four branched inner hair. The plate receives muscular insertions and there are large basal cells well displayed in the transparent newly ecdysed fourth instar larva. The hairs of the ventral fan a short distance from their bases divide into about three branches which for the most part continue without division.

The anal papillae, as shown by Wigglesworth (1933), are not gills but an adaptation to meet the physiological need for osmotic regulation of the haemolymph as described in chapter XXXI when describing the physiology of the larva. In the mature larva the papillae measure about 0·75 mm. in length. They taper towards their free end and are constricted at the base, where there is often a narrow ring of dark chitin and at which point they are very readily detached in manipulations.

(*d*) THE LARVAL CHAETOTAXY

The larval chaetotaxy has been much studied and modifications in numbering of the hairs by different systems are apt to be confusing. A few words therefore of explanation of these systems is desirable before giving the notation here used for *A. aegypti*.

The numbers given by Martini (1923) to the larval hairs of *Anopheles* have been found also applicable to those of mosquitoes generally and with some modification by Puri and others have been widely used. A particularly clear statement of this modified numbering has been given by Marshall, which will be useful when discussing more recent nomenclature. The work of Knight and Chamberlain (1948) on the pupal chaetotaxy of a large number of genera, and that of Baisas and Pagayon (1949) and of Belkin (1951–3) in determining the homologies between the larval and pupal chaetotaxy, has, however, raised the question of possible changes in Martini's system to fit the homologies shown. Such determinations have been based chiefly on the position of the developing pupal hairs as seen beneath the cuticle of the fourth instar larva before pupation, as also to some extent on the character of particular hairs and other considerations.

That the larval and pupal hairs are largely homologous has been clearly shown by Belkin (*loc. cit*) and this author has given a notation that will serve for both the larva and pupa. This notation, so far as it refers to the larva, seems to differ only in some details from the modified Martini system. Belkin (1952) indeed would appear from his statement on p. 116, to have adopted, for the sake of uniformity, the modified Martini notation for the larva, making the necessary changes for conformity in the notation as applied to the pupa. The pupal notation, based on that of a large number of genera, had been very fully worked out by Knight and Chamberlain independent of the larva. But Belkin has shown that this is very easily and simply made to correspond with the notation as in use for the larva. Thus the numbers given by Belkin for the pupa for such a purpose start with 0 (corresponding to 1 of Knight and Chamberlain) and are with a few exceptions consistently one less than the notation of these authors (see list of pupal hairs given later in chapter xv).

It will be shown later when describing the setae of the head, thorax and abdomen that very small modifications in the Martini notation will suffice to bring this into agreement with the Belkin numbering. Such agreement will be clear from the lists of hairs given in the following sections where the Belkin numbers are given in heavy type. (See also the accompanying figures where the Belkin numbers are indicated.)

The larval hairs in *Aëdes aegypti* are in general poorly developed as compared with those of many Culicines. They can be studied in the cast skin of the fourth instar (or of other instars if the changes in the instars are being studied). Or the mounted whole larva may be used. In a dorsal or ventral view the larger lateral hairs are not well displayed, appearing as tufts, the characters of the individual hairs not being well shown (Fig. 15 (1)). In a lateral view the hairs and their relations are often better shown, especially if a suitably mounted bisected larva be used (see under 'Technique', p. 117). In *A. aegypti* branching of most of the larger hairs is by the hair at a short distance from its base splitting up simultaneously in a flat series of branches like the prongs of a table fork compressed laterally (Fig. 15 (3), (4)). Though an impression of multiple branching is sometimes given, even the largest hairs rarely divide into more than four branches. Feathering as seen in *Anopheles* does not occur. The smaller hairs are simple, bifid or 3–4 branched, the branches in the last case usually arising at a large angle from near the base. A certain number of larger lateral hairs arise from basal thickenings, but except for those from which the pleural hairs take origin these are mostly small and few in number compared with the condition in many Culicines. The hairs on the dorsal and ventral surface are mostly small, possess no basal plate and are inconspicuous.

CHAETOTAXY OF THE HEAD

In all, twenty hairs are numbered by Martini and given in Marshall's list. But omitting those on the antenna and maxilla which are best treated in connection with these appendages the number becomes sixteen. If these sixteen are now numbered consecutively and some adjustment made in the sequence of the hairs 10–13, it will be seen from the accompanying list of head hairs (Table 9) that the Martini numbers are in complete agreement with the Belkin numbers here adopted.

Table 9. *List of head hairs*

Original* Martini	Modified† Martini	Belkin	Edwards and Given	Pupal hair	Name of hair
1	1	1	—	—	Inner preclypeal
1*a*	1*a*	0	—	—	Outer preclypeal
2	2	2	—	—	Inner clypeal (*Anopheles*)
3	3	3	—	—	Outer clypeal (*Anopheles*)‡
4	4	4	*d*	—	Post clypeal
5	5	5	*C*	—	Inner frontal§
6	6	6	*B*	—	Middle frontal
7	7	7	*A*	—	Outer frontal
8	8	8	*e*	—	Sutural
9	9	9	*f*	—	Transsutural
14	12	10	—	1	Supra-orbital
12	10	11	—	—	Basal (*Anopheles*)
15	13	12	—	2	Infra-orbital
—	—	(12*a*)	—	—	Subgenal¶
13	11	13	—	3	Subbasal or postmandibular
18	14	14	—	—	Postmaxillary
20	15	15	—	—	Submental

* As given by Marshall (1938) (hairs on antenna and mouth-parts omitted).
† Renumbered as explained in the text.
‡ Possibly the small hair on the clypeus posterior to hair 1*a* (Fig. 14 (2)).
§ For remarks regarding the order of these hairs, see text.
¶ See text.

There is a further notation for the head hairs which ought to be noted upon, namely, that by Edwards and Given (1928). These authors have designated certain of the head hairs that are important in systematic work by the letters *A*, *B*, *C*, *d*, *e* and *f*, a notation that has been very widely used. This notation is also indicated in the list of head hairs here given.

In regard to pupal homologies there is little to be said, as only three of the head hairs are represented in the pupal chaetotaxy, namely, nos. 10, 12 and 13 of Belkin's series corresponding respectively to hairs 1, 2 and 3 of the pupal cephalic hairs. These three hairs in the pupa are situated on the exposed portion of the head cuticle (ocular plate) and between the compound eye and the hollow of the curving pupal antenna, that is, on the genal area as in the larva (Fig. 42 (3) 1, 2, 3, p. 357).

The head hairs are readily identified and besides being given numbers they can be precisely indicated by names, which indicate their position and on this account, wherever this can be done, are helpful. Hair 1*a*, though minute, is present in *A. aegypti*. Actually on each side in this situation there are two hairs, namely, one on the preclypeus, which has been taken to be hair 1*a* (*outer preclypeal* of Puri and Marshall), and one just behind this on the clypeus (Fig. 12 (1)) the nature of which seems doubtful, but which may represent the *outer clypeal* present in *Anopheles* (hair 3 of Belkin). Hair 4 (*d* of Edwards and Given), *post-clypeal*, in *Aëdes aegypti* occupies a forward position which gives to the hairs of this species a superficial resemblance to the clypeal hairs of *Anopheles* (see drawings in Fig. 14 (3) where the hairs as given by Hopkins for certain genera are shown, the figure on the right being after this author's figure for *Aëdes aegypti*). Hairs 5, 6 and 7 are respectively the *inner*, *middle* and *outer* frontal hairs. When these are displaced it is difficult to homologise with certainty the two inner hairs and it is often considered that the middle hair *B* is the one which has advanced forwards closer to hair *d* (as in Hopkins's figure for *A. aegypti* referred to). It seems here, however, to be more correct to number as though it was hair 1 (*C* of Edwards and Given) which has moved forwards, as would appear to be the case from the position of the hairs in the earlier instars.

Hair 8 just internal to the suture and hair 9 just external to the suture, respectively the *sutural* and *transutural*, are well named and defined. Hair 10, *supraorbital*, stands out in both dorsal and ventral view in profile and can scarcely be mistaken. Hair 11, basal, arising at the base of the antenna and present in *Anopheles* is not here represented. Hair 12, *infra-orbital*, would appear to be a small unbranched hair situated ventral to the developing compound eye and a little distance behind the posterior end of the submaxillary apodeme (Fig. 14 (2) hair 12). Hair 13, *subbasal*, or named by Puri, *postmandibular*, arises from the bulge of the head capsule beyond which the mandible is situated (Fig. 14 (2) hair 13). Hair 14, *postmaxillary* is on the hypostomal area and hair 15, *submental*, on the labial area.

Whilst the above accounts for all fifteen hairs of the recognised head chaetotaxy it still would appear to leave one hair on the ventral surface unnoted. The hair here taken to be the infra-orbital is a simple hair situated just ventral to the developing compound eye (Fig. 14 (2) hair 12). In *A. aegypti*, however, there is a small trifid hair a little posterior to this (Fig. 14 (2) hair 12*a*). Which of these is the infra-orbital would therefore seem not certain, though the anterior simple hair fits the description better. Curiously enough it has been difficult to find any information on this point. Very few authors describe the hairs on the ventral aspect of the head. Martini's small original figures in this respect are not very helpful. Marshall's clear figure of the head with numbered hairs gives only one hair (hair 15) situated a little anterior to the supra-orbital hair seen in profile. Pending further consideration the second more posterior hair has been termed the *subgenal* and entered in the list of head hairs as no. 12*a*.

CHAETOTAXY OF THE THORAX

Each segment of the larval thorax carries on each side a number of hairs arranged for the most part in a line encircling the segment laterally, those in the series situated dorsally and ventrally being usually small, those situated more laterally being longer and reaching a

---

Figure 14. Chaetotaxy of head.

1 Dorsal view of head showing head hairs. On the right side the numbers relate to Martini's and Marshall's numbering of the hairs. On the left the lettering is according to Edwards and Given.

2 Ventral view of head showing head hairs. These are numbered after Martini and Marshall.

3 Hairs of dorsal surface of head of: A, *Anopheles*; B, *Culex fatigans*; C, *Aëdes caspius*; and D, *A. aegypti*, indicating changes in relative position of hairs. The hair **d** of Edwards and Given is shown throughout in solid black. Adapted from figures given by Hopkins.

4 Dorsal view of abdominal segments I and II to show numbering of the hairs.

5 Ventral view of the same. The letters *st* indicate the position of the spiracular puncta of the segments.

For explanation of numbers see the text and list of head hairs on p. 211.

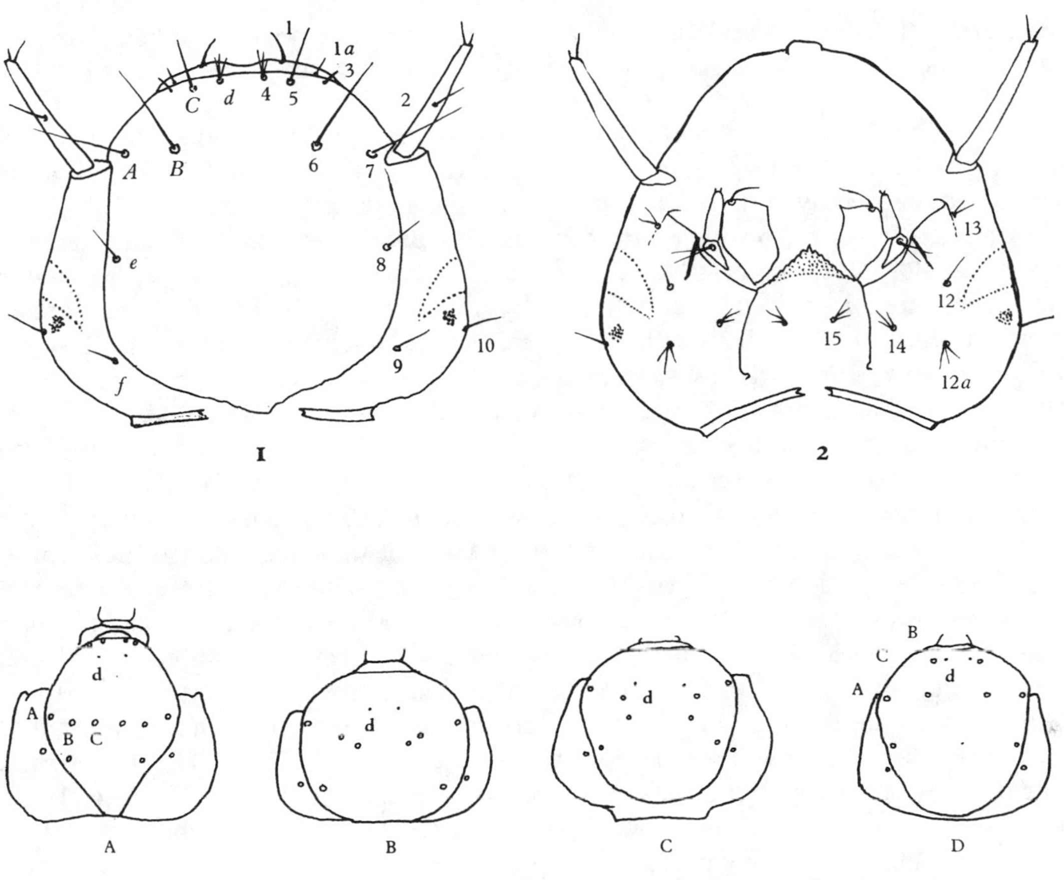
1
1a
3
C
d
4
5
2
A
B
6
7
e
8
9
10
f
1
13
12
15
14
12a
2
d
A
B
C
D
3

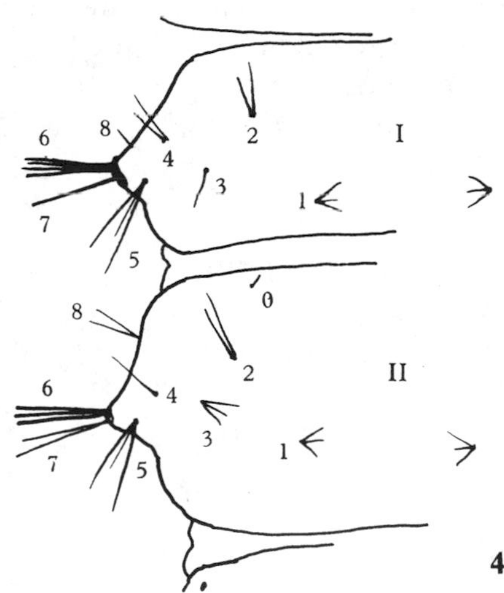
6
8
4
2
I
7
3
1
5
0
II
4
3
4

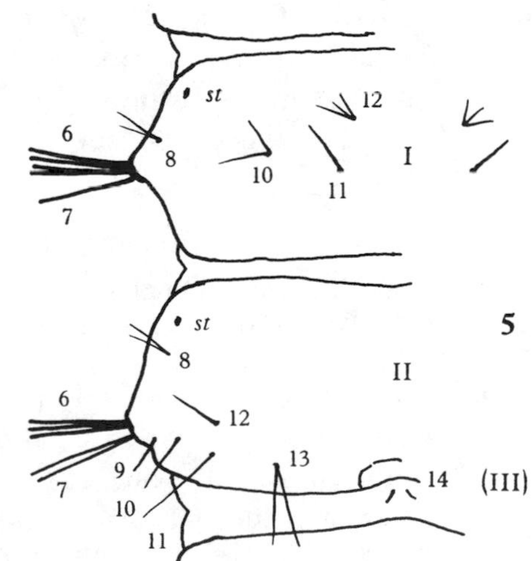
st
12
6
8
10
11
I
7
5
II
12
9
13
14
(III)
10
11

Figure 14

maximum in the *pleural hairs*. The number of hairs in the fourth instar larva on the prothorax is 14 (0–13), on the mesothorax 14 (1–14) and on the metathorax 13 (1–13) (Fig. 15 (1), (2)).

To simplify the numbering in Martini's original notation Puri has made a small modification by which each of the pleural hair tufts is formed of the hairs 9–12, a modification that has been generally adopted. This has the advantage that the pleural hairs which are important structures and obviously homologous are similarly numbered. It fails, however, to give homology to the hairs 8, 14 and 8 on the respective three thoracic segments, though these appear homologous and were all originally given as 8 (see Fig. 15 (2)). Pending more precise information on the homologies of the thoracic hairs some compromise would seem in this case inevitable. Belkin's numbers appear to be in accordance with Puri's modification and all that seems necessary here is to indicate the position of the hairs.

In Fig. 15 the thoracic hairs of *A. aegypti* fourth instar larva are given and numbered according to the modified scheme, which as noted is substantially also that of Belkin (1952). The arrangement and characters of the smaller hairs will be sufficiently evident from the figures. A brief description may, however, be given of the pleural hairs. These correspond with the position of the genital buds of the three pairs of legs. In both the mesothorax and metathorax the hairs (9–12) arise from basal plaques which are continued into strongly developed spines very characteristic of the species. On the prothorax the pleural hairs, though also four in number, are much smaller and arise only from a simple papilliform base without a spine, which is also situated more on the ventral surface.

The characters of the basal plaques and their hairs are shown in Fig. 15. The prothoracic hairs consist of three medium-sized simple hairs with one small bifid hair (Fig. 15 (2)). The mesothoracic hairs 9–12, passing from dorsal to ventral, are (1) a large triple forked hair; (2) a large simple hair arising in front of a serrated ridge on the basal plaque; (3) a large double forked hair; and (4) a very small hair on the posterior portion of the plaque (Fig. 15 (3)). The pleural hairs of the metathorax are almost exactly similar (Fig. 15 (4)). Hair 12 is, however, smaller and simple and hair 11 is minute but present.

As in the case of the head hairs there are but few pupal hairs to be considered in respect to homologies, namely, hairs 4–7 for the prothorax, hairs 8 and 9 for the mesothorax and hairs 10, 11 and 12 for the metathorax. The pupal hairs 4–7 are arranged in two groups, an anterior (4 and 5) and a posterior (6 and 7), situated just dorsal of the basal portion of the curved pupal antenna. According to Belkin (1952) they are probably homologous with hairs 1–5, 4 and 5 being equivalent to 1 and 2, 1 and 3 or 2 and 3, and 6 and 7 with 3 and 4,

---

Figure 15. Chaetotaxy of thoracic and abdominal segments I–VII.

1 Thorax and first abdominal segment, showing hairs of *Aëdes aegypti* larva numbered in accordance with Martini's system as slightly modified by Puri and Barraud. Left side dorsal. Right side ventral.

2 Lateral view of thorax of *A. aegypti*, showing position of hairs numbered as in 1.

3 Basal plaque of mesothoracic pleural hairs.

4 Basal plaque of metathoracic pleural hairs.

5 Lateral view of abdominal segments I–VII, showing position of hairs numbered after Martini. Hairs on segments III–VI are only shown numbered where the numbering is not obvious from that given on segments I and II.

Lettering: *st*, spiracular puncta.

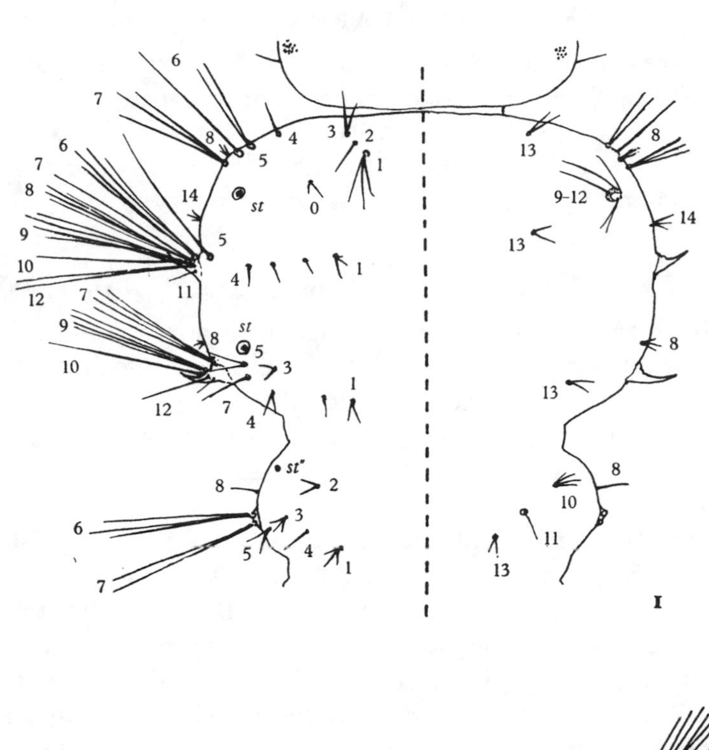

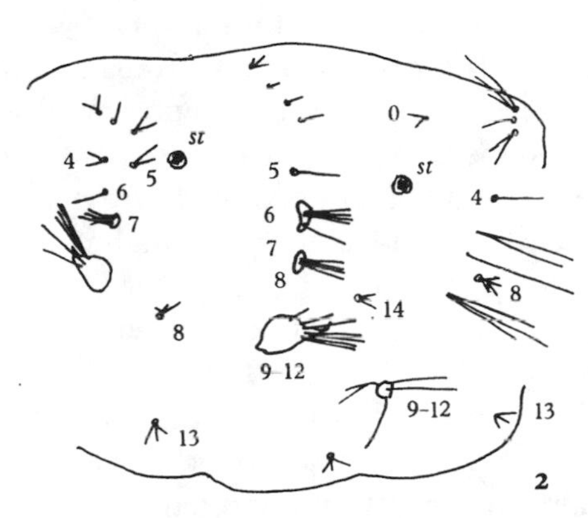

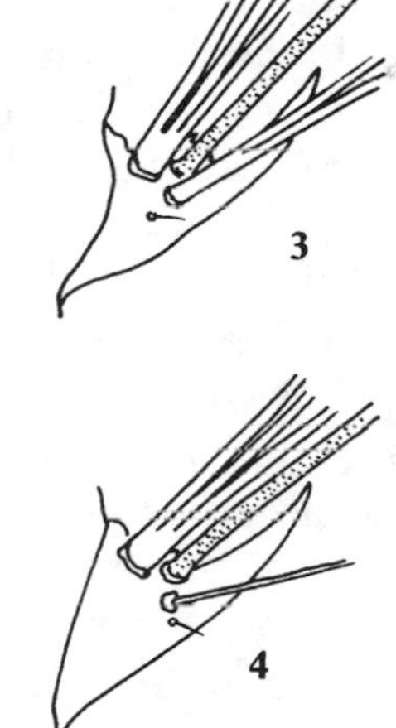

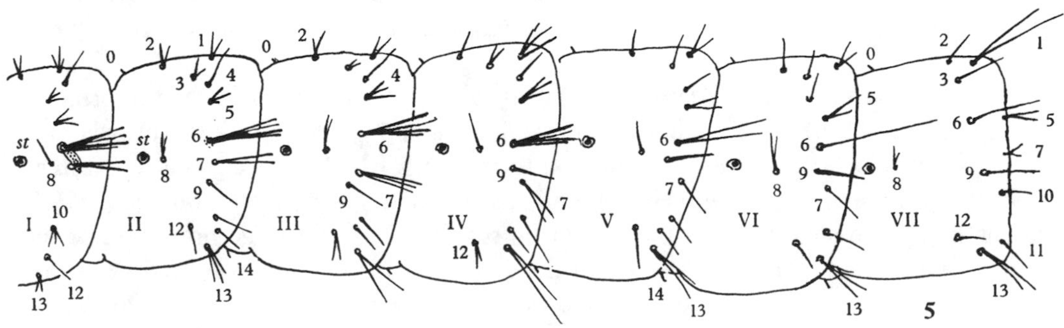

Figure 15

4 and 0 or 4 and 5, the exact homologies being uncertain. The mesothoracic pupal hairs 8 and 9 are situated far back on the mesonotum posterior to the trumpet, 8 being the more mesal hair, possibly homologous with larval hair 1, whilst 9 is 4 or 5. The metathorax (metanotal plate) of the pupa has three hairs on each side, namely, from within outwards 10, 11 and 12, corresponding to the larval metathoracic hairs 1, 2 and 3.

CHAETOTAXY OF THE ABDOMEN (Fig. 15 (5); Fig. 16 (4), (5))

The hairs of the abdomen in *A. aegypti* larva, though many are very small, accord very well with Martini's and Belkin's notations and with the pupal hairs. So far as I have been able to ascertain, however, the larval chaetotaxy for *A. aegypti* or any closely related species of *Stegomyia* has not yet been determined by actual comparison with the developing pupal hairs and the interpretation here given is therefore based on hair positions and characters only.

The maximum possible number of hairs for a segment is 15 (hairs 0–14). These are set out provisionally numbered in the figures noted above, with their numbers in Belkin's notation. An important point drawn attention to by Belkin (1953) is that two hairs, termed *transitory hairs*, present in the fourth instar are not represented in the first instar or in the pupa. These were first considered to be hairs 9 and 11 (Belkin, 1952), but later amended to 9 and 13 (Belkin, 1953). In *Anopheles freeborni* hair 9 first appears in the third instar and 13 in the second instar, neither being present in the pupa. Both in this species, however, are shown present in the first segment of the abdomen of the fourth instar. In *A. aegypti* these two hairs, in the interpretation here given, are also unrepresented in abdomen segment I of the fourth instar, as further commented upon below.

All fifteen hairs are normally developed and as a rule easily identified in *A. aegypti* on segments II–VI. On segment I only 11 hairs are present and allowing for the small hairs 0 and 14 this leaves two hairs to be accounted for. These appear to be 9 and 13 (see figuring on 15 (1) and (5)). On segment VII most of the hairs are very small so that with some exceptions identification is not very certain.

The arrangement of the hairs on the different segments is mostly very similar. Hairs 0–5 are situated dorsally; hairs 6 and 7 are usually large conspicuous lateral hairs; hairs 8 and 9 are slightly ventral to the mid dorso-ventral line, hair 8 being anterior and hair 9 close to hairs 6 and 7; hairs 10–13 are on the ventral aspect and the position of hair 14 is described later. The following are brief descriptions of the hairs.

Hair 0 is a minute simple hair situated dorsally on the anterior portion of the segment on each side. It is most readily seen in the mounted cast skin. It is present on segments II–VIII.

Hairs 1–5 show a characteristic pattern arrangement, that is, hair 1 is most internal and posterior, hair 2 is anterior to hair 1 and hairs 3 and 4 more or less in line towards hair 5 which is larger and branched and situated on the outer border of the segment just internal and posterior to hairs 6 and 7. Hair 1 on most segments is a short branched hair. It increases in size on segments IV and V and on segment VII is a large bifid hair dorsally overhanging segment VIII and the base of the siphon. In *Anopheles* it forms the palmate hairs. Hairs 2–4 are all small and on the more anterior segments are short, bifid, trifid or more branched except hair 4 which is characteristically a simple hair. Hair 5 becomes a well-marked branched hair increasing in size to segment IV and then becoming smaller.

Hair 6 is the largest of the hairs forming the main part of the large lateral hairs. It is usually four branched on segment I where it rises in common with the large single hair 7 from a small basal plate. It is much the same on segment II, but has only an indefinite thickening at its base. It is usually branched on III and IV, bifid on segment V and simple but large on segment VI. On segment VII it is difficult to identify with certainty. Hair 7 on segments I–III is a large simple, bifid or trifid large hair, but beyond becomes small, bifid or simple.

Table 10. *List of hairs of the terminal segments*

	Notation according to		
Situation and name of hair	Martini	Marshall	Belkin
Segment VIII*			
Pentad hair 1	VIII 6	α	1 (VIII 1)
Pentad hair 2	VIII 7	β	2 (VIII 5)
Pentad hair 3	VIII 9	γ	3 (VIII 7)
Pentad hair 4	VIII 11	λ	4 (VIII 10?)
Pentad hair 5	VIII 13	ρ	5 (VIII 12?)
Siphon tube			
Siphon hair	—	—	(IX 1)
Apico-dorsal hair	*h*	—	(IX 2)
Stigmal plate (segment IX)			
Anterior lobe (sensory pits)	VIII 1–3	—	(IX 3–5)
Lateral lobe			
Apical papillary	VIII 4	—	(IX 7)
Basal papillary	VIII 5	—	(IX 6)
Posterior lobe†			
Dorsal surface	*b*	—	(IX 9?)
	c	—	—
	d	—	(IX 8)
Ventral surface	*e*	—	—
	f	—	—
Anal segment (segment X)			
Saddle hair	—	—	(X 1)
Inner caudal hair	—	—	(X 2)
Outer caudal hair	—	—	(X 3)
Ventral fan	—	—	(X 4?)

* Also has hairs 0 and 14.

† Martini's lettering is shown in circles in Fig. 16 (3), (4).

Hair 8, the pilot hair, is easily identified on segments II–VII by its position a little posterior to the spiracular puncta of the segment. On segment I what is taken here to be hair 8 is closer than usual to the lateral hairs and appears to be given as hair 9 in *Anopheles* by Belkin (1953). Its relation to the puncta (which are present on segment I) still, however, holds good and its rather more posterior position may well be due to the narrower segment.

Hair 9 is a rather characteristic simple, but rather thick, hair much the same on all segments II–VII. It lies a little ventral to the lateral hairs.

Hairs 10–13 usually have somewhat the same position as hairs 1–4 dorsally, that is, one hair, taken to be hair 12, lies a little anterior to the hairs 10 and 11. Hair 13 is a conspicuous multi-branched hair that increases in size to a very large hair on segments IV and V.

Hair 14 is one of a pair of very minute hairs situated close together on the intersegmental membrane in the middle line ventrally. Like hair 0 it is best seen in the mounted cast skin. It is usually described as present on segments III–VIII, that is, situated on the anterior border of each segment on which it occurs, but in the larva it is situated on the intersegmental membrane nearer the anteriorly situated segment. Its situation suggests that the pair of little hairs might have some function in connection with the fold of membrane in this situation. In the pupa the hairs do actually arise on the very edge of the thickened cuticle of the hinder segment.

On segment VII the hairs are very small and mostly simple, except for the large hair 1.

On segment VIII the hairs constitute the pentad hairs of Marshall. In the accompanying list of hairs of the terminal segments (Table 10) the notation is given for these hairs, as also for the hairs of the siphon and for the anal segment as given by Martini (1923), and by Marshall with their probable numbering on the new system as given by Belkin. For description of the siphon hairs see (*e*), and for the pentad hairs and hairs of the anal segment description of the larva in (*a*).

(*e*) THE TERMINAL SPIRACLES AND ASSOCIATED STRUCTURES

THE RESPIRATORY SIPHON (Fig. 16)

The respiratory siphon of the mosquito larva was first described by Haller (1878), and a surprisingly complete description is given by Raschke (1887), a translation of which is included by Howard, Dyar and Knab in their description of the parts. The corresponding structures in the larva of *Anopheles* have been very thoroughly described by Imms (1907), and the structure of the siphon in the culicine larva with homologies of the parts has been dealt with by me in a previous communication (Christophers, 1922). The most complete

---

Figure 16. Showing structure of the respiratory siphon.

1 Outline of respiratory siphon and parts at its base to show muscles of the siphon. These are shown in black as they would appear if the near (left) terminal tracheal trunk had been removed leaving only the right-hand muscle of each pair except muscle *c* where both muscles of the pair are shown. *a–f*, the six muscles *a–f* as given in the list of these in the text. For other lettering see below.

2 Apical portion of siphon to show terminal portion of tracheal trunk of left side.

3 Dorsal view of stigmatic plate. Figures without attached dash relate to the inner plates respectively of anterior (1), lateral (2) and posterior (3) perispiracular lobes. Figures with attached dash relate to the outer plates of these lobes in the order given. 3″ indicates the soft membrane at the edge of the lobe. The small letters and figures in circles are hairs as named by Martini, this author's hairs 1–3 being those shown at *so* as sensory organs.

4 Apex of siphon showing closure of the terminal spiracles and muscles *a–e* of right side. Lobes and hairs as in no. 3.

5 The same with lobes (partially) expanded and muscle *f*.

6 Parts immediately after pupation showing median and lateral portions of the ninth tergite.

7 Apical view of peristigmatic lobes in retracted position and situation of the tips of the pupal paddles with paddle hair.

Lettering: *A*, anal segment; *al*, anterior perispiracular lobe; *ap*, spiracular apodeme; *ash*, apical hair of siphon tube; *cpt*, crypts opening on tracheal trunk; *fc*, felt chamber; *fc′*, upper edge of same; *ll*, lateral lobe of ninth tergite (paddle) of pupa; *ll′*, apical hair of pupal paddle (paddle hair); *ml*, median lobe of ninth tergite; *pl*, posterior perispiracular lobe; *pmp*, posterior median process; *tsp*, terminal spiracle; *tt*, tracheal trunk; *tt′*, terminal portion of tracheal trunk with fine taenidia; *w*, cells of postspiracular gland.

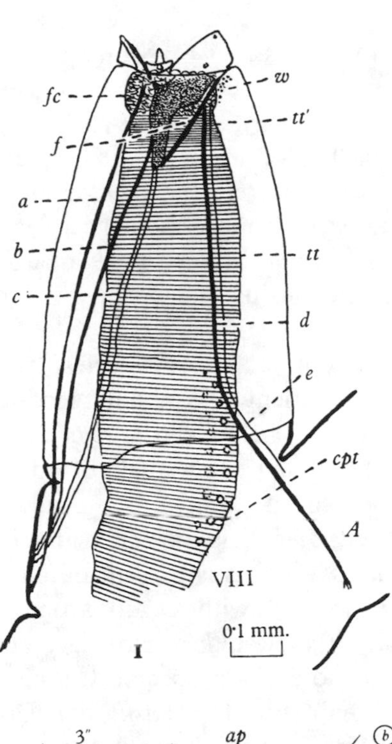

fc
w
tt'
f
a
b
tt
c
d
e
cpt
A
VIII
0·1 mm.
1

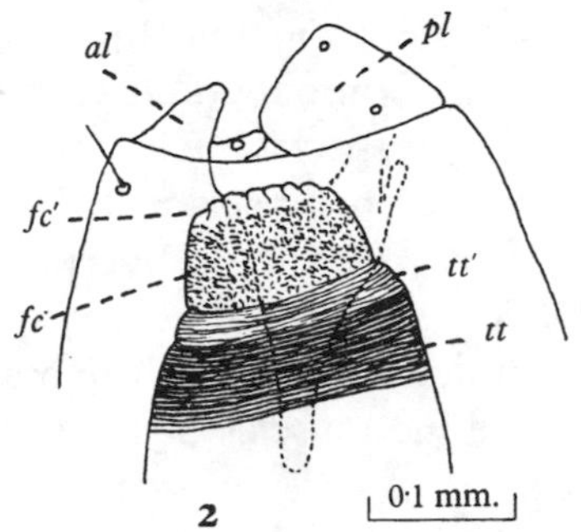

al
pl
fc'
fc
tt'
tt
2
0·1 mm.

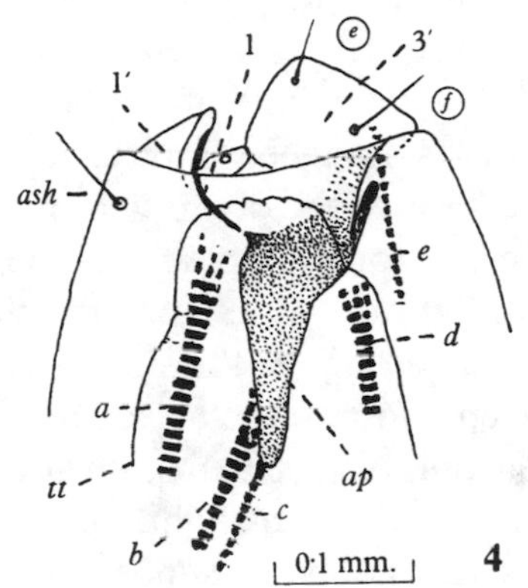

1
e
3'
1'
f
ash
e
d
a
tt
ap
b
c
0·1 mm.
4

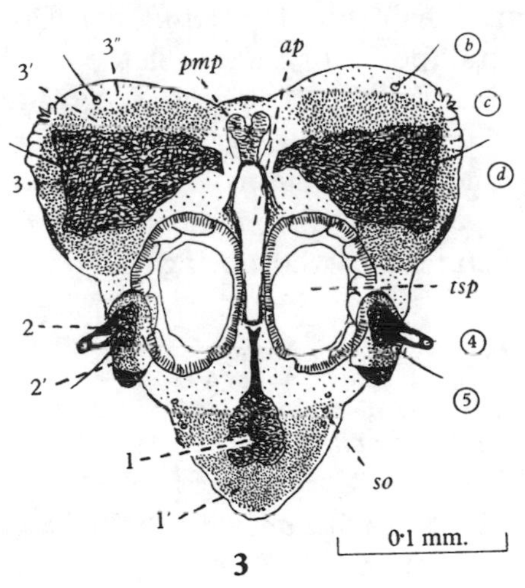

3"
3'
pmp
ap
b
c
d
3
tsp
2
4
2'
5
1
so
1'
0·1 mm.
3

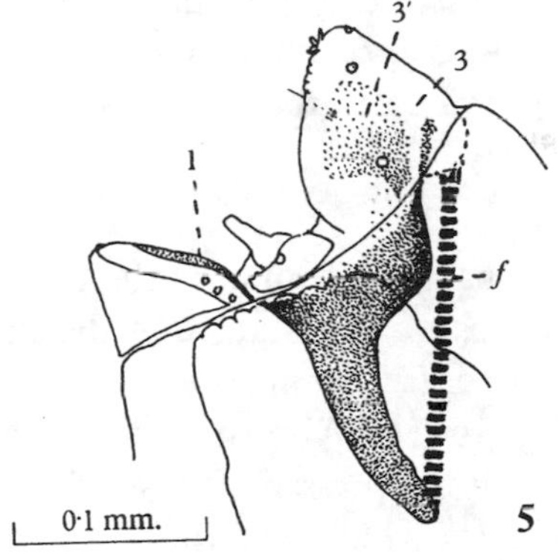

3'
3
1
f
0·1 mm.
5

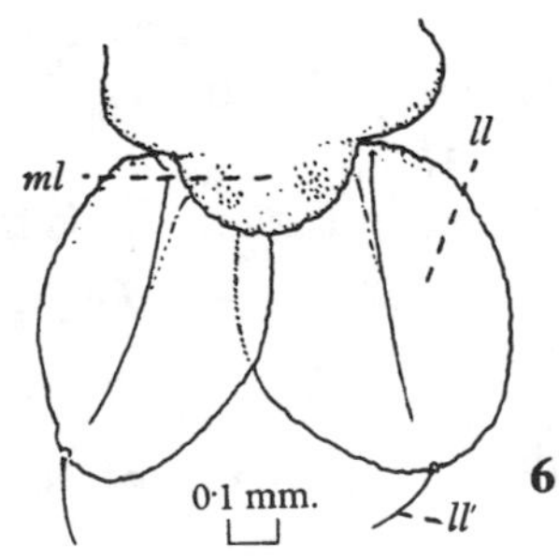

ll
ml
0·1 mm.
ll'
6

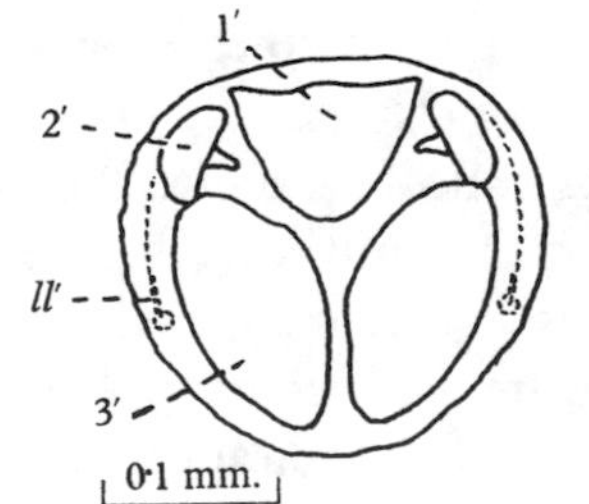

1'
2'
ll'
3'
0·1 mm.
7

Figure 16

description of the apical portions included in the 'stigmal plate' is that by Montchadsky (1930) in his monographic study in which he describes and figures these parts in a large number of species including *Aëdes aegypti* (see also Montchadsky, 1924, 1926, 1927). A more recent paper on the structure of the siphon is that by Keilin, Tate and Vincent (1935) in which they describe the perispiracular glands which secrete the oily or waxy substance which renders the parts about the spiracular openings hydrophobic and so assures their functioning. A still more recent paper is that by Sautet and Audibert (1946) who describe the muscles of the siphon and the mechanism of the closing and opening of the spiracles.

In *A. aegypti* fourth instar larva the main body of the siphon is a strongly sclerotised cylindrical tube tapering in its distal half, the *siphon tube*. Its length is about twice its breadth at the base, the siphonic index (length/diameter at base) being a little over 2. It arises from the posterior portion of the eighth abdominal segment at an angle of about 140° with the line of the body (see ch. XI, 'Hydrostatic balance', p. 253).

At its base the siphon tube terminates abruptly in a sharp edge which is continued around the base of the organ without any projecting lateral tag (*acus*) such as is present in many species, for example, in *Culex pipiens*. The edge ends in membrane without any articular connection. At its apex the siphon tube also ends abruptly in a circular sharp edge. The dorsal surface of the siphon tube (that facing anteriorly) is devoid of any special structures except close to the apex where there is on each side a small hair (Fig. 16 (4) *ash*), hair *h* of Martini, or *apico-dorsal hair* of Marshall.* On the ventral (posterior) surface in the basal half are the two slightly diverging lines of denticled spines forming the *pecten* as already described. These number in the fourth instar from 12–20 in each row, for the most part evenly spaced. A little beyond the most distal spine and a little internal to the line of spines is the *siphon hair*, usually 2–5 branched. At the apex of the siphon are the *terminal spiracles* surrounded by the *perispiracular lobes*, the structures as a whole being commonly referred to as the *stigmal plate*.

THE STIGMAL PLATE

The parts surrounding the terminal spiracles in the larva of *Anopheles* are represented, so to speak, in miniature in the stigmal plate of the culicine larva (Christophers, 1922). The structures forming the stigmal plate in the fourth instar larva of *Aëdes aegypti* are shown in Fig. 16 (3). Almost in the centre are the openings of the spiracles. These are approximately circular in shape with thickened rather transparent edges crossed by transverse crinkles which give them a somewhat scalloped appearance. The openings are frequently referred to as stigmata, each opening being a *stigma*, but there seems no reason for not using the term spiracle as defined by Snodgrass (1935). On their inner side the rim surrounding the openings is narrower and more sclerotised and is continuous with the edges of the hollow structure shortly to be described as the *apodeme*. This arrangement binds the openings firmly together and ensures that, when dragged upon by the apodeme in the closing in of the parts, the openings are pulled in and from being in a somewhat horizontal plane are dragged in apposition to a more vertical one (see also Sautet and Audibert, 1946).

Around the spiracular openings are certain more or less flap-like lobes capable of being opened up at the surface of the water when the spiracles are functioning and of being closed

* For homology of the hairs of the siphon see 'Lists of hairs of terminal segments', p. 217.

over these when the larva submerges. These lobes are five in number, one *anterior*, two *lateral* and two *posterior*. The lobes are rather fleshy, but carry certain plates. These are present on both surfaces leaving the edges rather soft and rounded. Following Marshall it is most convenient to refer to these plates as the *inner* and *outer plates* of the respective lobes. The outer plate of the anterior lobe (Fig. 16 (3) 1′) is broad and half-moon-shaped, the inner plate (Fig. 16 (3) 1) narrower, oval anteriorly and continued posteriorly as a narrow bar linking up with the apodeme. This portion is dragged well into the siphon tube when the parts are closed (Fig. 16 (4) and (5)). On each side at the base of the fleshy lateral edge of the lobe are three sensory hairs present in both anopheline and culicine mosquitoes (hairs 1–3 of Martini; Fig. 16 (3) *so*). The lateral lobes are papilliform in shape, the inner plate (2) carrying at its apex a minute, scarcely visible hair (hair no. 4 of Martini). The posterior lobes are somewhat triangular in shape. Their edges carry some small hairs (hairs *b*, *c* and, at the outer edge of the inner plate, hair *d* of Martini). The inner plate is rather densely sclerotised, rough and of an irregular shape (3). Some rather indefinite thickening connects it with the apodeme. The outer plate is larger and more indefinite (3′). It carries a small hair towards its apex and a larger hair more basally (hairs *e* and *f* respectively of Martini). There can be little doubt that the two plates represent, taken together, the scoop or median plate of the *Anopheles* larva. Between the two posterior lobes in the middle line is a small upright projection picked out by its bifid character, which in certain conditions can project as a small lobe. It may suitably be termed the *posterior median process* (*pmp*).

In the median line between the spiracles and extending back some distance between the inner plates of the posterior lobe is a slit-like opening the hardened edges of which connect up with all the other structures described and are continued into a deeply penetrating hollow chitinisation, the stirrup-shaped piece of Howard, Dyar and Knab and other authors, which will be referred to as the *spiracular apodeme* or shortly as the apodeme. This hollow projection receives the insertion of nearly all the muscles of the respiratory siphon and from this fact and its connections it is, so to speak, the operative structure of the respiratory mechanism. If dragged upon by muscles from below all the flaps are drawn inwards and closed over the tracheal openings, which are themselves also somewhat retracted into the siphon tube (Fig. 16 (4) and (7)). In the absence of such drag and at the water surface the flaps are pulled out by surface tension of the water into their extended position dragging with them the apodeme, a mechanism that can be very prettily shown as later described in a suitably put up detached siphon.

THE TRACHEAL TRUNKS AND THEIR TERMINATION (Fig. 16 (1) and (2))

In the spiracles of insects the trachea with its characteristic taenidia does not usually open directly at the surface of the body but into a secondary chamber, the *spiracular atrium*, so that, strictly, the external opening is that of the atrial orifice. Very commonly too the atrium is specially developed and has a closing apparatus and its walls may be provided with felting or other filtering device for the entering air. To this plan the terminal spiracles of the *A. aegypti* larva largely conform. The large tracheal trunks, after extending through the length of the abdomen, enter with little change into the siphon tube and continue side by side up this structure to within a short distance of the apex. Here just before opening to the exterior at the spiracles they enter a terminal chamber, the *felt chamber* of Keilin, Tate and Vincent, the walls of which are devoid of taenidia and under a low power give a granular effect due to minute microtrichia-like projections covering the inner surface. Their walls have also a certain rigidity and a rounded-off appearance which clearly distinguishes them from the uniform tracheal trunks. The trunks, as in the abdomen, show fairly coarse taenidian markings, but just before entering the felt chambers their walls become somewhat thinner and the taenidia much finer over what appears to be an annular ring of more pliable wall. As noted by Raschke at this point the felt chambers when retracted are to some extent invaginated into the ends of the tracheal trunks proper. This may be seen in preparations where in varying degrees the soft annular portion of tracheal wall is folded, like an intussusception, over the base of the felt chamber (Fig. 16 (2)) (see also Sautet and Audibert, 1946).

Associated with the terminal spiracles are the *postspiracular glands* (Fig. 16 (1) *w*). These have been described in *Culex pipiens* and *Anopheles* by Keilin, Tate and Vincent. As noted by these authors they are most readily seen in the living culicine larva, and especially in the just ecdysed fourth instar larva, due to the fat globules in the gland cells. They also stain with soudan red or soudan black using 75 per cent alcoholic solution (see later under fat-body). In the *Aëdes aegypti* larva these appear in sections as irregular agglomerations of cells with large nuclei enveloping the ventro-lateral region and to a less extent the distal dorsal parts of the felt chambers. In ventral view fat globules are also seen aggregated between the felt chambers along the apodeme. I have not succeeded in tracing any duct. The secretion of the glands is responsible for the hydrofuge character of the stigmal plates.

MUSCLES OF THE SIPHON (Fig. 16 (1) (4) (5))

It seems desirable to describe briefly here the rather specialised muscles of the respiratory siphon in relation to the parts that have just been described rather than to deal with these in the general description of the muscular system of the larva given later. There are five pairs of what may be termed extrinsic muscles of the siphon with attachments outside this organ and one small intrinsic muscle proper to the siphon parts only. These six pairs are as follows (*a*–*f* in figures):

Anterior series

a, A pair of muscles arising from the dorsum of the eighth segment towards its anterior end and inserted into the inner side of the felt chambers anteriorly where their wall links up with the apodeme and with the point of the inner plate of the anterior lobe.

b, A pair of muscles arising from the dorsum of the eighth segment towards its anterior border external to *a* and inserted into the apodeme towards its tip a little above muscle *c*.

c, A pair of muscles arising from the dorsum of the eighth segment and inserted into the tip of the apodeme.

Posterior series

d, A pair of muscles arising one on each side ventro-laterally from the intersegmental line between the eighth and anal segments, and inserted close together in the region of the base of the median posterior process. The muscles pass internal to the muscular bands *f*.

e, A pair of extremely fine thread-like muscles accompanying *d* for much of their course but arising from the dorsal part of the intersegment and inserted in the hypoderm of the inner plate of the posterior lobe. The muscles pass external to *f*.

f, A pair of short but relatively stout muscles intrinsic to the siphon. These take origin from the point of the apodeme and are inserted into the lowest point of the outer plate of the posterior lobe. They appear to be homologous with the longitudinal muscles described by Imms in *Anopheles*.

All except *f* are extremely long and thin muscles. Muscles *a–c* pass up in the angle left anteriorly by the tracheal trunks. Muscles *d–e* pass similarly up the siphon between the trunks posteriorly. It is scarcely possible to number these muscles in accordance with Samtleben's numbers for the muscles of the abdominal segments 1–7, but there seems little doubt that muscles *a–c* are median members of the dorsal longitudinal intersegmental series of the eighth segment passing from intersegmental line 7–8 to intersegmental line 8–9. Muscle *d* may be one of the dorso-ventral series and *e* possibly one of the dorsal longitudinal 9–10, but no certainty can be attached to either supposition. For reference to the muscle series see under 'Muscular system of the larva', p. 313.

THE CHORDOTONAL ORGANS OF THE SIPHON (Fig. 35 (10) (11))

A conspicuous feature of the siphon is the occurrence of a series of well-developed chordotonal organs. Demoll (1917) notes that there are three pairs of chordotonal organs in the siphon of the *Culex* larva, but does not describe these. Nor does any other author appear to have done so. There are three, possibly four, pairs in the siphon of *Aëdes aegypti*, three large passing through the whole length of the siphon and one small limited to the apical portion of the siphon. These are described later in the section on sensory organs of the larva. It seems desirable, however, briefly to record them here. The four pairs are as given below.

1. A posterior pair with the basal cell attached to the membrane just below the lower margin of the siphon tube at a position a little *internal* to the row of spines of their side and with the filaments passing up the siphon tube posteriorly to be attached by several branches to the hypoderm at the apex of the siphon tube below the posterior lobes. The filaments pass close to and are connected with the base of the siphon hair.
2. A lateral pair with the basal cell attached to the membrane just below the siphon edge in a position a little *external* to the row of spines of their side. Their filaments pass up the tube outside the tracheal trunk of its side and enter the lateral lobes of the spiracular plate.
3. An anterior pair passing from the anterior dorsal border of the eighth segment to the felt chamber of its side at the point of insertion of muscle *a*.
4. A small triangular cell with one filament attached to the wall of the siphon anteriorly a short distance below the apical hair and two passing respectively to the upper part of the felt chamber and the tracheal trunk where this ends in the ring of fine taenidia.

Two of these pairs, nos. 1 and 2, have well-marked and undoubted scolopale appearances at their base. The third seems undoubtedly of the same nature, but has a less distinct basal sense organ, and the fourth is a much smaller and more delicate structure, but appears to be of the chordotonal nature.

SEGMENTAL RELATIONS OF THE TERMINAL SPIRACLES AND ASSOCIATED STRUCTURES

The respiratory siphon is a peculiar structure the segmental relations of which are not at once apparent. In a previous communication (Christophers, 1922) I have dealt in some detail with the structure and development of the respiratory siphon showing that in the young fourth instar larva the hypoderm of this structure forms a conical protrusion of the shape of the siphon tube. As development proceeds the sides of this hypodermic cone are sheared off by fissures starting at the apex, the pieces so separated being the future pupal paddles. These extend through the whole length of the siphon, their apices lying beneath the posterior lobes of the stigmal plate. The median portion forms a separate cone with the spiracles at its apex. This position of the parts must not be confused with the growth of a genital bud arising from some point on the hypoderm and developing into an organ. The paddle rudiments are formed *in situ* from the hypoderm of the siphon tube, that is, the hypoderm of the sides of the siphon is what will become the paddle. The paddles of the pupa can be shown to be the lateral portions of the ninth tergite and the lateral wall of the siphon tube, so far as its hypoderm is concerned, is therefore the lateral lobe of the ninth tergite.

The central portion left by the fissures is at the time of pupation withdrawn from the siphon tube, but unlike the paddle rudiments, which harden and become pupal structures, shrinks rapidly, though not so rapidly that in the newly emerged pupa it cannot still be recognised as forming a semicircular plate lying behind the eighth tergite of the pupa (Fig. 16 (6) *ml*). It can eventually in fact be traced till it forms the median portion of the ninth tergite in the adult. If this median portion be examined shortly after pupation, it can readily be seen that on its lower surface are the pits left where the cast-off tracheal trunks have been drawn out, and in sections the pits show what has been the tracheal epithelium around the cast-off trunk linings.

It remains only to ascertain the path by which the spiracles have travelled to their abnormal position. They can scarcely be other than spiracles of the eighth segment which is the last segment to bear spiracles through the whole of the upper insects except the Diptera where it has been eliminated in the imago. There are no such structures in the higher insects as spiracles of a ninth segment. If this be so, then the eighth spiracles in the mosquito larva have been drawn up behind the ninth segment in whole or in part for, as has been seen, the pits from which their tracheae have been drawn lie behind the shrunken ninth tergite. In this connection it is of interest to compare the culicine siphon with the respiratory structures in the larva of *Anopheles*. It has already been noted that the stigmatic plate of the culicine siphon shows in miniature the parts surrounding the terminal spiracles of the *Anopheles* larva. Lying external to these parts in the *Anopheles* larva is, anteriorly on the eighth segment, a fold which in a previous communication (1922) I named the presiphonic fold, a fold which is continuous laterally with the areas delimited by the bar at the base of the comb. It is in these lateral areas that in *Anopheles* the pupal paddles develop, their tips being in the sides of the scoop, which correspond to the posterior lobes of the culicine stigmatic plate. It requires but little modification to see in these parts areas which have become the sides of the siphon in the culicine larva. The eighth spiracles would then appear to have migrated backwards between the ninth tergite and sternite with the result that the ninth sternite has been left in position at the base of the anal segment ventrally, where later in the adult stage it is seen as the narrow ninth sternite bearing in the male the gonocoxites, whilst the ninth tergite is shifted bodily dorsally to form in the culicine larva a part of the spiracular plate, or in the anopheline larva the presiphonic fold, its lateral parts becoming the pupal paddles and its central part the ninth tergite in the imago.

REFERENCES

Baisas, F. E. and Pagayon, A. V. (1949). Notes on the Philippine mosquitoes. XV. The chaetotaxy of the pupa and larva of *Tripteroides*. *Philipp. J. Sci.* **78**, 43–72.

Barraud, P. J. (1934). *The Fauna of British India. Diptera*. Vol. v. Taylor and Francis, London.

Belkin, J. N. (1951). A revised nomenclature for the chaetotaxy of the mosquito larva. *Amer. Midl. Nat.* **44**, 678–98.

REFERENCES

Belkin, J. N. (1952). The homology of the chaetotaxy of immature mosquitoes and a revised nomenclature for the chaetotaxy of the pupa. *Proc. Ent. Soc. Wash.* **54**, 115–30.

Belkin, J. N. (1953). Corrected interpretations of some elements of the abdominal chaetotaxy of the mosquito larva and pupa (Diptera: Culicidae). *Proc. Ent. Soc. Wash.* **55**, 318–24.

Blanchard, R. (1905). *Les moustiques: histoire naturelle et médicale.* Rudival, Paris.

Bonne-Wepster, J. and Brug, S. L. (1932). The subgenus *Stegomyia* in Netherlands-India. *Geneesk. Tijdschr. Ned.-Ind.* Byblad 2, 1–119.

Christophers, S. R. (1922). The development and structure of the terminal abdominal segments and hypopygium of the mosquito with observations on the homologies of the terminal segments of the larva. *Ind. J. Med. Res.* **10**, 530–72.

Cooling, L. E. (1924). The larval stages and biology of the commoner species of Australian mosquitoes, etc. *Aust. Dept. Hlth Serv. Publ.* (*Trop. Div.*), no. 8.

Crawford, R. (1933). The structure of the head of some anopheline larvae. *Malayan Med. J.* **8**, 25–38.

Demoll, R. (1917). *Die Sinnesorgane der Arthropoden ihr Bau und ihre Funktion.* Braunschweig.

Doane, R. A. (1910). *Insects and Disease.* Constable and Co., London.

Dyar, H. G. and Knab, F. (1906). The larvae of Culicidae classified as independent organisms. *J. N.Y. Ent. Soc.* **14**, 169–230.

Edwards, F. W. and Given, D. H. C. (1928). The early stages of some Singapore mosquitoes. *Bull. Ent. Res.* **18**, 337–57.

Eysell, A. (1913). *Die Stechmücken.* In Mense's *Handb. TropenKr.* Ed. 2, **1**, 97–183.

Goeldi, E. A. (1905). *Os mosquitos no Para.* Weigandt, Para.

Haller, G. (1878). Kleinere Bruchstücke zur vergleichenden Anatomie der Arthropoden. I. Ueber das Atmungsorgan der Stechmückenlarve. *Arch. Naturgesch.* Jahrb. 44, **1**, 91–5.

Hopkins, G. H. E. (1936). *Mosquitoes of the Ethiopian Region.* Vol. I, *Larval Bionomics and Taxonomy of Culicine Larvae.* Ed. 2. London, 1951, with notes by Mattingly. Brit. Mus. (Nat. Hist.).

Howard, L. O. (1901). *Mosquitoes: How they Live, How they Carry Disease, How they are Classified, How they may be Destroyed.* McClure Phillips and Co., New York.

Howard, L. O., Dyar, H. G. and Knab, F. (1912). See references in ch. I (*f*).

Imms, A. D. (1907). On the larval and pupal stages of *Anopheles maculipennis* Meigen. *J. Hyg., Camb.* **7**, 291–316.

James, S. P. (1899). Collection of mosquitoes and their larvae. *Indian Med. Gaz.* **34**, 431–4.

Keilin, D., Tate, P. and Vincent, M. (1935). The perispiracular glands of mosquito larvae. *Parasitology*, **27**, 257–62.

Knight, K. L. and Chamberlain, R. W. (1948). A new nomenclature for the chaetotaxy of the mosquito pupa, based on a comparative study of the genera. *Proc. Helm. Soc. Wash.* **15**, 1–10.

Kumm, H. W. (1931). Studies on *Aëdes* larvae in South Western Nigeria and the vicinity of Kano. *Bull. Ent. Res.* **22**, 65–74.

Macfie, J. W. S. (1917). Morphological changes observed during the development of the larva of *Stegomyia fasciata*. *Bull. Ent. Res.* **7**, 297–307.

Marshall, J. F. (1938). *The British Mosquitoes.* Brit. Mus. (Nat. Hist.). London.

Martini, E. (1923). Ueber einige für das System bedeutungsvolle Merkmale der Stechmücken. *Zool. Jb.* **46**, 517–90.

Martini, E. (1929). *Culicidae.* In Lindner's *Die Fliegen der Palaearktischen Region.* Parts 11, 12, 74–85.

Montchadsky, A. (1924). Neue Ergebnisse über Anatomie und Funktion des Atmungsapparats der Culicidenlarven. *Trav. Soc. Nat. Petrogr.* **54**, no. 1.

Montchadsky, A. (1926). Larva of *Aëdes* (*Ochlerotatus*) *pulchritarsis* Rond. var. *Stegomyina* Stack. and Monthc. nov. from Turkestan. *Bull. Ent. Res.* **17**, 151–7.

Montchadsky, A. (1927). Morphologische Analysis eines bisher nicht beachteten systematischen Merkmals der Culicidenlarven. *Bull. Acad. Sci. U.R.S.S.* nos. 5–6.

Montchadsky, A. (1930). Die Stigmalplatten der Culiciden Larven. *Zool. Jb.* **58**, 541–636.

PURI, I. M. (1931). Larvae of anopheline mosquitoes with full descriptions of those of the Indian species. *Indian Med. Res. Mem.* no. 21.

RASCHKE, E. W. (1887). Die Larve von *Culex nemorosus*. Ein Beitrag zur Kenntnis der Insekten Anatomie und Histologie. *Arch. Naturgesch.* **1**, 133–63.

SALEM, H. H. (1931). Some observations on the structure of the mouth parts and fore intestine of the fourth stage larva of *Aëdes* (*Stegomyia*) *fasciata* (Fab.). *Ann. Trop. Med. Parasit.* **25**, 393–419.

SAMTLEBEN, B. (1929). Anatomie und Histologie der Abdominal- und Thoraxmusculatur von Stechmückenlarven. *Z. wiss. Zool.* **134**, 179–269.

SAUTET, J. and AUDIBERT, Y. (1946). Études biologiques et morphologiques sur certaines larves de moustiques, etc. *Bull. Soc. Path. Exot.* **39**, 43–61.

SNODGRASS, R. E. (1935). *The Principles of Insect Morphology*. McGraw-Hill Book Co., New York and London.

THEOBALD, F. V. (1901). See references in ch. I (*e*).

THOMPSON, M. T. (1905). The alimentary canal of the mosquito. *Proc. Boston Soc. Nat. Hist.* **32**, 145–202.

WALKIERS, J. (1935). Contribution à l'étude des larves des Stegomyias de Maladi et Thysville. *Ann. Soc. Belge Méd. Trop.* **15**, 469–73.

WESCHE, W. (1910). On the larval and pupal stages of West African Culicidae. *Bull. Ent. Res.* **1**, 7–50.

WIGGLESWORTH, V. B. (1933). The function of the anal gills of the mosquito larva. *J. Exp. Biol.* **10**, 16–26.

X

THE LARVAL INSTARS

(*a*) GENERAL REMARKS ON THE INSTARS

No account of the larva would be complete that did not take into consideration the different instars and changes which the larva passes through in the course of its growth. The most complete account of these changes in *Aëdes aegypti*, including measurements of the parts instar by instar, is that by Macfie (1917).*

In *A. aegypti*, as in all mosquitoes, there are four larval instars, each instar terminating in a moult or *ecdysis*. The instars are periods of growth in which the larva starts at about the level it reached at the end of the previous instar and ends with that with which it starts in the next instar. At such time the soft parts increase in size whilst the hard cuticular structures such as the head, spines and hairs remain unchanged. The result is that whilst the hard parts in the different instars show throughout characters special to that instar the changes in the soft parts and relation of these to the hard give rise to a similarity in appearance of corresponding stages of growth of all instars. Thus each instar starts with a relatively large head, a short respiratory siphon and relatively very long lateral hairs, and ends with a small head, longer siphon and much less conspicuous lateral hairs, the general effect, except in size, depending more on the stage of growth than on the particular instar concerned (Fig. 17 (1), (2), (3); Fig. 18 (1), (2)).

IDENTIFICATION OF THE INSTARS

Except for the first instar, which has a number of special features, most changes in chaetotaxy and other external characters consist of an increased branching of the hairs, increase in the number and branching of spines and increase in complexity of parts such as the respiratory siphon, anal segment and mouth-parts. There are certain characters, however, which are specific to certain instars. The *first instar* at all stages is at once identified by the presence on the dorsum of the head of the *egg-breaker*. Further characters in which this instar differs will be given later. But the presence of the egg-breaker is so immediately diagnostic that no other characters are necessary for identification. The *second* and *third instars* are, with a little experience, to be identified by their size and general appearance. The most certain method of identification is by measurement of the *transverse diameter of the head*. In the successive instars I–IV this should be approximately 0·3, 0·45, 0·65 and 0·95 mm. In this respect the third instar in the later stages is apt to be somewhat variable. The dark head at this stage should, however, suffice to distinguish any well-grown third instar larva from the fourth instar larva in which the head, unlike all other instars, never fully darkens. The fourth instar also in all stages shows the rudiments of the pupal respiratory trumpets.

* For other descriptions of the instar characters see Howard, Dyar and Knab (1912), Tanzer (1921, *Anopheles Theobaldi, Culex* and *Corethra*), Cooling (1924, *Aëdes aegypti*), Puri (1931, *Anopheles*), Tate (1932, *Orthopodomyia*), Marshall (1938) who gives a good summary of the changes, Jobling (1938) especially changes in the respiratory siphon in *Culex*, Wigglesworth (1938) for description of the early first instar in *Aëdes aegypti*.

In cultures the cast skins appearing at the surface of the water also show the stage at which the majority of the larvae have arrived. At a little over 24 hours from hatching there appear in numbers the minute skins of the first instar, the dark heads visible to the naked eye only as tiny black specks. On the second day the skins of the second instar are readily seen as skins floating at the surface, as also on the third day are the large third instar skins. The skins at the fourth ecdysis are obviously those of the full-grown larva. They appear only some 3 days after the smaller third instar skins and when pupae are present.

RECOGNITION OF STAGES OF THE INSTARS

Special interest relates to the stage in each instar immediately following ecdysis and that immediately preceding change to the next instar. One reason is that these two stages constitute definite points in growth, which at other times during the instar proceeds without any structural landmarks. These are, therefore, the stages most suitable for measurements or other determinations relative to the instars. The two stages will be referred to respectively as the *immediate* and the *pre-ecdysis* stage, or, used in same sense, *post-ecdysis* may be substituted for immediate.

These stages have certain characteristics. Thus immediately following ecdysis, especially in the early instars, a most striking feature is the large ballooned, unpigmented and almost transparent head (Fig. 17 (1 A), (2 A), (3 A)). This later hardens and darkens, becoming in the first three instars almost black. It is thus easy by naked eye to pick out from other more normal larvae any which have recently undergone ecdysis. Where material is required for measurements such as are given later when dealing with growth, a useful technique is to collect in the first place such larvae in the instar preceding that required as show themselves on the point of ecdysis. This is done as explained below, and these larvae can then be kept under observation and removed immediately they leave their skins. It is advisable to wait a few minutes for the head structures to adjust themselves. Newly ecdysed forms are also very useful for making preparations to show larval structure since at this stage fat-body is at a minimum and the head, and even to a considerable extent the body, is so transparent that the internal organs and details of structure show to best advantage.

The most conspicuous feature of the pre-ecdysis stage is the appearance under the cuticle with approaching ecdysis of the dark hairs of the next instar wrapped round the body and clearly visible as dark bands across the thorax and abdominal segments (Fig. 17 (2 B), Fig. 18 (1 B)). Larvae in this state can usually be picked out by naked eye owing to their

---

Figure 17. General characters of instars.

1 First instar larva. A, newly hatched larva immediately after leaving the egg. A′, the same when the head has swollen to the normal condition for the instar. B, first instar larva ‘pre-ecdysis’.

2 Second instar larva. A, ‘immediate’; B, ‘pre-ecdysis’.

3 Third instar larva ‘immediate’.

4 First instar larva. *a*, *c*, lateral view of terminal segments at beginning and end of instar; *b*, *d*, posterior view of siphon at these times.

5 Second instar larva. Lettering as 4.

6 Third instar larva. Lettering as 4.

Figures 1–3 (also 1 and 2 of Fig. 18) are at the same magnification, and so approximately are Figs. 4–6.

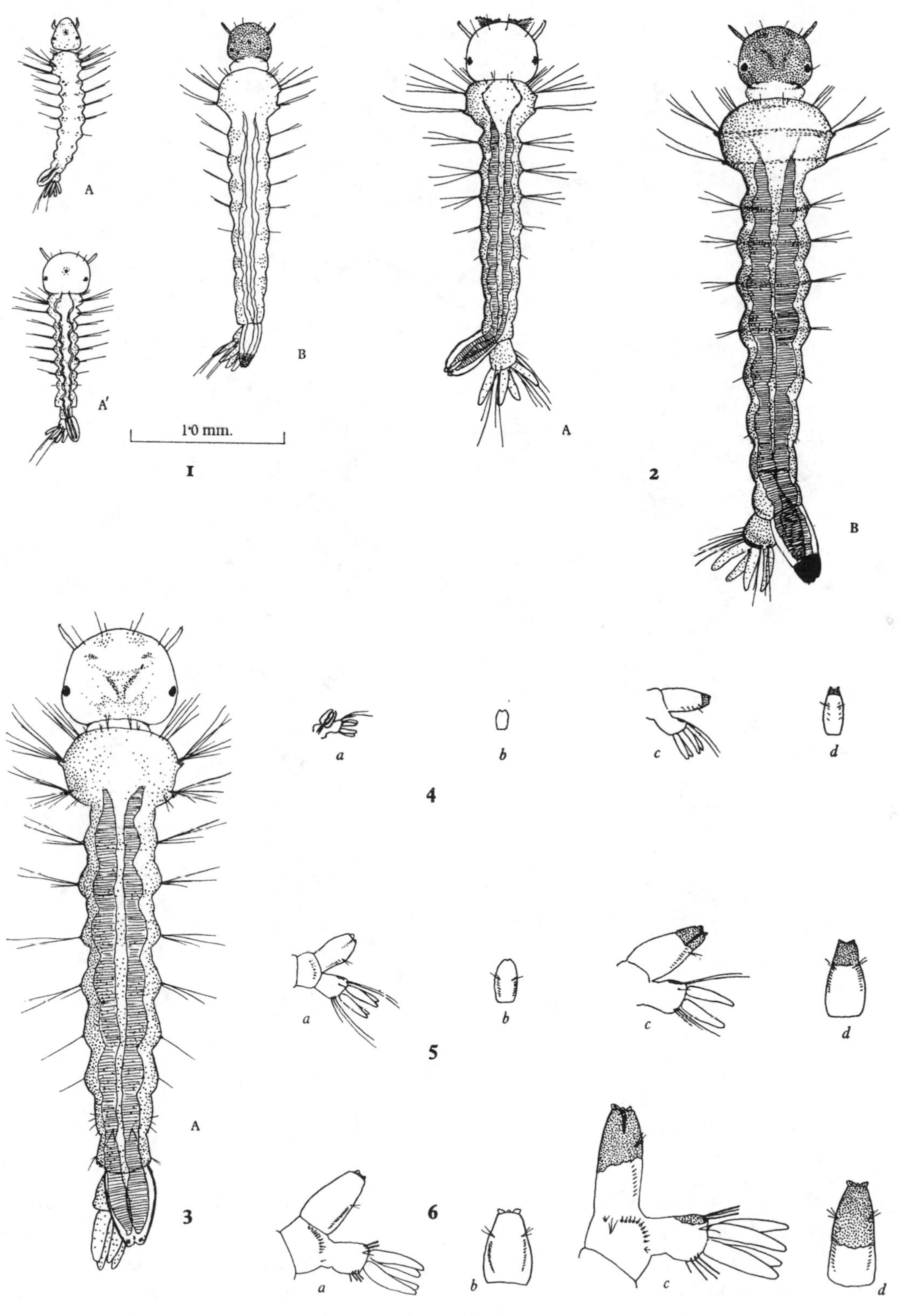

Figure 17

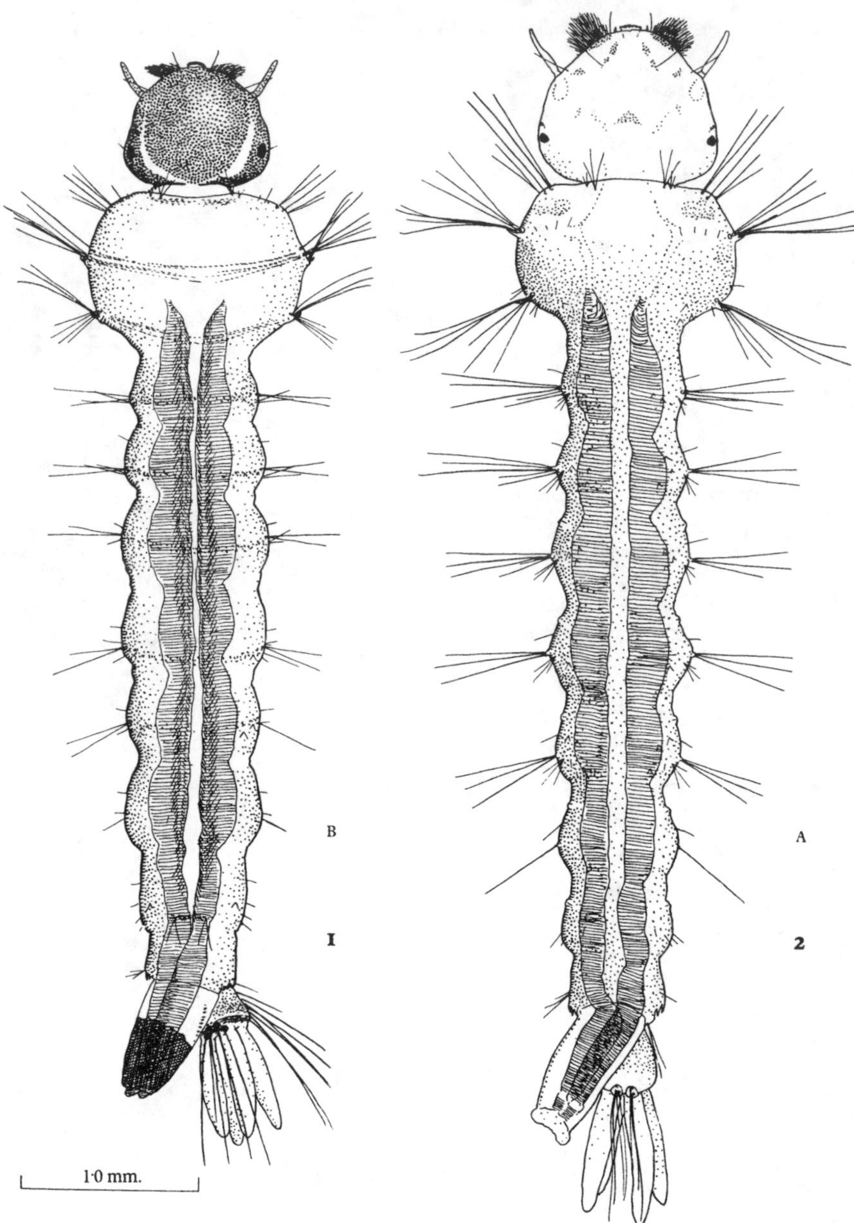

Figure 18. General characters of instars (cont.).

1 Third instar larva 'pre-ecdysis'.

2 Fourth instar larva 'immediate'.

The figures are at the same magnification as those of Fig. 17 (1)–(3).

slightly darker coloration produced by these bands. In the fourth instar, however, where there are no such hairs, since the oncoming pupal hairs are not conspicuous, the approach of ecdysis is readily judged by the degree of darkening of the pupal trumpet rudiments. These lie in the shoulders of the thorax and are very conspicuous in the later stages of this instar. At first pale they become as the larva matures first yellow and eventually brown, by which time the larva is on the point of pupation.

(*b*) LARVAL ECDYSIS

Of the physiological changes occurring at and responsible for ecdysis it will be sufficient for the present to note that the process begins with separation of the epidermal cells from the old and the laying down of a new cuticle, the space between the new and the old becoming filled with fluid, *ecdysial fluid*. Further the ecdysial fluid, by reason of its enzyme or enzymes, digests and dissolves the thick endocuticular inner layer of the old cuticle leaving only the thin outer layers. Hence the cast skin, as weighed for instance in the case of the mosquito larva, would not necessarily be that of the whole cuticle as originally laid down. Further, the cuticle of the larva over most of the body lacks any hardened layer (exocuticle) except for such parts as the head capsule, respiratory siphon and hair plaques. Of these hard parts it is only the head capsule that offers any bar to removal at ecdysis of the contained larval parts, though some tearing of the thin cuticle near the head may take place. In the case of the head capsule, dissolving of the endocuticle leaves certain lines of weakness where the exocuticle is deficient, *ecdysial lines*. Such are the lines that have the appearance, or actually are, the arms of the epicranial sutures and at ecdysis it is these that break to allow the new larval head to emerge. At ecdysis the head capsule is torn along these lines as far back as their termination at the neck and nearly as far forwards as the antennal prominences, thus allowing a broad wedge-shaped portion of the dorsum of the head to be pushed forwards as a flap and the cervical collar to part at this point. Apart from the head capsule and the delicate stigmatic cords there is little to prevent the body of the larva bathed in ecdysial fluid from being drawn out. Besides the external surface of the body similar processes take place in the cuticular lining of the tracheal trunks and vessels enabling these similarly to break at lines of weakness and so be drawn out as described below.

As ecdysis approaches, especially the final or pupal ecdysis, the larva tends to cease feeding and to remain at rest at the surface. In the early instars ecdysis is usually very rapid and the process is most readily followed in moulting of the third or fourth instar larva, though in the latter case it is somewhat modified by the fact that it is the pupa and not another larval instar to which it gives rise. One of the first signs that ecdysis is about to take place is the appearance of dark bands across the thorax due to the circularly wrapped lateral hairs of the next instar shining through the cuticle (Fig. 17 (2B); Fig. 18 (1B)). Under the microscope these bands are very conspicuous and are useful as enabling one to pick out at once larvae that are about to moult. The tracheal trunks are also now seen to have double contours, the original and still only functional trunk lying within the larger future trunk that is filled with fluid. It may now perhaps be observed that the ecdysial lines on the head capsule are parted, especially in the third instar where some widening of these lines may be present for some little time before active splitting occurs. With the active onset of moulting the pale transparent new instar head, covered with droplet-like marks on the cuticle is seen exposed as it forces forwards the flap-like dorsum of the head. Active pulsation movements are now seen in the interior of the head. These are due to contractions of the pharynx making swallowing movements as described by Wigglesworth (1938). Vermicular movements and twitchings also take place in the abdomen and there may be occasional lashing movements of this part. Twitchings are seen also in the occipital region where muscles of the mandibles and maxillae are assisting in the withdrawal of the new mouth-parts from the old. The tracheal trunks may now be seen to be broken at points distal to the cords passing from them to the puncta. Shruggings and flexures of the head accompanied by powerful muscular contractions eventually free the head, tearing the soft membrane of the neck and thorax and finally thrusting the old head capsule, now completely discarded, downwards. Almost imperceptibly now the thin cuticle of the body is slipped backwards dragging with it the broken-up pieces of the tracheal trunks that have been drawn out through traction on the cords. Probably also, to judge by appearances, the fragments of cord are partly squeezed out through the now open soft spiracular openings by movements of the body. As the thin wrinkling cuticle slips back, the long thoracic hairs are liberated and straighten. After the skin has come to lie in a wrinkled mass about the posterior extremity of the larva, the siphon and anal papillae are withdrawn and the larva wriggles free. Though the description is long the whole process, especially in the early instars, happens with surprising rapidity. With a healthy larva the time taken may be less than a minute. Where conditions are adverse it may take half an hour or more. The chief delay is the time taken to liberate the head. Once this is free full emergence usually quickly follows.

The discarded skin, at first closely compacted, after a time loosens and floats like a shadow larva with its siphon still attached to the surface film. In a good culture it may be seen so floating with hundreds of others after each ecdysis period.

The new larva now has an almost transparent head which slowly darkens, becoming in the first three instars almost black. In the fourth instar larva it never darkens so fully. The tracheal trunks, especially in the second and third instars are, after ecdysis, relatively enormous and the larva is extremely buoyant so that it floats even in spirit and may often be seen floating at the surface when first put into fixative when larvae of other stages sink.

Extraction of the tracheal trunks is according to a general plan as follows. With the siphon come away the tracheal trunks up to the level of the break in the trunks towards the anterior end of the seventh segment, bringing with them the branches supplying the anal segment. In the other abdominal segments the break in the trunk occurs behind the point of attachment of the cord, so that it is the portion of trunk from the preceding segment to that carrying the puncta that is drawn out. In the thorax the same plan holds good. The greater part of the trunks on each side are drawn out with the cords attached to the puncta of the first abdominal segment. A portion in front of this is drawn out by the cord from the metathoracic puncta and the lining of the trunks in the prothorax with that of the head tracheae by the cord from the puncta lying between the mesothoracic and metathoracic hair series, which has here been considered to be the mesothoracic puncta (Fig. 29, p. 301; Fig. 48 (1), p. 391).

The skins left after ecdysis, especially that of the fourth instar larva, have been much used for systematic and descriptive purposes since all the cutaneous structures are exceedingly well displayed and all possibility of error is avoided where the corresponding imago has been bred out and preserved. The skins also show with remarkable clearness the structures connected with the mouth-parts and pharynx. Such skins, when they have suitably relaxed, can be mounted flat after suitable treatment in canada balsam or Bhatia's medium, or for accurate systematic work or study of the chaetotaxy should be prepared by Puri's method. In this the skin is niggled under the dissecting microscope with needles to divide it along the ventral mid-line, floated on a slide and spread out with needles. It is then flooded with 90 per cent alcohol which causes the skin to flatten and adhere to the glass. Or if desired it can be detached and treated as a section.

Besides making use of the cast skin, the newly ecdysed larva is very suitable for morphological studies owing to its transparency. The head especially is so transparent that, examined fresh or mounted, the brain and other structures are very well displayed, especially if suitably stained, for example, by picro-carmine. Chordotonal organs and some other structures are also at this stage well seen, and owing to the fat-body being at a minimum at this stage in the fourth instar larva the newly ecdysed fourth instar larva forms excellent material for sectioning.

Still another useful purpose served by the newly ecdysed larva is that it forms a fixed point from which observations can be timed when studying growth, etc.

(*c*) THE FIRST INSTAR LARVA

The first instar larva when newly hatched measures about a millimetre in length, but grows during the instar to nearly twice that measurement. When first hatched it has a very characteristic appearance (Fig. 17 (1) A, A′). Good figures in this stage have been given by Roubaud (1929), and by Wigglesworth (1938). At this time the head is narrow and rather triangular in shape. The brushes lie impacted in the pre-oral cavity between the mandibles and it is only after some two hours, when the head has swollen, that they are liberated and begin to work intermittently. The swelling begins shortly after hatching and is well marked in about 2 hours when the head is much enlarged, globular and swollen out with fluid. This swelling is largely in the frontal region of the head, which remains bulged even when the head darkens and hardens, giving to this in the first, and to some extent in other early instars, an overhung shape differing from that of the fourth instar (Fig. 19 (1)). Other characteristic features of the early first instar are the elbowed antennae and the markedly annulated abdomen. Later the antennae straighten as the larva grows and the deeply indented abdomen becomes more uniformly cylindrical and worm-like. Other characteristics are given below.

THE EGG-BREAKER (Fig. 7 (4), (5); Fig. 19 (1), (2)

This is present only in the first instar larva and has already been described to some extent in the chapter dealing with eclosion. It is situated on the dorsum of the head on the most prominent part of the fronto-clypeus and consists of a strongly sclerotised, very sharp pointed, curved and somewhat flattened cone set in an oval area of soft membrane (Fig. 7 (4)). Passing backwards from the conical main body is a line of thickening joining this to the normal cuticle of the head and giving to the whole structure somewhat the appearance of a tennis racquet. Where this extension joins the head capsule is an area of thickening and into this on each side of the middle line is inserted the origin of the large median retractor muscles of the feeding brushes (head muscles no. 2).* Two smaller muscles, nos. 4 and 13, passing respectively to the roof of the mouth and the roof of the pharynx, also take origin, if not from the cone itself, at least from the hypoderm near this (Fig. 19 (1)). The egg-breaker in British mosquitoes has been described by Marshall (1938), who points out that in species of *Aëdes* the cone ends in a point, whereas in *Culex* it is chisel-shaped. In *Aëdes aegypti*, even under high magnification, the point is still round and very sharp.

* For numbering and description of muscles see ch. XIII (*c*), p. 306.

THE HEAD AND MOUTH-PARTS (Fig. 19 (1–5))

In the young larva the only sclerotised parts are the egg-breaker and the tiny mental sclerite. Later the head capsule darkens and becomes to the naked eye almost black. A curious feature of the cuticle of the dorsum of the head and especially in the region of the egg-breaker is its finely ribbed character giving an appearance very like the skin lines on the finger tips used in classifying finger prints. It also differs somewhat from that of later stages in being very widely open behind. Through this in the later stages of the instar a good portion of the brain is extruded into the swollen neck.

The labral parts are, in miniature, almost identical with those of the larva in later instars. The palatum, tessellated membrane, brush plates and their characteristic apodemes are all to be recognised (Fig. 19 (4), (5)). The feeding brushes have much the same characters as those of the mature larva and already show towards their inner ends the curved thickened brush filaments with fine comb-like ends as described in the mature larva. The preclypeus, though more overhung, is very similar to this structure in later instars and not only shows the preclypeal hairs but also the small outer hairs (hair 1*a*), or at least their basal marks.

The antennae, when they have straightened and hardened, differ little from these organs in the mature larva except that the antennal hair is relatively rather large and double. The parts at the apex are fully as complex showing a number of hairs and processes, namely, a dorsal hair, an outer small and a larger inner spine, a sensory papilla and two ventral hairs arising a little below the apex. Of these the inner hair is especially long, reaching about one-third of the length of the shaft of the antenna.

All the head hairs can be located if carefully looked for, though all are unbranched and the arrangement of the important hairs 4–7 different and seemingly more primitive (Fig. 19 (2), (3), (4)). Thus the three frontal hairs are now more or less in line, very like the three frontal hairs in the *Anopheles* larva, and the hair 4 is well in front. The position of

---

Figure 19. First instar larva.

1 Lateral view of head of first instar larva. *a*, overhanging frontal region; *b*, pale area showing position of pharynx.
2 Dorsal view of same to show chaetotaxy.
3 Frontal portion of same more highly magnified.
4 Ventral view of head of newly hatched larva.
5 Showing palatum and related structures.
6 Lateral view of thorax to show chaetotaxy. The prothoracic hairs are to the right.
7 Abdominal segments I–VIII to show chaetotaxy. Dorsal view. The asterisks mark a hair of the small dorsal group (hair 5?) that increases in length to segment V. *a*, hair 1; *b*, hair 2.
8 Lateral view of terminal segments of newly hatched larva. Only two of the four anal papillae are shown.
9 Lateral view of siphon of advanced first instar larva showing ballooned newly formed tracheal trunk of oncoming second instar surrounding pipe-like tracheal trunk of first instar.

Lettering: *A*, anal segment; *ap*, flabellar apodeme; *ash*, apical hair of siphon; *bmh*, basal maxillary hair; *dh*, dorsal hairs. One side only shown; *fb*, flabellum or feeding brush; *fb′*, flabellar plate; *h*, hypoderm of new forming brushes; *md*, mandible; *mx*, maxilla; *p*, palatum; *ph*, pilot hair; *sh*, siphon hair; *sh.f*, hair *f* of Martini; *tm*, tessellated membrane; *tt*, tracheal trunk.

Muscles are numbered as given in the list of head muscles in text (ch. XIII (*c*)).

Hairs are numbered as for the fourth instar larva (see Figs. 14, 15).

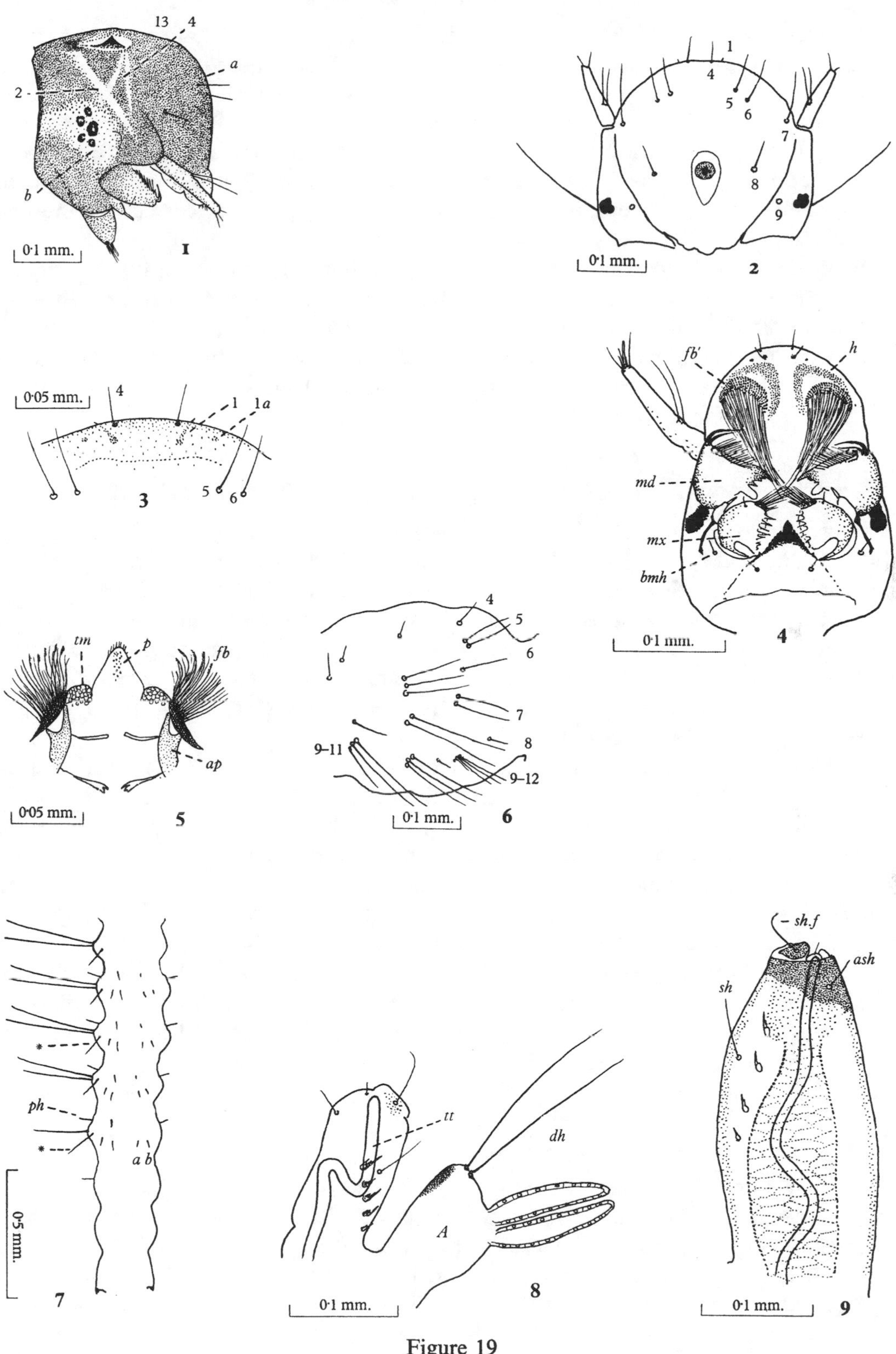
13
4
a
2
b
0·1 mm.
1
1
4
5
6
7
8
9
0·1 mm.
2
0·05 mm.
4
1
1a
5
6
3
fb′
h
md
mx
bmh
0·1 mm.
4
tm
p
fb
ap
0·05 mm.
5
4
5
6
7
8
9–11
9–12
0·1 mm.
6
*
ph
*
a b
0·5 mm.
7
tt
dh
A
0·1 mm.
8
sh.f
ash
sh
0·1 mm.
9

Figure 19

hair 8 also differs somewhat, being much nearer the middle line about half-way between the ocelli and the egg-breaker. Hair 9 is shown only as a pale spot near the posterior portion of the suture, which in the head of the first instar larva is very short. A point of some interest is that hair 17 of Martini (basal maxillary hair; Fig. 19 (4) *bmh*) does not arise from a separate basal sclerite of the maxilla, which has not yet been separated off, but from the head capsule on what will clearly later be this sclerite (Fig. 19 (4)). Thus the supposed error of some authors of depicting this hair on the head capsule is correct for the first instar larva.

The mandibles and maxillae are much as in the fourth instar, though somewhat simpler. The mandibles show the characteristic combs, teeth and a small molar process. The maxillae are rather conical with the terminal tuft less spread out than in the later instars. The maxilla has a terminal blunt spine about half the length of the organ. The palp is flattened dorso-ventrally with a crest bearing some smaller spines. On its side is a minute projection, but so far as could be made out no distinct hair.

THORACIC AND ABDOMINAL CHAETOTAXY (Fig. 19 (6), (7); Fig. 20 (5))

The long lateral hairs and all other hairs of the thorax and abdomen are simple and unbranched. Most, if not all, of the larger hairs of the fourth instar are represented. There are no strongly sclerotised basal plaques as seen in the later instars, though some indication where these will appear is sometimes given. The most notable missing hairs are the prothoracic hairs 1–3. Four hairs of the pleural series (hairs 9–12) are present on the prothorax and three on both the meso- and metathorax. The minute hair 12 on the mesothorax is not present until the fourth instar (or third according to Macfie). The same applies to hair 12 on the metathorax.

On abdominal segments I–IV the large lateral hairs (6 and 7) arise close together from separate bases and are simple. Next to these the most conspicuous hair is the pilot hair (hair 8) which arises a little in front of hairs 6 and 7. It is present on all segments except the first and eighth. As indicated on the figure, there are four small hairs only on the dorsal surface of the segments much in the relative positions of the small dorsal hairs in the fourth instar. The outermost of these hairs increases in length to segment V and would appear to be hair 5. In this case hair 4 is the probable missing hair (see also Mattingly (1917), who gives only four hairs for instars I and II (five for instar III)). Three hairs only are present on the ventral surface of segments II–V instead of five (hairs 9–13). This accords with Belkin's finding that hairs 9 and 13 only appear later. On segments VI and VII all hairs are relatively small. At least eight are present on each side of VI and ten on VII. The five pentad hairs are all simple.

THE SIPHON (Fig. 19 (8), (9))

In the newly hatched larva the siphon is almost entirely soft. Later the terminal portion only becomes darker and hardened, that is about one-fifth of its length, the sclerotised portion stopping, as in later instars except the fourth, at about the level of the distal end of the pecten. The pecten spines number 4–5 arising on soft membrane. The siphon hair is simple. As the parts harden towards the end of the instar it becomes possible to trace most of the structures of the spiracular plate. The tracheal trunks, which as described later are narrow and tubular, open by small orifices close together towards the anterior border of the siphon apex. Behind these are the two posterior lobes with their inner and outer plates. The inner plates are clearly visible from an early stage, the outer only later. From the outer surface of the lobe on each side, and from the outer plate when this becomes apparent through hardening of the cuticle, arises a largish curved hair (hair *f* of Martini (Fig. 19 (9) *sh.f*). This is figured by Wigglesworth and apparently takes a part in the early contacting of the surface of the water when the siphon is first brought against this. Usually in the advanced larva indications of the anterior lobe with its chitinised outer plate can be seen. The lateral lobes are not clearly shown at this stage, but their position is indicated by a small hair.

The most striking differences from the siphon in later instars relate to the tracheal trunks. As also throughout the abdomen these are narrow and tubular and even under the oil immersion show no indication of taenidia. They are soon surrounded by the developing, dilated and fluid-filled trunks of the succeeding instar within which they lie loose and usually thrown into bends as though allowing for expansion of the body with growth (Fig. (19 (8) *tt*, (9)). At their terminations they are rounded and appear to open only by a small pore or slit. They show no indication at any stage of the formation of the felt chambers.

THE ANAL LOBE (Fig. 17 (4) *c*; Fig. 19 (8))

The *anal lobe* carries dorsally at its apical extremity four well-developed *dorsal hairs*. These rise from a well-developed basal plate and are unbranched. The *saddle plate* is restricted to the extreme dorsal aspect of the lobe in its distal half. The *saddle hair* is simple and arises laterally. There is no *ventral fan* or any indication of this. The *anal papillae* are banana shaped and about the same length as the segment. They show in the early stages of the instar an outer layer of large flat, but thick, cells with little or no indication of any cuticle. The branching tracheae so conspicuous in later stages are undeveloped.

INTERNAL STRUCTURE (Fig. 20 (1), (2), (3))

The alimentary canal consists of the same parts and has much the appearance of that in the mature larva, the chief difference being in the disproportionate size of the proventriculus and salivary glands and the smaller gastric caeca. The *pharynx* is of the characteristic and dilated form seen in the mature larva with much of its structure including lines of fringe already marked out. The *fringes*, though rather delicate, are very similar to those in the fourth instar, the filaments and spaces being in proportion, in this case measuring together only about 1·0 μ. As shown in the figure, the lateral oral apodemes are present and show their characteristic fish-hook appearance. The *proventriculus* shows the same general structure as in the later stages, but is relatively very large, occupying much of the thorax. The intussuscepted portion of the oesophagus is long and narrow with a narrow lumen and there is reason to believe that it plays an active part as a sphincter in the early stages of the instar. The *gastric caeca* are small and at first almost globular. Later they become more oval, but remain small. The *salivary glands* extend almost to the thoracic wall and in size almost equal the gastric caeca. They show the bulla-like vesicle at their ends as in the mature larva. The *mid-gut* and *intestine*, as also the *Malpighian tubules*, do not differ greatly from these organs in the mature larva.

The *tracheal system* shows the same general plan as in the mature larva, except that the tracheal trunks are narrow and tubular and seem devoid of taenidia and that many future tracheal branches are as yet indicated only by strings of formative cells. The same system of commissures linking the trunks is present and rather more conspicuous than in the later instars. The sharp angular bend of the trunks in the thorax and the ending of these in the dorsal and ventral branches to the head are all as in the mature larva. What is characteristic is that owing to the narrowness of the tracheal vessels proper to the instar and the disproportionate size of those of the second instar the tracheal trunks and larger tracheae are everywhere in the later stages of the instar surrounded by the hugely dilated fluid-filled developing tracheae that will eventually replace them. At no other stage is this appearance so marked.

The *muscles* appear very similar in arrangement, though on a minute scale, to those of the fourth instar larva, but have not been investigated in detail. The same oblique arrangement of the chief muscular bundles is present in the abdominal segments as in the mature larva and the number of strands is not greatly less.

The *brain* differs notably from this organ in the fourth instar by the absence of the large optic lobes and the relative prominence of the deutocerebrum (Fig. 20 (3)). The protocerebrum is globular and increases greatly in size during the instar, becoming towards the end of this period partly pushed into the neck. A very conspicuous feature of the first instar larva is the chain of *ventral ganglia* which extends across the middle third of the segments as oval plaques of nerve ganglionic tissue lying close up against the cuticle (Fig. 20 (3)). As in the mature larva, no indication is given of any ganglion posterior to that of the eighth abdominal segment.

The fat-body is restricted chiefly to small masses occupying the prominent angles of the abdominal segments. In each lobule there is a small characteristic cluster of *large oenocytes,* these being of a size appropriate to the larva, namely about 10 μ in diameter as against 40 μ or more in the fourth instar.

(*d*) THE LATER INSTARS

The second instar larva after ecdysis is much the same length as the fully grown first instar larva, but is bulkier and the swollen head is enormous, an effect added to by its almost glass-like transparency when first liberated. The tracheal trunks are now enlarged and

---

Figure 20. Instar characters.

1 Dorsal view of thorax of first instar larva.
2 Lateral view of portion of head of same showing pharyngeal fringes.
3 Early first instar larva stained with picrocarmine showing brain and ventral nerve cord.
4 Schematic representation of pleural hair base to show positions of hair origins. The base is shown as though viewed from in front. *a, b, c, d,* hairs 9, 10, 11, 12 respectively; *e,* position of spine.
5 Pleural hairs of instars I–III. A, prothoracic; B, mesothoracic; C, metathoracic.
6 Comb scales (A) and pecten teeth (B). *a,* first instar; *b,* second instar; *c,* third instar.
7 Mental sclerite of first instar larva.
8 Mental sclerites to same scale of instars I–IV.

Lettering: *Cs,* gastric caeca; *Dc,* deutocerebrum; *f″,* pharyngeal fringes; *H,* hypostomal area; *L′,* labial area; *lpp,* lateral plates of pharynx; *md,* mandible; *mx,* maxilla; *Oe,* oesophagus; *P,* pharynx; *Pc,* protocerebrum; *Pv,* proventriculus; *sg,* salivary gland; *ss,* suspensorium; *ss′,* lateral oral apodeme.

For hair numbers see text, p. 214.

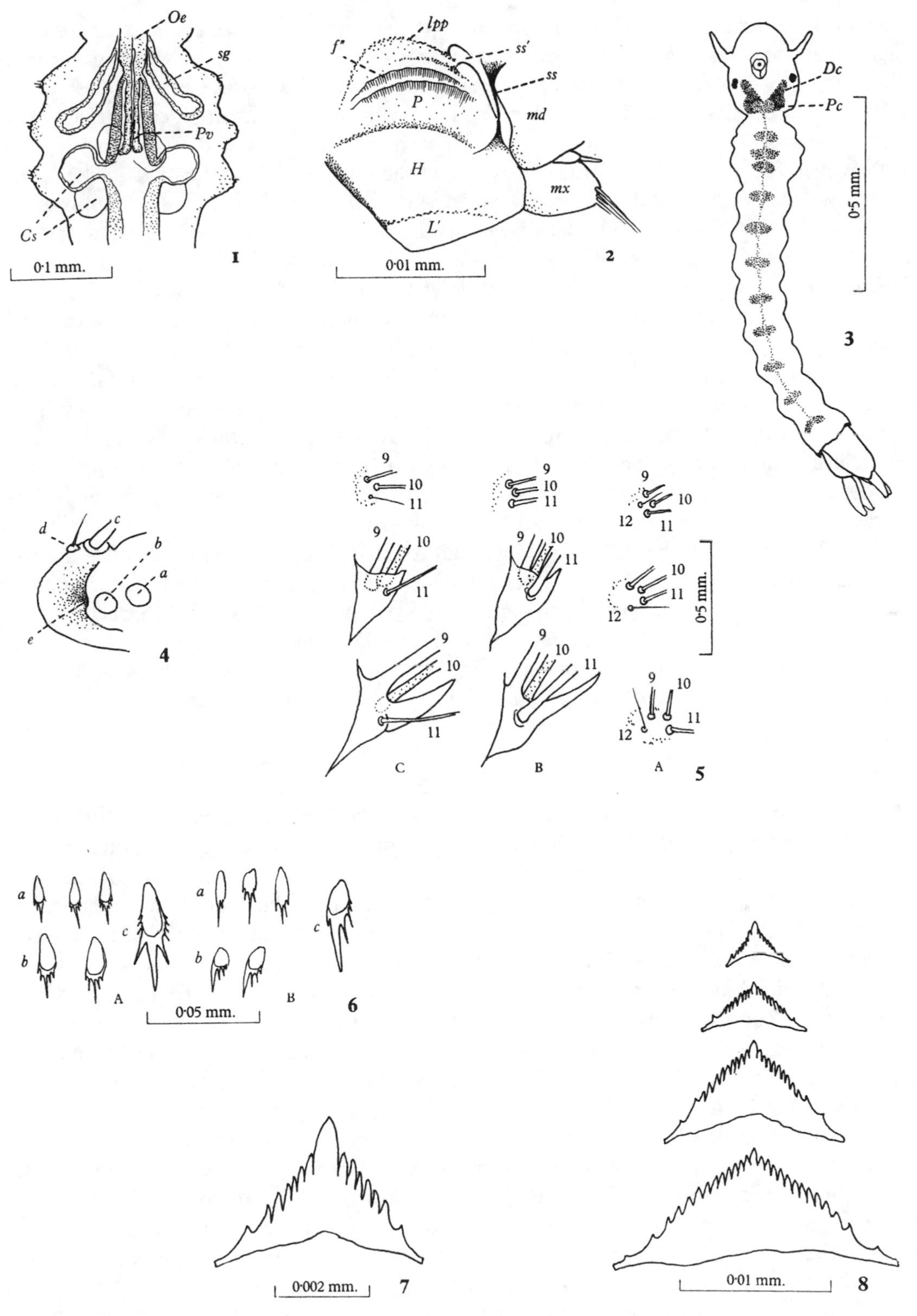
Oe
sg
Pv
Cs
0·1 mm.
1
lpp
f''
ss'
ss
P
md
H
mx
L'
0·01 mm.
2
Dc
Pc
0·5 mm.
3
d
c
b
a
e
4
9
10
11
12
0·5 mm.
C
B
A
5
a
b
c
A
B
0·05 mm.
6
0·002 mm.
7
0·01 mm.
8

Figure 20

lined with taenidia and the terminal portions in the siphon are ballooned. The larva of this stage when newly ecdysed is for some reason conspicuously light and when along with other larvae it is placed in 70 per cent alcohol it floats at the surface whilst other stages sink. Unlike the first instar larva and in common with the later instars it cannot, as noted by Wigglesworth (1938), take air into the tracheae if it is prevented from reaching the surface even for a short time from ecdysis. As in the previous instar the head darkens and the body becomes long and cylindrical, whilst with increase in size as ecdysis approaches the head, at first disproportionately large, again becomes disproportionately small. It almost repeats in fact the change in appearance of the first instar but on a larger scale. During the instar the larva grows in length from about 2 to about 3 mm.

The third instar larva, whilst it has certain characters of its own, is to a large extent a repetition on a still larger scale of the second instar larva. For some reason it is much more variable in head measurement than any other stage, there being large and small headed forms. It is very apt to have the ecdysial lines on the head capsule open for some time before ecdysis is definitely in progress. The pre-ecdysis stage is rather long and thin and being larger than the previous instars it shows this character more conspicuously. The fourth instar at a corresponding size has become much stouter due to development of the thoracic imaginal buds and accumulation of fat-body.

Little further need to be said regarding details of structure in these instars since in most respects they either conform to the description already given for the fourth instar or show only small departure from this. Such departure largely relates to external characters. For a very full description of these changes see Macfie (1917), and for measurements of the body parts this author's account as also the chapter in the present work on growth. The following is a brief account of changes undergone.

CHAETOTAXY

By the second instar the small head hair 4 has become dendritic as in the fourth instar and the arrangement of the frontal hairs has by this or the next instar also become as in the fourth instar. The antennal hair has become single and remains so. The prothoracic hairs 1–3, apparently lacking in the first instar, have in the second become quite conspicuous. The hairs 9–12 (pleural hairs) on the meso- and metathorax have by the second instar developed sclerotised basal papillae and these by the third instar have developed spines approaching in character those that are so conspicuous a feature in the fourth instar larva (Fig. 20 (5)). The exact relation to the pleural hairs to their basal papillae is in need of some brief explanation. On the prothorax there are three large and one small hair arranged as shown in Fig. 20 (5) A. These alter very little, even in size, in the successive instars and the basal papilla remains to the end little more than a soft tubercle. On the meso- and metathorax there are on each from the first instar onwards three large hairs and these remain recognisable throughout all the larval instars in spite of changed appearance due to branching. The basal papillae take the form of strongly sclerotised scoop-shaped structures with a smooth convex surface directed posteriorly. From the mesothoracic and metathoracic papilla there arise in each case, as will be seen from the figure, three large hairs, three equally large from the mesothoracic, two large and one somewhat smaller from the metathoracic papilla, the same arrangement of hairs as is clearly shown on the first instar larva. Two of these large hairs arise from the more hollow side of the scoop, one nearer the spine, the *posterior* hair, the other anterior to this, the *anterior* hair. A third hair, the *dorsal* hair, takes origin from the dorsal aspect of the convex outer surface of the scoop. This hair on the metathorax is the somewhat smaller hair of the three. The exact time of appearance and position of a fourth and minute hair is less certain. Such a hair is said by Macfie to be present on the metathoracic papilla from the third instar. In the scheme of culicine chaetotaxy as given by recent authors, four hairs, including a minute hair (no. 11), are given as present on the mesothorax group, but three only (no. 11 omitted) on the metathoracic papilla. In *Aëdes aegypti* there is what is obviously the same small hair on both the meso- and metathoracic papilla. This arises just posterior to the dorsal hair in both cases. It may be called the *accessory dorsal* hair. I have found it in both segments only in the fourth instar. The anterior hair tends to become bifurcate or multi-branched as development proceeds. The other hairs tend to remain unbranched. The anterior hair appears to be hair 9 of the accepted chaetotaxy, the posterior hair no. 10 and the dorsal hair no. 12.

Space does not permit of a detailed consideration of the smaller hairs of the thorax and abdomen or of the degree of branching of the larger hairs. Generally speaking, whilst all the large hairs, including the lateral hairs (notably the lateral hair 6 on segments I–III of the abdomen) in the second instar become bifid, in the third they are trifid and in the fourth three- or four-branched.

The pentad hairs are bifid in instar 2, and usually so are the siphon hair and the saddle hair. In this instar there are usually 4 cratal tufts and 2 precratal hairs in the ventral fan. In the third instar those pentad hairs that are branched in the fourth instar are trifid, as also is the siphon hair. The cratal hairs are increased in number usually to six with two pre-cratal hairs.

OTHER EXTERNAL STRUCTURES

The mental sclerite (Fig. 20 (7), (8)) has in the four instars respectively about 7, 9, 11 and 13 teeth on each side of the median tooth. A point of interest is that the median tooth, which is proportionately very large and broad in the first instar, remains practically the same size up to the fourth instar though the transverse diameter of the sclerite as a whole increases by four or five times.

The *comb spines* which are very delicate and number only about 5 in the first instar increase to 8 or 10 in the second instar with some further increase in number later, the new spines appearing beneath the cuticle in the position of the old.

The early condition of the *siphon* has already been described. In each instar period to the fourth only an increasing fraction of the whole siphon in its distal portion becomes dark and sclerotised. The proportion so hardened in the respective instars is approximately at the time preceding ecdysis 0·15, 0·25, 0·67, 0·70 and in the fourth instar 1·0. The *pecten spines* in the early instars are wholly or mainly on the soft membrane basal to the hard portion (Fig. 17 (4), (5), (6); Fig. 19 (9)), some 4 or 5 spines being present, however, on the hard portion in the third instar. The number of pecten teeth in the respective instars is about 6, 7–10, 12–15, 12–20. As noted by Macfie, both the comb scales and pecten teeth change somewhat in the successive instars. This change is mainly in the increased lengthening and strengthening of the main tooth (Fig. 20 (6)).

The *saddle plate* remains much the same proportionate size until at the fourth instar it increases from being restricted to the dorsal surface until it nearly covers the anal segment dorsally and laterally. The *dorsal hairs* also usually remain unbranched until the final instar. A common condition, however, in the third instar is for one, or more rarely both inner hairs, that is the more dorsal of the hairs, to have one thin branch.

INTERNAL STRUCTURE

Little change in internal structure occurs until, in the fourth instar, development of the future pupal structures begins. The compound eyes and the optic lobes begin to form in the late third instar, but appreciable development in these structures takes place only in the fourth instar. The imaginal buds at the bases of the pleural hairs become apparent in the second instar and are quite conspicuous in the third, though their main development is a feature of the fourth instar. The gonads are already evident in the sixth segment in the third instar, when also the buds of the developing gonocoxites become apparent.

(*e*) THE PREPUPA

During the fourth instar period the leg, wing and some other rudiments which have so far been enclosed in sacs as germinal buds become everted to appear as rough models of their final form. Especially the long cylindrical columns of the leg rudiments, towards the end of the instar, extend in loops over the lateral and ventral aspects of the thorax, concealing much of the real larval body (Fig. 21 (2), (3) *a*, *b*). Further, in this stage the hypodermis, as is very obvious in sections, becomes more and more separated from the larval cuticle and also different in form, leading ultimately to a condition in which the larva is little more than the roughly sculptured body of the future imago disguised in a larval skin. Such a condition in the mosquito indeed so closely resembles in many respects that which has been termed the prepupa in the higher Diptera (Wigglesworth, 1942) that this term may usefully be applied to the later stages of the fourth instar in the mosquito if only to emphasise how important in the post-embryonic development these changes are. It is, in fact, in the later stages of the fourth instar, not in the pupa, that the main outlines of the future imago are moulded. The pupa, as will be seen, with some exceptions to be noted, is merely the parts as they exist in this stage cemented together to form a simulacrum of an actual body proper to that stage. In this connection attention may be drawn to the paper by Giles (1903), entitled 'On the prepupal changes in the larvae of Culicidae', one of the few contributions dealing with this important stage in mosquito development. A connected description of post-embryonic development will be given later. In the meantime some description of the changes taking place in this respect in the fourth instar will be helpful.

DEVELOPMENT OF APPENDAGES

The leg rudiments. In the opening sentence of the subsection the word 'everted' was used to describe the appearance of the limb-like stage of the leg and wing rudiments. This word does not, however, truly describe the process here involved and some explanation seems desirable. The germinal buds of the legs are typically present already in the third instar where they have all the characters of the classical 'germinal bud'. In this stage and at the early fourth instar the buds form globular or pear-shaped sacs attached to the base of the pleural hairs (hairs 9–12) in each thoracic segment. They begin as a thickening of the hypoderm beneath these hairs. The edges of this thickening dip deeply into the subjacent tissue carrying with them a thin layer of hypoderm which forms the sac. The portion excavated forms the rudiment proper (Fig. 21 (5), (6), (7)). The apex of the rudiment up to a late stage maintains its connection with the point of origin, that is, in this case the pleural hair base. The rest of the rudiment elongates and becomes bent upon itself. Its

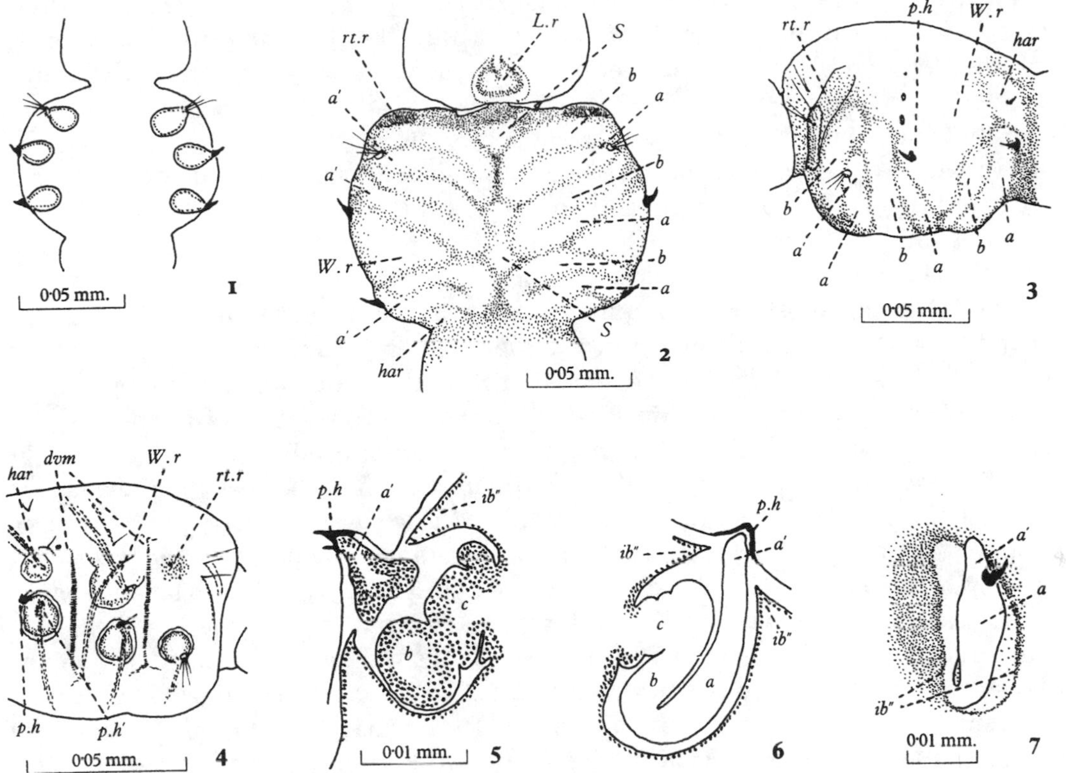

Figure 21. Showing imaginal buds of pre-pupa.

1 Thorax of mature third instar larva showing imaginal leg buds.

2 Ventral view of thorax of larva approaching pupation. *a′*, tip of tarsus (at or near pleural hair base); *a*, tarsus; *b*, tibia.

3 The same lateral view. Lettering as in 2.

4 Lateral view of thorax of third instar larva approaching ecdysis.

5 Section of imaginal leg bud showing maintenance of connection of tip of tarsus with pleural hair base. Lettering as in 6.

6 Schematic representation of imaginal leg bud. *a*, *a′*, tip and shaft of tarsus; *b*, tibia; *c*, femur.

7 Showing recession of edges of opening of sac (*ib″*) due to moulding of thorax.

Lettering: *dvm*, dorso-ventral muscle indicating limits of thoracic segments; *har*, halter rudiment; *ib″*, edge of imaginal sac opening; *L.r*, labial rudiment; *p.h*, pleural hair base; *p.h′*, pleural hair base of oncoming instar; *rt.r*, pupal trumpet rudiment; *S*, sternite; *W.r*, wing rudiment.

base, at the point where the excavating invagination has left it still in connection with the body of the larva, forms the eventual place of origin of the limb. The bent portion gives rise in its respective bends to femur, tibia and tarsus, the tip still at the pleural hair base being the tip of the tarsus or near this. In the later stages of the fourth instar the bending of the rudiment constitutes a loop which grows ventrally and continues to do so until it reaches the mid-line of the ventral surface of the larval thorax. The descending limb (*a*) is in all the rudiments situated anteriorly and forms the future tibia. The ascending limb (*b*), still situated at its tip at the pleural hair base or near this, eventually develops into the five tarsal segments. The femur (*c*) is formed from a more basal portion which is not visible externally. The general construction will be plain from Fig. 21 (6). Where the basal portion springs from the wall of the sac there develops a ring-like thickening which eventually forms the coxa and a second less marked ring indicates the position of the future trochanter.

It still remains to explain how the rudiment comes into the position previously referred to as everted. Actually there is no inversion in the sense of turning inside out and what happens to bring the rudiment eventually to a free position is as follows. The apex of the rudiment is still lying near the pleural hair base. Around this is the circular fissure formed by the invagination which has excavated the rudiment, in other words the narrow opening of the sac. Originally this opening is quite small, but as the rudiment increases in size and the hypoderm of the general surface of the thorax is dragged upon and shrinks in the process of the moulding of the thorax as development goes on, this narrow opening becomes increasingly enlarged. Eventually the loop of the rudiment consisting of the parts which will eventually be tibia and tarsus is extruded more and more and eventually comes to lie free on the side of the thorax (Fig. 21 (7)). The original edges of the opening of the sac are now widely extended and eventually merely form a ridge around the rudiment or even cease to be apparent. The tip of the rudiment which will become the tip of the tarsus may eventually become detached and even dragged away from the pleural base beneath which it was situated. But even so it is not usually far away as can be seen from the figures showing the limb rudiments in the mature larva. Further, on the wall of the flattened sac destined in the adult to form a portion of the lateral wall of the thorax, there appear thickenings which represent the beginnings of the future pleural sclerites. In the eventual history of these thickenings lies the ontogenetic problem to be worked out of their homologies in the imago.

The wing and halter rudiments. These differ from the leg rudiments in that they are more flap-like. In the course of the fourth instar the wing rudiment develops into a triangular flap covering the median portion of the lateral aspect of the thorax. Its further history will be dealt with in a later section. The halter rudiment is very similar in shape but smaller. The wing and halter rudiments lie respectively dorsal to the mid and hind leg rudiments and like these in their early stage are enclosed in sacs.

The pupal trumpet rudiments. The nature of these organs has been noted when dealing with the tracheal system of the larva. They first appear as relatively small areas of thickening about the mesothoracic spiracular puncta and they develop rapidly as organs formed around the spiracular cords leading from these. They lie with their openings directed somewhat downwards in the shoulders of the larva.

The antennal and labial rudiments. These develop as large buds occupying considerable space in the head. The antennal rudiments lie in elongated sacs passing back as far as the compound eyes. The labial rudiments occupy a single sac which when fully developed forms a considerable mass lying behind the labial area of the head capsule. The tips of the rudiment, which will become the tip of the proboscis, lie in the lateral papillae of the labial plate of the larva.

OTHER DEVELOPMENTAL CHANGES IN THE PREPUPA

Some other changes relating to development of the imaginal body in the prepupa will be mentioned when describing the different larval systems. Among these the most conspicuous is the early appearance and rapid increase of the future great thoracic imaginal muscles of the wings which complete their development in the pupa as also do the later changes in the alimentary system which bring this to the condition in the imago. Important changes also take place in the tracheal system, but these are most suitably dealt with in the chapter on the pupa.

(*f*) DURATION OF THE LARVAL STAGES

The times taken in the different instars added together should be that of the complete larval period. In actual practice determination of such data offers certain difficulties, notably that not only are the times largely dependent on temperature and other conditions, but also in cultures not all the larvae behave precisely similarly. Hence the results must usually be given on a statistical basis. For this reason it will be desirable to treat separately determination of the total larval period and of that passed in the different instars.

Many authors have given estimates, often of a rather general nature, as to the duration of the complete larval period. Francis (1907) gives the minimum period at constant temperature 80° F. (27° C.) as 7 days. Mitchel (1907) gives 8 to 13 days in fairly warm weather. Newstead and Thomas (1910) give 9 days at 23° C. Howard, Dyar and Knab (1912) quote the following: Reed and Carroll, 6 days; Agramonte, 10 days; Taylor, 9 days; Dupree and Morgan, 6–8 days. Howlett (1913), gives the shortest period as 5 days, the longest 22 days, on the average a week. Macfie (1915) gives for the larval period 7–13 days and (1916) in a tropical climate with sufficient food, 7 days. Many authors also refer to great extension of the period under certain conditions, for example Macfie (1915, 1916) gives up to 100 days, Bacot (1916) up to 70 days, and Bonne-Wepster and Brug (1932) up to 60 days. Usually such periods are the result of starvation. Shannon and Putnam (1934) give mean periods based on careful statistical data of 6·4 days (154 hours) from eclosion to pupation at 27° C. and 7 days for this period at 23–26° C.

Some very detailed figures are given by Trager (1937) and by De Meillon, Golberg and Lavoipierre (1945) in connection with larvae reared at 28° C. under sterile conditions on different essential food fractions. To express numerical rate of development Trager uses a figure obtained by multiplying the percentage of larvae reaching the fourth instar within 9 days by the reciprocal of the average time in days required by these larvae to reach the fourth instar ($N \times 1/t$, where N is the first item and t the second). The higher the figure the more rapid the development. Such a figure in his tables ranges from about 3, corresponding to a 16·5 day period, to 32, or one of 7·5 days. De Meillon, Golberg and Lavoipierre give very similar figures for larvae reared on liver extract, yeast and vitamin B, namely from 6–8 days from first instar to pupation at 28° C.

Putting aside those cases where the longer periods probably relate to adverse conditions the most usual period given lies between 6–8 days. Bacot (1916), giving 4 days for the minimum, however, continues by saying that males first appear on the fifth day, which would accord with our experience of cultures at 27–28° C., and probably is about the shortest time mentioned by any author. Though, however, this refers to the first males it is not the mean or usual period for pupation, which at 28° C. is about 144 hours or 6 days, though subject to variation within certain limits as discussed later.

The most detailed account of the duration of the different instar periods is that given by Macfie (1917), though both Trager and De Meillon, Golberg and Lavoipierre give information in considerable detail in this respect in their tables. Macfie's determination, made on 12 isolated larvae, gives for the time occupied by the four instar periods, 24, 24,

Table 11. *Duration of instar periods*

	Instars					
Hours	I	II	III	IV	*P*	Timing of stage
20	100	—	—	—	—	All first instar up to 20 hours
22	100	+	—	—	—	—
23	89	11	—	—	—	—
24	67	33	—	—	—	—
25	56	44	—	—	—	50 per cent ecdysis I–II 25–26 hours
26	44	56	—	—	—	
27	31	69	—	—	—	—
28	3	97	—	—	—	—
30	—	100	—	—	—	—
						All second instar 30–42 hours
42	—	100	—	—	—	—
45	—	74	26	—	—	—
46	—	37	63	—	—	50 per cent ecdysis II–III 46 hours
47	—	17	83	—	—	—
48	—	27	73	—	—	—
49	—	+	100	—	—	—
50	—	15	85	—	—	—
52	—	+	100	—	—	—
54	—	2	98	—	—	—
						All third instar 54–68 hours
68	—	—	100	—	—	—
69	—	—	73	27	—	—
70	—	—	60	40	—	—
71	—	—	58	42	—	50 per cent ecdysis III–IV 71–72 hours
72	—	—	32	68	—	
73	—	—	24	76	—	—
76	—	—	17	83	—	—
78	—	—	—	100	—	—
92	—	—	9	91	—	—
94	—	—	—	100	—	—
96	—	—	6	94	—	—
98	—	—	—	100	—	—
102	—	—	—	100	+	First male pupa 102 hours
117	—	—	9	91	—	—
122	—	—	—	50	50	Bulk of males pupated 122 hours
139	—	—	—	+	100	—
141	—	—	—	12	88	—
144	—	—	—	+	100	Bulk of females pupated 144 hours

A plus sign indicates that a few of this stage were present in the sample taken but too few to show as a percentage. The cultures were from eggs hatched within 15 minutes.

24 and 96–240 hours respectively. These times, except the last, correspond very closely to the times as given by me later. Since the length of time occupied by the instars is very closely linked with the study of growth dealt with in a later chapter a word may be said about the method used to determine this.

To determine satisfactorily in a statistical fashion the duration of the instars and the times of ecdysis it is necessary to take into account the fact that in a culture in spite of every care that the larvae should be hatched as nearly as possible simultaneously (in a number of cases within 15 minutes and in one within 5 minutes) spread still occurs becoming more and more pronounced with each successive instar. Two ways of ascertaining statistically the duration of the stages are possible. One is to follow the development of individual larvae. This has the disadvantage that it is difficult to ensure that the conditions of the isolated larva are those occurring in the culture and that it also necessitates a great many individual trials to cover the quite considerable range of variation. An alternative method is to make counts by sampling of the percentage of instars at different hours of culture, a method which does not in any way interfere with the progress of development of the culture from which the samples are taken. The results of such a series obtained from massed counts made from a number of cultures at 28° C. in which larvae were those hatched within 15 minutes or less are given in Table 11. From the table it is easy to see at any hour of culture what the proportion of instars is likely to be.

Taking 50 per cent change as the mean time of ecdysis, it will be seen that each of the first three instars lasts almost exactly 24 hours. The first instar is very slightly longer as a rule, and the second especially is liable to be rather shorter. The fourth instar lasts approximately from 68 hours, when the first ecdyses to the instar begin to take place, to 144 hours when, with a few abnormal exceptions, all larvae have pupated. From 50 per cent ecdysed from the third to the fourth instar to 50 per cent pupation was 50 hours. To 100 per cent pupation was 144 hours or exactly 6 days. The periods occupied by the different ecdyses, that is the time during which at least some forms will be found undergoing a particular ecdysis ending with the last larval ecdysis or pupation was respectively for the four ecdyses 6, 10, 20 and 40 hours.*

Even under standardised conditions slight variations in the time taken in different cultures did occur. Most noticeable were occasional instances of a more rapid complete maturation than usual. It was suspected that one cause of this was the previous history of the eggs though this requires confirmation. It would, however, be in accordance with Shannon and Putnam's finding that time of conditioning affects not only the hatching time, but also the course of subsequent culture. In order to obtain as simultaneous hatching as possible the present results were obtained with eggs kept less than a week after collection and the degree of variation amounted at most to a few hours.

(*g*) PUPATION

THE ACT OF PUPATION (Fig. 21; Fig. 45)

Ecdysis of the fourth instar larva, or pupation, differs from ecdysis in previous instars in that somewhat more extensive changes are involved since it results not merely in a larger and somewhat more developed larva but in the very different structure, the pupa. The process in *Culex* has been very fully described by Hurst (1890) and a great deal of information on internal changes given by Thompson (1905). In *Aëdes aegypti* the process is as follows.

As the larva approaches the time for pupation it develops a noticeable plump appearance. The thorax particularly is swollen due to increasing size of the wing and leg rudiments and also to greater accumulation of fat, the lobes of the fat-body being now almost contiguous.

* Möllring (1956) has made a similar study of the instar periods in *Culex* (autogenous and anautogenous) showing the period for each instar with overlap as shown here in Table 11.

The leg rudiments are particularly responsible. These form loops directed ventrally and as pupation approaches they approach and reach the mid ventral line leaving only a small area of sternum showing (Fig. 21 (2)). Coincident with these changes the respiratory siphons lying under the cuticle in the anterior corners of the thorax become more and more conspicuous. These, at first colourless, take on a yellow colour deepening to brown and so give a good indication how far the larva is from pupation, brown siphons indicating an almost immediate approach to ecdysis. At this time too the float hair of the oncoming pupa becomes very obvious under the cuticle of the first abdominal segment. The larva further now rests at the surface, ceasing to use the feeding brushes. Hence whilst at the surface it does not glide forwards as do younger larvae. It still swims away, however, if disturbed.

Just before the act begins, the larva watched from above shows a curious change in appearance as the compound eyes, previously not visible, are suddenly seen to be present as two black spots seemingly alertly watching over the anterior bulge of the thorax. This effect is the result of the head capsule splitting and the pupal head carrying the compound eyes being to some extent displaced backwards.

Suddenly now the body gives a jerk and the abdomen comes to lie horizontally just beneath the surface. Within a minute or so the whole aspect of the insect is changed. The changes taking place are as follows. At the bases of the breathing trumpets, which are still under the cuticle, a small bubble of air appears. This is air entering for the first time the length of tracheal trunk linking the pupal respiratory trumpets to the position of the future imaginal mesothoracic spiracle as described later in the pupa. Air also now begins to collect within and at the distal ends of the trumpets as yet still directed downwards and under the cuticle. This air spreads, forming a band across the front of the thorax behind the neck (Fig. 45 (3) *a*). Suddenly other collections of air make their appearance. A bubble appears in the space on the ventral aspect of the thorax between the lower bends of the leg rudiments, that is over the small area of sternum left uncovered. Extensive air collections are also seen behind and under the pupal wings. These last extend up under the wings to beneath the region of the halteres of the future pupa and will after pupation become the extensions of the ventral air space as described in the pupa. A band of air now forms across the base of the abdomen and air may even be seen elsewhere in the abdominal segments. All these collections of air occur before there is any obvious entry of air from outside due to tearing of the pupal skin. This, however, now takes place and the dorsum of the thorax is now seen to be formed by the exposed pupal thorax, the edges of the torn outer cuticle being pushed aside. This soon liberates the pupal trumpets, which spring out and make firm contact with the surface film.

Meanwhile changes are taking place which end in the old head capsule being pushed ventrally and the whole pelt slipped backwards. By this stage the tracheal trunks have parted at their lines of weakness and have mysteriously been drawn out of the body so that they lie end to end in a line along each side of the abdomen. Being dark, this gives the naked-eye appearance of dark lines marking out the lateral borders of the stretched-out abdomen (Fig. 45 (1)). It remains only for the displaced cuticle to slip backwards and the paddles and other structures to be finally withdrawn for the pupa to be free. The whole process normally takes only a minute or so. Thompson gives 3 minutes for *Culex*. It is sometimes, however, delayed, usually from some disturbing cause. The liberated pupa, after discarding the skin usually remains for a short time with the abdomen horizontal. After a few minutes it retracts this part under the body and assumes the normal pupal appearance. It is, however, at the moment of liberation very vulnerable and disturbance at this time may lead to it losing contact with the surface film and possibly difficulty in re-establishing this with the result that it drowns.

When first emerged the pupa is white, but in a short time shows pigment changes as later described.

THE MECHANICS OF PUPAL ECDYSIS

As already noted at the time preceding ecdysis, ecdysial fluid has been secreted which has dissolved the endocuticle of the cast-off integument, making this an extremely thin shell usually appearing only as a fine line in sections with a considerable space between it and the underlying new cuticle. Whilst this is one function of such fluid there is, in this case, another which is important, namely that it is presumably the ecdysial fluid which is responsible in the mosquito for cementing the pupal parts together. An account of ecdysial fluid in this connection has been given by Tower (1902), who describes its cementing function in the pupa of *Leptinotarsa* (Col.). This author notes that ecdysial fluid is present at all ecdyses, but is especially abundant at the ecdysis leading to pupation. At the time of pupation material from the fluid is gradually precipitated upon the surface of the oncoming pupa, leaving a more fluid portion in contact with the old cuticle. When the cuticle ruptures and the fluid is exposed to the air it hardens and cements the appendages to the body. It seems very probable that in *Aëdes aegypti* a similar cementing action is brought about. But this does not explain all that takes place to enable the pupa to become at emergence the form bound together in the fashion it is.

As already noted, the imaginal buds of the wings and legs are extended in the prepupal stage whilst still in the larva. The buds of the labral, labial and other mouth-parts are, however, at the time of commencing pupation still involuted and they are extended in the act of pupation. According to Thompson the extrusion of the long pupal mouth-parts may even be an important factor in rupturing the head capsule. Exactly how all the parts come to be cemented accurately in position as seen in the emerged pupa is not very clear. According to Thompson the labral rudiment is bent forwards from its position over the dorsum of the head. This author also notes that there is a short period of delay after extrusion of the more anterior parts (labrum, mandibles and maxillae) before the labial rudiment is extruded, thus possibly allowing time for the transverse fold of the hypoderm carrying the mental sclerite to be withdrawn.

The mechanism by which air as described appears in various parts under the larval cuticle is perhaps not so difficult to understand as might at first appear. Considerable volumes of air are ready to hand in the large tracheal trunks and as these are disrupted air may well follow or precede their fragments as these are dragged or squeezed out, for example, through the temporarily opened spiracular passages. Such air could then find locations in various spaces under the cuticle already provided by shrinkage or movements of the emerging pupal parts.

REFERENCES

Bacot, A. W. (1916). *Report of the Yellow Fever Commission.* London. Vol. III.

Bhatia, M. L. (1948). A simple medium for mounting mosquito larvae. *Indian J. Malar.* **2**, 283–4.

Bonne-Wepster, J. and Brug, S. L. (1932). See references in ch. II (*a–b*), p. 45.

COOLING, L. E. (1924). The larval stages and biology of the commoner species of Australian mosquitoes, etc. *Aust. Dept. Hlth Serv. Publ.* (*Trop. Div.*), no. 8.

DE MEILLON, B., GOLBERG, L. and LAVOIPIERRE, M. (1945). The nutrition of the larva of *Aëdes aegypti* L. *J. Exp. Biol.* **21**, 84–9.

FRANCIS, E. (1907). Observations on the life cycle of *Stegomyia calopus*. *Publ. Hlth Rep. Wash.* **22**, 381–3.

GILES, G. M. (1903). On prepupal changes in the larva of Culicidae. *J. Trop. Med.* (*Hyg.*) **6**, 185–7.

HOWARD, L. O., DYAR, H. G. and KNAB, F. (1912). See references in ch. I (*f*), p. 19.

HOWLETT, F. M. (1913). *Stegomyia fasciata*. *Proc. 3rd Meet. Gen. Mal. Comm., Madras*, 1912.

HURST, C. H. (1890). *On the Life History and Development of a Gnat* (*Culex*). Guardian Press, Manchester.

JOBLING, B. (1938). On two subspecies of *Culex pipiens* L. (Diptera). *Trans. R. Ent. Soc.* **87**, 193–216.

MACFIE, J. W. S. (1915). Observations on the bionomics of *Stegomyia fasciata*. *Bull. Ent. Res.* **6**, 205–29.

MACFIE, J. W. S. (1916). Rep. Accra Lab. for year 1915, pp. 71–9. Abstract in *Rev. Appl. Ent.* **5**, 47.

MACFIE, J. W. S. (1917). Morphological changes observed during the development of the larva of *Stegomyia fasciata*. *Bull. Ent. Res.* **7**, 297–307.

MARSHALL, J. F. (1938). *The British Mosquitoes*. Brit. Mus. (Nat. Hist.). London.

MITCHELL, E. G. (1907). *Mosquitoe Life*. G. P. Putnam's Sons, New York and London.

MÖLLRING, F. K. (1956). Autogene und anautogene Eibildung bei *Culex* L. zugleich ein Beitrag zur Frage der Unterscheidung autogener Weibchen an Hand von Eiröhrenzahl und Flügellänge. *Z. Tropenmed. u. Parasit.* **7**, 15–48.

NEWSTEAD, R. and THOMAS, H. W. (1910). The mosquitoes of the Amazon Region. *Ann. Trop. Med. Parasit.* **4**, 141–50.

PURI, I. M. (1931). Larvae of Anopheline mosquitoes with full description of those of the Indian species. *Ind. Med. Res. Mem.* no. 21.

ROUBAUD, E. (1929). Recherches biologiques sur le moustique de la fièvre jaune. *Aëdes argenteus* Poiret, etc. *Ann. Inst. Pasteur*, **43**, 1093–209.

SHANNON, R. C. and PUTNAM, P. (1934). The biology of *Stegomyia* under laboratory conditions. I. The analysis of factors which influence larval development. *Proc. Ent. Soc. Wash.* **36**, 185–216.

TANZER, E. (1921). (In Tanzer, E. and Osterwald, H.) Morphogenetische Untersuchungen und Beobachtungen an Culiciden-larven. I. Morphogenitische Beobachtungen. *Arch. Naturgesch., Abt. A*, 7. teil, 136–74.

TATE, P. (1932). The larval instars of *Orthopodomyia pulchripalpus* Rond. (Diptera: Nematocera). *Parasitology*, **24**, 111.

THOMPSON, M. T. (1905). The alimentary canal of the mosquito. *Proc. Boston Soc. Nat. Hist.* **32**, 145–202.

TOWER, W. L. (1902). Observations on the structure of the exuvial glands and the formation of exuvial fluid in insects. *Zool. Anz.* **25**, 466–72.

TRAGER, W. (1937). A growth factor required by mosquito larvae. *J. Exp. Biol.* **14**, 240–51.

WIGGLESWORTH, V. B. (1938). The absorption of fluid from the tracheal system of mosquito larvae at hatching and moulting. *J. Exp. Biol.* **15**, 248–54.

WIGGLESWORTH, V. B. (1942). *The Principles of Insect Physiology*. Methuen and Co., London. Ed. 2 (and subsequent editions).

XI
THE LARVA AND ITS ENVIRONMENT

(*a*) PHYSICAL CHARACTERS OF THE LARVA

Fundamental to behaviour of the larva are the physical characters not only of the environment, but of its own body in relation to these. Some knowledge of physical data relating to the larva is, therefore, essential to a proper understanding of many features in behaviour.

LENGTH AND WEIGHT

Determination of body measurements and weight has been very fully dealt with in the chapter on growth. It is here only necessary to give a brief summary of these results. As first emerged from the egg the larva measures about 1·0 mm. in length and weighs about 0·01 mg. At full size it measures to the base of the siphon about 7·0 mm. and weighs in the full-grown well-nourished female larva about 5·0 mg. The length and weight at the end of each instar are approximately as follows.

Instar	Length in mm.	Weight in mg.
I	1·97	0·0875
II	3·24	0·313
III	5·17	1·71
IV	6·80; 7·33	3·29; 4·92

The double figures given under instar IV relate to the male and female respectively. At the beginning of the instar both length and weight are approximately identical with the length and weight of the previous instar at the end of its period.

SPECIFIC GRAVITY

Living larvae of *Aëdes aegypti* are, due to the air spaces in their body, very nearly of the specific gravity of water. This is shown by the fact that at the point of the respiratory siphon when this is in contact with the surface film there is little or no funnel of depression. Actually there is a tiny portion of the spiracular parts projecting. How nearly the living larva is at about the specific gravity of water is also shown by observing larvae, for example, in a cylinder of water and noting how, when not actively swimming, they are at times to be seen slowly rising and at others slowly sinking. In general the large nearly mature fourth instar larvae sink when in the body of the fluid and rise only by swimming upwards. The younger forms under similar circumstances usually rise. In such observations it is necessary to use boiled or distilled water; otherwise minute bubbles of O_2 are liable to form on the hairs, mouth parts, etc. Even the smallest bubble has an obvious effect, again showing how nearly balanced is the weight of the larva and the specific gravity of water. Larvae also, though not actually swimming, may nevertheless be in motion due to the working of the mouth brushes. This may cause a larva to sink, sometimes very rapidly. Such action of the brushes never, or very rarely, causes a larva to rise. The body of the larva at rest is always oriented caudal end upwards whilst the action of the brushes is to propel the larva forward. The resulting effect is therefore for the larva to be propelled downwards. Only by using its body in swimming can it reverse this.

The above remarks apply to the living larva. The dead or long-submerged larva, on the other hand, is usually heavier than water and sinks. If larvae be very carefully chloroformed and the siphon then detached from contact with the surface film the same result, however, as with the living larva is obtained, that is the dead larva in this case is usually just a little lighter or heavier than water. On the other hand, larvae chloroformed, say on a glass plate, and transferred to water, sink, even if they are of the earlier instars. Further, if larvae killed by careful chloroforming when at rest at the surface are submerged several times they sink, though at first they may have floated. In the same way if some second instar larvae are placed at the bottom of a glass tube, say 4 feet long and closed at the bottom, and the tube then filled with water, or alternatively when filled after adding the larvae it is corked and reversed, the larvae will at once begin to swim upwards. But from time to time some will cease from swimming and then they will for the most part at such time slowly sink. These facts show how delicate is the adjustment of air space to weight of the larva. They also show what is presumably the effect of gaseous absorption by the tissues and that this is sufficient to tilt the balance from floating to sinking.

Trials with solutions of sodium sulphate of known specific gravity give results much as above. Living larvae are of about the specific gravity of water. The larger forms chloroformed at the surface mostly sink in sp. gr. 1·005 and 1·010 and rise in 1·020. In 1·005 and 1·010 younger forms rise and cannot at first be made to sink, but later they equilibrate and do so. The larger forms chloroformed on a glass plate in the great majority sink in 1·020; in 1·025 the majority sink but some rise; in 1·030 the majority rise and some sink; at 1·040 the majority rise and but few sink.

The importance of specific gravity lies, however, not only in relation to the larva as a whole, but even more strikingly in respect to its separate parts. When describing the tracheal trunks of the larva mention was made of the large size of these in the abdomen and their sudden narrowing on entering the thorax. The effect of this will clearly be to increase the specific gravity of the thorax and lessen that of the abdomen as compared with the larva as a whole. That this is so is easily shown by separating with a sharp razor the head and thorax from the abdomen and placing these separately in water; the former promptly sinks, the latter floats. The actual specific gravity as determined in sodium sulphate solutions of known density is for the head and thorax 1·040 and for the abdomen 0·982. This difference in specific gravity must have the effect on the body of the larva suspended in water of an upward and downward force forming a couple and it is not improbable that it is in the nice adjustment of the specific gravities of the parts as determined by the shape and size of the tracheal trunks that the balanced stance of the larva is brought about. It seems probable too that the characteristic attitude of the larvae of other species of mosquito may similarly be due to such adjustments (see Fig. 22).

In favour of such a view is the fact that in a medium of specific gravity 1·040 the front part of the body no longer tends to sink and the larva, living or dead, orients horizontally. Further in such a case the living larva clearly finds the position objectionable and one in which it is unable to attach its siphon to the surface film. Thus the tapering of the tracheal trunks, which might seem a mere arbitrary anatomical feature, is one of vital importance to the larva and one which is related to other physical features of the larval body.

VOLUME

Since the specific gravity of the living larva is near to that of water its volume stated in cubic millimetres should be approximately that of its weight in milligrams. Some determinations made on the living larva have, however, always given a somewhat greater volume than that so calculated. The method used for determination has been a modification of that used for determining the volume of the egg based on the method of Johnson (1937). The following are the results obtained:

Material	Observed Weight (mg.)	Observed Volume (mm.^{3})	Calculated specific gravity as weight/volume
Five mature female larvae with respiratory trumpets showing but not yet yellow	29·9	36·7	0·816
Five larvae somewhat earlier than the above	26·7	33·3	0·802
Seven larvae instar IV, still earlier stage	29·0	37·5	0·773

This would seem to make larvae much more buoyant than they appear to be and suggests that the volume figures obtained are too high. Though of interest, the matter has not been further explored.

The only observations in the literature bearing upon the subject of this section appears to be an observation by Wesenberg-Lund (1921), quoted by Senior White (1926), stating that young larvae are frequently over compensated hydrostatically, older larvae being more often under compensated. This would be in accordance with our observations on *Aëdes aegypti*.

(*b*) HYDROSTATIC BALANCE

ATTITUDE (Fig. 22 (1))

As with larvae of all mosquitoes, except in those few species that take oxygen from submerged vegetation, larvae of *A. aegypti* at all stages of growth spend much of their time at rest or slowly moving with their respiratory siphon in contact with the surface film. Whilst mosquito larvae are thus supported at the surface their body assumes a particular attitude which varies with, and is often characteristic of, the species. With some species the angle made by the body with the vertical is considerable and may occasionally approach 90°. That the larva of *A. aegypti* hangs almost vertically is a recognised characteristic of the species. The actual angle is about 20° from the vertical, varying somewhat at different stages and also whether the larva is at rest or in motion and the speed of such motion. In the sections dealing with specific gravity and volume such an attitude has been shown to be an expression of anatomical structure.

USE OF THE SPIRACULAR PARTS

Making and maintaining contact with the outer air and protection of the spiracular openings from entrance of water when so in contact, or when submerged, are the essential functions of the structures at the siphon apex. These functions are largely due to the strongly hydrophobic character of the inner aspects of the parts due to the waxy secretion from the peri-spiracular glands. When these parts are exposed by being opened up at the surface they therefore form a non-wettable area interposed into the surface film in the midst of which open the two large spiracular openings. Closing of the flaps is brought about by the action of the muscles, especially those inserted into the apodeme, traction on which brings all the parts of the spiracular plate together.

But whilst closure of the parts is brought about by the action of muscles there seems to be little or no action by muscles required in their opening which is normally capable of being brought about through action of the surface film alone. Watson (1941) when describing the spiracular parts says that the valves are held together by a relatively weak spring easily overcome by surface tension and even by the weaker force exercised at an oil–water interface, though with well-spread oil the tension may be too weak to do so.

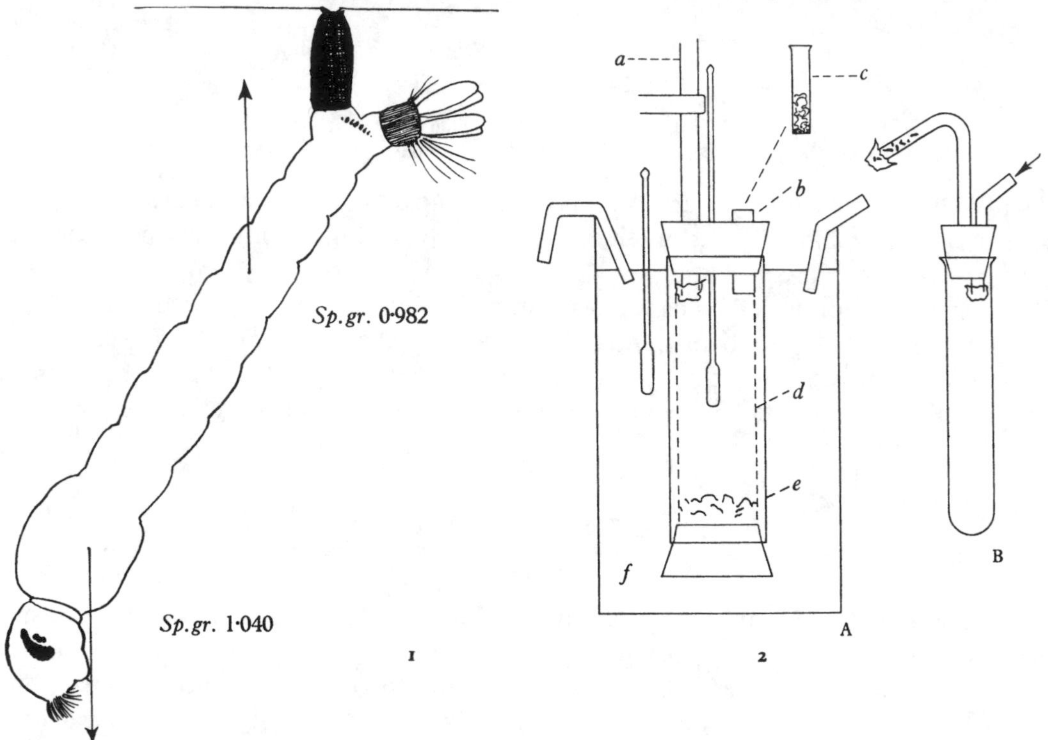

Figure 22. Physical characters of the larva.

1 Illustrating hydrostatic balance.

2 Apparatus used for determination of lethal temperature of the imago (see ch. XXV (*c*)). A somewhat similar apparatus but with a central vessel of water was used for lethal temperature for the larva. *a*, glass tube for support and air outlet; *b*, inlet tube for introducing material; *c*, slip-in tube to block inlet when not in use; *d*, wet or dry filter paper; *e*, wool; *f*, circulating fluid for constant temperature; B, pipette for manipulating adults which are gently blown into end of the pipette tube before blowing into inlet at *b*.

The mechanism is exactly as described. It may be studied and demonstrated very conveniently with a glass tube of 1½–2 mm. bore bent twice at right angles and fixed on a slide with plasticine so that one opening can be brought under the objective of a binocular microscope whilst the level of the water in it can be regulated by pipetting water in or out at the other end. The siphon with the eighth segment of a well-grown fourth instar larva is severed with a sharp razor and placed carefully in an upright position just inside the opening of the tube. Water is now added to bring the level under observation to the opening. When this is achieved the siphon as it engages the surface film will open up automatically. Or if the flaps remain closed a touch on the siphon tube will cause these to open in a manner as described by Watson such as to suggest the action of a weak spring.

Though contact with the surface film is over a small area it nevertheless gives rise to a supporting force the strength of which will be evident on attempting to sink a chloroformed larva with its siphon tip in position. An important point too is that whilst surface tension thus supports the larva at the surface, it allows complete freedom of movement in the horizontal plane since it in no way hinders such movement. Should the larva desire to sever the connection it has only to close the moveable flaps of the stigmal apparatus to cut out surface activity entirely.

Russell and Rao (1941) dealing with *Anopheles* larvae found the normal surface tension of the water to be about 70 dynes. (For pure water the value lies between 74·9 at 5° C. and 71·18 at 30° C. (Childs, 1934).) When the tension was reduced by soap solution to 27–36 dynes the larvae failed to cling to the surface. A similar effect is seen with the larvae of *Aëdes aegypti* in a suitable concentration of soap, the larvae becoming very excited attempting, but failing, to make adequate contact with the surface. The physical definition of surface tension is the force in dynes acting perpendicularly to a section of the surface 1·0 cm. in length. The area of the expanded spiracular plate of a full-grown larva of *A. aegypti* was found by measurement to be approximately 0·0514 mm.^{2}. Taking the area as circular, this would give as the length of the periphery about 0·41 mm., or, by measurement of the outline of a camera lucida drawing of known magnification, 0·43 mm. The proportion of 70 dynes (this being the force over 1·0 cm.) would therefore be, over this distance, about 3 dynes. Or, taking the critical point as found by Russell and Rao as 30 dynes for the fluid, about 1·3 dynes. This would be the force resisting withdrawal of the unclosed parts from contact with the film.

In this connection, however, Wigglesworth (1942) refers to the possibility that the effect noted by Russell and Rao is the result of angle of contact (see section (*j*)).

(*c*) MOVEMENTS

Movements of the larva are very characteristic and Shannon (1931) draws attention to the extreme restlessness of the larva of *A. aegypti* continually roaming through the container and distinguishable by such behaviour from all other Brazilian species except *A. fulvithorax*. This author also describes the characteristic looping movements and the small actual progress made when swimming. Actually translatory movements are of two kinds, namely *swimming* and *gliding*, the latter being brought about by action of the feeding brushes.

The essential movement in swimming is similar to that commonly seen in larvae of mosquitoes, namely a side-to-side lashing movement of the whole body. This, apart from such bending as occurs mainly in the abdomen, causes a partial rotation in the horizontal plane of the body about a point in the thorax, the head being driven in one direction and the longer caudal portion in an opposite direction. The body in the larva of *A. aegypti* being rather long in proportion, such lashing movements give the impression of looping. Progression in this type of locomotion is invariably tail first. Thus larvae swimming downwards from being at rest at the surface have first to manœuvre the body to get this into a suitable position. In swimming upwards the body is already correctly oriented and the larva swims upwards tail first.

When in a fluid of density greater than water progression up and down may become difficult and the larva tends to go round in circles at the same level.

The form of swimming by lashing of the body from side to side can be very clearly studied in the case of *Anopheles* larvae as they skim along the surface. Not only is the side-to-side movement of the head very evident as a result of the body lash, but also the backward propulsion. Movement is carried out here with much greater speed and precision than by the larva of *Aëdes aegypti.* This is probably partly accounted for by the fact that the energy expended by the larva is anchored strictly to the plane of movement. Further, the long body of the *A. aegypti* larva appears to be less favourable for such movement, hence the impression of poor progress. In contrast to lateral movement, upward and downward swimming appears to be more effective.

Besides swimming in the manner described the larva is able to move at a quite considerable rate, at times up to a centimetre or more per second, through the action of its feeding brushes. When working, these cause the larva to move forward smoothly and apparently without effort. Such movement may be seen when larvae are at the surface and their siphons in contact with the surface film, or when browsing at the bottom or on the sides of the vessel, the brushes in the latter case evidently dislodging and sweeping up particles to be swept into the pharynx. When so gliding the body of the larva lies nearly parallel to the surface over which it is passing, thus differing from its normal body attitude. This, however, is not a static condition and is due to the forward movement, as is shown by the fact that the angle made is largely dependent on the speed at which the larva is moving. Moving slowly the body has much the ordinary resting attitude. As speed increases, the line of the body becomes more and more horizontal. At times the additional urge from increased speed causes the body to bend before the movement is fully transmitted.

In the larvae of *A. communis* Hocking (1953) found that wriggling locomotion was rarely used in natural circumstances, most movements being by the feeding brushes.

(*d*) FEEDING

Feeding may be carried out in three entirely different ways, namely (1) by pharyngeal filtration of minute particles from currents produced by the brushes in the body of the medium, (2) by the gnawing and swallowing of solid particles of food using the mandibles, and (3) by browsing.

The first method which is characteristic of all mosquito larvae, except those of purely carnivorous habits, is naturally much in evidence with larvae of *A. aegypti.* Nevertheless, it is very largely supplemented in this case by the other methods, the relatively small head of the *A. aegypti* larva probably being an indication that the species is not one that specialises on feeding on very fine matter in suspension as are some open-water feeders such as *Culex vishnui*, where the head is very large, presumably to accommodate a correspondingly large pharynx (see p. 289). It is true that the larvae of *A. aegypti* are commonly found in clean drinking water, for example in native water pots, cisterns, etc. But in such situations there is usually a copious deposit of fragments of food and other organic debris. Further, it is doubtful how far the species could thrive in such a medium as clean water in the absence of fairly heavy organic contamination, as will be apparent when dealing with starvation effects.

On the other hand, feeding on particulate matter present in the sediment at the bottom, of containers is a rather characteristic feature of the larva of the species, which may commonly be seen moving along the bottom worrying particle after particle in a manner very reminiscent of a dog. In this second method the larva can be seen using its mandibles, and the brushes are not concerned.

The third method which may be described as browsing is also very commonly made use of where conditions are favourable. In this method the larva glides along the bottom or sides of the vessel propelled by its feeding brushes, which at the same time are used to detach matter from the surface over which the larva is gliding. Such detached matter is no doubt carried by the current produced by the brushes into the pharynx as in free feeding. This possibly explains why larvae so engaged usually stop abruptly every few seconds and start again on a new run, the bolus collected perhaps having to be disposed of. Larvae may browse in this way also on the surface film, though this is usually an indication of starvation conditions. Shannon and Putnam (1934) describe two methods of feeding at the surface. In one the larva retains its hold on the surface film by means of its siphon and bends the body up to use its brushes against this; in the other the larva swims ventral side upwards browsing on the film. The first is a common feature in starvation conditions.

(*e*) REACTION TO LIGHT AND OTHER STIMULI

REACTION TO LIGHT

Of various reactions of the larva to stimuli that of negative phototropism is especially characteristic of the larva of *A. aegypti*. Shannon (1931), when describing the behaviour of this larva, notes as especially characteristic, over and above its extreme restlessness, (1) strong negative phototropism, and (2) extreme sensitiveness to vibrations and to light.

Invariably in all circumstances and in all stages the larva at once begins to swim away from the light and to accumulate in that part of the container which is most remote from its source. When many larvae are present in a culture placed in front of a window or facing an electric light an extremely striking mass reaction takes place, the larvae swimming away from the illuminated side until brought up by the curved side of the vessel, which in turn directs them towards one spot, thus causing them to accumulate in a mass in a kind of focus at the point directly opposite the greatest intensity of light. Having to rise to the surface for respiration they form an ascending column of larvae which in turn creates a current. Moving up and driven forwards again towards the light by this current they again dive, only, on travelling a certain distance, again to join the throng. Such movement will continue unabated for hours. So delicate is the reaction to light intensity that where there are multiple sources of light the resultant column will act as an indicator of the greatest resultant illumination. Such intense negative phototropism, when first observed, was thought possibly to be due to the fact that the larvae were reared in the incubator and therefore in the dark. But larvae reared in daylight exhibit the same phenomenon and when cultures have been left undisturbed facing a fixed light a massive accumulation of larval faeces at a point opposite the light source shows how great a part of the larva's time has been spent under the permanent influence of this taxis.

A special study of the reaction of *Culex* larvae to light has been made by Folger (1946). He distinguishes an excitant and a directive effect. An interesting point is that with a light below the vessel the larvae at the surface dive and so in this case swim towards the light. In respect to such a condition, however, it should be noted that the larva at the surface cannot swim upwards and the only reaction open to it when disturbed and seeking safety is to swim downwards. The excitant effect would thus appear distinct from any directive action and stronger than an opposite negative phototropic effect. Folger also observes that a decrease in light intensity is also effective as an excitant. The excitant effect is largely dependent on the rapidity with which the degree of illumination changes, so that a sudden light or a sudden darkening (shadow) both affect the larvae in the same way, that is they dive.

As an index of negative phototropic effect Folger uses a figure obtained by dividing the number of larvae in the half nearer to the light by the number in the further half. To facilitate counting he uses black paper with a white line placed beneath the receptacle, namely a museum jar, 18 cm. by 6 cm. by 15 cm. deep, supported on a sheet of glass on a shelf in a dark room. An index of 1·0 indicates indifference to light, any figure less than 1·0 the existence of negative phototaxis. At 90–120 cm. (3–4 feet) the effect of an electric light gave as an average an index of 0·75–0·77. At 10 cm. the index varied between 0·06 and 0·71 (mean 0·31). Similar determinations for the larva of *Aëdes aegypti* would almost certainly be much higher, that is, a very low index, judging by the relative behaviour of the larvae of *A. aegypti* and those of *Culex molestus*.

The visual stimuli which provoke diving and the visual acuity of the larvae of *Aëdes communis* have been investigated by Hocking (1953). According to Omardeen (1957) negative phototactic reaction increases with age to the fourth instar and pupa, possibly related to growth of the compound eye.

An observation which may perhaps be considered in this relation is that of Jobling (1937) who records that *Aëdes aegypti*, *Culex fatigans*, *C. pipiens* and *C. molestus* can all develop from the egg to the adult in complete darkness. Of these species *Aëdes aegypti* was able in complete darkness to give rise to a second generation. The breeding of this species by the author has similarly been carried out in darkness except for occasional exposure during manipulations. Exposure to light may on the contrary be prejudicial in that this tends to cause the larvae to congregate at one spot in the culture. Even continuous exposure during the whole course of the culture does not abolish the reaction.

REACTION TO TEMPERATURE GRADIENT

Micha Bar-Zeev (1957), confirming results of Ivanova (1940) with *Anopheles maculipennis*, found that sensitiveness of larvae of *Aëdes aegypti* increased with later instars and pupa, the second instar congregating between 23° and 32° C. whilst later stages and pupae did so between 28° and 32° C.

REACTION TO VIBRATIONS AND MECHANICAL STIMULI

A prominent feature in the behaviour of larvae of *A. aegypti* emphasised by Shannon (1931) and other writers is great sensitiveness to vibration and mechanical shocks. A similar reaction is noted by Folger for the larvae of *Culex pipiens*. Sensitiveness is, however, very much greater with *Aëdes aegypti* larva than with that of *Culex pipiens*. Larvae of *Aëdes aegypti* accumulated at the surface will instantly dive if the side of the vessel is tapped to cause even the slightest jar. They behave in the same way if two hard objects are knocked together even at a distance of feet or yards from the culture. In this respect the larvae, especially the nearly full-grown forms, are more sensitive than the pupae. By grading the intensity of the tap on the glass the larvae can be caused to dive, leaving the pupae still at the surface, a procedure sometimes useful in collecting the latter free from the former.

The depth to which larvae under excitation by tapping will descend has been studied by Macfie (1923). At a tap on the tube containing larvae some descended 2½–3½ feet. Repeated tapping caused some to descend 8 feet.

CLINGING OF LARVAE TO EMPTIED RECEPTACLES

Though strictly speaking not a form of behaviour, there may be mentioned here a feature referred to by Howard, Dyar and Knab (1912) and other writers which will be familiar to anyone dealing with the species. This is the tendency of larvae to remain apparently clinging to the bottom when a vessel containing them is emptied. Such a tendency to cling seems to be due to the long lateral hairs of the larva becoming bound down by the surface film of the thin layer of fluid then left behind. The effect is no doubt greatly assisted by the action of the larvae in diving to the bottom of the vessel when this is handled and so largely remaining in the last remnants of fluid. It is doubtful whether beyond this the larvae take any active part in clinging.

(*f*) RESISTANCE TO SUBMERSION

Mitchell (1907) found that young larvae survived three hours' submergence and could be resuscitated after five hours. Older larvae survived only 1½–2 hours. Da Costa Lima (1914) found that larvae of *A. aegypti* survived after considerable periods of submergence. One such larva lived submerged for 53 days. The younger the larvae the more easily they live without air contact. Very advanced larvae will die as a rule in less than a day. Deprived of their papillae they remained alive for some time in aerated water, but died in a few hours in boiled water. Sen (1915) found that larvae of *A. albopictus* could withstand only 6–7 hours without coming to the surface. Bacot (1916) noted that larvae of *A. aegypti* were unable to withstand complete submergence for 20 hours. Macfie (1917) placed mature larvae in tubes of 8 c.c. fluid sealed with a centimetre-thick layer of melted paraffin of suitable consistence. After 2 hours three *Culex* larvae were dead. After 6 hours three, and after 7 hours four *Aëdes aegypti* larvae were dead. Younger larvae survived longer. Lowering the temperature prolonged resistance. Presence of organic matter and boiling the water shortened it. The same author notes that young larvae may live for many days, increase in size, moult and develop fully submerged in running water. Da Costa Lima similarly records that larvae of *A. aegypti* can live submerged for long periods in running water. Such larvae after a time ceased to search for air, but on exposure to air quickly attached themselves to the surface film in the usual manner. Carter (1923) found that larvae of *A. aegypti* will live 2–3 days under a film of petroleum, but not the pupae. Kalandadse (1933) found that fourth instar larvae of this species in a receptacle closed to air access with a glass sheet begin to die at 7 hours and were all dead after 22 hours. Young larvae began to die after 22 hours. The larvae were more resistant than the pupae which began to die in 2 hours, all being dead in 20 hours. In all such experiments it is very important that formation of oxygen bubbles liberated by the least rise of temperature should not take place.

It is, however, larvae of the first instar which most strikingly and uniformly exhibit resistance to submergence. As shown by Wigglesworth (1938) larvae of *A. aegypti* when first hatched from the egg are capable of living even without air entering the tracheae for several days. In the present author's experience larvae hatched from eggs in boiled water under a thick layer of liquid paraffin remain alive and active for days. Nevertheless, dissolved oxygen in some amount is evidently necessary, since if special steps beyond boiling be taken to free water from oxygen such larvae quickly die. Thus, when using a small tube with boiled water (Johnson's volume determinating apparatus) first instar larvae in the closed limb frequently died during manipulations whilst those in the open limb still remained alive. Similarly under strictly anaerobic conditions (Bulloch's apparatus as for anaerobic growth) all are usually dead overnight.

(*g*) FOOD REQUIREMENTS OF THE LARVA

The food requirements of the larva is a subject that has received much attention and may be briefly summarised. There is first the question of the physical nature of the food. A number of authors have attempted unsuccessfully to rear larvae on food in solution or in the colloidal state. Matheson and Hinman (1928), after many trials, obtained some development of *Aëdes aegypti* larvae in pond-water filtrate, but came to the conclusion that solutes and colloids are not a source of nourishment to mosquito larvae. De Meillon, Golberg and Lavoipierre (1945) also note that larvae from sterilised eggs in neutral broth, though they may remain alive several days, do not grow. Trager (1935*b*) found that larvae did not develop in 1 per cent Loeffler's serum, but later (1936) showed that larvae could be reared at least to the fourth instar if a certain amount of $CaCl_2$, and an organic growth factor shown to be a solute, were present. It seems clear, however, from the author's results that such a diet did little more than continue existence under starvation conditions and only a proportion of the larvae it would seem reached even to the fourth instar. It might be as the author suggests, that though the brushes would not be effective, the larvae could still swallow and so obtain some food matter in this way.

Quite apart from the question of food requirements it would seem very unlikely that larvae with their special mechanism for feeding, that is filtration of particles by the fringes of the pharynx and the accumulation of such matter in a bolus to be swallowed, would be able to obtain adequate nourishment by swallowing fluid medium. It may safely be said therefore that solutes and colloids are not normally the source from which larvae obtain food.

Bacot (1916) and Atkin and Bacot (1917) pointed out that bacteria and yeasts are an esssential food requirement of the larva and reared larvae on pure cultures of yeast. Dead bacteria or dead yeast they found had little value. Since these early researches there have been numerous observations which appeared to show that the food of mosquito larvae was inadequate unless this contained living bacteria and yeasts. Hinman (1933) failed to rear larvae on autoclaved material and considered living bacteria necessary. Rozeboom (1935) found that sterilisation rendered media unsatisfactory for development and that, whilst no development took place in sterile blood medium, contamination rendered this and other media satisfactory. MacGregor (1929) reared larvae on sterile bread, but his results have not been confirmed. Neither Hinman nor Rozeboom were able to rear larvae beyond the second instar using MacGregor's sterile bread technique. There seems no doubt that bacterial organisms and yeasts are an important, if not essential, food requirement and all satisfactory methods of culture clearly provide this in the form of fluids containing organic matter in which organisms are freely growing. Lewis (1933) sprinkles bread crumbs on the surface once in three days. If too much is used, scum forms and the larvae are killed; if too little is used, pupation ceases. Johnson (1937) uses powdered breakfast food as more convenient than yeast. MacGregor (1915) adds every alternate day a pellet of guinea-pig's faeces. Bonne-Wepster and Brug (1932) also found guinea-pig's faeces the most suitable food. Weyer (1934) used dried blood and powdered calf's liver and obtained vigorous adults, but not apparently under sterile conditions. MacGregor (1915) records cannibalism among larvae of *A. aegypti*, but this would not appear to be a usual feature of the species which does not exhibit carnivorous propensities.

The above would seem to bring the larvae of *A. aegypti* into the class of phytophagous insects. In such a connection Uvarov (1928) notes that a form of food taken very commonly by phytophagous insects is decaying vegetable matter, but that the actual food may not be the substance most obviously taken in. Thus in the food of *Drosophila* yeasts are always present and in their absence *Drosophila* develops slowly. The Cecidomyid *Miastor* living on bark feeds on the micro-organisms present and can be bred on agar culture but not on sterile bark.

Whether protozoa can provide an adequate dietary has been investigated by several authors. Barber (1928) reared larvae on pure cultures of *Colpidium* sp. and of *Paramecium* sp., but failed to produce adult insects. Mathis and Baffett (1934), however, found that though larvae died in 7–8 days in nutrient broth, full development occurred when a sterile culture of *Euglena* was added.

So far nothing has been said of the chemical nature of substances essential to satisfactory growth. For this it is necessary to turn to the work more especially of Trager (1935–7) and of De Meillon and his collaborators. Briefly Trager has shown that mosquito larvae require at least two growth-promoting substances. One is present in large amount in yeast, is alkali stable and like vitamin B_2 is heat stable. The heat labile vitamin B_1 is not required by *Aëdes aegypti* larvae. The second factor is present in substances effective against secondary anaemia. It is heat stable in neutral or slightly acid solutions and sensitive to prolonged alkali action even at room temperature. Buddington (1941) found that B_1 (thiamin hydrochloride) was ineffective, as also that B_1 and B_2 separately or together with yeast failed to bring larvae beyond instar IV, though with autoclaved yeast and liver extract 80 per cent reached the adult stage.

De Meillon, Golberg and Lavoipierre (1945) found that filtered liver extract improved growth and survival rates. Furthermore, they called attention to a factor which is necessary for emergence of vigorous adults. This factor is largely absent from many artificial media such as brewer's yeast and liver extract. They note that in nature it is unusual to find a weak adult, that is, one which cannot rise from the water. But in yeast and liver extract growth this is common. Later Golberg, de Meillon and Lavoipierre (1945) identified this substance with folic acid. Lichtenstein (1948) also found folic acid to be an essential food requirement for larvae of *Culex molestus*.

Folic acid (pteroylglutamic acid) is identical with vitamin M necessary for monkey growth, B_c vitamin essential to growth of the chick and the Norite eluate factor required for the growth of certain bacteria. It occurs in liver, yeast and many articles of diet (Subba Row 1946).*

* See also Golberg and De Meillon (1947, 1948). An important paper dealing with amino-acid requirements and giving references is that by Dimond, Lea, Hahnert and De Long (1956).

The procedure recommended for rearing *Aëdes aegypti* has been described in the chapter on technique. The culture medium there described clearly provides all that is necessary for optimum development, namely, food material in fine suspension and in coarser particles as sediment with yeasts and bacteria growing freely in an organic medium. Addition of a pharmaceutical liver extract has been tried, but with no appreciable difference in growth. Two cultures made under strictly comparable conditions side by side gave as the mean weight for females with liver extract and without respectively 2·43 and 2·44 mg. and for males 1·24 and 1·23 mg. The opportunity may be taken here to emphasise the points: (1) that by dog biscuits a certain type of biscuit has been indicated; (2) that the ingredients include bemax; and (3) that the solid ingredients are finely ground. It is believed from considerable experience that growth in such a medium is optimal. Whether addition of any substance could increase growth is not known. In this, as in all questions of nutrition of the species, the use of weight and wing length as a criterion is essential (see chapter XVII).

Table 12. *Data recorded in the literature regarding lethal temperature for larvae of different species of mosquito*

Species	Author	Lethal temperature in ° C. after exposure of		Remarks
		5 minutes	1 hour	
Anopheles gambiae	De Meillon (1934)	—	Over 45	Breeding in nature at 39° C.
Anopheles funestus	De Meillon (1934)	—	Under 45	—
Anopheles insulae-florum	Muirhead Thomson (1940)	40	—	—
Anopheles minimus	Muirhead Thomson (1940)	41	—	—
Anopheles hyrcanus	Muirhead Thomson (1940)	43–43·5	—	Tropical species
Anopheles barbirostris	Muirhead Thomson (1940)	43·5	—	—
Anopheles culicifacies	Muirhead Thomson (1940)	44	—	—
Anopheles vagus	Muirhead Thomson (1940)	44·5–45	—	—
Anopheles subpictus or *Anopheles ludlowi*	Brues (1939)	—	—	Breeding in nature at 39–40° C.
Anopheles bifurcatus	Wright (1923)	37	35	—
Aëdes detritus	Wright (1923)	42	37	Temperate species
Theobaldia annulata	Wright (1923)	36	35	—
Culex pipiens	Wright (1923)	40–41	—	—
Anopheles pharoensis	Bates (1949)	—	40	Subtropical species
Aëdes aegypti	Marchoux *et al.*	—	39	Tropical and subtropical species
Aëdes aegypti	Macfie (1920)	43	—	Tropical and subtropical species

(*h*) EFFECT OF TEMPERATURE

LETHAL EFFECT OF HIGH TEMPERATURE

The temperatures found lethal to larvae of a number of species of mosquitoes as determined by different observers are given in Table 12. The results show that for short exposures this is usually about 40°–45° C. For longer exposure it may in some cases be as low as 35° C. It is difficult to give a precise thermal death point since the result is liable to vary with the conditions of the test and that within a considerable range the lethal effect has to be expressed as a percentage. Macfie (1920) found that larvae apparently dead may recover. Muirhead Thomson (1940) notes that his results have been obtained after raising the temperature gradually.

Most of the results given in Table 12 refer to different species of *Anopheles* and Culicines from temperate regions. A very careful and detailed account of the effect of high temperature on the larva of *Aëdes aegypti* is, however, given by Macfie (1920) whose general results have been given as an average figure of 43° C. in Table 12. This author, however, found that a temperature of about 47° C. was necessary in an exposure of a few minutes to give complete lethality with no recoveries. For exposure of 15 minutes 45° C. was necessary for complete effect, though a temperature of 41° C. gave considerable mortality. For 30 minutes exposure 43° C. was completely effective, which is the figure given in the table. For more prolonged exposure a temperature of 39° C. produced considerable mortality and this temperature is that given as lethal by Marchoux, Salimbeni and Simond (1903). Headlee (1942) also gives 39° C. as lethal, in that larvae were dead in 5 days at 102° F. (38·9° C.).

Table 13. *Lethal effect of high temperature on larvae of* Aëdes aegypti

Temperature (° C.)	Periods of exposure: 3 minutes		15 minutes		30 minutes		60 minutes	
	a	*b*	*a*	*b*	*a*	*b*	*a*	*b*
47	100	100	100	100	100	100	—	—
46	90	100	100	100	—	—	—	—
45	70	30	100	100	—	—	—	—
44	11	44	100	88	—	—	—	—
43	27	27	100	67	100	100	—	—
42	8	7	73	53	100	—	—	—
41	—	—	60	50	25	45	—	—
40	—	—	0	0	25	42	34	38
39	—	—	—	—	—	—	8	42

a Number apparently dead at end of experiment.
b Number dead overnight.

In Table 13 are given the results of 3, 15, 30 and 60 minutes' exposure in which larvae filtered off on filter paper were added to fluid at the temperatures noted. About twenty three-quarters-grown fourth instar larvae were used for each determination and the numbers immobilised at the end of the experiment noted (first column of figures), as also the percentage found dead after exposure and being kept overnight at room temperature (second column of figures). The results are very close to those obtained by Macfie. The temperature required for complete effect was 46° C. for 3 minutes' exposure, 45° C. for 15 minutes, 43° C. for 30 minutes. Lethal effect ceased at 41° C. for 3 minutes' exposure, 40° C. for 15 minutes and there was 42 per cent eventual mortality with 39° C. for an hour, the number apparently dead at the end of the experiment being only 8 per cent.

LETHAL EFFECT OF LOW TEMPERATURE

Trofimov (1942) records that the larvae of *Anopheles pulcherrimus* and *A. bifurcatus* (*A. claviger*) resisted temperatures of −5° C. to −9° C. from 20 to 210 minutes. This observer also makes the observation that the haemolymph of the larvae freezes at −0·5° to −0·7° C. Records of *Anopheles* larvae in nature surviving when frozen in ice have not infrequently been made. M. G. Wright (1901) in England found *Culex* larvae at 29° F. (−1·7° C.) alive unless frozen in ice.

Marchoux, Salimbeni and Simond (1903) note that larvae of *Aëdes aegypti* are not killed by short exposure to 0° C. Bliss and Gill (1933), who give a good résumé of early work on the effect of low temperature on the larvae of *A. aegypti* record that the larvae will recover after being encrusted in ice. They found that larvae frozen for 2–10 hours at − 2° C. recovered in 2–3 hours. Those frozen for 11 hours or more did not recover. Bacot (1916) found that exposure to 4·4° C. for 2–3 hours caused larvae of *A. aegypti* to become dormant at the bottom of the vessel. They resumed activity at 27° C. Ramsay and Carpenter (1932) found that *A. aegypti* larvae survived freezing over 8 hours, but not 24 hours

As noted by Bacot the effect of exposure to low temperature is to cause larvae to become sluggish in their movements and, if sufficiently low, to become motionless and apparently dead. In observations by the present writer first instar larvae when immobilised were for the most part found lying inert and submerged. Older larvae are usually found floating inert or entirely motionless at the surface with their siphon in contact with the surface film. The following are observations on different stages of larvae at temperatures which might be considered lethal or on the border line. The temperatures are those of the water.

First instar

7° C. Larvae sluggish or motionless. Remained so throughout early instar I. Some dead by fifth day. Many dead by seventh day and majority dead by ninth day. A few still alive (early instar I) on twenty-third day.

1° C. After 24 hours' exposure appeared dead, but 25 per cent recovered on placing at room temperature. After 48 hours there were no recoveries.

Second instar

7° C. Newly ecdysed second instar larvae placed at 7° C. At 24 hours very sluggish, just able to move, many floating with the siphon attached to surface film. On third day many recovered at room temperature. On sixth day larvae as first described and many recovered. On ninth day all larvae dead with no recoveries.

1° C. At 24 hours all motionless and apparently dead. At 72 hours no recoveries.

Third instar

8° C. At 48 hours larvae sluggish, but made spasmodic use of brushes; tend to lie motionless around edges; some exhibited negative phototropism.

7° C. At 24 hours as described for instar II. At 72 hours many recovered at room temperature. At 6 days many recoveries.

1° C. At 24 hours all motionless and apparently dead. At 72 hours no recoveries.

Fourth instar

7° C. At 24 hours all larvae motionless, some showing sluggish movements on being disturbed. At 48 hours majority are motionless or dead floating with siphon at surface. 25 per cent recovered motility at room temperature, but were feeble and later development was effected. Fourth day some showed feeble movement when placed at room temperature. Sixth day no recoveries.

A temperature of 1° C. was lethal in most cases within 24 hours. A temperature of 7° C. was still lethal in all cases, but only after from the sixth to the ninth day. Though final death was not observed in the experiment with third instar larvae, it probably occurred. The only partial exception was with the first instar larvae some of which were still alive on the twenty-third day. The experiments show that a temperature of 7° C. is well below that at which the species could survive.

THE TEMPERATURE LIMITS OF EFFECTIVE DEVELOPMENT

As distinct from the temperature which is lethal for short periods of exposure is the question of the temperature limits within which the species can exist, that is undergo growth and complete development. Howard, Dyar and Knab (1912) quote Reed and Carroll as stating that at 10° C. young larvae of *A. aegypti* failed to pupate though growth took place. Gutzevich (1931) gives as the lower limit for *A. aegypti* 17° C. Headlee (1940–2) gives temperature limits for the species' continuance as 60–94° F., that is approximately 16°–34° C. At 54° F. and 102° F. (12°–39° C.) larvae do not mature.

The following are some observations made at water temperatures of 17°–13° C.

17° C. Larvae hatched from eggs remained active and developed slowly. Ecdysis to instar II took place on the fifth day; the larvae were half-grown instar III by the seventh day and pupae began emergence on the twenty-third day.

14–16° C. Newly hatched larvae at 96 hours were still instar I, and at 144 hours were still only partly instar II.

13° C. Larvae hatched from eggs at 96 hours were still half-grown instar I. All were dead at 68 hours.

The lower threshold for development seems therefore very precisely given by 16° C. An upper limit of 34° C. receives some support from the statement by MacGregor that adults hatched above 30·8° C. were small and feeble and from the data given above showing that exposure to 39° C. for one hour caused appreciable mortality.

THE TIME-TEMPERATURE CURVE

A considerable literature exists relating to the mathematical representation of the effect of temperature on development by the time-temperature curve. For information on this subject see Uvarov (1931), Janisch (1932), Wigglesworth (1942), also earlier papers by Peairs (1914) and Ludwig (1928). For application to development in mosquitoes see Headlee (1940–2), Huffaker (1944), Bates (1949).

Certain formulae, notably that of Blunk (1923), and of Janisch (1925), have been given for this curve. Such formulae are based on the assumption that between the limits of temperature at which development takes place this is proportional to the temperature. Blunk's formula is

$$t(v-c)=k,$$

where t is the time of development, v the given temperature, c the critical temperature at which development ceases and k the thermal constant. Thus Bodenheimer (1924), taking optimum temperature as 27° C., the lower limit of development as at 17° C. and the time of development as 10 days gives as the thermal constant for *A. aegypti* $10(27-17)=100$.

Janisch's formula is

$$t=m/2(a^{T}+a^{-T}),$$

where t is time of development, m time of development under optimum temperature, a a constant, and T and $-T$ the limits in degrees above and below the optimum. It is chiefly interesting in the present connection from the use made of it by Huffaker (1944) in following up the effect of variable temperature and the observations of Headlee (1942) in applying this to development of *A. aegypti*. For practical application it would seem from Headlee's results that with variable temperatures ranging through some six degrees the time to emergence is very close to that for the mean temperatures. Thus for 16–22° variable temperature the days to emergence were 22 as against 22·5 for a constant temperature of 19° C. For variable temperature 20–27° C. the time was 13 days as against 14·5 for a constant temperature of 24° C.

A time-temperature curve for *A. aegypti* and for some other species of mosquito is given by Bodenheimer (1924). Gutzevich (1931) states that the curve for *A. aegypti* follows that given by Peairs for insects in general, and later (1932) gives a curve for the species under laboratory conditions of culture.

EFFECT OF TEMPERATURE ON SIZE OF ADULT

Whilst speed of development increases with temperature up to and possibly beyond 28° C. this does not apply to the size of adults produced. MacGregor (1915) notes that reared below 19·5° C. or above 30·8° C. adult *A. aegypti* were undersized. At 23·8° and 25·9° C. adults were equally fine specimens. Martini (1923) notes that mosquito larvae subjected to warm temperature produce smaller-sized adults than at cooler temperatures. This effect, whether as a direct result of temperature or working through food supply, is very commonly seen in nature with species of *Anopheles.* Not only are alpine and temperate species apt to be larger than those of hot climates, but individual specimens of the same species bred in cool and shady places are usually finer (and darker) specimens than those bred in open situations with higher water temperature. The same effect appears to be present in *Aëdes aegypti* in that specimens bred at somewhat low temperatures are commonly of larger size than those bred at optimum temperatures. As noted in a later chapter the heaviest and largest specimens were those bred at a water temperature of 23° C. At 17° C., though the time to emergence was three times that at 28° C., the specimens were at least as large as those bred at 28° C. Yet, as noted above, a few degrees lower and the species does not develop at all.

(*i*) EFFECT OF STARVATION AND OVERCROWDING

If larvae of *A. aegypti* are reared in absence of sufficient food the period of development is increased and the size of the resultant adults reduced. If starvation is nearly or quite complete, as when larvae are placed in clean tap-water, they may remain long periods up to a week or more with little or no development and may or may not hatch out as adults. If they do complete their development, they form the small or minute pupae and adults characteristic of starvation forms.

Such forms are referred to by a number of authors, though most who do so give but few details. A special study of these forms has, however, been made by Weidling (1928). He describes the result of deficient food during the larval stage as leading to prolonged development, deficient size of adults, fewer ovarian tubes in the female, smaller egg measurement and fewer eggs laid. In regard to the larva, this author in a graph gives the normal well-fed larva as increasing in length during a period of 8 days to slightly over 7 mm. The starved forms increase at a lower rate, becoming stationary as regards growth after 10 days with a maximum length of 4 mm.

The larvae under such conditions after a time, if starvation is severe, remain motionless at the surface supported by contact of their respiratory siphons with the surface film, an appearance quite different from the normal. Or if less severe they may be seen characteristically using their brushes to sweep the under surface of the film, bending up to do so whilst similarly supported and stationary. In very advanced starvation a curious feature after some time is that the larvae come to lie flat at the surface and not only so but their bodies come to be liable to be more or less extensively involved in the surface film showing that there has been some effect making the cuticle hydrophobic. As time goes on they lose all trace of fat and become almost transparent, so that structures within the body show with extraordinary clearness and the most minute ramifications of the tracheae are vividly shown up. Imaginal buds, even in the fourth instar, may be inconspicuous or apparently absent. Ecdysis may take place, but if starvation conditions are sufficiently long maintained development beyond this is almost, if not quite, at a standstill. It would appear that even early fourth instar larvae placed in clean tap-water will pupate and do so more or less at the usual time, but producing minute pupae. With strict starvation, however, larvae in the earlier instars rarely survive to pupate. The minute pupae resulting show little difference in the size of the sexes, which may be scarcely distinguishable without examining the terminalia. Usually after a time the originally clean tap-water becomes faintly turbid and unless changed from time to time may eventually become a medium capable of allowing improved conditions. Third instar larvae starved in this way for 10–14 days have been made use of by Wigglesworth (1942) in his observations on reserve substances of the larva. The following are observations on different instars placed at 25° C. under starvation conditions (clean but not changed tap-water):

Early instar II. 48 hours. Some circling at surface.
6 days. Many still swimming. Two instars present heads measuring 0·4 and 0·6 mm. in diameter respectively.
14 days. Larvae mainly at surface inactive. Water clear. No further ecdysis.
30 days. Larvae largely dead.

Early instar III. 24 hours. Majority swimming.
6 days. Some swimming, some at surface. No skins, that is, no ecdysis.
8 days. About half have ecdysed. The skins have been eaten with the exception of the head capsule and siphon.
14 days. Larvae swimming and looking in fair condition. Water turbid.
30 days. Larvae still alive, a good deal of debris in water.
40 days. Some minute pupae.

Early instar IV. 6 days. Pupated to minute pupae (14), nine larvae still present. Head diameter 1·08 mm. Weight of a female pupa 2·19 mg., of two males 1·83, 1·77 mg.
8 days. All but two pupated.
12 days, Adults hatched out, but only one female and two males alive 5 days later. Female offered food took a long time to gorge and did not do so fully. Weak and fell on back in water on being transferred to tube. Died without oviposition or digestion of blood.

The following are measurements made on larvae which 24 hours from hatching were transferred to clean tap-water and examined 6 days (145 hours) later. Three instars were present, the largest presumably IV, since the head diameter was 0·89. The larvae were hanging motionless at the surface, obviously thin and transparent. Under the microscope they appeared a mass of tracheae owing to the absence of all fat. There was no trace of imaginal buds in the fourth instar larvae.

Measurement	Instar II (mm.)	Instar III (mm.)	Instar IV (mm.)
Body length to base of siphon	2·68 (2·60)	3·77 (4·15)	4·49 (7·05)
Head diameter	0·58 (0·54)	0·76 (0·74)	0·89 (0·97)
Thorax breadth	0·54 (0·57)	0·76 (1·00)	0·94 (1·47)
Siphon length	0·43 (0·38)	0·51 (0·56)	0·69 (0·82)
Anal papillae	0·36 (0·33)	0·51 (0·57)	0·62 (0·86)

The figures in brackets are the mean for normal pre-ecdysis measurements. There is clearly a progressive falling-off from the normal towards the fourth instar, especially noticeable in the body length and thorax measurements (62 and 64 per cent of the normal). The head diameter is 92 per cent, the siphon length 84 per cent and the anal papillae 72 per cent of the normal measurement for pre-ecdysis fourth instar.

A point that should be mentioned is that the first instar larva may be kept some days without food without apparently affecting future development. Bacot (1916) states that first instar larvae remain as such until organic matter is added. De Meillon, Golberg and Lavoipierre (1945) make use of this property by leaving first instar larvae hatched under sterile conditions in nutrient medium, in which they do not grow, for some days if necessary before transferring them to the food medium being tested.

Overcrowding tends to, and may seriously, reduce the size of the mature larva, pupa and resulting adult. The limits at which overcrowding becomes apparent are difficult to determine, but the numbers must be very considerable if conditions of culture are good, for example at over 1000 per litre. There is the possibility that too few larvae in a large culture may also give some reduction in size.

(*j*) EFFECT OF SALT, pH, SOAP AND CHEMICALS

Under certain conditions mosquitoes may breed in brackish water and the concentration of salt and sea-water that the larvae are able to withstand has been studied by a number of observers. For general accounts of mosquitoes breeding in brackish and saline waters see Balfour (1921), Wigglesworth (1933*a*, *b*. *c*). The following are data relating to *Aëdes aegypti*.

Macfie (1914) 0·5 per cent salt. Little effect.
2·0 per cent salt. Many dead at 6 hours; all dead at 12 hours.
2·3 per cent salt. Larvae rapidly killed.

Macfie (1922) Pure sea-water. Larvae killed in 2–4 hours.
50 per cent sea-water (1·6 salt). Larvae killed in 24 hours.

Wigglesworth (1933*c*) 0·9 per cent salt or less. Larvae not affected.
1·0 per cent salt. Larvae die in about a week.
1·3 per cent salt. Larvae die in about 72 hours.
1·4 per cent salt. Larvae die in about 48 hours.

Pantazis (1935) 0·5–1·0 salt per mille. Optimum salinity.
5·0 salt per mille. No development over this concentration.

By gradually increasing the salinity Wigglesworth (1933*c*) found that larvae of *A. aegypti* can become adapted to live in up to 1·75 per cent salt or sea-water equivalent. In 5·0 per cent the anal papillae drop off (1933*b*). If larvae reared in fresh water are transferred to salt water, the mid-gut cells swell and become detached. In larvae adapted to salt the gut cells absorb fluid but do not swell. If fed on stained particles, the cells in the salt-adapted larvae become crowded with granules; in the unadapted they are swollen and clear. Martini (1922) pointed out that larvae bred in saline water have short anal papillae and those bred in dilute acid long ones. The former statement applies also to certain mosquito species which habitually breed in nature in saline waters. The shortening is not so marked in *A. aegypti* as in *Culex* larvae, except as pointed out by Pagast (1936) in larvae brought from distilled water into very dilute salt, such as tap-water. *A. aegypti* larvae also, as shown by Wigglesworth, have a greater power than the larvae of *Culex* to obtain necessary salt from very dilute solutions; see also Gibbins (1932).

There is little reason to suppose that pH has any marked direct effect upon the larva. MacGregor (1921, 1929) found that pH of the medium had little direct importance. Senior White (1926) found larvae of *A. aegypti* in nature in waters with pH varying from 5·8 to 8·6. Buchman (1931) found that larvae of *Culex pipiens* could live in water at pH from 4·4 to 8·5, but died beyond these values. Rudolfs and Lackey (1929) found pH to have little effect. Woodhill (1942) notes that at pH 3·6–4·2 the percentage of emergence is considerably reduced and the period of development increased for certain species including *Aëdes aegypti*. The time of development at pH 9·2–9·5 was only slightly longer than at pH 6·8–7·2.*

Soap in suitable concentration has a lethal effect through so affecting the surface tension of the medium that the normal mechanism by which larvae contact the surface to obtain air is interfered with and the larvae eventually drown (Russell and Rao, 1941). Strickland (1929) found 4 per cent of ordinary soap an effective larvicide. This was much less effective, however, in the field. Shannon and Frobisher (1931) found that 1·0 per cent coconut oil soap killed larvae in an hour.

The effect of various chemical substances on larvae has been widely studied with a view to obtaining effective larvicides (see section on control). Various forms and grades of mineral oil have been extensively used as larvicides and the various theories as to how they act are discussed by Freeborn and Alsatt (1918). To some extent such oils are lethal by reason of forming a film which cuts contact of the larva with air and kills by asphyxiation. But this effect is relatively slight unless in addition the oil has other effects. One action of such oils is that by wetting the hydrophobic, but oil wettable, areas at the apex of the siphon they pass on from this to wetting and entering the tracheal trunks. The effect is further enhanced if the oil exerts in addition a toxic effect. Films of non-toxic oils such as liquid petroleum have very little immediate effect.

A study of the toxicity of kerosene has been made by Moore and Graham (1918). See also Freeborn and Alsatt (1918), who consider that it is largely the vapour which is toxic, oils being effective in proportion to the lowness of their flash point. A very good account of the action of oils and the factors concerned in their efficacy is given by Murray (1936) (see also Hacker, 1925, and Watson, 1937, 1941). The last-mentioned author notes that larvae will often themselves ensure contact of the oil with the hydrophobe surface at the tip of the siphon by nibbling there when they find obstruction in breathing.

Wigglesworth (1930) has described the process of entry of kerosene into the tracheal trunks and finest tracheal capillaries. Within half an hour gas had completely disappeared from the tracheal system. The author notes that the oil has a strong affinity for the lining membrane of the tracheae as judged by the highly concave meniscus at the free surface of the liquid as it moves along the tubes. The surface tension of the oil will therefore exert pressure on the gas in the tracheal system which is driven into solution in the tissue fluids. Watson (1941), possibly with a different oil, found the advancing meniscus convex and considers that the oil is drawn in by the using up of oxygen by the larval tissues. He notes that larvae affected by oils in the tracheal vessels show paralysis of the heart action when water-soluble toxins are present in oils that are fat solvents.

Though not in common use, if at all, as larvicides in practice a great variety of inorganic

* For lethal effect of pH see Tanimura Seizo and Sonoji Hine (1952).

and organic compounds have been found experimentally lethal to larvae. Matheson and Hinman (1928) found borax, 15 or more grams to the litre, a very efficient larvicide that might be used for water collections other than those used for drinking. The effects were permanent. Macfie (1923) found 1·2 per cent lithium chloride lethal to larvae. Frobisher and Shannon (1931) and Shannon and Frobisher (1931) have tested a large number of substances to ascertain the extent to which they were effective. Cresols and related substances killed larvae in 1 per cent concentration in 25 seconds. Some substances which it was suspected would be very lethal were surprisingly not so. In 1 per cent KCN full-grown larvae survived 1 hour. Ten per cent was fatal in 10 minutes, that is about the same as a corresponding KOH solution, though in greater dilution KCN was more lethal than KOH. Strong acids were very little lethal, 1 per cent sulphuric and nitric acid taking 75 minutes to kill and HCL 100 minutes. Ammonium sulphate was found by Brink and Chowdry (1939) in 0·75 per cent solution to prevent *Culex fatigans* breeding and to kill all larvae in 2 days, as well as being a fertiliser. Full-grown larvae, however, pupated and pupae emerged in 2·0 per cent. Yates (1946) found phenothiazine effective in one in a million, but the effect was erratic.

Barber (1944) tested the effect of *vapours* of certain compounds against larvae. Carbon bisulphide was most rapid, toluene, carbon tetrachloride and benzene almost as rapid (fatal in 2–3 minutes). Phenol and kerosene vapour required 12–30 minutes to cause larvae to become stationary.

The effect of *gases* and *vapours* has been tested by Williamson (1924), Barber (1944) and others. According to the former author first instar larvae of *Aëdes aegypti* are an exception to the general susceptibility of larvae to gases and vapours. SH_2 in saturated solution killed larvae of *Culex pipiens* in 1 minute. Chloropicrin was also effective in great dilution (1–200,000). Chlorine in high but undetermined amount maintained for 3 minutes killed all larvae and pupae. Macfie (1916) found chlorine 1: 10,000 lethal.

The effect of some pure gases has been described by Wigglesworth (1930). Larvae in an atmosphere of CO_2 within 4 or 5 minutes are completely narcotised. They recover when removed to a vessel of fresh water. Hydrogen has not this hypnotic effect, but through deprivation of oxygen the larva is rendered inactive in 10 or 15 minutes. Oxygen restores activity in less than thirty seconds. With narcotisation (CO_2) or hydrogen asphyxia the gas does not extend to the finest tracheal capillaries as it does in simple asphyxiation. This is due to the absence of violent muscular contractions which in simple asphyxiation cause absorption of fluid in the tracheoles. Chloroform vapour acted comparatively slowly, the muscles contracting for about 2 minutes. CS_2 acted similarly. HCN was more rapid in action with little muscular movement. With ammonia the larvae were dead in about 10 minutes.

The plant fish-poison, *Tephrosia vogelii*, was found by Worsley (1934) effective against a species of *Aëdes*, an alcoholic extract of the seeds 1–300 being used. Chopra, Roy and Ghosh (1940) also found *Tephrosia* effective in a 1–10 dilution of an acetone extract, all larvae being killed within 24 hours. A number of other vegetable poisons have been found actively insecticidal. A very valuable monograph on this is published by the Imperial Institute (1940).

The use of larvicides applied as dusts, for example Paris green, DDT and other insecticides of the chlorinated synthetic type are more adapted for use against the adult mosquito and as larvicides in antimalaria work since the larvae of *Aëdes aegypti* do not usually feed at the surface film. Also the breeding places of the species are not often suitable for such methods of application. Chopra, Roy and Ghosh (1940), however, found pyrethrum powder (a little scattered on the water) to be effective against larvae of *Aëdes aegypti*. Camphor or paradichlorobenzene used in this way has been suggested by Muir (1920). For an account of the chlorinated synthetic insecticides, see under 'Insecticides', pp. 555–6 in the chapter on viability (chapter XXV (*e*)).

Lewis (1955) found benzene hexachloride used in cement pellets as described by Bruce-Chwatt effective in some of the coast villages in the Sudan where water was precious.

In general larvae are little affected in short response to solutes in watery solution. Lipoid solvents in sufficient concentration are commonly rapidly lethal. Pupae will remain alive for long periods in many strong chemicals as a result of their waxy coat (see chapter XVI). But if given a preliminary short exposure to chloroform or other lipoid solvent they are rapidly killed under such treatment, the protective waxy layer with which they are coated being removed. In the pupa also the cuticular envelope is completely closed. In the larva not only is this envelope much thinner, but there are two points of entry not present in the pupa, namely by swallowing and by passage through the permeable cuticle of the anal papillae.

Under the present heading may be included the observations of MacGregor (1926) upon the effect of electric current on larvae. These are violently affected by 100 volts passed through swamp water, showing muscular twitchings and loss of muscular control. They are not affected by 2–6 volts but respond to make and break. Entanglement of bubbles in the mouth-parts due to electrolysis causes the larvae to swim downwards and attempt to attach their siphons to the bottom. The explanation of the last observation is probably the following. As previously pointed out in the section on movements, progression in swimming is always backwards and that to swim downwards the larva had of necessity to reverse its normal orientation. With a large bubble making the head end the lighter, any efforts unable to turn the head downwards must cause the larva to swim down and at the bottom to continue, as it would appear, attempting to attach its siphon to this.

REFERENCES

(*a–f*) PHYSICAL CHARACTERS AND RELATED SUBJECTS

BACOT, A. W. (1916). *Report of the Yellow Fever Commission* (*West Africa*). London. Vol. III.

CARTER, H. R. (1923). Comments on paper by Connor and Monroe, 1923. *Amer. J. Trop. Med.* **3**, 16–19.

CHILDS, W. H. J. (1934). *Physical Constants*. Methuen and Co., London.

DA COSTA LIMA, S. (1914). Contribution to the biology of the Culicidae. Observations on the respiratory process of the larva. *Mem. Inst. Osw. Cruz*, **6**, 18–34.

FOLGER, H. T. (1946). The reactions of *Culex* larvae and pupae to gravity, light and mechanical shock. *Physiol. Zool.* **19**, 190–202.

HOCKING, B. (1953). Notes on the activities of *Aëdes* larvae. *Mosquito News*, **13**, 62–4.

HOWARD, DYAR and KNAB (1912). See references in ch. I (*f*), p. 19.

IVANOVA, L. V. (1940). The influence of temperature on the behaviour of *Anopheles maculipennis*. (In Russian.) *Med. Parasit.* **9**, 58–70.

JOBLING, B. (1937). The development of mosquitoes in complete darkness. *Trans. R. Soc. Trop. Med. Hyg.* **30**, 467–74.

JOHNSON, C. G. (1937). The absorption of water and associated volume changes in the eggs of *Notostira erratica* L. etc. *J. Exp. Biol.* **14**, 413–21.

KALANDADSE, L. (1933). Materialen zum Studium der Atmungsprozesse der Mückenlarven und -puppen etc. *Arch. Schiffs.- u. Tropenhyg.* **37**, 88–103.

MACFIE, J. W. S. (1917). The limitations of kerosene as a larvicide with some observations on the respiration of mosquito larvae. *Bull. Ent. Res.* **7**, 277–93.

MACFIE, J. W. S. (1923). Depth and the larvae and pupae of *Stegomyia fasciata*. *Ann. Trop. Med. Parasit.* **17**, 5–8.

MICHA BAR-ZEEV (1957). The effect of external temperature on different stages of *Aëdes aegypti* (L.). *Bull. Ent. Res.* **48**, 593–9.

MITCHELL, E. G. (1907). *Mosquito Life*. G. P. Putnam's Sons, New York and London.

OMARDEEN, T. A. (1957). The behaviour of larvae and pupae of *Aëdes aegypti* (L.) in light and temperature gradients. *Bull. Ent. Res.* **48**, 349.

RUSSELL, P. F. and RAO, T. R. (1941). On the surface tension of water in relation to the behaviour of *Anopheles* larvae. *Amer. J. Trop. Med.* **21**, 767–77.

SEN, S. K. (1915). Observations on respiration of Culicidae. *Ind. J. Med. Res.* **2**, 681–97.

SENIOR WHITE, R. (1926). Physical factors in mosquito oecology. *Bull. Ent. Res.* **16**, 167–248.

SHANNON, R. C. (1931). The environment and behaviour of some Brazilian mosquitoes. *Proc. Ent. Soc. Wash.* **33**, 1–27.

SHANNON, R. C. and Putnam, P. (1934). See references in ch. x, p. 250.

WATSON, G. I. (1941). A physiological study of mosquito larvae which were treated with anti-malarial oils. *Bull. Ent. Res.* **31**, 319–30.

WESENBERG-LUND, C. (1921). Contributions to the biology of the Danish Culicidae. *Mem. Acad. Copenhagen*, (8), **7**, 1–210.

WIGGLESWORTH, V. B. (1938). The absorption of fluid from the tracheal system of mosquito larvae at hatching and moulting. *J. Exp. Biol.* **15**, 248–54.

WIGGLESWORTH, V. B. (1942). In review of Russell and Rao, 1941. *Trop. Dis. Bull.* **39**, 427–8.

(*g*) FOOD REQUIREMENTS OF THE LARVA

ATKIN, E. E. and BACOT, A. W. (1917). The relation between the hatching of the egg and the development of the larva of *Stegomyia fasciata* and the presence of bacteria and yeasts. *Parasitology*, **9**, 482–536.

BACOT, A. W. (1916). See under section (*a–f*).

BARBER, M. A. (1928). The food of culicine larvae. *Publ. Hlth Rep. Wash.* **43**, 11–17.

BONNE-WEPSTER, J. and BRUG, S. L. (1932). The subgenus *Stegomyia* in Netherlands-India. *Geneesk. Tijdschr. Ned.-Ind.* Byblad 2, 1–119.

BUDDINGTON, A. R. (1941). The nutrition of mosquito larvae. *J. Econ. Ent.* **34**, 275–81.

DE MEILLON, B., GOLBERG, L. and LAVOIPIERRE, M. (1945). The nutrition of the larva of *Aëdes aegypti* L. *J. Exp. Biol.* **21**, 84–9.

DIMOND, J. B., LEA, A. O., HAYNERT, W. F. and DE LONG, D. M. (1956). The amino-acids required for egg production in *Aëdes aegypti*. *Canad. Ent.* **88**, 57–62.

GOLBERG, L., DE MEILLON, B. and LAVOIPIERRE, M. (1945). The nutrition of the larva of *Aëdes aegypti* L. II. The essential water-soluble factors from yeast. *J. Exp. Biol.* **21**, 90–6.

GOLBERG, L. and DE MEILLON, B. (1947). Further observations on the nutritional requirements of *Aëdes aegypti* L. *Nature, Lond.* **160**, 582–3.

GOLBERG, L. and DE MEILLON, B. (1948). The nutrition of the larva of *Aëdes aegypti* Linnaeus. *Biochem. J.* **43**, 372–9, 379–87.

HINMAN, E. H. (1933). The role of bacteria in the nutrition of the mosquito larva. The growth-sustaining factor. *Amer. J. Hyg.* **18**, 224–36.

JOHNSON, H. A. (1937). Notes on the continuous rearing of *Aëdes aegypti* in the laboratory. *Publ. Hlth Rep. Wash.* **52**, 1177–9.

LEWIS, D. J. (1933). Observations on *Aëdes aegypti* L. (Diptera: Culicidae) under controlled atmospheric conditions. *Bull. Ent. Res.* **24**, 362–72.

LICHTENSTEIN, E. P. (1948). Growth of *Culex molestus* under sterile conditions. *Nature, Lond.* **162**, 227.

REFERENCES

MacGregor, M. E. (1915). Notes on the rearing of *Stegomyia fasciata* in London. *J. Trop. Med.* (*Hyg.*) **18**, 193–6.

MacGregor, M. E. (1929). The significance of the pH in the development of mosquito larvae. *Parasitology*, **21**, 132–57.

Matheson, R. and Hinman, E. H. (1928). A new larvicide for mosquitoes. *Amer. J. Hyg.* **8**, 293–6.

Mathis, M. and Baffet, O. (1934). Développement larvaire du moustique de la fièvre jaune en culture pure d'Euglenes. *C.E. Soc. Biol., Paris*, **116**, 317–19.

Rozeboom, L. E. (1935). The relation of bacteria and bacterial filtrates to the development of mosquito larvae. *Amer. J. Hyg.* **21**, 167–79.

Subba Row, Y. (1946). Folic acid. *Ann. N.Y. Acad. Sci.* **48**, 255–350.

Trager, W. (1935*a*). The culture of mosquito larvae free from living organisms. *Amer. J. Hyg.* **22**, 18–25.

Trager, W. (1935*b*). On the nutritional requirements of mosquito larvae (*Aëdes aegypti*). *Amer. J. Hyg.* **22**, 475–93.

Trager, W. (1936). The utilisation of solutes by mosquito larvae. *Biol. Bull.* **71**, 343–52.

Trager, W. (1937). A growth factor required for mosquito larvae. *J. Exp. Biol.* **14**, 240–51.

Uvarov, B. P. (1928). Insect nutrition and metabolism. A summary of the literature. *Trans. Ent. Soc. London.* **76**, 255–343.

Weyer, F. (1934). Der Einfluss der Larvalernährung auf die Fortpflanzungsphysiologie verschiedener Stechmücken. *Arch. Schiffs.- u. Tropenhyg.* **38**, 394–8.

(*h*) THE EFFECT OF TEMPERATURE

Bacot, A. W. (1916). See under section (*a–f*).

Bates, M. (1949). *The Natural History of Mosquitoes.* Macmillan Co., New York.

Bliss, A. R. Jr. and Gill, J. M. (1933). The effects of freezing on the larvae of *Aëdes aegypti. Amer. J. Trop. Med.* **13**, 583–8.

Blunk, H. (1923). Die Entwicklung der *Dytiscus marginalis* L. vom Ei bis zur Imago. *Z. wiss. Zool.* **121**, 171–391.

Bodenheimer, F. S. (1924). Ueber die Voraussage der Generationzahl von Insekten. II. Die Temperatur-Entwicklungskurve bei medizinisch wichtigen Insekten. *Zbl. Bakt.* (*Abt. 1, Orig.*) **93**, 474–80.

Brues, C. T. (1939). Studies on the fauna of some thermal springs in the Dutch East Indies. *Proc. Amer. Acad. Arts Sci.* **73**, 71–95.

De Meillon, B. (1934). Observations on *Anopheles funestus* and *Anopheles gambiae* in the Transvaal. *Publ. S. Afr. Inst. Med. Res.* no. 32, 195–248.

Gutzevich, A. V. (1931). The reproduction and development of the yellow fever mosquito under experimental conditions. *Mag. Parasit. Leningrad*, **2**, 35–54. Abstract in *Rev. Appl. Ent.* **21**, 2.

Gutzevich, A. V. (1932). The graphic representation of certain data in the biology of mosquitoes. *Mag. Parasit. Leningrad*, **3**, 5–16. Abstract in *Rev. Appl. Ent.* **23**, 46.

Headlee, T. J. (1940). The relative effects on insect metabolism of temperature derived from constant and variable sources. *J. Econ. Ent.* **33**, 361–4.

Headlee, T. J. (1941). Further studies of the relative effects in insect metabolism of temperatures derived from constant and variable sources. *J. Econ. Ent.* **34**, 171–4.

Headlee, T. J. (1942). A continuation of the studies on the relative effects on insect metabolism of temperatures derived from constant and variable sources. *J. Econ. Ent.* **35**, 785–6.

Howard, Dyar and Knab. See references in ch. I (*f*), p. 19.

Huffaker, C. B. (1944). The temperature relations of the immature stages of the malarial mosquito, etc. *Ann. Ent. Soc. Amer.* **37**, 1–27.

Janisch, E. (1925). Über die Temperaturabhängigkeit biologischer Vorgänge und ihre kurvenmässige Analyse. *Pflüg. Arch. Ges. Physiol.* **209**, 414–36.

Janisch, E. (1932). The influence of temperature on the life history of insects. *Trans. R. Ent. Soc. Lond.* **80**, 137–68.

LUDWIG, D. (1928). The effect of temperature on the development of an insect (*Popillia japonica* Newman). *Physiol. Zool.* **1**, 358–89.

MACFIE, J. W. S. (1920). Heat and *Stegomyia fasciata*: short exposures to raised temperatures. *Ann. Trop. Med. Parasit.* **14**, 73–82.

MACGREGOR, M. E. (1915). See under section (*g*).

MARCHOUX, E., SALIMBENI, A. and SIMOND, P. L. (1903). See references in ch. I (*f*), p. 18.

MARTINI, E. (1923). Theoretisches zur Bestimmung der Lebensdauer von Schädlingen, etc. *Z. angew. Ent.* **9**, 133–46.

MUIRHEAD THOMSON, R. C. (1940). Studies on the behaviour of *Anopheles minimus*. Part III. Thermal death point. *J. Malar. Inst. India*, **3**, 323–48.

PEAIRS, L. M. (1914). The relation of temperature to insect development. *J. Econ. Ent.* **7**, 174–81.

RAMSAY, G. C. and CARPENTER, J. A. (1932). An investigation on petroleum oils for malaria control purposes (freezing). *Rec. Malar. Surv. India*, **3**, 203–18.

TROFIMOV, G. K. (1942). Experiments on cold resistance of the larva of *Anopheles pulcherrimus* and *A. bifurcatus*. *Mag. Parasit. Moscow*, **11**, 79–81.

UVAROV, B. P. (1931). Insects and climate. *Trans. R. Ent. Soc. Lond.* **79**, 1–247.

WIGGLESWORTH, V. P. (1942). *The Principles of Insect Physiology*. Methuen and Co., London. Ed. 2 (and subsequent editions).

WRIGHT, M. G. (1901). The resistance of the larvae of mosquitoes to cold. Notes on the habits and life history of mosquitoes in Aberdeenshire. *Brit. Med. J.* **1**, 882.

WRIGHT, W. R. (1923). On the effects of exposure to raised temperatures upon the larvae of certain British mosquitoes. *Bull. Ent. Res.* **18**, 91–4.

(*i*) EFFECT OF STARVATION AND OVERCROWDING

BACOT, A. W. (1916). See under section (*a–f*).

DE MEILLON, B., GOLBERG, L. and LAVOIPIERRE, M. (1945). See under section (*g*).

WEIDLING, K. (1928). Die Beeinflussung von Eiröhrenzahl und -grösse einiger Dipteren durch Hunger im Larvenstadium, etc. *Z. angew. Ent.* **14**, 69–85.

WIGGLESWORTH, V. B. (1942). The storage of protein, fat, glycogen and uric acid in the fat-body and other tissues of mosquito larvae. *J. Exp. Biol.* **19**, 56–77.

(*j*) EFFECT OF SALT, pH, SOAP AND CHEMICALS

BALFOUR, A. (1921). Mosquito breeding in saline waters. *Bull. Ent. Res.* **12**, 27–34.

BARBER, M. A. (1944). A measurement of the toxicity to mosquito larvae of the vapours of certain larvicides. *Publ. Hlth Rep. Wash.* **59**, 1275–8.

BRINK, C. J. H. and CHOWDHURY, D. K. D. (1939). Ammonium sulphate as a combined fertiliser and mosquito larvicide. *J. Malar. Inst. India*, **2**, 111–12.

BUCHMANN, W. (1931). Untersuchungen über die Bedeutung der Wasserstoffionenkonzentration für die Entwicklung der Mückenlarven. *Z. Angew. Ent.* **18**, 404–16.

CHOPRA, R. N., ROY, D. N. and GHOSH, S. M. (1940). Insecticidal and larvicidal action of *Tephrosia vogelii*. *J. Malar. Inst. India*, **3**, 185–9.

FREEBORN, S. B. and ALSATT, R. F. (1918). The effect of petroleum oils on mosquito larvae. *J. Econ. Ent.* **11**, 299–307.

FROBISHER, M. JR. and SHANNON, R. C. (1931). The effects of certain poisons upon mosquito larvae. *Amer. J. Hyg.* **13**, 614–22.

GIBBINS, E. G. (1932). A note on the relative size of the anal gills of mosquito larvae breeding in salt and fresh water. *Ann. Trop. Med. Parasit.* **26**, 551–4.

HACKER, H. P. (1925). How oil kills Anopheline larvae. *F.M.S. Malar. Bur. Rep.* **3**.

IMPERIAL INSTITUTE (1940). *A Survey of Insecticide Materials of Vegetable Origin*. Imperial Institute, London.

LEWIS, D. J. (1955). The *Aëdes* mosquitoes of the Sudan. *Ann. Trop. Med. Parasit.* **49**, 164–73.

MACFIE, J. W. S. (1914). A note on the action of common salt on the larvae of *Stegomyia fasciata*. *Bull. Ent. Res.* **4**, 339–44.

REFERENCES

MACFIE, J. W. S. (1916). Chlorine as a larvicide. *Rep. Accra Lab. for* 1915, p. 71. Abstract in *Rev. Appl. Ent.* **5**, 47.

MACFIE, J. W. S. (1922). The effect of saline solutions and seawater on *Stegomyia fasciata*. *Ann. Trop. Med. Parasit.* **15**, 377–80.

MACFIE, J. W. S. (1923). A note on the action of lithium chloride on mosquito larvae. *Ann. Trop. Med. Parasit.* **17**, 9–11.

MACGREGOR, M. E. (1921). The influence of hydrogen-ion concentration in the development of mosquito larvae. *Parasitology*, **13**, 348–51.

MACGREGOR, M. E. (1926). Some effects of electric current on mosquito development. *Bull. Ent. Res.* **16**, 315–17.

MACGREGOR, M. E. (1929). The significance of the pH in the development of mosquito larvae. *Parasitology*, **12**, 132–57.

MARTINI, E. (1922). Über den Einfluss der Wasserzusammensetzung auf die Kiemenlänge bei den Mückenlarven. *Arch. Schiffs.- u. Tropenhyg.* **26**, 82.

MATHESON, R. and HINMAN, E. H. (1928). See under section (*g*).

MOORE, W. and GRAHAM, S. A. (1918). A study of the toxicity of kerosene. *J. Econ. Ent.* **11**, 70–5.

MUIR, F. (1920). A convenient mosquito poison. *Hawaiian Planter's Record.* Abstract in *Rev. Appl. Ent.* **9**, 38.

MURRAY, D. R. P. (1936). Mineral oil in mosquito larvicides. *Bull. Ent. Res.* **27**, 289–305.

PAGAST, F. (1936). Über Bau und Funktion der Analpapillen bei *Aëdes aegypti* L. (*fasciatus* Fabr.). *Zool. Jahrb., Zool.* **56**, 184–218.

PANTAZIS, G. (1935). Les effets de la salinité de l'eau sur les larves des culicines. *Prakt. Acad. Athens*, **10**, 348–56.

RUDOLFS, W. and LACKEY, J. B. (1929). The composition of water and mosquito breeding. *Amer. J. Hyg.* **9**, 160–80.

RUSSELL, P. F. and RAO, T. R. (1941). See under section (*a–f*).

SENIOR WHITE, R. (1928). Algae and the food of Anopheline larvae. *Ind. J. Med. Res.* **15**, 969–88.

SHANNON, R. C. and FROBISHER, JR. M. (1931). A comparison of the effect of various substances upon larvae of *Aëdes aegypti*. *Amer. J. Hyg.* **14**, 426–32.

STRICKLAND, C. (1929). Soap as a mosquito larvicide. *Trans. R. Soc. Trop. Med. Hyg.* **22**, 509–10.

TANIMURA SEIZO and SONOJI HINE (1952). An experimental study on the function of dilute solutions of strong inorganic acids upon larval mosquitoes, etc. *Shikoku Acta Med.* **3**, 36–40.

WATSON, G. (1937). Some observations on mosquito larvae dying in anti-malarial oils and other substances. *Ann. Trop. Med. Parasit.* **31**, 417–26.

WATSON, G. (1941). See under section (*a–f*).

WIGGLESWORTH, V. B. (1930). A theory of tracheal respiration in insects. *Proc. R. Soc. London* B, **106**, 229–50.

WIGGLESWORTH, V. B. (1933*a*). The effect of salts on the anal gills of mosquito larvae. *J. Exp. Biol.* **10**, 1–15.

WIGGLESWORTH, V. B. (1933*b*). The function of the anal gills of the mosquito larva. *J. Exp. Biol.* **10**, 16–26.

WIGGLESWORTH, V. B. (1933*c*). The adaptation of mosquito larvae to salt water. *J. Exp. Biol.* **10**, 27–37.

WILLIAMSON, K. B. (1924). The uses of gases and vapours for killing mosquitoes breeding in wells. *Trans. R. Soc. Trop. Med. Hyg.* **17**, 485–519.

WOODHILL, A. R. (1942). A comparison of factors affecting the development of three species of mosquitoes, etc. *Proc. Linn. Soc. N.S.W.* **67**, 95–7.

WORSLEY, R. R. LE G. (1934). The insecticidal properties of some East African plants. *Ann. Appl. Biol.* **21**, 649–69.

YATES, W. W. (1946). Time required for *Aëdes vexans* and *A. lateralis* larvae to obtain a lethal dose of several larvicides. *J. Econ. Ent.* **39**, 468–71.

XII
GROWTH

(a) GENERAL FEATURES OF LARVAL GROWTH

Certain 'laws' or generalisations regarding the growth of insects have been put forward. Dyar's rule (Dyar, 1890) was based on the fact that measurement of the head capsule of caterpillars increases in geometrical progression at each moult by a ratio (usually 1·4) which is constant for a given species. Other authors have found a similar relation in other groups. Przibram's rule, based on measurements of *Sphodromantis*, is that at each instar every cell divides once and grows to its original size giving an increase therefore in weight or volume at each instar of twice. Linear increase will therefore be at the rate of the cube root of two, that is 1·27. Though the hypothetical explanation of the rule as first given is not of general application, an increase of about twice the body weight at ecdysis has been shown by Bodenheimer (1933) to occur in many cases. On the contrary, in many cases there is no accurate doubling of weight (Wigglesworth, 1942).

The rate of growth is, however, not always, or even usually, the same for different instars. Ripley (1923) notes that increase from the first to the second instar is usually greater, and that from the penultimate to the last instar less, than that for other moults.

Measurements also of particular parts or organs may show harmonious growth, that is, in the same ratio as that of the body, or disharmonic, the parts growing at rates peculiar to themselves (Huxley, 1932).

Of authors who have given measurements of growth of the larva of mosquitoes in detail, are Kettle (1948), who gives measurements of the length and breadth of the head capsule and of certain soft parts in the larva of *Anopheles sergenti*, and Abdel Malek and Goulding (1948), who have studied the rate of growth of two sclerotised parts, namely, the head capsule and siphon, in four species of mosquito including *Aëdes aegypti*. The first-mentioned authors note that measurements of the length and breadth of the head capsule obeyed Dyar's law with a ratio of 1·524. Abdel Malek and Goulding found measurements of the width of the head capsule and length of the siphon for the different instars in μ to be respectively 246, 402, 635, 854 and 231, 388, 601, 756 giving respectively ratios of 1·63, 1·58, 1·34 and 1·68, 1·55, 1·27, results which are almost identical with results of the present author (see Tables 14 and 15).

A very full and detailed account of tissue growth in relation to metamorphosis in *Aëdes aegypti*, including growth measurements, is given by Trager (1937). Measurements of larval growth in this species are also given by Gander (1951) and for *Anopheles* by Jones (1953).

CONDITIONS AFFECTING GROWTH

In the life cycle of *Aëdes aegypti* growth as manifested by general increase in size and weight only occurs in the larval stage, no food being taken in during the egg or pupal stage and no appreciable growth in size or weight taking place. And though the imago may take in food and develop eggs no increase due to growth in the ordinary accepted sense occurs. It is through observations on the larval stage therefore that the rate of growth for the species as manifested by overall increase in weight has mainly to be ascertained.

Influences chiefly affecting growth are: (1) *temperature*, and (2) *nutrition*.

Under optimal food and other conditions the times taken for the larval stage of *A. aegypti* as a whole and for each instar have been found to be very constant for any given temperature. The optimal temperature appears to be round about 28° C. Above this there is some slight increase in rate of development, but with some decrease in size. From 28° to about 25° there is a gradual lengthening of the period. Below 25° there is a very great and increasing lengthening which reaches the limit of complete cessation of growth at about 10° C. There is, however, no corresponding decrease in size of insect until below about 18° C. when the period of growth is already extremely prolonged.

The effect of deficient food, up to a certain limit, appears chiefly as a decrease in total growth with little effect on the time taken for development so that it is chiefly evidenced by a smaller size of the adults. Such an effect applies not only to the amount (and suitability) of food present in the medium, but upon that actually taken, since this may be restricted due to interference with feeding by overcrowding, restlessness from exposure to light and other disturbing influences. There may also possibly be intrinsic factors in the insect itself which retard or hasten development.

With starvation conditions, however, deprivation of food has a profound effect in lengthening the period of development and decreasing total growth as indicated in the mature larva and the resulting pupa and adult. This may amount to an almost indefinite prolongation of the larval stage, subject only to viability, with reduction in size and weight to half the normal or less (see under 'Starvation', p. 266).

Interest in growth, however, chiefly relates to the laws concerning normal growth and particularly on the rate of growth as shown in different stages of growth and in different parts of the body. In these respects certain considerations would seem to be worth referring to.

In the first place, certain parts of the larva are soft and others hard and sclerotised, conditions of growth clearly differing in the two cases. Usually it is the sclerotised parts that are selected for measurement since these alter little if at all during the course of a particular instar, growth being, so to speak, hidden until at ecdysis the contained soft parts are liberated again to become fixed and sclerotised for the period of the new instar. This is not so in the case of the soft parts where growth is continuous. Here absence of any fixed point in the course of the instar at which measurements are to be made introduces a difficulty. Thus Abdel Malek and Goulding say, speaking of such difficulty, 'Determination of the nature of the rate of growth during successive moults cannot be accurately assayed without an accurate determination of the time spent by each larva in each stadium.'

METHOD OF INVESTIGATION

In the data here given an attempt has been made to determine growth rate in both soft and hard parts, and the relation of these to each other, in linear and weight measurements by (1) utilising as fixed points in development the stage immediately preceding and immediately following upon ecdysis, and (2) observations made at different hours on the culture as a whole. From such procedure as (1) the growth that has taken place can be determined accurately during each instar, that is from post-ecdysis when the instar begins to pre-ecdysis when it ends, and from (2) as will be shown later, very interesting information regarding growth in culture can be ascertained.

In relation to these data a point should be stressed that growth (as occurring in a larva or parts of its body) is essentially one of increase in volume or weight, these being equivalent where there is no change in shape or density. Linear measurement can be regarded as a linear expression of this. The importance of this will be clear later.

For observations at post- and pre-ecdysis samples of larvae from a culture at times when the different ecdyses are immanent have been kept under observation and larvae seen to be about to undergo change, or seen to have just completed the change, used for the required measurements. Observations given in the tables will show that linear and weight measurements in *A. aegypti* at the two times (post- and pre-ecdysis, or, to be more precise, to use the terms 'immediate' and 'pre-ecdysis') show that ratios of linear measurements and weight are almost identical for the soft parts. Thus for a soft part the ratio for growth during an instar is given by that of the measurement at pre-ecdysis to that at immediate. For a hard part it is that given by the immediate measurement of an instar to the post-ecdysis measurement of the preceding instar. Or in the last case with but little modification of that for any instar to that for the preceding instar, since in the sclerotised parts there is in most cases but little change once the parts have hardened.

For the second type of observation, that is the following-up of growth in culture, standard cultures as described by me elsewhere (Christophers, 1947) have been used. Samples have been taken at intervals by pipetting out in bulk (see p. 115) a mass of larvae, all of which, or a suitable proportion of which, have been measured or weighed. In all cases cultures used for such a purpose have been started with eggs that have hatched out within 15 or in some cases at as short an interval as 5 minutes. The results show: (1) the growth curve for the culture as a whole; (2) the degree of spread in the case of each instar, for though the larvae in a culture may start even, each successive instar becomes more and more spread out in time, the growth curve for the culture as a whole being a composite one formed of the overlapping frequencies of successive instars; (3) information regarding time passed in successive instars and much information about various other aspects of growth as will be apparent from succeeding sections and the tabular statements and figures.

(*b*) LINEAR MEASUREMENTS OF INSTARS AT PRE- AND POST-ECDYSIS

In Table 14 are given the mean values found in different instar stages for linear measurements of body length and for such parts as seemed suitable for measurement. In Table 15 are the factors of increase shown by these figures for each instar period or ecdysis as the case may be.

Under the heading 'soft parts' are those parts not enclosed in hard cuticle, including 'body length' which is mainly of this nature. It will be seen, comparing pre-ecdysis measurements with those for post-ecdysis, that there is no significant increase, and often a slight decrease, at the time of ecdysis. For this reason factors of increase at ecdysis for soft parts have not been entered in the table of ratios.

Under the heading 'hard parts' are those parts like the head that cannot expand except at ecdysis. In this case little or no change takes place during the instar period, and for reasons given above factors of increase for the instar period have not been given.

A third heading relates to 'special cases'. These are parts that behave not altogether as hard or soft parts or have characters that require special consideration. They will be discussed later.

Table 14. *Means of measurements of instars in millimetres at post- and pre-ecdysis stages*

	Instar I		Instar II		Instar III		Instar IV		
								Pre-	
	Eclosion	Pre-	Post-	Pre-	Post-	Pre-	Post-	Male	Female
Soft parts									
Body length	1·01	1·97	1·97	3·27	3·13	5·06	4·85	6·17	7·33
Thorax length	0·137	0·305	0·283	0·503	0·51	0·92	0·83	0·88	1·41
Thorax width	0·216	0·410	0·437	0·714	0·76	1·23	1·22	1·67	1·83
Abdomen length	0·60	1·40	1·36	2·24	2·03	3·52	3·28	4·10	5·14
Abdomen width*	0·147	0·260	0·280	0·471	0·46	0·76	0·75	0·99	1·24
Hard parts									
Head length†	(0·23)	0·255	0·420	0·41	0·56	0·65	0·82	—	—
Head width†	(0·18)	0·278	0·463	0·46	0·71	0·74	0·98	—	—
Antenna length†	(0·08)	0·108	0·169	0·162	0·237	0·245	0·348	—	—
Special cases									
Siphon length	0·128	0·280	0·287	0·448	0·44	0·67	0·75	0·80	0·94
Siphon width	0·08	0·149	0·167	0·268	0·276	0·392	0·40	0·41	0·45
Anal papillae	0·119	0·212	0·263	0·335	0·49	0·60	0·79	0·84	1·00

* Width of third abdominal segment.

† The data relate to before and after ecdysis. The first column (in brackets) gives the measurements at hatching before the head is expanded. They are not considered in relation to growth.

Table 15. *Growth rates calculated from data given in Table* 14.

	Instar periods				
				IV	
	I	II	III	Male	Female
Soft parts					
Body length	1·95	1·66	1·62	1·27	1·51
Thorax length	2·23	1·78	1·80	1·06	1·70
Thorax width	1·90	1·63	1·62	1·37	1·50
Abdomen length	2·33	1·65	1·73	1·25	1·57
Abdomen width	1·77	1·68	1·65	1·32	1·65
Mean rate	2·03	1·68	1·68	1·25	1·59
Hard parts*					
Head length	1·65	1·37	1·26	—	—
Head width	1·69	1·54	1·32	—	—
Antenna length	1·57	1·46	1·42	—	—
Mean rate	1·63	1·46	1·33	—	—
Special cases					
Siphon length	2·20	1·61	1·48	1·07	1·25
Siphon width	1·86	1·61	1·42	1·02	1·12
Anal papillae†	2·21	1·86	1·61	1·06	1·27

* Change at ecdysis at end of instar.

† Including rate for instar and at ecdysis.

LARVAL GROWTH

BODY LENGTH

Though the total body length consists of both 'hard' (head capsule) and 'soft' parts (thorax and abdomen) it is a desirable measurement to have and in the main it is a soft part. Since the respiratory siphon and the anal segment are set at an angle to the rest of the body they do not form a very suitable end-point for measurement and 'body length' has been taken as that from the tip of the head (preclypeus) to the point where the siphon begins on the eighth segment dorsally. In this measurement particularly the greatest care was taken to maintain the natural state of the body. After a number of trials the best results were obtained when a number of larvae were very gently chloroformed on a glass sheet covered with wet blotting paper, larvae which had died in a straight position being selected for measurement.

As with other measurements those of total body length show a graded decreasing growth ratio reaching a low figure in the fourth instar, especially in the male. The larva increases from a length of 1·0 mm. on emerging from the egg to a mean of 7·33 mm. for the female, or an increase factor of 7·33. The male increases only some six times, which, however, is very considerable seeing that the male pupa is only about half the weight of the female pupa. With each instar the growth factor steadily decreases from approximately 2·0 in the first instar to 1·25 in the male and 1·5 in the female. At ecdysis there is little change, usually a slight reduction.

THORAX AND ABDOMEN

The thorax and abdomen taken together form the greater part of the body length and represent growth as a soft part without the complication of the head capsule. This measurement increases during the larval period in the female from 0·737 mm. to 6·55 mm. or 8·9 times. In width the abdomen increases 8·4 and the thorax 8·5 times. The body as a whole, therefore, so far as the soft parts are concerned, maintains very closely its original form. As in all other measurements those for the thorax and for the abdomen show a steadily diminishing increase factor as growth proceeds.

THE HEAD CAPSULE

Apart from some slight increase which may precede ecdysis, the dimensions of the head capsule remain unchanged throughout the instars, increase being seen only at ecdysis. The mean diameter at hatching, allowing 2 hours for expansion at this time, and at the respective ecdyses was in the female 0·278, 0·463, 0·71 and 0·98 giving increase factors of 1·67, 1·54 and 1·32. The factors of increase for length are a little lower, being respectively 1·65, 1·37 and 1·26, corresponding to the fact that the shape of the head broadens slightly with growth.

THE ANTENNA

The antenna is interesting since the larval antenna takes no part in the final imaginal structure, the imaginal buds of the antennae being entirely within the head capsule. Further the length of the larval antenna would seem to be correlated with the working of the mouth parts forming part of an outer circle of sensory or protective hairs in advance of these parts. Being a hard part increase in size is entirely confined to the change at ecdysis. The growth factors are 1·57, 1·46 and 1·42.

THE RESPIRATORY SIPHON

The siphon in the fourth instar is a typical 'hard' structure, since so far as growth is concerned it is a completely closed chitinous box. In the earlier instars, however, it behaves like a 'soft' structure, since the chitinised portion is pushed up like a cap by the growing hypodermal siphon of the succeeding instar. It has about the same growth indices in the first three instars as the soft parts, though somewhat less in the third instar and, for reasons given, very little increase in the fourth instar.

THE ANAL PAPILLAE

These increase in size both during the instar and at ecdysis, as shown in the tables. To obtain the total growth in this case the two rates (for instar period and at ecdysis) must not be added but multiplied. The total growth is about the same as that for the soft parts.

Table 16. *Weight of instars at post- and pre-ecdysis*

Instar	Stage	Number of weighings	Number weighed	Weight (mg.)	Weight per larva (mg.)	Ratio increase in instar	Ratio increase at ecdysis
First	0 hours	1	1067	14·6	0·0137	—	—
First	2 hours	3	2201	46·5	0·0211	—	—
First	4 hours	1	908	20·9	0·0230	—	—
First	Pre-ecdysis	2	35	2·8	0·0800	5·8	—
Second	Post-ecdysis	2	29	2·4	0·083	—	1·03
Second	Pre-ecdysis	3	59	23·5	0·398	4·8	—
Third	Post-ecdysis	2	11	4·2	0·382	—	0·96
Third	Pre-ecdysis	4	34	61·0	1·794	4·7	—
Fourth	Post-ecdysis	4	26	42·4	1·615	—	0·90
	Pre-ecdysis	—	—	—	—	—	—
	Male	3	5	16·0	3·200	1·98	—
	Female	5	36	176·0	4·900	3·07	—
Pupa	Male	5	873	2275·1	2·606	—	0·81
	Female	4	822	3906·5	4·752	—	0·97

GENERAL

The above indicates growth ratios for the earlier instars approaching, and in the first instar reaching, doubling of the dimensions. The ratio of increase further shows a definite decrease in successive instars, especially so in the fourth instar. Not only is the ratio in the fourth instar reduced, but whilst the first three instars last each approximately 24 hours, the fourth instar is nearly three times as long. The measurements therefore are in accord with the findings of Ripley (1923), that increase from first to second instar is usually greater and that for the penultimate to the last instar less than for other instars.

(c) WEIGHT OF INSTARS AT PRE- AND POST-ECDYSIS

In Table 16 are given the results of weighing larvae at pre- and post-ecdysis stages. It will be seen that increase in weight is entirely that gained during the period of instar growth. At ecdysis there is no increase in weight and there may even be a decrease. This is shown not only by weighing larvae before and after, but by the fact as shown later that the weight of massed larvae in a culture at times of ecdysis is little influenced by the proportion of instars present, showing that at such times the two instars in the culture do not differ appreciably in weight.

Increase in weight of successive instars, as with linear measurements, shows a decreased ratio as development proceeds, the ratios for the four instars being respectively, 5·8, 4·8, 4·7 and 3·07. The last figure refers to the female. In the male the increase for the fourth instar is only 1·98. The mean weight of the larva increases as a whole from 0·0137 mg. at hatching to 4·90 mg. in the female just prior to pupation, an increase of 358 times.

The ecdyses occur at approximately 24, 48, and 72 hours and pupation at 144 hours in the case of the female. Pupation in the male is about 24 hours earlier. If the weights of the instars are compared with the graph given in Fig. 24 (p. 284) for the curve of growth in a culture it will be seen that the instar weight does not differ appreciably from the weight for the culture, in spite of the fact that all larvae do not pass through their stages of development simultaneously.

Table 17. *Cubes of linear increase ratios for body length for instars and observed weight ratios for these instars.*

				Instar IV		Mean		Total	
	Instar I	Instar II	Instar III	Male	Female	Male	Female	Male	Female
Linear increase	1·95	1·66	1·62	1·27	1·51	1·62	1·68	6·1	7·3
Cube of linear increase	7·4	4·6	4·3	2·0	3·4	4·6	4·9	226	389
Observed weight increase	5·8	4·8	4·7	2·0	3·0	4·3	4·6	234	358

COMPARISON OF LINEAR AND WEIGHT MEASUREMENTS

Since linear increase, other things being equal, would be represented by the cube of this in measurement made in volume, and, if no change in density occurred, in weight, it is of interest to see how far such a relation applies to the data obtained. In Table 17 are given the cubes of the ratios for linear increase in body length for each instar and in contrast the ratios obtained for weight increase (for source of data see Tables 15 and 16). It is obvious that such a relationship does hold good.

(*d*) MEASUREMENT OF GROWTH IN CULTURES

The course of development of *Aëdes aegypti* in cultures has already been indicated. Owing to the 'spread' caused by unequal rate of growth of individual larvae it is never possible, except in the very early stages, to find all larvae at the same stage of development. Even if only one instar is considered, the individual larvae are still in different stages of growth in that instar. It is necessary, therefore, in following up development in a culture, to obtain more information than is given by measurement of post- and pre-ecdysis stages.

The cultures made use of for this purpose have all been of considerable size, that is some hundreds or more, bred at 28° C. under conditions that experience had shown to be optimal and from hatchings that have taken place in a short interval of time, 15 minutes or less. Though it may be thought that such conditions are unduly artificial, this is not so to the extent that might at first sight appear. Under conditions in nature or in the laboratory favourable to the life of the insect, eggs laid by a single female almost entirely hatch over a very brief range of time and the cultures now considered do not differ to any marked degree in general characters from those put up in the course of ordinary routine work. Moreover, origin from simultaneous hatching was an essential requirement to avoid what might otherwise have been larvae starting development at different unknown times. In all, twenty-six cultures of this kind have been followed up during a period of about ten months.

As in the previous section, where measurement was concerned with particular stages in larval development, both linear measurements and weight have been taken, but in this case in relation to hours of culture. The method adopted has been to take samples at stated times by pipetting out indiscriminately a suitable mass of larvae, filtering these off after a preliminary wash, removing a suitable portion of the mass so yielded for measurement or

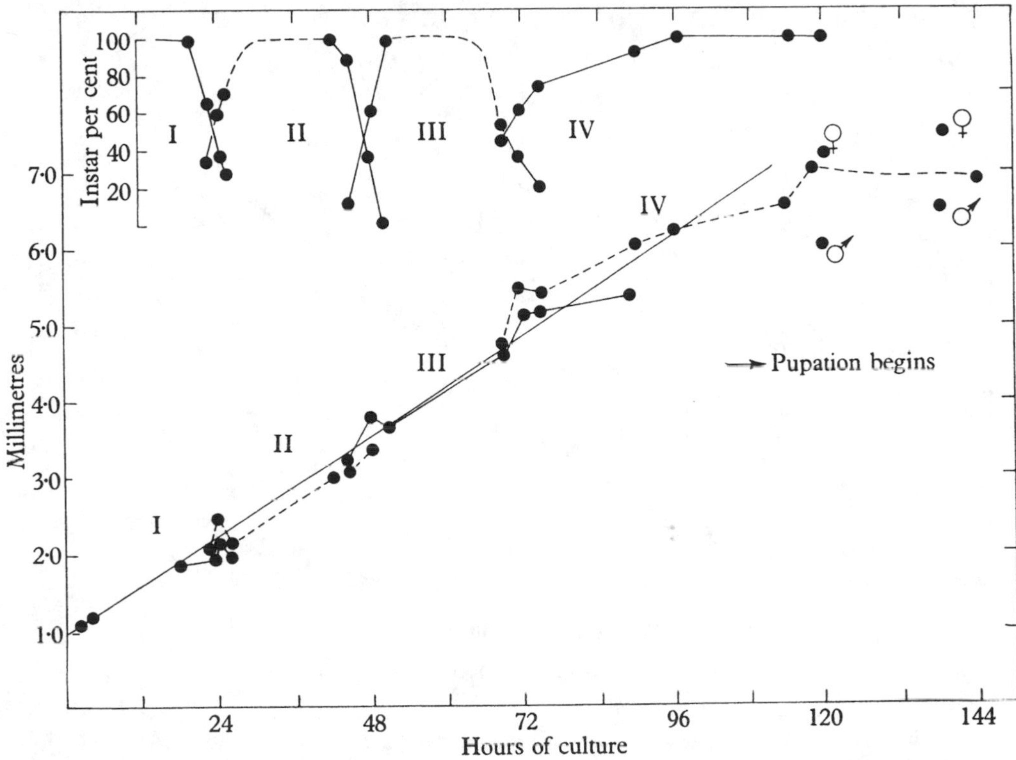

Figure 23. Chart giving graph for measurements of body length in culture at 28° C.

Observed values are shown as dots. In the body of the chart a line is drawn through these, giving the graph for linear increase. The dots for each instar are linked up alternately by unbroken and by broken lines.

At the top of the chart are given the percentages of the different instars at hours corresponding to those of the growth curve.

weight and at once killing the remainder by dropping into water at 80° C. for determination of the percentage of instars. This last procedure was necessary since near ecdysis time considerable change in the percentage may take place in a short time from ecdysis, especially in the earlier stages where this takes but a short time and goes on even among massed larvae. For linear measurements larvae have been treated as in the previous section. The sample for weighing was allowed to rest as a mass on a porous tile, which rapidly freed it from extraneous water. Care was always taken to avoid damaging the larvae by undue pressure or manipulation, since damage to the anal papillae causing these to be detached results in loss of haemolymph.

The data for body length and corresponding counts of instars are too voluminous for publication, but are shown in graph form for the cultures as a whole in Fig. 23, with the graphs for count of instars set along the top of the figure. It is very clear that growth as indicated by linear measurement plotted against time gives a straight line up to about 96 hours. After 96 hours there is some falling off in the rate. There are, however, by this time a number of complicating factors which would have to be allowed for before concluding that a reduction in growth rate is responsible. In the first place it is now that differentiation of sex begins to be apparent and as male larvae do not grow as large as the

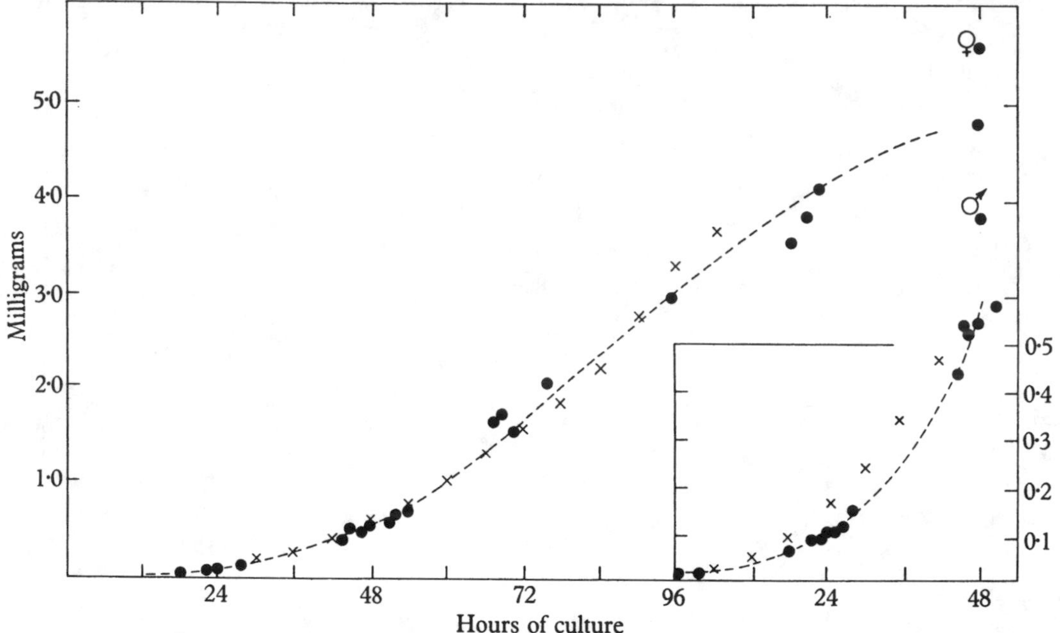

Figure 24. Graph of increase in body weight.

Giving graph for observed measurements of body weight during culture. Observed weights are shown as dots and what appears to be the course of increase as shown by the observations is indicated by the interrupted line. Plottings from the formula given in the text are shown by crosses. For data see Table 17 and text.

Inset. The first 48 hours shown on an increased vertical scale.

female this would be a cause of reduced body length in the culture as a whole. Also it is now that pupation is beginning and, as male pupae are considerably smaller than the female, this would remove more of the smaller male larvae than of the larger female and have a reverse effect. Towards the end of the culture period another effect is at work, namely that the relative number of larvae showing delayed development increases. These would be too few at an earlier period to have any appreciable effect. But as pupation removes more and more of the total larvae any retarded and usually poorly developed individuals begin to have an undue effect. In regard to linear measurement there seems no doubt that growth in a culture where larvae are started, so to speak, at scratch gives against time a straight line graph.

The result of weighing samples of larvae at different hours of culture is given in Table 18. Plotted, these data give the curve shown in Fig. 24. In the early stages, up to about 50 hours, that is about the end of the second instar, the form of the curve suggests an exponential function (see inset to graph). Later the curve is more uncertain, but may still be of this character. Such deductions as can be made regarding the nature of the growth curve will be considered in the next subsection.

Table 18. *Weight of larva at different hours of culture*

Hour	Number of larvae weighed	Weight (mg.)	Mean weight per larva	Hour	Number of larvae weighed	Weight (mg.)	Mean weight per larva
0	—	—	0·0137	53	326	223·7	0·69
18	1026	66·3	0·065	54	35	25·7	0·73
23	2081	175·6	0·089	69	146	249·4	1·71
24	2728	238·2	0·090	70	219	384·6	1·76
25	88	10·4	0·118	71	345	517·3	1·50
26	2900	336·5	0·119	76	178	367·7	2·07
28	2468	368·7	0·146	95	25	74·0	2·96
44	903	374·6	0·414	115	209	751·1	3·59
45	181	97·7	0·54	117	153	590·6	3·86
46	493	257·6	0·52	119	183	755·7	4·13
48	1087	594·1	0·55	144	6	20·7	3·90 (male pupae)
51	105	61·1	0·58	—	8	44·8	5·60 (female pupae)

(*e*) CHARACTERS OF THE GROWTH CURVE

The data that have been given show clearly that linear increase plotted against time gives a straight line. This at first sight seems to contradict what has been said as to the reduced ratio for instars as development proceeds. Actually the two facts are mutually explanatory. That the graph for linear increase is a straight line indicates that increase is not exponential, that is, proportional to the increase in size, but is a fixed increment added throughout. Imagine now a larva 1 mm. in length which increases in length per unit time by a fixed increment of 1 mm. The ratio of increase would then work out as $2/1=2$, $3/2=1{\cdot}5$, $4/3=1{\cdot}33$, ..., giving a reduced ratio, although the increase has been a regular one. The larva, however, is not a Euclidian line, but a solid body, and in such a case, provided the body maintains its shape, linear increase is only an indication of a three-dimensional increase, that is of one that would be represented by the cube of the linear increase and would be exponential. Furthermore, the reverse would hold good. If a mass is to grow evenly throughout, its increase must be exponential and if this were so then linear increase would be by even increments.

Consider, as a conveniently shaped mass, a sphere with a radius R which is increasing in size by a steady increment to the radius of r. Its volume, and equally its weight if its consistence remains unaltered, will increase according to the formula

$$1/3\pi(R+rt)^3,$$

where $1/3\pi$ is a constant characteristic of the sphere and could be given as a constant k.

The body of the larva is not a sphere, but it is of very constant shape at least in the earlier instars and it would be reasonable to think of a constant k that might characterise this in a similar formula. It might be reasonable therefore to consider a formula

$$W=k(L+lt)^3,$$

where W is weight, L the original length of the larva, l the increment, t time in hours and k a constant. Knowing that L at hatching is 1·0 mm. and that from the straight line graph given by body length the increment l can be found to be approximately 0·054, the formula could be written

$$W = k(1 + 0{\cdot}054t)^3.$$

Taking k as the proportion of weight to length of the larva at hatching, that is 0·0137, a formula was given the results from which are indicated in Fig. 24 by crosses. Though far from a perfect fit, it seems worth recording as a rational formula though it still does not give the essential factor of growth, utilising the value 0·054 as observed but without proceeding further in mathematical analysis.

Table 19. *Observed and calculated measurements of instars*

	Time in hours	Body length (mm.)		Weight (mg.)	
		Obs.	Calc.	Obs.	Calc.
Instar I at hatching	—	1·01	—	0·0137	—
Instar I at pre-ecdysis	24	1·97	2·26	0·08	0·164
Instar II at pre-ecdysis	48	3·27	3·59	0·40	0·63
Instar III at pre-ecdysis	72	5·06	5·88	1·79	1·56
Instar IV at pre-ecdysis (male)	96	6·17	6·50	3·20	3·20
Instar IV at pre-ecdysis (female)	120	7·33	7·48	4·90	4·18

In Table 19 is given the application of the formula to the instar values for body length and weight as determined at pre-ecdysis, that is as distinct from determinations on the cultures as a whole.

All observations relate to growth at 28° C. determined as described in the early portion of the chapter.

REFERENCES

Abdel Malek, A. and Goulding, R. L. (1948). A study of the rate of growth of two sclerotised regions within the larvae of four species of mosquitoes. *Ohio J. Sci.* **48**, 119–28.

Bodenheimer, F. S. (1933). The progressive factor in insect growth. *Quart. J. Biol.* **8**, 92–5.

Christophers, S. R. (1947). Mosquito repellents. *J. Hyg., Camb.* **45**, 176–231.

Dyar, H. G. (1890). The number of moults of lepidopterous larvae. *Psyche*, **5**, 420–2.

Gander, R. (1951). Experimentalle und oekologische Untersuchungen über das Schlüpfvermögen der larven von *Aëdes aegypti* L. *Rev. Suisse Zool.* **58**, 215–78.

Huxley, J. S. (1932). *Problems of Relative Growth.* Methuen and Co., London.

Jones, J. C. (1953). Some biometrical constants for *Anopheles quadrimaculatus* Say larvae in relation to age within stadia. *Mosquito News*, **13**, 243–7.

Kettle, D. S. (1948). The growth of *Anopheles sergenti* Theobald (Diptera: Culicidae) with special reference to the growth of the anal papillae in varying salinities. *Ann. Trop. Med. Parasit.* **42**, 5–29.

Przibram, H. and Megusar, F. (1912). Wachstumsmessungen an *Sphodromantis bioculata* Burm. I. Länge und Masse. *Arch. Entw. Mech.* **34**, 680–741.

Ripley, L. B. (1923). The external morphology and post embryology of noctuid larvae. *Illinois Biol. Monogr.* **8**, no. 4, 1–103.

Teissier, G. (1936). (Dyar's law.) *Livre Jubilaire E. L. Bouvier*, pp. 334–42.

Trager, W. (1937). Cell size in relation to the growth and metamorphosis of the mosquito *Aëdes aegypti. J. Exp. Zool.* **76**, 467–89.

Wigglesworth, V. B. (1942). *The Principles of Insect Physiology.* Methuen and Co., London. Ed. 2 (and subsequent editions).

XIII
THE LARVA: INTERNAL STRUCTURE

(*a*) THE ALIMENTARY CANAL

The alimentary canal of *Culex* larva has been very fully described by Thompson (1905) and that of *Anopheles* larva by Imms (1907).*

Passing backwards, the canal consists of *preoral cavity*, *mouth*, *pharynx*, *oesophagus*, *proventriculus*, *mid-gut* or *ventriculus* with its diverticula or *gastric caeca* and *intestine*, this last consisting of a thin-walled portion or *ileum*, a short thick-walled *colon* and a terminal portion, the *rectum*. The portion of mid-gut associated with the proventriculus is sometimes termed the *cardia*. In addition there are the *larval salivary glands* and the *Malpighian tubules*. The relations of these parts to each other and to the body are shown in Fig. 25 (1).

The pharynx, oesophagus and proventriculus (in part), collectively forming the *fore-gut*, are developed from the stomodaeum, the intestine constituting the *hind-gut*, as also seemingly in the mosquito the Malpighian tubules, from the proctodaeum, whilst the mid-gut and the gastric caeca are derived from the mesenteron. In keeping with this the pharynx and oesophagus, as also the intestine and rectum, have a chitinous intima and their epithelium is usually more or less of the flattened type. An exception, however, is the colon which has very large cells. The epithelium of the mid-gut and its caeca, on the contrary, consists of very large and massive cells. The cells of the Malpighian tubules are also very large, not unlike those of the mid-gut and colon.

THE PREORAL CAVITY AND MOUTH (Figs. 25, 26)

The preoral cavity is the functional mouth. Its roof is formed by the largely membranous structures on the ventral aspect of the labral and clypeal regions, including the epipharynx and especially the large membranous post-epipharyngeal lobe (Fig. 26 (3) *epl*). Its floor is formed mainly by the dorsal surface of the mandibles and to some extent of the maxillae. When these appendages are abducted, the floor is largely open. As shown by Christophers and Puri (1929), the emerging current produced by the brushes in the culicine larva when feeding is directed backwards from such opening thereby causing the larva to glide along in its characteristic way. When the brushes are retracted for combing or protection they are received in a compacted form into the preoral cavity and lie largely covered by the

* See also Raschke (1887, *Aëdes nemorosus*), Hurst (1890, *Culex*), Nuttall and Shipley (1901, *Anopheles*), Metalnikoff (1902), Wesche (1910), Schonichen (1921), Martini (1929), de Boissezon (1930), and various text-books on medical entomology. For special description of the pharynx see Kulagin (1905) and Salem (1931) and for description of the mid-gut Federici (1922, *Anopheles*) and Samtleben (1929). There are also many references to structure in the works of Wigglesworth, notably the very full description of the proventriculus of *Culex*, *Aëdes* and *Anopheles* larva (Wigglesworth, 1930) and the paper dealing with storage of fats, glycogen, etc. (Wigglesworth, 1942).

mandibles and maxillae. For a small extent in the middle line the floor is formed by the labial and hypopharyngeal rudiments and the sharp-toothed mental sclerite (Fig. 25 (2)).

Any true *buccal cavity*, if present, is very restricted in extent. Snodgrass (1935, p. 387) defines the buccal cavity as 'The first part of the stomodaeum, lying just within the mouth;

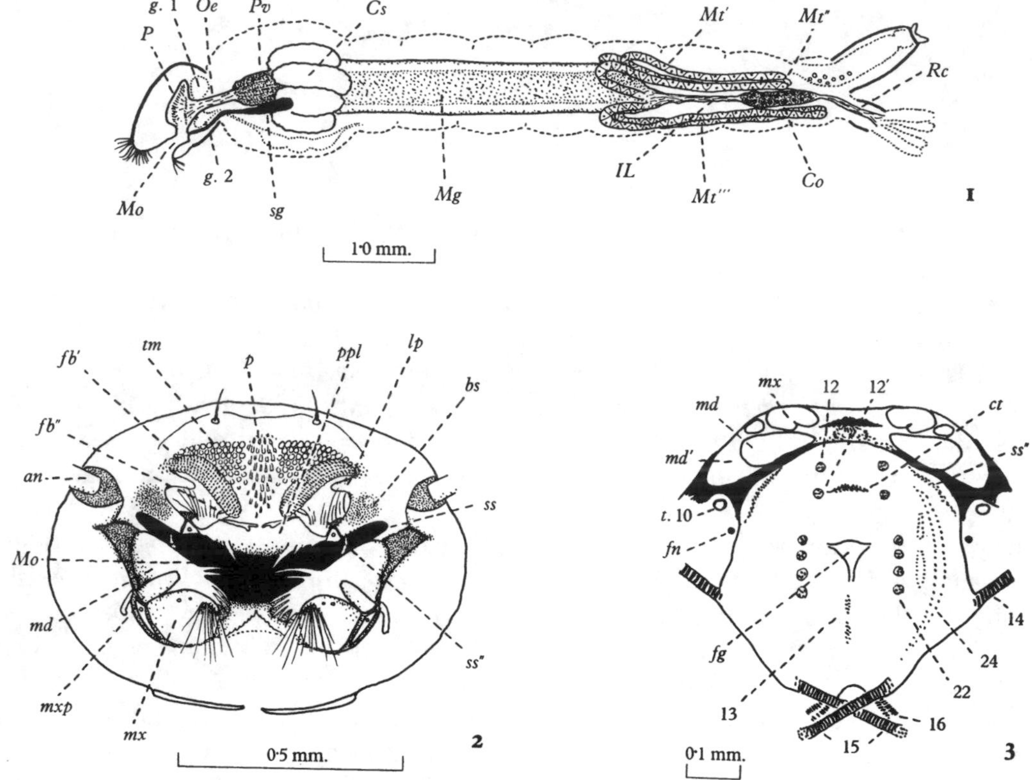

Figure 25. The alimentary canal.

1 Lateral view of larva showing general anatomy of the alimentary canal.

2 Anterior view of head showing preoral cavity and surrounding structures.

3 Pharynx and related parts. Dorsal view.

Lettering: *an*, antenna; *bs*, black spot area; *Co*, colon; *Cs*, gastric caeca; *ct*, the crescent; *fb'*, flabellar plate; *fb''*, scallop; *fg*, frontal ganglion; *fn*, frontal connective; *g.* 1, supra-oesophageal ganglion; *g.* 2, sub-oesophageal ganglion; *IL*, ileum; *lp*, process of lateral plate of clypeus; *Mg*, mid-gut; *Mo*, mouth; *Mt'*, Malpighian tubule (dorsal); *Mt''*, Malpighian tubule (dorso-lateral); *Mt'''*, Malpighian tubule (ventro-lateral); *md*, mandible; *md'*, membrane covering fenestra; *mx*, maxilla; *mxp*, maxillary palp; *Oe*, oesophagus; *P*, pharynx; *Pv*, proventriculus; *p*, palatum; *ppl*, post-palatal lobe; *Rc*, rectum; *sg*, salivary gland; *ss*, suspensorium; *ss''*, position of lateral oral apodeme; *t.* 10, ventral cephalic trachea; 12, 12', anterior elevators of roof of pharynx; 13, median elevators of roof of pharynx; 14, lateral elevators of roof of pharynx; 15, posterior decussating muscles; 16, dorsal suspensors of the pharynx; 22, dorsal compressors of the pharynx; 24, lateral compressors of the pharynx.

its dilator muscles arising on the clypeus, and inserted before the frontal ganglion and its connectives'. In *Aëdes aegypti* the frontal ganglion lies on the roof of the pharynx a little behind the level of the chitinisation described later as the crescent, and at about the same level on each side of the pharynx is a slight constriction where the frontal ganglion connectives pass up to the ganglion. Into the roof of the pharynx anterior to this are inserted two small muscles, 12 and 12′ (Fig. 25 (3)), taking origin from the falciform apodeme in a position that might well be clypeus. It would appear, therefore, that in this case the buccal cavity is the front portion of the pharynx. It will be simpler, however, to include such portion in the pharynx of which in this case it obviously forms a part.

THE PHARYNX (Fig. 26)

This has been described in some detail by Thompson in *Culex* and by Imms in *Anopheles*, but curiously enough the remarkable *fringes* are not referred to as such by either author. In sections these are quite inconspicuous and have been figured and considered merely as hairs. They are much more conspicuous in, for example, the cast skin of the larva, where their real nature as an important functional feature of the organ is obvious. It would seem that it is upon the fringes that the ability of the larva to filter out minute particles from the water depends. Further, it is probable that the necessity for such fringes has been the reason for the large pharynx and in turn for the rather large head of the mosquito larva, especially in some species which live in open clean water. Apart from the fringes it is difficult to see how the pharynx could filter particles from the current produced by the brushes. It is doubtful if the brushes themselves act to any extent as a filtering agency in the culicine mosquito, their main function being the production of a current directed into the pharynx.

The general appearance and structure of the pharynx, as also its relation to surrounding parts, can be best seen from the figures. Essentially it is a funnel-shaped muscular organ capped by a flat roof the lateral halves of which are capable of being folded up face to face in the middle line, thereby closing up and flattening the organ. The lateral edges of this roof form sharp longitudinal ridges referred to as *crests* of the pharynx. In life these crests are in constant in-and-out movement, expanding the organ when feeding is in progress, contracting it in the act of swallowing the bolus of food that gradually collects. Ventrally the pharynx opens by a broad slit-like *mouth* into the preoral cavity, this being, on the analogy already given, situated between the roof and the edge of the funnel in its most anterior and ventral part. The lateral limits of the opening are indicated by peculiar narrow curved chitinisations curiously like fish-hooks (*ss″*), especially in the *Culex* larva, which may be termed the *lateral oral apodemes*. In the middle line dorsally and ventrally just inside the opening of the mouth are the plates later described respectively as the *crescent* and *oval plate*. Though slit-like at rest the mouth by the lifting of its roof can be widely dilated. When the mouth and pharynx are so dilated they form, with the preoral cavity, a deep cavernous opening occupying much of the head space (Fig. 26 (1) *f*). Into this opening is directed the current produced by the feeding brushes.

Where the pharynx ends posteriorly it forms a dorsal backward projection of the organ reaching almost to the dorsum of the head. This *posterior angle* of the pharynx, as will be seen, has important relations with the opening of the aorta and with the sympathetic nerve system.

The pharyngeal roof consists of several plates which, taken together, form a roughly oval plaque. Anteriorly in the middle line is the small dark crescentic *crescent* (*ct*). Behind this the median portion of the roof is formed by the laterally symmetrical semilunar plates, the *dorsal plates* (*dp*). External to these and forming the lateral edges of the roof on each side are rather indefinite thickenings, the *lateral plates* (*lpp*). These are diffusely thickened towards their anterior ends and also posteriorly. The anterior thickenings receive the attachment of the lateral oral apodemes, narrow strips of sclerotisation which link the pharynx to the suspensoria and probably give support to it (Figs. 25 (2); 26 (3), (4), *ss″*). The posterior thickenings receive the insertions of the largest of the extrinsic muscles of the pharynx, the lateral dilators (no. 14 of head muscles), which help to sling the pharynx in its position in the head and to dilate it by drawing open the crests. It is the two posterior thickenings which, narrowing behind, come together and largely form the posterior angle of the pharynx, constituting a notch through which pass the aorta and the recurrent pharyngeal nerve from the frontal ganglion. Inserted into the posterior angle and its neighbourhood are various muscles as described later coming from the dorsum of the head. Whilst these and the lateral dilators sling the pharynx in the head, the organ receives support below from the lateral oral apodemes. In their natural position these lie more or less vertically springing from about the middle of the suspensoria. Though they have been termed and have the appearance of apodemes, they are largely surface thickenings of the membrane at the mouth angles.

Between the dorsal and the lateral plates of each side are the *dorsal fringes* (*pf′*). These consist of narrow ribbons of thickening crossed by fine transverse ridges which are continued beyond the ribbon's edge as stiff straight filaments, the whole forming a structure very like a small-toothed comb. The actual number of fringes in the series is not easy to be sure of, but there are at least three on each side, the outer ribbon carrying one, whilst the inner ribbon carries at least two, one arising from each edge. The filaments, excluding the ridge portion of the ribbon, are about 50μ in length and $0{\cdot}7\mu$ in width, the space between being very regularly only about $1{\cdot}0\mu$ or one-fifth of the diameter of a red blood corpuscle. Terminally some of the filaments end in fine straight branches, though the majority seem simple, continuing in strictly parallel course to their termination. In *Culex molestus* the branching is more marked. In sections food masses are frequently seen lying in the pharynx in close connection with the fringes. In the dilated pharynx they project as ridges along the roof of the cavity. As noted later a further fringe series lies along the ventral floor on either side.

On the floor of the pharynx anteriorly and at about the level of the crescent is an oval plate, *oval plate* (*op*), which forms a kind of threshold to the cavity. Behind this are a number of fine, backwardly directed hairs. This portion of the floor often appears in sections somewhat sunk to form a gutter and the hairs standing erect look as if they might be functioning as a valve, allowing entry to a current, but opposing its exit. The *ventral fringes*

Figure 26. Pharynx.

1 Sagittal section of head showing preoral cavity and pharynx.
2 Dorsal view of head showing pharyngeal connections.
3 Roof of pharynx and preoral cavity (cast skin preparation).
4 Floor of pharynx and hypopharynx (cast skin preparation).
5 Portion of a pharyngeal fringe highly magnified.
6 Basal portion of a flabellar hair at same magnification as 5.

Lettering: *ap*, flabellar apodeme; *ct*, the crescent; *dp*, dorsal plate of pharynx; *ep*, epipharynx; *epl*, post-epipharyngeal lobe; *f*, pharynx; *f″*, fringes; *hp*, hypopharynx; *L.r*, labial rudiment; *lpp*, lateral plate of pharynx; *ms*, mental sclerite; *op*, oval plate; *pe*, post-epipharyngeal bar; *pf′*, pharyngeal fringes (dorsal); *pf″*, pharyngeal fringes (ventral); *ppl*, post-palatal lobe; 4, elevators of the pharynx (two muscles); 17, accessory dorsal suspensors of the pharynx.

For other lettering and numbers see under explanation of Fig. 25.

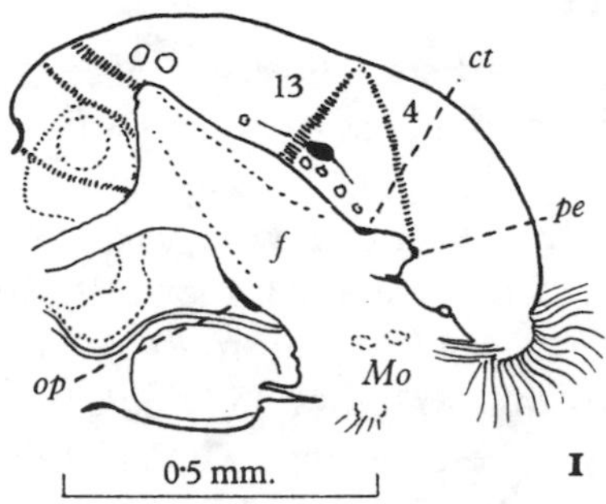
ct
13
4
pe
f
op
Mo
0·5 mm.
1

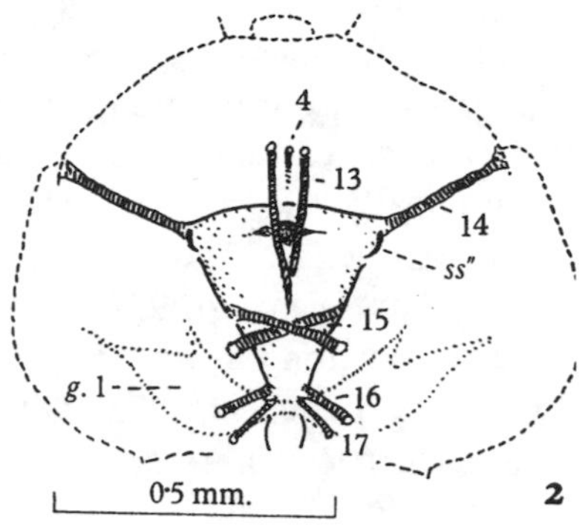
4
13
14
ss″
15
g. 1
16
17
0·5 mm.
2

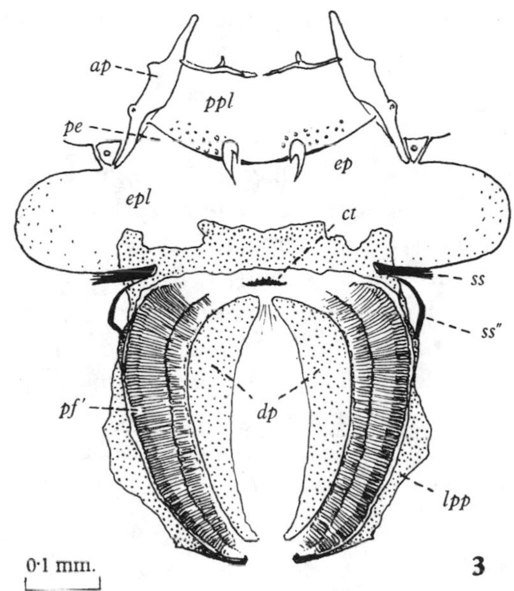
ap
ppl
pe
ep
epl
ct
ss
ss″
pf′
dp
lpp
0·1 mm.
3

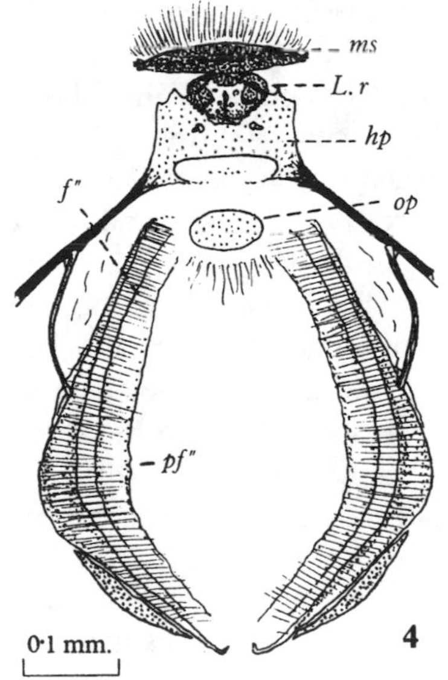
ms
L. r
hp
f″
op
pf″
0·1 mm.
4

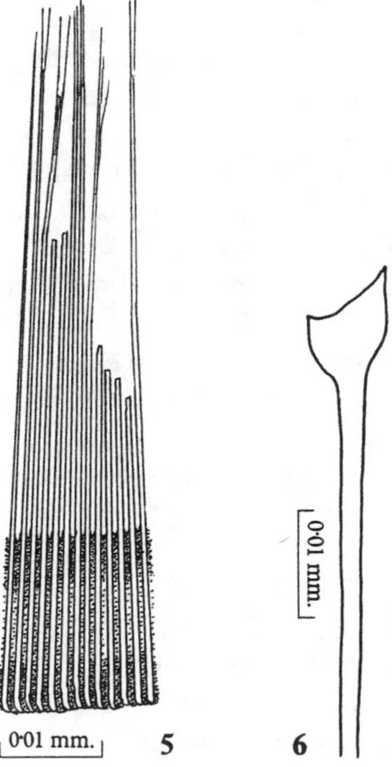
0·01 mm.
5
0·01 mm.
6

Figure 26

lying along the sides of the pharynx below the lateral plates here converge. Medianly, strictly speaking, there is no floor since ventrally the cavity of the pharynx forms a funnel-shaped passage leading to the oesophagus.

Histologically the pharynx has an inner lining of chitinous intima with small-celled pavement epithelium and an outer muscular coat. Apart from the small discreet dorsal compressor muscles (no. 22 of the head muscles) the dorsal roof has no muscular coat. The ventral portion has an inner coat of circular muscle with a strong band lying external to this beneath each crest laterally, termed by Thompson the *cingulum*. Though taking origin and insertion only from the pharynx this muscle has much the character of an extrinsic bundle (see Fig. 30 (1) *m*. 21). The muscular coat proper much resembles that of the oesophagus with an inner circular coat and some loosely arranged external longitudinal fibres. The extrinsic muscles of the pharynx and the nerve connections are dealt with later.

THE OESOPHAGUS

This is a short tube passing from the pharynx through the nerve ring to end in the proventriculus. It consists of an inner lining of flattened epithelium with chitinous intima, usually much thrown into folds, and an outer coat of circular muscle. Towards the base of the neck the muscular coat makes connection by strands of tissue with surrounding structures including the aorta which runs closely along its dorsal surface. As noted by Salem it has a ventral and a dorsal bulge in this position. Actually these effects are due to the oesophagus here passing through the cervical muscular diaphragm, which will be described later when the rather complicated parts in this region are dealt with.

THE PROVENTRICULUS (Fig. 27 (1))

The proventriculus has been described by Thompson in *Culex*, by Imms in *Anopheles* and in the three representative species *Anopheles plumbeus*, *Culex pipiens* and *Aëdes aegypti*, as also in a number of other Diptera and orders of insects, by Wigglesworth (1931 *b*). In all three forms it is roughly similar and consists essentially of an intussusception of the end of the oesophagus into the cardia of the mid-gut. Its structure in *Aëdes aegypti* will be evident from the figure showing it in longitudinal section. Its walls consist of three layers, namely an inner, or direct, face and an outer, or reflected, face of the intussuscepted oesophagus closely surrounded by a layer formed by the cardia (*c*). At the lower end of the intussusception is a conspicuous annular blood sinus. The cells of the outer wall formed from the cardia are large and resemble in type those of the mid-gut, though evidently specialised, being more columnar and deep staining. Where at the bend the small cells of the oesophageal type give place to the large cardia cells there is what Thompson describes as a 'break', a small gap where there appear to be no cells. At this point too there is a ligament-like extension of the connective tissue coat of the oesophagus on to the cardia. The lower end of the oesophageal intussusception differs somewhat in detail in the different mosquitoes. In *Anopheles* there is a strongly developed chitinous ring carrying fine curved spines and a special ring of longitudinal muscle above this not present in other forms (Wigglesworth, 1930). In *Culex pipiens* there is a general chitinous thickening of the reflected layer with an everted rim at the edge. In *Aëdes aegypti* the chitinous margin of the free edge is still more thickened (Fig. 27 (1) *a*).

In regard to the function of the organ it has been shown by Wigglesworth that not only does the layer of cardia cells secrete the peritrophic membrane but that the structure acts as a press, moulding the secretion as it passes out (see also under digestion in the chapter on physiology of the larva). The peritrophic membrane, as shown by Wigglesworth (1930), is composed of chitin, though the outer layer of secreting cells appears almost certainly a part

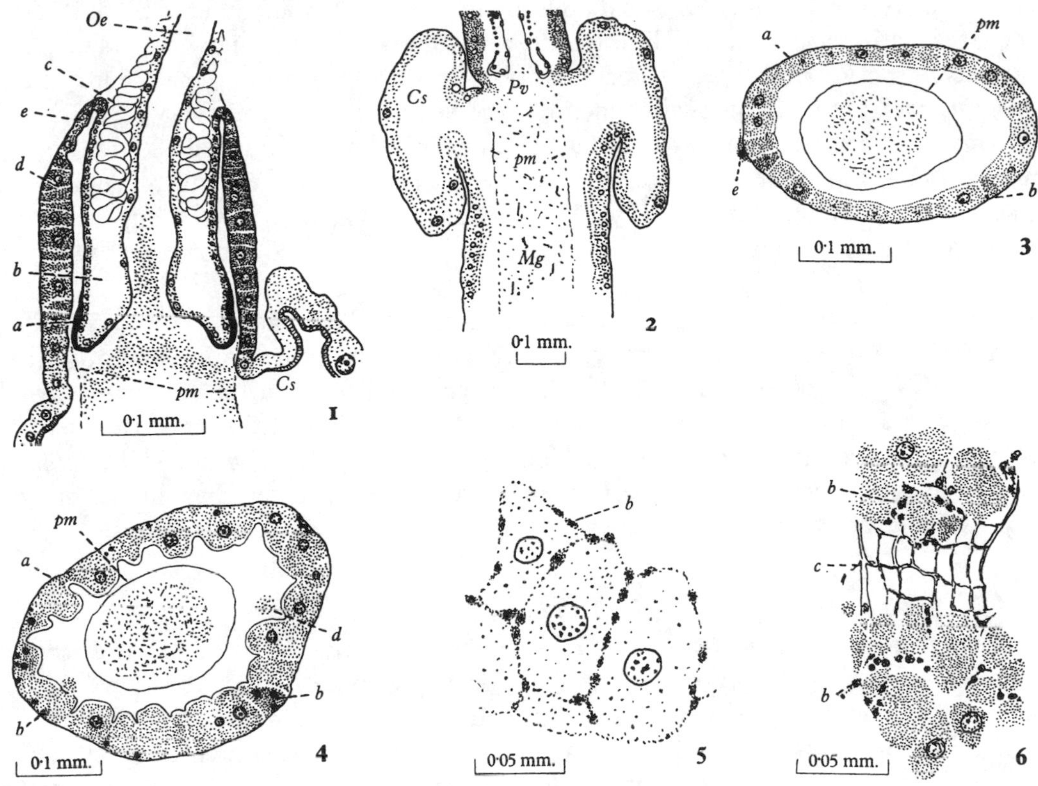

Figure 27. Proventriculus, gastric caeca and mid-gut.

1 Longitudinal section of proventriculus.
2 Section through anterior end of mid-gut and gastric caeca.
3 Transverse section of anterior portion of mid-gut of early fourth instar larva.
4 Transverse section of posterior portion of same.
5 Tangential section of mid-gut wall showing early infiltration of small basal cells.
6 Tangential section of extreme outer layer of mid-gut wall in a pre-ecdysis fourth instar larva, showing heavy infiltration with small basal cells and early formation of muscle fibres of future adult mid-gut.

Lettering: 1, *a*, annular chitinous thickening acting as moulding press; *b*, blood sinus; *c*, annular muscle; *d*, epithelium of cardia; *e*, ring of regenerative epithelium. 2–6, *a*, basement membrane; *b*, basal cell; *c*, developing muscle fibres; *d*, extrusion from cell; *e*, tracheal cell.

Other lettering: *Cs*, gastric caecus; *Mg*, mid-gut; *Oe*, oesophagus; *pm*, peritrophic membrane; *Pv*, proventriculus.

of the mid-gut and of mesenteral origin. The cells not only have a close resemblance to those of the mid-gut, but show the same inner clear margin or even a striated border. External to this border may often be seen a considerable thickness of clear substance, presumably secretion which will eventually become peritrophic membrane. Around the upper margin where the oesophageal cellular lining changes to the large cells of the cardia is a ring of epithelium which as Thompson notes forms a regenerative area in the pupal stage.

THE SALIVARY GLANDS AND THEIR DUCTS (Figs. 25 (1); 39 (3) *sg*)

On either side of the proventriculus and abutting upon the lower lateral caecum of its side are the larval salivary glands. Unlike the *Chironomus* larva, which in many species uses the salivary secretion to bind together particles to form a protective tube, the glands of the mosquito larva have no such special function to serve and are relatively very small. Each gland consists of a single tubular acinus having an open lumen surrounded by large deeply staining granular cells. The distal end of the acinus is usually somewhat dilated, sometimes considerably so, forming a spherical cyst-like portion which as noted by Thompson in the living larva is highly refractile. In sections very little indication is given of secretion in the lumen.

From each gland there passes forwards a delicate thin-walled duct, *salivary duct*. Each duct on the ventral aspect of the nerve connectives in the neck joins with its fellow of the opposite side a little behind the sub-oesophageal ganglion to form the *common salivary duct*. The common duct passes forwards along the dorsal surface of the labial rudiment to open by a transverse slit just dorsal to the labial plate (Fig. 13 (6)). The site of the opening figured by Salem as at the ventral part of the plate and immediately behind the mental sclerite is incorrect. None of the ducts is provided with taenidia as in the adult.

THE MID-GUT OR VENTRICULUS (Fig. 27 (2–6))

Immediately beyond the proventriculus is the portion of mid-gut from which arise the balloon-like diverticula or *gastric caeca*. These occupy a large part of the space of the thorax. They are eight in number arranged around the canal, two dorsal, two ventral and two lateral on each side. They all bulge somewhat forward, but to a greater extent backwards, the largest and longest being the lateral, which reach to the beginning of the abdomen, then the dorsal, the smallest being the ventral. They are lined with cells resembling those of the mid-gut, but of rather larger area. They have a broad striated border and like the cells of the mid-gut have rather pale clear cytoplasm staining strongly with eosin. There appears to be no muscular coat or indication of muscle fibres.

Beyond the origin of the caeca the mid-gut extends back as a long uniformly cylindrical organ to the fifth abdominal segment, the exact position to which it reaches being a little variable. Within it lies the long column of food debris enclosed in the tube of peritrophic membrane, a space being everywhere present between the membrane and the cells lining the gut.

The histology of the mid-gut wall is somewhat complicated by the changes which take place in the fourth instar leading to the condition present in the pupa. Essentially in the younger stages the gut is lined by uniform large cells with large globular nuclei and clear or finely granular cytoplasm. In vertical section the cells tend to appear cubical or rectangular and of more or less even depth. Cut tangentially they appear as large polygonal cells. Those of the anterior two-thirds or so of the gut are flatter and of more regular shape than those in the posterior portion where individually they project more freely into the lumen, especially in the later stages of development. There is a well-marked striated border which is thicker and more conspicuous posteriorly. This is of the 'brush' form, consisting of a palisade-like arrangement of separate filaments which in dissections may sometimes be seen sprayed out around fragments of cytoplasm like a coating of motionless cilia. In the advanced larva the gut cells, especially in the hinder portion of the gut, loose their regular form, show vacuoles and project far into the lumen which eventually becomes crowded by droplets and extruded masses of cellular material, a preliminary to the casting off of the whole old epithelial layer in the pupal stage.

In addition to the large cells there are a variable number of small darkly staining cells situated about the bases of the larger cells. These increase greatly in number towards the later stages forming lines or clusters of small cells along the lines of division of the larger cells (Fig. 27 (5)). These are regeneration cells which eventually in the pupa develop into the imaginal epithelium. The exact nature and source of the small basal cells are not clear. Whilst some are destined to form the imaginal epithelium, others appear to be forerunners of the imaginal muscular coat of the gut. There are also cells which appear to be of the nature of the normal replacement type. These may be of all sizes up to that approaching the normal cells. Whilst proliferation of the basal cells is very striking towards the end of the fourth instar a somewhat similar though less marked increase is present at the end of the third instar, suggesting that there is some degree of replacement of the gut epithelium at ecdysis.

Externally there is a well-marked basement membrane and the presence of numerous tracheal cells with their tracheole branches. How far there is a muscular coat seems somewhat doubtful. Thompson states that scattered muscle fibres are visible in sections. Samtleben (1929) notes that muscle in this case is inconsiderable and figures a section of the gut of the first instar larva of *Aëdes aegypti* with scattered small longitudinal muscle fibres. A curious feature is the presence in the fourth instar of two parallel strands passing along the whole length of the mid-gut on its mid-ventral surface and a similar pair dorsally. These bands have occasional nuclei and appear to be muscular in nature. Apart from these I have been unable in the early instar to detect definite muscle fibres. Later, however, there appears associated with abundant basal cells a regular network of fibres at right angles as in the adult mid-gut (Fig. 27 (6)). This would seem to be a prepupal development and if a muscular coat exists in the mid-gut apart from this it is of a very inconspicuous character. At its termination, the mid-gut with its large-celled epithelium gives place abruptly to the thin-walled ileum.

THE ILEUM (Figs. 25 (1); 28 (1–3 *IL*)

In marked contrast to the inert tubular mid-gut is the thin-walled muscular and actively contracting ileum. At its anterior end the ileum is somewhat dilated in a funnel-shaped manner, forming as in the adult a pyloric ampulla into which open at its upper end the five Malpighian tubules. Though anatomically the ampulla may be considered as part of the ileum, it shows some independent structural features. As noted by Thompson there is, for some distance beyond the termination of the mid-gut, a complete absence of the muscular coat of the ileum, which ends abruptly at a little distance from the mid-gut, bridged only by the delicate basement membrane and the thinned-out epithelium of the gut. The epithelium here too shows special characters, there being for a short distance beyond the mid-gut a zone of small rather cubical cells in contrast to both the large mid-gut epithelium and the much-flattened epithelium of the ileum (Fig. 28 (1)).

The remainder of the ileum consists of a length of thin-walled gut extending from the pyloric ampulla in the fifth or sixth segment to its termination in the colon in the seventh segment. In life it is usually seen exhibiting active vermicular movements. In fixed material it may appear as a thin-walled dilated tube of even diameter, or as a narrow portion of the gut with its walls thrown into folds. In the latter case it may be empty of any food column. The ileum has a lining of flattened epithelium with a thin cuticular lining and doubtful striated border. External to this is a thick outer coat of circular muscle. This consists of a single layer of regularly disposed muscle fibres with round nuclei in contrast with the oval nuclei of the epithelium (Fig. 28 (2), (3)). There do not appear to be any longitudinal fibres. In the seventh segment the ileum enters the colon, usually with a slight valvular-like protrusion of its walls into that organ.

THE COLON (Fig. 28 (4), (5))

The colon is a curiously thickened and seemingly rigid portion of the gut. It is thrown into longitudinal folds and lined by large cells not unlike those of the mid-gut or even of the Malpighian tubules. There is a thick well-marked striated border. Spaced along its outer wall at short intervals are circular muscle fibres linked up by membrane. In fresh preparations these spaced fibres, by constricting the organ, give it a curious beaded annulated appearance. It seems doubtful if there are any longitudinal fibres and none could be demonstrated by Thompson. Probably staining by Heidenhein's method might show such fibres where otherwise they have been missed. Towards the end of the eighth segment the colon gives place with no very clear line of demarcation to the rectum. It would be interesting to know what function is served by this peculiar portion of the gut. It is just possible that it may be concerned with breaking up of the tube of peritrophic membrane, which in the faeces of the larva is passed in short, rather regular lengths. See under 'Digestion', p. 704.

THE RECTUM (Fig. 28 (7), (8))

This is again a relatively thin-walled structure with its lining thrown into longitudinal folds. In histological structure it is not unlike the ileum with rather small-celled flattened epi-

---

Figure 28. Intestine, colon and rectum.

1 Section through junction of mid-gut and ileum, showing entrance of Malpighian tubules, gap in muscular wall and ring of cells at pylorus.

2 Longitudinal section of ileum in dilated condition showing regular arrangement of circular muscle fibres.

3 A, transverse section of contracted ileum on same scale as the mid-gut and colon in 1 and 4. B, portion of wall of same more highly magnified. C, showing muscular and flat epithelial layer and characteristic nuclei.

4 Longitudinal section through full length of the colon.

5 Transverse section of colon.

6 Section through junction of ileum and colon.

7 Continuation of colon into rectum showing continuation of the circular muscle fibres of the former into the latter.

8 Transverse section of rectum with diaphragm formed by tracheae and muscles of anal segment.

Lettering: *a*, entry of Malpighian tubule into pyloric ampulla; *b*, circular muscle fibres; *b'*, nuclei of same; *c*, gap following abrupt termination of muscular coat of ileum; *d*, ring of special cells; *e*, longitudinal muscular bundles of anal segment; *IL*, ileum; *Mg*, mid-gut; *tv*, tracheal vessel.

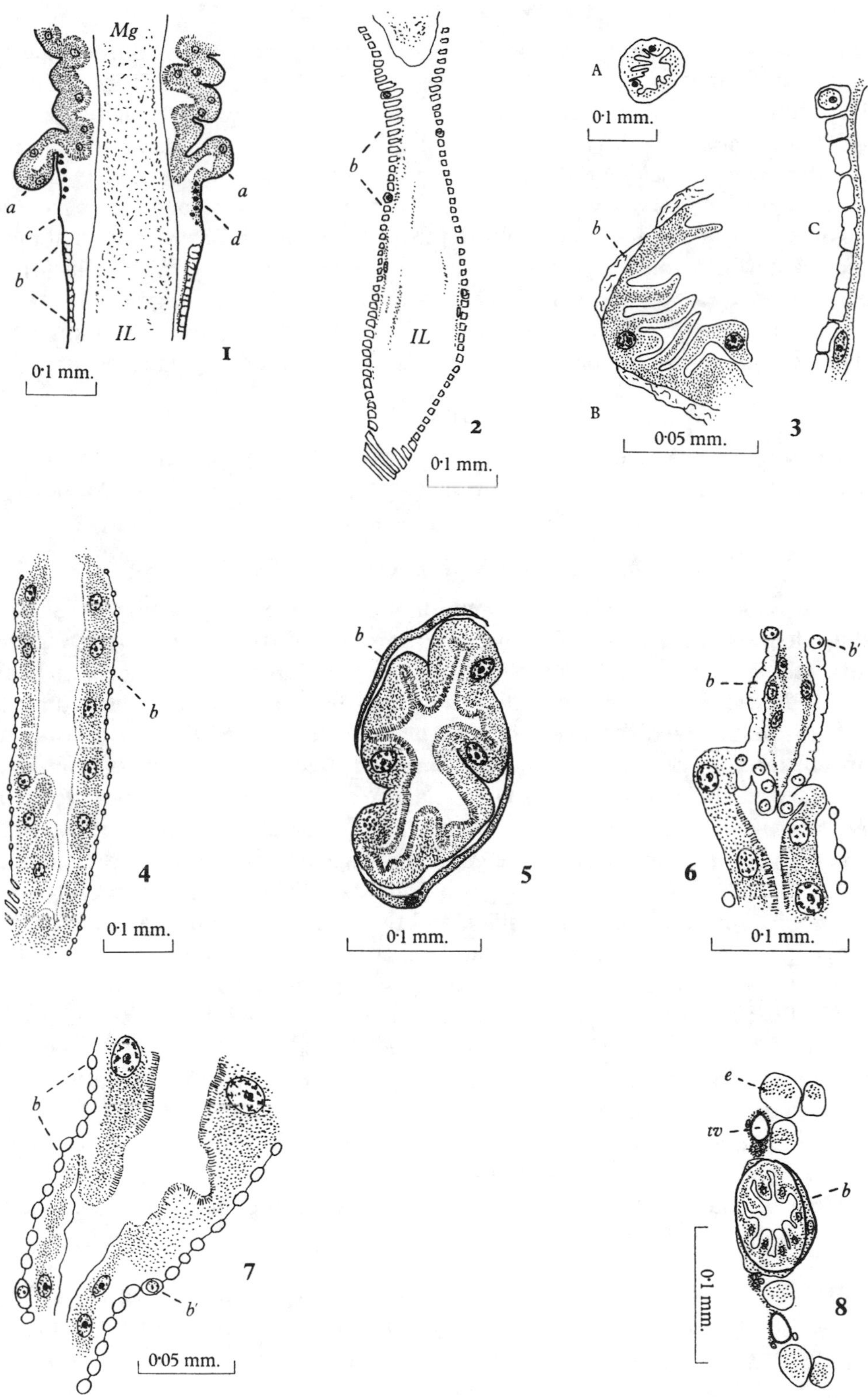
Mg
a
c
b
a
d
IL
I
0·1 mm.
b
IL
2
0·1 mm.
A
0·1 mm.
b
B
0·05 mm.
C
3
b
4
0·1 mm.
b
5
0·1 mm.
b
b'
6
0·1 mm.
b
b'
7
0·05 mm.
e
tv
b
0·1 mm.
8

Figure 28

thelium and an outer single layer of circular muscle fibres. These fibres continue the series of the colon, but with the fibres closer together (Fig. 28 (7)). Like the ileum the lining is cuticular without striated border. There do not appear to be any longitudinal fibres intrinsic to the organ. Towards its posterior end, however, the rectum receives the insertion of some segmental muscles of the anal segment (see under muscular system). In its passage through the anal segment the rectum is accompanied on each side by the large tracheae passing from the tracheal trunks to the anal papillae and by longitudinal muscles of the segment which together form a kind of supporting diaphragm (Fig. 28 (8)). The rectum terminates in the anus, which opens at the end of the anal segment between the bases of the four anal papillae. The lining at this point is thrown into many longitudinal folds. No structures corresponding to the rectal papillae of the adult are present in the larva.

THE MALPIGHIAN TUBULES

These are described under the excretory system. They are five in number and open into the ampulla, one dorsal and the other four arranged dorsally and ventrally at the sides.

(*b*) THE TRACHEAL SYSTEM

The tracheal system, besides playing an important part in the life of the insect, undergoes interesting changes in the course of the passage from larva, through pupa, to the imago, and without some knowledge of its details of structure much that happens in these changes cannot well be understood. It is desirable therefore to give a somewhat detailed account. The system is best studied in larvae which have been placed, when they have just reached the fourth instar, for 5 or 6 days or more in clean water (see under starvation forms). Such larvae may be examined alive under a cover-glass suitably supported, or mounted temporarily in glycerine or glycerine jelly. For the latter purpose a small opening must always be made to allow penetration, or otherwise gross shrinkage and collapse will result. A good plan is to place the larva, after puncturing some non-essential part, in 50 per cent glycerine, which may advantageously contain a stain, and then transfer to pure glycerine or melted glycerine jelly. Very good preparations can be made by killing the starved larvae by the formalin method (see under 'Technique', p. 116) and leaving them to harden for a day or two still floating. After puncture and treatment with 50 per cent glycerine and pure glycerine, they can be mounted by dropping melted glycerine jelly on them and covering. Air remains in the larger tracheae, which show up vividly, for a surprising length of time, at least for days or weeks. For a more permanent and thorough method, that described by Wigglesworth in a recent publication can be used (Wigglesworth, 1950). This shows up the ultimate fine tracheal branches in a very striking way.

As in other mosquitoes the larva of *A. aegypti* is metapneustic, that is the only functional spiracles are those opening at the apex of the siphon (post-abdominal spiracles of Keilin, 1944).

But though the terminal spiracles of the mosquito larva are the only functional ones, other spiracles normal to insects are represented in the *A. aegypti* larva by tiny black sclerotisations in the cuticle from which cords (actually tracheae without a lumen) pass inwards representing the spiracular tracheae of insects with open spiracles. The cords have been described as early as 1890 by Hurst and by Giles (1902). The small black sclerotisations, which may be termed *spiracular puncta*, are very clearly seen in the cast skin, as also in the cuticle of the pale living larva (Fig. 10 (1) *st*). They mark very clearly the position of the future spiracles in the adult and are present towards the anterior margin of abdominal segments I–VII and on the mesothoracic and metathoracic segments just in front of the line of hairs indicating these segments. There are no puncta in front of the prothoracic dorsal hairs or on the eighth abdominal segment.

Normally the spiracles of insects are ten in number on each side, two thoracic and eight abdominal, namely those on segments I–VIII, the eighth segment being the last segment on which spiracles are present in any of the orders. That they are clearly indicated in the mosquito larva in normal position towards the anterior end of the first abdominal segments, but not on the eighth, gives strong evidence, apart from other considerations pointing the same way, that the highly specialised posterior abdominal spiracles of the larva (and incidentally of other dipterous larvae so provided) are the spiracles of the eighth segment displaced and drawn up behind their proper segment. With change of the larva to the pupa these structures, as will be seen later, close up and not only cease to be functional, but for all practical purposes disappear, suggesting that their early specialisation is the reason why in the Diptera there are only seven abdominal spiracles in the imago instead of the usual eight in most other orders of insects.

The puncta also enable the segmentation of the two spiracles in the thorax to be determined with accuracy, for their relation to the thoracic hairs is quite definite. It may be noted that, according to Snodgrass (1935), the anterior of the two pairs of thoracic spiracles in insects is always mesothoracic, though it may in many cases appear to be prothoracic. In the mosquito it has by most authors been considered prothoracic. That it is here also mesothoracic is, however, clearly shown, as already noted, in the larva, and as their identity is traceable through the pupa to the imaginal stage the homology determined for the larva fixes that for the spiracles in the imago.

THE TRACHEAL TRUNKS

The most conspicuous feature of the tracheal system are the two large *tracheal trunks* (Fig. 29 (2)–(4) *d*) which extend from the terminal spiracles on the siphon through the siphon and abdomen to the thorax. In the mosquito larva they are relatively of enormous size. Giles (1902), who has given an early description of the tracheal system of the larva, suggests that they may serve as a store of air during submergence. A more obvious reason is that given by Hurst (1890) that they act as flotation chambers, reducing the specific gravity of the larva to that of the water. As has been seen they play an even more precise role in the hydrostatic balance of the larva. The trunks form in the abdomen a broad conspicuous band on each side of the narrow median space between them occupied by the pulsating heart. They are lined with fine taenidia, which, probably due to interference effect, give to them under the low power the appearance of watered silk. Especially just before ecdysis the trunks may have the appearance of being double-walled, due to the presence of the air-filled tracheae of the old instar lying surrounded by the much larger fluid-filled tracheae of the oncoming instar.

Each trunk is constricted to form a bulged portion in the metathorax and in each abdominal segment I–VII. Towards the posterior end of each bulge there is a thin line crossing the trunk which is free from taenidia and indicates a line of weakness. It is at this point that the intima is broken at ecdysis to enable segments of the trunks to be drawn out by traction of the cords attached to the puncta. Anterior to the bulge in the metathorax the trunks suddenly narrow and, diverging widely, pass round on either side of the thorax to terminate in branches to the head and other parts. At their posterior ends in the eighth segment before passing into the siphon there are on the ventral aspect of each trunk a number of crypts or depressions from which arises a felt-like mass of tracheoles passing to the space behind the terminal chamber of the heart. Lastly, after passing through the siphon and just before reaching the terminal spiracles, there is a short specialised portion of the trunks forming the felt chamber of Keilin as already described in the section on the siphon.

Across the middle line of the body the trunks are united by a number of *commissures*. One large and important commissure, *cervical commissure*, is situated just behind the neck. A narrow cord-like, but patent, commissure is situated in the thorax (Fig. 29 (2)) and there is one in each segment passing ventral to the heart a little anterior to each constriction in the abdomen (Fig. 29 (3)). On the outer side of each trunk, at the point where the cords from the puncta join the trunks, is a small marked-out area of the trunk wall, *insertion plaque*, from which arise the tracheal branches of the segment. Linking up the branches of one segment with those of the segment before and behind are the *connectives*, usually long uniform tracheae taking a wavy course from one tuft of branches to the other. Arising from the trunks or its branches, commissures and connectives are tracheal branches supplying all parts of the body.

According to Snodgrass (1935) there is in the generalised insect plan in each segment a trachea passing in from the spiracle, *spiracular trachea*, which sends off a dorsal and a ventral branch. Joining up the spiracular tracheae are connecting tracheae which collectively form a lateral longitudinal trunk. Similarly connectives may join up the dorsal branches to form a dorsal longitudinal trunk on each side and there may be also a ventral trunk formed by the joining up of the ventral tracheae of successive segments. On this plan the tracheal trunks of the mosquito larva are the dorsal trunks, the lateral trunks being represented only by the relatively small connectives.

Landa (1948) has described in great detail the tracheal system in the Ephemerid larva and has given a system of nomenclature, numbering and naming the different tracheae. It is a curious fact that though the Ephemeridae are a very primitive group and the Diptera far removed in zoological position, the characters of the tracheal system are remarkably similar in the two, so much so that in regard to the larger vessels and even many of the

---

Figure 29. The tracheal system.

1 Showing tracheation of head. On the left are shown branches of the dorsal cephalic trachea only and on the right branches of the ventral cephalic.

2 Showing tracheation of the thorax. Branches to the dorsal series of imaginal buds (trumpet, wing and halter) are shown on the left, those to the ventral buds (fore-, mid- and hind-leg) on the right. Similarly dorsal branches are shown on left and ventral on the right as indicated by the numbers.

3 Showing tracheation of an abdominal segment (segments I and II).

4 Showing tracheal supply to the terminal segments.

Lettering: *a*, spiracular puncta; *b*, site of insertion plaque; *c*, tracheal 'knot'; *d*, node line of tracheal trunk indicating position where the intima is broken at ecdysis for removal of the trunk segments.

Numbers indicate the tracheal branches as given in the list of tracheae in the text (pp. 304–5).

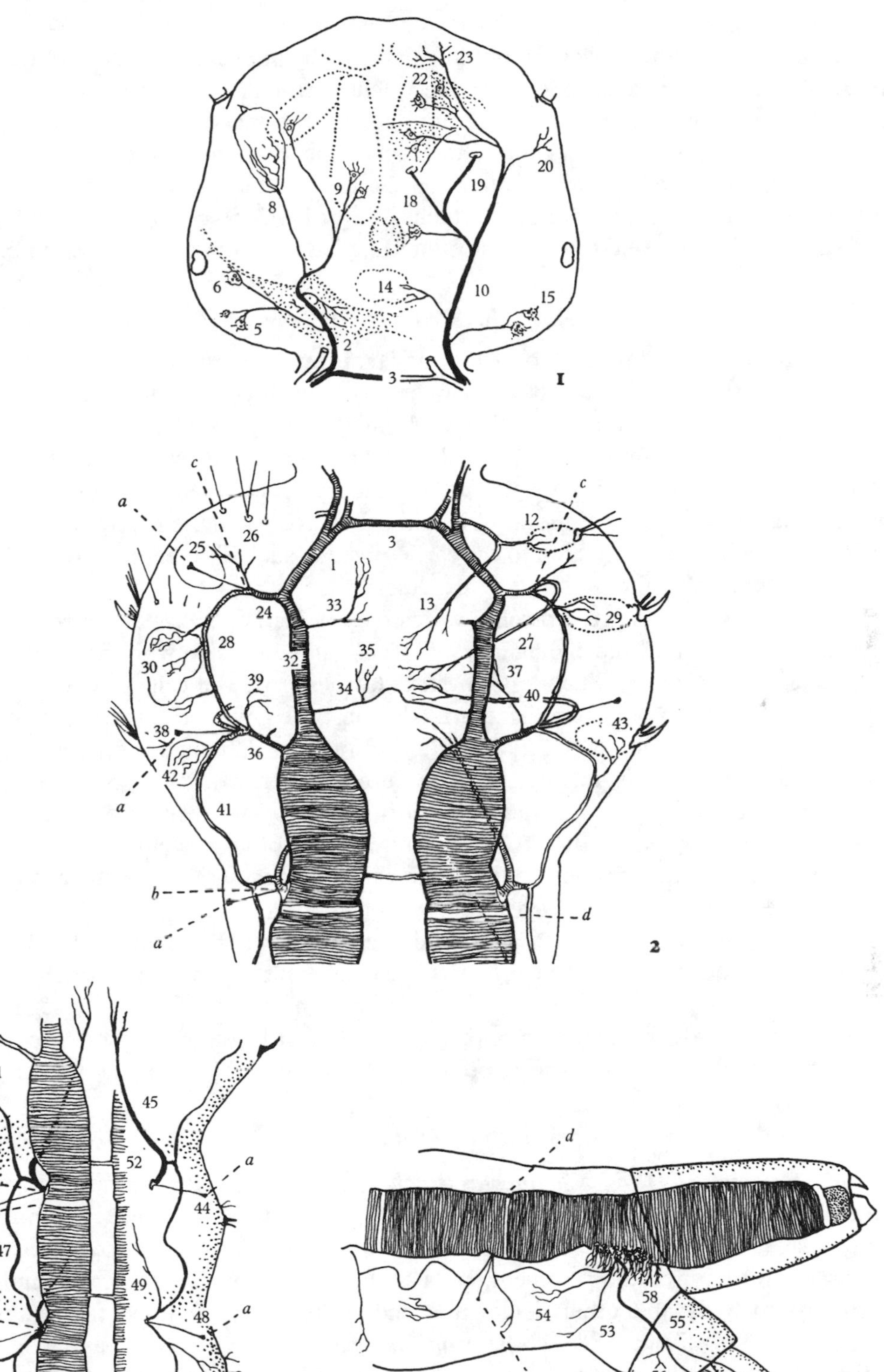
23
22
20
19
18
9
8
6
14
10
15
5
2
3
I
c
a
26
12
c
25
3
1
33
13
24
29
28
27
30
32
35
37
34
39
40
38
43
36
42
a
41
b
a
d
2
41
45
52
b
a
d
44
47
49
b
48
a
50
51
3
d
58
54
55
53
a
56
57
4

Figure 29

smaller branches the names given by Landa are clearly applicable to those of the mosquito larva. Landa does not consider that the tracheal trunks are merely connectives in the sense of Snodgrass and regards them, not as ordinary anastomoses, but as quite special organs superior to anastomoses and other tracheation. It is easier to describe the condition in the mosquito larva on this basis than on the view that the tracheal trunks are enlarged connectives, and in the description of tracheal branches here given Landa's nomenclature has been largely followed, though equivalent English names have been used in place of the Latin.

TRACHEAL BRANCHING

Though the arrangement of branches in the larva, pupa and adult superficially seem very dissimilar, fundamentally the basic plan remains throughout and a knowledge of the system in the larva will greatly help in describing the changes undergone in the later stages. To simplify the description an attached tabular statement gives the names of the chief tracheae with numbers which are also used in the figures or when necessary when referring to particular branches. This, and what is said below, relates to the fourth instar. It applies, however, with little modification to the earlier instars, such differences as occur being indicated later when the different instars are described.

The table and figures give branches which may be termed *tracheal vessels*, that is those furnished with taenidia and ending in *tracheal cells*, as contrasted with the ultimate tracheal capillaries or *tracheoles*, which pass in large leashes from the cells to enter the different tissues and which can only be satisfactorily demonstrated by special methods, being devoid of taenidia and often in part filled with fluid. For their demonstration see Wigglesworth (1950) and also earlier papers by the same author (Wigglesworth, 1930, 1931*a*, *b*, 1938). The tracheal cells from which tracheoles normally arise are relatively large, and usually stellate, cells seen lying on the surface of muscles and other structures. To a large extent they have fixed anatomical positions. As an example those present in the head are shown in the figure giving the head tracheation.

In the arrangement of branches a certain segmental plan is evident, though liable to considerable modification in certain parts. It is most clearly shown in tracheation of the abdominal segments, but also fairly clearly in tracheation of the thoracic segments. It is less evident in the head and some other parts. It will be simplest therefore to describe tracheation of the abdomen and to follow this in order by that of the thorax and head respectively.

TRACHEAL SUPPLY OF THE ABDOMEN

In each segment I–VII is a *spiracular trachea*, in this case the cord coming from the *spiracular puncta*, which joins the tracheal trunk at an *insertion plaque*. From this plaque there arise, (1) a *visceral branch* passing forwards and inwards under the tracheal trunk to the viscera in the preceding segment, (2) a *ventral branch* passing outwards and ventrally to supply fat-body and other tissues and ending in the ganglion of the segment, (3) a *connective passing forwards*, and (4) a *connective passing backwards*. The attachment of the connectives is a little curious. Each forward connective arises from the visceral branch near its origin and runs in a wavy fashion to join the ventral branch of the segment in front, usually some little distance from its origin, being then the backward connective of that segment. The arrangement will be clear from Fig. 29 (3) in the lower of the two abdominal segments shown.

The above arrangement holds good for all segments of the abdomen, but with some modification in segments I and VIII, exclusive that is of any segments posterior to this. In segment I the visceral branch is unusually large, passing forwards into the thorax to supply the large mass of the gastric caeca. In this case the forward and backward connective branches both arise from a short common vessel leading off from the large visceral branch (Fig. 29 (3)). In segment VIII the connective from VII has a common insertion with the visceral and ventral branch and a long stout branch which passes backwards into the anal segment. There is no spiracular cord. The branch passing to the anal segment comes to lie lateral to the rectum where it divides into a dorsal and ventral branch supplying the dorsal and ventral anal papilla of its side respectively (Fig. 29 (4)).

In connection with the eighth segment should also be mentioned the rather remarkable tracheole plexus already referred to as arising from the tracheal trunks in this position. The crypts from which the tracheoles arise are situated on the inner ventral aspect of the trunks. They are short tubular depressions with which conspicuous nuclei are associated, from which an intricate meshwork of branches passes ventrally to form a dense plexus. This is sometimes described as branches passing to the heart. Actually the plexus lies mainly posterior to the terminal chamber of the heart occupying a space beneath the base of the siphon and above the colon (Fig. 33 (2) *b*). It forms a mass of tracheoles of somewhat reversed pyramidal form limited, if not by membrane, by some form of limiting layer and, into the point of the pyramid, are inserted four tiny converging muscles, two on each side, an upper and a lower. The upper muscle takes origin from the membrane just below the edge of the siphon towards its posterior border and dips downwards and inwards to the point of the complex. The lower muscle takes origin from the dorsal knot VIII and passes upwards and inwards (Fig. 33 (2) *m*.35). The posterior openings of the heart open into the space occupied by the tracheoles, which is clearly an important functional accessory organ of circulation and as such will be further described when dealing with the heart.

TRACHEAL SUPPLY OF THE THORAX (Fig. 29 (2))

The tracheal system of the thoracic segments can best be seen from the figure. Here the trunks, after entering the mesothorax, somewhat diverge and become much narrower and, though still anchored by cords from the thoracic puncta, these cords are not inserted into the trunks directly, but into short branches from these from the ends of which other branches also arise, such points being termed by Landa *tracheal knots*. It is rather as if in the broad thorax and in their narrowed condition the trunks found it awkward to adhere to the proper plan and put out a branch to receive the insertion plaque. Otherwise much the same plan is seen in the thorax as in the abdomen. There are the visceral branches and ventral branches and also connectives. There are also here some small dorsal branches, branches which become much more important later in the pupa. The mesothoracic spiracle has, however, to cope with the supply both to the prothorax and head. There are also now the growing rudiments of the legs, wings and halteres to be supplied. The branches have been given in the tabular statement and are shown in Fig. 29 (1), (2). Only some of the more important points in the tracheal supply to the thorax need therefore be noted.

Beyond the level of the mesothoracic spiracle the trunks are continued on either side as a large common trunk, the *common cephalic* (1). This divides into two branches, a *dorsal cephalic* (2) and a somewhat larger *ventral cephalic* (10), both passing into the head. The ventral cephalic branch, however, before it enters the head sends off a recurrent branch which divides into two, one branch going to the rudiment of the first leg, the other sweeping round as the ventral branch of the prothorax to end in the first thoracic ganglion. The tracheal supply for the future pupal trumpet comes, however, from the mesothoracic knot. Further, the trumpet rudiment has an altogether different relation to the tracheation from that of the other rudiments, for the pupal trumpet merely forms around the mesothoracic stigmatic cord and is evidently a product of the hypodermal cells forming part of the tracheal system and not a generative bud such as are the rudiments of the legs, wings and halteres. These latter receive their tracheal supply, mainly at least, from branches arising from the connectives, the importance of which fact becomes more evident in the pupa.

TRACHEAL SUPPLY OF THE HEAD

The head (Fig. 29 (1)) receives its tracheal supply from the dorsal and ventral cephalic tracheae. Just before entering the head the dorsal cephalic trachea is linked to its fellow of the opposite side by a large and important commissure, the *cervical commissure*. Shortly after entering the head it lies in a deep groove in the brain. It supplies especially the brain (supra-oesophageal ganglion and optic lobe), the developing compound eye and the antennal rudiment, ending up in branches to the posterior portions of the large retractor muscles of the feeding brushes of its side. The ventral cephalic branch is a larger trachea supplying the mouth-parts and related structures. It extends forwards into the labral area and supplies also the labial rudiment. For further details see list below.

LIST OF TRACHEAL BRANCHES

Mesothoracic spiracle.

- 1. Common cephalic.
  - 2. Dorsal cephalic.
    - 3. Commissural. Linking with opposite dorsal cephalic.
    - 4. Lateral, branching into
      - 5. Branch to adductors of mandible.
      - 6. Ocular branch to region of eye.
    - 7. Ganglionic. Supplying brain and optic lobe.
    - 8. Antennal. Supplying antennal rudiment.
    - 9. Muscular to internal retractor of flabella, 9*a*, and to external retractor of flabella, 9*b*, towards origins.
  - 10. Ventral cephalic.
    - 11. Prothoracic. In thorax and dividing into
      - 12. Branch to rudiment of leg I.
      - 13. Prothoracic ventral. Ending in thoracic ganglion I.
    - 14. Ganglionic. To sub-oesophageal ganglion.
    - 15. Muscular. To adductor muscles in occipital region.
    - 16. Mandibulo-maxillary, branching to supply
      - 17. Labial rudiment.
      - 18. Maxilla.
      - 19. Mandible.
    - 20. Antennal. To antennal prominence.
    - 21. Labral, with branches to
      - 22. Internal retractor of flabella near insertion, 22*a*, and external retractor ditto, 22*b*.
      - 23. Labral membrane, flabella, etc.
- 24. Common mesothoracic. Leading to tracheal knot and
  - 25. Mesothoracic spiracular (cord to mesothoracic puncta).

26. Mesothoracic dorsal, with branch to rudiment of nymphal trumpet (forming around 25), and supplying shoulder.
27. Mesothoracic ventral, ending in thoracic ganglion 2.
28. Connective, meso- and metathoracic, sending branches to
    29. Rudiment of leg 2 (ventrally).
    30. Wing rudiment (dorsally). Anterior branch, 30*a*; posterior branch, 30*b*.
31. Dorsal-lateral, to thorax laterally.

32. Mesothoracic trunk; narrow portion of trunk linking common cephalic with ballooned metathoracic portion of trunk, giving off medianly
    33. Oesophageal branch, directed anteriorly by side of oesophagus, etc.
    34. Commissural, long thin commissure joining 32 towards posterior ends and giving off
        35. Branches to aorta.

Metathoracic spiracle.

36. Common metathoracic, linking trunk to tracheal knot, with branches
    37. Visceral; from 36 in its course.
    38. Metathoracic spiracular, cord leading to puncta.
    39. Metathoracic dorsal, to dorsum of thorax.
    40. Metathoracic ventral, ending in thoracic ganglion 3.
    41. Connective, metathoracic–abdomen I, giving off
        42. Branch to halter rudiment (dorsally).
        43. Branch to rudiment of leg 3 (ventrally).

Abdominal spiracle I.

44. Spiracular (cord leading to puncta).
45. Visceral. Large trunk supplying caeca and giving off
    46. Short branch dividing into 41 and 47.
47. Connective abdomen, segments I–II. Links up posteriorly with visceral branch of segment II.

Abdominal spiracles II to VII.

48 (II) to 48 (VII). Spiracular cords leading to puncta II–VII.

49 (II) to 49 (VII). Visceral, turning forwards under trunk and receiving connective from segment in front at or near its origin.

50 (II) to 50 (VII). Ventral, passing directly outwards and ventrally and giving origin shortly along its length to

51 (II) to 51 (VII). Connectives passing from near or at base of visceral branch of anterior segment to ventral branch of more posterior segment at some little distance along its course.

52 (II) to 52 (VII). Commissural. Short cords connecting the trunks in each segment medianly towards posterior ends of the trunk segments.

53. Ventral branch to segment VIII. Arising from trunk.
54. Visceral branch to segment VIII. Arising from trunk.
55. Anal segment branch. Large trachea on each side arising in common with 53 and 54 and passing into anal segment where it gives rise to
    56. Branch to eighth segment.
    57. Papillary branch dividing in segment VIII into branches supplying respectively the dorsal and ventral anal papilla of its side.
58. Feltwork of tracheae arising from ventral aspect of the trunks in the eighth segment and supplying terminal chamber of the heart.

(*c*) THE MUSCULAR SYSTEM

The larval musculature, though complicated and consisting of some hundreds of muscular bundles, in general arrangement is still broadly based on the general insect plan. The larval muscles, in contrast to those of the adult, are mostly narrow bands, many minute and thread-like, the only ones that could be described as massive in character being the four large muscles operating the feeding brushes and possibly one or two of the larger muscles in the thorax. Though all, even those of the visceral coats, are of striated type, the striation is usually much less marked than in the adult, and, in contrast also with the adult muscles, those of the larva consist of narrow strands of muscular tissue surrounded by very bulky sarcolemma (see under 'Muscle types', pp. 376–7). The muscular system has not only interest as an anatomical feature, but also is important as a help in tracing the homologies of skeletal parts. Some reference to it will also be required when dealing with the changes occurring during development from the larva through the pupa to the adult. It will be convenient to deal first with the muscles of the head and neck and later with those of the thorax and abdomen.

MUSCLES OF THE HEAD

The muscles of the larval head have been to some extent described by Thompson (1905), and later in more detail by Imms (1908). It seems desirable, however, to recapitulate and supplement these descriptions and for precision to number the head muscles as has been done by Samtleben (1929) for the thorax and abdomen. It would be unwise to attempt, however, to relate the many muscles of the complex head structures to any system such as that given by Samtleben for the abdominal and thorax muscles, and the twenty-four muscles of the head are given numbered consecutively in the following tabular statement (see also Fig. 30 (1), (2)).

TABULAR STATEMENT OF MUSCLES OF THE LARVAL HEAD

Antennal

1. *Antennal muscles.* Very small muscles arising from the anterior arm of the tentorium and inserted into the membrane at the inner side of the base of the antenna. Much larger in *Culex pipiens* where the antenna is larger and in which the muscle takes origin from the posterior arm.

Clypeal

2. *Median retractors of the flabella.* The largest muscles in the head taking origin on each side of the middle line of the dorsum of the head and inserted into the head of the stirrup apodeme. (Inner retractors of the flabellae of Thompson.)
3. *Lateral retractors of the flabella.* Almost equally large muscles arising laterally towards the front of the clypeus internal to the falciform apodeme and inserted into the shaft and point of the stirrup apodeme. (Outer retractors of the flabella of Thompson.)
4. *Elevators of the epipharynx.* Small, but long, twin muscles arising from the middle line of the dorsum of the clypeus between the origins of muscles 13 and inserted into the middle of the post-epipharyngeal bar. (Epipharyngeal muscle of Thompson.)
5. *Lateral retractors of the post-epipharyngeal lobe* (*dorsal lip of the pharynx*). Very small muscles arising from the inner ventral aspect of the falciform apodeme and inserted into membrane near the middle line posterior to the epipharynx.

Figure 30. Muscles of the head.

1 Left side of head viewed from the medial sagittal plane.

2 Left side of head viewed from dorsal aspect as though the cuticle were transparent.

Lettering: *Br*, brain; *md*, mandible; *mx*, maxilla; *Oe*, oesophagus; *P*, pharynx.
Numbers 1–24 are as given in the list of head muscles in the text.
The effect is that given when muscles are viewed by polarised light.

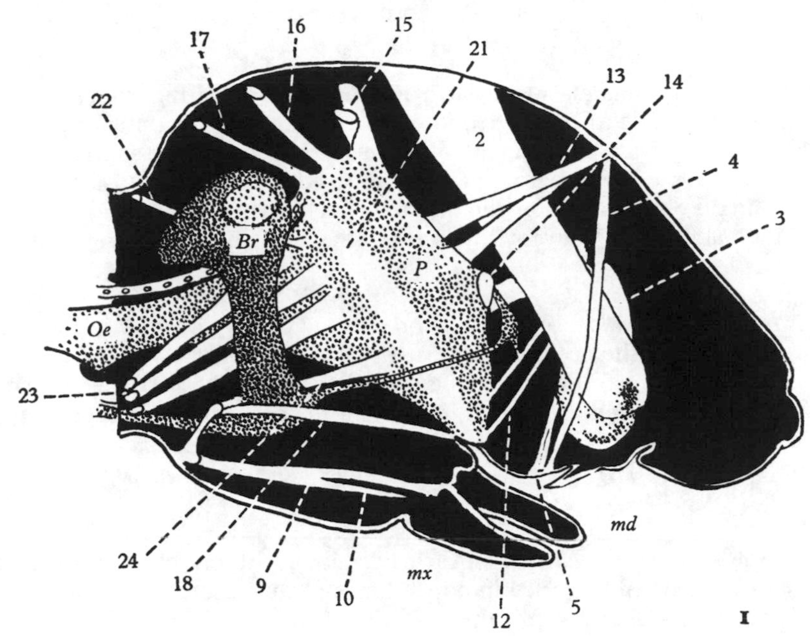
17
16
15
21
13
14
22
2
4
3
Br
P
Oe
23
md
24
18
9
mx
10
12
5
I

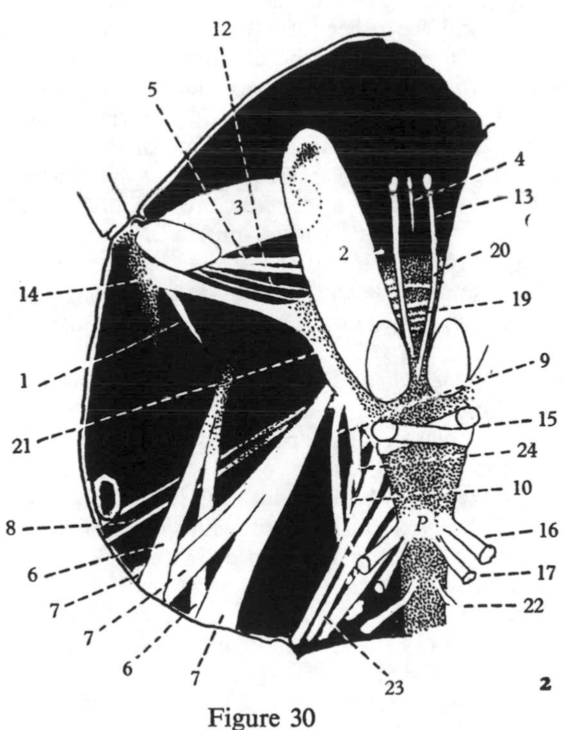
12
5
4
13
3
2
20
14
19
1
9
21
15
24
10
8
P
16
6
17
7
22
7
6
7
23
2

Figure 30

Mandibular

6. *Abductors of the mandibles.* Muscular bands, 6*a* and 6*b*, arising from the lateral occipital region and converging to be inserted into the membrane at the outer angles of the base of the mandibles. (Divaricator muscles of the mandibles of Thompson.)
7. *Adductors of the mandibles.* Large muscles composed of five bands, 7*a*–7*e*, arising separately from the lateral occipital region and inserted by a single tendon at the inner margin of the base of the mandible. (Converger muscles of the mandible of Thompson.)

Maxillary

8. *Depressors of the maxillae.* Twin bands on each side arising from the lateral aspect of the head a little behind the ocelli and inserted along with muscle 9 into a projection on the dorsal border of the base of the maxilla. (Depressors of the maxillae of Thompson.)
9. *Retractors of the maxillae.* Arising from the ventral part of the head capsule at the tentorial pit and inserted along with muscle 8 into the base of the maxilla. (Retractors of the maxillae of Thompson.)

Labial

10. *Retractors of the labium.* Arising on each side along with muscle 9, passing up the side of the labial rudiment and inserted into membrane at the base of the pouch on each side of the hypopharynx. (Depressor of the labium of Thompson.)
11. *Dilators of the salivary opening.* Two minute muscles lying in the hypopharyngeal rudiment and inserted into the dorsal surface of the opening of the salivary duct. (Fibres in hypopharynx of Thompson.)

Pharyngeal

12. *Anterior elevators of the roof of the pharynx.* Two small muscles on each side arising close together and along with muscle 5 on the inner ventral surface of the falciform apodeme and inserted one behind the other into the roof of the pharynx anterior to the frontal ganglion. (Lesser lateral muscles of the pharynx of Thompson.)
13. *Posterior elevators of the roof of the pharynx.* Two small muscles arising on each side of, and close to, muscle 4 in the median line of the clypeus and passing backwards to be inserted one behind the other in the median line of the roof of the pharynx behind the frontal ganglion and the small dorsal compressors of the pharynx. (Elevators of the dorsal plate of the pharynx of Thompson; elevator muscles of Imms.)
14. *Lateral pharyngeal.* Two moderately large muscles arising on each side just external to the origins of the lateral retractors of the flabellae and at the inner side of the falciform apodeme near the antennal base and inserted into the lateral plates of the pharynx at their chief convexity. (Lateral muscles of the pharynx of Thompson; lateral pharyngeal muscles of Imms.)
15. *Decussating muscles of the pharynx.* A small muscle on each side arising from the dorsum of the head just behind the origins of the median flabellar muscles, decussating and inserted into the lateral plate of the opposite side towards its posterior end. (Diagonal muscles of Thompson and of Imms.)
16. *Dorsal retractors of the pharynx.* A small muscle arising on each side of the middle line behind muscle 15 and inserted into the posterior angles of the pharynx. (Retractors of the pharynx of Thompson; retractor pharyngeals of Imms.)
17. *Accessory dorsal retractors of the pharynx.* Very small muscles arising behind muscle 16 and passing in front of the frontal ganglion to be inserted at the posterior angles of the pharynx a little behind and ventral to muscle 16. (Anterior dilators of the oesophagus of Thompson.)
18. *Ventral retractors of the pharynx.* Long thin strands arising from the inner ends of the posterior arms of the tentorium and passing external to the cerebral connectives to be inserted into the ventral lip of the pharynx anterior to the oval plate. (Ventral retractors of the pharynx of Thompson.)

19. *Dorsal compressors of the pharynx.* Four small intrinsic muscles of the pharynx passing from side to side across the middle line of the roof of the pharynx in the region of the frontal ganglion. The two most posterior muscles decussate and blend. (Dorsal muscles of the pharynx of Thompson and of Imms.)
20. *Frontal ganglion muscle.* A curious thin and thread-like, but long, muscle which arches over the pharynx accompanying the connectives of the frontal ganglion.
21. *The cingulum.* Thick muscle bundles situated along the sides of the pharynx just below the crests. They take origin on each side from the posterior angle of the pharynx and are inserted anteriorly at the sides of the lower lip of the pharynx. (Cingulum or girdle muscle of Thompson.)

Oesophageal.*

22. *Dorsal dilators of the oesophagus.* Small muscles arising dorsally on each side of the cervical collar and passing behind the cerebral connectives to be inserted into the wall of the oesophagus dorsally. (Dorsal, or sometimes given as posterior, dilators of the oesophagus of Thompson.)
23. *Lateral dilators of the oesophagus.* A row of muscular bands taking origin from the cervical collar ventrally on each side and passing up through the nerve ring at the sides of the oesophagus to be inserted in a line caudal of the cingulum. (Lateral dilators of the oesophagus of Thompson; lateral dilators of Imms.)
24. *Tentorial dilators of the oesophagus.* Muscles on each side taking much the same course as muscles 23, but taking origin from the inner end of the posterior arm of the tentorium.

The largest muscles of the larval head are the retractors of the feeding brushes. These are not inserted directly into the brushes or even into the main apodemes, but into the stirrups, which are separate apodemes and only indirectly linked to the main apodemes. The action of these muscles in retracting the brushes is easy to understand as they can be seen to draw down and compress together the whole of the labral structures (Fig. 34 *m.*2, *m.*3). This does not explain, however, how these muscles bring about the sweeping movements of these organs. Thompson notes that during the sweeping (flickering) movement of the brushes these muscles remain tense but still. Yet it seems certain that they must be concerned, since there are no other muscles that can be adequate for such a purpose and their great size would be explained if they were required for this, the one most vital action of the larva, namely the action necessary to obtain its food. The next largest muscles are the adductors of the mandibles, each arising by five rather widely separated heads, which along with other gnathal muscles in the occipital region all converge to the mandible and maxilla of their side (Fig. 30 (2) *m.*7).

A surprisingly large proportion of the head muscles relate to the pharynx, which is an extremely mobile organ and one by which again the larva mainly obtains its food, since it functions as the filtering apparatus of the currents brought by the brushes. Its exact orientation and correct participation in this function are probably nicely adjusted (see under feeding). The small muscles 4 and 5 probably also take an essential part in raising the membranous upper lip of the pharyngeal opening when feeding is in progress.

Certain segmental relations may be noted. The minute antennal muscle in *Aëdes aegypti* arises from the anterior arm of the tentorium. In *Culex pipiens* with much larger antenna it is more conspicuous and rises far back from the apex of the tentorial spur. The labral muscles all arise from what has been taken to be the clypeus, 2 and 3 taking origin from the dorsum of the head in the clypeal region and 4 and 5 from the inner aspect of the falci-

* The oesophagus is taken as extending from a line a little caudal of the cingulum muscle and excluding muscles 17 and 18.

form apodeme, that is that part corresponding to the clypeal surface. The mandibular muscles all take origin from the gena or parietal region of the head, the maxillary abductors from the same region. The maxillary retractor (9), however, arises from the front of the tentorial pit, that is from a part of the head capsule which, as already shown, is an extension backwards of the basal parts of the maxilla, the larval maxilla as ordinarily understood being the parts only of this structure distal to the origin of the maxillary palps. The labial parts in the larva are entirely rudimentary, as also is the hypopharynx, and the only muscles connected with these are the retractors of the labium arising from the tentorial pits and the tiny muscles at the salivary opening (no. 11).

MUSCLES OF THE NECK

The muscles of the neck are difficult to study owing to the way this structure is normally retracted and thrown into folds and also to the presence of the muscular diaphragm formed in this region by fibres from the oesophageal coat and other parts (see description of neck). Few muscles appear actually to be proper to this region, though the cervical collar gives origin to muscles passing into the head (notably no. 18) and insertion to muscles of the thoracic series, one of which (given as *c* in Fig. 31 (1)) is inserted into the basal fold of the neck membrane. Some muscles also pass from the thorax through the neck to be inserted into the tentorial spur (*vl* 1′ and *dl* 12′). The most nearly to be regarded as muscles proper to the neck are the two small muscles listed below as nos. 25 and 26 and shown in Fig. 32 (4) as *b* and *b*′.

25. A small muscle arising from the collar dorso-laterally and inserted into the neck membrane laterally.

26. A small muscle arising from the collar laterally along with the muscle given as *a* in the figures and inserted into the neck membrane on its ventro-lateral aspect.

No. 25 decussates with muscle *a* as this passes into the thorax, the decussation lying external to the large tracheal vessels in the neck. There is a further decussation of muscles in this position a little posterior to this where two muscles, *vl* 2′ and *dl* 12′ (see later), decussate at the base of the neck.

Curious features of the larval musculature are the number of long thin muscles and the occurrence not infrequently of twin strands. Two such strands arise in connection with muscle *a*, decussating in the middle line dorsally behind the neck. These are not mentioned by Samtleben and as there is some doubt as to their exact nature they have not been shown in the figures. Two long twin strands also pass through the neck to the outer knot. Again their nature is somewhat obscure.

MUSCLES OF THE THORAX AND ABDOMEN (Figs. 31, 32)

The muscles of the thorax and abdomen have been most minutely described by Samtleben (1929) in *Aëdes meigenensis* and some other species. The following up of these in serial sections is difficult and tedious and they are best studied in whole mounted preparations. Very useful mounts can be made by slicing hardened larvae on a paraffin block as described in the section on technique. See also method of dissection of fresh or fixed material given by Wigglesworth (1942) and by Abul-Nasr (1950, p. 344). The muscles can also be studied in cleared specimens observed under polarised light as described by Imms (1939) or by the method of Kramer (1948). Kramer's method is as follows: fix at 30° C. in Bouin's fluid,

8–10 hours; followed by 50 per cent alcohol, 10 minutes; 70 per cent, 60 minutes; 95 per cent, 10 minutes; 0·5 per cent eosin in 95 per cent alcohol. Return to 95 per cent alcohol and oil of wintergreen added dropwise at hourly intervals for 4–5 days. Transfer to oil of wintergreen. Care is necessary when adding oil of wintergreen to prevent collapse. The effect is much like that obtained with polarised light. By focusing the light and moving the larva about, the muscles stand out in the different depths of the specimen as these are examined. For technique of using polarised light see chapter on technique.

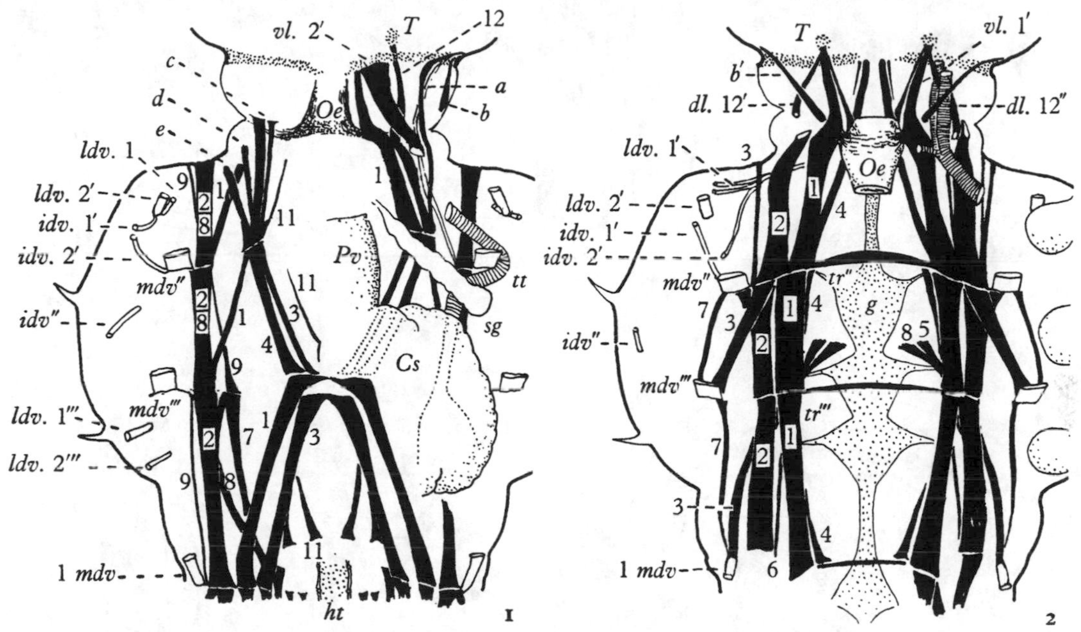

Figure 31. Muscles of the neck and thorax.

1 View of dorsal musculature of neck and thorax of fourth instar larva of *Aëdes aegypti*.

2 View of ventral musculature of neck and thorax of same.

Lettering: *a*, muscle arising along with *b'* (muscle 26) from collar laterally and passing to inner knot. Accompanying it are some thread-like twin muscles passing to the external knot; *b*, muscle 25 of neck; *b'*, muscle 26 of neck; *c*, muscle (double) inserted into neck fold; *Cs*, gastric caeca; *d–e*, muscles from inner knot to dorsal thoracic wall; *g*, thoracic ganglia; *ht*, heart; *Oe*, oesophagus with cervical diaphragm; *Pv*, proventriculus; *sg*, salivary gland; *T*, tentorium; *tt*, tracheal trunk.

On the left-hand side of 1 only the muscles inserted into the lower fold of the neck are shown. On the right only those passing through the neck are depicted. Muscles are numbered according to Samtleben's system (see text).

Samtleben's system of nomenclature enables the many muscles of the larva to be conveniently dealt with and the numbers given by him greatly simplify reference. Passing from the relatively simple arrangement in the abdominal segments this author finds the general plan similar, though more complicated, in the three thoracic segments. He distinguishes (1) a *dorsal longitudinal series*, (2) a *ventral longitudinal series*, and (3) a *dorso-ventral series*. There is also ventrally in each segment a *transverse intersegmental muscle* which takes its notation from the segment lying behind it. Both dorsal and ventral longitudinal muscles normally number up to nine in each quadrant in abdominal segments or somewhat more in some thoracic segments. In the abdomen nos. 1–4 in both dorsal and ventral series are usually relatively large muscles. Nos. 1 and 2 are normally directed outwards and forwards and lie deeper in the body than 3 and 4 which slant in an opposite direction. The arrangement, however, is less regular in the thorax. The remaining numbers are usually small or even quite thread-like. The dorso-ventral series are distinguished as (1) *median dorso-ventral*, of which there may be three, nos. 1–3 from within outwards, towards the front of the

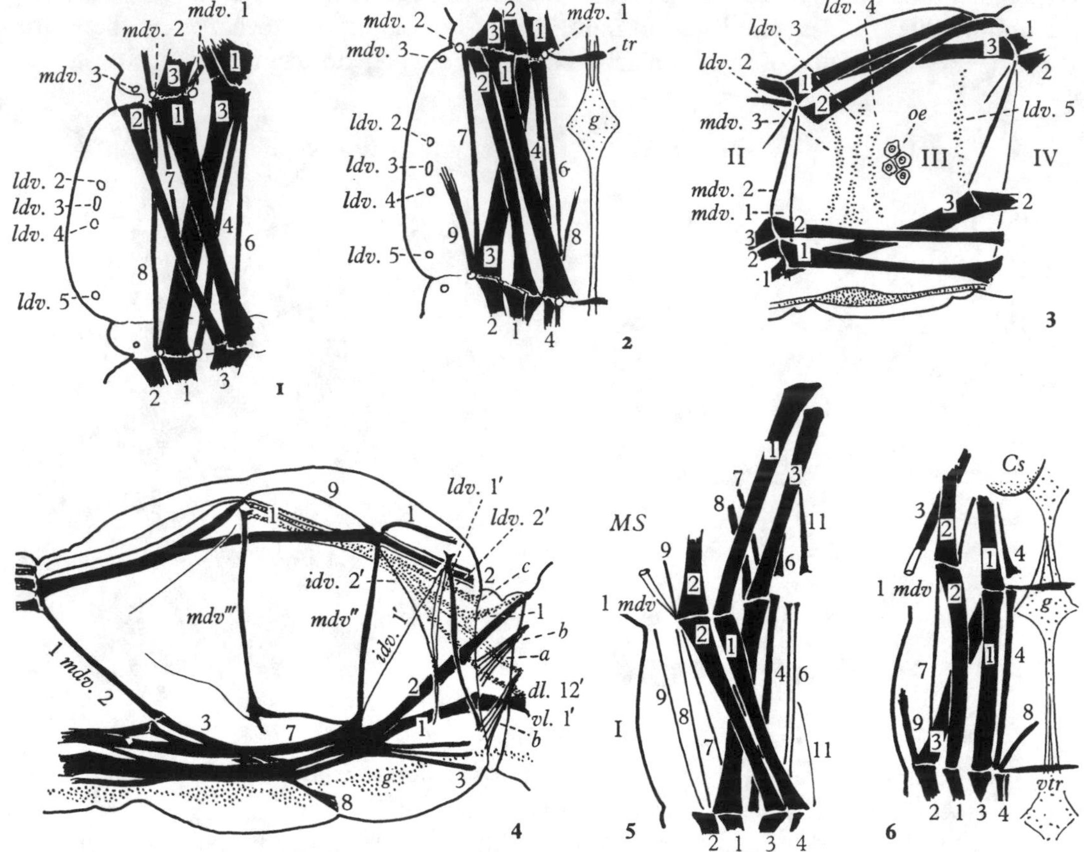

Figure 32. Muscles of thorax and abdomen.

1 One side of third abdominal segment showing dorsal musculature as seen from inside.

2 The same showing ventral musculature.

3 Lateral view of the same as seen from outside, showing also the dorso-ventral series and position of large oenocytes.

4 Lateral view of musculature of neck and thorax as seen from outside.

5, 6 Dorsal and ventral musculature of first abdominal segment with adjoining muscles of metathorax to show continuation of series.

Lettering: *a*, *b*, *b′*, *c*, as in Fig. 31; *Cs*, gastric caeca; *g*, ganglia of ventral chain; *oe*, large oenocytes; *MS*, mesothorax; *vtr*, transverse muscle of second abdominal segment crossing in front of ganglion of segment; I–IV, abdominal segments.

Muscles are numbered in accordance with Samtleben's system, the dorsal and ventral longitudinal series being indicated by their numbers only.

segment (Fig. 32 (1), (2), (3)); and (2) *lateral dorso-ventral*, of which there may be five, nos. 2–4 situated laterally about the middle of the segment, and no. 5 towards its posterior end, no. 1 being present only in the first abdominal segment; and (3) *intersegmental dorso-ventral* (in thorax only) where the muscle, usually oblique, passes dorso-ventrally from one segment to the next (Fig. 32 (4)). Abdominal segments II–VII, except for a few small details, have the typical musculature as shown in Fig. 32 (1), (2). Segment I, as also segment VIII, conform to the general plan, but show some differences (see Fig. 32 (5), (6)). Roughly speaking the arrangement as regards the main classes holds for the thorax.

By using the shortened forms *dlm* and *vlm* respectively for the dorsal and ventral longitudinal series, *mdv* and *ldv* for the median and lateral dorso-ventral and other appropriate lettering preceded by the number of the segment, any desired muscle can be indicated. To shorten such designations as much as possible I shall, when referring to muscles in this way, omit the letter *m* indicating muscle. The following therefore will be the abbreviations used in the figures and text:

dl dorsal longitudinal;
vl ventral longitudinal;
mdv median dorso-ventral;
ldv lateral dorso-ventral;
idv intersegmental dorso-ventral;
vtr transverse intersegmental.

Samtleben also gives the muscles names, for example *musculus pronoti secundus* (of the dorsal longitudinal series) for I *dl* 2, and *musculus dorsoventralis intersegmentalis mesothoracis* for II *idv*. The shortened form, however, appears to be all that is necessary here.

After working for some time on the musculature of *Aëdes aegypti* larva I have been unable to note any significant departure from Samtleben's description for *A. meigenensis*. Space does not permit of a detailed description and all that can be done in this respect has been to give figures for the appearance of the muscles in a typical abdominal segment and in the thorax. Where no doubt exists as to the number of a muscle in Samtleben's system, it is so numbered in the figures, the segment to which it relates being left to the reader to supply. Where, as in the prothorax, it has not always been possible to identify a muscle in this way it has been indicated by a letter with a reference in the legend. Where two numbers are given to the same muscle, the higher number or numbers refer to smaller muscles hidden under the larger muscle shown. Figures giving dorsal and ventral views show the muscles as they are seen in a bisected larva mounted and viewed from inside. In the lateral views, the better to show relations, muscles are shown as though viewed from outside.

A few general remarks are called for. The longitudinal series take origin and are inserted at the intersegmental lines, the cuticle at such situations usually being to some extent indented or even folded inwards. The muscles, however, are commonly continued with but little interruption into muscles of the adjoining segment, being mutually inserted and taking origin largely from each other. In the thorax muscles tend to arise in groups from what may be termed 'knots', much as in the tracheal system. Such 'knots' for a given segment may not always be at quite the same level, for example in the prothorax (Fig. 32 (4)) where those from an outer knot are shown in full black and those from an inner knot by dotting.

In addition to muscles proper to the larva there are, in the fourth instar and especially nearing pupation, the developing imaginal muscles. These newly forming muscles are very conspicuous owing to their numerous nuclei and dark staining which makes them quite distinct from the pale larval muscles. They relate almost entirely to the future wing and leg muscles and are restricted to the thorax. Some of the larval muscles may, however, undergo a form of reconstruction with accumulation of nuclei which makes them conspicuous. Both forms of muscle will be dealt with later when describing changes in metamorphosis.

Little can be said regarding functions of this complicated musculature. Synchronised action by the abdominal series could obviously result in the sculling-like movements of the abdomen so characteristic of the larva when swimming. Of the thoracic muscles a certain number in the prothorax are concerned with movements of the head and one at least (*c*) is inserted into the basal fold of the neck. It seems probable that to a large extent the thoracic muscles may be concerned mainly in maintaining the correct shape of the parts. It is of interest that not a single muscle appears to be directly in connection with the large lateral hairs of the thorax or abdomen.

MUSCLES OF THE EIGHTH AND ANAL SEGMENTS (Fig. 33)

Owing to changes in the general structure of the parts due to the presence of the siphon and other features, application of Samtleben's numbers to the different bands is, except in a general way, difficult. Beyond noting that the musculature is abnormal Samtleben does not include these segments in his survey. They have been described, however, to some extent by me in *Anopheles* (Christophers, 1922, 1923) and also by Imms (1908) in that genus. In *Aëdes aegypti* a notable feature is the number and complexity of the muscles present. In all there are some thirty-five pairs of muscles. Space does not permit of a detailed description, but it seems desirable that they should at least be indicated and the attached list and figures with the numbered muscles will enable them to be identified.

There are represented in the terminal parts almost certainly at least three and possibly four segments, namely segments VIII, IX, X and XI. Segment VIII appears to be mainly that segment. It possesses a dorsal and a ventral series of muscles as in previous segments. The dorsal series is represented by three small muscles, the siphon muscles *a*, *b* and *c* (nos. 1–3), and one large muscle (no. 4). The siphon muscles noted may be regarded as eighth segment muscles of the dorsal series carried up to intersegmental line VIII–IX at the anterior border of the stigmatic plate (ninth tergite). Muscle 4, the largest muscle in these parts, passes from intersegmental line VII–VIII external to the siphon muscles to be inserted at the posterior border of segment VIII into a point situated mid-laterally where a number of other muscles are also inserted or take origin. For convenience I have termed this *dorsal knot VIII*. Except that in the dorsal series there is a reduced number of muscles as compared with preceding segments, there is no very marked change from the normal arrangement other than the carrying up of the siphon muscles into the siphon.

The ventral series consists of four muscles on each side (nos. 5–8) arising as in previous segments at intersegmental line VII–VIII ventro-laterally. Instead, however, of all

---

Figure 33. Muscles and other structures of terminal segments.

1 Lateral view of terminal segments showing musculature as seen from outside.

2 The same of the eighth abdominal segment viewed from the medial sagittal plane, showing especially the tracheal plexus and the muscles 34, 35 associated with this.

Lettering: *a*, dorsal anal papilla; *a′*, ventral anal papilla; *b*, tracheal plexus; *Co*, colon; *ht*, terminal chamber of heart; *k*, lateral and dorsal muscular knots; *s*, siphon; *tt*, tracheal trunk.

Muscles are numbered in accordance with list of muscles of the terminal segments given in the text. The appearance is that given when muscles are viewed by polarised light.

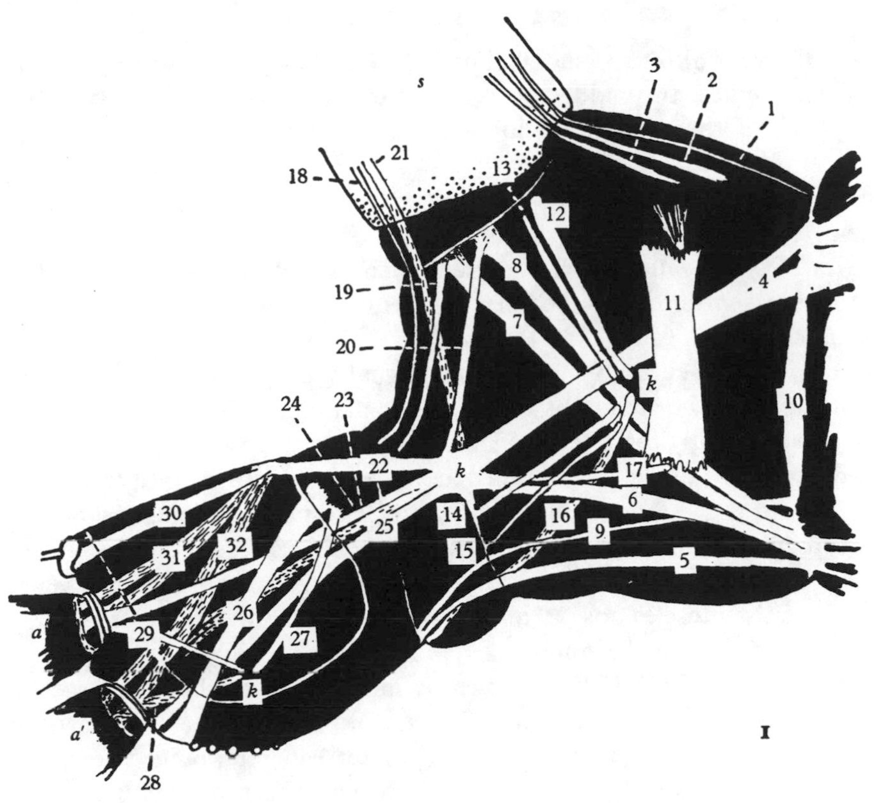
s
3
2
1
21
18
13
12
8
19
4
11
7
20
k
10
24
23
22
k
17
30
6
25
14
16
9
31
32
15
5
26
a
29
27
k
a'
28
I

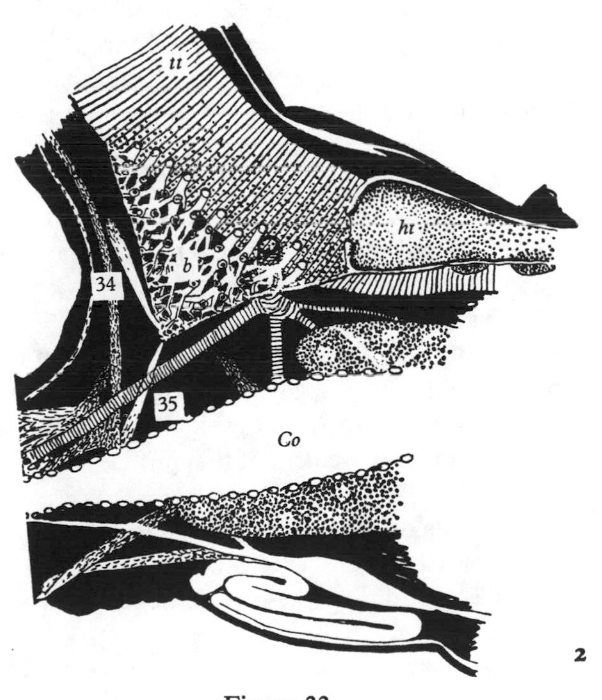
tt
ht
b
34
35
Co
2

Figure 33

proceeding to the ventral intersegmental line VIII–IX, they fan out widely to be inserted at points ranging from near the middle line ventrally to the dorsal aspect of the segment along the lower border of the siphon. In doing so they pass beneath muscle 4 of the dorsal series.

Besides the dorsal and ventral series there are two muscles of the dorso-ventral series, which appear to be respectively *ldv*. 2 and *ldv*. 3 (nos. 10 and 11). In addition there is a peculiar series of small muscles arising laterally from the segment and also fanning out from near the mid-ventral line to the dorsal aspect of the segment. I have entered these in the table as *lateral series* (nos. 12–17).

Segment IX is represented by tergal and sternal parts. The tergal parts are largely carried to the apex of the siphon as the stigmatic plate. The whole siphon from base to apex appears to be mainly ninth tergite which in the later stages in the hypoderm is slit by developing fissures into the lateral and median parts of the future ninth tergite of the pupa and adult. As with the anterior siphon muscles the two posterior muscles *d* and *e* (nos. 18 and 21) are drawn up to the stigmatic plate. These should represent the more median tergal series IX–X, though there is some indication that no. 21 may be from a ventral series. Besides these are the muscles nos. 19 and 20 which may be of this series. They pass from behind towards the edge of the siphon.

The sternal parts of segment IX as shown by the muscle attachments largely remain in their normal situation. As shown by muscles nos. 5, 9 and 16, the anterior border of the ninth sternite lies posterior to a rather prominent intersegmental bulged portion and is not at the same level as the line of dorsal muscle insertions. This area is indicated by both the male and female genital rudiments.

The quite numerous muscles of the anal segment belong to the dorsal series. There appears to be no ventral series. They take origin mainly from the dorsal knot and pass to the bases of the anal papillae (external aspect) and ventral fan area. Others arise from the saddle plate a little distance posterior to its anterior border. These are muscles 30, 31 and 32. No. 30 passes to the sides of the dorsal hair plaque. Nos. 31 and 32, each formed of twin bands, are inserted into the bases respectively of the dorsal and ventral anal papilla of its side on the internal aspect close to the anal margin. These would appear to be muscles pertaining to that part of the anal segment that later forms the cerci, that is pertaining to tergite XI. There is also a small lateral series. These arise from a point marked *k* in Fig. 33 (1) near the ventral edge of the saddle plate.

Apart from other points of interest the remarkable distortion of the parts indicated by the arrangement of the ventral series of segment VIII gives strong support to the view that the terminal spiracles of the larva are those of the eighth segment drawn up in some fashion behind this segment. Besides mere movement of the spiracles there is evidently some distortion of the hinder parts of the segment.

Mention should here be made of the curious little muscles 34 and 35 which pass into the capillary plexus of tracheoles behind the terminal chamber of the heart (Fig. 33 (2); 36 (5)). For a further description of these and related parts see under circulatory system.

Belonging to the muscular system are two alary muscles. These are the two most posterior of the alary series, being situated at the points corresponding to the two pairs of pericardial cells present at the anterior end of the eighth segment. The origin of one of these, the more posterior, is shown in Fig. 33 (1) just above the dorso-ventral muscle 11.

MUSCLES OF THE EIGHTH AND ANAL SEGMENTS

Segment VIII.

Dorsal series.

1–3. Respectively from within outwards siphon muscles *c*, *a*, *b*, arising from tergal area on each side of median line towards anterior end of the segment and passing into siphon anteriorly (see under siphon).

4. Largest muscle of segment. Arises dorso-laterally at intersegmental line VII–VIII external to 3 and inserted into the dorsal knot VIII (see text).

Ventral series.

5–8. Strong bands arising from intersegmental line VII–VIII on each side of middle line ventrally in order given from within outwards, and inserted:

5. Near median line into base of sternite IX anterior to rudiments of caecus and genital buds in female and male respectively.
6. Into dorsal knot VIII.
7. On line of thickening below edge of base of siphon about level of pecten spines.
8. As 7, but about level of middle of siphon.

9. Smaller muscle arising from the intersegmental line VII–VIII external to 8 and inserted just dorsal to insertion of 5.

Dorso-ventral series.

10. Muscle at anterior border of segment, possibly *ldv*. 2.
11. Very broad short muscle about middle of segment corresponding to *ldv*. 3 of previous segments.

Lateral series (see text).

12, 13. Small bands arising from lateral knot VIII, running parallel and inserted anterior to insertion of 8.

14–16. Small bands arising close together from lateral knot VIII and inserted:

14. Intersegmental line a little ventral to dorsal knot VIII.
15. As 14 but a little ventral.
16. Crossing beyond line between segment VIII and intersegmental bulge to be inserted external to 5 after passing ventral to that muscle and 9.

17. Arising a little ventral to lateral knot and inserted into dorsal knot VIII after passing internal to 14 and 15.

Segment IX (tergal series only represented: see text).

18. Arising from the intersegmental area between segment VIII and saddle plate and passing into siphon as the small siphon muscle.
19. Arising external to 18, but after short passage inserted below basal margin of siphon internal to insertion of 7.
20. Arising from dorsal knot VIII and inserted anterior to 19 near insertion of 8.
21. Siphon muscle *d*. Arises internal to and ventral to dorsal knot VIII and passes into siphon along with 18 posteriorly.

Anal segment.

22–5. Arising from dorsal knot VIII and inserted:

22. Near anterior border of saddle plate a little lateral to median line and muscles 32.
23. Base of dorsal anal papilla of its side externally.
24. Into sides of rectum towards its termination.
25. Side of ventral fan posteriorly and near base of ventral anal papilla of its side.

26–7. Arising just anterior to anterior edge of saddle plate and inserted:

26. Side of ventral fan posteriorly.
27. Small muscle arising near 26 and inserted into the lateral knot anal segment (see text).

28–9. Arising lateral knot anal segment and inserted:
  28. Base of dorsal anal papilla of its side externally.
  29. Base of ventral anal papilla of its side internally.

30–2. Arising close to middle line from saddle plate dorsally and inserted:
  30. Outer side of dorsal hair plaque.
  31. Inner side of base of ventral anal papilla of its side. Twin bundles.
  32. Inner side of base of dorsal anal papilla of its side. Twin bundles.

In addition there are the alary muscles (33, 33′) (see Fig. 36 (8) *alm*) and the small muscles passing into the capillary tracheal complex posterior to the terminal chamber of the heart (34, 35) (see Fig. 33 (2)).

REFERENCES

(*a*) THE ALIMENTARY CANAL

CHRISTOPHERS, S. R. and PURI, I. M. (1929). Why do *Anopheles* larvae feed at the surface and how? *Trans. Far East. Ass. Trop. Med. 7th Congr. India 1927*, **2**, 736–8.

DE BOISSEZON, P. (1930). Contribution à l'étude de la biologie et de l'histophysiologie de *Culex pipiens*. *Arch. Zool. Exp. Gen.* **70**, 281–431.

FEDERICI, E. (1922). Lo stomaco della larva di *Anopheles claviger* Fabr. *R. C. Accad. Lincei* (5), **31**, 264–8, 394–7.

HURST, C. H. (1890). The post-embryonic development of a gnat (*Culex*). *Proc. Trans. Liverp. Biol. Soc.* Reprinted in vol. II of *Studies from the Biological Laboratories of the Owens College*. Guardian Press, Manchester.

IMMS, A. D. (1907). On the larval and pupal stages of *Anopheles maculipennis* Meig. *J. Hyg., Camb.*, **7**, 291–316.

KULAGIN, N. (1905). Der Kopfbau bei *Culex* und *Anopheles*. *Z. wiss. Zool.* **83**, 285–335.

MARTINI, E. (1929). Culicidae. In Lindner's *Die Fliegen der palaearktischen Region.* Stuttgart. Part XI, 74–85.

METALNIKOFF, S. (1902). Beiträge zur Anatomie und Physiologie der Mückenlarve. *Bull. Acad. Sci. St-Pétersb.* **17**, 49–58.

NUTTALL, G. H. F. and SHIPLEY, A. E. (1901). Studies in relation to malaria. *J. Hyg., Camb.*, **1**, 51–73.

RASCHKE, E. W. (1887). Die Larve von *Culex nemorosus*. *Arch. Naturgesch.* Jahrb. 53, **1**, 133–63.

SALEM, H. H. (1931). Some observations on the structure of the mouth-parts and fore-intestine of the fourth stage larva of *Aëdes (Stegomyia) fasciata*. *Ann. Trop. Med. Parasit.* **25**, 393–419.

SAMTLEBEN, B. (1929). Zur Kenntnis der Histologie und Metamorphose des Mitteldarms der Stechmückenlarven. *Zool. Anz.* **81**, 97–109.

SCHONICHEN, W. (1921). *Prakticum der Insektenkunde*. Ed. 2. Gustav Fischer, Jena.

SNODGRASS, R. E. (1935). *The Principles of Insect Morphology*. McGraw-Hill Book Co., New York and London.

THOMPSON, M. T. (1905). The alimentary canal of the mosquito. *Proc. Boston Soc. Nat. Hist.* **32**, 145–202.

WESCHE, W. (1910). On the larval and pupal stages of West African Culicidae. *Bull. Ent. Res.* **1**, 7–50.

WIGGLESWORTH, V. B. (1930). The formation of the peritrophic membrane in insects with special reference to the larvae of mosquitoes. *Quart. J. Micr. Sci.* **73**, 593–616.

WIGGLESWORTH, V. B. (1931*b*). See under (*b*).

WIGGLESWORTH, V. B. (1942). The storage of protein, fat, glycogen and uric acid in the fat-body and other tissues of mosquitoes. *J. Exp. Biol.* **19**, 56–77.

(*b*) THE TRACHEAL SYSTEM

GILES, G. M. (1902). *A Handbook of Gnats or Mosquitoes*. Ed. 2. John Bale Sons and Danielsson, London.

HURST, C. M. (1890). On the life history and development of a gnat (*Culex*). *Ann. Rep. Trans. Manchester Micr. Soc.* Guardian Press, Manchester.

KEILIN, D. (1944). Respiratory systems and respiratory adaptations in larvae and pupae of Diptera. *Parasitology*, **36**, 1–68.

LANDA, V. (1948). Contributions to the anatomy of Ephemerid larvae. I. Topography and anatomy of the tracheal system. *Mém. Soc. Zool. tchécosl.* no. 12.

SNODGRASS, R. E. See under section (*a*).

WIGGLESWORTH, V. B. (1930). A theory of tracheal respiration in insects. *Proc. R. Soc.* B, **106**, 229–50.

WIGGLESWORTH, V. B. (1931*a*). The respiration of insects. *Biol. Rev.* **6**, 181–220.

WIGGLESWORTH, V. B. (1931*b*). The extent of air in the tracheoles of some terrestrial insects. *Proc. R. Soc.* B, **109**, 354–9.

WIGGLESWORTH, V. B. (1938). The absorption of fluid from the tracheal system of mosquito larvae at hatching and moulting. *J. Exp. Biol.* **15**, 248–54.

WIGGLESWORTH, V. B. (1950). A new method for injecting the tracheae and tracheoles of insects. *Quart. J. Micr. Sci.* **91**, 217–23.

(*c*) THE MUSCULAR SYSTEM

ABUL-NASR, S. E. (1950). See references in ch. XIV (i), p. 354.

CHRISTOPHERS, S. R. (1922). The development and structure of the terminal abdominal segments and hypopygium of the mosquito with observations on the homologies of the terminal segments of the larva. *Ind. J. Med. Res.* **10**, 530–72.

CHRISTOPHERS, S. R. (1923). The structure and development of the female genital organs and hypopygium of the mosquito. *Ind. J. Med. Res.* **10**, 698–719.

IMMS, A. D. (1908). On the larval and pupal stages of *Anopheles maculipennis*. *Parasitology*, **1**, 103–32.

IMMS, A. D. (1939). On the antennal musculature in insects and other arthropods. *Quart. J. Micr. Sci.* **81**, 274.

KRAMER, S. (1948). A staining procedure for the study of insect musculature. *Science*, **108**, 141–2.

SAMTLEBEN, B. (1929). Anatomie und Histologie der Abdominal- und Thoraxmuskulatur von Stechmückenlarven. *Z. wiss. Zool.* **134**, 179–269.

THOMPSON, M. T. (1905). See under section (*a*).

WIGGLESWORTH, V. B. (1942). See under section (*a*).

XIV

THE LARVA: INTERNAL STRUCTURE (cont.)

(*d*) THE NERVOUS SYSTEM

The nervous system of the larva consists of the *brain*, the *ventral nerve cord*, certain accessory ganglia belonging to the *stomogastric* or *visceral system* and nerves issuing from or connecting these structures. Also coming under this head are the *special sense organs*, notably the ocelli and chordotonal organs.

THE BRAIN (Fig. 34 (1), (2))

Apart from the general description of the brain of *Anopheles* larva by Imms (1908) little has been recorded of this organ in the mosquito larva or adult. For an account of what is known of the structure of the insect brain Snodgrass (1935), Imms (1938) and Wigglesworth (1942) should be consulted together with references given by these authors. The brain or supra-oesophageal ganglion lies in the posterior portion of the head cavity, lying entirely behind the level of the ocelli and forming a narrow band stretching almost across the head. It consists of two lateral masses, the *cerebral lobes*, linked across the middle line by a narrow commissure, *cerebral commissure*. It is also linked with the first ganglion of the ventral nerve cord or *sub-oesophageal ganglion* (*g*. 1) on each side by a connective, *circum-oesophageal connective*, passing from each cerebral lobe around the oesophagus. In structure it consists of an outer cellular cortical layer and a light staining medullary portion of nerve branchings, *neuropile*, which, besides occupying the central portions of the

---

Figure 34. Brain and ventral nerve cord.

1 Dorsal view of head of fourth instar larva showing the pharynx (shaded) and chief muscles with the position and appearance of the brain.

2 Anterior view of right half of the brain and sub-oesophageal ganglion of fourth instar larva with the position of issuing nerves. Reconstruction.

3 Dorsal view of the thoracic ganglia showing position of large nerve cells.

4 Ventral view of the same.

5 Transverse section of thoracic ganglion.

6 The same of an abdominal ganglion.

Lettering: *a*, protocerebrum; *a′*, neuropile of protocerebrum; *b*, deutocerebrum; *c*, tritocerebrum; *d*, cerebral commissure; *an.r*, germinal bud of antenna; *fg*, frontal ganglion; *g*, sub-oesophageal ganglion; *gc*, ganglionic connectives; *gc′*, intra-ganglionic connective tracts; *m*.2, median retractor of flabellum; *m*.3, lateral retractor of flabellum; *m*.24, tentorial dilator muscle of the oesophagus; *n*, issuing nerve; *nc*, large nerve cells; *ne*, nucleated sheath of ganglion; *np*, neuropile; *O*, optic lobe; *o′*, *o″*, *o‴*, neuropile of optic chiasma; *P*, pharynx; *snc*, ground mass of small nerve cells; *t*.2, dorsal cephalic trachea; *t*.10, ventral cephalic trachea; 1–8, cerebral nerves as given in the list of head nerves in the text.

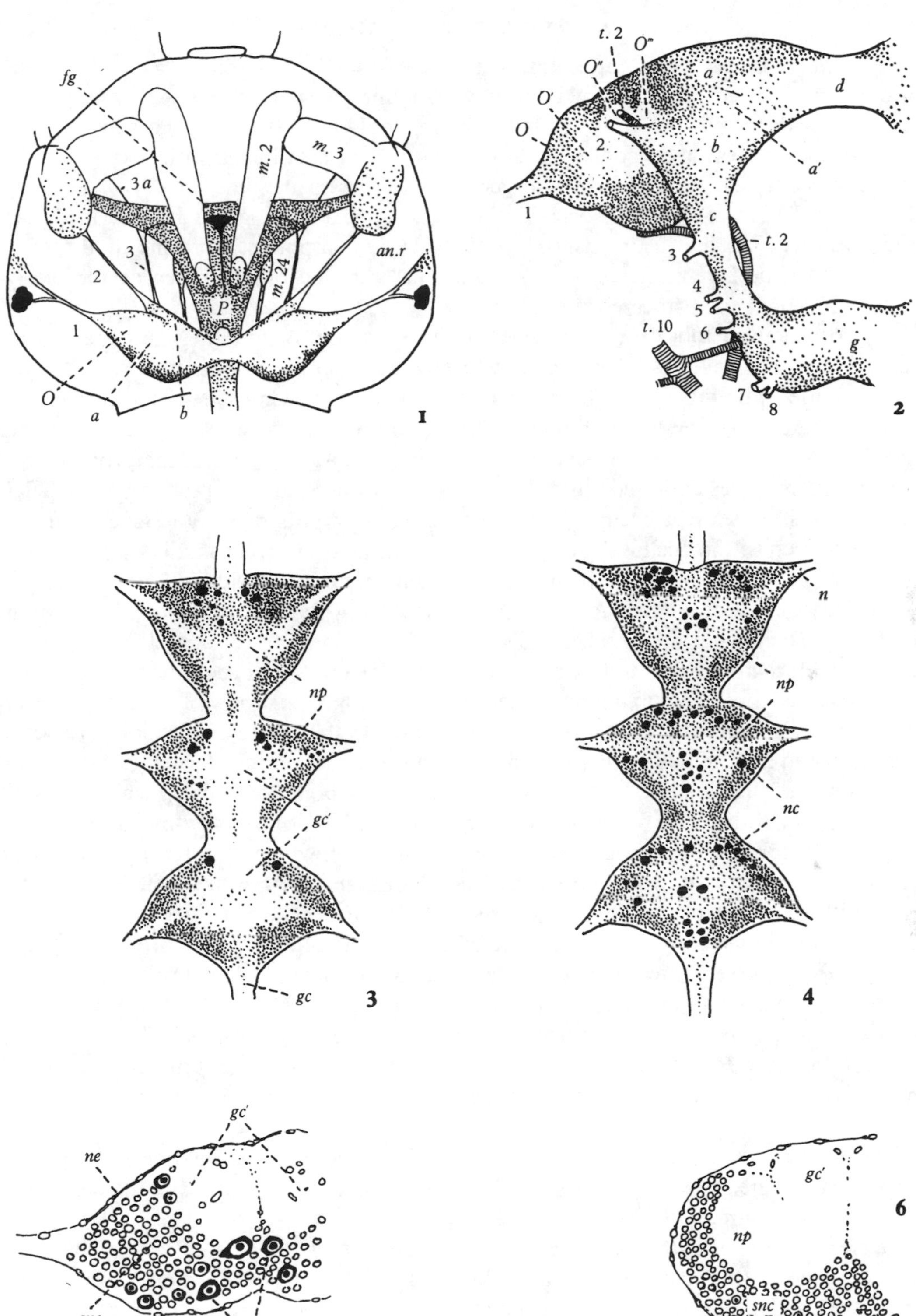
fg
3 a
m. 2
m. 3
3
2
1
m. 24
an.r
P
O
a
b
I
t. 2
O'''
O''
O'
O
a
d
a'
2
b
1
c
t. 2
3
4
5
6
t. 10
g
7
8
2
n
np
np
gc'
nc
gc
3
4
gc'
ne
snc
nc
5
gc'
np
snc
6

Figure 34

different lobes, forms tracts connecting the different parts of the brain as well as being continued into the commissure and connectives and into the nerves issuing from the brain. Compared with that of the adult the larval brain would appear to be relatively simple in structure, the cortical layer consisting almost entirely of small cells with deeply staining nuclei and scanty cytoplasm, though larger cells are present in some areas.

From its outer form, and especially from the appearances given by the medullary neuropile, there can be distinguished the three basic component portions of the insect brain, namely the fore-brain or protocerebrum (*a*), the mid-brain or deutocerebrum (*b*) and the hind-brain or tritocerebrum (*c*).

The protocerebral lobes form the median and hind portions of the main brain mass. Internally they are linked across the middle line by the cerebral commissure (*d*). Externally they carry the optic lobes (*o*). The commissure is composed almost entirely of neuropile and is quite narrow. There are present within it, however, some small regularly arranged nuclei. The optic lobes are pear-shaped with their bases abutting upon the protocerebral lobes and their apices continued into the optic tracts leading to the optic nerves (*l*). These pass outwards to the areas occupied by the developing compound eyes and give off in their course a branch to the ocellus. The optic lobes are at first relatively small, when the compound eyes have only begun to form, but later they increase considerably in size so that they extend outwards almost to meet the eyes fanning out to the area of developing ommatidia. The protocerebral lobes show a cortex composed of very densely packed small cells. There does not appear to be any appearance of distinct mushroom bodies or *corpora pedunculata* as described in many insect brains, though portions of surface showing aggregated cells referred to later may represent these. In the centre of each lobe is a somewhat lobulated area of neuropile (a') and between this and a more externally situated area of neuropile (o') in the optic lobe are two masses of neuropile (o'' and o''') which possibly represent parts of the optic chiasma. Arising from the posterior and ventral aspect of each protocerebral lobe is the curious small nerve (nerve no. 9) which may represent the tegumentary nerve described in some insect brains. There is no obvious neuropile extension to this as with the other nerves, but some large cells are present near its origin.

The *deutocerebrum* consists on each side of a lobe lying somewhat ventral to the protocerebral lobe. It is separated from the latter by a deep groove in which the dorsal cephalic tracheal trunk passes in its course to the dorsal region of the head. The deutocerebrum is produced into a conical projection from which the *antennal nerve* (2) arises. The larger cells previously referred to are chiefly present scattered over the surface of this lobe and neighbouring parts of the tritocerebrum. Their number seems to vary considerably in different preparations.

The *tritocerebrum* is an inconspicuous portion of the brain lying at the root of the circum-oesophageal connective. It gives rise to the nerve on each side proceeding to the frontal ganglion and labrum, *labrofrontal nerve* (3).

The *circum-oesophageal connectives* are stout structures which, in addition to neuropile, contain on their external aspects a fair thickness of cellular cortical substance. From near their lower end there arise from a common origin the *accessory mandibular nerve* (4) and *mandibular nerve* (5). Where they terminate in the cornua of the sub-oesophageal ganglion the *maxillary nerve* (6) takes origin.

The *sub-oesophageal* ganglion is a heart-shaped structure with its lateral angles drawn out to merge into the connectives. Posteriorly it is linked to the first thoracic ganglion by long connectives, *interganglionic connectives.* From the ventral aspect of its lateral angles there takes origin on each side a stout nerve passing to the labial rudiment, *labial* nerve (7) and from almost the same spot but passing more ventrally is a nerve, *hypostomal nerve* (8) passing to an enlargement near the cuticle just internal to the hypostomal suture.

The *frontal ganglion* (*fg*) is the chief ganglion of the stomogastric system. It lies in the median line above the anterior portion of the roof of the pharynx. It is pear-shaped with its narrow end pointing posteriorly. At its lateral angles it receives the frontal connectives. Posteriorly it gives off the *recurrent pharyngeal nerve* (10) which passes backwards in the middle line over the roof of the pharynx to the notch at the posterior end of this organ and beyond connecting up with the stomogastric system. A small nerve is also given off from the ganglion anteriorly passing forwards to the region of the upper lip of the pharynx.

In the accompanying table (Table 20) the nerves arising from the brain and sub-oesophageal ganglion are named and given numbers with a brief statement of their origin and distribution.

Table 20. *Tabular statement of nerves of the head*

No.	Name	Origin	Course
n. 1	Optic tract	Optic lobe	Fans out to area of developing imaginal eye
n. 2	Antennal	Deutocerebrum	Passes forwards to enter base of antennal rudiment
n. 3	Labrofrontal	Base of circum-oesophageal commissure (tritocerebrum)	Passes forwards by side of pharynx giving branch to 4 and dividing into 3*a* and 3*b*
n. 3*a*	Labral	Branch of 3 at crest of pharynx	Continues forwards towards falciform apodeme and passes beneath lateral retractor of flabella to clypeal and labral region
n. 3*b*	Frontal connective	Branch of 3 turning inwards over crest	Over roof of pharynx to lateral angle of pharynx
n. 4	Accessory mandibular	Junction of circum-oesophageal connective with cornua of sub-oesophageal ganglion in common with 5	Sends branch to plexus joining up with branch from 4 and continues to angle of mouth. Gives off 4*a*
n. 4*a*	Infra-orbital	Branch of 4 near origin	Passes ventral to ventral cephalic trachea to end in thickening close to cuticle in region of infra-orbital hairs. Supplies integument and adductor muscles
n. 5	Mandibular	Takes origin in common with 4 just above cornua of sub-oesophageal ganglion	Forms thickened trunk. Enters mandible at inner angle of base proximal to molar lobe
n. 6	Maxillary	Junction of connective with cornua of sub-oesophageal ganglion	Enters maxilla giving off branch to basal part of maxillary rudiment just external to hypostomal suture
n. 7	Labial	Takes origin from ventral surface of sub-oesophageal ganglion towards base of cornua	Divides into two branches, one passing ventral to labial rudiment to lip of invagination, the other passing dorsally to enter base of rudiment
n. 8	Hypostomal	Arises in common with 7, but passes external and ventral to this	Passes to hypodermis just internal to the hypostomal suture about its middle where it forms an enlargement
n. 9	Tegumentary	Takes origin from posterolateral aspect of brain	Crossing dorsal cephalic trachea has enlargement giving off branch to dorsum of occiput. Can be traced to region of tentorial pit
n. 10	Recurrent pharyngeal	Hinder end of frontal ganglion	Passes backwards over roof of pharynx to notch and beyond

The *ventral nerve chain* consists, besides the sub-oesophageal ganglion, which may be considered the first ganglion of the chain, of eleven further ganglia, namely three thoracic corresponding to the three thoracic segments and eight abdominal ganglia corresponding to the abdominal segments one to eight. Each ganglion is double consisting of two closely compacted lateral ganglia. The connectives linking the sub-oesophageal and first thoracic ganglia are very long whilst the three thoracic ganglia lie close together with very short connectives. The abdominal ganglia lie towards the anterior end of their respective segments, except the first which as development proceeds becomes drawn towards the metathoracic ganglion with which in the pupa it is fused. The ganglion of the eighth segment does not in the larva differ noticeably in the male or female and gives no evidence of composite nature.

The ganglia of the chain, like the brain, have a cellular cortical layer and medullary neuropile. In the cortex of the thoracic ganglia there are, in addition to the small cells resembling those of the brain, many larger cells, some of these very large and conspicuous in sections. These latter are especially numerous in certain areas of the ganglion, see Fig. 34 (3), (4), and appear to be large unipolar cells with their pointed end directed inwards towards the neuropile. The neuropile consists of two central tracts continuing the connectives and in each lateral half a central neuropile mass proper to the ganglion. The abdominal ganglia are smaller and possess few or none of the very large cells seen in the thoracic ganglia. They show two lateral masses of neuropile proper to each half of the ganglion, the connectives appearing to be continued into or abutting upon this without forming through tracts (*gc′*).

From both thoracic and abdominal ganglia nerves issue on each side from about the middle point of the ganglion, except in the case of the eighth segment where the issuing nerves arise posteriorly. The distribution of these can only here be very generally indicated, since a special study would be necessary for such a purpose and for further information the authorities already mentioned should be consulted. In the thorax a leash of nerves passes outwards from each ganglion towards the developing coxal regions which as development in the larva proceeds are becoming more and more defined by constrictions in the hypodermal walls. Similarly from each abdominal ganglion a nerve passes outwards beneath the ventral muscles to pass up the side of the segment in the region of the ventral trachea until it reaches the outer margin of the large tracheal trunk. From thence it turns inwards and is distributed to the dorsal muscles and other structures. Near its origin a branch turns posteriorly to end among the ventral muscles. In the eighth segment two large nerves pass from the ganglion of that segment, one on each side of the proximal portion of the rectum. Each then divides into two, one branch going to the posterior region of the siphon, the other passing into the anal segment.

THE VENTRAL SYMPATHETIC SYSTEM (Fig. 38 (4), (5))

This consists of a line of nerves situated in the median line dorsal to the ventral nerve chain. Whilst such a series appears to correspond to the *median nerves* of many insects it does not in *Aëdes aegypti* larva so far as can be made out follow the usual course as described for this series. Normally in insects a median nerve arises in each segment from the posterior part of the ganglion of the ventral nerve cord, taking origin between the connectives, and passes backwards to the transverse intersegmental muscle at the anterior border of the segment behind. At this muscle with which it is closely associated it gives rise to lateral branches proceeding outwards to the region of the spiracles. In *A. aegypti* larva the nerves pass from one transverse muscle to the next. In their course they pass over the ganglion but they do not enter or arise from this. The line of nerve forming the series may be clearly separate from the ventral chain or more closely approximated to this especially over the ganglia. But in the latter case they can be traced without any break past this structure. The nerves may possess only relatively few nuclei, but in the later stages of the larva are conspicuous for the numerous nuclei along their course. The series can be readily followed from the transverse muscle at the anterior border of the first abdominal segment to that at the anterior border of the eighth segment where the chain ends. The transverse muscle of segment I crosses the ganglion, but those of succeeding segments except the eighth lie a little forward of the ganglion. That at segment VIII passes over the front of the ganglion. The chain, if present at all, is not very clearly traceable in the thorax, though a nerve seems to pass from the third thoracic ganglion to the first transverse muscle. Small nuclei are usually present arranged about the transverse muscle in the region of the nerve and at their lateral terminations, but no lateral nerves have been made out. The appearance of the system will be clear from the figure 38.

A system of sympathetic nerves also forms part of the complex known as the retrocerebral complex. As, however, this system is very closely associated with the aorta and corpora allata, its description is deferred to a later section under the circulatory system.

(*e*) SPECIAL SENSE ORGANS

Under this head may be considered the *larval ocelli*, the developing *compound eyes* as these are seen in the larva, certain cuticular *sensory papillae and pits* and the structures commonly referred to as *chordotonal organs*.

THE LARVAL OCELLI

These come under the designation of *lateral ocelli*, or as sometimes termed, *stemmata*, as distinct from the dorsal ocelli of many adult insects. In the mosquito larva they form a close cluster of dark pigmented 'eye-spots' situated on the most prominent portion of the lateral aspect of the head, lying behind and in the mid-concavity of the developing compound eyes when these appear (Fig. 35 (1), (2)). A very complete description of the larval ocelli of *Culex pipiens* has been given by Constantineanu (1930).* In *Culex pipiens* according to Constantineanu the ocelli of each side are five in number arranged as shown in Fig. 35 (3). In *C. molestus* the arrangement is much as in *C. pipiens*, forming a compact group of closely aggregated spots. In *Aëdes aegypti* there are also five ocelli, but the appearance and arrangement are somewhat different. The general appearance is of three ocelli placed one above the other, the top and bottom ocelli being slightly posterior to the middle one (Fig. 35 (1), (2)). Closer inspection, especially in the earlier instars, however, shows that whilst the middle ocellus is single each of the other two is double, thus giving the five ocelli as in *Culex pipiens*. Further the large middle ocellus is more or less circular, not lengthened as in *C. pipiens*.

The general structure of an ocellus as seen in sections is shown in Fig. 35 (4), (5). Each ocellus forms a cone with its base directed outwards. Under the but little modified cuticle is a clear area, *lens*. Beneath this is a layer of elongate oval eosin-staining clear bodies corresponding to the pneumocones of the compound eye (crystalline bodies of authors).

* See also Zavrel (1907), Demoll (1917), and for a general account of ocelli in insects, Snodgrass (1935), Imms (1938), Wigglesworth (1942).

The remainder of the ocellus consists of elongate retinula columns containing rhabdomes, the retinula cells which compose these being heavily laden in the upper halves especially with pigment granules entirely obscuring their outlines (Fig. 35 (4)).*

THE COMPOUND EYES

These first appear in the early fourth instar larva as eyebrow-like dark streaks on the side of the head anterior to the ocelli. These areas increase progressively in an anterior direction by the appearance of successive rows of developing ommatidia until they occupy a large part of the lateral surface of the head. The ommatidia first appear as modified hypodermal cells which develop pigment and become arranged to form columns containing rhabdoms. At their base is a plexus of nerve fibres with nuclei continuous with the optic tracts. For further description of these changes see the account of the development of the compound eyes in *Vanessa* by Johansen and some further remarks given later under pupal changes.

SENSORY PAPILLAE

Of sensory papillae may be mentioned those at the apex of the antennae and of the maxillary palps. Two very complicated eminences with characteristic papillae constitute the central portion of each side of the labial plate. Sensory papillae are also present on the lateral portions of the stigmata at the apex of the respiratory siphon.

CHORDOTONAL ORGANS (Fig. 35 ((7)–11))

The chordotonal organs (*chordotonal sensilla* or *scolopidia*) are a simple type of sense organ in the form of a ligament or ligaments, attached to the integument by one extremity and by

* Details of structure of the insect eye, including various forms of ocelli and the compound eye, are given in the very full accounts, in addition to those already mentioned, by Grenacher (1879), Hickson (1885), Johansen (1893), Redikorzew (1900), Radl (1905), Bugnion and Popoff (1914); see also under sense organs of the imago (chapter XXIX).

Figure 35. Sense organs.

1 Ocelli and developing compound eye of fourth instar larva of *Aëdes aegypti.*
2 Outline of ocelli.
3 Outline (enlarged) from figure by Constantianu of ocelli of *Culex pipiens.*
4 Longitudinal section through ocellus.
5 Transverse section at level of crystalline bodies.
6 Transverse section through developing ommatidia of compound eye.
7 Chordotonal organ of first and second abdominal segment of fourth instar larva of *Aëdes aegypti.*
8 Tangential section of first abdominal segment showing the chordotonal organ passing from ventral to dorsal aspect.
9 Basal portion of chordotonal organ *d* of siphon.
10 Insertion of chordotonal organ *c* into lateral lobe of the stigmatic plate of siphon and the small chordotonal organ *a* at apex of siphon.
11 Lateral view of siphon to show course of the chordotonal organs *a, b, c, d.*

Lettering: *a, b, c, d,* chordotonal organs as listed in text; *cb*, crystalline bodies; *g*, ganglion of first abdominal segment; *le*, lens; *lm*, ventral longitudinal muscles of segment; *oc*, ocellus; *oe*, large oenocytes; *om*, developing compound eye; *pg*, pigment; *rb*, rhabdom; *rc*, retinula cells; *scp*, scolopidium; *scpn*, nerve to same; *sp*, spiracle; I, II, abdominal segments.

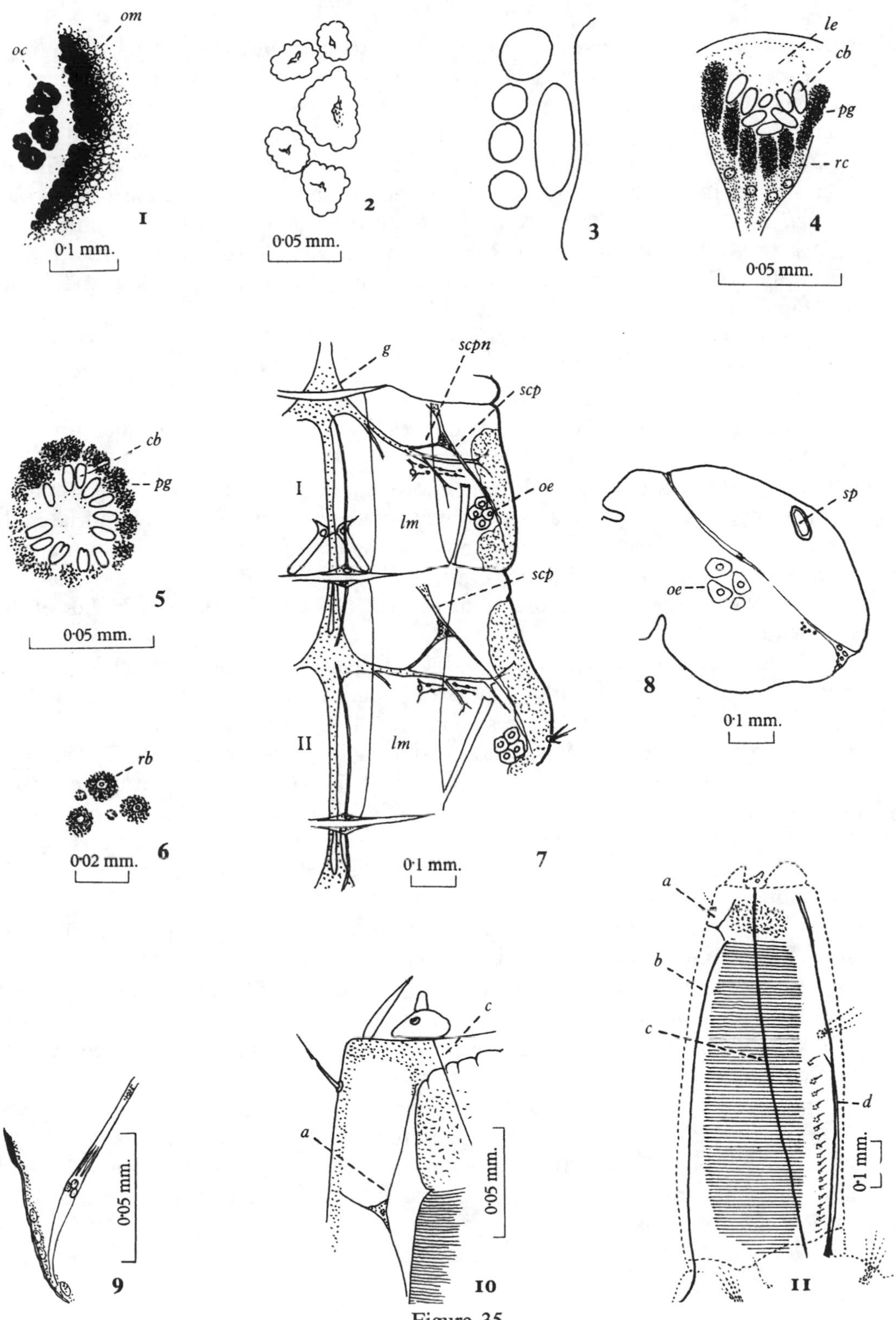
om
oc
1
0·1 mm.
2
0·05 mm.
3
le
cb
pg
rc
4
0·05 mm.
cb
pg
5
0·05 mm.
rb
0·02 mm.
6
g
scpn
scp
I
lm
oe
scp
II
lm
0·1 mm.
7
sp
oe
8
0·1 mm.
0·05 mm.
9
c
a
0·05 mm.
10
a
b
c
d
0·1 mm.
11
Figure 35

a short strand of connective tissue at the other, containing characteristic nerve endings, *scolopales*, the organs being considered as sensitive to mechanical stimuli or changes in tension. Such organs may be present in various parts of the body in different insects and have been described particularly in the abdominal segments of the larvae of Nematocera where they are most readily seen. Their structure has been minutely described by Graber (1882), as seen in the young *Chironomus* larva and in the transparent larva of *Corethra* (*Chaoborus*). In the latter they are present in all but the first and last body segments. In *Aëdes aegypti* larva they are most readily seen in the transparent stages of the living larva or in preparations of such stages. The most conspicuous is the organ in the first abdominal segment (Fig. 35 (7)) where the resemblance to Graber's figure is very evident. The cords in each segment pass diagonally from before backwards and from the ventral to the dorsal surface (Fig. 35 (8)). Demoll (1917) notes that three pairs of sensilla are present in the siphon of the *Culex* larva. These are very clearly to be seen in the siphon of *Aëdes aegypti*, passing up with the muscles of this organ, but readily distinguished from these by their narrower and less coloured appearance.

One such organ (*d*) arises on each side from a base showing distinct scolopale appearances (Fig. 35 (11)) attached just below the lower edge of the siphon posteriorly and laterally somewhat dorsal to the pentad hair and about in line with the pecten spines of its side. Its tendon passes up the siphon posterior to the tracheal trunk and close to the siphon hair of its side to terminate in one or more branches at the base of the posterior lobes of the stigmatic plate.

A second organ (*c*) with a similar base showing scolopale appearance is attached close to, but a little external and anterior to, organ *d* and about in line with the second or third scale of the comb. The tendon of this passes up external to the tracheal trunk crossing this obliquely to end in the lateral papilla of its side at the apex of the siphon. Dr Tate informs me that in *Culex* larva, where a tag, as in many mosquito larvae, is present on the lower margin of the siphon, this lies over a sensilla, which would appear to be that now being described. In *Aëdes aegypti* larva, however, no tag is present and the sensilla is unprotected, a fact in keeping with other evidence that the larva of this mosquito is poorly adapted to life other than in comparatively sheltered conditions.

A third cord (*b*) has an attachment at the anterior end of the eighth segment dorsally, passing backwards through this segment to enter the siphon on its anterior aspect. This passes up the siphon along with the muscles *a*, *b*, *c* to be inserted into the base of the felt chamber. Here the scolopale appearances are less distinct than in the two previous cases. There is a long basal cord in the eighth segment with some suggestive nuclei shortly before entering the siphon.

What appears to be a fourth organ distinct from (*b*) more resembles those in the abdominal segments. This is a much smaller organ situated anterior to the felt chamber in the apex of the siphon (Fig. 35 (10)). It has three cords, one passing towards the apical rim of the siphon, one going to the tracheal trunk and one short cord to the side of the siphon tube a little below the siphon hair.

(*f*) CIRCULATORY SYSTEM AND ASSOCIATED TISSUES

THE CIRCULATORY SYSTEM (Figs. 36–9)

For a general account of the circulatory system in insects with bibliography see especially that by Wigglesworth (1942*a*).

As in insects generally the blood, *haemolymph*, in the mosquito larva is a relatively colourless fluid occupying extensive spaces throughout the body and maintained in circulation through pulsation of the *dorsal vessel* with certain membranous and other guiding structures. The dorsal vessel as usual in insects consists of an abdominal portion, *heart*, and a thoracic portion passing forwards to the head, *aorta*. In the larva, except that ostia (as also alary muscles and pericardial cells) are not present in the aorta, there is no great contrast in the two portions. Associated with the heart are the *alary muscles*, rather poorly developed in the larva, and the characteristic groups of *pericardial cells* (dorsal nephrocytes) set along its course. The dorsal diaphragm, which is such a prominent feature in many insects and in the adult mosquito, is little developed in the mosquito larva, possibly as a result of the great size of the large tracheal trunks. Certain membranes exist, notably at the anterior opening of the aorta, as also at the neck where there is a muscular diaphragm. In the developing imaginal buds of the advanced larva may also be seen the forerunners of the longitudinal partitions conspicuous in the legs of the imago. No accessory pulsating organs additional to the heart and aorta have been described in the larva.

Other tissues convenient to describe along with the heart are: the haemolymph with its cellular constituents; certain free cells usually referred to as phagocytes; the oenocytes; and the fat-body.

Closely associated with the aorta is the retrocerebral nerve complex described later. For physiology of the circulation and behaviour of the heart see section (*g*) in chapter XXXI (Physiology).

The heart (Fig. 36). Among early descriptions of the heart in the larva or pupa of the mosquito see those by Raschke (1887, *Culex nemorosus* larva); Hurst (1890, *Anopheles* larva and *Culex* pupa); Metalnikoff (1902, *Culex* larva); Imms (1908, *Anopheles* larva and pupa). Several authors have studied the heart as seen in the transparent larva of *Chaoborus* (*Corethra*), namely Dogiel (1877); Lebrun (1926); Tzonis (1936). Lebrun, in the living larva suitably mounted after *intra vitam* staining, demonstrated in a very clear manner the ostia of the heart and the alary muscles.

The anterior opening of the aorta in the head of *Chironomus* larva has been described by Holmgren (1904) and Imms gives a careful description of this complex part in the larva of *Anopheles*. Very helpful in this respect too is the description by Wigglesworth (1934) of the aortic sinus in *Rhodnius*.

Recently a very complete study of the heart and associated tissues in the larva, pupa and imago of *Anopheles quadrimaculatus*, with references to other species, including *Culex pipiens* and *Aëdes aegypti*, has been made by Jones (1952, 1953, 1954) with, in addition to the morphology, much information given on the physiology of the heart and circulation in the mosquito as indicated in the chapter on physiology.

The heart in *A. aegypti* larva does not differ materially from the description by Imms and others of this structure in *Anopheles*. In the living insect it can be seen pulsating between the two tracheal trunks which lie lateral to it. In sections it appears as a delicate tube with laterally placed nuclei which project into the lumen and are characteristically spaced at intervals (about three pairs to a segment). In each of the abdominal segments I–VII, situated somewhat forward in the segment, is a pair of lateral *ostia*. These are recognisable by the presence of twin nuclei, smaller than those of the heart wall, of the cells guarding the

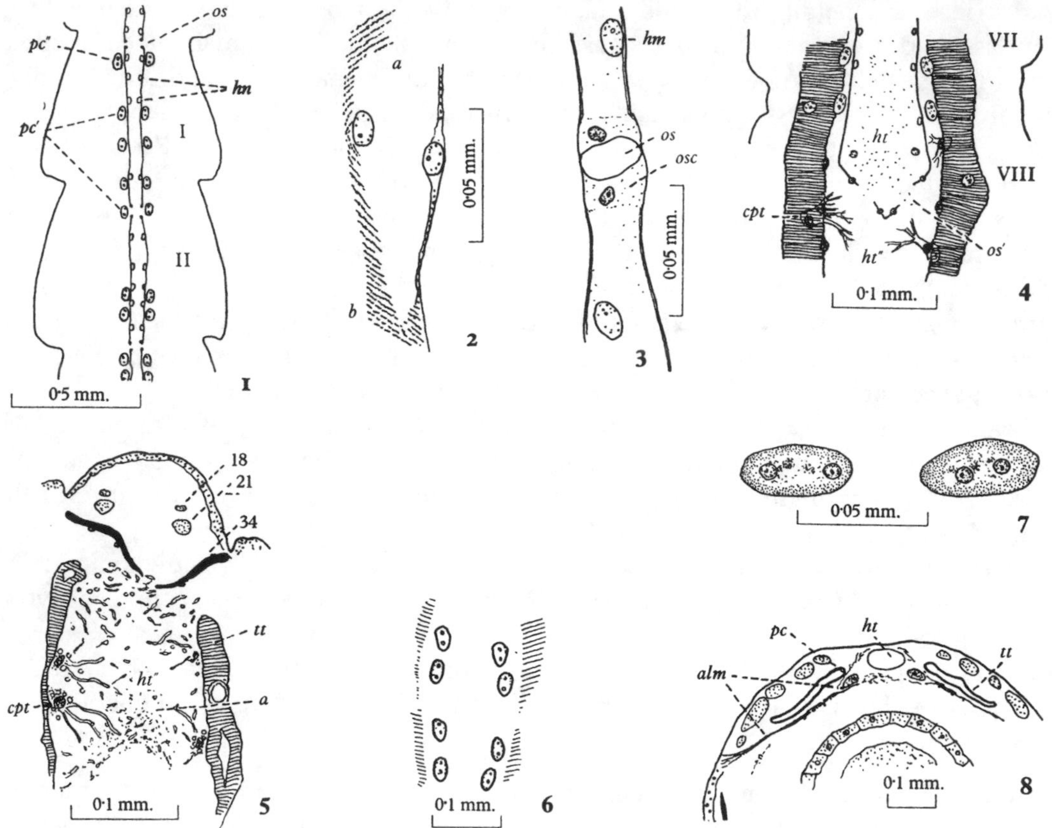

Figure 36. The heart and related structures.

1 Showing heart and arrangement of nuclei, ostia and pericardial cells.

2 Coronal section through heart showing lateral nuclei and fibrillae. *a*, floor of heart; *b*, roof of heart.

3 Sagittal section showing ostia and ostial cells.

4 Coronal section of terminal chamber of heart.

5 Similar section through tracheal plexus. *a*, terminal branches of tracheae forming limiting membrane of tracheal plexus.

6 Group of eight pericardial cells at an intersegment.

7 Pericardial cells showing cortical layer of cytoplasm and two nuclei.

8 Transverse section through abdominal segment showing alary muscles.

Lettering: *alm*, alary muscle; *cpt*, crypt with tracheal cell; *hn*, nuclei of heart wall; *ht*, heart; *ht′*, terminal chamber of heart; *ht″*, chamber posterior to heart occupied by tracheal plexus; *os*, ostium; *os′*, terminal ostium; *osc*, ostial cell; *pc*, pericardial cell; *pc′*, group of same at intersegment; *pc″*, single pair of same at anterior end of first abdominal segment; *tt*, tracheal trunk; 18, 21, muscles of siphon; 34, more dorsal of double pair of muscles inserted into apex of tracheal plexus; VII, VIII, abdominal segments.

ostia, one anterior and one posterior to the opening (Fig. 36 (3)). The openings are simple and slit-like. In each segment, just anterior to each pair of ostia, there is an ill-defined, and in the larva not very marked dilatation of the tube, pulsating chamber. Apart from the lateral openings of the ostia there are no valves in the heart.

Anteriorly the heart proper passes without much change into the aorta. Posteriorly it terminates in a somewhat swollen *terminal chamber*. This chamber in the larva ends somewhat abruptly about the middle of the eighth segment where it is closed posteriorly by a flattish membranous wall. This has an opening on each side of the middle line and there are small nuclei, resembling those at the ostia, in the intervening portion of wall and external to the openings, which therefore may be regarded as ostia, *terminal ostia*.

The walls of the heart are thin, delicate and transparent. According to Imms there is a delicate membrane (sarcolemma) externally and internally. Fine oblique or transverse parallel and closely set striae are visable on close inspection (about $0{\cdot}5\mu$ in thickness with intervals about twice this). According to Jones these fine muscular fibrillae are double and arranged spirally. Though muscular in nature the heart wall as a whole docs not show up conspicuously under polarised light, though the fine fibrils are faintly illuminated.

The tracheal plexus (Fig. 36 (5)). Beyond the terminal chamber the floor of the heart is continuous with a delicate membrane forming the floor of the structure previously referred to as the tracheal plexus (see under the tracheal system). It is from the chamber formed by the plexus, as shortly to be described, that the terminal ostia open into the heart and the plexus clearly forms an important adjunct to the organs of circulation being in effect a collecting chamber and an oxygenating organ. The plexus is formed by a mass of branching tracheole-like vessels which arise from short tubular crypts taking origin from the inner portions of the ventral walls of the tracheal trunks where these lie in the eighth segment and base of the siphon. The crypts number about fifteen to twenty on each tracheal trunk and are arranged in a staggered series extending from a little behind the anterior border of the eighth segment to some little distance within the base of the siphon. A few of the more anterior crypts lie over the terminal chamber of the heart, but the larger number are posterior to this structure. At the bottom of each crypt is a cell with a conspicuous large dark-staining nucleus and through the cytoplasm of these cells, as in a tracheal cell, there pass the tracheole-like vessels referred to. As with ordinary tracheoles these are devoid of taenidia.

Branching in various directions, but in the main directed ventrally, the tracheoles in mass form a defined cone-shaped structure coming to a point just dorsal to the colon. Here it receives the insertion of the four small converging muscles as described in the section dealing with the muscles of the terminal segments. The cone on its lower surface is definitely delimited by a delicate membrane previously mentioned as continuous with the ventral wall of the heart. The cone lies within, but does not fully occupy, a considerable space between the colon ventrally and the large tracheal trunks dorsally. The four small muscles coming from different directions to be attached to the point of the plexus presumably keep this structure suitably stretched in its correct position (Fig. 36 ((5)).

The aorta (Fig. 37). The aorta is very similar in appearance to the heart, its walls showing the characteristic lateral nuclei at intervals. It is, however, unprovided with ostia and neither alary muscles nor pericardial cells are present along its course. The two most anterior ostia are at the anterior border of the first abdominal segment, where also the pericardial cells end. The aorta may, therefore, be considered as that portion of the dorsal vessel anterior to this. From thence it passes forwards in the thorax in the mid-dorsal line lying over the groove between the two median dorsal gastric caeca and then dorsal to the proventriculus and oesophagus to the neck. Just before reaching the neck it lies close beneath the tracheal commissure with on either side the corpora allata. A little further on it pierces, and its walls are blended with, the muscular diaphragm in the neck. It now passes forwards close to the dorsal wall of the oesophagus and pharynx to the cerebral commissure linking the two halves of the brain. Its relations in the final portion of its course are complex and important and merit special consideration.

The prothoracic aortic sinus. Under this name Jones (1952, 1954) has described a specialised portion of the aorta in the prothorax which is dilated and can be seen in the living condition, especially in the first instar larva, actively pulsating. The author describes and figures it as a relatively large bulb-like sinus or sac lying between the corpora allata and attached to these structures by tissue strands. A single pair of intra-aortic thickenings is present at the posterior extremity of the sinus which opens through a narrow valve anteriorly into a narrow continuation of the aorta where this terminates behind the supra-oesophageal ganglion. In the living *Aëdes aegypti* larva it is seen as a swollen pulsating portion of the aorta. But in sections the organ is not so conspicuous and beyond the fact that it dips down sharply under the commissure there may be little to notice, though in some cases there is some lateral dilatation and some folding of the walls. At about the junction of the anterior two-thirds with the posterior third, however, where the aorta is crossed and closely contacted on each side by a small trachea passing dorsally from the lateral trunk, the wall of the aorta is thickened and there is a cluster of nuclei distinct from the spaced nuclei of the rest of the organ (see also description of the aorta in the imago).

The anterior termination of the aorta (Fig. 37 (1)–(3); 38 (1)). After passing the tracheal commissure and as the aorta approaches its anterior end it lies close to the dorsal wall of the pharynx, its ventral wall being closely adherent to, if not actually formed by, pharyngeal elements. At the pharyngeal notch it lies in the groove between the projections formed by the posterior ends of the pharyngeal plates, that is the groove in which the recurrent pharyngeal nerve also lies in its passage backwards. This groove is closed in dorsally by the cerebral commissure and so formed into a narrow passage-way. This passage-way is further restricted by two largish muscles coming from the dorsum of the head to be inserted into the posterior angles of the pharynx where these form the lateral boundaries of the

Figure 37. Aortic opening and related structures.

1 Sagittal section in median line of head giving lateral view of structures at anterior opening of aorta.

2 Dorsal view of structures at anterior opening of aorta. Reconstruction. For explanation see under lettering.

3 Posterior view of aortic opening. Reconstruction.

Lettering: *ao*, aorta; *Br*, brain; *Cc*, cerebral commissure; *P*, pharynx; *rn*, recurrent nerve passing through gap at posterior angle of the pharynx over muscle 24 (cingulum) to floor of aorta where it continues into structures described under retrocerebral system; *sn*, blood sinus; *sn′*, membrane continued from sarcolemma of muscle 16 to that of 15 to form roof of sinus (*sn*); *sn″*, membrane passing from cerebral commissure to dorsum of head; 15–24, muscles as listed in text under muscles of head.

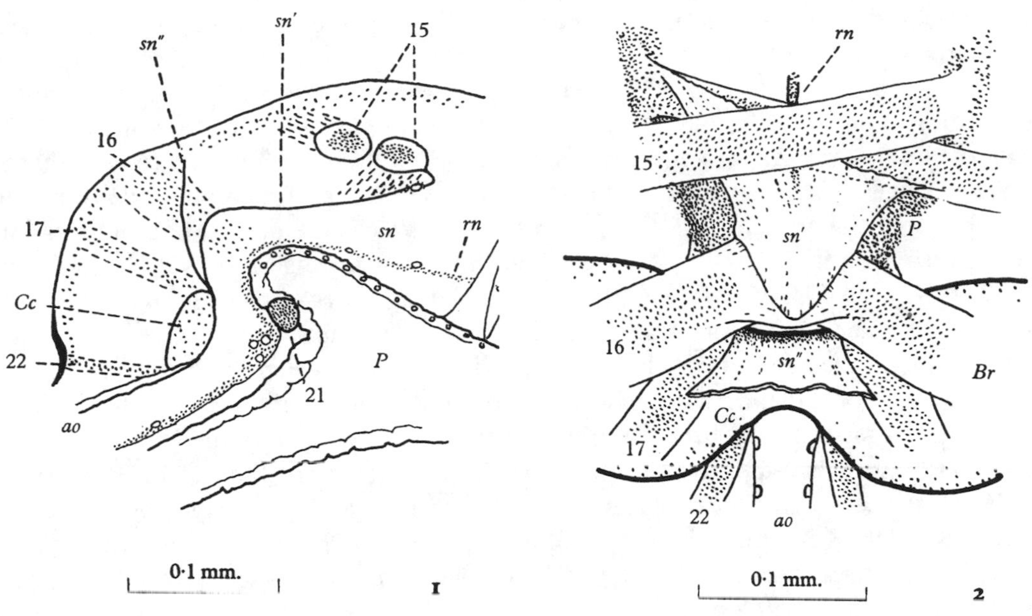
sn″
sn′
15
16
17
Cc
22
ao
sn
rn
P
21
0·1 mm.
1
rn
15
sn′
P
16
sn″
Br
Cc
17
22
ao
0·1 mm.
2

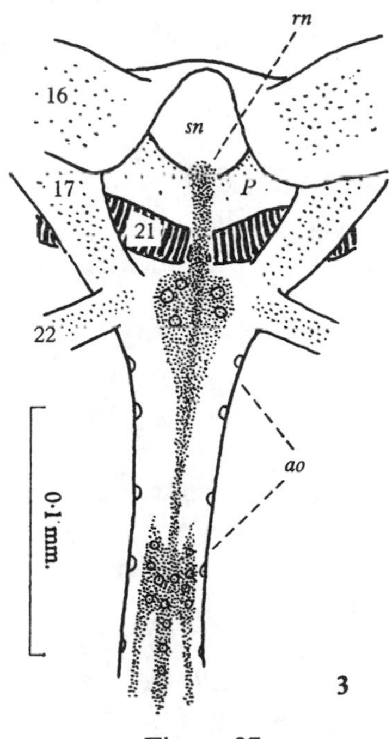
rn
16
sn
17
P
21
22
ao
0·1 mm.
3

Figure 37

notch (dorsal suspensors of the pharynx; muscles no. 16). With the walls of this narrow way those of the aorta are blended, that is the dorsal wall is fused with the connective tissue of the commissure and the ventral wall with the pharyngeal wall and the sarcolemma of the two muscles referred to (Fig. 37 (1)). Here the aorta proper terminates. Passing out of this opening, however, the blood still finds itself restricted by a roof formed of thin membrane linking the voluminous sarcolemma of the two muscles referred to with that of the decussating muscles situated some distance forwards. This membrane converts the trough between the crests of the pharynx as far forwards as the decussating muscles into a covered way continuing the aorta as a blood sinus. The arrangement will be clear from Fig. 37 (1), (2). In Fig. 37 (1) the aorta passes in from the neck on the left.

There is another membrane in this region not directly connected with the blood sinus, but which should be mentioned. This is a supporting membrane (*sn″*) which rises from the front of the cerebral commissure and passes to the dorsum of the head in the occipital region. This forms a pocket between itself and the sinus roof membrane (*sn′*).

There are also some important relations on the ventral side of the notch. Besides the muscles 16 there is a smaller pair (muscles no. 17) which also pass from the dorsum in front of the cerebral commissure along with the muscles 16 to be inserted into the angles of the pharynx just ventral to the insertion of these. These smaller muscles serve still further to close in the flanks of the blood passage. Still another pair of muscles (no. 21) pass behind the commissure and are inserted to each side of the floor of the aorta where this enters the notch. Between this pair and nos. 16 and 17 there is, so far as I have been able to ascertain, a passage-way for blood directed laterally and downwards. The similarity of the arrangement to that given by Wigglesworth (1934) for *Rhodnius* will be evident.

Where the aorta lies over the pharynx before reaching the knot it comes into close association with the retrocerebral nerve complex and is modified thereby. For further details see under the retrocerebral complex.

The alary muscles (Fig. 36 (8)). Intimately associated with the heart and its accompanying pericardial cells are the *alary muscles*. These are small muscles ending in delicate fibrillar branches which form a network closing in a space, *dorsal sinus*, the network forming the *dorsal diaphragm*. The muscles arise from the cuticular hypoderm of the segments laterally and just external to the outer edge of the longitudinal series of muscles and pass under these and the large tracheal trunks. Their terminal fibrillae pass above and below the pericardial cells enclosing these between the two layers. There is one such muscle in *Aëdes aegypti* to each pericardial cell and since these cells lie in groups of four pairs, two on each side of the intersegmental line the alary muscles have a similar distribution. There is only one pair at the anterior end of the first abdominal segment and two in the eighth segment. There are none connected with the aorta in the thorax. According to Jones (1954) the alary muscles maintain the heart wall taut and if cut the dilatation at the pulsating chamber disappears. They are not, however, concerned with the heart's pulsation otherwise than to keep this organ under some control. According to Lebrun (1926) the alary fibres in *Corethra* are not muscular but of an elastic nature. In *Aëdes aegypti* the fibrillae do not show up brilliantly when viewed by polarised light. This, however, would not seem to preclude their being of muscular nature since the heart itself is not vividly illuminated when so viewed. The same applies to a number of visceral muscles as compared with the brilliant effect with the usual systemic muscles.

The dorsal nephrocytes (Fig. 36 (1), (6), (7), (8)). In close relation with the heart are the *pericardial cells*, now commonly referred to as the *dorsal nephrocytes*, somewhat similar cells lying in the prothorax beneath the oesophagus being known as the *ventral nephrocytes*. The pericardial cells are large characteristic cells occurring in insects in close association with the heart. For a general review regarding their occurrence and characters see Hollande (1922). In the Diptera they have been described in the larva of Muscidae by Graber (1873) and by Kowalevsky (1889, 1892); in the larva of Anthomyidae by Keilin (1917); in Mycetophilidae larva by Madwar (1937); in *Chironomus* larva by Graber (1873). In the mosquito they have been described in *Culex pipiens* larva by Wielowiejski (1886) and by Metalnikoff (1902) (also referred to by Kowalevsky (1889)); in *Anopheles* briefly by Imms (1908); more recently in the larva, pupa and imago of *A. stephensi*, *C. fatigans* and *Aëdes aegypti* by Rajindar Pal (1944) and in several species of *Anopheles*, in *Culex pipiens* and *Aëdes aegypti* by Jones.

According to Metalnikoff there are two large masses in the thorax situated ventrally anterior to the caeca, two large cells in the first abdominal segment and eight smaller cells arranged in four pairs in each abdominal segment from II to VIII. The large masses in the thorax refer to the ventral nephrocytes. The remaining description agrees with that given by Rajindar Pal in the three species noted. Jones, however, finds in several species of *Anopheles* examined, including *A. stephensi*, many more than the twenty-eight pairs described. He distinguishes three sizes of pericardial cells, namely, large (av. 18·5μ by 28·1μ); medium (av. 12·3μ by 18·0μ); and small (av. 8·2μ by 15·7μ). The large occur principally in the region of the ostia, the medium and small all along the sides of the heart proper. Including the medium and small cells he finds in *Anopheles* as many as twenty-eight or more at each segment.

In *Aëdes aegypti* larva the number and arrangement of the pericardial cells with the exceptions noted below is as described by Metalnikoff and by Rajindar Pal. These are in groups of four pairs, two situated in the hinder part of the segment in front and two in the forward portion of the segment behind (Fig. 36 (1); 39 (9)). In segment I only one pair is present anteriorly (possibly in the metathorax as described by Rajindar Pal) and in the group VII–VIII there are two pairs only, both of which lie in the fore-part of the eighth segment. There are no cells in the hinder portion of the eighth segment, that is twenty-seven pairs in all. No cells that could be classed as small or medium cells have been seen in *A. aegypti* in the material examined by me, the intervening portions of heart between the groups of four pairs being quite free from any such type of cell. In mosquitoes taken in nature Rajindar Pal has found a certain small variation. In *A. aegypti* in the group VII–VIII there might be two, or three, pairs, two lying in the latter case in segment VII and one in VIII. In *Anopheles* there was some variation with season. It seems possible that conditions of culture might affect the numbers.

The cells are large, oval in shape and measuring in the full-grown larva some 40μ in length by 20μ in transverse diameter. They usually have two, but sometimes three, nuclei arranged in the length of the cell. There is an outer denser cortical zone of cytoplasm and an inner more reticular cytoplasm, commonly with vacuoles and inclusions. They lie lateral to and somewhat ventral to the heart and between the large tracheal trunks. They are surrounded and enmeshed by the fine branching fibrillae of the alary muscles, but owing to the inconspicuous nature of these branchings in the larva they often appear in sections to lie more or less free in the pericardial sinus. As noted by Metalnikoff and subsequent authors they tend to take up carmine particles when this substance is introduced into the body by injection or feeding. They are now accepted as concerned in the taking up of colloidal particles and along with cells of a similar nature have been termed *nephrocytes* (Keilin, 1917).

The ventral nephrocytes (Fig. 38 (1) *pc″*). In the mosquito larva the ventral nephrocytes are cells with a somewhat similar appearance to the pericardial cells and possessing the same power to take up carmine particles. In *Anopheles quadrimaculatus* they lie, as described by Jones (1954), below the oesophagus in the median line over the connective linking the sub-oesophageal ganglion with the first thoracic ganglion and attached to the middle portion of the larval salivary gland (at the isthmus). They are connected to each other by thin filaments

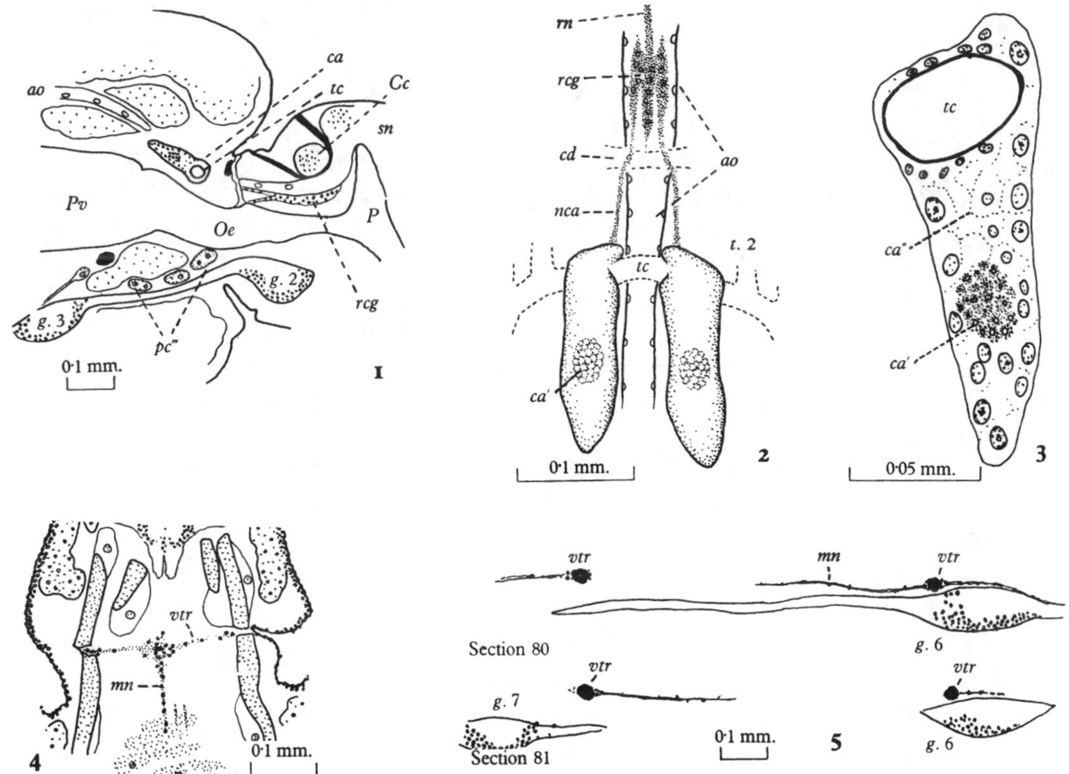

Figure 38. The corpora allata and the median nerve.

1 Sagittal section near middle plane of head showing corpus allatum complex and other structures.

2 The corpora allata and related structures of *Aëdes aegypti*. Reconstruction.

3 Sagittal section through a corpus allatum complex.

4 Coronal section through an intersegment showing ventral transverse muscle and median nerve.

5 Two successive sections showing median nerve.

Lettering: *ao*, aorta; *Cc*, cerebral commissure; *ca*, corpus allatum complex; *ca′*, corpus allatum; *ca″*, supporting cells of complex; *g.*2, sub-oesophageal ganglion; *g.*3, first thoracic ganglion; *g.*6, 7, abdominal ganglia VI, VII; *mn*, median nerve; *nca*, nerve to corpus allatum; *Oe*, oesophagus; *P*, pharynx; *Pv*, proventriculus; *pc‴*, ventral nephrocytes; *rcg*, hypocerebral ganglion; *rn*, recurrent nerve; *sn*, aortic sinus; *t.*2, dorsal cephalic trachea; *tc*, cervical tracheal commissure; *vtr*, ventral transverse muscle.

of cytoplasm. They are not conspicuous in sections, but are best displayed in larvae fed on carmine in dissections, being attached to the salivary glands when these are dissected by the method of Jensen (1955).

In *Aëdes aegypti* the ventral nephrocytes form a group of from one to three large pericardial-like cells in the median line lying over the nerve connectives.

THE BLOOD OR HAEMOLYMPH (Fig. 39 (8))

The blood (*haemolymph*) of insects is a more or less colourless fluid. It forms a considerable proportion of the total body substance of the mosquito larva and very considerable spaces occupied by the haemolymph occur in various parts of the body. The physical and chemical characters of the haemolymph in *A. aegypti* larva have been closely studied, notably by Wigglesworth (see under 'Physiology', p. 709).

Contained in and in connection with the haemolymph are certain cellular elements, *haemocytes*. For a general review of the subject of the haemocytes of insect blood with plates showing the chief forms see Rooseboom (1937).* A bibliography is given by Rooseboom and by Wigglesworth (1942*a*).

All authors agree in recognising small reproductive forms (*proleucocytes*) with deeply staining nucleus, scanty cytoplasm and showing active karyokinesis. In Rooseboom's plates these are shown in *Calliphora* measuring about 10μ in diameter with nucleus 6μ. Type 2 of Rooseboom are much larger, measuring according to his figure up to as much as 30μ in diameter with nucleus of 6μ and lightly staining. Among other forms described are *oenocytoids*. These have uniform acidophil cytoplasm resembling that of oenocytes, but are not (in *Rhodnius*) derived from these cells (Wigglesworth, 1942*a*).

Insect haemocytes are for the most part sedentary, occurring especially in many insects lateral to the heart. They are commonly pear-shaped or fusiform with pointed ends. In the late larval stage phagocytes are commonly seen in numbers attached to muscles in process of removal and such may contain inclusions of muscle substance (*sarcolytes*). Haemocytes may accumulate to form phagocytic tissue of a permanent character (Cuenot (1891, 1893); Lange (1932), and previously Pause (1919), describes in the terminal segments of *Chironomus* larva reticular membrane formed by phagocytic cells lining and enclosing fat-body).

In *Aëdes aegypti* larva blood cells, other than the small cells commonly seen accumulated about muscle fibres and other elements and presumably phagocytic, are not conspicuous. The commonest forms of detached cell seen in the haemolymph are round or oval cells with relatively small nucleus and clear eosinophil cytoplasm measuring about 10μ in diameter (Fig. 39 (8)). These are found as single isolated cells in many positions including the pericardial space about the heart. The cells clustered about muscles are small spherical cells with round deeply staining nuclei and will be referred to later.

OENOCYTES (Fig. 39 (4), (7), (9))

These are characteristic globular or polygonal cells with large central nucleus readily distinguished from other tissue cells by their appearance and dark eosin staining cytoplasm. For a general review see Hollande (1914) and for a more recent account Wigglesworth (1933, 1942*a*). They have been described in the larva of *Culex pipiens* by Wielowiejski

* See also Kollmann (1908), Hollande (1909, 1911), Mutkowski (1924), Wigglesworth (1933, 1937, 1942*a*).

(1886); in *Anopheles* by Imms (1908); and very fully by Hosselet (1925) in *Culex (Theobaldia) annulata*. In the advanced larva oenocytes occur in two forms, namely *small* and *large*. Small oenocytes measure about 20 μ in diameter. They occur characteristically set on the inner surface of the ventral parietal fat-body in segments 2–8. They may be quite numerous extending almost to the lateral border. They may also be found isolated or in scattered groups elsewhere. Large oenocytes resemble the small, but measure 50–60 μ in diameter. They form groups of 5 or 6 closely packed cells on each side of abdominal segments 1–8, lying in a pocket of the lateral parietal fat-body. In mounted preparations of the whole larva these lateral patches of dark-staining cells form a conspicuous and characteristic feature of the larval abdomen.

According to Hosselet oenocytes are secretory cells. In an early stage they contain abundant chondriome, which disappears as secretion accumulates in the cytoplasm. Eventually they discharge their contents into the general body cavity. In the small oenocytes discharge of secretion is continuous through larval and pupal stages. With the large oenocytes accumulation of the products of secretion takes place at a slower rate and full development and discharge of secretion results only in the pupa. The complete disappearance of oenocytes from the adult shows that these cells are concerned only with metamorphosis. They have been shown by Wigglesworth (1937) in *Rhodnius* not to initiate ecdysis, but to provide material for the building up of the new cuticle.

In *Aëdes aegypti* larva the lateral patches of about five large oenocytes on each side of abdominal segments I–VII form a striking feature (Fig. 39 (4), (9)). The cells lie close together in a roughly oval plaque in the mid-lateral line a little posterior to the middle of the segment and between lateral dorso-ventral muscles 4 and 5. The plaques lie usually in a recess on the inner aspect of the parietal layer of fat-body. The cells are very large, measuring up to 40 or 50 μ in diameter with large globular nuclei and uniform strongly

---

Figure 39. Malpighian tubules, fat-body, oenocytes and haemocytes.

1 Sliced preparation to show ventral parietal fat-body of thorax and first abdominal segment. Effect of staining with sudan black.

2 Thoracic visceral fat-body. *a*, cape-like prothoracic parietal fat-body. Ventral portion; *b*, median lobe of visceral fat-body lying on each side of the aorta and partly covering the corpora allata; *c*, extension of dorsal visceral fat-body lying over salivary gland; *d*, lateral extensions of ventral visceral fat-body.

3 Transverse section of thorax of early fourth instar larva to show lobes of visceral fat-body. *a* and *b* as for no. 2.

4 Transverse section of abdominal segment to show position of large and small oenocytes.

5 Malpighian tubules. Unstained preparations. A, tubule of early fourth instar larva; B, the same of pre-ecdysis fourth instar larva.

6 Transverse section of Malpighian tubule showing striated border.

7 Oenocytes. A, large oenocytes; B, small oenocytes. Same magnification.

8 Haemocytes. A, as seen in various situations; B, lying near heart.

9 Schematic tangential section to show distribution of large and small oenocytes, also pericardial cells and the dorso-ventral segmental muscles.

Lettering: *ao*, aorta; *Cs*, gastric caecus; *ca*, corpus allatum; *gb*, imaginal bud of first leg; *ldv*. 2–5, lateral dorso-ventral muscles. *ldv*. 1 is present only in segment I; *oe*, large oenocytes; *oe'*, small oenocytes; *Pv*, proventriculus; *pc'*, group of four pairs pericardial cells; *pc''*, group of single pair; *pc''''*, group of two pairs; *sg*, salivary gland; *sp*, spiracle; *t*, trachea (*t*. 32).

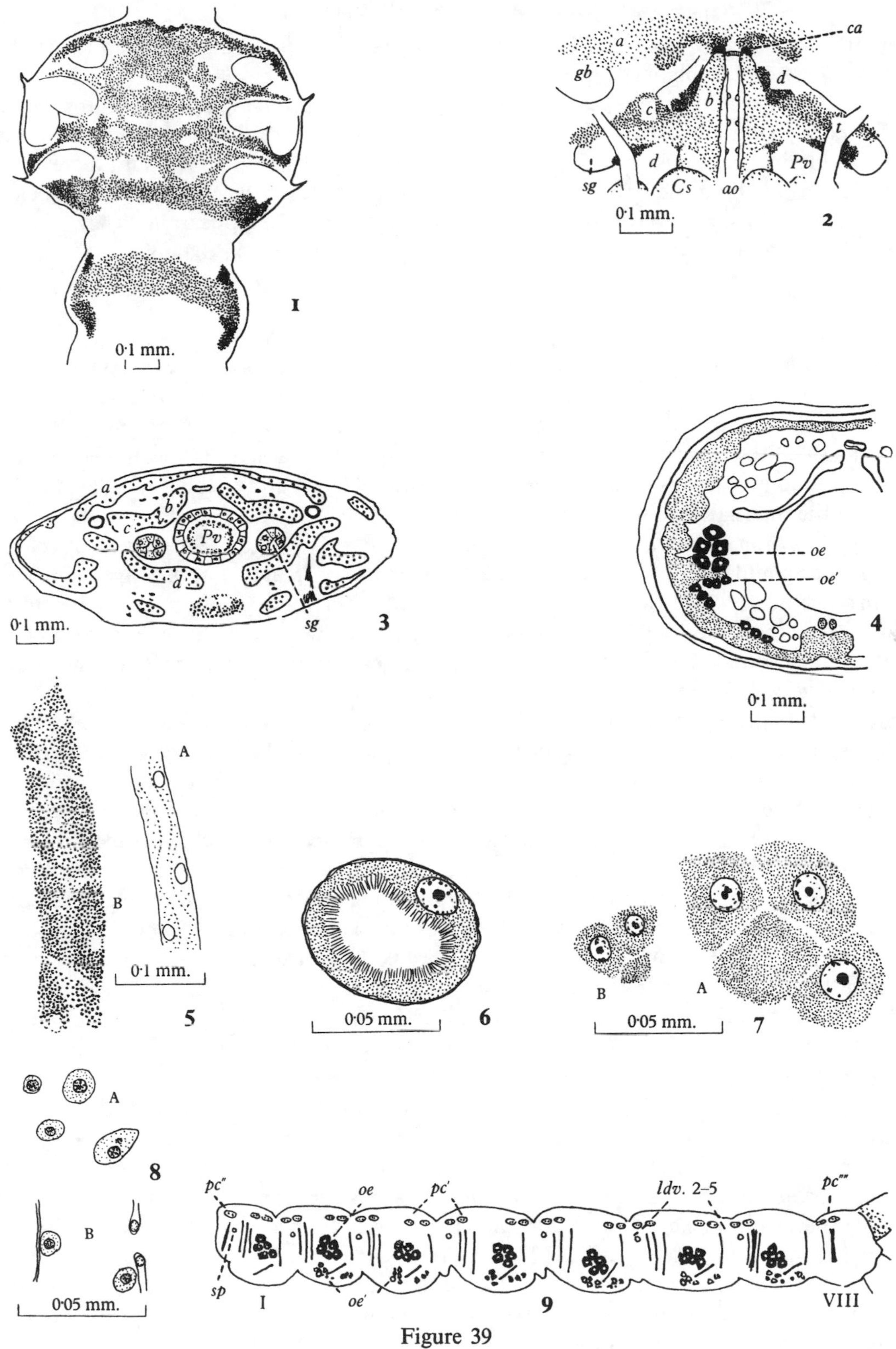
0·1 mm.
I
ca
a
gb
d
b
c
t
d
Pv
sg
Cs
ao
0·1 mm.
2
a
b
c
Pv
d
sg
0·1 mm.
3
oe
oe'
4
0·1 mm.
A
B
0·1 mm.
5
0·05 mm.
6
B
A
0·05 mm.
7
A
8
B
0·05 mm.
pc''
oe
pc'
ldv. 2–5
pc''''
sp
I
oe'
9
VIII

Figure 39

eosinophil cytoplasm (Fig. 39 (7)). The only segment not showing the plaques is the eighth. There are no plaques in the thorax.

Small oenocytes are another constant and conspicuous feature of larval structure. These lie embedded, or more strictly recessed, on the inner surface of the ventral parietal fat-body in all abdominal segments except the first and eighth (Fig. 39 (4), (9)). Usually there is an irregular group about the level of and ventral to the plaque of large oenocytes. Others are present more towards the posterior end of the segment. In appearance they closely resemble the large oenocytes, but measure only up to about 20μ in diameter.

THE FAT-BODY (Fig. 39 (1)–(4))

This normally consists of layers or lobes composed of cells of more or less uniform character, 30–40μ in diameter, whose cytoplasm is swollen with oil globules and watery vacuoles and may contain both reserve and secretory substances. As in insects in general (Weber, 1938) the fat-body in the mosquito larva consists of a layer immediately beneath and largely conterminous with the hypodermis, *parietal layer*, and of lobes lying around or between the internal organs, *visceral layer*.

The disposition of the fat-body is best studied in the early stages of larval growth before excessive accumulation of fatty tissue and development of imaginal appendages obscures the arrangement. Even so, though in the mature larva there may appear to be little evidence of lobes from these becoming pressed together, their limits can still generally be made out. Further, the distribution has a definite relation to the metamerism and other features and so has some interest and importance.

Study of the fat-body can often be facilitated by suitably staining fixed material with a fat-soluble stain, for example sudan black. Dissections or sliced larvae may be overstained in soudan black in 90 per cent alcohol, transferred to 90 per cent alcohol to decolorise if necessary and then to 70 per cent alcohol followed by 50 per cent alcohol, equal parts glycerine and 50 per cent alcohol and pure glycerine with final mounting in glycerine jelly. As oil-soluble stains are very rapidly removed in alcohol and clearing fluids, mounting in balsam is unsuitable. The fixative used is important. Good results have been given with Tower's method. The lobes stand out very vividly (see Fig. 39 (1)).

The arrangement of fat-body is seen in its simplest form in the abdomen where each segment has its own fat-body with a wide intersegmental gap, where in the living larva the pulsating heart and tracheal trunks are clearly displayed. The dorsal layer consists of a thin sheet lining the tergum and reaching anteriorly to the level of the spiracular cord and posteriorly to the end of the segment proper. It is this dorsal sheet, only about one cell in thickness, which is referred to by Wigglesworth (1942*b*) as being very convenient for the study of reserve materials under the microscope using the living larva under a cover-glass. The ventral sheet is less uniform. It lines the ventral surface of the segment leaving a gap at the intersegment. It has a median ridge on which the ventral nerve cord rests. Laterally the two sheets are thickened, turned in and largely fused to form a thick mass of fat tissue occupying the side of the segment. This mass is grooved anteriorly by the spiracular cord passing from the spiracular punctae to the tracheal trunks and perforated by the dorso-ventral muscles of the segment. Occupying a pocket on the inner aspect of the lateral mass at about the middle of the segment or a little posterior to this is the characteristic group of large oenocytes. Lying on the inner surface of the ventral sheet on either side of the ventral nerve cord are the small oenocytes. In the eighth segment the ventral layer is deficient posteriorly and the dorsal sheet does not pass the base line of the respiratory siphon. The anal segment has a separate sheet of parietal layer only on each side. Apart from lobules of the parietal layer the abdomen is devoid of any visceral layer.

In the thorax the fat-body occurs both as parietal layer and extensions of this or separate lobes forming a visceral layer. As in the abdomen the parietal layer is divided corresponding to the three thoracic segments, though the gaps between are narrower and more bridged across than in the abdomen. Fenestrae at the insertion of muscles are also more conspicuous (Fig. 39 (1)). Laterally the dorsal and ventral sheets are linked up by extensions of fat-body passing internal to and between the genital buds of the legs and wings. A conspicuous feature is the continuous cape-like layer of prothoracic fat-body. This covers completely the anterior aspect of the thorax, sending lateral extensions over the genital buds of the first legs and abutting upon the rudiments of the pupal trumpets lying just dorsal and external to these (not shown in the figure).

In addition to the thoracic dorsal and ventral parietal layers there are lobes of fat-body situated around and between the internal organs. The most important of these consist of dorsal and ventral masses filling the space anterior to the gastric caeca. The dorsal mass consists on each side of a lobe which has an inner fusiform portion (Fig. 39 (2), (3) *b*) abutting upon the aorta and partially covering over the corpora allata of its side. Externaly this lobe forms an extension lying over the salivary gland (*c*). It has been figured by Wigglesworth (1942*b*) in his paper dealing with reserve materials in the mosquito larva. The ventral mass forms a pad at the level of the anterior end of the proventriculus resting upon the ventral nerve cord connectives (Fig. 38 (1)). From this extensions pass backwards on to the first thoracic ganglion and along the under side of the salivary gland (*d*). This lobe has also been figured by Wigglesworth. The arrangement is liable to considerable variation in details, though on the whole the general plan is adhered to. Lateral connecting strips joining the dorsal and ventral parietal sheets and junctions of these with the visceral lobes may resemble lobes. In the late larval stages also gross accumulation of fat tissue tends to obliterate details.

The head is almost entirely devoid of fat-body and the spaces, which are considerable, are occupied by haemolymph.

(*g*) THE RETROCEREBRAL COMPLEX AND THE CORPORA ALLATA

Under the name retrocerebral complex is included a rather complicated system of ganglia and nerves constituting what was formerly often termed the *sympathetic* or *stomatogastric* nervous system. Closely associated with this are certain bodies of a glandular nature the most important of which are the *corpora allata* or endocrine glands responsible for hormones concerned in metamorphosis (see ch. XXXI (*b*)).

THE RETROCEREBRAL COMPLEX

The nervous structures composing the complex have in the past been very variously homologised and named by different authors. Recently an important comparative study, including these parts in the mosquito (*Culex pipiens*) has been made by Cazal (1948) with a suggested nomenclature.

According to Cazal the generalised scheme for the arrangement in insects is as follows. The recurrent nerve from the frontal ganglion passing backwards between the aorta and the oesophagus enlarges to form an oval ganglion, *hypocerebral ganglion*. From this there passes backwards a median, or two lateral, *oesophageal nerves*. These end at the fore part of the gut on each side in a *ventricular* or *gastric ganglion*. This constitutes the *stomatogastric system*. The *cephalic neuro-endocrine* or *aortic system* consists of a pair of glands, *corpora paracardiaca* (*corpora cardiaca* of many authors), situated laterally to the aorta, or in some forms fused in the middle line beneath the aorta, and two further glands lying posterior to these, the *corpora allata*. Nerves from the protocerebral lobes (an inner from the cerebral commissure and an outer from the protocerebral lobe), which may be fused, pass to the corpora paracardiaca and from these in turn nerves pass to the corpora allata. There are also some collateral nerves among which may be noted one from the nerve to the corpus paracardiacum going to the muscles of the tentorium or mastigator muscles; one from the corpus paracardiacum anastomosing with the maxillary nerve forming a ganglion under the oesophagus and others anastomosing with nerves in the prothorax. There is a nerve from the corpus paracardiacum to the hypocerebral ganglion and one from the corpus paracardiacum to the aorta.

According to Cazal in *Culex pipiens* pupa the hypocerebral ganglion and the corpora paracardiaca are fused into one structure, namely a composite ganglion lying ventral to the aorta. This receives the recurrent pharyngeal nerve and from it proceed two lateral oesophageal nerves terminating in two ventricular ganglia, thus completing the normal stomatogastric system. From the ganglion also arise on each side a nerve to the corpus allatum. Cazal notes, however, that in the larva the composite gland is very poorly developed, whilst the paracardiaca nerves are fused to the hypocerebral ganglion and end in some cells of doubtful nature (see also Possompes, 1946, 1948, who has recently described the above parts in *Chironomus*. A very clear account with figure is given by Imms in his fourth edition (1938), p. 65, embodying much of what is said above).

In *Aëdes aegypti* the parts are much as described by Cazal in *Culex*. The recurrent nerve passes backwards over the roof of the pharynx to the posterior notch in that organ. Here it turns sharply downwards, passing through a median groove in the cingulum muscle to enter the curious T-shaped body forming the first part of the floor of the aorta and which is thought to be the structure described by Imms in *Anopheles* larva as the 'supporting collar'. Further back it enters a rather diffuse ganglionic structure on the ventral floor of the aorta. In some late larvae this appears to consist of a dorsal and a ventral mass. From the ventral mass arise two lateral nerves which at the level of the muscular neck diaphragm end in small collections of cells possibly representing the ventricular ganglia. The dorsal mass gives off stout nerves which pierce the diaphragm and pass to the corpora allata complexes. It has not been found possible to demonstrate the presence of the lateral nerves passing to the hypocerebral ganglion from the brain.

THE CORPORA ALLATA

The corpora allata in *Aëdes aegypti* are included and form a part only of the structures which have been termed by Bodenstein (1945) the *corpus allatum complex*. A very complete description has been given by Bodenstein, including the variation seen in a number of species and genera of mosquitoes, including *A. aegypti*, and the changes taking place in the pupa and adult.* The complex, which in the advanced larva is a conspicuous and easily identified structure, is situated just behind the neck on the tracheal commissure which connects the two dorsal cephalic tracheal trunks (Fig. 38 (2)). It consists on each side of the median line of a backwardly directed somewhat elongated conical structure composed of large cells with large oval nuclei containing, some distance down its length, a round or oval body composed of much smaller cells, the actual corpus allatum. The bodies enclosing the corpora allata are not only attached to but appear in continuity with the tracheal wall. The cells composing them are large with rather clear voluminous cytoplasm and large conspicuous oval nuclei. They are considered by recent authors (Possompes, Cazal) to be of the nature of pericardial cells. The cells composing the corpora allata are small with limited cytoplasm, clearly marked off from the surrounding larger cells. On their inner aspects a nerve enters each complex anteriorly and passes through their substance to the neighbourhood of the corpora allata. As shown by Bodenstein, the outer cells disappear in the adult making the complex much less conspicuous. The characteristic small cells of the corpus allatum are, however, surrounded by a single layer of cells of intermediate size which have first appeared in the young pupa. The differences noted by Bodenstein in the species and genera of mosquitoes examined, including *Anopheles*, though appreciable are not great.

In their figure showing the stomatogastric system in *Chironomus* Miall and Hammond, as does Possompes, indicate peculiar flange-like structures arising from the trachea and extending to the corpora allata. These flanges are clearly homologous with the supporting part of the corpora allata complex in mosquitoes, though in *Chironomus* they arise, not from the tracheal commissure, but from the dorsal cephalic tracheae just caudad of the commissure. As already noted, the cells of the supporting structure have been considered to be of the nature of pericardial cells. But very similar cells with large oval nuclei may be present where large new tracheal vessels are being initiated, for example, the large trachea that forms in connection with the abdominal spiracle of the pupa, or large new tracheae forming to supply the developing wing muscles of the adult. There would appear to be more than a merely casual relation of the complex with the tracheal system.†

(*h*) THE EXCRETORY SYSTEM

THE MALPIGHIAN TUBULES (Fig. 39 (5), (6))

The Malpighian tubules of *Culex pipiens* larva have been described by Metalnikoff (1902); Roubaud (1923); and de Boissezon (1930). Those of the larva of *Anopheles maculipennis* have been described by Nuttall and Shipley (1903); Imms (1907); Roubaud (1923); Missiroli (1925, 1927). Many points regarding their structure and physiology in *Aëdes aegypti* have been given by Wigglesworth (1933, 1942*a*). Studies on the Malpighian tubules of the same species have been made by Ramsay (1951).

* A paper in Russian by Mednikova (1952) also describes the endocrine organs, corpora allata and corpora cardiaca, of mosquitoes (see under references), but I have not seen any translation of this.

† Of authors other than those mentioned who have dealt with the retrocerebral complex in the larvae of related forms to mosquitoes may be noted; Dogiel (1877, who in *Corethra* first described the corpora allata), Miall and Hammond (1900, *Chironomus*), Holmgren (1904, *Chironomus*), Nabert (1913, Tipulid), Frew, (1923, *Forcipomyia*), Selke (1936, Tipulid), Madwar (1937, Mycetophiliidae), Burtt (1937, *Chironomus* and Tipulid), Hanstrom (1942, Diptera), Thomsen (1943, *Calliphora*, with short general review in English), Arve and Gabe (1947, *Chironomus*), Zee and Pai (1944, *Chironomus*).

The tubules of *A. aegypti* have the same general anatomical characters as described in other mosquitoes. Both in the larva and in the imago they are five in number, a peculiarity of Culicidae shared only by *Psychoda*, *Ptychopters* and Blepharocidae (Pentanephridia of Muller, 1881). They are best studied dissected out and examined fresh in Ringer's solution or after special fixation on a cover-glass. As noted by Wigglesworth (1931) in connection with *Rhodnius* they are not usually well fixed in sections. The striated border, as also noted by Wigglesworth, *loc. cit.*, is well shown by dark-ground illumination. For observing the course of the tubules they can be examined in transparent mounted larvae or better in fixed material, the detached abdomen being brought through alcohol and glycerine equal parts to glycerine. The soft cuticle is then readily removed, leaving the hardened tubes and gut in their natural position.

The tubules open into the intestine just behind the characteristic cells of the mid-gut into a slight dilatation of the intestine corresponding to the pyloric ampulla of the imago, one dorsal, *dorsal tubule*, and two on each side, one dorsal and one ventral, that is a *dorso-lateral* and a *ventro-lateral* pair. From their point of entry, which is at about the posterior end of the fifth segment, each tube loops forward to about the anterior border of the segment and, bending backwards, so continues to terminate in a blind end in the eighth segment. In the larva, after the initial forward loop, the course of the tubules is fairly direct. The two ventro-lateral tubules extend to the end of the segment, the others stopping slightly short of this. Though only loosely held in place by tracheal branches the different tubules have a more or less constant position. The ventro-lateral pair pass back on each side in a wavy fashion ventral to the intestine to terminate with their blind ends lying close together near the middle line or more or less separated. The dorso-lateral pair lie dorsal to the intestine and about half-way along their course come together in the middle line, afterwards diverging to lie along each side of the sausage-shaped colon (Fig. 25 (1)). Sometimes they cross each other and come to lie on opposite sides of the colon to their origin. The dorsal tubule in its terminal portion usually comes to lie alongside one or other of the dorso-lateral pair, usually the left.

In the early stages, up to the early fourth instar, the tubules are narrow, of even thickness throughout and translucent. During the fourth instar the cells of the tubules become so charged with granules that the tubules are considerably swollen, opaque and milky white. The portion of tubule forming the ascending part of the anterior loop differs somewhat from the rest of the tubule, being somewhat narrower and remaining relatively free from granular deposit so that it is more difficult to see in the whole mounted larva when the rest of the tubule, being opaque and white, stands out markedly.

In structure each tubule consists of an outer thin membranous coat devoid of muscle within which are large characteristic cells with a striated border surrounding a relatively narrow lumen. The cells are aligned along the tubule on opposite sides alternately, giving to the tubule a somewhat zigzag effect. Each cell occupies the greater part of the circumference of the tube, bulging in its thicker central portion in which the nucleus lies into the lumen and thinning out towards its edges. They may be visualised as more or less square, wrapped round the tube with two opposite angles meeting, or nearly meeting, on the opposite side to the nucleus. The adjoining cells being oppositely placed make good the deficiency where the narrow points meet. There are well-marked lines of demarcation between adjoining cells, which, when the cells are opaque with granules, appear as clear transparent lines. The cytoplasm is uniform in character, finely granular or loaded with coarse granules depending on circumstances affecting excretion. The nucleus is large, clear and spherical and situated centrally in the cell. The lumen in the early larval stages is fairly wide and clearly outlined. As a result of the cells projecting on alternate sides of the tube it has usually a wavy course. When the cells are heavily loaded with granules it tends to be narrow and ill-defined unless distended with fluid or solid matter. The presence of solid secretion in the lumen is commonly described, usually associated with cells transparent and clear from granular matter. This would appear to depend to some extent at least upon conditions affecting the larva. Wigglesworth (1933) notes that if larvae are grown in isotonic or slightly hypertonic salt the lumen of the Malpighian tubules becomes clogged with granules. After transfer to fresh water the tubules in a few minutes are clear and transparent. A similar filling of the lumen by solid excretory matter is described by Roubaud in larvae grown under crowded conditions.

In that portion of the tubule forming the proximal limb of the forward loop which was noted as differing somewhat in character from the remainder of the tubule the cells are smaller and usually more flattened. Also, as noted by Imms, the line of demarcation between the cells is little apparent.

An important point is the nature of the striated border. A well-marked and defined striated border can usually be made out except when the cells are heavily loaded with granules. As seen under dark-ground illumination the border appears as a dark clear zone about $1{\cdot}0\mu$ in width situated next to the more granular cytoplasm and above this, bordering the lumen, is a brightly illuminated line about $2{\cdot}0\mu$ in width. In the early larval stages this clearly defined line can be traced throughout the whole lumen including that in the proximal part of the forward loop. In the larva of *Anopheles maculipennis* Imms also figures the ascending portion of the forward loop as being lined by a uniform narrow layer which has all the appearance of being striated border of the honeycomb type. The question to what extent the tubules in *Aëdes aegypti* show striated border of the brush type is later discussed under the excretory system of the imago.

As shown by Wigglesworth the Malpighian tubules of the larva play an important role in the maintenance of the osmotic control of the larval haemolymph (see under 'Physiology', p. 709 and also further remarks regarding the Malpighian tubules and their functions given when discussing the excretory system of the imago).

Unlike the mid-gut the epithelium of the Malpighian tubules is not replaced during metamorphosis.

OTHER TISSUES CONCERNED IN EXCRETION

Besides the Malpighian tubules certain other tissues also play a part in excretion usually by the storage of waste products. In the mosquito the only tissues of importance in this respect are the cells of the fat-body and the nephrocytes, including the ventral nephrocytes and the pericardial cells. For information regarding these structures in the larva see appropriate headings in the section dealing with the circulatory system and associated tissues.

(*i*) THE REPRODUCTIVE SYSTEM

Certain stages in the development of the reproductive system take place in the body cavity and hypoderm of the larva. There are no external indications of these such as are present in the pupa. These changes have been described in some detail by Christophers (1922, 1923) and reference to some of the structures has been given by Imms (1908). In a recent publication Abul-Nasr (1950) has studied the development of the reproductive system in three representative species of the Nematocera, including *Chironomus dorsalis*, and as a result has employed a common nomenclature for the parts based on the homologies as ascertained by the observed development. The stages present in the mosquito larva are very similar to those described by Abul-Nasr in *Chironomus* and it will be convenient to use Nasr's nomenclature. Before describing the appearances seen in the larva of *Aëdes aegypti* it will facilitate matters to describe very briefly and in outline the general plan on which this nomenclature is based.

Following Abul-Nasr the organs of generation include (*a*) the primary reproductive system consisting of the sexual glands or gonads with associated mesodermal structures, (*b*) the secondary reproductive system comprising those parts of ectodermal origin other than the segmental parts, and (*c*) modified segmental parts including the genital appendages.

Little need be said regarding the gonads and mesodermal structures. These are respectively in the female and male the *ovaries* and the *testes* with their *terminal filaments*, when present, and their *mesodermal strands.* According to Abul-Nasr the last-mentioned structures in *Chironomus* in both sexes are branched, one branch proceeding to the seventh segment and the other to the anterior border of the ninth sternite. The branch which is thickest and eventually develops depends on the sex, namely the strand to the seventh segment in the female and that to the ninth segment in the male, the other strand disappearing before completion of larval life.

The secondary systems differ in the two sexes. In the female the structures are derived chiefly from invaginations at the posterior border of the eighth sternite. Two imaginal buds appear first as separate disks of hypodermal thickening on each side of the mid-ventral line and later coalesce to form the rudiment of the *vagina.* On the dorsal surface of each of these buds there arise the *spermathecal rudiments.* Later a median invagination between and slightly anterior to those forming the spermathecal rudiments forms the rudiment of the *common oviduct.* At a later stage an invagination arising from a thickening on the dorsal wall of the vagina forms the *vaginal apodeme* rudiment. Besides these developments at the posterior border of the eighth segment there is formed in the female a further pair of hypodermal buds on each side of the median line at the base of the ninth segment, which later join and form the rudiment of the *caecus.* Later these two sets of rudiments, those on the eighth and those on the ninth segment, are brought together so that they open into a common cavity, the *atrium.* This latter is a development from the intersegmental membrane between the eighth and ninth segments, enclosed as these segments are brought into closer apposition. It develops certain sclerotisations which will be referred to later when describing the parts in the adult. Meanwhile the common oviduct invagination progresses forwards and links up at the posterior border of the seventh segment with the mesodermal strands either still as an undivided median structure or after bifurcation, thus giving rise to the *oviduct* rudiments. The oviducts, right and left, may in different species be variously formed, therefore, from mesodermal or ectodermal elements, but are liable to eventual replacement of the former by the latter due to the ectodermal tissues replacing or creeping over the mesodermal. The segmental modifications need not here be dealt with as they will be fully described later when dealing with the pupal and adult parts. It is only necessary to say that in the female sex there are no genital appendages.

THE REPRODUCTIVE SYSTEM

In the male the first indications of the reproductive system other than the mesodermic elements appear as hypodermal disks on each side of the middle line at the base of the ninth sternite ventrally. These develop into conspicuous imaginal buds which are the rudiments of the future *gonocoxites* (male forceps, claspers or coxites). Later two further invaginations appear at the bases of the gonocoxite rudiments which are the rudiments of the *ejaculatory ducts*. From each of the ejaculatory duct rudiments there develops an outgrowth which will eventually link up with the mesodermal strand of its side. The ejaculatory duct rudiments and their outgrowths, together with the mesodermal strands, will eventually develop into the vesiculae seminales, vas deferens and other structures which need not concern us for the present. There now occurs a still further and independent invagination which carries forwards with it the ejaculatory duct rudiments and will eventually form the terminal undivided median seminal passage. This is termed by Abul-Nasr the *penis tube*. The name is a little apt to cause some confusion, due to other terms made use of in connection with the extremely complicated reproductive organs in some of the Nematocera, for example in the Tipulidae. For the present it will be sufficient to note that it commonly happens that the penis tube with the tissue surrounding it is excavated by a circumferential fissure to form a projecting organ, the *penis*, and that this may lie within the hollow formed by the encircling fissure which constitutes the *genital cavity*. These later changes need not be further pursued for the present. It is necessary, however, to say something regarding the genital appendages in the male. Internal to the gonocoxite rudiments, arising independently or as lobes divided from the gonocoxite rudiments, are the *paramere lobes* of Pruthi (1925). In many forms these further divide into an outer pair of lobes, the *parameres*, and an inner pair which may form or take part in the formation of the intromittent organ.

With this résumé of the nature of the structures present in the two sexes the reproductive system as present in the larval stage in *Aëdes aegypti* can be described in due perspective.

THE MALE SYSTEM

The testes are situated in the sixth abdominal segment where they are recognisable in sections at least as early as the second instar and quite conspicuous in the third. In the fourth instar they form stout sausage-shaped structures measuring about 0·3 mm. in length, that is nearly the length of the segment. They lie on each side of the segment close to the lateral parietal wall and ventral and external to the large tracheal trunks. Each is closely invested by a special coat of fat-body, which, however, is not pigmented or otherwise different from other lobes of this tissue. The anterior extremity is narrowed, but is not continued into a terminal filament as is that of the ovary. Distally there is a well-marked mesodermal strand which can be traced to the ventral aspect of the posterior border of the seventh segment. I have not been able to demonstrate in *A. aegypti* larva any strand extending to the ninth sternite. Imms, too, in describing the larval testis in *Anopheles*, mentions only a strand passing to the seventh segment. Nor, in the larva, does development appear to proceed as far as the linking up of the strands with the ejaculatory duct rudiments. In structure the testis in the fourth instar has a very characteristic appearance, being largely composed of transverse layers of spermatogonia and spermatocytes from two to more cells in thickness as distinct from the polygonal loculi in many insects. There is a membranous coat with nuclei within which the contained spermatocytes lie apparently without special support. The apical portion, as does also the very young organ, consists of small round cells with dark nuclei. There does not appear to be any large 'apical cell' as present in some insects (Snodgrass, 1935, p. 571). Following upon the apical portion are layers of different stages in the formation of spermatozoa. No fully developed spermatozoa have been seen in the larval stage.

The secondary reproductive system in the male is entirely confined to structures originating at the base of the anal segment. Already in the third instar are present imaginal buds in the position of the imaginal buds of the gonocoxites in the fourth instar. These appear widely separated on either side of the ninth segment close to its anterior border on the ventral aspect. They are very conspicuous and may enable the sex of a larva to be determined even in a fresh specimen. During the fourth instar these buds develop into conical lobes which are clearly the future gonocoxites. These come to occupy the whole length of the anal segment, their apices reaching as far as the ventral fan. Owing to the thickening of the hypoderm over a considerable area at the base of the segment the gonocoxite rudiments when extruded are linked together basally by thickened hypoderm which will eventually form the ninth sternite in the pupa and adult (Fig. 40 (3)–(5)).

Lying basal and internal to the gonocoxite rudiments after these have evaginated are two conspicuous more or less globular masses with central cavities. Each of these is extended outwards as a fine projection (Fig. 40 (5)). There can be no doubt, from the resemblance of these structures to the description and figures of Abul-Nasr for *Chironomus*, that they are the ejaculatory duct rudiments as described by that author. They first appear in the late third instar as thickenings on the gonocoxite imaginal buds. Their cavities are continued into the extensions which arise from them and will become the *vasa deferentia*. These extensions have not been traced in the larva to the mesodermal strands and my previous statement that they were the swollen ends of the strands (Christophers, 1922) requires correction. Late in the fourth instar the position where the invagination of the penis tube will take place is very clearly shown as a dimple in the middle line between the bases of the gonocoxite rudiments (Fig. 40 (5) *pi'*). At this stage also the commencing fissure that will divide off the paramere lobes from the gonocoxites makes its appearance. Later changes are described under the pupa and discussed when dealing with the reproductive system in the adult.

Figure 40. Male reproductive system.

1 Showing position of male gonads and secondary reproductive system. Fourth instar larva.

2 Longitudinal section of male gonad.

3 Anal segment of late third instar larva, showing gonocoxite genital buds not everted and beginning formation of the ejaculatory duct rudiments.

4 The same in an early fourth instar larva, showing everted gonocoxite rudiments and appearance of rudiments of vas deferens on ejaculatory duct rudiments.

5 The same in a late stage fourth instar larva. The vas deferens rudiments have elongated and position of penis cavity invagination indicated.

Lettering: *A*, anal lobe; *gcx.r*, rudiment of gonocoxite; *ejd.r*, rudiment of ejaculatory duct; *ft*, parietal fat-body of segment; *gm*, spermatogonia; *m.go*, male gonad; *mst*, mesodermal strand; *pi'*, site of penis invagination; *spc*, layers of developing spermatocytes; *t.ft*, layer of fat-body surrounding the male gonad; *vd.r*, rudiment of vas deferens.

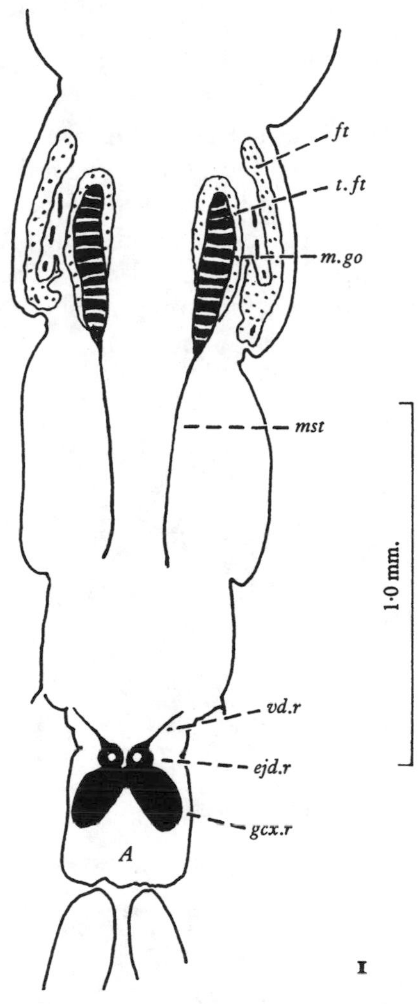

ft
t.ft
m.go
mst
1·0 mm.
vd.r
ejd.r
gcx.r
A
1

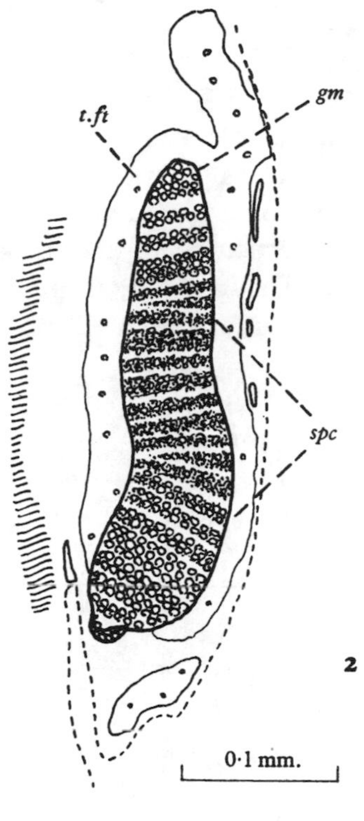

t.ft
gm
spc
2
0·1 mm.

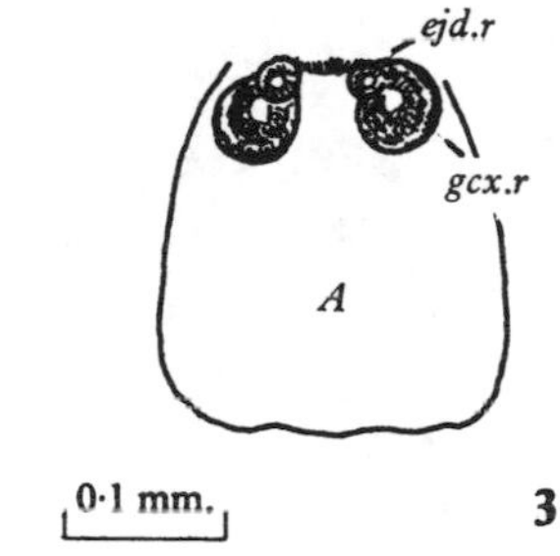

ejd.r
gcx.r
A
0·1 mm.
3

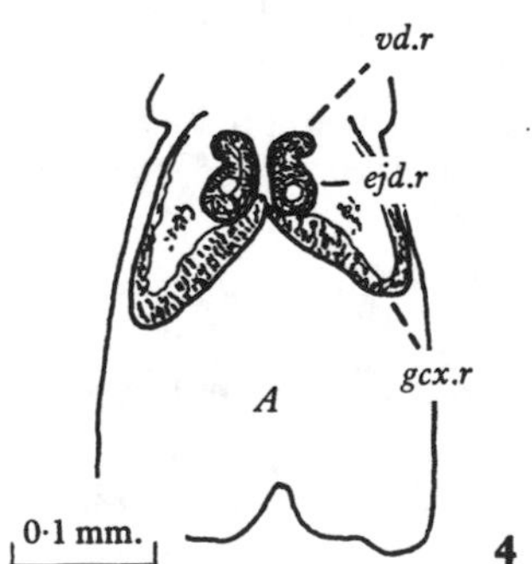

vd.r
ejd.r
gcx.r
A
0·1 mm.
4

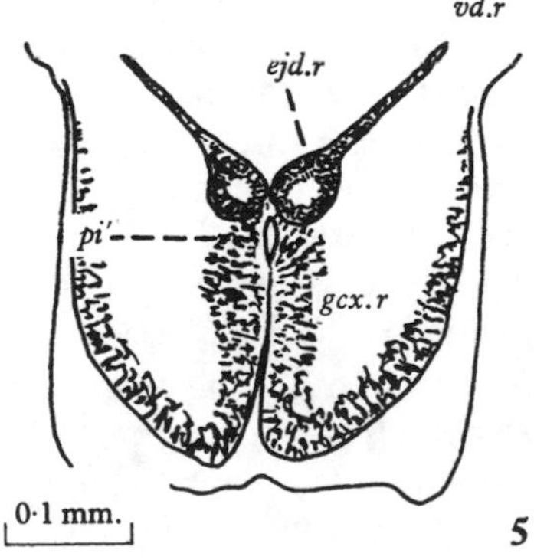

vd.r
ejd.r
pi
gcx.r
0·1 mm.
5

Figure 40

THE FEMALE SYSTEM (Fig. 41)

The female rudiments in *Aëdes aegypti* conform to the usual type of development in the Nematocera as described by Abul-Nasr. The ovaries are situated in the sixth abdominal segment, being already present in the third instar. In the fourth instar they are rather elongate structures of about the length of a segment and extend somewhat into the fifth segment. They are not, as are the testes, provided with a special coating of fat-body. They possess an outer nucleated containing membrane and indications in the arrangement of the

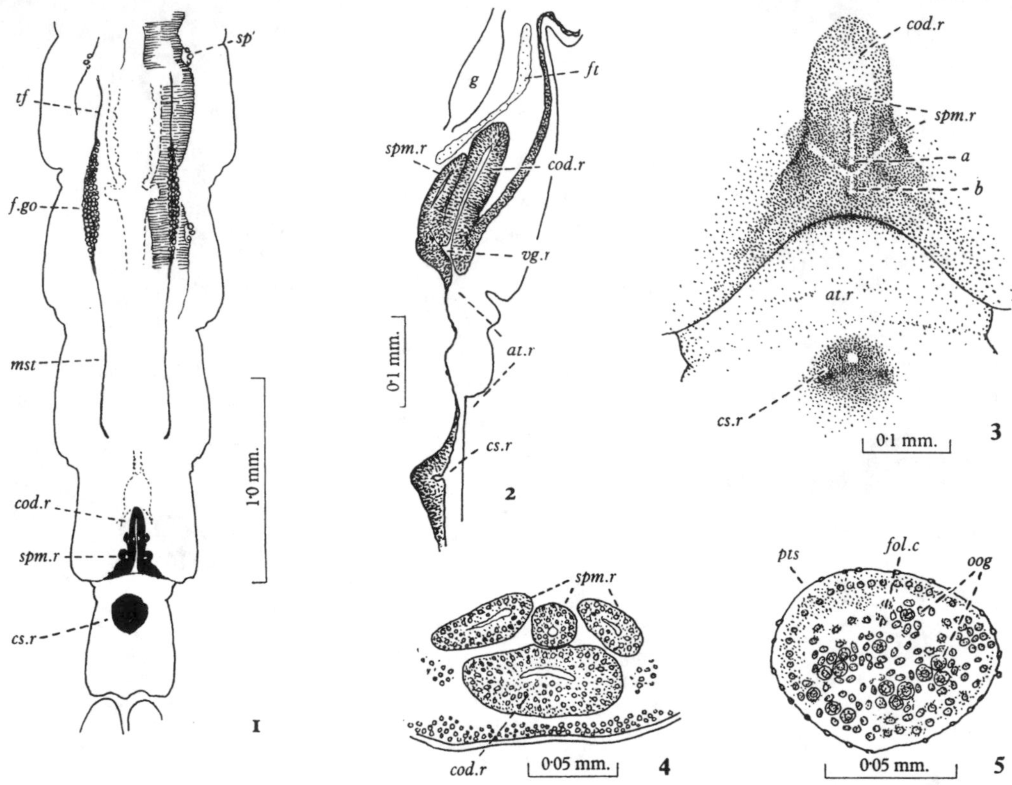

Figure 41. Female reproductive system.

1 Showing position of gonads and secondary reproductive system of female. Fourth instar larva.

2 Sagittal section, showing hypodermal thickenings and invaginations of developing common oviduct, spermathecae and caecus.

3 Ventral view of the same parts. Reconstruction. *a*, opening of duct of median spermathecal rudiment into the vaginal rudiment. *b*, opening of the conjoined ducts of the lateral spermathecae.

4 Transverse section across the common oviduct rudiment and the rudiments of the median and lateral spermathecae.

5 Section of ovary in young larva, showing early formation of follicles and nucleated peritoneal sheath of ovary.

Lettering: *at.r*, area becoming atrium on closing in of parts; *cod.r*, common oviduct rudiment; *cs.r*, caecus rudiment; *f.go*, female gonad; *fol.c*, follicular cells; *ft*, lobe of fat-body; *g*, ganglion of eighth abdominal segment; *mst*, mesodermal strand; *oog*, oogonium; *pts*, nucleated peritoneal sheath; *sp'*, spiracular plaque on tracheal trunk; *spm.r*, spermathecal rudiments; *tf*, terminal filament; *vg.r*, vaginal rudiment.

cells throughout their substance of early follicle formation (Fig. 41 (5)). A conspicuous terminal filament is present, passing forwards from the apex of the organ nearly to the anterior border of the fifth segment in the neighbourhood of the tracheal plaque of that segment where it ends on the surface of the tracheal trunk. A well-marked mesodermal strand extends from the posterior extremity to the hinder border of the seventh segment.

The secondary reproductive system consists of the rudiments of the caecus on the ventral aspect of the base of the anal segment (ninth sternite) and those of the common oviduct and spermathecae at the posterior border of the eighth segment. The caecus rudiment is already present in the third instar, where, as also early in the fourth instar, it consists of two areas of thickened hypoderm with a central depression in each. Later these join to form a single centrally situated rudiment with a central dimple, in which condition it remains throughout the larval period.

The rudiments of the common oviduct and spermathecae are later in appearing. In the early fourth instar there appear on the eighth segment towards its posterior ventral border, and widely separated, two small hypodermal thickenings. These later extend and fuse, forming a median rudiment consisting of a common oviduct rudiment and three spermathecal rudiments. The common oviduct rudiment has a wide posterior opening and a rapidly advancing invagination which at the time of full larval development has extended to the hinder border of the seventh segment. The spermathecal rudiments are three in number. They lie one on each side and one in the middle behind the common oviduct rudiment, the ducts from their cavities joining and entering the cavity of the oviduct rudiment some little distance from its opening. From the resemblance of the early appearance of the two widely separated hypodermal disks to the condition described by Abul-Nasr in *Chironomus* and other features in the development, it is almost certain that the portion of invagination distal to the entry of the ducts is really a separate rudiment, namely that of the vagina. This would be formed by the junction of the original two hypodermal disks coming together as described in *Chironomus*. Unfortunately these early changes have not been sufficiently observed in detail to enable this to be definitely stated. The resemblance, however, of the development in the two forms leaves little doubt of the fact that development in the mosquito and *Chironomus* is closely similar. In *Chironomus* there are two spermathecae, those corresponding to the lateral rudiments in the mosquito. The median spermatheca of the mosquito is therefore not represented, unless its homology be the vaginal apodeme as figured by Abul-Nasr.

A point that should be mentioned is that in the later stages the lateral thickenings of the area on the eighth sternite become much thickened and form projecting ridges. They represent the beginning of the process by which the parts on the eighth and ninth sternites are brought together to form the atrium.

REFERENCES

(*d–e*) THE NERVOUS SYSTEM AND SPECIAL SENSE ORGANS

Bugnion, E. and Popoff, N. (1914). Les yeux des insectes nocturnes. *Arch. Anat. Microsc.* **16**, 261–304.

Constantineanu, M. J. (1930). Der Aufbau der Sehorgane bei den in Süsswasser lebenden Dipterenlarven und bei Puppen und Imagines von *Culex*. *Zool. Jahrb., Anat.* **52**, 251–346.

Demoll, R. (1917). Die Sinnesorgane der Arthropoden, ihr Bau und ihre Funktion. Braunschweig.

GRABER, V. (1882). Die chordotonalen Sinnesorgane und das Gehör der Insekten. *Arch. wiss. Anat.* **20**, 506–640; **21**, 65–145.

GRENACHER, H. (1879). *Untersuchungen über das Sehorgan der Arthropoden.* Göttingen.

HICKSON, S. J. (1885). The eye and optic tract of insects. *Quart. J. Micr. Sci.* **25**, 215–51.

IMMS, A. D. (1908). On the larval and pupal stages of *Anopheles maculipennis* Meigen. *Parasitology,* **1**, 103–32.

IMMS, A. D. (1938). *A General Textbook of Entomology.* Methuen and Co., London.

JOHANSEN, O. A. (1893). Die Entwicklung des Imagoauges von *Vanessa urticae. Zool. Jahrb., Anat.* **6**, 446–80.

RADL, E. (1905). Über das Gehör der Insekten. *Biol. Zbl.* **25**, 1–5.

REDIKORZEW, W. (1900). Untersuchungen über den Bau der Ocellen der Insekten. *Z. wiss. Zool.* **68**, 581–624.

SNODGRASS, R. E. (1935). See references in ch. XIII (*a*).

WIGGLESWORTH, V. B. (1942). *The Principles of Insect Physiology.* Methuen and Co., London (see also many later editions).

ZAVREL, J. (1907). Die Augen einiger Dipterenlarven und Puppen. *Zool. Anz.* **31**, 247–55.

(*f*) THE CIRCULATORY SYSTEM AND ASSOCIATED TISSUES

CUENOT, L. (1891). Étude sur le sang et les glandes lymphatiques dans la série animale. *Arch. Zool. exp.* (2) **9**, 365–99.

CUENOT, L. (1893). Études physiologiques sur les Orthoptères. *Arch. Biol.* **14**, 293–341.

DOGIEL, J. (1877). Anatomie und Physiologie des Herzens der Larve von *Corethra punctipennis. Mém. Acad. Sci. St-Pétersb.* (7), **24**, no. 10.

GRABER, V. (1873). Ueber den propulsatorischen Apparat der Insekten. *Arch. mikr. Anat.* **9**, 129–96.

GRABER, V. (1876). Ueber den pulsierenden Bauchsinus der Insekten. *Arch. mikr. Anat.* **12**, 575–82.

HOLLANDE, A. C. (1909). Contribution à l'étude du sang des Coléoptères. *Arch. Zool. exp. gen.* (5), **2**, 271–94.

HOLLANDE, A. C. (1911). Étude histologique comparée du sang des insectes etc. *Arch. Zool. exp. gen.* (5), **6**, 283–323.

HOLLANDE, A. C. (1914). Oenocytes (review and orig.). *Arch. Anat. Micr.* **16**, 1–64.

HOLLANDE, A. C. (1922). La cellule péricardiale des insectes. *Arch. Anat. Micr.* **18**, 85–307.

HOLMGREN, W. (1904). Zur Morphologie des Insektenkopfes. II. Zum metameren Aufbau des Kopfes der *Chironomus* Larve. *Z. wiss. Zool.* **76**, 439–77.

HOSSELET, C. (1925). Les oenocytes de *Culex annulatus* et l'étude de leur chondriome au cours de la sécrétion. *C.R. Acad. Sci., Paris,* **180**, 399–401.

HURST, C. M. (1890). See references in ch. XIII (*b*).

IMMS, A. D. (1908). See section (*d–e*).

JENSEN, D. V. (1955). Method for dissecting salivary glands in mosquito larvae and pupae. *Mosquito News,* **15**, 215–16.

JONES, J. C. (1952). Prothoracic aortic sinuses in *Anopheles, Culex* and *Aëdes. Proc. Ent. Soc. Wash.* **54**, 244–6.

JONES, J. C. (1953). On the heart in relation to circulation of haemocytes in insects. *Ann. Ent. Soc. Amer.* **46**, 366–72.

JONES, J. C. (1954). The heart and associated tissues of *Anopheles quadrimaculatus* Say (Diptera: Culicidae). *J. Morph.* **94**, 71–124.

KEILIN, D. (1917). Recherches sur les Anthomyides à larve carnivores. *Parasitology,* **9**, 325–450.

KOLLMANN, M. (1908). Recherches sur les leucocytes et le tissu lymphoide des invertébrés. *Ann. Soc. Nat. Zool.* (9), **8**, 1–238.

KOWALEVSKY, A. (1889). Ein Beitrag zur Kenntnis der Excretionsorgane. *Biol. Zbl.* **9**, 33–47, 65–76, 127–8.

KOWALEVSKY, A. (1892). Sur les organes excréteurs chez les Arthropodes terrestres. *Congr. Internat. Zool., Moscow,* Part 1, 187–234.

LANGE, H. H. (1932). Die Phagocytose bei Chironomiden. *Z. Zellforsch.* **16**, 753–805.
LEBRUN, H. (1926). L'appareil circulatoire de *Corethra plumicornis*. *Cellule*, **37**, 183–200.
MADWAR, S. (1937). Biology and morphology of the immature stages of Mycetophilidae (Diptera: Nematocera). *Phil. Trans. R. Soc.* B, **227**, 1–110.
METALNIKOFF, S. (1902). Beiträge zur Anatomie und Physiologie der Mückenlarve. *Bull. Acad. Imp. Sci. St-Pétersb.* **17**, 49–58.
MUTTKOWSKI, R. A. (1924). Studies on the blood of insects. II. The structural elements of the blood. *Bull. Brooklyn Ent. Soc.* **19**, 4–19.
PAUSE, J. (1919). Beitrag zur Biologie und Physiologie der Larve von *Chironomus gregarius*. *Zool. Jahrb., Physiol.* **36**, 339–452.
RAJINDAR PAL (1944). Nephrocytes in some Culicidae. Diptera. *Ind. J. Ent.* **6**, 143–8.
RASCHKE, E. W. (1887). See references in ch. XIII (*a*).
ROOSEBOOM, M. (1937). Contribution à l'étude de la cytologie du sang de certaines insectes etc. *Arch. Néerl. Zool.* **2**, 432–551.
TOWER, W. L. (1902). Observations on the structure of the exuvial glands and the formation of the exuvial fluid in insects. *Zool. Anz.* **25**, 466–72.
TZONIS, K. (1936). Beitrag zur Kenntnis des Herzens der *Corethra plumicornis* Larve (*Chaoborus crystallinus* Geer.). *Zool. Anz.* **116**, 81–90.
WEBER, H. (1938). *Grundriss der Insektenkunde*. Gustav Fischer, Jena.
WIELOWIEJSKI, H. R. (1886). Über das Blutgewebe der Insekten. *Z. wiss. Zool.* **43**, 512–36.
WIGGLESWORTH, V. B. (1933). Physiology of the cuticle and of ecdysis in *Rhodnius prolixus* (Triatomidae, Hemiptera), with special reference to the function of the oenocytes and of the dermal glands. *Quart. J. Micr. Sci.* **76**, 269–318.
WIGGLESWORTH, V. B. (1934). The physiology of ecdysis in *Rhodnius prolixus*. II. Factors controlling moulting and metamorphosis. *Quart. J. Micr. Sci.* **77**, 191–222.
WIGGLESWORTH, V. B. (1937). Wound healing in an insect (*Rhodnius prolixus*, Hemiptera). *J. Exp. Biol.* **14**, 364–81.
WIGGLESWORTH, V. B. (1942*a*). *The Principles of Insect Physiology*. Ed. 2. London.
WIGGLESWORTH, V. B. (1942*b*). The storage of protein, fat, glycogen and uric acid in the fat-body and other tissues of mosquito larvae. *J. Exp. Biol.* **19**, 56–77.

(*g*) THE RETROCEREBRAL COMPLEX AND CORPORA ALLATA

ARVY, L. and GABE, M. (1947). Contribution à l'étude cytologique et histochimique des formations endocrines rétro-cérébrales de la larve de *Chironomus plumosus* L. *Rev. Canad. Biol.* **6**, 777–96.
BODENSTEIN, D. (1945). The corpora allata of mosquitoes. *44th Rep. Connect. State Entom. Connect. Agric. Exp. Sta.* Bull. 488, 396–405.
BURTT, E. T. (1937). On the corpora allata of dipterous insects. *Proc. R. Soc.* B, **124**, 13–23.
CAZAL, P. (1948). Les glandes endocrines rétrocérébrales des insectes (Étude morphologique). *Bull. biol.* Suppl. 32.
DOGIEL, J. (1877). See under section (*f*).
FREW, J. G. H. (1923). On the larval and pupal stages of *Forcipomyia piceus*. *Ann. Appl. Biol.* **10**, 409–41.
HANSTROM, B. (1942). Die corpora cardiaca und corpora allata der Insekten. *Biolog. generalis*, **15**, 485–531.
HOLMGREN, N. (1904). See under section (*f*).
IMMS, A. D. (1938). See under sections (*d*)–(*e*).
MADWAR, S. (1937). See under section (*f*).
MEDNIKOVA, M. V. (1952). The endocrine glands, corpora allata and corpora cardiaca of mosquitoes (Fam. Culicidae). (In Russian.) *Zool. Zhurnal*, **31**, 676–85.
MIALL, L. C. and HAMMOND, A. P. (1900). *The Structure and Life History of the Harlequin Fly* (Chironomus). Clarendon Press, London.
NABERT, A. (1913). Die corpora allata der Insekten. *Z. wiss. Zool.* **104**, 181–358.

Possompes, B. (1946). Les glandes endocrines post-cérébrales des Diptères. I. Études chez la larve de *Chironomus plumosus* L. *Bull. Soc. Zool. France*, **71**, 99–109.

Possompes, B. (1948). Les corpora cardiaca de la larve de *Chironomus plumosus* L. *Bull. Soc. zool. Fr.* **73**, 202–6.

Sellke, K. (1936). Biologische und morphologische Studien an schädlichen Wiesenschnaken (Tipulidae: Dipt.). *Z. wiss. Zool.* **148**, 465–555.

Thomsen, E. (1943). An experimental and anatomical study of the corpus allatum in the blow-fly, *Calliphora erythrocephala* Meig. *Vidensk. Medd. dansk naturh. Foren. Kbh.* **106**, 320–405.

Thomsen, E. (1952). Functional significance of the neurosecretory brain cells and the corpus cardiacum in the female blow-fly, *Calliphora erythrocephala* Meig. *J. Exp. Biol.* **29**, 137–72.

Zee, H. C. and Pai, S. (1944). Corpus allatum and corpus cardiacum in *Chironomus* sp. *Amer. Nat.* **78**, 472–7.

(*h*) THE EXCRETORY SYSTEM

De Boissezon, P. (1930). Contributions à l'étude de la biologie et de la histophysiologie de *Culex pipiens*. *Arch. Zool. Exp. Gen.* **70**, 281–431.

Imms, A. D. (1907). On the larval and pupal stages of *Anopheles maculipennis* Meigen. *J. Hyg., Camb.* **7**, 291–316.

Metalnikoff, S. (1902). See under section (*f*).

Missiroli, A. (1925). I tubuli del Malpighii nell' *Anopheles claviger*. *Ann. Igiene*, **35**, 113–22.

Missiroli, A. (1927). I tubuli del Malpighii nell' *Anopheles claviger*. *Riv. di Malariol.* **7**, 1–7.

Muller, F. (1881). Wissenschaftliche Mitteilungen. I. Verwandlung und Verwandtschaft der Blepharociden. *Zool. Anz.* **4**, 499–502.

Nuttall, G. H. F. and Shipley, A. E. (1903). Studies in relation to malaria. *J. Hyg., Camb.* **3**, 166–215.

Ramsay, J. A. (1951). Osmotic regulation in mosquito larvae. The role of the Malpighian tubules. *J. Exp. Biol.* **28**, 62–73.

Roubaud, E. (1923). Les désharmonies de la fonction rénale et leur conséquences biologiques chez les moustiques. *Ann. Inst. Pasteur*, **27**, 627–79.

Wigglesworth, V. B. (1931). The physiology of excretion in a blood-sucking insect, *Rhodnius prolixus* (Hemiptera, Reduviidae). II. Anatomy and histology of the excretory system. *J. Exp. Biol.* **8**, 428–42.

Wigglesworth, V. B. (1933). The adaptation of mosquito larvae to salt water. *J. Exp. Biol.* **10**, 27–37.

Wigglesworth, V. B. (1942*a*). See under section (*f*).

(*i*) THE REPRODUCTIVE SYSTEM

Abul-Nasr, S. E. (1950). Structure and development of the reproductive system of some species of Nematocera (Diptera: Nematocera). *Phil. Trans.* B, **234**, 339–96.

Christophers, S. R. (1922). The development and structure of the terminal abdominal segments and hypopygium of the mosquito with observations on the homologies of the terminal segments of the larva. *Ind. J. Med. Res.* **10**, 530–72.

Christophers, S. R. (1923). The structure and development of the female genital organs and hypopygium of the mosquito. *Ind. J. Med. Res.* **10**, 698–719.

Imms, A. D. (1908). See under section (*f*).

Pruthi, H. Singh (1925). Development of the male genitalia in Homoptera and preliminary remarks on the nature of these organs in other insects. *Quart. J. Micr. Sci.* **69**, 59–96.

Snodgrass, R. E. (1935). See references in ch. XIII (*a*).

XV
THE PUPA

(*a*) GENERAL DESCRIPTION AND EXTERNAL CHARACTERS

GENERAL DESCRIPTION (Fig. 42)

The mosquito pupa, unlike the pupa of most insects, is quite an active creature. It is not, however, as it is sometimes termed, a nymph, a name which is properly reserved for the immature stages of insects which pass through a gradual metamorphosis. Three types of pupa are distinguished by Comstock (1920), namely those in which the legs and wings are free as in the Hymenoptera and Coleoptera (exarate pupac); those in which the legs and wings are glued to the body as in the Lepidoptera (obtected pupae); and those enclosed in the hardened larval skin as in many Diptera (coarctate pupae). The legs and wings of the mosquito pupa are glued to the body and in Comstock's classification it is an *active obtected pupa.*

In general appearance the pupa of *Aëdes aegypti* resembles that of most other culicine mosquitoes. It differs from the anopheline pupa in having a more rounded and plump body and breathing trumpets which are cylindrical, not broadly flap-like as in that genus. The chief structures used in identification from other culicine pupae are: (1) the form of the respiratory trumpets; (2) the arrangement and characters of the hairs, especially those of the abdominal segments; and (3) the characters of the tail-fins or paddles.

The body of the pupa consists of a large globular anterior portion, *cephalothorax*, and a narrower articulated *abdomen.*, which is normally kept flexed under the cephalothorax and is used to propel the insect when swimming. This outward appearance, however, is largely a simulacrum, the result of glueing together of head, body, wings and legs to form a seeming whole. This can be clearly demonstrated if a newly emerged pupa is placed in some fixative, for example formalin acetic alcohol, which dissolves the cement bathing the parts before this hardens. Pupae so treated show the head and mouth-parts free, the wings opened out, the legs free and loosely displayed and the true body (thorax) made evident.

Outstanding among descriptions of the pupa and its structure are the classical papers by Hurst (1890), the very detailed description of the changes taking place in this stage by Thompson (1905), and the very useful short account of pupal characters used in systematic work by Edwards (1941). Much information on pupal structure is also given in the papers by Marshall and Staley (1932), and by Brumpt (1941), dealing with the mechanism of emergence (see also Imms (1907, 1908); Eysell (1911); Séguy (1923); Sen (1924); and the early work of Causard (1898), drawing attention to the part played by air in emergence of aquatic pupae with mention of the mosquito on p. 260).

On pupal chaetotaxy are: Ingram and Macfie (1917, 1919); Macfie (1920) whose description of the chaetotaxy of the pupa of *A. aegypti* still forms the basis of work on this subject, and more recent papers by Senevet (1930–40); Christophers (1933); Baisas (1934–8) Crawford (1938); Belkin Knight and Rozeboom (1945); Rozeboom and Knight (1946); Knight and Chamberlain (1948); Mattingly (1949); Baisas and Pagayon (1949); Belkin (1951–3).

THE PUPA

Contributions dealing with special features are: Giles (1903), on the prepupa (see ch. x (*e*)); Kulagin (1905), on the pupal head and mouth-parts; Theodor (1924), on the respiratory trumpets of *A. aegypti* pupa; Adie (1912), Cantrell (1939) and Trembley (1944), on sex differentiation; Zavrell (1907) and Constantinianu (1930), on pupal ocelli; Hosselet (1925), on oenocytes in the pupa.

THE HEAD AND MOUTH-PARTS (Figs. 42, 43)

At the front of the cephalothorax is the flattened *head* with the trunk-like *mouth-parts*, forerunners of the proboscis, curled round beneath the cephalothorax like a keel. If the head, preferably of a young pupa, be carefully detached from the rest of the cephalothorax and laid on the flat facing the observer and so mounted without undue pressure, most of the structures composing these parts will be well displayed (Fig. 42 (2), (3), (4)). Conspicuous on the sides of the head are the large *compound eyes* and behind these the five *ocelli* much as in the larva. In front of the compound eyes in their upper portions and approaching each other in the median line are the large globular *basal segments of the antennae*, somewhat larger in the male than in the female. From these the *antennae* sweep back in a curved

---

Figure 42. The pupa.

1 Lateral view of female pupa of *Aëdes aegypti.*

2 Anterior view of same. The left wing is shown slightly separated from the cephalothorax. *a*, outer hard cuticle forming part of cephalothoracic cuticle; *b*, thin membrane of inner surface of wing not normally exposed.

3 Anterior view of detached head of *A. aegypti* pupa (female).

4 The same of the male pupa. In both cases the much retracted hypoderm forming the imaginal parts is shown.

5 Bilobed membranous sac of tip of pupal labium showing thickenings and papillae. *a*, *b*, thickenings of the membrane; *c*, papilla of left lobe; *d*, developing labella.

6 Transverse section of respiratory trumpet at level of opening. *a*, outer cuticular layer; *b*, felt-like inner lining composed of thin filamentous or sheet-like extensions of cells. Note double layer of hypoderm.

7 Dorsal view of cephalothorax and first two abdominal segments. *a*, inset membranous area in first abdominal tergite; *b*, rod-like thickening on membrane extending from near base of float hair.

8 Lateral view of first abdominal segment showing extensive membranous area with inset large functional spiracle opening into the ventral air cavity.

9 Ventral view of same. Only a thin line of thickening is present. Segment II ventrally has cuticle resembling that of other segments.

Lettering: *an*, antenna; *an′*, basal lobe of antenna; *ats*, antero-thoracic setae; *CL*, clypeus; *E*, compound eye with 5 ocelli; *G*, general plate; *ha*, pupal halter; *Lr*, labrum; *md*, mandible; *mt*, metanotum; *mx*, maxilla; *mxp*, maxillary palp, with developing imaginal palp; *psc*, post-scutellar area; *psc′*, torn edge of same; *rt*, respiratory trumpet; *rt′*, anterior thickening at base of same; *rt″*, posterior thickening from wing base; *sc*, scutum; *sc′*, torn edge of same; *so*, sensillum ('hairless ring'); *sp*, spiracle; *st*, mark of functionless spiracle; *T*, tentorium with posterior end torn from head membrane; *T′*, anterior opening of tentorium; *tb*, tibiae (outer surface); *tb′*, *tb″*, tibiae of fore and mid legs; *ts*, tarsi; *vp*, vertical plate; *W*, pupal wing.

Roman numerals indicate segments.

Arabic figures indicate the setae as given under Belkin's (1953) system in Table 21.

For items in above lettering not shown on Fig. 42 see Fig. 43.

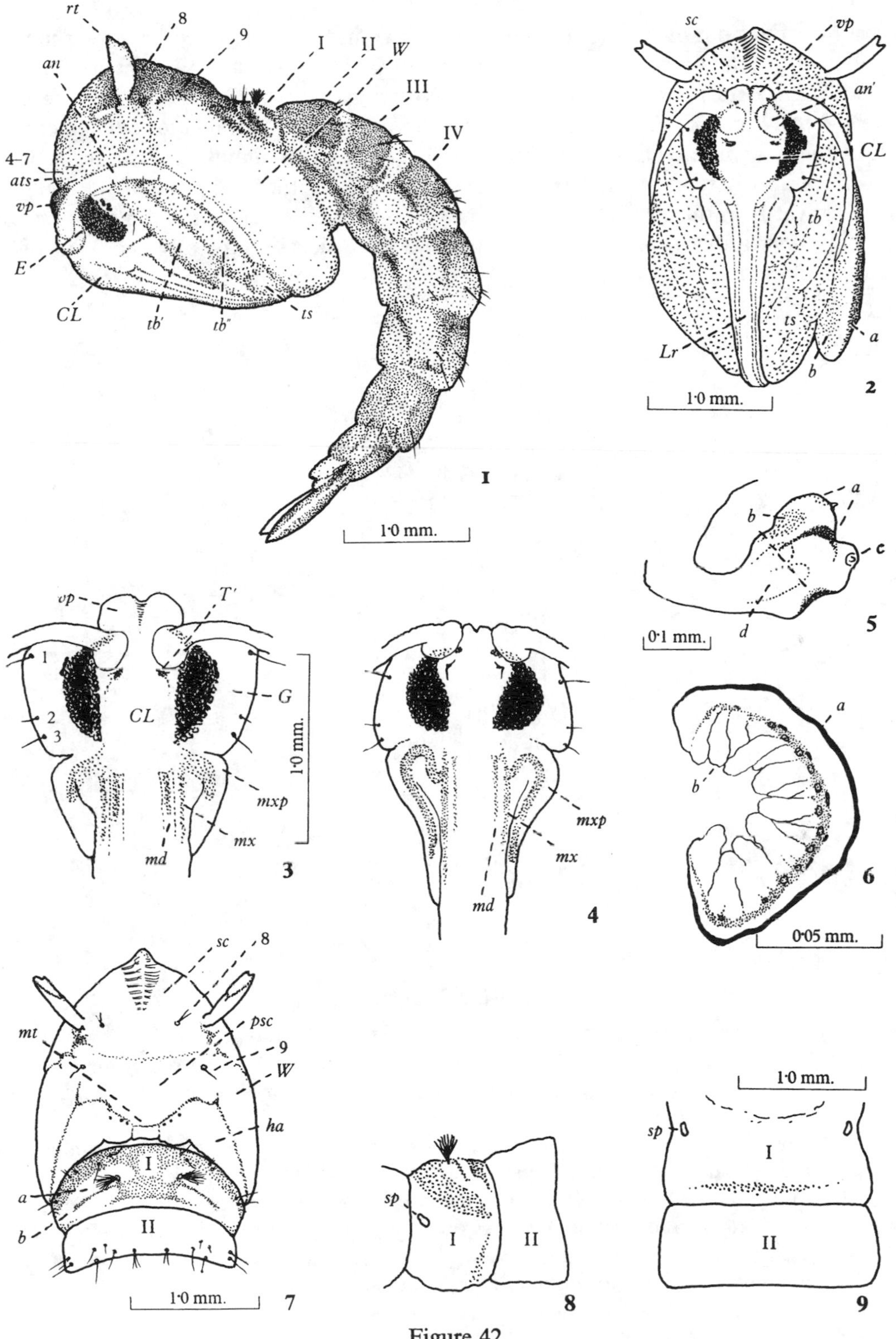
rt
8
9
I
II
W
III
IV
an
4–7
ats
vp
E
CL
tb'
tb"
ts
1·0 mm.
sc
vp
an'
CL
tb
ts
Lr
a
b
1·0 mm.
2
a
b
c
d
0·1 mm.
5
vp
T'
1
2
3
G
CL
1·0 mm.
mxp
mx
md
3
mxp
mx
md
4
a
b
0·05 mm.
6
sc
8
psc
mt
9
W
ha
I
a
b
II
1·0 mm.
7
sp
I
II
8
1·0 mm.
sp
I
II
9

Figure 42

fashion over the sides of the thorax. Passing back from the narrow gap between the basal lobes onto the dorsum of the head are two small plates separated by a shallow median groove. These correspond to the vertex of the future imaginal head and may be termed the *vertical plates* (*head shields* of Belkin, Knight and Rozeboom). During the act of emergence the longitudinal split down the back of the pupa ends on reaching the posterior border of these plates and is replaced by a transverse slit. It is here that the head of the imago is pushed forwards in the early stages of emergence. After emergence is complete the plates are pushed forwards to become two conspicuous oval pieces at the extreme front of the cast pupal skin (Fig. 43 (6) *vp*).

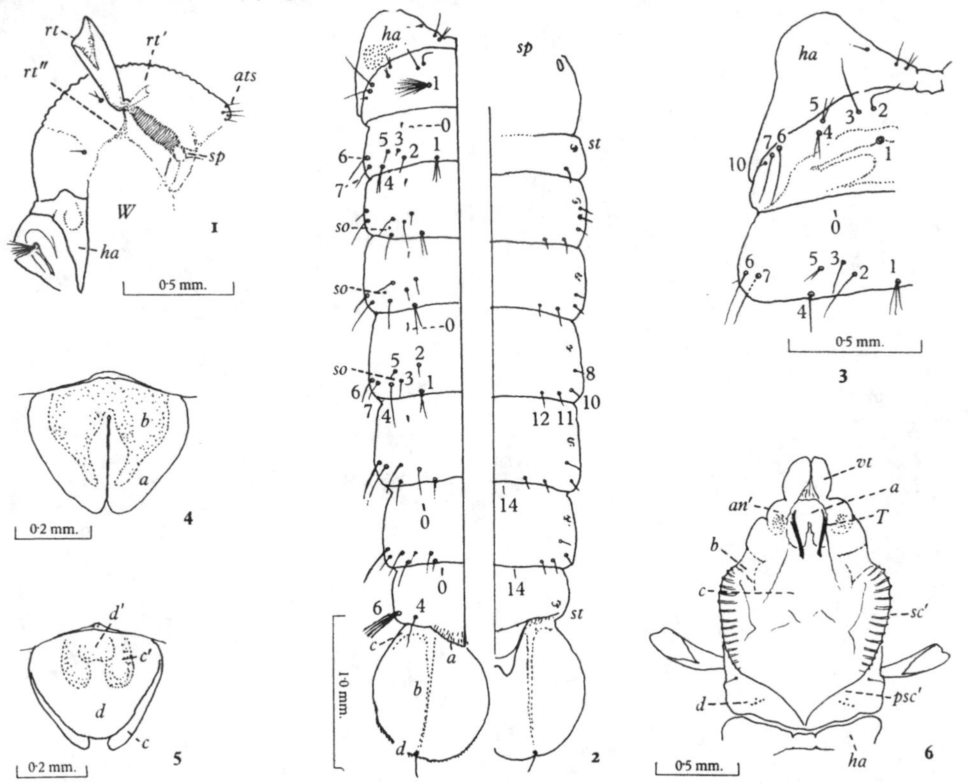

Figure 43. The pupa.

1 Lateral view of scutum showing connections of respiratory trumpets.

2 Dorsal (left) and ventral (right) view of abdominal segments showing setae numbered according to Belkin (1953). *a*, medium portion ninth tergite; *b*, midrib; *c*, buttress; *d*, paddle hair.

3 Dorsal view of metanotum and left halter. Also dorsal view of left side of abdominal segments I and II. Hairs as under 2 but at higher magnification.

4 Pupal hypopygium of male. Within the pupal gonocoxites are the retracted developing gonocoxites of the imago. *a*, pupal gonocoxite; *b*, developing gonocoxite of imago.

5 Pupal hypopygium of female. Ventral view. *c*, cercus partly hidden by *d*, ninth sternite; *c'*, developing cercus of imago; *d'*, developing ninth sternite of imago.

6 Dorsal view of pupal skin as left floating after emergence. Only cephalothorax is shown. See also Fig. 48. *a*, opening into sheath of clypeus and mouth-parts; *b*, membrane of back of head now opened out; *c*, hollow occupied by anterior coxae; *d*, sensory pits.

For lettering see under Fig. 42.

Lateral to the compound eyes there is, on each side of the head, a smooth flat area of cuticle which carries the three *postocular setae* and corresponds to a part of the genal area of the larval head. At its posterior border this is reflected inwards as soft membrane forming the back of the head and apparent only on dissection or in sections. At a little distance below the level of the bottom of the compound eyes this surface terminates on each side at a cleft or groove beyond which is the large flattened maxillary palp of its side. This area forming the head laterally may be termed the *genal plate* of the pupa (Fig. 42 (3) *G*).

On the front of the head between the compound eyes and below the gap between the basal lobes of the antennae is a conspicuous bulged area, the *clypeus* (Fig. 42 (3) *CL*). On each side of this in the angle between the compound eye and the basal lobe of the antenna are the *anterior openings of the tentorium*. As in the larva the hollow tentorial bars pass through the head, in this case to the end in the soft membrane already referred to as forming the back of the head. Passing downwards the clypeus continues without a break into the trunk-like *labrum*. Glued on each side of the labrum in both male and female are the long thin *pupal mandibles* and external to these the *pupal maxillae*. External to the maxillae in their basal portions are the broad flat areas of the *pupal maxillary palps* (Fig. 42 (3), (4) *mxp*). In the female these are somewhat shorter than in the male and within them can be seen the short rather club-shaped female imaginal palps modelled in the hypoderm. In the male the contained developing imaginal palps are long and thin and bent into an S-shaped curve (Fig. 42 (4) *mxp*). Behind, and largely hidden by the parts so far described, is the stout, but largely membranous, *labium*. In the female, lying in the same pupal sheath as the labium is the future imaginal *hypopharynx* identifiable by its sharp-pointed end. The hypopharynx develops as a ridge on the dorsal aspect of the labium and in the female later separates as a distinct organ. In the male it fails to separate. This organ is one of the few which is formed entirely in the pupal stage.

All these elements continue in close approximation along the ventral aspect of the cephalothorax to its posterior part where they curve upwards to end in close association with the membranous ventral aspects of the first and second abdominal segments. Here the labium ends in a bilobed bag-like swelling furnished with some whispy sclerotisations and at the tip of each lobe a minute papilla (Fig. 42 (5)). Within this membranous bag lie the much shrunken-away labella of the imaginal labium. Closely applied to the sac ventrally is the turned-up spatulate end of the pupal labrum marked by conspicuous transverse ridges and at its sides are the closely applied undifferentiated ends of the pupal mandibles and maxillae.

Owing to the early retraction of the hypoderm all these structures in the pupa are in the form of cuticular sheaths and already, even in the young pupa, there is a wide space between this pupal cuticle and the developing hypoderm forming the corresponding imaginal parts. The pupal wings and legs have already been extended in the formation of the prepupa within the body of the larva. But the mouth-parts up to the time of pupation have been still in the form of invaginated rudiments, namely: a large plicated backwardly projecting mass in the frontal region very conspicuous in sections of the later stages of the pupa, which when projected forwards constitutes the pupal labrum; a large invaginated double germinal bud lying in and fully occupying the labial area, which when extruded at pupation forms the pupal labium; and still other rudiments compacted in the stumpy larval mandibles and maxillae, which during pupation are pushed out to form the long slender organs of the pupa. More will be said about the extrusion of these various structures when describing the act of emergence in which as shown by Thompson they play a distinct role.

THE THORAX (Figs. 42, 43)

When the head has been removed there comes into view that part of the pupal surface corresponding to the prothorax and above this a convex surface corresponding to the anterior promontory of the imaginal thorax. On the sides of the pupal thorax just above the curved antennae towards their base are the *antero-thoracic setae* described later and further back is the broad convex *scutum* of the mesothorax. Along the median dorsal line of the scutum for the greater part of its extent is a sharp crest-like ridge, *median keel* of Belkin, Knight and Rozeboom. On each side of this crest is a band of smooth cuticle crossed by regularly spaced transverse thickenings. The central ridge forms the line of weakness along which the cuticle splits in the act of emergence. The bands of ridged cuticle at its sides are pushed over to the sides during emergence and form supporting edges giving stability to the pupal skin and helping to make this a safe platform for the emerging imago (Fig. 43 (6) *sc'*). Projecting from the sides of the scutum are the *respiratory trumpets* and dorsal to these towards the median line is on each side the single small *dorsal seta.* A little posterior to the trumpets are the *wing roots* from which the *pupal wings* pass downwards on the sides of the thorax. The pupal wings show little or no indication of venation, which, however, is already showing up in the contained hypoderm forming the developing wings of the imago. Across the scutum at a level a little behind the wing roots is a transverse line behind which the surface is smooth and just before which the median crest and its accompanying striated areas cease. Beneath the cuticle in this position and stretching from side to side can be seen the developing scutellum of the imago and behind this the postscutellum. On each side just dorsal to the wing root is the single small *supra-alar seta.* Behind the extensive scutum is a short transverse bar, the *metanotum*, linking together two flat triangular areas constituting the *pupal halteres.* These lie behind the pupal wings with their apices directed posteriorly along the sides of the first abdominal segment. Within them can be seen the much smaller club-shaped developing imaginal halteres. On the bases of the halteres are the *metanotal hairs*, three on each side (Figs. 42 (7) *ha*; 43 (1), (2), (3) *ha*).

Occupying the remaining parts of the lateral aspects of the pupa are the outer sides of the tibiae of the three pairs of legs, those of the fore- and mid-legs in front of the wings and those of the hind legs lying just under the wing edge (Fig. 42 (1) *tb*). In the prepupa it has been noted that the tarsi are turned up and lie behind and parallel to their respective tibiae. But during pupation they have become straightened out and in the pupa they lie in line with the tibiae and are curled up under the wing apex. The tarsi of the fore- and mid-legs make single loops in this position and those of the hind legs a double S-shaped loop.

The respiratory trumpets (Fig. 43 (1) *rt*). The respiratory trumpets of the Culicini, in contrast to those of the Anophelini, are more or less cylindrical in shape, those of the Anophelini being more flap-like. A summary of the characters used in identification is given by Edwards (1941). Following Ingram and Macfie (1919), the basal closed portion is termed the *meatus* and the open portion the *pinna.* The meatus is further subdivided into a proximal *tracheoid* portion marked with transverse striae and a distal *reticulated* portion in which the surface is covered with a fine network. The respiratory trumpets of *Aëdes aegypti*, as noted by Theodor (1924) possess no tracheoid portion. They are tubular, somewhat expanded from base to apex with a fairly oblique opening and when flattened appear rather broadly triangular. The open portion occupies one-third to one-half of the whole with a rounded notch in the apical lip. The basal end of the opening is narrow and terminates as a fine slit. The base is very loosely articulated allowing the trumpet to be easily swung into position by surface forces when brought to the surface. At the level of this articulation is a sharp bend which may possibly act as a valve closing the air passage when the trumpet is in certain positions. Giving support to the basal articulation is a cuticular fold anteriorly and an area of thickening extending from the wing base posteriorly (Fig. 43 (1) *rt′*, *rt″*).

The trumpets arise from a triangular area formed by an extension of the scutum into the angle between the wing root behind and the curve of the antenna in front, an area that in the imago will form a somewhat depressed region of the pleura surrounding the anterior thoracic spiracle of the imago. Shining through the cuticle here is seen a considerable length of ballooned tracheal trunk. This leads from the trumpet to the developing anterior spiracle of the imago and is cast off with the pupal skin at emergence (see under tracheal system).

The ventral air space (Fig. 48 (5) *a*). When the parts forming the cephalothorax become cemented, any spaces where the parts have not been in contact will become sealed off within the apparent body of the pupa. An important space of this nature is a large cavity lying ventral to the thorax. It is walled in laterally by the wings and legs and below by the mouth-parts. Within this space, which is strictly outside the real body of the pupa, is a collection of air forming a large bubble which can be readily seen in the young pupa before the parts darken. It is to a large extent due to this bubble, as pointed out by Hurst, that the pupa, besides being made very buoyant, is properly aligned at the surface in relation to the respiratory trumpets. Also as pointed out by Hurst there is a most important point in connection with this space, namely that into it open the large patent first abdominal spiracles. Thus the air in the cavity communicates via the tracheal system of the pupa with the external air. An interesting point in this connection, which does not appear to have been so far noted, is that pressure within the sac is apparently under control. Looking down upon a young (white) pupa the presence of the bubble where this presents in the region of the halteres is very evident. From time to time as this area is watched something very like a wink occurs. At first this was thought to be due to some movement of the halteres, but was afterwards seen to be due to the opening and closing of the large first abdominal spiracle.

THE ABDOMEN

The abdomen of the pupa has a curious resemblance to that of a lobster, consisting as it does of heavily sclerotised segments which are freely moveable on each other in flexion and extension with little or no lateral movement, the resemblance being further added to by the large flat terminal paddles or tail fins. Disregarding the tail fins and parts derived from more posterior segments there are eight fairly equally sized abdominal segments. The first lies posterior to the metanotum and carries on each side a conspicuous large branched hair, the *dendritic tuft* or *float hair* of Edwards. These hairs, from their size and black colour, are already conspicuous under the cuticle of the larva as it approaches pupation and stand out prominently to the naked eye in the newly emerged white pupa. Each arises at the edge of an oval area of soft membrane set in the tergite. Extending across the membrane, one end in close connection with the hair, is a long straight thickening strongly suggesting the stylus of a drum membrane. When the pupa is at the surface the float hairs lie close beneath the surface film. As suggested by Hurst they are probably sensory in function. Ventrally and laterally the first abdominal segment is largely membranous. On it, situated laterally and a little dorsally, are the large oval first abdominal spiracles.

The second to the seventh segments are approximately of the same size and character. They consist of completely sclerotised rings, there being no area of pleural membrane laterally to indicate division between tergite and sternite, though markings on the cuticle give an appearance of this (Fig. 42 (1)). The segments carry rather inconspicuous hairs as described later. At least two of these hairs, hairs 8 and 10, arise from the pleural region of the segment.

The eighth segment carries the paddles and other structures which are portions of the ninth segment. It carries a reduced complement of hairs, the most conspicuous being a large branched hair on each posterior angle.

Dorsally the eighth tergite is continued backwards into a semilunar median projection with crinkled edges, the *median lobe of the ninth tergite*. This is the form finally taken by the median portion of the hypoderm of the larval siphon after the tracheal trunks and other stigmatic structures have been withdrawn from it at pupation. Beneath this plate can still be seen in sections the shrunken remains of the cellular coat of the tracheal trunks left to degenerate after the cuticular intima has been withdrawn. On each side of the posterior margin of the segment are articulated the *pupal paddles*. These are withdrawn at pupation from the lateral portions of the larval siphon and are the *lateral lobes of the ninth tergite*. Lying ventral to the paddles are the parts withdrawn from the anal segment of the larva that will form the proctiger of the hypopygium of the imago with characters depending upon the sex.

The pupal paddles (Fig. 43 (2) (*b*)). In *A. aegypti* the paddles are almost circular in shape, being only slightly longer than broad (index 1·2–1·3). They possess a median longitudinal thickening, the *midrib*, and at their base a thickening termed by Belkin, Knight and Rozeboom the *buttress*. At the termination of the midrib is the *paddle hair* or *apical seta*, which in *A. aegypti* is about one-quarter the length of the paddle. In the *Anopheles* pupa a small accessory paddle hair is present near the main hair arising ventrally. In *Culex* an accessory hair is also present arising dorsally. In *Aëdes aegypti* there is no accessory hair. On the distal margin of the paddle there are present in *A. aegypti* fine microtrichiae-like spines, somewhat larger on the portion external to the paddle hair. The paddles normally overlap by about half their width. They are the chief organs of pupal motility.

The pupal hypopygium (Fig. 43 (4), (5)). Ventral to the paddles in the male are the relatively large *gonocoxites*. These form conical projections with a deep median fissure between them extending almost to their bases. The ninth sternite is a narrow inconspicuous transverse band at their base. The gonocoxites are already formed in the larva on the ventral aspect of the anal segment. The parts withdrawn from the apical dorsal region of the anal segment do not figure conspicuously in the male pupa where they lie behind the gonocoxites. They become the anal lobe in the male hypopygium.

In the female there are also two conical projections, but these are smaller and blunter than in the male and are not so deeply separated by the median fissure. Moreover, they are not genital appendages but are the cerci being formed from the hypoderm in the apical dorsal part of the anal segment of the larva. Further in the female there is present, lying ventral to the projections and extending well up their length, a conspicuous semicircular plate (Fig. 43 (5) *d'*). This is the ninth sternite and its presence or absence enables a final decision as to the sex to be made. It is easily visible in lateral view, thus enabling final proof of sexing to be made even in the living pupa with little or no difficulty. There are no parts representing the gonapophyses.

THE PUPAL SKIN

The cast skin of the pupa has been extensively used for purposes of systematic description. As noted by Edwards (1941), for examination it should be suitably dissected, as otherwise when mounted whole flat the parts are much confused. The method of dissection used by Edwards is as follows. First the whole abdomen should be carefully detached, taking care not to injure the first abdominal segment; next the two sides of the thorax should be separated by means of needles inserted into the dorsal slit, separating the head with the antennae and proboscis which readily comes away, and cutting through the bar of the metanotum; the four parts, namely the abdomen, the two sides of the thorax and the head, should then be mounted flat, avoiding undue pressure. Knight and Chamberlain separate the metanotum with the abdomen.

The pupal skin differs from the cast skin of the larva in that, besides the cuticle of such parts as are exposed at the surface and are more or less firm and sclerotised, there are extensive areas not exposed at the surface where the cuticle is thin and membranous and difficult to follow except in sections. Thus the back of the head, which in the pupa is hidden by the head being flattened against the thorax, is still represented by thin membrane to which the ends of the tentorial bars are attached before they are torn away in the act of emergence. The pupal skin left floating after emergence is on this account rather more complicated than might be expected and a word may be said in this connection. Pushed out well in front are the displaced small oval vertical plates (Fig. 43 (6) *vp*). Along the sides of the now widely open empty skin are the edges of the dorsal split with their characteristic striated strips. These strips, resting on the water, amount almost to lateral outboards giving to the floating skin considerable extra stability. Between the striated strips and the vertical plates are the now empty globular antennal bases, and projecting below these are the tentorial bars with their torn-away free ends. Behind the antennal bases are areas of membrane that have been covered in at the back of the head and front of the thorax and with these is a portion of harder cuticle carrying the anterior thoracic hairs. Posterior to the striated edges are the smoothly torn edges of the postnotal area extending back to the metanotum, which is left intact. On the floor of the empty skin are the openings into the sheath of the clypeus and depressions in which the anterior coxae have lain. Laterally are the openings into the wing sheaths. Furthest back of all is the widely distended empty skin of the abdomen.

DIFFERENTIATION OF THE SEXES

Differentiation of the sexes, which in the larva is only obscurely evident, becomes in the pupa obvious. The outstanding distinction is size. This is most marked where the rearing has been optimal. But even in poor cultures the sex is usually obvious. Only in starvation forms are the two sexes almost equal in size. Even so the female is usually slightly larger. The only authors, so far as I am aware, who emphasise, or even mention, this useful point are Cantrell (1939) and Trembley (1944). The first-mentioned author gives a table of six series of measurements of the length of the cephalothorax in female and male pupae of

A. aegypti. In five of the series the difference is as between 10 and 19 per cent, which in volume or weight would be equivalent to the cube of this.

Besides difference in size there are slight differences in shape, the female pupa being usually more bulged at the sides. The most certain distinction is of course the character of the genitalia which are situated ventral to the tail paddles and are readily examined in the dead pupa or still more readily in the cast skin. In the case of the living pupa, however, there is little difficulty in determining the sex in this way if the pupa is allowed to lie upon its side, as it naturally does, on the slide. The critical feature is then the presence or absence of the projecting plate of the ninth sternite (see under description of the pupal hypopygium).

Ability to distinguish the sex of pupae is extremely useful, almost essential in critical experimental work. Thus, for example, 100 pupae of a particular sex are quite easily pipetted out for emergence. This is not only much less laborious than collecting 100 of a given sex from the adults in a cage with suction tube, but has the further advantage that there is no handling of the adult insect. Even apart from the hypopygeal characters the pupal skins of the two sexes are readily distinguished at a glance both by size and by reason of the greater thinness and delicacy of those of the male. It is easy therefore to check up after hatching out a batch of pupae what number of each sex has emerged.

(*b*) CHAETOTAXY OF THE PUPA

As already noted when dealing with the larval chaetotaxy (ch. IX (*d*), there have been a number of different systems of notation for the larval and pupal hairs. The first fully to describe the pupal hairs was Macfie (1920), who gave a description of the hairs of the pupa of *A. aegypti* with a nomenclature for hairs of the cephalothorax and abdomen. Later a very complete study was made of the pupal abdomen by Baisas (1934–8), who adopted with some modifications the scheme of notation introduced by Macfie and developed by Senevet (1930) and Christophers (1933). It was this notation that was adopted by Edwards (1941) and that, following this author, has been widely used. As a result, however, of observations on a large number of genera, more especially by Knight and Chamberlain (1948), and of the study of the homologies between larval and pupal chaetotaxy by Rozeboom and Knight (1946); Baisas and Pagayon (1949); and by Belkin (1951–3), a new notation had become necessary, that put forward by Belkin (1952) with corrections by Belkin (1953) being now in general acceptance, and giving a serial notation applicable to both the larva and pupa (see also remarks under the larva in ch. IX).

Alterations beyond a certain point, however, are inevitably confusing and there are important works dealing with the pupa, such as the extensive work of Senevet on North African Culicines and of Edwards on Ethiopian species, as also of Knight and Chamberlain and others on eastern forms where notation has been that used by Edwards or earlier systems. In Table 21 there has therefore been given for convenient cross reference the original notation by Macfie; the modification by Baisas as used by Edwards; the more consecutive numbers used by Knight and Chamberlain and adopted by Belkin modified to apply both to larva and pupa; and the final consecutive numbering by Belkin (1953).

Mention should here be made of the item termed by authors the 'hairless socket', a small hairless ring present on abdominal segments III–V of the pupa and situated between hairs 5 and 4. This is not included in Table 21, and is now considered to be a sensillum and not a hair.

The reader should also be reminded that in the first larval instar and in the pupa the hairs 9 and 13, the transitory hairs, do not appear in the corrected Belkin notation (see ch. IX (*d*)).

Table 21. *Designation of pupal hairs according to authors*

Position of hair		Macfie (1920)	Edwards (1941)	Knight and Chamberlain (1948)	Belkin (1953)	Notes
Cephalothorax						
Post ocular		1–3	1–3	3–1	1–3	
Antero-thoracic		4–7	4–7	4–7	4–7	
Mesothoracic		8–9	8–9	8–9	8–9	
Metathoracic		8	*O*	10	10	
		9	*P*	11	11	
		10	*R*	12	12	
Abdomen						
Segment I	Dorsal	*C*	*t*	2	1	
		1	*H*	3	2	
		2	*K*	4	3	
		4	*M*	5	4	
		3	*L*	6	6	
		5	*S*	7	6	
		6	*T*	8	7	
		7	*U*	10	10	(*a*)
	Ventral		Nil			
Segments II–VII	Dorsal	*D*	5	1	0	(*b*)
		C	*C*	2	1	
		C″	*C′*	3	2	
		C′	4	4	3	
		B	*B*	5	4	
		B′	2	6	5	
		A	1	7	6	
		A′	*A*	8	7	
	Ventral	*E*	6	9	8	
		D	7	10	10	
		C	*D*	11	11	
		B	8	12	12	
		A	—	13	14	
Segment VIII	Dorsal	*D*	5	1	0	
		P	*A′*	5	4	
		A	*A*	8	7	
	Ventral	—	—	13	14	
Segment IX	Paddle	—	—	8	7	(*c*)
		—	—	7	6	(*d*)

(*a*) Hair 10 displaced dorsally by presence of pupal haltere.
(*b*) Sensillum (*O* of Knight and Chamberlain) between 5 and 4.
(*c*) Paddle is the lateral portion of tergite IX.
(*d*) Not present in *Aëdes aegypti*.

A point that may be of some help in identification of abdominal hairs and explains to some extent the somewhat confusing use of letters to signify some of the hairs is that originally made by Macfie in giving his system. Looking at an abdominal segment of the pupa in dorsal view there are usually three hairs arising on or near the posterior margin of the segment. From without inwards these are given by Macfie as *A*, *B* and *C* (6, 4 and 1 of Belkin). Near *A*, and a little anterior, is *A′* (7 of Belkin). A little anterior to *B* is *B′* (5 of Belkin) and anterior to *C* are *C′* and *C″*. The changes undergone in this quite helpful early notation will be seen from the table. In the present text and in Figs. 42 and 43 the numbering of hairs is that of Belkin (1953), as given in Table 21.

THE PUPA

CHAETOTAXY OF THE PUPAL CEPHALOTHORAX

Hairs 1–3 of the cephalothorax are situated on each side on the plate behind the compound eye and ventral to the curved basal portion of the antenna. Hair 1 is the most dorsal and is situated above the ocellus. Hairs 2 and 3 lie below the ocellus and towards the point of origin of the maxillary palps from the plate. The plate, usually termed the 'post-ocular plate', represents part of the gena of the imago, the hairs 1–3 being considered respectively to represent head hairs 10, 12 and 13 of the imago.

Hairs 4–7 are situated on each side above the curved antenna on the anterior lateral angles of the thorax on a portion of the dorsal shield corresponding to the prothoracic lobe of the imago (Fig. 43 (1) *at*), hairs 4 and 5 being a little anterior to 6 and 7. They represent the prothoracic hairs of the imago.

The mesothoracic hairs 8 and 9 are situated on the convex scutum, hair 8 being a little dorsal and posterior to the trumpet, and 9 just mesad of the line of origin of the wing (Fig. 43 (1), (8), (9)). The metathoracic hairs 10, 11 and 12 are situated on each side in a line on the metathoracic plate internal to the more expanded portion representing the haltere (Fig. 43 (3), (10), (11), (12)). Hair 10 is mesad and commonly bifid, hair 11 is longer than the others and single. They are homologous with the larval metathoracic hairs 1, 2 and 3 respectively.

CHAETOTAXY OF THE PUPAL ABDOMEN (Fig. 42 (2), (3))

Except for segments I and VIII the hairs of the pupal abdomen resemble in their number and arrangement those of the larva. Their notation by different authors is given in Table 21. As in the larva, those for each segment are numbered 0–14 (omitting the transitory hairs 9 and 13), 0–5 being dorsal, 6–7 lateral and 8–14 ventral.

The small hair 0 is present on each segment II–VIII much as in the larva towards the anterior portion of the segment. The remaining dorsal hairs also have much the arrangement seen in the larva (Fig. 43 (2), (3)). Hair 1 is the most mesad and like hair 4 arises on or near the posterior margin of the segment. Hairs 2, 3 and 5 lie more on the segment and are usually small and simple hairs. Hair 4 is Macfie's hair *B*. Approximately between 5 and 4 on segments III–V is the sensillum (hairless ring of authors). Hairs 6 and 7 are conspicuous as being at the outer posterior angle of the segment. Hair 6 is simple, long on segment VI and very small on segment VII. Hair 7 increases gradually in size, becoming thick and more thorn-like to segment VI. Ventrally are hairs 8–14 (Fig. 43 (2)). Hair 8, as in the larva, is situated laterally and more forward than the others and, as in the larva, posterior to the pupal spiracular marks (pupal puncta). Hairs 10, 11 and 12, on the outer half of the posterior end of the segment. The minute hair 14, as in the larva, is present in the intersegmental line anterior to segments VII and VIII and is stated by Macfie to be present on the remaining segments, though I have not been able to verify this. The two hairs are much more conspicuous on the anterior margin of segment VIII than on other segments.

On segment I hair 1 forms the conspicuous *dendritic tuft* hair. The remaining hairs are in a line towards the anterior margin of the segment, 2 and 3 being most mesal, then another pair 5 and 4, and towards the outer border 6, 7 and 10, the last being displaced from its normal position by the large haltere rudiments (Fig. 43 (3)). There are no hairs on the ventral aspect of segment I.

On segment VIII the only hairs on the dorsal surface are, on each side, the small anteriorly situated hair *O* and a fairly developed simple hair on the posterior border (hair 4). Laterally is the large branched hair 7, which on this segment consists of 4 or 5 large thick branches of about equal length carrying very fine lateral filamentous branches. There are no hairs on the ventral surface.

The paddles carry only the single unbranched main paddle hair. The paddle hair (hair 1) and the accessory paddle hair (hair 2) are considered by Belkin (1953) to be probably the two lateral hairs of the ventral valve of the larval siphon. The paddles are the lateral portions of the ninth tergite and it would seem very likely that the paddle hair and accessory paddle hair represent the hairs 6 and 7 conspicuous at the lateral angles of the other tergites.

The chief characteristics of the pupal hairs of *Aëdes aegypti* are given by Edwards (1941) as follows:

(1) The dorsal thoracic hair (hair 8) is very short and usually double.

(2) The middle metathoracic hair is markedly longer than the other two.

(3) The two inner and the two outer hairs on segment I (hairs 2, 3 and 6, 7) are subequal in length and single (an unusual condition) (Fig. 43 (3)).

(4) Hair 1 on segment II is usually double or triple.

(5) On segment III hair 1 is usually double placed obliquely internal to hair 4 (not behind it).

(6) Hair 6 is rather spine-like even on segment II, progressively longer on III–V, but not longer on VI than V. It is single and strong and much longer on VII than on VI. On VIII it has 2–5 subplumose branches reaching to about the middle of the paddle.

(7) Hair 4 on segments II–VI is single, not more than half as long as the segments.

(*c*) THE TRACHEAL, MUSCULAR AND OTHER SYSTEMS

GENERAL CONSIDERATIONS ON INTERNAL STRUCTURE OF THE PUPA

The pupa is essentially an intermediate form between the larva and the imago, both of which lead an active life of their own, the larva as an aquatic and the imago as a terrestrial winged insect. Hence almost any stage between the structures required for the larva and those for the imago will be found in the pupa, making any description of the systems a study of development rather than one relevant to the pupa as such. One system, however, namely the tracheal system, plays so prominent a role in the life of the pupa in both the beginning and the end of this stage that it merits special consideration. Another system which cannot be passed over lightly is the muscular, in that the enormously developed imaginal musculature is imposed anew upon that of the larval. The circulatory and nervous systems are less profoundly altered though in these as in the alimentary canal there are stages in the pupa which are helpful in deciding some homologies of imaginal structures.

THE TRACHEAL SYSTEM (Fig. 44)

The tracheal system of the pupa is formed around the larval system as present in the prepupa and ends as that of the adult insect lying ready for emergence swathed in the pupal cuticle. From one point of view the characters most properly pertaining to the pupa are those in the young pupa and later changes might be regarded as in increasing degree those pertaining to the imago. Such a view, here adopted, will simplify description and enable changes more properly belonging to adult structure to be dealt with more satisfactorily later. To a large extent the main features of the pupal tracheal system have already been laid down at pupation when, the intima of the larval tracheae being withdrawn, the newly formed system behind this has come into use. According to Hurst no fresh intima to the tracheae proper to the pupa is developed with the exception of that lining the length of trachea linking the respiratory trumpets to the position where the imaginal spiracle is eventually formed and the lining of the branch joining the tracheal trunk to the spiracle of the first abdomal segment, both of which are cast off with the pupal skin. It is true that in the cast pupal skin no such wealth of cast intima is to be seen as in the cast larval skin. Careful examination will, however, show that there is a considerable extent of very thin intima cast which is devoid of any taenidia, indicating that the pupal stage does not, with the exceptions noted, produce a taenidial lining proper to itself.

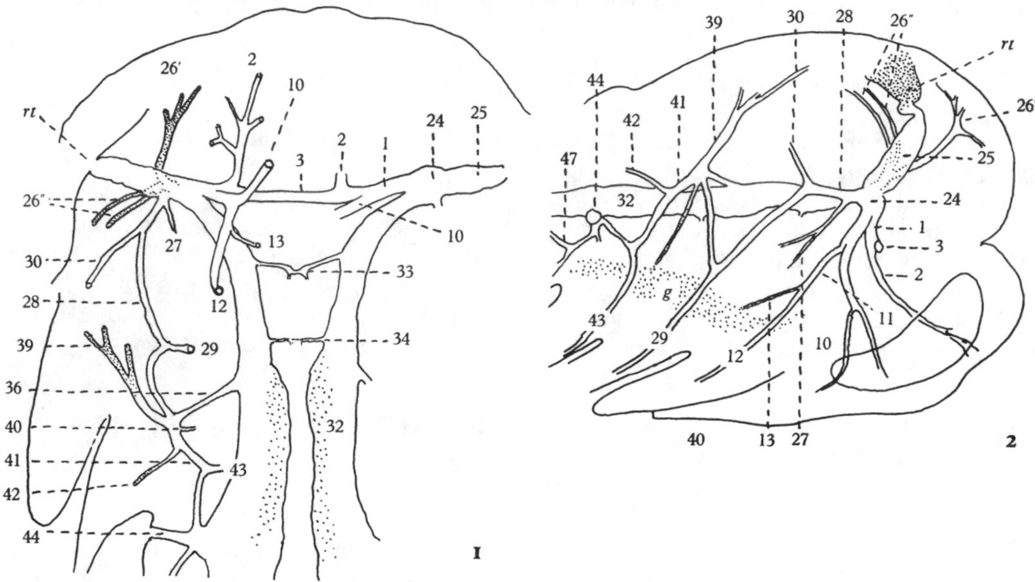

Figure 44. The pupa. Tracheal system.

1 Tracheae of the cephalothorax viewed from ventral aspect.

2 The same viewed from the side.

Lettering: *rt*, base of respiratory trumpet; *g*, thoracic ganglia; 1–47, numbers of tracheae as given in tabular statement on p. 369 of pupal tracheae.

The following are the chief changes that occur at pupation:

(1) The structures about the terminal spiracles of the larva with the cuticular lining of the tracheal trunks up to the first line of weakness in the seventh segment are withdrawn and not replaced. The cellular layer representing these shrinks and forms solid cords extending from the terminations of the pupal tracheal trunks to the under side of the ninth pupal tergite.

(2) The spiracular cords leading from the mesothoracic punctae to the mesothoracic knot are replaced by large patent tracheal trunks linking the mesothoracic spiracles with the main system.

(3) The spiracular cords and puncta of the first abdominal segment are replaced by large patent tracheae and functional spiracles opening into the ventral air space.

(4) With the general flattening of the pupal head and neighbouring parts the more anterior portions of the trunks in the thorax are compressed and pushed back making a

sharp angle or kink on each side where the fore and hind parts of the tracheal trunk system meet at the breathing trumpets (Fig. 44 (2)).

Changes other than those noted mainly occur in the hypoderm and are hence to be regarded as imaginal rather than pupal or in providing new vessels for the increased mass of wing muscle in the thorax. Thus in the pupa strictly speaking there is no metathoracic functional spiracle. But at an early stage a large spiracle is developed in this position in the hypoderm. In the pupal abdomen there are no functional spiracles (except for the first segment). But marks on the cuticle (best seen in the pupal skin) indicate the position of spiracles on segments II–VIII, those on segment VIII being near the posterior border.

In face of the considerable changes noted above, the tracheal system in the pupa nevertheless still maintains in a remarkable manner the essential features of the system in the larva. Thus the connectives which in the larva take a more or less direct course giving off quite small branches only to the germinal bud rudiments of leg, wing and haltere still give the same, but much enlarged, branches to the now extended legs, wings and halteres of the pupa passing in long loops down the sides of the coxae to do so. This identity of the pupal and larval tracheal systems will be very apparent from the numbers given in the tabular statement of tracheae where the homologous pupal branches have been given the numbers previously applied to those of the larva. Almost all the pupal branches are capable of being thus numbered.

As in the larva the main tracheal trunks in the pupa are much dilated and ballooned. In addition to the large anterior commissure small commissural branches are present as in the larva, two in the thorax and one in each segment II–VII. The other tracheal vessels and even their smaller branches are very much as in the larva. Thus in the head the dorsal cephalic still supplies the brain and antennal region and the ventral cephalic the mouthparts and labrum with branches much as in the larva.

Tracheation of the abdominal segments, except for the eighth and anal segments, is very similar to that of the larva. For each segment II–VII there is on each tracheal trunk towards the anterior end of the segment an *insertion plaque*. From this there take origin: (1) a spiracular trachea; (2) a connective passing forwards; (3) a visceral trachea passing directly inwards; (4) a ventral trachea passing round the segment to its ventral aspect; (5) a connective passing backwards. This last usually takes origin a little distance from the base of the ventral trachea (as in the larva). From the backward connective about the middle of the segment there arises a small dorsal branch.

The tracheal trunks, as patent vessels, now end at the seventh segment, being continued only as solid cords composed of cells remaining over from the epithelial coats of the trunks after withdrawal at pupation of their intimal lining. These cords, as already noted, continue backwards to the under side of the semicircular median lobe of the ninth tergite where they may still show some indication of their original tubular structure. The eighth and anal segments receive their tracheal supply from lateral tracheae taking off from the trunks where their tracheal character terminates.

TABULAR STATEMENT OF PUPAL TRACHEAE

1. *Common cephalic.* A short trunk directed medianly and ventrally. Gives off the following branches at its inner end:
   2. *Dorsal cephalic.* To brain and antennal region (branches 4–9).
   3. *Commissural.* The main anterior commissure as in the larva but stouter.

10. *Ventral cephalic*. To mouth-parts and other head structures (branches 14–23). In thorax gives off:
    11. *Prothoracic* dividing into:
        12. Trachea to first leg.
        13. Branch ending in region of thoracic ganglion 1.

24. *Common mesothoracic*. In the pupa contracted to a bulbous dilatation at junction of the common cephalic with the main thoracic trunk. From it the following branches arise:
    25. *Mesothoracic spiracular*. A sausage-shaped tracheal trunk linking the respiratory trumpet of its side with the main system and joining 24 at position of future imaginal spiracle.
    26. *Mesothoracic dorsal*. Tracheae arising from common origin or short basal trunk, namely, 26′ passing forwards and inwards to front of dorsum, and 26″ twin tracheae passing backwards and outwards to lateral area of dorsum.
    27. *Mesothoracic ventral*. Arises along with or as branch near base of 28. Divides into branch to pleura and one ending near thoracic ganglion.
    28. *Meso-meta connective*. Long looped trachea directed ventrally. Gives off:
        30. *Branch to wing*. Taking origin towards anterior end.
        29. *Trachea of middle leg* arising from apex of loop.

32. *Thoracic tracheal trunk*. Dilated trunk with inner wall thin and membranous. Extends from junction with 24 to level of open spiracular trachea of first abdominal segment. Gives off:
    33. *Commissural* (mesothoracic). Linking trunks towards anterior end and giving branches to heart, etc. A small trachea.
    34. *Commissural* (metathoracic). A small trachea linking trunks in the metathoracic region.
    36. *Common metathoracic*. Trachea from trunk to metathoracic knot and branches radiating from that, namely
        38. *Spiracular*. Becomes open trachea in hypoderm.
        39. *Metathoracic dorsal*. Newly formed or enlarged branches passing into posterior portion of dorsum to supply wing muscles.
        40. *Metathoracic ventral*. A small branch to pleura.
        41. *Connective* (meta-I abd.). Forming ventral loop and giving branches:
            42. *Branch to haltere*. From near anterior end of connective.
            43. *Branch to hind leg*. Arising from apex of loop.

44. *Spiracular* (I abd.). A patent and wide trachea passing from the trunk to the open spiracle. From its inner end (insertion plaque), which receives the connective 41, arise also the following:
    45. *Visceral branch*. Passes directly inwards to end in branches among the remnants of the gastric caeca.
    47. *Connective* (I–II). About its middle gives off:
        50. *Ventral branch* for the segment.

(For description of the abdominal tracheae see text.)

THE ALIMENTARY CANAL (Fig. 45)

The alimentary canal during the pupal stage undergoes changes that convert the larval into the imaginal canal. These changes begin in the late (prepupal) stage of the larva by the withdrawal of the invaginated portion of oesophagus from the proventriculus. At pupation this is followed by other changes. The entire elaborate cuticular structures lining the mouth and pharynx are withdrawn, leaving these structures more or less featureless with a simple cellular lining. At the point that was previously a recessed angle between the cardia and the oesophageal invagination there develops an annular thickening which eventually projects far into the lumen (Fig. 45 (4), (7) *ar*). There follow a number of other changes which are mainly degenerative. The cells of the outer wall of the proventriculus (cardia), and also those lining the gastric caeca, show marked nuclear and other changes and are eventually cast off into the lumen of their respective parts, leaving these structures with thin walls formed only by small regenerative cells. Similar changes occur in the mid-gut, the whole of the alimentary canal from the annular thickening to the point of entry of the Malpighian tubules being reduced to a much narrowed simple tube at first filled with massive cast-off epithelium and later, when this is absorbed, almost free of contents. Both cardia and the gastric caeca are included in these changes and later cease to be differentiated from the rest of the canal (Fig. 45 (6), (7)).

Posterior to the pyloric ampulla changes in the ileum consist chiefly in the absorption and replacement of the muscular coat. The colon loses its characteristic larval appearance and its walls degenerate, its epithelium being replaced according to Thompson by encroachment of cells growing backwards from the ileum and forwards from the rectum.

Following this general degeneration and absorption of larval elements reconstructive changes take place which eventually result in formation of the imaginal structures. The now much narrowed portion of canal anterior to the brain (pharynx and oral cavity) develops a strong cuticular lining and becomes the 'buccal cavity' of many authors (Fig. 45 (1), (2)). That part of the oesophagus lying in the head immediately behind the cerebral commissure likewise develops a strong cuticular lining and forms the 'sucking pump' as described later in the imago. The portion of canal behind this as far back as the annular ring remains as the 'oesophagus' of the imago.

These parts have been differently named by authors and their homologies may be briefly considered. Sinton and Covell (1927) and Barraud and Covell (1928), following Nuttal and Shipley's nomenclature, have used the terms pharynx and buccal cavity for the two parts in the imago. It will be clear, however, that the sucking pump of the imago is not homologous with the pharynx of the larva, whilst the 'buccal cavity' is. Snodgrass (1938), p. 388, defines the pharynx as that part between the mouth and the oesophagus usually not extending beyond the nerve ring. The pharynx of the larva corresponds accurately to this description and complies with other requirements and would appear therefore to be correctly so named. The 'pump' of the imago lies behind the nerve ring and corresponds to what in the larva is the anterior portion of the oesophagus. Snodgrass, however, notes that in some insects there is a 'posterior pharynx' behind the brain, and to such the sucking pump of the mosquito may be considered to conform. Morphologically, therefore, this organ may conveniently be so termed. It consists of that portion only of the oesophagus of the larva that lies within the head extending back to about the position occupied by the cervical diaphragm. The pharynx lies anterior to the cerebral commissure. Its posterior angle is clearly defined by the insertions of the larval muscles 15, 16 and 17 as seen in the early pupal stage (Fig. 45 (1), (2)). In the late pupa this becomes the roof of the anterior end of the posterior pharynx of the imago. At this point, that is at the meeting point of pharynx and posterior pharynx, there is a small strongly sclerotised lateral bar which receives the insertions of the larval head muscle 13 (adult head muscle 19) as also strong lateral muscular bundles apparently corresponding to larval head muscle 14.* This appears to represent the much contracted lateral angles of the larval pharynx just in front of which the connective to the frontal ganglion passes in the imago as in the larva. It appears to be the *lateral flange* of the pharyngeal armature of Sinton and Covell.

* For numbering of the head muscles see tabulations under muscular system for the larva and for the adult; also cross homologies given in Table 22.

The anterior limit of the imaginal parts corresponding to the larval pharynx is shown by the position in the early pupa of those structures which in the hypoderm have underlain the epipharynx and other labral parts including the feeding brushes. These form a much shrunken bilobed cellular mass lying above the roof of the oral passage a little posterior to the entry of the salivary duct (salivary pump). At this time head muscle 4 (elevator of the epipharynx) is still to be seen inserted at its posterior border (Fig. 45 (1)). In spite of its conspicuous appearance this cellular mass is later wholly absorbed, the only trace left in the imago of all the organs it represents being the 'palatal papillae' of Sinton and Covell ('small conical setae' of Thompson, p. 183). The roof of the larval pharynx forms the anterior and posterior hard palate of Sinton and Covell.

That portion of the oesophagus behind the posterior pharynx undergoes little change. The inner epithelial coat remains intact. The muscular coat, however, in accordance with what appears to be a general rule for all the viscera, undergoes lysis and in the imago the walls of this portion, that is what is usually regarded as oesophagus, are thin and poorly supplied with muscles. At the posterior end, just cephalad to the annular ring, there develops at each lateral angle a small pouch with characteristic walls, the *dorsal diverticula* of the imago. From its median ventral wall there develops the much larger rudiment of the *ventral diverticulum* (Fig. 45 (7) *vdl*). This appears as a long rod-like growth of characteristic tissue which lies ventral to the now greatly narrowed mid-gut as far back as the

---

Figure 45. The pupa. Alimentary system.

1 Median sagittal section of head of early pupa.

2 The same of a pupa approaching emergence. The two figures show the homologies as between the imaginal and larval pharynx and related parts. The changes in musculature are also indicated by the muscle numbers. Larval muscles (Fig. 1) 4–23. See Table 22. Imaginal muscles (Fig. 2) 4–24. See Table 22. Larval muscles 4, 15, 18 are not carried through; nos. 13, 16, 17, 22, 23 become respectively imaginal muscles 19, 21, 22 (not shown in figure), 23, 24, of the imaginal series. *a*, pupal cuticle now with a large space between it and the retracted hypoderm forming the imaginal parts.

3 Transverse section through the mouth-parts of the pupa showing pupal sheaths and contained developing imaginal parts.

4 Longitudinal section through the proventriculus and related parts in a larva approaching pupation. *a*, epithelium of invaginated portion of oesophagus now withdrawn from the proventricular invagination; *b*, muscle of muscular coat of oesophagus; *c*, cellular coat of cardia.

5 Longitudinal section through portion of mid-gut of an early pupa showing: (*a*) cast-off and disorganised epithelial coat and (*b*) formation of new coat of small cells with beginnings of the imaginal muscular coat (*c*).

6 Longitudinal section through the region of the proventriculus and gastric caeca of the early pupa. *a*, cast-off epithelial cells; *b*, disorganised epithelium of gastric caeca; *c*, cast-off mid-gut epithelium.

7 The same parts of an advanced pupa. Same magnification.

Lettering: *ann*, antennal nerve; *ar*, annular ring; *Cc*, cerebral commissure; *ca*, corpus allatum; *cd*, cervical diaphragm; *cg*, cingulum muscle; *Dd'*, level at which the dorsal diverticula develop; *epr*, developing epipharynx; *fg*, frontal ganglion; *hp*, hypopharynx; *L*, labium; *lr'*, pupal cuticle of labral sheath; *lr''*, developing imaginal labrum; *Mg*, mid-gut; *md*, mandible; *mx*, maxilla; *mxp*, maxillary palp; *Oe*, oesophagus; *P*, pharynx; *P'*, posterior pharynx; *sg*, salivary gland; *slm*, dilator muscle of salivary pump; *slp*, salivary pump; *slp'*, developing salivary pump; *vdl*, developing ventral diverticulum; *vtr*, ventral transverse muscle.

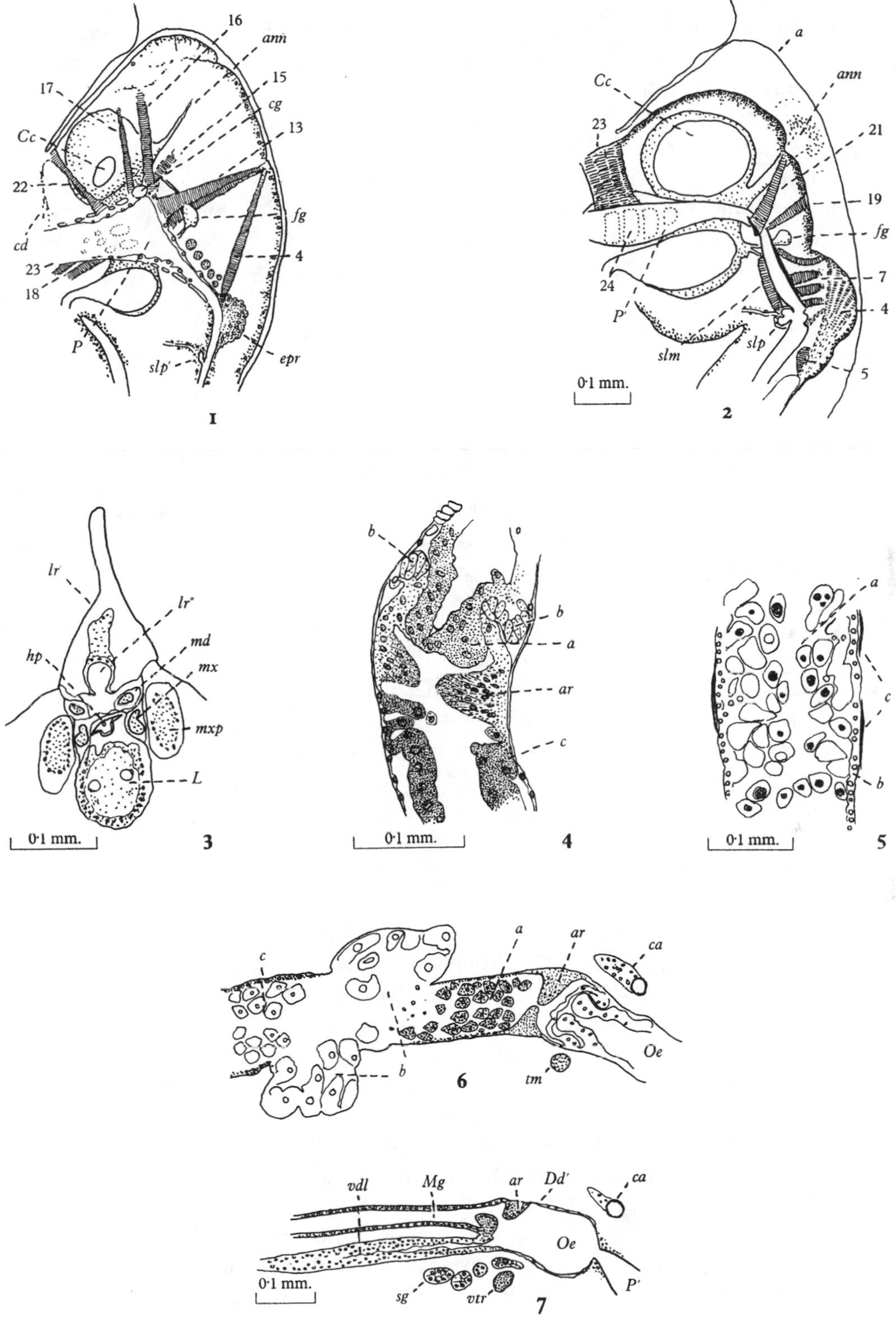
16
ann
15
cg
13
17
Cc
22
fg
cd
4
23
18
P
slp′
epr
1
a
Cc
ann
23
21
19
fg
24
7
P′
4
slm
slp
5
0·1 mm.
2
lr′
lr″
md
hp
mx
mxp
L
0·1 mm.
3
b
b
a
ar
c
0·1 mm.
4
a
c
b
0·1 mm.
5
c
a
ar
ca
b
6
tm
Oe
vdl
Mg
ar
Dd′
ca
Oe
0·1 mm.
sg
vtr
7
P′

Figure 45

first abdominal segment. The annular ring becomes the so-called proventriculus of the imago, a structure clearly not wholly homologous with the more complex organ of the larva and better termed the *oesophageal valve*.

As already noted, the cellular lining of the larval cardia and of the gastric caeca is cast off and absorbed. With this the organs become merged into and form part of the new mid-gut, now an extremely long and narrow empty tube (Fig. 45 (7)). Apart from the disappearance of the colon as a distinct organ the most notable happening in the hinder part of the canal is the appearance of the rectal glands of the imago.

Of accessory structures the single salivary gland on each side of the larva early shows degeneration and vacuolisation of its cells and is eventually lysed and absorbed. The acini of the imaginal glands develop as new growths from the distal end of the salivary glands. These lie in the pupa as a number of dark-staining tubules on each side of and ventral to that portion of the gut corresponding to the cardia (Fig. 45 (7)). The Malpighian tubules remain intact from larva to imago.

THE MUSCULAR SYSTEM (Fig. 46)

During the pupal period an almost complete change occurs from the larval to the imaginal musculature. With a few exceptions the larval muscles undergo histolysis and disappear, whilst the majority of the imaginal muscles are formed independently of those of the larva. In certain cases muscles are carried through from the larva to the imago, but even in such cases the appearance and character of the muscles and even its anatomical relations and functions are usually more or less changed.

The change from one musculature to the other takes place almost entirely in the pupal stage, the musculature in the early pupa being that of the larva, whilst in the late pupa it is that of the adult. Differing as the two system do it becomes scarcely practicable to use the same numbered list of muscles for the two series and it has been thought preferable to give separate numbered lists of larval and imaginal muscles as has been done under the muscular system of the larva in chapter XIII and of the imago in chapter XXVI. Questions of homology can, however, be discussed in the present chapter under the pupa, in which stage the change from one to the other system takes place.

Disappearance of the larval muscles, as noted by Thompson (1905), does not appear to be associated with marked phagocytosis, though accumulations of small cells with dark nuclei do occur. Most of the imaginal muscles, however, appear first as collections of such small cells (myoblasts), sometimes associated with the histolysing larval muscles but often seemingly independent of such. Some difference in the degree of change and the period at which such change takes place depends upon the part of the body concerned, that is whether it is the musculature of the head, the thorax or abdomen that is being considered.

The head muscles are almost entirely new, not only in respect to separate origin, but, with the great changes in the head, even the same muscle where it has been carried through from larva to adult has changed relations to the general structure of the parts (see Fig. 45 (1) and (2), where the parts in an early and a late stage of the pupa are shown). The changes in the head musculature have been very fully dealt with by Thompson in *Culex*, the changes in *Aëdes aegypti* being not noticeably different. Thompson notes that all the muscles of the head continue intact as in the larva up to the eighth or tenth hour when histolysis occurs. This brings about the final disappearance of the majority of the larval muscles, but certain muscles are reformed into imaginal muscles. In this case the large reticular nuclei

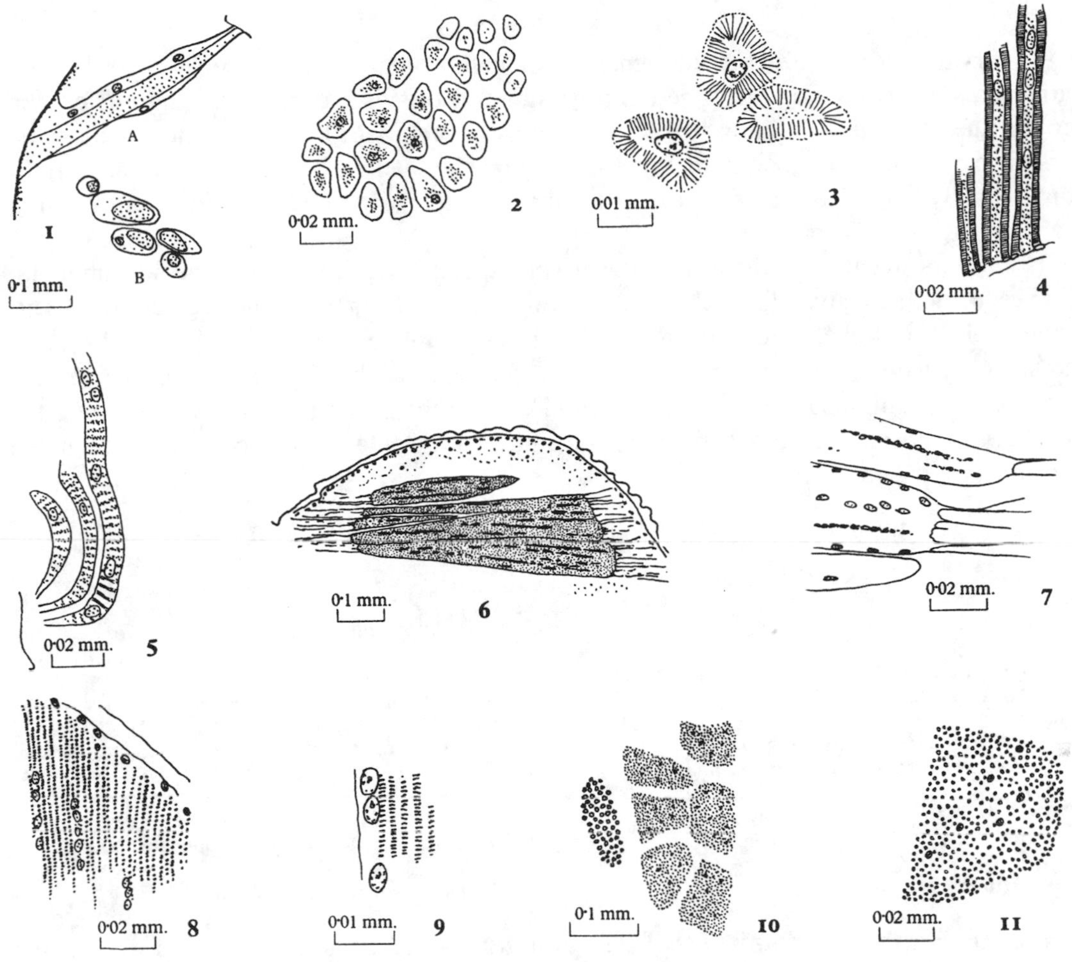

Figure 46. The pupa. Muscular system.

1 Larval muscle type. A, longitudinal view of a muscle band. B, transverse section of a group of larval muscles.

2 Transverse section of a tubular muscle showing peripheral ring of muscle substance and centrally situated nuclei.

3 Three fibres of same more highly magnified, showing arrangement of fibrillae in radiating lines.

4 Longitudinal section of three muscle fibres of tubular type.

5 Three fibres of another type of imaginal limb muscle with peripherally placed nuclei.

6 Sagittal section of dorsal area of thorax of early pupa showing developing masses of indirect wing muscle. The masses have not yet directly attached themselves to the hypoderm but are indirectly attached to this by areas of tonofibrillae.

7 Portion of same more highly magnified, showing ending of muscle fibres in tonofibrillae.

8 Attachment area of portion of a fully matured bundle of wing muscle type.

9 A small portion of same, more highly magnified. This is the same magnification as no. 3, showing fibres of tubular muscle. Appearance shown at the surface of the muscle band suggests sarcolemma, but care has to be taken to avoid confusing tracheal nuclei, for those of the sarcolemma.

10 Transverse section of muscular bundles of indirect wing muscle. A tubular leg muscle type band is shown lying in juxtaposition.

11 A portion of 10 more highly magnified.

of the larval muscle fibres become smaller, stain more darkly and lose their reticulated character, some becoming histolysed. Large numbers of small, darkly staining nuclei then appear, the two sets remaining side by side until the twentieth hour when new contractile substance staining with haematoxylin appears and the nuclei take on the imaginal condition. Finally the amount of contractile substance increases and the imaginal staining reactions are shortly reached.

The muscles so carried over, as also the larval muscles disappearing and the new imaginal muscles formed, are given in Table 22, the numbers and names being those given in the lists respectively of larval and adult head muscles. Those numbers in the table given in brackets and without accompanying name are larval muscles that are finally histolysed or imaginal muscles formed independent of larval muscles. Numbers not in brackets and with the name of the muscle are, under the conditions noted above, larval muscles that are carried over as imaginal muscles.

Table 22. *Relation of larval to imaginal musculature of the head*

Larval	Imaginal
(1–10)	(1–17)
11 Dilators of the salivary opening	18 Salivary muscle
(12)	
13 Posterior elevators of the roof of the pharynx	19 Valvular muscles
14 Lateral pharyngeal	20 Lateral retractors of the buccal cavity
(15)	
16 Dorsal retractors of the pharynx	21 Dorsal retractors of the buccal cavity
17 Accessory dorsal retractors of the pharynx	22 Accessory dorsal retractors of the buccal cavity
(18–21)	
22 Dorsal dilators of the oesophagus	23 Dorsal dilators of the oesophagus pump
23 Lateral dilators of the oesophagus	24 Lateral dilators of the oesophageal pump
(24)	
25 *vl l′* (thoracic)?	25 Depressor muscle of the head

In the thorax the great bulk of the imaginal musculature is wholly new consisting of the large indirect wing muscles and those connected with the limbs that have largely formed in the imaginal buds. The original larval muscles, however, are still represented in the less conspicuous visceral system as described under the imago.

In the abdomen the relatively massive muscles are retained throughout the pupal stage and are only finally replaced by the true imaginal muscles in the imago after emergence.

Besides replacement of the larval muscles by those of the adult there is also a great change in the character of the muscles in the two systems. Further remarks on these systems will be found later when the adult musculature is being described. Since, however, it is in the pupal stage that the change in character takes place, it will be appropriate to give in the present connection some account of the types of muscle fibres. Good accounts of insect muscle describing the different types are given by Snodgrass (1925) in his *Anatomy and Physiology of the Honey Bee*; see also Hürthle (1909); Kielich (1918); Jordan (1919, 1920); Weber (1933); Studnitz (1935); Tiegs (1955). In the pupa the following types are distinguishable in the somatic musculature.

The larval type (Fig. 46 (1)). All somatic muscles of the larva are of this type. Each muscle is commonly a single fibre, often long, thin and even thread-like, rarely massive. Such muscle fibres consist of a centrally situated very faintly fibrillated and striated strand of muscle substance surrounded by a distinct sarcolemma and between this and the strand a clear, often quite voluminous, zone of transparent sarcoplasm in which are spaced at intervals large nuclei. The muscle fibre substance stains lightly with eosin and most other stains.

Imaginal limb type (Fig. 46 (2)–(5), (10)). This type is seen most characteristically in the leg muscles, including those of the coxa and sternal parts. It is also seen in the direct wing muscles. It occurs as muscles formed of massive bundles of moderate-sized discrete fibres ranging from 10 to 20μ in diameter. Each fibre consists of an outer zone of muscle substance with a central sarcoplasmic core in which is embedded a close series of moderate-sized nuclei. The fibres in cross-section appear very characteristically as rings with clear centre, or, if the section passes through this, with a central nucleus. From this appearance they have been termed 'tubular muscles' (Morison after Röhrenmuskeln of Brünnich). They are usually distinctly or even conspicuously striated and show no sarcolemma. The component fibrillae of the muscle substance are very minute, forming radially arranged lamellae.

Along with this form of muscle are others with fibres of about the same diameter but with peripherally situated nuclei. These are often very strongly striated (Fig. 46 (5)).

Imaginal indirect wing type (Fig. 46 (6)–(11)). This form of muscle is entirely different from those so far described. It appears early in the prepupal stage as sausage-shaped masses of very dark-staining substance, so much so that they appear almost as foreign bodies. In this early stage they are crowded with lines of very dark-staining bodies resembling flattened nuclei. Later these masses increase greatly in size and become dark-staining bands stretching across the thorax and attached to the hypoderm by broad zones of fine fibrils (tonofibrillae). Eventually the dark bands become the imaginal indirect wing muscles. These are massive muscles formed of hundreds of fibrillae in the region of 2μ in diameter and markedly striated embedded in sarcoplasm. Among the fibrillae are rows of nuclei and granular material. When fully developed such muscle fibres are inserted directly into the cuticle. In cross-section they show large polygonal areas formed from cross-section of numerous fibrillae, contrasting with the bundles of larger single fibres such as form the leg muscles (Fig. 46 (10)). They have been termed fibrous muscles (Morison, 1927) and are considered to be each a giant muscle fibre as described in *Calliphora* (Lowne, 1895; Tiegs, 1955).

The imaginal muscles of the head are mostly of tubular type, but some, for example those of the pharynx and post-pharynx, consist of fine fibrillae and seem to approach in character the indirect wing muscle type. The main mass of indirect wing muscle type occurs in, and largely occupies, the dorsum of the large mesothoracic segment, but a much smaller mass passes between phragmata in the metathorax.

The muscles in the thorax connected with the legs, as also those in the leg segments, are of tubular type. The muscles of the imaginal abdomen still remain of the larval type, that is with sarcolemma and large peripheral nuclei, though a few muscles of tubular type are developed in the pupal stage.

THE NERVOUS SYSTEM (Fig. 47)

Changes in the nervous system taking place in the pupa and leading to the imaginal condition are chiefly those in the brain. Even so the main lines in structure are largely followed and the brain of the imago, as will be seen from the figures, is not very markedly different to that of the larva. The chief changes are:

(1) Increase in size in proportion to the head capsule and especially in the volume of white medullary substance.

(2) Great increase in size of the optic lobes and chiasmata as the compound eyes grow to their full maturity. Actually the total increase in size of the brain is largely due to this development of the optic lobes.

(3) Increased size of the antennal nerves arising from the deutocerebrum. At the beginning of the pupal period these nerves are relatively small, but later they become massively developed as large trunks passing to Johnston's organ in the basal lobe of the antenna.

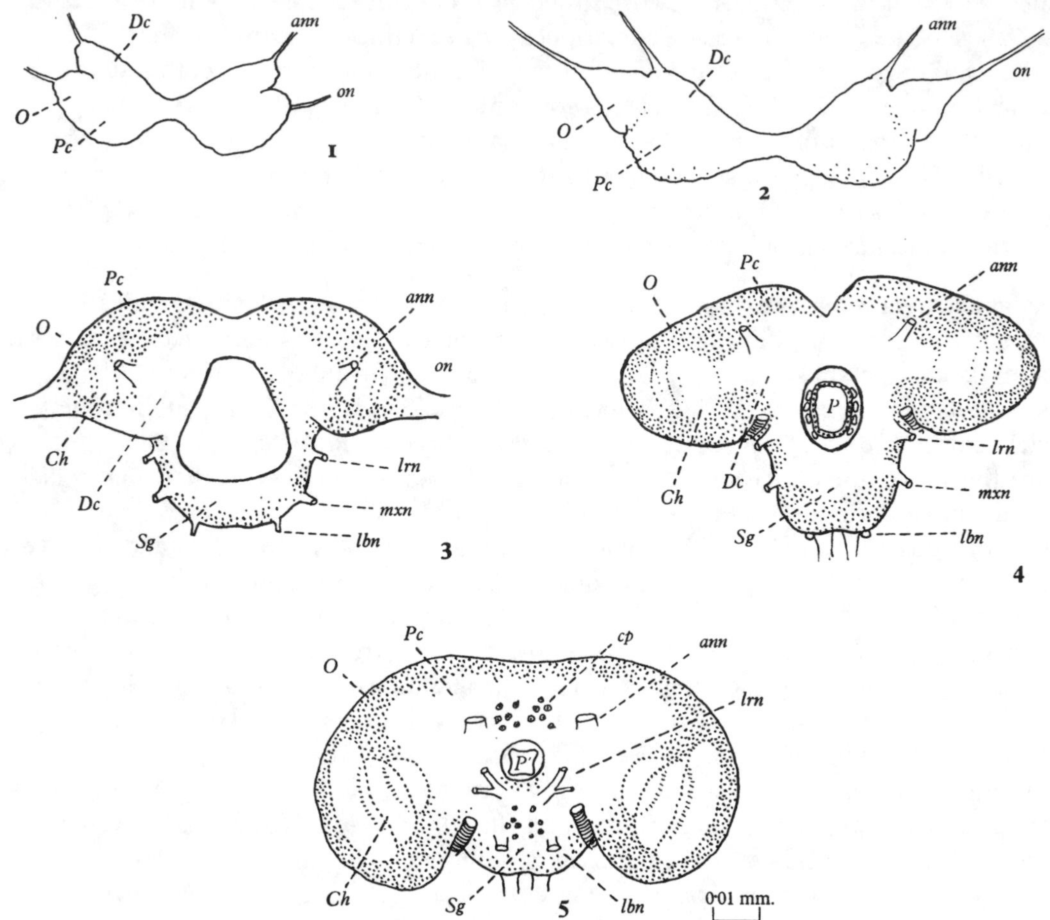

Figure 47. The pupa. Nervous system.

1 Dorsal view of brain of pre-ecdysis third instar larva.

2 The same of an immediate post-ecdysis fourth instar larva before the compound eyes have shown any degree of development. Same magnification.

3 Anterior view of brain of advanced fourth instar larva. Same magnification. Reconstruction.

4 Anterior view of brain of early pupal stage. Same magnification. Reconstruction.

5 Anterior view of brain of pupa approaching emergence. Same magnification. Reconstruction.

Lettering: *ann*, antennal nerve; *Ch*, optic chiasma; *cp*, area of large cells probably representing corpora pedunculata of many insects; *Dc*, deutocerebral lobe; *lbn*, labial nerve; *lrn*, labral nerve; *mxn*, maxillary nerve; *O*, optic lobe; *on*, optic nerve; *P*, pharynx; *P'*, posterior pharynx; *Pc*, protocerebral lobe; *Sg*, sub-oesophageal ganglion.

The shaded and pale areas in 3, 4 and 5 give a general representation of the distribution of cortical neurocellular tissue and medullary white matter (neuropile).

(4) Closing in of the foramen between the cerebral connectives owing to massive thickening of the latter and to reduced size of the pharyngeal canal.

Disproportionate increase in size of the brain to that of the insect takes place only at the beginning of the fourth instar when the first beginnings of the compound eyes occur. In the pupal stage further increase takes place as the compound eyes grow, so that from stretching as a narrow band across the posterior region of the head the organ in the imago largely fills this structure.

A peculiar point appears to be the small development of the *corpora pedunculata* as these are described in many insects. In *Aëdes aegypti* up to the end of the pupal stage only a small number of large cells in the protocerebral lobes appear to represent these bodies.

Changes in the ventral nerve chain are not very great. There is some increase in size of the thoracic ganglia with enlargement of the nerves to the now greatly increased leg rudiments. The first abdominal ganglion which in the advanced larva has moved up somewhat towards the thoracic ganglia becomes in the pupal stage more or less incorporated with these. A similar change takes place terminally, the ganglion of the eighth segment, both in the male and in the female, moving forwards during the pupal period to fuse with that of the seventh, so forming a double ganglion.

The frontal ganglion remains present throughout the pupal period and still receives nerves from the tritocerebrum, but these are now much shorter. A visceral nerve chain is present much as in the larva. These and related structures as finally present in the imago will be described later.

OTHER STRUCTURES

The heart, except for its posterior terminal portion, is not essentially different in the pupal stage to the description already given in the larva. The *corpora allata* are still present in the late pupa, but are much less conspicuous and appear to be undergoing some degeneration. Oenocytes are present in the same situations in the pupa as in the larva. In the advanced pupa, however, the small oenocytes are seemingly reduced in size and prior to emergence even the large oenocytes appear reduced and many degenerated. According to Hosselet (1925) oenocytes occur in the pupa of *Culex annulatus* in the wing roots.

REFERENCES

Adie, H. A. (1912). Note on the sex of mosquito larvae. *Ann. Trop. Med. Parasit.* **6**, 463–6.

Baisas, F. E. (1934). Notes on Philippine mosquitoes. IV. The pupal and certain adult characters etc. *Philipp. J. Sci.* **59**, 65–84.

Baisas, F. E. (1935). Notes on Philippine mosquitoes. V. The pupal characters of Anophelines under the *Myzorhynchus* series, etc. *Mon. Bull. Bureau Hlth Philipp. Islds*, **15**, 291–339.

Baisas, F. E. (1936). Notes on Philippine mosquitoes. VI. The pupal characters of Anophelines of the subgenus *Myzomyia*. *Philipp. J. Sci.* **61**, 205–20.

Baisas, F. E. (1938). Notes on Philippine mosquitoes. *Mon. Bull. Bureau Hlth. Philipp. Islds*, **18**, 175–232.

Baisas, F. E. and Pagayon, A. U. (1949). Notes on Philippine mosquitoes. XV. The chaetotaxy of the pupa and larva of *Trypteroides*. *Philipp. J. Sci.* **78**, 44–72.

Barraud, P. J. and Covell, G. (1928). The morphology of the buccal cavity in Anopheline and Culicine mosquitoes. *Ind. J. Med. Res.* **15**, 671–9.

Belkin, J. N. (1951). A revised nomenclature for the chaetotaxy of the mosquito larva. *Amer. Midl. Nat.* **44**, 678–98.

BELKIN, J. N. (1952). The homology of the chaetotaxy of immature mosquitoes and a revised nomenclature for the chaetotaxy of the pupa. *Proc. Ent. Soc. Wash.* **54**, 115–30.

BELKIN, J. N. (1953). Corrected interpretations of some elements of the abdominal chaetotaxy of the mosquito larva and pupa (Diptera: Culicidae). *Proc. Ent. Soc. Wash.* **55**, 318–24.

BELKIN, J. N. (1954). The dorsal hairless setal ring of mosquito pupae (Diptera: Culicidae). *Pan-Pacif. Ent.* **30**, 227–30.

BELKIN, J. N., KNIGHT, K. L. and ROZEBOOM, L. E. (1945). Anopheles mosquitoes of the Solomon Islands and New Hebrides. *J. Parasit.* **31**, 241–65.

BRUMPT, E. (1941). Méchanisme d'éclosion des moustiques. *Ann. Parasit.* **18**, 75–8.

CANTRELL, W. (1939). Relation of size to sex in pupae of *Aëdes aegypti* (Linn.), etc. *J. Parasit.* **25**, 448–9.

CAUSARD, M. (1898). Sur le rôle de l'air dans la dernière mue des nymphs aquatiques. *Bull. Soc. Ent. Fr.* 1898, 258–61.

CHRISTOPHERS, S. R. (1933). *Fauna of British India: Diptera*, vol. IV. *The Anophelini.* Taylor and Francis, London.

COMSTOCK, J. H. (1920). *An Introduction to Entomology*. Ed. 2. Comstock Publ. Co., Ithaca, New York.

CONSTANTINIANU, M. J. (1930). See references in ch. XIV (*d–e*), p. 351.

CRAWFORD, R. (1938). Some anopheline pupae of Malaya with a note on pupal structure. *Gov. Straits Settlem. Mal. Adv. Bd. F.M.S.* Singapore.

EDWARDS, F. W. (1941). *Mosquitoes of the Ethiopian Region.* Vol. III. *Culicine Adults and Pupae.* Brit. Mus. (Nat. Hist.) London.

EYSELL, A. (1911). Das Schlüpfen der Stechmücken. *Arch. Schiffs.- u. Tropenhyg.* **15**, 273–82.

GILES, G. M. (1903). On prepupal changes in the larvae of Culicidae. *J. Trop. Med.* (*Hyg.*), **6**, 185–7.

HOSSELET, C. (1925). Les oenocytes de *Culex annulatus*, etc. *C.R. Acad. Sci., Paris*, **180**, 399–401.

HURST, C. H. (1890). *The Pupal Stage of* Culex. Guardian Printing Works, Manchester.

HURTHLE, K. (1909). (Structure of striated muscle; *Hydrophilus.*) *Arch. ges. Physiol.* **126**, 1–164.

IMMS, A. D. (1907), (1908). On the larval and pupal stages of *Anopheles maculipennis* Meig. *J. Hyg., Camb.*, **7**, 291–316; *Parasitology*, **1**, 103–32.

INGRAM, A. and MACFIE, J. W. S. (1917). Notes on some distinctive points in the pupae of West African mosquitoes. *Bull. Ent. Res.* **8**, 73–91.

INGRAM, A. and MACFIE, J. W. S. (1919). The early stages of West African mosquitoes. *Bull. Ent. Res.* **10**, 59–69.

JORDAN, H. E. (1919), (1920). (Structure of insect muscle.) *Anat. Rec.* **16**, 217–45; **19**, 97–124.

JORDAN, H. E. (1920). Studies on striped muscle structure. VI. The comparative histology of the leg and wing muscle of the wasp, etc. *Amer. J. Anat.* **27**, 1–67.

KIELICH, J. (1918). Beiträge zur Kenntnis der Insektenmuskeln. *Zool. Jahrb., Anat.* **40**, 515–36.

KNIGHT, K. L. and CHAMBERLAIN, R. W. (1948). A new nomenclature for the chaetotaxy of the mosquito pupa based on a comparative study of the genera. *Proc. Helm. Soc. Wash.* **15**, 1–10.

KULAGIN, N. (1905). Der Kopfbau bei *Culex* und *Anopheles*. *Z. wiss. Zool.* **83**, 285–335.

LOWNE, B. T. (1895). *The Anatomy, Physiology, Morphology and Development of the Blow-fly.* Porter, London.

MACFIE, J. W. S. (1920). The chaetotaxy of the pupa of *Stegomyia fasciata*. *Bull. Ent. Res.* **10**, 161–9.

MARSHALL, J. F. and STALEY, J. (1932). On the distribution of air in the oesophageal diverticula and intestine of mosquitoes; its relation to emergence, feeding and hypopygeal rotation. *Parasitology*, **24**, 368–81.

MATTINGLY, P. F. (1949). Notes on some oriental mosquitoes. *Proc. R. Ent. Soc., Lond.*, **18**, 219–28.

MORISON, G. D. (1927). The muscles of the adult honey bee (*Apis mellifica* L.). *Quart. J. Micr. Sci.* **71**, 395–463, 563–651.

ROZEBOOM, L. E. and KNIGHT, K. L. (1946). The *punctulatus* complex of *Anopheles* (Diptera: Culicidae). *J. Parasit.* **32**, 95–131.

REFERENCES

SEGUY, E. (1923). *Histoire naturelle des moustiques de France.* Paul Lechavalie, Paris.

SEN, S. K. (1924). Observations on the bionomics of *Aëdes (Stegomyia) albopictus* Skuse. *Rep. Proc. 5th Entom. Meet. Pusa,* 1923, 215–25. Abstract in *Rev. Appl. Ent.* **13,** 135.

SENEVET, G. (1930). Contribution à l'étude des nymphs de Culicides. *Arch. Inst. Pasteur Algér.* **8,** 297–382.

SENEVET, G. (1931). Contribution à l'étude des nymphs d'Anophelines (2nd. Mem.). *Arch. Inst. Pasteur Algér.* **9,** 17–112.

SENEVET, G. (1932). Contribution à l'étude des nymphs d'Anophelines. *Arch. Inst. Pasteur Algér.* **10,** 204–54.

SENEVET, G. (1934). Contributions à l'étude des nymphs d'Anophelines. *Arch. Inst. Pasteur Algér.* **12,** 29–76.

SENEVET, G. (1940). Contributions à l'étude des nymphs d'Anophelines. *Arch. Inst. Pasteur Algér.* **18,** 443–7.

SINTON, J. A. and COVELL, G. (1927). The relation of the morphology of the buccal cavity to the classification of anopheline mosquitoes. *Ind. J. Med. Res.* **15,** 301–8.

SNODGRASS, R. E. (1925). *Anatomy and Physiology of the Honey Bee.* McGraw-Hill Book Co., New York and London.

SNODGRASS, R. E. (1938). *The Principles of Insect Morphology.* McGraw-Hill Book Co., New York and London, p. 388.

STUDNITZ, G. VON (1935). Über die Feinstruktur, Kontraktion und Färbbarkeit quergestreifter Arthropodenmuskeln. *Z. Zellforsch.* **23,** 1–23.

THEODOR, O. (1924). Pupae of some Palestinian Culicidae. *Bull. Ent. Res.* **14,** 341–5.

THOMPSON, M. T. (1905). The alimentary canal of the mosquito. *Proc. Boston Nat. Hist.* **32,** 145–202.

TIEGS, O. W. (1955). The flight muscles of insects; their anatomy and histology etc. *Phil. Trans.* B, **238,** 221–348.

TREMBLEY, H. L. (1944). Mosquito culture technique. *Mosquito News,* **4,** 103–19.

WEBER, H. (1933). *Lehrbuch der Entomologie.* Gustav Fischer, Jena.

ZAVREL, J. (1907). Die Augen einiger Dipterenlarven und Puppen. *Zool. Anz.* **31,** 247–55.

XVI

THE PUPA (cont.)

(*a*) PHYSICAL CHARACTERS

DIMENSIONS AND WEIGHT

Measurements of the length of the cephalothorax of *Aëdes aegypti* pupa are given by Cantrell (1939). For the female these were from 1·80–3·00 mm. and for the male 1·45–2·65 mm. with one exception which measured 2·95 mm.

Very similar figures were obtained from measurement of the cephalothorax of fifty female and fifty male pupae of *A. aegypti* taken at random on a number of occasions. The distance measured was that from the prominent anterior point of the head to the posterior bulge of the cephalothorax taken with the pupa lying on its side. This gave for the female from 2·32 to 2·75 mm. (mean 2·55 mm.) and for the male from 1·92 to 2·17 mm. (mean 2·10 mm.), the proportion of female to male being 1·21. The maximum transverse diameter of the thorax in a well-nourished female pupa was about 1·8 mm., whilst that for the male was about 1·4 mm.

The mean weight of some hundreds of pupae taken on different occasions was 4·73 mg. for the female and 2·65 mg. for the male, or a ratio of female to male of 1·78. This shows the pupa to be about the same weight as the mature larva and considerably heavier than the adult mosquito some time after emergence. Starvation pupae may be only about half the weight or less of the well-nourished female pupa. Departure from the mean weight in the male is not so great.

The toughness of the outer pupal skin gave an opportunity to make an estimate of the surface area of the pupa. For this a female and a male pupal skin were carefully broken up to give fragments which when mounted in gum would lie flat. Camera lucida drawings of the fragments were then made at a magnification of × 50 and tracings of these placed over squared paper and the areas determined. The result gave an area of 17·29 mm.^{2} for the female and of 12·36 mm.^{2} for the male, or a proportion of female to male of 1·40. For the female therefore the proportion of area in mm.^{2} to weght in mg. was 17·29 to 4·73 or approximately 3·6 to 1, as against 6 to 1 for a cube or 3 to 1 for a sphere. For the male the proportion was 4·5 to 1.

Some interest attached to the weight of the pupal cuticle, which in the female particularly is of considerable thickness. The weight of 100 female pupal cast skins dried overnight at room temperature after blotting off excess fluid was 27 mg. or a mean weight per pupa of 0·27 mg. The cast pupal skin therefore weighs about 4 per cent of the pupal weight. The weight is also about ten times that of the cast skin of the fourth instar larva.

PHYSICAL CHARACTERS

SPECIFIC GRAVITY AND VOLUME

In contrast to the larva the pupa is hydrostatically lighter than water and can only maintain itself below the surface by active diving movements. Observations on carefully chloroformed pupae made at 14° C. with different mixtures of alcohol and water gave the following values for specific gravity:

Female (dark)	0·982
Female (white pupa)	0·982
Male (dark)	0·977

The specific gravity as determined from weight and volume as determined by displacement (see ch. VI) was somewhat lower as shown below, the mean volume being 5·60 mm.^{3} and the mean weight of the pupae tested 4·96 mg., giving the mean sp.gr. as 0·905.

	Weight (mg.)	Volume (mm.^{3})	Specific gravity
Three white female pupae	15·1	18·3	0·825
Five grey female pupae	23·3	25·3	0·921
Four advanced female pupae	21·1	23·6	0·894

ORIENTATION AND HYDROSTATIC BALANCE

The respiratory trumpets of the pupa of *A. aegypti* are tubular in their basal portion and rather broadly trough-shaped in their apical third. The edges of the trough when in contact with the surface film are expanded, leaving a considerable area open. The inner surface is also markedly hydrophobic and has the power when submerged to retain a comparatively large bubble of air. In addition to the trumpets the float hairs on the first abdominal segment also contact the surface film. Thus the film is broken by the comparatively large hydrophobic surface of the trumpet openings and behind this by the two small contacts by the float hairs.

As already noted it is the air cavity beneath the cephalothorax which is mainly responsible for the pupa's buoyancy and this also plays a part in giving a correct fore and aft orientation to the pupa by which when at the surface its trumpets are suitably disposed. It would appear, however, that with this flotation centre placed low down an unstable condition must exist and according to Manzelli (1941) it is the tips of the trumpets held by surface tension which enables the pupa to keep its balance. If the surface tension is reduced, the pupae turn on their side, the top of the pupa being the heavier part. Manzelli also notes that the trumpets normally hold a bubble of air and that they are folded back by muscular action to prevent this being displaced. As already noted, the trumpets are much constricted at their base and have a kink in this position that may in certain positions act as a valve cutting off connection between their cavities and the general tracheal system.

The effect of lowered surface tension is the reason why certain substances have a lethal effect on pupae. It is due also to the attachment of their respiratory trumpets to the surface film that pupae in a vessel tend to collect and lie up at the edge of the water against the sides of the receptacle. The reason for such a position is that when a pupa comes within a certain distance of the edge its trumpets are drawn up the slope of the meniscus by flotation until the pupa touches the glass. Pupae rising in the body of the fluid are not so affected. But as pupae are constantly liable to dive sooner or later, most get drawn to the edge where they tend to rest more quietly.

(*b*) THE PUPAL CUTICLE

In contra-distinction to the thin and delicate cuticle of the larva, other than that of the head and the siphon, the hardened pupal cuticle is thick and strong and highly resistant to loss of water by desiccation and to the action of chemicals. That this is due to some cause over and above that dependent on a thick and strongly tanned cuticle is shown by the fact that pupae, while extremely resistant to bleaching with chlorine, are readily bleached after dipping for a few seconds in chloroform. This applies also to diaphanol. Untreated pupae will remain for long periods in diaphanol without bleaching, but are bleached at once after dipping in chloroform, thus differing from most insect cuticle which is rapidly bleached in this reagent.

That this property of resistance is due to a waxy substance within or on the cuticle receives confirmation from the fact that pupal skins washed in distilled water, dried and allowed to remain in chloroform yield on evaporation of this a fatty or waxy substance the melting point of which is approximately 52–53° C. This substance forms 2·67 per cent of the weight of the dried cast pupal skin. In a thin layer after evaporation from chloroform it shows clumps of material which give bifringence with polarised light.

The newly emerged pupa is a pure dead white, the only dark parts being the float hairs on the first abdominal segment, which on the white pupa show up to the naked eye as two minute black puncta towards the base of the abdomen. The following are changes of colour which take place by the times noted:

30 minutes	No longer pure dead white
45 minutes	Distinctly a greyish tinge
60 minutes	Grey
90 minutes	Dark grey with pale abdominal bands
120 minutes	The usual dark appearance of the pupa

A more precise determination on twenty-three pupae observed at 28° C. from time of emerging from the larval skin is as follows:

	Colour change				
Time in minutes	Dead white	Not quite white	Light grey	Dark grey	Dorsum black
5	22	1	—	—	—
15	20	3	—	—	—
30	12	9	2	—	—
45	—	19	4	—	—
60	—	—	—	23	—
120	—	—	—	—	23

At 28° C. any pure white pupa may safely therefore be taken as on the average not more than 15 minutes from casting of the larval skin, a fact which may often be useful in making observations on pupae at stated times from emergence.

The colour of the darkening pupa cannot, like that of the egg, be described as blue, though the grey is of a character rather suggestive of that colour and has no element of yellow or red in it. The respiratory trumpets are, however, in conspicuous contrast, being yellow turning brown. The large compound eyes are red, but the ocelli are from the beginning black. The parts first showing darkening are the scutum, postnotum and halteres with an ornamental pattern on the first abdominal segment (Fig. 42 (7)), the basal portions of the abdominal segments II–VI and fainter effects on the wings and some leg segments.

White pupae broken up in water with a few crystals of pyrocatechol show the characteristic tomato red, indicating presence of polyphenase. Under such treatment the paddles soon show up pink and also the pupal wings. Later the same tint is seen in various parts of the pupal cuticle, head, thorax, antennal bases, etc. This colour reaction indicates that polyphenase is the enzyme concerned in the reaction leading to the darkening of the pupal cuticle, the ferment present in the insect's haemolymph acting as shown by Pryor (1940) through the liberation of polyphenols, which acting upon the protein in the cuticle give rise to darkening and tanning characteristic of the sclerotisation (so-called chitinisation) of insect cuticle. The darkening in this case is independent of light as it takes place in much the same time whether the pupa is exposed to light or kept in the dark.

As the time of emergence approaches the pupa becomes of a dull sooty black in contrast to the younger pupae, which retain throughout a certain amount of light coloration in the abdominal banding and under-surface of the body. This final darkening is not due to change in the pupal cuticle, but to the black scaling of the contained imago showing through. Such pupae within a short time will undergo emergence, the colour change being a useful guide to the choice of specimens for observing emergence.

(*c*) BEHAVIOUR AND VIABILITY

BEHAVIOUR

When disturbed the pupa, like the larva, dives. But owing to their great buoyancy they can normally only maintain themselves below the surface by a succession of dives produced by strokes of the abdomen with its terminal paddles. The result of a stroke by the paddles is to tilt the cephalothorax forwards and downwards and progression consists largely of a series of U-shaped dives which carry the insect some distance horizontally as well as downwards. Whether the power of the stroke is obtained by extension or flexion of the abdomen is not easy to ascertain by observation. The former, however, would seem most likely since movement is forwards. Rising is mainly passive.

Owing to the great buoyancy of the pupae it is only by swimming in this fashion that they can keep beneath the surface. Nevertheless, by a succession of small strokes they can remain down for some minutes and often more or less in a stationary position. Very occasionally the pupa becomes anchored by its paddles or trumpets to some particle on the bottom or inequality on the side of the vessel. This, however, is unusual in *A. aegypti* in captivity though common in nature with the pupae of some species.

Very noticeable features in the behaviour of the pupa are, as with the larva, negative phototropism and sensitiveness to vibrations. The pupa is, however, less responsive to all these forms of stimulus than the larva, and it is possible by regulating the intensity of the tap given to the vessel in which they are contained to cause the larvae to dive whilst the pupae remain at the surface, a fact often made use of when collecting pupae free from larvae for experiment.

RESISTANCE TO SUBMERSION AND DESICCATION

Da Costa Lima (1914) found that pupae showed little resistance to prolonged submersion. Owing to the relatively thick and impervious cuticle the pupa is unlikely to be able to make any use of dissolved oxygen. Further, pupae when submerged tend to lose buoyancy owing to the using up of oxygen by their tissues, and when submerged beyond a certain time they may become too heavy to rise to the surface, efforts to do so only driving them further downwards. Carter (1923) notes that larvae of *A. aegypti* will live 2 or 3 days under a film of petroleum, but not the pupae. Kalandadse (1933) found pupae began to die under a glass sheet excluding air from the water in 2 hours, whereas fourth instar larvae began to die only after 7 hours. As previously noted first instar larvae under such circumstances may live for days. Macfie (1923*a*) found that, in a tube, larvae will descend on tapping the tube to a distance of as much as 8 feet. Pupae, on the other hand, showed anxiety to return at beyond 3–3½ feet, and one at 7 feet sank to the bottom.

Pupae, however, have considerable powers of resistance to desiccation and when removed from water may remain alive and active for some hours. Larvae, on the contrary, unless kept moist, quickly begin to dry and die. Howard (1913) notes that pupae of *A. aegypti* can live and eventually emerge on moist blotting paper. Young (1922) found that pupae still hatched when dry twenty-four hours from pupation. Gillett (1955) utilises this property in isolating adults for experiment.

LETHAL EFFECT OF HIGH AND LOW TEMPERATURE

Bacot (1916) notes that at 40° F. (4·4° C.) pupae of *A. aegypti* become dormant at the surface of the water. They resume activity at 60° F. (15·6° C.).

My own observations on the lethal effect of low temperature on the pupa are but few. At 7° C. pupae are immobilised or capable of very feeble movement. Emergence does not take place. At 17° C. pupae are active, but time of emergence is greatly prolonged or does not take place. At 18° C. emergence takes place with a pupal period of from 116 to 120 hours.

For the lethal effect of high temperature Macfie (1920) gives the following:

Immediate effect	
44° C. or higher	Pupae inert and lying at surface
42–43° C.	Similar effect but slighter.
41° C.	No deleterious effect.
Final effect	
50° C.	No recoveries.
45° C.	Seldom survived above this temperature.
44° C.	Seldom showed permanent ill effects.

Table 23 gives the results of exposing female and male pupae of *A. aegypti* to different temperatures for various periods and observing the immediate effect and eventual mortality as shown after 24 hours.

It will be seen from the table that the temporary immobilising effect from which many pupae may recover, as also the completely lethal effect, is practically identical for the two sexes. Another interesting feature is that, for immobilising effect, time of exposure is not very important, the effect for different periods of exposure beginning to be pronounced always at a temperature of 43° C. Time of exposure is more important with higher temperatures. The critical lethal temperature for 3 minutes exposure is 48° C., that for 15 minutes 45° C. and that for 30 minutes 44° C., though from the data obtained there would seem to be little difference for exposures over this. 43° C. was still not completely lethal, even at 60 minutes exposure. For a short exposure, therefore, that is for periods from a few minutes to an hour, lethal temperature for the pupa, taking this as producing 100 per cent mortality, is from 44 to 48° C. Lethal temperature for the pupa is therefore within a few degrees of that for the larva. In this case pupae from a culture were added directly to water at the temperature noted. For periods much over those noted a lower temperature would probably be lethal. According to Davis (1932) a temperature of 36° C. greatly shortens life.

Table 23. *Lethal effect of high temperature on pupae of* Aëdes aegypti

		Females			Males		
Exposure (min.)	Temperature (° C.)	No. exposed	Immobile at end of exposure	Dead at 24 hours	No. exposed	Immobile at end of exposure	Dead at 24 hours
3	42	16	—	—	11	2	2
3	43	7	—	—	14	3	—
3	44	17*	13	4	20*	15	4
3	45	12*	12	1	12*	12	2
3	46	11*	11	3	25*	23	9
3	47	13*	13	7	18*	18	10
3	48	14†	14	14	11†	11	11
3	49	12†	12	12	19†	19	19
15	40	11	—	—	13	1	1
15	41	11	1	1	15	1	2
15	42	12	1	1	11	4	2
15	43	8*	8	3	10*	10	1
15	44	24*	14	11	34*	34	11
15	45	13†	13	13	8†	8	8
15	46	7†	7	7	16†	16	16
15	47	11†	11	11	18†	18	18
30	40	4	—	—	14	—	—
30	43	30*	30	2	30*	22	2
30	44	48†	48	48	50†	50	48
60	39	11	—	—	27	2	—
60	40	10	2	—	21	2	4
60	41	5	—	—	9	2	—
60	43	41*	38	30	36*	33	19

* Indicates over 50 per cent immobilised. † 100 per cent killed.

EFFECT OF SALT AND CHEMICALS

The pupa, unlike the larva, is almost entirely closed in an impervious cuticle and is remarkably resistant to chemicals other than vapours and gases. Nevertheless, as previously noted, this resistance is not entirely a matter of thickness and hardness of cuticle and seems to be due to a considerable component of waxy substance on or within the cuticle.

Macfie (1914) notes that in 2 per cent salt (that is, about two-thirds the concentration of sea water) pupae of *A. aegypti* continue to hatch out for the first 2 days, after which all are dead. The same author (1916) noted that pupae of the same age always hatched out earlier in saline solutions than in tap water and that development is hastened by a low concentration of salt, for example 0·49 per cent. Woodhill (1942) notes that pupae of *A. aegypti* were not affected by salinity of 3·5 or 7 per cent. Curiously enough, Macfie (1923*b*) notes that 1·2 per cent lithium chloride not only killed larvae but also pupae within 4 hours.

Pupae may remain alive for many hours in 70 per cent alcohol, 10 per cent formalin or other usual fixatives. Chopra, Roy and Ghosh (1940) found that pupae of *A. aegypti* were less readily killed by pyrethrum scattered in powder on the water than were larvae. Hoskins

(1932) found pupae of *Culex tarsalis* much less susceptible than the larvae to acid and basic solutions of sodium arsenite. 50 per cent of pupae died in an exposure of 26·5 minutes to 0·03 molar solution of arsenite at pH 5, but continuous exposure for 402 minutes was necessary for full effect.

In contra-distinction to chemicals which act, if at all, by penetration of the cuticle are those lethal through their effect in lowering the surface tension of the fluid and so, as already noted, causing the pupae to lose attachment of their respiratory trumpets to the surface film. Conspicuous among such substances is soap. Senior White (1943), working with saturated bar soap solution, found a dilution of 1–50 in tap-water lethal to pupae in 45 minutes. 1–100 dilution took over 2 hours and at 1–400 pupae progressed to emergence. In distilled water 1–500 killed in 135 minutes, emergence taking place at 1–750 or over. Manzelli found soap, 1–100, killed in 3 minutes. 1–400 did not kill in 4 hours. A patent wetting substance, sanlomerse, at 1–4000 reduced surface tension to half. Other effects of the lowered surface tension were that adults alighting on the water sank, as also did egg-rafts. Russell and Rao (1941), working with the pupae of *Anopheles*, also found soap lethal to pupae.

There would seem no reason why pupae should be especially resistant to gases and vapours. Williamson (1924) found that chlorine in high but undetermined amount maintained for 3 minutes killed all larvae and pupae. See also Macfie (1916).

(*d*) EMERGENCE

DURATION OF THE PUPAL PERIOD

For the duration of the pupal stage of *Aëdes aegypti* Marchoux, Salimbeni and Simond (1903) give 30–50 hours ordinarily and 3–4 days at 22° C., Mitchell (1907) 1–5 days, Newstead and Thomas (1910) 2–3 days at 23° C., Howlett (1913) 2–3 days, Bonne-Wepster and Brug (1932) 2–3 days in a tropical climate. De Meillon, Golberg and Lavoipierre (1945) found the mean time for emergence of 1000 pupae was 2·4 days (57 hours), 97 per cent emerging within a range of 47–69 hours. Shannon and Putnam (1934) give for the time between pupation and emergence at 23–27° C. 45 hours for males and 60 hours for females.

The data given in the accompanying table (Table 24) were obtained for pupae which had been either timed from observed pupation (39 pupae) or used as pure white pupae (241 pupae). As already noted, pupae remain pure white only for an average period of

Table 24. *Time from pupation to emergence at different temperatures*

Temperature (° C.)	Sex	Number observed	Number emerging at different hours											Mean in hours
			43	44	45	46	47	48	49	50	51	52–69	116–120	
29	F.	43	40	3	—	—	—	—	—	—	—	—	—	43
28	F.	55	—	—	—	14	30	11	—	—	—	—	—	47
26	F.	31	—	—	—	—	1	5	12	10	3	—	—	49
23	F.	12	—	—	—	—	—	—	—	—	—	12	—	60
18	F.	6	—	—	—	—	—	—	—	—	—	—	6	118
7	F.		Failed to emerge											
29	M.	38	37	1	—	—	—	—	—	—	—	—	—	43
28	M.	32	—	—	—	13	15	4	—	—	—	—	—	47
26	M.	9	—	—	—	1	8	—	—	—	—	—	—	47
23	M.	31	—	—	—	—	—	—	—	—	—	31	—	60
18	M.	7	—	—	—	—	—	—	—	—	—	—	7	118
7	M.		Failed to emerge											

15 minutes, a period which may be neglected without much error when giving time to emergence in hours.

The curious point is that though the period of larval development in the male is about one day shorter than in the female the much smaller male pupa takes almost exactly the same time to reach emergence as the female. It has already been observed that males emerge in cultures before the females. This, however, appears to be the result of a shorter period of larval development; the pupal period is not shorter. At 17° C. pupae do not develop further; or if emergence takes place at all it is greatly delayed. At temperatures below 17° C. it does not occur. At some temperature between 17° C. and 7° C. pupae are immobilised and in time killed.

THE ACT OF EMERGENCE (Fig. 48)

The emergence of the imaginal mosquito from the pupal skin has been a subject of interest as far back as the time of Reaumur (1738) who describes the act. See also Meinert (1886); Causard (1898), whose reference to the part played by air in emergence of aquatic nymphs has already been noted; Hurst (1890), whose classical papers on pupal structure have also been noted; Eysell (1905, 1911), who appears to have been the first to record that air enters under the pupal cuticle at separation of the pupal trumpets and a translation of whose account is given by Howard, Dyar and Knab (1912); Brocher (1913), who describes the act in *Culex* as one of the object lessons of the small aquarium; Seguy (1923), who gives three pages of description of the act; Sen (1924), who describes the swallowing of air and movements of the pharyngeal pump during the act; MacGregor (1927), who gives a detailed account of emergence of *A. aegypti*; and lastly the important papers by Marshall and Staley (1932) and Brumpt (1941) on which detailed knowledge of the mechanics of the processes concerned largely rests.

The general features of the act of emergence are well described in the following translation of a description of emergence of *A. aegypti* by Eysell (1911) given by Howard, Dyar and Knab.

> The abdomen straightens steadily and is completed in 10 to 15 minutes. Air meanwhile appears between the imago and the pupal skin. The anterior parts of the cephalo-thorax are then raised by a powerful jerk so that the pupa is now touching the surface of the water in front with the dorsal part of the pro- and meso-thorax, in the middle with the trumpets and behind with the end of the body. A split now occurs in the middle line of the cephalo-thorax exposing the dry surface of the imaginal dorsum. By jerks the split is widened and two transverse slits are added on each side and after the pupal skin has receded about ten times the thorax in its entire width is exposed. The scutellum, the stretched neck and vertex of the strongly bent down head next appear. Jerks now cease and the insect emerges with steady ghost-like motion. When the head is born the antennae first become free from their sheaths, then the palpi and proboscis are drawn from their sheaths. The last segment of the pupal skin is now filled with air. The head is raised, the antennae, palpi and proboscis extended and the first legs withdrawn by bending the knee-joints and later the joints of the feet in a plane with the body axis. After the first legs the middle legs are similarly extracted. The front and mid legs are now extended and the tarsal joints laid gently on the water. The body is now firmly supported and the hind legs are withdrawn. The legs are disengaged independently of movements of the body, being extracted from their sheaths in pairs by repeated very short movements. Finally the wing tips and abdomen become free. The entire process is completed within a few minutes.

The following are some further observations on emergence in *A. aegypti* as seen under the low-power microscope. The first indication that emergence is approaching is change in the colour of the pupa from a greyish black to a deep sooty black, the result of the black scales of the adult showing through the pupal cuticle and adding to the already dark colour of the pupa. Light patches on the abdomen also become visible from the pale scales on the abdomen of the imago showing through, and also the dark-scaled veins on the wings are very noticeable. The bubble at the angles behind the bases of the wings becomes more conspicuous and air is seen collected about the base of the breathing trumpets. At this time there are often seen some small, but definite, jerky movements of the body, apparently unconnected with movements of the abdomen, the nature of which is uncertain.

The exact sequence of the events that follow at this time seems to vary somewhat, but an early first indication that emergence is imminent is that the abdomen from being bent under the cephalothorax gradually becomes extended so that the paddles become visible from above. Air may be seen at this stage not only at the bases of the trumpets but below this. It may even sometimes be seen at the sides of the abdomen before further happenings occur. Usually, however, before air under the cuticle is at all extensive a white line suddenly appears along the top of the crest on the scutum. This is caused by air apparently entering into small fissures either within the substance of, or under, the cuticle of the ridge. The line at this time is not an actual split. Without further warning a sudden movement of violent

---

Figure 48. The pupa. Pupation and emergence.

1 Fourth instar larva of *Culex pipiens* killed during pupation at time of withdrawal of tracheal trunk sections. Camera lucida drawing. *a*, median flap of head capsule pushed forwards by the head of the emerging pupa. Behind the head are indications of membranous cuticle slipping backwards and just about to expose the respiratory trumpets; *b*, portion of tracheal trunk dragged from metathorax; *c*, portions of tracheal trunks respectively from the first, second and third abdominal segments, that is the section anterior to their present position; *d*, wrinkled cuticle and sections of tracheal trunks from segments four, five and six collected at terminal portion of abdomen; *e*, tracheal trunks drawn out from seventh and eighth abdominal segments attached to siphon.

2 Head and thorax of fourth instar larva of *Aëdes aegypti* at commencement of pupation showing air collected about the respiratory trumpets. Dorsal view. Lettering as for 3.

3 The same at a later stage showing air collections under the cuticle at various places. *a*, ventrally at front of thorax; *b*, in space left between limb loops on ventral aspect of thorax; *c*, air collections under pupal wings. Ventral air space; *d*, ventrally at front of abdominal segments.

4 Horizontal section of pupa at level of first abdominal spiracles showing relation of pupal sheaths of mouth-parts, legs, wings and halteres. The dotted lines indicate imaginal parts lying within the sheaths. The lateral extensions of the ventral air space up to the region of the halteres, etc. are shown in black.

5 The same at a more ventral level and passing through the main portion of the ventral air space.

6 Showing the way in which the sheaths of the mouth-parts and legs terminate to form the floor of the ventral air space.

7 Lateral view of pupal skin with emerging imago, showing the boat-like character of the pupal cuticle and position of the imago just prior to withdrawing the proboscis and legs. The water line is arbitrarily indicated. Camera lucida drawing.

Lettering: *an*, antenna; *cx′*, *cx″*, *cx‴*, coxae of first, second and third legs; *fe′*, *fe″*, *fe‴*, femora of first, second and third legs; *ha*, imaginal haltere in pupal sheath; I, cut-across base of abdomen; *L*, labium and other mouth-parts in their pupal sheaths; *sp*, spiracle of first abdominal segment opening into the ventral air space; *tb′*, *tb″*, *tb‴*, tibiae of first, second and third legs; *ts′*, *ts″*, *ts‴*, tarsi of the same; *va*, ventral air space; *va′*, lateral extensions of same; *W*, pupal wing sheath with developing imaginal wing.

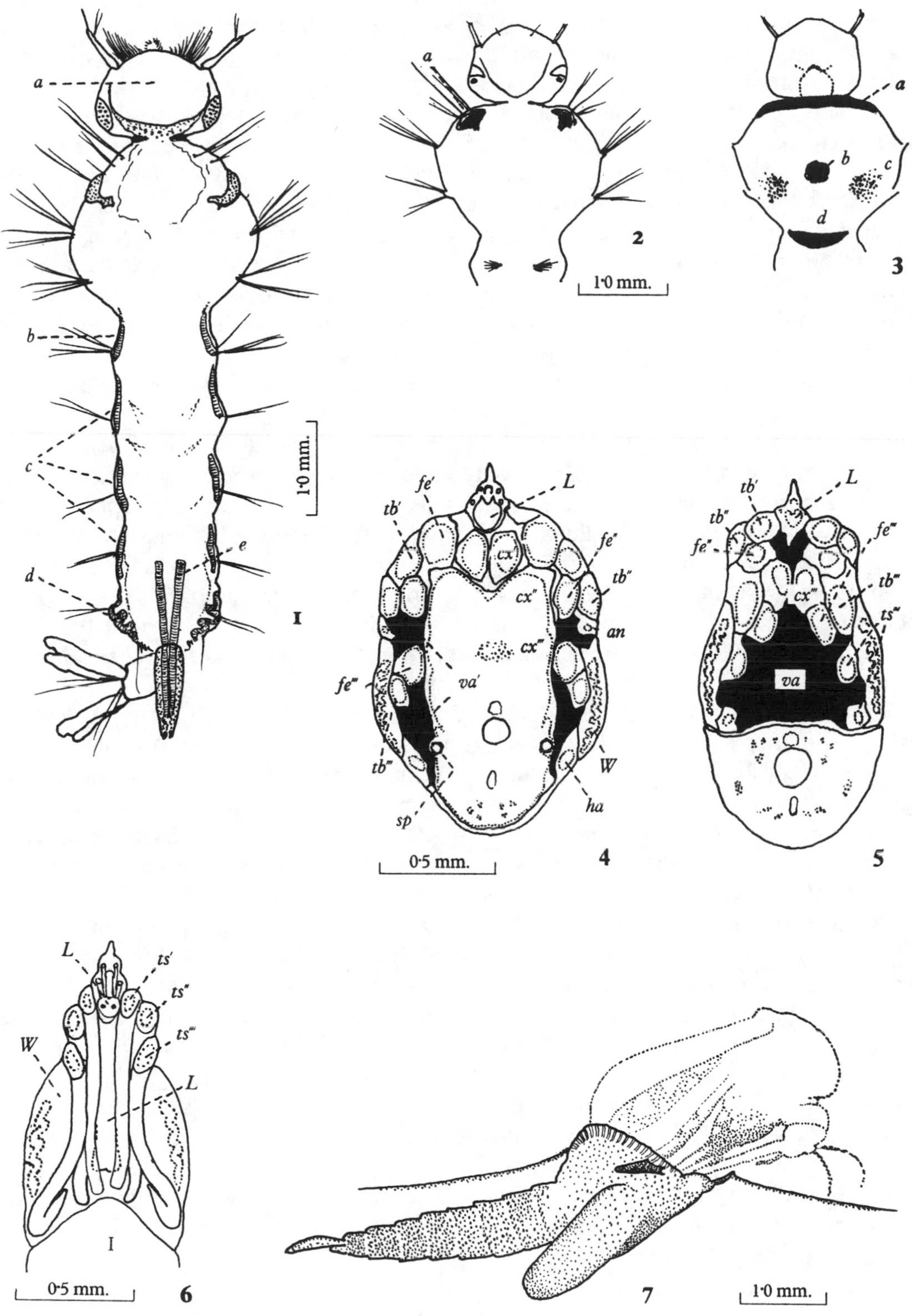
a
b
c
d
e
1·0 mm.
I
a
2
1·0 mm.
a
b
c
d
3
fe'
L
tb'
cx'
fe''
cx''
tb''
an
cx'''
fe'''
va'
tb'''
W
sp
ha
0·5 mm.
4
tb'
L
tb''
fe'''
fe''
cx'''
tb'''
ts'''
va
5
L
ts'
ts''
ts'''
W
L
I
0·5 mm.
6
7
1·0 mm.

Figure 48

extension of the pupal body now occurs bringing the abdomen completely level with, or even concave to, the surface, the cephalothorax at the same time being tilted backwards so that the antennae now become conspicuous at the front of the pupa as this is viewed from above. This almost convulsive movement of extension would appear to be an important stage in emergence, but its exact significance or implications are not clear. It is usually almost immediately after this that a median crack appears along the white line on the crest. Once this happens emergence is usually quickly completed.

The crack rapidly widens, exposing the dry surface of the striped imaginal thorax. Then the stretched back of the neck and finally the occipital surface of the bent-down head are exposed. Over the lateral regions of the now exposed occiput curious twitching movements occur, evidently due to contraction of muscles taking origin in this situation. These may be lateral or dorsal dilators of the sucking pump or, as is thought possible, muscles connected with the mouth-parts and engaged in liberating these from their sheaths. Or both may be in action.

Meanwhile air has appeared at the sides of the eighth segment and subsequently other segments, and the imaginal abdomen is seen to be exhibiting vermicular movements and to be retreating from the pupal cuticle, leaving this distended with air which has rapidly spread over the whole abdomen and allowed the scales of the imaginal abdomen to show up clearly. At the same time the widening longitudinal split is further supplemented by short transverse splits behind the vertical plates and also by the opening up of folds in the pupal skin behind the head as previously described when discussing the cast skin of the pupa. As the pupal skin more and more retracts, the thorax, scutellum, head and prothoracic lobes all emerge to view.

Now the head is suddenly raised, the antennae, palps and proboscis lifted out and the first legs withdrawn as described by Eysell. At this stage the front and mid-legs are more or less simultaneously extended and the tarsal joints laid gently on the water. In *A. aegypti*, according to Brumpt, the liberation of the three pairs of legs is almost simultaneous, though in some other species the legs are withdrawn successively. The wings, as the insect emerges, are drawn out of their sheaths and come to lie folded on the swollen abdomen as this leaves the now rigid air-filled pupal skin of the abdomen.

The pupal skin has now become a rigid floating base on which the insect continues its emergence very much as if making use of a small boat. Its body directed upwards like a spectre rising higher and higher without apparently any struggle or motive force has been a familiar subject for illustration and its slow emergence is certainly one of the curious sights in nature. At one stage, that is just before the mosquito places its fore-legs on the water, its body projects so far forwards that it appears at any moment to be in danger of toppling over (Fig. 48 (7)). But the critical moment passes once the legs are withdrawn and placed on the water. The insect, now stably supported, requires only to draw out the hind tarsi to be free.

That the simile of the pupal skin as a small boat, a simile noted by Réaumur, is not after all so fanciful is shown by the following observations giving the volume of displacement during the act of emergence.

Volume occupied by the intact pupa	5·48 $mm.^3$
Volume shortly after emergence began	5·21 $mm.^3$
Volume during emergence	4·95 $mm.^3$
Volume when emergence completed	3·00 $mm.^3$

Since 2·48 mm.^{3} of water (the difference when the insect moved away) would represent about 2·48 mg. weight, this should be about the weight of the emerging adult. The mean weight found for the newly emerged adult has been given in another section as 2·91 mg.

The time taken over emergence varies as does that of the different stages. Timed for a number of pupae on a warm stage at 30° C. emergence up to the appearance of the wing tips was most commonly completed in from 10 to 12 minutes, but in some cases as soon as 5 minutes. From the time of the sharp extension of the body to emergence as above was rarely more than 2 minutes. These times relate to the female pupa.

The newly emerged mosquito is at first pale with strongly inflated abdomen. It is at first somewhat incoordinate in its movements and weak. But after about 10 minutes it can make short flights. Normally, however, it remains on the water. After about an hour it is able to fly more or less normally and usually then leaves the water and settles on the wool plug or netting at the top of the container or on the sides of this.

The question has been raised whether mosquitoes following upon emergence drink water. Experiments in which the mosquitoes had emerged from pupae in water coloured with fuchsin and had been left undisturbed until they left the water voluntarily appeared to show that this was not a normal procedure following upon emergence, since in no case were traces of colour ever found in any part of the alimentary canal. It is usual for mosquitoes that have died, or become moribund, on the water surface to distend the diverticula with fluid. This, however, occurs at any age as an ante-mortem condition and has no special relation to emergence.

Owing to its swollen abdomen the mosquito has much the appearance of one that has fed or become gravid. But after a short time the abdomen takes on its usual appearance in the unfed insect. A small quantity of faecal matter may be passed at this time and has been termed *meconium*. It represents such remains as there may be of the cast off and undigested gut epithelium of the larval stage.

As with the discarded skin of the larva at pupation that of the pupa after emergence forms important material for the systematist owing to the clearness with which the hairs and other cuticular structures are displayed. Methods of preparing the pupal skin for examination have been given in chapter XV.

THE MECHANISMS INVOLVED IN EMERGENCE

Briefly the mechanisms employed in emergence are as follows. In the first place before emergence begins the usual steps leading to ecdysis as already described have been in operation, that is ecdysial fluid has been excreted, has dissolved the endocuticle of the pupal cuticle and has been absorbed. In some insects, including some aquatic Nematocera, absorption of the ecdysial fluid may in itself be a cause of entry of air under the cuticle (Wigglesworth, 1942, p. 27). Brumpt does not consider that this is so in the mosquito since the imago within the pupal cuticle is already dry before entry of air begins. It is still possible that some degree of negative pressure due to absorption of fluid might persist and assist entry of air when an opening for this occurs. The fact remains that early in the act air can be seen in some situations under the cuticle. Various views of the source of such air have been held. Brumpt, who has given special attention to the question, points out that the air of the large bubble held between the wings and legs of the pupa (that is the ventral air space) lies outside the pupal cuticle and hence cannot be the source of air observed under the cuticle. (This may require a little modification as will be seen later.) The swallowed air according to this author must come from that which enters under the pupal skin, as suggested by Eysell, at the time when the connections of the pupal trumpets are broken. It will be remembered that leading from the bases of the respiratory trumpets on each side is a length of dilated tracheal trunk leading to the position of the future mesothoracic spiracle of the imago, and that this portion of trunk seems to be entirely a pupal structure and is cast off with the pupal skin. Consideration will show that at this time the connections where these trunks are attached at the imaginal spiracles must be the only points at which the included imago is still attached to the pupal skin, and there would seem to be no difficulty in these connections being torn through at the commencement of emergence as appears to be the case. If they are so torn, air would certainly be able to enter under the pupal cuticle, whether from the air in the tracheal trunks as considered by Eysell, or from the external air via the respiratory trumpets as maintained by Brumpt. What does not follow is that such air is necessarily at once available for swallowing. Air entering at the bases of the respiratory trumpets would, to reach the mouth, require to pass through the relatively narrow pupal neck, and though this passage-way is not obvious externally it none-the-less exists, even if only represented in membrane. The structure of the parts in such questions requires to be borne in mind. Air, for example, could not be swallowed, as it would be in the emerged mosquito, through the proboscis, since the mouth-parts that will eventually constitute the channel for blood or air are at this stage in separate pupal sheaths and could not unite to form a channel. The mouth, however, situated high up in the region of the clypeus, is so available and if air were present in this region it could conceivably be swallowed. Its entry might be facilitated by the tearing of the ventral head membrane by dragging of the tentorial bars from their posterior attachments as previously described. It cannot even be said with certainty that air could not be obtained from the ventral air space. Though the presence of air in this space may appear as rounded bubbles, the space itself has many ramifications (Fig. 48 (4), (5)) among the cemented parts, and if membranes are torn air might well be made available in unlimited amount from this space. This space too is in direct communication with the outer air via the open first abdominal spiracles and the tracheal trunks. Even air in the abdomen is not necessarily derived from rupture at the base of the respiratory trumpets, since it might well escape from the tracheal trunks through the open imaginal spiracles when these are freed from pupal tracheal lining. This might explain the early appearance of air in the abdominal segments. That air is eventually swallowed and is largely responsible for emergence of the imago is undoubted. There remains the issue, however, whether at this early stage it is the swallowing of air which is responsible for the split on the thorax and if so what exactly is the source of the swallowed air. Marshall and Staley found no air in the diverticula or mid-gut at the time when the pupa straightens out, that is a very short time before the split occurs.

Clearly the crest carries a special line of weakness in the cuticle as is evidenced by the admission of air into cracks within or beneath the cuticle at the time the white streak appears and often some time before the split follows. According to Brumpt the splitting is initiated by a cluster of erect hairs, which at the first appearance of the split appear to be in the median line of the imaginal notum, but which in reality, as is seen when the imaginal thorax is fully expanded, are situated to the side of this, there being before the thorax expands a sulcus between the hairs of the two sides. In *Theobaldia* and *Culex* the hairs form a tuft. In *Aëdes aegypti* they lie well back on the notum arranged in the sagittal plane.

EMERGENCE

Once the split on the thorax has appeared emergence is rapidly completed, now mainly the result of swallowing air which by causing the abdomen to swell gradually forces the imago from the pupal skin. As pointed out by Brumpt it is the dorsal concavity of the swollen imaginal abdomen pressed against the resistance offered by the back of the slit (incidentally strengthened by the strong metathoracic bar) that causes the emerging mosquito to adopt the vertical attitude characteristic of the later stages of emergence. In the liberation of the mouth-parts and legs muscular action seems to be the chief motive force, the wings being passively lifted out of their sheaths, though they also lengthen, presumably from blood pressure.

The distribution of the swallowed air has been especially investigated by Marshall and Staley. In describing the swallowing of air during emergence Howard, Dyar and Knab state that this air passes into the diverticula. This, however, is not immediately the case. In the freshly emerged insect what was found by Marshall and Staley was a large bubble in the mid-gut and no air in the diverticula, as has also been confirmed by me on many occasions. Only later is the air passed as small bubbles into the diverticula. Pupae dissected in the act of straightening out and with air beneath the pupal cuticle had no air either in the diverticula or mid-gut. At a somewhat later stage small bubbles were present in the oesophagus and mid-gut. Later still the mid-gut was swollen with a large elongate bubble extending to the third and fourth and later to the fifth and sixth abdominal segments, but there was still no air in the diverticula. During the time following emergence, however, air is gradually passed from the gut into the diverticula, which take on the characteristic appearance of these organs distended with numerous small air bubbles.

REFERENCES*

BACOT, A. W. (1916). See references in ch. VI, p. 155.

BONNE-WEPSTER, J. and BRUG, S. L. (1932). See references in ch. II (*a–b*), p. 45.

BROCHER, F. (1913). *L'aquarium de chambre.* Payos, Paris.

CARTER, H. R. (1923). (Appendix to Connor and Monroe.) *Amer. J. Trop. Med.* **3**, 19.

CHOPRA, R. N., ROY, D. N. and GHOSH, S. M. (1940). Action of pyrethrum on mosquito larvae. *J. Malar. Inst. India*, **3**, 457–63.

DA COSTA LIMA, S. (1914). Contributions to the biology of the Culicidae. Observations on the respiratory process of the larva. *Mem. Inst. Osw. Cruz*, **6**, 18–34.

DAVIS, N. C. (1932). The effect of heat and cold upon *Aëdes* (*Stegomyia*) *aegypti. Amer. J. Hyg.* **16**, 177–91.

DE MEILLON, B., GOLBERG, L. and LAVOIPIERRE, M. (1945). The nutrition of the larva of *Aëdes aegypti* L. *J. Exp. Biol.* **21**, 84–9.

EYSELL, A. (1905). *Die Stechmücken.* In Mense's *Handb. TropenKr.* II, 44–94.

EYSELL, A. (1911). Das Schlüpfen der Stechmücken. *Arch. Schiffs.- u. Tropenhyg.* **15**, 273–82.

GILLETT, J. D. (1955). Mosquito handling. A recent development in technique for handling pupae of *Aëdes aegypti. Rep. Virus Res. Inst. E. Afr. High Comm.* 1954–5, p. 24.

HOSKINS, W. M. (1932). Toxicity and permeability. I. The toxicity of acid and basic solutions of sodium arsenite to mosquito pupae. *J. Econ. Ent.* **25**, 1212–24.

HOWARD, L. O. (1913). The yellow fever mosquito. *Fmrs' Bull. U.S. Dept. Agric.* no. 547. Superseded by *Fmrs' Bull.* no. 1354, 1923.

HOWARD, L. O., DYAR, H. D. and KNAB, F. (1912). See references in ch. I (*f*), p. 19.

HOWLETT, F. M. (1913). *Proc. Gen. Malar. Comm., Madras*, 1912, p. 205.

* For authors not given here see under references to previous chapter.

KALANDADSE, L. (1933). Materialen zum Studium der Atmungsprozesse der Mückenlarven und -puppen und der Einwirkung von Petroleum, Beschattung und Verunreinigung des Wassers auf dieselben. *Arch. Schiffs.- u. Tropenhyg.* **37**, 88–103.

MACFIE, J. W. S. (1914). A note on the action of common salt on the larvae of *Stegomyia fasciata. Bull. Ent. Res.* **4**, 339–44.

MACFIE, J. W. S. (1916). Chlorine as a larvicide. *Rep. Accra Lab. for 1915*, pp. 71–9. Abstract in *Rev. Appl. Ent.* **5**, 47.

MACFIE, J. W. S. (1920). Heat and *Stegomyia fasciata.* Short exposures to raised temperatures. *Ann. Trop. Med. Parasit.* **14**, 73–82.

MACFIE, J. W. S. (1923*a*). Depth and the larvae and pupae of *Stegomyia fasciata. Ann. Trop. Med. Parasit.* **17**, 5–8.

MACFIE, J. W. S. (1923*b*). A note on the action of lithium chloride on mosquito larvae. *Ann. Trop. Med. Parasit.* **17**, 9–11.

MACGREGOR, M. E. (1927). *Mosquito Surveys.* Baillière, Tindall and Cox, London.

MANZELLI, M. A. (1941). Studies on the effect of reduction of surface tension on mosquito larvae. *Proc. N.J. Mosq. Ext. Ass.* **28**, 19–23.

MARCHOUX, E., SALIMBENI, A. and SIMOND, P. L. (1903). See references in ch. I (*f*), p.

MEINERT, F. (1886). De eucephale myggallarver. *K. Danske Viedensk. Selsk.* **3**, 373–493.

MITCHELL, E. G. (1907). *Mosquito Life.* G. P. Putnam's Sons, New York and London.

NEWSTEAD, R. and THOMAS, H. W. (1910). The mosquitoes of the Amazon Region. *Ann. Trop. Med. Parasit.* **4**, 141–50.

PRYOR, M. G. M. (1940). On the hardening of the ootheca of *Blatta orientalis. Proc. Roy. Soc.* B, **128**, 378–93.

REAUMUR, R. A. F. DE (1738). *Mémoires pour servir à l'histoire des insectes*—. Mém. XIII. *Histoire des cousins.* L'Imprimerie Royale, Paris. Vol. IV, pp. 573–636.

RUSSELL, P. F. and RAO, T. B. (1941). On the surface tension of water in relation to the behaviour of *Anopheles* larvae. *Amer. J. Trop. Med.* **21**, 767–77.

SENIOR WHITE, R. (1943). Effect of reduction of surface tension on mosquito pupae. *Indian Med. Gaz.* **78**, 342.

SHANNON, R. C. and PUTNAM, P. (1934). The biology of *Stegomyia* under laboratory conditions. I. The analysis of factors which influence larval development. *Proc. Ent. Soc. Wash.* **36**, 185–216.

WIGGLESWORTH, V. B. (1942). *The Principles of Insect Physiology.* Ed. 2. (See also many later editions.) Methuen and Co., London.

WILLIAMSON, K. B. (1924). The use of gases and vapours for killing mosquitoes breeding in wells. *Trans. R. Soc. Trop. Med. Hyg.* **17**, 485–519.

WOODHILL, A. R. (1942). A comparison of factors affecting the development of three species of mosquitoes (*Aëdes* (*Pseudoskusea*) *concolor* Taylor; *Aëdes* (*Stegomyia*) *aegypti* Linnaeus and *Culex* (*Culex*) *fatigans* Wied.). *Proc. Linn. Soc. N.S.W.* **67**, 95–7.

YOUNG, C. J. (1922). Notes on the bionomics of *Stegomyia calopus* Meig. in Brazil. *Ann. Trop. Med. Parasit.* **16**, 389–406.

XVII
THE IMAGO

(*a*) GENERAL DESCRIPTION

SYSTEMATIC

Descriptions of the imago of *Aëdes aegypti* from the systematic point of view are numerous (see ch. II). One of the most detailed of these, including all early published descriptions of the species under its different synonyms, is that by Howard, Dyar and Knab (1912, 1917). A more general account with much information is that by Howard (1913) under the title 'The yellow fever mosquito', republished in a later bulletin of the same series (1923).* Illustrations are also numerous, but many, though accurate as regards markings, do not well show the general aspect of the species. One of the most realistic and accurate in every respect is the large-sized lateral view of the female given by Soper, Wilson, Lima and Antunes (1943). Other good figures are by Martini (1920), who gives lateral and dorsal views; by Goeldi (1905) and by Edwards (1932, 1941), both authors giving figures in colour; and by Barraud (1934), who gives photographs of the species and of related forms. Good figures are also given by Neumann and Mayer (1914) in their Atlas of human parasites and their vectors.

Recent synoptic tables giving identification characters are given by: Edwards (1941) and Mattingly (1952) for African species; Barraud (1934) for Indian species; Bonne-Wepster and Brug (1932) for Indonesian species; and Dyar (1928) for American species.

EXTERNAL CHARACTERS (Fig. 49 (1), (2))

In general appearance and body structure the imago of *A. aegypti* closely resembles many other Culicine species, differing from these chiefly in ornamentation and scale structure. Along with most other members of the genus *Aëdes* it differs from mosquitoes of the large genus *Culex* in having the abdomen somewhat pointed and the eighth segment reduced in size and largely retracted within the seventh, no pulvilli in connection with the tarsal claws and the fore- and mid-claws in the female toothed.

The *head* is rather small, globular and largely occupied by the *compound eyes*, which almost meet in both sexes above and below. Springing in front from between these are the *antennae* (*an*) moderately hairy in the female, plumose in the male. Below these are the *maxillary palps* (*mxp*) long in the male, short in the female, usually referred to in systematic works merely as the palps. Below the palps springing from the snout-like *rostrum* (*ro*) is the characteristic *proboscis* (*pb*) containing the highly modified *mouth-parts*. The *neck* is largely membranous but with strong lateral sclerites giving it ample support and allowing the head to be extended or retracted into the hollow of the prothorax. The *thorax* is relatively very large, accommodating the enormously massive wing muscles and consists of

* See also descriptions by Theobald (1901), Blanchard (1905), Martini (1920), Bonne-Wepster and Brug (1932) Barraud (1934), Edwards (1941).

pro-, meso- and metathorax with their appendages. These comprise the fore-, mid- and hind-*legs*, the *wings* and the *halteres*. The *abdomen* is composed of the largely unmodified abdominal segments I–VIII and the modified terminal segments forming the *hypopygium* (*Hp*) according to sex. All these parts will be described in detail later. But before undertaking this considerable task there are certain characters of the species which may suitably be first considered.

As is usual in mosquitoes the male is distinguished from the female by the plumose antennae and the longer and more developed palps, which in the male are about as long as the proboscis, whilst in the female they are only about one-quarter as long. Though to some extent hairy, the male palps in *Aëdes aegypti* are much less conspicuously so than in the males of most species of *Culex*, where they are commonly longer than the proboscis and heavily clothed with hairs apically. Contrasted also with the narrow pointed termination of the abdomen in the female carrying the small lobose *cerci* is the conspicuous male hypopygium with the large claw-like *gonocoxites* (claspers). The male also differs from the female in having only rudimentary mandibles, no distinct and separate hypopharynx, narrower wings in proportion to their length and a more developed fringe apically, larger tarsal claws and a different ungual formula as noted later.

Of interest from the biological point of view is the marked and constant difference in the size of the sexes, the female being nearly twice the weight of the male and much more bulky and robust. The sexes differ also in many points of behaviour, in the absence in the male of the blood-sucking habit, in the speed of flight, in the note emitted in the hum and notably in the normal length of life. In all data regarding the imago it is necessary therefore to give separate valuations for the sexes. The female being so much more important, however, as a vector of disease and as an insect suitable for experimental work in the laboratory, more attention has naturally been given to its study. Usually, therefore, where no reference is made to the sex, an observation may be taken to apply to the female, unless from its nature it may be assumed to apply to both sexes.

ORNAMENTATION (Fig. 49 (1); 55 (1); 59 (4))

The general colour effect of the normal form of *A. aegypti*, as with many related species of *Aëdes*, especially in the subgenus *Stegomyia*, is to the naked eye dark, amounting, especially in the darker varieties, almost to black ornamented with white. A good deal of variation in this respect occurs, though not usually so conspicuously in the ordinary laboratory strains, and darker and paler forms are described (see under varieties, strains and hybrids in chapter II).

The dark coloration is due mainly to the scaling, but the cuticle of both the head and thorax and to a large extent that of the legs and wings is dark (fuscous), that of the abdomen in the female being, however, light-coloured, though in the male the last few segments of the abdomen with the claspers have the cuticle dark-coloured. The cuticle of the legs is dark where the scales are dark and light at the white bands. Vividness of the markings, both dark and pale, is, however, due mainly to the scaling. Certain of the pale markings are silvery, gleaming white in some lights, duller and less conspicuous in others. These silvery markings occupy fairly fixed positions and are not usually much subject to variation. They are characterised in many cases by a peculiar form of scale described later. Other areas of pale scaling, and especially the extent of yellow scaling, are more subject to variation and are not usually so vivid.

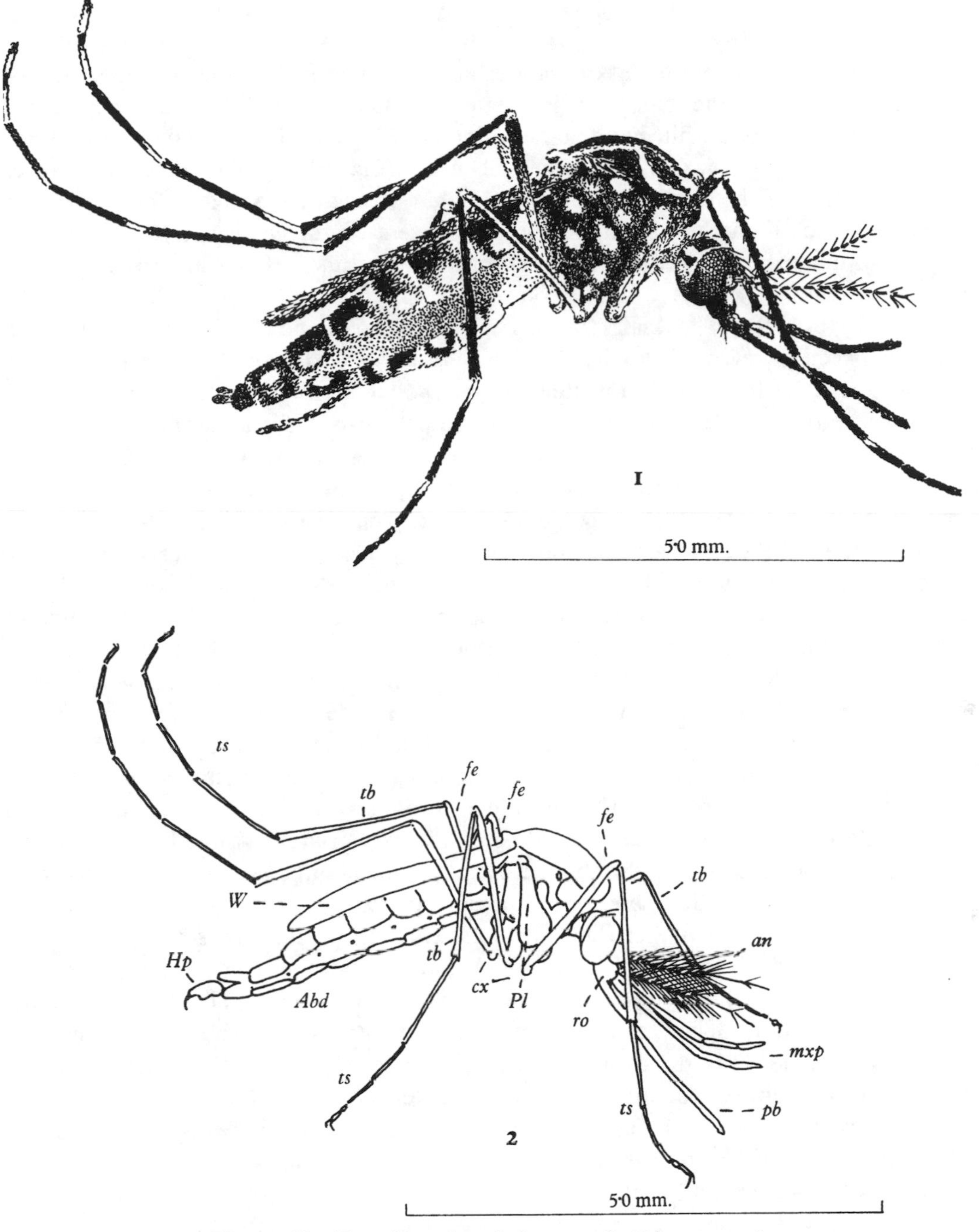

Figure 49. Female and male imago of *Aëdes aegypti.*

1 Female to show ornamentation. See also Fig. 55 (1) (thorax) and Fig. 59 (4) (abdomen). The female has fed some time previously and shows partially developed ovaries (pale area towards terminal segments). Note also pale area basally due to diverticulum.

2 Male to show male characters and general imaginal parts.

Note the plumose antennae, long palps, normal-sized eighth abdominal segment and male hypopygium.

Lettering: *Abd*, abdomen; *an*, antenna; *cx*, coxae; *fe*, femora; *Hp*, hypopygium; *mxp*, maxillary palps; *pb*, proboscis; *Pl*, pleura; *ro*, rostrum; *tb*, tibia; *ts*, tarsus; *W*, wing.

Silvery white patches are present on the tips and dorsal surface of the apical half of the palps in the female. In the male there is a patch of white on the dorsal surface near the base of the palps, a broad white band in the region of the pseudo-joint and patches of white scales at the bases of the two apical segments beneath (Fig. 54 (4)). There are white scales on the basal lobes of the antennae in both sexes, but especially marked in the female. In the female two small patches of white scales are present on the clypeus, a feature present in only one other species of *Stegomyia* (*A. vittatus*). Sometimes a few white scales are also seen in the male. The proboscis is all dark.

The head is clothed with flat scales forming a white median band on the dorsum extending forwards between the eyes. On each side of this is an area of dark scales covering most of the dorsal surface of the head, but white scales extend round the margin of the eyes to the ventral surface and cover the lateral aspects of the head except for a lateral dark band.

Brilliant silvery broad scales form conspicuous patches on the prothoracic lobes and below these on each side are white markings forming a double V, one above the other, formed by white scales on the propleurae and on the front of the anterior coxae. On the dorsal surface of the thorax are the characteristic 'lyre-shaped' markings (Fig. 55 (1)). The most conspicuous parts of these are the two semilunar white areas on the 'shoulders' extending back for about half the length of the thorax where each ends in a thin line of white or yellow scales passing back to the region of the lateral lobes of the scutellum. Between these markings and close to the middle line are two fine lines starting at a small white spot on the front of the scutum and passing back nearly to the scutellum where, after fusing, they continue along each side of the small triangular bare space which lies in front of the scutellum. Laterally on the scutum are patches of diffuse pale scaling external to the crescentic marks and a prominent white area further back above the wing root. In addition an area of flat white scaling on the paratergite is visible from the dorsal aspect. The general surface of the scutum, especially in the lighter varieties, is covered with diffusely scattered light yellowish scales. But the dark cuticle showing through may still give a general dark effect. On each lobe of the scutellum are particularly striking patches of silvery white scales,* which with those on the pleurae and sides of the abdomen are largely responsible for the striking silvery ornamentation of the species. The location of the patches on the pleurae will be described later when the pleural sclerites have been indicated. Other similar patches are on the coxae. Besides the conspicuous patches on the anterior coxae already mentioned, there are two on the outer surface of the mid-coxa and one larger area covers much of the outer surface of the hind-coxa (Fig. 49 (1)).

All the femora are dark on their upper surface, especially towards their apices, and have pale knee spots. All are dully pale towards their base and for most of their anterior and posterior surfaces. On the anterior surface of the middle femur there is a narrow brighter white-scaled line not present on the other legs. The tibiae of all the legs are dark throughout. On the fore- and mid-legs there are pale spots at the bases of the first and second tarsal segments, usually on the upper aspect only. The hind tarsi have broad basal pale bands on all the segments, the pale areas on the last two segments including most of the segment, whilst some pale scaling may extend from the pale area at the base of the third segment for some distance along the outer aspect of the segment. Sometimes the fifth segment is almost or entirely white (Fig. 49 (1)).

* As in nearly all *Stegomyia* there is a small dark spot at the apex of the mid-lobe of the scutellum (Mattingly *in lit.*).

GENERAL DESCRIPTION

With the exception of a minute pale spot near the base of the costa the wing veins are entirely dark-scaled. The halteres are entirely pale.

The markings on the abdominal tergites and sternites are liable to some variation, giving rise to the different named varieties when such variation is very marked. The markings are formed by white, pale or yellow scaling shown up on a black background of black scaling, the cuticle being pale and entirely covered by the scaling. The white scaling gives rise to basal bands on the tergites of segments II–VI and to silvery spots on the sides of the tergites. A small pale spot is present in the same situation on tergite VII which is otherwise usually all black-scaled, though a small spot may be present in addition on the sternite. The sternites are largely pale-scaled with dark scales laterally from segment III onwards, increasing in extent to the ventral aspect of segment VII, which is usually all dark. Within these lateral areas of dark scaling are silvery spots matching those on the tergites. Basal banding is pale but not silvery and is rather diffuse. On each side of tergite I is a particularly prominent white-scale patch. The sternite is pale throughout as is also usually sternite II. These appearances are best seen in the distended abdomen, that is in a gorged or gravid specimen, the sternal lateral spots often being hidden in the unexpanded abdomen by curling in of the tergites, so much so that in the dried specimen even the lateral silvery spots on the tergites may not be visible in dorsal view. The pleurae are not scaled and in the distended abdomen, being transparent, allow the organs and underlying silvery tracheal vessels to be seen.

Ornamentation in the male is very similar to that of the female, except for the palps and the termination of the abdomen. The eighth segment is dark, the sternite (in dorsal position due to rotation) having some conspicuous white scales, while the claspers are black.

SCALE CHARACTERS (Fig. 50)

The scales of mosquitoes are figured as a microscopic curiosity by Hogg as early as 1854 and more fully described by this author in a later paper (Hogg, 1871). Hogg refers to Muller and Delpino as previously describing them in a German periodical, but I have been unable to trace this reference. Scale characters have been extensively used by Theobald in his early classification of mosquitoes, and the genus *Stegomyia* was first separated on such characters and it is still distinguished as a subgenus of *Aëdes* by the flat scaling of the head and scutellum.

Scales arise from more or less regularly arranged pit-like depressions on the cuticle, rather similar to, but smaller than, those from which hairs and spines arise. Each scale consists of a stalk, *pedicel*, and an expanded flattened apical portion, *squame*, which is invariably marked with striations. Those occurring in mosquitoes are for the most part to be classified as (1) flat scales, (2) narrow curved scales, and (3) upright forked scales. Those of *A. aegypti* can be mainly included under these heads.

Flat scales (Fig. 50 (1), (4)) are most commonly bat-shaped, but may be oval, orbicular or triangular. Except in the special silvery flat scales described later the striations are strongly marked and parallel, following a straight course independently of the shape of the scale. The striations are mostly spaced at about the same interval, usually about 2μ, whatever the width of the scale. In the vestiture scaling of the legs and some other parts up to about 9 striations may be present, about 5 or 6 being common. In the broader flat scales of the head, abdomen, scutellum and pleurae up to 12–15 striations are usual. For the most part flat scales are abruptly truncated, the cut across the striations giving the terminal margin a serrated appearance. Especially where the vestiture scales are small they may be more elongate-oval in shape (Fig. 50 (1) A). The pedicel of vestiture flat scales is usually curved or bent to allow the squame to lie parallel to the surface from which the scale arises. Such scales are spaced at such intervals as to overlap somewhat, the squames forming a sheet-like layer with a space between it and the underlying cuticle. Flat scales may be dark, giving rise to black ornamentation, or white. Dark scales mounted in balsam retain their colour; white scales become colourless and transparent, and seen *in situ* in balsam preparations of parts are almost invisible except for the pedicels which then have the appearance of microtrichae.

Many of the flat scales giving rise to the more brilliant white spots such as those on the pleurae and scutellum and the lateral white spots on the abdomen are peculiar. These scales are very broad and when mounted are even more transparent than usual. Their striations are finer and branched, and when examined dry under a high power they show characteristic dark interference banding (Fig. 50 (4)). Not only do they differ from the ordinary white scales in showing branching of the striations, but appear to have striations on both surfaces as lines are seen obliquely crossing others. It seems to be clusters of these scales that are responsible for the brilliancy of some of the silvery markings that give so characteristic an ornamentation to the species.

Narrow curved scales (Fig. 50 (2)) are crescentic in shape with longitudinal striations which conform to the shape of the scale. The number of striations is usually five or six, but some broader scales occur in some of the white markings. The scales terminate in a sharp point, and the pedicel is not bent but merges gradually into the squame so that an even crescent is formed (Fig. 50 (2)). Narrow curved scales may be dark or light, but are commonly cream or yellowish.

Upright forked scales (Fig. 50 (3)) are usually narrow and rather solid in appearance. They expand gradually towards the tip and usually show a deep median cleft giving them a forked appearance. They are practically restricted in distribution to the dorsum of the head.

Figure 50. Integumentary structures.

1 Flat scales. A, ordinary vestiture scales. B, broader type from head, abdomen or other situations specially ornamented.

2 Narrow curved scales.

3 Upright forked scales.

4 Silvery ornamental scales. A, very large scales from pleural silvery patch, showing fine striations with branching and crossing; B, the same showing banding due to interference.

5 Transverse sections of scales. A, ordinary flat scales; B, different types of ornamental scales; C, upright forked scales.

6 Pleural and other membranes. A, section of pleural membrane of abdominal segment showing small plates of exocuticle bearing microtrichae. *a*, plates of exocuticle; *b*, endocuticle; *c*, epidermis; B, plates as above showing stellate arrangement of microtrichae : abdominal pleura; C, portion of abdominal pleura with plates as above usually carrying 4 microtrichae; D, neck membrane showing single stout microtrichae arising from the membrane.

7 Portion of abdominal sclerite showing usual structures. *a*, hair-like microtrichae arising direct from exocuticle; *b*, scales; *c*, spine-like hairs.

Nos. 1–4 are on same scale as shown.

For text regarding nos. 6 and 7 see section on the integument in ch. XXXI (*d*).

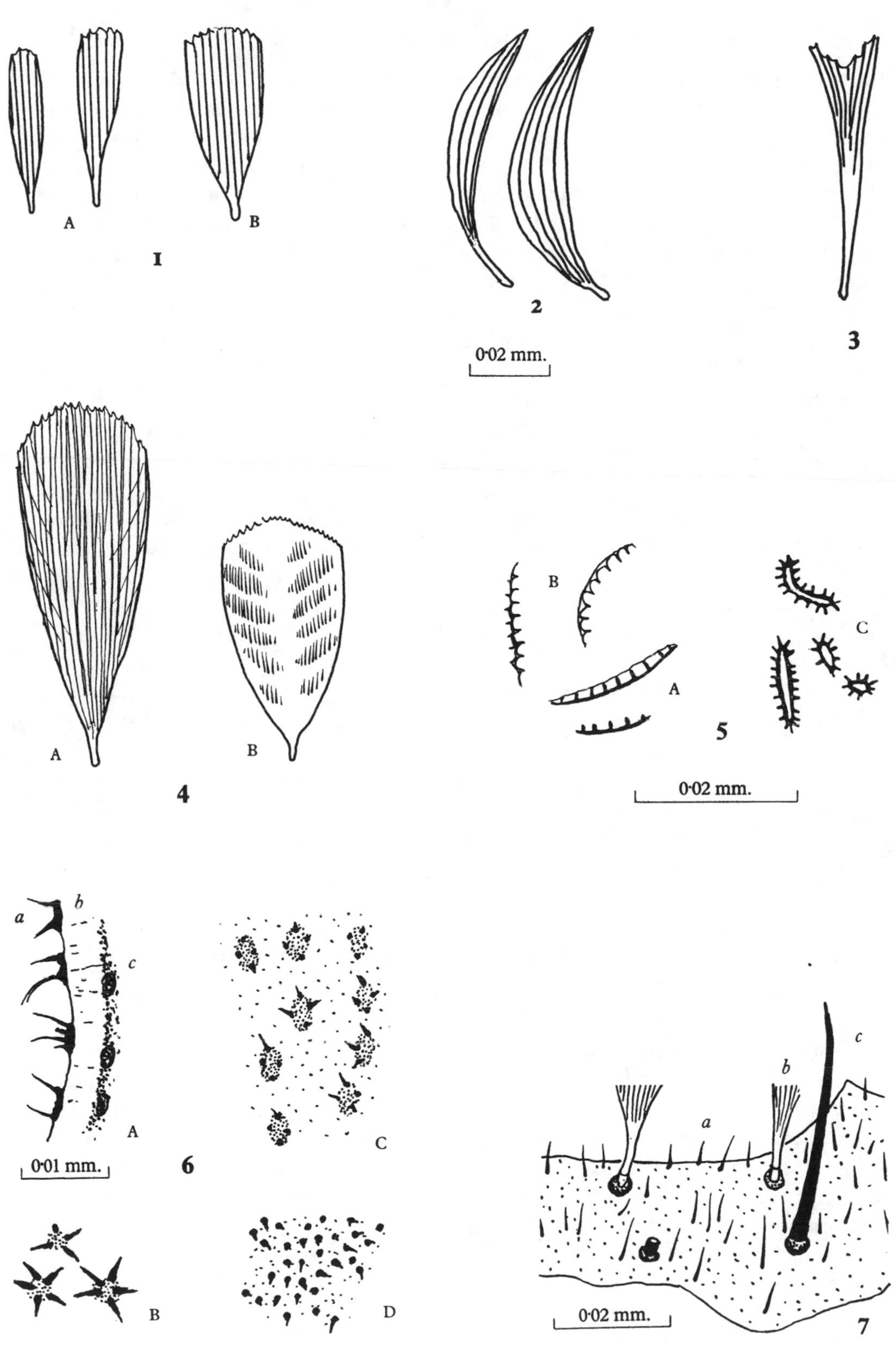

Figure 50

In *A. aegypti* the scales of the proboscis, legs, head, pleurae, wings and abdomen are for the most part flat scales, those forming the more conspicuous white spots such as those on the thoracic pleurae and sides of the abdomen and on the head and scutellum being usually broader than the usual vestiture scales. The more vividly white and silvery spots are largely composed of very broad scales of the peculiar type referred to which would appear to have the effect of giving extra brilliance. Unlike other parts scaling on the mesonotum consists mainly of narrow curved scales. The crescents of the lyre are formed of broader scales than usual of this type. The groundwork scales of the mesonotum are not dark but yellow. As, however, they are narrow and do not overlap, a good deal of the dark-coloured cuticle shows through and gives a general darkish effect to the scutum as a whole. The only situation where forked scales occur is forming a row some five or six deep on the occiput behind the flat scaling covering most of the head. They show longitudinal striations in their upper portion (Fig. 50 (3)). The scaling of the wings is in some degree special and will be described later when dealing with these organs. Scaling of the male is not significantly different to that of the female.

In structure the scales show considerable variation as seen in transverse section. The ordinary flat scales show a thick lower layer from which even ridges extend. Sometimes a delicate upper membrane can be made out, but usually the ridges are free (Fig. 50 (5) A). Ridges are sometimes present on both surfaces and various other effects are seen (Fig. 50 (5) B). Scales from the white areas may show oval bodies embedded in the ridges giving them a beaded appearance. Upright forked scales differ considerably from the other scales. They show a hollow central cavity surrounded by thick walls with ridges (Fig. 50 (5) C). Many hairs on the head and in other situations are ridged very like the upright forked scales.

There is a large literature on scales mostly relating to the wing scales of Lepidoptera and concerned with colour effects. Short descriptions of scale structure in mosquitoes are given by Imms (1938) and by Wigglesworth (1942). A paper by Onslow (1921) gives sections of many kinds of butterfly scales, some of them resembling those depicted from *A. aegypti*. Mayer (1897), referring to silvery-white scales, says that these must have polished surfaces directed towards the observer, as by reflected light they appear merely milk-white. This author also gives four or five pages of bibliography.

CHAETOTAXY

Little systematic importance attaches to the chaetotaxy of the imaginal head. Such chaetae as are present will be described later when dealing with the head capsule and mouth-parts. The chaetotaxy of the thorax and especially of the pleural sclerites is, on the contrary, important and has been made use of by Edwards (1932) in classification. But since their description requires reference to the thoracic parts these also will be dealt with later when these parts are being described. Little importance attaches to the chaetae of the limbs or abdomen and these also will be given later.

(*b*) ATTITUDE

ATTITUDE AT REST (Figs. 49, 51)

The resting attitude of the living insect varies somewhat with its condition, especially in the female where gorging completely alters its whole appearance and stance. To a large extent attitude is dependent on the body shape. It is this which gives to the insect the 'hunch-back' appearance often ascribed to it. Any change in the stance is, however, chiefly due to varying positions taken by the joints of the legs. In the normal resting position the three femora of each side are directed upwards and outwards at a slope of about 45° to the vertical, the fore- and hind-femora each at about this angle to the longitudinal axis of the body respectively forwards and backwards, the femur of the mid-leg being approximately at right angles to the body axis, though usually directed a little backward of this. In the fore- and hind-legs the tibiae are held at an angle of about 60° to the femora and the tarsi at about an angle of 120° to the tibiae so that they lie almost parallel to the surface rested upon, though sloping slightly to the clawed tips, which on a vertical surface appear to be the only parts in actual contact with the surface.

Support is normally given mainly, if not entirely, by the fore- and mid-legs, the hind-legs being either held in the air, or if held touching the surface make little or no change in the attitude when they are lifted. The tips of the fore-legs reach forward to points slightly in front of the level of the tip of the proboscis and at some distance lateral to this, whilst the tips of the mid-legs complete with these the points of a quadrilateral area somewhat wider behind than in front. By this arrangement, and the fact that the legs at their bases arise close together, what amounts to an adjustable tripod (or more correctly quadrupod) is formed on which the body is slung. In such a position the fore-femora pass up close to the antero-lateral margin of the thorax and the tibiae viewed from the side cross the line of the proboscis (Fig. 49). The mid-femora in side view cross the line of the wing edge a short distance from the base. The hind-femora with the legs raised are brought far forwards and commonly near their origins cross the line of the mid-femora to lie internal and anterior to these, the tibiae being raised above and held roughly parallel to the wing margins, whilst the tarsi are directed upwards with a sweep and often waved about or moved up and down by movement of the femora as though from a sense of well-being.

The body of the insect meanwhile is poised so that the tips of the coxae lie at a distance of only about half the depth of the thorax from the supporting surface. The tip of the abdomen, as likewise the tip of the proboscis, almost touch the supporting surface when this is vertical or horizontal. When hanging from a horizontal surface the hinder end of the body is usually further away. By narrowing the femoro-tibial angles, thus drawing the tibial apices and tarsi inwards and so lowering the body, or widening the angle by straightening the legs and raising the body, or tilting the body by changing position of the legs, a considerable latitude in stance can be brought about. Nevertheless, when resting on an even surface the attitude of the unfed insect is nearly always that described. Change is associated only with some form of activity, such as feeding or probing, ovipositing, etc.

Almost invariably on a vertical surface the insect rests with the head directed accurately upwards, and if such surface be rotated so as to alter the line of the insect's body, it will at first make small adjustments to counteract this and, if the displacement continues, will re-establish its correct orientation.

In the normal insect the wings when resting are closed and locked in position in relation to the thorax and more or less in line with the abdomen, lying over but not touching it.

When gorged the whole appearance of the insect is changed by the relatively enormous distension of the abdomen by the blood meal, the dark blood showing through the transparent pleural membrane. The abdomen is also swollen when the insect is gravid, though not to the same extent.

Various aspects of the insect as depicted by sketches from life are shown in figure 51. Some of these have been drawn under the camera lucida, but others, for example the female in the act of ovipositing, have been sketched as accurately as possible whilst observing the insect.

The male is much more slender and less robust-looking than the female. But except that it tends to rest with the abdomen more raised and the hypopygium turned up, its position at rest is much as in the female.

MOVEMENT

Though both sexes are usually either flying or in one position at rest, they do at times traverse small distances with a rather awkward crab-like walk. The manner of walking appears to be as described for most insects (see Imms, 1938; Wigglesworth, 1942), namely that the middle leg of one side is moved forwards simultaneously with the fore- and hind-leg of the opposite side, the other legs meanwhile supporting the body. A sideways movement is also commonly seen, the mid-legs being advanced and a grip taken with the claws, the other legs then following. When prodding in search for a place to puncture, the insect's attitude is alert, with the body well raised and movements quick and nervous (Fig. 51 (5)).

(*c*) BODY SHAPE AND PROPORTIONS

Curiously enough very few authors, except occasionally as a rough measurement in the systematic description, have given any indication as to measurements of the specimens of *A. aegypti* with which they have been dealing. Peryassu (1908) gives length as 3·5–5·0 mm. for the female and 3·0–4·5 for the male. Bonne and Bonne-Wepster (1925) give about 4 mm. for the male and female. The last-mentioned authors also give as wing measurements 3·5 mm. for the female and 3·0 for the male. Such approximations are, however, of little value in comparing for example the effect of various conditions on size of the adult, and some more critical form of measurement for recording size is very desirable.

It is not easy in an insect shaped like a culicine mosquito to find a good measurement for body length, and before discussing what is the most desirable criterion for size and what

---

Figure 51. Attitude.

1 Female resting. Dorsal view.

2 The same. Viewed from in front.

3 Unfed female, resting. Lateral view.

4 Recently gorged female. Note the distended abdomen with white lines of stretched tracheae over the dark area of ingested blood and the pushing of the pale viscera (Malpighian tubules and hind-gut) into the sixth and seventh abdominal segments.

5 Alert female intent on feeding.

6 Female ovipositing on upright filter paper slip. Note the curiously mobile abdomen. An egg will shortly appear extruded at the tip and will be deposited on the paper.

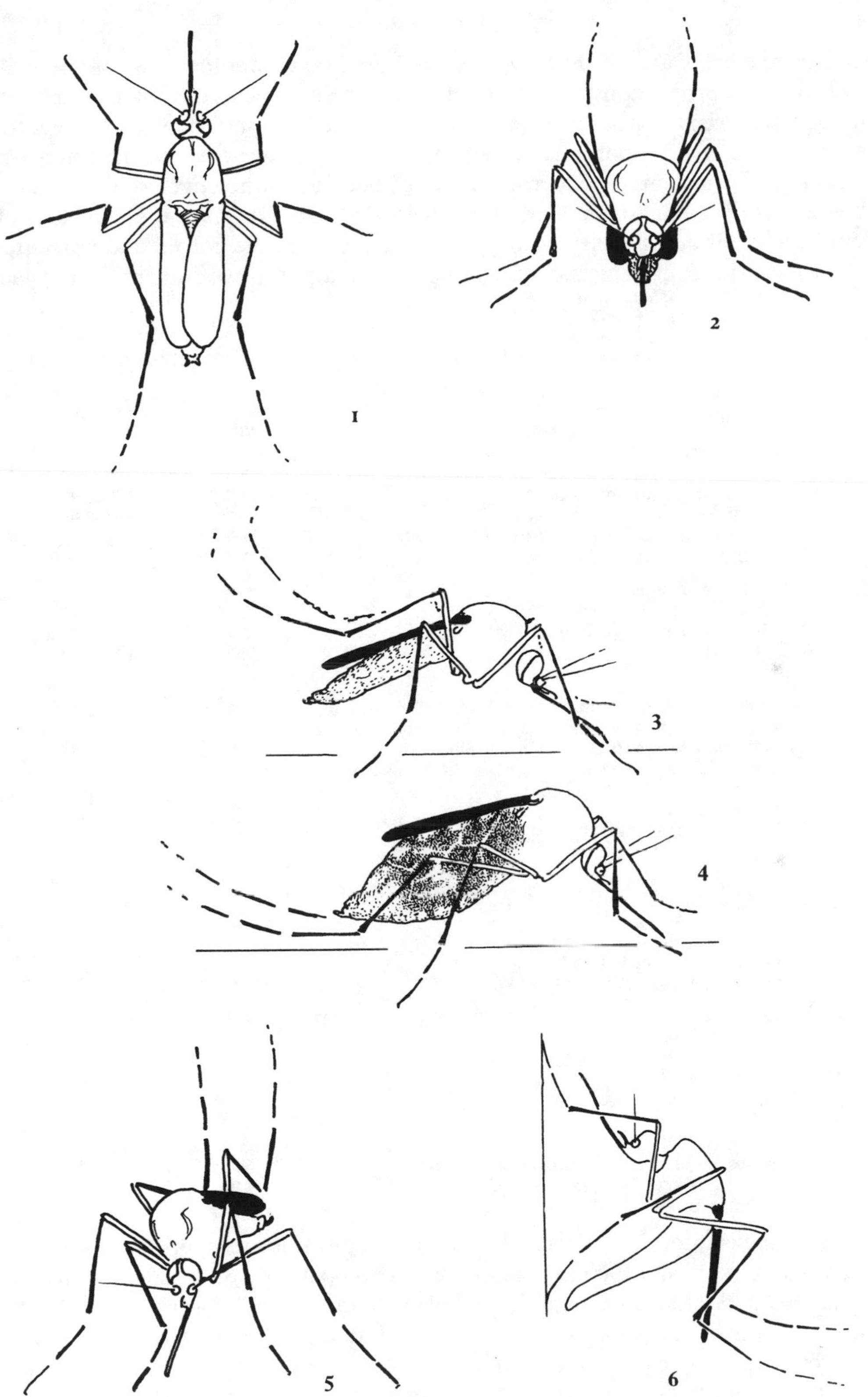

Figure 51

linear measurements are satisfactory and significant, consideration must be given to the shape and bodily proportions of the insect as a whole. In Table 25 certain measurements of body parts are given. A word may be said on the method adopted in taking those measurements relating to the proportions of the whole body. One very noticeable feature of the insect is that, apart from the moveable legs, the body as a whole including the head and proboscis has a certain rigidity. Nor are the parts liable to change greatly except the abdomen, which is of different dimensions at emergence, in the gorged and gravid female and to a less extent with age and starvation. Advantage has been taken of these facts in making certain measurements.

Table 25. *Measurements in mm. of body parts of normal-sized fifth-day unfed female* Aëdes aegypti

Measurement	Mean	Approximate range
Body proportions		
Base line (tip of proboscis to tip of abdomen)	6·57	6·45–6·67
Perpendicular (top of thorax to base line)	2·59	2·32–2·75
Tip of coxae to base line	0·79	0·54–0·94
Tip of proboscis to scale patch	3·48	3·33–3·55
Scale patch to tip of abdomen	3·71	3·48–4·06
Head and proboscis		
Tip of proboscis to clypeus	2·13	2·10–2·17
Tip of proboscis to occiput	2·78	2·68–2·90
Width of head	0·78	0·75–0·80
Thorax		
Anterior promontory to posterior border of scutellum	1·49	1·46–1·52
Base of neck to base of abdomen	1·09	—
Dorsum to coxae (on perpendicular line)	1·81	1·78–1·81
Breadth at scutal angles	1·02	1·00–1·04
Breadth at prealar prominence	1·18	1·12–1·20
Breadth at wing roots	1·14	1·10–1·16
Width of scutellum	0·74	0·70–0·80
Abdomen		
Base to tip	3·26	3·12–3·48
Greatest width	0·78	—
Wings		
Length (excluding fringe)	3·41	3·26–3·77
Greatest breadth (excluding fringe)	0·86	0·82–0·90
Legs		
Femur. Fore	2·07	1·88–2·17
Mid	2·15	1·96–2·25
Hind	2·18	2·03–2·25
Tibia. Fore	2·05	1·81–2·17
Mid	2·25	2·10–2·32
Hind	2·33	2·06–2·46
Tarsus. Fore (1·12; 0·57; 0·33; 0·21; 0·22)	6·54	5·94–6·88
Mid (1·21; 0·65; 0·40; 0·22; 0·22)	7·03	6·60–7·32
Hind (1·39; 1·04; 0·85; 0·51; 0·36)	8·56	8·00–8·77

In a carefully chloroformed insect all three legs of one side are snipped off near the base with a sharp scissors and the insect allowed to lie on that side on a slide. A tiny strip of paper slipped under the remaining legs and made to touch the tip of the proboscis and the tip of the abdomen then gives a convenient base line which can be measured, along with other proportions, by means of an ocular micrometer scale or camera lucida. Measurements have in all cases been made on females taken at random from routine standard cultures on the fifth day from emergence with access to water but no food.

BODY SHAPE AND PROPORTIONS

If, with the insect viewed in profile as described above, a perpendicular be dropped from the highest point of the dorsum of the thorax on to the above-mentioned base line, it will pass through a particularly prominent patch of white scales on the sternopleuron and continuing on will pass through or near to the tip of the anterior coxa enabling the distance from the base of the legs to the base line to be measured. Using the scale patch referred to, a measurement can also be made from the tip of the proboscis to this spot and from the spot to the tip of the abdomen. The second measurement divided by the first will give an index or ratio indicating degree of abdominal lengthening. Thus in the unfed insect the two measurements are approximately equal, giving an index near 1. Immediately after gorging, the index is 1·56. Also the angle formed by the two lines to fix the bodily configuration of the insect can be determined. The angle is about 120°. The following are remarks on the measurements of the different proportions and parts.

BODY LENGTH

The measurement given is that along a base line as described above. It is the same measurement as would be made between the same two points looking down on the insect. It is not a satisfactory measurement for general use in recording size owing to the hunch-back shape of the insect, its liability to change as the insect dries and it is very dependent on the condition of the abdomen. It is useful, however, for working out the body shape and proportions.

HEAD AND PROBOSCIS

The proboscis measured from the tip of the labella to the sulcus separating it from the clypeus dorsally is about 2·0 mm., which is therefore about the maximum possible thickness of woven fabric which the insect could penetrate (Christophers, 1947). The head and proboscis taken together measure about 3·0 mm.

THE THORAX

The thorax is in general shape an inverted pyramid with the base (mesonotum) rounded off, the closely approximated coxae forming the apex. The two pleurae with the anterior and posterior surfaces of the thorax, all of which are roughly flat, form the four sides. The wings spring from points about one-sixth of the length of the thorax from its posterior end. The legs all rise close together round a point midway between the two mid-coxae, the site of selection for the pin point when mounting a mosquito. Some measurements of the thorax are given in the table.

THE WINGS

The length of the wing has been taken from the base of the costa where this has a pale spot and makes a sharp bend or knuckle at the wing root, to the tip of the wing at the wing margin, that is excluding the fringe, for which 4–5 per cent of the wing length can be allowed if desired. The mean length of the female wing is 3·5 mm. and for the male 2·5 mm. The male wing is also narrower in proportion than that of the female, the ratio of length to width being about 4 to 1 in the female and 4·3 to 1 in the male.

The wing length appears to be the simplest and most reliable measure for size of mosquito in experimental work. As shown later it is roughly correlated with weight, but unlike weight is not liable to vary with the condition of the insect. It does not alter appreciably with drying or mounting in balsam in which states it can be kept very conveniently for record.

The wing is most readily detached using a dry specimen. If to be mounted in balsam it should be dropped in the first place into xylol and pipetted on to the slide, as otherwise a troublesome air bubble is apt to form. It will also prevent detachment of scales.

The area of the wing has some importance in relation to studies on flight. It has been given by Sotavalta (1947) and others for some species of mosquito, and by Sotavalta (1952) for *A. aegypti*, the area for two wings male and female being 2·4 and 5·0 mm.^{2} Data obtained by the present author for dimensions of the wing are given in Table 26.

Table 26. *Length, breadth and area of the wing**

		Normal forms		Large forms†		Starvation forms	
Sex	Measurement	No. obs.	Mean	No. obs.	Mean	No. obs.	Mean
Female	Length mm.	42	3·50	14	3·85	10	2·52
Female	Breadth mm.	19	0·88	—	—	10	0·58
Female	Area mm.^{2}	6	2·28	—	—	10	0·90
Male	Length mm.	34	2·55	—	—	6	2·22
Male	Breadth mm.	33	0·59	—	—	6	0·42
Male	Area mm.^{2}	2	0·98	—	—	6	0·71

* Exclusive of fringe. † Bred at 17° C.

THE LEGS

The femora are all almost equal in length, increasing only very slightly in length from before backwards, but especially so for the hind-legs. The same applies to the tibiae, which are about the same length as the femora. The tarsi of the mid-legs are only slightly longer than those of the fore-legs, but in the hind-legs the tarsi are about half as long again, the increase being chiefly in the third and fourth segments.

THE ABDOMEN

The abdomen in the unfed female some days from emergence is about the same length as the wings which just reach to about the level of the cerci. At emergence the abdomen is to some extent distended owing to the large air bubble in the mid-gut, and probably also to the presence of more fluid and reserves than later. The abdomen is also distended and lengthened after feeding and as a result of enlargement of the ovaries, under which conditions it extends to a greater or lesser degree beyond the tips of the wings when these are in the folded position.

(*d*) WEIGHT

WEIGHT OF IMAGO UNDER DIFFERENT CONDITIONS (Tables 27, 28 and 29)

The only author giving data regarding the weight of *A. aegypti* imago under different conditions appears to be Roy (1936) in connection with the role of blood in ovulation in this species. He gives the following results for the weight of the 3-day-old female mosquito based on a total of 137 mosquitoes weighed. Maximum 2·32 mg., average 1·32 mg., minimum 0·86 mg. As noted by Roy, the weight of the unfed female undergoes considerable diminution in the course of the first 2 days.

WEIGHT

Table 27. *Reduction in weight of adult* Aëdes aegypti *following upon emergence*

	Culture of 15. viii. 46 (females)		Culture of 7. v. 48 (females)		Culture of 7. v. 48 (males)	
Time from emergence	No. of mosquitoes	Mean weight (mg.)	No. of mosquitoes	Mean weight (mg.)	No. of mosquitoes	Mean weight (mg.)
0–4 hours	—	—	24	3·04	28	1·67
0–24 hours	8	2·91	12	3·16	24	1·65
19–24 hours	57	2·31	53	2·98	28	1·41
Second day	—	—	108	2·52	60	1·27
Fourth day	—	—	31	2·29	79	1·18
Fifth day	190	2·14	104	2·34	91	1·14

Table 28. *Weight and wing length frequency*

Weight (mg.)	Wing length (mm.)	Weight (mg.)	Wing length (mm.)	Weight (mg.)	Wing length (mm.)	Weight (mg.)	Wing length (mm.)	Weight (mg.)	Wing length (mm.)
		2·23	3·48	2·41	3·33	1·98	3·26		
Culture A*		2·24	3·48	2·42	3·48	2·05	3·33	Culture G*	
1·67	3·19	2·26	3·29	2·44	3·41	2·07	3·33	2·26	3·70
1·74	3·15	2·26 (2)	—	2·45	3·41	2·09	3·33	2·41	3·77
1·76	3·08	2·28	—	2·46	3·48	2·14	3·26	2·49	3·62
1·84	3·12	2·29	3·44	2·49	3·41	2·15	3·22	2·49	3·70
1·86	3·26	2·31	—	2·50	3·41	2·15	3·33	2·63	3·84
1·89 (2)	3·19 (2)	2·32	3·29	2·50	3·36	2·17	3·26	2·75	3·84
1·90	3·19	2·32	3·48	2·51	3·48	2·19	3·33	2·87	3·91
1·90	—	2·33	—	2·52	3·48	2·20	3·33	2·88	3·84
1·92	3·26	2·34	—	2·53	3·41	2·20	3·48	2·91	3·99
1·93 (2)	—			2·54 (2)	3·51 (2)	2·22	3·29	2·93	3·84
1·95	—	Culture B*		2·55	3·41	2·23	—	2·97	3·88
1·97	3·19	1·99	3·33	2·60	3·51	2·24	3·44	3·14	4·13
1·97	—	2·12	3·33	2·61	3·44	2·25	3·26	3·40	3·99
1·98	—	2·20	3·36	2·61	3·48	2·29	3·29	3·44	3·91
2·00	3·41	2·24	3·41	2·63	3·44	2·29	3·41		
2·00 (2)	—	2·25	3·19	2·65	3·33	2·31	3·41	Culture F*	
2·01	—	2·25	3·36	2·65	3·41	2·32	3·33	(Males)	
2·02	3·26	2·29	3·36	2·65	3·48	2·35	3·29		
2·02	3·33	2·30	3·33	2·66	3·51	2·37	3·37	0·91	2·46
2·03	3·26	2·30	—	2·67	3·41	2·39	3·37	1·09	2·54
2·03	—	2·31	3·22	2·67 (2)	3·55 (2)	2·41	3·33	1·13 (2)	2·54 (2)
2·04	3·37	2·31	—	2·68	3·58	2·41	3·37	1·14	2·57
2·05	—	2·33	—	2·70	3·51	2·43	3·33	1·17	2·68
2·06	3·33	2·34	—	2·71	3·44	2·46	3·48	1·18	2·57
2·06 (2)	—	2·36	—	2·72	3·48	2·48	3·44	1·25	2·50
2·07	3·26	2·46	—	2·73	3·48	2·48	3·55	1·26	2·54
2·08	3·26	2·47 (2)	—	2·75	3·48	2·50	3·62	1·26	2·64
2·08	—	2·48 (4)	—	2·75	3·51	2·57	3·51	1·27	2·68
2·09	3·29	2·50 (2)	—	2·75	3·55	2·60	3·33	1·28	2·54
2·09	—	2·51	—	2·78	3·48	2·60	3·44	1·30	2·54
2·10	—	2·52	—	2·79	3·55	2·60	3·51	1·40	2·75
2·12	3·41	2·54 (2)	—	2·85	3·48	2·63	3·48	1·53	2·68
2·13	3·22	2·56 (2)	—	2·90	3·58	2·77	3·48		
2·13	3·41	2·57	—	2·91	3·77	2·78	3·26	Culture H.	
2·13	—	2·59	—	2·95	3·66	2·81	3·55	Starvation	
2·15	3·26	2·60	—			2·85	3·55	forms*	
2·15	3·33			Culture D*		2·89	3·58	(Females)	
2·15	—	Culture C*		1·13	2·90	—	3·19	0·97	—
2·16	3·26	2·14	3·36	1·22	2·75	—	3·22	1·06	2·75
2·16	3·36	2·20	3·33	1·29	2·93	—	3·26 (5)	1·43	2·68
2·16 (2)	3·41 (2)	2·24	3·36	1·59	3·19	—	3·29	1·49	2·86
2·16	—	2·32	3·33	1·65	3·04	—	3·33 (2)	1·59	2·86
2·17	3·37	2·32	3·36	1·68	3·15	—	3·37	1·61	2·83
2·17	—	2·32	3·41	1·75	3·12	—	3·41 (5)		
2·18	—	2·36	3·44	1·76	3·12	—	3·44	(Males)	
2·19 (2)	—	2·39	3·29	1·82	3·26	—	3·48 (2)	0·74	2·12
2·20	—	2·40	3·41	1·97	3·33	—	3·51 (2)	1·11	2·35
2·21	—								

* For particulars of cultures see Table 29.

In Table 27 is given the weight of female and male *A. aegypti* mosquitoes kept at 70 per cent relative humidity without food but with access to water. It will be seen that the mean weight of the female in these series has become reduced by the third day from about 3·0 mg., shortly after emergence, to 2·5 mg., and by the fourth day to about 2·3 mg. The male weight has also decreased, there being a steady reduction from 1·6 mg. to under 1·2 mg. Hereafter there appears to be little further loss of weight up to, in the female, the tenth day when as a rule considerable mortality has occurred.

Table 29. *Index to cultures*

Culture	Date	Temperature (° C.)	Description	Mean weight (mg.)	Mean wing length (mm.)	Index
A	10. v. 48	28	A rather poor culture (females)	2·08	3·30	0·63
B	12. v. 48	28	A good normal culture (females)	2·40	3·32	0·72
C	2. vi. 48	28	A good normal culture (females)	2·57	3·46	0·74
D	10. vi. 48	28	Some overcrowding (females)	2·22	3·33	0·67
G	23. i. 48	17	Fine specimens (females)	2·83	3·85	0·74
F	16. vi. 48	28	A good normal culture (males)	1·22	2·58	0·47
H	—	25	Starvation forms			
			Females	1·36	2·80	0·49
			Males	0·92	2·23	0·41

Apart from variation in weight due to difference in age there is considerable difference in weight depending upon the cultural conditions under which the adults have been reared (Fig. 52 (2)). In Table 28 are given the results of weighing adult *A. aegypti* reared under different conditions, including some routine standard cultures. The mean weight for the four routine cultures (195 females weighing 446·77 mg.) is 2·29 mg. The heaviest female in the series was 2·95 mg. and the smallest starvation form weighed 0·97 mg. In the culture of 10. vi. 48, a somewhat overcrowded and rather poor culture, the smallest female weighed 1·13 mg. Otherwise the smallest form in normal culture weighed 1·67 mg. The usual run of females under normal cultural conditions (for which culture of 2. vi. 48 is a fair example) range between 2·0 and 2·6 mm. All the figures relate to 5-day-old mosquitoes given access to water but without food. The weight of the male is only about half that of the female. In starvation forms, though the difference in size of the two sexes is much less, the female is still recognisedly slightly heavier than the male.

(*e*) CORRELATION OF WING LENGTH AND WEIGHT

For recording the size of mosquitoes experience would suggest some linear measurement or weight. Whilst weight might be regarded as the final arbiter of size, it has to be remembered that it will differ greatly following upon feeding or in the gravid condition, and that it is liable to smaller changes as a result of age or other possible circumstances such as the taking in of water where this is provided in the cage. Nevertheless, weight, if taken under suitably specified conditions, is clearly a most desirable index, and if there is a close correlation with weight under given conditions and some easily taken linear measurement it would be useful to know this. Body length, as already noted, is unsatisfactory. If, therefore, a linear measurement is to be used that of the wing length would appear to be the most suitable for the purpose, since it is a relatively large part of the mosquito, is subject to little or no change and is easily measured.

In Fig. 52 (1) is shown the result of plotting wing measurements against weight, as great a variety as possible in size of mosquito being made use of from the largest forms available to the smallest starvation forms. To obtain the necessary material, cages of adults on the fifth day from emergence were selected and the mosquitoes removed and weighed individually until all, or nearly all, in the cage had been dealt with. After weighing, each mosquito was put away in a numbered small tube until it was convenient to take the wing measurement. The results are given in detail in Tables 28 and 29, the latter table giving a brief description of the cultures.

It is evident from the plottings (Fig. 52 (1)) that a fairly high correlation exists and that a curve can be drawn about which the plottings are fairly evenly distributed. On general principles it was thought that the most likely relation of wing length to weight would be that between the linear measurement of the wing and the cube root of the weight, weight being largely equivalent to the third dimensional character of volume. Trial was therefore made on the basis of the formula:

$$L = k\sqrt[3]{W} + a,$$

where L is the wing length in mm., W weight in mg. and k and a constants. The curve given in the graph is one using this formula with $k = 2{\cdot}54$. It appeared unnecessary to consider the constant a which for all practical purposes could be written off as equal to 0. It is reasonable, therefore, to believe that the wing measurement and weight under the conditions noted are related in a simple manner and that as a gauge of size wing measurement is a very suitable and practical index.

Marked by crosses, plottings for a small number of males are included in the graph. These lie a little off the female curve, as is to be expected, since the male wing does not in its characters entirely conform to that of the female.

The graph gives some evidence, however, that relations of size and weight may be more complex than at first sight might appear. The plottings for Culture G (low temperature rearing) appear to suggest a curve more to the right than that relating to the species bred at 28° C. These very large specimens in fact require a curve with a higher value of k, for example $k = 2{\cdot}67$. Similarly, males should be best represented by a lower value.

Apart from weight, volume or other measurement is the somewhat indeterminate conception of *size*. But though lacking any precise form by which it can be expressed size often lies at the back of what it is desired to express by giving weight, length or other measurement. *A. aegypti* is being used in various experimental investigations and especially in the testing of aerosols and other products where precise results are required. For this purpose standardisation is essential, not only of the conditions of tests employed, but of the mosquitoes used in the tests. If the technique of rearing is faulty, that is does not provide conditions to produce the species in a satisfactory and standardised state, the most obvious evidence of this is reduced size of the mosquitoes reared. Of the criteria of size two have here been considered, namely weight and wing length. In Table 29 the ratios of weight to wing length for a number of cultures are set out in the column headed 'Index'. This has varied from about 0·63 to 0·74 for females, low if the culture is poor, higher if the culture is good. Such a relation does not appear to be accidental. It is evident that if the relation of wing length to weight is as assumed in the equation, namely that of a linear to a three-dimensional value, then such relation will ensure, as a physical necessity, other things being equal, that the lower the index the smaller the mosquito. Hence the 'index' given above may also be an indicator of size that might be useful on occasion.

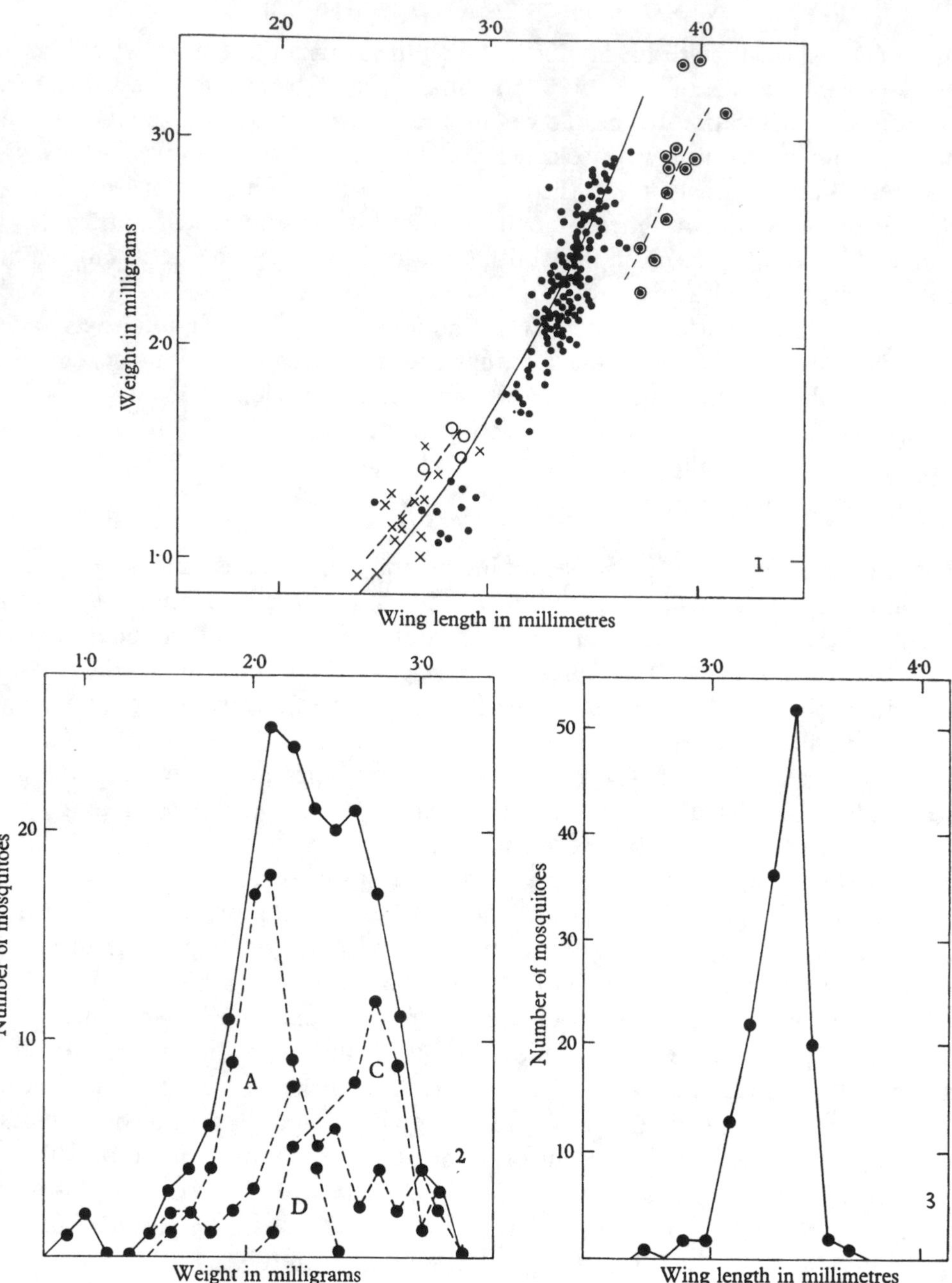

Figure 52. Graphs of size and weight.

1 Graph showing plottings of weight against wing length. The unbroken line is drawn at $k=2{\cdot}54$. The broken line curves are for $k=1{\cdot}89$ and $k=2{\cdot}67$ respectively and relate to starvation forms (crosses) and specimens bred at low temperature (circles). For nature of k see text.

2 Frequency distribution curves for weight. The unbroken line is for the weights of 195 females from the four cultures A–D. The broken lines are for the cultures A, C, and D separately. For details of the cultures see Tables 28 and 29 and text.

3 Frequency distribution curve for wing lengths of 161 females of the four cultures A–D. For measurements see Table 28.

Note. All observations relate to females, except where males are indicated.

WING LENGTH AND WEIGHT

The frequency curve for weight of females from a single culture is usually fairly symmetrical as shown in Fig. 52 (2). With massed cultures, where the means for the different cultures may vary, the resulting curve is liable to be more drawn out and irregular. With wing length the frequency curve is asymmetrical, there being a more abrupt fall in size with the larger forms than at the small end of the curve (Fig. 52 (3)).

(*f*) OTHER PHYSICAL DATA RELATING TO THE IMAGO

Whilst volume and specific gravity do not suggest themselves as of much importance in relation to the adult mosquito, there are some points not without interest.

VOLUME

Volume was ascertained by observing the rise of the meniscus in a small test-tube by means of the binocular microscope arranged horizontally as for determining the volume of the larva and pupa, but using butyl alcohol to ensure wetting and to minimise effect of evaporation. Five 5-day females with a mean weight of 2·30 mg. gave a mean value of 2·50 mm.^{3}, which would indicate a specific gravity of 2·30/2·50 or 0·920.

SPECIFIC GRAVITY

The specific gravity was also ascertained directly by submerging carefully chloroformed adults in different strengths of alcohol of which the specific gravity was determined. It was found necessary to dip the insect momentarily into 80 per cent alcohol first to ensure wetting and freedom from bubbles. By moving the mosquito up and down several times while held in a forceps and leaving it in the middle of the tube to rise or sink as the case might be, reasonably reliable readings were possible. Newly emerged females rose in all dilutions, even slightly in absolute alcohol (sp. gr. 0·780). This was evidently due to the presence of the large air bubble in the mid-gut as described under emergence. Seven-day-old females showed considerable variation, being in equilibrium (that is neither rising nor sinking) in the case of different individuals in dilutions ranging from 10 to 60 per cent (sp. gr. 0·985–0·900). Seven-day-old males were much lighter and gave more uniform results. The following gives the number of mosquitoes just sinking at the specific gravity given, a mosquito rising in 10 per cent and sinking in 20, for example, being given under sp. gr. 20.

Percentage of alcohol	Specific gravity	Females	Males
20	0·960	2	—
30	0·955	2	—
40	0·940	4	—
50	0·920	5	—
60	0·900	1	—
70	0·870	—	5

When rising in the fluid females tended to rise evenly right away up (as in flying position). Even if disturbed they regained this position. When in fluid of such specific gravity that they sank, they sank upside down. This rather suggests that in flight it is important, as in the larva, that the body should balance correctly.

DRY WEIGHT

A number of authors (notably Buxton, 1932, 1935; Theodor, 1936; Hochrainer, 1942) have dealt with the percentage of weight due to water in insects. Buxton found that this might range from under 50 to over 90. Hochrainer gives the range as from 55 to 85. The amount is to a considerable extent related to the fat present, water content being greater as that of fat is less. Buxton (1935), in hibernating *Culex pipiens* females, found that from October to March as during hibernation fat decreased from 27·9 to 6·3 per cent, water increased from 54 to 64·9 per cent. The following results show the percentage of water as shown by dry weight of *A. aegypti* adults on the fifth and sixth days from emergence kept at 25° C. and 70 per cent relative humidity with access to water but no food.

	Original weight (g.)	Dry weight (g.)	Percentage of water
Fifth day			
40 females	0·0922	0·0278	69·8
21 males	0·0235	0·0066	71·9
Sixth day			
136 females	0·2930	0·0981	66·5
150 males	0·1546	0·0458	70·4

Hibernation has not been described in *A. aegypti*. The susceptibility of the species to mortality where kept at higher temperatures without access to water has already been noted.

REFERENCES

BARRAUD, P. J. (1934). See references in ch. II (*a–b*), p. 45.

BLANCHARD, R. (1905). See references in ch. II (*a–b*), p. 45.

BONNE, C. and BONNE-WEPSTER, J. (1925). Mosquitoes of Surinam: a study of Neotropical mosquitoes. *Kolon. Instit. Amsterdam*, Meded 21. *Afd. Trop. Hyg.* no. 13.

BONNE-WEPSTER, J. and BRUG, S. L. (1932). See references in ch. II (*a–b*), p. 45.

BUXTON, P. A. (1932). Terrestrial insects and the humidity of the environment. *Biol. Rev.* **7**, 275–320.

BUXTON, P. A. (1935). Changes in the composition of adult *Culex pipiens* during hibernation. *Parasitology*, **27**, 263–5.

CHRISTOPHERS, S. R. (1947). Mosquito repellents. *J. Hyg., Camb.*, **45**, 176–231.

DYAR, H. G. (1928). See references in ch. II (*a–b*), p. 45.

EDWARDS, F. W. (1932). See references in ch. II (*a–b*), p. 45.

EDWARDS, F. W. (1941). See references in ch. II (*a–b*), p. 45.

GOELDI, E. A. (1905). *Os mosquitos no Para.* Weigandt, Paris.

HOCHRAINER, H. (1942). Der Wassergehalt bei Insekten und die Factoren die denselben bestimmen. *Zool. Jahrb., Zool.* **60**, 387–436.

HOGG, J. (1854). *The Microscope: its History, Construction and Applications.* W. S. Orr and Co., London.

HOGG, J. (1871). On gnat's scales. *Mon. Micr. J.* **6**, 192–4.

HOWARD, L. O. (1913). The yellow fever mosquito. *U.S. Dep. Agric., Farmers' Bull.* no. 547. Superseded by *Fmrs' Bull.* no. 1354, published 1923.

HOWARD, L. O., DYAR, H. G. and KNAB, F. (1912, 1917). See references in ch. I (*f*), p. 19.

IMMS, A. D. (1938). *A General Textbook of Entomology.* Methuen and Co., London.

MARTINI, E. (1920). Über Stechmücken besonders deren Europäische Arten und ihre Bekämpfung. *Arch. Schiffs.- u. Tropenhyg.* **24**, Beih. 1, 1–267.

MATTINGLY, P. F. (1952). The sub-genus *Stegomyia* (Diptera: Culicidae) in the Ethiopian Region. Part I. *Bull. Brit. Mus.* (*Nat. Hist.*), **2**, no. 5.

REFERENCES

MAYER, A. G. (1897). On the colour and colour patterns of moths and butterflies. *Bull. Mus. Comp. Zool. Harv.* **30**, 169, 256.

NEUMANN, R. O. and MAYER, M. (1914). Atlas und Lehrbuch wichtiger tierischer Parasiten und ihrer Überträger. In Lehmann's *Mediz. Atlanten*, **11**, München.

ONSLOW, H. (1921). On a periodic structure in many insect scales and the cause of their iridescent colours. *Phil. Trans. R. Soc.* **211**, 1–74.

PERYASSU, A. G. (1908). Os Culicideao do Brazil. Typog. Leuzinger, Rio de Janeiro.

ROY, D. N. (1936). On the role of blood in ovulation in *Aëdes aegypti* Linn. *Bull. Ent. Res.* **27**, 423–9.

SOPER, F. L., WILSON, D. B., LIMA, S. and ANTUNES, W. SA. (1943). *The Organisation of Permanent Nation-wide Anti-*Aëdes aegypti *Measures in Brazil.* The Rockefeller Foundation. New York.

SOTAVALTA, O. (1947). The flight-tone (wing stroke frequency) of insects. *Acta Entom. Fennica*, **4**, 117 pp.

SOTAVALTA, O. (1952). The essential factor regulating the wing stroke frequency of insects in wing mutilation and loading experiments, etc. *Ann. Zool. Soc. Vanamo*, **15**, no. 2, 1–67.

THEOBALD, F. V. (1901). See references in ch. II (*a–b*), p. 46.

THEODOR, O. (1936). On the relation of *Phlebotomus papatasii* to the temperature and humidity of the environment. *Bull. Ent. Res.* **27**, 653–71.

WIGGLESWORTH, V. B. (1942). *The Principles of Insect Physiology*. Ed. 2. (see also many later editions). Methuen and Co., London, pp. 354–5.

XVIII
THE IMAGO: EXTERNAL CHARACTERS

(*a*) THE HEAD CAPSULE

GENERAL DESCRIPTION (Figs. 53, 54)

For accounts of the head capsule in insects in general see Comstock (1920); Crampton (1921); Snodgrass (1935, 1943, 1953). For a very complete study of the head capsule in a large number of Dipteran types, including the mosquito *Psorophora* and a standardised nomenclature of parts see Petersen (1916). A very complete account of the head capsule and its appendages in the mosquito is that by Robinson (1939) on these parts in *Anopheles*. Another recent helpful paper is that by Jobling (1928) on the head capsule of *Culicoides*, which has many resemblances to that of the mosquito; see also Christophers, Shortt and Barraud (1926) on the head and mouth-parts of *Phlebotomus*.

The homologies of the mosquito head capsule have been studied mainly from the comparative point of view; but consideration may also be given to ontological evidence, namely to the changes in passage from the larva, where in many respects the nature of the parts is more evident, to the imago. These changes largely take place in the prepupa and are made further evident in the pupa. Besides extrusion of the antennal and mouth-part rudiments there are changes in the head capsule itself. These are largely accounted for by

---

Figure 53. The head capsule.

1 Dorsal view of head.

2 Ventral view of head.

3 Lateral view of head. *a*, extension from region of the anterior opening of the tentorium to mandible.

4 Lateral view of clypeal apodeme and furcal plate. *a*, clypeal fossa; *b*, point of attachment of clypeo-labral muscles; *c*, joint-like break in wall of clypeus; *d*, dorsal labral sclerite; *e*, ventral wall of labrum and roof of blood channel; *f*, blood channel.

5 Enlarged view of left half of frontal area showing the pedicel displaced to display the scape and its attachments. *a*, articulating points of scape with pedicel; *b*, articulating points of scape with the head capsule.

6 Showing apical portion of clypeus everted and relation to the clypeal apodemes.

7 Tentorial bar of male imago. *a*, posterior more tubular portion; *b*, anterior ballooned portion.

Lettering: *apn*, anterior pronotal lobe; *BP*, bucco-pharynx; *CL*, clypeus; *CLa*, clypeal apodeme; *CLa'*, everted *CLa*; *csc*, cervical sclerite; *E*, compound eye; *E'*, ocular sclerite; *eps*, epicranial suture; *fp*, furcal plate; *fsc*, sockets of forked scales; *hp*, hypopharynx; *is*, inter-antennal suture; *L*, labium; *L'*, labial area; *Lr*, labrum; *md*, mandible; *mxa*, maxillary apodeme; *mxp'*, base of maxillary palp; *Oc*, occiput; *Oc'*, paraocciput; *oc'*, paired ocellus; *of*, occipital foramen; *oh*, orbital hairs; *P'*, post-pharynx; *pas*, pre-antennal suture; *pd*, pedicel; *pfs*, post-frontal suture; *pge*, postgena; *sca*, scape; *sd*, salivary duct; *sd*, salivary pump; *slh*, sublabial hairs; *soh*, suborbital hairs; *T*, tentorium; *T'*, anterior opening of tentorium (position of); *V*, vertex.

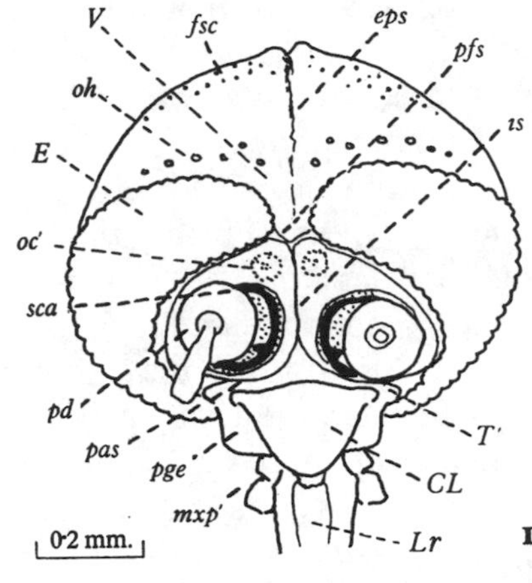
V
fsc
eps
pfs
oh
ts
E
oc'
sca
pd
pas
pge
mxp'
T'
CL
Lr
0·2 mm.
1

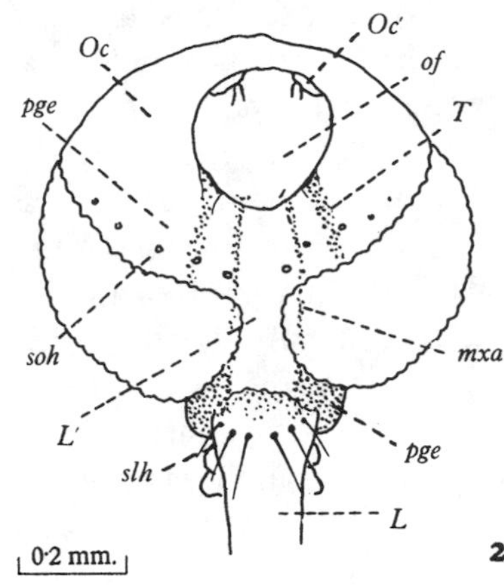
Oc'
Oc
of
pge
T
soh
mxa
L'
slh
pge
L
0·2 mm.
2

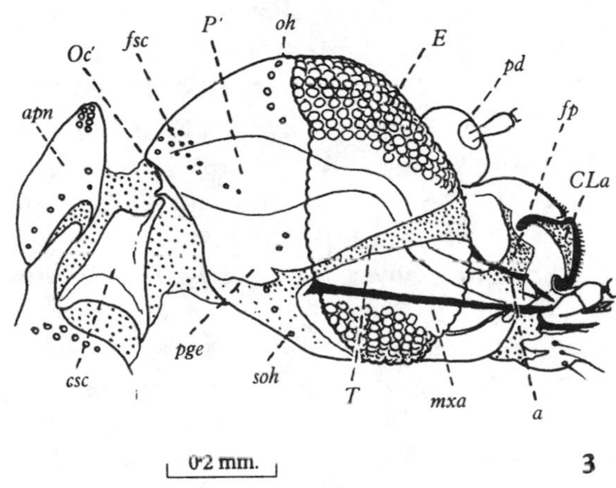
Oc'
fsc
P'
oh
E
pd
apn
fp
CLa
pge
soh
T
mxa
a
csc
0·2 mm.
3

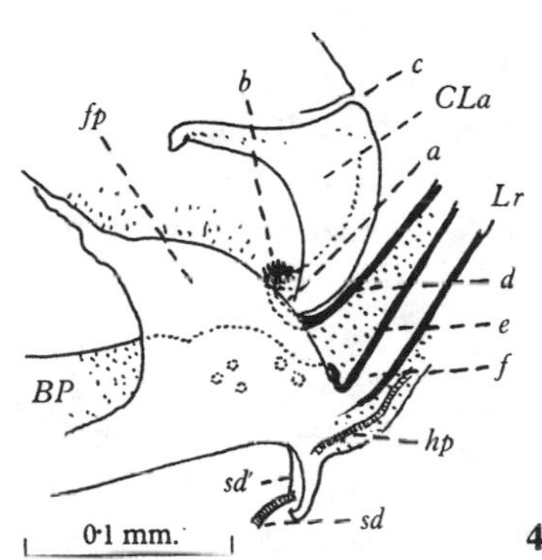
b
c
CLa
fp
a
Lr
d
e
f
BP
hp
sd'
sd
0·1 mm.
4

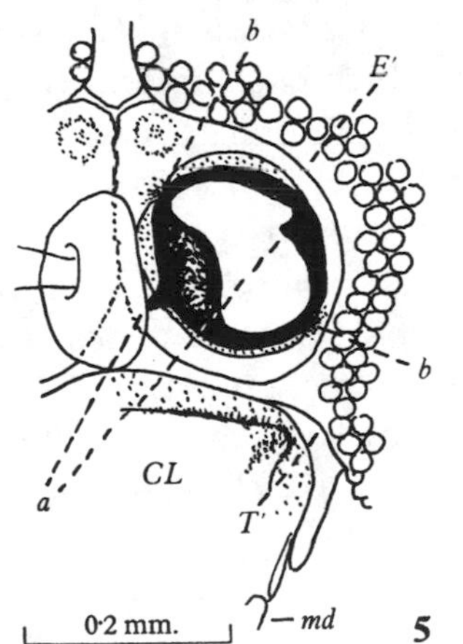
b
E'
b
CL
a
T'
md
0·2 mm.
5

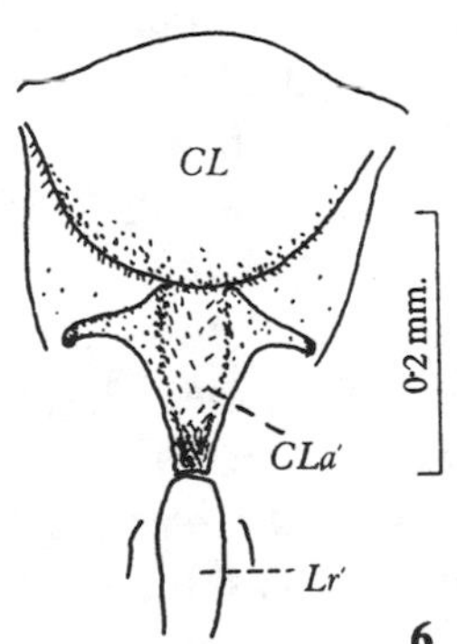
CL
0·2 mm.
CLa'
Lr'
6

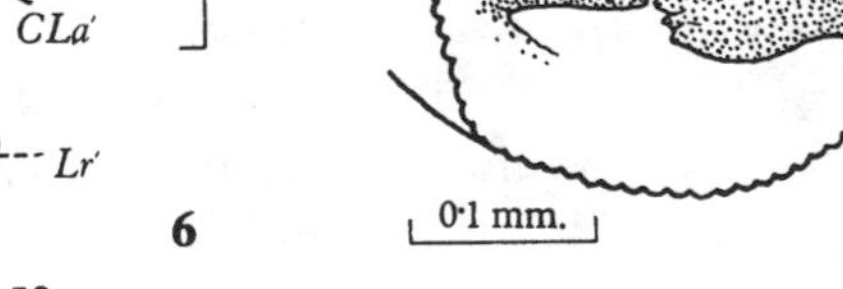
a
b
0·1 mm.
7

Figure 53

the great reduction, almost amounting to collapse and disappearance, of those portions of the larval head that have been adapted to meet the method of feeding in the larval stage, namely the great expanse of the labral and clypeal regions necessary to accommodate the relatively enormous pharynx and the elaborate feeding brush mechanism. By this collapse the whole of the front portion of the larval head, amounting to perhaps two-thirds of the head, is reduced in the imago to the relatively insignificant clypeus. Further the reduction of the median parts of the head capsule has brought inwards, nearly to the median line, the antennal prominences with their contained antennal base rudiments and with these the genal areas of the head bearing the large compound eyes. Also through the shift of the parts due to these changes the anterior openings of the tentorium are brought from the ventral position as in the larva to the dorsal aspect, that is in the natural position of the head, to the anterior aspect of the head. Also the shrinkage of the median parts and the increase in the area of the compound eyes at the expense of the genal areas results in the genal areas in the imago being restricted to the relatively small tracts named by Petersen the postgenae.

With these general considerations a brief description of the head capsule in *Aëdes aegypti* may be given. In both sexes the head is noticeably globular, the anterior half being largely occupied by the large convex kidney-shaped *compound eyes* (*E*), whilst the posterior half is formed mainly of the rounded occiput (*Oc*) and, ventrally, of the postgenae (*pge*). Between the compound eyes, one on each side of the middle line, are the *antennae* (Fig. 54 A, B), plumose in the male, less conspicuously hairy in the female, with their large conspicuous globular basal second segments (*pd*). Below these in the middle line there projects a snout-like prolongation of the head sometimes termed the *rostrum* (Fig. 49 (2) *ro*), its dorsal portion formed by the characteristic convex *fronto-clypeus*, or as it will here be designated *clypeus* (*CL*). Extending forwards from below the clypeus is the cylindrical *proboscis* (Fig. 54 (2)), with on each side the *maxillary palps*, usually referred to in systematic works on the mosquito as the palps, short in the female, as long as the proboscis in the male (Fig. 54 (3), (4)).

Posterior to the occiput is the large *occipital foramen* (*of*) and passing forwards from the upper margin of this in the median line is a deep median suture, partly apodematous but becoming shallower anteriorly, which is considered by Petersen to be the *epicranial suture* (*eps*), also termed the coronal suture (Robinson). Anteriorly the occiput extends forwards between the eyes as a narrow strip-like area, the *vertex* (*V*), which ends anteriorly at a short V-shaped line of thickening called by Robinson the *prefrontal suture* (*pfs*), and which at this point links together the marginal framework of the two compound eyes. Still further forwards is an area lying between the hollowed-out anterior margins of the two compound eyes and from which the antennae take origin. This is traversed in the median line by a strong suture, the nature of which is discussed later, but which Robinson very suitably terms the *interantennal suture* (*is*). Just distal to the postfrontal suture are two small circular areas, one on each side of the median suture, that are strongly suggestive of, and have all the appearance of, degenerate ocelli, including a lens-like thickening of the cuticle (Fig. 53 (1) *oc′*; 56 (10)). Whether these are, as they appear to be, the dorsal ocelli commonly present in this situation or not is still a matter of discussion: see under 'Special sense organs', p. 648 and under 'Propulsatory organs', p. 619.

After extending to a level a little anterior to the bases of the antennae the interantennal suture bifurcates, the branches (*pas*) passing to the anterior openings of the tentorial bars (Fig. 53 *T′*), that is to the lateral angles of the area now being considered, which, for reasons given later, will be termed the *frontal area*. According to Petersen these branches as seen in *Psorophora* are the arms of the epicranial suture (frontal suture of Snodgrass). This view is that accepted by most recent authors, but, for reasons given later, it is open to some doubt and the name *pre-antennal sutures* (*pas*) will provisionally be used for them, being less committal as to their nature.

There still remain to be considered the posterior and ventral aspects of the head and the intra-capsular structures. Posteriorly and ventrally a considerable extent of the head capsule is occupied by the large occipital foramen. On its upper margin on each side of the middle line is a thickening carrying a projection (*Oc'*) which articulates with the anterior arm of the neck sclerite and which Petersen terms *paraocciput*. Ventrally on the margins of the occipital foramen are thickenings where the posterior arms of the tentorium are attached and internal to these the attachment of the maxillary apodemes, as described later. Lateral to the foramen the occiput merges without break on each side into the smooth cheek-like areas termed by Petersen the *postgenae* (*pge*), again a term that will be open to some discussion later. The postgenae are continued without break on to the ventral aspect of the head and may be regarded as extending forwards between the lower margins of the compound eyes to the under surface of the rostrum. In *Psorophora* Petersen notes the presence of a median depression passing from the ventral margin of the occipital foramen in the median line to the base of the proboscis where this takes origin from the rostrum. An indication of such a line as shown by a clear interval in the stained cuticle is present in *Aëdes aegypti*.

Passing through the head on each side are the imaginal *tentorial bars* (Fig. 53 (3) *T*). In the imago these are large open tubes, patent throughout and much greater in diameter than in the larva. Anteriorly they open by a wide funnel-shaped opening in the position already indicated. Posteriorly they are attached to thickenings on the ventral margin of the occipital foramen. A little behind the middle point of each tube is a line of junction, or slight membranous gap, between the posterior portion, which is narrower and of more even diameter, and the anterior portion, which is wider and more uneven in shape. In the male the anterior portion is ballooned (Fig. 53 (7) *b*) possibly as a balancing device during flight. Towards the anterior end of the tube in *Anopheles* Robinson describes a spur representing the dorsal arm of the tentorium. This is not usually present in the female *Aëdes aegypti*, but in the male a somewhat plate-like expansion takes off towards the posterior end of the ballooned portion. Near the posterior end of the bar in the female is a short spur very similar to that present in the larva, inconspicuous or absent in the male.

Internal and ventral to the tentorial bars are two conspicuous rods, thicker and denser in the female, narrower and less dense in the male, which represent the basal portions of the maxillae. Posteriorly these are attached to membrane at the ventral edge of the occipital foramen a little internal to the posterior arms of the tentorium (Fig. 53 (2) *mxa*). Anteriorly they articulate with the free portion of the maxilla (galea) and give support to the maxillary palps. They are considered by Petersen in *Psorophora* to be the *stipes*. The basal segment or *cardo* is not distinguishable.

THE COMPOUND EYES (Fig. 53 (1), (2) *E*)

These are approximately of equal size in the two sexes, nearly meeting on the dorsum of the head and ventrally. In some cases, both in the male and in the female, they may actually meet below. Each eye is surrounded by a rim of thickened cuticle, *ocular sclerite* (*E'*), united on the vertex as already described by the postfrontal suture (*pfs*).

The eyes are roughly oval in shape with a hollowed-out anterior border, deeper in the male than in the female, caused by the presence of the large basal antennal segments. They are composed of *ommatidia* (see 'Special sense organs', p. 652) there being from 400 to 500 in each eye. The facets of the ommatidia as seen in preparations of the eye cuticle are circular, not hexagonal as described in some insects, often with a small triangular space left where the circles are not in contact. They measure from 21–24μ in diameter and in *Aëdes aegypti* are about equal in size in all parts of the eye and in both sexes, though on careful determination of the size there is some variation in different parts of the eye and those of the male are somewhat smaller, though not fully in proportion to the smaller size of this sex. For further information see under special sense organs.

About the middle of the posterior border of the compound eye there are still present traces of the larval ocellus as described by Constantineanu (1930) in *Culex pipiens*. The extent to which such traces are present varies. In some individuals they are large and well separated from the margin of the compound eye. In others they are chiefly evident as a thickening or deformation of the border of the eye. They leave no distinct facets on the cuticle after treatment with potash as do the ommatidia of the compound eye.

THE ANTENNAE (Fig. 54 (1) A, B *an*)

The antenna in both sexes consists of a ring-like basal segment, *scape* (*sca*), a globular second segment, *pedicel* (*pd*), and a long, thirteen-segmented *flagellum* (*fil*). The two basal segments differ but little in general structure in the two sexes, though the scape is possibly more distinctly separated from the head capsule and the pedicel is larger and more strictly globular in the male. On the inner aspect of the scape is a short projection which articulates with the base of the pedicel. A less marked thickening is present on the outer portion of the ring. These two attachments tend to restrict movement to a hinge-like up and down and somewhat oblique direction of movement of the antennae. The scape is further surrounded by a ring-like thickening of the head capsule separated from it by a narrow area of membrane. This thickening has two small projections between those by which the scape articulates with the pedicel so that there is altogether a more or less developed double hinge joint with the axes approximately at right angles. The pedicel in both sexes contains the special sense organ known as Johnston's organ, but rather more developed in the male.

In the typical insect antenna the scape is usually long, and, in addition, there is a ring-like thickening on the head capsule which carries a process, *antennifer*, on its lower part. In the present case the basal ring is the scape. The fact that the globular segment contains Johnston's organ leaves no doubt as to its being the pedicel.

The flagellum in the female consists of thirteen distinct segments, the second of which is the shortest, after which the segments increase in length gradually to the thirteenth, which is about one-tenth of the length of the whole flagellum (Fig. 54 (1) A *b*). Each segment of the flagellum, except the first, is formed of a narrow basal ring followed by a clear membranous area on which is borne a whorl of five or six hairs arranged round the segment. This is followed by a long apical cylindrical sclerotised portion forming the main length of the segment. This apical portion is somewhat tuberculated and carries small hairs with some rather larger ones apically, making a subsidiary series to the main hairs borne on the membranous area. This subsidiary series is most developed towards the base of the antenna and becomes less conspicuous towards the apex. The first segment of the flagellum, like the other segments, has a basal ring where it joins the pedicel and a clear membranous area following this, but with no hairs. The long hairs in this case spring from the body of the segment.

The flagellum in the male (Fig. 54 (1) B; 7) also consists of thirteen segments. Of these the two terminal ones are of about equal length, but disproportionately long as compared with the remaining segments, the two together forming about half the length of the whole flagellum. The terminal segment, like that in the female, has a basal clear area from which springs a circlet of about 8 hairs. Each of the remaining segments is peculiarly modified and of quite complex structure. Each carries towards its apex conspicuous crescentic outgrowths situated opposite one another on the segment and carrying whorls of about up to 20 long hairs which spring in a curious manner from the lower outside edge of the projecting outgrowths (Fig. 54 (7)). Apical to the crescents is a short length of the segment carrying small hairs. What appears at first sight to be a narrow length of segment basal to the hair-bearing crescents is seen on closer examination to be a dark cylindrical core surrounded by a thin transparent membrane bearing small hairs and widely separated from the core by a clear space (Fig. 54 (7)). The segments are separated by narrow, clear bands. The long penultimate segment, unlike the long apical segment, carries crescentic outgrowths and whorls of hairs like the more basal segments.

THE CLYPEUS (Fig. 53 (1) *CL*)

The reasons for terming this structure clypeus, rather than fronto-clypeus, is partly that it has long been so designated in a very large systematic literature dealing with the mosquito and partly for reasons of homology based on the present studies. The clypeus forms a rounded snout-like projection constituting the dorsal portion of the rostrum and largely forming the part of the head from which the mouth-parts appear to spring. Anteriorly in the middle line its wall is sharply bent in over the base of the labrum, forming a small invagination termed by Robinson the *clypeal fossa*. At its apex in the middle line is a thickening to which the base of the labrum is attached and into which is inserted the fan-shaped clypeo-labral muscle (Fig. 53 (4) *b*). Dorsal to this thickening laterally there project backwards into the cavity of the clypeus a pair of hook-shaped apodemes which receive the insertion of the muscles termed by Robinson the elevators of the labrum and by Thompson the epipharyngeal muscles, and which arise from the lateral portions of the roof of the so-called buccal cavity (bucco-pharyngeal cavity). These are the *clypeal apodemes* of Robinson, the *alae clypeales* and *laminae frontales* of other authors (Fig. 53 (3), (4) *CLa*).

The real nature of these parts does not seem to have been fully appreciated. The clypeus is not so wholly a rigid structure as its appearance might suggest and considerable areas on its surface are membranous. Further, just above the clypeal apodemes is a break in the thick anterior wall very suggestive of a partial joint. Very often in a preparation of the parts the front portion of clypeus with the apodemes is seen everted, forming a plate of which the lateral angles are the clypeal apodemes (Fig. 53 (6) *CLa'*). Muscles pulling the apodemes downwards would tend to project the apical portion of the clypeus with the labrum forwards, that is the apodemes appear to be levers which would tend to project the labrum. Elevators of the labrum would be retractors of this structure to return it to position, should it have been projected forwards.

External to the clypeal apodemes and occupying the middle portion of the sides of the clypeus are two apodematous sheets termed by Robinson the *fulcral plates* (*fp*). These pass upwards from the sides of the bucco-pharynx to blend with the walls of the clypeus. Their posterior borders form a conspicuous feature when the parts are viewed laterally as transparent mounted preparations (Fig. 53 (3), (4) *fp*; 61 (2) *ap*).

Where the sides of the clypeus merge ventrally into the postgenae there are on each side rod-like extensions continued forwards from the floor of the tentorial openings which are referred to later when discussing the basal connections of the mandibles (Fig. 53 (3) *a*).

THE FRONTAL AREA (Fig. 53 (1), (5))

This is the portion of head capsule situated caudad to the clypeus and which lies between the compound eyes and anterior to the post-frontal suture. It has very generally been accepted as a portion of the vertex, largely because of the interpretation placed on the median suture as the epicranial suture, and its branches as the branches of this suture. Such a view requires either that the frons is missing or incorporated without any revealing suture in a fronto-clypeus. It also hypothecates that the epistomal suture which normally passes between the two tentorial openings is absent, the suture taking this position being already ear-marked as the branches of the epicranial suture. A view based on what has been said regarding the crowding out of the median areas of the head would, on the other hand, allow of a much more natural explanation of the homologies. Thus the antennae of insects normally arise from the parietal plates of the head capsule, never, according to Snodgrass, from the frons. The paired ocelli are also borne on these plates. In the standard insect head the single ocellus is commonly borne on the frons. If now it be supposed that the two genae bearing the large compound eyes have overridden the central parts, carrying in the antennae and forming by their contact the deep central suture (in which the frons is engulfed) and by their anterior edges, where they press up against the clypeus, forming the branches of this suture, that is no longer regarding these as branches of the epicranial suture, but secondary structures, there are few difficulties in giving a natural explanation to all the parts. Thus the antennae and paired ocelli are as they should be. The frons and the single median ocellus are eliminated. The single ocellus, instead of having to be given a position well down on the fronto-clypeus away from the paired ocelli, can be regarded as merely buried in its natural place near the paired ocelli. The epistomal suture is not eliminated but merely backed up against by the overriding parts and still passes, as it should do, behind the clypeus and in front of the frons from one tentorial opening to the other.

It might be held that even in the larva the dorsum of the head shows only a combined fronto-clypeus and no separate frons. But this is by no means certain, for something could be said for regarding the greatly swollen dorsum of the head between the sutures passing from the neck to the neighbourhood of the antenna not as fronto-clypeus, but as frons, and it is not impossible that the so-called pre-clypeus is the rightful clypeus (see under 'Embryology', p. 191). There clearly remains here an interesting point for determination by careful ontological investigation. Meanwhile, in the undue development of the median parts of the dorsum of the head in the larva to allow for the special mode of feeding, there is clearly a tendency for these parts, following upon such development, to become much reduced or even to be eliminated, much as the eighth spiracle of the larva has disappeared, in the imago.

It has been suggested earlier in this chapter that the term *postgena* requires some consideration. The *gena* in the larva has been accepted as the side of the head bearing the developing compound eyes and lying between the frontal suture dorsally and the hypostomal suture ventrally. Within these limits there are no appearances to suggest any separate postgenal area, since posteriorly the gena merges gradually into the occipital area. Occupying the very middle of this area in the fourth instar larva are the rapidly developing compound eyes. In the imago these have still further progressed, leaving merely remnants of the original gena about their periphery. If the shrinkage of the median parts and the overriding dorsally (and probably ventrally) of the lateral portions of the head be accepted, there would seem no reason to introduce a new term, postgena, for these parts. On such a view, therefore, the genae in the imago extend dorsally to the so-called epicranial suture. Seeing that the maxillary region, which in the larva lies between the hypostomal sutures, has in the imago become deeply buried and apodematous the genae also have met ventrally and the small area on the ventral aspect of the head not occupied by the compound eyes is again merely a vestige of the gena.

CHAETAE OF THE HEAD CAPSULE (Fig. 53 (1), (2))

The most conspicuous, indeed almost the only, chaetae of any size on the head consist of a row of stout bristles arising behind the margins of the compound eyes, *orbital chaetae* of Edwards (ocular of Christophers, 1933). In *Aëdes aegypti* these form a continuous or almost continuous line, there being about eight to twelve large bristles on each side dorsally and about six or more smaller hairs ventrally (*oh*, *soh*). The dorsal series do not extend forwards onto the small vertex. There may be somewhat of a gap between the series laterally. The series is present in both sexes. Ventrally in the female there may be a further line or staggered row of small hairs behind the ventral orbital row, *postgenal hairs*. This series is doubtfully present in the male where the ventral postgenal area is much less extensive than in the female.

The only other hairs that might be considered as relating to the head, other than to the mouth-parts, are those on the antennal segments described under that organ and the six hairs at the base of the labium.

(*b*) THE MOUTH-PARTS

The mouth-parts of the mosquito have been described by many authors. See Thompson (1905) and later accounts by Robinson (1939) of the parts in *Anopheles* and the description by Gordon and Lumsden (1939) of the parts and their function in *Aëdes aegypti.* These last two papers leave little more to be done in description of the parts and the mechanism by which they function. The mouth-parts in the male in a number of species have been described by Marshall and Staley (1935).

All the structures, except the maxillary palps, lie in what is commonly referred to as the proboscis (Fig. 54 (2)). As seen externally this is constituted by the *labium* (*L*), which forms a sheath-like trough in which the other parts are contained except during the act of puncturing the skin and sucking blood, when they are in part exposed as described later. These inner parts consist of six stylets, namely a median dorsally situated *labrum* (*Lr*), a part lying ventral to this, the *hypopharynx* (*hp*), a pair of *mandibles* (*md*) and a pair of *maxillae* (*mx*). All these parts, as shown by Robinson, are held together by surface tension of the oily secretion with which they are bathed. As such they form a single perforating organ, the *fascicle*. The extreme tenuity of these stylets is one of the outstanding marvels of natural history, so much so that, until their method of use in the form of the fascicle had been shown by Gordon and Lumsden, it was difficult to understand how such delicate and filamentous organs could perform the function of piercing the human skin. The general appearance of the parts as they are seen in a low-power preparation under the microscope is shown in Fig. 54 (2) and in section in Fig. 62.

THE LABIUM (Fig. 54 (2) *L*)

The labium externally is covered with vestitural scales like other external parts. The main portion of the organ is a long even cylindrical structure corresponding to what in some other diptera is termed the theca. Its walls, especially when stained, show fine irregular transverse striations formed by darker and lighter alternating bands. This appearance has been remarked upon by Dimmock as early as 1881, 1883 in *Culex*. It has been attributed by Vogel to the ability of the labium to adjust itself after feeding and by Patton and Cragg to its property of acutely bending during feeding, both actions probably indicating considerable elasticity in its walls. In *Anopheles*, as noted by Robinson, this appearance is present only on the proximal and distal portions. In *Aëdes aegypti* the striae are present throughout the length of the organ.

Though often spoken of as forming a gutter for the other mouth-parts, the margins of the slit along the upper surface of the labium very nearly approach one another except in about the basal third where the dorsally situated labrum is somewhat exposed. Towards the apex the borders of the slit close in and entirely enclose the stylets. During the act of feeding, as described later, the fascicle may be thrust out by bending of the labium except at the extreme end. Even if forcibly removed from this terminal portion of the groove the insect with some little trouble is able to return the parts to their sheath.

At its base ventrally the labium forms a slight bulge from which in the female arise six

---

Figure 54. Appendages of the head.

1 Antenna. A, male; B, female.

2 General view of mouth-parts of female as they appear in a mounted preparation.

3 Maxilla of female showing galea, maxillary palp and stipes.

4 Maxillary palp of male. Ventral side is to the right.

5 Tip of galea of female. A, as seen fully on the flat; B, as commonly seen partly on edge. Half scale of A.

6 Two segments of female antenna. *a*, main hair series arising from clear area at base of segment; *b*, subsidiary series at apex of segment; *c*, intersegmental membrane.

7 Third antennal segment of male. *a*, intersegmental membrane; *b*, one of pair of hair-bearing crescents; *c*, dark core of segment; *d*, outer thin membrane bearing microtrichae; *e*, basal sclerotisation of *d*, giving appearance of an additional dark band.

8 A, tip of labrum of female; B, tip of male labrum; C, tip of hypopharynx.

9 Dorsal view of tip of labium, showing labella and median organ. *a*, plates on inner aspect of labella.

10 Ventral view of same.

Lettering: *fil*, flagellum; *ga*, galea; *hp*, hypopharynx; *L*, labium; *Lr*, labrum; *lb*, labellum; *lg*, ligula; *md*, mandible; *mx*, maxilla; *mxp*, maxillary palp; *pd*, pedicel; *sca*, scape; *sti*, stipes.

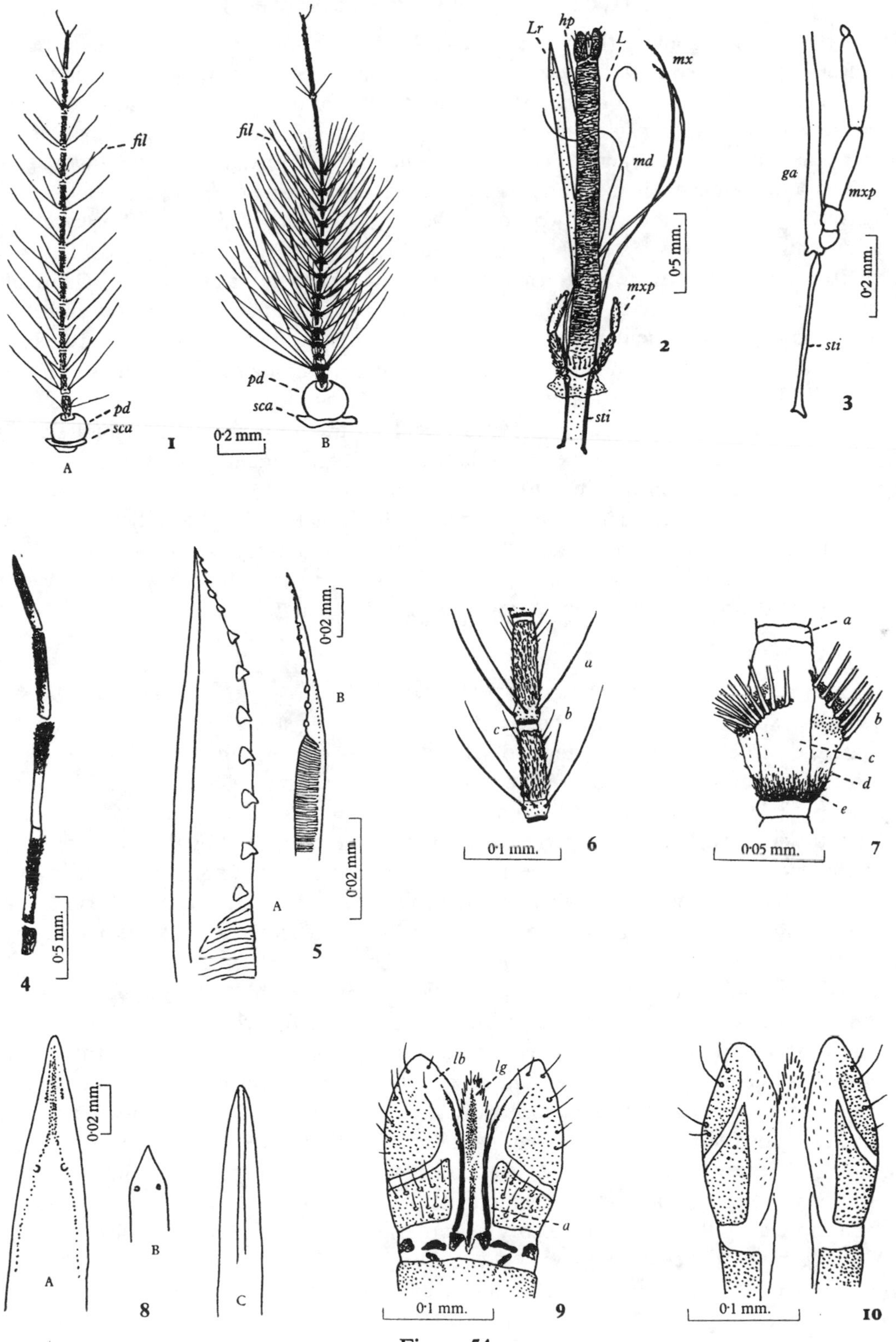
fil
pd
sca
A
I
0·2 mm.
B
Lr
hp
L
mx
md
0·5 mm.
mxp
2
sti
ga
mxp
0·2 mm.
sti
3
0·5 mm.
4
0·02 mm.
B
0·02 mm.
A
5
a
b
c
0·1 mm.
6
a
b
c
d
e
0·05 mm.
7
0·02 mm.
A
B
C
8
lb
lg
a
0·1 mm.
9
0·1 mm.
10

Figure 54

very regularly placed hairs (Fig. 53 (2) *slh*). Hairs are also present in the male in this situation, but are not so constantly regular in number and arrangement. The junction of the base with the rostrum ventrally is not specially thickened or articulated and is separated from the ventral head capsule only by a thin membranous line. Laterally, however, the edges of the groove are thickened and closely bound to the maxillary bases where the galea and the apodematous rods meet and where the maxillary palps take origin.

Terminally the labium carries two small lobes, the *labella* (Fig. 54 (9), (10) *lb*), between which in the middle line is a sharp-pointed median structure, the *ligula* (*lg*). In *A. aegypti* each labellum consists of two segments, a proximal and a distal, as indicated by cuticular plates. The ligula terminates in a hairy point (particularly hairy in the male) and has a median raphe which articulates at its base with the body of the labium. Between the labella and the labium is a membranous area in which are some small very dense sclerites apparently forming basal articulations to the labella and on the inner aspect of each labellum is a thickening which articulates basally with the inner of the small sclerites (Fig. 54 (9)). On their inner aspects the labella are largely membranous and hollowed, forming a space in which the tips of the stylets normally lie.

The labella are generally considered to represent the labial palps. The body of the labium has been very variously interpreted. It is considered by Crampton to consist of the fused palpifers of the labial appendages. According to Snodgrass the labium of the mosquito is probably the prementum, other more basal parts being unrepresented.

THE LABRUM (Fig. 54 (8) A, B)

The labrum is the stoutest of the stylets. It resembles a long straight sword with a bevelled pointed end. It consists of a median dorsal plate and two lateral thickenings all linked by membrane except apically where they become for a short distance continuous. The dorsal plate (Fig. 53 *Lr′*) is attached basally by a narrow point to the median process of the turned-in tip of the clypeus. The lateral thickenings with the membrane linking them are continued from the roof of the bucco-pharyngeal cavity. The organ as a whole forms an inverted gutter, the thickened edges of which with their flanges by meeting ventrally form a tube, the *food canal*, through which blood or other fluids are conveyed to the mouth and pharynx. For sections see Fig. 62.

Apically the labrum ends in a sharp bevelled point resembling very much that of a hypodermic needle, except that it is slightly curved at the tip when seen in lateral view. A little way back from the point on each side at the termination of the lateral thickenings is a small papilla believed to be a gustatory sense organ. There seem to be no other organs which could be concerned in taste, at least in *A. aegypti*. But in *Anopheles* there are other structures in this situation which may have that function. According to Robinson the labrum is the only one of the stylets provided with sensory organs. In the female the labrum has a long tapering point. In the male the open part is shorter and the end is cut off at a more abrupt angle. The sensory pegs in this sex are quite conspicuous.

It was formerly believed that the food canal was completed below by apposition of the hypopharynx. It has been shown, however, by Vogel and others that the sides of the labrum themselves complete the canal except for a short distance towards the base. Formerly also the labrum was generally referred to as the labrum-epipharynx. As noted, however, by Snodgrass, there seems no reason for regarding the ventral surface of the labrum in insects where this is elongated as being epipharynx. The complicated larval epipharynx in the mosquito is absorbed in the pupal stage and even if a relic of it still remains in the imago it plays no part in the building up of the labrum in this stage. At the base of the lateral rods of the labrum in the midge *Culicoides* Jobling (1928) describes small sclerites noted by Petersen as present in certain Diptera and referred to as *tormae*. Such are not mentioned by Robinson in *Anopheles* and I have not been able to identify such in *Aëdes aegypti*.

THE HYPOPHARYNX (Fig. 53 (4); 54 (2); 62)

Ventral to the labrum is the next stoutest of the stylets, the *hypopharynx* (*hp*). The hypopharynx is a thin flat stylet with a midrib along its full length within which lies the terminal portion of the *salivary duct* (*sd*). At its base is the *salivary pump* (*sd'*). Its termination is shaped rather like a long pen-nib (Fig. 54 (8) C). At or near the point is the opening of the salivary duct. In *Anopheles* (Macloskie, 1887, 1888; Robinson, 1939) the opening is a trace proximal to the actual point. In *Aëdes aegypti* the opening so far as could be ascertained is terminal. Basally the hypopharynx is continuous with the floor of the buccopharynx. In the male it is represented only by a ridge on the labium, the salivary duct or groove passing along the floor of the labial gutter to the tip of the ligula.

THE MANDIBLES (Fig. 54 (2) *md*)

The mandibles are much the least conspicuous of the stylets and in *Aëdes aegypti* may require a little trouble to display. They lie close along the labrum and are apt to remain so in a dissection. The basal half or so is fairly stout and readily seen, but the terminal half is extremely thin and transparent. The mandibles of the mosquito are commonly depicted as they are seen in *Anopheles*, that is as having a triangular expansion at the tip furnished with fine teeth. In *Aëdes aegypti* they appear to terminate in a smooth lancet-like tip with no teeth. Gordon and Lumsden, who worked with *A. aegypti*, do not describe the mandibles which they say were never observed when watching penetration by the fascicle. Dimmock (1881) figures the mandibles in his '*Culex rufus*' much as described above, that is lancet-shaped with no teeth. There are two species given by Dyar (1928) under the synonym of *C. rufus*, namely *Aëdes cinereus* and *Culex pipiens*. Since Dimmock gives *C. pipiens* as another of the species he was working with it seems probable that the *Culex rufus* referred to was an *Aëdes*. It is possible, therefore, that in this genus the mandibles of the female are of this character.

According to Robinson the mandibles cover the opening of the labrum when this organ is not in use, which would appear to be their main, if not only, function.

Marshall and Staley (1935), also Marshall (1938), note that in British mosquitoes mandibles are present in the male, though usually reduced, in a number of genera, but are absent in *Aëdes*. They are absent in *A. aegypti*.

The mandibles are articulated at the sides of the rostrum with a small sclerite linking them to an extension passing forwards from the floor of the anterior opening of the tentorium. This small sclerite is evidently the 'mandibular condyle' first described by Jobling in *Culicoides*. Robinson describes a similar condition in *Anopheles* when he describes the mandibles as articulating at their base with a thickening on the genal wall termed by the author the 'genal shelf'. In *A. aegypti* the parts referred to can be correlated with structures in the larva the nature of which is more evident. The mandible in the larva, as already described, takes origin from the edges of a large fenestra on the ventral aspect of the head, but has a definite articulation at one point only on the ventral edge of the fenestra, the rest of its connections being membranous. External to this articulation the ventral rim of the fenestra, it may be remembered, is formed by the edge of the gena, which edge is continued without a break outwards to the region of the anterior opening of the tentorium, where it links up with the outer end of the suspensorium. In the imago the portion of the ventral rim of the fenestra linking up the point of articulation of the mandible with the suspensorium has remained as the now isolated small suspensory sclerite, other parts, except what remains of the suspensorium, having ceased to be represented. A small outer portion of the suspensorium, sufficient to give attachment to the suspensory sclerite, has remained as the genal shelf (Fig. 53 (5) *md*). It is again a matter of the changes in the imaginal parts brought about by the great recession in the whole clypeal region with the great reduction in the diameter of the bases of the mouth appendages.

THE MAXILLAE (Fig. 54 (2), (3), (5))

The maxillae of the imago, like those of the larva, are present as free structures only as the portions distal to the origin of the maxillary palps, the basal portions being in the form of the apodematous rods already described. The free portion consists of an inner stylet, the *maxilla* of authors, more correctly the *galea* of the generalised insect maxilla, and an outer five-jointed *maxillary palp*. The galea (*ga*) is a long, rather stiff ribbon-like stylet, the inner edge of which is formed by a rod-like thickening and its outer portion by a thin even flange marked by transverse striations. At its base the galea is thickened and lies along the lateral edge of the labium after articulating with the maxillary apodeme of its side. In the proboscis the galeae lie along the ventral side of the hypopharynx, one on each side of the midrib. Apically the galea ends in a characteristic curved point carrying teeth. The teeth are borne on the external edge, that is on the side of the flange. The striations of the flange, however, cease at the level of the most basal tooth and from this point to the termination the stylet is rigid (Fig. 54 (5)). In *A. aegypti* the teeth are very regularly twelve in number. The more proximal seven or eight are large and blunt and only very slightly recurved at the point. They rather call to mind some form of mammalian molar teeth and are implanted in the substance of the maxillary tip. The terminal four teeth or so are more sharp pointed and increasingly recurved so that the last tooth or two are very fine and strongly curved with sharp points.

The five-jointed female maxillary palp consists of two small segments at the base, then two longer segments forming the greater part of the length of the organ and a minute globular apical segment.

In the male the maxillae are short, only about a third of the length of the proboscis, very slightly sclerotised and devoid of teeth and the characteristic striation of the female organ.

As will be seen from the section dealing with the mechanism of feeding, the maxillary stylets are extremely important structures as it is by their forward and backward rapid movements that penetration by the fascicle is brought about.

THE FASCICLE

The fascicle is formed by all the six stylets held together by surface tension acting as a single organ. At rest as described by Gordon and Lumsden the labrum occupies the front of the fascicle, the ends of the maxillary blades lying side by side beneath it with their teeth projecting at its sides. In action the point of the fascicle advances into the tissues in a series of minute thrusts in the manner of a pneumatic drill, the maxillae in rapid movement of small magnitude cutting into the tissues but never advanced in front of the labrum. For further information see the section on the mechanism of feeding, p. 486.

Some importance attaches to the measurements of the mouth-parts, since upon such largely depends the ability of the insect to penetrate fabrics. The proboscis excluding the labella measures in length in an average-sized female *A. aegypti* 2·32 mm. and with the labella 2·5 mm. At the base it is 0·12 mm. in width and depth; at the apex 0·10 and 0·08 mm. respectively in width and depth. The fascicle measures in width 0·06 mm. at the base and 0·04 at the apex. At the apex its depth is 0·03 mm. The blood channel is approximately circular in outline and measures 0·025 mm. (25 μ) in diameter.

(*c*) THE NECK AND THORAX

THE NECK (Figs. 53 (3); 55 (4))

The neck links the head with the thorax. Though it is to a considerable extent membranous it receives very substantial rigid support from the *cervical sclerite* (*csc*). The cervical sclerite is shaped something like the letter **H**, the cross piece of the **H** forming a bar across the ventral aspect of the neck, the upper arms articulating with the margin of the occipital foramen in its dorsal portion and the lower arms with the prothorax ventrally. This gives a double-hinged strut on each side of the neck which, while allowing the head to be extended or brought backwards, gives complete rigidity. In *A. aegypti* what may be likened to the sides of the **H** are expanded into substantial broad plates covering much of the lateral aspects of the neck. Seen from in front the sclerite has much the appearance of a broad collar. Such parts of the neck as are not occupied by the cervical sclerites are membranous, the membrane being evenly covered with regularly situated tiny spots formed by circles of minute microtrichae-like elevations.*

THE THORAX (Fig. 55; 56 (1))

The thorax of the mosquito is formed of numerous sclerites the nature and nomenclature of which have been the subject of study by a number of authors. According to Crampton (1909) the first to give nomenclature to the parts in insects was Audouin in 1820. Certain of the parts in the mosquito have been variously interpreted and named, see Snodgrass (1912); Christophers (1915); Prashad (1918); Patton and Evans (1929); Edwards (1921, 1932); Christophers (1933); Barraud (1934); Komp (1937). The nomenclature now very generally accepted is that given by Edwards (1941) in his treatise on Ethiopian mosquitoes; see also Natvig (1948). For thoracic parts in insects see Crampton (1909, 1914, 1942) and especially Snodgrass (1927, 1935).

The thorax in the mosquito is large and massive, with an extensive convex dorsal surface and extensive flat pleurae which converge ventrally to the narrow coxal origins, giving to the structure as a whole a wedge shape, or, since the coxae arise from a relatively small area and both the anterior and posterior surfaces of the thorax are rather flat, the shape is more correctly that of an inverted pyramid with the base broadly rounded and compressed somewhat laterally.

* For information on the cervical sclerites in insects see Crampton (1917, 1926), Martin (1916), Snodgrass (1935).

It consists of pro-, meso- and metathorax, the greater bulk being mesothorax. Dorsolaterally it carries the wings and halteres and ventrally the three pairs of legs. Laterally on the pleurae are the large meso- and metathoracic spiracles. It is highly ornamented with white and silvery white markings, giving on the dorsum the well-known lyre-shaped effect (Fig. 55 (1)), which except for a few related species with rather similar markings is characteristic.

THE PROTHORAX

If the fore part of the thorax bearing the anterior pair of legs be detached and mounted without undue pressure, an adequate picture of the prothoracic segment will be obtained (Fig. 55 (4)). On each side of the neck space in its dorsal half are the conspicuous prothoracic lobes, or *anterior pronotal lobes* (*apn*), and between these a median area where pronotal structure appears to be absent, the large overhanging anterior promontory of the mesonotum (Fig. 55 (4)) occupying the gap. Dissection will show, however, that there is in this situation an apodeme projecting inwards from the suture at the base of the mesonotum not so far noted by any author, but well displayed in stained potash preparations. This would appear to be the *first phragma* by the notation of Snodgrass, and strictly speaking belongs to the mesothoracic acrotergite, primitively part of the prothoracic segment. It would seem in this case sufficiently correct to term it the *pronotal apodeme* (Fig. 55 (4), (5) A *pna*). From this laterally there proceeds backwards on each side a long apodematous rod referred to below. The anterior pronotal lobes in *A. aegypti* carry conspicuous patches of flat white scales. They also carry a line and apical clump of bristles, the *anterior pronotal chaetae*.

Ventral to the anterior pronotal lobes the neck opening is encircled by the arms of a large median shield-shaped sclerite which pass up on each side to articulate with the lobes. The shield-shaped sclerite is the prothoracic sternum, *prosternum* (*S.* 1), and the arms are part at least of the prothoracic pleura, *propleura* (*Pl.* 1). The prosternum in most species of Culicini, including *A. aegypti*, is bare of scales. It has a strong median suture. Lateral to it, linked by membrane, lie the anterior coxae, with the lower inner border of which its

---

Figure 55. The thorax.

1 Dorsal view of head and thorax to show ornamentation.

2 The same to show structure and chaetotaxy. *a*, scutal angle; *b*, bare area.

3 Lateral view of thorax to show pleural sclerites and chaetotaxy.

4 Anterior view of neck and thorax.

5 A, pronotal apodeme; B, sternopleural bridge and furca 1. *a*, the sternopleural bridge; *b*, apodematous ring behind prosternum; *c*, thickened anterior margin of sternopleuron; *d*, branches of furca extending towards neck.

6 Anterior view of hind coxae and metasternum.

7 Posterior view of hind coxae showing furca 3. *a*, membranous area at back of coxae.

Lettering: *apn*, anterior pronotal lobe; *csc*, cervical sclerite; *cx.* 1–3, coxae 1–3; *fu.* 3, furca 3; *ha*, haltere; *k′*, small clear space; *me*, mesepimeron; *mes*, mesepimeral suture; *mr*, meron; *mt*, metanotum; *mtr*, metameron; *N*, neck; *Pl.* 1, propleuron; *pa*, prealar knob; *pna*, pronotal apodeme; *ppn*, posterior pronotal area; *psp*, postspiracular area; *ptg*, paratergite; *ptn*, postnotum; *S.* 1–3, sternites 1–3; *Sp.* 1–2, spiracles of meso- and metathorax; *sc*, scutum; *sc.p*, anterior promontory; *sct*, scutellum; *so*, sensory pits; *ssp*, subspiracular area; *stp*, sternopleuron; *tr*, trochanter; *W*, wing.

0·5 mm.
1

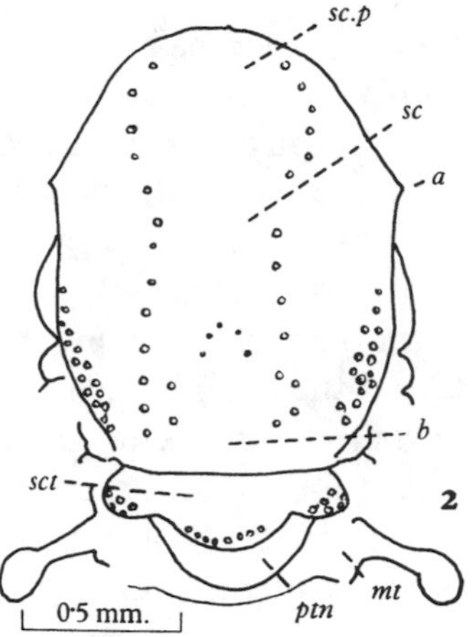

sc.p
sc
a
b
sct
2
0·5 mm.
ptn
mt

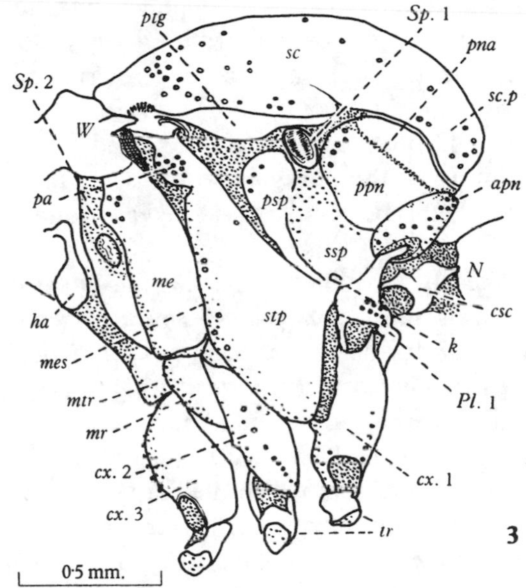

ptg
Sp. 1
pna
sc
Sp. 2
sc.p
W
pa
psp
ppn
apn
ssp
me
N
ha
stp
csc
mes
k
Pl. 1
mtr
mr
cx. 2
cx. 1
cx. 3
tr
3
0·5 mm.

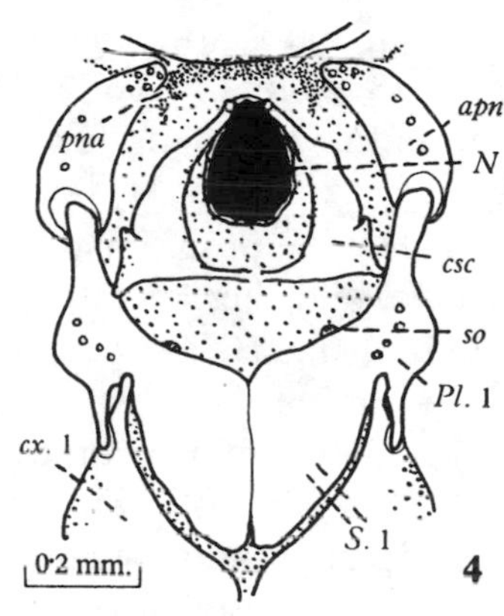

apn
pna
N
csc
so
Pl. 1
cx. 1
S. 1
0·2 mm.
4

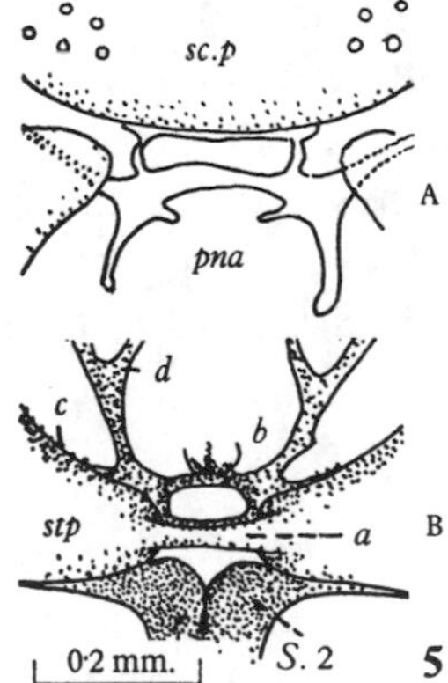

sc.p
A
pna
d
c
b
stp
a
B
0·2 mm.
S. 2
5

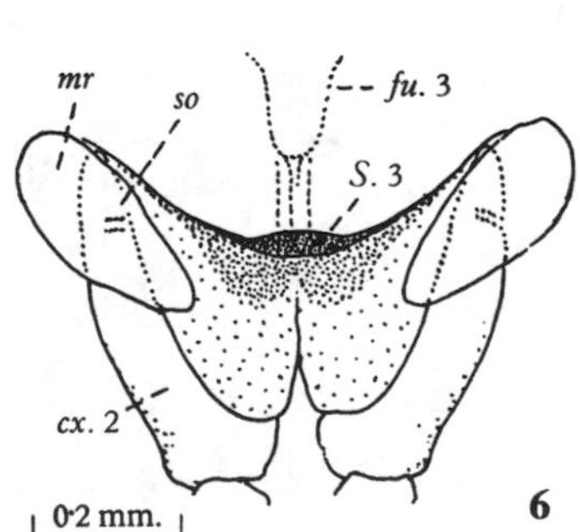

mr
so
fu. 3
S. 3
cx. 2
0·2 mm.
6

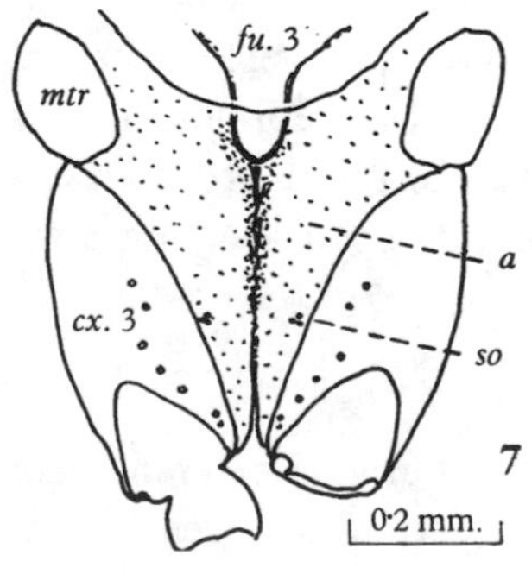

fu. 3
mtr
a
cx. 3
so
7
0·2 mm.

Figure 55

posterior border near the middle line articulates (Fig. 55 (3), (4) *cx.* 1). On each side anteriorly it is prolonged into the propleurae. At the base of each propleura is a prominent boss carrying a conspicuous patch of flat white scales and dark chaetae (*propleural chaetae*). From the region of the boss there projects ventrally a long spur-like process which articulates with the anterior coxa near its upper margin. Extending from the boss dorsally is the arm that articulates with the anterior pronotal lobe of its side. With the inner margin of this arm the ventral cornu of the cervical sclerite articulates.

Behind the anterior pronotal lobe on each side and forming part of the pleura is a flat somewhat circular sclerite carrying on its posterior border a ridge and a line of chaetae situated a little anterior to the mesothoracic spiracle. The dorsal border lies close along the lateral border of the anterior promontory separated by a membranous suture. The sclerite is considered by Edwards to be pronotal in origin and has been termed the *posterior pronotal area* (Fig. 55 (3) *ppn*). It is bare of scales, but readily picked out by its characteristic shape. The line of chaetae at its posterior border are the *posterior pronotal chaetae*, not to be confused with a line of chaetae present in some genera on a narrow area beyond the ridge and nearer to the spiracle termed the *spiracular chaetae*. Passing internal to its upper portion and attached at its end to the posterior border of the posterior pronotal area is the thin rod-shaped apodeme coming from the pronotal apodeme referred to above (Fig. 55 (3) *apn*). The presence of this extension from the phragma would seem to be in favour of the view that the area now being described is correctly considered as pronotal in origin.

Ventrally the posterior pronotal area is limited by two lines of thickening, but is continued beyond these without break into another smooth area of the pleura rather noncommittally named the *subspiracular area* (*ssp*) and a further extension backwards from this to lie in front of the large sternopleuron of the mesothorax is the *postspiracular area* (*psp*). Whether these are prothoracic or mesothoracic is perhaps doubtful. They are both given by Edwards as parts of the mesothoracic pleura. They are, however, more connected with the posterior pronotal area than with the sternopleuron from which they are separated by a strong suture. Both areas, however, are intimately connected with the sternopleuron at a sort of meeting place or knot in the region of the propleural boss and coxal junction, where also the furcal system of the prothorax has connections. On the subspiracular area at this point is a curious small clear spot (Fig. 55 (3) *k′*). The postspiracular area in *A. aegypti* carries a small patch of white scales and some chaetae (*postspiracular chaetae*).

The names given above are those now very generally accepted. But certain other terms have been in use which it is desirable to mention. Thus the propleuron and the posterior pronotal area were considered to represent the usual two pleural sclerites, namely, the episternum and epimeron of the prothorax, and were therefore termed respectively proepisternum and pro-epimeron. Other names, notably by Peus (1933) (see Natvig, 1948), have also been used for some of the parts.

THE MESOTHORAX AND METATHORAX

The scutum and related parts (Fig. 55 (2), (3)). Dorsally almost the whole of the thorax is composed of the greatly developed *mesonotum*, consisting of scutum, scutellum and postnotum, the metathorax being here represented only by an insignificant narrow band behind the postnotum. The greater portion is formed by the *scutum* (*sc*). This is a large domed structure providing space for the greatly developed indirect wing muscles. There is no suture dividing it into prescutum and scutum as in some Diptera and in the mosquito it is usual to apply the term scutum to the whole. Anteriorly the scutum projects forwards over the neck as the *anterior promontory* (*sc.p*) and laterally, at about the junction of the anterior and middle third, it is produced into somewhat pointed angles, the *scutal angles* (Fig. 55 (2) *a*). For some distance behind the scutal angles the lateral borders are continued downwards at an angle into the pleurae as small triangular pieces, the *paratergites* (Fig. 55 (3) *ptg*). Some way further back the sides are expanded above the wing roots where they are heavily clothed with scales and chaetae. Posteriorly the scutum is separated from the bar-like scutellum by a deep transverse groove and just anterior to this groove in the middle line is a small triangular area free from scales known as the *bare space* (Fig. 55 (2) *b*). Apart from the bare space the whole surface of the scutum is clothed with narrow curved scales which produce the ornamentation already described. Patches of white flat scales are present, however, on the paratergites and less markedly on the sides of the anterior promontory. The chaetae carried by the scutum will be described later in the section on chaetotaxy.

Posterior to the scutum is the *scutellum* (*sct*). As is general in the Culicini the scutellum is three-lobed, as distinct from the more uniformly convex part in the Anophelini. Each lobe bears a particularly conspicuous patch of flat white scales and its own cluster of black chaetae. From the under aspect of each lateral lobe there rises a long thin ridge which is continued into the wing as the *posterior wing root* (Fig. 58 (4) *pwr*).

Posterior to the scutellum is a characteristic smooth and bare area, the *postnotum* (*ptn*). This is considered to be phragmal in nature and represents the second phragma of Snodgrass. On its inner surface it receives the attachment of the longitudinal indirect wing muscles, which pass the scutellum to be inserted into the postnotum.

Dorsally the *metathorax* forms on each side of the postnotum a somewhat triangular area from the outer border of which the halteres take origin (Fig. 55 (2) *mt*). Linking these behind the postnotum and separated by a suture from this structure, is a narrow uniform band which represents the metathorax dorsally in the middle line. It has no setae or other special features. Behind it is membrane with, posterior to this, the anterior line of thickening (acrotergite) of the first abdominal tergite.

The meso- and metathoracic pleurae (Fig. 55 (3)). The mesothorax forms the greater part of the pleural areas of the thorax posterior to the parts already described as prothoracic. In the upper anterior angle is the large mesothoracic spiracle (*Sp.* 1) and posterior to this is a small depressed membranous area lying ventral to the small triangular paratergite (*ptg*) with its conspicuous white scale patch. Still further back in front of the wing root is a very prominent knob carrying a conspicuous tuft of hairs, the *pre-alar knob* (*pa*). Continued obliquely downwards to between the mid and hind coxae from this knob is the large conspicuous sclerite, the *sternopleuron* (*stp*). Its upper part is narrow, ending at the pre-alar knob, but below it forms a wide smooth area eventually passing between the coxae to the ventral surface of the thorax. It carries conspicuous patches of flat white scales, notably a large patch about its middle already referred to as a useful landmark in ascertaining body proportions. In addition, there is a smaller patch on its lower border above the coxa. Besides the *pre-alar chaetae* on the pre-alar knob it carries a line of small chaetae along its length near its posterior border, *sternopleural chaetae*. Posterior to the sternopleuron is the almost equally large quadrangular shaped *mes-epimeron* (*me*). Dorsally the mes-epimeron extends up to below the wing root where it makes important connections with certain of the wing root sclerites (see subsection on the wings). At its lower border there is a peculiar and unmistakable fold and suture separating it from a small convex area called the *meron* (*mr*). The mes-epimeron carries two patches of flat white scales, one above the other and on its upper part a number of chaetae, *upper mes-epimeral chaetae*. Lower mes-epimeral chaetae are wanting in *Aëdes aegypti*. Between it and the sternopleuron is an important suture, extending between the middle coxa and the wing, the two sclerites on either side of it and the suture forming the usual arrangement of the parts in the standardised insect pleura of a wing segment. The suture in this case is the *mes-epimeral* (*mes*).

The meron in insect taxonomy is a sclerite originally forming part of the coxa which in some forms, including the mosquito, has become attached to the thorax and become a part of this structure (Snodgrass, 1935, p. 196). The meron is connected to the middle coxa. A small sclerite at the base of the hind coxa has been variously interpreted, but has been termed by Edwards the *metameron* (*mtr*).

Posterior to the mes-epimeron is a recessed triangular area on the upper part of which is the *metathoracic spiracle* (*Sp.* 2). This area on its more dorsal portion carries the *haltere* (*ha*). At its lower end is the metameron, which appears as a small, almost isolated structure above the hind coxa apparently appended to the back of the metathorax on each side (Fig. 55 (7)).

The sternal area and coxae (Fig. 55 (4), (7); 56 (1), (5)). The coxae form so intimate a part of the thorax that, though strictly speaking they are basal joints of the legs, it is most convenient to consider them when dealing with the sternal areas of the thorax. The thorax being wedge-shaped and very narrow ventrally, the lower borders of the coxae of the two sides are articulated almost to the middle line leaving only the narrowest space for the sternal parts between them. Each coxa may be considered as an elongate cylinder the upper part of which has been sliced off at a low angle leaving a long inner surface by which it is articulated to the thorax. The upper end of this surface is narrow and is applied to the pleura. The lower end is broad and circular and applied to the sternal area. The edges are linked to the thorax largely by membrane, but, especially in the fore coxae, there is an articular junction with the pleural sclerites dorsally and with the sternal elements ventrally. Thus in order of first, second and third coxa they articulate dorsally respectively with the propleura, with the lower ends of the sternopleuron and mes-epimeron and with the metapleuron. Ventrally all come in contact with their respective sternites near the middle line. The middle coxa is peculiar in that it is hollowed out basally and below this hollow forms a prominent bulge (Fig. 55 (6) *cx.* 2).

The sternal parts are difficult to examine as they are largely overlaid by the coxae and being much reduced and modified are not easy to define. With the exception of the more obvious prosternum they have not been described. Even if of no systematic importance, however, it seems desirable that some description of them should be put on record. The following method has been found useful in their study.

A spirit-hardened mosquito is placed on its side on a slab of paraffin under the dissecting microscope and rigidly fixed by thrusting a fine stainless steel pin through the thorax into the wax. With a sharp razor or mounted fragment of a Gillette blade the specimen is now cut down upon in such direction as it is thought will best display the parts. The detached portion is then passed through any necessary alcohols to 10 per cent potash where it is left, preferably at a warm temperature, for a day or more and removed to distilled water for a like time. After deeply staining with fuchsin in 70 per cent alcohol it is passed through alcohols, carbon xylol to a drop of canada balsam on a slide and after examination and suitable arrangement mounted without pressure. Some remarkable preparations can be made in this way.

Taken in order from before backwards the sternal structures are as follows. Between the anterior surfaces of the two fore coxae is the relatively broad prosternum already described with its deep median suture and double pointed posterior end (Fig. 55 (4)). Lying internally to the prosternum at its posterior end is a rather complex ring of apodeme from which there projects forwards on each side what may be termed the *first furca* (Fig. 55 (5) B). Just posterior to this point is the peculiar bar joining the lower ends of the sternopleura across the middle line as noted by Nuttall and Shipley in *Anopheles* (Fig. 55 (5) B *a*). Behind the bar is an area formed by two small plates separated by a median suture and largely folded together in a V-shaped fashion. Laterally these plates send out extensions which pass anterior to the bases of the middle coxae. They also largely encircle the coxae on their inner side and appear to represent the mesothoracic sternite or some portion of this (Fig. 55 (5) B). From the region of these plates there extends into the thorax a complex apodematous extension which may be considered to be the *second furca* (Fig. 56 (1)). Still more posteriorly there is a deep depression largely separating off the hind coxae and their associated basal structures from the parts anterior to this and in this depression is a further double plate from which lateral extensions pass to the region of the meron (Fig. 55 (6)). From now on the sternal area is membranous, forming a V-shaped area between the hind coxae anteriorly and a similar area posteriorly, as is well seen examining the thorax from the posterior aspect (Fig. 55 (7)). On the more posterior of these membranous areas in the middle line is a line of apodematous suture continued at its upper end into the thorax as a characteristically shaped fork which may be termed the *third furca* (*fu.* 3).

The furcae when suitably exhibited are striking structures. They may be briefly described as follows.

Furca 1. Rises at the posterior end of the prothoracic sternum as an angular rather ring-like apodeme from which extensions pass out laterally, namely one continuous with the greatly thickened anterior border of the sternopleuron and one, a free apodeme, passing forwards towards the neck and bifurcating.

Furca 2. Rises from the region of the mesosternum as a median thickening from which arise two arms forming a flattened U, each of which is expanded at the tip in a bell-shaped manner (Fig. 56 (1)).

Furca 3. Enters the thorax posteriorly just anterior to the ventral wall of the abdomen where this joins the thorax. It resembles a two pronged hay-fork.

The furcae give attachment to muscles, most of which are massive and fully occupy all available surface given by the furcal extensions.

Before concluding the description of the thoracic parts, mention may be made of the relation of the meron and metameron to the sternal structures. The meron terminates after passing for a short distance onto the anterior wall of the depression between the bases of the middle and hind coxae (Fig. 55 (6)). The metameron of each side, when seen in posterior view, is perched above the base of the hind coxa on the lateral parts of the posterior surface of the thorax below the attachment of the abdomen. It has no particular attachments or extensions on its inner side (Fig. 55 (7)).

THE THORACIC CHAETAE

The chaetae of the thorax have already been largely indicated and named in the foregoing sections. The following is a seriatim list:

PROTHORACIC

1. *Anterior pronotal.* Usually an apical double row and a line around the base of the lobe.
2. *Posterior pronotal.* A line of about eight hairs along the posterior margin of the posterior pronotal sclerite (anterior to the marginal crest).
3. *Propleural.* A line of about eight hairs on the propleural process.
4. *Posterior spiracular.* Usually about four hairs on the postspiracular plate.

MESONOTAL

5. *Lateral anterior mesonotal.* A cluster of hairs on each lateral angle of the anterior promontory continuing posteriorly into 7.
6. *Median anterior mesonotal.* A small cluster anteriorly in the middle line on the anterior promontory.
7. *Dorso-central.* A single line of spaced stout chaetae passing backwards from 5 to a short distance in front of the lateral lobe of the scutellum.*
8. *Postero-medial.* A V-shaped line of small hairs in the middle line, a little in front of the posterior bare space.
9. *Antero-marginal.* A few hairs on the edge of the scutum above the posterior pronotal area.
10. *Postero-marginal.* A conspicuous line and cluster of small and medium hairs on the margin of the scutum above and behind the paratergite.
11. *Supra-alar.* A dense cluster of large hairs above the wing roots. This is the largest cluster of chaetae on the body. It sends forwards a line of some large chaetae between 10 and 7. A single hair isolated from the group lies always above the scutal flange (*scutal flange hair*).
12. *Median scutellar.* A circumscribed bunch of large chaetae on the middle lobe of the scutellum.
13. *Lateral scutellar.* A similar circumscribed bunch on each lateral lobe.

MESOPLEURAL

14. *Sternopleural.* A line of about six hairs along the posterior border of the sternopleuron about its middle. The most dorsal hair situated more anteriorly.
15. *Pre-alar.* A cluster of stout chaetae (usually about twelve hairs) on the pre-alar knob.
16. *Upper mes-epimeral.* About six small hairs on the upper part of the mes-epimeron below the subalar sclerite. The lower mesepimeral chaetae are unrepresented.

Spiracular chaetae, and also the lower mesepimeral are absent. No chaetae are present on the metathorax.

REFERENCES

BARRAUD, P. J. (1934). *Fauna of British India. Diptera.* Vol. v, *Culicini*, p. 29.

CHRISTOPHERS, S. R. (1915). Pilotaxy of *Anopheles. Ind. J. Med. Res.* **3**, 362–70.

CHRISTOPHERS, S. R. (1933). *Fauna of British India. Diptera.* Vol. IV, *Anophelini*, p. 15.

CHRISTOPHERS, S. R., SHORT, H. E. and BARRAUD, P. J. (1926). The anatomy of the sandfly *Phlebotomus argentipes*. I. The head and mouth parts of the imago. *Ind. Med. Res. Mem.* no. 4, 177–204.

COMSTOCK, J. H. (1920). *An Introduction to Entomology.* Ed. 2. Comstock Publ. Co., Ithaca, New York.

CONSTANTINEANU, M. J. (1930). See references in ch. XIV (*d–e*), p. 351.

* A line of hairs in the middle line of the scutum, *acrostichal setae*, are absent in *Aëdes aegypti* and most *Stegomyia*. Both the dorso-central and the acrostichal setae are important taxonomically (Mattingly, *in lit.*).

REFERENCES

CRAMPTON, G. C. (1909). A contribution to the comparative morphology of the thoracic sclerites of insects. *Proc. Acad. Nat. Sci. Philad.* 1909, 1–53.

CRAMPTON, G. C. (1914). The ground plan of a typical thoracic segment in winged insects. *Zool. Anz.* **44**, 56–67.

CRAMPTON, G. C. (1917). The nature of the veracervix or neck region in insects. *Ann. Ent. Soc. Amer.* **10**, 187–97.

CRAMPTON, G. C. (1921). The sclerites of the head and the mouthparts of certain immature and adult insects. *Ann. Ent. Soc. Amer.* **14**, 65–103.

CRAMPTON, G. C. (1926). A comparison of the neck and prothoracic sclerites throughout the orders of insects etc. *Trans. Amer. Ent. Soc.* **52**, 199–248.

CRAMPTON, G. C. (1942). External morphology of the Diptera. A guide to the insects of Connecticut. Part VI. *Connect. State Geo. and Nat. Hist. Survey.* Bull. 64, 10–165.

DIMMOCK, G. (1881). *Anatomy of the Mouthparts and of the Sucking Apparatus of some Diptera.* Thesis, Leipzig. A. Williams and Co., Boston.

DIMMOCK, G. (1883). Anatomy of the mouth-parts and suctorial apparatus of *Culex*. *Psyche*, **3**, 231–41.

DYAR, H. G. (1928). See references in ch. II (*a–b*), p. 45.

EDWARDS, F. W. (1921, 1932, 1941). See references in ch. II (*a–b*), p. 45.

GORDON, R. M. and LUMSDEN, W. H. R. (1939). A study of the behaviour of the mouth-parts of mosquitoes etc. *Ann. Trop. Med. Parasit.* **33**, 259–78.

JOBLING, B. (1928). The structure of the head and mouth-parts in *Culicoides pulicaris* L. (Diptera: Nematocera). *Bull. Ent. Res.* **18**, 211–36.

KOMP, W. H. W. (1937). The nomenclature of the thoracic sclerites in the Culicidae and their setae. *Proc. Ent. Soc. Wash.* **39**, 241–52.

MACLOSKIE, G. (1887). Poison fangs and glands of the mosquito. *Science*, **10**, 106–7.

MACLOSKIE, G. (1888). The poison apparatus of the mosquito. *Amer. Nat.* **22**, 884–8.

MARSHALL, J. F. (1938). *The British Mosquitoes.* Brit. Mus. (Nat. Hist.), London.

MARSHALL, J. F. and STALEY, J. (1935). Generic and subgeneric differences in the mouth-parts of male mosquitoes. *Bull. Ent. Res.* **26**, 531–2.

MARTIN, J. F. (1916). The thoracic and cervical sclerites of insects. *Ann. Ent. Soc. Amer.* **9**, 35–83.

NATVIG, L. R. (1948). See references in ch. II (*a–b*), p. 46.

NUTTALL, G. H. F. and SHIPLEY, A. E. (1901–3). Studies in relation to malaria. *J. Hyg., Camb.*, **1**, 459–67; **3**, 167–70.

PATTON, W. S. and CRAGG, F. W. (1913). *Textbook of Medical Entomology.* London.

PATTON, W. S. and EVANS, A. M. (1929). *Insects, ticks, mites, etc.* Lpool Sch. Trop. Medicine.

PETERSEN, A. (1916). The head capsule and mouthparts of Diptera. *Illinois Biol. Monogr.* **3**, no. 2, 171–284.

PEUS, F. (1933). Zur Kenntnis der *Aëdes*-arten des deutschen Faunengebietes, etc. *Konowia*, **12**, 145–59. Quoted by Natvig, p. 10.

PRASHAD, B. (1918). The thorax and wing of the mosquito. *Ind. J. Med. Res.* **5**, 610–40.

ROBINSON, G. G. (1939). The mouth parts and their function in the female mosquito *Anopheles maculipennis*. *Parasitology*, **31**, 212–42.

SNODGRASS, R. E. (1912). The thorax of *Psorophora ciliata*. In Howard, Dyar and Knab, **1**, 55–9.

SNODGRASS, R. E. (1927). Morphology and mechanism of the insect thorax. *Smithson. Misc. Coll.* **80**, no. 1.

SNODGRASS, R. E. (1935). *The Principles of Insect Morphology.* McGraw-Hill Book Co., New York and London.

SNODGRASS, R. E. (1943). The feeding apparatus of biting and sucking insects affecting man and animals. *Smithson. Misc. Coll.* **104**, no. 1.

SNODGRASS, R. E. (1953). The metamorphosis of a fly's head. *Smithson. Misc. Coll.* **122**, no. 3.

THOMPSON, M. T. (1905). The alimentary canal of the mosquito. *Proc. Boston Soc. Nat. Hist.* **32**, 145–202.

VOGEL, R. (1921). Kritische und ergänzende Mitteilungen zur Anatomie des Stechapparats der Culiciden und Tabaniden. *Zool. Jb., Abt. Anat.* **42**, 259–82.

XIX

THE IMAGO: EXTERNAL CHARACTERS (cont.)

(*d*) THE THORACIC APPENDAGES

The appendages of the thorax are the three pairs of legs pertaining respectively to the pro-, meso- and metathoracic segments and named respectively first, second and third, or fore, mid and hind pairs, the wings pertaining to the mesothorax and the halteres, pertaining to the metathorax.

THE LEGS (Fig. 49)

Each leg consists of a basal segment, the *coxa*, a small linking segment, the *trochanter*, with *femur*, *tibia* and five-jointed *tarsus*, the last tarsal segment carrying the tarsal claws or *ungues*.

The coxae (Fig. 55 (3), (6), (7); 56 (1)). The relations of the coxae with the thorax have been described in the previous section. At their distal ends they articulate with the small second segment, the trochanter, the lower margin of each coxa being deeply excavated to allow for the presence of the trochanter when the legs are drawn up, that is, anteriorly for the fore-legs, externally for the mid-legs and posteriorly for the hind-legs. The three coxae carry a spaced row of setae respectively on their anterior, external and posterior aspects, and also a close row or cluster on their inner apices. They also all carry conspicuous patches of flat white scales, a V-shaped patch on the fore-coxa, two small patches externally on the mid-coxa and a large patch both externally and internally on the hind-coxa. Each of the coxae also carries one or more groups of sensory pits (*so*). One such is present on the first coxa anteriorly about the middle of its inner margin. On the middle coxa there is a double line of about four sensory pits each on the edge of its anterior border just below its attachment to the sternopleuron. On the hind-coxa there is a group of sensory pits on its anterior border near its base and a tiny group on the edge of the membrane about two-thirds of the way down posteriorly (Fig. 55 (6), (7); 56 (1) *so*). There is also a small group of sensory pits on the membrane behind the mid-coxa near the termination of the meron.

The trochanters. These are short cylindrical segments which articulate by a hinge joint with the coxa and by a peculiar ball joint with the femur, the rounded end of which projects into the trochanter. Each trochanter is more or less bent so that the planes of its articular surfaces are at an angle, thus assisting the femur in making the acute angle which it does with the coxa when in flexion by taking up some of the necessary arc. Near the upper margin of each trochanter is a small ridge on each side of which is a group of sensory pits (Fig. 56 (4), (9) A, *a*, *b*).

The femora. Each femur at its articulation with the trochanter has a somewhat contracted neck with a smooth convex, almost hemispherical, projecting head composed of conspicuously dark sclerotised cuticle, the result of the articulation almost amounting to a ball and socket joint (Fig. 56 (4) *e*). Just distal to the neck on the ventral surface of all the femora is a rounded swelling on which are two longitudinally arranged lines of sensory pits of a larger size than those so far mentioned (Fig. 56 (5) *b*). At their distal ends the femora articulate with the tibiae, a row of spines dorsally projecting over the joint. They also have sparsely spaced inconspicuous spines along their length projecting from among the scales, most marked in the apical portion and in the hind femora.

The tibiae. The tibiae at their articulation with the femora are somewhat constricted with two lateral convex darkly sclerotised expansions which are received into the hollow of the femur end and constitute what might be termed a dicondylic ball and socket hinge joint (Fig. 56 (6)). At their apices the tibiae, especially those of the fore- and hind-legs, are expanded and on the inner aspect of the fore- and hind-tibiae there is a triangular area of fine parallel hairs giving a comb-like effect and suggesting the cleaning brushes of some lemurs (Fig. 56 (8)). The tibiae also carry apical spines and small spines along their length.

Since the above was written and the drawing given in the figure made, Mattingly (*in lit.*) informs me that distal to the triangular area referred to there exists at the apex of the fore-tibia an extremely regular 'tibial rake' consisting of a row of straight closely packed spines strongly attached at their base. These are best seen after scraping away the triangular patch of spines which to some extent overlies them. Mattingly and Hamon (1955) also describe at the apex of the fore-tibia in many groups including some *Aëdes*, but not so far in *Stegomyia*, a small spine that may resemble (in some *Ficalbia*) a tarsal claw. A study of the presence of the claw in various genera has been made by Mattingly and Grjebine (1957).

The tarsal segments. These are five in number, decreasing in size from the first to the fifth, except in the male where in the fore- and mid-legs the fourth segment is unduly short, little more than half the length of the fifth (Fig. 56 (2) D). The segments are for the most part long, narrow and evenly cylindrical and, especially in the hind-legs, carry fairly numerous spines along their length. The fifth segment carries the *tarsal claws* or *ungues*. The tarsal joints are of the hinge type with a long anteriorly projecting spur on the proximal end of the distal segment which extends some distance into the preceding segment and receives the long tendon to the segment (Fig. 56 (7) A, B).

The tarsal claws (Fig. 56 (2), (3)). These differ in the male and female and on the different legs. Their characters have considerable systematic importance as well as indicating the sex. Thus the claws in the female on some of the legs at least are usually toothed in *Aëdes*, but simple in *Culex*. The development of pulvilli into pubescent lobes characteristic of *Culex* is also restricted to this genus. The following is the arrangement in the two sexes in *Aëdes aegypti*.

	Fore-leg	Mid-leg	Hind-leg
Male	One very large outer. Uniserrate	One very large outer. Simple	Two small simple
	One large inner. Simple	One large inner. Simple	—
Female	Two large. Uniserrate	Two medium sized. Uniserrate	Two small simple

The claws, one inner and one outer, are borne on membrane at the end of the terminal tarsal segment, which is hollowed out to some extent with a thickened rim where the membrane is attached. Between the claws in the middle line of the membrane ventrally is a thickening marked with transverse ridges which forms a pad and continues as a tuft of hairs of considerable length in the fore- and mid-ungues of the male. This constitutes the *empodium* (Fig. 56 (2) A *b*). There does not appear to be any structure that can be taken as representing the *pulvilli*. At the base of the tarsal claws on the dorsal margin of the tarsal segment is a thickening to which the claw is attached, the *unguifer* (Fig. 56 (2) A *a*). The ventral margin of the segment is also more or less thickened forming, in the fore-tarsus of the male what amounts to a blunt spur (Fig. 56 (2) A (6)).

Ornamentation of the legs (see under 'Ornamentation' in ch. XVII, p. 400) is of importance in identification of species, especially the presence and character of tarsal banding and to a less extent of markings on the femora and tibiae. The markings are due to scaling, but the pale and dark colour of the underlying cuticle often accords with this. Figures of the leg ornamentation of *A. aegypti* and some other species of *Stegomyia* are given by Edwards (1941).*

THE WING (Fig. 57)

For good short general accounts of the wing and its attachments in insects see Comstock (1920); Snodgrass (1935); Imms (1938). For description of the mosquito wing and its venation see Nuttall and Shipley (1901); Shipley and Wilson (1902) (base of wing); Prashad (1918); Christophers and Barraud (1924) (pupal wing tracheation); Christophers (1933); Barraud (1934); Edwards (1941); Natvig (1948). For description of the basal sclerites and attachments of the mosquito wing see Prashad (1918) who is the only author who has made a detailed study of the mosquito wing in this respect.

As in other mosquitoes the wings in *A. aegypti* are rather narrow structures about four times as long as their greatest breadth and shaped not unlike an aeroplane propeller blade. They consist of thin membrane, *wing membrane*, strengthened by lines of thickening, nervures or *veins*, the arrangement of which constitutes the *venation*. The membrane is free from scales, but is closely covered with minute hairs, *microtrichae*. The veins are

* For the structure of the insect leg in general see Snodgrass (1935), pp. 193–209; also Dahl (1884), Crampton and Hasey (1915), Gillett and Wigglesworth (1932), Imms (1938). For the general plan of the arthropod leg and nomenclature of the parts see Ewing (1928), Snodgrass (1935), pp. 83–99. A simple account of the parts in the crayfish is given by Huxley (1906). On the tarsal claws in mosquitoes see Edwards (1912).

Figure 56. Thoracic appendages: the legs.

1 Posterior view of the mesosternal area and mid-coxae. *a*, furca 2; *b*, mesosternum; *c*, transverse apodeme.

2 Male ungues. A, fore-leg; B, mid-leg; C, hind-leg; D, fore-tarsus (segments 4–5). *a*, unguifer process; *b*, empodium; *c*, spur.

3 Female ungues. Lettering as for 2.

4 Trochanter and head of femur of fore-leg. Anterior view. *a–d*, different groups of sensory pits; *e*, sclerotised extension of femur.

5 Head of femur of hind-leg. Ventral view. *a*, sclerotised extension of femur; *b*, sensory organ shown at 4*c* more highly magnified.

6 Femoro-tibial joint. Fore-leg.

7 Tarsal joint. A, lateral view; B, dorsal view.

8 Inner aspect of apex of fore-tibia showing hair organ.

9 Sensory organs. A, of trochanter as shown at 4*a*, *b*; B, of coxa as shown at 1*so*; C, of coxa as shown at Fig. 55 (6).

10 Coronal section through paired ocelli of frontal area. Newly emerged female.

Lettering: *cx.*2, mid-coxa; *oc'*, cuticular thickening resembling lens; *so*, sensory pits.

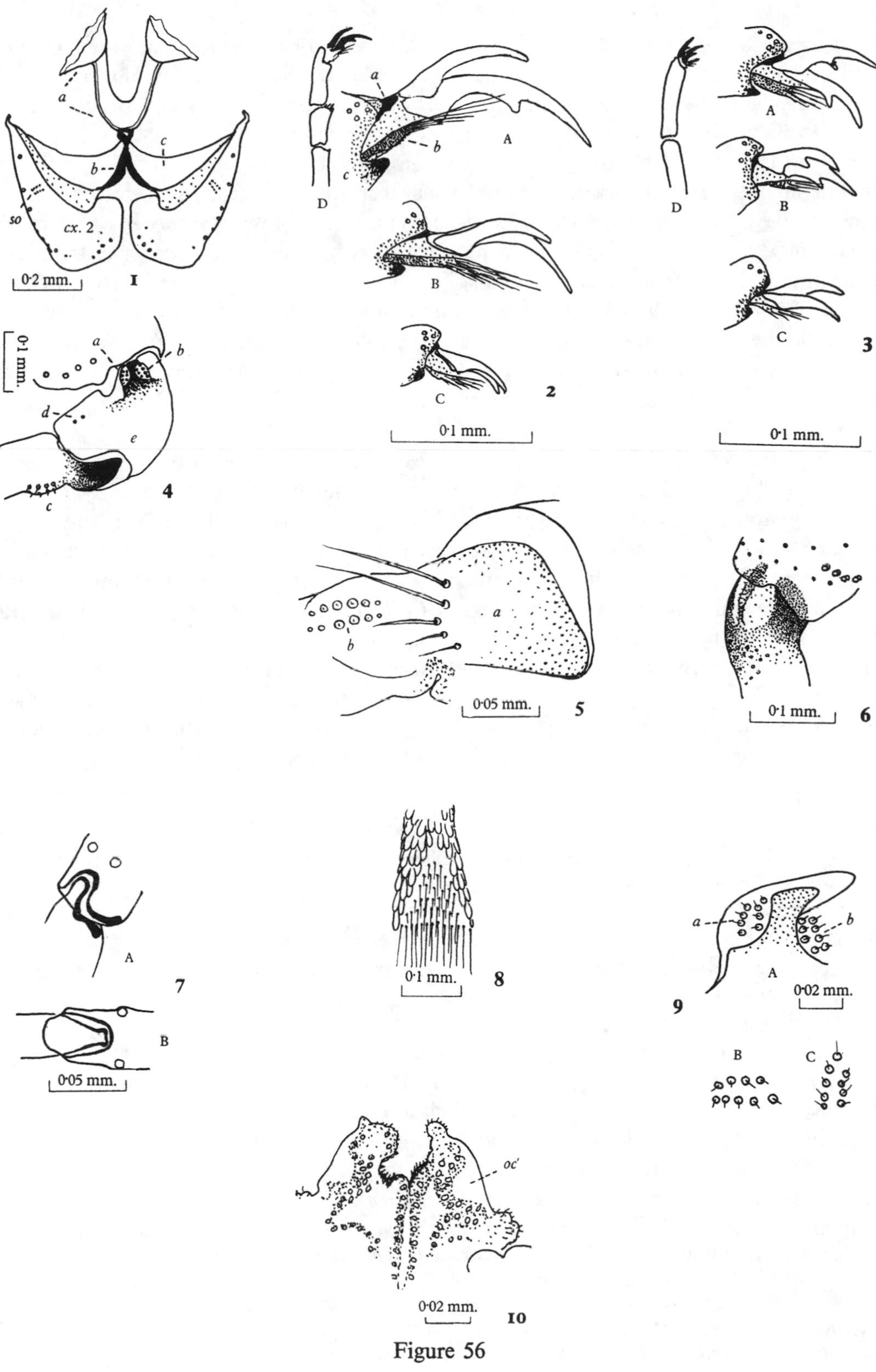
a
b
c
so
cx. 2
0·2 mm.
1
a
b
c
D
A
B
C
2
0·1 mm.
A
D
B
C
3
0·1 mm.
0·1 mm.
a
b
d
e
c
4
a
b
0·05 mm.
5
0·1 mm.
6
A
7
B
0·05 mm.
0·1 mm.
8
a
b
A
0·02 mm.
9
B
C
oc'
0·02 mm.
10
Figure 56

clothed with scales forming the *wing vestiture*, which is often characteristic for particular genera or species. The male wing is in general similar to that of the female, but besides being smaller is somewhat narrower in proportion to its length and has a broader apical fringe. It differs also in some other respects as mentioned later.

The wing is attached to the thorax largely through the intervention of a complicated system of small sclerites as described later. It has a straight thickened scaled anterior edge formed by the *costa* (*C*), a rounded *apex* (*apx*) and a convex *posterior border* which carries a broad *fringe* (*fr*). Towards the base of the wing the convex posterior border terminates at a notch, internal to which is a short, almost bare and only slightly convex portion of edge, the *alula* (*aa*), and internal to this a much more convex small lobe fringed with long hairs, which in the flexed position of the wing is sharply folded, the *squama* (*sq*). The terms here given to these two structures are as used by Edwards (1941) and as there has been in the literature some uncertainty as to which term applies to which a few words may be said on this point.

The term squama was first used by Sharp (1899) and first applied to the mosquito by Nuttall and Shipley (1901). The latter authors state that in doing so they are following Sharp and give a reference to this author as 1895 (Cambridge Natural History). In vol. VI of this work Sharp (1899) gives a figure after Schneider where the single basal lobe in the wing of an acalypterate fly is named alula. Squama was a new name here given by Sharp to a lobe internal to this which he figures in the wing of *Calliphora*. This, as in the wing of the mosquito, is characterised by a fringe of conspicuous hairs.

The fringe in *A. aegypti* consists of a single row of long fusiform erect scales, *fringe scales*, set along the dorsal surface of the wing edge, and in the intervals between these smaller shorter scales of the same type, *secondary fringe scales*, set along the ventral edge. In addition in the female, set along the edge external to these, one line ventral and one dorsal,

---

Figure 57. The wing.

1 Wing of female showing venation and scaling. Nomenclature as in systematic works. *a*, humeral cross-vein; *b*, subcostal cross-vein; *c*, cross-vein 1–2; *d*, supernumerary cross-vein; *e*, mid cross-vein; *f*, posterior cross-vein.

2 Wing of male on same scale. Nomenclature on generalised insect plan.

3 Portion of fringe of female wing. *a*, fringe scale; *b*, secondary fringe scale; *c*, ventral basal scale; *d*, dorsal basal scale.

4 The same of male wing. Ventral.

5 Portion of subcosta showing broad scaling of both aspects.

6 Portion of vein 2 showing scaling of ventral aspect. *a*, median squame scale; *b*, lateral squame scale; *c*, plume scale (showing through membrane).

7 Portion of vein 3. Ventral aspect. Lettering as in 6.

8 Base of wing showing remigium and other basal structures.

9 Base of remigium more highly magnified.

Lettering: *An*, anal vein; *aa*, alula; *apx*, apex of wing; *art*, articular sclerite; *C*, costa; *Cu*, cubitus vein; *cn*, conoid; *cn′*, ulnoid; *crc*, coracoid; *epu*, epaulet; *fk*, anterior forked cell; *fk′*, posterior forked cell; *fr*, wing fringe; *inc*, incrassation; *M*, media vein; M_1, M_2, M_3, branches of *M*; *m-cu*, median cubital cross-vein; *pwr*, posterior wing root; *R*, radial vein; R_1, R_2, R_3, R_4, R_5, branches of *R*; *Rg*, remigium; *Rs*, radial sector; *r-m*, radio-median cross-vein; *Sc*, subcosta; *Sc′*, lobe at base of subcosta; *sab*, sabroid; *sc*, scutum; *sep*, subepaulet; *sq*, squama; *tg*, tegula.

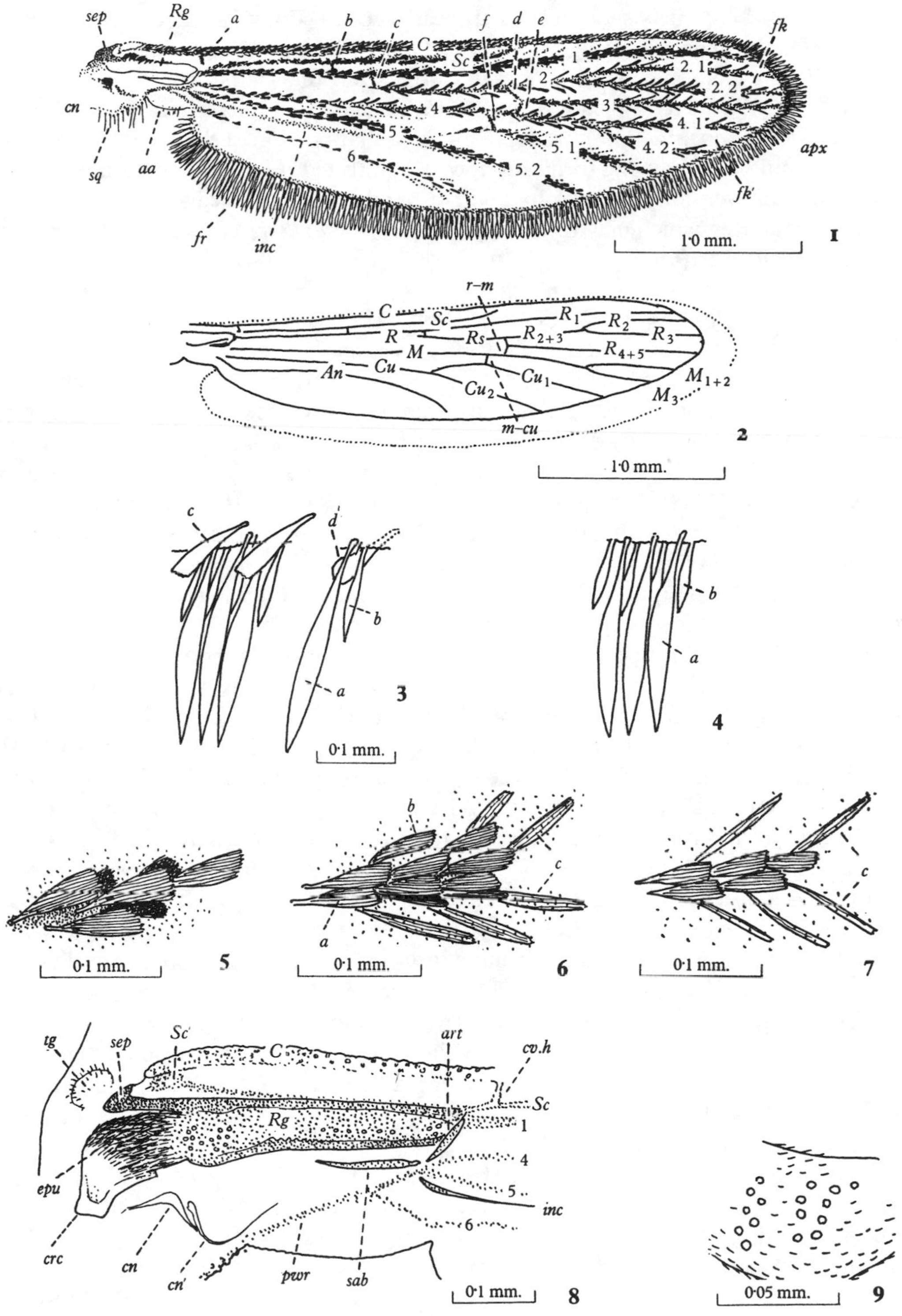
sep
Rg
a
b
c
f
d
e
fk
C
Sc
1
2
2. 1
2. 2
3
4
4. 1
4. 2
5
5. 1
5. 2
6
cn
sq
aa
apx
fk'
fr
inc
1·0 mm.
I
r–m
C
Sc
R1
R2
R
Rs
R2+3
R3
M
R4+5
An
Cu
Cu1
Cu2
M1+2
M3
m–cu
2
1·0 mm.
c
d
b
a
3
0·1 mm.
b
a
4
b
c
a
c
0·1 mm.
5
0·1 mm.
6
0·1 mm.
7
tg
sep
Sc
C
art
cv.h
Sc
Rg
1
4
5
epu
inc
6
crc
cn
cn'
pwr
sab
0·1 mm.
8
0·05 mm.
9

Figure 57

is a single line of obliquely set small flat truncated scales (Fig. 57 (3) *c*, *d*). These last are not present on the male wing (Fig. 57 (4)). The alula, like the posterior wing border, carries some small and inconspicuous fringe scales with a few fine hairs. The squama, on the contrary, has a line of conspicuous large hairs arising from small basal papillae. This description of the squama, however, applies only to that portion of the basal fold which in the flexed position of the wing forms an upwardly directed outer fold. Internal to this is a long length of ventrally directed fold with a peculiarly marked edge resembling somewhat a coiled spring which may be distinguished as the *axillary cord* and will be referred to again later (Fig. 58 (4)).

The wing venation (Fig. 57 (1), (2)). The venation of the wing in *A. aegypti* is of the usual Culicid type and only in small details differs from that in the great majority of mosquitoes. Compared with the venation in many Diptera the mosquito venation is of relatively simple character and whilst a notation following that of Comstock and Needham for the generalised insect wing was early put forward by Nuttall and Shipley and has with some minor modifications been accepted by later authors, see Edwards (1941), and still more recently by Natvig (1948), it is more usual in systematic works on the mosquito to employ a simpler notation in which the six main longitudinal veins are indicated by numbers in the order of occurrence of the veins from before backwards. The method of numbering is shown in Fig. 57 (1) and may be briefly indicated, leaving the question of the nomenclature of the veins in accordance with the generalised scheme to follow.

The *costa* (*C*), forming the anterior edge of the wing, extends the whole length of the wing to the apex where it is continued into a thickening along its posterior border. The *subcosta* (*Sc*) joins the costa a little external to its middle. The *first longitudinal* vein (1) lies parallel to the costa and extends without break to the wing apex. The *second longitudinal* (2) begins about the middle of the wing where it is linked by a cross-vein to the first longitudinal. Towards the apex it branches into 2·1 and 2·2, forming what is commonly referred to as the *anterior forked cell*, the relative length of which to its stem may be of diagnostic importance in some groups. The *stem* is that portion of vein 2 which lies between its bifurcation and the cross-vein linking it to vein 3 (*cv*. 2–3). In *A. aegypti* this ratio is about 2·5. Vein 4, though it extends from the base of the wing, in some respects resembles vein 2. At about the same level it bifurcates, forming the *posterior forked cell*. Between these two forks lies the short straight vein 3, linked at its inner end by cross-veins to veins 2 and 4, the general picture of two simple forked cells with a short vein 3 lying between these being almost diagnostic of the mosquito wing. Vein 5 also branches in a characteristic way, forming 5·1 and 5·2, and behind this is the unbranched vein 6 joining the wing edge at about its middle and marking off posteriorly and basally a portion of wing sometimes referred to as the anal area.

Besides the true veins which are scaled there are certain thickenings which are not scaled and have not the appearance of true veins. Three such are noted in the wing of *Anopheles* by Prashad, namely a *medial fold* along the upper edge of vein 5, a *cubital fold* between 5 and 6 and an *anal fold* posterior to 6. The last two and traces of the first can usually be seen in the wing of *Aëdes aegypti*, especially if stained. The thickening behind 5 is the most conspicuous and runs close behind vein 5 and its posterior branch 5·2 (Fig. 57 (1), (8) *inc*).

Of some importance too are the *cross-veins*. These are best shown in the dry wing suitably flattened or in the scale-denuded stained mounted wing. In the canada balsam mounted wing unstained they are only with difficulty made out. In the mosquito wing these are six in number, namely the humeral (Fig. 57 (1) *a*) near the base of the wing and linking the costa with the subcosta; the subcostal (*b*) linking the subcosta about its middle to vein 1; the cross-vein linking vein 2 at its base to vein 1; and cross-veins *cv.* 2–3, *cv.* 3–4 and *cv.* 4–5. The last three mentioned are much the most conspicuous of the cross-veins and their relative positions have been given some systematic importance. They are respectively the *supernumerary*, the *mid* and the *posterior* cross-veins.

The subcostal cross-vein is not easily visible since it is liable to be concealed by the relative positions of the subcosta and vein 1 and is often not very conspicuously marked. Yet, though holding no recognised position in the generalised scheme as at present applied, it is always present and in the developing wing is the largest of all the cross-veins (see Christophers and Barraud, 1924).

Application of the generalised scheme of wing venation to that of the mosquito. In the generalised plan of insect wing venation there is recognised, in addition to the costa and subcosta and a variable number of anal veins, the following three veins and their branches, namely the *radius* (*R*), the *media* (*M*) and the *cubitus* (*Cu*). The radius is regarded as having one branch, R_1, which takes off anterior to the main vein, and a branch, *radial-sector*, (R_s), which takes off posteriorly and terminates typically by dividing dichotomously into R_{2+3} and R_{4+5} and finally into the four branches, R_2, R_3, R_4 and R_5. The media similarly, after giving off an anterior branch dividing into MA_1 and MA_2, divides into M_{1+2} and M_{3+4}, with M_1–M_4 if there is further division. The cubitus divides into Cu_1 and Cu_2. There may be anal veins, An_1, An_2, etc. This notation as it has been applied to the mosquito wing, following Edwards (1941) and Natvig (1948), is shown in Fig. 57 (2).

It will be observed that only three of the six cross-veins actually present appear in the generalised scheme, namely the humeral cross-vein, *cv. R–M* and *cv. M–Cu*. The others are considered (except *cv. Sc–R* which is not included) merely as cross-veins developed near where the veins take origin, their original connection being lost.

The generalised scheme is based on development, and especially early tracheation, of the wing in primitive insects. Its application to the mosquito from such a point of view has been dealt with by Christophers and Barraud (1924), who have studied developmental changes and tracheation as seen in the wing in the larva and pupa. In the mature larva the wing is a thick flap-shaped organ in which the course of the forming veins and the tracheation accompanying these is clearly traceable (Fig. 59 (1)). The whole wing at this stage is surrounded by a cord-like thickening with a channel similar to those which will eventually become veins and which links up the terminations of veins where these reach the wing periphery. On the wing field are tracts indicating the formation of the future veins and cross-veins. The arrangement differs from that in the imaginal wing in that in the place of longitudinal veins linked by narrow cross-veins there is a broad arc from which the terminal portions of the future veins pass to the marginal channel. The anterior portion of this arc clearly represents the future radius and from it there pass to the marginal channel vein 1 (the portion beyond the cross-vein area), vein 2 (as represented by the fork) and vein 3. The posterior portion of the arc forms a broad hand-shaped area occupying the hinder basal half of the wing (Fig. 59 (3) *b*). From it strands of future vein tissue pass to the marginal channel representing vein 4 (the fork), vein 5 with its bifurcation and vein 6, also two shorter arms basal to this. The condition therefore represents very clearly the generalised plan and shows that the cross-veins are not secondary structures but the shrunken remains of the original broad arc.

Evidence from the tracheation is less straightforward. The tracheal leashes in the wing in the larva consist of an anterior bundle of fine tracheae in the subcostal area (*a.tr*) and one in the cubito-anal area (*b.tr*). The first of these gives a branch to the costa, but this fails to reach even as far as the subcostal junction. The branch to the subcosta, however, extends to the margin of the wing apex. Other branches pass through the wide *cv.-Sc* cross-vein area to veins 1 and 2. The cubito-anal leash sends branches to veins 4, 5 and 6, but also may send others, crossing branches from the subcostal leash, to veins 2 or 3. This last feature is not so remarkable as might at first sight appear since the wing at this stage is bag-like with the dorsal and ventral surfaces widely separated. There is even some indication that tracheal leashes may be dorsal or ventral in nature. Thus the subcostal leash is rather ventral and tends to supply the ventral surface, whilst the cubito-anal is dorsal. The most striking feature in later development is the formation in the pupa of a new tracheal branch from the thorax from which there develops an entirely new leash, the radial. Representation of the media seems doubtful.

The wing scaling (Fig. 57 (5), (6), (7)). The wing scaling in *A. aegypti* shows few characteristics other than those commonly present in many species of mosquito. All the longitudinal veins are scaled. In this respect two types of vein occur which have relation to some degree of plication of the wing, namely convex or *direct* veins which run along the crests of folds directed dorsally and concave or *reverse* veins having an opposite relationship. On the upper surface of direct veins and the lower surface of reverse veins the scaling consists of flat, usually rather broad and abruptly truncated overlapping scales, *squame scales*. In most mosquitoes these are typically of two kinds, namely those arranged longitudinally on the vein, *median squames*, usually short and overlapping, and those projecting laterally over the membrane, *lateral squames*, which are longer but still lie relatively flat. On the lower surface of direct veins and upper surface of reverse veins the scales are longer, narrower and more pointed and project away from the vein, *plume scales*. The direct veins in mosquitoes are: the costa (up to the subcostal junction) and veins 1, 3, 5 and 6. Veins 2 and 4 are reverse veins. The subcosta in position appears as a reverse vein, but the scaling, in *A. aegypti* more resembles that of a direct vein. Usually the scaling tends to become narrower and longer and the distinction of forms less marked towards the apex.

The scaling in *A. aegypti* is rather narrow and without any outstanding specific characters. The median squame scales over most of the wing have about 6–9 striations, reaching about 10 towards the front of the wing. Plume scales have about 3–6 striations. Squame scales measure in the female 80–100μ in length and the plume scales 150μ or more. Some of the veins tend to show particular characters. Thus the subcosta has broad short scales on both aspects (Fig. 57 (5)) and vein 1 has the lateral squames well developed (Fig. 57 (6) *b*) which are not very marked on many veins in the species. The scales are uniformly dark showing no wing ornamentation except for a minute white spot at the extreme base of the costa.

Scaling in the male is on the same plan, but much less profuse and the scales are smaller, the squame cells having at most about eight striations and measuring about 60μ in length, whilst plume scales do not much exceed three striations and measure only up to about 100μ. The pale spot at the base of the costa is also present in the male.

The wing base (Figs. 57 (8); 58). Apart from the neuration of the wing as seen in the distal portions of this organ and already described, there are important structures at the base of the wing still requiring notice. Structurally the wing consists of two somewhat independent parts, an anterior more rigid costo-radial portion and a posterior more labile cubito-anal region. Movements of the former are largely controlled by those basal parts forming what may be termed the *anterior wing root*. The latter is intimately connected with a complex line of membranous wing and thickenings forming the *posterior wing root*. The

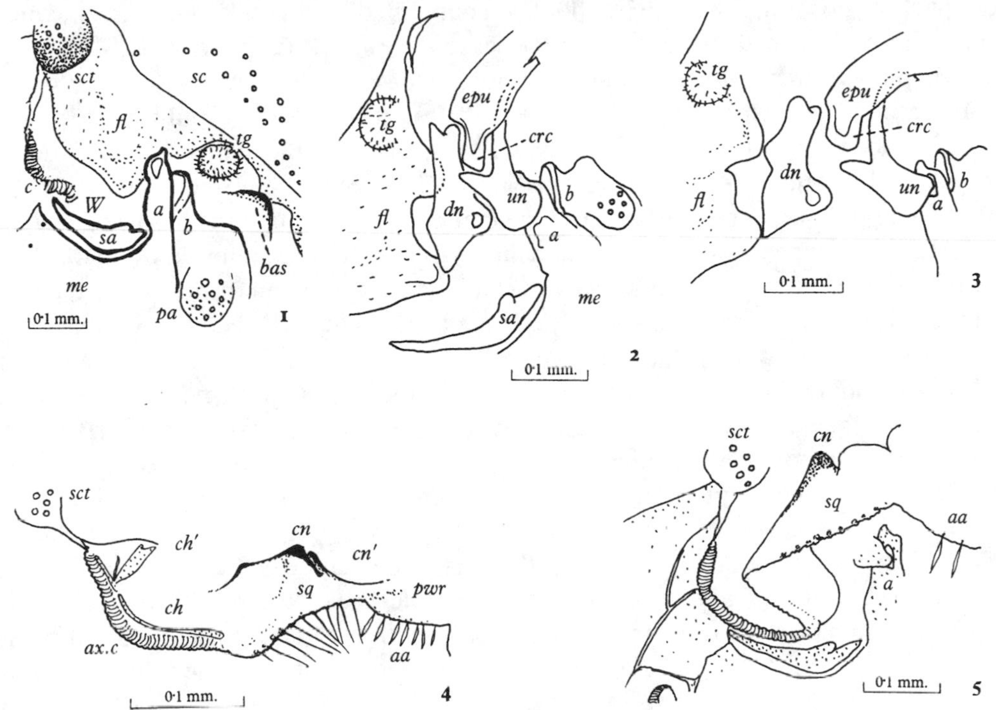

Figure 58. The wing root.

1 Showing pleural attachment of the wing. The thicker lines show the line of ventral attachment in a plane nearer to the observer than the thinner line of dorsal attachment. *a*, hamuloid process; *b*, unguoid process; *c*, point at which the axillary cord becomes the free wing edge.

2 Showing basal sclerites. Lettering as in 1.

3 The same preparation as 2 drawn apart.

4 Membranous posterior wing root as drawn out on the flat.

5 The same in natural position.

Lettering; *aa*, alula; *ax.c*, axillary cord; *bas*, basalare; *ch*, thickening towards edge of axillary cord portion; *ch'*, thickening at extreme base of fold; *Cn*, conoid; *Cn'*, ulnoid; *crc*, coracoid; *dn*, dens; *epu*, epaulet; *fl*, flange of scutum; *me*, mesepimeron; *pa*, prealar knob; *pwr*, posterior wing root; *sa*, subalar; *sc*, scutum; *sct*, scutellum; *sq*, squama; *tg*, tegula; *un*, unguiculus; *W*, area occupied by wing attachment.

latter term is applied by Prashad to the line of thickening formed by the basal junction of veins 4, 5 and 6, but is here used also in a wider sense to include other structures in this region.

Although the *costa* is seemingly the most rigid vein in the wing it plays little part in giving basal support to this organ, since at the base of the wing it is neither continued into, nor articulates with, any further basal structure, but ends in membrane close to, though not in contact with, the small sclerite named by Prashad the *subepaulet* and which appears to be the *humeral plate* of Snodgrass (Fig. 57 (8) *sep*). Nevertheless, though there is no actual articulation in this position, the subepaulet is hollowed where it faces the end of the costa and may act as a check to this structure during extreme extension of the wing. Thus the costa is largely free basally so far as rotation or movements within certain limits are concerned. The anterior wing root, which forms as it were a handle by which the movements of the fore part of the wing are brought about, is constituted by a conspicuous thickening of the wing lying posterior to the costa and termed the *remigium* (Fig. 57 (8) *Rg*).

The remigium is essentially a strengthened portion of wing. It appears in dorsal view as a double structure consisting of two bar-like thickenings, an anterior and a posterior, the general effect being not unlike a long brooch, the posterior thickening forming the pin. Actually it is a bent-up area of wing convex dorsally and deeply hollowed ventrally. It has, therefore, an upper and under surface. The two bar-like thickenings are the thickened edges of this hollow. The space between the thickenings is the roof of the hollow. Along the dorsal surface of the anterior bar, besides scales, which on its basal portion form a marked cluster, there are a number of circular sensory pits. These form a cluster basally and a line of pits more distally (Fig. 57 (8), (9)). The posterior bar is chiefly noticeable as having on its inner aspect, that is facing the hollow, about its middle third, a row of rounded ridges, which, depending how they are viewed, appear as rounded teeth or parallel grooves. It was the peculiar appearance of these regular ridges that led Shipley and Wilson, who very thoroughly investigated these parts in *Anopheles*, to consider that they might form a stridulating organ. In *Aëdes aegypti* there are some fifteen or so of these rounded ridges confined to the posterior bar. Sometimes in viewing the parts it may appear that there are ridges on the anterior bar. But critical examination has always negatived this and there appear to be no ridges in opposition. It is doubtful what their function is, but Prashad considers that they prevent sliding of the parts upon each other in a longitudinal direction during rotation of the wing.

The appearance of the remigium suggests that its bars represent two wing veins, possibly the subcosta and first longitudinal. At its distal end the remigium is joined by both these veins, but the subcosta, after passing close alongside the anterior bar, leaves this to end in a small thickening on the ventral aspect of the wing just below the inner end of the costa (Fig. 57 (8) *Sc′*).

Closely associated with the remigium are two further thickenings, namely one running parallel to the remigium a little posterior to it and a small sclerite linking this to the remigium, respectively the *sabroid* (*sab*) and the *articular* (*art*) of Prashad.

The term *posterior wing root* has been applied by Prashad to the line of thickening which results at the base of the wing from the linked up bases of veins 4, 5 and 6. This thickening, which resembles a vein without scales, passes to the wing border at the junction of the alula and squama (Fig. 57 (8) *pwr*).

Functionally the posterior wing root is formed by a long line of membranous wing formed from within outwards by the alula, squama and a region internal to the squama bounded on its external margin by a curious cord-like structure having somewhat the appearance of a length of spiral spring, the *axillary cord* (Fig. 58 (4) *ax.c*). The margins of these three portions are characterised respectively by small fringe scales (alula), long hairs with conspicuous bases (squama) and a long length of axillary cord. When stretched out this basal membranous area is surprisingly long (Fig. 58 (4)). But even when the wing is extended it is thrown into folds which are rigidly and exactly maintained by thickenings in the fold to be mentioned shortly. The alula is not affected, but the squama, defining this part as above, forms a flat, upwardly directed fold and internal to this is a deep ventral fold formed by the portion margined by the axillary cord (Fig. 58 (5)). The thickenings in the area in *Anopheles* have been termed by Prashad from within outwards, the *intermediary*, the *conoid* and the *ulnoid*. In *Aëdes aegypti* the conoid (*cn*) is very conspicuous owing to its almost black coloration, usually an indication that the part is dense and elastic. External to the conoid is a line of thickening (*cn'*) which corresponds fairly well to Prashad's description of the ulnoid. This forms a line from just external to the conoid to the wing margin near the junction of the alula and squama. There is a further line of thickening internal to the conoid which demarcates the inner edge of the folded squama and also a long line of thickening lying just internal to the axillary cord which serves to maintain the downward projecting bulge of membrane internal to this. There is also a rather plate-like thickening which arises near the posterior end of the subalar plate and further gives rigidity to the fold (Fig. 58 (3), (4) *ch*, *ch'*). There can be little doubt but that these different thickenings act as an elastic control drawing this membranous portion of wing into precisely determined folds when the wing is flexed and even assisting in the complete flexion of the wing.

The basal sclerites (Fig. 58 (1), (2), (3)). The wing is articulated with the thorax largely through a complex of small basal sclerites (*pterales* of Snodgrass) which very much call to mind the carpal bones of the human wrist. They have been described and individually named in *Anopheles* by Prashad and whilst some may be homologised with the axillary sclerites of the generalised insect wing there are others which are not readily so homologised and for which the names as given by Prashad must provisionally be used.

The most readily identified and largest of these sclerites lies at the base of the remigium with which structure it is largely continuous, namely a dark flattish shield-shaped hairy plate, which from its position and relations is the *epaulet* (Figs. 57, 58 *epu*), and possibly the *second axillary* of the generalised insect plan. Fused with the anterior border of the epaulet is a small sclerite which is hollowed facing the end of the costa and can be identified as the *subepaulet* (*sep*), and probably the *humeral plate* of Snodgrass. Situated where the inner margin of the epaulet bends deeply down against the thorax and articulates with the flange-like extension of the scutum described later is the *dens*, possibly the *first axillary* of the generalised plan. Though the dark hairy surface of the epaulet ceases at the apparent inner margin of this sclerite its substance is more deeply continued as a snout-like process, the *coracoid* of Prashad (*crc*). This process lies between the diverging fang-like arms of a large sclerite which has in certain positions a remarkable resemblance to an inverted molar tooth, the *unguiculus*, probably the *third axillary*. Whilst the main articulation of the dens is with the scutal flange, that of the unguiculus is with the hamuloid process of the mesepimeron, which as described later with the unguoid process of the mesosternum forms the pleural wing process or main ventral articulation point of the wing with the pleura. Whilst the head of the unguiculus articulates as described, its two fang-like processes pass dorsally, the more external continuing under and fusing with the posterior border of the epaulet. The irregular shape of these sclerites and their complicated articulations make it difficult to follow or describe them in detail, but much information will be found in the paper by Prashad. It is clear, however, that in essence they are thickenings of the upper and lower cuticular surface of the wing and their complete relationships are most likely to be revealed by tracing their formation in the pupa.

The pleural attachments of the wing (Fig. 58 (1), (2), (3)). The wing takes origin from the thorax at the meeting place of a number of thoracic sclerites, namely dorsally the scutum towards it posterior end, anteriorly the upper portion of the sternopleuron in the region of the pre-alar knob, ventrally the mes-epimeron, and posteriorly, through the axillary cord, the scutellum (Fig. 58 (1)).

The most conspicuous feature dorsally is the broad smooth *flange* (*fl*) projecting downwards almost vertically from the margin of the scutum. This flange is devoid of hairs or scales and is normally hidden under overhanging chaetae arising from the scutal prominence lying over the wing root. Its lower border articulates with the bent down inner end of the wing base forming the main dorsal articulation of the wing. Ventrally the main feature is the upwardly projecting pleural wing process. This is formed anteriorly by a somewhat claw-like process from the parts in the region of the pre-alar knob, the *unguoid process* (*b*) and posteriorly by the *hamuloid process* (*a*). The hamuloid process is a projection from the upper anterior edge of the mes-epimeron. It is a very characteristic structure in appearance and useful in working out relationships of the basal sclerites. Lying in the membrane posterior to the two conjoined processes is an isolated crescentic sclerite situated above the upper edge of the mes-epimeron, but separated from this by a narrow membranous gap. This is one of the accessory sclerites, epipleurites or paraptera and being situated posterior to the pleural wing process can be identified as the *subalar* of Snodgrass (*sa*). What may be the *basalare epipleurite* is an isolated sclerite situated anterior to the unguoid process and a little dorsal to the pre-alar knob. It is shaped rather like a limpet shell and faces rather inwards (*bas*). A little posterior to it at the edge of the tergite and near the subepaulet is a curious little hair lobe, the tegula (*tg*).

Briefly the wing root dorsally, through the dens, is articulated with the downwardly projecting scutal flange representing the notal wing processes of the generalised insect plan, and incidentally forming a deep gutter-like depression between the wing base and the thorax. Ventrally the wing articulates with the upwardly projecting strong pleural wing process reaching to about the level of the scutal flange, but at some distance external to this. The result when the wing is extended is a lever system with what may be regarded as the fulcrum at the pleural process and the force operating at the end of the short arm of the lever at the scutal flange, the power being provided by the indirect wing muscles dragging downwards the scutum. This, however, is by no means the whole story of the wing mechanism which will be further dealt with in the chapter on flight.

THE HALTERES

The halteres of Diptera represent the hinder wings of other insect orders adapted as sense organs concerned with co-ordination in flight (see Pringle, 1948, and chapter XXIII on flight). In general structure the halteres of mosquitoes resemble those of other Diptera. They consist of a knob-like head, *capitellum* (Fig. 59 (3) *cap*), a narrow stalk or neck, *scape* or *pedicel* (*sca*) and a dilated basal portion, *scabellum* (*scb*). They are articulated along the line of junction of the metathoracic pleuron and the lateral portion of the metanotum a little posterior to the metathoracic spiracle by a hinge-like joint allowing movement in a dorso-ventral elliptical curve. The sensory structures they carry consist of areas of sensilla arranged in rows, namely one dorsally at the base of the scape, *dorsal scapal plate* (*scala superior*) (*scs*), one situated dorsally proximal to this, the *basal plate* (*cupola*) (*cup*) and one on the ventral surface opposite to the dorsal scapal plate, *ventral scapal plate* (*scala inferior*) (Fig. 80 (4)). The names in brackets are those employed by Prashad (1916) adopting Lowne's nomenclature, those not so enclosed are more recent terms used by Pringle in his description of the halteres of *Calliphora*. The dorsal and ventral scapal plates consist of rhomboidal sensilla, the basal plate of circular sensilla. For further information on these structures see section given later on sense organs. In addition to these, Pringle in *Calliphora* records two further areas of sensilla of smaller size, one situated on the dorsal surface close to the basal plate on its anterior side, *dorsal Hick's papillae*, and a still smaller group on the ventral surface of the scabellum, *ventral Hick's papillae*. There is also a simple large papilla just anterior to the distal portion of the dorsal scapal plate, *undifferentiated papilla*. Contained in the haltere but not appearing externally is probably a large and a small chordotonal organ (see under 'Special sense organs', p. 663).

In the mosquito the halteres have been described by Prashad (1916) in *Ochlerotatus pseudotaeniatus* Giles (*Aëdes* (*Finlaya*) *pseudotaeniatus* Giles); see also Nuttall and Shipley (1903); Shipley and Wilson (1902); Shipley (1915); who give a brief description of these organs in *Anopheles maculipennis*.

In *Aëdes aegypti* the halteres of a number of specimens of normal size gave as mean length for the female 0·37 mm., with range 0·34–0·4 mm. In the male the mean was 0·3 mm. with range 0·29–0·33. This compares very well with a length of 0·4 mm. given by Nuttall and Shipley for the rather large species of *Anopheles*, *A. maculipennis*, and 0·56–0·6 mm. for the rather large Culicid studied by Prashad. It is also not very different in proportion to the halteres in *Tipula*. As given in the figure of natural size given by Pringle these measure about 3·0 mm. to a wing length of about 28·0 mm., a ratio to wing length of about one-ninth (0·107). For *Aëdes aegypti* the mean wing length has been previously given as 3·4 mm. (see under 'Body shape and Proportions', p. 408), which, taking the figure for the halteres given above, would give an index of 0·109. As pointed out by Prashad the halteres of the mosquito are relatively large as compared with those of the higher Diptera, the length of the haltere for *Calliphora* as given by Pringle being only 0·7 mm.

In shape the halteres of *Aëdes aegypti* are broad and stumpy as compared with the more slender and elongate organs of *Calliphora* and *Tipula*, but closely resemble in shape those of other mosquitoes so far as these have been figured. The capitellum is nearly half the total length of the haltere (2·0–2·5 mm.). It is globose and bulged posteriorly with a slight kink in this direction towards its junction with the scape. The scape is short and stout, forming about one-third of the length of the haltere. It consists of a thickened and smooth anterior border passing from the capitellum to the scabellum the distal portion of which, like the capitellum, carries scales, those on the capitellum being light coloured, those on the scape darker. Behind the thickened anterior border is a soft membranous portion with a dorsal and a ventral flattish surface and a rather narrow membranous edge. Dorsally this carries the area of sensilla forming the dorsal scapal plate and ventrally the somewhat narrow bulged area on which is situated the ventral scapal plate. On the posterior border are two rather flat papillae (*ps*). The scabellum consists of a smooth sclerotised portion at the base of the scape which, forming a rounded angle, is continued along the base of the haltere to form the free distal portion of the hinge-like joint by which the organ is articulated. In the angle between these two portions is a membranous area forming a rather complex system of lobes. Proximal to the dorsal scapal plate the area is depressed, forming a hollow in which is situated the basal lobe. If structures corresponding to the two Hicks papillae are present in *Aëdes aegypti*, they are quite inconspicuous and their presence still requires confirmation.

The whole haltere is hinged approximately horizontally, but not in the direct line of the body and diverging from this obliquely in a forward direction at an angle of a little under 30°. The hinge does not extend for the whole length of the scabellum leaving a portion at the anterior angle to project freely without attachment. To this projecting angle and the oblique hinge line the thick anterior edge appears to be external whether the haltere is raised to the vertical or flexed, an effect that is at first rather puzzling. The general effect of the haltere as a moving structure can be likened to a short length of rod with a swollen knob at its free end and the other end bent at an angle a little less than a right angle, this bent portion rotating on its axis as a hinge (Fig. 59 (2) *a–b*). Movement of the haltere is mainly in an up and down direction through 180° in a plane approximately at 60° to the median vertical plane of the body. It cuts this latter at a vertical line passing through the middle point of the postnotum. Further information about movements are given in chapter XXIII in connection with flight.

(*e*) THE ABDOMEN

Apart from the terminal segments forming the terminalia the abdomen, in the male, consists of seven normally developed segments and an eighth segment more or less normal in character but rotated following emergence through 180°. In the female the eighth segment is much reduced in size and largely retracted within the seventh.

Each unmodified segment consists of tergite, sternite and membranous pleurae. Both tergites and sternites, with the exception of sternite I which is largely membranous, consist of roughly quadrangular plates with no special thickenings or division into parts. Normally both tergites and sternites overlap for a short distance and no intersegmental membrane is visible, but in the gorged abdomen the plates become widely separated with a considerable extent of membrane showing. Both tergites and sternites are uniformly covered with scales giving rise to the ornamentation already described (see Figs. 48 and 59). The scales are fairly broad flat scales with up to about ten striations, except those forming the silvery spots which are as previously described. Apart from a line or so of inconspicuous hairs largely confined to the posterior and lateral borders and especially notice-

---

Figure 59. Larval wing, halteres and abdomen.

1 Larval wing rudiment (after Christophers and Barraud).

2 Metanotum and halteres, showing hinge line *a–b*.

3 Haltere. A, dorsal view of left haltere of female; B, portion of scala superior more highly magnified; C, optical section of B; D, portion of cupola more highly magnified.

4 Female abdomen. A, as seen in dorsal view in ungorged female; B, dorsal view of tergites as seen in gorged female; C, as in B but showing sternites; *a*, lateral silvery spot on tergite I; *b*, basal white band; *c*, silvery spot on sides of tergite and sternite; *d*, apical pale band.

Lettering: *An*, anal vein; *C*, costa; *Cu*, cubitus vein; *cap*, capitellum; *cup*, cupola; *cv′*, area of cross-veins; *M*, media vein; *mt*, metanotum; *ps*, posterior papillae; *ptn*, postnotum; *R*, radius; *Sc*, subcosta; *sca*, scape; *scb*, scabellum; *scs*, scala superior; I–VIII, abdominal segments.

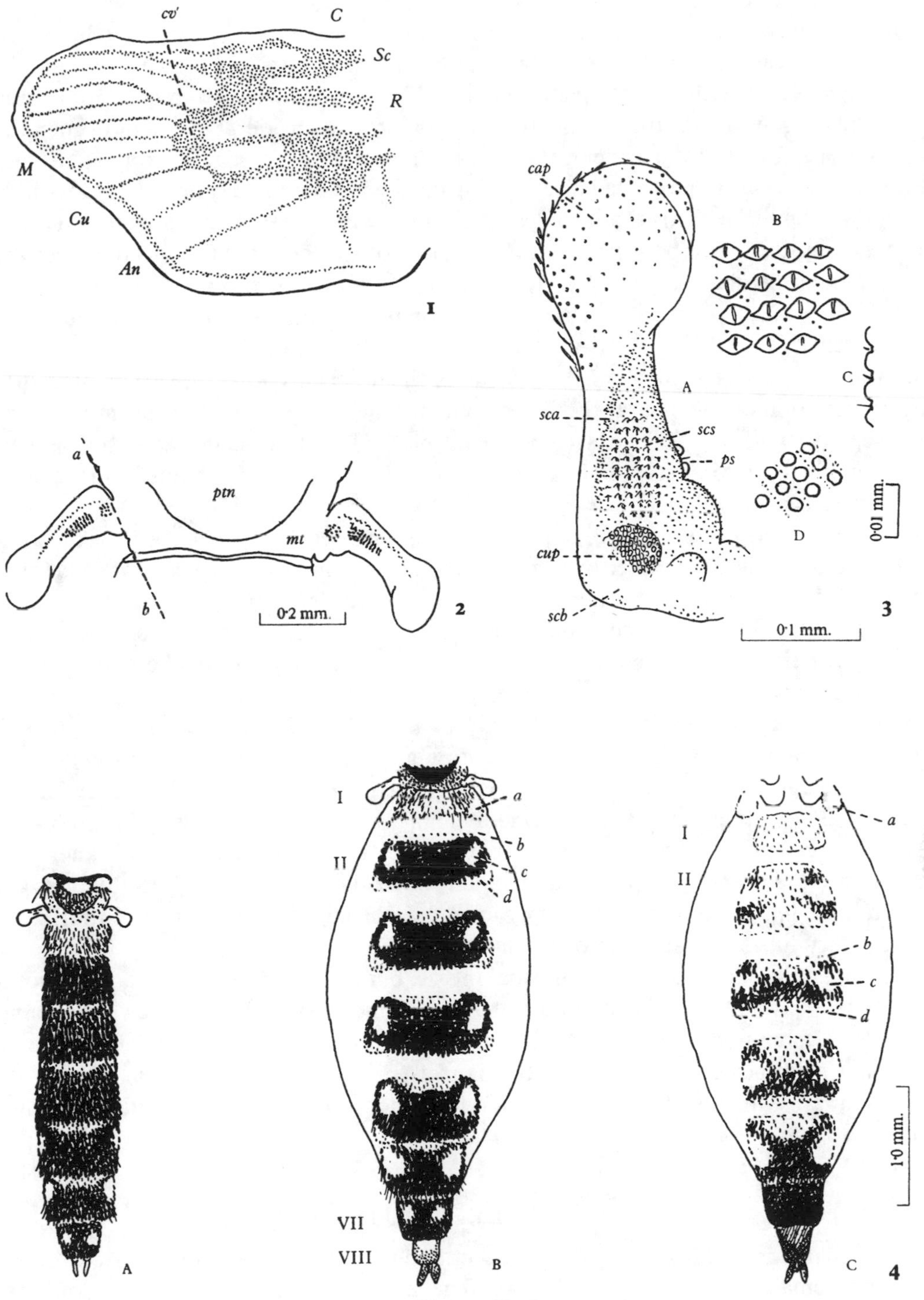

cv'
C
Sc
R
M
Cu
An
1
a
ptn
mt
b
0·2 mm.
2
cap
B
A
C
sca
scs
ps
D
0·01 mm.
cup
scb
0·1 mm.
3
I
a
b
II
c
d
VII
VIII
A
B
C
1·0 mm.
4

Figure 59

able on the sternites, there are no special setae, but segments VI and VII carry numerous hairs among the scales.

The pleurae consist of membrane only and are entirely devoid of scales and hairs. As is commonly the case with membranes, for example the neck, the wing membrane, etc., they are closely covered with microtrichae, which in this case are arranged in groups of three or four. Whilst, especially in the female, the pleurae form a considerable extent of membrane they are normally but little evident, the edges of the sternites, except when the abdomen is greatly distended as in the gorged insect, being inserted under the tergite carrying with them the pleural membrane. This is not folded or elastically collapsed, but is spread out and flat under the tergite, as can be seen by lifting the sides of the tergite with a needle in an unfed female. When the abdomen is fully expanded, as after a blood meal, the full extent of the pleura in the middle segments of the abdomen may be as great as that of the tergite and sternite, but it is doubtful if it is stretched rather than merely fully opened out. On the membrane of each of segments I–VII, towards the anterior end of the segment and near the upper margin of the pleura, there is a spiracular opening. No spiracles are present in either sex on segment VIII. The last spiracle normal to insects as a class is on segment VIII, but in imaginal Diptera, seemingly as a result of premature development in the larva, the last spiracle-bearing segment is VII.

Segment I differs in some respects from the succeeding segments. The tergite is only about half the length of the more anterior segments and, in addition to scales, is uniformly set with hairs. Laterally it carries a particularly conspicuous silvery white scale patch and is here extended in a somewhat lobose manner ventrally. The sternal area is largely membranous with only a narrow strip of sternite carrying scales posteriorly; or, in the male, it may be entirely membranous and devoid of scales.

In the female the seventh segment has only a narrow line of pleurae and in the distended abdomen remains relatively undistended, forming a cuff-like end to the abdomen, within which is the almost wholly invaginated small reduced eighth segment. The eighth segment in the female is rather smooth with some fine hairs but devoid of scales. Proximally it is joined to segment VII with a considerable extent of intersegmental membrane. At its terminal end are the flap-like cerci and median subgenital plate of the female hypopygium. In the middle line posteriorly it is deeply indented, the edges of the sternite forming semi-circular hairy lobes on each side of the median fissure.

In the male the eighth segment is normal in size, but after emergence is gradually rotated through 180° so that the sternite and claspers become dorsal. It is somewhat expanded apically to accommodate the large claspers. During life the segment with the claspers is commonly held raised in a characteristic manner (Fig. 49 (2)).

Abdominal segments distal to the eighth in both sexes are modified to form the terminalia.

(*f*) THE MALE TERMINALIA

The modified terminal segments of the insect abdomen are commonly referred to as the male and female genitalia, or male and female hypopygium respectively. But many authors prefer the name *terminalia* as more correctly designating the complex to which the term refers.

The parts show a great variety of structure in the different insect orders and no adequate general account appears to exist. For the general plan in Nematocera may be specially mentioned the work of Crampton (1942), on the parts in Diptera including the primitive form *Panorpa*, and that of Abul-Nasr (1950), describing the structure and development of the parts in three families of Nematocera.

On the male terminalia of mosquitoes there is a large literature, including among papers dealing with structure and terminology the following: Howard, Dyar and Knab (1912, figures of American species); Edwards (1920, 1941, structure and terminology); Christophers (1922, development); Christophers and Barraud (1923, terminology); Root (1923, 1924, structure); Freeborn (1924, an important contribution on structure, more especially of the anopheline terminalia); Martini (1922, 1928, structure and terminology); Matheson (1929, technique); Barraud (1934, terminology and figures of culicine species with named parts); Marshall (1938, British species); Natvig (1948, an historical summary and account with many figures of Fenno-Scandian species); Marks (1949, on the *Aëdes scutellaris* group and information on the basal lobe in *Aëdes*).

On the parts in *A. aegypti* are more especially the descriptions and figures given by Edwards (1941) and Barraud (1934). See also Iyengar and Menon (1955), and references given by these authors.

TERMINOLOGY

The nomenclature of the parts now generally adopted is mainly that of Edwards and may be gathered from the figure given by Edwards (1941), on p. 15 of his monograph on Ethiopian mosquitoes. A useful list of terms with corresponding lettered figures is also given by Barraud (1934), p. 4. With one or two exceptions the nomenclature of Edwards is that here used.

Where the segmental nature of the parts is clear, these are named accordingly. This applies to the tergite and sternite of the ninth abdominal segment, the homologies of which are not in doubt. It is also now generally accepted that the side-pieces, claspers, forcipes of the older literature are appendages of this segment and the similarity to the coxae of the limbs justifies the name *coxite* which has been applied to them. Possibly the term *gonocoxite* recently in use for the parts in other Nematocera would be even better as also indicating their genitalic character.* The terminal blade-like structure at their apex is the *style* and the small terminal segment would seem most simply termed the *appendage of the style*. Such a term would still be suitable where, as in some cases, the appendage is large and complicated. The appendage is given by Edwards as the *articulated spine of the style*, but this name is not repeated in Barraud's list.

From analogy with the parts in more primitive insects the median lobe bearing the anus would appear to be homologous with the cercus-bearing complex of the tenth tergite and eleventh sternite of many forms and is now generally referred to as the *proctiger*, its lateral plates being the *paraprocts*. The plates on its dorsal aspect appear to be, and are named by Edwards, the *tenth tergite*; or alternatively the *dorsal plates of the proctiger* or *epiprocts*, as also used by this author.

In regard to structures of non-segmental nature the homologies are still somewhat uncertain. According to Abul-Nasr the terminal portion of the male genital duct in Nematocera is formed from a later and separate invagination than that forming the ejaculatory ducts and the portion of duct so formed he terms the *penis tube*. The penis tube may, as in

* The coxite (gonocoxite) is also commonly termed the *basistyle*, the terminal piece being the *dististyle*.

Chironomus, merely open directly on the intersegmental membrane, or it may with its surrounding tissues be excavated to form a projecting intromittent organ carrying the tube. Or the intromittent organ may be otherwise formed by outgrowths at the inner base of the gonocoxites, the *paramere lobes* of Pruthi (1924). These lobes may divide, the inner pair uniting to form the intromittent organ, while the outer pair become the parameres. Or the paramere lobes may not divide but remain as single prominences (paramerophores of Abul-Nasr).

In the first case, where the intromittent organ is formed from sclerotisations of the intersegmental membrane it may, following Abul-Nasr, be termed *theca*, a term used by Wesche. It is also termed *penis sheath* by Abul-Nasr. When the intromittent organ is formed as a result of excavation, as is the filamentous organ in Tipulidae, it may be termed, following Abul-Nasr, *penis* (penis tube and surrounding tissue). Where formed from the inner parameral lobes it would, following Pruthi, be *aedeagus*. An advantage of a terminology based on correct homology is that there may be present more than one organ concerned with intromission, for example in the Tipulidae where, besides the coiled filamentous penis, there may be a sclerotised organ, penis sheath or theca, formed by intersegmental membrane. Or a penis may be present associated with an aedeagus.

The name now most usually applied to the intromittent organ in the mosquito is *phallosome*, though *aedeagus* is sometimes used. The name aedeagus was originally applied to the intromittent organ in Coleoptera. How far it is applicable to the mosquito is uncertain. Edwards uses phallosome for the median organ and aedeagus for the whole complex including the parameres. Though appearing in potash specimens as projections the parameres are actually sclerotisations of the sides of the depression in which the phallosome lies, that is the genital cavity or penis sac of Abul-Nasr. Until more is known of the details of development in the mosquito it would seem best at present to retain the termes phallosome and parameres as now in common use.

A characteristic of the mosquito terminalia in the male is that, as first pointed out by Christophers (1922), all the parts are rotated through 180°. This rotation involves segment VIII. It takes place subsequent to emergence, shortly after which the parts may be found unrotated or in an intermediate position. To avoid confusion any reference in what follows will be to the morphological position, that is, to the parts as they were before rotation.

THE MALE TERMINALIA OF *AËDES AEGYPTI*

The male terminalia of *A. aegypti* conform in general to the description given above when discussing the nomenclature of parts. As in other mosquitoes the parts including segment VIII are rotated. Unlike the condition in the female, segment VIII in the male is not unduly small or telescoped within segment VII. Though rotated with the other parts it is not otherwise modified and it retains the scale characters of its respective surfaces. Tergite IX is peculiar in having two densely sclerotised conical processes divided by a deep median emargination (Fig. 60 (1) IX*t*). Laterally the tergite narrows abruptly to stout rib-like bars that extend ventrally to meet the sternite. A peculiar feature is the termination of these arms in flat triangular expansions which end largely in membrane but articulate by one corner with the narrow sclerotised portion of the ninth sternite (Fig. 60 (2) *a*). Similar triangular expansions are referred to by Martini (1928) and are figured by Natvig in *A. communis* (see Fig. 12 of his monograph). Sternite IX is peculiar. Anteriorly it consists of a narrow crescent-shaped sclerotised plate (Fig. 60 (2) IX*s*″) which appears to be apodematous and gives rise along its length to muscular bands (muscle no. 4 of the series given later). This portion posteriorly merges gradually into the extensive area of membrane that covers the parts ventrally proximal to the bases of the gonocoxites ((2) IX*s*′). In some cases this membrane is slightly thicker and gives the appearance of a thin sclerite. If sternite, it would also account for the apparent lack of relation of the expanded terminal portion of the tergite arms since these would then, if the membrane were sternite, show a normal relation to this.

The gonocoxites are rather stumpy and so shaped and aligned that their inner margins at their bases approach each other ventrally, whilst their outer and dorsal margins are widely separated. Their ventral surface ((2) *Gc*″) is uniformly sclerotised and carries hairs and scales. Along their basal edge is a line of thickening beyond which there is a narrow zone of sclerotised membrane. There is no similar line of thickening on the dorsal surface. The dorsal surface is largely membranous ((1) *Gc*′). About one-third of the surface externally only is sclerotised, and like the ventral surface, with which it is continuous over the outer aspect of the gonocoxite, carries hairs ((1) *Gc*). It is this narrow sclerotised area which is continued as the basal apodeme. Set in the membranous portion towards the apex of the gonocoxite is a conspicuous spinose plate, the *basal lobe* ((1) *bl*).

The *basal lobe*, though situated in *A. aegypti* more apically than basally, seems clearly homologous with a similar structure which in many species of *Aëdes* is basal in position and in other at various levels. It is a strongly sclerotised plate carrying stout spinose hairs arising from sockets. Along the inner edge of the plate is a row of specially stout hairs with extra large sockets some of which have characteristic flat blade-like angled tips (Fig. 61 (1) *a*; (3)). Iyengar and Menon (1955), who first drew attention to these specialised hairs in *A. aegypti*, give the number of such hairs in the material studied by them as 3–4. But in the material at my disposal there seem very regularly to be just two such hairs showing the full characters, though some of the other large hairs are sometimes somewhat bent. At the base of the large hairs the sockets are 7–10μ in diameter and it is possible the hairs are sensory in nature. Proximally the basal lobes are continued into smooth folds with regularly arranged small hairs that give them a characteristic appearance. On each of these folds are two medium-sized hairs with sockets which may possibly represent the harpago of many species (Fig. 61 (1) *b*). The folds on each side pass to the gap between the gonocoxites where they are connected by a rectangular-shaped thickening of the membrane (Fig. 61 (1) *c*). The two basal lobes with the folds into which they are continued and the membrane linking them across the median line are readily detached in dissection as a single structure which would appear to represent the *harpagonal fold* of many species.

The *proctiger* is largely membranous with conspicuous paraprocts. These each consist of an apical and a basal portion separated by a break or joint (Fig. 61 (5)). The apical portions are continued into projecting curved processes (*a*) and carry ventrally smooth knob-like projections (*b*). They appear to represent the cerci. The basal portions pass dorsally to connect with the conical lobes of tergite IX towards their bases internally. They possibly represent sternite XI. On the dorsal aspect of the proctiger are two rather indefinite plates which appear to represent the tenth tergite.

Situated ventrally to the proctiger is the heavily sclerotised *phallosome* (Fig. 60 (2), (4) *Ph*). This is shaped rather like a slightly open cockle shell with its open flanges directed ventrally and bearing on the distal portions of their edges short dark spinous processes.

Lateral to the phallosome are the *parameres*. These are almond-shaped sclerites attached at their bases to the lateral plates of the phallosome and connected about the middle of their outer edge to a process of the basal apodeme of their side.

In the usual conditions under which the terminalia are studied the phallosome in *A. aegypti* would appear as if it opened directly on the ventral surface proximal to the bases of the gonocoxites. Actually in this direction the whole area is completely closed over by membrane (the membranous portion of sternite IX) and there is no opening in this direction (Fig. 60 (2)). The opening of the pouch in which the phallosome lies is at the summit of the membranous fold between the bases of the gonocoxites and a little dorsally. The opening is, therefore, posterior to the ninth sternite. It is in fact close to the anus separated only by a narrow fold (Fig. 61 (7)). Situated just at this level are the two ventral processes of the paraprocts. It seems possible from the appearances seen in various preparations that these processes may function as the guardians of, if not the openers of, the membranous opening into the phallosome sac, for in sections they may be seen embedded in folds of membrane in this situation (Fig. 61 (6) *b′*).

In the mosquito male terminalia three apodematous structures have been named, namely, the *external apodeme*, the *apodeme of the side-piece* and the *basal apodeme*. In *A. aegypti*, and possibly in all mosquitoes, the apodeme of the side-piece and the basal apodeme are two parts of what appears to be a single structure, namely the basal apodeme, of which the apodeme of the side piece is the more superficial part and the basal apodeme of some authors the deeper part. In *A. aegypti* both the external and basal apodeme originate as extensions of the edge of the gonocoxite, the external apodeme at its outer aspect and the basal apodeme from the sclerotised portion of the dorsal surface, the two apodemes in this species being continuous (Fig. 60 (2)). In many species the two apodemes are separated by an interval, the basal apodeme taking off more towards the middle line from the dorsal edge. In *A. aegypti* its position is apparently determined by the position and extent of the only sclerotised portion of this surface of the gonocoxite, that is, the narrow strip previously referred to. This narrowness of the sclerotised portion of the dorsal surface of the gonocoxite is a character rather special to *Aëdes*, the result of the large extent of membrane on the dorsal surface of the gonocoxite in this genus. In *Anopheles* where the whole gonocoxite surface is sclerotised the basal apodeme takes off from the extreme inner portion of this structure.

The superficial portion of the basal apodeme is a rather flat band continuing the narrow dorsal portion of the gonocoxite into the tissues (Fig. 60 (2)). It then expands into the

---

Figure 60. Male terminalia.

1 Dorsal view of male terminalia.

2 Ventral view of same. *a*, expanded terminal portion of arm of ninth tergite; IX*s′*, membranous portion of ninth sternite; IX*s″*, sclerotised apodematous portion of the same.

3 Lateral external view. Lettering as for 2.

4 Lateral internal view as with the near gonocoxite removed.

Lettering: *ap.b*, basal apodeme; *ap.e*, external apodeme; *bl*, basal lobe; *bl′*, proximal portion of fold carrying two hairs; *Cr*, terminal portion of ventral sclerite of proctiger (probably cercus); *Gc*, gonocoxite; *Gc′*, membranous area of dorsal surface of same; *Gc″*, sclerotised ventral surface of gonocoxite; *Ph*, phallosome; *Ph′*, the same seen behind membrane; *Pr*, proctiger; *pr*, paramere; *sl*, style; *sl′*, appendage of style; IX*s*, X*s*, respective sternites; IX*t*, X*t*, respective tergites.

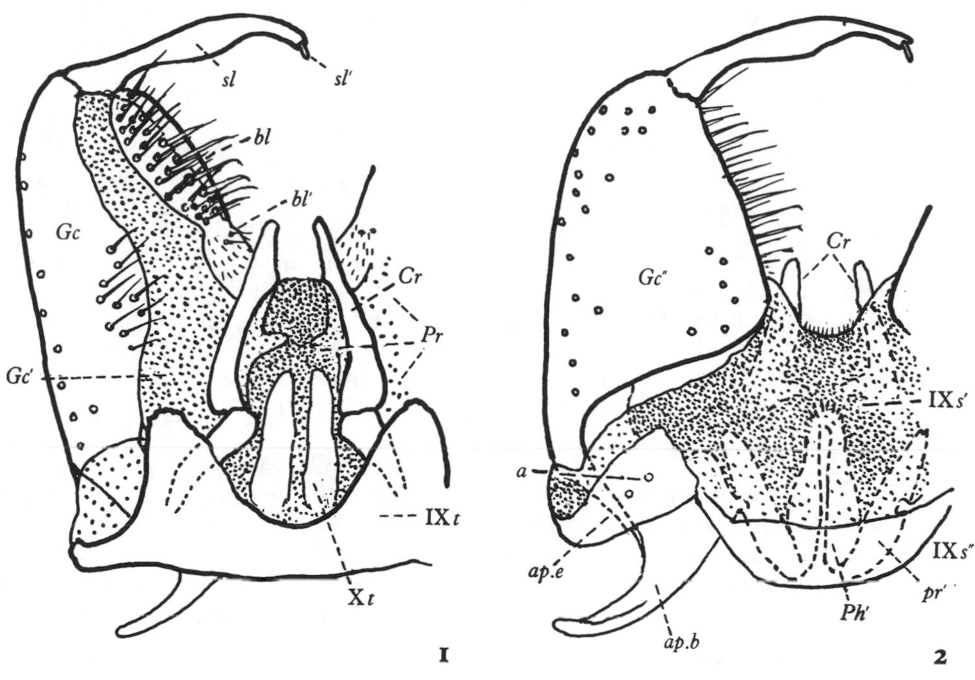
sl
sl'
bl
bl'
Gc
Cr
Pr
Gc'
IXt
Xt
1
Cr
Gc"
IXs'
a
ap.e
IXs"
pr'
Ph'
ap.b
2

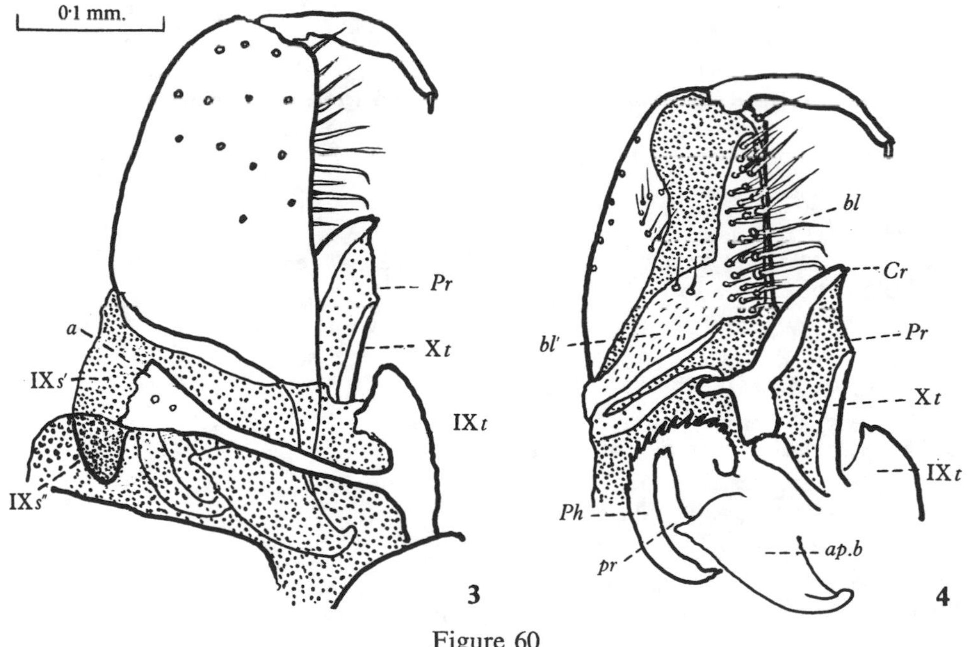
0·1 mm.
Pr
a
Xt
IXs'
IXt
IXs"
3
bl
Cr
bl'
Pr
Xt
IXt
Ph
ap.b
pr
4

Figure 60

more basal portion which forms an expanded plate at somewhat of an angle to the first part (Fig. 60 (2) *ape*). Internally and ventrally the plate connects with the paramere and externally it is continued into a long spur-like process giving origin to a number of important muscles as described later.

There remains only to give a brief description of the musculature of the parts. A complete study of this has not been made, but the following is a numbered list of the more important muscles (Fig. 61 (8)). The musculature has also been dealt with by Martini (1922). See also Natvig (1948).

1. A muscle arising from the outer aspect of the gonocoxite and inserted into the outer tubercle of the base of the style (extensor of the style).
2. A massive muscle consisting of converging bands arising from an extensive area of the gonocoxite and inserted into the region of the inner tubercle of the base of the style (flexor of the style).
3. A long muscle arising from the spur of the basal apodeme and passing up the whole length of the gonocoxite to be inserted in the region of the inner tubercle of the base of the style (adductor of the style).
4. Muscular bands arising along the length of the apodematous portion of segment IX and passing up on each side of the phallosome to be inserted into the proximal portion of the basal lobe (retractor of the basal lobe).
5. Muscle bands arising from the tenth tergite and inserted into the basal apodeme posteriorly.
6. Massive muscles arising from the spur of the basal apodeme internal to no. 3 and inserted into the paramere distal to its articulation with the basal apodeme (protractor of the phallosome).

The action of the muscles would appear to be adduction of the gonocoxites with flexion of the styles, the dragging downwards and inwards of the basal lobes and eversion of the phallosome by the powerful basal apodeme-paramere muscles.

Figure 61. Male and female terminalia.

1 Harpagonal fold with the two basal lobes. *a*, specialised hairs with angled tips; *b*, two hairs on fold possibly representing the harpago; *c*, thickened portion of membrane forming small plate.

2 Basal lobe more highly magnified with hairs removed to show large sockets of specialised hairs and hairy portion of fold. *a*, sockets of large hairs including two with angled tips; *b*, hairs possibly representing the harpago.

3 Specialised hair with angled tip.

4 Appendage of style.

5 Lateral view of proctiger. *a*, terminal cornu; *b*, process of ventral plate; *c*, basal sclerite of ventral plate (possibly XI sternite).

6 Sagittal section a little lateral of median line to show genital opening. *a*, apex of ninth sternite; *b*, apex of proctiger; *b′*, process of ventral plate; *c*, sclerotisations at base of phallosome; IX*s′*, membranous portion of ninth sternite; IX*s″*, sclerotised apodematous portion of same.

7 Another section of same series in middle line. *a*, *b*, as in no. 6.

8 Showing chief muscles numbered as given in list of muscles in the text.

9 Dorsal view of female terminalia.

10 Ventral view of same.

11 Lateral view of same. *a*, cloaca-like hollow formed by overlapping sternal portion of segment VIII. This is quite distinct from the more deeply situated atrium.

Lettering: *Ao*, opening of anus; *ap.b*, basal apodeme; *Cr*, cercus; *cw*, cowl; *Go*, genital opening; *hg*, hinge; *in*, insula; *Ph*, phallosome; *pgp*, postgenital plate; *Rc*, rectum; *spd*, opening of spermathecal duct; *spm*, spermathecae.

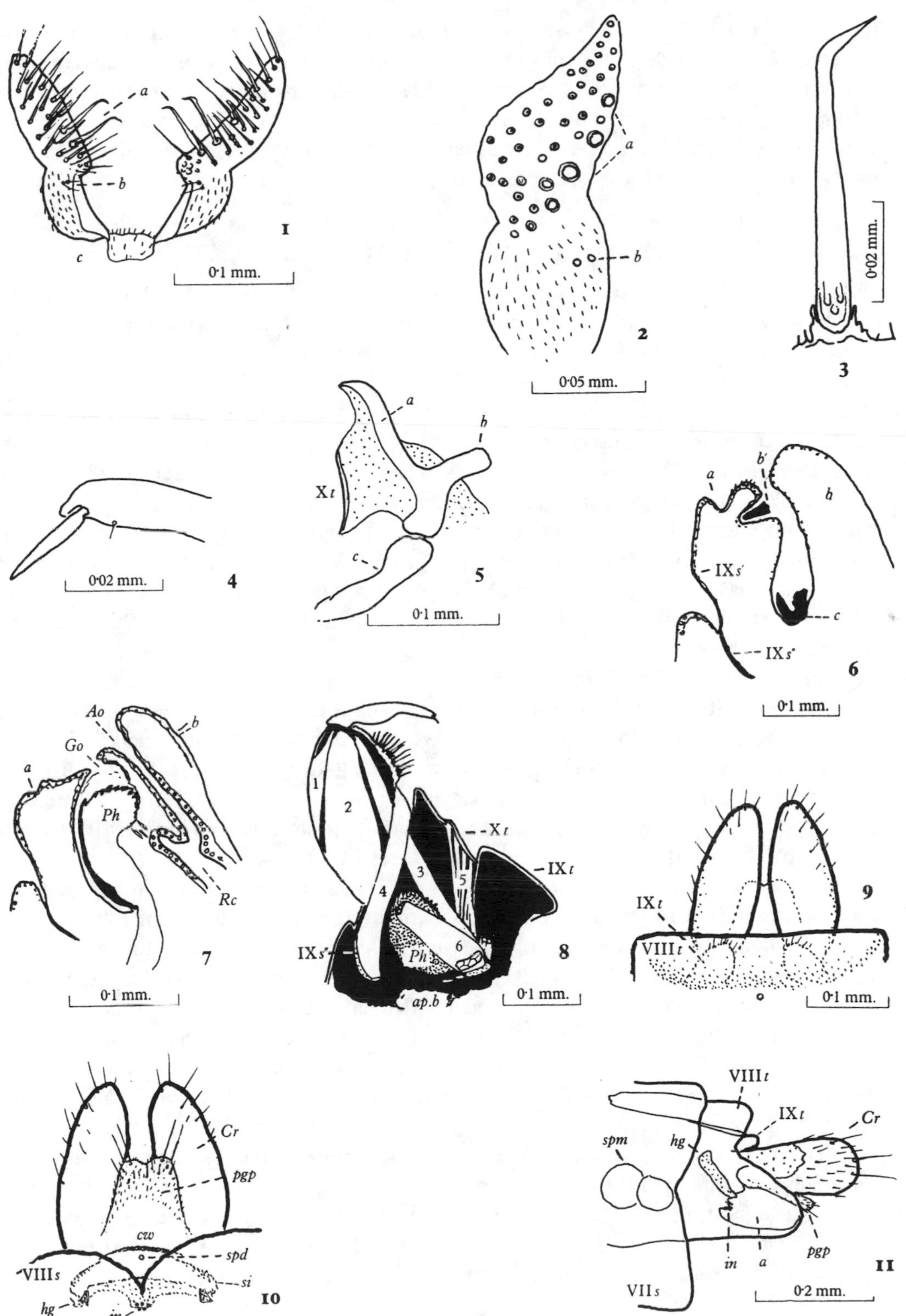

a
b
c
1
0·1 mm.
a
b
2
0·05 mm.
0·02 mm.
3
0·02 mm.
4
a
b
Xt
c
5
0·1 mm.
a
b′
b
IXs′
c
IXs″
6
0·1 mm.
Ao
b
Go
a
Ph
Rc
7
0·1 mm.
1
2
3
4
5
Xt
IXt
IXs″
Ph
6
8
ap.b
0·1 mm.
IXt
VIIIt
9
0·1 mm.
Cr
pgp
cw
VIIIs
spd
si
hg
in
10
VIIIt
IXt
Cr
spm
hg
in
a
pgp
11
VIIs
0·2 mm.

Figure 61

From the systematic point of view the male terminalia of *Aëdes aegypti* are peculiar in several respects. Edwards notes that the conical lobes of tergite IX are peculiar among African species and Barraud that *A. aegypti* differs from all other Indian species, except *A.* (*Stegomyia*) *desmotes*, in the well-developed ventral arms of the paraprocts. The character of the basal lobe and its specialised hairs is also of systematic value. As also according to Mattingly (*in lit.*) is the shape of the style.

(*g*) THE FEMALE TERMINALIA

The female genitalia or hypopygium, or, as many prefer, the *female terminalia*, of the mosquito have been studied by a number of observers: Macfie and Ingram (1922); Christophers (1923, whose description of the parts and terminology has been generally followed); Davis (1926, *Anopheles*); Freeborn (1926, *Aëdes*); Gerry (1932, morphology); Gjullin (1937, 1946, *Aëdes*); Edwards (1941, who gives a good summary of structure and terminology with a useful plate giving characters of the different culicine genera); Bohart (1945, *Aëdes*); Roth (1946, *Wyeomyia*); Ross (1947, 28 species); Coher (1948, a good recent account with generic characters). Useful papers are also Crampton (1942), on *Panorpa*; Abul-Nasr (1950), on the development of the parts in Nematocera.

The segments composing the female terminalia in different orders of insects are in the main as in the male, that is, segments IX–XI. But segment VIII is commonly also included, its appendages with those of segment IX forming the ovipositor when this is present. In the mosquito there is no ovipositor or indication of the parts forming such a structure. Nor is segment VIII, apart from its reduction in size in certain forms, much modified. Usually, however, its posterior border is indented or notched in the median line of the sternite. The tergite, other than reduction in size in certain forms, is not modified.

Segment IX, especially the sternite, is almost eliminated, the segmental parts distal to segment VIII being little more than the postgenital plate (usually considered sternite XI), reduced tergite IX and possibly tergite X and the cerci. If sternite IX is present, it may be represented by the small hairy plate, the insula, described below.

The chief clue to the nature of the parts other than those mentioned lies in the existence of a deep invagination in the intersegmental membrane posterior to sternite VIII. As a result of this invagination there is approximation of the postgenital plate to sternite VIII closing in the parts so that a cloaca-like cavity is formed, the *atrium*. Into this cavity open three separate ducts, two relating to segment VIII, namely the *common oviduct* and the *spermathecal duct*, and one pertaining to segment IX, the opening of the *caecus* or *mucus gland*. It should perhaps be mentioned that the common oviduct does not directly open into the cavity but through a small terminal part, the *vagina*, as described when dealing later with the generative organs.

The rim of the atrium is sclerotised, forming a narrow ribbon-like thickening surrounding a transversely elongate opening at the lateral angles of which it is sharply bent in, the whole strip strongly suggesting the metal framework of a clasp purse. The anterior portion of the rim is the *sigma* (Fig. 61 (10) *si*), the posterior portion the *cowl* (*cw*). At each corner where the sigma and cowl join there is usually a small thickening termed by Coher the *hinge* (*hg*). There may be further thickenings extending inwards from the hinges on to the roof of the atrium, *atrial plates*. In the median line, just anterior to the sigma and sometimes continuous with this structure, is a small hairy sclerite, the *insula* (*in*).

In *Aëdes aegypti* the parts conform to the above description, though, associated with the general reduction in size of the terminalia as a whole, the structures are small and rather ill-defined. Sternite VIII is bilobed with a median cleft in its posterior edge (Fig. 61 (10)). Tergite VIII, except for reduction in size, is not modified. Posterior to tergite VIII are two small hairy plates linked together basally by a rather indefinite thickening of the cuticle. These have been generally regarded as tergite IX. Beyond these small plates are the *cerci* (*Cr*). These are elongate flap-like structures well covered with hairs, but without scales. Ventrally, lying below the cerci, is the *postgenital plate* (*pgp*). This has a median notch in its posterior border, giving it a somewhat bilobed appearance. It is largely freely projecting, its ventral surface and the distal portion of the dorsal surface being covered with small hairs and in addition there are some larger hairs arising from small basal papillae towards the apex, one of which just ventral and one just dorsal to the apex of each lobe are longer than the others (Fig. 61 (10)).

The appearance of the insula suggests that it is of segmental nature and it has been considered by some authors to be possibly sternite IX. It lies anterior to the atrial opening, so that if the insula is sternite IX the opening of the oviduct and vagina is posterior to sternite IX, a position that is not usually accorded to it. Edwards (1941), p. 19, however, discussing this, considers that the female opening in the Diptera is, like that of the male, posterior to sternite IX, where it is situated in *Panorpa*, an ancestral form of the Diptera. A point in favour of such a view is that in many insects sternite IX is deficient in its central parts (see Christophers and Cragg (1922) on the female parts in *Cimex*). Edwards is also opposed to the assigning by Gerry (1932), and by Gjullin (1946), of the cowl to the ninth sternite. If such were the case, the opening of the caecus would be anterior to this sternite, which would not be in accordance with its position as shown by development. The structure termed insula by Abul-Nasr (1950) in several species of Nematocera lies between the anterior portion of the atrium which receives the oviduct and spermathecal openings and the posterior portion which receives the opening of the caecus. This would make it distinct from the insula as here described.

REFERENCES

(*d*) THE THORACIC APPENDAGES

BARRAUD, P. J. (1934). See references in ch. II (*a–b*), p. 45.

CHRISTOPHERS, S. R. (1933). See references in previous chapter.

CHRISTOPHERS, S. R. and BARRAUD, P. J. (1924). The tracheation and venation of the wing of the mosquito. *Ind. J. Med. Res.* **11**, 1103–17.

COMSTOCK, J. H. (1920). See references in previous chapter.

COMSTOCK, J. H. and NEEDHAM, W. M. (1923). *The Wings of Insects*. The Comstock Publ. Co., Ithaca, New York.

CRAMPTON, G. C. and HASEY, W. H. (1915). The basal sclerites of the leg in insects. *Zool. J., Anat.* **39**, 1–26.

DAHL, F. (1884). Beiträge zur Kenntnis des Baues der Funktionen der Insektenbeine. *Arch. Naturgesch.* Jahrg. **50**, 1, pp. 146–193.

EDWARDS, F. W. (1912). African Culicidae other than *Anopheles* (p. 7 male claws). *Bull. Ent. Res.* **3**, 1–53.

EDWARDS, F. W. (1941). See references in ch. II (*a–b*), p. 45.

EWING, H. E. (1928). The legs and leg-bearing segments of some primitive arthropod groups etc. *Smithson. Misc. Coll.* **80**, no. 11, 41 pp.

IMAGO: EXTERNAL CHARACTERS

GILLETT, J. D. and WIGGLESWORTH, V. B. (1932). The climbing organ of an insect *Rhodnius prolixus* (Hemiptera: Reduvidae). *Proc. R. Soc.* B, **111**, 364–75.

HUXLEY, T. H. (1906). *The Crayfish.* Ed. 6 (Ed. 1, 1879), Kegan Paul, Trench, Trubner and Co., London.

IMMS, A. D. (1938). *A General Textbook of Entomology.* Methuen and Co., London.

LOWNE, B. T. (1895). *The Anatomy, Physiology and Development of the Blowfly.* Vol. II. Porter, London.

MATTINGLY, P. F. and GRJEBINE, A. (1957). (Tibial claw.) In Press. *Mem. Inst. Sci. Madagascar.*

MATTINGLY, P. F. and HAMON, J. (1955). Position taxonomique et synonymie de quelques *Ficalbia* de la région Ethiopian (Dipt. Culicidae). *Ann. Parasit. hum. comp.* **30**, 488–96.

NATVIG, L. R. (1948). See references in ch. II (*a–b*), p. 46.

NUTTALL, G. H. F. and SHIPLEY, A. E. (1901–1903). See references in ch. I (*d*), p. 17.

PRASHAD, B. (1916). The halteres of mosquitoes and their function. *Ind. J. Med. Res.* **3**, 503–9.

PRASHAD, B. (1918). The thorax and wing of the mosquito. *Ind. J. Med. Res.* **5**, 610–40.

PRINGLE, J. W. S. (1948). The gyroscopic mechanism of the halteres of Diptera. *Phil. Trans. R. Soc.* **233**, 347–84.

SHARP, D. (1899). *Cambridge Natural History.* Vol. VI, *Insects.* Macmillan and Co., London.

SHIPLEY, A. E. (1915). Insects and war. XII. The mosquito *Anopheles maculipennis. Brit. Med. J.* **1**, 330–1.

SHIPLEY, A. E. and WILSON, E. (1902). On a possible stridulating organ in the mosquito (*Anopheles maculipennis*). *Trans. R. Soc. Edinb.* **11**, 387–72.

SNODGRASS, R. E. (1935). See references in previous chapter.

(*f–g*) MALE AND FEMALE TERMINALIA

ABUL-NASR, S. E. (1950). Structure and development of the reproductive system of some species of Nematocera (order Diptera: suborder Nematocera). *Phil. Trans.* B, **234**, 339–96.

BARRAUD, P. J. (1934). See references in ch. II (*a–b*).

BOHART, R. M. (1945). A synopsis of the Philippine mosquitoes. *Nav. Med. Bull.* **580**, 1–88.

CHRISTOPHERS, S. R. (1922). The development and structure of the terminal abdominal segments and hypopygium of the mosquito with observations on the homologies of the terminal segments of the larva. *Ind. J. Med. Res.* **10**, 530–72.

CHRISTOPHERS, S. R. (1923). The structure and development of the female genital organs and hypopygium of the mosquito. *Ind. J. Med. Res.* **10**, 698–719.

CHRISTOPHERS, S. R. and BARRAUD, P. J. (1923). Descriptive terminology of male genitalic characters of mosquitoes. *Ind. J. Med. Res.* **10**, 827–35.

CHRISTOPHERS, S. R. and CRAGG, F. W. (1922). On the so-called 'penis' of the bed-bug and the homologies generally of the male and female genitalia of this insect. *Ind. J. Med. Res.* **9**, 445–63.

COHER, E. J. (1948). A study of the female genitalia of Culicidae: with particular reference to characters of generic value. *Entom. Amer.* **28**, 75–112.

CRAMPTON, G. C. (1942). See references in previous chapter.

DAVIS, N. C. (1926). Notes on the female hypopygia of anopheline mosquitoes with special reference to some Brazilian species. *Amer. J. Hyg.* **6**, 1-22.

EDWARDS, F. W. (1920). The nomenclature of the parts of the male hypopygium of Diptera Nematocera with special reference to mosquitoes. *Ann. Trop. Med. Parasit.* **14**, 23–40.

EDWARDS, F. W. (1941). See references in ch. II (*a–b*), p. 45.

FREEBORN, S. B. (1924). The terminal abdominal segments of male mosquitoes. *Amer. J. Hyg.* **4**, 188–212.

FREEBORN, S. B. (1926). The mosquitoes of California. *Univ. Calif. Publ. Ent.* **3**, 333–460.

GERRY, B. I. (1932). Morphological studies of the female genitalia of Cuban mosquitoes. *Ann. Ent. Soc. Amer.* **25**, 31–75.

GJULLIN, C. M. (1937). The female genitalia of *Aëdes* mosquitoes of the Pacific Coast States. *Proc. Ent. Soc. Wash.* **39**, 252–66.

REFERENCES

GJULLIN, C. M. (1946). A key to the *Aëdes* females of America north of Mexico. *Proc. Ent. Soc. Wash.* **48**, 215–36.

HOWARD, DYAR and KNAB (1912). See references in ch. I (*f*), p. 19.

IYENGAR, M. O. T. and MENON, M. A. U. (1955). Mosquitoes of the Maldive Islands. *Bull. Ent. Res.* **46**, 1–9 (with appendix by Mattingly, P. F., 9–10).

MACFIE, J. W. S. and INGRAM, A. (1922). On the genital armature of the female mosquito. *Ann. Trop. Med. Parasit.* **16**, 157–88.

MARKS, E. N. (1949). Studies of Queensland mosquitoes. Part IV. *Univ. Queensland Papers. Dep. Biol.* **2**, no. 11, 1–41.

MARSHALL, J. F. (1938). See references in previous chapter.

MARTINI, E. (1922). Über den Bau der äusseren männlichen Geschlechtsorgane bei den Stechmücken. *Arch. Naturgesch.* Jahrg. **88**, Abt. A, Heft 1, 134–42.

MARTINI, E. (1928). Über die segmentale Gliederung Nematoceran Dipteren. IV. Die Terminalia der Culiciden und Psychodiden. *Zool. Anz.* **76**, 147–62.

MATHESON, R. (1929). *A Handbook of the Mosquitoes of North America.* Baillière, Tindall and Cox, London.

NATVIG, L. R. (1948). See references in ch. II (*a–b*), p. 46.

PRUTHI, H. SINGH (1924). On the post-embryonic development and homologies of the male generative organs of *Tenebrio molitor* (Coleoptera). *Proc. Zool. Soc. Lond.* 1924, 857–68.

ROOT, F. (1923). The male genitalia of some American *Anopheles* mosquitoes. *Amer. J. Hyg.* **3**, 264–79.

ROOT, F. (1924). Further notes on the male genitalia of American *Anopheles*. *Amer. J. Hyg.* **4**, 456–65.

ROSS, H. H. (1947). The mosquitoes of Illinois. *Illinois Nat. Hist. Surv.* Bull. 24, 1–16.

ROTH, L. M. (1946). The female genitalia of *Wyeomyia* of North America. *Ann. Ent. Soc. Amer.* **39**, 292–7.

WESCHE, W. (1906). The genitalia of both sexes in Diptera and their relation to the armature of the mouth. *Trans. Linn. Soc. Lond.* (*Zool.*) (2), **9**, Part x, 339–86.

XX
THE IMAGO: FOOD AND FEEDING

(*a*) NATURE OF FOOD

That the natural food of the female mosquito in the majority of species is blood from some vertebrate host and that the male does not suck blood is the result of general experience. This applies in full measure to *Aëdes aegypti*, the female of which, whether in nature or in the laboratory, is a particularly active biter and feeds readily and consistently under most circumstances when given the opportunity. The male, though it may annoy by hovering about and even in some circumstances settling on the skin, does not attempt to puncture. The female may, however, maintain existence for long periods on food other than blood.

Goeldi (1905) kept females alive 31–102 days fed on honey. Macfie (1915) notes that females take honey for the first day or two and that males feed only on honey. Bonne-Wepster and Brug (1932) note that sugar-containing fluids are readily taken. Gordon (1922*b*) observed both females and males sucking fluid from flowers. Many observers have observed that sugary fluids, raisins, banana, etc., are sucked by both sexes and it is a common practice to put such material in the mosquito cage to maintain the insects alive. In the case of food of this kind the fluid does not, however, pass into the mid-gut but into the diverticula as described later, and apart from experimental feeding (through membranes or in other artificial ways) the taking of other food than blood by the female is in an altogether different category from, and unaccompanied by many characteristic features in the insect's behaviour when taking the main object of her existence, the *blood meal*.

Further, without such a meal eggs are not developed and the insect does not oviposit. Finlay (1886) found that *A. calopus* (*aegypti*) could not develop its eggs without blood. Goeldi (1905), after numerous experiments with fruit, sugar, honey, etc., came to the conclusion that blood was necessary for the production of eggs. Fielding (1919) kept females alive for 25 to 142 days fed on a variety of foods (sugar, peptone, apples, etc.), but failed to obtain eggs. Sen (1918), dealing with a closely related species, *A. albopictus*, records, however, that eggs were laid on a diet of milk and protein and on two occasions on cane sugar only. Such experiences are, however, very unusual and not only does oviposition not occur in the absence of a blood meal but development of the ovarian follicles beyond a certain early stage does not take place.

In the case of *A. aegypti* the blood meal is very commonly, but not necessarily, human blood. Of blood other than human, Boyce (1907) states that the species feeds normally on warm-blooded vertebrates. Howard, Dyar and Knab (1912) note that, while human blood is preferred, *A. aegypti* will also feed on mammals and birds. Other authors (Connor, 1922; Gordon, 1922*b*) have had the same experience and Davis and Philip (1931) found occasionally in nature that the precipitin test was positive for chicken blood. Gordon (1922*b*) found *A. aegypti* feeding with great voracity on bats. Kumm (1932) records the same. The species is now commonly fed in the laboratory with full gorging and good egg development on chicken's blood (James and Tate, 1938; Bishop and Gilchrist, 1946; and others). It is also commonly fed on the guinea-pig and rabbit. But whilst it can be fed on experimental animals these do not always form a good substitute in the laboratory for human blood. Fielding (1919) found feeding with small animals as a rule unsuccessful even if left overnight. Lewis (1933) notes that *A. aegypti* will feed on the guinea-pig only if the animal is shaved. Mathis (1934) found that for oviposition several blood meals are necessary if fed on the rabbit, whilst it is exceptional for more than one feed to be required when fed on human blood. Much depends upon the technique used and feeding on small animals usually requires immobilisation and shaving of some part. With suitable technique many authors use the guinea-pig for routine feeding for oviposition and the writer has found the rabbit an exceptionally useful animal for the purpose, the insects feeding readily and fully on the shaved abdomen using the attachment already described. With suitable technique Bishop and Gilchrist (1946) and other authors have similarly found the chicken a very suitable animal for routine use with the species.

How far *A. aegypti* will feed upon cold-blooded vertebrates is less certain. Goeldi failed to get the species to feed on a lizard and Gordon (1922*b*) similarly failed with the gecko. Woke (1937*a*), however, succeeded in getting the species to take a full blood meal from the frog and from the turtle. He found (1937*b*) that more eggs were produced per cubic millimetre from the blood of turtle, frog, canary, rabbit and guinea-pig, in this order, than from that of man or monkey. From other data given, however, it was clear that much less blood was taken from cold blooded animals per feed, so that in effect production from feeding on mammals was greater. Gordon and Lumsden (1939) found difficulty in getting *A. aegypti* to feed on the frog's foot and that the insects only did so when excited in the first instance by an interrupted feed on man.

Whatever source of blood *A. aegypti* may at times utilise there is little doubt from the domestic habits of the species as ordinarily encountered and its close association with man that its normal blood food is that of man.

Whether there is any selective preference for particular races or individuals is a question often raised. Connor (1922) found white races preferred to coloured and that the young and healthy were most often bitten. Marchoux, Salimbeni and Simond (1903) as also Gordon (1922*a*) found, however, no difference between sweating and dry, hairy or smooth skin, or young or old persons, or nationality. Some statements in this respect may be based more upon the immediate or after effects of the bites than upon actual observation of the number of bites. An old resident, probably to a considerable extent immunised to bites, may take little note of such bites as he receives, since there is a minimum of reaction. But to the new-comer the immediate and after effects of even a single bite usually attracts attention. Further, since the biting urge is largely excited by the temperature of the object bitten, it is very probable that young persons with highly vascular skins would be especially attractive. On the whole, however, it may be doubted whether a hungry *A. aegypti* female would take much note of the delicate distinction between one nationality and another or between individuals. The question how far smell may be concerned is dealt with later. In general it may be said that evidence is against the view that smell enters largely into the urge to attack and that any effect it might have is almost certainly overshadowed by the species' reaction to temperature.

Feeding on the human corpse has been recorded by several observers. Durham (quoted by Theobald, 1903) observed a specimen of *A. aegypti* feeding on the inner side of the thigh of a Chinaman dead 3 hours where it was allowed nearly to fill itself. Rosenau, Francis and Goldberger (quoted by Howard, Dyar and Knab) record the species as sucking blood from the human corpse 6 or even 12 hours after death. Howard, Dyar and Knab found female *A. aegypti* feeding on a corpse 12 hours after death, but only one out of a number apparently obtained blood. Three mosquitoes in another case succeeded in doing so half an hour after death. Gordon (1922*b*) states that *A. aegypti* will bite corpses as long as 18 hours after death and will draw blood as long as 6 hours after. Though feeding on corpses has an importance in connection with the spread of yellow fever, it does not raise any very special point as to food selection, since, provided the corpse is still warm, as may well be the case under tropical conditions, or even as a result of postmortem changes, there is nothing unusual in *A. aegypti* being attracted to and attempting to prod any warm object and at some hypostatically congested part obtaining blood.

Female *A. aegypti*, as shown by Bishop and Gilchrist (1946), will feed through a membrane upon whole blood, but not upon plasma, even though presented at 40–42° C. Under such circumstances a suspension of twice-washed red cells in citrate in equal volume to the plasma removed was fully gorged upon, though the percentage doing so was lower than with full blood. Blood haemolysed by freezing and thawing gave a still smaller percentage feeding, whilst the majority of those that did feed imbibed only a small amount. Dilution of the haemolysed blood to reduce the high viscosity led to fewer still feeding. On a suspension of rice starch granules in a solution of haemoglobin they refused to feed.

The effect of the red cells in modifying the physical properties of plasma will be very apparent to anyone attempting to make a thin film of plasma or very anaemic blood and contrasting the experience with that obtained from whole blood or red cells in citrate, both of which film perfectly.

Much depends also upon the form in which blood or fluids are presented. The above observers found that fluid presented in open drops often give rise to a reaction different to that to the same fluid presented behind a membrane. A much smaller percentage fed on heparinised blood presented openly than when this was offered behind a membrane. Haemoglobin in plasma was not imbibed when presented in drops, though it was taken through a membrane. A small number, however, imbibed haemoglobin in water from drops. Plasma was not taken either in drops or through a membrane. Saturated glucose solution was taken by a large proportion of the mosquitoes, the fluid passing to the diverticula. It seems possible that the physical properties of the fluids, to some extent at least, may determine whether they are imbibed or not, especially as such fluids have to be passed very rapidly through the excessively fine blood channel, approximately only 0·03 mm (30μ) in diameter.

Briefly summarising, it may be concluded that there are two forms of food of quite different nature taken by the female mosquito, namely (*a*) blood from a living host, the act being accompanied by all the vivid reactions of the insect to such an episode and leading to eventual oviposition; and (*b*) various fluids (including water), but especially sweet fluids such as honey, glucose, etc., taken much as by any creature, useful for maintaining life, but not leading to oviposition and not associated with the violent impulses and zest that characterise desire for the first type.

Of interest as to the effect of sweetness in stimulating feeding are the observations by Lang and Wallis (1956). These authors note that sugars act as a stimulant to feeding, sucrose being the most effective of a number of sugars experimented with (lactose, glucose, sucrose, and fructose). An interesting point noted is that the attractiveness to the mosquito differs somewhat from their sweetness to man, and very noticeably that glycerine which is relatively sweet to man had little attractiveness for *Aëdes aegypti*. The following figures are given for 10 per cent concentrations of different sugars and glycerine in attractiveness to *A. aegypti*.

	Man	*A. aegypti*
Water	0	0
Lactose	16	12
Glucose	74	56
Sucrose	100	100
Glycerine	108	4
Fructose	173	52

The nutritive value of different bloods, for example, mammalian or avian, or even different mammalian bloods, in relation to oviposition appears to vary somewhat, as also does the amount imbibed at a meal. Such questions have been investigated chiefly by Woke (1937*a*, *b*). The amount of blood taken in relation to the development and number of eggs laid is discussed in chapter XXII.

(*b*) TIME FROM EMERGENCE TO FIRST BLOOD MEAL

That a certain time elapses after emergence before the female will attempt to feed is well recognised. Marchoux, Salimbeni and Simond (1903) state that the female is ready to feed in about 24 hours after emergence. Drake-Brockman (1913) was never able to get *A. aegypti* to feed on the same day as emergence. Howard, Dyar and Knab (1912) refer to Peryassu as saying that both fertilised and virgin females can be induced to suck blood 18 hours after emergence, and Howard (1923) states that the female will bite 18–24 hours after emergence. Macfie (1915) found that, when given opportunity for feeding at least twice daily, females fed for the first time on the second day, more often on the third day. H. A. Johnson (1937) says that the first blood meal is taken 20–40 hours from emergence.

Seaton and Lumsden (1941), investigating this question, found that out of 100 mosquitoes only seven fed in the first 24 hours from emergence, twenty-seven 24 and 48 hours and forty-four in the next 24 hours. Between 72 and 96 hours sixty-two fed and between 96 and 120 hours seventy fed. After the fifth day there was a marked decline associated with high mortality. Bishop and Gilchrist (1946) also give data for three strains of *A. aegypti* fed: (1) on the chicken; (2) through a membrane on chicken blood. The mosquitoes in the first-mentioned series were applied to the breast of the chicken in a biting box in the dark at 28° C. in a moist atmosphere and 30 minutes allowed for the feed. Only a few (0–6 per cent) fed in the first 24 hours. Somewhat under 50 per cent fed between 24 and 48 hours and in two strains 76 and 95 per cent between 48 and 72 hours. Through membrane the number feeding was considerably lower, the three strains on the fourth day giving only 54, 73 and 30 per cent feeds respectively.

In routine standard testing extending over many months the present author found that females reared and allowed to emerge at 28° C. and hereafter kept at 25° C. and relative humidity of 80 per cent in the dark without other food than water reached maximum biting potential on the human arm on the fifth to seventh day (Christophers, 1947). Under good conditions (well-reared mosquitoes, warm arm and other suitable conditions) feeding within an allowed 15 minutes should approach or reach 100 per cent. The majority of females will, however, feed even by the third day or even second day from emergence and a variable proportion may feed even on the day following emergence. In this last case gorging may be deficient. Thus in a cage of 100 female and 100 male mosquitoes offered the arm for 15 minutes on the day following emergence forty-one females fed of which twenty-eight were fully gorged and thirteen only partially so.

Though males were usually included along with the females in the cages, their presence or absence appeared to make little difference to feeding for the first blood meal. In a cage containing 175 females reared from female pupae and without access to males offered the arm for 26 minutes, 116 were considered gorged to the fullest extent (mean weight after feed 6·60 mg.) and fifty-nine very nearly so (mean weight 5·90 mg.). This does not apply to subsequent feeds by unfertilized females as noted later.

(*c*) TIME OF SUBSEQUENT BLOOD FEEDS

Once fully gorged, which is the normal condition following an undisturbed human blood meal, the female does not feed again until after oviposition. Macfie (1915) offered females that had fed a blood feed twice daily, but they did not feed again until eggs had been laid. The same author notes that after oviposition the female is ready, almost at once, for the next feed and that it sets out to search for food as soon as the eggs are laid. Bacot (1916) says that vigorous females feed within 24 hours of egg laying. David, Bracey and Harvey (1944) state that feeding takes place on the day of oviposition. Since oviposition normally is completed about 3 days from the blood meal, the female with the necessary opportunity feeds subsequently about every third or fourth day (Macfie, 1915). H. A. Johnson (1937) gives the intervals as short as 2 days. According to Macfie (1915) unfertilised females do not feed at regular intervals.

Under artificial conditions the number of feeds in the life of the mosquito may be very considerable. Macfie (1915) records instances of fifteen and ten feeds in mosquitoes living respectively 50 and 60 days. Howard (1923) records that in 31 days a female fed twelve times at mean intervals of 2·6 days. The great majority of females under ordinary laboratory conditions, however, do not survive for such long periods. Following the first feed the numbers in a cage considerably diminish and after a second feed mortality is very heavy.

(*d*) TIME OF DAY OF FEEDING

Aëdes aegypti is of active diurnal habit and is widely known in the tropics as the day-biting mosquito. Durham and Myers (1902), speaking of Manaos, say that *A. aegypti* is very active in the early afternoon, but after sundown there is not one. Dutton (1903), speaking of West Africa, says it only bites during the day. Goeldi (1905) says that at Para it bites from sunrise to sunset. Doane (1914) says it ceases to feed at night when *Culex fatigans* takes over. Howlett (1913) says it hardly ever bites after dark. Howard, Dyar and Knab (1912) say that if it bites by night it does so in lamp-lighted rooms. Marchoux, Salimbeni and Simond (1903), however, held that *Aëdes aegypti* bit by day and night during the first 6–8 days from emergence, but later at night only. In general, authors have not supported this view (Carter, 1904; Siler, Hall and Hichens, 1926; Bonne and Bonne-Wepster, 1925). That the species may feed at night is, however, shown by the observations of Gordon and Young (1921). These authors liberated thirty-one marked females in a room at night not less than 14 days from their first blood meal. During the succeeding 4 days five marked and three unmarked fed during daylight. None were observed to feed at night. Of fifty similarly liberated, ten marked and ten unmarked fed in daylight, six marked and eight unmarked at night. Schwetz (1915) describes it as biting towards evening and as very active after sunset; later, 7–8 p.m., it ceases to attack. There seems little doubt that *A. aegypti* as commonly encountered is definitely diurnal in its habits, though it very characteristically clings to dark corners and shady places and may under certain circumstances attack at night. How far this may cease to apply under some conditions is further discussed below.

In the laboratory the species feeds readily at all times of the day, in light or in the dark. Bishop and Gilchrist (1946) performed their feeding experiments on chickens, or through membranes, in the dark, obtaining 94 per cent gorged on the chicken in 30 minutes exposure. The present author, using the same standard technique as when feeding in the normal routine, but with the arm introduced and allowed to remain in the cage in complete darkness, observed no marked difference either in the number of wheals produced in 15 minutes exposure, or so far as could be ascertained in the rate of settling on the arm.

Recently there have been references to *A. aegypti* in native huts in tropical Africa feeding by night. Lumsden (1957), referring to conditions in Southern Province, Tanganyika, and investigating the time of feeding of *A. aegypti* as shown by hourly baited catches, notes that, though the activity of the *A. aegypti* female is diurnal, the night activity is by no means insignificant. He gives the following figures as the result of a 48 hours baited catch in a hut.

	Hut interior	Hut verandah
By day	1280	416
By night	669	189

Day and night were taken as between sunrise and sunset, in this case about 5 a.m. to 5 p.m. for the day catches. A noticeable feature was the occurrence of periods of special activity. From the charts given, maximal catches in the hut interior for females are notable at 8 a.m. (159), 2 p.m. (199), 9 p.m. (107) and 3 a.m. (102). In the verandah catches of fifty or over were at 5 a.m. (50), 8 a.m. (52), 1 p.m. (72), 5 p.m. (50). During the day the number of females caught in the verandah, allowing an hour before and after sunrise and sunset for probable daylight, was 466 as against 134 (77·6 per cent). In the hut with the same provision the numbers were 1398 and 574 (70·9 per cent) or very little different. The hut interiors are noted by the author to be practically dark even by day, and a point of possible importance in reference to what is said below is that in making the catches it was necessary to use lamps. Besides females a considerable number of males were captured on the bait, a total of 249 and 354 in the hut and verandah respectively as compared with 1973 and 621 for females. Much other information is given in this very interesting paper which space does not allow further to enter into.

The choice of daytime so commonly associated with this species in its annoying activities would seem therefore to be dependent on other factors than a natural instinct to feed only by day. Whilst much no doubt remains to be ascertained regarding habits of the species in nature, it is thought that the following may give some explanation of contradictory experiences described. In searching for its food, as noted in the section dealing with reactions of the imago (ch. XXIV), *A. aegypti* appears to be influenced by two major considerations, namely attraction to dark moving objects, operative therefore at a distance from the food source, and attraction by warm air convection currents, obviously most marked near the source. In an experimental cage the effect is demonstrably almost, if not entirely, the second form of attraction. In an open room, however, there is reason to believe that the first form of attraction is that chiefly operative. Under many circumstances this might be sufficient to give the general effect, as commonly seen in better class houses, of mosquitoes active and troublesome in rooms by day and ceasing to be so by night. In a dark hut the circumstances might more nearly resemble those seen in the laboratory where the species is being studied in a cage.

Apart from special circumstances that may affect behaviour of the species there can be no doubt that compared with many mosquitoes, and including almost all anophelines and the majority of culicines, *A. aegypti* is in the main of diurnal habits.

(*e*) EFFECT OF TEMPERATURE AND OF SOME OTHER CONDITIONS ON FEEDING

EFFECT OF ENVIRONMENTAL TEMPERATURE

The effect of temperature on feeding is rather complex. It includes the effect on the insect of the temperature of the air forming its environment, the temperature of the potential food source, the relation between this and the temperature of the environment and the question of convection currents set up by the food source. For the present we are concerned only with the direct effect of low and high temperature of the environment on feeding. Environmental temperature appears to affect feeding mainly through the general effect upon the activity of the insect, which will in the main attempt to bite at any temperature which leaves it physically able to attack.

The lower temperature limit. At low temperatures somewhat above the lethal point *A. aegypti* becomes torpid, unable to fly or even move its limbs except slowly and with difficulty. In such condition it is rendered incapable of attacking. Otto and Neumann (1905) record that exposed out of doors to freezing it is killed. At 4° C. it could be revived after an hour, but with longer exposures it was killed. At 7–9° C. females were rendered stiff and moved slowly and were nearly always on the bottom of the jar, presumably being unable to cling to the glass sides of the vessel. Lewis (1933) found *A. aegypti* at 10° C. dormant and only stimulated to move when warmed. The temperature at which the insects cease to bite has been very commonly given as about 15° C. (Marchoux *et al.* 1903; Howard *et al.* 1912; Connor, 1924). Reed and Carroll note that it will bite at 63° F. (17° C.) and above. They have never succeeded in getting the species to feed below this temperature. Cossio (1931) records *A. aegypti* as biting in nature at 15° C. and experimentally at 14° C. Gutzevich (1931) notes that *A. aegypti* on the Black Sea Coast does not feed or oviposit below 16° C. In the author's laboratory at Cambridge escaped free *A. aegypti* attacked in an attempt to feed during some cold weather at a temperature in the room of 15° C. This would appear to be the lower temperature limit for such activity.

Optimal temperature. What exact temperature is optimal for biting can only be very generally stated. Marchoux *et al.* (1903) note that between 19° and 25° C. females are slow to bite, but that from 26° to 35° C. they will largely feed. Connor (1924) states that the species is most active at 28° C., and this temperature is very generally considered by authors optimal for the species. Aragao (1939) gives between 27° and 32° C. as the vital optimum, all vital activities being retarded between 25° and 17°, whilst below 17° C. they cease. Seaton and Lumsden (1941) for their controlled experiments used a fixed temperature of 24–26° C. Our own experience is that the biting activity at 28° C. with high humidity is somewhat greater than at 25° C. At both temperatures and relative humidity of 70 per cent or over they settle to feed rapidly and as a rule gorge fully within 5 minutes (Christophers, 1947). Under favourable conditions even lower environmental temperatures seem quite compatible with active attacking and gorging. We have found that in cages they will settle rapidly on the arm and gorge well, provided the arm is warm, at 18° C.

Higher temperature limit. Regarding biting at air temperatures above 28° C. there is little information. Lewis (1933) at 35° C. obtained 11 per cent feeding at 96 per cent relative humidity and 17 per cent at 77 per cent relative humidity, the figures not being very different to those obtained at 30° C. None fed at 40° C. and relative humidity 83 per cent. As will be shown later, 40° C. is practically a lethal temperature for the adult.

EFFECT OF HUMIDITY

That a high humidity is favourable to the life of the species is well recognised and in nature attack by the insect is most common in relatively high humidities. But biting may take place quite freely at lower humidities, for example, at ordinary English summer temperature and humidity. For routine testing a relative humidity of 70–90 per cent at 25° C. was found to ensure maximum and standard settling and biting effects. As noted by Lumsden (1947) relative humidity at time of biting did not appear to be of importance. A very high humidity has, however, been noted as having a deterrent effect on feeding at higher temperatures.

EFFECT OF LIGHT

Light, if subdued, does not appear to have a deterrent effect. But during feeding under an electric light it was noticeable that the insects when settling to feed tended to avoid the illuminated area for those in shadow. The question has been studied under controlled conditions by Seaton and Lumsden (1941). These authors found that light (0·5 metre candles at the skin surface) reduced the number of fourth-day mosquitoes feeding at 24–26° C. and Hg. sat. def. 4·5–5·5 mm. by almost half. Of 100 mosquitoes fed in lots of ten, sixty fed in darkness and thirty-four in light. This is in keeping with the known habit of the species in nature of attacking in the subdued light of dark corners and on the shady side of the victim.

EFFECT OF FERTILISATION

That fertilisation is not necessary to ensure the female feeding has been noted by a number of observers and the point has been put to careful test by Seaton and Lumsden (1941). Experiments by these authors showed that the presence of males up to the third day from emergence made no difference in the number feeding. The present author's experience in the same direction has already been given. This applies, however, only to the first blood meal, the course of subsequent feeds being much modified by the irregular ovarian development as shown later.

OTHER GENERAL CONDITIONS AFFECTING FEEDING

Some effect upon biting potential was thought to result when the insects had been previously allowed to feed on sugar, fruit and even water shortly before experiment. Thus deprivation from water for an hour combined with exposure to light was found greatly to increase and standardise biting activity and was used as a routine. The use of sugar solution or fruit has been widely used by observers with the idea of maintaining the life of the insects. Except for the purpose of maintaining insects alive for special purpose over long periods such usage would not seem desirable where critical experimentation on biting activity is being carried out and where the condition of the insects used should be as standardised as possible. Up to the seventh day such feeding is, with suitable technique, unnecessary and introduces an unknown factor where every endeavour should be to reduce conditions to the simplest possible.

Moistening the skin when mosquitoes were sluggish has been thought by some observers to be conducive to a more rapid settling rate. Lumsden and Bertram (1940) noted that mosquitoes showing reluctance to feed could often be induced to do so if saliva were applied to the skin when feeding on the fowl. This, however, was not found to have any effect by Bishop and Gilchrist (1946) in their feeding experiments using the chicken.

Sweating was not found by Gordon (1922*a*) to affect biting. Out of thirty individuals sweating and fifty-eight not sweating the average bites per individual was in both cases seven. Our own experience has been that sweating, if it did not increase the settling and biting rate, did not interfere with it. In this case the arm is bound to be warm and thus favourable to biting.

Presence of a heavy growth of hair was to some extent deterrent to rate of settling, but, unless combined with presence of repellents, was not eventually protective. Gordon (1922*a*) found the mean number of bites for 'hairiness showing' and 'hairiness not showing' to be respectively 7·1 and 6·9.

Of all influences adversely affecting biting (short of sublethal temperatures and other abnormal conditions) the most obvious was lack of warmth of the observer's skin. On a cold day introduction of a cold arm into a cage gave delayed feeding. Warming up the arm before introduction had an entirely different effect. Few results in respect to feeding urge are more illuminating than those of Howlett (1910) who, using a warm test-tube, showed that this when held upright gave reduced attraction, thus showing that it was warm convection currents rather than radiant heat that was effective in bringing about attack (see ch. XXIV, p. 540). The species in this case was *A. albopictus* the behaviour of which is very similar to that of *A. aegypti*. Crumb (1922) has observed a similar effect with *Culex pipiens*.

(*f*) CHEMICAL ATTRACTANTS AND REPELLENTS

Chemical substances having an attractive effect appear to be rare and such as have been shown to have such an effect possess it in very moderate degree. Crumb, though he found warmth powerfully attractive to *Culex pipiens*, notes that the components of perspiration and blood produced only faint erratic response. Rudolfs (1922, 1930) with *Aëdes sollicitans* and other species also found sweat, sebaceous secretions and many other substances little attractive. But CO_2 and weak ammonia were 'strongly activating'. Van Thiel (1935), as also Reuter (1936), found with *Anopheles atroparvus* that acids in human sweat did not attract. Bates (1949) was never able to trap mosquitoes using absorbent cotton that had been rubbed over sweating animals as bait, though it was easy to trap mosquitoes with vegetable bait such as slices of apple or sugar solutions. Parker (1948), as also Willis (1947), found human arm odour attractive. It seems very doubtful if olfactory stimuli are important or necessary in causing *Aëdes aegypti* to attack with the object of feeding. There is clearly too a distinction between the urge with which a female *A. aegypti* is attracted, say to fruit or sugar solution, and the excited behaviour of the insect intent on attacking for the purpose of a blood meal. This is evident not only in the time taken to produce the result, but in the characteristic behaviour of the insect.

Chemical substances having a repellent effect are, on the contrary, many and various. During the last war very considerable research was undertaken in testing and evaluating large numbers of chemical substances with a view to selecting those that could most effectively be employed in protection of troops from the bites of mosquitoes. As a result three substances became especially recognised as suitable for such use, namely indalone, a proprietary preparation; Rutger's 612 (2-ethylhexane-1:3-diol) synthesised at Rutger's University and dimethyl phthalate. Of these dimethyl phthalate (commonly referred to as DMP) has now come chiefly into use. For an account of the history, method of testing and chemical nature and properties of repellent substances with a bibliography see Christophers (1947) and for some points regarding repellents for civilian use, Christophers (1945). An account of the work done at Orlando, 1942–7, is given by Travis *et al.* (1949). For other work of more recent date see Knipling and Dove (1944); Dethier (1947); Travis and Smith (1951); Sarkaria and Brown (1951); King (1951); Roadhouse (1953). For tests in the field against northern *Aëdes* in Canada see Applethwaite and Smith (1950); Applethwaite and Cross (1951); Altman and Smith (1955).

Three points are of special importance in the selection of compounds suitable for use as repellents; these are: (1) the degree of repellent effect; (2) the permanence of this effect when the substance is spread upon human skin; and (3) absence of irritant or toxic effect. It is also useful to know its boiling point and viscosity (see Christophers, 1947).

Whilst tests of efficacy in the field have played an important part in determining the value of different compounds, accurate information has mainly resulted from testing in the laboratory. In this latter respect *A. aegypti* has been very largely used. Such tests are valuable largely in proportion to the use of standardised conditions employed. The following are substances that with the author have yielded high degrees of repellency. They are given with a rating value which requires some explanation. A substance may be very effective but its effect short-lasting. Or it may be only moderately effective, but very lasting. All grades may be found in regard to these two points. In the rating value given in the list complete repellency from 10 c.mm. spread on the arm per square inch is indicated by A, A' means an equivalent effect but at half the dose, and A'' the same with a still further halving of the dose. The figure after the stroke indicates the number of hours that the substance in that dosage under standard conditions remains completely effective as shown overleaf.

The degree of permanence is largely dependent upon the boiling point of the compound. Thus, to give a longer protection in the standard dose than 2 hours the boiling point must be at least about 200° C. For this reason many powerful repellents, such as citronella, lack the permanence required for military purposes, that is at least 6 hours, and usually give protection for about 2 hours. With boiling point over 300° C. there is a tendency for the substance to give lasting effect but of relatively poor intensity. Thus dimethyl phthalate with boiling point of 282° C. gives about the optimum effect for the 6 hours required duration, but dibutyl phthalate with boiling point 325° C. gives a poorer repellency. On account of its higher boiling point, however, dibutyl phthalate has been much used for rendering clothing protective, since articles impregnated with this substance retain repellent effect for long periods and after washing.

Substance	Boiling point (° C.)	Rating
Oil of citronella	—	*A*/2
Citronellal	80·7	*A*/4
Citronellol	—	*A*/4
Hydroxycitronellal	—	*A'*/6
Oil of cassia (cinnamic aldehyde)	246	*A*/4
Rutger's 612	—	*A'*/6
Another related diol	328	*A*/6
An unsaturated long chain dihydric alcohol	314	*A"*/6
Dimethyl phthalate	282–5	*A'*/6
Di-butyl phthalate	325	*B*/2
Di-ethyl cinnamate	271–5	*A'*/6
Isopropyl cinnamate	342	*A*/2
Trichloroacetyl aminoethyl chloride (15 per cent in ointment)	—	*A"*/6

Whilst a high degree of permanence has been regarded as an essential requirement for protection of troops in the field, this is less essential as a rule for civilian use and some of the lower boiling point repellents may here be more useful, for besides being often more effective for a period of, say, 2 hours, they have the great advantage of being more effective at a distance. Thus very often in place of all skin areas requiring to be covered, it may be necessary only to apply the substance here and there on the clothing, for example on the socks or about the wrists or behind the collar. The most commonly used repellent of this type is oil of citronella, but there are still more effective substances that could be used, such as citronellal, the active principle, or hydroxycitronellal, a very effective repellent with a pleasant smell. The chief precaution necessary is that the substance should not be irritant or poisonous.

What the essential reason may be for a substance to be repellent is not known. Many such compounds have citrous or other quite pleasant smells. The high boiling point esters of the lower methyl series and high molecular weight organic acids (phthalic, adipic, cinnamic, citric, etc.) appear to owe their repellent effect to methyl radicles in suitable combination. In the long chain alcohols it is apparently the position of the OH groups which ensures repellency, the methyl radicles being little concerned.

Repellent substances may be used 'neat' by being spread thinly by smearing over exposed skin, but may also be used in the form of some pharmaceutical preparation, such as a lotion, cream or ointment, a cream being that for the most part adopted. In general the neat application is the most effective but may be unpleasant for delicate skins. Dimethyl phthalate (DMP) is not very readily combined in any considerable strength as a cream owing to its low viscosity, a favourable condition for the 'neat' use. The following, however, is a permanent cream spreading readily and containing 25 per cent of the compound:

Dimethyl phthalate	12·5 c.c.
White wax B.P.	9·0 g.
Arachis oil	27·5 c.c.

Melt the wax in the arachis oil over a water-bath and stir in the dimethyl phthalate. If necessary filter through a little wool. Put up as an ointment in a wide-mouthed bottle or pot. Though it contains wax it is not waxy or unpleasant on this account and may be varied slightly in composition to meet climatic requirements. Or an excellent cream can be made using 70 per cent DMP and 30 per cent magnesium stearate.

Oil of citronella is liable to be adulterated and a good Java or other guaranteed oil should be obtained, or if cost is not very material the essential compound citronellal or citronellol used. Such low boiling point compounds can usually be put up as creams or more simply applied here and there on clothing, etc. Some preparations on sale for use against mosquito bites appear to be based on the idea that more of the substance can be included by use of zinc oxide as an adsorbent, but even if the adsorbed compound is really active the cosmetic effect of zinc oxide makes such preparations very undesirable. For information about impregnation of clothing and other facts connected with repellents see Christophers (1947). It is understood that Dr Busvine is shortly publishing a paper on repellents.

(*g*) BITING RATE

Various methods have been used to indicate the number and intensity of attacks by mosquitoes, either in the open under natural conditions or in cages in the laboratory. One such that has been in common use is 'time to first bite' expressed in minutes. Another method makes use of the number of settlings or bites per minute and has been termed 'the biting rate'.

In tests carried out in nature these indices, determined on an untreated subject, give an indication of the number and biting urge of mosquitoes present at the time, usually a very variable factor depending upon climatic conditions, time of day and other factors. It gives, however, a basis for evaluating the degree of effectiveness of the preparations tested.

Table 30. *Frequency distribution of number of females settling in 15 seconds exposure on the untreated arm*

			Number settling and wheals in groups of 5						
Group	Result	Number of observations	0–5	6–10	11–15	16–20	21–25	26–30	Total settlings or wheals
A	Settlings	209	0	24	78	78	29	0	3390
	Wheals*	95	4	46	23	19	1	1	1046
B	Settlings	74	1	11	26	19	18	—	1191
	Wheals	74	1	33	20	19	1	—	886

* Relates only to feedings in this series for which wheals were recorded.

Table 31. *Number of females settling and leaving in a gorged condition in an exposure of* 15 *minutes*

			Time intervals						
Group	Number of females used	Result	15″	30″	60″	2′	5′	10′	Totals
A	100	On arm	8	10	38	50	67	67	—
		New settlings	8	2	28	26	23	2	89
		Left gorged	—	—	—	14	38	22	74
B	200	On arm	21	40	105	161	171	42	—
		New settlings	21	19	65	70	57	0	232
		Left gorged	—	—	—	14	61	93	158

In the case of cage experiments response to the untreated arm, provided the mosquitoes are in a suitable condition for use in testing, is very rapid and some modification of the above use of the test is necessary. For this purpose the author uses a preliminary short exposure of the untreated arm immediately before carrying out a test for repellency. The arm being inserted only for 15 seconds does not interfere with the adequacy of the test cage, the result serving to show that the mosquitoes are normal in their reaction. As this test has been carried out and the results recorded over a long period, a considerable mass of data has been accumulated in this respect serving to show how extremely rapid is the attacking reaction under such circumstances. In Table 30 is given the distribution of the number of females settling in exposures of the arm for 15 seconds tests. The area of arm exposed is 30 square inches, the cages those described as type *B* cages and the mosquitoes 5–7 days from emergence kept in small type *A* cages at 25° C. and 70 per cent relative humidity until, shortly before the repellency test, they are transferred to the larger cage and exposed for 1 hour, without access to water, to illumination from an electric lamp. Each experimental cage contains mosquitoes hatched from 100 female and 100 male pupae. The data relate to two observers, *A* and *B* respectively.

The observations show under *A* that out of 209 tests from ten to twenty females settled in the 15 seconds in no less than 156 of these and that in a fair number of cases up to twenty-five had settled to feed in that short time. Under *B* a very similar result is shown.

The table shows also the recorded number of wheals that followed on these exposures. Some interest attaches to these since the presence of a wheal had been found to show that a mosquito had remained at least some 6 seconds from the time of settling to puncture. From series *B* in which observations on wheal formation were recorded for all tests it will be seen that wheals formed only 74 per cent of settlings, those mosquitoes settling within less than 6 seconds from termination of the test having probably not had time to puncture, or punctured long enough to leave a wheal. In this brief period of the preliminary test 4581 settlings or 16·3 per cent are accounted for out of a possible total of 283×100 if all the 28,300 females used had settled.

In Table 31 are given the results of counts made by two observers each recording settlings and leaving in a gorged condition of mosquitoes on their side of a line down the middle of the exposed area of 30 square inches of skin with a time interval of 15 minutes. Some discrepancies are due to the difficulty of keeping accurate record under such conditions. Also a certain number of the mosquitoes, usually amounting to about 20 per cent, may leave the situation on which they have first alighted for the eventual feed. The table, however, gives a good idea of the settling rate as also for the time of engorgement and of leaving the arm in a gorged condition. Many, however, remain on the arm in a fully gorged condition, having withdrawn their mouth-parts for some considerable time unless disturbed.

When the test is performed with a repellent, in place of almost immediate settling and gorging, the 'time to first bite' may be prolonged for hours in the case of a strong repellent. Where the repellent is one acting at a distance, mosquitoes usually remain resting on the roof or sides of the cage and make little or no attempt to feed. Substances that have little or no action at a distance may show the curious phenomenon that may be described as 'hopping'. The mosquitoes settle, but quickly leave, only again to alight and leave. Another frequent condition when the repellent is not too strong is for the mosquito to fly to the arm as if to alight but fail to do so and to leave the arm at a tangent. Under such circumstances, though they do not alight, a faint touch may be felt by the observer.

REFERENCES

ALTMAN, R. M. and SMITH, C. N. (1955). Investigation of repellents in protection against mosquitoes in Alaska 1953. *J. Econ. Ent.* **48**, 67–72.

APPLETHWAITE, K. H. and CROSS, H. F. (1951). Further studies of repellents in Alaska. *J. Econ. Ent.* **44**, 19–22.

APPLETHWAITE, K. H. and SMITH, C. N. (1950). Field tests with mosquitoes and sandfly repellents in Alaska. *J. Econ. Ent.* **43**, 353–7.

ARAGAO, DE B. H. (1939). Mosquitoes and yellow fever virus. *Mem. Inst. Osw. Cruz*, **34**, 565–81.

BACOT, A. W. (1916). *Rep. Yell. Fev. Comm.* (*West Africa*). Vol. III. London.

BATES, M. (1949). *The Natural History of Mosquitoes.* Macmillan Co., New York.

BISHOP, A. and GILCHRIST, B. M. (1946). Experiments upon the feeding of *Aëdes aegypti* through animal membranes, etc. *Parasitology*, **37**, 85–100.

BONNE, C. and BONNE-WEPSTER, J. (1925). *Mosquitoes of Surinam: a Study of Neotropical Mosquitoes.* Kolonial Institute, Amsterdam.

BONNE-WEPSTER, J. and BRUG, S. L. (1932). See references in ch. II (*a–b*), p. 45.

BOYCE, R. (1907). Quoted by Howard, Dyar and Knab (1912).

CARTER, H. R. (1904). Some characteristics of *Stegomyia fasciata* which affect its conveyance of yellow fever. *Med. Rec. N.Y.* **65**, 761–6.

CHRISTOPHERS, S. R. (1945). Insect repellents. *Brit. Med. Bull.* **3**, 222–4.

CHRISTOPHERS, S. R. (1947). Mosquito repellents: being a report of the work of the mosquito repellent enquiry, Cambridge 1943–5. *J. Hyg., Camb.*, **45**, 176–231.

CONNOR, M. E. (1922). Renseignements sur l'*Aëdes aegypti* (*Stegomyia*). Mexico. Abstract in *Bull. Off. Int. Hyg. Publ.* **15**, 506.

CONNOR, M. E. (1924). Suggestions for developing a campaign to control yellow fever. *Amer. J. Trop. Med.* **4**, 277–307.

COSSIO, V. (1931). Observaciones sobre al *Aëdes aegypti*, etc. *Bol. Cons. Nac. Hig. Uruguay.* Abstract in *Rev. Appl. Ent.* **19**, 230.

CRUMB, S. E. (1922). A mosquito attractant. *Science*, **55**, 446–7.

DAVID, W. A. L., BRACEY, P. and HARVEY, A. (1944). Equipment and method employed in breeding *Aëdes aegypti* in the biological assay of insecticides. *Bull. Ent. Res.* **35**, 227–30.

DAVIS, G. E. and PHILIP, C. B. (1931). The identification of the blood meal in West African mosquitoes by means of the precipitation test. *Amer. J. Hyg.* **14**, 130–41.

DETHIER, V. G. (1947). *Chemical Insect Attractants and Repellents.* H. K. Lewis and Co., London.

DOANE, R. W. (1914). Disease-bearing insects in Samoa. *Bull. Ent. Res.* **4**, 265–9.

DRAKE-BROCKMAN, R. E. (1913). Some notes on *Stegomyia fasciata* in the coast towns of British Somaliland. *J. Lond. Sch. Trop. Med.* **2**, 166–9.

DURHAM, H. E. Quoted by Theobald, *Mono. Cul.* **3**, 1903.

DURHAM, H. E. and MYERS, W. (1902). Report of the yellow fever expedition to Para. *Lpool Sch. Trop. Med.* Mem. VII.

DUTTON, J. E. (1903). Report of the malaria expedition to Gambia. *Lpool Sch. Trop. Med.* Mem. X.

FIELDING, J. W. (1919). Notes on the bionomics of *Stegomyia fasciata* Fabr. *Ann. Trop. Med. Parasit.* **13**, 259–96.

FINLAY, C. (1886). Yellow fever transmission by means of the *Culex* mosquito. *Amer. J. Med. Sci.* **92**, 395–409.

GOELDI, E. A. (1905). *Os mosquitos no Para.* Weigandt, Paris.

GORDON, R. M. (1922*a*). The susceptibility of the individual to the bites of *Stegomyia calopus. Ann. Trop. Med. Parasit.* **16**, 229–34.

GORDON, R. M. (1922*b*). Notes on the bionomics of *Stegomyia calopus* Meig. in Brazil. *Ann. Trop. Med. Parasit.* **16**, 425–39.

GORDON, R. M. and LUMSDEN, W. H. R. (1939). A study of the behaviour of the mouth-parts of mosquitoes when taking up food from living tissues etc. *Ann. Trop. Med. Parasit.* **33**, 259–78.

REFERENCES

GORDON, R. M. and YOUNG, C. J. (1921). The feeding habits of *Stegomyia calopus* Meig. *Ann. Trop. Med. Parasit.* **15**, 265–8.

GUTZEVICH, A. V. (1931). The reproduction and development of the yellow fever mosquito under experimental conditions (in Russian). *Mag. Parasit. Leningrad,* **2**, 35–54. Abstract in *Rev. Appl. Ent.* **21**, 2.

HOWARD, L. O. (1923). The yellow fever mosquito. *U.S. Dep. Agric. Fmrs' Bull.* no. 1354.

HOWARD, DYAR and KNAB. See references in ch. I (*f*), p. 19.

HOWLETT, F. M. (1910). The influence of temperature upon the biting of mosquitoes. *Parasitology*, **3**, 479–84.

HOWLETT, F. M. (1913). *Stegomyia fasciata. Proc. 3rd Meet. Gen. Malar. Comm. Madras, 1912*, p. 205.

JAMES, S. P. and TATE, P. (1938). Exoerythrocytic schizogony in *Plasmodium gallinaceum* Brumpt. *Parasitology*, **30**, 128–39.

JOHNSON, H. A. (1937). Note on the continuous rearing of *Aëdes aegypti* in the laboratory. *Publ. Hlth Rep. Wash.* **52**, 1177-9.

KING, W. V. (1951). Repellents and insecticides for use against insects of medical importance. *J. Econ. Ent.* **44**, 338–43.

KNIPLING, E. F. and DOVE, W. E. (1944). Recent investigations of insecticides and repellents for the armed forces. *J. Econ. Ent.* **37**, 477–80.

KUMM, H. W. (1932). Yellow fever transmission experiments with South American bats. *Ann. Trop. Med. Parasit.* **26**, 207–13.

LANG, C. A. and WALLIS, R. C. (1956). An artificial feeding procedure for *Aëdes aegypti* L. using sucrose. *Amer. J. Trop. Med. Hyg.* **5**, 915–20.

LEWIS, D. J. (1933). Observations upon *Aëdes aegypti* L. (Diptera: Culicidae) under controlled atmospheric conditions. *Bull. Ent. Res.* **24**, 363–72.

LUMSDEN, W. H. R. (1947). Observations on the effect of micro-climate on the biting of *Aëdes aegypti* (L.) (Diptera: Culicidae). *J. Exp. Biol.* **24**, 361–73.

LUMSDEN, W. H. R. (1957). The activity cycle of domestic *Aëdes* (*Stegomyia*) *aegypti* (L.) (Dipt., Culicid.) in Southern Province, Tanganyika. *Bull. Ent. Res.* **48**, 769–82.

LUMSDEN, W. H. R. and BERTRAM, D. S. (1940). Observations on the biology of *Plasmodium gallinaceum* Brumpt 1935 in the domestic fowl with special reference to the production of gametocytes and the development in *Aëdes aegypti*. *Ann. Trop. Med. Parasit.* **34**, 135–60.

MACFIE, J. W. S. (1915). Observations on the bionomics of *Stegomyia fasciata*. *Bull. Ent. Res.* **6**, 205–29.

MARCHOUX, E., SALIMBENI, A. and SIMOND, P. L. (1903). See references inch. I (*f*), p. 19.

MATHIS, M. (1934). Agressivité et ponte comparées du moustique de la fièvre jaune en conditions expérimentales. *C.R. Soc. Biol., Paris*, **115**, 1624–6.

OTTO, M. and NEUMANN, R. O. (1905). Studien über Gelbfieber in Brasilien. *Z. Hyg. InfektKr.* **51**, 357–506.

PARKER, A. H. (1948). Stimuli involved in the attraction of *Aëdes aegypti* L. to man. *Bull. Ent. Res.* **39**, 387–97.

PERYASSU, A. G. (1908). *Os Culicideos do Brazil.* Leuzinger, Rio de Janeiro.

REUTER, J. (1936). *Orienteerend onderzoek naar de oorzaak van het gedrag van* Anopheles maculipennis *Meigen bij de voedselkeuze* (summary in English). 118 pp. Proefschr, Rijksuniv, Leiden. Abstract in *Rev. Appl. Ent.* **24**, 223.

ROADHOUSE, L. A. O. (1953). Laboratory studies on insect repellency. *Canad. J. Zool.* **31**, 535–46.

RUDOLFS, W. (1922). Chemotropism of mosquitoes. *N.J. Agric. Exp. Sta. Bull.* no. 367.

RUDOLFS, W. (1930). Effects of chemicals upon the behaviour of mosquitoes. *N.J. Agric. Exp. Sta. Bull.* no. 496.

SARKARIA, D. S. and BROWN, A. W. A. (1951). Studies of the responses of the female *Aëdes* mosquito. Parts I and II. *Bull. Ent. Res.* **42**, 115–22.

SCHWETZ, J. (1915). Preliminary notes on the mosquitoes of Kabinda (Lomani), Belgian Congo. *Ann. Trop. Med. Parasit.* **9**, 163–8.

REFERENCES

SEATON, D. R. and LUMSDEN, W. H. R. (1941). Observations on the effect of age, fertilization and light on biting by *Aëdes aegypti* (L.) in a controlled micro-climate. *Ann. Trop. Med. Parasit.* **35**, 23–36.

SEN, S. K. (1918). Beginnings in insect physiology and their economic significance. *Agric. J. India*, **13**, 620–7.

SILER, J. F., HALL, M. W. and HICHENS, A. D. (1926). Dengue: its history, epidemiology, mechanism of transmission, etc. *Philipp. J. Sci.* **29**, 1–304.

THEOBALD, F. V. (1903). See references in ch. I (*e*), p. 18.

TRAVIS, B. V., MORTON, F. A., JONES, H. A. and ROBINSON, J. H. (1949). The most effective mosquito repellents tested at the Orlando Florida Laboratory 1942–7. *J. Econ. Ent.* **42**, 686–94.

TRAVIS, B. V. and SMITH, C. N. (1951). Mosquito repellents selected for use on man. *J. Econ. Ent.* **44**, 428–9.

VAN THIEL, P. H. (1935). Onderzoekingen omtrent den gedrag van Anopheles ten opzichte van Mensch, etc. *Geneesk. Tijdschr. Ned.-Ind.* **75**, 2101–18.

WILLIS, E. R. (1947). The olfactory response of female mosquitoes. *J. Econ. Ent.* **40**, 769–78.

WOKE, P. A. (1937*a*). Cold-blooded vertebrates as hosts for *Aëdes aegypti* Linn. *J. Parasit.* **23**, 310–13.

WOKE, P. A. (1937*b*). Comparative effects of the blood of different species of vertebrates on egg production of *Aëdes aegypti* Linn. *Amer. J. Trop. Med.* **17**, 729–45. See also further references given under ch. XXII.

XXI
THE BLOOD MEAL

(*a*) DESCRIPTION OF THE ACT OF FEEDING

Few biting insects are more apparently intelligent, cunning and cautious than *Aëdes aegypti* in its behaviour when attacking. It rarely attacks blindly as some mosquitoes do or makes a frontal attack. Usually it approaches from the shady side and from behind, and it is difficult to get the attacking insect plainly visible against a suitable background. Commonly after failure to achieve its object at a first trial it makes a renewed attack from a quarter different to that which is engaging the victim's attention. Its approach is silent. According to Howard, Dyar and Knab (1912) a good description of the stealthy attack has been given by Parker, Beyer and Pothier (1903). The male, though it does not bite, behaves very much as the female, hovering and darting about the seemingly intended victim apparently without any object, since it rarely settles. Only under one condition does the female mosquito appear to disregard all caution, namely when it has been interrupted in the act of feeding. It may then literally precipitate itself upon the victim regardless of consequences.

On alighting on the skin it quickly punctures, usually if hungry without hesitation and at the spot at which it first alights. In a certain proportion of cases more than one puncture is made, though normally the insect does not easily leave the first-selected site and if blood is not tapped within the first minute it will often remain several minutes until it is ultimately successful. On the other hand, it may after a time withdraw its mouth-parts and make one or more further punctures. Gorging is complete with the majority of insects in good condition within from 2 to 5 minutes. Evidence of puncture in the normal subject is given by the subsequent appearance of a wheal.

A number of authors have described with various degrees of detail the act of feeding, a good general description for *Culex* being given by Reaumur as early as 1738. The following is a description of the act as it can be seen by the naked eye or under a lens.

The insect alights, folds its wings and takes up a position much as when resting. The tip of the proboscis is then naturally close to the skin at a point roughly central to the four points of support provided by the fore- and mid-legs. If strong and hungry the insect straightens its fore-legs and raises the front part of its body to bring the proboscis into a more vertical position before sinking it in the tissues. In many cases, however, the mosquito simply remains as if resting until it is seen that the labium is buckling and the mouth-parts evidently sinking in. As the mouth-parts sink in, their direction usually, however, becomes more vertical. At first the hind-legs may be held, as they commonly are at rest, in the air giving no support to the body. But with commencing penetration the hind tarsi are brought down on to the skin. The femoro-tibial angles of all the legs are made more acute and the tarsi are drawn in and brought flat on the skin. In other words the insect appears to settle to work and take a grip, a fact supporting the contention of Robinson (1939) that the force for initial penetration of the epidermis is supplied from the legs through the neck (see section (*b*) on mechanism of feeding). It is this drawing in of the legs which gives the feeding insect its characteristic crouched appearance.

Usually without any sign of effort the mouth-parts appear to sink into the skin, whilst the ensheathing proboscis buckles to allow of this. Though the palps may be raised, the act is not conspicuous as with the long palps of *Anopheles* and usually they appear much as when at rest. During sinking of the mouth-parts, however, they exhibit a distinct tremor as described by MacGregor (1930), due to the action of the maxillary muscles. A similar tremor is often seen on withdrawing the parts. It soon becomes evident, especially in a lateral view, that the labium has bent sufficiently to expose some length of fascicle which is forced out of the groove it normally occupies. Bending of the labium, as it buckles more and more, is mainly at a point at the junction of the basal with the middle third of the organ. The bending at this point is quite sharp and becomes sharper as penetration of the fascicle progresses. When the fascicle has penetrated to about half its length blood may begin to enter as is shown by the sudden appearance of a bright red streak in the fascicle. The labium now has the appearance of being sharply bent, the basal third sloping downwards to the bend and the distal two-thirds passing to the site of puncture. Sometimes the distal third remains approximated to the fascicle, at others the labium retains its hold only at the extreme end, being bent like a bow with a kink where it is more sharply bent. With deeper penetration the kink in the labium from being an angular bend becomes like the bend in a hairpin, the basal third of the labium being brought into apposition eventually with the middle third should the fascicle be sunk to its full depth. By this time the bent labium has become directed backwards and upwards into the hollow between the head and prothorax. The head is depressed in such a position that the occiput viewed from above only just shows in front of the anterior promontory of the thorax. Sometimes after being sunk more deeply the fascicle is somewhat withdrawn or it may be moved up and down several times.

At some stage a bright red streak of blood suddenly appears in the fascicle and following upon this the abdomen slowly swells; at first rather slowly, later more rapidly. The insect now remains passive as the abdomen enlarges more and more. As it does so it ceases to remain parallel to the wings as in the unfed condition and sags away from these which no longer suffice to cover it. As engorgement proceeds the cerci are extruded and a drop of clear fluid is shot out on to the skin. This may be repeated several times.

As full engorgement approaches the abdominal pleurae are greatly stretched so that they become at least as extensive as the scaled portions. The membrane of the intersegmental areas is also stretched out and exposed so that these and the pleurae as far down as the fifth abdominal segment show up bright red from the ingested blood. Unless there is some delay in striking blood, completion of gorging usually takes from 2 to 3 minutes. Often once started it is completed within a minute.

If undisturbed, the insect now remains for some little time with the fascicle lying in the tissues apparently no longer engaged in conveying blood. Then the fascicle is withdrawn. Before this is done it may be partially withdrawn and resunk, often several times, seeming to slide up and down with no effort, as though in a bore hole. Final withdrawal, however, is often accompanied with considerable effort, and a marked tremor of the head may occur and be transmitted to the palpi and even the antennae. Or the fascicle may be withdrawn unobtrusively and the proboscis observed to be now no longer active but just resting on or near the skin.

Having completed gorging, the insect, unless disturbed, does not leave its position but remains at rest with its proboscis laid along or near the skin for as much as 15 minutes or more, though disturbance causes it at once to fly away, or as described by MacGregor (1915) hop away, for it now flies with difficulty and rapidly comes to rest if only on the floor of the cage.

Having found a suitable place nearby to settle the insect now remains quiet, and if in a tube with damp filter paper it will remain for days resting towards the top of the strip or on the netting covering the tube until oviposition approaches.

A very curious phenomenon is commonly seen shortly after the fascicle has been withdrawn, namely a strange writhing movement of the proboscis. This has been described by MacGregor as vibratory, but the movement is more like a man settling his arm in a sleeve, and the appearance is obviously the insect settling the fascicle again satisfactorily in its sheath. What makes it so surprising to watch is that the proboscis, which one is apt to look upon as a more or less rigid, or at least inert, organ literally squirms, there being no doubt, watching it, that intrinsic muscles must be concerned (see under 'Muscles of the labium', p. 585).

Another striking feature of feeding is that the insect, once it has begun to suck blood, appears to become oblivious of all danger and considerable physical force is required to make it give up its hold. Moreover, even when dislodged, it will often refuse to leave the spot and in spite of being pushed about will endeavour to bite again. This feature is referred to by Gordon and Lumsden (1939) when they note that they were only able to get *A. aegypti* to feed on the frog's foot by employing mosquitoes which had been allowed to start feeding on the human arm. When nearing repletion, however, the insect usually leaves readily if disturbed.

(*b*) MECHANISM OF FEEDING

The mechanism of feeding has been very thoroughly worked out by MacGregor (1931), Robinson (1939) and Gordon and Lumsden (1939) (see also Griffiths and Gordon, 1952). A number of points have been further clarified by the work of Bishop and Gilchrist (1944, 1946). It is interesting to note that the bending back of the labium was observed by Finlay as far back as 1886. He says that in the case of man the stylets penetrate 1½–2 mm. before a blood vessel of sufficient calibre is met with. He gives 1–7 minutes for completion of the act.

MacGregor, following up the observation of Hertig and Hertig (1927) and of Kadletz and Kusmina (1929) that mosquitoes can be fed artificially with a capillary tube, detached altogether the fascicle from the labium before the operation and was thus enabled to study the induced automatic imbibition of fluid. He described two methods of feeding—the continuous and the discontinuous. When feeding on blood in the normal way the insect used the continuous method and the blood passed into the mid-gut. Fluids such as sugar mixed with blood are taken in by discontinuous aspiration and pass into the diverticula. When blood sucking is interrupted, any fluid left in the proboscis is diverted to the diverticula. These organs according to the author act not only as food reservoirs for fluids other than blood, but also as air separation chambers against airlocks.

Robinson (1939) has further followed the mechanism of feeding and has described very fully the action of the different mouth structures in this process. For the details of the very complete exposition the original paper should be consulted. Briefly the initial maxillary thrust is due to the contraction of the maxillary protractor muscle. The maxillary galeae (maxillae of authors), being thus bent by impact with the skin, tend to bend and separate from the labrum and other stylets, but are held in place by surface tension and so are forced into the tissues. Penetration of the fascicle is thus begun. Two phases in penetration are to be distinguished, namely puncturing the epidermis and penetration into the dermis.

During the initial thrust Robinson considers that the thrusting force is exerted on the thorax by the clinging action of the legs. This is transmitted through the neck to the head and steadies the area of origin of the maxillary protractor muscles. The author notes that though the neck is membranous it may act in this way by reason of turgidity caused by blood pressure. The neck, however, is provided through the cervical sclerites with broad chitinous plates passing from articulation with the prothorax to the occipital portion of the head, which should assist in giving the necessary support. There seems no doubt from the stance of the insect when commencing to puncture that the legs are important in providing at least a fulcrum for action (see under description of the act of feeding). Later by alternate forward movements of the maxillae, using anchorage of the opposite maxilla for purchase, the maxillae in rapid rhythmical movement literally draw the fascicle into the tissues through the pull of the maxillary retractor muscles. Eventually the fascicle is sunk often for nearly its full length into the tissues.

The third stage, which follows when blood has been tapped, is one of passivity, the movements of the mouth-parts ceasing and the insect sitting passively on the skin until engorgement is completed. Engorgement, once blood has begun to flow, is very rapid, seeing how fine is the capillary blood channel through which the blood has to pass. The channel is about 2 mm. long and has a diameter of about 0·03 mm. Through it commonly in 2 minutes passes from 2 to 4 c.mm. of blood. The volume of the channel space from the above data works out as 0·0014 c.mm. To pass the volume of blood given it would therefore have to be filled from 1430 to 2860 times in 120 seconds, which would give the rate of flow as from 2 to 4 centimetres per second.

During withdrawal of the mouth-parts movements of the maxillae again take place, usually assisted to a greater or less extent by traction exerted by the legs transmitted through the neck and head.

Gordon and Lumsden (1939) observed the feeding of *A. aegypti* in the web of the frog's foot viewed from behind. The first sign of intention to feed was the repeated application of the tip of the labium to the skin, the palps being often raised during the process. The labella were always kept pressed close together as described by Robinson and not separated as has been stated by Vogel (1921) and other authors. Before the skin was pierced by the fascicle the labium ceased to wander and remained stationary for several seconds at the selected spot. The labella seen through the web appeared as a dark circular spot from the centre of which the fascicle was protruded. At this stage an oscillatory movement was seen for a few seconds presumably due to cutting of the superficial layers of the skin by the galeae of the maxillae, followed by appearance of the sharp point of the labrum, on each side of which could be seen the rapidly moving galeae. The fascicle penetrating the tissues did so in a purposive manner. The tip was actively bent and the flexible remainder followed. When blood was taken from an extravasation into the tissues from an injured vessel, feeding might take ten minutes for completion (pool feeding). When occurring as a result of the fascicle entering the lumen of a capillary, full engorgement took only 3 minutes (capillary feeding).

These observations by Gordon and Lumsden were made on the transparent web of the frog's foot. In 1952 Griffiths and Gordon describe an apparatus (see also Griffiths and Gordon, 1951) which enables the process of feeding to be observed in the tissues of a live rodent. By this apparatus microscopic observation can be carried out by transillumination of the ear of an anaesthetised laboratory white mouse. Here as in the web of the frog the remarkable movements of the fascicle can be followed. The authors note that the fascicle enters the skin at right angles to the surface and may sometimes continue in such a direction. Usually, however, immediately after penetrating, it bends nearly at right angles and then proceeds to tunnel through the tissues parallel to the skin surface. Usually before the fascicle becomes fully extended the tip of the labium is seen to bend sharply and the forward movement is continued in a new direction. The authors comment upon the rapidity and apparent ease with which the fascicle penetrates and the effortless manner in which it alters its direction almost as if moving in a liquid. The shaft, however, shows that there is complete absence of lateral movement and the curvature assumed by the fascicle is dependent on its adaptation to the tunnel cut by the galeae. During the process of searching for blood, discharges of clear fluid, evidently salivary secretion, take place from time to time. As in the web of the frog's foot blood may be obtained by pool feeding from an extravasation or from a capillary. In the case of capillary feeding the fascicle may pass some distance along the capillary lumen. A peculiar feature in the latter case is the common occurrence of a pulsation possibly due to suction, collapse of the vessel wall and an increased velocity imparted to the contained red cells.

(*c*) THE GORGED INSECT

Immediately after feeding the abdomen is enormously swollen, measuring in a fully gorged female up to over 4·0 mm. in length and nearly 2·0 mm. in greatest breadth. The wings which when folded formerly completely covered and reached to about the end of the abdomen now only partially cover the enlarged mass which extends beyond them for a third of its length as well as laterally beyond their borders. Occupying the greater part of the mass is the enormously swollen mid-gut extending posteriorly, as shown by its vivid red colour shining through the membranous pleurae, to the posterior border of the fifth abdominal segment. Formerly entirely hidden or indicated only as a line of meeting of the tergites and sternites the pleurae now form extensive areas on either side of greatly stretched thin membrane, under which can be seen by the binocular thin white lines formed by stretched tracheal vessels and posteriorly some coils of Malpighian tubules shown up on the red background. The intersegmental membrane between the tergites and that between the sternites are similarly stretched so that these structures are widely separated by membrane.

Anteriorly at the base of the abdomen and occupying the first abdominal segment can be seen a pale area where the end of the ventral diverticulum extends into the abdomen. Here the sternite is represented only by thin membrane, and the characteristic bubbles of the diverticulum can be seen shining through.

Compressed into the sixth segment and distending this are now all the organs of the abdomen other than the mid-gut. The seventh segment (the last except for the small eighth normally retracted within the seventh for most of its extent) has only narrow pleurae and is not distended, the tergite and sternite remaining in contact. Beyond the cuff-like seventh segment in the anaesthetised or moribund insect there projects in constant uneasy movement part of the tiny eighth segment with the cerci.

During engorgement there is projected from time to time from the anus a tiny drop of clear fluid. These drops have commonly been assumed to be plasma or serum exuded from the blood mass in the gut (for example, Bonne-Wepster and Brug, 1932), in support of which view was held to be the fact noted by Fulleborn (1908) that he found the haemoglobin content much higher in the blood from the gut than in normal human blood. Allowed to dry on a slide such drops show crystals and give exactly the appearance shown by small evaporated drops of normal saline. They do not coagulate on heating or give the Millon reaction. Diluted they give a copious precipitate with $AgNO_3$. From the work of Wigglesworth on excretion in insects (Wigglesworth, 1942) it is clear that these droplets are urine secreted by the Malphighian tubules and not directly derived from the ingested blood. Very often after some time, droplets left on the glass of tubes in which fed mosquitoes have been confined for oviposition show a white deposit at the bottom of the drop, which under the microscope is seen to consist of the characteristic globular crystals of urates. Passage of such droplets is not confined to the time when the insect is engorging but continues copiously for some time after.

Dissection of the freshly gorged insect shows a mass of partially clotted blood distending the bag-like mid-gut, the walls of which are now so thin and delicate that they are ruptured at the least touch. The contained blood has a rather tarry consistence, but is not at this stage apparently haemolysed, since mixed with normal saline the fluid is not coloured and the red cells are still largely intact. The ovaries are tiny organs measuring about a millimetre when not stretched out and composed of follicles in an early stage of development. The ventral diverticulum forms a large saccular organ with bubbles and measuring 3–4 mm. in length and about 2 mm. broad when flattened by its own weight.

(*d*) THE FUNCTIONS OF THE DIVERTICULA

These organs in the mosquito were observed as long ago as 1851 by Dufour. As in other mosquitoes those of *A. aegypti* are three in number, namely two quite small dorsal and one large ventral, all arising from the thin-walled dilated oesophagus in front of the proventriculus. Normally these delicate thin-walled structures contain at least some air bubbles, and the large ventral diverticulum may be distended by a mass of such. The diverticula may also contain fluid, especially after feeding on certain substances such as sugary fluids or in moribund mosquitoes that have become entangled with the water surface. These very commonly swallow water and air until the ventral diverticulum becomes enormously distended. Even as a normal result of engorgement and digestion, as the gut is reduced in size and with oviposition the ovaries are emptied, the ventral diverticulum comes to occupy a large part of the abdomen, extending to segment 3 or 4 or beyond.

The functions of the diverticula were not at first very evident and considerably divergent views on the subject have been given. That when mosquitoes are fed on sugary fluids these, unlike blood, pass, not into the mid-gut, but into the diverticula was first shown by Nuttall and Shipley (1903), and is now generally accepted. Also the result of the observations by Wright (1924), Kadletz and Kusmina (1929), MacGregor (1930, 1931), Bishop and Gilchrist (1944), and more recently of Trembley (1952) and of Day (1954), has been to establish that when the mosquito feeds normally on blood this passes direct to the mid-gut. Whether blood is ever taken into the diverticula is less certain and recorded findings in this respect are sometimes conflicting. Roy (1927) states that when the mid-gut is full, blood may be passed into the diverticulum. This, however, is against usual experience. Fifty females were allowed to gorge for 5 minutes and killed and examined within a further 3 minutes. All stages of gorging were found including many that had gorged and left the arm. In none was blood found in the diverticulum and in all fully gorged specimens the clear space at the base of the abdomen occupied by the air-filled ventral diverticulum was evident to external examination.

MacGregor (1930, 1931), however, found that when fed artificially by the method of Kadletz and Kasmina (1929) with blood mixed with honey the fluid passed almost entirely to the diverticula, if fully fed both to the dorsal and ventral diverticula, otherwise to the ventral diverticulum. In normal feeding blood passed directly to the mid-gut, but with interrupted feeds it may go to the diverticula.

Bishop and Gilchrist (1944, 1946) have studied the imbibition of different fluids in the case of mosquitoes fed through an animal membrane and directly from open drops. In these experiments it was found that whole blood and red blood corpuscles in saline, when ingested through a membrane, go directly into the mid-gut which becomes fully distended. Haemoglobin in plasma or distilled water is ingested to a lesser degree and, in feeding through a membrane, plasma alone rarely. But if they are ingested these pass to the mid-gut. Solutions containing glucose or honey are seldom imbibed through a membrane. But if they are, they pass to the mid-gut or diverticula, only the diverticula, however, being distended.

Feeding from an open drop, blood is rarely ingested, but if it is it passes to the mid-gut. Haemoglobin in plasma or water is rarely ingested, but if it is it passes to the mid-gut. Solutions containing glucose or honey are readily ingested and pass mainly to the diverticula which become fully distended, though traces may be found in the mid-gut. In this last respect it may be recalled that it was noted by MacGregor (1930) that traces of fluid when the diverticula are full may pass into the mid-gut, the point at which they do so being shown, if formalin be added to the fluid, by immediate death of the insect. Summarising, Bishop and Gilchrist conclude that it is the nature of the food, not the method of feeding, which determines its destination.

Day (1954), who found that sugars are diverted to the diverticula and blood to the mid-gut, suggests that such discrimination is carried out by reason of the sensory papillae of the buccal cavity (see description of the buccal cavity given in ch. XXVI, p. 571 and description of the sensory papillae in ch. XXIX, p. 664) transmitting impulses through the stomogastric system causing relaxation of the cardiac sphincter in the case of blood and of sphincter fibres at the neck of the ventral diverticulum for sugar.

Wright (1924) suggests that the diverticula in Diptera were originally food reservoirs, this function being largely suspended in the case of blood-sucking females. To act as storage receptacles for certain forms of food is not, however, the only function these curious thin-walled yet muscular organs appear to serve. Mention has been made of the fact that in the act of engorgement the ventral diverticulum passes well into the abdomen. It seems very possible that this is a protective device. The newly gorged insect with an amount of blood that may be equal to or even greater than its own weight is in a position where trauma may readily occur. When, too, as digestion proceeds and the gut shrinks and eventually the ovaries are emptied, the ventral diverticulum comes to occupy a large part of the abdomen.

Thus it appears to play the part of an adjustable air cushion in the abdomen that might well be purposive in effect.

Even the small dorsal diverticula may not be without some function. These lie in a very narrow space between the massive dorso-ventral wing muscles of the two sides and are further restricted by the anterior-posterior wing muscles above and the thoracic ganglia below. It is possible, therefore, that some cushioning may here also be present.

Wright, commenting on the view that the diverticula may act in a fashion to lighten the body for flying, turns such a view down since the diverticula being filled with air cannot lighten the body. They might, however, increase the ratio of resistance to the air and weight in the same way that birds possess air cavities in the bones.

Reference has been made to males and females found lying moribund on the water with their diverticula swollen with fluid. It would appear that the insects are unable to prevent fluid passing up the blood channel as a purely mechanical process or that some form of sucking action by the moribund insect leads to this result. The cause of the condition which is very commonly seen is not known.

(*e*) SALIVARY GLAND SECRETION

A number of observations have been made on the salivary glands of mosquitoes and their secretion. The first to describe the glands appears to have been Macloskie (1887). Since then they have been further described, especially as regards the glands of *Anopheles*, by many authors.* The glands consist on each side of three acini, two lateral and one median, the lateral acini in the body lying one above and one below the median. In *Anopheles* the median acinus is shorter and more saccular than the lateral acini and differs from these in appearance. In *Culex* and *Aëdes* the difference is little marked (Fig. 66 (6)). Like the glands of *Culex* the central duct of each acinus in *Aëdes aegypti* is not enlarged towards the end of the acinus but remains uniformly narrow to its termination. According to Metcalf (1945) the intra-acinar salivary ducts are lined with chitin. Compared with the large glands of *Anopheles* those of *Aëdes aegypti* are small. The most recent method of dissection of the glands is described by Shute (1940).

The nature of the salivary secretion differs in different species. Nuttall and Shipley (1903), experimenting with an emulsion of the glands of *Culex pipiens*, found that this did not prevent the clotting of human blood. Cornwall and Patton (1914), trying the action on blood of an emulsion of the salivary glands of two species of *Anopheles* (*A. rossi* and *A. jamesi*—probably *A. subpictus* or *A. vagus* and *A. annularis*), found that these gave strong and immediate agglutination and that the blood failed to clot. A very complete study was made by Yorke and Macfie (1924) of the salivary secretion of *Anopheles maculipennis*, *Culex pipiens*, *Theobaldia annulata* and *Aëdes aegypti*. With *Anopheles* glands they obtained complete and immediate agglutination. With *Culex*, *Theobaldia* and *Aëdes aegypti* there was no trace of agglutination against human blood or that of a number of animals. In no case was there haemolysis, but in *Anopheles* there was a definite anti-coagulation effect on human blood. The glands of *A. aegypti* gave neither agglutination nor anti-coagulation effect. Two hours after feeding, the blood in the mid-gut of *Anopheles* was agglutinated and unclotted. In *Aëdes aegypti* it was coagulated and not agglutinated. Shute (1935) found that broken up in saline, citrate or defibrinated blood the glands of

* Christophers (1901), Grassi (1901), Nuttall and Shipley (1903), Goldi (1913), Perfilev (1930), Shute (1940) and others.

A. maculipennis (vars. *atroparvus*, *messeae* and *type* form) caused agglutination of the red cells of man and of most animals and birds tested. The male glands did not agglutinate, nor did the glands of *A. claviger* (*bifurcatus*), *A. plumbeus*, *Culex pipiens*, *C. fatigans*, several species of *Aëdes*, including *Aëdes aegypti*, *Theobaldia annulata* or *Taeniorrhynchus richiardii*. Later (1948) he found that the glands of *A. stephensi* did not agglutinate. Due to the agglutinating action of the salivary secretion in *A. maculipennis*, oocysts in infected specimens tend to be congregated in the posterior portion of the gut. If, however, the mosquitoes be kept head downwards for 24 hours after the feed, the oocysts are in the anterior portion of the mid-gut. In *A. stephensi*, the salivary secretion of which does not agglutinate, the oocysts are scattered over the gut.

Metcalf (1945) in *A. maculipennis* found the pH of the median acinus to be 6·8–7·0, whilst that of the lateral acini was 6·0–6·2. Agglutination was caused by the median acinus only. Tests for amylase, protease and lipase were negative. Both lateral and median acini, but especially the latter, contained a thermostable anti-coagulin; the median acinus also contained a thermolabile agglutinin, lacking or in very small amount in the lateral acini. Blood in the stomach of *Aëdes aegypti* was coagulated but not agglutinated 15 minutes after a feed.

Gordon and Lumsden (1939) allowed *A. aegypti* to bite for interrupted periods of 10 seconds and found the reaction to the tenth bite as marked as that following the first. Since Boyd and Kitchen (1939) found sporozoites in the tissues in the neighbourhood of the bite they consider it probable that injection of the salivary secretion takes place into the tissues and not into the blood stream. This has since been observed to take place by Griffiths and Gordon (1951) (see section *b*).

(*f*) AMOUNT OF BLOOD TAKEN

Fulleborn (1908) gave the blood taken by a gorged female *A. aegypti* as 1·1 mg. Bonne-Wepster and Brug state that this is probably too small. Darling (1910) weighed *Anopheles albimanus* bred in the laboratory 24 hours old with the mid-gut empty and with a moderate blood feed, the weight being respectively 0·8 and 1·6 mg., evidently a very small mosquito, but taking its own weight of blood. Caught specimens with much blood in the gut but undeveloped ova weighed 1·8 and 2·1 mg. With some blood in the gut and half-developed ovaries the weight was 1·9 mg. and with much blood and developed ovaries 3·5 mg. Taking the specific gravity as 1·050 he calculated that the volume of blood was 0·76 c.mm.

A considerable number of observations on the weight of blood taken by *Aëdes aegypti* have been given by Roy (1936) in his work on the relation of the blood meal to development of the ovaries. The weight of a 3-day-old *A. aegypti* as determined on 137 mosquitoes before and after a blood meal in mg. was as follows.

	Mean	Maximum	Minimum
Unfed females	1·32	2·32	0·86
Same fed	2·07	3·16	1·06
Amount of blood taken	0·75	0·84	0·20

He also gives a detailed statement of the mean weight of blood taken by mosquitoes of different weight. The majority weighed 1·34 mg. and took as blood meals a mean of under 2 mg. On the average the mosquitoes took a blood feed less than their own weight, but 53 out of 137 took one greater than their own weight.

In the present author's material the mosquitoes were evidently larger, as also the amount of blood taken, but about the same proportion of weight of mosquito to blood taken holds good as in the fifty-three more active feeders in Roy's series as shown in Table 32.

In the series given in the table the mosquitoes were weighed immediately following the feed, that is as soon as they could be anaesthetised and placed in the weighing bottle. In every case all the fed mosquitoes in the cage were taken. The operation served to show that the passing of drops of fluid during and after feeding was by no means restricted to the few drops seen when the mosquito was actually feeding, as during weighing droplets of this fluid were always found covering the glass of the weighing bottle and in one sample weighed the weight of fluid amounted to 0·1 mg. per mosquito. The control unfed mosquitoes were from parallel series reared under identical conditions to those fed. It is clear, therefore, that a fifth-day *A. aegypti* female when fully gorging takes more than its own weight of blood and may take nearly twice this.

Table 32. *Weight of mosquitoes and of blood taken*

Date	Age in days	Degree of gorging	Number fed	Mean weight fed	Control unfed	Weight of blood taken	Proportion, weight of blood to that of mosquito
16. viii. 46	0	Full	28	5·07	2·31	2·66	1·13
16. viii. 46	0	Partial	13	3·43	2·31	1·12	0·48
11. vii. 46	1	Full	90	5·57	2·20	3·27	1·49
11. vii. 46	1	Full	90	5·58	2·20	3·48	1·58
19. viii. 46	4	Full	84	5·80	2·67	3·13	1·17
5. vi. 46	5	Full	159	6·35	2·14	4·21	1·97

(*g*) RATE OF DIGESTION

The blood meal is very rapidly digested. Some of the fluid of the ingested blood is absorbed and evacuated during and shortly after the feed in the shape of droplets of clear urine. The weight of such droplets is quite considerable. Those left on a weighing bottle after weighing a number of recently fed females weighed about 0·1 mg. per mosquito. An equal amount or more is probably evacuated during the feed, and copious secretion of urine continues for some time after.

By the second or third day a considerable amount of brown faeces is passed and the mosquito has already lost much of its gorged appearance and much of its extra weight.* Nine mosquitoes on the second day after engorgement gave a weight of only 2·33 mg. mean weight which is about the weight of the newly emerged mosquito. After 48 hours at 25° C. the gut usually still contains a much reduced quantity of altered blood, but this disappears by about 52 hours from the feed. The mid-gut is always completely free from all trace of blood before oviposition occurs. During this time in addition to the dark faeces there are to be seen on the tube in which the mosquito is confined droplets containing white deposit. This deposit commonly settles to the bottom of the drop leaving clear fluid above. Under the microscope characteristic ball-shaped crystals of urates are seen.

For further information on digestion and changes following a blood meal see under 'Digestion', ch. XXXI, p. 707.

* Gillett (1956) has recently shown that elimination of the blood meal by defaecation usually occurs from 28 to 32 hours after feeding in mosquitoes which develop oocysts to stage 3 only, whereas elimination of the blood meal is delayed for a further 7–10 hours in mosquitoes which develop oocysts to maturity.

(*h*) BEHAVIOUR OF MALES DURING THE BLOOD FEED

When offering the arm to a cage containing several hundred females and a like quantity of males the former have for the greater part settled on the skin and are actively engorging within some 2 minutes. The males do not settle at this time, or only an occasional one does so. The males, however, become extremely active, flying in a dancing-like fashion to and fro, backwards and forwards and sideways in an irregular lurching manner. The most conspicuous movement is backwards and forwards flight at an angle of about 30° to the horizontal through a distance seemingly of some 4 or 5 inches, reversing without turning the direction of the body. When glimpsed in this flight the insect's body appears to be held at a slope from head to tip of abdomen of about 30° to the horizontal, that is much the same as the angle of flight. Whilst thus flying they give rise to a shrill humming noise. This activity is kept up almost indefinitely so long as the arm is present and even when the majority of the females have left. When the arm is withdrawn the dance ceases. When after some 15 minutes many females have left the arm the males often settle on the skin and walk about on this. They may be seen applying their proboscis tip with the long palpi to the skin, possibly taking up sweat or fluid voided by the females when feeding.

Flight of the males, though it appears continuous, seems to be taken up in relays, there being always a number of males settled on the roof of the cage, but on watching the individuals they remain resting only for a time; others on the contrary are settling to rest. The excitement does not seem to be especially associated with copulation and there is no conspicuous appearance of females entering the crowd as described for swarms in the open. A certain number of males do, however, when on the arm solicit females who are still sucking, but are usually rebuffed.

(*i*) REACTION FOLLOWING BITES

Where there is nothing to cause interference with feeding *A. aegypti* for the most part completes gorging where it first alights and without making any second puncture. Gordon (1922) makes the observation that unless disturbed *A. aegypti* never bit twice. A certain proportion, however, even if not disturbed, do puncture more than once. Usually the total number of wheals after a large feed is about 25 per cent in excess of the number biting. In Table 33 are given examples where counts have been made. In all cases the temperature at which feeds were made was 25° C. and the exposure 15 minutes. Care was taken that no movement of the arm or other disturbing factor was present.

It might seem that the results here given, showing an excess of wheals over the number of mosquitoes, were in contradiction to previous results given where the reverse was shown.

Table 33. *Number of wheals following feeds by mosquitoes*

Date	Age in days	Number fed	Number of wheals	Proportion of wheals to mosquitoes
16. viii. 46	0	41	47	1·14
11. vii. 46	1	92	105	1·14
17. viii. 46	1	82	102	1·24
19. viii. 46	3	84	111	1·32
13. iii. 46	6	84	111	1·32
Total		383	476	1·24

The difference, however, is owing to the fact that in the test for biting rate only 15 seconds was allowed for the biting, whereas the present figures relate to conditions where the mosquitoes were given full time to complete their feeding.

As already noted it is necessary before a wheal is produced on a susceptible individual for a certain time to elapse from the first commencing to puncture. This appears to be about 6 seconds. Bites in which the mosquito has left under 6 seconds in the individuals under observation have not usually left a wheal. The size of the wheal, other things being equal, is proportional to some extent to the time taken in the feed. Thus bites of short duration in which the insect is disturbed or prevented from remaining due to presence of a repellent may be very small, whilst a full-time bite gives a good wheal. A certain time is taken before the wheal develops. This may be from 5 to 15 minutes depending upon the individual and other circumstances. After expiry of this time a tingling sensation is felt and an urticarial-like wheal appears as a small discrete papule. The tingling felt is distinct from that due to the action of the mouth-parts in puncturing, which may or may not be felt at the time of the bite. The papule increases in size with typical urticarial appearance and may reach a centimetre or more in diameter. In some individuals the wheal is more diffuse, forming a soft indefinite swelling. This urticarial wheal usually subsides after an hour or so. What has been described may be called the 'initial reaction'. At some later time, usually 12–24 hours, a more diffuse type of papule develops, which may have a tiny superficial bleb and is itchy. This, which may be termed the 'secondary reaction', may last a variable time up to 48 hours or more and is liable to exacerbation from time to time, as, for example, when the bitten person is warm in bed. This stage, especially if a bleb forms, is liable to become excoriated or may even become septic.

Immunity to the secondary reaction is soon developed. Such immunity is to a large extent local to the area frequently bitten, but apparently not entirely so. In the case of the writer, who after some years' constant biting by the species has become entirely immune to the secondary reaction, there still remains a large degree of susceptibility to the initial wheal.

A description of the reaction of man to the bites of *A. aegypti* has been given by Gordon and Crewe (1948). These authors state that the reaction in the majority of individuals does not develop until some considerable time after the bite, such reaction being termed by the authors 'the delayed reaction'. If persons showing a delayed reaction are irregularly exposed to further bites, a proportion become 'sensitised', that is they show an 'immediate reaction' and the delayed reaction ceases to occur. If regularly bitten at short intervals, they do not become sensitised and the delayed reaction becomes progressively less intense and of shorter duration. If persons showing an immediate reaction are irregularly exposed to further bites, the sensitivity persists for an indefinite period. If regularly bitten at short intervals, however, the immediate reaction becomes progressively, though irregularly, less. The desensitisation is highly specific, and persons sensitised to *Aëdes*, *Culex* and *Anopheles* and subsequently desensitised to *Aëdes* still gave an immediate reaction to *Culex* and *Anopheles*.

Reaction and immunity to mosquito bites have also been investigated by Brown, Griffiths, Erwin and Dyrenforth (1938), who describe Arthus's phenomenon, namely wheals followed by blebs with subsequent necrosis and scarring following introduction of glandular substance into the skin.

The natural assumption would be that these effects following the bite were due to the salivary secretion injected during the act. Schaudinn (1904), however, ascribed them to the products of yeasts and other micro-organisms which are always present in the diverticula. Leon (1924) came to the same conclusion, since he obtained no effect from the glands but did from the diverticula. Roy (1927) punctured the skin with a fine needle and rubbed gland substance and diverticula suspension over the puncture. He also obtained results (though small) from the diverticula but none from the glands. Hecht (1928), however, has found yeasts present in the diverticula only when mosquitoes are fed on sugary fluids, whilst secondary papules follow implantation of the glands, the effects of the bite being due, not to contents of the diverticula, but to the salivary secretion.

The 'initial reaction' is here referred to as a 'wheal', this describing its characters precisely. Wheals are of great value in quantitative testing work with *Aëdes aegypti*, and in the testing of repellents their number and distribution on the experimental skin area in relation to the number of females used has been the most precise of all methods of assessing protection. No difficulty in the author's experience has been found in those partaking in the experiments not being 'sensitised' or in loss of sensitisation following biting, and the extent of the latter after working with the species for more than a year with almost constant biting was only a matter of degree.

It seems possible that there may be two distinct processes operating to produce the 'initial' and 'secondary' reaction. The secondary reaction is presumably due to the injection into the tissues of the salivary secretion. In view, however, of the mechanism of biting by what must amount to a considerable degree of cellular laceration through the action of the maxillae, it is thought the initial reaction might be the result of such trauma with the usual liberation of histamine. There would not seem to be anything contradictory in this to the views put forward by Gordon and Crewe, desensitisation then being development of immunity to the products of cellular trauma, not to an effect of the salivary secretion.*

For the treatment of ordinary mosquito bites little has been advocated beyond the usual methods of allaying irritation by cooling lotions, application of ammonia, rubbing with grass or other household remedies. The subject has been dealt with to some extent by Howard (1917). W. R. Shannon (1943) has advocated administration by mouth of thiamin chloride (80–100 mg. on the first day, followed by 10 mg. daily) to allay irritation and to diminish tendency to be bitten. But thiamin chloride (vitamin B_1 hydrochloride) was not found protective by Wilson, Matheson and Jackowskie (1944). See also Goldman, Thompson and Trica (1952) on cortisone acetate. A preparation, 'anthisan cream', by May and Baker, is on sale for irritation from mosquito bites. I have been informed by Mr Mattingly that he understands that a very dramatic reduction of swelling and irritation can be obtained by the application of certain antihistamines, for example Pyranisamine maleate, and that this and other more recent compounds are available in proprietary form, for example anthisan.

REFERENCES

Allen, A. C. (1949). Persistent 'insect bites' (dermal eosinophilic granulomas) simulating lympho-blastomas, histiocytosis and squamous cell carcinoma. *Amer. J. Path.* **24**, 367.

Bishop, A. and Gilchrist, B. M. (1944). A method for collecting sporozoites of *Plasmodium gallinaceum* by feeding infected *Aëdes aegypti* through animal membranes. *Nature, Lond.*, **153**, 713.

* See also Bristowe (1946), Goldman, Johnson and Ramsay (1952). Allen (1949) refers to persistent 'insect bites' (dermal eosinophic granulomas) simulating lymph blastomas, histiocytosis and squamous cell carcinoma.

REFERENCES

BISHOP, A. and GILCHRIST, B. M. (1946). Experiments upon the feeding of *Aëdes aegypti* through animal membranes, etc. *Parasitology*, **37**, 85–100.

BONNE-WEPSTER, J. and BRUG, S. L. (1932). See references in ch. II (*a–b*), p. 45.

BOYD, M. F. and KITCHEN, M. F. (1939). The demonstration of sporozoites in human tissues. *Amer. J. Trop. Med.* **19**, 27–31.

BRISTOWE, W. S. (1946). Man's reaction to mosquito bites. *Nature, Lond.*, **158**, 750.

BROWN, A., GRIFFITHS, T. H. D., ERWIN, S. and DYRENFORTH, L. Y. (1938). Arthus's phenomenon from mosquito bites. *Sth. Med. J., Nashville*, **31**, 590–6.

CHRISTOPHERS, S. R. (1901). The anatomy and histology of the adult female mosquito. *Rep. Malar. Comm. R. Soc.* (4), 1–20. London.

CORNWALL, J. W. and PATTON, W. S. (1914). Some observations on the salivary secretion of the common blood-sucking insects and ticks. *Ind. J. Med. Res.* **2**, 569–91.

DARLING, S. T. (1910). Factors in the transmission and prevention of malaria in the Panama Canal Zone. *Ann. Trop. Med. Parasit.* **4**, 179–223.

DAY, M. F. (1954). The mechanism of food distribution to midgut or diverticula in the mosquito. *Aust. J. Biol. Sci.* **7**, 515–24.

DUFOUR, L. (1851). Recherches anatomiques et physiologiques sur les Diptères. *Mém. Acad. Sci., Paris*, **11**, 171–360.

FINLAY, C. (1886). Yellow fever, its transmission by means of the *Culex* mosquito. *Amer. J. Med. Sci.* **92**, 395–409.

FULLEBORN, F. (1908). Über Versuche an Hundefilarien und deren Übertragung. *Arch. Schiffs.- u. Tropenhyg.* **12**, Beih. 8, 5–43.

GILLETT, J. D. (1956). Initiation and promotion of ovarian development in the mosquito *Aëdes (Stegomyia) aegypti* (Linnaeus). *Ann. Trop. Med. Parasit.* **50**, 375–80.

GOLDI, A. (1913). *Die sanitärisch-pathologische Bedeutung der Insekten und verwandter Gliedertiere.* Berlin, pp. 43–5.

GOLDMAN, L., JOHNSON, P. and RAMSAY, J. (1952). The insect bite reaction. I. The mechanism. *J. Invest. Derm.* **18**, 403.

GOLDMAN, L., THOMPSON, R. G. and TRICA, R. E. (1952). Cortisone acetate in skin diseases. *Arch. Derm. Syph., N.Y.*, **65**, 177.

GORDON, R. M. (1922). The susceptibility of the individual to the bites of *Stegomyia calopus. Ann. Trop. Med. Parasit.* **16**, 229–34.

GORDON, R. M. and CREWE, W. (1948). Man's reaction to the bites of *Aëdes aegypti. Trans. R. Soc. Trop. Med. Hyg.* **42**, 7–8.

GORDON, R. M. and LUMSDEN, W. H. R. (1939). A study of the behaviour of the mouth-parts of mosquitoes when taking up food from living tissue etc. *Ann. Trop. Med. Parasit.* **33**, 259–78.

GRASSI, B. (1901). *Studi di uno zoologo sulla malaria.* Rome.

GRIFFITHS, R. B. and GORDON, R. M. (1951). A simple apparatus designed in order to observe insects feeding on living tissue or the penetration of helminth larvae. *Trans. R. Soc. Trop. Med. Hyg.* **44**, 366.

GRIFFITHS, R. B. and GORDON, R. M. (1952). An apparatus which enables the process of feeding by mosquitoes to be observed in the tissue of a live rodent together with an account of the ejection of saliva and its significance in malaria. *Ann. Trop. Med. Parasit.* **46**, 311–19.

HECHT, O. (1928). Über die Sprosspilze, der Oesophagusausstulpungen und über der Giftwirkung der Speicheldrüsen von Stechmücken. *Arch. Schiffs.- u. Tropenhyg.* **32**, 561–75.

HERTIG, A. T. and HERTIG, M. (1927). A technique for artificial feeding of sand-flies and mosquitoes. *Science*, **65**, 328–9.

HOWARD, L. O. (1917). Remedies and preventives against mosquitoes. *U.S. Dep. Agric., Frms' Bull.* no. 444, 1–15.

HOWARD, L. O., DYAR, H. G. and KNAB, F. See references in ch. I (*f*), p. 19.

KADLETZ, N. A. and KUSMINA, L. A. (1929). Experimentale Studien über den Saugprozess bei *Anopheles* mittels einer zwangsweisen Methode. *Arch. Schiffs.- u. Tropenhyg.* **33**, 335–50.

LEON, N. (1924). *Contributions à l'étude de parasites animaux de Roumanie 1898–1924.* Bucharest.

REFERENCES

MacGregor, M. E. (1915). Notes on the rearing of *Stegomyia fasciata* in London. *J. Trop. Med. (Hyg.)*, **18**, 193–6.

MacGregor, M. E. (1930). The artificial feeding of mosquitoes by a new method which demonstrates certain functions of the diverticula. *Trans. R. Soc. Trop. Med. Hyg.* **23**, 329–31.

MacGregor, M. E. (1931). The nutrition of adult mosquitoes. Preliminary contribution. *Trans. R. Soc. Trop. Med. Hyg.* **24**, 465–72.

Macloskie, G. (1887). Poison fangs and glands of the mosquito. *Science*, **10**, 106–7.

Metcalf, R. L. (1945). The physiology of the salivary glands of *Anopheles quadrimaculatus*. *J. Nat. Malar. Soc.* **4**, 271–8.

Nuttall, G. H. F. and Shipley, A. E. (1903). Studies in relation to malaria. *J. Hyg., Camb.*, **3**, 167–70, 186–96.

Parker, H. B., Beyer, G. E. and Pothier, O. L. (1903). Report on the working party. *Yell. Fev. Inst. Bull.* no. 13, 21–7.

Perfilev, P. P. (1930). Zur Frage über die vergleichende Anatomie van *Anopheles*. *Mag. Parasit., Leningrad*, **1**, 75–96. Abstract in *Rev. Appl. Ent.* **19**, 230.

Reaumur, R. A. F. (1738). *Mémoires pour servir à l'histoire des insectes. Mém. XIII. Histoire des cousins*, vol. IV, pp. 573–636. L'Imprimerie Royale, Paris.

Robinson, G. G. (1939). The mouth-parts and their function in the female mosquito *Anopheles maculipennis*. *Parasitology*, **31**, 212–42.

Roy, D. N. (1927). The physiology and function of the oesophageal diverticula and of the salivary glands in mosquitoes. *Ind. J. Med. Res.* **14**, 995–1004.

Roy, D. N. (1936). On the role of blood in ovulation in *Aëdes aegypti* Linn. *Bull. Ent. Res.* **27**, 423–9.

Schaudinn, F. (1904). Generations und Wirtswechsel bei *Trypanosoma* und *Spirochaete*. *Arbeit. Kaiserl. GesundheitsAmt* **20**, 387–439. Berlin.

Shannon, W. R. (1943). An aid in the solution of the mosquito problem. *Minn. Med.* **26**, 799.

Shute, P. G. (1935). Agglutination of the red blood corpuscles of man, animals and birds by the salivary glands of *Anopheles maculipennis*. *J. Trop. Med. (Hyg.)*, **38**, 277–8.

Shute, P. G. (1940). Supernumerary and bifurcated acini of the salivary glands of *Anopheles maculipennis*. *Riv. Malariol.* **19**, sez. 1, 16–19.

Shute, P. G. (1948). The comparative distribution of oocysts of human malaria parasites on the stomach wall of *Anopheles* mosquitoes. *Trans. R. Soc. Trop. Med. Hyg.* **42**, 324.

Trembley, H. L. (1952). The distribution of certain liquids in the oesophageal diverticula and stomach of mosquitoes. *Amer. J. Trop. Med. Hyg.* **1**, 693–710.

Vogel, R. (1921). Kritische und ergänzende Mitteilungen zur Anatomie des Stechapparats der Culiciden. *Zool. Jahrb., Anat.* **42**, 259–82.

Wigglesworth, V. B. (1942). *The Principles of Insect Physiology*. Ed. 2 (see also many later editions). Methuen and Co., London.

Wilson, C. S., Matheson, D. R. and Jackowskie, L. A. (1944). Ingested thiamin chloride as a mosquito repellent. *Science*, **100**, 147.

Wright, W. R. (1924). On the function of the oesophageal diverticula in the adult female mosquito. *Ann. Trop. Med. Parasit.* **18**, 77–82.

Yorke, W. and Macfie, J. W. S. (1924). The action of the salivary secretion of the mosquito and *Glossina tachinoides* on human blood. *Ann. Trop. Med. Parasit.* **18**, 103–8.

XXII
MATING AND OVIPOSITION

(*a*) THE SEXES

DISTINCTION BETWEEN THE SEXES

The male, as in the majority of mosquitoes, is distinguished from the female by the plumose antennae and the large hairy palpi, which in *Aëdes aegypti* are about as long as the proboscis in marked contrast to the very short female palpi only about a quarter of the length of this organ. In addition the male differs from the female in the absence of mandibles, narrower wings in proportion to their length and narrower scaling of these organs, a more developed fringe apically and absence of scales on the fringe margin, the presence of larger ungues on the fore- and mid-legs and a different ungual formula. The male differs also, not only in possessing the large gonocoxites (claspers) and other characters of the male hypopygium, but in the large and well-developed eighth abdominal segment contrasting with the small retracted part in the female.

Of interest from the biological point of view is the marked and constant difference in size, the female being nearly twice the weight of the male and much more bulky and robust. The difference in size begins to be apparent in the fourth instar larva, is marked in the pupa and is present in about the same degree in the imago.

The sexes differ also in many points of behaviour, in the blood-sucking habit, in length of life, in the speed of flight and in the note emitted in the hum.

In all data regarding the imago it is necessary therefore to give separate valuation for the sexes. Since, however, the female is more important as a blood-sucking insect, as a carrier of disease and as an insect useful for experiment, it has received more attention in the literature than has the male, most observations where sex is not mentioned relating to the female.

PROPORTION OF THE SEXES

A characteristic of *A. aegypti* to which attention has been drawn by a number of observers is preponderance of the number of males over that of females hatching out. Rees (1901) is quoted by some authors as referring to *A. aegypti* when he notes that males hatch out first and in greatest numbers. Actually in his paper he refers to 'mosquitoes', but it is probable that he was here dealing with this species. Gordon (1922*b*), however, definitely gives figures for the species, namely a total of 142 males to ninety-eight females (59·2 per cent males) with maximum food supply and 105 males to ninety-one females (53·6 per cent males) with a minimum of food. Young (1922) gives the relative number of males to females as 323 to 194 (62·5 per cent males). He notes that preponderance of males appears to be a constant feature under conditions employed in the laboratory.

Teesdale (1955), in two experiments with ninety-six and 134 eggs, hatched out respectively seventy-seven and 121 adults, sixty males and seventeen females and seventy-six males and forty-five females in the two series, giving for the eggs hatched out a percentage for males of 77·9 and 62·8 in the two series respectively.

Mattingly (1956) found a proportion of five males to three females (62·5 per cent males) and a similar proportion for all published records for *aegypti-albopictus* hybrids.

That an increased percentage of males over females is not, however, confined to *A. aegypti* or even to *Aëdes* would appear from data given by Qutubuddin (1952). Out of a total of 4353 *Culex fatigans* bred out 2427 were males and 1926 females (56·1 per cent males).

Table 34. *Proportion of females to males in thirteen cultures*

					Total			
Culture no.	Date	Male	Female	Left over	Male	Female	Left over	Percentage females*
1	27. iv. 43	865	542	23	865	542	23	39·5
2	8. v. 43	1,050	140					
	9. v. 43	177	700	269	1,227	840	269	47·5
3	26. v. 43	324	100					
	27. v. 43	73	106					
	28. v. 43	8	91	28	405	297	28	44·5
4	25. v. 43	1,142	846	111	1,142	846	111	45·6
5	18. v. 43	530	395	30	530	395	30	44·5
6		835	558	126	835	558	126	45·0
7	5. vii. 43	1,634	607					
	6. vii. 43	71	504					
	7. vii. 43	8	84	56	1,713	1,195	56	42·2
8	22. vii. 43	1,021	432					
	24. vii. 43	66	478	75	1,087	910	75	47·5
9	2. viii. 43	1,686	316					
	3. viii. 43	252	422					
	4. viii. 43	23	401	289	1,961	1,139	289	42·1
10	6. ix. 43	1,056	65					
	7. ix. 43	719	554					
	8. ix. 43	97	566					
	9. ix. 43	12	346	118	1,884	1,531	118	46·7
11	14. ix. 43	1,081	26					
	15. ix. 43	350	270					
	16. ix. 43	0	700					
	17. ix. 43	30	328	73	1,461	1,324	73	48·9
12	25. x. 43	946	309					
	26. x. 43	14	537					
	27. x. 43	1	23	45	961	869	45	48·7
13	24. i. 44	1,749	135					
	25. i. 44	54	1,180					
	26. i. 44	6	83	21	1,809	1,398	21	44·0
			Grand total		15,880	11,844	1,264	45·0

* In calculating the percentage for reasons given in the text the number of left-over larvae and pupae (including undiagnosed dead) has been added to the females.

In Table 34 are given the results of counts carried out in 1943–4 as a routine in setting out standardised tests in connection with work on repellents (Christophers, 1947). It will be seen from the table that in all thirteen counts recorded, amounting in all to 27,724 mosquitoes, males were consistently more numerous than females, though the percentage was not so high as counts already noted. These counts were carefully made from cultures obtained from standard egg-pots (see under chapter v describing technique), the number of pupae and any unchanged larvae or dead larvae being also counted after making up the day's experimental cages. The eggs in the pots were massed eggs, the larvae being reared from overnight hatching. Previous repeated trials had shown that with the technique adopted very few eggs failed to hatch as shown after replacing pots to hatch.

Taking all the cultures 27,724 pupae gave 57·3 per cent males. Dead or unpupated larvae numbered in all 1264 and as noted elsewhere the majority of such larvae would almost certainly be females. Adding these to the females gives a percentage of 54·7 males. The lowest percentage of males was in experiment 11 (51·1 per cent) and the highest in experiment 1 (60·5 per cent).

As will be seen later in chapter XXXI such a preponderance of males has considerable interest from the genetical point of view and an accurate figure is very desirable. If, however, a really accurate determination is to be made, some important precautions are necessary. Rearing conditions should be such that practically all eggs used will safely reach the adult stage or be capable of having the sex determined. If the proportions in the offspring of single females are being determined, it would be necessary to see that all the eggs of a given ovulation are laid. Such a careful appraisal does not appear to have so far been made. For the present it can be said that males predominate to the extent of from approximately 55 to 65 per cent.

ORDER OF HATCHING OF THE SEXES

Mention has been made that Rees (1901) not only noted that males were in excess, but that they hatched out first. Bacot (1916) also states that males are usually a day quicker in development than the females, and the same is noted by Gordon (1922*b*) and by Putnam and Shannon (1934), as also probably by others. Such observations are in accordance with the counts made of male and female pupae on the first, second and subsequent days of collection as given in Table 35, where the results of routine counts made in connection with the testing of repellents during 1943–5 are given.

Table 35. *Counts of female and male pupae collected on successive days*

Day of collection	Females	Males
First day	2,130	10,547
Second day	4,476	1,452
Third day	2,223	167
Fourth day	674	42
Left over	997	—
Total	10,500	12,208

But though the male larva reaches full development and pupates a day earlier than the female larva, the male pupa as will be seen from figures given under the pupa requires almost the same time from pupation to emergence as the much larger female pupa.

BEHAVIOUR OF THE SEXES

Behaviour of the male and female imago differs in a number of respects other than in those directly connected with sexual activity, for example in their behaviour in the presence of the host during feeding, in the character of their flight and in their general behaviour in a room. Thus males will hover about a person sitting in a room quite openly and in a different manner to the surreptitious approach of the female.

A careful experimental study of the sexual behaviour of *A. aegypti* is given by Roth (1948). The swarming of males so conspicuously evident in many Chironomidae, Tipulidae and some mosquitoes, such as *Culex pipiens*, is not a feature with *Aëdes aegypti*, though a modified form of swarming has been described, for example on a small scale in a corner of the laboratory. As in some other species males may show sensitivity to the wing note of the female. Roth and Willis (1952) note that males will respond to a tuning fork held close to the cage, clinging to the cloth so that females can be more conveniently captured. The effective vibration is between 320 and 512.

(*b*) COPULATION AND FERTILISATION

THE ACT OF COPULATION

Translation of a passage in which Godeheu de Riville (1760) describes the act of copulation in what was presumably *A. aegypti* is given by Howard, Dyar and Knab. Another early description is given by Low (in Theobald, 1903) who states that as soon as the insects emerge from the pupae, even on the first day, they begin to copulate, the males flying after and chasing the females until they catch them, the pair then flying about joined together. A male was observed to copulate with a female several times. The time occupied was a minute or less.

Marchoux, Salimbeni and Simond (1903), describing copulation in *A. aegypti*, say that the male seizes the female in flight—belly to belly—and keeps in this position by means of legs hooked to the thorax and by fixing its claspers near the vulva, the act lasting about a minute, the mosquitoes continuing in flight.

MacGregor (1915) states that almost as soon as they can fly copulation occurs, usually in mid-air, either completely so, or more usually the female settled on the muslin, the male with the back and legs pressed against the support and the abdomen arched. He states that males do not alight on resting females. Bonne-Wepster and Brug (1932) state that in captivity copulation always began with both parties on the wing. It may be finished in the air or against the wall of the cage. In the latter case the female presses the male against the cage. These authors never saw a male alight on a resting female. Hovanitz (1946) states that in *A. aegypti* copulation is most frequent at rest after flight, common partly in flight and partly at rest and least common in flight.

Our own experience is that copulation takes place quite commonly with the female at rest. It also takes place very commonly on the wing, especially when a number of the mixed sexes are disturbed and take suddenly to flight. Seemingly in the latter case it occurs at random as the insects fly about, a couple appearing to meet in air and then often tumbling together on to the floor of the cage. It is common at such times to see two, or even three, males clinging to the same female, the agglomeration tumbling about or flying. When coupling occurs with the female at rest the pair usually shortly take to flight. The act lasts a very short time and often seems more in the nature of a 'pass' than a true pairing. It never lasts more than some seconds.

In copulation the bodies are opposed. Howard, Dyar and Knab (1912) state that in mosquitoes with simple claws in the female (*Culex*, *Anopheles*) the position in copulation is end to end, the pair facing in opposite directions, whilst in those forms in which the female claws are toothed copulation takes place face to face, the pair clasping each other with their claws. In *Aëdes aegypti* the bodies are opposed, the legs of the male clasping the female and the abdomen of the male arched upwards and clasping the tip of the female abdomen. A good picture of a couple in copulation is given by Wesenburg-Lund (1943).

It is difficult to see what takes place when copulation begins in flight. When the female is at rest the male alights alongside and inserts himself beneath her, usually passing in from the side, and clasps her with his body opposed at a little distance. The couple then usually fly off. A very full description of copulation in *A. aegypti* is given by Roth (1948) in his paper on the sexual behaviour of *A. aegypti.*

TIME AT WHICH COPULATION OCCURS

Most authors note that copulation occurs early and independent of a blood meal. Macfie (1915) says that the species in captivity pair very soon after emergence, sometimes as soon as rested. In our experience copulation is at a maximum in the first few days after emergence. It may be so active on the second or third day that it is difficult to capture a female with a test-tube without including one or more males. At such times it occurs in bursts whenever the insects are disturbed and take to flight. Later it becomes less noticeable, though liable to occur so long as males are present.

Copulation is not particularly noticeable at feeding time when the females on, say, the fifth day from emergence collect at once on the arm and the males become excited and active, flying backwards and forwards with a shrill hum in the upper parts of the cage. At most, towards the end of the operation, when males begin wandering over the arm, some may molest a female that is still feeding, often to be rebuffed. Even when nearly all, or all the females have left the arm to settle about the cage the males still maintain their to-and-fro flight without copulation to any extent taking place. Whether this would happen with females kept apart from males to feeding day is not known. On removal of the arm the males cease their activity.

FERTILISATION

The usual test of fertilisation is that the spermatheca contains spermatozoa. Little is known as to details of the process by which the spermatozoa become packed in the organ. Very rarely a mass of spermatozoa on sectioning may be found in the atrium and most probably spermatozoa pass up the spermathecal ducts by their own activity.

Once fertilised *A. aegypti* is capable of laying fertile eggs for some considerable time. Macfie obtained fertile eggs 37 days after the last possible date for copulation and Bacot after 62 days. In Bacot's observation, eggs laid subsequent to this were not fertile. Fielding records that a female which laid 752 eggs during 72 days of life, feeding sixteen times, originally segregated with two males and with no other chance of copulation, gave eggs that were 80–90 per cent fertile.

Roy (1940) has shown that in *Anopheles* the presence of sperms in the spermatheca may act as a stimulus to ovarian development. More recently Gillett (1955) has shown that in a certain strain of *A. aegypti* mating may be a necessary stimulus to full development of the eggs and oviposition (see chapter II under 'strains').

At 20–25° C. Marchoux *et al.* (1903) found fertilisation to be the rule. Below 20° C. the proportion unfertilised was considerable.

(c) OVIPOSITION

THE ACT OF OVIPOSITION

Mitchell (1907) and also Howard, Dyar and Knab (1912) give a translation of a description of the act of oviposition by Agramonte.

> The mosquito alighted upon the water, which was in a small beaker inside the jar, with legs spread apart. The abdominal segments being bent forwards and downwards, she dipped her whole body until the last segment touched the surface of the water; then she rose, walked on a few steps, and dipped again. This she would do repeatedly (14 to 22 times), when she would remain for a slightly longer time with the last abdominal segment touching the water, and would allow a minute egg to issue forth upon the surface. In this way she laid at the rate of 3 eggs per minute, resting quietly after every sixth or eighth egg for about 30 seconds when she would resume the process.

Fielding (1919) states that the favourite position is resting on the side of the receptacle with the tip of the abdomen just touching the water. Eggs were often laid on the water as females moved about on it. Up to fifteen eggs may be laid in a batch.

Owing to the habit of *A. aegypti* laying eggs a few at a time at intervals and the inconspicuousness of the newly laid eggs, which are almost invisible on white filter paper, it is not easy to observe females in the act of oviposition. It has, however, been observed on a number of occasions. On most of these the female was one isolated in a specimen tube. After remaining for some days, following upon feeding, almost entirely at rest near the top of the filter paper slip or on the muslin cover of the tube the insect shortly before ovipositing begins to move about in a restless way. Shortly after this ovipositing begins.

The insect during the act has much the same attitude as when resting, but the abdomen is from time to time bent downwards in a curved manner and is moved about to touch the wet surface. Each time the tip touched the surface an egg was seen to have been deposited. The action was a rather unexpected one from the deliberate manner in which the abdomen was used first to touch one place and then another, much as an elephant might use its trunk. Without the body moving, eggs were laid at spots 2 or 3 mm. apart, the abdomen moving laterally quite distinctly to do this. After laying four or five eggs the body was often raised and the abdomen straightened out, whilst the apex was vigorously rubbed by the hind-legs. The abdomen was then lowered as before and the depositing of eggs continued. After laying a group of about twenty eggs the female in each case moved to another part of the tube. The eggs were found to be adherent to the glass when the film of liquid was dried off.

SELECTION OF A SITE FOR OVIPOSITION

Two issues are here concerned, namely (1) choice of the type of site, and (2) choice of the position where the eggs are to be laid. The first relates to the type of breeding place selected by the insect and has already been discussed when dealing with the nature of the breeding places made use of by the species. The second is concerned with exactly where in the chosen receptacle the female places her eggs. It is mainly the second issue which can be studied in the laboratory and with which we are here concerned.

In the laboratory *A. aegypti* at full maturation of the ovaries will, under almost any conditions, deposit its eggs freely on any water or wet surface that may be available. It appears never to lay on a dry surface and in the absence of water or a wet surface will hold its eggs.* The wet surface may be glass, earthenware, filter paper, sponge or the floor of the cage wet with seeped or spilled water.

They may be laid (1) on the surface of the water; (2) at the edge of the water, almost invariably on the wet surface just above the edge. Which of these sites is chosen depends largely upon the nature of the receptacle. If this is of glass, the majority of the eggs are laid on the water. If of unglazed earthenware, they are laid along the edge. Hence if floating eggs are required, a glass pot should be placed in the cage and vice versa. Howard, Dyar and Knab say that *A. aegypti* lays on the surface of the water only in captivity, glass being an unsuitable surface. Probably glass is too smooth to give the foothold that earthenware gives. That in the case of the earthenware pot both water and wet earthenware surface are available points to the fact that the insects prefer the latter. The accumulation of a line of eggs at the water's edge is not as sometimes thought the result of eggs being drawn into this position by capillarity. Actually *Aëdes* eggs are not very apt to be drawn to the edge by capillarity, though empty shells and collapsed eggs are. The eggs are so laid *in situ* by the female and are retained in position by their sticky chorionic pads until when dry they are firmly cemented in place by this same structure.

Obtaining eggs on wet filter paper may be a convenient method for enabling small samples of eggs to be used in experimentation. Floating eggs collected on cover-glasses or slides may also be useful for some purposes. But as noted in the section on technique the most suitable method and one with many advantages for obtaining and storing eggs is by the use of unglazed earthenware pots. Also, it may be noted from what has been said about hatching that filter paper, where eggs are employed for critical hatching data, requires to be used with much caution. Recently the effect of the physical characters of the container on oviposition and the desirability of using a rough porous surface have been emphasised by O'Gower (1955) in respect to *A. scutellaris*.

Eggs are invariably laid with the ventral surface, that is the surface furnished with the peculiar refractile bodies, upwards. When not over-crowded they are mostly laid in lines. Very often inspection shows that the line has followed a streak in the glass, or that the eggs have been laid along the edge of the filter paper strip where this meets the glass. In this case the eggs are almost invariably cemented to the glass, not to the paper. Usually the eggs are laid lenthwise in the direction of the line, but not necessarily all pointing in the same direction.

A point frequently referred to in the literature is the question of the choice for oviposition of the suitable fluid. It is commonly stated that this is by preference water containing organic matter, as against clean water. Beattie (1932) found the largest number of eggs on test bottles containing leaf infusion (8309 eggs from forty-two layings, as against 1241 eggs with twenty-nine layings with distilled water). In the laboratory, however, females appear to lay without any hesitation on clean tap-water and as a routine our pots have always been boiled and well scrubbed and placed in the cage with fresh clean tap-water. There are certain advantages in cleanliness in this respect and, to avoid unnecessary fouling by pre-oviposition defaecation, the pots for oviposition are freshly changed after this has taken place.

It is doubtful indeed if there is any very marked selection of the character of the water if other conditions are suitable, except in respect to strong salinity or other objectionable feature. Fielding, for example, notes that eggs may be laid on 40 per cent sea-water. Wallis (1954), studying the oviposition activity of mosquitoes including *A. aegypti*, found

* Incidentally if it did so the eggs would certainly perish; see description of changes following oviposition in ch. VI (*d*).

that the female could still detect an objectionable amount of salt when the movements of the abdomen were restricted and the palps, proboscis and antennae were coated with wax, and considered that, in general, sensitivity was located in the tarsal segments (see ch. XXIX).

RELATION OF BLOOD MEAL TO FERTILISATION AND OVARIAN DEVELOPMENT

The blood meal has little or no relation to copulation or fertilisation. Marchoux, Salimbeni and Simond state that a blood meal is not necessary to fertilisation. Macfie (1915) states that fertilisation can take place without feeding. Howard (1923) says that virgin females will feed, but that fertilised females are more greedy. Brug (1928) states that unfertilised females suck blood as readily as fertilised and, contrary to the statement by Macfie, (1915) they also lay eggs (unfertilised).

That for ovulation a blood meal is necessary is now generally accepted. It seems to be immaterial at what stage of life the blood meal is taken. Bacot records that a female fed 56 days on honey and white of egg laid fertile eggs 4 days after the first blood meal. Fielding, who failed to obtain eggs from females fed on a variety of foods other than blood, also found that egg production ceased on substitution of banana for blood. Gordon (1922) offered fifty-four females serum, washed red cells and whole blood and found eggs produced only on the last-mentioned.

That more than one blood meal is necessary before *A. aegypti* can develop its eggs has been sometimes stated. Whilst this is obviously incorrect where a full blood meal has been taken it raises the question how much blood is necessary. This has been very fully investigated by Roy (1936) whose work on the weight of blood taken at a meal has already been referred to. This author compared the number of eggs which were laid following imbibition of different amounts of blood. The data given in Table 36 are taken from his paper.

Table 36. *Number of eggs laid following upon different amounts of blood feed (from data given by Roy)*

Range blood (mg.)	Number of females	Mean weight blood (mg.)	Total number of eggs laid	Mean number of eggs laid	Number of times no eggs laid	Number of eggs per mg.
0·0–0·5	21	0·38	0	0	21	—
0·6–1·0	28	0·77	197	7·04	21	9·14
1·1–1·5	35	1·22	1197	34·2	2	28·06
1·6–2·0	34	1·79	1699	47·03	1	27·77
2·1–2·5	11	2·17	544	49·45	1	22·71
2·6–3·0	6	2·72	404	67·33	0	24·77
3·1–3·5	2	3·2	171	85·5	0	26·72

The table shows very clearly that to obtain a mean of 85·5 eggs, which it will later be shown is about the normal, a full meal of 3·0–3·5 mg. blood is required and that for any eggs at all at least 0·5–1·0 mg., or on the average 0·8 mg., of blood is necessary.

Further, small amounts of blood in successive feeds were capable of producing eggs only when the weight of blood taken at the third feed exceeded 0·5 mg. The number of eggs under such conditions was, however, as a rule small. In one batch mosquitoes given feeds from 0·13 to 0·45 mg. produced no eggs. But seven females given over 0·47 all produced some eggs. The author considered that these minimal amounts were necessary to stimulate the ovaries and that excess over this goes to form eggs. In chapter XXXI in the section on hormones are recorded recent findings that a hormone secreted in response to nervous stimuli aroused by blood entering the gut, or stretching the gut wall, is necessary for initiation and promotion of ovarian development and Roy's minimum dose may be such a stimulus. For the present it will suffice to note that successful oviposition is dependent on sufficient blood to enable the ovarian follicles to develop to full maturity and that this entails for a normal oviposition 3·0–3·5 mg. of blood, or in other words a full meal.

Among recent papers dealing with nutritive requirements for development of the ovaries and oviposition are: Greenberg (1951), Dimond *et al.* (1955, 1956), Lea *et al.* (1956), Woke (1955), Woke *et al.* (1956). All deal with *A. aegypti.*

TIME FROM BLOOD MEAL TO OVIPOSITION

Most references on this point are of a rather general nature. One of the most detailed of early statements is that of Marchoux *et al.* (1903) who gave the following data for different temperatures.

29–30° C. 48 hours, almost always by third day, sometimes fourth.
25–27° C. Fourth to fifth day.
20–25° C. Fourth day, sometimes seventh to ninth.
Under 20° C. May be retarded 26–7 days.

A complicating fact is that, unlike many culicines, *A. aegypti* does not usually lay all its eggs at one act of oviposition. Buxton and Hopkins (1927) note that in nature it usually deposits about twenty eggs at any one time. In captivity, too, it is not unusual for a female to take several days to lay all the eggs of one batch. Normally, however, after a full gorge on human or rabbit blood, mature fertilised insects under laboratory conditions, once oviposition is started, commonly lay all their eggs within the space of at most a few hours.

The time after a blood meal when eggs are laid depends upon the temperature and to some extent on the conditions under which the gravid female is kept. Two methods of observation are open to be used, namely (1) when large numbers of the insects are fed in cages; and (2) when single females after feed are isolated in tubes with a strip of filter paper and a little water.

In the first case, at a temperature of 25° C. and 70 per cent relative humidity, egg-laying in the author's experience usually begins on the afternoon of the third day from feed, counting the day of feed as 0. Feeding was usually carried out between 2 and 4 p.m. on the human arm with cages of some hundreds of females and males on the fifth to the seventh day from emergence, some 90 per cent of the females gorging. A large number of such layings were carried out as a routine in the course of some years' work on repellents and records kept show that at the temperature and humidity noted at most one or two females could have laid before 68 hours, the usual time for appearance of the first eggs being about 70 hours. The great bulk of eggs were laid between 72 and 112 hours. Comparatively few were laid from the sixth day onwards. A few observations of this kind have been made with cages in an incubator at 28° C. In some cases oviposition began as early as 48 hours and was practically completed by 96 hours.

Single females in tubes mostly began to oviposit some time during the third day (68–75 hours) when kept at 25° C. At 23° C. there was a marked delay, making the time of commencing oviposition commonly a little under 4 days (80–90 hours), and at 18° C.

the time was about doubled, retained eggs (see chapter VII) being for the first time noted.

In all the above observations the females were fed on fifth to seventh day from emergence. When females are fed on the second day from emergence, as previously noted, they do not engorge so fully and as was to be expected oviposition was delayed. Not only was there delay but considerable irregularity, which was very noticeable compared with the very constant results from later-fed females.

In all the above the mosquitoes were largely in the dark, being in a constant-temperature room with artificial light left on only when in occupation. Since these observations were made, an important research on oviposition time has been undertaken by Haddow and Gillett (1957) showing that the time of day, and thus the time from the blood meal, is largely determined by the effect of light. With normal daylight and darkness these authors found a marked peak of oviposition in the late afternoon, that is taking 4 hourly periods starting at 6·0 a.m. (sunrise) the peak was in the period 2.0–6.0 p.m. When subject to continuous artificial light (closed room with 4-foot fluorescent daylight-type tubes) oviposition was without any marked peak period, and when illumination was made opposite to the normal day and night the peak occurred at 2.0–6.0 a.m.

Regarding the time from blood meal to oviposition in these experiments cages with females (20) and males (10), emerged on the same day, were fed at 8.0 a.m., 4.0 p.m. and 12.0 midnight forming two series each of three cages so fed. The first series were kept at 27° C. and 70 per cent relative humidity under continuous artificial daylight as noted above. Oviposition began respectively in the three timed feedings at 58, 55 and 52 hours. The second series was kept at a temperature of 23° C. and a humidity much as in the first series, but under normal daylight and darkness. The time of beginning oviposition was respectively in the three timed feedings at 77, 73 and 83 hours from the blood feed, that is at 1.0 p.m., 1.0 a.m. and 11.0 a.m. respectively. The authors summarise their conclusions as follows:

> Under conditions of almost constant temperature and humidity, but with normally fluctuating daylight, *Aëdes aegypti* shows a regular cycle of oviposition, with one peak of major activity, which occurs in the afternoon. Under controlled conditions with constant light, this pattern breaks down and laying becomes aperiodic. Where the hours of light and darkness are artificially reversed (by means of daylight-type fluorescent lamps), the cycle also is reversed. Darkness delays the onset of oviposition. Light is the master factor, and there is a strong suggestion that the peak of oviposition is associated with the onset of light, and, furthermore, that the underlying mechanism may be hormonal.

Gillett (1956*a*) has distinguished as between 'ovarian development' or development of the oocytes in the ovarioles; 'ovulation' or passing of a ripe oocyst into the oviduct; and 'oviposition' or egg-laying. The times of initiation and completion of these, especially of the first and third, are each liable to be dependent on circumstances, for example ovarian development as recently shown is initiated by certain stimuli and hormonal effects, and the period to completion is dependent on temperature, amount of blood taken, age of the mosquito when fed and other conditions. Ovulation does not necessarily mean oviposition; and oviposition, besides being dependent on full ovarian development, may quite well be affected by many conditions, including as shown above by 'light', but such also may well include ability or inclination of the insect to find a satisfactory site for the act. Regarding the terms used it might be preferable to make 'ovulation' equivalent to completion of egg formation, which would of course be indicated by passage of at least one egg into the oviduct, but might be arrived at without this happening. From the data given by Haddow and Gillett's observations an estimate of the time to completion of ovarian development might be made as 55 hours at 27° C. (mean of the times in continuous light) and 73 hours at 23° C. (time when those fed at 4.0 p.m. laid at optimal time 4–5 p.m.) in this number of hours. Those fed at midnight (83 hours) it may be presumed were waiting overnight for appropriate light conditions (11–12 noon).

Under normal favourable conditions oviposition once started is commonly completed in the laboratory in 5 or 6 hours. At 28° C. thirty-nine out of forty-four (89 per cent) of females completed oviposition in 6 hours or less. At 25° C. thirty-six out of fifty-four (66 per cent) and at 23° C. eight out of twenty (40 per cent) did so in the same time. At 18° C. the time taken for completion of oviposition from the start was a full day or more.

Usually oviposition has been found to be complete on dissecting the mosquito after a reasonable time had been given for completion of the act. Short of dissection a guide to completion is the condition of the abdomen, which on completion of egg-laying is thin and cylindrical, with the ventral sclerites deeply folded under the tergites, so that the prominent lateral silvery spots on the sternites, so conspicuous in the distended abdomen, are quite hidden. A useful guide to the impending commencement of oviposition in a cage of fed mosquitoes is the passing of dark faeces staining the filter paper fans or soiling the receptacle. Gillett (1956*b*) has noted that mosquitoes that do not develop oocytes beyond stage 3 usually eliminate the blood meal by defaecation at 28–32 hours from the feed, whereas those that develop eggs to maturity do so 7–10 hours later.

DATA REGARDING SUBSEQUENT EGG BATCHES

Since more than one batch of eggs is normally laid in the lifetime of the female, provided there are further blood meals, the total number of eggs laid by one female may be considerable. Macfie (1915) records instances of fifteen and ten ovipositions, the first at intervals of 3 days by a female that lived 50 days, the second at longer intervals by a female that lived 60 days. Fielding records a female which laid 752 eggs in 72 days. In the first 31 days it fed eight times and laid 437 eggs (55 eggs per feed), in the next 31 days it fed six times and laid 260 eggs (43 eggs per feed) and in the next 10 days it fed twice and laid 55 eggs. Owing, however, to heavy mortality at oviposition the number of females with such numerous ovipositions is small and even after the first oviposition there is a considerable reduction in numbers. Howard, Dyar and Knab note that the French Commission (Marchoux *et al.*) found that out of 100 females only thirty laid eggs a second time and twenty-one from three to seven times. The largest number of ovipositions observed in a single female was seven. H. A. Johnson (1937) records females producing viable eggs to an age of 6 weeks.

The data given in the previous subsection relate to the time of oviposition following upon the first blood meal. Owing to the fact that after oviposition the follicles next following on the shed follicles are to a certain extent advanced in development, the time for these to reach maturity is, in the case of *Anopheles*, somewhat shortened as compared with that for the first oviposition. Whether this occurs in *Aëdes aegypti* has not been ascertained.

NUMBER OF EGGS LAID

Very varied estimates of the number of eggs laid at an oviposition by *A. aegypti* are given in the literature. Among earlier records are the following.

Marchoux *et al.* (1903)	First batch up to 100. Later batches below 30
Otto and Neumann (1905)	20–40 or more
Howard, Dyar and Knab (1912)	
J. R. Taylor (Havana)	35–114
American Commission	40–150
French Commission	First batch 70–95; max. 144; later batches max. 30
Goeldi	50–100
Boyce (1911)	27–97
Howlett (1913)	About 50
Macfie (1915)	100 to as few as 7
Buxton and Hopkins (1927)	20 on each occasion in nature
Connor (1924)	35–150
F. H. Taylor	Up to 80
Roy (1936)	Average max. 76

What at first seems very puzzling is that so many of these determinations are far below the number of follicular tubes in the ovaries. The ovaries of *A. aegypti*, like those of other mosquitoes, consist of a number of ovarian tubules in which successive follicles develop into eggs, the lowest in the tubes simultaneously undergoing full development whilst the next follicle behind remains undeveloped until after oviposition and another blood meal. In normally fed females many dissections following a blood meal at different intervals of time have invariably shown all the lower follicles simultaneously developed. The number of such follicles in the two ovaries is approximately 120. The number of eggs following a full blood meal should therefore be about one hundred. In *Culex molestus* a different condition may be found, namely that only a proportion of the lower follicles show development. This has not been seen in *Aëdes aegypti*, but may possibly occur under some conditions. In most of the above estimates the number of eggs at an oviposition clearly relates to the number laid under observation, no indication being given whether this might differ from the total number of oocysts developing to maturity, as might be shown by dissection.

In Table 38 is shown the number of females out of thirty observed that laid eggs in numbers between 85 (the lowest) and 125 (the highest). The mean was 115 eggs. The results were verified by dissection, the ovaries being found empty of developed follicles (eggs) except in a few instances where one or two retained eggs were found. To explain the very low figures for the number of eggs sometimes given it must be supposed either that the eggs counted did not include all the eggs formed, or that only a proportion of the lower follicles underwent development, or more probably that eggs were retained (see later).

There does not appear to be any close relation between the number of eggs and the size of the female. This would appear from the figures by Roy given in Table 39 and my own results as given in Table 37.*

* The number of eggs laid in relation to the amount of blood ingested has recently been the subject of investigation by Woke *et al.* (1956). A single first blood meal was taken at different days from emergence between the fifth and twenty-eighth days. The blood taken ranged from 0·1 to 4·8 mg. and the eggs laid from 0 to 136. The means for feeds on fifth to fourteenth day were 89, 85, 95 and 86. The number of eggs rose with 0·5 increment of blood from 0 in a correlated curve to 99 beyond which the curve levelled off. Postponement of the first feed to 28 days led to a lower number (mean 56). Size of mosquito did not affect the number.

Table 37. *Number of eggs laid by females of different weight and wing length*

Weight of female after oviposition (mg.)	Wing length (mm.)	Number of eggs laid*
2·01	3·33	126
2·04	3·37	100
2·08	3·55	126
2·30	3·48	116
2·30	3·37	117
2·32	3·48	132
2·40	3·48	114
2·43	3·41	133
2·49	3·62	123
2·64	3·48	120

* In all cases ovaries examined. An occasional retained egg in a few cases included.

Table 38. *Frequency distribution of number of eggs laid*

Number of eggs per batch	80–9	90–9	100–9	110–19	120–9	130–9
Number of mosquitoes laying	2	2	4	14	8	0

Table 39. *Number of eggs laid by females of different weight (from data given by Roy)*

Class (mg.)	Mean weight unfed (mg.)	Mean number of eggs laid	Total number of females
Below 1·0	0·85	42	3
1·1–1·5	1·35	44	16
1·6–2·0	1·82	57	6
2·1–2·5	2·17	36	2

EGG LAYING BY UNFERTILISED FEMALES

The laying of eggs by unfertilised females was stated by Macfie (1915) not to occur. That unfertilised females do lay eggs, however, is now generally recognised, the points at issue being the degree to which they do so and circumstances associated with this. A very complete study in this respect has been made by Lang (1956) and by Wallis and Lang (1956) who have compared egg production in continuously mated females with that by virgin females. The following are the chief conclusions arrived at:

1. There is no difference in the engorgement rate at the feed.
2. Continuously mated females produce more eggs than virgin females, one-third of which failed to lay.
3. The average time for oviposition for virgin females was longer than that for mated females and more variable. Of forty-six mated females 89 per cent oviposited within 7 days. Of 128 virgin females only 55 per cent oviposited within this period. With partial or continuously mated females maximum oviposition was within 7 days, but only 60 per cent of virgin females gave maximum oviposition in this period.
4. Dissection of ovaries 2–3 days after feed provides a rigid assay for presence of an ovarian stimulation factor in an experiment.
5. Feeding on sucrose did not provide a stimulus, showing that it is unlikely that distension of the abdomen acts as a stimulus, which is more likely to be from some chemical in blood.

MATING AND OVIPOSITION

Many interesting facts regarding the effect of mating and other stimuli connected with oviposition have been recently given by Clements (1956) and by Gillett (1955, 1956) (see section on hormones in chapter XXXI).

RETAINED EGGS

That eggs may be retained beyond their natural time is commonly shown by females that have not had access to water. Goeldi (1905) notes that eggs may be retained 23–102 days and still laid. Woke (1955) records that female *A. aegypti* forced to withhold oviposition by absence of water for from 5 to 90 days from the blood meal may still oviposit when given moist filter paper.

In the author's experience dissection of females isolated in tubes that have been considered as having completed oviposition has rarely shown any considerable number of retained eggs. Under 2 per cent usually show one or two eggs still retained in the common oviduct or cavity of the ovary. That such eggs are mature is shown by their appearance and the fact that on exposure they darken and become quite black as in the normally laid egg. In females from cages the presence of a few retained eggs has been more frequent (up to 21 per cent).

Among a small number of females fed and kept at 18° C. two were found which had laid respectively 20 and 50 eggs, but in which 94 and 73 fully matured eggs were found still largely in position in the ovaries. On exposure the majority of these darkened only slowly, showing that they were mature.

It should perhaps be mentioned here that all eggs when first laid (see ch. VI (*d*)), and whilst still white, measure only about 0·5–0·6 mm., whereas the same eggs when darkened give the usual measurement for eggs of about 0·6–0·7 mm. Failure of retained eggs to meet normal egg measurement does not therefore invalidate the fact that they are fully developed. In many cases retained eggs have already within the body partly or fully darkened.

OVARIAN CHANGES

In the unfed 5-day female the ovaries, when dissected out and allowed to lie in salt solution, measure about a millimetre in length and in the insect are situated in the hinder parts of the fifth and sixth segments. On the fourth day after the blood meal, just before oviposition, they measure up to about 3 mm. and almost completely fill the abdomen from the first to the seventh segment, though the swollen abdomen is never so distended as after a full blood meal. At this time the gut is empty and even the ventral diverticulum occupies a very small space.

At oviposition the follicles, now in the form of eggs, occupying the lower ends of the follicular tubes, pass one by one into the oviducts with the micropilar end directed posteriorly and the rather concave ventral surface directed inwards. They are fertilised as they pass the opening of the spermathecal ducts, some dissections suggesting that as they pass this point a single spermatozoon with its head directed forwards lies at the opening ready to perform this function.

REFERENCES

BACOT, A. W. (1916). *Rep. Yell. Fev. Comm.* (*West Africa*), London.

BEATTIE, V. F. (1932). The physico-chemical factors of water in relation to mosquitoes breeding in Trinidad. *Bull. Ent. Res.* **23**, 477–96.

REFERENCES

BONNE-WEPSTER, J. and BRUG, S. L. (1932). See references in ch. II (*a–b*), p. 45.

BOYCE, R. (1911). The prevalence, distribution and significance of *Stegomyia fasciata* (*calopus* Mg.) in West Africa. *Bull. Ent. Res.* **1**, 233–63.

BRUG, S. L. (1928). Remarks on Prof. Hoffmann's paper. *Meded. Volks. Ned.-Ind.* Foreign ed. **17**, 184–5.

BUXTON, P. A. and HOPKINS, G. H. E. (1927). Researches in Polynesia and Melanesia. Parts I–V. *Mem. Lond. Sch. Hyg. Trop. Med.* no. 1.

CHRISTOPHERS, S. R. (1947). Mosquito repellents. *J. Hyg., Camb.*, **45**, 176–231.

CLEMENTS, A. N. (1956). Hormonal control of ovary development in mosquitoes. *J. Exp. Biol.* **33**, 211–23.

CONNOR, M. E. (1924). Suggestions for developing a campaign to control yellow fever. *Amer. J. Trop. Med.* **4**, 277–307.

DIMOND, J. B., LEA, A. A., BROOKS, R. F. and DE LONG, D. M. (1955). A preliminary note on some nutritional requirements for reproduction in female *Aëdes aegypti*. *Ohio J. Sci.* **55**, 209–11.

DIMOND, J. B., LEA, A. A., HAHNERT, W. F. and DE LONG, D. M. (1956). The amino-acids required for egg production in *Aëdes aegypti*. *Canad. Entom.* **88**, 57–62.

FIELDING, J. W. (1919). Notes on the bionomics of *Stegomyia fasciata* Fabr. *Ann. Trop. Med. Parasit.* **13**, 259–96.

GILLETT, J. D. (1955). Behaviour differences in two strains of *Aëdes aegypti*. *Nature, Lond.*, **176**, 124.

GILLETT, J. D. (1956*a*). Genetic differences affecting egg-laying in the mosquito *Aëdes* (*Stegomyia*) *aegypti* (Linnaeus). *Ann. Trop. Med. Parasit.* **50**, 362–74.

GILLETT, J. D. (1956*b*). Initiation and promotion of ovarian development in the mosquito *Aëdes* (*Stegomyia*) *aegypti* (Linnaeus). *Ann. Trop. Med. Parasit.* **50**, 375–80.

GODEHEU DE RIVILLE (1760). Mémoire sur l'accouplement des cousins. *Mém. Acad. Sci., Paris*, **3**, 617–22.

GOELDI, E. A. (1905). *Os mosquitos no Para*. Weigandt, Para.

GORDON, R. M. (1922*b*). See references in ch. XX, p. 481.

GREENBERG, J. (1951). Some nutritional requirements of adult mosquitoes (*Aëdes aegypti*) for oviposition. *J. Nutr.* **43**, 27–35.

HADDOW, A. J. and GILLETT, J. D. (1957). Observations on the oviposition-cycle of *Aëdes* (*Stegomyia*) *aegypti* (Linnaeus). *Ann. Trop. Med. Parasit.* **51**, 159–69.

HOVANITZ, W. (1946). Comparison of mating behaviour, growth rate and factors influencing egg-hatching in South American *Haemogogus* mosquitoes. *Physiol. Zool.* **19**, 35–53.

HOWARD, L. O. (1923). The yellow fever mosquito. *U.S. Dep. Agric. Frms' Bull.* no. 1354.

HOWARD, L. O., DYAR, H. G. and KNAB, F. (1912). See references in ch. I (*f*), p. 19.

HOWLETT, F. M. (1913). *Stegomyia fasciata*. *Proc.* 3*rd Meet. Gen. Mal. Comm. Madras*, 1912, pp. 205.

JOHNSON, H. A. (1937). See references in ch. XX, p. 482.

LANG, C. A. (1956). The influence of mating on egg production by *Aëdes aegypti*. *Amer. J. Trop. Med. Hyg.* **5**, 909–14.

LEA, A. O., DIMOND, J. B. and DE LONG, D. M. (1956). Role of diet in egg development of mosquitoes (*Aëdes aegypti*). *Science*, **123**, 890.

MACFIE, J. W. S. (1915). See references in ch. XX, p. 482.

MACGREGOR, M. E. (1915). Notes on the rearing of *Stegomyia fasciata* in London. *J. Trop. Med.* (*Hyg.*), **18**, 193–6.

MARCHOUX, E., SALIMBENI, A. and SIMOND, P. L. (1903). See references in ch. I (*f*), p. 19.

MATTINGLEY, P. F. (1956). See references in ch. II (*c*), p. 47.

MITCHELL, E. G. (1907). *Mosquito Life*. G. P. Putnam's Sons, New York and London.

O'GOWER, A. K. 1955. The influence of physical properties of a water container surface upon its selection by the gravid females of *Aëdes scutellaris scutellaris* (Walker) for oviposition (Diptera: Culicidae). *Proc. Linn. Soc. N.S.W.* **79**, 211–18.

OTTO, M. and NEUMANN, R. O. (1905). See references in ch. I (*f*), p. 19.

Putnam, P. and Shannon, R. A. (1934). The biology of *Stegomyia* under laboratory conditions. 2. Egg-laying capacity and longevity of adults. *Proc. Ent. Soc. Wash.* **36**, 217–42.

Qutubuddin, M. (1952). The emergence and sex ratio of *Culex fatigens* Wied. (Diptera: Culicidae) in laboratory experiment. *Bull. Ent. Res.* **43**, 549.

Rees, D. C. (1901). Malaria: its parasitology, with a description of methods of demonstrating the organisms in man and animals. *Practitioner* (new series), **13**, 271–300.

Roth, L. M. (1948). A study of mosquito behaviour. An experimental laboratory study of the sexual behaviour of *Aëdes aegypti* (L.). *Amer. Midl. Nat.* **40**, 265–352.

Roth, L. M. and Willis, E. R. (1952). Method of isolating males and females in laboratory colonies of *Aëdes aegypti*. *J. Econ. Ent.* **45**, 344–5.

Roy, D. N. (1936). On the role of blood in ovulation in *Aëdes aegypti* Linn. *Bull. Ent. Res.* **27**, 423–9.

Roy, D. N. (1940). Influence of spermathecal stimulation on the physiological activities of *Anopheles subpictus*. *Nature, Lond.*, **145**, 747–8.

Taylor, F. H. (1942). Contributions to a knowledge of Australian Culcidae. No. V. *Proc. Linn. Soc. N.S.W.* **67**, 277–8.

Teesdale, C. (1955). Studies on the bionomics of *Aëdes aegypti* (L.) in its natural habitats in a coastal region of Kenya. *Bull. Ent. Res.* **40**, 711–42.

Theobald, F. V. (1903). See references in ch. I (*e*), p. 18.

Wallis, R. C. (1954). A study of oviposition activity of mosquitoes. *Amer. J. Hyg.* **60**, 135–68.

Wallis, R. C. and Lang, C. A. (1956). Egg formation and oviposition in blood-fed *Aëdes aegypti*. *Mosquito News*, **16**, 283–6.

Wesenberg-Lund, C. (1943). Biologie der Süsswasserinsekten. Copenhagen, p. 433.

Woke, P. A. (1955). Deferred oviposition in *Aëdes aegypti* (Linn.). *Ann. Ent. Soc. Amer.* **48**, 39–46.

Woke, P. A., Ally, Mona S. and Rosenberger, C. R. Jr. (1956). The number of eggs developed related to the quantity of blood ingested in *Aëdes aegypti* (Linnaeus). *Ann. Ent. Soc. Amer.* **49**, 435–441.

Young, C. J. (1922). Notes on the bionomics of *Stegomyia calopus* Meig. in Brazil. *Ann. Trop. Med. Parasit.* **16**, 389–406.

XXIII
FLIGHT

(*a*) SPEED AND RANGE OF FLIGHT

SPEED OF FLIGHT

The speed of flight of a number of insects as recorded in the literature is given by Wigglesworth (1942). This varies in the list given from 0·6 (*Chrysopa*) to 15·0 (Sphingids) metres per second. Speeds given for the bee vary from 2·5 to 6·0 metres per second and for Diptera from 2·0 (*Musca*) to 14·0 (*Tabanus*). For further examples see especially Magnan (1934), also Demoll (1918). In many cases such estimates of speeds have been arrived at by observation alone.

Seen clearly in a beam of light or against a suitable background, estimation of the cruising speed of the unfed female *Aëdes aegypti* has given values between 0·5 and 1·0 metre per second (50 and 100 cm./sec.). It is thought that the male has in general a more rapid flight than the female. Accurate determination by such method is, however, difficult.

Observations based on experimental conditions have been given by Kennedy (1940). As shown by this author the flight of the female is largely controlled by visual response to the background. In a wind tunnel in still air over a background of moving shadow stripes the mosquitoes flew in the same direction as the movement, often faster but never more slowly than the stripes. In a gentle wind of 40 cm./sec. over a stationary background the mosquitoes made headway and were never carried backwards. A minority of individuals temporarily swept in the direction of the wind sometimes turned in mid air and flew up wind again. When disturbed the mosquitoes (females) flew directly against a 40 cm./sec. wind to settle on the netting on the upper end of the wind tunnel. As the wind grew stronger, more and more mosquitoes alighted, those still flying in 100 cm./sec. wind maintaining their position until they suddenly alighted on the floor without being carried backwards. With winds over 150 cm./sec. they were made to alight and could not take off. Kennedy notes that Afridi and Abdul Majid (1938) found that *Culex fatigans* refused to take off in a wind of 10 miles an hour (about 450 cm./sec.). These results indicate an average speed of about 50 and a maximum speed of 150 cm./sec.

The effect of such visual control of speed according to Kennedy would be likely to cause the speed to be greater when flying at a distance from the 'ground'. When close to the 'ground' (not above 10 cm.) and with quite gentle winds the speed did not exceed 30–40 cm./sec. When well up in the air, however, the mosquitoes should be able to fly to the limit of their capacity (150 cm./sec.). When flying the length of the tunnel without air or stripes movement the average speed was about 17 cm./sec., the highest being 37 and the lowest 9 cm./sec. The author terms speed so controlled the 'tolerated image movement'.

The flight of males during feeds has some features of interest. The rate of flight is such that at 10 inches distance the insects in flight can be glimpsed, but little or no detail made out. Flight is backwards and forwards in a plane at an angle of about 20° to the horizontal, the head facing upwards in the plane. There does not appear to be any turning for the backward movement, though this requires confirmation. If so, this must mean that the male can fly at an almost equal rate in reverse. Though the flight is kept up continuously for half an hour, should the arm still be in the cage, not all individuals are in flight for such a period, as numbers are constantly dropping out, settling on the sides of the cage and after a time taking wing again.

RANGE OF FLIGHT

Though *Aëdes aegypti* has considerable powers of rapid and directive flight, it is not, like certain salt-water breeding mosquitoes and many anophelines, recognised as a species having a tendency to travel considerable distances and is largely restricted in its activities to haunting rooms of dwellings and such-like places, normally making short flights and frequently settling. Soper (1935) notes that though the species may fly long distances in exceptional circumstances, its flight is normally limited to within 25–30 yards of its breeding places. Hinman (1932) states that the species is incapable of flying more than a few hundred yards. Cumming (1931), dealing with the species in connection with control of yellow fever, gives flight (dispersal) as 400–1000 metres. Shannon and Davis (1930) also state that *A. aegypti* is capable of sustained flight over water of 1 kilometre. A good summary is given by Whitfield (1939).

Dispersal as shown by capture of stained liberated mosquitoes is much less than some of these estimates. Shannon and Davis released 20,000 stained specimens in a small village. Two were recovered in houses more than 300 metres away and ninety-five at smaller distances. Less than 0·4 per cent were recovered, the latest capture being 13 days from time of liberation. Shannon, Burke and Davis (1930) liberated 3500 stained mosquitoes in one house. Dispersion to other houses began within 24 hours. By the end of a week over 90 per cent had disappeared from the house of release. At the end of two weeks less than 1 per cent were found. Bonnet and Worcester (1946), in Hawaii, give the result of capturing marked *A. albopictus*, a species of somewhat similar habits to *A. aegypti*. They found the maximum distance covered in 21 days to be 232 yards, the mean distance for all captures being 68·7 yards. In 5 days the mean distance was 36·3 yards, in 10 days 113·8 yards and in 15 days 103 yards. Wind had little effect, the insects flying into the wind when velocity was low; otherwise they clung to vegetation. Wolfinsohn and Galun (1953) liberated 28,000 laboratory bred gravid females towards evening in an area in Palestine free from wild *A. aegypti* or of dwellings likely to cause attraction and searched for eggs in 150 jars. They found eggs laid up to 2·5 kilometres from the liberation point, all within 45° of the direction of the wind.

METHODS USED IN MARKING MOSQUITOES FOR DISPERSAL STUDIES

The chief difficulty in determination based on liberation of marked mosquitoes has always been (1) the very small numbers recaptured, and (2) the unknown extent of possible lethal action of the means taken for marking. The most commonly used method of marking has been that first used by Zetek, namely dusting of the mosquitoes with aniline dyes, some of which were clearly harmful. Shannon and Davis found spraying with 1 per cent gentian violet caused death, but 1 per cent methylene blue was used successfully. Griffiths and Griffitts (1931) used a 2 per cent water solution of yellowish eosin distributed by an atomiser. Bonnet and Worcester have experimented with auramine 0, pyrazol red, gentian violet, malachite green, methyl orange, toluidine blue, chloramine red and pyrazol violet.

Chloramine red was the only one that proved successful. See also Clarke (1943*a*, *b*), Eyles and Bishop (1943).

To avoid toxic effects Majid (1937) used a metallic marking. Mosquitoes not exceeding 200 are transferred to a small lamp shade (6½ inches high by 3 inches diameter opening at base), the top and bottom closed by netting. 'Gold' powder, as used in Press work for decorative writing was then insufflated by a dusting tube made from a corked 3 × 1 inch specimen tube fitted with a rubber ball bellows and inlet and outlet tubing. The marked mosquitoes are readily recognised after capture.

Hassett and Jenkins (1949) use radio-active phosphorus, third and fourth instar larvae being transferred to culture medium containing 10 μc./ml. radio-active phosphorus (P^{32}). Second instar larvae reared in culture fluid containing $Na_2HP^{32}O$ in from 0·05 to 10 μc./ml. were all greatly retarded in development, but at 1 μc./ml. adults emerged after 15 days and varied in radioactivity from 6761 to 10,132 cpm. Third and fourth instar larvae in 10 μc./ml. begin to emerge in 6 days. For converting curies to grams, etc., see Morgan (1947). Bugher and Taylor (1949) also used radiophosphorus and radiostrontium (Sr^{89}) as Na_2PO_4 and $SrCl_2$. Use of 10–20 radioactive mμc. per c.c. in the larval bath gave a radioactivity of 2–16 mμc., using 5 larvae per c.c. There was no absorption by pupae. In the mosquitoes 40 per cent of radioactive substance was in the legs, practically none in the wings.

For dispersal of *A. aegypti* by man see under 'Geographical distribution', p. 41.

(*b*) MECHANISM OF FLIGHT

GENERAL NATURE OF FLIGHT MECHANISM*

Movements of the wings in flight have been studied, especially in the bee and wasp, by (1) observation of the movements of the wing as seen in the dead insect (Stellwaag); (2) observations made on the stationary insect with the ends of the wings gilded (Marey); (3) study of cinematograph records (especially Bull, Magnan and Voss); (4) observing the movements under the stroboscope as dealt with in later subsections, and other recent methods (Sotavalta, Hocking).

To a large extent the mechanism concerned in flight is very similar in the majority of insects including the mosquito. The main action is the up and down movements of the wing brought about through the powerful indirect wing muscles in the thorax acting indirectly on the wing by leverage. In addition, to ensure forward progress, the wing is also to some degree rotated on its long axis and/or so composed with a thickened anterior and more flexible posterior portion that some degree of propeller-like action results. Some arrangement has also usually to be made to bring the wings into position for action and for flexing and disposing of them when they are not required.

The wings of the mosquito and the structure of the parts concerned in flight have already been described in chapter XIX. It is necessary here only to note how the parts are used.

What is required of the wings is (1) the main up and down movement, (2) some rotation of the wing in its long axis or other adaptation required for progressive flight, and (3)

* A considerable literature exists dealing with insect flight, see especially Marey (1895), Bull (1904, 1909, 1910), Stellwaag (1910, 1916), Ritter (1911), Voss (1913, 1914), Buddenbrock (1930, review), Snodgrass (1930, how insects fly), Magnan (1934), Wigglesworth (1942). Very complete monographs with bibliographies on insect flight are given by Sotavalta (1947) and by Hocking (1953).

extension and flexion with adjustment of the wings to the resting position. The part played by the halteres and other considerations connected with flight are dealt with later.

UP AND DOWN MOVEMENT OF THE WING

Through the structures at its base the wing has the arrangement of a lever, the long arm being the wing itself and the short arm the bent down root of the wing. The power is applied at the articulation of the wing with the edge of the scutal flange (Fig. 58 (1), (2), (3) *fl*). With the wing extended, thus bringing the lever system into action, pressure downwards by the scutal flange will cause the wing to be lifted in the upstroke. Similarly any raising of the flange by return to its original position will produce the downstroke. It is easy to show this effect of the flange by pressing with a needle on the tergum when the wing is extended.

The wing, however, is not free to move in all positions equally. The base of the wing on its upper surface ends at its inner margin in an irregular knuckle-like ridge. Between this and the edge of the scutum is a deep sulcus, the inner wall formed by the smooth surface of the scutal flange, the outer by the downward-projecting base of the second axillary. This arrangement limits very largely the wing movement to that on a hinge-like line. It also largely prevents movement of the wing upwards beyond a certain point, as will be very evident on attempting to raise the extended wing in a freshly killed insect much beyond the horizontal plane. It will also be found difficult to extend the wing forwards beyond a certain point. This is because as the wing is pushed forwards the free inner end of the costa will be seen to engage in the hollow of the tooth-shaped humeral plate (Fig. 57 (8) *sep*) and this in turn to abut against the tegula, movement being further restricted by the limpet-shaped sclerite referred to in the description of the parts as the basalare (Fig. 58 (1) *bas*). Only in a backward direction and to some extent downwards after some anti-clockwise rotation is the wing free to move in any marked degree. Clockwise rotation, that is rotation forwards, is much more restricted. Nor is the lever action of the up and down movement entirely a good simile, for the movement of the wing in flight is somewhat vibratory, that is the terminal portion of the wing has the appearance in flight of moving through a considerable arc, whilst the basal parts show only very slight movement, suggesting rather what is seen with the vibrating limb of a tuning fork.

ROTATION OF THE WING

The parts concerned in rotation of the wing on its longitudinal axis are largely the second axillary and the remigium. The remigium is a very complex structure. As already noted it seems essentially to consist of two partially fused veins, a dorsal radial element more or less superposed on a subcostal element, though the parts are far from being merely an unmodified continuation of the respective veins in the wing field. These continue basally in a normal manner up to the point where the splinter-like articular sclerite of Prashad links on the sabroid of the remigium. It is on the two elements of the remigium that the series of ridges and hollows described by Shipley and Wilson (1918) as a possible stridulating organ in *Anopheles maculipennis* occur. These are considered by Prashad to be concerned in rotation, the ridges and hollows preventing the two parts of the remigium from moving longitudinally on each other. If this be so, the poor development of the ridges in *Aëdes aegypti* as compared with that in *A. maculipennis* may be related to the much larger wing and greater powers of flight in the latter species. That some rotation of the parts in the base of the wing takes place is shown by the shift in position of the second axillary (epaulet) on manipulation of the wing.

EXTENSION AND FLEXION

The folding mechanism of the wing, as indeed its full extension, is in some considerable degree at least automatic and due to the structure and elastic nature of the parts. If the extended wing be gently pushed back, there is, after moving it to a certain point, a sudden jerking back of the wing towards, and even completely up to, its resting position. Similarly a touch on the half-closed wings may throw both wings into the completely extended position, an action very familiar when mounting mosquitoes. The main 'joint' of the wing in this movement seems to be the flexible articulation of the remigium with the second axillary (epaulet). In the action the connexion is wrenched round and the epaulet rotated forwards. When flexing the wing from the fully extended position the free basal end of the costa lies anterior to the base of the remigium. But as the wing moves backwards there comes a point where the membrane linking the end of the costa to the humeral plate (sub-epaulet) slips over the bend made by the remigial joint, thus removing the costa as a bar to further bending. The mechanism indeed suggests that did the costa not end freely in membrane it might well embarrass working of this movement. When the wing is fully bent back, that is, fully flexed, the second axillary-remigial joint (epaulet) projects forwards forming a kind of knuckle. The free end of the costa then forms an additional small knuckle external to this (Fig. 57 (8)).

The horizontal backward movement of the wing at full flexion is also terminated by a kind of locking device. This is largely brought about by the squama. As the wing closes, the outer squama merely follows the movement of the posterior border of the wing and so comes to lie over the bent-in inner squama. The movements of the inner squama are more difficult to follow. In the extended wing the inner squama is bent down in its inner half to form a flat depressed area linking on to the axillary cord (Fig. 58 (4), (5) *sq*). With flexion of the wing the greater part of the inner squama is folded back and completely hidden beneath the outer squama. So folded, it lies packed away in a hollow behind the posterior border of the scutal flange. The folding is seemingly not passive, but accurately regulated by the elasticity of the parts to which the sclerites of the inner squama give precision in the folding, the elastic border and ribbed axillary cord keeping the edges firm and rigid.

When the wings are fully flexed there is left behind the scutellum a triangular uncovered area consisting of part of the postnotum and some portion of the first abdominal segment. Over this space the rather long fringe scales at the base of the wing project. On the outer squama there are no long fringe scales, thus leaving room for the fringe scales of the posterior wing border and the scutellar chaetae.

As the wings close in they pass very neatly one over the other, slipping into place without any interference. It was thought interesting to know whether one wing was always the upper one or not, if not, what was the proportion in which right or left wing overrode. In thirty-eight females examined at random eighteen gave the left wing over the right and twenty the right wing over the left. These mosquitoes were then individually re-examined after being allowed to recover from the anaesthetic and fly for a time. This gave almost equal numbers in which the overriding wing was changed or was still the right or left as the case might be. It was also found that it was not necessary that the wing first closed should be the covered one. As often as not the second wing to close slipped beneath the already closed wing; this without the least hesitation or interference.

(*c*) THE HALTERES

Well behind the wings lie the halteres. The structure of these organs in the mosquito has been minutely described by Prashad (Prashad, 1916). They consist of a round knob-like head and narrow stalk and a somewhat pyramidal base attached by a linear hinge-like joint which allows movement only in a fixed plane. The direction of the hinge-line is very close to that of the sulcus and articular ridge of the base of the wing. It follows that the halteres, which vibrate through a considerable arc, do so almost exactly in parallel movement with the wings. When the wings vibrate the halteres do so too, stopping immediately the wings stop. This can be readily demonstrated on a female mounted as for the stroboscope or merely pinned through the tip of the abdomen. The nature of the sense organs on the base of the halteres is described in chapter XIX when dealing with structure of these organs.

That the halteres are essential to flight in that they allow of co-ordinate movement of the insect is now well known. It is not difficult to touch the halteres of an anaesthetised insect with a fine needle wet with liquid glue and to fasten them down to the abdomen without injury to the insect. Such an insect on recovering from the anaesthetic will fly but in a wholly incoordinate manner as will be well demonstrated on a large sheet of paper, the insect flying up only to dash itself violently to the ground.

For information on the function of the halteres and bibliography see especially Pringle (1948).

(*d*) WING-STROKE FREQUENCY

THE FLIGHT NOTE

The wing-stroke frequency in insects has been studied by a number of observers, some of whom have given data regarding mosquitoes. In respect to mosquitoes determination of the frequency has for the most part been deduced from the note emitted by the insect, or as it is commonly called the 'hum' of the mosquito.

Recently a very complete study of flight tone has been made by Sotavalta (1947). The method used has been mainly the acoustic method, that is, matching the note by ear. For this the author uses a covered glass cylinder 10 by 15 cm. high, through the cover of which are fixed the cup of a binaural stethoscope, a thermometer and two glass tubes by which temperature and humidity can be adjusted, the latter being read by a small hygrometer on the bottom of the cylinder. Two notes may be given by insects under certain circumstances, namely the true frequency note (flight tone) and the 'whining tone' on the average an octave higher. The latter can be elicited by cutting the wings short. It is interesting to note that Rudolfs (1922) records a shrill note when certain mosquitoes are subjected to some essential oils (angry note). The highest flight notes are given by small midges (up to c''' equivalent to 1046 vib./sec.). They have a proportionately large thorax to allow for the necessary muscles. For Culicidae spp. the author gives about 300 vib./sec. for the female and 500 for the male. The rate alters with the temperature. The note is also changed by load, by fixing the insect and by fatigue. Goeldi (1905) refers to alteration of the note as a result of dilatation of the abdomen by food. In *Anopheles* Nuttall and Shipley found the tone raised in proportion to the extent of gorging.

Table 40. *Recorded flight note and wing vibration per second of different mosquitoes*

Species	Note with approximate corresponding vibrations per second															
	d 143	*e* 161	*f* 171	*g* 192	*a* 213	*b* 241	*c′* 256	*d′* 287	*e′* 322	*f′* 342	*g′* 384	*a′* 427	*b′* 483	*c″* 512	*d″* 574	*e″* 644
								Female								
Anopheles																
maculipennis (2)	—	—	—	—	—	×	×	×	×	—	—	—	—	—	—	—
maculipennis (4)	—	—	—	—	—	—	—	—	×	—	—	—	—	—	—	—
maculipennis (7)	—	×	×	×	×	×	—	—	—	—	—	—	—	—	—	—
punctipennis (4)	—	—	—	—	—	—	—	—	—	—	—	—	—	—	—	—
Theobaldia																
annulata (1)	—	—	—	—	—	—	—	—	—	—	—	—	—	—	—	—
spp. (7)	—	—	×	×	×	×	—	—	—	—	—	—	—	—	—	—
Aëdes																
aegypti (3)	—	—	—	—	—	—	×	—	—	—	—	—	—	—	—	—
spp. (7)	—	—	—	—	—	×	×	×	×	×	—	—	—	—	—	—
Culex																
pipiens (1)	—	—	—	—	—	—	—	—	—	—	—	—	—	—	×	—
pipiens (7)	—	×	×	×	—	—	—	—	—	—	—	×	×	—	—	—
spp. (6)	—	—	—	—	—	×	×	×	×	—	—	—	—	—	—	—
Culicidae																
spp. (7)*	×	×	×	×	×	×	×	×	×	×	—	—	—	—	—	—
spp. (7)†	—	—	—	×	×	×	×	×	×	×	×	—	—	—	—	—
spp. (5)	—	—	—	—	—	—	—	×	—	—	—	—	—	—	—	—
								Male								
Anopheles																
maculipennis (2)	—	—	—	—	—	—	—	—	—	—	—	×	—	—	—	—
maculipennis (7)	—	—	—	—	—	—	—	—	×	—	—	—	—	—	—	—
Theobaldia																
annulata (1)	—	—	—	—	—	—	—	—	—	—	—	—	—	—	—	×
spp. (7)	—	—	—	—	—	×	×	×	×	×	—	—	—	—	—	—
Aëdes																
aegypti (3)	—	—	—	—	—	—	—	—	—	—	—	×	—	—	—	—
spp. (7)	—	—	—	—	—	—	—	—	×	×	×	×	×	×	×	—
Culicidae																
spp. (7)*	—	—	—	—	—	—	—	—	—	—	—	—	×	×	—	—
spp. (7)†	—	—	—	—	—	—	—	×	×	×	×	×	×	×	×	—

(1) Landois (1867). (2) Nuttall and Shipley (1902). (3) Goeldi (1905). (4) Weber (1906). (5) Howard *et al.* (1912). (6) Voss (1913, 1914). (7) Sotavalta (1947). * In nature. † Under different temperatures, 7–39° C.

In Table 40 are given the notes recorded by different observers for various species of mosquitoes. In the table, in order to give a general figure for Culicidae, the frequencies, whether given by the author as notes or as vib./sec. are entered as crosses covering the range recorded. For exact details, references should be consulted in original. The allocation of frequencies to the notes is on the basis of the scale given below. It will be evident from the table that, excluding the very high notes given by Landois (possibly the whining notes) the extreme range for the female mosquito extends through about the two octaves *c* to *c″* (128–512 vib./sec.), the more usual records being from *f* to *f′* (170–340 vib./sec.).

For the male the range is between *b* and *d″* (240–570). The large species (*Theobaldia*) tend to give a lower note. The only author giving data for *Aëdes aegypti* is Goeldi (1905) who records *c′* for the female and *a′* for the male, or about 256 and 430 per second respectively.

It is perhaps desirable to give a brief explanation of the notation system which to a non-musician is apt to be confusing. For an account of the different systems see Catchpool and Satterly (1944). Middle octave notes are recorded as *c, d, e, f, g, a, b*, those of the next higher octave as *c′* — *b′* and for the still higher octave *c″* — *b″*, the octave below the *c* — *b* being given as capitals. In relation to the written clefs the note for the first line of the treble clef is *e*, that for the third space from the bottom being *c′*. In the bass clef the top line is *A*, one line above being *c* (Fig. 62 (5)). The exact number of vibrations corresponding to these notes differs according to the standard of pitch in use, *c″* in English concert pitch being 546, in the Old Philharmonic 540 and in the tonic sol-fa 507. For ordinary scientific purposes *c″* is commonly taken as 512 which gives convenient numbers (powers of 2). The notes intermediate between the octave notes are also somewhat dependent upon the scale in use. But the usual equal-tempered scale is based on the series 1, $2^{2/12}$, $2^{4/12}$, $2^{5/12}$, $2^{7/12}$, $2^{9/12}$, $2^{11/12}$, 2, or in actual figures in the proportions, 1, 1·122, 1·260, 1·388, 1·498, 1·682, 1·888, 2. Using these proportions and approximate figures the value of the notes likely to be required in the present connection will be as follows:

	c	*d*	*e*	*f*	*g*	*a*	*b*
Lower octave (*c*)	128	143	161	171	192	213	241
Middle octave (*c′*)	256	287	322	342	384	427	483
Upper octave (*c″*)	512	574	644				

These figures have been used in the table, any notes recorded as flats or sharps being given under the note.

Notes can be obtained, sometimes of great strength, by confining mosquitoes in tubes about an inch in diameter and of different lengths closed with a piece of large mesh mosquito netting held in place with a rubber band. The note is best taken when the mosquito is flying near the netting, that is at the mouth of the tube. The tubes are advantageously placed on a sounding board. A powerful note, for instance, is given by the newly gorged female *A. aegypti* when segregated in this way in a 3 by 1 inch specimen tube. The note calculated at a wavelength of four times the length of the tube,* the rule for a closed tube, should be at 25° C. 1151 vib./sec. Actually it was thought to be about middle *g* or about 380 vib./sec. In such cases the note heard may not be the fundamental but some harmonic of this, indicating a possible serious source of error in deducing vibrations from the note. See also Nuttall and Shipley (1902), who while giving the male note as 880 consider this possibly an overtone and that the real vibration rate was 440. See also Hannes (1926) who found in the bee that the wing gave rise to two pressure waves, thus doubling the note frequency but not the wing beats. It should be noted that by frequency, when speaking of wing beats, is meant the double vibration, that is down and up stroke counted as one.

Other methods than that of the note emitted are available for determining the wing beat, namely (1) in the case of a large insect recording on a blackened drum, (2) observation with the stroboscope, and (3) use of the oscillographic method.

* The number of vibrations per second is given by velocity of sound divided by length of wave (in this case 12 inches). The velocity of sound at 0° C. is 33,170 cm./sec., 54 cm. (1·8 feet) being allowed for each degree of rise in temperature. One-third of the diameter of the tube is added to the length of wave (four times the tube length) as a correction.

WING-STROKE FREQUENCY

STROBOSCOPIC OBSERVATION

Very valuable observations on flight mechanism and wing stroke frequency can be made by the use of the stroboscope (see Chadwick, 1939; Williams and Chadwick, 1943; Laird, 1948).

When the instrument is adjusted to near the wing vibration (or its harmonics) the wings appear in slow motion, every movement under the binocular being clearly displayed. The stroboscope can also be used to determine the wing stroke frequency, since the points at which the wing is seen stationary or in slow movement as registered by the instrument gives the frequency, or harmonics of this, at those breaks in the light corresponding to movements of the wing. For such observations Laird (1948) uses a mosquito which, after being lightly anaesthetised, has been mounted on a suitably bent piece of wire with a minute portion of quick-drying cement at its end pressed against the front of the mesonotum (Fig. 62 (1), (2)). It is important that the wire be so placed that it does not enter the field of vibration of the wings or affect the position assumed by the legs during flight. When under observation the legs must not be allowed to touch any support or the wings will cease to vibrate. Where observations are to be made on the effect of different gases, temperatures, etc., Williams and Chadwick mount such insects on a supported strip inserted into a length of pyrex tube provided with a tap at each end.

The instrument available to the present author has been the Stroboflash, type 1200 A, made by the Daws Instruments, Ltd., Harlequin Avenue, Great West Road, Brentford, Middlesex. The duration of the flash is given as 5–10 micro-seconds. The maximum number of vibrations on the scale is 14,600 per minute (about 240 per sec.), readings being given in hundreds. For any higher rate it is therefore necessary to use the lower harmonics of such rate, this explaining why in what follows it has been necessary to deduce the number of vibrations from the harmonics instead of arriving at these directly from the readings. Whilst in use the mosquito can be studied under the low-power binocular placed near the face of the instrument in a darkish room or even roughly shaded from too much extraneous light by a dark cloth or other device. I am much indebted to Mr C. H. Moore of the Zoological Laboratory, Cambridge, for instruction in its use, as also for help in respect to the observations on the oscillograph given in the next subsection.

For mounting, an ordinary quick-drying cement as sold for mending glass, china, etc., considerably diluted (1–10) with solvent (ethyl acetate) worked very well. But it was found that on a thin wire the diluted cement tended to collect, due to capillarity, in a drop above, instead of at, the actual end, whilst a thicker wire was clumsy. Much better results were obtained by using a long, thin bent entomological pin, such as is used for mounting Lepidoptera. The rounded head of the pin when touched in the diluted cement maintained a terminal film that adhered readily to the mesonotum giving a minimum but effective area of cemented contact. Insects so mounted would frequently continue to vibrate the wings without cessation for an hour or more, provided their tarsi did not contact any solid body. The best results were obtained when the insects were originally very lightly anaesthetised using the collecting tube with funnel for giving a graduated dose as described in the chapter on technique. Table 41 gives the results obtained, the readings being repeated approximately on several occasions.

The readings given in the table are those of the scale on the instruments, namely vibrations per minute/100. The figures for ratios give in each case the ratio to the figure next below, for example, the first entry in the table is $110/73 = 1{\cdot}51$. Now the true number of vibrations with its harmonics are in the proportions 1, 1/2, 1/3, etc. and these treated as above for ratios give 2, 1·5, 1·33, 1·25, etc., the first harmonic being 1·5. The figures in the first column of the table are therefore clearly not the actual vibrations, but the first harmonics of these and the real number of vibrations, as given in the last column of the table, are twice this. The number of vibrations for the female is therefore 367 at 18° C. and 427 at 25° C. That for the male at 18° C. is 467. The corresponding notes would be *g′*, *a′* and *b′*. Since the determinations have been made on the fixed insect, they might be expected to be somewhat higher than for the free flying mosquito, the rate for the fixed insect according to Sotavalta being that for an insect with maximum load (see also Roch, 1922).

Table 41. *Readings (vibrations per minute/100) when wings appeared stationary or nearly so with ratios as explained in the text*

Sex	Temperature (° C.)		Readings and ratios					First harmonic vib./sec.	Vib./sec. wing
Female	18	Readings	110	73	53	40	33	183·5	367
		Ratios	1·51	1·33	1·25	1·20	—		
Female	21	Readings	115	80	—	—	—	191·7	383
		Ratios	1·44	—	—	—	—		
Female	25	Readings	128	85	60	48	40	213·3	427
		Ratios	1·51	1·42	1·25	1·20	—		
Male	18	Readings	140	95	70	56	46	233·3	467
		Ratios	1·47	1·36	1·25	1·22	—		

During such observations it was noticeable that as a rule the vibration rate was not throughout always entirely uniform, there being periods of 'blur' indicating temporary changes in the rate.

At those points at which the wings appeared stationary, or nearly so, the halteres also came into view moving synchronously in slow movement with the wings. They moved through a considerable arc, nearly touching the abdominal surface at each end of the swing.

OSCILLOGRAPHIC OBSERVATION

Another method open to the study of frequency is the use of a mosquito mounted as before and set with the pin carrying the mount against a crystal as used for wireless, the vibrations being projected and studied on the screen of the oscillograph. When so connected in circuit a female *A. aegypti* showed smaller waves imposed on the 50 cycle per second waves of the normal supply. These smaller waves numbered about 8 to the cycle, indicating a value of about 400 vib./sec. Oscillographic observation therefore confirms the order of about 350–400 vib./sec. for female *A. aegypti* independent of any question of harmonics. For a useful small guide to the oscillograph see *The Cathode-Ray Tube Handbook* by S. K. Lewer, Pitman and Sons, Ltd., London.

MATHEMATICS OF FLIGHT

That there is a relation between the weight of the insect and the number of wing strokes and also between these and the area of the wing has been noted by several observers. Thus for *Culex pipiens* Prochnow (1907) gives

Weight	4·0 mg.
Wing area	11·0 mm.^{2}
Wing length	5·1 mm.

Magnan (1934) gives an index, usually about 1·5 (varying between 0·7 and 2·2) for length of wing (in metres) multiplied by vibrations per second. If this held good for *Aëdes aegypti*, and taking the wing length as 3·40 mm. and vibrations as 360, the index would work out as 1·22 which would conform to the relation.

A very complete study of the mathematics of flight in insects is given in the monograph by Sotavalta (1947) and subsequent papers (1952–1955).

(*e*) OTHER OBSERVATIONS ON THE WINGS IN FLIGHT

Interesting observations on the movements of the wings can be made on mosquitoes mounted as described. In such a mosquito raised so that the tarsi are not in contact and examined by unaided eye or under the low-power binocular the extent and direction of movement of the wings can be to a large extent determined by the 'blur' made as the wings vibrate. Viewed from in front or behind, this blur is triangular. It is at once clear that the extent of movement in the mosquito is far less than that figured in text-books for such insects as the wasp. By means of a micrometer scale in the eye-piece, the length of the wing and the extent of the vibratory movement can be measured. The extent varies somewhat in different insects and at different times, but is on the whole reasonably constant. Taking the precaution to measure the excursion accurately in its plane of movement this for a normal sized female in good condition is about 2·5 mm. Since the wing length is about 3·5 mm., this gives an angle of about 45° (sine about 0·7), or possibly slightly less since the extent of the blur may be slightly increased as the effect of some rotation of the wing on its long axis.

Since the wings are articulated not very far below the level of the dorsum of the thorax the insect with vibrating wings viewed from in front or behind gives the appearance shown in Fig. 62 (4), that is rather like an aeroplane with high-level wings. The extent of the stroke appears to be about the same above and below the horizontal line through the bases of the wings.

Viewed from above it is seen that the wings when vibrating are not set accurately at right angles to the body but are sloped somewhat backwards, the angle which the two wings make with each other being about 140°.

Viewed from the side this latter fact, as also the relatively high position of the wings, is very evident (Fig. 62 (2)), the blur of the wings in movement being directed backwards and upwards in much the direction that the proboscis is directed forwards and downwards.

In estimating the degree to which the wings are directed forwards at the end of the stroke, much depends upon the attitude adopted by the insect when flying. If the insect is poised with the abdomen horizontal, the wings would be working almost exactly up and down. But with the body tilted as it appears to be when flying there would be considerable forward movement during the down stroke, much as is depicted by Stellwaag for the bee (see also Wigglesworth, 1942, p. 89).

This raises a curious point, which at first might seem to point to some error in the observations, in that if the downstroke is directed forwards this should, like a man rowing, propel the insect backwards. After noting this apparent discrepancy I have been informed by Dr Sotavalta that several observers have noted this fact and that the downstroke acts in sustaining the insect, it being the upstroke that propels the insect forwards.

Much information on subjects touched upon above and others relative to insect flight, including references to the mosquito and midges and to recent authors' works, will be found in later papers by Sotavalta than that given in the text above. These are given with the titles in the list of references. They include those concerned with wing-stroke frequency and various factors, such as temperature, air pressure and other matters in relation to this (1952*a*, *b*; 1953); a very useful paper on reserves and energy output (1954*a*) with one giving a description of a simple energy mill (1955); two on muscle structure and thoracic temperature (1954*b*, *c*); and one on the source of sensory impulses (1954*d*). See also Haufe (1954).

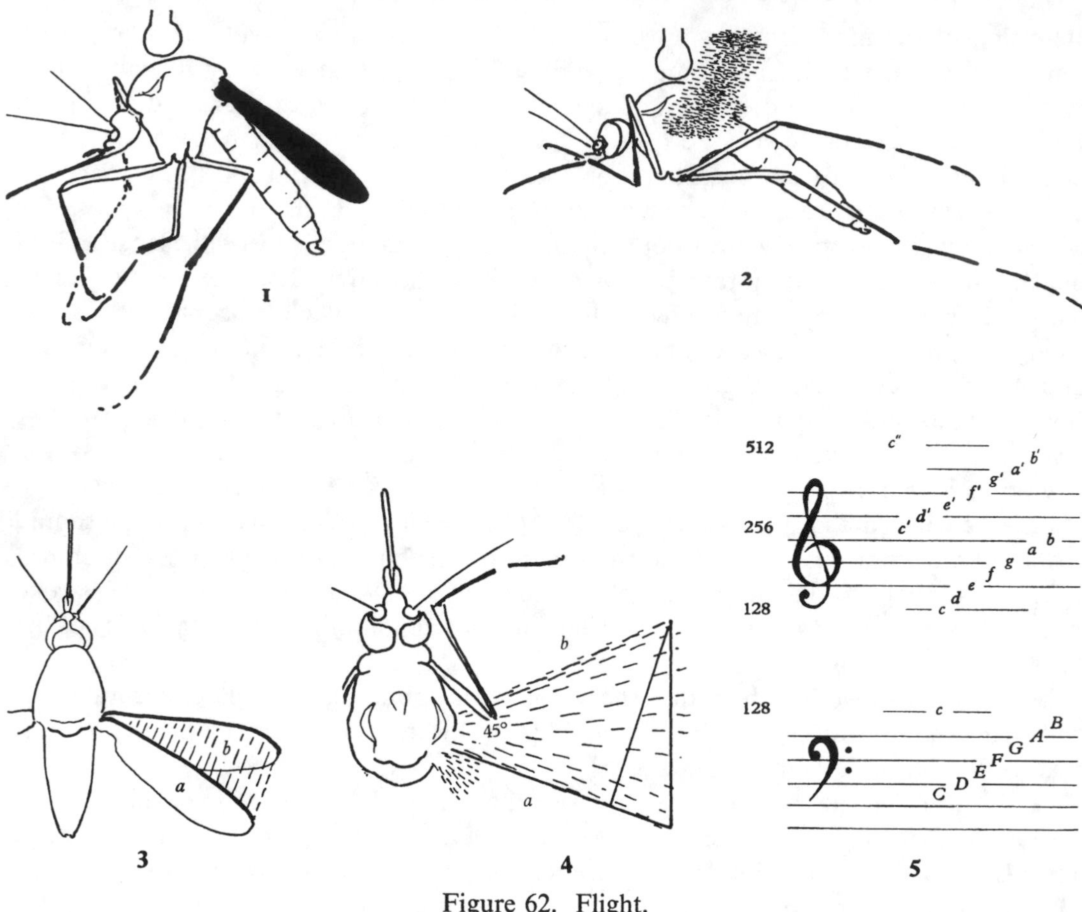

Figure 62. Flight.

1 A female *Aëdes aegypti* as seen under the stroboscope showing attitude when, from the tarsi making contact or other cause, flight movements have ceased. Freehand drawing.

2 The same specimen when in the act of flight. The area of the moving wings is only approximately indicated.

3 The same mosquito viewed from the dorsal aspect showing: *a*, upper limit of movement of the wing; *b*, lower limit of same.

4 Wing movement with the mosquito tilted to bring the plane of movement horizontal, that is showing maximum excursion. *a*, upper limit of movement of wing; *b*, lower limit of same. The angle between *a* and *b* as determined from camera lucida drawing is 45°. The lines indicate the sine (approx. 0·7).

5 Treble and bass clefs showing notation and corresponding vibrations of *c*, *c′* and *c″*. For intermediate values see text.

(*f*) ATTITUDE IN FLIGHT

When an insect, mounted as for stroboscopic observation, is placed so that the tarsi do not touch any support, it frequently happens that the wings will vibrate for short periods followed by others in which they are folded as at rest. In a normal insect any intermediate state is unusual or at the most of momentary duration. In the two states the arrangement of the legs is entirely different. When the wings are closed the legs usually are held more or less hanging down, either motionless or to a certain extent moving about. The halteres like the wings are motionless, their tips usually just visible beneath the costal edge at about the humeral cross vein. When the wings are erected and vibrate the whole appearance of the insect changes. The fore-legs are brought up in a very characteristic way and the mid- and hind-legs raised and held straight out more or less approximated to the abdomen so that their tips lie beyond and behind this structure. In other words the insect adopts what amounts to a stream-line attitude. This is particularly striking in the case of the fore-legs. These are sharply bent with the tibiae closely approximated to the femur which latter is held close against the thorax, the tarsus being raised somewhat in a praying attitude, being directed upwards and forwards in a curve roughly parallel with the antennae. As the wings stop, all this changes, the legs dropping, to be raised again when the wings restart vibrating. It is not necessary that the stroboscope should be used to see these changes which can be readily seen under the binocular or with the unaided eye.

If an insect, mounted as for the stroboscope, be held, not in the usual position, but markedly sideways, it usually stops vibrating the wings and scrambles with its legs, and this even if the change be made quite slowly. It has already been noted that *A. aegypti* when at rest is sensitive to changes in orientation and tends to re-orient itself if the displacement is at all appreciable. The same sensitiveness is associated with attitude during flight. Some trials with complete reversal, that is with the insect held completely upside down, showed that the wings in this case might sometimes continue to vibrate. In all other cases there was strong objection to vibrating the wings in a position at a marked angle to gravity. For reference to attitude of insects in flight see Poujade (1884), Stellwaag (1916), also chapter XVII (*f*) on specific gravity of the imago.

REFERENCES

(*a*) SPEED AND RANGE OF FLIGHT

AFRIDI, M. K. and ABDUL MAJID, S. (1938). Dispersion of *Culex fatigans*. *J. Malar. Inst. India*, **1**, 155–68.

BONNET, D. D. and WORCESTER, D. J. (1946). The dispersal of *Aëdes albopictus* in the territory of Hawaii. *Amer. J. Trop. Med.* **26**, 465–76.

BUGHER, J. C. and TAYLOR, M. (1949). Radiophosphorus and radiostrontium in mosquitoes. Preliminary report. *Science*, **110**, 146–7.

CLARKE, J. L. (1943*a*). Studies on the flight range of mosquitoes. *J. Econ. Ent.* **36**, 121–2.

CLARKE, J. L. (1943*b*). Flight range and longevity of mosquitoes dusted with aniline dye. *N.J. Mosq. Exterm. Ass. Proc.* **30**, 227–34.

CUMMING, H. S. (1931). Present day problems of yellow fever. *Publ. Hlth Rep. Wash.* **46**, 361–6.

DEMOLL, R. (1918). *Der Flug der Insekten und der Vögel.* Jena.

EYLES, D. E. and BISHOP, L. D. (1943). An experiment on the range of dispersion of *Anopheles quadrimaculatus*. *Amer. J. Hyg.* **37**, 239–45.

GRIFFITHS, T. H. D. and GRIFFITTS, J. J. (1931). Mosquitoes transported by airplanes. Staining methods used in determining their importation. *Publ. Hlth Rep. Wash.* **46**, 2775–82.

HASSETT, C. C. and JENKINS, D. W. (1949). Production of radioactive mosquitoes. *Science*, **110**, 109–10.

HINMAN, E. H. (1932). The water breeding and activity of Culicine mosquitoes at New Orleans. (30° N. Lat.) *Amer. J. Trop. Med.* **12**, 263–71.

KENNEDY, J. S. (1940). The visual response of flying mosquitoes. *Proc. Zool. Soc. Lond.* A, **109**, 221–42.

MAGNAN, A. (1934). *Le vol des insectes.* Hermann et Cie, Paris.

MAJID, S. ADBUL (1937). An improved technique for marking and catching mosquitoes. *Rec. Malar. Surv. India*, **7**, 105–7.

MORGAN, K. Z. (1947). Tolerance concentrations of radio-active substances. *J. Phys. Chem.* **51**, 984–1003.

SHANNON, R. C., BURKE, A. W. and DAVIS, N. C. (1930). Observations on released *Stegomyia aegypti* (L.) with special reference to dispersion. *Amer. J. Trop. Med.* **10**, 145–50.

SHANNON, R. C. and DAVIS, N. C. (1930). The flight of *Stegomyia aegypti* (L.). *Amer. J. Trop. Med.* **10**, 151–6.

SOPER, F. L. (1935). El probleme de la fiebre amarille en America. *Bol. Ofic. Sanit. Pan-Amer.* **14**, no. 3, 203–13.

WHITFIELD, F. G. S. (1939). Air transport. Insects and disease. *Bull. Ent. Res.* **30**, 365–442.

WIGGLESWORTH, V. B. (1942). *Principles of Insect Physiology.* Ed. 2 (see also many later editions). Methuen and Co., London.

WOLFINSOHN, M. and GALUN, R. (1953). A method for determining the flight range of *Aëdes aegypti* (Linn.). *Bull. Res. Coun. Israel*, **2**, 433–6.

ZETEK, J. (1913). Determining the flight of mosquitoes. *Ann. Ent. Soc. Amer.* **6**, 5–21.

(*b–f*) FLIGHT MECHANISM

BUDDENBROCK, W. VON (1930). Der Flug der Insekten. In *Handbuch der normalen und pathologischen Physiologie*, **15**, 349–61.

BULL, L. (1904). Méchanisme du mouvement de l'aile des insectes. *C.R. Acad. Sci., Paris*, **138**, 590–2.

BULL, L. (1909). Recherches sur le vol de l'insecte. *C.R. Acad. Sci., Paris*, **149**, 942–4.

BULL, L. (1910). Sur les inclinaisons du voile de l'aile de l'insect pendent le vol. *C.R. Acad. Sci., Paris*, **150**, 129–31.

CATCHPOOL E. and SATTERLY J. (1944). *Text Book of Sound.* Ed. 6. Univ. Tutorial Press, London.

CHADWICK, L. E. (1939). A simple stroboscopic method for the study of insect flight. *Psyche*, **46**, 1–8.

GOELDI, E. A. (1905). *Os mosquitos no Para.* Weigandt, Para.

HANNES, F. (1926). Bienenflugton und Flügelschagzahl. *Biol. Zbl.* **46**, 129–42.

HAUFE, W. O. (1954). The effects of atmospheric pressure on the flight responses of *Aëdes aegypti* (L.). *Bull. Ent. Res.* **45**, 507–26.

HOCKING, B. (1953). The intrinsic range and speed of flight of insects. *Trans. R. Ent. Soc. Lond.* **104**, 223–345.

HOWARD, L. O., DYAR, H. G. and KNAB, F. (1912). See references in ch. II (*a–b*), p. 45.

LAIRD, M. (1948). A method of securing living mosquitoes to mounts. *Science*, **107**, 656.

LANDOIS, H. (1867). Die Ton- und Stimm-apparate der Insekten. *Z. wiss. Zool.* **17**, 107–186.

MAGNAN, A. (1934). See under section (*a*).

MAREY, E. J. (1895). *Le movement.* W. Heinemann, London. Many other publications on flight, including that of insects. See bibliography given by Sotavalta (1947).

NUTTALL, G. H. F. and SHIPLEY, A. E. (1902). *J. Hyg., Camb.*, **2**, 77–80.

POUJADE, G. A. (1884). Note sur les attitudes des insectes pendant le vol. *Ann. Soc. Ent. Fr.* (6), **4**, 197–200.

REFERENCES

PRASHAD, B. (1916). The halteres of mosquitoes and their function. *Ind. J. Med. Res.* **3**, 503–9.

PRASHAD, B. (1918). The thorax and wing of the mosquito. *Ind. J. Med. Res.* **5**, 610–40.

PRINGLE, J. W. S. (1948). The gyroscopic mechanism of the halteres of Diptera. *Phil. Trans.* **233**, 347–84.

PROCHNOW, O. (1920). Mechanik des Insektenfluges. In Schröder's *Handb. d. Entomologie*, **1**, 534–69.

PROCHNOW, O. (1907). Die Lautapparate der Insekten. *Intern. ent. Zeitschrift* (Berlin).

RITTER, W. (1911). The flying apparatus of the blowfly. *Smithson Misc. Coll.* **56**, no. 12, 1–77.

ROCH, F. (1922). Beiträge zur Physiologie der Flugmuskulatur der Insekten. *Biol. Zbl.* **42**, 359–64.

RUDOLFS, W. (1922). Chemotropism of mosquitoes. *N.J. Agric. Exp. Sta. Bull.* 367.

SHIPLEY, A. E. and WILSON, E. (1902). On a possible stridulating organ in the mosquito (*Anopheles maculipennis*). *Trans. R. Soc. Edinb.* **11**, 367–72.

SNODGRASS, R. E. (1930). How insects fly. *Ann. Rep. Smithson. Inst. for* 1929, pp. 383–421.

SOTAVALTA, O. (1947). The flight tone (wing-stroke frequency) of insects. *Acta Ent. Fenn.* **4**, 1–117 (with 12 pp. bibliography.)

SOTAVALTA, O. (1952*a*). The essential factor regulating the wing-stroke frequency etc. *Ann. Zool. Soc. Vanamo*, **15**, no. 2, 1–65.

SOTAVALTA, O. (1952*b*). Flight-tone and wing-stroke frequency in insects and the dynamics of insect flight. *Nature, Lond.*, **170**, 1057.

SOTAVALTA, O. (1953). Recordings of high wing-stroke and thoracic vibration frequency in some midges. *Biol. Bull.* **104**, 439–44.

SOTAVALTA, O. (1954*a*). On the fuel consumption of the honey bee (*Apis mellifica* L.) in flight experiments. *Ann. Zool. Soc. Vanamo*, **16**, no. 5, 1–27.

SOTAVALTA, O. (1954*b*). On the thoracic temperature of insects in flight. *Ann. Zool. Soc. Vanamo*, **16**, no. 8, 1–22.

SOTAVALTA, O. (1954*c*). Sarcosomes from the flight muscles of certain insects. *Ann. Ent. Fenn.* **20**, no. 3, 145–7.

SOTAVALTA, O. (1954*d*). Preliminary observations on sensory impulses from insect wing nerves. *Ann. Ent. Fenn.* **20**, no. 3, 148–50.

SOTAVALTA, O. (1955). A simple method of studying and demonstrating the energy consumption in flying insects. *Nature, Lond.*, **175**, 543.

STELLWAAG, F. (1910). Bau und Mechanik des Flugapparates der Biene. *Z. wiss. Zool.* **95**, 518–50.

STELLWAAG, F. (1916). Wie Steuern die Insekten während des Fluges? *Biol. Zbl.* **36**, 30–44.

VOSS, F. (1913/14). Vergleichende Untersuchungen über die Flugwerkzeuge der Insekten. *Verh. Dtsch. Zool. Ges.* **23**, 113–42; **24**, 59–90.

WEBER, S. E. (1906). Notes on mosquitoes. *Ent. News*, **17**, 214–17.

WIGGLESWORTH, V. B. (1942). See under section (*a*).

WILLIAMS, C. M. and CHADWICK, L. B. (1943). Technique for stroboscopic studies of insect flight. *Science*, **98**, 522–4.

XXIV

THE IMAGO: SPECIAL SENSES AND BEHAVIOUR

(a) REACTION TO LIGHT AND VISUAL RESPONSE

Though the lateral ocelli of the larva and pupa remain present in the imago they are clearly obsolescent and it is improbable that they are functional. The dorsal ocelli, assuming that the bodies described in the frontal region of the imago are these, would also appear not to be functional. The effective visual organs consist therefore solely of the large compound eyes. As previously noted, in *Aëdes aegypti* these are not noticeably larger in comparison with the size of the body in the male than in the female, nor do they extend dorsally and ventrally more than in this sex. Though, as shown by Kennedy (1940), different portions of the compound eyes respond somewhat differently to visual stimuli, there is little anatomical evidence of difference and the difference in size of the facets noted by Shôzô Satô (1950) in *Culex pipiens* var. *pallens* appears to be much less in *Aëdes aegypti* (see chapter XXVIII on measurement of facets).

A. aegypti in habit is more diurnal than nocturnal and is widely recognised in the tropics as the 'day biting mosquito'. Nevertheless, it shows a marked preference for shade and dark corners and in general avoids bright light and open spaces. It will feed, however, readily in moderate or bright daylight at any time of the day. In the laboratory it will feed readily in complete darkness. This has been found to be so by Bishop and Gilchrist (1946) when feeding on the chicken and we have found that it will do so on the human arm and so far as could be ascertained with no delay in numbers settling or gorging.

Response of mosquitoes in general to light is commonly negatively phototactic. Yet Holmes (1911) found with *Culex pipiens* and *C. territans* that they kept on the side of the cage nearest the window or in a dark room followed the light and that tubes can be held with the opening away from the light without the mosquitoes escaping. They were still positively phototactic after being kept some weeks in confinement. The light reaction, however, was mixed, that is, they selected dark corners but at the same time were photo-positive. Some species of *Anopheles* are at night definitely positively phototactic in so far as they will settle on an illuminated sheet, for example *A. hyrcanus* (Stephens and Christophers, 1902), or are activated and fly towards a light (Christophers, 1914; Bentley, 1914), or are seen in the morning flying at window panes. Yet the majority of the species are powerfully attracted by dark doorways which they enter for day resting places and thus in this respect would appear to be negatively phototactic. De Meillon (1937) notes that *A. gambiae* gave both positive and negative reactions to light. Hundertmark (1938) found that by day 45 per cent of *A. maculipennis* chose black to various shades of grey and grey to white, whereas at night the response was reversed, 62 per cent choosing white.

The response of *Aëdes aegypti* to light has been carefully studied by Kennedy (1940), who has given the reactions of suspended females to a horizontal beam of light varying in intensity from 3 to 3100 metre candles. The insects exhibited negative phototaxis, the most frequent readings being orientation at 180° from the source of light. Departures from this exact angle suggested that the insects did not receive rays falling on the postero-lateral parts of the eyes. Very similar results were obtained by Rao (1947) with mosquitoes (*Culex molestus* and *Anopheles maculipennis*) rendered flightless by clipping the wings. In all cases the insects moved away from the light. Observing cages of *Aëdes aegypti* at different temperatures between 19° and 28° C. the writer has found any reaction to light, whether diffuse or a beam of artificial light, to be mildly negative phototactic, that is, the insects tended to settle on the sides away from the light. But when flying there was no obvious drift. Yet in a test tube held as described by Holmes, *A. aegypti* of either sex will keep flying towards the light, and though it may escape if the tube is left open too long, in the main it attempts to fly through the closed end of the tube. Further, if it does escape it then flies towards the window. The reactions of the adult *A. aegypti* to light are therefore mixed and seem more related to some objective of the insect than to an invariable taxis such as is so conspicuous with the larva and pupa. The facts might, to some extent, be summarised by saying that in seeking to escape the insect flies towards the light, but in seeking a resting position it selects situations in the shade. There is another important reaction which will be dealt with later, namely orientation and movement towards a dark object or area, especially if this be in motion. It was of interest to note on one ocasion that when a beam of winter sunlight entered a cage situated well back in a room the mosquitoes, which for some days had been normally resting, kept up a continuous flight showing no particular relation to the direction of light, that is, they were merely activated.

That dark objects such as clothes hanging on a wall are attractive to *A. aegypti*, so that in nature they are found often in numbers resting by preference in such situations, is a common observation in the tropics. A similar response is displayed in the laboratory when a dark object, for example a dark stripe on a pale ground is presented to them. Kennedy (1940), using a single stationary vertical stripe, found that the majority of suspended females oriented themselves to face the stripe. This occurred with all widths of stripe subtending visual angles from 45° to 315°. With two stripes the insects oriented themselves to one or other, not to between the stripes. Rao (1947) obtained very similar results with mosquitoes (*Anopheles maculipennis* and *Culex molestus*) rendered flightless by removal or fixation of the wings. Such mosquitoes moved towards a single dark stripe. When two stripes were present movement was to one, not between. Reaction was invariably positive, no matter how feeble the illumination, and was effective for stripes as narrow as a visual angle of 8°, but not narrower. Stripes directly behind were not reacted to unless they subtended an angle of 135° from the front. When one eye was blinded, circus movement took place towards the blind eye. If a stripe comes into the field of vision of the functional eye, the mosquito, however, turned sharply towards this. The author stresses that there was never any indication of positive phototaxis. The mechanics of orientation seemed to be in conformity with current ideas of negative phototaxis. There was no evidence of photohorotaxis (Kalmus, 1937; orientation to light-dark margin).

Seemingly somewhat opposite to the above Sippell and Brown (1953) found a mirroring surface more attractive than a dull one, a black enamelled surface more attractive than a flat black one, a silvered mirror more attractive than polished metal.

COLOUR

How far does *Aëdes aegypti* distinguish colour? Preference for particular colours has usually been determined on coloured cloths. Nuttall (1901) found that with *Anopheles maculipennis* dark blue was preferred to green, yellow, red or white. Le Prince (see Herms, 1940) found no clear preference with *A. quadrimaculatus*. The colour preferences of *Aëdes aegypti* using coloured cloths has been investigated by Brett (1938). Adult females 2½ days from emergence were used at temperatures between 21·7° and 23·3° C. and relative humidity 55–84 per cent. The mosquitoes were in cages about one foot cube with glass tops and netting on three sides. The fabrics were introduced stretched over the investigator's hand at a distance of about 0·8 cm. Fifty females were used in each test and the proportions settling on each colour as against black and white respectively determined. The author's results arranged in order of contrast to black are given in Table 42 (see also Brown, 1954).

Table 42. *Colour preferences of* Aedes aegypti *as determined by Brett re-arranged in order of contrast to black*

Colour	Reflection factor	Proportion to black (percentage)	Proportion to white (percentage)
Pure white (MgO)	100	—	—
Yellow	48·2	27	50
Light khaki	40·3	28	45
Green	9·3	31	59
Blue	7·5	32	58·5
Khaki	14·6	32	60
Grey	17·8	35	56
Heliotrope	10·0	40	57
Brown	7·6	40	62
Red	10·4	43	69
Pure black	2·08	—	—

It would seem that in Nuttall's observations what was being tested was colour preference for 'settling to rest'. In Brett's case the presence of the human hand close beneath the cloth must clearly have acted as a stimulus to feed and what was being tested therefore was colour preference for 'settling to feed', possibly a different reaction. In such a case it seems certain that the colour choice would be secondary to the urge to feed, the main urge being to settle with some predilection for the exact situation modifying this. From the table the general trend would appear very largely preference for the colour with the greatest proportion of black and least reflection factor, or in other words the greater amount of shadow. A somewhat similar condition is commonly observed when feeding on the human arm, more mosquitoes settling on the more shaded parts (see also Smart and Brown, 1956). The choice of dark blue in Nuttall's series could be explained in the same way. Without more evidence of colour selection as apart from the amount of light reflected, it would not seem that either of the above series of observations offers definite proof that mosquitoes possess colour sense.

The effect of colour as displayed in different coloured lights has been studied by Headlee (1937). Traps with white light (25 watt frosted globe), red light (neon), greenish yellow (mercury) and blue (mercury argon) caught respectively 1621, 1096, 582 and 599 mosquitoes, or taking white light as one, in the proportions respectively of 1·0, 0·676, 0·359 and 0·369. In terms of energy received the relative attractiveness would have been in the same order 1·0, 6·1, 12·3 and 21·5. Preference for the traps was, therefore, in reverse order to the amount of energy given out. The author, however, found that different species behaved very differently in their preference for certain colours of light and the reactions of *A. aegypti* are not given. Recently a study has been made of the comparative attractiveness of coloured lights of equal intensity to specific species of mosquitoes by Bargren and Nibley (1956), but I have not as yet been able to see these authors' paper. On the whole, it would seem uncertain whether mosquitoes, or *A. aegypti* in particular, possess a sense of colour as usually understood.

MOVEMENT

With moving stripes Kennedy found that the mosquitoes oriented themselves towards the following edge and were affected by moving stripes up to 90°, following a retreating stripe and turning towards an advancing stripe when it passed this angle. Such observations when extended indicated the importance of the visual background in the orientation and behaviour of the insects when in flight. With free-flying mosquitoes in a wind tunnel Kennedy found that females flew consistently up wind, keeping position or making headway so long as a background of stripes moving below was present, but not in darkness. In the absence of wind with a background of moving stripes they flew in the same direction as the stripes or ahead of the movement. Kennedy considered that the lateral ommatidia initiate the responses to light and to moving objects working in a different way from the dorsal and ventral ommatidia which are concerned in upwind orientation. Sippell and Brown (1953), studying the responses of the female *Aëdes* spp., note that movement nearly doubles attractiveness of animals or inanimate objects.

The importance of visual stimuli from movement, especially of dark objects, will be seen from what is said in section (*f*) of this chapter regarding the means by which *A. aegypti* locates its victim.

(*h*) REACTION TO SMELL

That smell is an important guide to mosquitoes in searching out their prey has been very widely assumed. Smell undoubtedly plays a large part in reactions of many insects, for example, the blowfly is attracted by putrid odours and fruit-flies by various chemical compounds. It is somewhat surprising, therefore, to find that a number of workers who have experimented with mosquitoes in this respect have obtained either negative or indefinite results with materials that might have been expected to give marked attraction through smell. Thus Rudolfs, dealing with species of *Aëdes*, found that sweat and sebaceous secretions were little attractive. Sen (1918) found the blood of the goat after shedding was not attractive to *A. albopictus*. Howlett (1910) also found fresh blood did not attract *A. scutellaris*. Reuter (1936) found that acids in human sweat did not attract *Anopheles maculipennis*. Crumb (1922) found mixtures of the components of perspiration and of blood produced only faint erratic response from *Culex*. Van Thiel (1935) found acids of sweat unattractive.

Hackett (1937) considered, however, that *Anopheles maculipennis* was attracted by the smell of cow-dung. Haddow (1942), experimenting in Kenya, by counting the bites of *Anopheles* received by five washed and five unwashed native children sleeping naked in clean blankets in similar huts found that the unwashed children received more bites. He also thought that dirty clothing acted as an attraction. Some evidence that the olfactory sense plays some part with *Aëdes aegypti* is given by Lumsden (1947) using closed receptacles with air at known temperature and humidity. Goeldi (1905) found sweat to be attractive, but did not exclude the possible effect of warmth and humidity. Human breath has been noted to be very attractive which can be very readily demonstrated by breathing on the gauze of a cage containing *A. aegypti*, but the same objection lies that exhaled breath is both warm and humid. An olfactory sense in *A. aegypti* is noted as possibly present by MacGregor (1931) in that the insect is able to determine the nature of a fluid immediately the tip of the unsheathed proboscis (fascicle) either touches or comes into close proximity to the surface of such fluid, the author concluding that the perceptive sense is therefore of an olfactory nature. Even so this would appear to be a reaction connected with the actual act of feeding rather than one by which the insect is attracted to its host.

Willis (1947), using a special form of olfactometer designed to avoid the effect of temperature, carried out critical experiments to test the reaction of *A. aegypti*, among other things, to odour of the human arm. The olfactometer was a modification of that used by Hoskins and Craig (1934) and by Wieting and Hoskins (1939), in which air is passed through warming and humidifying parts and is presented to the mosquitoes as two equal and similar gauze covered funnels, the number of mosquitoes clustering on the respective gauzes indicating the attractiveness of any odour as compared with a control. In this case an arm was inserted in one of two thin celluloid cylinders inserted into the respective circuits. Two species of mosquito were used, namely *Anopheles quadrimaculatus* and *Aëdes aegypti*. The authors obtained what were considered to be positive results from the counts, in that the numbers on the gauze receiving air coming from the arm was greater than that on the control gauze receiving only air of the same temperature and humidity. The counts are given in Table 43 (see remarks, however, later).

Table 43. *Results by Willis of observations relating to* Aëdes aegypti *and arm odour*

Temperature air-stream (° C.)	Air flow	Air exhaust	Number of tests	Total response		Attraction to odour (percentage)
				Test	Control	
34	100	300	40	2498	947	73
24–26	100	300	10	195	180	52
36·5	100	300	10	463	369	56
34	100	300	10	546	353	61
34	100	200	10	534	380	58
34	100	250	10	420	369	53

Schaerffenberg and Kupka (1951, 1952) claim that blood gives off a volatile substance which attracts *Culex* to feed. This they claim attracts the mosquitoes from a distance and can act through a membrane.

Apart from human scent certain gases and substances have been recorded as attracting mosquitoes. Several observers have considered that CO_2 had an attractive effect and might be responsible for making human skin attractive. As noted by Willis, however, Rudolfs does not describe an attractive response to CO_2 but states that it makes mosquitoes restless. He found CO_2 not attractive to *A. aegypti* at one per cent in air and at 10 per cent it repelled. Crumb found admixtures of various amounts of CO_2 with air bubbled through warm water and directed into a cage containing *Culex pipiens* had no observable effect. The attractive effect of CO_2 to *Anopheles maculipennis atroparvus* has been studied by

Van Thiel (1937, 1947), and by Van Thiel and Weurman (1947), and in respect to different species of mosquitoes by Reeves (1951, 1953), some attractive effect being noted (see also Dethier, 1954). On *Aëdes aegypti* the effect of CO_2 has been determined by Willis and Roth (1952) using a small and a large olfactometer. In the small olfactometer a stream of air with from 0·1 to 500 per cent CO_2 repelled most individuals. In the large olfactometer only a small number of the mosquitoes participated, but some were attracted. The authors concluded that at least some females of *A. aegypti* respond positively to stimulation by CO_2. It would seem that recently CO_2 has been regarded as being more of the nature of an excitant than an attractant and the idea that it is the cause of females being attracted to human skin seems improbable. Another substance found to have some attraction by Wieting and Hoskins (1939) was ammonia in concentration of 0·012 per cent. At concentration 0·03 it strongly repelled (see also Laarman, 1955).

Very numerous volatile substances, on the other hand, repel mosquitoes. How far this is through smell or otherwise is not at present known. Some of such substances are not unpleasant to human smell, for example the citronella-like compounds, which nervertheless are strongly repellent.

On the whole the evidence that smell is an important stimulus in the attraction of *A. aegypti* to feed is not very strong. The careful experiments with this species by Willis (1947), though they showed some attraction from the human arm ascribed to smell, did not indicate any powerful effect such as clearly activates *A. aegypti* in its normal reaction to human skin. Thus it will be seen from Table 43 giving this observer's results that, of the six recordings, two gave only 52 and 53 per cent favouring the arm screen and three only 56–61 per cent, results quite unlike what might be expected in any experiment where an arm, or even a warm object, is offered and where under reasonably good conditions some 90 per cent of the females in a cage would have been on the attractive object within 15–30 seconds.

(*c*) REACTION TO SOUND

Hearing in mosquitoes, using the term as indicating perception of air vibrations, so far as described, has been mostly in relation to the antennae with their large globular basal segment containing Johnston's organ. Mayer (1874) states that the fibrils of the male antenna vibrate to certain notes when the vibrations are transverse to their length, but not in the direction of their length. Rotating the stage of the microscope until the fibrils ceased to vibrate located the tuning fork to within 5°. The mechanism of locating the female was thought to be as follows. The antennae of the male have a range of motion in a horizontal direction so that the angle between them can be varied considerably. The song of the female vibrates the fibrillae of one antenna more forcibly than those of the other. The insect turns his body in the direction of the antenna whose fibrils are most affected and thus gives greater intensity to the fibrils of the other antenna. When the effect on the two antennae is equal, the insect is pointing in the direction of the female. According to Child (1894) vibrations are transmitted by movement of the antennal shaft to the membrane closing in Johnston's organ. For a description of Johnston's organ and related parts see under sense organs (ch. XXIX).

Assuming the above to be correct, it is not clear why the female should possess a Johnston's organ almost as developed as the male, though it is true that the antennal fibrils are far less developed. The answer might be that, whilst the male antenna may well be concerned in detecting the presence of the female, in certain circumstances Johnston's organ has other functions of an auditory nature applicable to both the male and the female.

Many accounts are given of mosquitoes being trapped or killed through use of a buzzing or musical note. Shipley (1915) refers to an account by Howard of an engineer who found mosquitoes being attracted in large numbers by the buzzing of a particular electric machine and designed a method of destroying the attracted insects by electric contact. The same author also refers to an account by Sir Hiram Maxim of a lamp, which, through buzzing, attracted mosquitoes, all males. See also Kahn, Celestin and Offenhauser (1945).

That hearing may play some part in attracting the female mosquito to feed has been suggested by Grassi (1900) who noted that people talking were more bitten by *Anopheles* than those keeping silent. Ordinary noises, however, which it was thought might attract *Anopheles* from breeding places to villages, failed to attract in an experiment undertaken by the present author and Dr Bentley (Christophers, 1914; Bentley, 1914); these authors used a long tunnel with netting closing each end. When a light was shown at one end the mosquitoes rapidly flew the length of the tunnel (some yards) and congregated on the netting at the lamp end. Noises such as might be expected to originate from habitations failed to cause any such response.

(*d*) REACTION TO HEAT

THE EFFECT OF WARMTH

Howlett (1910) found female *Stegomyia* to be attracted and excited by a warm test tube. Other species of mosquito showed a similar reaction, and he gave his opinion that temperature was a dominant factor in the attack of mosquitoes. A similar reaction was later described for *Anopheles* by Marchand (1918) who showed attraction by glass plates heated to 1° C. above body temperature. Howlett's observations, however, went further than this, for he noted that if the tube was inclined the effect was much greater than if held vertically, from which he deduced that the effect was not due to radiant heat but to heated currents of air. Crumb (1922), under the title of 'a mosquito attractant', described the effect of air bubbled through warm water and directed through a funnel upon the muslin of cages containing *Culex pipiens*. As the temperature rose slightly above that of the surrounding air a beard-like appearance was produced due to the mosquitoes being attracted and pushing their proboscis through the cloth of the cage so long as the moist current of air flowed. There was no specific optimum temperature, but response occurred between 90° and 110° F. (15–30° F. above the temperature of the surrounding air). At 120° F. (49° C.) less interest was displayed, below 85° F. (29° C.) there was very little, if any, response.

Petersen and Brown (1951), studying the attractive effect of warmth, found that the number of female *Aëdes aegypti* attracted to a billiard ball warmed to 100–110° F. (38–43° C.) was twice that attracted to the same ball 20° F. cooler. Smart and Brown (1956), studying the attractiveness to *A. aegypti* females of the human hand, found the hand of warm-skinned Caucasians significantly more attractive than those of cool skinned ones. Warmer skins, even if of lighter hue, were more attractive than cool and the artificially cooled hand less attractive than the normal hand. Experimenting in the open with wild *Aëdes* spp. Brown (1951) found that warmth increased the attractiveness of a robot so that at 98° F. (36° C.) three times as many mosquitoes were attracted as when at 50–65° F. (10–18° C.). Van Thiel (1935, 1937), when testing the attractiveness of sweat to *A. aegypti*, found none of the sweat acids attractive but warmth was so, and though CO_2 increased attractiveness of warm moist air it did not bring about reaction in unwarmed air.

Another direction in which the effect of warmth is evident is in the need of warmth in blood or other fluid as a stimulus to feed. Bishop and Gilchrist (1946), when feeding *A. aegypti* through a membrane, found that it was necessary for effective feeding that the fluids behind the membrane should be warm. Further, for effective feeding it was necessary to have a difference in the temperature of the fluid and the environmental temperature. Their results in this respect are shown in Table 44.

Table 44. *Effect of difference in temperature of environment and fluid in attraction to feed of* Aëdes aegypti (*Bishop and Gilchrist*, 1946)

Temperature of environment (° C.)	Temperature of fluid (° C.)	Difference (° C.)	Number used	Number gorged	Percentage gorged
24	24	0	430	26	6
28	28	0	439	35	8
37	37	0	393	74	19
24	42	18	436	224	51
28	42	14	403	285	71

Lea *et al.* (1955) found that temperature played a considerable part in causing *A. aegypti* to feed on artificial media such as citrated blood. For this purpose they used a U-shaped tube the two ends of which were outside the cage so that the tube could be kept warm by circulating water at 115–120° F. (40–49° C), the food material being given on cotton (without sugar) placed on the tube within the cage.

That it is the effect of warmth which dominates the behaviour of *A. aegypti* in being attracted to feed is familiar to those using the species in many ways. If into a cage of resting *A. aegypti* is introduced a container warmed to about 40° C. it does not take very long before the majority of the females will be clustered about the object in an excited manner thrusting with their proboscis as though attempting to puncture and behaving generally as they would have done had the object been a human arm. It is easy to show that in the case of an arm placed in a cage of the mosquitoes visual attraction is not what is in action. If the arm is placed just below a sheet of glass, there is no settling until the glass has had time to warm. Then, too, it is easy to demonstrate with a cage of the mosquitoes that there is a considerable difference in the rate of settling whether the operators arm be warm or cold. When a rubber glove is worn with some of the arm skin also exposed, the mosquitoes at first when the arm is introduced settle on the glove as well as the arm. If the glove is thin often as many will settle on this as on the skin, only leaving the glove when they find it is not giving what is required.

Not only is there very strong evidence that warmth attracts *A. aegypti* in a very pronounced way in the urge to feed, but there has not been described in the literature any other attractive influence so active and characteristic in its results. Certain requirements must be met such as that the mosquitoes must be of an age to be ready to feed, are not already fed on blood or sugar and other common-sense precautions and the conditions regarding temperature must be those, as shortly to be described, actually involved in bringing about the necessary stimulus.

THE FORM IN WHICH WARMTH ACTS AS AN ATTRACTANT

In the last section it has been noted that for temperature to be effective as a stimulus to feed there must be a difference in the temperature of the attractive object and that of the general environment. Further than this Petersen and Brown (1951), studying the effectiveness of warm bodies to *A. aegypti*, found that such attractiveness was eliminated by the interposition of an air-tight window that allowed almost all the radiant heat to pass through but prevented heat convection. These authors thus concluded that radiant heat in itself was not the attractive factor and that heat was effective through the formation of warm convection currents, curiously enough what Howlett had shown in his simple experiment with a warm test tube. There can indeed be no doubt that the effective factor (disregarding for the moment certain visual reactions at a distance discussed later) causing *A. aegypti* to attack to feed are warm convection currents set up by the host. Experience with the species over a number of years indeed has shown that this sensitiveness to convection currents is very highly developed in the species and amounts almost to a special sense. What sensilla are responsible is not at present known, but the abundance of such sensory organs about the coxae and associated structures (see ch. XIX (*g*)) suggests that these might be concerned.

It is easy to give demonstrations of the sensitiveness to such currents. Thus, if a hand be held against the glass side of a cage of hungry *A. aegypti*, females will soon begin to collect in flight, not opposite the hand, but a little way above this where the warm glass has evidently set up ascending convection currents of warm air. If in a cage the arm be exposed enclosed in a box with gauze top and sides, it will be noted that the females congregate above rather than at the sides, even if the arm be placed much nearer the sides than the top. Again the females settle rapidly on a white cloth stretched a little distance over the arm since this gets warmed or allows warm air to pass through. The mosquitoes, however, take very little interest in such a cloth without the arm, even if it has been well rubbed over the arm, so that it can scarcely be smell that has affected them. The possibility of CO_2 being the attractive factor is not ruled out, but from what is said later this is unlikely to be the main reason for the behaviour described.

Very interesting in the above connection are the results obtained by Mer *et al.* (1947) on the attraction of *Anopheles elutus* by human beings. These authors summarise their findings as follows: (1) the air expired by human beings attracts mosquitoes; (2) not only expired air, but also factors emanating from the rest of the human body (though probably with lesser intensity) attract; (3) in absence of horizontal air movement attraction is exerted in a vertical direction, too, the attracting factor probably being carried upwards with the column of warm air rising from the human body; (4) the mosquitoes were attracted from distances of from 30–330 cm. in an exposure time of 15 minutes.

Whilst the above authors credit the attractiveness to some content of the air currents it clearly might have been the air currents that were the cause of the attraction. Very similar experiences are described by Haddow in *Aëdes simpsoni* attacking human beings (see (*f*)).*

Whilst, therefore, *A. aegypti* displays such behaviour in perhaps an unusually pronounced fashion, the same effect seems to be very general in mosquitoes. The effect of moisture by itself and in combination with warmth is dealt with in the next section.

It should perhaps be mentioned here that Parker (1948, 1952), unlike most observers,

* A very comprehensive and thorough study of the common European *Anopheles*, *A. maculipennis atroparvus*, dealing with host behaviour, is given by Laarman (1955), with an extensive bibliography.

found in his experiments no attraction from a warm dry object and that a cold moist surface was attractive. Since some importance has been attached to these results some remarks seem called for. The author's tests were carried out in an apparatus consisting of a tube 15 cm. long by 5 cm. in diameter. The experiments were carried out in a surrounding environment of 25° and one of 28° C., the latter a rather high temperature for such trials. The mosquitoes used in each experiment were of mixed ages and included in almost all cases those of ages from 2 days, or even in some cases one day, from emergence. It is possible that the small narrow tube acted as a closed space and taken along with what was possibly not good material gave these unusual results.

(*e*) REACTION TO MOISTURE

Crumb (1922), in his experiments with *Culex pipiens*, used air bubbled through heated water. The attraction observed was, therefore, caused by the warmth or by the moisture or by a combination of both. Human breath, which was also observed to attract, similarly has moisture and warmth. Crumb used boxes with a hole 2 inches square cut in the lid and covered with cheese-cloth with wool behind. When the cloth was moistened with warm water mosquitoes were attracted. They were not attracted to an identical box moistened with cold water.

Ordinarily a cold damp object causes no attraction to *Aëdes aegypti* in the sense of attraction to feed. Obviolusly it might attract it to obtain water and it seems desirable here to call attention to the need when using the word 'attraction' to define clearly what is meant. Here the term is applied to a condition in which the mosquitoes are activated and actively attracted, characteristically exhibiting hurried prospecting combined with prodding when on the attractive surface, that is, briefly 'attraction to feed'. In a cage provided with a moist pot or moist filter paper fan, a proportion of the mosquitoes will always be found resting on such an object and in a sense have been attracted. Or mosquitoes may settle on a darker part of a cage in preference to a lighter part or on one colour rather than another. In none of these conditions do they exhibit the characteristic behaviour of mosquitoes 'attracted to feed'. Yet in giving the result of experiments on colour, smell or other conditions the word attraction, often suggesting attraction to attack, is commonly used for any condition where more mosquitoes are found in one place rather than another and where possibly something quite different to attraction to feed is involved. With recognition of a difference between attraction to feed and other influences bringing about settling it would seem that moisture unassociated with warmth plays little or no part in the first. Associated with warmth and especially considered as a condition favouring warm convection currents of air, moisture clearly may be expected to play a considerable role. How far such may depend on a separate sensory stimulus for warm humid air and warm dry air is a matter for determination by carefully designed experiment.

The extent to which humidity at different temperatures affects the biting of *A. aegypti* has been investigated by Lewis (1933) whose results are given in condensed form in Table 45. Here both warmth and humidity are concerned. But the mosquitoes in the experiments were in a small enclosed atmosphere and the conditions were therefore akin to those obtained by Bishop and Gilchrist when there was no difference of temperature between the environment and the object and they relate therefore more to the effect of a humid environment on the mosquitoes than on the effect of moisture in the present sense as an attractive agent to feed in the open.

Brown (1951), in field experiments with female *Aëdes* spp., found that moisture on clothing increased attractiveness of a warm object 2–4 times when air temperature exceeded 60° F. (15·6° C.) but decreased at lower temperatures. Thompson and Brown (1955), studying the attractiveness of human sweat, found sweat from the armpits of Canadian males superficially attractive when tested in the olfactometer if in sufficient concentration. Attractiveness of moist clothing was not increased by addition of simple sweat. Volatile acids from sweat had a repellent effect. Smart and Brown (1956) found the hands of individuals with low moisture output more attractive than those with high and hands with free perspiration less attractive than normal hands. Some determinations by the present author (Christophers, 1947) using an apparatus designed to measure attraction and repellency are given in Table 46. In these there was little effect in substituting wet for dry filter paper pads over the attractive warm object (warm container). Reference may also be made to the findings of observers that moistening the skin or a membrane with saliva increased, or did not increase, attractiveness to feed. The present author has found that moistening the arm sometimes appeared to increase the rapidity of settling, but no constant effect could be shown.

Table 45. *Number of* Aëdes aegypti *females feeding at different temperatures and humidities* (*Lewis*, 1933)*

Temperature (° C.)	Relative humidity (percentage) 0	60–70	77–88	91–95
15	—	—	0 (27)	—
20	—	—	—	5 (37)
25	6 (34)	12 (33)	—	33 (43)
30	11 (36)	17 (35)	—	38 (39)
35	—	—	11 (36)	17 (36)
40	—	—	0 (24)	—

* The figures in brackets are the numbers of mosquitoes used, those not in brackets the numbers that fed.

Table 46. *Index for attraction of* Aëdes aegypti *using wet or dry filter paper over warm container*

Alignment of container	Position of filter paper to container	Index wet	Index dry
Horizontal	Lateral	2·8	2·5
Vertical	Below	2·4	2·0
Vertical	Above	2·2	2·4

For nature of index see Christophers (1947).

If, as here suggested, convection currents arising from the object are the chief stimulus to attack to feed, it is probable on purely physical grounds that moisture would have some, but possibly a variable, effect. It is a well-recognised fact in meteorology that it is the presence of aqueous vapour in the air which largely causes columns of air to continue ascending, dry air through adiabatic adjustment soon coming to equilibrium. Presumably what happens on a large scale may also occur to some extent with such a source as the human body. Being invisible, such currents are probably much more extensive than is generally appreciated. Haddow (1945), dealing with *A. simpsoni*, a diurnal species very like *A. aegypti* in its habits, notes that it attacks the head and shoulders by preference and people standing to those sitting, as also persons sitting to those lying. This is compatible with stimulation by convection currents from the warm human body, the bulk of which is towards the upper part and currents of which would be more concentrated in an upward direction standing than in any other situation. Some further observations on such currents will be found in the next section where some observations on *A. aegypti* in a free state are given.

Perhaps here it may be desirable to note that whilst mosquitoes in general appear from the literature to be largely attracted to feed by warmth, it would not be legitimate to assume that all species are alike in the degree to which this occurs or in absence of other sources of attraction or degrees of such. In many respects *A. aegypti* is a peculiar species in its markedly domestic and in its diurnal habits and even in the way in which in freedom it attacks. Whilst therefore the importance of convection currents in its behaviour has been emphasized, it does not necessarily follow that this applies to all mosquitoes. Nor contrariwise does proof of other factors being more dominant in other species negative conclusions drawn regarding *A. aegypti*.

(*f*) STIMULI CONNECTED WITH FEEDING

It seems probable that a number of stimuli and their responses are concerned with *A. aegypti* attacks and feeds and there may be distinguished as possibly distinct: (1) stimuli leading to the location of the victim at a distance; (2) stimuli leading to attack and settling to feed; (3) stimulus to puncture; (4) stimulus to suck and pass to the gut.

From observation of the free insect and what has previously been said regarding its reactions it seems probable that *A. aegypti*, in locating its victim, may be guided in the first place by visual response, namely attraction to dark and especially moving dark objects. It does not take long for an escaped insect in an ordinary-sized room to be found hovering in the initial stages of attack and it is thought that the insect in such circumstances makes prospecting flights with periods of rest until, from visual stimuli or the encountering of attractive currents of warm air, it comes to recognise the presence of a prospective host. The same behaviour is characteristic of males which locate and remain to hover about any occupant though they do not often settle and do not bite.

It is not impossible, however, that apart from any visual stimulus random flight may bring the insect within range of warm convection currents. Females liberated one by one at a lower corner of a net 6 feet by 3 feet by 6 feet high in a room warmed approximately to 28° C. were observed to fly upwards at a slant until they approached the netting. They then usually coasted at a height of about 5 feet from the ground slowly along the netting or made a diversion across the net. Eventually, usually flying at about 5 feet from the ground, they came into the neighbourhood of an observer seated in the net and at once when this occurred they made preliminary attack manœuvres, their rapid movements then being more difficult to follow. The impression gained was that the first indication of the existence of a victim was the presence of an aura compatible with the existence of convection currents arising from the body as a whole. Usually under these circumstances the mosquitoes located the observer within from 4 to 20 seconds. They did so equally when he was enveloped in a white cloth (see the very similar account given for *Anopheles* by Mer *et al.* referred to in section (*d*)).

Having located its victim the insect from this time onwards is concerned with 'settling to feed', taking flights, however, for safety if disturbed and reapproaching over and over again if necessary, usually from the dark side. There would appear to be a good deal more in the behaviour at this time than merely automatic response to a simple stimulus. If seriously threatened by movements of the victim it will usually retire for a time and may commonly be found resting on some nearby surface, a feature in its behaviour which often enables a particularly evasive pest to be captured, a towel or handkerchief being flapped vigorously and the insect then looked for on the nearest wall. It is interesting to contrast this caution with the behaviour of a female that has begun to feed and has been interrupted in an early stage of gorging. Such an insect then cares nothing for danger and will precipitate itself recklessly and without caution upon its victim. It was by using such mosquitoes that Gordon and Lumsden (1939) were able to get *Aëdes aegypti* to feed on the web of a frog's foot under the conditions of their experiments. A contrast in the opposite direction is seen when the skin is smeared with a repellent. The insect may, if a strong repellent is used, remain resting on the cage with no attempt to investigate. Or with a less strong repellent it may 'prospect', giving rise to the curious soft 'touches' which it has been thought are due to the trailing hind tarsi (Christophers, 1947). With a weak repellent it may essay to feed but leave the arm before gorging, or it may approach from a vantage point, inserting its proboscis over the protective edge of clothing or other object. Under normal conditions, however, sooner or later the free insect, as it does in a feeding cage, 'settles to feed'. It may do so on the clothing, much as it does often on the glove in the feeding cage, leaving this, if it is not satisfactory, for a place where it can find skin or sense warmth through thin clothing. As already noted stimulus to 'settle to feed' in a cage is mainly, if not entirely, thermal and there is no reason to doubt that it is the same stimulus which causes the free insect to attack and attempt to settle where it can feed.

Bishop and Gilchrist (1946) ascribe the effect of attraction by a warm object to a heat gradient. It is not very clear what physical condition would correspond to such a term. There would appear to be two conditions relative to heat set up by such a body, namely radiant heat and air warmed by contact. The human body is a considerable mass at a temperature of about 37° C. usually some degrees warmer than the surrounding air and convection currents must inevitably be set up. From facts already given it would seem that it is these currents, rather than radiant heat, which form the stimulus. If such a conclusion is correct, and having regard to the rapidity and certainty with which the response is evoked, the insect must possess what practically amounts to a special sense in this respect.

In regard to puncturing one might provisionally assume that, having reacted to the stimuli which bring the hungry mosquito to settle on its prospective host, there is still a further stimulus required to cause the female to puncture. Such an insect will, however, even if it cannot reach the skin to puncture it, that is, if stopped from doing so by netting a few millimetres distant, will vigorously thrust its proboscis through the netting with all the vigour of an insect on the skin though it has not touched this. Further, *A. aegypti* females will persistently prod, as if attempting to puncture, warm glass or other material where there can be no other stimulus than a thermal one. It would seem, therefore, that the attempt to puncture is part of the same chain of behaviour set up by the original thermal stimulus, though this may be completed or reinforced possibly by actual contact with the warm body.

Having punctured, however, there is evidence as already described in ch. XXI (*d*) that the taking in or refusal of fluids encountered and their despatch respectively to the mid-gut or to the diverticula lies with the insect now to decide. Sen (1918), dealing with *A. albopictus*, a closely related species, considered that though warmth actuates the mosquito to bite it does not encourage it to suck. Probably the author came to this conclusion because he found that shed blood had no attraction. MacGregor found that in artificial feeding the nature of the fluid could be detected through some sense possibly of an olfactory nature when it touched or approached close to the end of the hypopharynx, suggesting that the two small sensory pits on the termination of this organ have an olfactory or gustatory function, there being no other sense organs in this situation.

(*g*) STIMULI CONNECTED WITH OVIPOSITION

The essential requirement for oviposition in the case of *A. aegypti* is water. Even in captivity in tubes females of the species will die before they will lay eggs except on water or a wet surface. The behaviour of the female in respect to water finding for this purpose has been investigated by Kennedy (1941). As with *Anopheles* and *Culex* the only evidence obtained of a directed reaction (taxis) contributing to water finding in a small space such as a cage was random movement. The behaviour in the three cases, however, differed somewhat. *Anopheles* on encountering water breaks into a hovering flight laying its eggs during the 'oviposition dance' or while settled on the water. *Culex* stops dead on the water and lays its eggs there. *Aëdes aegypti* acts like *Culex* but with some fidgeting about. As previously described, *Aëdes aegypti*, when confined in tubes, remains quiet and at rest until oviposition approaches when it becomes restless and wanders about the tube. According to Kennedy the only direct form of attraction was a contrasting dark appearance (black paper on a white background) or a reflecting surface (mirror, or covering a dark target with glass). Water itself in dishes caused no attraction to *Culex* at a distance of over 10 cm., either above or from the side. Actual contact with water was the chief stimulus to oviposition. No eggs were laid by *Aëdes aegypti* in absence of such contact.

It is doubtful, if water is available, whether in the laboratory the fully gravid female *A. aegypti* pays much attention to any particular time of day or temperature for oviposition. Macfie (1915) found oviposition had usually taken place overnight before 6–7 a.m., over 50 per cent of females having laid before 8 a.m. It is to be noted, however, that the night hours are very long in proportion to the short hours of the working day, being usually double or more. Muirhead Thomson (1940) refers to Buxton as saying that *A. aegypti* oviposits (in nature) at dusk in the morning or evening, or during a dull day and that experiments showed that darkness was more attractive than light. Jobling (1937) found that *A. aegypti* would oviposit in complete darkness.

Haddow (*in lit.*) has recently given the present author figures obtained as a result of hourly counts of eggs laid by *A. aegypti* in cages in the laboratory of the Virus Research Institute, Entebe. The heaviest laying was in a peak period in the late afternoon at a time of rather low and rapidly falling intensity of light, 1675 eggs out of a total of 1922 being laid between the seventh and thirteenth hour from sunrise. In the present author's experience oviposition has chiefly been in relation to the time at which the mosquitoes were fed. No particular difference was noted in the case of isolated females kept in the dark or exposed to artificial light. Pots in cages in positions least exposed to light as a rule, however, showed heavier egg laying than those in the same cage in a less protected position.

In regard to temperature Hecht (1930) found that *A. aegypti* laid most eggs between 20° and 30° C., but the temperature choice was less sharply defined than with *Anopheles*. Oviposition, according to Gutzevich (1931), does not occur below 16° C. With the present author it has taken place at all temperatures between 19° and 28° C., the only difference noted being some spreading of the time taken at the lower temperatures.

There are many references in the literature to the varying attractiveness of different media for oviposition by *Aëdes aegypti*. Buxton and Hopkins (1927) found infusions of hay, rice or bran very attractive for females in nature. There has never been in our case any hesitation of females in captivity in laying on or around clean tap-water. The masses of eggs, amounting to 10,000 or more in a pot, were laid on clean water and the pots for technical reasons were perfectly clean and free from organic matter and newly placed in position after fouling by the pre-oviposition defaecation. More important than the nature of the fluid, provided this is not objectionable, is the nature of the receptacle (see under 'Selection of site for oviposition', ch. XXII (*c*), p. 504).

In regard to salinity Macfie (1915) notes that *A. aegypti* shows a selective choice for tap-water over salt solutions and that females will not oviposit at all on 2 per cent salt, or will do so only after much delay. Fielding (1919) found eggs laid on 70 per cent sea-water (about 2 per cent salt), but in higher concentrations none were laid. Woodhill (1941) states that *A. aegypti* distinguishes between water containing 5, 10, 17·5 and 35 parts per thousand of salt and will not lay on water containing 35 parts. Roubaud and Colas Belcour (1945) note that all types of water, tap, pure, distilled, or diluted sea-water are utilised with no particular preference, though eggs on sea-water remained white and the chorion did not harden (see ch. VI (*d*)). Of 17,940 eggs laid 28·9 per cent were on tap-water, 28 per cent on 31 per cent sea-water, 22·2 per cent on 42·6 per cent sea-water, 17·8 on 62 per cent sea-water and 3·1 per cent on undiluted sea-water. These authors consider that the only influence affecting oviposition is hygrotropism, which affects both males and females and apparently is governed by a special sensibility of the tip of the abdomen in respect to unfavourable substances in the medium. Howlett (1919) found sodium citrate and tartrate had an attractive effect for oviposition of *A. albopictus*.

REFERENCES

BARGREN, W. C. and NIBLEY, C. (1956). Comparative attractiveness of coloured lights of equal intensity to specific species of mosquitoes. *Res. Dep. 3rd Army Area Med. Lab.* Fort McPherson, Ga.

BENTLEY, C. A. (1914). Notes on experiments to determine the reaction of mosquitoes to artificial light. *Ind. J. Med. Res.* Suppl. 5, 9–11.

BISHOP, A. and GILCHRIST, B. M. (1946). Experiments upon the feeding of *Aëdes aegypti* through animal membranes, etc. *Parasitology*, **37**, 85–100.

BRETT, G. A. (1938). On the relative attractiveness to *Aëdes aegypti* of certain coloured cloths. *Trans. R. Soc. Trop. Med. Hyg.* **32**, 113–24.

BROWN, A. W. A. (1951). Studies in the response of female *Aëdes* mosquitoes. Part IV. Field experiments on Canadian species. *Bull. Ent. Res.* **42**, 575–82.

BROWN, A. W. A. (1954). Studies on the responses of the female *Aëdes* mosquito. Part VI. The attractiveness of coloured clothing to Canadian species. *Bull. Ent. Res.* **45**, 67–78.

BUXTON, P. A. and HOPKINS, G. H. E. (1927). Researches in Polynesia and Melanesia. Parts I–IV. *Mem. Lond. Sch. Hyg. Trop. Med.* no. 1.

REFERENCES

CHILD, C. M. (1894). Beiträge zur Kenntnis der antennalen Sinnesorgane der Insekten. *Z. wiss. Zool.* **58**, 475–528.

CHRISTOPHERS, S. R. (1914). (In discussion.) *Ind. J. Med. Res.* Suppl. 1, 255.

CHRISTOPHERS, S. R. (1947). Mosquito repellents. *J. Hyg., Camb.*, **45**, 176–231.

CRUMB, S. E. (1922). A mosquito attractant. *Science*, **55**, 446–7.

DE MEILLON, B. (1937). Some reactions of *Anopheles gambiae* and *Anopheles funestus* to environmental factors. *Publ. S. Afr. Inst. Med. Res.* no. 40, 313–27.

DETHIER, V. G. (1954). The physiology of olfaction in insects. *Ann. N.Y. Acad. Sci.* **58**, 139–55.

FIELDING, J. W. (1919). Notes on the bionomics of *Stegomyia fasciata* Fabr. *Ann. Trop. Med. Parasit.* **13**, 259–96.

GOELDI, E. A. (1905). *Os mosquitos no Para.* Weigandt, Para.

GORDON, R. M. and LUMSDEN, W. H. R. (1939). A study of the behaviour of the mouth-parts of mosquitoes when taking up food from living tissues, etc. *Ann. Trop. Med. Parasit.* **33**, 259–78.

GRASSI, B. (1900). Studi di uno zoologo sulla malaria. *Accad. Lincei* (5), **3**, 299–516.

GUTZEVICH, A. V. (1931). The reproduction and development of the yellow fever mosquito under experimental conditions. *Mag. Parasit.* **2**, 35–54. (In Russian; abstract in *Rev. Appl. Ent.* **21**, 2.)

HACKETT, L. W. (1937). *Malaria in Europe.* Oxford Univ. Press, London.

HADDOW, A. J. (1942). The mosquito fauna and climate of native huts at Kisumu, Kenya. *Bull. Ent. Res.* **33**, 91–142.

HADDOW, A. J. (1945). The mosquitoes of Bwamba County, Uganda. II. Biting activity with special reference to the influence of microclimate. *Bull. Ent. Res.* **36**, 33–73.

HEADLEE, T. J. (1937). Some facts underlying the attraction of mosquitoes to sources of radiant energy. *J. Econ. Ent.* **30**, 309–12.

HECHT, O. (1930). Ueber den Wärmesinn der Stechmücken bei der Eiablage. *Riv. Malariol.* **9**, 706–24.

HERMS, W. B. (1940). In Herms and Gray's *Mosquito Control.* New York and London, p. 92.

HOLMES, S. J. (1911). The reaction of mosquitoes to light in different periods of their life history. *J. Anim. Behav.* **1**, 29–32.

HOSKINS, W. M. and Craig, R. (1934). Olfactory responses of flies in a new type of olfactometer. *J. Econ. Ent.* **27**, 1029–36.

HOWLETT, F. M. (1910). The influence of temperature upon the biting of mosquitoes. *Parasitology*, **3**, 479–84.

HOWLETT, F. M. (1919). Report of the Imperial Pathological Entomologist. *Sci. Rep. Agric. Res. Inst., Pusa*, 1917–18, pp. 117–20.

HUNDERTMARK, A. (1938). Über das Helligkeitsunterscheidungs-Vermögen von *Anopheles maculipennis. Anz. Schädlingsk.* **14**, 25–30.

JOBLING, B. (1937). The development of mosquitoes in complete darkness. *Trans. R. Soc. Trop. Med. Hyg.* **30**, 467–74.

KAHN, M. C., CELESTIN, W. and OFFENHAUSER, W. (1945). Recording of sounds produced by certain disease-carrying mosquitoes. *Science*, **101**, 335–6.

KALMUS, H. (1937). Photohorotaxis, eine neue Reaktionsart gefunden an den Eilarven von Dixippus. *Z. vergl. Physiol.* **24**, 644–55.

KENNEDY, J. S. (1940). The visual response of flying mosquitoes. *Proc. Zool. Soc. Lond.* (A), **109**, 221–42.

KENNEDY, J. S. (1941). On water finding and oviposition by captive mosquitoes. *Bull. Ent. Res.* **32**, 279–301.

LAARMAN, J. J. (1955). *The Host Seeking Behaviour of the Malaria Mosquito* Anopheles maculipennis atroparvus. Thesis: Utrecht.

LEA, A. O., KNIERIM, J. A., DIMOND, J. B. and DE LONG, D. M. (1955). A preliminary note on egg production from milk-fed mosquitoes. *Ohio J. Sci.* **55**, 21–2.

LEWIS, D. J. (1933). Observations on *Aëdes aegypti* L. (Diptera: Culicidae) under controlled atmospheric conditions. *Bull. Ent. Res.* **24**, 363–72.

LUMSDEN, W. H. R. (1947). Observations on the effect of microclimate on the biting of *Aëdes aegypti* (L.) (Diptera: Culicidae). *J. Exp. Biol.* **24**, 361–73.

MACFIE, J. W. S. (1915). Observations on the bionomics of *Stegomyia fasciata*. *Bull. Ent. Res.* **6**, 205–29.

MACGREGOR, M. E. (1931). The nutrition of adult mosquitoes: preliminary contribution. *Trans. R. Soc. Trop. Med. Hyg.* **24**, 465–72.

MARCHAND, W. (1918). First account of thermotropism in *Anopheles punctipennis* with bionomic observations. *Psyche*, **25**, 130–5.

MAYER, A. M. (1874). Experiments on the supposed auditory apparatus of the mosquito. *Amer. Nat.* **8**, 577–92.

MER, G., BIRNBAUM, D. and AIOUB, A. (1947). The attraction of mosquitoes by human beings. *Parasitology*, **38**, 1–9.

MUIRHEAD THOMPSON, R. C. (1940). Studies on the behaviour of *Anopheles minimus*. I. The selection of the breeding place and the influence of light and shade. *J. Malar. Inst. India*, **3**, 265–94.

NUTTALL, G. H. F. (1901). The influence of colour upon *Anopheles*. *Brit. Med. J.* **2**, 668–9.

PARKER, A. H. (1948). Stimuli involved in the attraction of *Aëdes aegypti* L. to man. *Bull. Ent. Res.* **39**, 387–97.

PARKER, A. H. (1952). The effect of a difference in temperature and humidity on certain reactions of female *Aëdes aegypti* (L.). *Bull. Ent. Res.* **43**, 221–9.

PETERSEN, D. G. and BROWN, A. W. A. (1951). Studies on the responses of the female *Aëdes* mosquito. Part IV. The responses of *Aëdes aegypti* (L.) to a warm body and its radiation. *Bull. Ent. Res.* **42**, 535–41.

RAO, T. R. (1947). Visual responses of mosquitoes artificially rendered flightless. *J. Exp. Biol.* **24**, 64–78.

REEVES, W. C. (1951). Field studies on carbon dioxide as a possible host stimulant to mosquitoes. *Proc. Soc. Exp. Biol. Med.* **77**, 64–6.

REEVES, W. C. (1953). Quantitative field studies on a carbon dioxide chemotropism of mosquitoes. *Amer. J. Trop. Med. Hyg.* **2**, 225–31.

REUTER, J. (1936). Orienteerend onderzoek naar de oorzaak van het gedrag van *Anopheles maculipennis* Meigen bij de voedselkeuze. *Proefschr. Rijksuniv. Leiden.* Abstract in *Rev. Appl. Ent.* **24**, 223.

ROUBAUD, E. and COLAS-BELCOUR, J. (1945). Influence de la salure des eaux sur le développement de l'*Aëdes aegypti*. *Bull. Soc. Path. Exot.* **38**, 136–45.

RUDOLFS, W. (1922). See references, ch. XXIII, p. 529.

SCHAERFFENBERG, B. and KUPKA, E. (1951). Untersuchungen über die geruchliche Orienterung blutsäugender Insekten. I. Über die Wirkung eines Blutduftstoffes auf *Stomoxys* und *Culex*. *Österr. zool. Z.* **3**, 410–24. Also 1952 *Trans. Internat. Congr. Entom.* 9th Congr. Amsterdam. **1**, 359–61.

SEN, S. K. (1918). Beginnings in insect physiology and their economic significance. *Agric. J. India*, **13**, 620–7.

SHIPLEY, A. E. (1915). Insects and war. The mosquito (*Anopheles maculipennis*). *Brit. Med. J.* **1**, 331.

SHÔZÔ SATÔ, S. (1950). Compound eyes of *Culex pipiens* var. *pallens* Coquillett. *Sci. Rep. Tohôku Univ.* (4) (Biology), **18**, 331–41.

SIPPELL, W. L. and BROWN, A. W. A. (1953). Studies of the responses of the female *Aëdes* mosquito. Part V. The role of visual factors. *Bull. Ent. Res.* **43**, 567–74.

SMART, M. R. and BROWN, A. W. A. (1956). Studies on the responses of the female *Aëdes* mosquito. Part VII. The effect of skin temperature, hue and moisture on the attractiveness of the human hand. *Bull. Ent. Res.* **47**, 89–100.

STEPHENS, J. W. S. and CHRISTOPHERS, S. R. (1902). Some points in the biology of species of *Anopheles* found in Bengal. *Rep. Malar. Comm. Roy. Soc.* (6), 11–19.

THOMPSON, R. P. and BROWN, A. W. A. (1955). The attractiveness of human sweat to mosquitoes and the role of carbon dioxide. *Mosquito News*, **15**, 80–4.

REFERENCES

VAN THIEL, P. H. (1935). Onderzoekingen omtrent den gedrag von *Anopheles* ten opzichte van mensch en dier etc. *Geneesk. Tijdschr. Ned.-Ind.* **75**, 2101–18.

VAN THIEL, P. H. (1937). Quelles sont les excitations incitant l'*Anopheles maculipennis atroparvus* à visiter et à piquer l'homme ou le betail. *Bull. Soc. Path. Exot.* **30**, 193–203.

VAN THIEL, P. H. (1947). Attraction exercée sur *Anopheles maculipennis atroparvus* par l'acide carbonique dans un olfactomètre. *Acta tropica*, **4**, 10–20.

VAN THIEL, P. H. and WEURMAN, C. (1947). L'attraction exercée sur *Anopheles maculipennis atroparvus* par l'acide carbonique dans l'appareil de choix, II. *Acta trop., Basel*, **4**, 1–9.

WIETING, J. O. G. and HOSKINS, W. M. (1939). The olfactory response of flies in a new type of insect olfactometer. II. Responses of the housefly to ammonia, carbon dioxide and ethyl alcohol. *J. Econ. Ent.* **32**, 24–9.

WILLIS, E. R. (1947). The olfactory responses of female mosquitoes. *J. Econ. Ent.* **40**, 769–78.

WILLIS, E. R. and ROTH, L. M. (1952). Reactions of *Aëdes aegypti* (L.) to carbon dioxide. *J. Exp. Zool.* **121**, 149–79.

WOODHILL, A. R. (1941). The oviposition responses of three species of mosquitoes (*Aëdes* (*Stegomyia*) *aegypti* Linn., *Culex* (*Culex*) *fatigans* Wied., *Aëdes* (*Pseudoskusea*) *concolor* Taylor) in relation to the salinity of water. *Proc. Linn. Soc. N.S.W.* **66**, 287–92.

XXV
VIABILITY UNDER DIFFERENT ENVIRONMENTAL CONDITIONS

(a) NORMAL DURATION OF LIFE

The conditions which mainly determine length of life of *Aëdes aegypti* in the laboratory are: access or not to water; low or high humidity; whether fed or not; whether fed on blood with resulting oviposition; whether given only raisins, banana or sugar solution.

The presence of water for the insects to drink is important as otherwise extreme mortality may readily occur, even in high humidity. A cage containing about 600 mosquitoes was accidentally left at 25° C. and 70 per cent relative humidity, but without water. In this cage all mosquitoes were dead in 4 days, whereas there was negligible mortality in two similar cages with water.

A high humidity is favourable even up to approximate saturation (80 per cent relative humidity or more) at 28° C. It was found that placing porous pots of water in the incubator to bring about an almost saturated atmosphere reduced mortality at time of emergence almost to nil and insects at high humidities are very active provided the temperature is 25° or over. In the absence of high humidity life is much shortened if the temperature is at all high. At relatively low temperatures, for example 18–20° C., duration of life, however, may often be very considerable even with low humidity and insects may be kept alive if given water, even without food, under such circumstances for weeks. At a temperature of 25° C. and 70 per cent relative humidity with water provided there is relatively little mortality among females before the seventh day whether fed or not. Bacot (1916) gives the duration of life under these circumstances without food as averaging 6–8 days, maximum 12 days. There is usually, however, considerable mortality in males even shortly after emergence, especially at relatively high temperatures (28° C.) and in the absence of high humidity. After the seventh day from emergence, mortality among unfed females and in males increases rapidly and there is heavy mortality by the tenth day, few surviving the fourteenth day at 25° C. even at high humidity and with water provided. A few males may be present up to the end.

If given a blood meal within the first 7 days there is usually little mortality until after oviposition, when it may be considerable as at subsequent ovipositions.

When given regular blood meals or food such as fruit or glucose solution life may be very considerably prolonged. Fielding (1919) gives the following:

Not fed	Lived up to 7 days
Fed on sugar	Lived up to 20 days
Fed on milk and sugar	Lived up to 19 days
Fed on banana	Lived up to 68 days
Fed on blood	Lived up to 93 days

Bonne-Wepster and Brug (1932), however, state that life is longer where the insects are given sugar and water as food instead of blood. In this case there is not the mortality following oviposition. Connor (1924) states that 70 per cent of females die after the first batch of eggs has been laid.

Most accounts of very prolonged life relate to females that have been allowed to feed on some form of sugary substance or fruit, with or without occasional blood feeds. Such records probably almost always relate to some few individual mosquitoes that have survived and become conditioned to the environment. Marchoux, Salimbeni and Simond give for females up to 106 days, males 50 days; survival of one mosquito for 154 days is recorded by Guiteras, as quoted by Howard (1923); MacGregor (1915) gives females up to 4–6 weeks, males 10–21 days; Macfie (1915) found no female living beyond the sixty-second day, maximum for males 28 days; Bacot gives a maximum of 74 days if water is not available for oviposition. Beeuwkes, Kerr, Wetherbee and Taylor (1933) give the maximum for females as 131–225 days, with average of 70–116 days, and for males maximum of 82–135 days with average of 40–61 days. The most detailed data of a statistical kind are given by Putnam and Shannon (1934). These authors give two series, namely (*A*) one of 118 females offered frequent and regular blood meals, and (*B*) one of 190 females given only honey and water. The longest lived female, was among those not fed on blood. The weekly mortality was as follows:

Weeks	0	1	2	3	4	5	6	7	8	9	10	11	12	13	14	15	16	17
A	0	3	6	1	10	10	9	12	8	8	12	10	13	7	5	3	1	—
B	0	1	0	0	1	0	3	0	1	37	41	20	32	12	31	10	0	1

From these figures it would appear that there was very little mortality among the honey-fed mosquitoes until the ninth week when heavy mortality began. In the blood-fed series mortality was more uniformly distributed. Since there were more than half as many again mosquitoes in the second series, the final result in longevity was not very different.

Macdonald (1952 *a*, *b*, 1957), making an analysis of available data, gives the expectation of life of a mosquito (*Anopheles*) as $p^n/-\log_e p$ or $p^n/-2{\cdot}303 \log p$, using the ordinary form of logarithm, where p is the expectation of life through n days, together with a table of data as calculated for a number of values of p from which graphs may readily be drawn and interesting effects studied. Some data on survival for *Anopheles maculipennis atroparvus* are also given by Laarman (1955), p. 31. Kershaw, Chalmers and Lavoipierre (1954) found that for *Aëdes aegypti* under laboratory conditions there is a general pattern of survival, namely, that the rate of mortality increases with age in such a manner that its logarithm is directly proportional to age. This, the authors point out, is in accordance with the rate in man in adult life and some mammals, the rate of mortality following the Gompertz function (Gompertz, 1825; Gaddum, 1945; Brues and Sacher, 1952). *A. aegypti* were maintained by them at a temperature of 77° F. (25° C.) and relative humidity of 71 per cent and offered a blood meal daily, as also glucose and water. In two series of observations dealing respectively with ninety-two and thirty females 50 per cent survived in both series to the sixtieth day and a female in the first series survived over 100 days. Curves given by plotting the logarithm of the percentage of survival against time in days were convex to the origin with increasing declination. The Gompertz function was derived from the curves by fitting tangents to points on these and measuring their slope. This gave the rate of mortality at the points of contact by using the equation:

$$\mu = \frac{2{\cdot}303\,(\log_{10} N_o - \log_{10} N_t)}{t},$$

where μ is the mortality, N_o and N_t the numbers at times 0 and t respectively and t the time. If the survival curve is a straight line with an unchanging rate of mortality, the Gompertz function is a horizontal straight line; if the survival curve is convex to the origin with the rate of mortality increasing with age (and provided the rate increases regularly with age), then the function is a straight line at an inclination. Curves for *Anopheles hyrcanus*, *A. gambiae* and *A. funestus* are also given, based on data available from the literature.

The long periods given above show how very considerable the length of life of *Aëdes aegypti* under carefully controlled conditions may be. They are, however, much beyond ordinary laboratory experience. They show well, however, that if it is desired merely to keep insects alive the simplest method is to provide some form of sugary food, thus avoiding the rather heavy mortality that is apt to occur at oviposition. This is usually done by providing raisins or glucose in water. But a very clean and satisfactory method in practice is to maintain a supply of pieces of apple on which both sexes feed greedily. For reasons given elsewhere, however, it is usually undesirable, except for some special purpose, in critical experiments or testing to use insects thus kept alive and standardised conditions are much more precisely attained by employnig recently reared insects of known age from emergence (which should always be given) and fed only on water (Christophers, 1947).

The length of life of *A. aegypti* in nature has been little investigated. This is given by Korovitskii and Artemenko (1933) at Odessa out of doors for females as on the average 15 days (maximum 42 days), in basements 18 days (maximum 62 days) and in warm rooms as 10 and 24 days respectively for the mean and maximum. Here, however, the species is at or near the limit of distribution as largely determined by temperature. A point that is borne out by experience in the laboratory is that longevity tends to be greater at some suitable lower temperature than at that which is optimum for activity of the species.

(*b*) EFFECT OF LOW TEMPERATURE

Because of its medical importance many observers have investigated the effect of low temperatures upon adult *A. aegypti*. Temperatures below 0° C. appear to be fatal to the species if continued even for a short time. Flu (1920) found temperatures below freezing certain to kill the adult in 24 hours. At such temperatures the insect is completely immobilised and it is only a question after exposure whether recovery takes place at a warmer temperature. At 4° C. insects exposed for 1 hour survived; longer exposures killed (Otto and Neumann, 1905). At 7–9° C., however, these authors kept *A. aegypti* alive for 82 days. After 30 days half the females were still alive.

The point at which *A. aegypti* is rendered quite inert and incapable of movement is about 10° C. At various temperatures between this and 15° C. they are in greater or less degree torpid. Howard (1923) says that between 54° and 57° F. (13–14° C.) they are torpid, fly with difficulty and are no longer firm on their legs. Lewis (1933) notes that at 10° C. the mosquitoes become dormant after a few days. Touch will only stimulate them to move if warmed. At higher temperatures they can always be induced to fly until within a few hours of death.

A good deal of the immediate effect depends upon the suddenness with which the temperature reduction takes place. Mosquitoes taken out of the incubator at 28° C. into a cold room cease flying at once and cling to the sides and roof of the cage, or if the temperature is under 15° C. they fall completely inert to the bottom of the cage. At 13° C. the present author has found that *A. aegypti* maintained for days round about this temperature still maintain their normal resting position and if seriously disturbed move their legs slowly in an attempt to walk. If detached from their position they fall. At 14° C. if detached they flew but quickly resettled. At 14·5° C. they flew on minimal disturbance or even voluntarily and without being disturbed they often walked slowly and awkwardly and raised themselves high on their legs as if stretching their limbs. At 15° C. behaviour in respect to flight and attitude was more or less normal.

The point at which females cease to feed or oviposit has been given by Gutzevich (1931) as 16° C. Howard says that below 62° F. (17° C.) *A. aegypti* is sluggish and will not feed. Marchoux, Salimbeni and Simond (1903) state that between 19° and 25° C. they are slow to bite and below 15° C. they do not bite. Reed and Carroll (quoted by Howard, Dyar and Knab) state that *A. aegypti* will bite at 62° F. (17° C.) and above, but that they never succeeded in getting it to bite below this temperature.

Pairing, according to Gutzevich, does not occur below 18–19° C. The same author notes that above 25° C. males are active and females rarely escape fertilisation. At 20–25° C. fertilisation is still the rule. Below this temperature the proportion fertilised sinks rapidly.

The effect of low temperature is well summed up by Howard, Dyar and Knab (1912), who state that adult *A. aegypti* exhibit greatest activity around 82° F. (28° C.). Below 62° F. (17° C.) they are sluggish. From 54–57° F. (12–14° C.) they become torpid, fly with difficulty and are no longer firm on their legs. At 0° C. they die quickly. Below 62° F. (17° C.) they will not feed and below 68° F. (20° C.) they are seldom fertilised.

Davis (1932), testing the ability of the species to maintain itself at low temperatures, found that colonies successfully maintained themselves at 18–19° C., but not below this temperature.

(*c*) EFFECT OF HIGH TEMPERATURE AND THE THERMAL DEATH POINT

The optimum temperature for the species is generally taken to be 28° C. Any increase above this appears to be increasingly unfavourable, and over 40° C. it is quickly fatal. Mellanby (1932), speaking of insects generally, states that at over 40° insects die from the effects of heat, below 36° C. all insects survived short exposures. These limits apply well to *A. aegypti*. Finlay (1886) gives the following effects of high temperature on the species.

95–100° F. (35–38° C.)	Uncomfortable
102–105° F. (39–40° C.)	Remains motionless in apparent death
105–110° F. (41–43° C.)	Apparent or actual death

Macfie (1920) found that adults at 37° C. were not affected as regards activity, but life was shortened. There was reduced blood-sucking urge and fertility. After 5 minutes at 39° C. they were fairly active, at 40° C. profoundly affected and at 41° C. or higher all insects were almost instantaneously rendered inert. There ware no eventual recoveries after 44° C. Davis (1932) says that temperatures above 36° C. greatly shorten the life of adults. H. A. Johnson (1935) gives somewhat higher figures. He found *A. aegypti* relative to other species of mosquito rather resistant. 106° F. (41° C.) did not cause rapid shortening of life and the species could survive 30 minutes at 113° F. (45° C.) and 10 minutes at 117° F. (47° C.).

In determining the thermal death point the present author has used the apparatus as shown in Fig. 22 (2). This consisted of a cylindrical lamp glass of thin glass about 4·5 cm. internal diameter and 15 cm. in length, closed below with a rubber bung and above with a bung bored to take a short length of tubing a little over a centimetre in internal diameter, a long piece of stout tubing and a thermometer. The long tube, closed below by netting, was intended as an outlet and also as a convenient handle to enable the whole to be set in a stand and sunk in a water-jacket with controlled temperature. The short tube was for introduction of the mosquitoes. This was fitted with a smaller-sized tube plugged at the lower end with wool to drop in and so, when mosquitoes had been introduced, prevent any entering the tube during the experiment.

Mosquitoes, roughly to the required number, were drawn by suction into the receiver of the collecting pipette shown in the figure. The collecting tube of the pipette was then exchanged for a suitably bent tube. When transferring the mosquitoes they were first gently blown into the upper end of the tube, temporarily closed with netting, and then blown gently into the lamp glass through the short tube. In most of the exposures the inner surface of the lamp glass was covered with damp filter paper so that humidity was at about saturation. Some observations were, however, made with the apparatus in a dry condition, the effect as shown in the table being rather less lethal. The results are given in Table 47. As will be seen, the lethal effect varied with time of exposure. Full lethal effect (100 per cent deaths as shown by no recoveries overnight) for an exposure of 3 minutes required a temperature of 46–47° C. With 30 minutes' exposure 43° C. sufficed. Lethal effect began to be shown at 41° C. which temperature gave about 50 per cent deaths with females after one hour's exposure. The effect at 40° C. with the exposures used was slight.

Knock down (heat stupor) with females began at 45° C. with 3 minutes' exposure and was complete at 46–47° C., the insects being either motionless or showing only some slight movements of the legs. Usually after short exposures some or even a large proportion would later, at least partially, recover. Males appeared more susceptible to knock out than females. For dry heat the effects were very similar, though possibly slightly less for short exposures.

The thermal death point, basing this upon one hour's exposure might be placed at 41–42° C. For rapid lethal effect in 15 minutes with no recoveries it was 44° C.

The temperature limit at which the species was able to maintain itself in a colony has been studied by Davis (1932). At 36° C. colonies ceased to maintain themselves.

(*d*) EFFECT OF HUMIDITY

A high relative humidity not only favours duration of life, but also the general activity of the insects. This is especially so when high humidity is combined with temperature approaching 28 °C. Thus in cages of insects kept at temperatures of 25° C. or over and relative humidity 90 per cent the increased activity is very marked and made evident by the loud hum and the increased settling rate when food is offered. A very high humidity has been found unfavourable by some authors. At the temperatures mentioned, however, it has not

Table 47. *Effect of high temperature on the imago*

Temperature (° C.)	Time exposed in minutes	No. of mosquitoes	Knock down*			Final result		Minimal lethal temperature
			Partial	Complete	Still active	Affected	Killed	
				Females				
Sat. humidity								
47	3	10	—	10	—	—	10	—
46	3	9	2	7	—	5	4	—
45	3	7	2	5	—	7	—	45
44	3	5	—	5	1	4	—	—
43	3	12	4	8	6	3	2	—
42	3	9	9	—	9	—	—	—
40	3	10	10	—	10	—	—	—
45	15	9	—	9	—	—	9	—
44	15	13	—	13	—	—	13	—
43	15	10	2	8	—	1	9	43
42	15	10	m	—	3	2	5	—
41	15	9	m	—	—	m	—	—
40	15	13	13	—	13	—	—	—
43	30	10	—	10	—	3	7	—
42	30	5	1	4	—	—	5	42
41	30	11	11	—	—	10	1	—
41	60	17	5	12	—	10	7	—
40	60	12	12	—	5	7	—	—
39	60	12	11	1	10	2	—	—
Low humidity (dry container)								
47	3	13	—	13	—	—	13	—
43	15	12	—	12	—	4	8	43
42	15	10	m	—	10	—	—	—
				Males				
Sat. humidity								
47	3	13	—	13	—	—	13	—
46	3	10	—	10	—	9	1	—
45	3	8	—	8	—	2	6	—
44	3	3	—	3	—	1	2	—
43	3	8	—	8	—	1	7	43
42	3	8	6	2	8	—	—	—
40	3	6	6	—	6	—	—	—
45	15	7	—	7	—	—	7	—
44	15	8	—	8	—	—	8	—
43	15	17	2	15	—	2	15	43
42	15	14	m	—	1	2	11	—
41	15	17	m	—	—	m	—	—
40	15	12	12	—	12	—	—	—
43	30	7	—	7	—	1	6	—
42	30	3	—	3	—	2	1	42
41	30	5	2	3	—	4	1	—
41	60	4	—	4	—	—	4	41
40	60	17	15	2	—	13	4	—
39	60	10	9	1	—	7	2	—
Low humidity (dry container)								
47	3	10	—	10	—	—	10	—
43	15	14	—	14	—	—	14	43
42	15	8	6	2	8	—	—	—

* m = Majority.

been found prejudicial to life and is especially favourable to emergence. Lewis (1933) notes that at 23° C. the mean life of the unfed adult at 0 per cent relative humidity is 1·5 days and at 100 per cent relative humidity 7 days, and the data given for other percentages by this author shows a steady increase in length of life as relative humidity is raised towards 100 per cent.

A high humidity appears to be in general very favourable to the biting of mosquitoes in nature. Rudolfs (1925) found a steady increase in the numbers settling in outdoor tests from relative humidities below 60 per cent to a relative humidity of 96 per cent.

Accurate determination of the humidity in the open or in a room is best carried out using the sling psychrometer. Where this is not possible, as in a cage, the wet and dry bulb is used. For methods of using these and full tables for the purpose, see Psychromatic Tables by Marvin (1941), published for the U.S. Department of Commerce Weather Bureau and obtainable from C. F. Casella and Co. Ltd., Regent House, Fitzroy Square, London, W. 1. For papers dealing with humidity in relation to insects and methods of producing fixed humidities under experimental conditions see Buxton (1931, 1932); Buxton and Mellanby (1934); Bertram and Gordon (1939).

(*e*) INSECTICIDES

Unlike the pupa and larva the imago is rapidly killed when brought into contact with such fluids as fixatives, being killed almost instantaneously if submerged in alcohol. The adult is also very susceptible to the effects of volatile substances and gases. A peculiar effect seen on exposure of some adult mosquitoes to certain vapours, notably those of some of the essential oils, described by Rudolfs (1922) is the casting off of limb segments (autotomy). This author also notes a peculiar shrill hum given by mosquitoes when exposed to irritant vapours, etc. Further the adult is very susceptible to various insecticides, including not only gases and vapours but also such as act on contact with the cuticle. It is this last form of agent that is chiefly used in operations against the adult insect.*

Insecticides may be classed as:

(1) Those of vegetable origin and especially in the present case pyrethrum and the pyrethrins.

(2) Miscellaneous poisonous substances that may be lethal to various kinds of insects, such as arsenicals, fluorine compounds, mercury salts and such substances as creosote, carbolic acid, etc.

(3) Certain chlorinated synthetic compounds including DDT, gammexane and other contact insecticides of a like nature.

Under the first heading are a number of effective insecticides widely used in agriculture, such as nicotine, anabasin, rotenone and derris. But the only one now used to any extent against the adult mosquito is pyrethrum. In the form of the powdered flowers this has long had a reputation as 'insect powder'. It is now commonly used as the extract or in the form of, or in terms of, its active principles, pyrethrins I and II. These are respectively the mono- and diesters of a carboxylic acid (chrysanthemic acid) with a base (pyrethralone). For an account of its history, cultivation and other facts about pyrethrum see Gnadinger (1933); also on this and other insecticides of vegetable origin the booklet issued by the

* For good general accounts of insecticides and their use see Brown (1951), Busvine (1951, 1952), King (1950, 1951), Buxton (1945, 1952), and for accounts of their use against mosquitoes: Bates (1949); Muirhead-Thomson (1951). Much of the older literature has been made largely obsolete by the discovery and use of recent types of insecticides. References to a number of the more important are, however, given in the text later.

Imperial Institute (1940). For recent views on the chemical nature of the pyrethrins see Busvine (1951, 1952).

The pyrethrins are also used as an adjuvant in sprays both for knock-down effect and as an excitant to cause insects to take to flight and so increase the dose of insecticide received (see under aerosols). Formerly pyrethrum was one of the commonest substances burnt for fumigation to act as a repellent to mosquitoes in the house.

Of the second class of substances that have been used against adult mosquitoes mention may be made of an effective proprietary synthetic insecticide much used in sprays in America, namely lethane (lethane 384 and lethane 60, see Plumb, 1944), also of another proprietary substance, thanite (see editorial, *Science*, 95, no. 2473, Suppl.; p. 10) which is strongly lethal. Another well-known proprietary spray insecticide is 'Flit', much used in households against domestic flies. Kerosene, though not itself markedly insecticidal, is much used as a basis for insecticidal sprays, a role for which it is specially fitted by reason of its great spreading qualities. Formerly sulphur, creosote, carbolic were burnt or volatilised as fumigants.

Insecticides of the third class are remarkable especially in three respects, namely (1) they are intensely lethal on contact to insects, (2) being solids and but very slightly vaporisable they remain unchanged for long periods during which they retain their lethal character, and (3) they are relatively non-poisonous to man and mammals.

The first to be recognised as possessing these qualities and to be employed as a mosquito insecticide was DDT. Its history has been frequently told (see West and Campbell, 1946). Like most other modern contact insecticides it is a chlorinated synthetic compound (2,2-bis(*p*-chlorphenyl)1,1,1-trichlorethane). Its discovery as a substance with remarkable properties was followed by that of another chlorinated synthetic compound, gammexane (BHC, or more correctly gamma BHC, this being the only one of five isomers having such properties). Its chemical name is gamma hexachlorcyclohexane.

Gammexane is some ten times more powerful in its effect than DDT, but slightly more fumigant. Both are to some extent repellent, that is, cause mosquitoes to leave after a short contact, gammexane more so than DDT. But whereas after such contact with DDT mosquitoes often escape a fatal dose, with gammexane, owing to its higher toxicity, the dose received is usually fatal (Hadaway and Barlow, 1952). Gamma BHC may also, owing to its fumigant effect, act lethally when it has been absorbed into a mud or clay surface.

Other powerful insecticides of this type are: chlordane, dieldrin, aldrin, DDD, methoxychlor, all chlorinated hydrocarbons; toxaphene, a chlorinated camphene; parathion, an organo-phosphorus compound; allethrin, a synthetic pyrethrin; and others. The following is a list of the new insecticides discovered later than DDT and BHC with a brief summary of their chief characteristics taken from Busvine (1952).*

Chlordane. A dark brown viscous liquid soluble in oils. Effectiveness against *Aëdes aegypti* compared with DDT as one, 0·66. Oral lethal dose to rats, 200–500 mg./kg. Persistence in residual film low. Chiefly used against the domestic fly.

Aldrin. Wettable powder. Effectiveness against *A. aegypti*, 2·6. Lethal dose to rats, 45 mg./kg. Used also as larvicide.

Dieldrin. Wettable powder. Effectiveness against *A. aegypti*, 4·0. Lethal dose to rats, 50–55 mg./kg. Used also as larvicide, considered a promising insecticide against flies and mosquitoes by Rajindar Pal (1952).

* A review of the chemistry of insecticides with bibliography is given by Martin (1956); see also Brown (1951), Metcalf (1955*a*).

DDD. Similar to DDT, but toxicity to mammals low: lethal dose to rats, 2500 mg./kg. Useful where toxicity is important, for example, cattle spraying.

Methoxychlor. Toxicity still less: lethal dose to rats 6000 mg./kg. Used as larvicide as less toxic to fish.

Parathion. Strongly toxic to mammals: lethal dose to rats, 3·5–10 mg./kg. Used against bed-bug. Also as larvicide and against adult.

Toxaphene. Waxy solid. Low insecticidal properties, but prolonged persistence in film.

Allethrin. Similar to natural pyrethrins in action. Knock-down effect less rapid, lethality about the same. Effectiveness against *A. aegypti*, 4·0. Also used as larvicide.

WAYS IN WHICH INSECTICIDES ARE USED

Insecticides directed against the adult mosquito may be used as: (1) fumigants; (2) sprays; (3) mists or aerosols; (4) residual films. There are also to be considered substances that, without necessarily being fatal, protect against the bites of mosquitoes, namely (5) repellents.

At one time *fumigants* were the usual, if not almost the only, type of insecticidal action taken against mosquitoes. They include: smoke; the burning or volatilisation of such substances as sulphur, creosote, etc. and especially the burning of pyrethrum, usually in the form of cones of pyrethrum powder made up with nitre. Most of such measures are more correctly described as *culicifuges*. Though not very powerful under ordinary circumstances they have a certain usefulness as a household remedy. For further particulars see Howard (1911), Howard and Bishopp (1928), also Covell (1941), who gives much information with recipes regarding all forms of anti-mosquito measures.

The term fumigant has, however, been used recently to indicate the effect where insecticides act at a distance, presumably by means of their vapour. Also the term 'smokes' has been applied to certain fumigants driven off by heat in the form of solid particles (see Busvine and Kennedy, 1949).

Most usually insecticides against the adult mosquito have been used as *sprays*, either as a direct means of attacking the insects, or for the purpose of laying down residual films. For the first purpose a very commonly used medium has been kerosene in which the active insecticide is dissolved. Where danger of fire exists, or for other purposes, other substances may be used as the medium, for example, carbon tetrachloride. For automatic propulsion the substance 'Freon' is used as, or added to, the medium, or compressed gas is used, giving the 'insecticidal bomb'. For small automatic sprays 'sparklets' (solid CO_2) such as may be used for making soda-water have been utilised.

The insecticide formerly mostly used was pyrethrum in the form of the extract dissolved in kerosene (see Ginsburg, 1935; Symes, 1935; Hicks and Diwan Chand, 1936). Another important insecticide used in this way is lethane, as also DDT or other new insecticide. Or for reasons given later a small amount of pyrethrum extract may be added to the medium. The following are given by Busvine (1951) as representative sprays:

0·1 per cent pyrethrins, or
5·0 per cent lethane 384, or
10·0 per cent lethane (special), or
5·0 per cent thanite, or
0·35 per cent gamma BHC,
or
0·05 pyrethrins with 0·3 DDT, or
0·03 pyrethrins with 0·5 DDT.

A modification of the spray is the recently developed mist or *aerosol*, the droplets of the sprayed fluid being made sufficiently small to remain floating for some time in the air. Examples of such given by Busvine are:

3·0 per cent DDT.
5·0 per cent lubricating oil.
0·3 per cent pyrethrins.
dissolved in Frenon,
or
3·0 per cent DDT.
15–20 per cent DDT solvent.
0·4 per cent pyrethrins.
1–2 per cent non-volatile oil.
75–80 per cent Freon.

For an account of the principles involved in the use of sprays and aerosols see especially David (1946*a*, *b*), David and Bracey (1946), Goodhue (1942, 1946), May (1945), Brown (1951), Kruse *et al.* (1951), Mackerras *et al.* (1950).

The degree of atomisation depends upon the nozzle design, the propulsive force, the relative amount of air and fluid passing the nozzle and the viscosity and other physical properties of the fluid. For optimal effect of an aerosol the droplets should be of such a size that they will remain for a time suspended. They should not be so small that they rebound or fail to contact the insect. Velocity of the droplets and their uniform distribution is important. Owing to the movement of the wings, etc., a mosquito during flight receives a larger dosc than when stationary. Since pyrethrum has an exciting action on mosquitoes causing them to take to flight, this substance may be used to increase the effectiveness of other insecticides employed. Pyrethrum may also be added to give rapid knock-down effect, its action in this respect being much more rapid than that of most other insecticides.

One of the most extensively used anti-malarial measures is the employment of *residual films* of DDT, gammexane or other insecticide of this type. A large literature relates to the use and testing of such films, their physical characters, especially the nature of the deposit, whether crystallinc, size of granule, etc., their effectiveness on various backgrounds and other points relating to their use (see especially the article by Hadaway and Barlow (1952), and the references given by these authors; also Gahan *et al.* (1948), who give data on *A. aegypti*; Field (1950); Cutkomp (1947); Fay *et al.* (1947); Peffley *et al.* 1949); Reid (1951)). Besides films laid down by spraying, the new insecticides can also be used as insecticidal paints (Gilmour, 1946). See also ch. IV under 'Control and protection', p. 84.

MODE OF ACTION OF INSECTICIDES

Entry of contact insecticides through the cuticle has been especially studied by Hurst (1940) and by Wigglesworth (1942, 1948), the latter with special reference to the structure of insect cuticle. According to Wigglesworth oils passing through the lipoid layer of the cuticle come into contact with water. Water is then liberated in the form of droplets which appear on the surface of the cuticle (see chapter on physiology under 'The integument', p. 698). Light oils pass in more readily than heavy and entry is very slow with vegetable oils. Addition of 5 per cent oleic acid to refined petroleum greatly increases its rate of entry.

Hurst (1940) states that feebly dissociating compounds of high dielectric constant penetrate the cuticle more readily in presence of relatively apolar substances of low dielectric constant. The toxicity of unsaturated compounds, such as are many insecticides, may for this reason be increased by the addition of non-toxic paraffins and cycloparaffins. Thus alcohol and kerosene separately have only a very slow effect on the larva of *Calliphora*. But together they are rapidly fatal.

Poisoning of the adult mosquito by DDT is chiefly displayed on the nervous system. Mosquitoes that have received a dose when resting on a surface impregnated with DDT soon become restless and fly off. Later they show increasing incoordinate movements of the legs with eventual stupor. In poisoning by pyrethrins the insects appear to be brought into a spastic condition, the legs, especially the fore-legs, being held in a characteristic rigid fashion and when attempting to recover their position when overcome they use movements that strongly suggest a man suffering from spastic paralysis.

A review of the mode of action of insecticides with four pages of bibliography is given by Kearns (1956); see also Metcalf (1955*a*).

ACQUIRED RESISTANCE TO INSECTICIDES

Mention has already been made when dealing with control measures (ch. IV (*b*)) of acquired resistance against insecticides with a brief account of this condition. It remains to give very briefly what is known of such resistance.

Acquired resistance to insecticides is not confined to mosquitoes or only recently known, but has been encountered as far back as 1914 in the control of aphides and was known in some other insects (Harrison, 1952). Nor is it confined to compounds of the hydrochlorinated type, though it is in respect to such that it has become recently more fully recognised and important from the widespread control operations now undertaken against flies and mosquitoes. As previously noted it was first reported in mosquitoes from Italy for a resistant strain of *Culex* and more recently for such a condition in *Aëdes aegypti*. It has also more recently been recorded as present in some degree for several species of *Anopheles*. The degree of resistance developed varies under different conditions but may amount to complete innocuousness to spraying with DDT and other chlorinated compounds. Brown (1956) notes that dieldrin-resistant flies may be 800 times more resistant than the susceptible species.

Such resistance is now regarded, not as the result of acquirement of resistance by the individual, but due to the development of a resistant population through elimination of susceptible individuals and the hereditary passage of the character of resistance by those remaining, the regular killing off of susceptible strains resulting in time in a population of resistant strains. This original proportion of resistant individuals may be small but is commonly present. Thus Brown (1956) notes that 5 per cent of flies in the Gambia are naturally resistant and can by careful dosage be made the basis of a resistant strain. It is not certain, however, that all populations contain such a resistant component. Acquired resistance with such an origin becomes therefore very largely a genetical problem and its study from this point of view is now being actively carried on in a number of laboratories. Three chances of development of resistance are hypothecated by Busvine (1957), namely

(*a*) The frequency and importance of genes conferring resistance in the original population;

(*b*) the intensity of selection (that is the size of the population exposed to the insecticide and the proportion killed);

(*c*) The number of generations per year.

The method by which resistance is produced in the case of DDT is thought to be through possession by the resistant insect of an enzyme (dehydrochlorinase) which converts DDT into a harmless compound DDE ((2,2-bis-(*p*-chlorophenyl)-1,dichlorethylene), the process being known as dehydrochlorination. The enzyme is present in all tissues of the fly and especially the integument. DDT does not induce mutation, but those with resistance become the form of the insect left by elimination of non-resistant forms (Brown and Perry, 1956).*

(*f*) REPELLENTS

Repellents are substances that can be smeared on the skin without being irritant or poisonous and which prevent mosquitoes biting. They are usually oily fluids and may be used 'neat', or in some form of cream. They are also used to impregnate clothing to afford protection from bites and also wide-mesh netting used as a protection to the face and hands. They include a very large number of substances of varied chemical nature and different degrees of effectiveness depending upon their essential repellent properties, the time for which they remain effective when applied and whether they act at a distance or only on contact by the mosquito.

Substances especially effective are (1) certain natural essential oils, notably oil of citronella, or their essential principles, or synthetic substances of this type (for example the very powerfully repellent substance hydroxycitronellal); (2) a number of the dihydric alcohols (diols) among which is Rutger's 612 (ethylhexane diol-1:3); and (3) certain esters, for example dimethyl phthlate or other less well known esters of this type, such as isopropyl cinnamate or diethyl cinnamate or many others.

An important character largely determining their suitability for use is the boiling point. Organic compounds with boiling points below 200° C. are of little practical use as they evaporate from the human skin too rapidly. Substances with boiling point much above 300° C., whilst they give lasting effect usually lose in effective repellency. A high boiling point may, however, be useful for a special purpose, for example dibutyl phthalate (b.p. 325° C.) has been used for impregnating clothing since it stands up to washing.

Probably the two most generally useful repellents are dimethyl phthalate for military or similar purposes where a strong and lasting effect is essential and a good oil of citronella for civilian use where a milder and less permanent effect may serve all that is required. The former can be used 'neat' and applied after anointing the palms over all exposed skin (avoiding the near neighbourhood of the eyes or mucus membrane) or as a cream, a very satisfactory formula being the following:

Dr Hamill's cream (see Christophers, 1947)	
Dimethyl phthalate	12·5 c.c.
White wax (cera alba)	9 g.
Arachis oil	27·5 c.c.

* For further information see reviews by Metcalf (1955 *b*), Hoskins and Gordon (1956), and papers by Busvine (1957) and later. See also Brown (1956), Hadaway and Barlow (1956). On genetics of insect resistance see Crow (1956). For instances of resistance developed during operations against yellow fever see Jenkins and West (1954), Severo (1956).

Melt the wax in the arachis oil on a water bath and stir in the dimethyl phthalate. It can be modified by slightly changing the proportions of wax and arachis oil to suit a hotter or colder climate. It spreads very readily and is pleasant to use.

Oil of citronella if of good quality (a good Java oil) has the advantage of acting very strongly at a distance so that smearing of all exposed skin is unnecessary and a little on the clothing, socks, etc., may be sufficient to afford all that is required for comfort where danger of infection is not present.

Little is known as to the reason why particular substances have repellent properties or what determines the degree of effectiveness. For an account of repellents and their testing see Christophers (1947). See also references there given and some recent papers on mosquito repellents given here under references.

REFERENCES

(*a–d*) VIABILITY

BACOT, A. W. (1916). See references in ch. VI, p. 155.

BEEUWKES, H., KERR, J. A., WETHERBEE, A. A. and TAYLOR, A. W. (1933). Observations on the bionomics and comparative prevalence of the vectors of yellow fever, etc. *Trans. R. Soc. Trop. Med. Hyg.* **26**, 425–47.

BERTRAM, D. S. and GORDON, R. M. (1939). An insectarium with constant temperature and humidity control, etc. *Trans. R. Soc. Trop. Med. Hyg.* **33**, 279–88.

BONNE-WEPSTER, J. and BRUG, S. L. (1932). See references in ch. II (*a–b*), p. 45.

BRUES, A. M. and SACHER, G. A. (1952). In Nickson's *Symposium on Radiobiology*. Chapman and Hall, London, p. 441.

BUXTON, P. A. (1931). The measurement and control of atmospheric humidity in relation to entomological problems. *Bull. Ent. Res.* **22**, 431–47.

BUXTON, P. A. (1932). Terrestrial insects and the humidity of the environment. *Biol. Rev.* **7**, 275–320.

BUXTON, P. A. and MELLANBY, K. (1934). The measurement and control of humidity. *Bull. Ent. Res.* **25**, 161–5.

CHRISTOPHERS, S. R. (1947). Mosquito repellents. *J. Hyg., Camb.*, **45**, 176–231.

CONNOR, M. E. (1924). Suggestions for developing a campaign to control yellow fever. *Amer. J. Trop. Med.* **4**, 277–307.

DAVIS, N. C. (1932). The effects of heat and cold upon *Aëdes* (*Stegomyia*) *aegypti*. *Amer. J. Hyg.* **16**, 177–91.

FIELDING, J. W. (1919). Notes on the bionomics of *Stegomyia fasciata* Fabr. *Ann. Trop. Med. Parasit.* **13**, 259–96.

FINLAY, C. (1886). Yellow fever, its transmission by means of the *Culex* mosquito. *Amer. J. Med. Sci.* **92**, 395–409.

FLU, P. C. (1920). Onderzoek naar de levensduur van *Stegomyia fasciata* bij lage temperaturen. *Meded. Burg. Geneesk. Ned.-Ind.* **7**, 99–105.

GADDUM, J. H. (1945). Lognormal distributions. *Nature, Lond.*, **156**, 463–6.

GOMPERTZ (1825). Quoted by Gaddum (1945).

GUTZEVICH, A. V. (1931). See references in ch. XXIV, p. 545. Abstract in *Rev. Appl. Ent.* **21**, 2.

HOWARD, L. O. (1923). The yellow fever mosquito. *U.S. Dep. Agric. Fmr's Bull.* no. 1354.

HOWARD, L. O., DYAR, H. G. and KNAB, F. (1912). See references in ch. I (*f*), p. 19.

JOHNSON, H. A. (1935). The effect of high temperatures on the length of life of certain species of mosquito. *J. Tenn. Acad. Sci.* **10**, 225–7.

REFERENCES

KERSHAW, W. E., CHALMERS, T. A. and LAVOIPIERRE, M. M. J. (1954). Studies on arthropod survival. I. The pattern of mosquito survival in laboratory conditions. *Ann. Trop. Med. Parasit.* **48**, 442–50.

KOROVITSKII, L. K. and ARTEMENKO, V. D. (1933). Zur Biologie des *Aëdes aegypti.* (In Russian.) *Mag. Leningrad,* **2**, 400–6. Abstract in *Rev. Appl. Ent.* **22**, 78.

LAARMAN, J. J. (1955). See references, ch. XXIV, p. 545

LEWIS, D. J. (1933). Observations on *Aëdes aegypti* L. (Diptera: Culicidae) under controlled atmospheric conditions. *Bull. Ent. Res.* **24**, 363–72.

MACDONALD, G. (1952*a*). The analysis of the sporozoite rate. *Trop. Dis. Bull.* **49**, 569–86.

MACDONALD, G. (1952*b*). The analysis of equilibrium in malaria. *Trop. Dis. Bull.* **49**, 813–29.

MACDONALD, G. (1957). *The Epidemiology and Control of Malaria.* Oxford Univ. Press, London.

MACFIE, J. W. S. (1915). Observations on the bionomics of *Stegomyia fasciata. Bull. Ent. Res.* **6**, 205–29.

MACFIE, J. W. S. (1920). Heat and *Stegomyia fasciata*: short exposures to raised temperatures. *Ann. Trop. Med. Parasit.* **14**, 73–82.

MACGREGOR, M. E. (1915). Notes on the rearing of *Stegomyia fasciata* in London. *J. Trop. Med.* (*Hyg.*), **18**, 193–6.

MARCHOUX, E., SALIMBENI, A. and SIMOND, P. L. (1903). See references in ch. I (*f*).

MARVIN, C. S. (1941). *Psychrometric Tables.* U.S. Dept. Commerc. Weather Bur. (see text).

MELLANBY, K. (1932). The influence of atmospheric humidity on the thermal death point of a number of insects. *J. Exp. Biol.* **9**, 222–31.

OTTO, M. and NEUMANN, R. O. (1905). Studien über Gelbfieber in Brasilien. *Z. Hyg. InfektKr.* **51**, 357–506.

PUTNAM, P. and SHANNON, R. A. (1934). The biology of *Stegomyia* under laboratory conditions. *Proc. Ent. Soc., Wash.,* **36**, 217–42.

RUDOLFS, W. (1925). Relation between temperature, humidity and activity of house mosquitoes. *J. N.Y. Ent. Soc.* **33**, 163–9.

(*e*) INSECTICIDES

BATES, M. (1949). *The Natural History of Mosquitoes.* Macmillan Co., New York.

BROWN, A. W. A. (1951). *Insect Control by Chemicals.* Chapman and Hall, London, and Wiley and Sons, New York.

BROWN, A. W. A. (1956). (Resistance to dieldrin.) *World Hlth Org.* X, no. 3.

BROWN, A. W. A. and PERRY, A. S. (1956). Dehydrochlorination of DDT by resistant houseflies and mosquitoes. *Nature, Lond.,* **178**, 368–9.

BUSVINE, J. (1951). *Insects and Hygiene.* Methuen and Co., London.

BUSVINE, J. (1952). The newer insecticides in relation to pests of medical importance. *Trans. R. Soc. Trop. Med. Hyg.* **46**, 245–52.

BUSVINE, J. (1957). Insecticide resistance. Strains of insects of public health importance. *Trans. R. Soc. Trop. Med. Hyg.* **51**, 11–31.

BUSVINE, J. and KENNEDY, J. S. (1949). Experiments with insecticidal smokes for indoor use. *Ann. Appl. Biol.* **36**, 20–85.

BUXTON, P. A. (1945). The use of the new insecticide DDT in relation to the problems of tropical medicine. *Trans. R. Soc. Trop. Med. Hyg.* **38**, 367–93.

BUXTON, P. A. (1952). The place of insecticides in tropical medicine: an introduction. *Trans. R. Soc. Trop. Med. Hyg.* **46**, 213–26.

COVELL, G. (1941). *Malaria Control by Antimosquito Measures.* Thacker and Co., London.

CROW, J. F. (1956). Genetics of insect resistance to chemicals. *Ann. Rev. Ent.* **2**, 227–46.

CUTKOMB, L. K. (1947). Residual sprays to control *Anopheles quadrimaculatus. J. Econ. Ent.* **40**, 328–33.

DAVID, W. A. L. (1946*a*). Factors influencing the interaction of insecticidal mists and flying insects. I. The design of a spray testing chamber and some of its properties. *Bull. Ent. Res.* **36**, 373–93.

DAVID, W. A. L. (1946*b*). Factors, etc. II. The production and behaviour of kerosene insecticidal spray mists and their relation to flying insects. *Bull. Ent. Res.* **37**, 1–28.

DAVID, W. A. L. and BRACEY, P. (1946). Factors, etc. III. Biological factors. *Bull. Ent. Res.* **37**, 177–90.

FAY, R. W., COLE, E. L. and BUCHNER, A. J. (1947). Comparative residual effectiveness of organic insecticides against houseflies and malaria mosquitoes. *J. Econ. Ent.* **40**, 635–40.

FIELD, J. W. (1950). Fumigant and repellent effects of BHC (gammexane) and DDT upon *Anopheles. Trans. R. Soc. Trop. Med. Hyg.* **43**, 547–8.

GAHAN, J. B., GILBERT, I. H., PIFFLEY, R. L. and WILSON, H. G. (1948). Comparative toxicity of four chlorinated organic compounds to mosquitoes, houseflies and cockroaches. *J. Econ. Ent.* **41**, 795–801.

GILMOUR, D. (1946). The toxicity to houseflies of paints containing DDT. *J. Coun. Sci. Indust. Res.* **19**, 225–32.

GINSBURG, J. N. (1935). Protection from mosquito bites in outdoor gatherings. *Science*, **82**, 490–1.

GNADINGER, C. B. (1933). *Pyrethrum Flowers.* McLaughlin, Gormley King and Co., Minneapolis.

GOODHUE, L. D. (1942). Insecticidal aerosol production. *Ind. Engng Chem.* **34**, 1456.

GOODHUE, L. D. (1946). Aerosols and their application. *J. Econ. Ent.* **39**, 506–9.

HADAWAY, A. B. and BARLOW, F. (1952). Some physical factors affecting the efficiency of insecticides. *Trans. R. Soc. Trop. Med. Hyg.* **46**, 236–44.

HADAWAY, A. B. and BARLOW, F. (1956). Effects of age, sex and feeding on the susceptibility of mosquitoes to insecticides. *Ann. Trop. Med. Parasit.* **50**, 438–43.

HARRISON, C. M. (1952). The resistance of insects to insecticides. *Trans. R. Soc. Trop. Med. Hyg.* **46**, 255–60.

HICKS, E. P. and DIWAN CHAND (1936). Transport and control of *Aëdes aegypti* in aeroplanes. *Rec. Malar. Surv. India*, **6**, 73–90.

HOSKINS, W. M. and GORDON, H. T. (1956). Arthropod resistance to chemicals. *Ann. Rev. Ent.* **1**, 89–122.

HOWARD, L. O. (1911). Remedies and preventives against mosquitoes. *U.S. Dep. Agric. Fmrs' Bull.* no. 444.

HOWARD, L. O. and BISHOPP, F. C. (1928). Mosquito remedies and preventives. *U.S. Dept. Agric. Fmrs' Bull.* no. 1570.

HURST, H. (1940). Permeability of insect cuticle. *Nature, Lond.*, **145**, 462–3.

IMPERIAL INSTITUTE (1940). A survey of *insecticide materials* of vegetable origin. Imperial Institute. London.

JENKINS, D. W. and WEST, A. S. (1954). Yellow fever in Trinidad: a brief review. *Mosquito News.*

KEARNS, C. W. (1956). The mode of action of insecticides. *Ann. Rev. Ent.* **1**, 123–48.

KING, W. V. (1950). DDT-resistant houseflies and mosquitoes. *J. Econ. Ent.* **43**, 527–32.

KING, W. V. (1951). Repellents and insecticides available for use against insects of medical importance. *J. Econ. Ent.* **44**, 338–43.

KRUSE, C. W., PHILEN, E. A. and LUDVIK, G. F. (1951). Characteristics of larvicidal sprays applied by aircraft for the control of *Anopheles quadrimaculatus. J. Nat. Malar. Soc.* **10**, 23–4.

MACKERRAS, I. M., RATCLIFFE, F. N., GILMOUR, D. and MULES, M. W. (1950). The dispersal of DDT from aircraft for mosquito control. *Bull. Commonw. Sci. Indust. Res. Org. Austral.* no. 257.

MARTIN, H. (1956). The chemistry of insecticides. *Ann. Rev. Ent.* **1**, 149–66.

MAY, K. R. (1945). The cascade impactor. An instrument for sampling coarse aerosols. *J. Sci. Instr.* **22**, 187–95.

METCALF, R. L. (1955*a*). *Organic Insecticides. Their Chemistry and Mode of Action.* Interscience Publishers Inc. New York.

METCALF, R. L. (1955*b*). (Review of resistance.) *Physiol. Rev.* **35**, 197–232.

MUIRHEAD THOMSON, R. C. (1951). Mosquito Behaviour in Relation to Malaria Transmission and Control in the Tropics. Ed. 2. Edward Arnold, London.

PEFFLEY, R. L. and GAHAN, J. B. (1949). Residual toxicity of DDT analogs and related chlorinated hydrocarbons to houseflies and mosquitoes. *J. Econ. Ent.* **42**, 113–16.

REFERENCES

PLUMB, G. H. (1944). Lethane 384 Special for control of the brown dog tick. *J. Econ. Ent.* **37,** 292–3.

RAJINDAR PAL (1952). Dieldrin for malaria control. *Indian J. Malariol.* **6,** 325–30.

REID, J. A. (1951). A laboratory method for testing residual insecticides against Anopheline mosquitoes. *Bull. Ent. Res.* **41,** 761–77.

RUDOLFS, W. (1922). Chemotropism of mosquitoes. *N.J. Agric. Exper. Sta. Bull.* no. 367.

SEVERO, O. P. (1956). Acquired resistance to DDT in the Dominican Republic. *Chron. World Hlth Org.* **10,** 347–54.

SYMES, C. B. (1935). Insects in aeroplanes. *Rec. Malar. Res. Lab. Kenya,* no. 6.

WEST, L. S. and CAMPBELL, G. A. (1946). *DDT the Synthetic Insecticide.* Chapman and Hill, London.

WIGGLESWORTH, V. B. (1942). Some notes on the integument of insects in relation to the entry of contact insecticides. *Bull. Ent. Res.* **33,** 205–18.

WIGGLESWORTH, V. B. (1948). Mode of action of new insecticides. *Rep. 5th Commonw. Entom. Conf.* 1948, 35–37.

(*f*) REPELLENTS

ALTMAN, R. M. and SMITH, C. N. (1955). Investigation of repellents for protection against mosquitoes in Alaska. *J. Econ. Ent.* **48,** 67–72.

APPLETHWAITE, K. H. and CROSS, H. F. (1951). Further studies of repellents in Alaska. *J. Econ. Ent.* **44,** 19–22.

APPLETHWAITE, K. H. and SMITH, C. N. (1950). Field tests with mosquito and sandflies in Alaska. *J. Econ. Ent.* **43,** 353–7.

CHRISTOPHERS, S. R. (1945). Insect repellents. *Brit. Med. Bull.* **3,** 222–4.

CHRISTOPHERS, S. R. (1947). Mosquito repellents. *J. Hyg., Camb.,* **45,** 176–231.

DETHIER, V. G. (1947). Chemical insect attractants and repellents. Blakiston Co., Philadelphia, and Lewis and Co., London.

DETHIER, V. G. (1956). Repellents. In *Ann. Rev. Ent.* **1,** 181–202, 144 references.

KING, W. V. (1951). Repellents and insecticides available for use against insects of medical importance. *J. Econ. Ent.* **44,** 338–43.

KNIPLING, E. F. and DOVE, W. E. (1944). Recent investigations of insecticides and repellents for the armed forces. *J. Econ. Ent.* **37,** 477–80.

XXVI
THE IMAGO: INTERNAL STRUCTURE

(*a*) THE ALIMENTARY CANAL

GENERAL DESCRIPTION (Figs. 63–6)

Many authors have described with more or less completeness the alimentary system or parts of this in the imaginal mosquito. Among very early descriptions are those by Pouchet (1847), who describes in *Culex* eight vesicular stomachs; Dufour (1851), who in *Theobaldia annulata* first described the diverticula; Apostolidon (1901), who very early figured the alimentary tract in *Anopheles*. Works of more modern type are those of Grandpré and Charmoy (1900), Christophers (1901); Grassi (1901); Nuttall and Shipley (1903); Thompson (1905); Leon (1923, 1924*a*, *b*); De Boissezon (1930); Perfiliev (1930). Of these accounts those by Nuttall and Shipley and by Thompson are particularly complete and detailed, that by Thompson being especially valuable as giving much information on the transition from larval to imaginal parts. On the changes in metamorphosis are also: Hurst (1890*a*, *b*); Holt (1917); Richins (1938). Some excellent figures giving the appearances of the alimentary canal as seen in dissections will be found in the Atlas by Neumann and Mayer (1914), some of which have been reproduced by Martini (1920).

On the intracranial portion of the canal are the very thorough studies by Sinton and Covell (1927) in *Anopheles* and by Barraud and Covell (1928) in other genera of Culicidae, also the contributions by Robinson (1939), and by Jobling (1928) (the latter dealing with *Culicoides*, where the parts in question have a close resemblance to those in the mosquito). Snodgrass (1943, 1944, 1953), in respect to the structures situated in the head, has dealt with their nature and given a terminology in accordance with the generalised insect plan.

On the visceral portions of the tract see Jagujhinskaia (1940) and Rajindar Pal (1943), who describe the peritrophic membrane following a blood feed; Trembley (1951), who, besides other observations, describes the pyloric spines; Engel (1924), who describes the rectum and rectal papillae, including those of the male. On the salivary glands, first described in the mosquito by Macloskie (1887, 1888), and whose figure of the sucking apparatus of the mosquito is reprinted in an editorial (1898) in the *British Medical Journal*, are Christophers (1901); Grassi (1901); Nuttall and Shipley (1903); Goldi (1913); Leon (1923); Shute (1940); Jensen (1956); Jensen and Jones (1957); Bhatia, Wattal and Kalra (1957). A photograph of the glands of *Aëdes aegypti* is given by Beyer, Pothier, Couret and Lemann (1902). A description of the salivary pump is given by Cornwall (1923). References to the Malpighian tubules will be found under the excretory system in ch. XXX; see also the section on digestion in ch. XXXI.

Though some parts of the canal in the imago are still recognisable as those present in the larva, many profound changes have taken place to adapt structures to the two entirely different methods of feeding and two entirely different kinds of food to be dealt with. This applies especially to those parts within the head, the homologies of which have been very variously interpreted by different authors. It is helpful here to appreciate the changes previously described which take place in passage from the larval to the imaginal head, namely the shrinking of the clypeofrontal portion, greatly ballooned in the larva to accommodate the enormous pharynx with its fringes, to the crowded-up contracted condition as seen in the imago.

In the imago the alimentary canal consists of:

(1) A prestomial portion represented by the *labral canal* (Fig 64 (1) *Lc*) in the proboscis and other structures anterior to the true mouth.

(2) A stomodaeal portion, which consists of two pump-like organs, the *buccal cavity* (*Bc*) and the *oesophageal pump* (*p'*) continued as a short muscular *oesophagus* from which arise three sac-like *diverticula* (Fig. 65 (1)), two small *dorsal* (*Dd*) and one much larger *ventral* (*Dv*).

(3) The part of the canal derived from the mesenteron, *ventriculus*, or as it is usually termed in the mosquito, mid-gut (Fig. 65 (1) *Mg*).

(4) A proctodaeal portion, the *hind-gut*, made up of a small proximal dilatation, *pyloric ampulla* (*am*) into which the Malpighian tubules open, a narrow intestine-like *ileum* (*IL*) sometimes called ileo-colon, and a terminal portion, *rectum* (*Rc*) that opens at the anus.

At the junction of the oesophagus with the mid-gut is a thickened portion with a valve-like intussusception of the former into the latter corresponding to the much more developed structure in the larva commonly termed the *proventriculus*, but more accurately named the *cardia* (*Ca*). In the rectum are the *rectal papillae* (Fig. 66 (1) *rp*).

Also connected with the alimentary canal are the *salivary glands* (Fig. 66 (6) *sg*) with their ducts, *salivary ducts* (*sd*), which combine beneath the sub-oesophageal ganglion to form the *common salivary duct* (*csd*) terminating in the *salivary pump* (*slp*). Opening into the canal, but described under the excretory system are the five *Malpighian tubules* (Fig. 65 (1) *Mt*).

The general character and arrangement of these parts will be clear from the accompanying figures.

The capacious pharynx of the larva, the large well-developed proventriculus (cardia), the voluminous gastric caeca and the swollen colon so conspicuous in the larva are none of them present in the imago. Nor in the mid-gut is there the conspicuous feature of a peritrophic membrane. If a peritrophic membrane exists, it is confined to the short period of digestion of a blood meal. On the other hand, the characteristic rectal papillae of the imago are lacking in the larval rectum.

THE LABRAL CANAL (Fig. 64 (1–3) *Lc*)

The first portion of the food canal lies in the proboscis, the parts composing which have already been described. It lies wholly in the labrum which is bent round ventrally to form a deep groove that for practical purposes is a closed canal. It may conveniently be named the *labral canal*. Its walls are formed entirely by the *ventral plate* of the labrum. This, as already noted when describing the parts of the proboscis, is a separate structure from the externally situated *dorsal plate* and has quite different connections basally from this. The dorsal plate is continued basally into the moveable front portion of the clypeus here provisionally termed the preclypeus (Fig. 63 (1–3) *PC*). The ventral plate is the sclerotised ventral aspect of the labrum and is continuous basally with the roof of the buccal cavity. It was formerly thought that the labrum formed only an open gutter and that to complete the canal this was closed in below by the hypopharynx (Fig. 64 (4) *hp*). The thin lateral edges of the ventral plate are, however, practically in contact through all except the most basal portion of the canal. Nevertheless, as observed by Thompson, who gives an excellent account of these structures, some support to these thin edges may still be necessary from the hard flat upper surface of the hypopharynx. What may be important also in this respect are two folds (*Lr'''*), one on each side of the under surface of the labrum, each of which forms a flat surface resting on the flat upper surface of the hypopharynx. Towards the base of the proboscis these folds become increasingly sclerotised and eventually flatten out into thin plates which take over the function of closing-in the canal in its basal portion where the edges of the ventral plate separate (Fig. 64 (4) *Lr''*). As these lateral folds and their basal continuations are important parts of the canal, they may for convenience of description be termed the *lateral plates* (of the labrum).

The basal connections of the parts of the proboscis, so far as these take part in the formation of the food channel, are shown in Fig. 63. The dorsal plate of the labrum is attached basally to the moveable portion of the clypeus (*PC*) and has no direct connection other than by membrane with the food channel (Fig. 63 (2) *a*). The ventral plate at the base of the labrum ends abruptly in membrane and is connected only by a thin median thickening with the sclerotised roof of the buccal cavity, its edges here diverging and ceasing to close in the food canal ventrally (Fig. 63 (5) *c*). At this level the lateral plates, after broadening out to close in the canal, abruptly thicken into rods and in doing so leave widely open the ventral aspect of the labral canal. It is here that the labral canal terminates

---

Figure 63. Alimentary canal.

1 Lateral view of buccal cavity and oesophageal pump.

2 Anterior portion of same to show region of mouth. *a*, dorsal plate of labrum; *b*, ventral plate of same; *c*, lateral plate of same; *d*, hypopharynx with salivary channel; *e*, labium.

3 Dorsal view of buccal cavity and oesophageal pump.

4 Dorsal view of roof of buccal cavity. *a*, lateral plate of labrum; *b*, membranous area of roof; *c*, hummocky area lying behind posterior hard palate.

5 Anterior hard palate and related structures (ventral view). *a*, lateral plates forming thin flanges at base of labrum; *b*, rod-shaped continuation of same; *c*, dorsal portion of ventral plate of labrum; *d*, thinned ventral edge of same.

6 Sagittal section through mouth to show opening of labral canal into alimentary canal proper. *a*, dorsal plate of labrum continued into preclypeus; *b*, ventral plate with flange-like sides forming lateral wall of labral canal; *c*, flange of lateral plate of labrum; *d*, hypopharynx forming floor of true alimentary canal.

The arrow indicates position where the labral canal enters the alimentary canal, that is the mouth.

7 A, section through a sensory pit on the anterior hard palate; B, a sensory pit on membranous roof of buccal cavity.

8 Transverse section of hypopharynx with salivary channel. *a*, position of lighter area in cuticle.

9 Apex of hypopharynx showing subapical opening.

10 Dorsal view of base of hypopharynx showing the salivary pump.

11 Portions of common salivary duct. Only the cuticular intima is shown. *a*, oblique lines of thickening.

For explanation of lettering see that for Fig. 64.

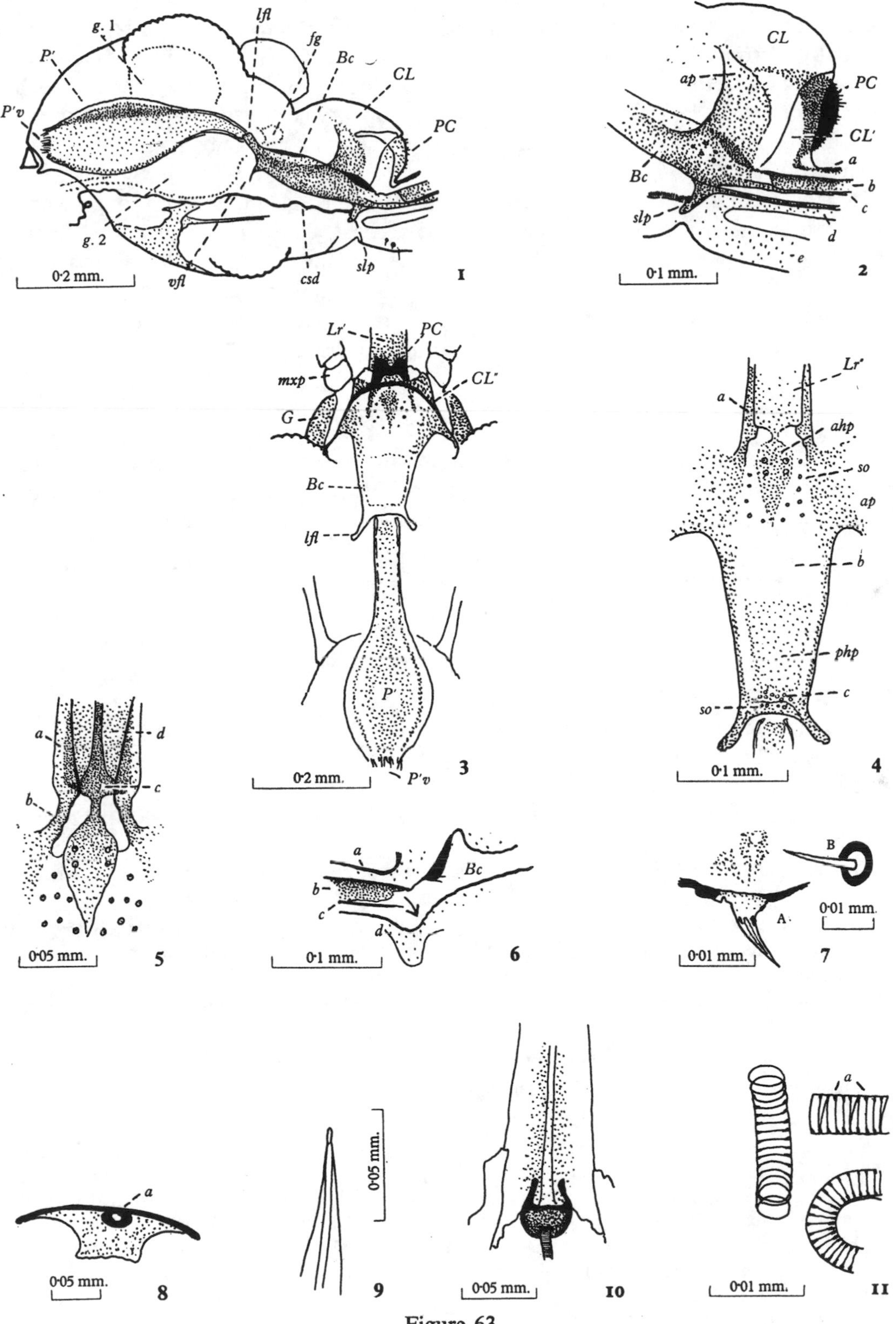
g. 1
lfl
fg
P′
Bc
CL
P′v
PC
g. 2
0·2 mm.
vfl
csd
slp
1
CL
ap
PC
CL′
Bc
a
b
c
slp
d
e
0·1 mm.
2
Lr′
PC
mxp
CL″
G
Bc
lfl
P′
0·2 mm.
P′v
3
Lr′
a
ahp
so
ap
b
php
c
so
0·1 mm.
4
a
d
c
b
0·05 mm.
5
a
Bc
b
c
d
0·1 mm.
6
B
A
0·01 mm.
0·01 mm.
7
a
0·05 mm.
8
0·05 mm.
9
0·05 mm.
10
a
0·01 mm.
11

Figure 63

in the alimentary canal proper. It should perhaps be pointed out that the labral canal is essentially a groove in the under surface of the labrum and that its floor is formed by the curled-in edges of this groove and not by the true floor of the alimentary canal, which in this position would be the dorsal aspect of the hypopharynx. At the gap between the bars, however, the floor is the dorsal aspect of the hypopharynx and the channel becomes the alimentary canal (Fig. 63 (6)).

The labral canal in most of its length is very nearly circular in section, though with the transverse diameter somewhat greater than the vertical, the diameters being respectively about 24μ and 18–20μ. Towards the apex of the labrum, where the dorsal and ventral plates have fused to form a hard point, the channel becomes somewhat narrower and basally it widens considerably reaching up to 30μ or more in transverse diameter (Fig. 64 (3)). The labral canal in the male is very similar to that in the female.*

THE MOUTH (Figs. 63 (6); 64 (7))

Where the labral canal terminates proximally would appear to be the true morphological *mouth*. Not only is this the level at which for the first time the alimentary canal becomes closed laterally, but the arrangement of the surrounding parts, allowing for the elongation,

* In a recent description (1957) of the mouth-parts M. L. Bhatia and B. L. Wattal (*Ind. J. Malar.* 11, 183–9) describe the presence of delicate septal rings projecting at intervals into the lumen of the labral canal (19–29 in the female; 26–33 in the male).

Figure 64. Alimentary canal.

1 Transverse section through region of the labella.
2 The same through labial portion of the proboscis.
3 The same through the extreme base of the proboscis.
4 The same through the tip of the clypeus.
5 Transverse section showing labral canal bounded below by the thin lateral plates of the labrum. *a*, labral canal; *b*, alimentary canal.
6 Transverse section a little posterior to 5 to show the lateral plates continued as rods with wide opening between. *a*, termination of labral canal; *b*, alimentary canal.
7 Transverse section at level of true mouth. *c*, anterior end of anterior hard palate; *e*, extensions of salivary pump and fold passing to region of mandibular socket (suspensorium).
8 Transverse section through anterior portion of buccal cavity. *d*, posterior end of anterior hard palate.
9 Transverse section through buccal cavity at level of membranous roof.
10 The same at level of posterior hard palate.
11 The same through posterior end of buccal cavity showing frontal ganglion.

Lettering: *ahp*, anterior hard palate; *ap*, apodeme; *Bc*, buccal cavity; *Bc'*, floor of buccal cavity; *CL*, clypeus; *CL'*, membranous area; *CL"*, line of clypeus; *csd*, common salivary duct; *fg*, frontal ganglion; *G*, gena; *g.*1, supra-oesophageal ganglion; *g.*2, sub-oesophageal ganglion; *hp*, hypopharynx; *L*, labium; *L"*, basal end of labium; *Lc*, labral canal; *Lr'*, dorsal plate of labrum; *Lr"*, ventral plate of labrum; *Lr'"*, lateral plate of labrum; *lb*, labellum; *lfl*, lateral flange of buccal cavity; *lg*, ligula; *ln*, labial nerve; *Mo*, mouth; *md*, mandible; *mx*, maxilla; *mxa*, maxillary apodeme; *mxp*, maxillary palp; *P'*, oesophageal pump; *P'v*, oesophageal valve; *PC*, preclypeus; *php*, posterior hard palate; *sd"*, salivary channel; *slp*, salivary pump; *sm*, salivary muscle; *so*, sensory pit; *t*, trachea; *vfl*, ventral flange of buccal cavity.

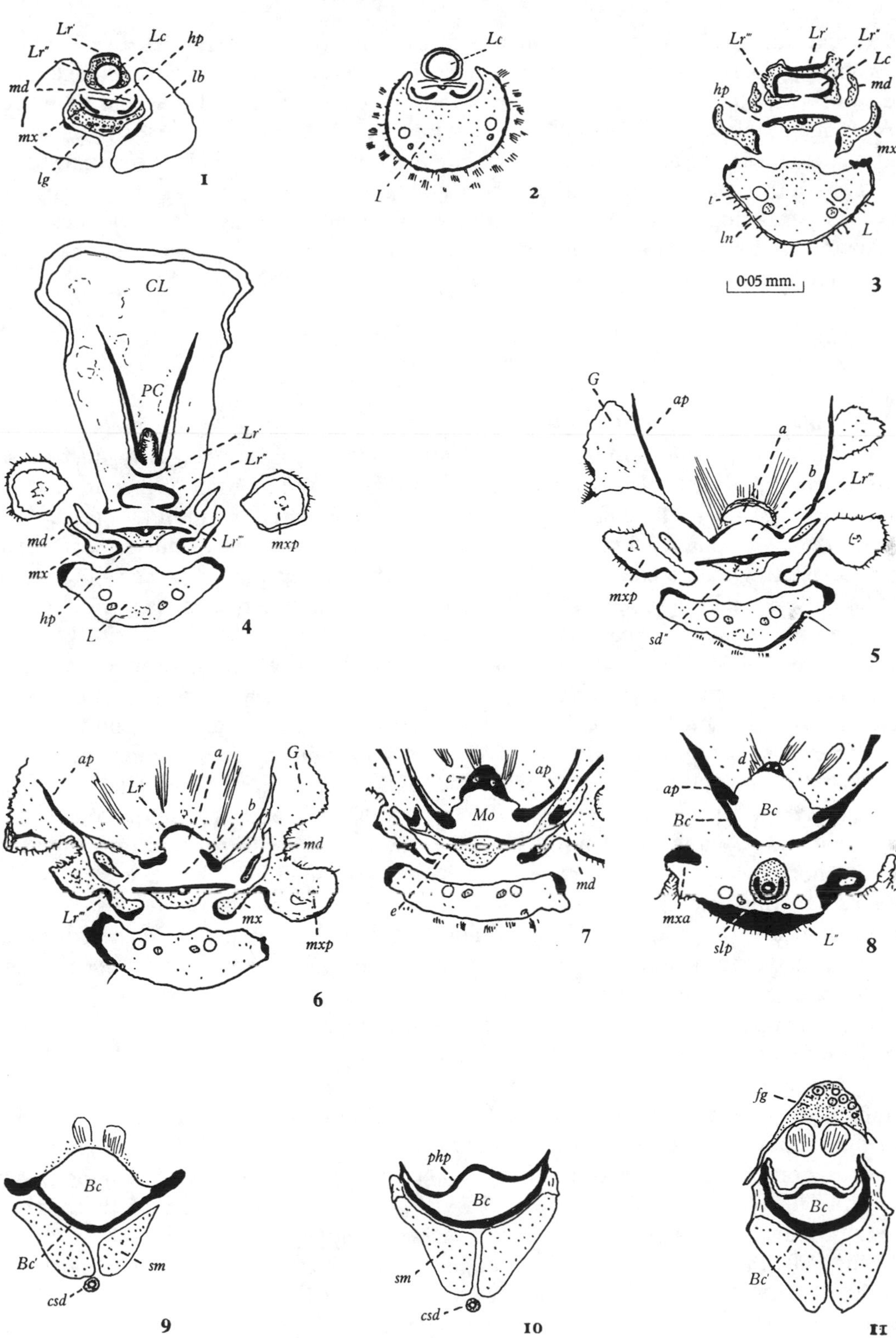
Lr'
Lr"
Lc
hp
lb
md
mx
lg
1
Lc
l
2
Lr'''
Lr'
Lr"
Lc
md
hp
mx
t
ln
L
0·05 mm.
3
CL
PC
Lr'
Lr"
md
Lr'''
mxp
mx
hp
L
4
G
ap
a
b
Lr'''
mxp
sd"
5
ap
a
G
Lr'
b
md
Lr'''
mx
mxp
6
c
ap
Mo
md
e
7
d
ap
Bc
Bc'
mxa
slp
L"
8
Bc
Bc'
sm
csd
9
php
Bc
sm
csd
10
fg
Bc
Bc'
11

Figure 64

has a great similarity to that already indicated as the mouth in the larva. There is the same transverse slit-like opening, dorsal to which lies the under surface of the base of the labrum and ventrally in the median line the base of the hypopharynx with what would have been the opening of the salivary duct in the larva were it not that changes have led to this being continued in the imago as a closed canal in the elongated hypopharynx. On each side of the slit-like mouth in their due positions the mandibles and maxillae take origin with the maxillary palps, no longer the tiny organs of the larva but relatively large structures. Very similar too to the condition in the larva is the fold passing from the hypopharynx to the now much reduced mandibles (Fig. 64 (7) *e*).

THE SALIVARY PUMP AND SALIVARY CHANNEL

(Figs. 63 (1), (2), (10); 64 (5))

The *common salivary duct* at the level of the base of the hypopharynx enters a heavily sclerotised structure, the *salivary pump* (*sd'*). From this a further length of duct passes down the median line of the hypopharynx to open at the termination of this organ. This latter portion of duct differs in character from that entering the pump in that it lacks the characteristic taenidial markings of the salivary and common salivary ducts. It is of late formation during the pupal stage and in some insects and possibly some mosquitoes may be wholly or partially an open gutter. To distinguish it from the true salivary ducts it may be termed the *salivary channel* (*sd"*).

The *salivary channel* in the female *Aëdes aegypti* is a thick-walled narrow duct extending from the salivary pump to the point of the hypopharynx where it opens subterminally, as described by Robinson in *Anopheles*, on the dorsal aspect. At its origin for a short distance it is somewhat dilated, narrowing gradually to a very uniform bore throughout. It lies immediately below the sclerotised upper surface of the hypopharynx, in contact and fused with this. A small, somewhat lighter, area may be seen in some sections in the cuticle lying over the duct, but so far as could be made out the duct is a closed canal throughout its length. Thompson speaks of the channel as the salivary gutter. But his figures show it as in *Aëdes aegypti*. Robinson's reference to a subterminal opening would also seem to suggest a closed canal in *Anopheles*. In female *Culicoides* Jobling (1928) describes the channel as a closed canal in the proximal third of the hypopharynx and an open groove for the remainder of the organ.

The *salivary pump* is situated at the base of the hypopharynx. It is a strongly sclerotised cup-shaped structure firmly fixed in the sclerotised base of the hypopharyngeal dorsal wall. The open end of the cup is directed posteriorly and closed by a thick but seemingly soft membrane through the middle of which the common salivary duct enters the cup's cavity. Into the membrane on each side is inserted the comparatively large *salivary muscle* of its side taking origin from the flange of the buccal cavity and lying beneath the floor of that cavity throughout much of its length. From the upper anterior aspect of the cup the salivary channel issues with on each side a thickened supporting spur (Fig. 63 (10)).

In the male there is an almost identical pump. But there is no separate hypopharynx and the salivary channel runs as a slightly sclerotised channel along the soft membrane of the labial gutter terminating near the tip of the ligula. Thompson figures the channel as an open groove without any special thickening of its walls. In *Culicoides* Jobling's figure appears to show it as a closed canal throughout.

ALIMENTARY CANAL

THE BUCCAL CAVITY (Figs. 63, 64 *Bc*)

The structure of the part here termed buccal cavity has been very fully described in *Anopheles* by Sinton and Covell (1927), whose nomenclature of its parts is here followed. The buccal cavity in a number of species and genera of Culicines, including that of *Aëdes* (*Stegomyia*) *albopicta*, has also been described and figured by Barraud and Covell (1928). The characters of the organ are of considerable systematic importance in *Anopheles*. But apart from such small differences as may distinguish some forms the general structure throughout the Culicini is very similar. The elaborate armature borne on the posterior border in many Anophelini is absent in the Culicini, including *Aëdes aegypti*.

This part of the intra-cranial alimentary canal has by many authors been termed pharynx (Jobling, 1928 (in Culicoides); Marshall, 1938; Martini, 1920). But though this term is clearly rightly applied to the large dilated pharynx of the larva, it is doubtful if it is a correct name for the part now under consideration in the imago. In the larva the chief bulk of the organ with its fringes lies posterior to the frontal ganglion. But in the imago only the most posterior portion is so situated (Fig. 63 (1)). According to Snodgrass (1935) the buccal cavity in insects is defined as the first part of the stomodaeum lying just within the mouth; its dilator muscles arising on the clypeus, and inserted before the frontal ganglion and its connectives. On this definition the greater part of the organ now being discussed would be buccal cavity. As will be seen later, muscles of the part are also in conformity with this view. On the other hand, Snodgrass (1943, 1944) homologises this organ in the mosquito as the *cibarium* (preoral cibarial pocket), a part of the canal situated anterior to the true mouth which he considers in the mosquito to have been carried forward by lateral extension forwards of the union of the clypeal borders with the edge of the ciberial surface of the hypopharynx. In respect to homology with the larval parts Thompson has very succinctly and clearly described the relation when he says: 'The valve and posterior hard palate regions are derived from the "pharynx" of the larva. The remainder of the pharynx and thc proximal parts of the proboscis canal are formed from the "buccal cavity" of this earlier stage.' It is clear that the parts in the mosquito imago have been very profoundly modified, making accurate homologies in relation to the organ as a whole difficult, and it has seemed most in accordance with the facts to retain the name buccal cavity as has been very largely done by systematists dealing with the Culicidae.

In general shape the buccal cavity is tubular, flattened dorso-ventrally, and, though in part membranous, is in thc main heavily sclerotised. It lies for most of its length ventral to, but at some distance from, the clypeus, extending from the mouth at the base of the proboscis to a short distance beyond the posterior border of the clypeus where it ends in two conspicuous diverging lateral processes, *lateral flanges* (Fig. 63 (3) *lfl*), and is linked between these by membrane to the anterior ends of the plates forming the oesophageal pump. Its roof is largely membranous with thickenings anteriorly and posteriorly. The anterior thickening, *anterior hard palate* (Fig. 63 (4) *ahp*) is situated in the middle line entirely surrounded by membranous roof, except anteriorly where it is linked by a narrow isthmus to the ventral plate of the labrum (*Lr″*). The anterior hard palate is rather irregularly fusiform, shaped, as described by Nuttall and Shipley, somewhat like a trowel with the pointed tip directed backwards. It is heavily sclerotised and carries on each side two sensory pits with short spines. Lateral to it on each side on the membrane are six more pits arranged as shown in the figure (Fig. 63 (4), (5)). Posterior to the anterior hard palate, is an extensive membranous area into which muscles are inserted and behind this, occupying about the posterior third of the organ, is a more indefinite thickening with its lateral margins forming indefinite bar-like thickenings, *posterior hard palate* (*php*). This normally lies somewhat sagged below the level of the lateral borders of the cavity. It terminates shortly before the end of the organ in membrane with small hummocky thickenings (Fig. 63 (4) *c*).

The floor is uniformly thickened extending from the base of the hypopharynx with the salivary pump to a thickened bar forming its posterior border, which in *Aëdes aegypti* carries on each side of the middle line two closely approximated small sensory pits (*so*). Laterally its thickened border, which is continued backwards from the thickened bars previously described at the base of the labrum, is continued into the lateral flanges and from across its ventral aspect towards its posterior end arises the flange-like projection, *ventral flange* (Fig. 63 (1), *vfl*), from which the salivary muscles take origin.

For about their anterior third the thickened lateral borders are continuous with the broad apodematous sheets that extend downwards from the sides of the clypeus. These pass between the partially membranous genae and the median rostrum. Their posterior borders are concave and well defined. But anteriorly they merge into the membranous area separating the preclypeal portion of the rostrum from the clypeal as described in the section on the head capsule (Fig. 63 (2)).

Numerous muscles take their insertion at different parts of the organ, notably along the length of the roof coming from the clypeus and on the lateral flanges coming from the maxillary apodemes. These will be described later when dealing with the muscles of the head. Lying beneath the floor of the cavity are the salivary muscles. Above the roof near the posterior border is the frontal ganglion.

The buccal cavity in the male does not differ materially from that in the female.

THE OESOPHAGEAL PUMP (Fig. 63 (1), (3) *P′*)

The oesophageal pump is shaped rather like the bulb of a Higginson's syringe with a bulbous posterior portion and a thin forward extension. The swollen portion is situated posteriorly to the brain, the narrow anterior portion occupies the narrow passage left in the imago between the cerebral lobes and the sub-oesophageal ganglion. The organ consists of three plates, one dorsal and two ventro-lateral. The plates are wide and fusiform in the bulbous portion, becoming narrow and elongated to form the narrow extension. They are all very similar in shape, with a thickened area medianly and thinner borders that are curved outwardly where they are connected by membrane. The curled edges are continued along the narrow portion, giving a curious effect in section to the anterior part of the organ. It is clear from the appearance of the plates in section that the organ must be capable of considerable expansion when these are dragged apart by the powerful muscles inserted into them. These are the dorsal and ventro-lateral dilator muscles taking origin respectively from the occipital and latero-ventral regions of the head capsule. At the anterior end of the narrow portion the area between the two ventro-lateral plates forming the floor is, as noted by Thompson, flat. It is also somewhat sclerotised so that it almost forms a narrow fourth plate.

Posteriorly each of the plates ends in a somewhat thinner manner just before continuing into the soft walls of the oesophagus. On this area there arise backwardly projecting hairs forming what amounts to a valve, the *oesophageal valve* (Fig. 65 (5)). As figured by Nuttall and Shipley in *Anopheles maculipennis* the hairs appear as tufts arising from circular pale depressions. In *Aëdes aegypti* a certain number of tufts arise from the raised margins of circular depressions. Others arise in tufts from less definite thickenings. The hairs are of considerable length and for the most part have a sinuous course. Just beyond the valve there is a marked annular muscle forming a sphincter.

The organ in the male is essentially similar, but with the anterior narrow portion longer and somewhat thicker relatively to the posterior part than in the female. The bulbous posterior part is also somewhat smaller even allowing for the smaller size of the male mosquito. Thompson has estimated that the female pump in *Culex stimulans* has a capacity of 0·002–0·004 mm.^{3} and that of the largest male a capacity of 0·0008 mm.^{3}. In *Aëdes aegypti* the female bulbous portion measures about 0·3–0·35 mm. in length and the narrow portion measures in a straight line about 0·18 mm. The corresponding measurements in the male are 0·18 and 0·20 mm.

The oesophageal pump has been termed the 'sucking pump' or antlia by Thompson; pharynx by Sinton and Covell; oesophageal pump by Jobling in *Culicoides*; and posterior pharynx by Snodgrass (1935), and by Robinson (1939) in *Anopheles*. Most authors recognise it as a modified portion of the oesophagus.

OESOPHAGUS, DIVERTICULA AND CARDIA (Fig. 65)

The oesophagus is a short soft-walled muscular tube passing between the oesophageal pump and the cardia. Anteriorly, following upon the oesophageal valve, it is narrow and encircled by a muscular band forming an annulus. Posteriorly it becomes more dilated and in sections appears either distended and somewhat cone-shaped with a wide base at the cardia or thrown into longitudinal folds. It may in the latter case give the appearance of forming a ventral pouch. Its walls consist of an inner layer of epithelium which may appear columnar (especially at the narrow anterior end), cubical or flattened depending on the degree of distension. There are spindle-shaped circular muscular fibres externally with some longitudinal fibres especially towards the posterior end. In dissections it rarely comes away with the other viscera.

At the posterior end of the oesophagus just anterior to its junction with the cardia are the openings of the diverticula. The dorsal diverticula are relatively small with narrow openings and directed anteriorly in the median space between the dorso-ventral and under the longitudinal indirect wing muscles. From their position it has been thought that they may yield up glucose from their contents to these muscles, so providing extra energy for flight (Hocking, 1953), though the impervious nature of their walls would seem against such a view. The ventral diverticulum is very much larger. It lies in the middle line ventral to the mid-gut and dorsal to the ventral nerve chain. It varies much in size, depending upon its air content, and when distended may reach as far back as the sixth abdominal segment and if excessively inflated may cause the whole abdomen to be swollen as after a blood meal. It plays an important role as already described in relation to the blood feed, but is equally large in the male where blood is not taken.

Usually both the ventral and dorsal diverticula appear as transparent lobulated sacs filled with bubbles of air of various sizes (Fig. 65 (1)) or, if collapsed, as solid wrinkled organs. Their walls consist of thin transparent membrane which is highly elastic and, except when fully distended, is thrown into innumerable small folds. In fresh preparations this plicated appearance is very characteristic and may simulate sporozoites seen in mass. In sections the walls are seen to consist of an inner layer of epithelium with small nuclei and an outer elasto-muscular layer. That this outer layer is at least in part muscular is shown by the fact that when distended with sugary fluid and examined in salt solution the diverticula may, as noted by Nuttall and Shipley, exhibit peristalsis. The microscopical appearances also indicate the presence of muscular fibres. These consist of rather broad flat bands passing circularly around the sac and ending in branches consisting of small isolated fibres. The bands are spaced at regular and rather considerable intervals and are arranged so that the branched ends and unbranched thicker parts alternate (Fig. 65 (3)). Nuttall and Shipley give an instructive figure showing a diverticulum dissected out after a full meal on sugary fluid, the organ allowed to dry externally and stained with fuchsin. The fibres noted above are slightly illuminated under polarised light, but not strikingly as are skeletal muscle fibres, a character they share with muscles of the mid-gut and some other structures, for example the heart. Besides the terminal branches there are numerous fine cross fibres linking the muscular bands. According to Thompson these are narrow folds in the membrane, but though such folds may be in part present these cross bands appear to be largely muscular in nature forming with the larger fibres an arrangement not unlike a spider's web (Fig. 65 (4)).

A peculiar character of the diverticula commented upon by Nuttall and Shipley is the extreme imperviousness of their walls to passage of water. Dissected out on a slide and allowed to dry at room temperature these authors found that the diverticula of *Anopheles maculipennis* still retained fluid after some months.

The diverticula are equally present in the male and appear if anything even proportionately somewhat larger.

At the junction of the oesophagus with the mid-gut is the structure here termed *cardia.*

Figure 65. Alimentary canal.

1 Alimentary canal as displayed when drawn out in dissection.

2 Mid-gut as flattened for examination, showing tracheal branches from the abdominal spiracles and scalloping of edges due to constriction by circular muscle fibres.

3 Portion of flattened diverticulum showing muscle fibres as shown by polarised light.

4 Thin portion of wall of diverticulum showing nuclei of epithelial layer and branched muscle fibres.

5 Showing arrangement of backwardly directed hair tufts of oesophageal valve.

6 Proventriculus (cardia) of male in coronal section. That of the female is essentially similar but somewhat larger.

7 A, section of wall of mid-gut in stretched condition (from newly emerged female with distended gut); B, epithelium of mid-gut as usually seen; C, the same in tangential section.

8 Pyloric valve and ampulla. *a*, thickening of wall of mid-gut around opening; *b*, muscular sphincter; *c*, fold in wall of ampulla; *d*, muscle fibres extending from wall of ampulla to mid-gut.

9 Spines on intima of ampulla.

10 A, Transverse section of ileum showing cubical epithelium. B, longitudinal section of junction of ileum with rectum.

Lettering: *am*, pyloric ampulla; *Ca*, cardia; *Dd*, dorsal diverticulum; *Dv*, ventral diverticulum; *IL*, ileum; *Mg*, mid-gut; *Mt*, Malpighian tubule; *Oe*, oesophagus; *ov*, ovary; *Rc*, rectum; *spm*, spermathecae.

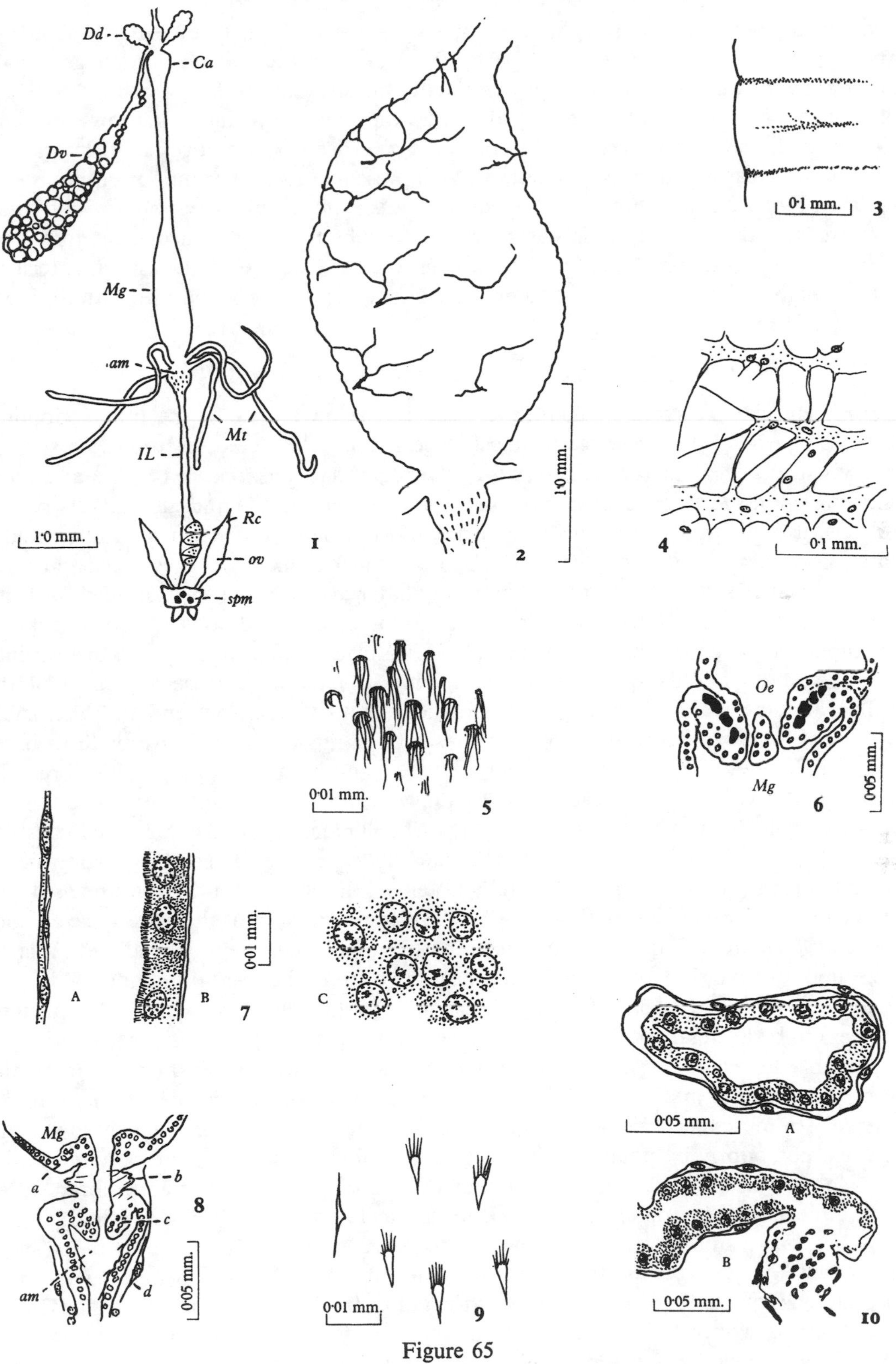
Dd
Ca
Dv
Mg
am
Mt
IL
Rc
ov
spm
1·0 mm.
I
1·0 mm.
2
0·1 mm.
3
0·1 mm.
4
0·01 mm.
5
Oe
Mg
0·05 mm.
6
A
B
0·01 mm.
7
C
Mg
a
b
c
am
d
0·05 mm.
8
0·01 mm.
9
0·05 mm.
A
B
0·05 mm.
10

Figure 65

This is a thickened portion at the anterior end of the mid-gut formed partly from the cardiac portion of the mid-gut and partly from invagination of the oesophageal walls (Fig. 65 (1), (6) *Ca*). It has, therefore, essentially the structure of the so-called proventriculus of the larva though less developed and lacking the blood sinus. It shows the same folds with the recessed angles which in the larva give rise to the peritrophic membrane, a structure which as noted in the next subsection if present at all in the adult mosquito is seen only after a blood feed. Though in the mosquito it is commonly termed the proventriculus it would seem that this organ should more correctly be termed the cardia, the term proventriculus being, according to Snodgrass, applied to a structure situated more anteriorly and especially well developed in mandibulate insects where it may be supplied with chitinous teeth and form a gizzard.

MID-GUT, PYLORIC AMPULLA AND ILEUM (Fig. 65)

The *mid-gut* extends from the cardia situated in the prothorax to the fifth abdominal segment, or when fully distended to the sixth segment. All trace of gastric caeca so conspicuous in the larva at the cardiac end of the organ have disappeared in the adult and following upon the small swollen portion forming the cardia the mid-gut consists of two portions, namely an anterior long and narrow portion lying mainly in the thorax and a posterior swollen portion situated in the abdomen which alone undergoes distension when a blood meal has been taken or, in the newly hatched insect, when it is filled with air (Fig. 65 (1) *Mg*). The anterior portion remains under all conditions a narrow tubular structure. The posterior portion when empty is fusiform in shape and usually thrown into some longitudinal folds. When distended it may be ballooned to occupy most of the swollen abdomen with its walls greatly thinned. Both the narrow and the distensible portions are very similar in structure and consist of an inner epithelial layer with its basement membrane on which there are sparse circular and longitudinal muscular fibres. In addition, externally there are tracheal cells from which numerous tracheoles pass on to the surface of the organ. The epithelium appears as columnar, cubical or flattened cells depending upon the condition of the organ, that lining the anterior narrow portion being usually uniformly cubical whilst when fully distended the epithelium may be stretched and flattened to an extreme degree (Fig. 65 (7)). On the inner aspect of the cells is the characteristic *striated border*, the degree to which the striations are evident depending to a large extent upon the fixation and staining, but possibly also on the condition under which the organ is examined (Fig. 65 (7) *b*). The muscle fibres of the outer layer are not conspicuous and seem to be less marked than in the mid-gut of *Anopheles*. They consist of circular and longitudinal fibres which form an open rectangular network. According to Grassi the muscle fibres in the mid-gut of *Anopheles* appear to be embedded in the basement membrane (elasto-muscular tunic of Grassi). They are striated but not conspicuously so and are demonstrated poorly by polarised light. No peritrophic membrane corresponding to the conspicuous cylinder in the larva is normally present; but formation of a membrane after a blood meal has been described by Jaguzhinskaya (1940) and by Rajindar Pal (1943), in *Anopheles*, and a similar membrane may be seen in *Aëdes aegypti* limiting the blood mass after a feed. Whether this is strictly comparable with the peritrophic membrane and how it is secreted requires further study. The mid-gut of the male is similar to that of the female except that the swollen portion is much less marked. It has a well-marked cardia (Fig. 65 (6)).

The *pyloric ampulla* is a short bulbous thin-walled portion of the hind-gut following immediately upon the mid-gut (Fig. 65 (1) *am*). Into it open the five Malpighian tubules, one dorsal and two on each side, one dorsal and one ventrally situated. The opening from the mid-gut is narrow and around it the epithelium of the mid-gut is thickened to form a fold. The ampulla shows a still more pronounced fold. Between these two folds is an area of circular muscle forming a strong sphincter (Fig. 65 (8) *b*). Just distal to the sphincter the Malpighian tubules enter. Where these tubules enter, the characteristic cells of the tubules are replaced by a mass of cells with small nuclei, the tissue at the point of entry forming a posteriorly projecting mass. The whole forms a very marked valve. The walls of the ampulla consist of an inner small-celled epithelial lining and in its more distal part a layer of circular muscle fibres continued from the ileum. Proximal and distal to the valve the walls are very thin and almost devoid of any muscular layer. Longitudinal fibres pass from the distal parts of the wall to the mid-gut often at some distance from the sphincter opening (Fig. 65 (8) *d*). The epithelium has a fine cuticular lining which carries backwardly projecting spinous processes (Fig. 65 (9)). These have been referred to by a number of observers (Eysell, 1905; De Boissezon, 1930*a*; Richins, 1938; Wigglesworth, 1942; Trembley, 1951). They give the ampulla as seen in dissections a characteristic appearance (Fig. 65 (2)). They extend also some little way along the ileum, the passage from ampulla to ileum not being well defined. The spines are not unlike those seen on the larval cuticle in some situations, namely a thorn-like base which is continued into from four to six fine spines projecting in a horizontal plane.

From the pyloric ampulla to the rectum is a portion of the hind-gut sometimes termed ileo-colon. It, however, no longer shows the marked distinction into ileum and colon as seen in the larva, and as noted by Thompson the colon of the larva is replaced by growth from the ileum and rectum during metamorphosis. The term *ileum* is therefore preferable, though some change in character does occur in its passage to the rectum, the epithelial cells tending to be more massive in its hinder portion. It is, throughout its course, a narrow tube and very heavily supplied with muscular bands frequently showing active peristalsis when dissected out in saline. After leaving the ampulla it is at first directed posteriorly and somewhat ventrally. It then forms a loop directed dorsally which terminates in the rectum close beneath the seventh tergite. Its walls consist of an inner layer of rather large cubical epithelium which is thrown into longitudinal folds and an outer muscular layer consisting of more or less circularly disposed broad and well-striated muscular fibres. Longitudinal fibres seem to be present, but the actual relation of the fibres is difficult to determine with certainty. Very frequently a circular fibre is seen tending to become longitudinal, and it is possible fibres may be more or less spirally arranged. At the junction of the ileum with the rectum the swollen epithelial cells of the former become replaced by the small, flattened epithelium of the latter, the contrast being very marked (Fig. 65 (10) *b*).

THE RECTUM AND RECTAL PAPILLAE (Fig. 66 (1–4))

The *rectum* consists of two portions of about equal length, a dilated proximal portion, *rectal sac* (*Rc′*) situated in the seventh abdominal segment and in which the rectal papillae are situated, and a distal narrow portion extending to the anus. The walls of the rectal sac consist of a thin inconspicuous epithelium and an outer layer of discrete broad bands of striated muscle fibres (Fig. 66 (1) *m*). The distal portion of the rectum has a thin epithelial layer with a much folded and wrinkled intima and an outer layer of muscle bands similar though shorter than those of the rectal sac. Towards its termination, besides numerous muscular fibres in its coat, there are fibres of skeletal type taking origin from surrounding parts and inserted into the tube. The anal opening is in the membrane between the bases of the cerci. Small spicules may be present on the intima.

The *rectal papillae*, sometimes called the rectal glands, are in the female six in number. They consist of conical papillae each formed of a number of large cells with clear cytoplasm and large nuclei arranged around a central core-like space. They take origin from the dorsal wall of the rectal sac, one posteriorly, two pairs side by side medianly and one situated anteriorly (Fig. 66 (2)). Their bases lie close beneath the posterior end of the heart and are freely exposed to the body cavity, the gut wall being attached around their base and not extending over this (Fig. 66 (3)). A large tracheal branch on each side comes from the spiracle complex of the seventh segment and, passing behind the rectal sac, gives off leashes of branches some of which enter and pass up the core-like centres of the papillae to break up into a network of tracheoles in the substance of the papillae (Fig. 66 (4)). The whole complex has a certain resemblance to the colon of the larva with its large clear cells and to the tracheal network of the larval anal papillae.

In the male the rectal parts are very similar to those in the female but, as noted by Engel (1924), only four rectal papillae are present. These are arranged as in the female, but there is only one median pair.

THE SALIVARY GLANDS AND DUCTS (Fig. 66 (5–9))

The correct name for such glands in general insect terminology should be *labial glands*. But the name salivary glands has been so largely employed that it seems undesirable to use another name, especially as they are the only glands in the mosquito that could merit the name salivary.

Figure 66. Rectum and salivary gland.

1 Sagittal section through rectum showing rectal papillae arising from dorsal wall of the rectal sac. The section shows relation to segmental parts and to pericardial cells and heart.

2 Coronal section through rectal sac showing position of the six rectal papillae in the female. In the male only one pair is present medianly.

3 Section through rectal papilla. *a*, Base of papilla exposed to body cavity; *b*, wall of rectal sac attached to periphery of the papillary base with a nucleus in the papilla at this point; *c*, position where tracheal vessels break up to form tracheoles passing into papilla.

4 Rectal papilla showing tracheoles. *a*, position where large tracheal branches break up to give off tracheoles supplying the papilla.

5 Transverse section through anterior portion of thorax to show anatomical relations of the salivary acini. The three acini on each side are shown in black.

6 Salivary glands of *Aëdes aegypti* as seen in dissection. *a*, proximal portion of lateral acinus; *b*, lateral acinus; *c*, median acinus; *d*, anterior portion of median acinus.

7 A, transverse section of acinus in newly emerged female; B, the same of male acinus.

8 Portion of acinus of 24-hour-old female showing beginning of formation of secretion.

9 Transverse section of mature acinus distended with secretion. From an ovipositing female.

Lettering: *asd*, intra-acinar duct; *Dv*, ventral diverticulum; *ht*, heart; *Mg*, mid-gut; *m*, muscle fibres; *nu*, nucleus; *pc*, pericardial cell; *Rc′*, rectal sac; *rp*, rectal papillae; *sd*, salivary duct; *sec*, secretion; *sg*, salivary gland; *t*, trachea; VII, VIII, abdominal segments.

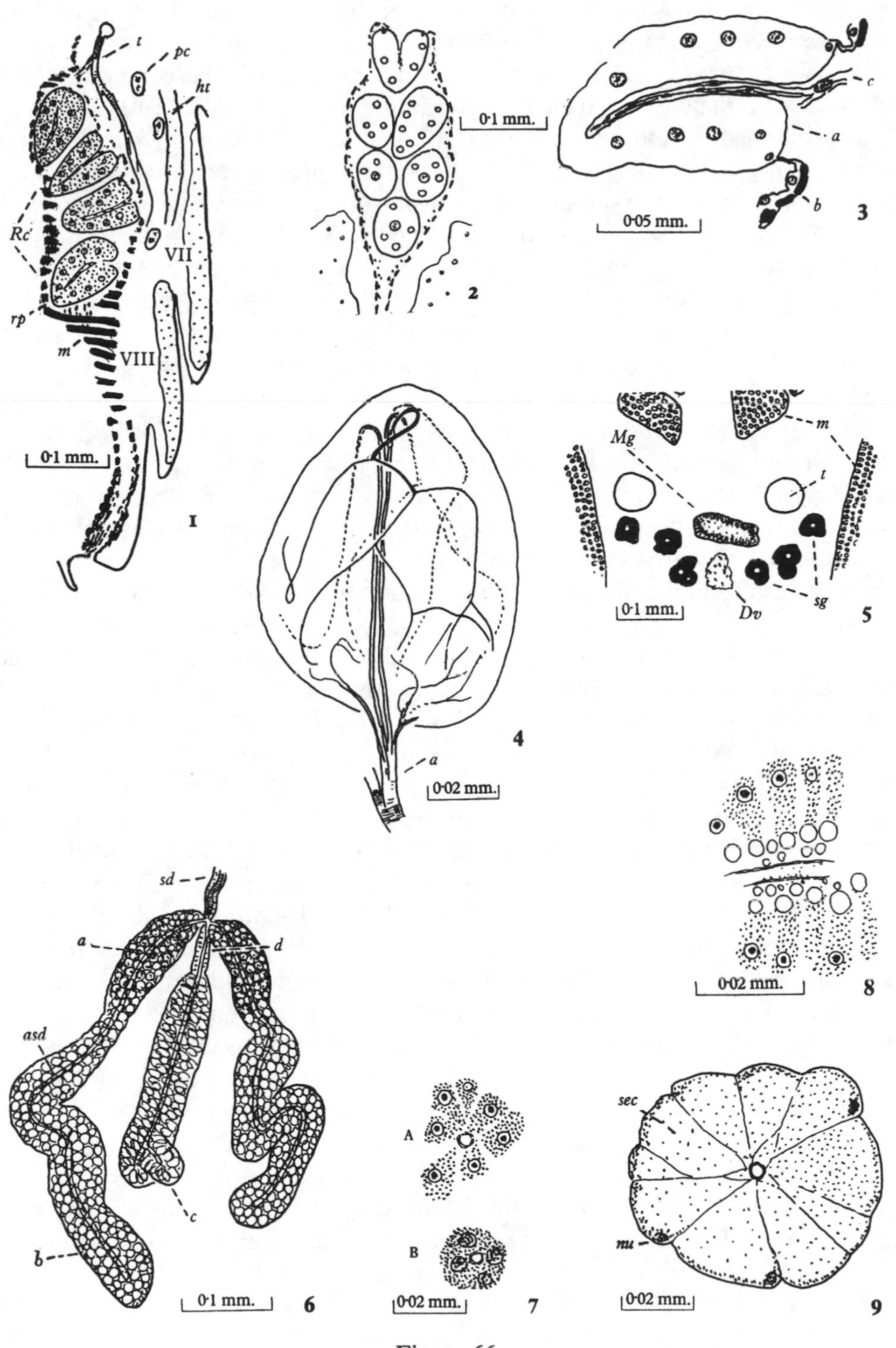
t
pc
ht
Rc
rp
m
VII
VIII
0·1 mm.
I
0·1 mm.
2
c
a
b
0·05 mm.
3
a
0·02 mm.
4
m
Mg
t
sg
Dv
0·1 mm.
5
0·02 mm.
8
sd
a
d
asd
c
b
0·1 mm.
6
A
B
0·02 mm.
7
sec
nu
0·02 mm.
9

Figure 66

The glands of *Aëdes aegypti* resemble in general those of *Anopheles*, but are smaller and rather more stumpy. They are readily dissected out by the technique in general use as previously described. Their appearance so dissected is shown in Fig. 66 (6). The glands of each side consist of three lobes or *acini*, a *median* and two *lateral*. These are linked at their anterior ends by the junction of their ducts forming the *salivary duct* of its side (*sd*). The glands lie surrounded by and largely embedded in lobes of the fat-body close behind the prosternum, their anterior ends close to the base of the neck through which the salivary ducts pass to unite beneath the sub-oesophageal ganglion to form the *common salivary duct* (*csd*). The acini, coiled somewhat irregularly, extend backwards to lie alongside and close to the hinder end of the oesophagus and cardia and for a short distance by the side of the anterior portion of the mid-gut. The three acini lie in dorso-ventral series with the upper lateral acinus somewhat external to the others and the lower lateral acinus closer to the middle line, where its termination comes to lie close up to the ventral diverticulum (Fig. 66 (5)).

Each acinus is in the form of an elongated cylinder in the centre of which is a duct with thick cuticular lining which may have taenidia much resembling those of tracheae, *intra-acinar duct* (*asd*). In *Anopheles* the median acinus is shorter, more sacculated and the intra-acinar duct markedly dilated. In *Aëdes aegypti* and most Culicini the median acinus, though shorter than the lateral acini, is not ampullated and, as usually described, has the intra-acinar duct of uniform diameter and extending to the end of the acinus. Recently Bhatia, Wattal and Kalra have studied the intra-acinar ducts in different species of mosquitoes including *Aëdes aegypti*. In *Anopheles*, in the female, the intra-acinar duct of the median acinus dilates immediately after entering the lobe, its walls losing their taenidia. In the lateral lobes the intra-acinar ducts after running for about half the length of the acinus dilate somewhat and towards the end of the lobe may become invisible. As in the median acinus the ducts in the lateral acini lose their taenidia on entering the lobe. In *Culex fatigans* the intra-acinar ducts in the female are narrow with taenidial lining throughout their length in all three acini. In the female in *Aëdes aegypti* the duct of the median acinus is somewhat dilated and without taenidia when it enters the lobe. In the lateral acini taenidial rings surround the ducts for about half their length, their distal half being slightly dilated and without taenidia. In the male *Anopheles* the intra-acinar duct of the median acinus passes to the end of the acinus without forming an ampulla. In the males of *Culex fatigans* and *Aëdes aegypti* the ducts in all three acini are narrow and with taenidia to their terminations. In *Culex fatigans* male the median acinus is shorter than the lateral acini. In *Aëdes aegypti* male the three acini are about of equal length. The authors point out that these and other differences may be of great value in their bearing on the systematics of mosquitoes.*

As usually depicted the three intra-acinar ducts are shown meeting and joining up at one point. Bhatia *et al.*, however, find that this is unusual and that commonly the duct from the median acinus joins with one of the lateral acinar ducts and this common portion shortly afterwards joins up with the other lateral duct.

Though in *Aëdes aegypti* the median acinus does not differ from the lateral acini to the same extent as in *Anopheles*, the distinction is quite apparent in that the median acinus is shorter and lacks the swollen anterior portion at the proximal end of the lateral acini. In

* The authors note that dissecting out the glands in 0·68 per cent saline and mounting them in 60–70 per cent glycerine gives better results than the conventional fixatives, stains and other mounting media.

Anopheles this swollen proximal portion differs considerably in structure from the rest of the acinus. In the freshly dissected gland it is usually more bulky than the remainder of the acinus, though in salt solution it is apt to become shrunken and narrower. It also shows other differences as noted by Christophers (1901), including the presence of small extensions from the duct into the cells. In *Aëdes aegypti* the differentiation of such an anterior portion is also apparent as indicated in Fig. 66 (6) *a* by greater bulk, by being separated by some degree of constriction from the rest of the acinus and by somewhat darker staining. In the median acinus this portion seems to be represented by a small and narrow rather conical portion of gland lying between the swollen ends of the two lateral acini (Fig. 66 (6) *d*).

In structure each acinus consists of a single layer of cells disposed around the central duct. The appearance of these differs depending on the age of the mosquito. Immediately following emergence the cells are small, deeply staining, coarsely granular and with a centrally situated nucleus. Usually there are considerable clefts between contiguous cells. The nucleus is noticeable as having an unusually large central deeply stained body. By 24 hours clear vacuoles have formed occupying the inner portions of the cells and in the lateral acini especially there may be numerous small globules disposed around the duct. By the second day the vacuoles of secretion have largely filled the cells, the cytoplasm and flattened nucleus being pushed to the cell periphery. In later stages the cells are greatly distended with secretion, the acini being now much swollen, strongly refringent and with a noticeable rigidity. If crushed the glands in the fresh state exude massive quantities of oily-looking secretion, leaving the gland substance relatively transparent and inconspicuous.

A feature of the glands noted by Shute (1940) is the frequent presence of small accessory outgrowths or duplication of the normally cylindrical acini. This seems to be particularly the case in *A. aegypti* where most specimens show some degree of this condition, the glands shown in the figure having been selected from a number to avoid a too excessive degree of aberration.

A study of the development of the salivary glands is given by Jensen and Jones (1957).

The *salivary* and *common salivary ducts* consist of a thick chitinous intima with an outer layer of somewhat flattened cells. The intima carries raised cuticular thickenings similar to, but more pronounced and thicker than, the taenidial markings of tracheal vessels. I have not been able to ascertain whether these are distinct rings or whether the thickenings are spiral. Usually, however, about every third or fourth band there is an oblique line of thickening joining two adjacent bands (Fig. 63 (11)).

(*b*) THE MUSCULAR SYSTEM

The musculature of imaginal insects is extremely complex and in this respect the mosquito is no exception. It is much more complex in the imaginal mosquito than in the larva, and in the imago it appears to be composed of two elements: (1) the original body muscles (of Snodgrass and others) as seen in the larva and continued or reconstituted in the imago; and (2) newly formed muscles, the result of development of elements in the imaginal buds or of other newly formed muscles in the imago. The muscles of these two systems differ markedly in a number of respects and may for convenience of description be referred to for want of better terms as those of the protomorphic and neomorphic muscular systems respectively.

Muscles of the protomorphic type closely resemble the larval body muscles in that they tend to be long and often thread-like, whereas the neomorphic type are most commonly massive or fan-like. Very marked too is the difference in the appearance of the two types in fixed and stained sections. Muscles of the protomorphic type are pale, tend to show indefinite structure with wide sarcolemma and few nuclei. They usually show poor striation. Muscles of the neomorphic type on the contrary usually stain much more deeply and show marked structural characters. Thus the muscles of the neomorphic type commonly consist of tubular-form fibres plentifully supplied with nuclei and markedly striated. The neomorphic type also for the most part take origin from points or areas on the various cuticular sclerites; the protomorphic strands on the other hand tend to pass from intersegmental line to intersegmental line or, where not at the surface, to pass between muscular knots as have been described in the larval musculature.

In the head there is little evidence of the early type, but one important muscle, the tentorially inserted flexor of the head, would appear to belong to the series. In the thorax this series is wholly subordinated to the new type of muscle, but is nevertheless still present as strands of muscle lying above and below the alimentary canal and associated with ventral transverse muscles as in the larva. The musculature of the abdomen in the imago is almost purely protomorphic and closely resembles that of the larval abdomen.

The musculature of the head has been that mostly studied, since it possesses a special interest in relation to the mechanism of feeding and in some other respects. Also the number of muscles is not too great to make a close study of each possible. In the present account therefore the muscles of the head have been numbered and described seriatim and named according to their functions. The thoracic muscles of the mosquito imago have not, so far as I am aware, been described as a whole and their number is very great, so that

---

Figure 67. Muscles of the head.

1 Longitudinal view of head of imago showing the musculature. *a*, frontal ganglion; *b*, scape of far side; *c*, annular muscle of oesophagus.

The muscles, indicated by their numbers as given in the text, are shown as though seen in a bisected head viewed from the right side. With a few exceptions they are therefore those of the left side. For clearness a few are shown as if to the right of the median line, namely nos. 8, 18 and 24. The tentorium and maxillary apodemes also are those of the right half of the head.

2 Intrinsic muscles of the antenna as seen in coronal section at the level of the base of the pedicel. *a*, scape; *b*, base of pedicel showing insertion of muscles; *c*, antennal nerve.

3 Coronal section through the clypeus to show musculature. *a*, invaginated preclypeal edges (preclypeal apodeme); *b*, position of frontal ganglion.

4 Coronal section at level of junction of buccal cavity and oesophageal pump. *a*, posterior end of buccal cavity; *b*, lateral flange of buccal cavity.

5 Reconstruction showing anterior end of the tentorium with arrangement of muscles taking origin from it. It is as would be seen on an enlarged scale in the first figure were it not hidden by the hind end of the buccal cavity.

6 Transverse section through apical portion of proboscis to show the labial musculature. *a*, sclerotised tendons. Both the extensor and flexor possess tendons.

7 Apical half of labium to show the extensor and flexor muscles. Only one of each is shown on their respective sides. The appearance of the muscular bands is that seen under polarised light. The tendons do not show up so illuminated.

Lettering: *Bc*, buccal cavity; *G*, gena; *Gs*, genal shelf; *md*, mandible; *mx*, maxilla; *mxa*, maxillary apodeme; *P′*, oesophageal pump; *ss*, suspensorium; *T*, tentorium; 1–26, muscles as given in list of head muscles in the text.

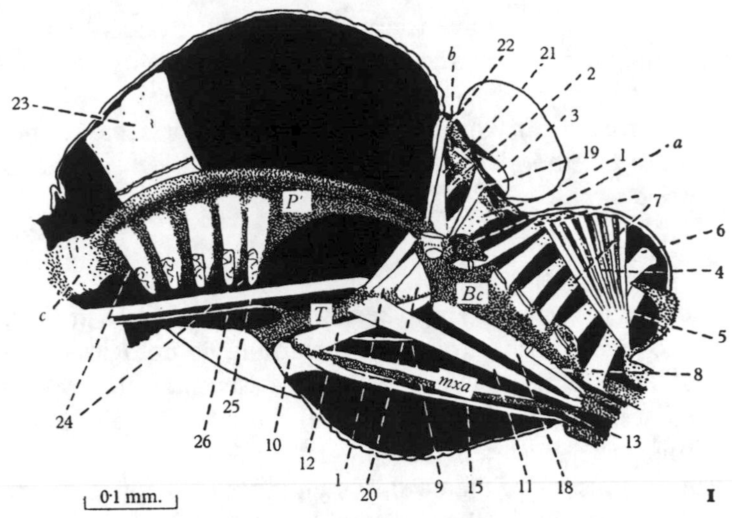

23
b
22
21
2
3
19
1
a
7
6
P'
4
Bc
T
c
5
8
mxa
24
25
26
10
12
1
20
9
15
11
18
13
0·1 mm.
1

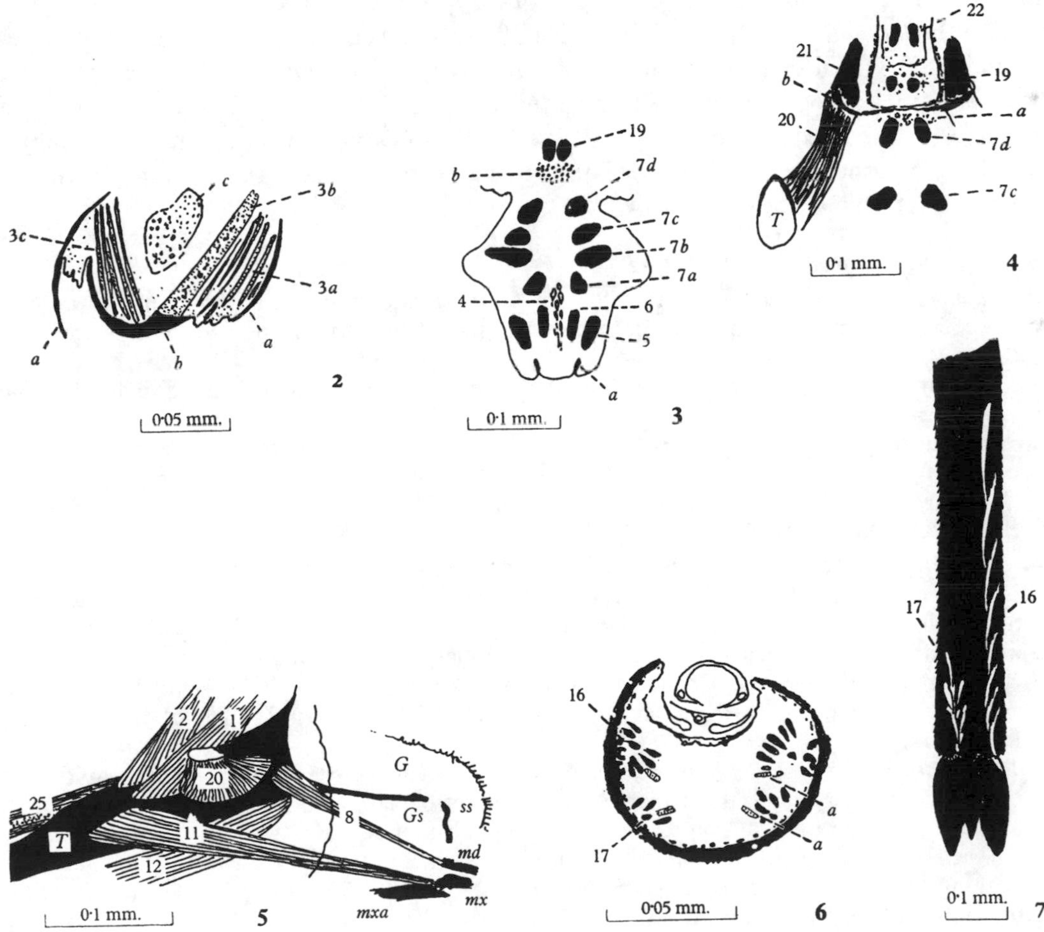

22
21
19
b
a
20
7d
T
7c
0·1 mm.
4
19
b
7d
7c
7b
7a
4
6
5
a
0·1 mm.
3
c
3b
3c
3a
a
b
a
2
0·05 mm.
17
16
16
a
a
17
0·05 mm.
6
0·1 mm.
7
2
1
20
G
25
8
Gs
ss
T
11
md
12
mx
mxa
0·1 mm.
5

Figure 67

some form of classification and nomenclature is essential to avoid confusion. The same applies to the abdominal muscles where, however, Samtleben's system serves for their naming.

It will be most convenient in what follows to deal under separate headings with the muscles of the head, neck, thorax and abdomen, using such descriptive methods as seem suitable in the particular case.

MUSCLES OF THE HEAD (Fig. 67)

The muscles of the head of the imago have been very fully described in *Culex* by Thompson (1905), who has also given much information regarding homologies of the imaginal with the larval muscles. The muscles in *Anopheles*, more especially those of the mouth-parts, have also been well described by Robinson (1939). Very helpful also is the description by Jobling (1928) of the head muscles in *Culicoides*.*

Below are given the muscles of the head as seen in *Aëdes aegypti*, the numbers given to them being those by which they will be referred to and by which they are indicated in the figures. In describing the origins and insertions the head is considered as arranged dorso-ventrally in line with the body so that the *frons* is posterior to the *clypeus* and the *labrum* dorsal to the *labium*. The site of origin is given first, followed by the site of insertion unless it is uncertain which end of the muscle is which, when it is given as 'between' the sites given. A name has been given where muscles have an obvious function, in most cases as they have been named by previous authors. Thompson's nomenclature has been largely used, though names by Robinson or Jobling or new names have been given where such seem more suitable.

Muscles of the antennae (Fig. 67 (1), (2))

1. *Adductor of the scape.* Dorsal surface of tentorium towards its anterior end to anterior portion of basal ring of antenna (scape). (Inner antennal of Thompson; adductor of the scape of Jobling.)

2. *Abductor of the scape.* Dorsal surface of the tentorium external to the adductor to outer portion of scape.

3. *Intrinsic muscles of the antennae* (3*a*–3*d*). Muscular fibres from scape to margins of pedicel (see note 1).

Muscles of the labrum and clypeus

4. *Retractors of the labrum.* Fan-shaped muscles composed of numerous separate bundles on each side of middle line of dorsum of clypeus converging to base of dorsal plate of labrum at its junction with the preclypeus in the preclypeal pouch. (Labral muscle of Thompson; retractor of the labrum of Robinson; elevator of the labrum epipharynx of Jobling.)

5. *Elevators of the labrum.* Cornua of preclypeal apodeme to sides of ventral plate of the labrum in its basal portion. (Epipharyngeal muscle of Thompson; elevator of the labrum of Robinson.)

6. *Elevators of the anterior hard palate.* Sides of clypeus a little external to the middle line passing between the preclypeal cornua to sides of anterior hard palate. (No. 1 elevators of the palate of Thompson; possibly the dilators of the pharynx of Jobling.)

7. *Elevators of the roof of the buccal cavity.* Four parallel muscles, 7*a*–7*d*, on each side from a line along the side of the clypeus to the sides of the roof of the buccal cavity from just behind the anterior hard palate to the posterior end of the organ. 7*a*–7*b* are inserted into the membranous

* See also earlier work by Annett, Dutton and Elliott (1901), Nuttall and Shipley (1901, 1903), Kulagin (1905), and the still earlier work of Dimmock (1881, 1883).

roof, 7*c* into the posterior hard palate and 7*d* into the hummocky area behind the posterior hard palate. 7*d* lies just in front of the frontal ganglion and passes beneath the ganglion to its insertion. (Elevators 2–5 of the palate of Thompson; dilators of the pharynx of Jobling in part.)

Muscles of the mandibles

8. *Mandibular muscle.* A small muscle on each side from the floor of the tentorium at its anterior end to the base of the mandible. These small muscles are conspicuous in transverse sections through the fore part of the head situated on each side of the buccal cavity and are so shown in the figure by Annett, Dutton and Elliott. (Mandibular muscles of Thompson; retractors of the mandibles of Robinson.) (See note 2.)

9. *Subocular muscles.* Minute muscles arising on each side from the postgenal area a little anterior to the foremost lateral dilator muscles of the oesophageal pump and passing forwards internal to the lower part of the compound eye of its side and according to Thompson inserted into the tentorium immediately caudal of the origin of the dorsal retractor of the maxilla (see note 3). (Subocular muscle of Thompson.)

Muscles of the maxillae

10. *Ventral retractors of the maxillae.* A large muscle on each side arising from the occipital area ventrally just external to the posterior end of the tentorium and inserted into the anterior end of the maxillary apodeme. It lies in most of its course along the outer side of the maxillary apodeme. (Retractors of the maxillae of Thompson; ventral retractors of the maxillae of Robinson; retractors of the galeae of Jobling.)

11. *Dorsal retractors of the maxillae.* A large muscle on each side from the inner surface of the tentorium about its middle to the spur of the maxilla. (Double retractors of the maxillae of Thompson; dorsal retractors of the maxillae of Robinson.)

12. *Protractors of the maxillae.* A large muscle on each side from the inner surface of the tentorium for much of its extent to the posterior end of the maxillary apodeme. (Protractor maximus of Thompson; protractor of the maxillae of Robinson; to some extent protractor of the stipes of Jobling.)

13. *Elevators of the maxillary palps.* Very small muscles present in the male of *Aëdes aegypti.* Anterior end of maxillary apodemes to base of palps dorsally. (Elevators of the palps of Robinson and of Jobling.)

14. *Accessory elevators of the palps.* Muscular fibres conspicuous in the male, present to a less extent in the female, in basal segment of palps. (Accessory elevators of the palps of Robinson.) (Not figured.)

Muscles of the labium

15. *Retractors of the labium.* Small muscles on each side from the posterior ends of the maxillary apodemes to posterior margins of the labium. (Maxillo-labial muscles of Thompson; retractors of the labium of Robinson and of Jobling.)

16. *Extensors of the labella.* Muscular fibres arising as series of strands from the lateral wall of the distal part of the labium and inserted into the base of the labella on their outer sides.

17. *Flexors of the labella.* Muscle fibres from ventro-lateral wall of labium on each side towards its distal end to base of labella at their inner ventral angles. (Flexors of the labella of Robinson.) (See note 4.)

Muscles of the salivary pump

18. *Salivary muscle.* A stout muscle on each side arising from the lateral and ventral flanges of the buccal cavity and inserted into the posterior face of the salivary pump. (Hypopharyngeal muscles of Thompson; dilators of the salivary syringe of Robinson; salivary pump muscles of Jobling.)

Muscles of the buccal cavity

19. *Valvular muscles.* Two small muscles from middle line of frons between the scapes of the antennae to membrane at junction of buccal cavity and oesophageal pump. (Valvular muscles of Thompson.)

20. *Lateral retractors of the buccal cavity.* Stout bulky muscles arising on each side from inner surface of the tentorium at its anterior end and inserted into the lateral flanges of the buccal cavity. (Lateral pharyngeal muscles of Thompson.)

21. *Dorsal retractors of the buccal cavity.* Median line of frons between the dorsal ocelli to lateral flanges of the buccal cavity. (Ascending pharyngeals of Thompson; retractors of the pharynx of Jobling.)

22. *Accessory dorsal retractors.* A small muscle on each side arising in common with 21 but inserted into the anterior end of the dorsal plate of the oesophageal pump. (Anterior dorsal dilator of the pump of Thompson and of Jobling.) (See under 'Anterior termination of the aorta' ch. XXVII, p. 618.)

Muscles of the oesophageal pump

23. *Dorsal dilators of the oesophageal pump.* Two large flat muscles arising close together on each side of the median line of the occiput and inserted into the dorsal plate of the pump. (Posterior dorsal dilator of the oesophageal pump of Thompson and of Jobling.)

24. *Lateral dilators of the oesophageal pump.* Five muscular masses on each side, one behind the other, arising in series from the postgenal areas and inserted into the lateral plates of the oesophageal pump. (Lateral dilators of Thompson and of Jobling.)

Muscles from the neck and thorax

25. *Depressor muscle of the head.* Enters the head from the neck on each side of the ventral nerve connectives and inserted towards the anterior end of the tentorium on its outer surface. A long muscle with distinctive characters and the only large single tubular muscle in the head. (See also under 'Muscles of the thorax p. 588)'. (Tentorial muscle of Thompson; depressor muscle of the head of Jobling.)

26. A muscle from the prothorax in the form of a thin tendon that closely accompanies muscle 25, but is inserted into the spur on the tentorium near its posterior end. For further particulars see under muscles of the thorax.

Note 1. The intrinsic muscles of the antenna, that is those passing from the scape to the margins of the pedicel, consist of numerous fibres arranged as shown in Fig. 67 (2), a fan-like arrangement passing from the inner and posterior part of the scape beneath the pedicel to be inserted along the inner half of its anterior border (3*a*). External to this fan is a single small stout muscle (3*b*). Externally in the same plane are fibres from the outer posterior part of the scape to the outer anterior border of the pedicel (3*c*). 3*d* is not strictly an intrinsic muscle but consists of fibres from the edge of the ocular sclerite near the dorsal ocellus passing to the outer posterior part of the scape.

Note 2. In the larva the mandibular muscles (adductor and abductor bundles) arise from the postgenal area behind the eyes. It is therefore strange to find the only mandibular muscle in the imago arising from the anterior end of the tentorium. In *Culicoides*, however, a muscle is described by Jobling additional to the large adductor and abductor muscles arising from the genal area. This has much of the character of the mandibular muscle in the mosquito. In *Culicoides* all three muscles are well developed, as also is the mandible. In the mosquito the mandible is obsolescent, especially so in *Aëdes aegypti*, where it terminates in a flat lanceolate tip devoid of any teeth.

Note 3. This small muscle has been traced to below the tentorium, but its insertion has not with certainty been traced by me. In *A. aegypti* it does not seem to be inserted into the tentorium and it is suspected that it may represent an obsolescent strand representing the larval adductor or abductor muscles (see note 2).

Note 4. These muscles in *A. aegypti* are as figured by Robinson for *Anopheles*. Robinson, however, gives as the origin of the flexor muscles the *dorsal* aspect of the labium. I think this may be a slip of the pen. They clearly arise ventro-laterally. Robinson also describes the muscle as ending in a long tendon passing into the labella. In *Aëdes aegypti* they appear to be inserted as given in the synopsis. The muscles are shown up by polarised light but less vividly than the usual systemic muscles.

A reference may be made to the relation of the larval to the imaginal musculature of the head. As a generalisation it may be said that the larval muscles are either completely histolysed and disappear, or in certain cases follow through from the larva to the imago but much modified. Yet even when new muscles are formed in the imago these often show an homology with those of the larva. With the disappearance of the massive labral structures of the larva with their feeding brushes and apodemes and associated muscles and the much contracted state of these parts in the imago, any larval muscles still represented in the imago must be looked for as small and insignificant ones crowded together in the clypeus. Contrariwise, as a result of the appearance of the large oesophageal pump small insignificant muscles in the rear parts of the larval head may be expected to be large and important muscles.

Of the twenty-six imaginal head muscles, eleven (10, 15, 18, 19, 20, 21, 22, 23, 24, 25 and 26) appear to be larval muscles carried through, five (1, 4, 5, 6 and 7) appear to be histolysed but represented by imaginal muscles, and the remaining ten new imaginal muscles not represented in the larva.

MUSCLES OF THE NECK (Fig. 68 (3), (4))

A large part of the neck laterally is covered by the broad flat cervical sclerites, hinged dorsally to the sides of the occipital prominence and ventrally to the propleura. Anterior and posterior to the cervical plates the neck is membranous and normally thrown into deep folds, one in front of and one behind the sclerite. The muscles of the neck are given in tabular form below with the numbers by which they will be referred to and by which they are indicated in the figures. Their origins and insertions and their general positions are indicated in Fig. 68 (3).

The largest muscles are those from the occiput to the cervical sclerites. Of special interest are the decussating muscles 29 recalling a similar condition in the larva. The muscles assigned to the protomorphic system are those connecting with this system in the prothorax as described later. Among them are the long muscles 25 from the tentorium in the head traversing the neck to pass to the ends of the prothoracic ventral transverse muscle in the thorax.

Though not strictly a cervical muscle it is convenient here to include muscle 38 (pronotal muscle) which takes origin from the walls of the prothoracic lobe (anterior pronotum) and converges to be inserted into the narrow neck linking this with the propleura. Its function is not very clear, though it is a fairly large muscle. Possibly it has to do with giving support to the head.

The following are the muscles of the neck:

	Number
Occipital protuberance	
to cervical sclerite: anterior	27
to cervical sclerite: posterior	28
to cornu of first phragma: decussating	29
to cornu of first phragma: lateral	30

	Number
Cervical sclerite	
anterior border to posterior fold of neck	31
ventral arm to prothoracic furca	32
ventral arm to propleura	33
ventral arm to dorsal knot	34
Posterior fold of neck	
to first phragma: dorso-lateral	35
to first phragma: ventro-lateral	36
to dorsal knot	37
Passing through neck from head*	
as muscle from tentorium anteriorly	25
as tendon from posterior spine of tentorium	26

* Already given under muscles of the head.

MUSCLES OF THE THORAX (Figs. 68, 69)

For a general description of the thoracic musculature of insects see especially Snodgrass (1912, 1927, 1935); Morison (1927–1929) on the muscles of the honey bee; Dirkes (1928) on the imaginal musculature of *Psychoda* and its development from the larval system; Maki (1938) on the thoracic muscles of various orders of insects including Diptera. Among the types given by the last-mentioned author is a very detailed study of a tipulid, *Ctenacroscelis mikado*, in which sixty-two numbered muscles are listed for the thorax. So far as I am aware no author has described the thoracic musculature of the imaginal mosquito. A reference to certain of the wing muscles has, however, been made by Prashad (1918).

For the mosquito *Aëdes aegypti* the number of distinct muscles for the thorax now listed is 64. These are set out in tabular form below in four series as they relate to: the wings and halteres; the spiracles; the legs; and the protomorphic series. The table also gives the categories as commonly used so far as these seem applicable, for example tergo-pleural, sterno-coxal, etc. In addition their situations, origins and insertions have as far as possible been indicated in the figures. Some further information is given in the form of notes.

One interesting point has been brought out by the study of the muscles, namely the extent to which characters of the sclerites are shown to be determined by the requirements for the attachment of certain muscles. Particularly noticeable in this respect is the shape of the sclerite named the posterior pronotum. This gives the necessary surface for the large trochanteral muscle of the fore-leg. Another instance is the peculiar-shaped sclerite termed the post-spiracular. This seems to be the shape it is to give attachment to the large fan-shaped posterior spiracular muscle. Other instances are the paratergite, which appears to exist to provide origin for the large quadrangular muscle 44, and the turned-in edge of the scutum anterior to this that gives origin to the dorsal spiracular muscle. The characteristic sternopleuron of the mosquito owes its large size and solid appearance to the requirement for attachment of the dorso-ventral indirect wing muscles. Even the whole shape of the thorax with its projecting anterior promontory and elongate wedge shape is evidently to give sufficient length for the indirect wing muscles.

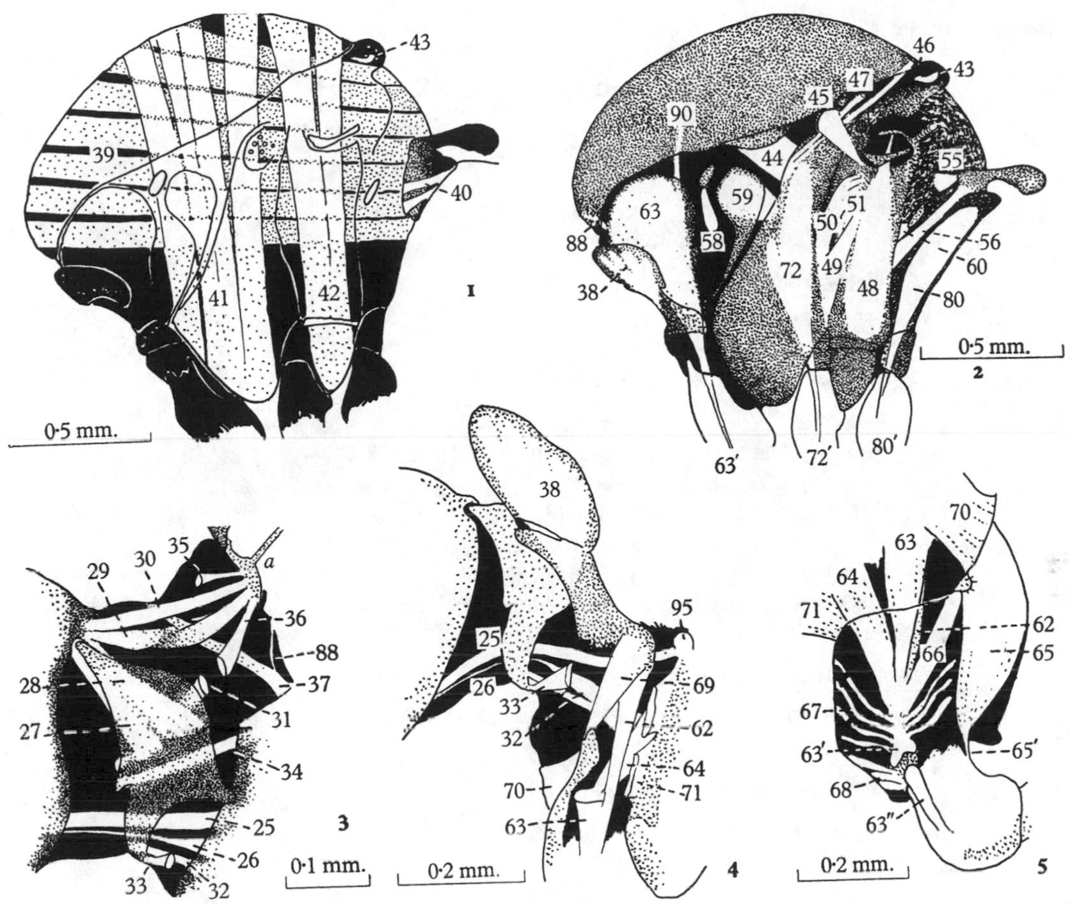

Figure 68. Muscles of the thorax.

1 Lateral view of thorax showing the indirect wing muscles.

2 The same showing other muscles as these would be seen showing through the transparent sclerites. The indirect wing muscles have been omitted.

3 Muscles of the neck. Lateral view as seen with the neck extended (reconstruction). In the middle of the figure the cervical sclerite is shown as though transparent. The posterior part of the head is on the left and the position of the prothoracic parts indicated on the right. *a*, first phragma. Extending backwards from it is seen the rod-like apodeme that extends backwards from the first phragma behind the posterior pronotum.

4 Muscles of the prothorax. The occipital region of the head is on the left, the anterior pronotal lobe and its stalk with parts of the thorax on the right.

5 The first coxa. Internal view to show muscles.

The numbers indicated on the figures are those given to the muscles in the tabular statements in the text.

Numbers marked with a dash indicate tendons of correspondingly numbered muscles. The number with two dashes (63″) refers to the strap-like thickening into which the tendon proper is inserted.

The following are the muscles of the thorax:

A. Muscles of the wings and halteres

	Number
Indirect wing muscles (note 1)	
Longitudinal (mesothoracic)	39
Longitudinal (metathoracic)	40
Dorso-ventral, anterior	41
Dorso-ventral, posterior	42
Transverse (scutellum)*	43
Tergo-pleural muscles (note 2)	
Anterior pronotal	38
Paratergite to sternopleuron	44
Basalar to mesepimeron	45
Angle of scutum to sternopleuron	46
Axillary, basalar and subalar muscles (note 3)	
Sternopleuron to dens (2nd axillary)	47
Mes-epimeron to posterior wing root	48
Pleural suture to posterior wing root	49
Pleural suture to hamular process	50
Pleural suture to mes-epimeral apodeme	51
Furco-pleural line to wing root	52, 53
Furco-entopleural (note 4)	
Cup of mesofurca to furco-pleural line	54
Muscles of the halteres (note 5)	
Elevator of the halter	55
Depressor of the halter	56

* See under note 1.

B. Muscles of the spiracles (note 6)

	Number
Mesothoracic spiracle	
Ventral: from subspiracular area	58
Posterior: from postspiracular sclerite	59
Metathoracic spiracle	
Ventral: from metapleuron	60

C. Thoracic muscles of the legs (note 7)

	Number		
	Pro-	Meso-	Meta-
Pleuro-trochanteral: with extensor tendon	62, 63	72	80
Furco-trochanteral: joining extensor tendon	64	73	81
Coxo-trochanteral			
External surface to coxa (flexor)	65	74	82†
Accessory extensor	66	(75)*	(83)*†
Extensor tendon adjustor	67	76	84†
Apical	68	77	85†

* See note 7.
† Not shown in the figures.

	Number		
	Pro-	Meso-	Meta-
Pleuro-coxal			
Propleural knob to first coxa	69	—	—
Sterno-coxal			
Anterior	70	78	86
Posterior	71	79	87

D. Muscles of the protomorphic series (note 8)

	Number
Dorsal series	
Cervical sclerite to prothoracic dorsal knot	34*
Dorsum of neck to prothoracic dorsal knot	37*
Anterior promontory joining muscle 37	88
Cornu of first phragma to prothoracic dorsal knot	89
Lateral region of scutum to prothoracic dorsal knot	90
Prothoracic dorsal transverse	91
Continuation of dorsal prothoracic strand	92
Ventral series	
Tentorium to prothoracic ventral knot	25†
Branch from muscle 90 to prothoracic ventral knot	93
Branch from muscle 93 to behind first coxa	94
Prothoracic ventral transverse	95
Continuation of strand to mesothoracic ventral knot	96
Mesothoracic ventral transverse	97‡
Continuation of ventral strand	98‡

* Already given under muscles of the neck.
† Already given under muscles of the head. The muscle of the tendon accompanying this muscle goes to the prothoracic furca, not to the ventral knot.
‡ Not shown in the figures.

Note 1. The indirect wing muscles act on the wings indirectly through changing the shape of the thorax. The longitudinal muscles act as *depressors* and the dorso-ventral as *elevators* of the wings. Both consist of massive muscular bundles more or less rectangular in cross section and separated by intervening spaces. These number about twelve on each side.

The *longitudinal* muscles occupy the median portion of the dome-shaped upper part of the thorax. They extend from the anterior promontory and the dorsum of the scutum for about the anterior two-thirds of its extent to the postnotum and its phragma. The series is continued beyond the phragma as small short muscles of the same type in the metathorax. These pass from the second to the third phragma.

The *dorso-ventral* muscles extend from the ventral portion of the thorax to the lateral aspects of the scutum. They consist of two separate series, an anterior and a posterior. The *anterior series* extend from the expanded lower portion of the sternopleuron to about the middle third of the lateral portions of the scutum. They consist on each side of about 8–10 muscular bundles. The *posterior series* pass from the meron to about the posterior third of the lateral portions of the scutum. They consist on each side of about 5–6 muscular bundles. According to Snodgrass the posterior series are originally tergal remotors (extensors) of the middle leg which in Diptera, with removal of the meron from the coxa to the thorax, have become indirect wing muscles.

The indirect wing muscles differ profoundly in their minute structure from all other thoracic muscles, consisting of massed fine fibres with numerous intervening nuclei, whilst the limb muscles are usually formed of much larger fibres commonly of tubular type with centrally situated nuclei. They differ also in origin, appearing during the late larval and pupal stages as new, isolated, sausage-shaped masses resembling foreign bodies which eventually attach themselves to the cuticle.

The small transverse muscle in the scutellum (43) has been placed in the present group as it may assist in the deformation of the thorax, but it is more probably an accessory pulsatory muscle (see under 'The circulatory system', ch. XXVII, p. 619).

Note 2. The muscles 44 and 45 (along with the spiracular muscle 59) are very characteristic thoracic muscles. Muscle 44 takes origin from the paratergite and is inserted into a line at the anterior edge of the sternopleuron in its upper part. This line is also made use of by the two small muscles 46 and 47. Muscle 45 arises from the limpet-shaped sclerite which has been here termed the basalar, though this designation requires confirmation. It is a striking fan-shaped muscle inserted into the top of the mes-epimeron just anterior to the subalar sclerite at a point below which muscle 50 is inserted. This is at the base of the hamular process of Prashad (posterior portion of the pleural process of Snodgrass).

Note 3. The tracing of the insertion of muscles into the basal sclerites of the wing root (axillary muscles) is difficult. But there appear to be three muscles so inserted, namely 47, 48 and 49. Muscle 47 is very small and seems to be inserted into the dens (second axillary). Three muscles are stated by Prashad to be so inserted, one into the dens, one into the unguiculus and one into the remigium. Apart from 47 the only other muscles so far observed by me have been the flexor muscles 48 and 49 inserted into or near the posterior root. Muscle 48 is much the largest direct wing muscle. It arises from nearly the whole of the surface of the mes-epimeron and terminates in a tendon that passes behind the subalar sclerite to be inserted into a thickening on the membrane at the posterior base of the wing near the point of junction of the axillary cord. It is clearly the flexor of the wing. Muscle 49 which arises from the posterior border of the sternopleuron appears to join 48 in a common tendon. Still another muscle, 50, rises from the posterior edge of the sternopleuron and is inserted into the mes-epimeron at the base of the hamular process, and a third muscle so arising, 51, is inserted into the long rod-like apodeme lying just internal to the mes-epimeron. Muscles 52 and 53 arise from an apodematous extension where the upper portion of the second coxa lies between the sternopleuron and the meron. In the table this extension has been termed the *furco-pleural line*. Both muscles are inserted in the region of the wing root.

Note 4. The muscle here termed furco-entopleural is named after a term used by Maki which appears to be applicable. It arises from the cup-shaped process of the mesofurca and passes outwards to the furco-pleural line. It is included among the wing muscles though its action is not very clear.

Note 5. The muscles of the halteres appear to be only two, namely a very small fan-shaped muscle inserted into the anterior portion of the base of the haltere and arising from the metathorax just anterior to the haltere and which may be an *elevator* of the organ, and a larger, longer and more conspicuous muscle, 56, arising from the posterior edge of the mesepimeron in its lower part which would appear to be a *depressor* muscle.

Note 6. The spiracular muscles are indicated in Fig. 68 (2). Muscle 59 is much the largest and is assumed to be a dilator muscle of the mesothoracic spiracle as its action would seem to be of this nature. It is a conspicuous fan-shaped muscle arising from the post-spiracular sclerite. The closing muscle for the mesothoracic spiracle is 58, situated ventral to the spiracle. For the metathoracic spiracle there is a small fan-shaped muscle, no. 60, arising from the posterior border of the mes-epimeron and inserted into the lower end of the spiracular opening.

Note 7. In the mosquito the coxae form a fixed and integral part of the thorax. Hence muscles concerned with movements of the legs are for the most part inserted into the more moveable trochanter. For each leg there is one large muscle (*pleuro-trochanteral*) which arises from an extensive area on the respective pleura. These pass through the coxae as long tendons to be inserted into triangular, bifid or heart-shaped small patella-like sclerites from which in turn a ligament extends to that part of the trochanter that is situated opposite to the femur attachment. Hence action of the muscle is to straighten out the femur and leg, a movement here given as extension (covering the terms suitable to some cases of depressor or promotor). In each leg, joining the tendon as it enters the coxa, there is a muscle arising from the corresponding sternal complex (furca) and given in the table as *furco-trochanteral*. The common tendon may be termed the *extensor tendon*.

The extensor tendon also receives fibres arising from the walls of the coxa. These tend to form a series of separate fibres which on reaching the tendon are inserted into it almost at right angles. For some reason these separate fibres in preparations very commonly show a wavy or curled character which serves to distinguish them. The muscle formed by them is given in the table as the *extensor tendon adjustor*. Another muscle of coxal origin entering the tendon has been termed the *accessory extensor*. In the coxa of the first leg this is a small distinct compact muscle arising from the antero-internal border of the coxa close to the small hairy node in this situation. It passes down the coxa and joins the extensor common tendon. Fibres in a somewhat similar position occur in the other coxae, but it is difficult to say whether they represent a separate muscle or form a part of the loose-fibred muscle, the adjustor. For the two hinder coxae therefore the numbers for this muscle have been given in brackets.

The *flexor* (remotor or elevator) muscles working through the trochanters arise from the coxae. These arise from the external aspects of the coxae and are inserted into the trochanter on the same side as the femur attachment by a simple tendon. Towards the apex of each coxa, but particularly large in the mid coxa where it occupies the inner bulge, is a fan-shaped arrangement of fibres which I have termed in the table the *apical muscle*.

In addition to the above each coxa receives anteriorly and posteriorly the insertion of a more or less sheet-like muscle arising from the apodematous median ridge of the sterna and inserted into the corresponding proximal margins of the coxae. The anterior muscles arise from the sterna proper (basi-sterna), the posterior from the furcal ridge. These sheet-like muscles, though largely transverse in direction, are to be distinguished from the true transverse muscles which cross the median line. In the table they are entered as *sterno-coxal*. Some idea of their character will be obtained from Fig. 69 (2).

A point that should be mentioned in respect to the leg musculature as here given is that, though the large extensor muscle for the first pair of legs has, in accordance with those for the other legs, been given as pleuro-trochanteral, there is in this case some doubt since the part from which it takes origin has been now generally known as the posterior pronotum. It is possible, on the other hand, that the posterior pronotum as now named is not pronotal in nature but a pleural sclerite, the pro-epimeron, as originally designated by Snodgrass (1912) and others (Christophers, 1915; Prashad, 1918).

Note 8. The protomorphic system of the thorax occurs as two layers of muscular strands, one dorsal and one ventral to the alimentary canal with some contributory muscles from the head, neck and thorax. On reaching the posterior end of the thorax these strands become continuous respectively with the dorsal and ventral abdominal muscles.

The dorsal series of each side commence anteriorly with the muscles as given in the table and Fig. 69 (9). These meet forming a knot-like plexus about the large tracheae entering the thorax from the spiracles. At the level of the dorsal diverticula there appears to be a small transverse muscle. From the knot the strands pass back beneath the longitudinal indirect wing muscles and above the narrow part of the mid-gut to join the dorsal system of the abdomen.

The ventral series commences anteriorly on each side with the tentorial muscle from the head. These pass to the ends of the small prothoracic ventral transverse muscle. Reinforced by a branch from the dorsal knot, the tract passes backwards beneath the ventral diverticulum and above the ventral nerve chain to join the ventral series of the abdomen. There is a second ventral transverse muscle situated a little anterior to the mesofurca.

The impression given by the system is that it has little relation to bodily movements, but may have visceral and perhaps tracheal functions.

Figure 69. Muscles of the thorax and abdomen.

1 Middle coxa and mesofurca to show muscles.

2 Horizontal section through the neck and thorax showing muscles. *a*, prothoracic furca; *b*, mesothoracic furca. On the left the cup-shaped process is shown. On the right the stem of same is shown as though cut across at a lower level; *c*, metathoracic furca.

3 Showing muscles of the protomorphic system in region of the prothorax. The tracheal vessel passing forwards and dividing into two (common cephalic) has been shown as though cut away.

4 Showing muscles at the trochantero-femoral joint.

5 The same at the femoro-tibial joint. *a*, spiny strap-like thickening into which the tendon of the extensor muscle is inserted; *b*, sclerotised area on ventral aspect of head of tibia.

6 The same at the tibio-tarsal joint.

7 Showing dorsal muscles of abdominal segment of the newly emerged imago. The numbers give the identification of the muscles with Samtleben's numbers for the larval muscles of the series (dorsal longitudinal). *a*, nos. 2–4 of the lateral dorso-ventral of Samtleben; *b*, no. 5 of the same.

The median dorso-ventrals are indicated only in Fig. 8.

8 The same for the ventral longitudinal series. *a*, nos. 2 and 3 of the dorso-ventral series.

9 Showing course of the ventral tracts of the protomorphic series of the thorax and their continuation as the ventral longitudinal series of the abdomen. *a*, prothoracic muscular knot and ventral transverse muscle; *b*, cup-shaped process of mesofurca; *c*, muscular strands lying median to the main strands.

10 Tangential section through portion of the pleura and adjacent tergite showing thread-like muscle fibres in the former. *a*, portion of section through the pleural membrane; *b*, portion through the tergite showing absence of fibres.

Lettering: *csc*, cervical sclerite; *Dd*, dorsal diverticulum; *Dv*, ventral diverticulum; *ht*, heart; *Mg*, mid-gut (anterior portion); *Oe*, oesophagus; *Sp*, mesothoracic spiracle.

The numbers (except those in Figs. 7 and 8) indicate muscles as tabulated in the text.

The numbers (large) in Figs. 7 and 8 are Samtleben's numbers for the abdominal muscles in the larva.

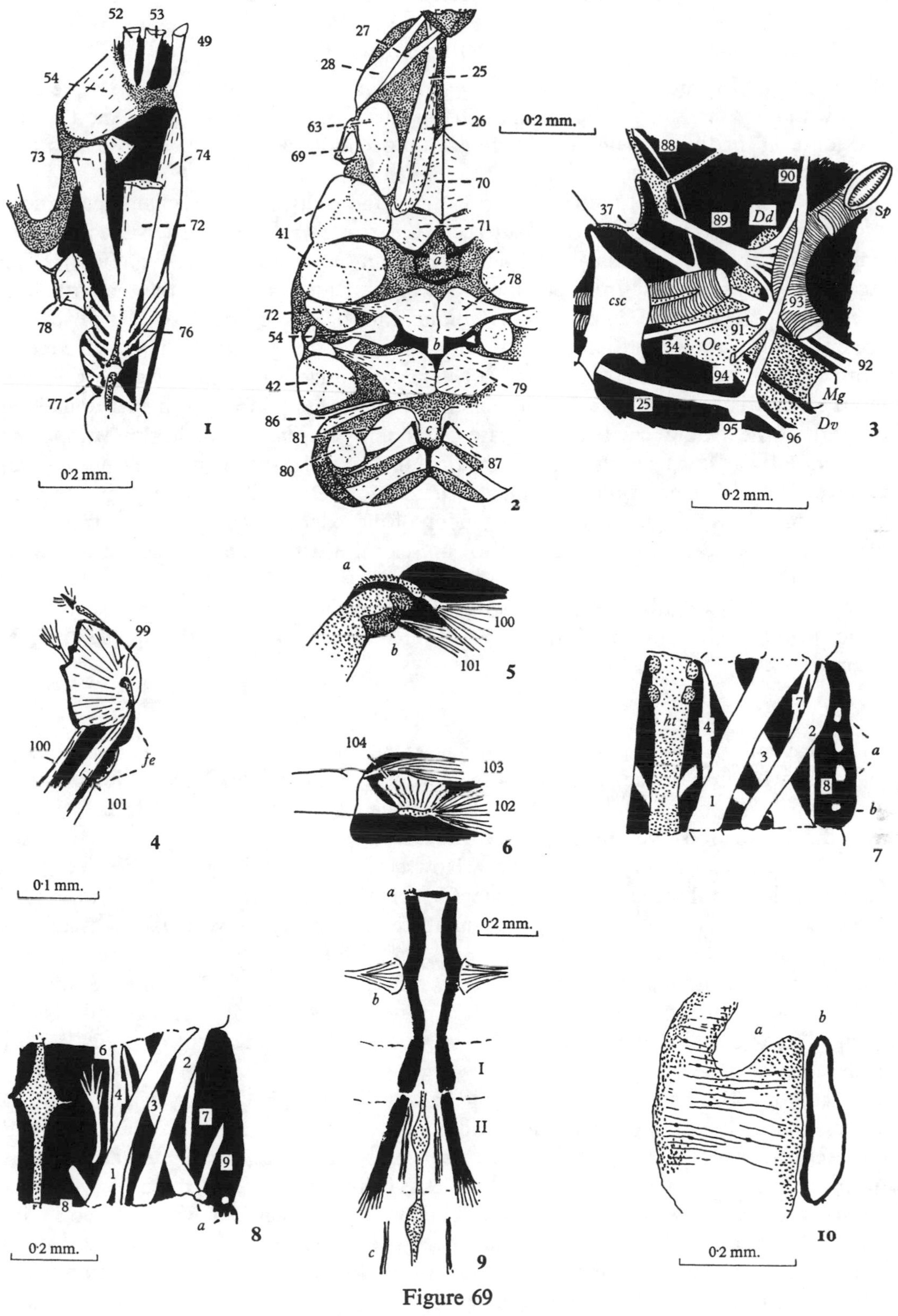
52
53
49
54
73
74
72
78
76
77
I
0·2 mm.
27
28
25
63
26
69
70
41
71
a
78
72
54
b
42
79
86
81
c
80
87
2
0·2 mm.
88
90
Sp
37
89
Dd
csc
93
34
Oe
91
94
92
Mg
25
95
96
Dv
3
0·2 mm.
99
100
fe
101
4
0·1 mm.
a
100
b
101
5
104
103
102
6
ht
4
7
2
3
1
8
a
b
7
a
0·2 mm.
b
I
II
c
9
6
2
4
3
7
9
1
8
a
8
0·2 mm.
a
b
10
0·2 mm.

Figure 69

MUSCLES OF THE LEGS (Fig. 69)

Following upon the muscles moving the trochanter on the coxa in each leg are those concerned with movement of the femur on the trochanter, the tibia on the femur and the first tarsal joint on the tibia. In addition there is a long tendon passing through the tibia and tarsal segments to the ungues.

The trochantero-femoral muscle (99) rises from the walls of the trochanter and converges to be inserted into the proximal surface of the strongly sclerotised spur-like processes which project from the proximal end of the femur into the trochanter (Fig. 69 (4)). The femoro-tibial muscles arise from the upper end of the femur and along the femoral shaft and consist respectively of an extensor (100) and a flexor of the tibia (101). The extensor muscle is inserted into a strap-like sclerite with fine spines that passes to the dorsal edge of the proximal end of the tibia (Fig. 69 (5)). The flexor muscle is inserted by a short tendon on to the ventral edge of the same. Between the two muscles and forming a partition along the length of the femoral cavity is a delicate membrane carrying the tracheal branch to the limb. Similarly arising from the walls of the tibia in its distal portion are the extensor (102) and flexor (103) of the first tarsal segment. In addition, especially noticeable in the hind leg, are some fibres at right-angles to the tendon of the extensor muscle (the ventral of the two muscles) that call to mind the adjustor muscles in the coxae and which may have a similar function (104).

According to Snodgrass (1935) the long tendon to the ungues arises in insects from muscles in the femur and tibia and is inserted into the basal plate of the ungues (unguetractor plate).

MUSCLES OF THE ABDOMEN

With the exception of some points mentioned later the abdominal musculature of the imago is identical with that of the larva. For this and other reasons given later a full account segment by segment as given for the larva seems uncalled for. It consists as in the larva of longitudinal and dorso-ventral segmental muscular bands. The longitudinal bands are arranged in a dorsal and a ventral series. How close the resemblance of these series to those of the larva will be clear from Fig. 69 (7), (8), which give *camera lucida* drawings of the two series in the imago, the muscles numbered according to Samtleben's system of nomenclature for the larva (see ch. XIII, p. 311). In the dorso-ventral series there are, as in the larva, in the normal segment four laterally situated muscles (*ldv*. 2–5), three about the middle of the segment (*ldv*. 2–4) and one at its posterior end (*ldv*. 5). There are also bands passing dorso-ventrally at the intersegmental line (median dorso-ventral, *mdv*. 1–3) of Samtleben.

All these muscles have the characters of the larval series; that is, in preparations they stain less markedly; do not stand out so vididly under polarised light as do the wing and leg muscles; consist of single fibres with sarcolemma, never bundles of tubular fibres as with most of the leg muscles and are much less vividly striated than these. Very characteristically too they pass from intersegmental line to intersegmental line and moreover are continuous in series with the protomorphic strands of the thorax, the dorsal longitudinal strands of the thorax continuing as the dorsal longitudinal series of the abdomen and the ventral strands as the ventral abdominal series.

What has been said applies to the system as seen in the newly emerged or young mosquito shortly after emergence. In mosquitoes several days old there is a general deterioration of the muscles. In an unfed mosquito 48 hours from emergence the bands can still largely be made out but are more attenuated and seem degenerated. In still older mosquitoes practically the only muscles showing up under polarised light are two thin thread-like fibres in each segment as described below. Hocking (1952) has described autolysis of the flight muscles of a Canadian mosquito as a means of prolonging life. If the above observations are confirmed, it would appear that in *Aëdes aegypti* the abdominal muscles, after being passed on from the larva and pupa, are in the imago made use of in a somewhat similar way during the few days in which the mosquito following upon emergence does not feed. Following upon emergence it has already been noted that there is a reduction in weight and in the size of the abdomen. It is possible that the using up of the muscles plays some part in this.

A feature very characteristic of the musculature of the imago is the presence in the first abdominal segment on each side of a long band of muscle passing from the metathoracic knot to beyond the second segment and ending in diffusely splayed out branches. Between these are the two thin muscles referred to above, similar muscles being repeated in the subsequent segments (Fig. 69 (9)). Another feature is the presence just beneath the soft pleural membrane of rows of fine muscular fibres arranged dorso-ventrally. Very similar fibres may be present extending from the ends of the lateral dorso-ventral muscles, but those referred to seem quite distinct from such. They arise and terminate entirely in the subcuticular tissue of the pleural membrane and are particularly conspicuous in the recently fed mosquito where the membrane is stretched and the fibres correspondingly long (Fig. 69 (10)).

No extrinsic muscles have been noted in connection with the abdominal spiracles. The alary muscles are described when dealing with the circulatory system and those of the terminal segments and hypopygium under the generative system.

REFERENCES

(*a*) THE ALIMENTARY CANAL

Apostolidon, N. X. (1901). *Anophelon konopon.* Athens. (Copy in the Shipley Collection, Molteno Institute, Cambridge.)

Barraud, P. J. and Covell, G. (1928). The morphology of the buccal cavity in Anopheline and Culicine mosquitoes. *Ind. J. Med. Res.* **15**, 671–9.

Beyer, G. E., Pothier, O. L., Couret, M. and Lemann, J. J. (1902). Bionomics, experimental investigations with *Bacillus sanarelli* and experimental investigations with malaria in connection with mosquitoes of New Orleans. *New Orleans Med. Surg. J.* 1902, 1–78.

Bhatia, M. L. and Wattal, B. L. (1957). Description of a hitherto unknown structure in the mouthparts of mosquitoes. *Ind. J. Malar.* **11**, 183–9.

Bhatia, M. L., Wattal, B. L. and Kalra, N. L. (1957). Structure of salivary glands in mosquitoes. A preliminary note. *Ind. J. Malar.* **11**, 55–9.

Christophers, S. R. (1901). The anatomy and histology of the adult female mosquito. *Rep. Malar. Comm. R. Soc.* (4), 1–20.

Cornwall, J. W. (1923). On the structure of the salivary pump in certain blood-sucking and other insects. *Ind. J. Med. Res.* **10**, 996–1007.

De Boissezon, P. (1930*a*). Contribution à l'étude de la biologie et de l'histophysiologie de *Culex pipiens*. *Arch. Zool. Exp. Gén.* **70**, 281–431.

De Boissezon, P. (1930*b*). Sur l'histologie et l'histophysiologie de l'intestine de *Culex pipiens* (imago) et en particulier sur la digestion du sang. *C.R. Soc. Biol., Paris*, **103**, 568–70.

Dufour, L. (1851). Recherches anatomiques et physiologiques sur les Diptères. *Mém. Acad. Sci., Paris*, **11**, 171–360.

Engel, E. O. (1924). Das Rectum der Dipteren in morphologischer und histologischer Hinsicht. *Z. wiss. Zool.* **122**, 503–33.

Eysell, A. (1905). Die Stechmücken. In *Mense's Handb. Tropenkrankh.* **2**, 44–94. Barth, Leipzig (and subsequent editions).

Goldi, A. (1913). *Die sanitärisch-pathologische Bedeutung der Insekten und verwandter Gliedertiere.* Berlin. (Culicidae, pp. 31–53.)

Grandpré, A. D. de and Charmoy, D. d'E. de (1900). *Les moustiques. Anatomie et biologie.* Port Louis, Mauritius.

Grassi, B. (1901). *Studi di uno zoologo sulla malaria.* Ed. 2. Rome.

Hocking, B. (1953). The intrinsic range and speed of flight of insects. *Trans. R. Ent. Soc. Lond.* **104**, 223–345.

Holt, C. M. (1917). Multiple complexes in the alimentary tract of *Culex pipiens. J. Morph.* **29**, 607–17.

Hurst, C. H. (1890*a*). *On the life history and development of a gnat* (Culex). Guardian Press, Manchester.

Hurst, C. H. (1890*b*). *The Pupal Stage of* Culex. Guardian Press, Manchester.

Jensen, D. V. (1953). The morphological and postembryonic development of the salivary glands of *Anopheles albimanus* Wiedemann (Diptera: Culicidae). M.S. Thesis. George Washington University. 56 pp.

Jensen, D. V. and Jones, J. C. (1956). A note on some variations in the structure of the salivary glands of *Anopheles albimanus* Wied. *Ann. Ent. Soc. Amer.*, **50**, 464–9.

Jobling, B. (1928). The structure of the head and mouthparts in *Culicoides pulicaris* L. (Diptera: Nematocera). *Bull. Ent. Res.* **18**, 211–36.

Leon, N. (1923). Ueber die Speichenpumpe der Culiciden. *Zbl. Bakt.* (*Abt. 1, Orig.*) **90**, 361–2.

Leon, N. (1924). *Contributions à l'étude de parasites animaux de Roumanie 1898–1924.* Bucharest.

Macloskie, G. (1887). Poison fangs and glands of the mosquito. *Science*, **10**, 106–7.

Macloskie, G. (1888). The poison apparatus of the mosquito. *Amer. Nat.* **22**, 884–8.

Macloskie, G. (1898). (Edit.) *Brit. Med. J.* **2**, 901–3.

Marshall, J. F. (1938). *The British Mosquitoes.* Brit. Mus. (Nat. Hist.), London.

Martini, E. (1920). Über Stechmücken besonders deren europäische Arten und ihre Bekampfung. *Arch. Schiffs.- u. Tropenhyg.* **24**, Beih. 1, 1–267.

Neumann, R. O. and Mayer, M. (1914). Atlas und Lehrbuch wichtiger tierischer Parasiten und ihrer Überträger. In Lehmann's *Mediz. Atlanten.* Vol. xi. München.

Nuttall, G. H. F. and Shipley, A. E. (1903). Studies in relation to malaria. The structure and biology of *Anopheles. J. Hyg., Camb.*, **3**, 166–215.

Perfiliev, P. P. (1930). Zur Frage über die vergleichende Anatomie von *Anopheles. Mag. Parasit., Leningrad*, **1**, 75–96. Abstract in *Rev. Appl. Ent.* **19**, 230.

Pouchet, F. (1847). Sur l'apparat digestif du Cousin (*Culex pipiens* Linn.). *C.R. Acad. Sci., Paris*, **25**, 589–91.

Rajindar Pal (1943). On the histological structure of the midgut of mosquitoes. *J. Malar. Inst. India*, **5**, 247–50.

Richins, C. A. (1938). The metamorphosis of the digestive tract of *Aëdes dorsalis* Meigen. *Ann. Ent. Soc. Amer.* **31**, 74–83.

Robinson, G. G. (1939). The mouthparts and their function in the female mosquito *Anopheles maculipennis. Parasitology*, **31**, 212–42.

Shute, P. G. (1940). Supernumerary and bifurcated acini of the salivary glands of *Anopheles maculipennis. Riv. d. Malariol.* **19**, Sez. 1, 16–19.

Sinton, J. A. and Covell, G. (1927). The relation of the morphology of the buccal cavity to the classification of Anopheline mosquitoes. *Ind. J. Med. Res.* **15**, 301–8.

REFERENCES

SNODGRASS, R. E. (1935). *The Principles of Insect Morphology*. McGraw-Hill Book Co., New York and London.

SNODGRASS, R. E. (1943). The feeding apparatus of biting and sucking insects affecting man and animals. *Smithson. Misc. Coll.* **104**, no. 1, 1–151.

SNODGRASS, R. E. (1944). The feeding apparatus of biting and sucking insects affecting man and animals. *Smithson. Misc. Coll.* **104**, no. 7, 37–50.

SNODGRASS, R. E. (1953). The metamorphosis of a fly's head. *Smithson. Misc. Coll.* **122**, no. 3, 1–25.

THOMPSON, M. T. (1905). The alimentary canal of the mosquito. *Proc. Boston Soc. Nat. Hist.* **32**, 145–202.

TREMBLEY, H. L. (1951). Pyloric spines in mosquitoes. *J. Nat. Malar. Soc.* **10**, 213–15.

WIGGLESWORTH, V. B. (1942). *The Principles of Insect Physiology*. Ed. 2 (and subsequent editions). Methuen and Co., London.

YAGUZHINSKAYA, L. W. (1940). Présence d'une membrane péritrophique dans l'estomac de la femelle adulte d'*Anopheles maculipennis*. *Mag. Parasit., Leningrad*, **9**, 601–3.

(*b*) THE MUSCULAR SYSTEM

ANNETT, H. E., DUTTON, J. E. and ELLIOTT, J. H. (1901). Report of the malaria expedition to Nigeria. Part II. *Lpool Sch. Trop. Med.* Mem. **4**, 73–89.

CHRISTOPHERS, S. R. (1915). Pilotaxy of *Anopheles*. *Ind. J. Med. Res.* **3**, 362–70.

DIMMOCK, G. (1881). *Anatomy of the Mouthparts and of the Sucking Apparatus of some* Diptera. Boston.

DIMMOCK, G. (1883). Anatomy of the mouthparts and suctorial apparatus of *Culex*. *Psyche*, **3**, 231–41.

DIRKES, L. (1928). Das larval Muskelsystem und die Entwicklung der imaginal Flugmuskulatur von *Psychoda*. *Z. Morph. Ökol. Tiere*, **11**, 182–228.

HOCKING, B. (1952). Autolysis of flight muscles in a mosquito. *Nature, Lond.*, **169**, 1101.

JOBLING, B. (1928). See under section (*a*).

KULAGIN, N. (1905). Der Kopfbau bei *Culex* und *Anopheles*. *Z. wiss. Zool.* **83**, 285–335.

MAKI, T. (1938). Studies of the thoracic musculature of insects. *Mem. Fac. Sci. Agric. Taihoku Imp. Univ.* **24**, no. 1, Entom. no. 10.

MORISON, G. D. (1927–29). The muscles of the adult honey-bee (*Apis mellifera* L.). *Quart. J. Micr. Sci.* **71**, 395–463, 527–651, **72**, 511–26.

NUTTALL, G. H. F. and SHIPLEY, A. E. (1901, 1903). Studies in relation to malaria. *J. Hyg., Camb.*, **1**, 458–80; **3**, 166–215.

PRASHAD, B. (1918). The thorax and wing of the mosquito. *Ind. J. Med. Res.* **5**, 610–40.

ROBINSON, G. G. (1939). See under section (*a*).

SAMTLEBEN, B. (1929). Anatomie und Histologie der Abdominal- und Thoraxmuskulatur von Stechmückenlarven. *Z. wiss. Zool.* **134**, 179–269.

SNODGRASS, R. E. (1912). The thorax (*Psorophora ciliata*). In Howard, Dyar and Knab, **1**, 55–9.

SNODGRASS, R. E. (1927). Morphology and mechanism of the insect thorax. *Smithson. Misc. Coll.* **80**, no. 1.

SNODGRASS, R. E. (1935). See under section (*a*).

THOMPSON, M. T. (1905). See under section (*a*).

XXVII
THE IMAGO: INTERNAL STRUCTURE (cont.)

(*c*) THE TRACHEAL SYSTEM

LITERATURE AND TECHNIQUE

A general account of the tracheal system is given by Snodgrass (1935) and a good short summary by Imms (1938) and by Weber (1938). See also especially Landa (1948) who gives a nomenclature based on the tracheal system of the ephemerid larva; Lehmann (1926) on the tracheal system and its development in the grasshopper with general remarks on homology; Kennedy (1922) on homology of the system in insects; Ripper (1931) on homology in arthropods.

The majority of descriptions of the tracheal system in Nematocera relate to the larva. Among authors who have described the system in the imago in Nematocera, or referred to this in describing that of the larva or pupa, are: Taylor (1902), *Simulium* (larva and imago); Bodenheimer (1923), *Tipula* (larva); Feuerborn (1927), *Psychoda* (larva). There does not appear to be any adequate description in the literature of the tracheal system of the imaginal mosquito. Recently a study of the tracheal system of a number of species of Diptera, including the mosquito, has been made by Miss J. Thomas using the injection technique of Wigglesworth referred to later, but at the time of writing these results have not been published. The early figure given by me (Christophers, 1901), whilst indicating the general arrangement of the system, is incorrect in several respects and the description of the system by Nuttall and Shipley (1901) is not very detailed. In this connection it may be noted that the paper by Pouchet (1872) entitled 'Développement du système trachéen de l'Anophèle' relates not to *Anopheles*, but to *Chironomus*.

Many attempts have been made to arrive at a satisfactory method of injecting the tracheal system. Recently a technique has been described by Wigglesworth (1950, 1953, 1954) which gives very remarkable results, enabling even the finer branches and tracheoles to be displayed. The following is a brief description. For fuller details the author's papers should be consulted.

The insect to be injected is placed alive in a small test tube with a hole drilled in the bottom (or alternatively closed with fine metal gauze) and the tube arranged in an outer receptacle in such a way that, when air has been drawn out of the receptacle, the tube can be pushed down, the insect submerged in the injection fluid and air then readmitted. This is done by means of a thin glass rod set in the cork of the tube and through a hole in the cork of the outer receptacle. To enable the rod to be easily manipulated it should be wetted with glycerol. There must be a slit or other opening in the cork of the tube to allow free communication between it and the outer vessel. As described by the author the vacuumising tube from the receptacle leads to a three-way tap by which connection can be made with a vacuum pump and manometer, or alternatively with a supply of hydrogen in a bladder or to the outside air. Only very moderate negative pressure, however, is necessary and this can if desired be easily obtained from mouth suction. It is uncertain too whether the use of hydrogen is necessary or adds to the effectiveness. Good results have been obtained with *Aëdes aegypti* using a negative pressure of 60–70 mm. Hg kept up for 10 minutes or more. This was not found to cause bubbling of the preparation used for injection (see below). But a negative pressure as low as 10–15 mm. Hg can be used, this being the pressure originally mentioned by the author. Newly emerged insects are to be preferred, especially if these can be obtained before pigmentation sets in, for example white pupae. Curiously enough *A. aegypti* appears to stand even greater air deprivation than that noted without very marked effect beyond some restlessness.

The injection medium consists of cobalt naphthenate, a tar-like substance, made up with one or two times its own volume of white spirit (a light petrol used by painters). After a brief stay in the injection fluid and readmission of air the insect is quickly rinsed in several changes of white spirit and transferred to white spirit saturated with H_2S through which H_2S is bubbled. The insect is then fixed in Carnoy and may be sectioned or mounted whole or in desired portions or slices. Exposure to H_2S can alternately be very conveniently made by the following procedure. To a tube half filled with white spirit add a few millilitres of ammonium sulphide solution. This sinks below the spirit and does not mix with it. Pipette in a few drops of acetic acid. This causes evolution of H_2S with which on shaking the tube the white spirit becomes saturated. The injected insect remains in the white spirit and does not sink into, or even make contact with, the ammonium sulphide solution. Alternatively the insect can be placed on wet filter paper on the cork of the tube. A few minutes' exposure seems all that is necessary. To minimise the time exposed to white spirit the insect can be transferred, directly after rinsing, to Carnoy that has been treated by the addition of a few drops of ammonium sulphide solution. All these methods give blackening of the tracheal vessels.

For whole mounted specimens the effect is greatly enhanced by bleaching. The method recommended by Wigglesworth in his first paper to avoid the effect of diaphanol or hydrogen peroxide upon the injection material is, before bleaching, to soak well first in saturated potassium ferricyanide. Care should be taken that no steel instruments or needles should come in contact with the specimen at any stage. In place of canada balsam and xylol, specimens, after bleaching and dehydration, may be cleared in tetralin and mounted in DPT medium [distrene 80 (10 g.) dissolved in tetralin (40 ml.) with the addition of tricresyl phosphate (7·5 ml.)] (see Giglioli, 1953; also Baker, 1945, p. 178). Various alternative methods of treating the injected insects are given by Wigglesworth in the papers noted regarding which the author should be consulted in original.

GENERAL REMARKS ON THE IMAGINAL SYSTEM

The tracheal system in the imago conforms in essence to the general plan as seen in the larva. It has, however, been profoundly modified to meet altered conditions, namely (1) the complete change in the site of entry of air to the system, (2) the vastly greater air supply needed for the now existing massive muscles of the wings and legs, necessitating new and enlarged tracheal vessels especially in the thorax.

The large terminal spiracles of the larva have disappeared and the main entry of air is now via the mesothoracic and metathoracic spiracles. In the abdomen, in place of the puncta and spiracular cords of the larva and pupa, there are now open spiracles with open spiracular tracheae on abdominal segments II–VII. The absence of spiracles on segment I has been noted by Nuttall and Shipley (1901). Remarkably, the pupa has large functional spiracles in this situation as described by Hurst in 1890. But in the imago nothing remains to indicate their presence. Normally too in insects the last pair of spiracles are those on the eighth segment. Even in Diptera there is a pair in the larva posterior to those on segment VII which it is arguable are those of segment VIII. In the imago of Diptera, however, the last pair, as in the mosquito, is on segment VII, suggesting that the absence of spiracles on segment VIII is due to their precocious development in the larva (see also later reference to the tracheal supply of segment VIII in which there is a suggestive indication of obliterated spiracles on this segment).

The tracheal vessels of the system have also undergone considerable changes. The great tracheal trunks that in the larva play such a dominant role are still recognisable, but no longer dominant. In the thorax they are still the largest tracheae, but now, except in their extreme posterior part, they are rigid tubes approximately circular in section, unlike the widely dilated trunks of the larva. Furthermore, they have become subordinated in function to numbers of newly developed large tracheae. As a result of these new tracheae being of large size, with much ballooning of the parts where they take origin from the spiracular atrium, the original relation of the trunks to the spiracles as represented by the puncta and connecting cords is obscured, the trunks appearing to open directly into the dilated atrial cavity.

Still greater changes have taken place in these trunks in the abdomen. Except for a narrow connecting passage in abdominal segment I there is now no functional connection of the trunks with the tracheal knots which in the larva were situated at their external borders and opened out from them. The trunks in the imago have lost all resemblance to their original character and if, in the older mosquito, not removed altogether are represented by tissue in process of active removal. With these changes the abdominal tracheal system has become quite independent of the trunks with some interesting consequences referred to later.

THE SPIRACLES (Figs. 70–2)

In the imago the spiracles are two pairs in the thorax, an anterior (mesothoracic) and a posterior (metathoracic) pair, and a pair on each of the abdominal segments II–VII. Their structure has been described by Hassan (1944) in *Anopheles maculipennis* and in *Theobaldia annulata*; see also Alessandrini and Missiroli (1926). Both thoracic and abdominal spiracles in *Aëdes aegypti* conform closely to Hassan's description.

The *mesothoracic spiracles* are situated on each side in a depressed triangular area of membrane close beneath the lateral edge of the scutum and between the posterior pronotum and the post-spiracular sclerite (Fig. 72 (3–5)). The opening, *atrial opening* (*ato*), is elongate-oval, or more slit-like, depending upon the extent to which the spiracle is open or closed. It is aligned obliquely, its dorsal end being posterior. On each side of the opening is a hairy bar-like thickening, *peritremes* (of Snodgrass), *peritremal sclerites* (of Hassan), the anterior a little shorter than the posterior. When the opening is narrow the peritremes lie approximately parallel, but with the opening widened they diverge correspondingly at their upper ends. Seen in section they have a certain depth and, as their inner surfaces are covered with small hairs, some approach to a hair-lined outer chamber of the spiracle is made (Fig. 72 (7)). At the bottom of this inner surface there arise thin, almost valve-like,

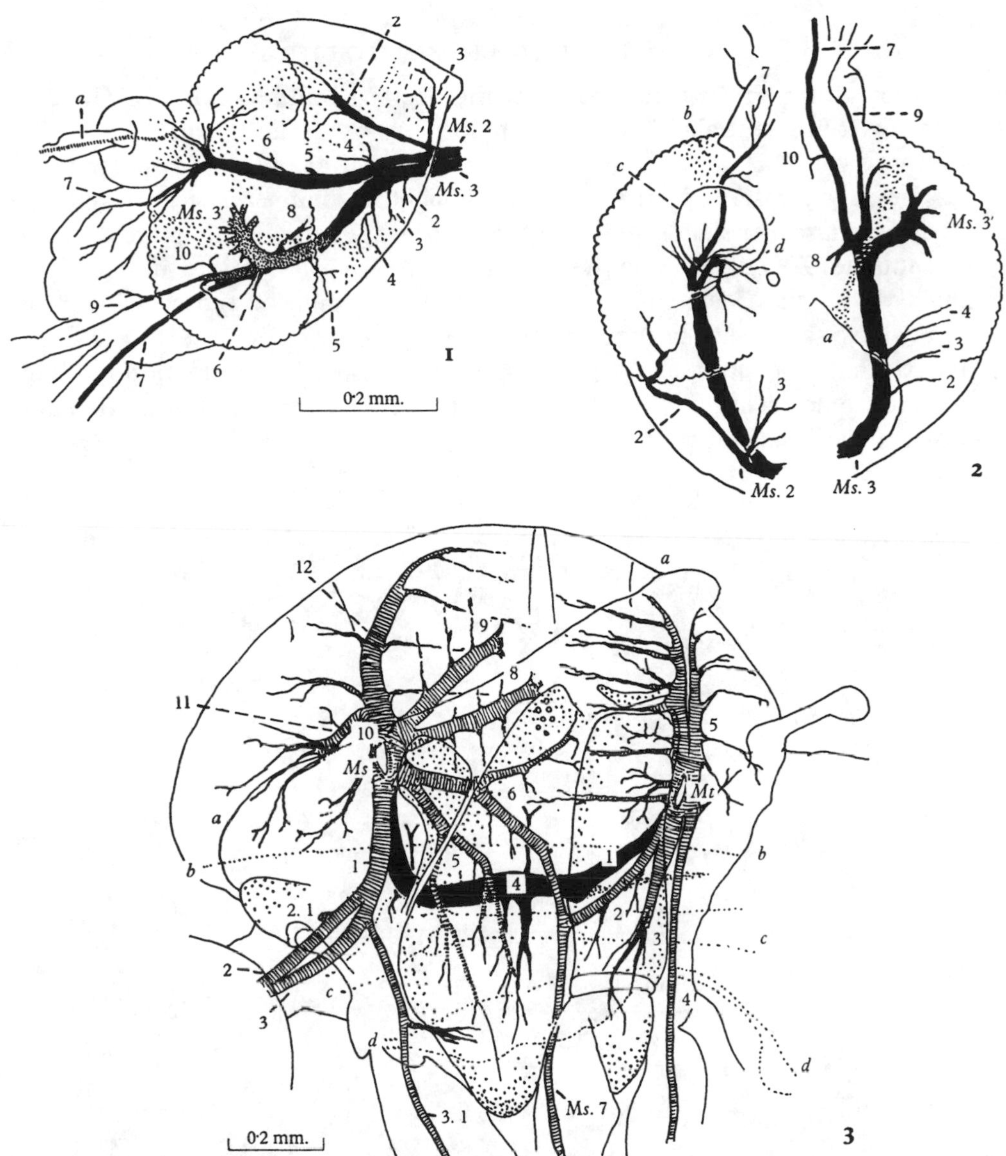

Figure 70. Tracheal system of head and thorax.

1 Lateral view of tracheation of head showing numbered branches of dorsal cephalic (*Ms.* 2) and ventral cephalic (*Ms.* 3). The numbers are those given in the text for the branches of the tracheae respectively but without their full notation, for example the figure 2 when shown as a branch of *Ms.* 2 would in full notation be *Ms.* 2. 2 and 3 would be *Ms.* 2. 3. Similarly if shown as a branch of *Ms.* 3 the figure 2 would indicate *Ms.* 3. 2. *a*, base of antennal flagellum showing a trachea-like blood vessel from the antennal pulsatile organ.

2 Dorsal view of tracheation of head showing on the left side the dorsal cephalic and on the right the ventral cephalic. The branches indicated as in the first figure. *a*, position of tentorium in relation to the ventral cephalic. The tentorium is dorsal to the trachea; *b*, position of anterior end of tentorium; *c*, outline of pedicel; *d*, position of frons. The organ of Clements is shown by a small circle.

3 Lateral view of tracheation of the thorax. The numbers are as given in the text for *Ms* and *Mt* branches. *a–a*, line of tergo-pleural suture; *b–b*, dotted line indicating lower limit of the longitudinal indirect wing muscle bundles; *c–c*, dotted lines indicating position of the midgut; *d–d*, dotted lines indicating position of the ventral nerve cord ganglia.

clear expansions forming the actual edges of the opening, *spiracular lids* of Hassan (*li*); being more developed ventrally they form a calliper-like border to the atrial opening. The atrial opening is closed by a single small muscle at the ventral end of the opening (muscle no. 58) as previously described. It was at first thought from appearances in sections that there was also a muscle arising from the tergite at the dorsal end of the spiracle (originally entered as muscle 57, but later omitted from the list). This seems to be merely some thickening with fibrous structure in the membrane in this region. A muscle which from its position may have some effect in enlarging the opening, however, is muscle 59 arising from a large part of the inner aspect of the post-spiracular sclerite. Its contraction by dragging on the sclerite would seem likely to enlarge the opening, even if it had other functions to perform as a pleural indirect wing muscle.

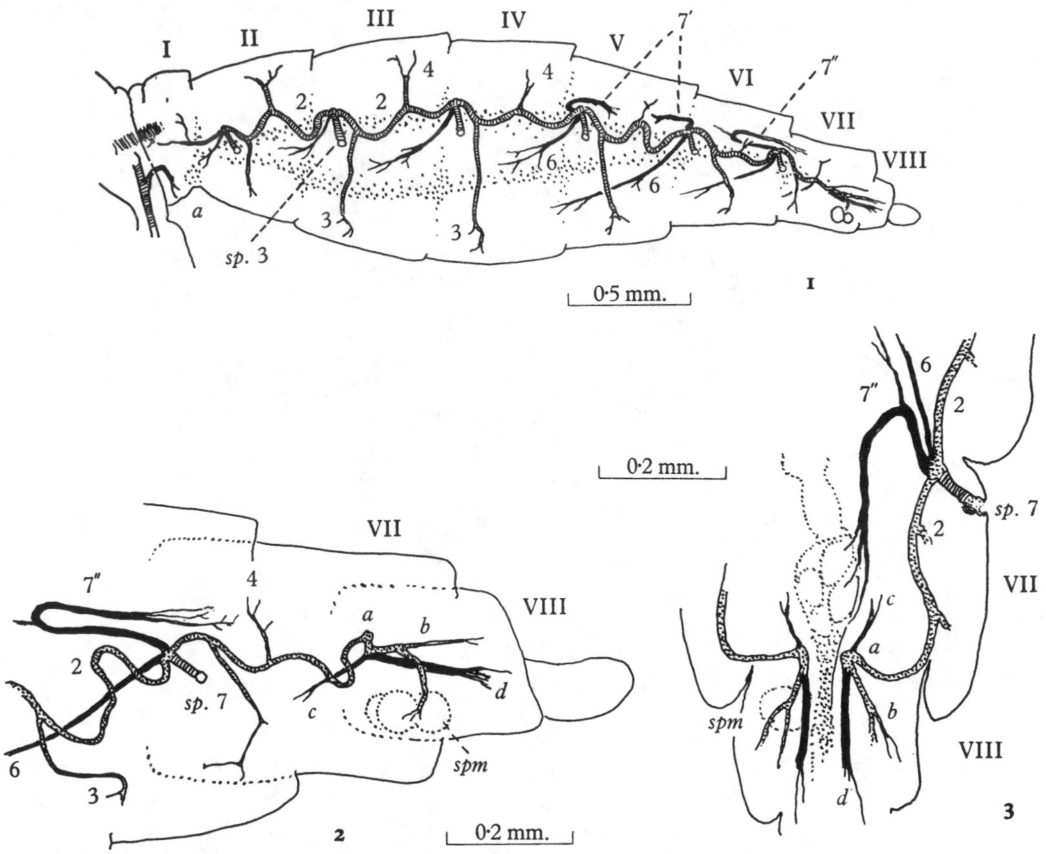

Figure 71. Tracheal system of the abdomen.

1 Showing tracheation of the abdomen. The figures indicate the tracheae as listed in the text, namely (1) spiracular (not numbered); (2) connective; (3) ventral; (4) dorsal; (5) accessory ventral (not figured but arising, when present, from the connective opposite to the dorsal trachea); (6) visceral; (7′) posterior visceral to ovary; (7″) the same to rectum. *a*, branch from trachea of hind leg into first abdominal segment.

2 Showing tracheation of segments VII and VIII; lateral view left side. *a*, appearance suggesting tracheal knot in segment VIII; *b*, connective of segment VIII giving off ventral branch; *c*, visceral branch of segment VIII; *d*, posterior visceral of segment VIII.

3 The same, dorsal view. Lettering as for 2.

Lettering: *sp.*3, spiracle of segment III; *sp.*7, spiracle of segment VII; *spm*, spermathecae.

The atrial opening passes into a shallow cavity of somewhat greater extent than the spiracle as marked by the peritremes (Fig. 72 (7)). Into this cavity open a number of large tracheae. Certain of these (tracheae nos. Ms. 5, 6, 8 and 9) open superficially into the sides of the cavity, but the thoracic trunks, the common cephalics and the large tracheal branches supplying the muscles in the dorsal portion of the thorax enter deeply into its base. The walls of the cavity are to a certain extent free from taenidia so that it may be regarded as the *atrium*, though greatly encroached upon by the massive tracheae opening into it. The spiracle in this case would appear to fall into the class of atrial spiracle with lips and peritremes as figured by Snodgrass (1935) in his Fig. 230 B, the lips and peritremes being more or less flush with the surface and the atrium with no closing apparatus at the entrance of the trachea.

The *metathoracic spiracle* is situated behind the upper part of the mes-epimeron a little anterior and ventral to the base of the haltere and on the rather flat and partially membranous metapleuron. It is of very similar construction to the mesothoracic spiracle, but the hair-bearing peritreme thickenings form a complete ring enclosing the narrowly oval atrial opening, and the lids are less conspicuous. As in the mesothoracic spiracle the atrial opening passes directly into a shallow cavity representing the atrium, but of which the walls are largely formed by the entering tracheae. There is one small muscle arising from the lower part of the mes-epimeron and inserted into the ventral end of the spiracular opening which acts as a closing muscle. As with the mesothoracic spiracle there is no closing apparatus at the entrance of the tracheae into the atrium.

The abdominal spiracles (Fig. 72 (9–11)). These are situated in the pleural membrane towards the anterior end of their respective segments. They are very similar in all the segments in which they occur. Though small compared with the thoracic spiracles they have a well-developed closing apparatus at the tracheal entrance to the atrium. At their external orifice there is a slight depression of the microtrichae-covered pleural membrane and at the bottom of this a circular 'lip'. This bears hairs and has a certain thickness so that a small superficial chamber is formed at the bottom of which is a thickened ring leading into a small but well-marked atrium (*a*), the walls of which are free from taenidia and thickened. From the atrium a narrow passage leads into the relatively wide *spiracular trachea*. At this point the atrial wall is specially thickened, forming a boss-like elevation (*c*) from which there extends a strongly sclerotised *bar* (*ap*). Associated with these structures are small, but typically striated, muscular bands enclosed in a capsule that forms a projecting bulge externally where atrium meets trachea (Fig. 72 (11)). The spiracular trachea is a straight length of trachea with well-marked taenidia which passes from the spiracle to the segmental knot where the chief tracheal branches of the segment meet or take origin. At this point, which may be termed the *segmental tracheal knot*, the component tracheae commonly form a triple knuckle-like structure which in mounted preparations is often more conspicuous than the spiracle and may cause the quite long spiracular trachea to be overlooked.

LIST OF MAIN TRACHEAE (Figs. 70, 71)

Below are given the larger tracheae and their main branches. The numbers given to them are those by which they are indicated in the figures and by which they can if necessary be referred to in the text. They are listed in relation to the spiracles with which they are connected, those connecting two spiracles being listed under the more anterior spiracle, unless there are reasons to the contrary. Those connected with the mesothoracic spiracle are listed as Ms. followed by a number. Those connected with the metathoracic spiracle are listed as Mt. and those of the abdomen under the numbered abdominal segment to which the spiracle belongs. The dorsal trunks in the thorax are treated as they take their place as tracheal vessels.

Ms. 1. *Common cephalic.* Takes origin from the spiracle ventrally and just external to the thoracic tracheal trunk. After passing forwards to the level of the anterior pronotal lobe each common cephalic divides into:

Ms. 2. *Dorsal cephalic.* Passes forwards through the neck to enter the head through the occipital foramen. At about the middle of its course to the foramen it is linked to its fellow of the opposite side by a short commissure, *anterior commissure* (Ms. 2. 1). For branches in the head see under 'Tracheation of the head' in a subsequent section.

Ms. 3. *Ventral cephalic.* Passes forwards along with Ms. 2 to enter the head through the occipital foramen a little ventral and external to this trachea. At about the same level as the anterior commissure it gives off the *trachea of the first leg* (Ms. 3. 1). For branches in the head see as for the dorsal cephalic.

Ms. 4. *Thoracic trunk.* A wide tracheal trunk arising from the lower part of the atrium, passing at first ventrally and then horizontally through the thorax below the medianly situated longitudinal indirect wing muscles to connect up by a branch with the lower part of the meta-

---

Figure 72. Tracheal trunks and spiracles.

1 Transverse section of trunks at level of mesofurca. The trunks show as rigid tubes circular in section. *a*, protomorphic muscular strands, dorsal series; *b*, the same, ventral series; *c*, lower limit of longitudinal indirect wing muscles.

2 Transverse section of trunks at entrance to abdomen. Lettering as for 1.

3 Mesothoracic spiracle. External view with anterior direction to left. The bar across the atrial opening is the ridge on the floor of the atrium separating the opening into the atrium of tracheae Ms. 10 above and Ms. 4 and Ms. 1 (trunk and common cephalic) below. *a*, membrane with dotted line showing limits of atrium beneath; *b*, closing muscle (muscle no. 58).

4 Showing relation of mesothoracic spiracle to thoracic sclerites.

5 Muscles in region of the mesothoracic spiracle.

6 Metathoracic spiracle. External view with anterior direction to left. About half magnification of 3. *a*, membrane with dotted line showing limits of atrium beneath; *b*, closing muscle (muscle no. 60); *c*, depressor muscle of haltere (muscle no. 56).

7 Horizontal section through mesothoracic spiracle. *a*, tracheae entering atrium laterally and superficially; *b*, trachea entering floor of atrium.

8 Atrial opening of abdominal spiracle as seen in tangential section just not cutting the opening. In the right-hand upper corner is left intact an area of cuticle with its microtrichae.

9 Horizontal section through abdominal spiracle. *a*, depression in pleural membrane leading to a circular 'lip'; *b*, atrium; *c*, opening into the spiracular trachea with boss-like thickening; *d*, spiracular trachea.

10 Section across abdominal spiracle.

11 Closing bar and muscle.

Lettering: *ap*, apodeme of closing apparatus; *ato*, atrial opening; *Dv*, ventral diverticulum; *fp*, furca; *g*, ganglion; *ht*, heart; *li*, lip of atrial opening; *m*, muscle of closing apparatus; *Mg*, mid-gut; *ppn*, posterior pronotal area; *psp*, posterior spiracular area (sclerite); *ptg*, paratergite; *ptn*, portion of postnotum just included in section; *ptr*, peritreme of atrial opening; *sc*, edge of scutum; *tt*, tracheal trunk.

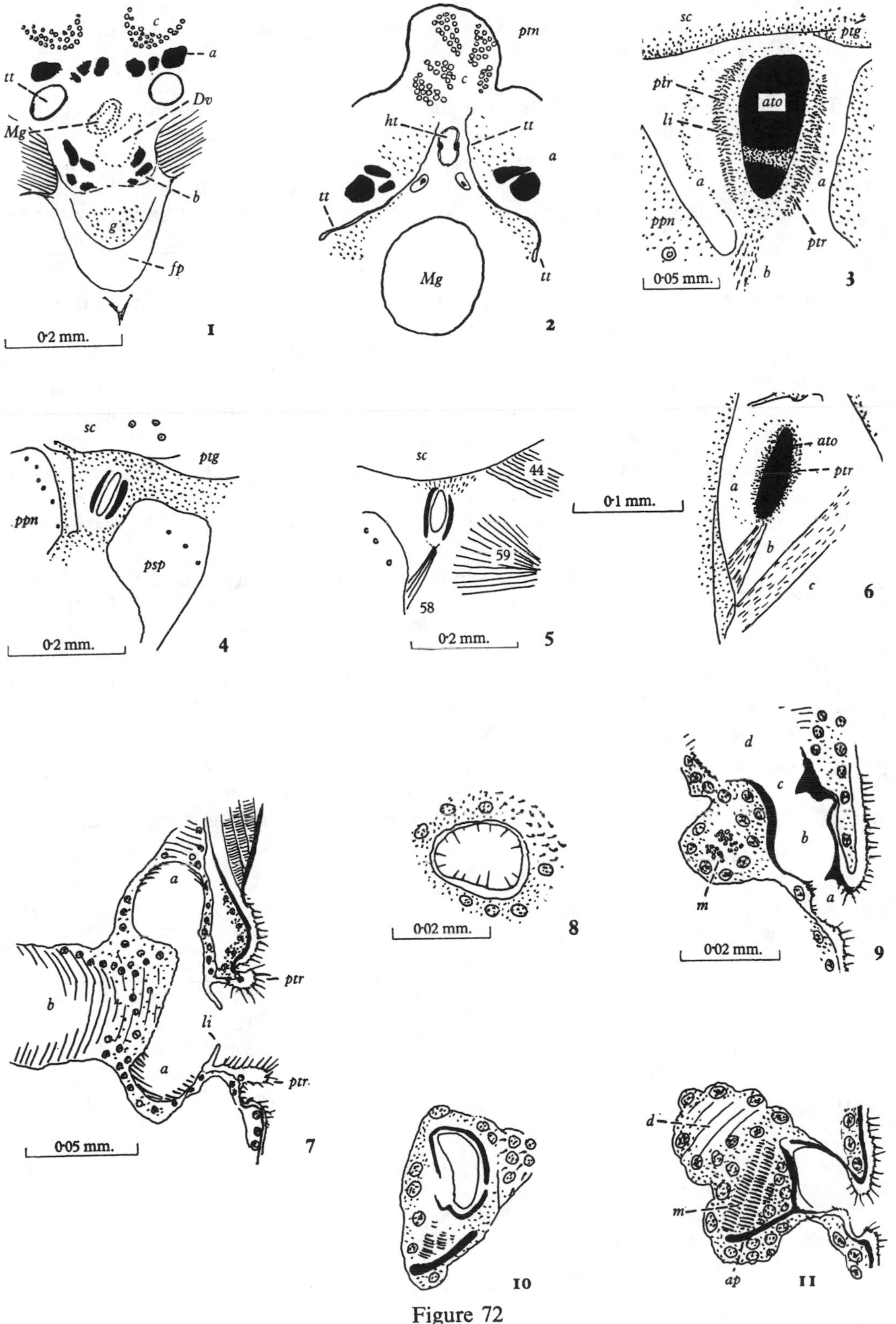
c
a
tt
Dv
Mg
b
g
fp
0·2 mm.
1
ptn
c
ht
tt
a
tt
tt
Mg
2
sc
ptg
ptr
ato
li
a
a
ppn
ptr
b
0·05 mm.
3
sc
ptg
ppn
psp
0·2 mm.
4
sc
44
59
58
0·2 mm.
5
ato
ptr
0·1 mm.
a
b
c
6
a
ptr
b
li
a
ptr.
0·05 mm.
7
0·02 mm.
8
d
c
b
m
a
0·02 mm.
9
10
d
m
ap
11
Figure 72

thoracic spiracular atrium. Continued into the abdomen in a modified form only. For branches and characters of the thoracic and abdominal portions see under 'Tracheation of the thorax' and 'Tracheation of the abdomen'.

Ms. 5. A trachea arising superficially from the ventral portion of the atrium posteriorly to Ms. 1 and in common with Ms. 6 and passing into the lower lateral portion of the sternopleuron of its side.

Ms. 6. A trachea arising from a short common portion with Ms. 5, giving off a dorsal branch to the area of the wing root and forming with Mt. 2 the mesothoracic-metathoracic connective from which is given off:

Ms. 7. *Trachea of the middle leg.*

Ms. 8. A trachea taking origin superficially from the posterior portion of the atrium and passing backwards just ventral to the tergo-pleural suture. Gives off branches dorsally and ventrally to the anterior dorso-ventral indirect wing muscles of its side.

Ms. 9. A trachea arising superficially from the dorsal portion of the atrium and passing backwards just dorsal to the tergo-pleural suture and giving off branches to the anterior dorso-ventral indirect wing muscles of its side.

Ms. 10. A short very stout trachea arising dorsally from the deepest part of the atrium and after giving off a branch anteriorly (Ms. 10. 1) dividing into:

Ms. 11. Short wide trachea passing forwards into the anterior portion of the dorsum of the thorax and branching freely to the muscles in this area.

Ms. 12. Similar but larger trachea passing posteriorly and supplying parallel branches among the bundles of the longitudinal indirect wing muscles anteriorly and posteriorly, the latter extending backwards to come into connection with corresponding branches from Mt. 5 passing forwards.

Mt. 1. Trachea linking up the metathoracic atrium with the thoracic trunk, Ms. 4, and having the appearance of the terminal portion of this, the true continuation being flattened and specially modified, see under 'Tracheation of the abdomen'.

Mt. 2. Trachea arising more superficially from the anterior ventral portion of the atrium, giving off branches to the posterior dorso-ventral indirect wing muscle of its side and joining with Ms. 6 to form a connective loop from which the *trachea of the middle leg* takes origin.

Mt. 3. A stout trachea passing ventrally and inwards from the ventral portion of the atrium to the coxal and sternal regions of the middle leg.

Mt. 4. The *trachea of the hind-leg*. Takes origin from the posterior ventral portion of the atrium and passes down behind the mes-epimeron to the hind-leg.

Mt. 5. A massive trachea arising from the dorsal portion of the atrium and passing, after dividing into two branches, to the region of the scutellum. It gives off numerous branches, notably a series of parallel branches passing between the bundles of the longitudinal indirect wing muscles and another series passing outwards to the posterior dorso-ventral indirect wing muscle bundles of its side.

I–VIII. 1.* *Spiracular trachea.* From each abdominal spiracle to tracheal knot. Not in segments I or VIII.

I–VIII. 2. *Connective.* The designation applies to the connective passing from any given segmental knot to that of the segment behind. From it are given off:

I–VIII. 3. *Ventral trachea.* Arises from the connective a little distance posterior to the knot and passes more or less directly to the ventral region of the segment.

I–VIII. 4. *Dorsal trachea.* Arises towards the posterior end of the connective and passes dorsally.

* For description and interpretation of tracheal branches in segments I and VIII see section on 'Tracheation of the abdomen', p. 612.

I–VIII. 5. *Accessory ventral.* A small trachea commonly arising near the dorsal trachea but passing ventrally.

I–VIII. 6. *Visceral trachea.* Arises from the anterior part of the knot and passes inwards and forwards to supply viscera some distance anterior to the segment.

V–VIII. 7. *Posterior visceral.* Large branches arising from inner side of knot and passing to ovary, testis, or rectum depending on segment.

TRACHEATION OF THE HEAD (Fig. 70 (1), (2))

The tracheal supply of the head is derived from the dorsal and ventral cephalic tracheae. After giving off respectively the anterior commissural trachea (Ms. 2. 1) and the trachea to the first leg (Ms. 3. 1) these enter the head on each side through the occipital foramen lying close together, the dorsal cephalic a little dorsal and internal to the ventral cephalic.

The *dorsal cephalic* (Ms. 2) immediately after passing through the foramen gives off a large branch, the *occipital trachea* (Ms. 2. 2). This passes outwards and upwards in the occipital area giving off branches behind the brain and extending to the posterior dorsal portion of the compound eye of its side with a branch to the imaginal ocellus. As with many of the large tracheae of the head it may show much variability in size and may be locally somewhat ballooned. From near its origin there passes directly dorsally a small trachea (Ms. 2. 3) which supplies the dorsal dilator muscle of the oesophageal pump. The main dorsal cephalic continues forward beneath the brain in the groove between the optic lobe externally and, in succession, internally the protocerebrum and the deutocerebrum (antennal lobe). Shortly after entering the groove it sends a branch upwards into the protocerebrum (Ms. 2. 4) and after passing some distance forwards one outwards into the optic lobe (Ms. 2. 5). Just before leaving the groove it gives off a branch in the region of the antennal lobe (Ms. 2. 6). At each of these points where the branches enter the brain substance there is present a conspicuous large nucleus (tracheal cell) which, embedded among the smaller nerve cells of the part, might be mistaken for a larger ganglion cell. Similar cells are present at other points where tracheae enter the brain and are useful as giving an indication of such entry.

Where the dorsal cephalic emerges from the groove at the front of the brain it lies just external to the broad nerve tract passing from the antennal lobe of the brain to Johnston's organ, here termed the antennal connective to distinguish it from the nerve to the filament. At this point the dorsal cephalic breaks up rather abruptly into branches coming mainly from two main divisions of the trunk, a dorsal and a lateral. The dorsal portion sends branches which largely supply muscles and other structures in the frontal area and others passing backwards over the anterior surface of the brain. It also sends branches into the base of the antenna, supplying structures in the pedicel from which twin tracheae pass into the flagellum. In the more basal flagellar segments there is, in addition to the two small tracheal branches, a conspicuous tube resembling, and which has been mistaken for, a trachea, but which is without taenidia and is the antennal blood vessel described later passing from the pulsatile organ near the base of the antenna into the flagellum of the antenna. The lateral branch of the dorsal cephalic continues forwards to enter the clypeus at its lateral angles. It largely supplies the muscles in this structure.

The *ventral cephalic* is somewhat larger than the dorsal cephalic. After entering the occipital foramen it passes on each side over the lateral dilators of the oesophageal pump giving off small branches that pass between the muscular bundles. Most usually there are two single branches followed by one breaking up into three (Fig. 70 (1), (2)) (Ms. 3. 2–4). A little further forwards is a small branch passing to the genal area behind the eye. In its forward course the ventral cephalic crosses the tentorium at about one-third of its length from its posterior end lying external to this structure and the depressor muscle of the head (muscle no. 25). Here a short branch arises externally (Ms. 3. 6) giving branches to the lower ocular area. On its inner side passing round the tentorium the trachea gives off the large branch that is continued to the proboscis as the labial trachea of its side (Ms. 3. 7). What appears as the continuation of the main trachea then passes forwards, outwards and dorsally as a short stumpy trachea giving off numerous branches to the central parts of the compound eye (Ms. 3′). The labial trachea at its origin gives off a branch on its inner side which enters the side of the sub-oesophageal ganglion (Ms. 3. 8), its point of entry being marked by the usual tracheal cell. Some distance further on the labial trachea gives off a branch on its outer side that continues forward to the genal area of the rostrum (Ms. 3. 9). One of its branches passes into the maxillary palps and the other ends in the region of the genal shelf and may possibly supply the mandible and maxilla. Some distance beyond the origin of Ms. 3. 9 the labial branch gives off a trachea to the salivary muscle of its side (Ms. 3. 10). From thence the labial trachea runs a conspicuous straight course forward and into the proboscis running parallel to the trachea of the opposite side. These two parallel large tracheae, which pass through the length of the proboscis in the labial sheath, are in great contrast to the tracheal supply to the other mouthparts which have failed to show any definite tracheal vessels.

In general, therefore, the dorsal cephalic provides tracheal supply to the dorsal areas of the head, the supra-oesophageal ganglion, the dorsal dilators of the oesophageal pump, the scape and pedicel, the frons and clypeus. The ventral cephalic provides tracheation for the lateral dilators of the oesophageal pump, the gnathal muscles and gnathal parts of the head, the sub-oesophageal ganglion, a large portion of the compound eyes, the salivary muscles and the genal, maxillary and labial areas including the proboscis sheath.

TRACHEATION OF THE THORAX

The general character of the tracheal supply to the thorax will be clear from Fig. 70 (3) in which the position of all the main tracheal branches is shown. Many of the tracheae are almost entirely concerned with tracheal supply to the indirect wing muscles which form the greater bulk of the contents of the thorax. Thus the large tracheae passing into the dorsal portion of the thorax, Ms. 10–12 anteriorly and Mt. 5 posteriorly, are mainly concerned with sending parallel branches as shown in the figure between the muscular bundles of the longitudinal indirect wing muscles which occupy the median portion of the dorsal part of the thorax. On each side Ms. 8 and 9 send similar branches between the muscular bundles of the dorsal parts of the anterior dorso-ventral, while Ms. 5 supplies the ventral portion of this muscle. The trachea Ms. 5 divides into two branches, the more posterior passing down the external aspect of the muscle, the more anterior passing inwards and ending in branches about the anterior border and inner aspect of the muscle. The posterior dorso-ventral indirect wing muscles are supplied on each side by Mt. 5 and Mt. 3, the first mentioned supplying the dorsal portion and the latter the ventral. In this case the branches from Mt. 5 are directed outwards, not anteriorly as are those supplying the longitudinal series, and seen fore-shortened in the figure. Mt. 3 is the middle of the three large tracheae passing ventrally from the metathoracic spiracle. It supplies branches to the ventral portions of the posterior dorso-ventral indirect wing muscle of its side extending into the meron. It also sends branches to the sternal muscles. The tracheal supply to the smaller thoracic muscles which are not of the indirect wing muscle type is of a miscellaneous character and seemingly relatively less abundant. Branches in the prothorax are largely derived from the ventral cephalic (Ms. 3). Those of the rest of the thorax are derived from Ms. 5, the thoracic trunk (Ms. 4) and Mt. 3.

Very conspicuous are the long tracheae to the legs. These pass down the whole length of the limb with little or no major branching. Those of the fore-legs take origin from the ventral cephalic. Some little distance from their origin they give off a rather large branch to the ganglion. They also give a branch almost at their origin to the salivary glands and surrounding tissues. The tracheae to the middle legs take origin from the connective loop formed by Ms. 6 and Mt. 2. This is in keeping with the generalised condition where the leg tracheae take off from the connectives. The tracheae of the hind-legs take origin from the posterior part of the atrium of the metathoracic spiracle. Near its origin it gives off a small branch that links up with the tracheal system of the abdomen and has a short tubular connection with a point on the scalloped edge of the modified abdominal trunk.

The tracheal supply to the wing in the imago is almost non-existent. In injected material I have not found the wing veins injected. In the larva and pupa the tracheal supply to the wings is derived from tracheae coming from the anterior portion of the connective loop from which the trachea of the mid-leg takes origin (Christophers and Barraud, 1924). In the imago a corresponding trachea, namely the upper branch of Ms. 6, can be traced to the root of the wing.

A tracheal vessel of special interest is the large thoracic trunk passing from the mesothoracic spiracle backwards into the metathorax. It does not itself pass to the metathoracic spiracle, but is joined to this by a length of trachea (Mt. 1), continuing in a modified form into the abdomen. Up to close to the junction with Mt. 1 the trunk is a rigid tube, more or less circular in section with well-marked taenidia. Just before it is joined by Mt. 1 its inner wall becomes partly membranous and beyond the junction only the outer portion has resemblance to a trachea and this rapidly becomes smaller and flatter. The inner portion is flattened and extended, becoming a flat membranous sheet which, as it passes into the abdomen, becomes increasingly difficult to distinguish from the dorsal membrane, if it is not actually that structure. Injected material has failed to penetrate this modified portion of the trunk.

The thoracic portion of trunk, after passing some distance ventrally from the mesothoracic atrium, makes a sharp bend and subsequently passes on a more or less level course to the metathorax. From each trunk at the bend a small branch passes inwards and then upwards passing close to the sides of the aorta and breaking up into branches in the fat-body in which this structure lies. Near their origin some branches pass from these to the proventriculus. In the level part of its course each trunk gives off branches which pass outwards and ventrally to the neighbouring dorso-ventral indirect wing muscle, and, about their middle dorsally, another branch on each side is given off which, passing upwards, lies close on each side of the aorta and breaks up into branches in the medium gap between the longitudinal muscles of the two sides. Where these branches contact the heart this organ has thickened walls and a number of closely aggregated nuclei (see 'Aorta', p. 616 and Fig. 73 (3)).

The tracheal supply of the conjoined ganglion is derived from (*a*) branches from the trachea of the fore-leg which supplies the prothoracic ganglion; (*b*) branches from the inner division of Ms. 5 which enter the mesothoracic ganglion; and (*c*) branches from Mt. 3 which pass to the posterior part of the conjoined ganglion behind the furca. Large conspicuous nuclei (tracheal cells) are present in the ganglia at the point of entry of branches as described for the brain.

TRACHEATION OF THE ABDOMEN (Fig. 71)

The tracheation of the abdominal segments shows in general considerable uniformity. In all the normal segments, that is excluding segments I and VIII which lack spiracles, a short straight *spiracular trachea* (no. 1 of the list of tracheae) passes from the spiracle to the tracheal knot at which it, the connective of the previous segment, and that of the segment in question meet to form a triple knuckle-like junction (*tracheal knot*). The connective (2) of the segment passes backwards from the knot to the knot of the following segment. It is usually thrown into a wavy or coiled course with a loop directed downwards followed by one directed upwards (Fig. 71 (1)). Towards the anterior end of the connective there is given off the *ventral trachea* (3) which passes ventrally and usually a little backwards to supply the ventral portion of the segment with a branch to the ganglion. From the more posterior bend a branch is given off dorsally, *dorsal trachea* (4), which supplies the dorsal portion of the segment. This branch is large in the more anterior segments, becoming increasingly smaller in the segments towards the posterior end of the abdomen. A small ventrally directed branch, *accessory ventral* (5), is commonly present taking off opposite the dorsal branch.

Where the connective from the previous section joints the knot a trachea takes origin on the inner side of the knot and passes inwards and very characteristically forwards to supply viscera, *visceral trachea* (6). The viscera supplied are usually situated from one to several segments anterior to that from which the trachea takes origin. The visceral tracheae from segments II–V largely supply branches to the mid-gut, those from VI and VII mainly supply the Malpighian tubules. In segments V–VII, in addition to those branches already noted, there is a large trachea taking origin from the posterior part of the knot on its inner aspect which tends to pass backwards to supply important organs, *posterior visceral* (7). In segments V and VI this passes to the ovary where it forms a close coiled or more open plexus depending on the state of development of this organ. In the male it supplies branches to the testis. In segment VII it provides tracheal supply to the anterior dilated portion of the rectum, including the network in the rectal papillae (see chapter XX and Fig. 55 (4)). This trachea often forms a loop, at first passing forwards and then backwards dorsal to the rectum. It sends a branch also forwards.

Tracheation of segments I and VIII departs somewhat from the above, though to some extent homologous tracheae are present. In segment I there is no spiracle and no spiracular trachea. A trachea passes forwards into the segment from the knot of segment II which gives off a dorsal branch and probably represents the connective. Anteriorly it has connection with the abdominal continuation of the thoracic trunk which at this level still shows some cavity. Another trachea takes origin from the trachea of the third leg and enters the segment anteriorly. It forms an upward loop and then passes into the ventral part of the segment. It possibly represents in its terminal portion the ventral trachea of the first segment, its first part being connective, that is the posterior part of the loop from which the trachea of the third leg takes origin.

In segment VIII there is again no spiracle. But the connective from segment VII passes to a point towards the anterior end of segment VIII from which a number of tracheae take origin (Fig. 71 (2), (3) *a*). One of these tracheae divides into a branch continuing laterally and one directed ventrally to the region of the spermathecae and common oviduct. Its appearance suggests that it may represent the connective for segment VIII giving off the ventral branch for this segment. A stout straight trachea passes deeper in along the side of the rectum ending in a brush of small branches supplying the parts in the rectal and anal area. It might correspond to the posterior visceral. A small branch directed anteriorly may be the visceral. There is in fact every indication that this point towards the anterior end of segment VIII represents the tracheal knot of that segment.

In the description of the imaginal circulatory system given later in this chapter there is a reference to the fact that the outer edges of the remnants of tissue which in the imago represent the larval abdominal trunks are characteristically scalloped. The scalloping is due to the trunks, even in this stage, retaining their connection by projecting points with the functional tracheal knots, the edge being concave between these points. It is there further noted that one such point connects with the tracheal knot of the eighth segment, a further support for the view that this is in fact a tracheal knot of the segmental series. It was at this stage that it occurred to me to re-examine the larval structures in this connection. At once it became evident that the supposed tracheal knot in the imago was exactly repeating conditions at the point at the base of the siphon tracheal trunk from which arose tracheae nos. 53–8 (see Fig. 29 (4)). The only difference was that the trunk had been done away with. The knot in the imago has actually the rather peculiar appearance of having been left, so to speak, hanging.

If the characters of the abdominal spiracles in the imago are borne in mind, that is the small outer taenidia-free chamber, the constriction at its base and the stout open trachea leading to the tracheal knot of the segment, it is further not unreasonable to correlate these structures with those of the terminal spiracles and the siphon trunks. On such a basis the larval respiratory siphon consists of spiracles and spiracular tracheae of the eighth segment that have been carried backwards taking with them the ninth tergite as already hypothecated on other grounds, the projecting mass being secondarily provided with sclerotised support, the siphon tube.

Before concluding, attention may be drawn to a notable feature of the imaginal tracheation, namely the almost negligible degree of communication between the thoracic and the abdominal systems. This virtual isolation of the two areas, already suggested by the great difference in the types of spiracle, would seem possibly to have some physiological reason. An explanation might be that the two systems in the imago have to meet different requirements, that of the abdomen to meet ordinary tissue requirements with the means of regulating entry of air in accordance with conditions of humidity, etc., and of a more or less permanent character, and that of the thorax to meet sudden calls for an ample oxygen supply for the great muscles in flight, with at other times little air interchange. It might be advantageous in such circumstances that the two systems did not communicate too freely.

(*d*) THE CIRCULATORY SYSTEM AND ASSOCIATED TISSUES

GENERAL DESCRIPTION

An account of the heart and associated tissues in the larva has already been given. Apart from some early references (Christophers, 1901; Eysell, 1913), there has been little in the literature regarding the heart and circulatory system in the imaginal mosquito. Recently, however, a very full account of the structure and physiology of the heart and associated organs of larva, pupa and imago in *Anopheles*, *Culex* and *Aëdes* has been given by Jones (1952, 1953, 1954) who also deals with the technique for examination of living material in the imago.

That the pulsations of the heart in the living condition are readily studied in the larva through the cuticle has already been noted. That this is also possible in the imago under the dissecting microscope up to a magnification of 250 has been shown by Jones, quoted above. For this the insects, preferably one-day mosquitoes fed on sugar solution, are glued to the slide by the wings and tip of the abdomen, or with the dorsal surface downwards and the slide examined reversed. A special resin, Resin Adhesive no. 502, manufactured by Southern Adhesives Corporation, was used for the purpose. The insects so put up may be kept alive in damp petri dishes. Observations are helped by removal of scales with a soft brush. In counting the heart beats it is desirable to record the time for a limited number of beats and to reduce this to beats per minute, since counting is liable to be interfered with by temporary cessation of beats.

Besides the study of serial sections useful information can be obtained from mounts made from sliced hardened material suitably stained. Such a preparation of the dorsum of the abdomen shows well the characters of the heart and its accessory structures and can be used for examination under polarised light.

As in the larva the chief propulsatory organ is the *dorsal vessel*. This is divisible into an abdominal portion, the *heart*, provided with ostia and associated pericardial cells and alary muscles, and a thoracic portion, the *aorta*, devoid of these structures.

THE HEART (Fig. 73 (1–3))

The heart is a thin-walled muscular tube with laterally placed nuclei at intervals along its course. Its walls show circular and oblique striae as in the larva. It extends from the posterior end of segment VII to the metathorax, lying close beneath the tergites. Towards the anterior end of each of segments I–VII there is a pair of lateral *ostia*. These appear as pouch-like clefts in the lateral wall and their presence is further indicated at each ostium by a pair of small nuclei, markedly smaller than those of the heart wall. These are the nuclei of the *ostial cells* bordering the openings as in the larva. The ostial cells in the imago appear to be less distinctly crescentic and more ballooned than are those in the larva, so that the ostia appear commonly as double-walled blebs in the wall of the heart. Anterior to each pair of ostia, that is, in the region of the group of pericardial cells as described below, the wall of the heart is somewhat bulged, giving rise to *pulsatory chambers* of which, counting the terminal chamber, there are eight.

At its anterior end, a little in front of the first pair of ostia, the heart enters the thorax to become the aorta. Posteriorly it becomes narrower and ends towards the posterior end of segment VII. At its termination it is somewhat blunt dorsally but is continued ventrally as a tapering extension, much as in the larva. Whether there are terminal ostia seems doubtful and the elaborate surrounding tracheal meshwork of the larva is entirely lacking.

Pericardial cells (Fig. 73 (2) *pc*). Ranged along each side of the heart are groups of *pericardial cells* (*dorsal nephrocytes*). These are present in groups of four pairs at each intersegment, two pairs lying in the hinder portion of the segment in front and two pairs in the fore portion of the segment behind, except in segment I, where anteriorly there is only a single pair, and intersegment VII–VIII, where there are two pairs in the hinder part of segment VII and none in segment VIII. Altogether there are, therefore, twenty-seven pairs. Some variation in numbers of cells at the anterior end of segment I and at the VII–VIII intersegment may occur as noted by Rajindar Pal (1944), who gives two pairs as usually present in the anterior group which are considered to be in the metathorax. The single pair in *Aëdes aegypti* imago has an alary muscle, however, which appears to be in abdominal segment I.

The cells closely resemble those in the larva, usually possessing two nuclei and other characters as described for them. The above description refers to large, characteristic pericardial cells of the older literature. Jones, *loc cit.*, however, describes additional *small* and *medium* nephrocytes as also present. Including such, the dorsal nephrocytes according to Jones, *loc. cit.*, may in some species be present in considerable numbers, some hundreds being present in *Anopheles quadrimaculatus*. In *Aëdes aegypti*, in material studied by me, no other cells comparable in size with the typical pericardial cells and resembling them in appearance have been present near the heart, the groups of four pairs, each cell with its alary muscle as described below being highly characteristic and, with the exceptions noted, repeated with great regularity in the segments. It is not known whether conditions of breeding or maintenance may affect the appearances seen.

Alary muscles (Fig. 73 (2) *alm*). At each set of four pairs of pericardial cells there are four pairs of branched *alary muscles*. These take origin from the hypodermis at the intersegment laterally as one muscular stem anterior to, and one posterior to, the intersegmental line, each stem then dividing into two, one for each pericardial cell. Each of these branches further divides into some four or five or more branches spreading out fan-wise above and below the pericardial cell and then ending in a fine network of fibrils, resembling a spider's web, beneath the heart. This meshwork extends for some little distance anterior and posterior to the area of the four pairs of pericardial cells, that is, approximately over the area of the pulsatory chamber, but does not equally cover the whole of the intervening area between the groups. The meshwork of branches and fibrillae divides off a chamber in which the heart lies, the *dorsal* or *pericardial sinus*, the meshwork itself forming the *dorsal diaphragm*. From the nature of its origin the dorsal diaphragm has a scalloped outer border, the concavities of the scallop corresponding to the intervals between the groups of four alary muscles. An alary muscle is present for each of the pericardial cells at the anterior part of segment I. No muscles appear to be present for the two pairs of cells at intersegment VII–VIII.

The tracheal trunk remnants (Fig. 73 (1) *c*). Lying on each side of the heart external to the dorsal diaphragm and lying over the stems of the alary muscles there is commonly a more or less distinct band of tissue formed by the disorganised and flattened remains of the abdominal tracheal trunks. These bands, their appearances depending probably on the age of the mosquito, show more or less distinctly tracheal tissue with large nuclei, but infiltrated with small cells containing deeply staining granules ('phagocytes'). On the inner side these bands are slightly scalloped, the projecting portions lying against the groups of pericardial cells and, where there have been tracheal commissures, tracheal tissue lies across and dorsal to the heart. Externally, as noted under the tracheal system, the bands are also scalloped, the points contacting the tracheal knots of the corresponding side. The bands, now narrowed, continue into the eighth segment approaching each other in the middle line and contacting the structures described as probably representing the tracheal knots of the eighth segment, an interesting relic in the imago of the larval terminal spiracles (see under tracheal system of the imago).

The aorta (Fig. 73 (3), (6), (7) *ao*). On entering the thorax the dorsal vessel takes on the characters of the aorta. This is somewhat narrower than the heart, lacks ostia and is not accompanied by pericardial cells or alary muscles. Throughout most of its course in the thorax the aorta no longer lies close beneath the tergites but is situated deeply just below the lower border of the longitudinal indirect wing muscles embedded in a flat lobe of fat-body situated in the median space between the longitudinal muscular bundles of the two sides. Lateral to it lie bundles of the dorsal protomorphic muscles (representing the

---

Figure 73. Circulatory system.

1 Dorsal view of heart and related tissues. Female imago. *a*, still patent thoracic tracheal trunk passing into abdomen; *b*, point where trunk makes contact with functional tracheal system of abdomen in segment I representing tracheal knot of segment I; *c*, showing extent of tissue representing obsolescent tracheal trunk; *d*, tracheal knots of functional system with which the obsolescent tracheal trunks still maintain contact; *e*, tracheal knot of segment VIII; *e'*, continuation of trunk into segment VIII.

2 Portion of heart showing pericardial cells and alary muscle. *a*, network of alary muscle branches; *b*, nuclei of heart wall.

3 Portion of thoracic aorta where crossed by tracheae arising from thoracic trunks. *a*, thickened wall of aorta; *b*, tracheae impinging on aorta; *c*, muscular bundles of protomorphic system alongside which the aorta passes.

4 Coronal section of scutellum showing pulsatory muscle. *a*, muscle no. 43. Dorsal transverse muscle of scutellum; *b*, lateral lobe of scutellum; *c*, median space between longitudinal indirect wing muscles of the two sides (seen in cross-section).

5 Portion of ventral diaphragm. *a*, lateral intersegmental point and origin of segmental muscles; *b*, portion of ventral diaphragm.

6 Sagittal section through frons and anterior termination of the aorta. *a*, bundle of muscular fibres passing from aortic wall to Clements' organ.

7 Coronal section of same. *a*, Clements' organ; *b*, muscular fibres passing from aorta.

Lettering: *alm*, alary muscles; *ao*, aorta; *Bc*, buccal cavity; *Br*, brain; *F*, frons; *fg*, frontal ganglion; *ft*, lobe of fat-body; *ht*, heart; *os*, position of ostium; *P'*, anterior end of oesophageal pump; *pc*, pericardial cells (dorsal nephrocytes); I–VIII, abdominal segments.

Muscles: *7d*, most posterior of elevators of the buccal cavity at insertion into membrane between oesophageal pump and buccal cavity. 19, valvular muscles: middle line of frons between scapes to roof of anterior end of oesophageal pump. 21, dorsal rectractors of buccal cavity: middle line of frons between dorsal ocelli (Clements' organ) to flanges of buccal cavity. 22, accessory dorsal retractors of buccal cavity: arising in common with 21 and inserted into the anterior end of dorsal plate of oesophageal pump.

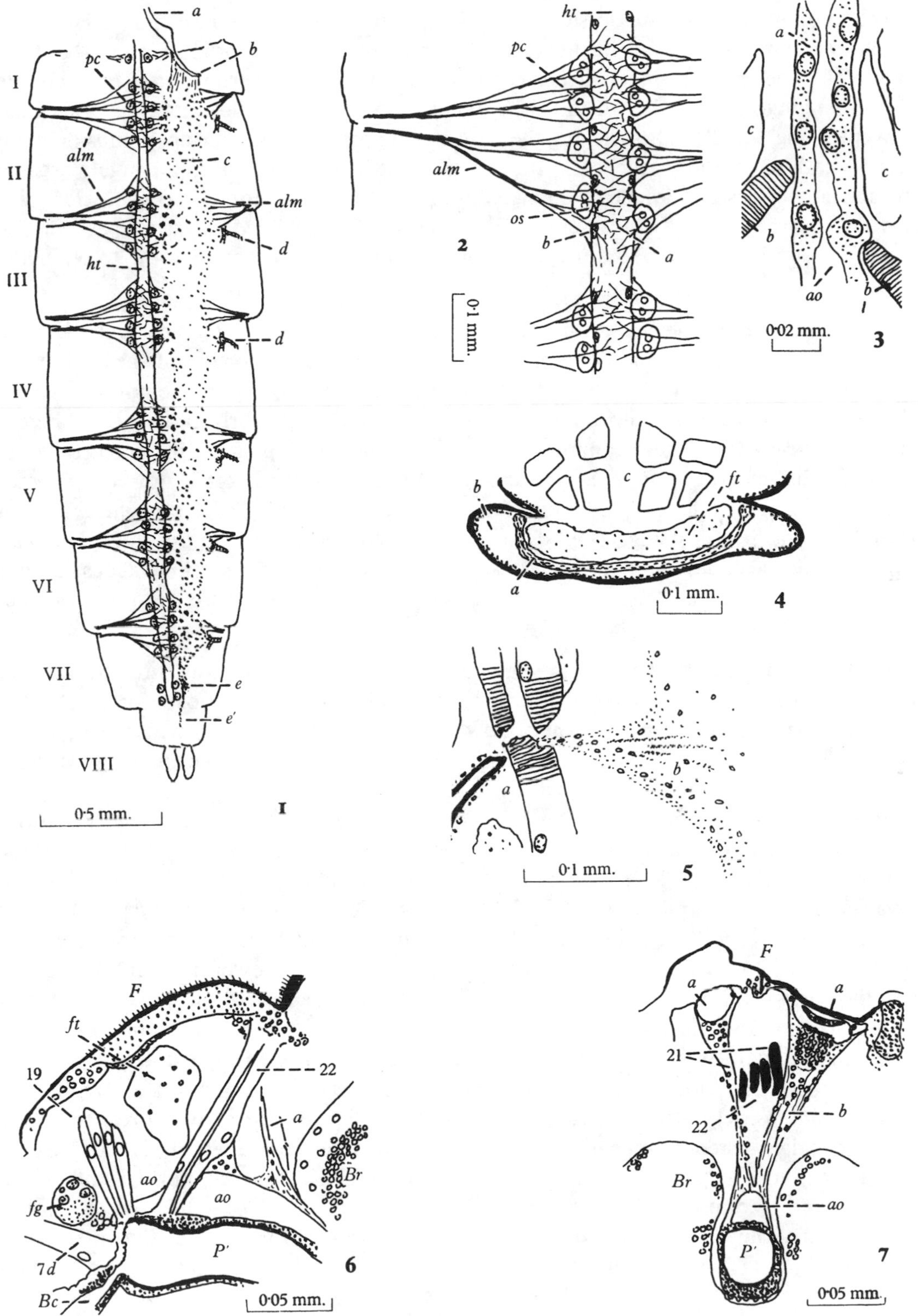
a
b
pc
I
II
alm
c
alm
d
III
ht
d
IV
V
VI
VII
e
e′
VIII
0·5 mm.
1
ht
pc
alm
os
b
a
2
0·1 mm.
a
c
c
b
ao
b
0·02 mm.
3
b
c
ft
a
0·1 mm.
4
a
b
0·1 mm.
5
F
ft
19
22
a
ao
ao
Br
fg
P′
7d
Bc
0·05 mm.
6
F
a
a
21
22
b
Br
ao
P′
0·05 mm.
7

Figure 73

thoracic continuation of the dorsal longitudinal muscles of the abdominal segments). Its structure is similar to that of the heart with laterally placed nuclei at intervals. About the middle of its course two small tracheae coming from the thoracic trunks pass on each side of it impinging on its walls which at this point show clustered nuclei and are somewhat thickened (Fig. 73 (3)). On reaching the level of the proventriculus the aorta dips down to pass beneath the anterior tracheal commissure. A little before doing so it is again crossed on each side by a small trachea coming from the angles of the thoracic trunks. Here also to a less degree the wall shows clustered nuclei and is thickened. Passing between the corpora allata and close beneath the commissure the aorta passes forward to enter the head.

The prothoracic aortic sinus. An important structural feature is described by Jones, *loc. cit.*, under the name of the *prothoracic aortic sinus*. This is a dilated portion of the aorta lying just behind the tracheal commissure where the aorta passes between the corpora allata. It is described as having an intra-aortic thickening (valve) posteriorly and a valve where it passes beneath the tracheal commissure. It is especially conspicuous in the early larval stage. In the imago of *Aëdes aegypti* the structure in sections is often not conspicuous. But in some cases a considerable expansion or even ballooning of the aorta is present behind the tracheal crossing. The posterior valve referred to may be the thickening and collection of nuclei described above where the aorta is crossed by a small trachea. I have not been able to detect any valve at the crossing of the commissure, though the commissure itself may act as such.

The anterior termination of the aorta (Fig. 73 (6), (7)). Passing forwards through the neck the aorta enters the head through the occipital foramen over the posterior end of the oesophageal pump and then along the hollow of the dorsal plate of this organ to the cerebral commissure. In doing so it lies over the hypocerebral ganglion with the entering recurrent pharyngeal nerve of the retro-cerebral system.

On emerging from beneath the cerebral commissure the aorta is still recognisable as a tube with laterally placed nuclei more or less semicircular in section with a flat floor where it lies upon the narrow anterior portion of the dorsal plate of the oesophageal pump towards its anterior termination. At this point a bundle of muscular fibres passes from its lateral walls to each of the dorsal ocelli (organs of Clements). These appear to be what Thompson (1905) took to be nerves from the brain. They are, however, clearly connected with the aorta and are muscular in nature. Beyond this the aorta is still traceable but with walls now thin and membranous and closely associated with the sarcolemma of certain small muscles, namely nos. 21, 22 and 19. The space now representing the aorta, probably to be regarded as a blood sinus, widens laterally over the anterior end of the oesophageal pump until it entirely surrounds this structure. In doing so its dorsal extent reaches to the frontal ganglion where it appears to end. Ventrally it widens considerably and appears to open downwards to the region beneath the floor of the buccal cavity, but it is difficult to ascertain with certainty its limits.

The walls of the sinus so far as described above are formed of thin but distinct membrane (see Fig. 73 (6), (7) *ao*). As with the termination of the aorta in the larva the membrane is closely connected with the sarcolemma of the muscles mentioned above. Muscle 21 passing to its insertion into the lateral cornu of the buccal cavity lies on each side external to it. Muscle 22 coming from the same origin as 21, that is the most posterior portion of the frons, passes as a twin muscle inside the two muscles no. 21 to be inserted into the dorsal plate of the pump towards its anterior end; and muscle 19 coming from the fore part of the frons and directed somewhat backwards is inserted just in front of muscle no. 22 at the junction of the dorsal plate with the membrane connecting the pump with the buccal cavity. Curiously both muscles nos. 19 and 21 perforate the membrane to reach their insertion. It is not possible without a more certain knowledge of the homologies of the imaginal and larval muscles concerned to say with certainty if the imaginal aorta terminates at the same point as the larval. But allowing for the great forward development of the brain in the imago and the pushing forwards of any muscles arising in front of it from the frons there would seem to be no great difference. It can, however, be said that with the contraction of the fore part of the head in passing from larva to imago, bringing the frons from a wide expanse occupying most of the dorsum of the head to a narrow and short median inverted apodeme between the now almost touching antennal bases, the scale of the structures concerned is greatly reduced.

ACCESSORY PULSATORY STRUCTURES (Fig. 73 (4), (5))

Certain accessory pulsatory organs are commonly present in insects. In the imaginal mosquito the following are noted by Jones, *loc cit.*, in addition to the prothoracic aortic sinus, which may be so considered, namely (1) a muscular diaphragm above the nerve cord; (2) a pulsatile muscle in the scutellum; (3) possible pulsatile structures in the legs; (4) Clement's organ.

The muscular diaphragm above the nerve cord is somewhat after the plan of the alary muscles dorsally, but without distinct muscular elements. It consists of fan-shaped sheets of delicate transparent membrane with striae which are presumably muscular in nature and scattered small nuclei (Fig. 73 (5) *b*). The sheets arise from the intersegmental line somewhat laterally and converge to form a layer, with scalloped outer edge, lying over the nerve cord and ventral fat-body.

What may be a pulsatile organ is muscle no. 43. This is a sheet of muscle of skeletal type extending across the scutellum from one lateral lobe to the other and in the middle arching nearly to the roof of the middle lobe (Fig. 73 (4) *a*). It rests upon a small lobe of fat-body and is situated behind the longitudinal indirect wing muscles. There is no appearance of any blood sinus structure, but the median gap between the longitudinal wing muscles of the two sides may afford passage for haemolymph so that contraction of the muscle may assist circulation of this fluid. A similar muscle is described by Maki (1938) in a tipulid and is thought to be a pulsatory organ.

Pulsatory organs are described in the legs of certain insects. Such have not so far been described in the mosquito. Septa are present in all the legs extending to the tarsal segments. These begin in the coxae and are present in trochanter, femur and tibia stretched across the limb with the leg trachea incorporated in their substance.

The nature of the curious organ situated on each side in the frontal region of the head described by Clements (1953, 1956*a*, *b*) and Day (1955), and which has been referred to in the section on the tracheal system, is not very certain. It seems certain, however (see Clements, 1956*b*), that it is not a tracheal sac, as might be suggested by the tracheal-like appearance of the passage leading into the antennal filament, but an accessory propulsatory organ as has been described in a number of insects (Pawlowa, 1895; Brocher, 1922; Freudenstein,

1928; see also Wigglesworth, 1942*a*, p. 224). As the structure of this curious organ is, however, somewhat complex and as it appears to be essentially of the nature of a modified dorsal ocellus, its full description may conveniently be deferred to the section on special sense organs even though its functions in this respect have presumably been lost.

Very briefly it consists on each side of a sac situated in the frontal region between the two compound eyes and a little dorsal and external to the bases of the antennae (Fig. 53 (1) *oc'*). From the sac a stout tube, very like a tracheal vessel but lacking taenidia, extends to the base of the antenna of its side and passing through the central gap in Johnston's organ along with the nerve to the antennal filament enters and passes along that structure. For further details see under 'Special sense organs', p. 648.

HAEMOCYTES, OENOCYTES AND VENTRAL NEPHROCYTES

For literature relating to haemocytes see under 'Circulatory system of the larva', p. 337. In the imago, except for the presence of phagocytic cells associated with structures undergoing absorption, haemocytes are not conspicuous and what has been said under the larva must suffice. The phagocytic cells as seen in fixed preparations are most noticeable on account of deeply stained globules. The cells are about 6–8μ in diameter with a round nucleus about 3μ in diameter. Their cytoplasm contains relatively large globules up to about 2μ in diameter that stain intensely in preparations stained with haematoxylin. They are especially numerous about the abdominal tracheal trunks undergoing absorption, but occur in other situations such as clustered about muscle fibres or free.

The conspicuous groups of *large oenocytes* seen in the abdominal segments of the larva are lacking or represented only by groups of distorted cells clearly in process of disappearing. According to Hosselet (1925) the large oenocytes are present throughout the larval period, attain their maximum during pupation, and disappear in the adult stage. Some small oenocytes bordering the fat-body in the ventral regions of the segments are present, but appear to be much less numerous than in the larva.

The *ventral nephrocytes* are cells very similar in appearance to the pericardial cells. They are present in the imago as a cluster of five or more cells on the lower surface of the lobe of fat-body in which the salivary glands lie and a little anterior to the ventral transverse muscle. Jensen (1953) describes a method of dissecting the salivary glands in *Anopheles* by which the ventral nephrocytes are displayed.

THE FAT-BODY (Fig. 74)

As in the larva the fat-body in the imago consists of a peripheral and a visceral layer. It is, however, for the most part much less massive and more broken up, especially in the thorax, owing to the greater development of muscles and the more complicated cuticular structure.

In the head there is a layer of peripheral fat-body over the occipital and genal areas extending to the posterior margins of the compound eyes (1). Other smaller masses are in the frontal area and basal lobes of the antennae (2). There is a long narrow strand on each side accompanying the tentorium (3), a lobe on each side at the junction of the buccal cavity and oesophageal pump (4), and one in the labial region (5).

The thoracic fat-body is relatively small in amount and much broken up. There is, however, an extensive layer of peripheral fat-body over such parts of the scutum as are not occupied by insertions of the indirect wing muscles (6). In the median line this layer sends a crescentic extension into the gap between the insertions of the longitudinal wing muscles of the two sides (6*a*). At its posterior end this vertical sheet is continued as a flat ribbon-like extension just short of the scutellum in which there is a small transverse lobe referred to when discussing accessory pulsatory organs (6*c*). Anteriorly in the space on each side not occupied by the dorso-ventral muscles the scutal fat-body forms a considerable mass, the largest unbroken area of fat-body in the thorax (6*d*). From these areas the scutal fat-body is continued down on each side of the anterior promontory to the dorsal aspect of the root of the neck where there is a somewhat distinct lobular mass at the base of the neck dorsally (7). Over the pleurae the peripheral fat layer is thin and broken; but there are thickened portions with extensions passing inwards making junction with deeper portions of fat substance behind and among the muscles (8). One of these is continued inwards in the anterior spiracular area accompanying the large tracheal vessels (8*a*). Another is between the anterior and posterior dorso-ventral muscle bands (8*b*) and still another posterior to the latter muscle in the metathoracic area and the posterior spiracular tracheal branches (8*c*).

The visceral fat-body is in the form of longitudinal strands associated with the thoracic viscera. A thin median vertical sheet passes backwards from just above the base of the neck between the lower bundles of the longitudinal indirect wing muscles of the two sides and extends back to the metathorax (9). In this the aorta lies embedded for much of its course. A second portion of visceral fat-body lies in the middle line of the thorax ventrally beneath the oesophagus, proventriculus and ventral diverticulum and above the thoracic ganglia. Its anterior portion supports the salivary gland acini and on its lower surface are the ventral nephrocytes (10*a*). Beyond the furca this portion of fat-body forms a layer surrounding the narrow portion of the mid-gut, the ventral diverticulum and the ventral series of protomorphic muscular strands, that is, it surrounds the thoracic viscera proper, as distinct from the large mass of secondary muscle in the thorax (10*b*). In this region it forms also a considerable vertical sheet on each side internal to the posterior dorso-ventral indirect wing muscle bundles (10*c*) and passing backwards eventually comes to lie beneath the flattened tracheal trunk of its side accompanying this into the first abdominal segment (10*d*).

In the abdomen the peripheral layer is to a large extent divided by gaps at the inter-segmental lines so that the fat-body of each segment is to a considerable extent distinct. It is further partially divided into lobe-like masses by the presence, or previous presence, before these have been absorbed, of the abdominal muscular bands. Dorsally there is a gap occupied by the heart. Ventrally it forms a median ridge continuous from segment to segment on which the nerve cord lies (11). Laterally it forms a thick sheet covering the side of the segment dorsal and ventral to the infolded pleural membrane. On the inner aspect of the sheet are lobular extensions which extend inwards and assist in supporting the viscera. The lobes show an intimate relation to tracheal branches around which they largely lie (Fig. 74 (2)).

The fat-body plays an important role in the life history of the insect and shows changes in relation to the different stages of development. These changes have been especially studied by De Boissezon (1930*a*, *b*, *c*; 1932) in *Culex pipiens* (see also Roubaud, 1932; Buxton, 1935; and Wigglesworth, 1942*b*). According to De Boissezon the fat-body is composed entirely of one type of cell, *trophocytes*. These show three forms of inclusions in their cytoplasm, namely *fat-globules* that stain black with osmic acid; *albuminoid granules* that stain pink with Millon's reagent; and *brown refractile granules* consisting of purine material (urates). In the two-day-old larva the trophocytes are small cells with irregular contour, measuring 10–15μ in diameter, with strongly basophil cytoplasm already containing fat globules 3–4μ in diameter. When the larva approaches full growth the trophocytes reach a size of 20–22μ and the cytoplasm has become clear and acidophile and loaded with inclusions. The fat globules have increased in size to 5–6μ and the albuminoid granules to 4–5μ. In the imago the fat-body tissue for the first 24 hours is as described. But if the mosquito is unfed the albuminoid granules entirely disappear, being used up in the maturation of the eggs.

The conditions are very similar in *Aëdes aegypti*. The fat-body consists of cells (trophocytes) with globular nuclei measuring in all stages from the first instar to the imago approximately the same size (4–6μ in diameter). Depending upon the stage of development and conditions the cytoplasm is variously loaded with inclusions and the cell outlines usually obscured. In the first instar immediately following hatching the cytoplasm already has its characteristic vacuolated appearance with occasional fat globules up to 5μ in diameter (Fig. 74 (4)). Usually it also shows the characteristic brown refractile granules such as are usually only seen in the later stages of development. It is possible that these granules, which are very characteristic of this stage, are the result of metabolism during the period of diapause in the egg, though this has not been verified. By the third instar the cells have become packed with fat globules up to about 3μ in diameter and small albuminoid spheres up to 2–3μ in diameter are also present (Fig. 74 (5) *a*). In the fourth instar still larger fat globules are present with albuminoid spheres up to 2–3μ in diameter with numerous smaller spheres. Just before pupation there is almost as much albuminoid material as fat globules, many of the spheres reaching a diameter of 6μ (Fig. 74 (6) *a*). Much the same appearances are present through the pupal stage and in the newly emerged imago.

Figure 74. The fat-body.

1 Showing the fat-body in the head and thorax (reconstruction). The figures are those given to the lobes in the text.

2 Preparation showing lateral view of the fat-body of an abdominal segment and relation to tracheal branches. *a*, portion of lateral lobe dorsal to pleural membrane; *b*, portion of same ventral to pleural membrane; *c*, ventral fat-body; *g*, ganglion; *sp*, spiracle; *vt*, ventral trachea.

3 Lateral portion of an abdominal segment of a newly hatched first instar larva, showing refractile brown granules.

4 Lobe of fat-body in newly hatched first instar larva more highly magnified, showing refractile brown granules (camera lucida). Section stained with haematoxylin and eosin.

5 Portion of fat-body of pre-ecdysis third instar larva, showing accumulation of moderate-sized fat globules and small albuminoid sphere stained black with Heidenhein's haematoxylin. Same magnification as 4.

6 Portion of fat-body of imago half an hour from emergence, showing cells packed with fat globules, many of large size, large albuminoid spheres and some purine bodies (pseudonuclei). Spheres stained black with Heidenhein's haematoxylin. Same magnification as 4. *a*, albuminoid spheres; *b*, purine bodies.

7 Portion of fat-body of imago three days from emergence without food, showing absence of all reserves other than reduced fat globules. Same magnification as 4.

8 Fat-body of imago 48 hours after blood feed, showing absence of albuminoid spheres and numerous purine granules (*c*).

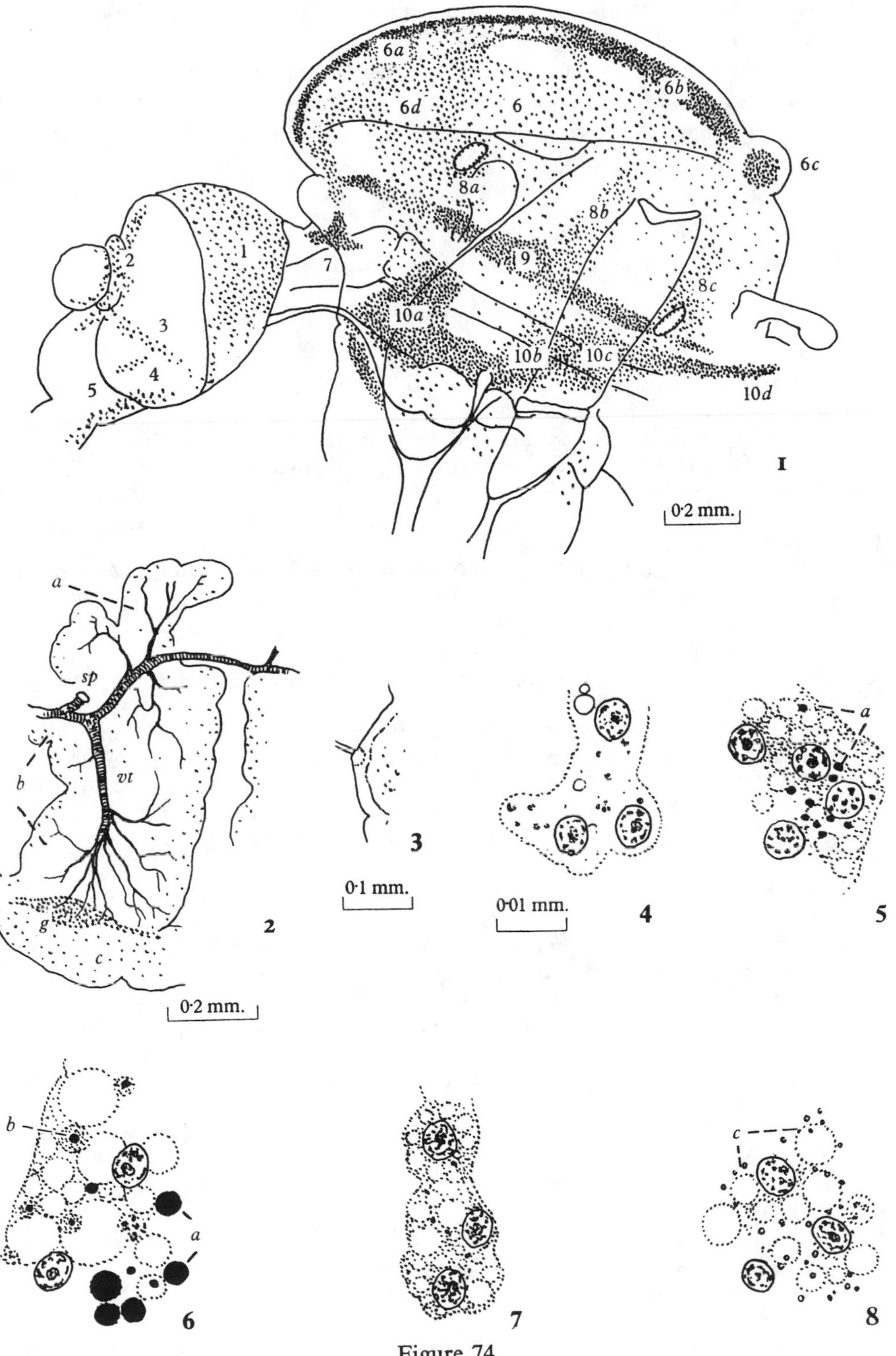
6a
6b
6d
6
6c
8a
8b
1
2
7
19
8c
3
10a
4
10b
10c
5
10d
I
0·2 mm.
a
sp
b
vt
g
c
2
0·2 mm.
3
0·1 mm.
4
0·01 mm.
a
5
b
a
6
7
c
8

Figure 74

Some days after emergence, though fat globules are still present, albuminoid spheres and granules have almost entirely disappeared and the most noticeable feature is the increasing number of brown refractile granules some of which, situated in the centre of spherical globules, give the appearance of nuclei (the so-called 'pseudonuclei'). Following a blood meal the fat-body does not appear to receive further reserve material which is apparently diverted to growth of the egg follicles.

REFERENCES

(c) THE TRACHEAL SYSTEM

ALESSANDRINI, G. and MISSIROLI, A. (1926). Sulla struttura dell'apparato respiratorio delle *Anopheles*. *Riv. di Malariol.* **5** (n.s. 1), 35–43.

BAKER, J. R. (1945). *Cytological technique.* Ed. 2. Methuen and Co., London.

BODENHEIMER, F. (1923). Beiträge zur Kenntnis der Kohlschnäke, *Tipula oleracea*. *Z. wiss. Zool.* **121**, 393–441.

CHRISTOPHERS, S. R. (1901). The anatomy and histology of the adult female mosquito. *Rep. Malar. Comm. R. Soc.* (4), 1–20.

CHRISTOPHERS, S. R. and BARRAUD, P. J. (1924). The tracheation and venation of the wing of the mosquito. *Ind. J. Med. Res.* **11**, 1103–17.

FEUERBORN, H. J. (1927). Die Metamorphose von *Psychoda alternata* Say. I. Die Umbildungsvorgänge am Kopf und Thorax. *Zool. Anz.* **70**, 315–28.

GIGLIOLI, M. E. C. (1953). A way of studying structures in insects by whole mounts. *Trans. R. Soc. Trop. Med. Hyg.* **47**, 266–7.

HASSAN, A. A. G. (1944). The structure and mechanism of the spiracular regulating apparatus in adult Diptera and certain other groups of insects. *Trans. R. Soc. Ent. Lond.* **94**, 103–53.

HURST, C. H. (1890). *The Pupal Stage of* Culex. Guardian Printing Works, Manchester.

IMMS, A. D. (1938). *Textbook of Entomology.* Ed. 4. Methuen and Co., London.

KENNEDY, C. H. (1922). The homologies of the tracheal branches in the respiratory system of insects. *Ohio J. Sci.* **22**, 84–9.

LANDA, V. (1948). Contributions to the anatomy of Ephemerid larvae. I. Topography and anatomy of tracheal systems. *Mém. Soc. Zool. tchécosl.* **12**.

LEHMANN, F. E. (1926). *Über die Entwicklung des Tracheensystems, etc.* In Leuzinger, Weismann and Lehmann, *Zur Kenntnis der Anatomie und Entwicklungsgeschichte der Stabheuschrecke, etc.* Gustav Fischer, Jena, pp. 330–414.

NUTTALL, G. H. F. and SHIPLEY, A. E. (1901). Studies in relation to malaria. *J. Hyg., Camb.*, **1**, 451–84.

POUCHET, G. (1872). Développement du système trachéen de l'Anophèle (*Corethra plumicornis*). *Arch. Zool. Exp.* **1**, 217–31.

RIPPER, W. (1931). Versuch einer Kritik der Homologiefrage der Arthropodentracheen. *Z. wiss. Zool.* **138**, 303–69.

SNODGRASS, R. E. (1935). *The Principles of Insect Morphology.* McGraw-Hill Book Co., New York and London.

TAYLOR, T. H. (1902). On the tracheal system of *Simulium*. *Trans. Ent. Soc. Lond.* 1902, 701–16.

WEBER, H. (1938). *Grundriss der Insektenkunde.* Gustav Fischer, Jena.

WIGGLESWORTH, V. B. (1950). A new method for injecting the tracheae and tracheoles of insects. *Quart. J. Micr. Sci.* **91**, 217–23.

WIGGLESWORTH, V. B. (1953). Surface forces in the tracheal system of insects. *Quart. J. Micr. Sci.* **94**, 507–22.

WIGGLESWORTH, V. B. (1954). Growth and regeneration in the tracheal system of an insect, *Rhodnius prolixus* (Hemiptera). *Quart. J. Micr. Sci.* **95**, 115–137.

REFERENCES

(*d*) THE CIRCULATORY SYSTEM AND ASSOCIATED TISSUES

BROCHER, F. (1922). Étude expérimentale sur le fonctionnement du vaisseau dorsal et sur la circulation du sang chez les insectes. 5. La *Periplaneta orientalis*. *Ann. Soc. Ent. Fr.* **91**, 156–64.

BUXTON, P. A. (1935). Changes in the composition of adult *Culex pipiens* during hibernation. *Parasitology*, **27**, 263–5.

CHRISTOPHERS, S. R. (1901). See under section (*c*).

CLEMENTS, A. N. (1953). Organ of unknown function in the head of mosquitoes. *Proc. R. Ent. Soc. Lond.* (C), **18**, 44.

CLEMENTS, A. N. (1956*a*). (Exhibit.) *Proc. R. Ent. Soc. Lond.* **21**, 7.

CLEMENTS, A. N. (1956*b*). The antennal pulsating organs of mosquitoes and other Diptera. *Quart. J. Micr. Sci.* **97**, 429–33.

DAY, M. F. (1955). A new sense organ in the head of the mosquito and other Nematocerous flies. *Aust. J. Zool.* **3**, 331–5.

DE BOISSEZON, P. (1930*a*). Contribution à l'étude de la biologie et de histophysiologie de *Culex pipiens*. *Arch. Zool. Exp. Gen.* **70**, 281–431.

DE BOISSEZON, P. (1930*b*). Les réserves dans le corps gras de *Culex pipiens* et leur rôle dans la maturation des œufs. *C.R. Soc. Biol., Paris*, **103**, 1232–3.

DE BOISSEZON, P. (1930*c*). Le rôle du corps gras comme rein d'accumulation chez *Culex pipiens* et chez *Theobaldia annulata*. *C.R. Soc. Biol., Paris*, **103**, 1233–5.

DE BOISSEZON, P. (1932). Localisation du glycogene et du fer chez *Culex pipiens*. *C.R. Soc. Biol., Paris*, **111**, 866–7.

EYSELL, A. (1913). Die Stechmücken. In Mense's *Handb. TropenKr.* Ed. 2, **1**, 97–183.

FREUDENSTEIN, K. (1928). Das Herz und das Circulationssystem der Honigbiene (*Apis mellifica* L.). *Z. wiss. Zool.* **132**, 404–75.

HOSSELET, C. (1925). Les oenocytes de *Culex annulatus* et l'étude de leur chondriome au cours de la sécrétion. *C.R. Acad. Sci., Paris*, **180**, 399–401.

JENSEN, D. V. (1953). See references in ch. XXVI (*a*).

JONES, J. C. (1952). Prothoracic aortic sinuses in *Anopheles, Culex* and *Aëdes*. *Proc. Ent. Soc. Wash.* **54**, 244–6.

JONES, J. C. (1953). On the heart in relation to circulation of haemocytes in insects. *Ann. Ent. Soc. Amer.* **46**, 366–72.

JONES, J. C. (1954). The heart and associated tissues of *Anopheles quadrimaculatus* Say (Diptera: Culicidae). *J. Morph.* **94**, 71–124.

MAKI, T. (1938). Studies of the thoracic musculature of insects, etc. See references in ch. XXVI (*b*).

PAWLOWA, M. (1895). Über ampullenartige Blutcirculationsorgane im Kopfe verschiedener Orthopteren. *Zool. Anz.* **18**, 7–13.

RAJINDAR PAL (1944). Nephrocytes in some Culicidae. Diptera. *Ind. J. Ent.* **6**, 143–8.

ROUBAUD, E. (1932). Des phénomenès d'histolyse larvaire postnymphale et d'alimentation autotrophe chez le moustique commun, *Culex pipiens*. *C.R. Acad. Sci., Paris*, **194**, 389–91.

THOMPSON, M. T. (1905). The alimentary canal of the mosquito. *Proc. Boston Soc. Nat. Hist.* **32**, 145–202.

WIGGLESWORTH, V. B. (1942*a*). *The Principles of Insect Physiology*. Ed. 2 (and subsequent editions). Methuen and Co., London.

WIGGLESWORTH, V. B. (1942*b*). The storage of protein, fat, glycogen and uric acid in the fat-body and other tissues of mosquito larvae. *J. Exp. Biol.* **19**, 56–77.

XXVIII
THE IMAGO: THE NERVOUS SYSTEM AND RETROCEREBRAL COMPLEX

(*e*) THE NERVOUS SYSTEM

The imaginal nervous system includes the brain and ventral nerve cord, the peripheral nerves, certain ganglia and nerves forming the stomodaeal or stomatogastric system, and various sensory nerve systems relating to the cuticle, muscles, viscera and sense organs. For literature see general accounts of the nervous system in insects by Snodgrass (1935); Weber (1933); Imms (1938); also further references given in the separate sections below, dealing with particular parts of the system.

Nervous structures are characterised by nerve cells, *neurocytes*, fibre extensions of these, *axons*, branch fibres from neurocytes, *dendrites*, or from the axons, *collateral* or *terminal arborisations*. For satisfactory demonstration of axon fibres silver impregnation methods have been much used. Some recent techniques are those of Holmes (1947); Romanes (1950); Wigglesworth (1953); and Samuel (1953). For mounted sections spread by the diluted glycerine albumen method the technique of Holmes has been found very constantly successful, giving interesting colour effects for some tissues as well as staining the axons of nervous tissue. Excellent preparations for anatomical structure of the brain and giving dark staining of peripheral nerves have been obtained, however, with iron haematoxylin using a suitable degree of differentiation. But such preparations do not show neurone fibres sharply stained as they are by impregnation methods.

The anatomical characters of the nervous system of the larva have already been described. The chief differences that have resulted in the imago are: (1) Increased size of the brain largely due to the great develoment of the optic tracts and antennal lobes, but also to the thickening of the circum-oesophageal connectives and other changes. Also owing to the contraction of the clypeal region in the imago, the brain occupies a more forward position in the head and there is some crowding of the cerebral nerves making these less easy to trace than in the larva. (2) Some increase in size and complexity of the thoracic ganglia and the appearance in the imago of large new nerves necessitated by parts formed from the imaginal disks such as the legs and halteres, the nerves to the halteres being among the largest in the body. (3) The apparent absence of the median nerve system.

THE BRAIN

Literature. So far as I have been able to ascertain no author has described in any detail the brain of the mosquito. Nevertheless, there exists an extensive literature dealing with the structure of the brain in other insects and without some reference to such work any adequate description of the brain of the mosquito could scarcely be given. A number of the earlier papers on the insect brain are by Viallanes (1882–93), including that by Viallanes (1893), in which there is given a diagram showing the generalised structure of the insect brain that includes much of what is known of the anatomy of this organ and which has been reproduced as illustrative of present-day knowledge in at least one important textbook. Many other early papers on the brain of different insects are also to be found in various zoological journals, for example Newton (1879), on the brain of the cockroach; Packard (1880) and Burgess (1880), on the brain of the locust; Kenyon (1896), on the brain of the bee with references to earlier papers. More recently there are: Haller (1905), on the brain of the bee, of the wasp and of *Musca* showing neurones stained by the Golgi method; Janet (1905) and Thompson (1913), on the brain of the ant; Böttger (1910), on the brain of *Lepisma*; Kuhnle (1913), on that of the earwig. Very helpful is the paper by Jonescu (1909), on the brain of the bee, giving a very full description accompanied by figures showing transverse sections of the brain at different levels. Still more recently are the papers by Baldus (1924), on the larva and imago of *Libellula*; of Bretschneider, (1913–24), on the brain of the cockroach and species of Coleoptera and Lepidoptera; of Hanström (1925–30), on the larva and imago of *Pieris*, of the white ant and of the leaf insect; of Holste (1923), on the brain of *Dytiscus*; of Jawlewski (1936), on the brain of beetles; and of Snodgrass (1925) on that of the honey-bee. Of special interest in the present connection is the description by Buxton (1917) of the brain of the small primitive moth, *Micropteryx*, where as will appear the structure seems nearer to that of *Aëdes aegypti* than that of most of the above.

It will be seen that little mention is made of description of the brain of Diptera. Here reference can be made to the very full, but not very recent, description of the brain of the blow-fly by Lowne (1895). Very few other authors have studied the Dipterous brain and none apparently that of any species of Nematocera. There is a paper by Brandt (1879), entitled 'Vergleichend-anatom. Untersuchungen über das Nervensystem der Zweiflügler', which I have not, however, so far seen. Flögel (1878), discussing structure of the brain in the different Orders of insects, has a section on Diptera of one page and mentions *Tabanus*.

The optic tracts have been studied by Berger (1878); Hickson (1885); Zawarzin (1914); and by Cajal (1918). For description of neurones and physiological research on relation of structure to function see Wigglesworth (1950) and recent papers there quoted.

Structure (Fig. 75, 76) The brain of the imaginal mosquito, as of the larva already described, consists of a complex ganglionic mass, the *supra-oesophageal ganglion*, two thickened *circum-oesophageal connectives*, or *cerebral crura*, and a composite ganglion, *sub-oesophageal ganglion*, in which are located the nerve centres for the gnathal and labial structures. External to the supra-oesophageal ganglion proper on each side are the large *optic lobes*, or as they are commonly termed *optic tracts*, each forming in the imago almost a third of the total brain mass. These parts, however, in the imago are so compacted that, with the much thickened crura, they form an almost globular mass perforated by a narrow tunnel which just suffices to allow passage of the anterior portion of the oesophageal pump accompanied by the aorta and the recurrent pharyngeal nerve (Fig. 75 (4), (14)).

As in the larva the supra-oesophageal ganglion consists of *protocerebrum*, *deutocerebrum* and *tritocerebrum*. The protocerebrum forms the greater part of the supra-oesophageal ganglion consisting of two lateral halves, *protocerebral lobes*. In practice the deutocerebrum and tritocerebrum are so little differentiated and their limits so uncertain that it is often convenient to include these with the respective protocerebral lobes which may then

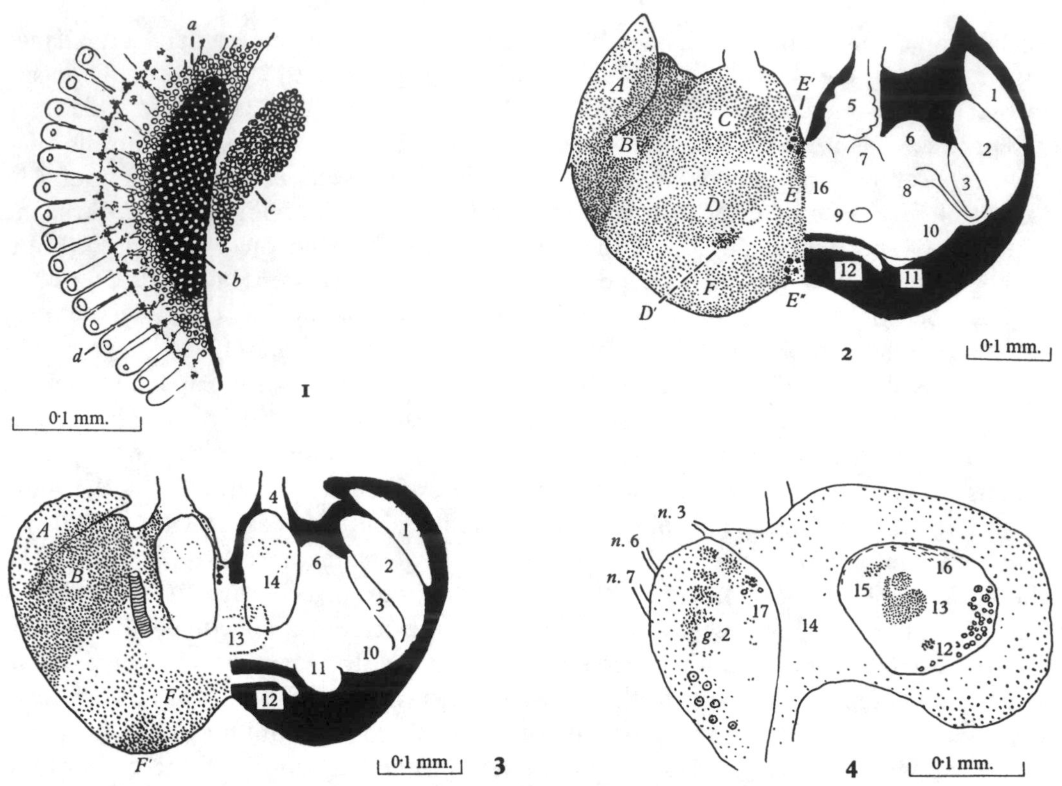

Figure 75. The brain.

1 Section through the ganglionic plate showing cortical layer giving off nerve fibres to the ommatidia and patterned neuropile. *a*, cortical layer; *b*, neuropile; *c*, cortical layer of external medullary mass. The section passes a little dorsal of the level at which the ganglionic plate is in contact with the medullary mass and includes only the tip of the latter; *d*, ommatidia with basal layer of pigmented cells and nerve fibres.

2 Dorsal view of supra-oesophageal ganglion, showing on the left side the cortical layer and on the right the neuropile.

3 Ventral view of the same. The cerebral crura are shown as if cut across and the sub-oesophageal ganglion removed. The portion of trachea (dorsal cephalic) marks a deep embayment (see Fig. 76 (5)) where this is shown in section.

4 Brain shown as though sliced in half along the median line. It shows the area over which the cerebral lobes are connected and the broad crus flanking the canal in which the anterior portion of the oesophageal pump lies with the aorta and recurrent pharyngeal nerve.

Lettering (Figs. 75 and 76): *A*, ganglionic plate; *B*, cortical layer of medullary masses; *C*, post-antennal cortical area; *D*, dorsal cortical area; *D'*, cell mass lying below level of cortical layer; *E*, cortical area bordering the median fissure with groups of large nerve cells *E'*, *E"*; *F*, occipital cortical area, with dark-staining area *F'*; *G*, cortical area of deutocerebrum (antennal lobes); *H*, position of tritocerebral lobe; *a–e*, parts of optic lobe; *g.* 2, sub-oesophageal ganglion: seen in section in Fig. 75 (4) are tracts linking the nerve neuropile masses across the median line; *nl*, neurilemma; *n.* 2, antennal nerve; *n.* 3, fronto-labral nerve; *n.* 6, maxillary nerve; *n.* 7, labial nerve; *tcl*, nucleus of tracheal cell; 1, neuropile of ganglionic plate; 2, neuropile of external medullary mass; 3, neuropile of internal medullary mass; 4, neuropile formed by continuation of antennal nerve within the brain; 5, antennal neuropile centre showing lobulation; 6–11, neuropile masses (see text); 12, suboccipital commissure; 13, central body; 14, circum-oesophageal connective (crura); 15, anterior commissural tract; 16, layer of neuropile forming dorsal surface layer of cerebral commissure; 17, tritocerebral commissure.

be conveniently termed the *cerebral lobes* as in later parts of this section. The two lateral halves of the supra-oesophageal ganglion are joined in the middle line over a limited extent of their inner surface to form what may still be regarded as the *cerebral commissure*. Elsewhere the lobes are separated by a deep median fissure, *median fissure*, into which the neurilemma covering the brain dips. In this considerable degree of separation of the two lobes the brain of the mosquito appears to differ from that described for many insects. The deutocerebrum is formed by the *antennal lobes* from which the stout antennal nerves pass to Johnston's organ in the base of the antenna. The tritocerebrum is an inconspicuous lobe situated on each side ventral and internal to the antennal lobes at the base anteriorly of the crus of each side. It is chiefly to be located as the situation from which the fronto-labral nerve of its side arises. This nerve shortly after leaving the brain turns up in front of the lateral flange of the buccal cavity to enter the frontal ganglion as a very short frontal ganglion connective and continues as the relatively small labral nerve.

Histologically the brain, and the optic tracts, consist of an outer layer of nerve cells, *cortical layer*, and an inner mass or masses formed by the ramifications of axons and their branches, here termed, following Snodgrass, *neuropile*. Surrounding the whole is the neurilemma, a delicate membrane with scattered lightly staining, small nuclei (Fig. 76 (1) *nl*). The supra-oesophageal ganglion is almost entirely sensory and associative in function and its cortical layer is composed of very small nerve cells with deeply staining nuclei, large in comparison with the cell and measuring as a rule about 6–8 μ in diameter, *sensory* or *association neurone cells*. Some large cells, *motor neurone cells*, are present, however, in the cortical layer flanking the median furrow. Motor neurone cells are more numerous in the sub-oesophageal ganglion from which muscles of the mouth-parts are innervated. Here and there embedded in the cortical layer where tracheal branches enter the brain substance are the large and conspicuous nuclei of tracheal cells (Fig. 76 (1) *tcl*).

The neuropile forms lobulated masses of non-medullated nerve fibres and usually, owing to the relative absence of nuclei, stains lightly in sharp contrast to the dark cortical layer. It may appear as somewhat indefinite lobe-like *masses* or as aggregated bundles of fibres forming *tracts*. In material stained by silver impregnation methods sharply defined axon fibres are seen traversing the masses in various directions. Special aggregations of these constitute the tracts connecting the different neuropile masses.

The optic lobes (Fig. 75). Each optic lobe consists of three ganglionic masses arranged in series, namely from without inwards, the *ganglionic plate* (*periopticon*), the *external medullary mass* (*epiopticon*) and the *internal medullary mass* (*opticon*).

The ganglionic plate is an oval somewhat mushroom-shaped structure consisting of an outer layer of cellular cortex and an inner neuropile mass which, cut in certain directions, shows a regular pattern resembling woven fabric (Fig. 75 (1)). Its outer surface and thinned-out edges are approximated to the inner surface of the compound eye of its side, but separated by a space. Across this space numerous bundles of nerve fibres pass outwards from the surface of the ganglionic plate to form a plexus under the basement membrane of the ommatidia (see under 'Special sense organs', p. 652). From the thinned-out posterior margin one such nerve tract passes to the pigmented, but now degenerate, lateral ocellus. On its inner side the neuropile mass of the ganglionic plate is in contact over its middle portion with the neuropile mass of the external medullary mass, the line of junction forming a conspicuous chiasma, *outer chiasma*. Between the external and internal medullary masses is another line of chiasma, but much narrower than that between the ganglionic plate and external medullary mass, *inner chiasma*. This commonly has the appearance in sections of a regular row of small circular spaces with intervening fibre substance (Fig. 76 (5) 2, 3). The external and internal medullary masses each have a central rather lens-shaped neuropile mass, that of the internal mass being somewhat the smaller. Surrounding these is a common layer of cortical cellular substance. This cortical layer is peculiar in being composed of smaller, more compacted and more deeply staining cells than is the cortical layer of the rest of the brain. This difference in appearance makes it possible in sections to determine where the thick cortical layer of the optic tract ends and that of the brain proper begins, otherwise the two cortical areas are largely continuous (Fig. 75 (2), (3) B). From the inner medullary mass fibres pass inwards to the central neuropile mass of the protocerebrum forming what is sometimes regarded as a third chiasma. The intimate structure of the tracts is complicated and for information the literature should be consulted. The structure in the mosquito is clearly that of a very general plan in insects.

The cerebral lobes (Fig. 75 (2–4)). The two lateral halves of the supra-oesophageal ganglion may conveniently be termed the *cerebral lobes* and it is convenient in the present case to include in this term the antennal and tritocerebral lobes. It will simplify description of these structures, which practically constitute the brain, to note very briefly what has been recorded in the literature regarding the brain in other insects.

Outstanding among named structures are the *corpora pedunculata* or 'mushroom bodies'. These are normally large and conspicuous structures. They consist in each hemisphere of one, or often two, masses of so-called globuli cells embedded in, or underlying, the cortical layer of the dorsal aspect of the brain and very often forming elevations on the surface. There may be cup-shaped fibrous structures, *calices*, underlying the globuli masses. From each globuli mass a thick fibrous tract, root, stalk or *pedunculus*, passes into the neuropile mass of the lobe. There is commonly an elbow-shaped median root (Balken) directed inwards beneath the central body where those of the two sides nearly meet and a posterior root, *cauliculus* or recurrent root, which extends posteriorly and dorsally anterior to the central body. The central body (*corpus centrale*) is a conspicuous, somewhat segregated portion of the neuropile occupying a central position between the two cerebral lobes. It is described and figured in the blow-fly by Lowne, who gives the name *nodulus* to a spherical body connected to the hilum of the *corpus centrale* by a slender median cord. Other structures named are the *pons cerebralis* or bridge, the *pars intercerebralis*, the *corpora*

Figure 76. The nervous system: the brain.

1 Camera lucida drawing of sagittal section through cerebral lobe at level of entry of antennal nerve (section no. 38 of series of 10 μ sections).

2 The same at level of fronto-labral nerve (section no. 42 of the series).

3 Coronal section through dorsal portion of brain (about level of plane through *D* to *n.*2 in Fig. no. 1; section no. 56 of the series).

4 The same about level of from *D′* to *G* in Fig. no. 2 (section no. 58 of the series).

5 Transverse section at level of central body (section no. 81 of the series).

6 The same through posterior part of the brain (section no. 88 of the series).

7 A, central body as seen in transverse section no. 80 of same series; B, the same at section no. 81; C, reconstruction as seen in lateral view; D, the same from ventral aspect.

For lettering see explanation of Fig. 75.

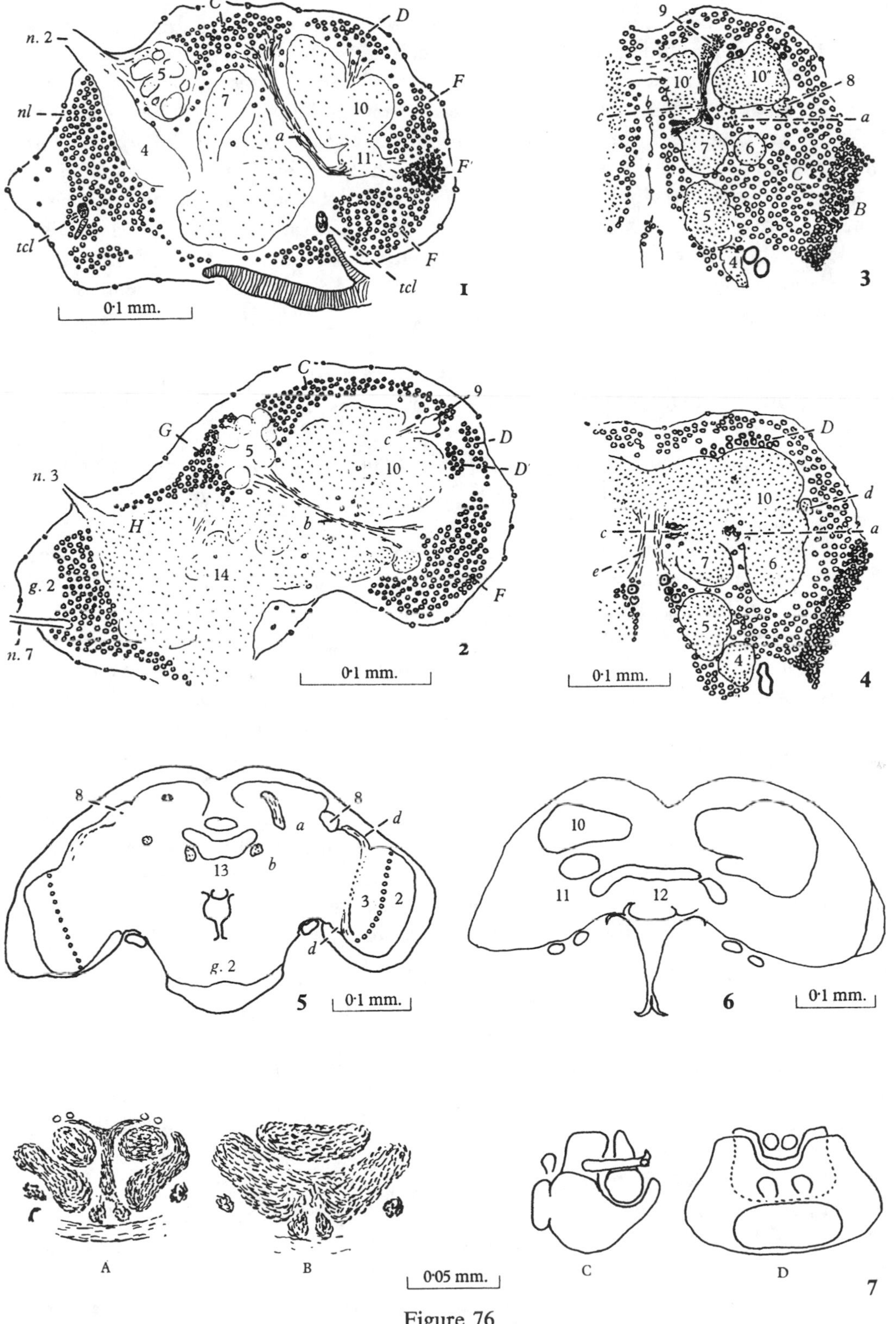
C
D
n. 2
5
nl
7
10
4
a
11
F
F′
tcl
F
tcl
1
0·1 mm.
9
10′
10″
8
c
a
7
6
C
B
5
4
3
C
9
G
5
c
D
10
D′
n. 3
H
b
14
g. 2
F
n. 7
2
0·1 mm.
D
d
10
c
a
e
7
6
5
4
0·1 mm.
4
8
8
d
a
13
b
3
2
d
g. 2
5
0·1 mm.
10
11
12
6
0·1 mm.
A
B
0·05 mm.
C
D
7

Figure 76

ventralia and the *corpora optica*. The *pons cerebralis* is situated in the upper and posterior part of the conjoined lobes. It is transversely elongated, horse-shoe shaped or dumb-bell shaped with the concavity downwards or forwards. The *pars intercerebralis* is that part of the brain where the two lobes form a common portion. It is usually described as that portion situated between the *corpora pedunculata* of the two sides. The *corpora ventralia* (Lebenlappen) are given by Snodgrass as situated in the ventro-lateral parts of the brain just above the antennal glomeruli of the deutocerebrum (this, depending on the view taken of the orientation of the head, might be posterior to the antennal glomeruli). They are regarded by some authors as belonging to the deutocerebrum. They are stated to be united by a transverse commissural tract that passes beneath the central body and the median roots of the *corpora pedunculata*. The *corpora optica* are chiefly present in the Apterygota. They lie dorsal to the pons.

Certain neuropile masses and tracts are also recognised. In the antennal lobes are the *antennal neuropile centres*. These are large globular masses of neuropile characterised by their lobulated appearance. *Protocerebral lobe* is a term applied by Jonescu to the neuropile mass lying internal to the optic tract, seemingly neuropile mass 6 as described by me in the mosquito. Of commissural tracts may be mentioned (1) the commissure linking the antennal neuropile centres, usually shown as passing above the oesophagus; (2) the tritocerebral commissure, sometimes a separate nerve trunk passing beneath the stomodaeum, but may be included in the crura; (3) the *pons* already alluded to; (4) commissures linking various parts of the cerebral lobes across the middle line.

It will be clear how difficult it would be to fix precisely some of these structures where they are not typically developed. It is certainly difficult to be sure of the homology with the above of many parts of the mosquito brain. In the mosquito *Aëdes aegypti* there are no large and unmistakable *corpora pedunculata*. Some tracts may represent these but how far they may do so must be for the present left uncertain. The central body is well represented. The *pars intercerebralis* must be of small extent since the cerebral lobes are divided by a deep median fissure. The bridge or *pons* may be the structure described later as the suboccipital commissure, but it is difficult to describe its position in the mosquito as dorsal. It seems, therefore, desirable to describe first what has been ascertained of the structure of the cerebral lobes in the mosquito, leaving the question of homology and nomenclature of the parts for subsequent consideration.

The cortical layer of the cerebral lobes (Figs. 75 (2), (3); 76 (1), (2)). A cortical layer of small association neurocytes covers the greater part of the surface of the cerebral lobes. It varies, however, in thickness and certain areas of the cortex can be distinguished as more or less distinct and such may even in places be separated by breaks in continuity of the layer. Two such lines of break occur on the dorsal surface and may be termed the anterior and posterior fissures respectively. By such and other structural features the following areas of cortical layer may be distinguished.

(1) The cortical layer of the antennal lobes. On each side this forms a fairly thick layer around the entering antennal nerves and the antennal neuropile centre (Fig. 76 (1), (2), (5)).

(2) A thick mass of cortical layer is situated posterior* to the antennal neuropile centres. It extends backwards to the anterior fissure. Internally it is continuous with the cortical

* References to orientation regard the head as extended, the frons dorsal to the labium and posterior to the clypeus. This is the usual method adopted by authors, but in some cases (for example, in the cockroach) the head has been regarded as oriented with the occiput dorsal.

layer bordering the median fissure and externally meets the cortical layer of the medullary masses of the optic tracts. It may be called the *postantennal cortical area* (Fig. 75 (2) *C*).

(3) A portion of cortical layer lying between the two fissures. It may be termed the *dorsal cortical area* (Fig. 75 (2) *D*).

(4) A particularly thick and massive portion of cortical layer occupying the occipital region, *occipital area* (Fig. 75 (2) *F*). A portion of this area consists of darker staining and more compact tissue resembling the cortical layer of the medullary masses of the optic tracts (Fig. 75 (3); 76 (1) *F'*).

(5) A portion of cortical layer bordering the median fissure, notable as containing clusters of large neurocytes, *median cortical area* (Fig. 75 (2) *E*).

Of a different character to the above there is, on the posterior portion of the dorsal surface of each cerebral lobe, a small mass of dark-staining cells situated beneath the general level of the cortical layer and partly exposed in the posterior fissure (Fig. 75 (2) *D'*).

Ventrally the cellular cortical layer is absent or very thin over a considerable area. This bare area in part forms the roof of the passage occupied by the oesophagus and in part lies upon the more anterior bundles of the lateral dilators of the oesophageal pump where these pass in to their insertion (Fig. 75 (3)).

Neuropile masses of the cerebral lobes (Figs. 75; 76 (1–15)). In the central neuropile mass certain portions (here termed *masses*) can be distinguished. These are as numbered in the figures and text:

1–3. Neuropile masses of the optic lobes. See under 'Optic lobes', p. 629.

4. A considerable mass formed by the entering antennal nerve as distinct from that portion passing to the antennal centre (Fig. 76 (1), (4)).

5. The antennal neuropile centres. These are the most distinct and outstanding of all the neuropile masses in the cerebral lobes. They are approximately globular in shape and consist of a number of lobules giving them a characteristic appearance as noted in a number of other insects (Fig. 75 (2); 76 (1–3), (5)).

6. A considerable mass forming a lobe situated external to the antennal neuropile centre and just internal to the internal medullary mass of the optic lobe of its side from which it receives fibres.

7. A neuropile mass situated internal to 6 and posterior to the antennal centre.

8. A peculiar small isolated neuropile mass on the dorsal aspect of each cerebral lobe and partly exposed in the anterior fissure of the cortex. It is peculiar as receiving a small but distinct tract which passes up the side of the cerebral lobe from the most ventral part of the internal medullary mass (Fig. 76 (5) *d*). Though it usually appears in sections as an isolated mass, it is a projecting spur from the antero-external portion of neuropile mass 10, that is, that portion shown pointing upwards in Fig. 76 (1).

9. Another small isolated neuropile mass situated in or near the posterior fissure. It is situated near the cortical mass *D'* though not directly below it. From it an important tract (*tract c*) passes forwards as described later.

10. A large neuropile mass situated dorsally and posteriorly. It shows indications of consisting of an inner posterior and an outer anterior portion. It forms a large part of the cerebral neuropile mass and receives fibres from the optic lobes. Numerous loosely arranged commissural fibres pass between the masses of the two sides.

11. A small but distinct and rather isolated neuropile mass situated posterior and ventral to 10. It lies beneath the occipital cortical area and receives fibres from it, especially from the dark-staining portion of the occipital cortical area (*F'*).

12. A conspicuous bar-shaped mass of neuropile situated posterior and ventral to the cerebral lobes across the median line. It appears to be of commissural nature and possibly connects the two occipital cortical areas. It has characters that suggest it might be the *pons*. It is provisionally named the *suboccipital commissure* (Fig. 75 (2), (3), (12)).

13. The *central body*. This is a mass of neuropile situated in the centre of the ganglion and forming a part of the connection between the cerebral lobes constituted by the cerebral commissure. It is to a considerable extent isolated from surrounding neuropile by clefts and the presence of small nuclei which appear to be those of neurilemma or neuroglia type cells. It consists of a larger mass partially enfolding a smaller mass, both of which extend dorsally in blunt papilla-like processes. There is also a forked rod-shaped structure and some smaller accessory masses. Its appearance in sections depends a good deal on the angle at which the brain has been cut. A reconstruction of the part is shown in the figure (Fig. 76 (7) *C*, *D*).

14. There are considerable masses of neuropile situated ventrally which are difficult to delimit and which are closely connected with, if not part of, the crura. They include on each side a considerable neuropile mass formed by the passage backwards within the brain of a large portion of the antennal nerves. The two neuropile masses 7 are also here united by a commissural tract linking the two. Here also on each side is the neuropile mass giving origin to the fronto-labral nerve and linked across the median line by the tritocerebral commissure.

Neuropile tracts of the cerebral lobes. In material stained by silver impregnation stained axons traverse the neuropile in almost every direction making an assessment of connection between different parts of the mass difficult. Certain aggregations of fibres, however, constitute characteristic tracts. The more conspicuous of these, given under the respective letters by which they are indicated in the figures and referred to in the text are:

Tract a. This is a thick compact tract passing from the posterior part of the postantennal cortical area to neuropile mass 11. It passes some distance lateral to the central body and ends in the neuropile mass 11 and its neighbourhood. It links the postantennal cortical area with the occipital cortex, especially the dark-stained area *F′*, which also sends fibres to the neuropile mass 11 (Fig. 76 (1) *a*).

Tract b. This is a similarly long and conspicuous tract passing from the antennal neuropile centre on each side to the occipital region and the same neuropile mass 11 or its neighbourhood (Fig. 76 (2) *b*).

Tract c. This is a short but thick tract passing forwards from the neuropile mass 9 to near the origin of tract *a* (Fig. 76 (3)). It is the only main tract that has any relation to the glomeruli-like cortical mass *D′* and this is only a somewhat indirect one as the cortical mass *D′* appears more related to the neuropile lobe 10 than to the small mass 9 though the masses are somewhat near each other (Fig. 76 (2)).

Tract d. This is the narrow but distinct tract passing from the ventral part of the internal medullary mass up the side of the cerebral lobe to the small neuropile mass 8 (Fig. 76 (5)).

Tract e. Small tracts on each side of the median fissure passing from the anterior group of large neurocytes on the sides of the median fissure backwards on each side of the fissure to the inner portions of the neuropile mass 10 (Fig. 76 (4) *e*).

Mention should also be made of commissural tracts. A commissural tract crosses in front of the central body linking the two neuropile masses 7 (Fig. 75 (4), 15). Numerous fibres also pass between the neuropile masses 10, mainly posterior to the central body. A connecting portion of neuropile between the two bodies 10 forms the roof of the cerebral commissure lying over the central body (Fig. 75 (4), 16). This does not, however, consist of transverse fibres but of fibres passing down into the crura. A commissural tract linking the two antennal neuropile masses is commonly present in insect brains. In the mosquito the deep median fissure makes any such tract difficult to follow, but tracts seem to pass down from these masses into the crura.

Homology of the cerebral lobe structures. Interest in this respect relates chiefly to the question how far the structures in the mosquito brain show the presence of the *corpora pedunculata*. In the tracts described there is a suggestion that the characteristic mushroom bodies of the insect brain may be represented if only in a modified and imperfect form. It is difficult, however, to put such a suggestion into any concrete conclusion. In the absence of typical *corpora pedunculata* there is a danger that almost any tracts might be fitted into the standard scheme. The findings have a general resemblance to those found by Buxton in *Micropteryx* and the mosquito brain would seem to be of primitive type. Further information on the brain of other Nematocera would be very desirable. The large size of the optic lobes would seem to be a feature in keeping with the enormous size of the compound eyes and the large tracts linking the antennal neuropile centres and antennal nerve tracts with the rest of the brain would seem to point to the importance of the antennal sensory mechanism in *Aëdes aegypti*.

The sub-oesophageal ganglion (Figs. 75 (4); 78 (3)). This is a club-shaped structure, swollen anteriorly and narrowing as it passes backwards to divide into the two long connectives linking it through the neck with the prothoracic ganglion. It is a composite ganglion comprising a number of fused ganglia of the anterior portion of the ventral nerve cord. From it normally issue the mandibular, maxillary and labial nerves. It consists of a neuropile mass which is exposed dorsally where it forms the floor of the canal in which the oesophagus lies and is surrounded ventrally and laterally by cortical substance containing both association and motor neurone cells.

Nerves pass out through the cortical layer from neuropile masses situated in the ventral portion of the ganglion, the respective neuropile masses being linked across the median line by transverse tracts. The more dorsal portion of the ganglion is largely occupied by fibres passing in from the brain through the crura. Anteriorly on each side is the neuropile mass from which the fronto-labral nerve arises (tritocerebral lobes). Whether there is a mandibular nerve is a little doubtful, but appearances suggest that this may be a very small nerve passing out a little ventral to the fronto-labral nerve or a branch of the labral nerve representing this (see note 4 following list of peripheral nerves, p. 642). Ventral to the frontolabral nerves are the maxillary nerves and still more ventral and posterior and closer to the median line there are the large labial nerves (Fig. 78 (3)).

Posterior to the points of issue of the labial nerves the cortical layer contains four groups of large neurocytes, one on each side of the middle line some distance behind the point of exit of the labial nerves and another pair behind these.

THE VENTRAL NERVE CORD

The ventral nerve cord consists of the sub-oesophageal ganglion (described above), the long double connectives passing through the neck, the fused ganglia of the three thoracic segments, the ganglion of the first abdominal segment now fused with the ganglia of the thoracic segments and the six ganglia of abdominal segments II–VII with their double connectives (Fig. 76 (1)).

The three large thoracic ganglia are situated ventrally in the thorax above the sternites and at about the level of the upper part of the coxae. They occupy the space in the median line between the lower ends of the dorso-ventral indirect wing muscles of the two sides and below the fore part of the mid-gut and proventriculus, but separated from these by the ventral diverticulum and the lobe of visceral fat-body in which lie the salivary glands and the ventral nephrocytes.

The conjoined ganglia form a roughly cylindrical mass in which, however, the component ganglia are still evident (Fig. 76 (1)). This is especially so in the case of the metathoracic ganglion which is sharply divided off by a constriction where the conjoined ganglia pass through the stirrup-shaped opening between the arms of the mesofurca (Fig. 76 (1) *mf*). The three ganglia are approximately equal in size in spite of the fact that the greater part of the thorax is mesothoracic. The small ganglion of the first abdominal segment is attached to the posterior and upper part of the metathoracic ganglion, rather like a stumpy dog's tail and from it the connectives are continued into the abdomen.

The neuropile of the conjoined ganglia is continuous medianly and dorsally. Laterally and ventrally it is in the form of distinct lobes which project ventrally and from the apices of which pass out the six large nerves to the respective legs. The cortical layer is thin and largely deficient dorsally, but thick ventrally and laterally where it occupies space left between the neuropile masses (Fig. 77 (3), (4)). The neuropile of the small first abdominal ganglion is in the form of two lateral masses as in the ganglia of the other abdominal segments, but basally these are continuous with the neuropile of the metathoracic ganglion. In material stained by silver impregnation methods the continuous dorsal portion of neuropile is seen to be traversed by numerous long axon fibres continued from the anterior connectives and passing various distances through the ganglion before terminating in the neuropile, some extending its whole length. These fibres mainly form tracts towards each side of the ganglion, one dorsal along its upper surface and a parallel tract that early separates and passes more ventrally at about the level of the bases of the conical neuropile masses (Fig. 77 (3) *a*, *b*).

The neuropile is seen to consist mainly of a confused mass of fine branching fibres with occasional thicker fibres, some of which form lines or tracts passing to issuing nerves, whilst others are seen passing into the neuropile from parts of the cortical layer. Fibres forming small commissural tracts are also present, one of which in the mesothoracic ganglion is particularly noticeable. The cortical layer is mainly formed of small association type neurocytes, but occasional larger cells are present being most noticeable in the mesothoracic ganglion especially posterior to the mesothoracic neuropile in the ventral portion of the cortical layer (Fig. 77 (6)).

Nerves issuing from the ganglion for the most part appear to take origin from the neuropile. The large nerves to the legs arise from the apices of the six conical neuropile masses. Next in importance in respect to size are the two nerves to the halteres arising from the lateral dorsal aspects of the metathoracic ganglion just posterior to the mesofurca. Their origin is peculiar in showing a conspicuous fan-shaped aggregation of what appear to be axon fibres coming largely from the dorsal axon tracts previously noted and which appear to terminate largely in these nerves. From the antero-lateral face of the prothoracic ganglion three rather small nerves pass to the sternal and other prothoracic regions. In addition to the large leg nerves a smaller nerve in the meso- and metathorax arises along with or near these and passes backwards into the posterior part of the coxae. Apart from the nerves to the halteres the only nerves of any size arising from the ganglionic complex dorsally are a group of moderate-sized nerves arising laterally from the upper part of the mesothoracic ganglion. On each side these are three in number, one arising anteriorly apparently from the anterior face of the neuropile mass at some depth (Fig. 77 (1) *b*; (5) *a*) and two rising together in a mid antero-posterior position from the dorsal part of the neuropile mass (Fig. 77 (1) *c*, *d*; (5) *b*, *c*). The more anterior of these latter two passes forwards and links up with, or accompanies, the first-mentioned nerve as described later, whilst the more posterior one passes backwards and upwards anterior to the projecting cups of the mesofurca.

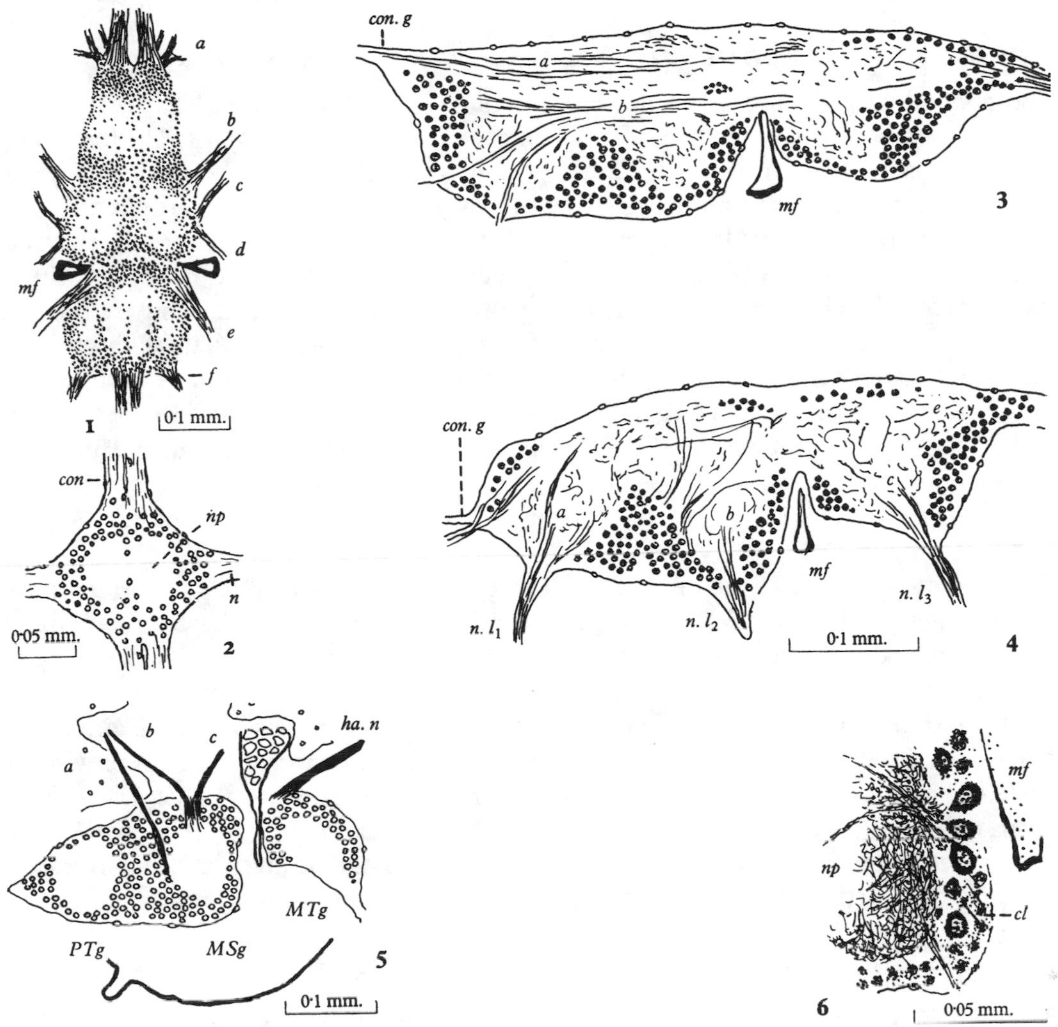

Figure 77. The ventral nerve cord.

1 Dorsal view of thoracic ganglia. *a*, prothoracic nerves; *b*, *c*, *d*, nerves arising dorsally from the mesothoracic ganglion; *e*, nerve to haltere; *f*, nerve to hind-leg.

2 Ganglion of an abdominal segment.

3 Sagittal section of thoracic ganglia towards lateral margin showing axon tracts. *a*, dorsal tract of side; *b*, ventral tract of side; *c*, position a little internal to where the nerve to the haltere takes off.

4 Sagittal section of thoracic ganglia passing through the lateral neuropile masses at level of origin of nerves to the legs. *a*, *b*, *c*, prothoracic, mesothoracic and metathoracic neuropile masses; *d*, dorsal continuous layer of neuropile; *e*, neuropile mass of lateral half of ganglion of the first abdominal segment.

5 Sagittal section of thoracic ganglia at level of origin of dorsal nerves. *a*, *b*, *c*, nerves corresponding to those labelled *b*, *c*, *d* in no. 1 of plate.

6 Portion of sagittal section showing cortical layer posterior to the mesothoracic neuropile mass at level of group of large pyriform neurocytes.

Lettering: *cl*, cortical layer; *con.*, ganglionic connectives; *con.g*, connectives from sub-oesophageal ganglion entering the prothoracic ganglion; *ha.n*, nerve to haltere; *mf*, mesofurca; *MSg*, mesothoracic ganglion; *MTg*, metathoracic ganglion; *n*, nerve issuing from lateral angle of ganglion; *n.*l_1, *n.*l_2, *n.*l_3, nerve to first, second and third leg respectively; *np*, neuropile; *PTg*, prothoracic ganglion.

The abdominal ganglia are much smaller than those of the thorax. That of the first segment is attached in the imago, as already noted, to the thoracic complex. The remaining ganglia lie close to the sternites in the anterior portion of their respective segments. Each ganglion is diamond-shaped and consists of two oval neuropile masses partially fused in the middle line and a cortical layer of small neurocytes. The cortical layer, as in the thoracic ganglia, is thickest ventrally and laterally and but slightly developed dorsally. A single nerve on each side takes origin from the lateral angle of the ganglion. Linking the ganglia are double connectives; those linking the ganglion of the first with that of the second segment are somewhat longer than the others, but otherwise the distances between the ganglia are not very different. The ganglia also are of approximately equal size, except that the ganglion of the seventh segment is somewhat larger than those in the segments immediately preceding and in both sexes shows a constriction marking off an anterior and a posterior portion and suggesting that the ganglion represents a fused seventh and eighth abdominal ganglion.

For intimate structure of the insect ventral cord ganglia see Binet (1894), on the ventral cord in insects and Zawarzin (1924) on the intimate structure of the cord ganglia in *Aeschna*. The abdominal ganglia in *Aëdes aegypti* are, however, small and appear to be of relatively simple construction.

There does not appear to be in the imago any median nerve system such as is present in the larva. In this connection may be mentioned the delicate membrane that has previously been described (see section on accessory pulsatile structures in ch. XXVII, p. 619) as overlying the ventral nerve cord. In sagittal sections this appears as a line of tissue with small nuclei spaced at regular intervals lying just above the cord and at first sight might suggest a median nerve system. Traced laterally, however, its membranous nature is apparent.

THE PERIPHERAL NERVOUS SYSTEM

More study is necessary before a full and adequate account can be given of the peripheral nerves and their distribution. This is the more so since there seem relatively few anatomical studies of the peripheral nerves and their distribution in insects other than those of a very general character. Among authors who have described the nervous system including the peripheral nerves are: Holste (1910), on the nervous system of *Dytiscus*; Hammar (1908), on the nervous system of the larva of *Corydalis*; Hanström (1928 *a*, *b*), on the nervous system in arthropods; Rogosina (1928), on segmental nerves in *Aeschna*. Though the muscles of the head of the mosquito have been minutely described by several authors, the nerves have received no attention. Jobling (1928), in his study of the head and mouth-parts in *Culicoides*, gives figures in which the mandibular, maxillary and labial nerves are shown, but does not describe these or give their origin or distribution. The following list of nerves of the imago of *Aëdes aegypti* must be considered provisional and may require modification and addition. The notes to which reference is made follow on the list, p. 642. For nerves of the stomodaeal system see section on the retrocerebral complex.

List of peripheral nerves

Head

Optic nerve and nerve to lateral ocellus (see under 'Special sense organs', p. 648).
(Nerves to dorsal ocelli: not represented, see note 1, p. 642).
Antennal nerve (see note 2).

Tegumentary nerve (see note 3).

Nerves of the stomatogastric system (see under 'Retrocerebral complex', p. 643).

Fronto-labral nerve. A short stout nerve on each side arising from the neuropile of the trito-cerebral lobe, but issuing antero-laterally from the sub-oesophageal ganglion. Divides almost at once into:

Frontal ganglion connective. A short stout nerve passing up in front of the lateral flange of the buccal cavity to join the side of the frontal ganglion (Fig. 78 (4)).

Labral nerve. Passes forwards and dorsally past the side of the buccal cavity into the lower part of the clypeal dome supplying branches to muscles and ending at a group of cells on each side in the region of the anterior hard palate (Fig. 78 (4) *n.* 3″).

Near its origin it gives off a branch passing forwards along the side of, and somewhat ventral to, the buccal cavity (see note 4). The muscles in the upper part of the dome are mostly supplied by the *frontal nerve* of the stomodaeal system as shown in Fig. 78 (4).

Mandibular nerve (see note 4).

Maxillary nerve. A relatively large nerve arising from the lateral anterior aspect of the sub-oesophageal ganglion. It passes forwards and outwards dorsal to the ventral cephalic trachea, giving off branches to maxillary muscles and ending as a well-marked nerve entering and continuing along the maxillary palp (Fig. 78 (2), (4) *n.* 6).

Labial nerve. The largest of the nerves arising from the sub-oesophageal ganglion. Arises ventrally on each side of the middle line some little distance back from the anterior end of the ganglion and passes forwards ventral to the ventral cephalic trachea to enter and continue along the labium to the labella forming with the nerve of the opposite side the characteristic twin nerves of the labium (Fig. 78 (3), (4) *n.* 7).

Prothoracic ganglion

Nerve arising just external to the entry of the anterior connectives and passing forwards to the neck.

Nerve arising with the next tabulated nerve a short distance posterior to the above and passing to the prosternal and propleural regions.

Nerve arising along with the above and passing into the first coxa accompanying the trachea to the first leg.

Nerve to the fore-leg. A large nerve arising ventrally from the apex of the lateral lobe of the prothoracic ganglion and passing down the posterior part of the fore coxa and beyond (Fig. 77 (4) *nl.* 1).

Mesothoracic ganglion

Anterior mesothoracic. Arises on each side laterally and dorsally from the fore part of the mesothoracic ganglion and passes forwards after linking with the next tabulated nerve in front of the anterior indirect wing muscle to the region of the anterior spiracle and the large tracheae in that region (Figs. 77 (5); 78 (1) *b*).

Middle mesothoracic. Arises posterior to the above from about the middle of the outer border of the ganglion along with the next tabulated nerve, but leaves this to pass forwards to accompany, or fuse with, the anterior mesothoracic nerve (Figs. 77 (5); 78 (1) *c*).

Posterior mesothoracic. Arises in common with the previous nerve and passes posteriorly and dorsally anterior to the furcal cup and to the general region of the wing root (Figs. 77 (5); 78 (1) *d*).

Nerve to the middle leg. A large nerve arising ventrally from the apex of the lateral lobe of the mesothoracic ganglion and passing down the coxa to the trochanter and beyond (Fig. 77 (4) *nl.* 2).

Posterior coxal nerve of the mid-leg. A nerve taking origin from the apex of the lateral lobe just posterior to the main nerve to the leg and passing into the posterior part of the coxa.

Metathoracic ganglion

Nerve to the haltere. A very large nerve arising on each side dorsally and laterally from the fore part of the ganglion just posterior to the stem of the mesofurca and passing posterior to the cup of that structure directly to the base of the haltere (see note 6).

Nerve to the hind-leg. A large nerve arising from the apex of the lateral lobe of the metathoracic ganglion and passing somewhat backwards and ventrally into the coxa and beyond.

Posterior coxal nerve of the hind-leg. Arises from the apex of the lateral lobe just posterior to the above and passes into the posterior part of the hind coxa.

First abdominal segment

Segmental lateral nerve. A small nerve arising on each side from the connective behind the ganglion and passing into the first abdominal segment to become the lateral nerve of that segment.

Abdominal segments II–VI

Segmental lateral nerves of the segments. From each side of each ganglion a nerve takes origin, passing at first outwards external to the segmental muscles and then up the side of the segment just internal to the parietal fat-body. In each segment it accompanies the ventral tracheal branch of the segment.

Abdominal segment VII in the female

Segmental lateral nerve. A nerve on each side takes origin from the anterior portion of the ganglion and passes outwards to the side of the segment as in other segments giving off branches near its origin to the fore and hind parts of the segment. The nerve and its branches pass ventral to the oviducts (Fig. 78 (5) *a*).

Postero-ventral nerve. On each side a nerve arises from the posterior pointed extremity of the ganglion and passes to the lateral parts of the eighth segment ventrally (Fig. 78 (5) *b*).

Postero-dorsal nerve. Arising along with the above but passing more medianly and dorsally this nerve passes through the terminal segments subjacent to the rectum and behind the common oviduct and spermathecae and sends a branch into the cercus of its side (Fig. 78 (5) *c*).

Abdominal segment VII in the male (Fig. 78 (6))

Segmental lateral nerve. A small nerve passing up the side of segment VII.

Posterior nerve. A rather large nerve arising on each side from the posterior end of each lateral half of the ganglion and passing backwards at the sides of the median genital ducts. It gives off a branch from which a nerve passes up the side of segment VIII and continues through this segment eventually to enter the gonocoxite of its side. It supplies the muscles connected with this structure and sends some fine branches to the parts connected with the median copulatory structures (Fig. 78 (6)).

Figure 78. Peripheral nerves.

1 General view of the ventral cord and its nerves. *a*, prothoracic nerves; *b*, *c*, *d*, dorsal nerves of mesothoracic ganglion; *e*, nerve to the halteres.

2 Dorsal view of clypeus, showing frontal and labral nerves and course of maxillary nerves (dotted).

3 Ventral view of sub-oesophageal ganglion, showing position of issuing nerves and other features. *a*, commissural tracts linking neuropile masses from which nerves take origin; *b*, collections of large neurocytes.

4 Lateral view of clypeus, showing frontal, labral, maxillary and labial nerves.

5 Dorsal view of terminal segments of female, showing the terminal ganglion and its nerves. *a*, lateral nerves of seventh abdominal segment; *b*, ventral posterior nerve; *c*, dorsal posterior nerve.

6 Lateral view of terminal segments of male, showing the terminal ganglion and its nerves.

Lettering: *ac.g*, accessory glands; *cod*, common oviduct; *fg*, frontal ganglion; *fn*, frontal nerve; *Gc*, gonocoxite; *g.*2, sub-oesophageal ganglion; *g.* II, *g.* VII, ganglia of second and seventh segment; *m.*41, *m.*42, indirect wing muscles; *m.*63, muscle to first leg; *mxp*, maxillary palps; *n.* 3, *n.*3′, *n.*3″, frontal-labral, frontal and labral nerves; *n.*6, maxillary nerve; *n.* 7, labial nerve; *Ph*, phallosome; *Rc*, rectum; *rn*, recurrent nerve; *vd*, fused vasa deferentia; VII, VIII, IX, abdominal segments.

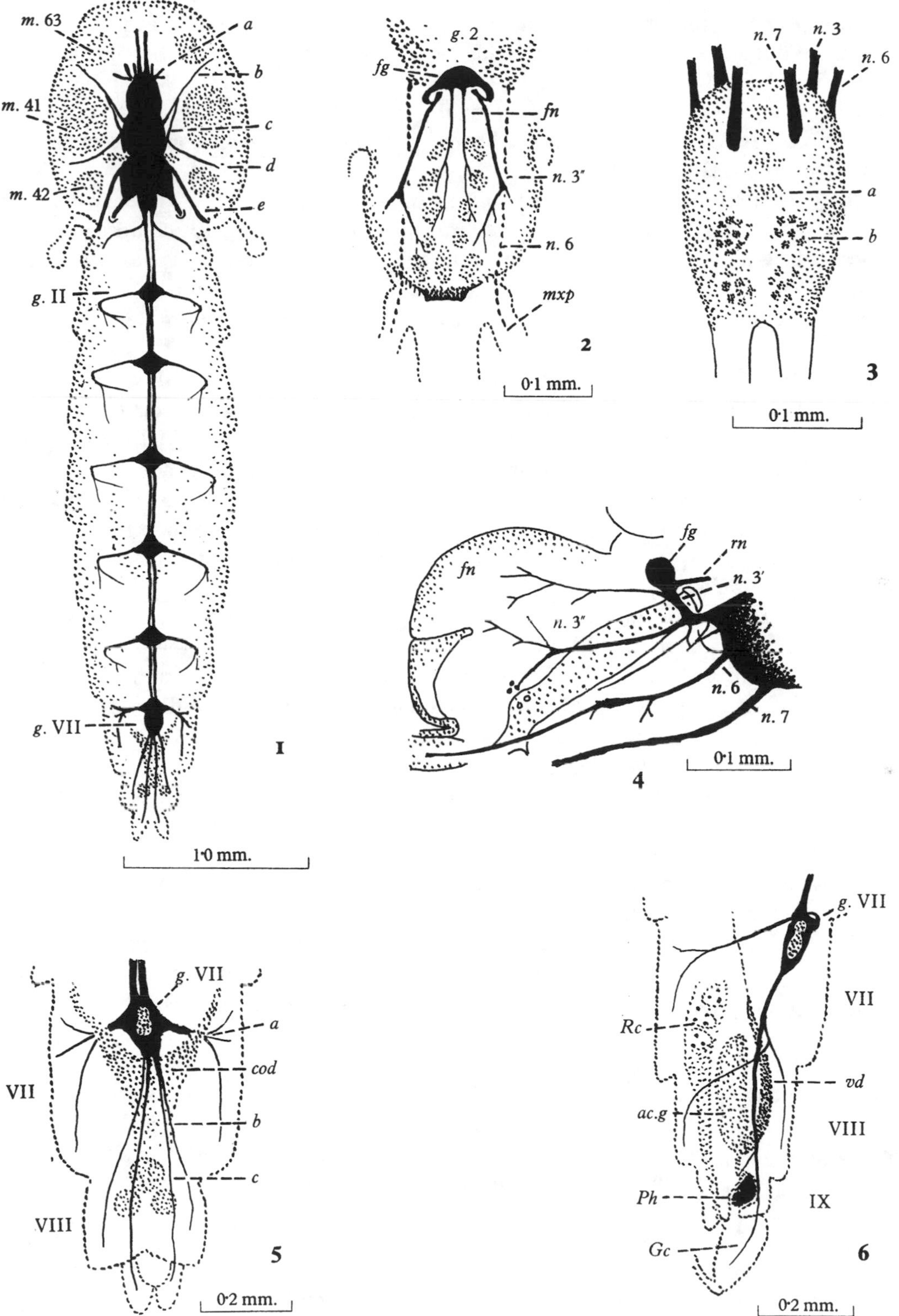
m. 63
a
b
m. 41
c
d
m. 42
e
g. II
g. VII
I
1·0 mm.
g. 2
fg
fn
n. 3″
n. 6
mxp
2
0·1 mm.
n. 7
n. 3
n. 6
a
b
3
0·1 mm.
fg
rn
n. 3′
fn
n. 3″
n. 6
n. 7
0·1 mm.
4
g. VII
a
cod
VII
b
c
VIII
5
0·2 mm.
g. VII
VII
Rc
vd
ac.g
VIII
Ph
IX
Gc
6
0·2 mm.

Figure 78

Note 1. The structure figured by Thompson (1905) as a nerve passing from the cerebral lobe to the dorsal ocellus (frontal pulsatile organ) is clearly not nerve but muscle fibres passing from the aorta to that organ.

Note 2. The major portion of the antennal nerve which is very massive terminates in Johnston's organ. A separate strand is, however, distinguishable as a small nerve passing on into the antennal filament.

Note 3. A small nerve can be made out in some preparations taking origin from the posterior part of the occipital lobe of the brain just internal to where the dorsal cephalic trachea enters the groove between the optic and cerebral lobes. It appears to send a branch forwards and ventrally over the lateral dilator muscles of the pump and one passing backwards. It may be the tegumentary nerve or possibly connected with the retrocerebral system.

Note 4. In keeping with the obsolescent character of the mandible in *Aëdes aegypti* a mandibular nerve, if present, is small. A small nerve bundle which may represent it passes out from the ganglion a little ventral to the fronto-labral nerve. Or a larger nerve that branches from the fronto-labral nerve and passes forwards to the general region of the tiny mandibular muscle may function as the mandibular nerve. At the point of issue of the above nerves there are crossing fibres forming a small plexus. It has been noted in the larva that there is a plexus connected with the nerves to the mouth structures and an accessory mandibular nerve.

Note 5. There are some points of interest in respect to the thoracic nerves which deserve comment. In the first place the contrast between the large nerves to the legs and the relatively insignificant nerves which presumably are to serve the far greater volume of muscle indirectly connected with the wings suggests something peculiar about the innervation of the latter. The indirect wing muscles it may be recalled differ from the leg muscles not only in their histological structure, but in that they are considered to be of the nature of giant fibres, that is, each bundle on such a view is a single fibre enormously greater than the fibres making up most other muscles of the body. Another peculiar feature is that the small nerves that one might expect to be concerned in innervation of these great muscular masses do not mainly proceed to them but seem, so to speak, to by-pass them. Thus the anterior dorsal mesothoracic nerve is mainly traceable as proceeding to the region of the anterior spiracle and the mass of large tracheal trunks in this region. A discussion on the innervation of muscles cannot here be entered upon, but attention may be drawn to the interesting and stimulating paper by Hoyle (1955) on this subject; see also Tiegs (1955).

Note 6. Mention has already been made of the large size of the nerves to the halteres. I have not seen any reference to the size of these nerves in other Diptera or to the unusual nature of their relations to the axon tracts in the ganglion. It suggests the importance of these organs in connection with the more anterior parts of the nervous system in the mosquito. Visual, antennal and haltere senses seem to be dominating features in the nervous system of this actively flying insect.

A description of the peripheral nervous system would be incomplete without some mention of the terminations of the nerves. Such may be in the subcuticular nervous system, in muscle-fibre endings, in the alimentary canal and other viscera and in special endings in sense organs. The last mentioned will be dealt with in a subsequent section. In regard to other methods of termination, only a very brief reference need be made. The subcuticular sensory system, apart from ectodermal sense organs such as sensitive hairs, sensory pits, etc., described later, is usually seen most developed in soft-skinned larval forms. It consists of a subcuticular network of multipolar neurocytes and their branches. So far as known no observations have been made on any such system in the imaginal mosquito where sensitive hairs and sensory pits may be expected mainly to form the cuticular sensory system. For literature see Snodgrass (1926, 1935), and references there given.

The ending of nerves in insect muscle are most usually by branching of the nerve fibres surrounding or penetrating the muscle fibres or by a form of motor end-plate, 'Doyère's cones', which are according to Marcu (1929) points where the fibres penetrate the muscle fibre as a brush-like group of branches. Claw-like endings of nerve fibres on muscle fibres are described by Hoyle (1955). No observations appear to have been made on the imaginal mosquito.

The termination of nerves to the alimentary canal in different insects has been studied by Zawarzin (1916); Orlov (1924); Rogosina (1928).

(*f*) THE RETROCEREBRAL COMPLEX

The stomodaeal, stomatogastric, sympathetic or visceral nervous system is intimately associated with certain endocrine glands having important functions in growth and metamorphosis. The two elements combined form what Cazal (1948) has termed the *retrocerebral complex*. This dual system has already been described with references to the literature in the chapters dealing with larval structure where it is both more developed and more easily studied than in the imago. It will here, therefore, only be necessary to note what changes have taken place in passage to the imago and any special features that may require mention in this stage.

The stomatogastric nervous system in the strict sense, that is the frontal ganglion with its connectives, recurrent nerve and the terminal ending of the recurrent nerve in a hypocerebral ganglion show little essential change. The frontal ganglion is still readily identifiable forming a most useful landmark in assigning homologies under the greatly changed conditions that have taken place during metamorphosis in the pharynx and oesophagus and accessory structures. The recurrent nerve has the same essential relations passing back in the middle line. But in place of the soft muscular pharynx and oesophagus on which it lay in the larva it now lies deeply enclosed in what is almost a closed canal formed by the deeply folded heavily sclerotised dorsal plate of the oesophageal pump. It is still accompanied by, and underlies, the aorta which is here greatly narrowed. Just before the posterior end of the plate the recurrent nerve ends abruptly in a not very large group of cells forming a flattened ganglion lying beneath, and seemingly in continuity with, the floor of the aorta. From this ganglion, which resembles in position the hypocerebral ganglion of the larva, certain nerves pass backwards. What these are may for the moment be left unspecified pending discussion of the glandular elements of the complex.

The component glands of the complex have already been indicated when describing the complex in the larva. They consist of the *corpora allata* and the mass of cells in which these bodies, more especially in the larva, lie embedded, the two components together forming the *corpora allata complex* of Bodenstein (1945). As Bodenstein notes, the *corpora allata* are still present unchanged in the imago. The surrounding element of the complex, which in the larva would appear to be the *peritracheal gland* of Possompes (1948, 1953) (*prothoracic* gland of Williams, 1949, and other authors) is, however, no longer present in its original form and consists of a different type of cell and is without the special characters of this part of the complex in the larva.

Cazal (1948) has described the retrocerebral complex in the pupa of *Culex pipiens* in which stage as he says it is most developed and most readily studied. In the posterior part of the neck lying by the side of the aorta there are two nerves, each with two roots, internal and external, which are fused at their point of emergence from the brain. These issue from the posterior aspect of the brain in the region of the root of the cerebral commissure and run parallel with the recurrent nerve. According to Cazal these pass to a group of cells

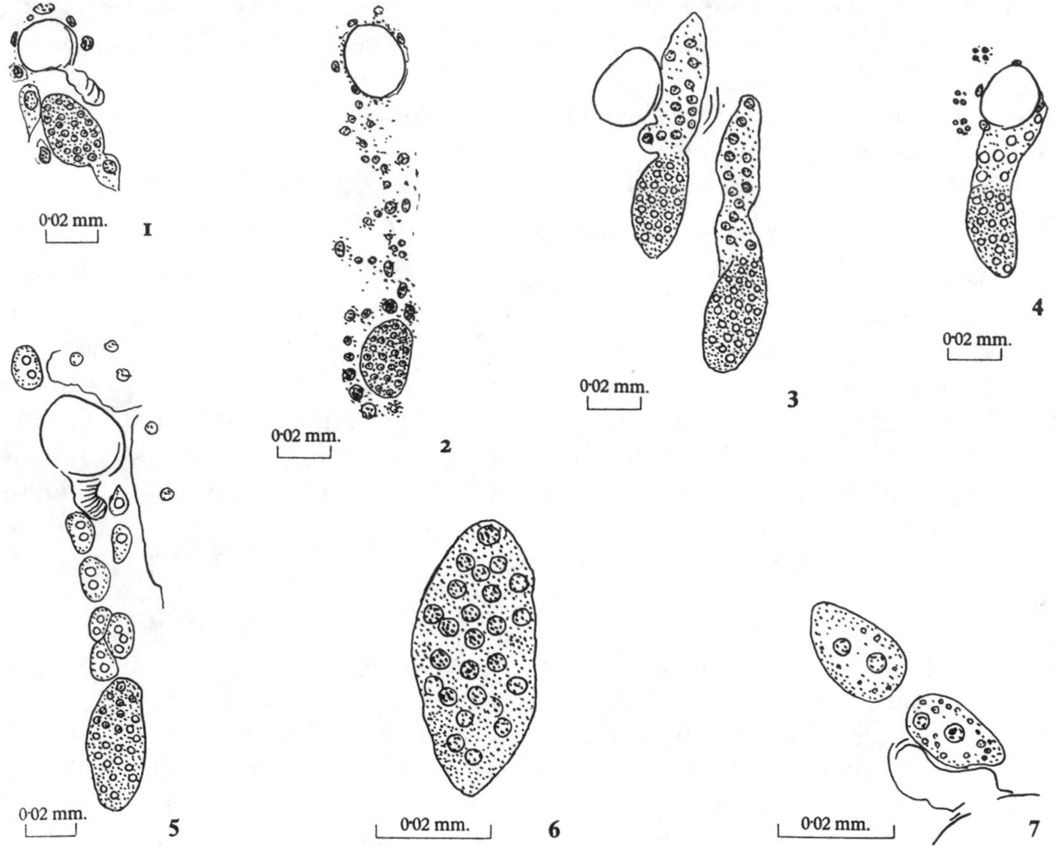

Figure 79. The retrocerebral complex.

1 Corpus allatum of a newly emerged female showing the gland situated close up to the anterior commissural trachea and with little indication of any peritracheal gland tissue.

2 The same in a newly emerged mosquito where the unchanged corpus allatum lies at a considerable distance from the trachea with intervening apparently disorganised peritracheal gland tissue between.

3 The corpora allata of a female (age uncertain) showing almost unchanged peritracheal glands.

4 The complex in a male.

5 The complex in a female mosquito that has had a blood feed and was ovipositing (age probably about 7 days from emergence). The corpora allata are unchanged and the site of the larval peritracheal gland is occupied by what appear to be typical pericardial cells.

6 The corpus allatum of the same more highly magnified.

7 Two of the pericardial-like cells more highly magnified.

incorporated in the hypocerebral ganglion that represent a single median *corpus paracardiacum* (*corpus cardiacum* of Possompes). From the single *corpus paracardiacum* included in the hypocerebral ganglion in which the recurrent nerve ends there pass back on each side a nerve to the *corpus allatum* and one to the oesophagus. At the level of the *corpora allata* the nerves and the aorta are crossed by a large tracheal vessel. In this region the trachea, with the *corpora allata* and the ends of the *corpora allata* nerves, is surrounded by a mass of cells which Cazal considers to be of the nature of pericardial cells. In the imago the pericardial cells disappear (see, however, later) so that the complex as a whole is much reduced. The anterior postcerebral gland of Possompes is also considered by Cazal to be a pericardial cell charged with waste products.

In the imago of *A. aegypti* the *corpora allata* with their characteristic appearance as small oval bodies with small regularly arranged nuclei have, as noted by Bodenstein in *Anopheles*, remained unchanged. But the regular compact appearance of the *corpora allata* complex in which these bodies in the larva lie embedded has been largely lost, the small oval *corpora allata* being no longer definitely included in a cellular mass but lying more or less free. The appearance of the complex was observed to vary a good deal in different specimens even of the same age. Shortly after emergence there was sometimes a distinct appearance of peritracheal gland still present (Fig. 79 (3)). In other cases the *corpora allata* were close up to the trachea with scarcely any associated tissue (Fig. 79 (1)). Very commonly the *corpora allata* were situated at some distance posterior to the trachea linked to this by what appeared to be disorganised tissue (Fig. 79 (2)). The *corpora allata* were also present unchanged in the male (Fig. 79 (4)).

As the *corpora allata* seemed to be still in a condition to function it was thought of interest to note whether in mosquitoes that had lived some time, had had a blood meal and had oviposited, the *corpora allata* were still unchanged. In such mosquitoes the *corpora allata* were unchanged, but in place of the peritracheal glands or cells replacing these the space between the *corpora allata* and the trachea was occupied by clusters of large cells undistinguishable from pericardial cells (Fig. 79 (5–7)). Before this observation it had been difficult to appreciate Cazal's reference to the pericardial cell nature of the peritracheal glands. It would appear, however, that with the passage to the imaginal state the outer portion of the *corpora allata* complex, representing the peritracheal gland, disorganises whilst some of its cells, or new cells, take on a typical nephrocyte condition. Lying, as the complexes do, close beside the aorta there would not seem anything impossible in these being pericardial cells.

In the imago the nerve to the *corpus allatum* of each side is much less conspicuous than in the larva and more difficult to follow to its termination. Two nerves, however, appear to continue from the hypocerebral ganglion on each side, one going to the neighbourhood of the *corpora allata* and one to the dilated portion of the oesophagus near the proventriculus.

REFERENCES

(*e*) THE NERVOUS SYSTEM

BALDUS, A. (1924). Untersuchungen über Bau und Funktion des Gehirnes der Larve und Imago von Libellen. *Z. wiss. Zool.* **121**, 557–620.

BERGER, C. A. (1878). Untersuchungen über den Bau des Gehirns und der Retina der Arthropoden. *Arb. Zool. Inst. Wien u. Trieste*, **1**, 173–220.

BINET, A. (1894). Contribution à l'étude du système nerveux sous-intestinal des insectes. *J. Anat. Phys.* **30**, 449–580.

BÖTTGER, O. (1910). Das Gehirn eines niederen Insekts, *Lepisma saccharina*. *Jena Z. Nat. Zool.* **46**, 801–44.

BRANDT, A. (1879). Vergleichend-anatom. Untersuchungen über das Nervensystem der Zweiflügler. *Hor. Soc. Ent. Rossic.* **15**.

BRETSCHNEIDER, F. (1913). Der Centralkörper und die Pilzformigenkörper im Gehirn der Insekten. *Zool. Anz.* **41**, 560–9.

BRETSCHNEIDER, F. (1914). Über die Gehirne des Goldkäfers und des Lederlaufkäfers. *Zool. Anz.* **43**, 490–7.

BRETSCHNEIDER, F. (1916). Über die Gehirne der Küchenschabe und des Mehlkäfers. *Jena Z. Naturwiss.* **52**, 269–362.

BRETSCHNEIDER, F. (1921). Über das Gehirn des Wolfsmilchschwärmers *Deilephila euphorbiae*. *Jena Z. Naturwiss.* **57**, 423–62.

BRETSCHNEIDER, F. (1924*a*). Über das Gehirn eines Bärenspinners *Callimorpha dominula*. *Jena Z. Naturwiss.* **60**, 147–73.

BRETSCHNEIDER, F. (1924*b*). Über die Gehirne des Eichenspinners und des Seidenspinners (*Lasiocampa quercus* L. and *Bombyx mori* L.). *Jena Z. Naturwiss.* **60**, 563–78.

BURGESS, A. F. (1880). *Second Rep. U.S. Entom. Comm.*

BUXTON, P. A. (1917). On the protocerebrum of *Micropteryx* (Lepidoptera). *Trans. Ent. Soc. Lond.* 1917, 112–53.

CAJAL S. R. (1918). Observaciones sobre la estrutura de los ocelos y vias nerviosas ocelares de algunos insectos. *Trab. Lab. Invest. Biol. Univ. Madr.* **16**, 109–39.

FLÖGEL, J. H. L. (1878). Ueber den einheitlichen Bau des Gehirns in den verschiedenen Insektenordnungen. *Z. wiss. Zool.* **30**, Suppl. 556–92.

HALLER, B. (1905). Ueber den allgemeinen Bauplan des tracheaten Syncerebrums. *Arch. micr. Anat.* **65**, 181–279.

HAMMAR, A. G. (1908). On the nervous system of the larva of *Corydalis cornuta*. *Ann. Ent. Soc. Amer.* **1**, 105-27.

HANSTRÖM, B. (1925). Comparison between the brain of the caterpillar and the imago in *Pieris brassicae*. *Ent. Tidskr.* **46**, 43–52.

HANSTRÖM, B. (1928*a*). Die Beziehungen zwischen dem Gehirn der Polychäten und dem der Arthropoden. *Z. Morph. Ökol. Tiere*, **11**, 152–60.

HANSTRÖM, B. (1928*b*). *Vergleichende Anatomie des Nervensystems der wirbellosen Tiere.* Berlin.

HANSTRÖM, B .(1930). Ueber das Gehirn von *Termopsis nevadensis* und *Phyllium pulchrifolium* nebst Beiträgen zur Phylogenie der *corpora pedunculata* der Arthropoden. *Z. Morph. Ökol. Tiere*, **19**, 732–73.

HICKSON, S. J. (1885). The eye and optic tract of insects. *Quart. J. Micr. Sci.* **25**, 215–51.

HOLMES, W. (1947). *Recent Advances in Clinical Pathology* (chapter on peripheral nerves). Churchill, London.

HOLSTE, G. (1910). Das Nervensystem von *Dytiscus marginalis*. *Z. wiss. Zool.* **96**, 419–76.

HOLSTE, G. (1923). Das Gehirn von *Dytiscus marginalis*. *Z. wiss. Zool.* **120**, 251–80.

HOYLE, G. (1955). The anatomy and innervation of locust skeletal muscle. *Proc. Roy. Soc.* B, **143**, 281–92.

IMMS, A. D. (1938). *Textbook of Entomology*. Ed. 4. Methuen and Co., London.

JANET, C. (1905). *Anatomie de la tête du* Lasius niger. Limoges.

JAWLEWSKI, H. (1936). Über den Gehirnbau der Käfer. *Z. Morph. Ökol. Tiere*, **32**, 67–91.

JOBLING, B. (1928). The structure of the head and mouthparts in *Culicoides pulicaris* L. (Diptera: Nematocera). *Bull. Ent. Res.* **18**, 211–36.

JONESCU, C. N. (1909). Vergleichende Untersuchungen über das Gehirn der Honigbiene. *Jena Z. Nat. Zool.* **45**, 111–80.

KENYON, F. C. (1896). The brain of the bee. *J. Comp. Neurol.* **6**, 133–210.

KUHNLE, K. F. (1913). Das Gehirn des gemeinen Ohrwurms. *Jena Z. Naturwiss.* **50**, 147–276.

REFERENCES

LOWNE, B. T. (1893–95). *The Anatomy, Physiology, Morphology and Development of the Blowfly*. London, **2**, 351–745.

MARCU, O. (1929). Nervenendigungen an den Muskelfäsern von Insekten. *Anat. Anz.* **67**, 369–80.

NEWTON, E. F. (1879). On the brain of the cockroach. *Quart. J. Micr. Sci.* **19**, 340–56.

ORLOV, J. (1924). Die Innervation des Darmes der Insekten. *Z. wiss. Zool.* **122**, 425–502.

PACKARD, A. S. (1880). *The Brain of the Locust*. Second Rep. U.S. Ent. Comm.

ROGOSINA, M. (1928). Über das periphere Nervensystem der *Aeschna*-Larve. *Z. Zellforsch.* **6**, 732–58.

ROMANES, G. J. (1950). The staining of nerve fibres in paraffin sections with silver. *J. Anat.* **84**, 104–15.

SAMUEL, E. P. (1953). Towards controllable silver staining. *Anat. Rec.* **116**, 511.

SNODGRASS, R. E. (1925). *Anatomy and Physiology of the Honeybee*. McGraw-Hill Book Co., New York and London.

SNODGRASS, R. E. (1926). The morphology of insect sense organs and the sensory nervous system. *Smithson. Misc. Coll.* **77**, no. 8.

SNODGRASS, R. E. (1935). *Principles of Insect Morphology*. McGraw-Hill Book Co., New York and London.

THOMPSON, C. B. (1913). A comparative study of the brains of three genera of ants with special reference to the mushroom bodies. *J. Comp. Neurol.* **23**, 515–71.

THOMPSON, M. T. (1905). The alimentary canal of the mosquito. *Proc. Boston Soc. Nat. Hist.* **32**, 145–202.

TIEGS, O. W. (1955). The flight muscles of insects. *Phil. Trans.* B, **238**, 221–348.

VIALLANES, H. (1882). Recherches sur l'histologie des insectes. *Ann. Soc. Nat. Zool.* (6), **14**, 1–348.

VIALLANES, H. (1885). Le ganglion optique de la Libellule (*Aeschna maculatissima*). *Ann. Soc. Nat. Zool.* (6), **18**, Article 4, 1–34.

VIALLANES, H. (1887*a*). Le cerveau de la guêpe (*Vespa crabro* et *V. vulgais*). *Ann. Sci. Nat. Zool.* (7) **2**, Article 1, 1–100.

VIALLANES, H. (1887*b*). Le cerveau du criquet. *Ann. Soc. Nat. Zool.* (7), **4**, 1–120.

VIALLANES, H. (1893). Centres nerveuses et les organes des sens des animaux articulés. *Ann. Sci. Nat. Zool.* (7), **14**, 419–56.

WEBER, H. (1933). *Lehrbuch der Entomology*. Gustav Fischer, Jena.

WIGGLESWORTH, V. B. (1950). *Principles of Insect Physiology*. Ed. 4. Methuen and Co., London.

WIGGLESWORTH, V. B. (1953). The origin of sensory neurones in an insect. *Rhodnius prolixus* (Hemiptera). *Quart. J. Micr. Sci.* **94**, 93–112.

ZAWARZIN, A. (1914). Histologische Studien über Insekten. IV. Die optischen Ganglien der *Aeschna*-Larve. *Z. wiss. Zool.* **108**, 175–257.

ZAWARZIN, A. (1916). Quelques données sur la structure du système nerveux intestinal des insectes. *Russk. zool. Zh.* **1**, 176–80.

ZAWARZIN, A. (1924). Histologische Studien über Insekten. VI. Das Bauchmark der Insekten. *Z. wiss. Zool.* **122**, 323–424.

(*f*) THE RETROCEREBRAL SYSTEM

BODENSTEIN, D. (1945). The corpora allata of mosquitoes. *44th Rep. Connect. State Entomologist, Connect. Agric. Exp. Sta. Bull.* 488, 396–405.

CAZAL, P. (1948). Les glandes endocrines rétrocérébrales des insectes (Étude morphologique). *Bull. Biol. Fr. Belg.* Suppl. 32, 227 pp.

POSSOMPES, B. (1948). Les corpora cardiaca de la larve de *Chironomus*. *Bull. Soc. Zool. Fr.* **73**, 202–6.

POSSOMPES, B. (1953). Recherches expérimentales sur la détermination de la métamorphose de *Calliphora erythrocephala* Meig. *Arch. Zool. Exp. Gén.* **89**, 203–64.

WILLIAMS, C. M. (1949). The prothoracic glands of insects in retrospect and in prospect. *Biol. Bull.* **97**, 111–14.

XXIX
THE IMAGO: SPECIAL SENSE ORGANS

(*g*) SPECIAL SENSE ORGANS

For general accounts of special sense organs in insects see: Snodgrass (1926, 1935); Eltringham (1933); Wigglesworth (1942). For location of different senses in mosquitoes, including *Aëdes aegypti*, see Frings and Frings (1949); Frings and Hamrum (1950); Roth (1951); Roth and Willis (1952). For accounts of particular organs see references given under headings below.

Special sense organs in the mosquito include: the ocelli and compound eyes; Johnston's organ in the pedicel of the antenna, commonly considered an organ of hearing; various forms of sensitive hairs, sensory pits, chemo-receptor, thermo-receptor and other sense receptors.

THE OCELLI

The lateral ocelli (Fig. 80 (1) *oc*). As shown by Constantineanu (1930) the lateral ocelli, which are the functioning eyes in the larval mosquito, are still present in the pupa and imago. In the imago they are to be seen on removing the head scales as pigmented areas behind and close to the posterior margin of the compound eyes. So viewed they show little evidence of organisation and they leave no mark on the cuticle in potash preparations. Nevertheless they receive a nerve from the ganglionic plate of the optic tract and in sections show some indication of ocellar structure.

The dorsal ocelli (Fig. 80 (2–4)). Under the circulatory system mention is made of the organs described by Clements (1953) description of which was postponed to the present chapter since they appeared to be of the nature of modified ocelli. The organs are situated one on each side of the middle line just in front of the thickening in the head capsule linking the two ocular sclerites and marking the line between vertex and frons (Fig. 80 (2)). They appear externally as small circular areas outlined by a somewhat indefinite depression in the cuticle of the frons and with a darker central spot.

In section each organ is seen to consist of an outer thick plate of cuticle covered with small microtrichae (3*a*). Underlying this is a cushion-like pad of tissue (*b*) consisting of columnar epithelial cells with large nuclei towards their inner ends (Clements, 1956*b*). This layer is continued as a flattened nucleated membrane forming the walls of the vesicle. According to Clements there is an opening in the wall posteriorly which is closed in a valve-like fashion by an extension of the wall of the vesicle into the sac. At the back of the vesicle in *A. aegypti* is a cone-shaped nucleated mass which might be described as somewhat resembling a ganglion (*d*). This mass forms the back of the vesicle the walls of which, at least in part, are attached to its margin. Its attenuated portion passing backwards is continued into a bundle of muscle fibres passing backwards to the aortic wall where this structure emerges from beneath the cerebral commissure. In *Culex pipiens* Clements describes the cellular mass behind the vesicle as a spherical mass with a very large number of nuclei narrowly separated by cytoplasm apparently without cell boundaries (syncytium). These cells are termed receptor cells by Day (1955), but according to Clements they bear no resemblance to typical sense cells and apparently are not innervated. In their position, however, they certainly call to mind the mass of cells seen in the dorsal ocelli. The bundle of muscle fibres passing from the aortic wall to the syncytial body are described by Clements as appearing to splay out over or to be inserted into this body. It is these fibres that have been figured by Thompson (1905) as nerves passing to what he took to be the dorsal ocelli. Clements, however, using the silver impregnation method (Samuel, 1953), has been unable to detect any nerve fibres in the bundles.

Passing from the outer wall of the vesicle by a well-marked somewhat narrowed opening is a stout-walled tube having the appearance of a trachea, but with no taenidial markings (Fig. 80 (4) *e*). This vessel can be traced to the base of the antenna where it passes through a gap in the antennal nerve to the central space of the organ of Johnston and from thence into the antennal flagellum. It is clearly this tube which is described by Eggers (1924), and by Risler (1953), as a blood vessel in the antennal flagellum, though neither author notes the existence of the organs now being described from which the vessel arises.

It is also quite evident that these organs in the mosquito are of a similar nature to organs that have been described as propulsatory organs in the cockroach and other insects (see figure after Pawlowa (1895), given by Wigglesworth (1942), p. 224, and accompanying text). The figure referred to shows quite clearly tubes leading into the antenna and muscular fibres passing from the aorta. Such organs appear to have been recorded from a wide range of insect orders. Wigglesworth, *loc. cit.*, notes them as recorded in *Periplaneta*, *Locusta* and Acridiidae and that similar vessels and ampullae supply the antennae of the honey-bee. Clements (1956*b*) records an antennal organ strikingly similar in *Panorpa* (Mecoptera). What appears to be the same organ has been recorded not only in *Anopheles*, *Culex*, *Theobaldia*, *Aëdes* and *Chaoborus* (Clements, 1956*b*) but in many Nematocera: Tipulidae (Dufour, 1851); Ptychoptera, Tricoceridae, Scatopsidae, Blepharocidae (Clements, 1956*b*); Chironomidae (Miall and Hammond, 1900); Ceratopogonidae (Jobling, 1928); Psychodidae (Christophers, *et al.* 1926); Rhyphidae, Cecidomyidae, Mycetophilidae and Sciaridae (Day, 1955). The organ has not been found by Clements in *Thaumalea testacea*, *Simulium ornatum*, or *Dilophus febrilis*. Day was unable to find the organ in a species of *Stratiomyidae* or in *Musca domestica* and considered it was probably absent in Diptera Brachycera. Clements, however, has confirmed the description by Miller (1950) in *Drosophila melanogaster* Mg. of a single vesicle below the ptilinum from which two muscles arise directly without a distinct syncytial organ and run to the aorta. Antennal vessels pass from each side of the vesicle to the antennae.

The whole appearance of the organs, situated just where the dorsal ocelli might be expected and with some indication of what might well be degenerate ocellar structure, suggests that they are ocelli modified as propulsatory organs. That such organs might easily be recorded as ocelli would seem likely; and contrariwise ocelli might sometimes be thought to be these organs. According to Miall and Denny (1886), the white fenestrae which in the cockroach lie internal to the antennal sockets may represent two simple eyes that have lost their dioptric apparatus (p. 100). Jobling (1928) gives a figure showing a longitudinal section of an organ in *Culicoides* (Pl. IX, 1) considered by him to be a dorsal ocellus which has some resemblance to the organ in the mosquito except that it does not possess a cavity. He notes that the chitinous integument is transparent, that the cellular layer of the anterior part of the ocellus apparently represents the modified corneagen cells and that below this layer is a cone formed by cells that are loaded with pigment granules and are in direct connection with the fibres of ocellar nerves arising from the protocerebral lobes.

The answer as to the nature of the organs would seem to be decided by whether, when such pulsatile structures are present, dorsal ocelli are absent. Day (1955) states that in certain families the presence of the organs together with ocelli shows that the two structures are distinct. Apparently, however, in the cockroach there are no structures other than the propulsatory organs that could be taken as dorsal ocelli and clearly in *Culicoides* as

---

Figure 80. Compound eye and ocelli.

1 Tangential section in region of lateral ocellus, showing nerve to ocellus from the ganglionic plate of the optic tract.

2 Frontal region of head, showing position of the dorsal ocelli.

3 Coronal section through the two antennal pulsatory organs (modified dorsal ocelli). *a*, cuticular plate; *b*, cellular mass below same; *c*, cavity (blood sinus); *d*, ganglion-like mass (possibly representing retinal cells); *e*, muscle fibres passing to *d* from wall of aorta.

4 Reconstruction from serial sections, showing antennal blood vessel passing from antennal pulsatory organ. *a–d*, as in no. 3; *e*, tubular vessel.

5 Logitudinal section of ommatidium of compound eye of *Aëdes aegypti*, as seen without removal of pigment. *e*, level at which the transverse section given in no. 12 is taken.

6 The same of two ommatidia after removal of pigment. *a*, nucleus of iris pigment cell; *b*, nucleus of accessory pigment cell; *c*, indicates level at which transverse section given in no. 10 has been taken; *d*, level at which no. 11 is taken; *f*, level at which no. 13 is taken.

7 Portion of tangential section of compound eye passing through the dark staining rim of the corneae (as shown at *a* and *b* in no. 8). *a*, position corresponding to *a* of no. 8; *b*, position *b* of no. 8; *c*, the four crystal cell nuclei situated close beneath the lens (shown dotted in ommatidium to the left); *d*, nuclei of the two iris pigment cells; *e*, portion of periphery where the section passes through the corneae just above the dark-staining (more sclerotised) lower rim.

8 Vertical section through rim of cornea, showing sclerotised rim and interval between the corneae (cuticular frame). *a*, corresponding to position *a* in no. 7; *b*, corresponding to position *b* in no. 7.

9 Transverse section through level corresponding to *b* in no. 6 (level of nuclei of accessory pigment cells). Shows iris pigment cells surrounding narrowed portion of cone-like body and surrounding nuclei of accessory pigment cells (pigment removed).

10 Transverse section at level *c* of no. 6 (pigment removed).

11 The same at level *d*. Rhabdome formed of rhabdomeres surrounded by cytoplasm of retinula cells (pigment removed).

12 The same at level *e* of no. 5. Rhabdome in centre with pigment surrounded by ring of pigment in inner portion of the retinula cells.

13 The same at level of *f* in no. 6. Showing nuclei of 7 retinula cells.

Lettering: *ao*, aorta; *apc*, accessory pigment cells; *cb*, cone-like body; *cbn*, nucleus of cone-like body; *co*, cornea; *E*, compound eye; *F*, frons; *fil*, flagellum (of antenna); *gp*, ganglionic plate of the optic tract; *hpc*, hypobasal pigment cell; *ipc*, iris pigment cell; *jo*, Johnston's organ; *le*, lens; *m*, muscle bundles (*m*.21 and 22); *n*, nerve to ocellus; *oc*, ocellus; *P'*, anterior portion of oesophageal pump; *pd*, pedicel; *rb*, rhabdome; *rc*, retinula cell; *rcn*, nucleus of retinula cell; *sca*, scape; *t*, trachea; *V*, vertex.

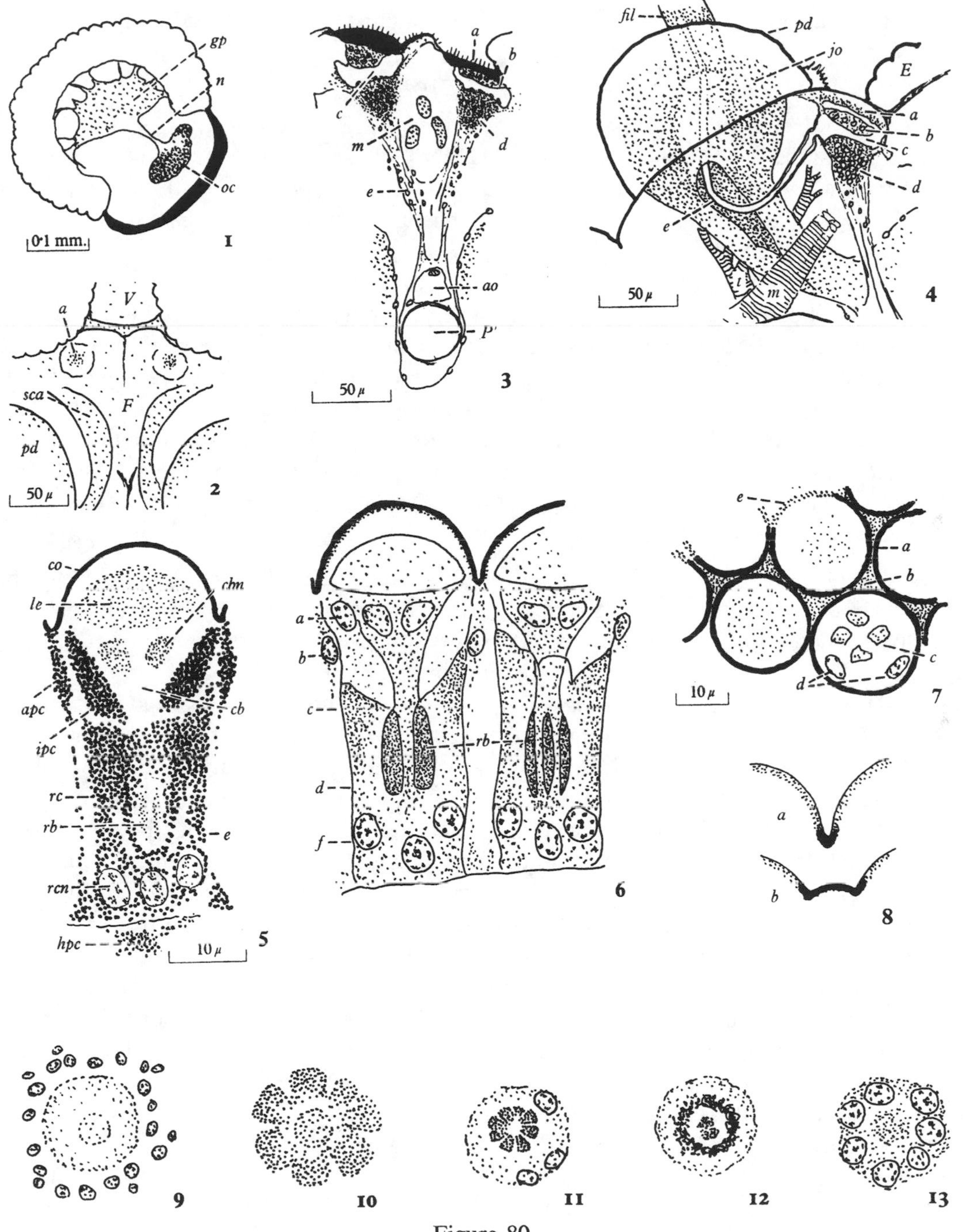
gp
n
oc
0·1 mm.
1
a
V
F
sca
pd
50 μ
2
a
b
c
d
m
e
ao
P
50 μ
3
fil
pd
jo
E
a
b
c
d
e
i
m
50 μ
4
co
le
chn
apc
ipc
cb
rc
rb
e
rcn
hpc
10 μ
5
a
b
c
d
f
rb
6
e
a
b
c
d
10 μ
7
a
b
8
9
10
11
12
13

Figure 80

described by Jobling only either the ocelli or the organs are represented. It seems possible that the organs have the function of maintaining due tension in the antenna, especially where these are long and filamentous. It is even further possible that there might be an alternative relationship as between the use of long antennae and the dorsal ocelli as perceptive organs.

THE COMPOUND EYES

The compound eyes of *Culex* have been described by Constantineanu (1930); see also Buddenbrock (1939). Recently a very detailed study of the structure and development of the compound eyes of a species of *Culex* (*Culex pipiens* var. *pallens* Coq.), of a species of *Aëdes* (*A.* (*Finlaya*) *japonicus* Theo.) and of *Anopheles* (*A. hyrcanus* var. *sinensis* Wied.) has been made by Shôzô Sato (1950; 1951*a*, *b*; 1953*a*, *b*).

In *Aëdes aegypti* the compound eyes occupy the greater part of the lateral aspects of the head and are approximately of the same relative size in both sexes. They are kidney-shaped with a hollowed-out border anteriorly, deeper in the male than in the female, to accommodate the basal antennal segments. Dorsally the eyes approach each other, leaving a narrow isthmus of vertex between them. Ventrally they nearly meet beneath the head. Each eye is surrounded by a rim of thickened cuticle which forms a shallow apodematous frame, the *ocular sclerites*. These are linked together dorsally by a thickened bar where the vertex meets the frons. Each eye is composed of some hundreds of *ommatidia*, which on the surface appear as regularly aligned facets.

The ommatidia (Fig. 80 (5–13)). The structure of the ommatidia of the compound eye in the mosquito has been described by Shôzô Sato in the species studied by him. His description closely accords with the findings in *Aëdes aegypti*. Each ommatidium consists of three parts, namely a corneal portion, an iris portion and a retinal portion. The corneal portion includes a thin outer cuticle, the *cornea* (Fig. 80 (5) *co*) and an underlying biconvex *corneal lens* (*le*). The cornea is a rigid strongly convex structure about a micron in thickness and almost hemispherical in form. The lens is also strongly convex distally, but somewhat flatter, though to a variable extent, proximally. When stained it commonly shows a darker portion towards its base or several such areas in a nest-like arrangement. Below the lens is the *cone-like body* (*cb*). In the mosquito this is formed of four *crystal cells*, the nuclei of which, being situated just below the lens, form a characteristic appearance in tangential sections through the eye (Fig. 80 (7) *c*). That the cone-shaped body is not a true crystalline body secreted by the crystal cells but constituted by these, places the mosquito eye in the class of *acone* eyes, as distinct from the *eucone* eyes of many insects. The crystal cells narrow proximally so that the cone-like body tapers towards the retinal portion of the ommatidium, finally meeting the central sense body or rhabdome. Surrounding the cone-like body are two densely pigmented cells, *iris pigment cells* (*ipc*), the nuclei of which appear in tangential sections situated to the side of the four crystal cell nuclei (Fig. 80 (7) *d*). The retinal portion of the ommatidium consists of eight pigmented sense cells, *retinula cells* (*rc*), the large nuclei of seven of which are situated close to the basement membrane (Fig. 80 (6), (13)). Formed by the inner portions of the retinula cells are the *rhabdomeres*, which together form the *rhabdome* or central sense organ (Fig. 80 (5), (6) *rb*). The retinula cells contain pigment granules, massed especially in the inner portions around the rhabdome, but leaving a clear area, *intermediate area*, consisting of neurofibrils (Fig. 80 (5–12)). Basally is the *basement membrane* through which fibres pass from the retinula cells to the plexus of nerve fibres passing between the retinal portions of the ommatidia and the ganglionic plate of the optic tract. Beneath the basement membrane opposite each ommatidium is a branched pigmented cell, *hypobasal pigment cell* of Shôzô Sato (*hpc*). Surrounding the ommatidium is a circle of elongate pigment cells passing between the external cuticle of the eye and the basement membrane, but with nuclei all at a level slightly below the crystal cells, *accessory pigment cells* (*apc*). The outer parts of these cells form a dense continuous pigment mass underlying the cuticular frame in which the corneae are set as described later in the section on the facets. The iris pigment cells, the retinula cells, the accessory pigment cells and also the basement membrane, hypobasal cells and nerve plexus are all heavily loaded with pigment. The cone-shaped body and to some extent the rhabdome are free from pigment and appear as clear areas in the otherwise densely pigmented ommatidium.

Certain changes taking place in the eyes of *Culex* under conditions of light and darkness are described by Shôzô Sato (1950). In the light adapted condition the parts are as described above for *Aëdes aegypti*. After some hours exposure to darkness, however, the rhabdome pushes outwards opening up the crystal cells and moving into the iris portion of the ommatidium, approaching and nearly touching the lens, which thickens proximally. Some change in the same direction is figured by Shôzô Sato (1953*a*) in *A. japonicus*, though to a less degree.

It is clear from the description given above that the mosquito eye is to be classed as of *appositional type*, that is, the retinula cells and rhabdom are placed immediately behind the cone and not separated by a long interval with segregation of pigment to allow lateral passage of light between neighbouring ommatidia as in the *superpositional type* of eye as seen in some nocturnal insects.

The facets (Figs. 80, 81). The facets of the compound eyes of insects, including those of the mosquito are commonly described as hexagonal. In the case of the mosquito as seen in *Aëdes aegypti* this requires some modification, or at least amplification. The appearance of the facets varies a good deal depending upon the conditions and magnification under which the eye membrane is viewed. If the eye membrane of *A. aegypti*, appropriately treated as described later, is viewed in water under fairly high magnification, each facet appears as a clear transparent circular area, representing the cornea with the subjacent lens, surrounded by what appears as a ring of less transparent cuticle (Fig. 81 (3)). The rings, like the corneae, are circular and hence leave areas where their contours fail to meet. Such areas appear as small equilateral triangles, six in number, set around each cornea. That these are not merely artefacts is shown by the fact that air may be trapped beneath them forming small triangular air bubbles (Fig. 81 (4)). When, however, the eye membrane is examined mounted in a medium such as canada balsam, the corneae usually appear as large clear circular areas filling or almost filling the facet and with their circumferences in contact. Being circular they still leave triangular areas where their circumferences fail to meet. Sometimes the small triangular areas described above may be made out within such spaces, but commonly the cuticle being now more transparent the appearance is merely of triangular spaces between the circular corneae (Fig. 81 (5)). Occasionally the corneae may themselves be more or less hexagonal, possibly as a result of pressure in mounting, but even so in no case do they form sharp-angled hexagons.

As shown by the study of mounted preparations and sections, the following conclusions appear to hold good. The corneae are circular in outline, leaving a certain amount of relatively flat cuticle between them. The lower rims of the corneae also consist of sclerotised cuticle as distinct from the corneal substance proper. The difference between the corneal substance proper and the darker-staining sclerotised margins may often be very clearly shown as in the section stained by Holmes' method shown in Fig. 80 (7). Thus, taken together, the cuticular substance forms a framework in which the corneae proper are set. The intercorneal cuticle further corresponds to and overlies space between the ommatidia in which the accessory pigment cells lie. In the undecolorised eye the accessory pigment cells underlying the circumferential area form a densely pigmented mass of cytoplasm immediately beneath the cuticle further emphasising the fact that the facets are essentially circular.

The facets, may, however, conveniently be considered as hexagonal in so far as in apportioning the total eye area between them the area allotted per facet would need to be hexagonal. The point, as will be seen later, arises in relation to measurement of the facet.

Number of facets. In *Culex pipiens* Shôzô Sato (1950) found the number of facets in each eye in the female to be 503–60, and in the male 440–62, the surface of the eye in the female being 0·285–0·295 mm.² and in the male 0·211–0·236 mm.². In *Aëdes japonicus* the numbers were 504–27 in the female and 409–28 in the male, the eye area being respectively 0·286–0·314 mm.² and 0·199–0·223 mm.².

Table 48. *Number of facets and area of the compound eye of female and male* Aëdes aegypti *with calculated data based on same*

Material	No. of facets	Area of eye (mm.²/1000)	Calculated area of hexagonal facet (μ^2)	Calculated diameter of inscribed circle (d) (μ)
Female no. 1	468	216	462	23·1
Female no. 2	492	236	480	23·6
Female no. 3	421	203	482	23·6
Female no. 4	447	211	472	23·4
Mean	457	216	474	23·4
Male no. 1	362	161	445	22·7
Male no. 2	358	140	391	21·2
Male no. 3	369	131	355	20·3
Mean	363	144	397	21·4

Figure 81. Facets of the compound eye.

1 Showing relation of area of hexagon to diameter of the inscribed circle. The figure shows that the area of the hexagon equals that of six equilateral triangles with sides equal to *h* and that *r* is readily calculated from *h* since the half base *ad* is equal to half *h* and the angle *adc* which is a right angle.

2 Camera lucida drawing of eye membrane divided and flattened. The facets are indicated by dots placed in their centre. The small crosses indicate lines that terminate (see text). A–D, areas of eye; A, dorsal; B, antero-lateral; C, lateral; D, postero-lateral; E, ventral; *a*, vertex; *b*, cuticle surrounding base of antenna; *c*, cuticle of occiput.

3 Showing appearance of facets as examined in water.

4 Showing trapping of air under inter-corneal cuticle.

5 Showing appearance of facets as mounted in balsam.

6 To show arrangement of facets in lines at 120°.

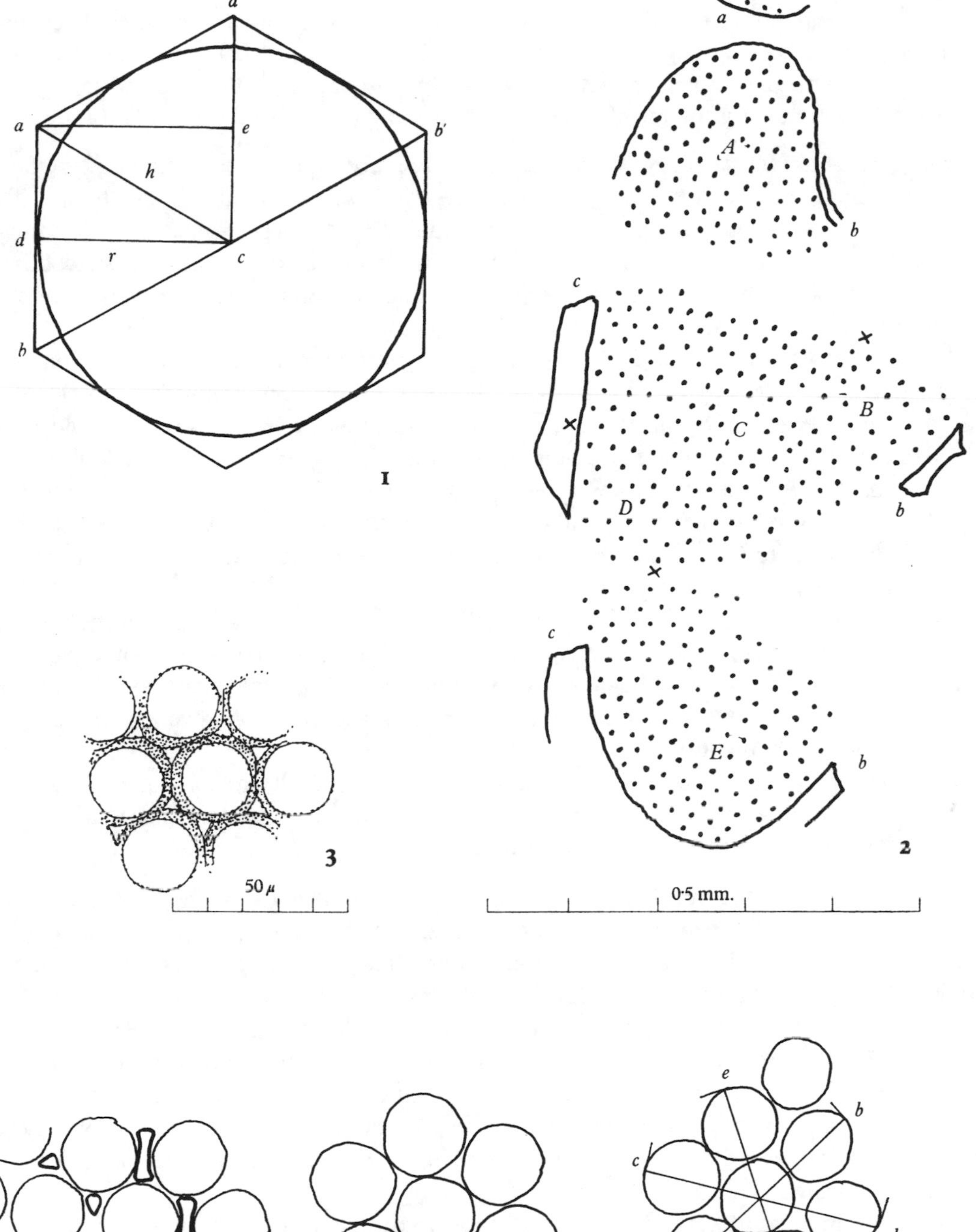
a'
a
e
b'
h
d
r
c
b
I
a
A
b
c
x
B
C
x
b
D
x
c
E
b
2
3
50 μ
0·5 mm.
50 μ
4
50 μ
5
e
b
c
d
a
f
6
50 μ

Figure 81

The area of the eye and number of facets as determined in four female and three male *A. egypti* are given in the accompanying table (Table 48). The counts have been made from camera lucida drawings of eye membrane suitably cut to lie flat. The mean number of facets for female and male respectively works out as 457 and 363. Allowing for some possible loss in cutting the membrane the actual numbers may be very slightly greater.

Arrangements of facets. A few words may be said regarding the arrangement of the facets. It is evident that these are arranged in very regular rows. It is also evident that, if the facets are circular or hexagonal, of equal size and fitted as closely as possible, they must of necessity on a flat or uniformly curved surface be arranged so that in three directions at 120° to each other they form lines in which diameters through their points of contact are continuous and in the same straight line (Fig. 81 (6)).

Except that the lines of facets in the eye are slightly curved, this is very obviously the arrangement (Fig. 81 (2)). The lines are quite long, running from one margin of the eye to another, one series nearly horizontal, one dorso-anterior to ventro-posterior and one at the same angle in an opposite direction. The surface of the eye not being wholly equally convex some adjustment is, however, necessary. One method by which this is effected is by the dropping of a line at a suitable point (Fig. 81 (2) *x*). Another is to make some increase in the inter-facetal intervals as is commonly seen in the antero-lateral portion of the eye as noted later.

Since rows of facets are often nearly straight, there are always in all parts of the eye many examples where three or four facets have the diameters through points of contact in a straight line so that they can be measured simultaneously and the mean taken, thus much reducing the labour involved and probably increasing accuracy. An even more important point is that with the three directions noted measurements can be made in the three directions. In measurements here given, unless otherwise stated, the diameters of the facets given have been the means of a considerable number of observations, nearly always fifty or more, made in the three directions noted.

Size of facet. In recording size of facets Shôzô Sato divides the kidney-shaped eye into the following areas: dorsal; antero-lateral; lateral; postero-lateral; and ventral. In *Culex pipiens* the facets in the postero-lateral area were smaller than in other parts of the eye (diam. 17·7–23·1 μ), and those in the ventral area larger (diam. 23·1–27·1 μ). In *Aëdes japonicus* the smallest facets were also in the postero-lateral area (diam. 19·0–29·5 μ), the facets in other parts of the eye ranging from 23·3–29·5 μ. In this case gradation in size showed a contour effect, the facets being smaller posteriorly and increasing in size towards the anterior margin with a maximum in the ventral area. In measuring the size of the facet, which he describes as hexagonal, Shôzô Sato uses the diameter of the inscribed circle within the hexagon.

A special interest attaches to the size of facet in different areas of the eye in *A. aegypti* since, while *Culex pipiens* is essentially a nocturnal or crepuscular feeder, *Aëdes aegypti* under natural conditions is strictly diurnal, though usually attacking in the shade. Another reason why size of facet in different parts of the eye is of interest is that Kennedy (1940), basing his conclusions on behaviour in flight of *A. aegypti*, considered that there was some difference in function of the different parts.

Before giving observations made on this species a word may be said on technique. One difficulty is the pigment which must be removed. Another is the markedly convex shape of the eye surface which prevents the eye membrane being spread flat without dividing it up in some way. The best results have been obtained by treating the bisected, or otherwise divided, head, of material fixed in Bles's fluid (formol acetic alcohol), overnight in 10 per cent KHO at 37° C. and after changing to water removing with a few touches of the needle any still adherent tissue. This leaves the eye membrane perfectly clear from pigment and retaining its original form. I have not found Grenacher's fluid so effective for this purpose. To ensure the eye membrane being suitably flat for examination it is necessary either to nick it, or better, to divide it up into recognisable portions (or the head before potashing may be suitably divided). The portions may then, after transfer to water, be mounted direct in a very small drop of water under a cover-glass. The same fragments can later, if desired, be mounted in the usual way. A useful medium for the purpose is Bhatia's medium (see under 'Technique', p. 116) to which the specimen can be transferred direct from absolute alcohol. By measuring the same specimens in water and again after mounting in Bhatia's medium it was ascertained that no appreciable difference in size of facet had occurred. The advantage of examining in water is that the material is softer and lies flatter. Measurements made from sectioned material also gave very similar results indicating that the eye membrane is a relatively rigid structure, as is also shown by the shape being retained after treatment with potash.

One method of determining the mean area of the facet is to determine the total eye area and divide this by the number of counted facets. Data in this respect are given in the accompanying table (Table 48). The mean area so determined from four females was 0·216 mm.^{2} and from three males 0·144 mm.^{2}.

The area of facet so given is that where the facet is considered as a hexagon and to obtain the diameter of the inscribed circle as used by Shôzô Sato it is necessary to calculate this from the data. This can readily be done.

By graphical methods it can be shown in regard to a regular hexagon and the inscribed circle that the area of the hexagon equals twelve triangles of the size of the triangle *ADC* (Fig. 81 (1)), two of which are equal to the parallelogram *AEDC* of which the sides are equal respectively to the radius of the inscribed circle and half the side of the hexagon. From such a relation it can be readily worked out that the area of the hexagon in terms of d $(2r)$ is given by the expression

$$d = 1{\cdot}075\sqrt{A},$$

where d is the diameter of the inscribed circle and A the area of the hexagon. Using this formula, the mean for the diameter of the inscribed circle in four female and three male *A. aegypti* has been given in column 4 of Table 48. The values give as the mean for the female 23·35μ and for the male 21·4μ.

To arrive at the size of facet for different areas of the eye, however, it is necessary to measure d directly. Data in this respect are given in Table 49.

The data given in the table show very definitely that the facets in the male are smaller than in the female, the mean values being 23·4μ for the female and 21·4μ for the male, incidentally almost exactly the figures obtained by the first method of arriving at these values.

It is evident also that in *A. aegypti* any difference in size of facet in different parts of the eye is very small. Such differences as do appear are the slightly smaller facets in the dorsal area and somewhat larger facets in the lateral and antero-lateral areas, the effect in the latter instance being, at least to some degree, due to an increase in the cuticular framework.

There is also some indication, as noted by Shôzô Sato in his species, of an increase in size passing forwards from the postero-lateral area to the antero-lateral, including to some extent the ventral area. It may be worth noting that in the larva the ommatidia of the compound eyes commence as a fine crescentic line just in front of the ocellus that will eventually form the posterior margin of the eye and that growth of the eye occurs by successive addition of lines of facets anterior to this.

Table 49. *The diameter of the inscribed circle of facets in different areas of the compound eye of female and male* Aëdes aegypti

Material	Dorsal (μ)	Antero-lateral (μ)	Lateral (μ)	Postero-lateral (μ)	Ventral (μ)	Examined in	Mean
Female *A*	21·3	23·8	23·9	22·4	23·1	Water	22·9
Female *B*	22·5	24·5	23·4	23·4	24·2	Water	23·6
Female *C*	22·4	24·4	22·9	22·6	23·1	Water	23·6
Female *D*	21·9	24·4	23·2	22·6	23·4	Water	23·2
Female *E*	23·4	23·8	23·8	23·4	23·4	Water	23·6
Female *F*	22·1	24·7	23·8	23·4	22·9	Water	23·4
Female *G*	21·6	24·2	22·9	22·2	22·9	Bhatia's medium	22·8
Female *H*	22·9	22·1	24·6	22·5	22·5	Serial sections	22·9
Mean	22·3	24·0	23·6	22·8	23·2	—	23·25
Male *K*	20·6	22·1	21·2	20·3	20·9	Water	21·0
Male *L*	20·1	21·2	21·9	21·3	21·8	Bhatia's medium	21·3
Male *M*	19·9	22·6	22·5	19·1	21·6	Serial sections	21·1
Male *N*	20·3	21·6	22·6	21·2	20·3	Serial sections	21·2
Mean	20·2	21·9	22·05	20·5	21·15	—	21·15

Note. *A, B, C, D* are normal-sized females.
E is a selected large female, measuring 7·2 mm. from tip of proboscis to tip of abdomen.
F is a selected small female giving 6·0 mm. for the same measurement.
G is the same female as *F* but measurements made on the eye parts after mounting.

Since there was a reduction in size in the male as compared with the much larger female insect it was thought of interest to see if reduction in size of facet occurred with reduction in size of the female. As shown in the table a small female, however, gave very similar measurements to those of a selected large female. To obtain a female as small as a male would have required special breeding and the question must be left unanswered for the present. It is clearly, however, mainly by reduction in number of facets that adjustment is made to the size of the eye in the respective sexes.

JOHNSTON'S ORGAN

The sense organ in the second segment of the antenna commonly known as Johnston's organ is present very generally in insects, but is especially developed in the Culicidae where it has received much attention. First drawn attention to by Johnston (1855) as the auditory apparatus of the *Culex* mosquito it was studied in respect to its functions by Mayer (1874) and its structure described by Child (1894). Later it has been described in great detail by Eggers (1924), in *Culex* and in *Anopheles stephensi* by Risler (1953, 1955) (see also Debauche, 1935; Tischner, 1953). I have seen a reference to P. Mayer, *Sopra certi organi di senso nelle antennae dei Dipteri* (Rome, 1879), but have been unable to see this work or find any note as to its contents.

Essentially the organ is a massive collection of sensilla of scolopophore type inserted into the region of the articulation of the antennal flagellum with the pedicel. It is considered to register movements or vibrations of the flagellum. To judge from its great size and complexity it almost certainly has some very important function and according to Mayer the vibrations of the antennal hairs enable it to function as a directive organ towards sounds, notably by the male, in which the antennal hairs and the organ are especially developed, in locating the female. The differences in the organ in the sexes in *A. aegypti*, however, seem insufficient to point to the organ as having a function restricted to the male for such a purpose.

The organ is located in and occupies the whole interior of the globular second antennal segment or pedicel. Apically this is flattened and its walls turned in and depressed to form a deep somewhat funnel-shaped depression or pit, *apical pit*, at the bottom of which is the articulation with the third segment of the antenna forming the base of the flagellum. The outer wall of the pedicel is strongly and uniformly sclerotised with an inner lining of cells of hypoderm type. There appears to be no special attachment of the organ to the walls. The apical pit is also lined with hard and rigid cuticle covered with numerous microtrichae and irregular circularly disposed linear thickenings. The lower margin of the wall of the pit, which is much thickened, forms a strongly sclerotised circular ridge covered with thorn-like microtrichae, *basal ridge*. The floor of the pit is flat formed by a circular plate in the middle of which is set the base of the flagellum. The plate, *basal plate*, articulates by a narrow line of articular membrane with the base of the flagellum and by its outer edge with the sides of the apical pit below the basal ridge (Fig. 82 (1)). The plate stains differently from the usual cuticular thickenings and may be more resilient in character. It is thickest where it articulates with the base of the flagellum and thins out towards its edge. Beneath it is a thick layer of cells with large nuclei resembling those of the subcuticular hypoderm. In connection with the plate is a rather complex arrangement for the insertion of the numerous scolopale elements or sense rods (*Stift* of German writers) of the mass of sensilla forming the organ. Briefly this consists of ridges which radiate out on the basal plate and are continued on to the lower wall of the apical pit below the basal ridge. Into these ridges (*Spangen* of the German writers) are inserted rows of the numerous fine sense rods coming from the main mass of scolopophores, that is, those coming from the sense cells of all the outer layer of the organ other than those of the extreme apical portion. Between the ridges are membranous *septa* which form a conspicuous feature in sections that pass through them (Fig. 82 (1), (2)).

Not all the scolopophores are inserted into the ridges. Three sets of scolopophores are distinguished by Eggers. The main bulk, *outer series*, are those referred to above as arising from the greater part of the outer layer of the organ and having sense rods inserted into the ridges described above. In the apical portion of the organ are scolopophores, *inner series*, the sense rods of which pass down between the septa to be inserted into the upper surface of the basal plate towards its outer edge (*scp'*). A third and still smaller series of scolopophores have rods inserted into the lower surface of the basal plate (*scp'''*).

The scolopophores as described by Debauche (1935) and figured by Eggers and others consist of an outer *sense cell* and two supporting cells, namely a proximal *sheath cell* and a distant *cap cell*. But though the sensilla of Johnston's organ are of scolopophore nature they do not according to Snodgrass possess the characteristic peg-shaped rod or *scolops* characteristic of the scolopophore sensilla of some other sense organs. The outer zone of the organ consists of the sense cells, the nuclei of which form a double or triple layer (Fig. 82 (1) *sn.c*). Internal to this outer zone is a zone free from nuclei that consists of nerve fibres. Internal to the zone of nerve fibres there are two zones characterised by elongate oval nuclei, the nuclei respectively of the sheath and cap cells. The nuclei of these cells and the bundles of scolopidial filaments are arranged, especially in the male organ, in radially aligned sheets, the nuclei situated externally and the mass of fine filaments internally.

In the centre of the organ is a space in which pass two nerves to the flagellum, the tubular blood vessel from the propulsatory organ and a small trachea. The central space is formed by the antennal nerve opening up as it enters the organ in a calyx-like manner. As it does so it leaves ventrally and internally a gap that acts as a passage-way into the central space, it being through this gap that the blood vessel of the propulsatory organ and the trachea pass. It is from here also that the two nerves to the flagellum arise from the inner aspect of the antennal nerve (Fig. 82 (1) *b*).

THE SENSE ORGANS OF THE HALTERES

A description of the halteres has already been given when dealing with the external structure of the imago. It is necessary to supplement this in the present connection by a more detailed account of the sensory organs with which these are so well supplied. A most detailed account, giving practically all that is known of the structure of the sensory organs

---

Figure 82. Sense organs of Johnston's organ and haltere.

1 Vertical section through second antennal segment to show main features of Johnston's organ. *a*, central space; *b*, showing position of opening through the antennal nerve and site of giving off of nerves to the filament.

2 Showing insertion of the three series of scolopopidia and nature of the septa. *a*, septum seen on the flat; *b*, septa as usually seen cut in transverse section.

3 Lower portion of right haltere showing dorsal surface and position of the dorsal scapal plate and basal plate. On the right border are the two undifferentiated papillae.

4 Portion of ventral surface of left haltere showing raised membranous protuberance on which is situated the ventral scapal plate. Note size and number of microtrichae.

5 A small portion of the dorsal scapal plate to show features of the rhomboidal sensillae and scattered microtrichae.

6 Cross-section of rhomboidal sensilla of scapal plates. A, dorsal scapal plate; B, ventral scapal plate.

7 Tangential and vertical section of sensilla of basal plate. A, superficial cut showing central sensory rod and absence of microtrichae; B, cut through base of sensilla showing central body and cuticular framework; C, vertical section showing processes of sense cells.

8 Sections through rhomboidal sensillum to show structure (after Pflugstaedt and Pringle). A, cut in direction of slit; B, cut transversely to slit.

9 Tangential section near tip of terminal segment of maxillary palp to show club-shaped and hair-like sensilla. *a*, thin-walled club-shaped sensitive hairs; *b*, thin-walled sensitive hairs; *c*, microtrichae; *d*, normal thick-walled hairs.

Lettering: *a.p*, apical pit; *b.p*, basal plate (of Johnston's organ); *b.r*, basal ridge; *b.v*, blood vessel (from propulsatile organ); *bp*, basal plate (of haltere); *dsp*, dorsal scapal plate; *nf*, one of the two nerves to antennal filament; *scc*, sclerotised anterior border of haltere; *scm*, membranous area; *scp*, zone of distal cap cells (inner) and enveloping cells (outer) of scolopophorous sensilla; *scp'*, inner series of scolopophorous sensilla; *scp"*, outer series of scolopophorous sensilla; *scp,‴* third series of scolopophorous sensilla inserted beneath the basal plate; *sn.c*, sense cell zone; *spn*, cuticular rods (Spangen); *up*, undifferentiated papillae.

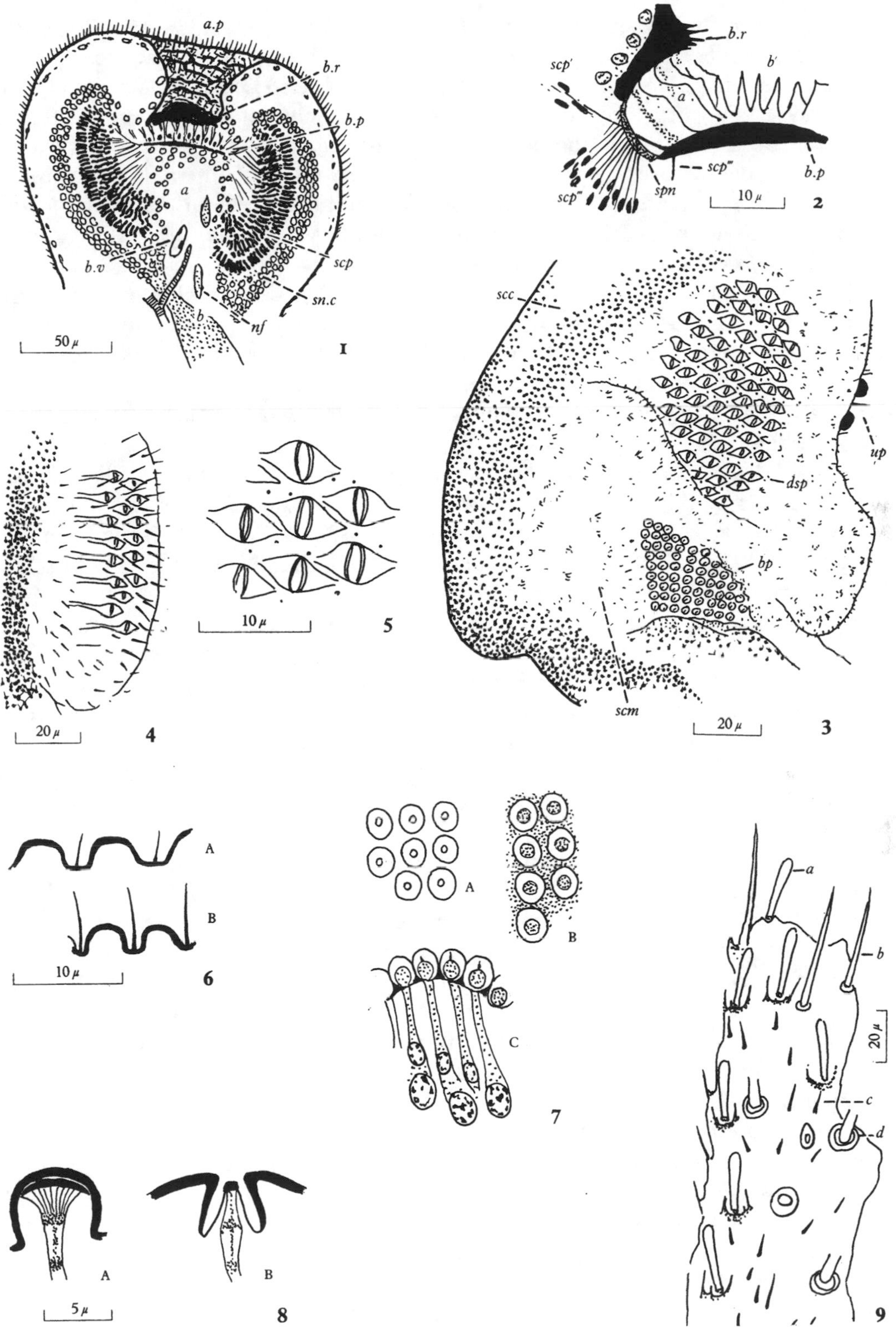

a.p
b.r
b.p
a
b.v
scp
sn.c
nf
b
50 μ
1
b.r
scp'
b'
a
scp'''
b.p
scp''
spn
10 μ
2
scc
up
dsp
bp
scm
20 μ
3
20 μ
4
10 μ
5
A
B
10 μ
6
A
B
C
7
A
B
5 μ
8
a
b
20 μ
c
d
9

Figure 82

of the halteres in the higher Diptera (*Sarcophaga*, *Calliphora*, *Syrphis*, and *Eristalis*), is that by Pflugstaedt (1912). A more recent account, giving a very clear and concise description of the organs in *Calliphora*, is that by Pringle (1948) (see also Lee, 1885; Weinland, 1891; Lowne, 1895). On these organs in the mosquito the only detailed account is that by Prashad (1916). From this account it is clear that the organs, both in anatomical features and minute structure, are very similar to those of the higher Diptera. They consist of (1) the dorsal scapal plate; (2) the ventral scapal plate; (3) the basal plate; and (4) two undifferentiated papillae on the posterior edge of the scape. How far other structures can be stated to be present will be discussed later.

In *Calliphora* there are also present: a group of sensilla situated close to the anterior margin of the basal plate, dorsal Hicks papillae; a smaller group on the ventral surface proximal to the ventral scapal plate, ventral Hicks papillae; an undifferentiated papilla, apparently distinct from (4) as given above, and situated on the anterior side of the dorsal scapal plate; and two chordotonal organs situated in the basal portion of the haltere, a small and a large chordotonal organ.

The *dorsal scapal plate* in *Aëdes aegypti* consists of about 12 rows of sensilla, the same number as noted by Prashad for *A. pseudotaeniatus*. In the middle rows of this series the number of sensilla is seven, the total number of sensilla in the group being from 60 to 65 (Fig. 82 (3)). The sensilla consist of rhomboidal papilla each composed of two opposed lateral elevations with thickened crests which form the edges of a cleft-like opening into a cavity in which the termination of the sensory cell is situated. The arrangement of the sensilla suggests ridges thrown into oblique corrugations, each corrugation forming a sensillum (Fig. 82 (5)). When cut transversely to the ridges the sensilla appear as mound-like elevations with flat intervening areas of cuticle from which arise scattered microtrichae (Fig. 82 (5)). Within the opening of the sensillum there is an elongate sunken cuticular body and deeper in than this a globular mass. The sensilla therefore conform to the structure given by Pflugstaedt for the rhomboidal sensilla in *Calliphora* and the higher Diptera as indicated in Fig. 82 (8).

The *ventral scapal plate* has about nine to ten rows of sensilla, but differs from the dorsal scapal plate in that each row is formed of two, or at most three sensilla. The sensilla are rhomboidal and resemble those of the dorsal plate, but are slightly smaller and the microtrichae arising from the cuticle are much longer and thicker, so much so that the sensilla are not so clearly displayed as are those of the dorsal plate. Also the sensilla of the ventral scapal plate, in place of being like those of the dorsal scapal plate on a flat surface, appear to be borne upon a rather prominent narrow membranous elevation situated just posterior to the hard scapal anterior margin. On this elevation the more anterior of the sensilla are contained as double furrows (Fig. 82 (4)).

The *basal plate* consists of nine rows of sensilla, the number of sensilla in the rows varying from nine anteriorly to two or three posteriorly, the whole forming a roughly triangular area (Fig. 82 (3) *bp*). The sensilla appear as hemispherical swellings smaller than the rhomboidal sensilla of the scapal plates and in close contact without any microtrichae. In section they show as rather globular structures set in a cuticular framework. Passing into each sensillum is a stout process from a sensory cell, the whole series of such processes forming in sections a fan-like arrangement (Fig. 82 (7) *C*). Accompanying the large nuclei of the sensory cells are smaller nuclei of the associated cells.

The two *undifferentiated papillae* are situated on the thin membranous edge of the scape towards its proximal end (Fig. 82 (3) *up*). They are on the opposite side of the dorsal scapal plate to the large single undifferentiated papilla as figured by Pringle in *Calliphora.*

In *Calliphora* the dorsal Hicks papillae are situated close to the basal plate on its anterior side, that is towards the curved-in lower end of the sclerotised anterior border of the haltere. In *Aëdes aegypti* there is no separate group of sensilla in this position, but the two most anterior rows of sensilla of the basal plate, which are slightly out of line with the rest, may represent this structure (Fig. 82 (3) *bp*). The ventral Hicks papillae in *Calliphora* are situated on a prominence proximal to the ventral scapal plate. I have not succeeded in finding a corresponding structure in *Aëdes aegypti.*

The two chordotonal organs in *Calliphora,* as given by Pringle after Pflugstaedt, are situated in the basal portion of the haltere. The large chordotonal organ consists of a large number of typical sensilla situated beneath a well-marked protuberance of the cuticle on the ventral side of the basal region of the haltere posterior to and proximal to the scapal plate. Its direction is across the base at an angle of 45° to the longitudinal axis of the haltere sloping distally towards the rear. The sensilla are not visible on the surface. The small chordotonal organ extends from the dorsal surface at the distal margin of the basal plate vertically to the ventral surface behind the proximal end of the scapal plate. The dome-shaped protuberance appears to be present in *Aëdes aegypti* as also a mass of sensory cells, but the actual structures present still require to be ascertained.

Entering the base of the haltere is the large haltere nerve. Before entering the haltere this gives off a lower branch which passes to the base of the organ. The main nerve divides into fan-shaped portions, passing to the sensory cells at the bases of the different sensory organs.

SENSE ORGANS OF THE ANTENNAE, PALPS AND MOUTH-PARTS

Trichoid sensilla consisting of thin-walled hairs on the antennae of both sexes in *A. aegypti* are described by Roth and Willis (1952) (see also Roth, 1951). In the male these are restricted to the last two flagellar segments (the other segments, apart from the large hairs, being smooth and hairless). But in the female such hairs are present on all the thirteen flagellar segments. They are 40–50μ in length, thin walled and without articulated base, arising from thin membrane over a pore canal surrounded distally by a semicircular ridge in the cuticle. Proximally a median ridge passes backwards. They are set at irregular intervals along the length of the segment and in the female number some 30–50 on a segment. They are considered on experimental evidence by Roth and Willis to be hygroreceptors. In addition to the thin-walled hair sensilla the above-mentioned authors also note the presence of a certain number of peg-like sensilla on the antennal segments. Smith (1919) has figured basiconic sensilla at the tip of the antenna in *Culex pipiens.* In the male of *Anopheles stephensi* Risler (1953) notes the presence of sense cells at the base of the long hairs on the first flagellar segment. In *Aëdes aegypti* in both male and female the tip of the antenna ends in a rather suddenly constricted spike at the end of which are two thin-walled papilla-like elevations.

The female palps of *A. aegypti* are abundantly supplied on the terminal segment with thin-walled club-shaped sensilla as described in this species by Roth and Willis (1952). This segment is free from scales on its ventral aspect where the sensilla are chiefly present and the surface is strongly convex, suggesting a definitely tactile organ. On the preceding segment few or none of these sensilla are present. The sensilla are characteristically club-shaped hairs about 20–30 μ in length arising from membrane covering a circular pore canal surrounded by a circular cuticular ridge (Fig. 82 (9) *a*). Pointed trichoid sensilla are also present (*b*). The terminal segment of the male palps is not swollen like that of the female and is uniformly covered with scales. It lacks the above-mentioned sensilla and has not the appearance of an organ used for tactile purposes. At the extreme tip of the female palps is a large circular area of membrane carrying a central short sclerotised peg.

SENSE ORGANS OF THE LABIUM AND BUCCAL CAVITY

Frings and Hamrum (1950) describe the sensory structures on the labella of *A. aegypti*. They note four types of hairs on the labella in both female and male, namely (1) short epicuticular hairs about 7 μ long (non-sensory); (2) long pointed hairs about 40 μ in length arranged in a fan-shaped pattern at the tip of the labella and scattered elsewhere, which appear to be tactile; (3) medium-sized curved hairs 20 μ in length at the tip and on the ventral surface similar to gustatory end-organs on labella of other Diptera (Frings and O'Neal, 1946; Frings and Frings, 1949) that may be chemoreceptors; (4) short blunt peg-like hairs, 6 μ in length, of which there are about fourteen on the dorsal surface of each labellum of unknown function.

Towards the tip of the labrum are the two small gustatory papillae (of authors) (Fig. 54 (8) *A*). They are present also on the modified labrum of the male (Fig. 54 (8) *B*).

In the buccal cavity are the sensory pits on, and at the sides of, the anterior hard palate and those on the posterior margin of the organ (Fig. 63 (4), (5), (8)). These appear as circular pore canals covered with membrane bearing usually a small central hair. Those on the hard palate itself are larger and more typically of basiconic type.

SENSE ORGANS OF THE LEGS, WINGS AND GENITALIA

One of the most striking features of the sensory equipment of *A. aegypti* apart from the major sense organs is the large number of 'sensory pits' with which the under surface of the thorax and bases of the legs are provided. A rather conspicuous sensillum is present on each side of the face of the prothorax at the edge of the propleural extensions from the prosternum (Fig. 55 (4) *so*). On the mid-coxae are the little groups of eight sensilla in two rows of four (Fig. 56 (1) *so*) and on the posterior aspect of the hind coxa is on each side a little group of three sensilla (Fig. 55 (7) *so*). On the trochanters, close to their articulation with the coxae are two groups, each of two rows of four sensilla (Fig. 56 (4), (9) *a*, *b*) and on the bulge near the base of each femur is again a double row of sensilla (Fig. 56 (4), (5)).

According to Frings and Hamrum there are on the tarsi of the fore- and mid-legs of *A. aegypti*, among non-sensory spines and scales, many slightly curved hairs probably tactile in function. At the distal end of each segment of the tarsus these authors also note the presence of groups of short curved hairs of approximately the same size and appearance as those they have described on the labella. These may be tactile receptors, though proof is wanting. Menon (1951) describes two rows of bud-like sensilla on the fore- and mid-tarsi in Culicines, on the fore-tarsi only in Anophelines and none in *Megarhinus*. Wallis (1954) in *Aëdes aegypti* found all tarsal segments provided with curved thin-walled spines corresponding with the sensory areas. There do not appear to be any sensory pits, so far as I have been able to ascertain, on the tarsi of *A. aegypti*.

Two small groups of scattered sensory pits are present on the remigium at the base of the wing (Fig. 57 (8) *Rg*). These are circular or slightly oval sensilla about 3 μ in diameter, hemispherical in shape and lacking any central hair. If they are proprioreceptors and of campaniform type, they would resemble the sensilla of the basal plate of the haltere and not the rhomboidal sensilla of the scapal plates.

In another communication (Christophers, 1951) I have described in *Culex pipiens* and *C. molestus* a sensory pit of the hair-bearing type at the side of the female hypopygium just posterior to the ninth tergite at its outer angles.

REFERENCES

BUDDENBROCK, W. V. (1939). *Grundriss der vergleichenden Physiologie*. Ed. 2. Berlin.

CHILD, C. M. (1894). Beiträge zur Kenntnis der antennalen Sinnesorgane der Insekten. *Z. wiss. Zool.* **58**, 494–500.

CHRISTOPHERS, S. R. (1951). Note on morphological characters differentiating *Culex pipiens* L. from *Culex molestus* Forskål and the status of these forms. *Trans. R. Ent. Soc. Lond.* **102**, 376 (figure).

CHRISTOPHERS, S. R., SHORTT, H. E. and BARRAUD, P. J. (1926). The anatomy of the sandfly *Phlebotomus argentipes* Ann. and Brun. (Diptera). I. The head and mouth parts of the imago. *Indian Med. Res. Mem.* no. 4.

CLEMENTS, A. N. (1953, 1956*a*, *b*). See references in ch. XXVII (*d*), p. 625.

CONSTANTINEANU, M. J. (1930). See references in ch. XIV (*d–e*), p. 351.

DAY, M. F. (1955). A new sense organ in the head of the mosquito and other Nematocerous insects. *Aust. J. Zool.* **3**, 331–5.

DEBAUCHE, H. (1935). Les organes sensoriels antennaires de *Hydropsyche longipennis*. *Cellule*, **44**, 45–83.

DUFOUR, L. (1851). See references in ch. XXVI, p. 598.

EGGERS, F. (1924). Zur Kenntnis antennaler stiftführender Sinnesorgane der Insekten. *Z. Morph. Ökol. Tiere*, **2**, 259–349.

ELTRINGHAM, H. (1933). *The Senses of Insects*. Methuen's Biological Monographs, London.

FRINGS, H. and FRINGS, M. (1949). The loci of contact chemoreceptors in insects. A review with new evidence. *Amer. Midl. Nat.* **41**, 602–58.

FRINGS, H. and HAMRUM, C. L. (1950). The contact chemoreceptors of adult yellow fever mosquitoes, *Aëdes aegypti*, *J. N.Y. Ent. Soc.* **58**, 133–42.

FRINGS, H. and O'NEAL, B. R. (1946). The loci and thresholds of contact chemoreceptors in females of the horsefly, *Tabanus sulcifrons* Meig. *J. Exp. Zool.* **103**, 61–79.

HOLMES, W. (1947). *Recent Advances in Clinical Pathology* (chapter on peripheral nerves). Churchill, London.

JOBLING, B. (1928). The structure of the head and mouthparts in *Culicoides pulicaris* L. (Diptera: Nematocera). *Bull. Ent. Res.* **18**, 211–36.

JOHNSTON, C. (1855). Auditory apparatus of the *Culex* mosquito. *Quart. J. Micr. Sci.* (old series), **3**, 97–102.

KENNEDY, J. S. (1940). The visual response of flying mosquitoes. *Proc. Zool. Soc. Lond.* A, **109**, 221–42.

LEE, A. B. (1885). Les balanciers des diptères, leur organes sensifères et leur histologie. *Rec. Zool. Suisse*, **2**.

LOWNE, B. T. (1895). *The Anatomy, Physiology, Morphology and Development of the Blowfly*. London.

MAYER, A. M. (1874). Experiments on the supposed auditory apparatus of the mosquito. *Amer. Nat.* **8**, 577–92.

MAYER, P. (1879). *Sopra organi di senso nelle antennae dei Ditteri*. Rome.

MENON, M. A. V. (1951). On certain little-known external characters of adult mosquitoes and their taxonomic significance. *Proc. R. Ent. Soc. London* (B) **20**, 63–71.

MIALL, L. C. and DENNY, A. (1886). *The Cockroach.* Lovell Reeve and Co., London.

MILLER, A. (1950). In Demerec, M., *Biology of* Drosophila. Wiley and Sons, New York.

PAWLAWA, M. (1895). Über ampullenartige Blutcirculationsorgane im Kopfe verschiedener Orthopteren. *Zool. Anz.* **18**, 7–13.

PFLUGSTAEDT, H. (1912). Die Halteren der Dipteren. *Z. wiss. Zool.* **100**, 1–59.

PRASHAD, B. (1916). The halteres of mosquitoes and their function. *Ind. J. Med. Res.* **3**, 503–9.

PRINGLE, J. W. S. (1948). The gyroscopic mechanism of the halteres of Diptera. *Phil. Trans.* B, **233**, 347–84.

RISLER, H. (1953). Das Gehörorgan der Männchen von *Anopheles stephensi* Liston (Culicidae). *Zool. Jb.* **73**, 163–6.

RISLER, H. (1955). Das Gehörorgan der Männchen von *Culex pipiens*, L. *Aëdes aegypti* L. und *Anopheles stephensi* Liston (Culicidae), eine vergleichend morphologische Untersuchung. *Zool. Jb.* **74**, 478.

ROTH, L. M. (1951). Loci of sensory end organs used by mosquitoes (*Aëdes aegypti* (L.) and *Anopheles quadrimaculatus* Say) in receiving host stimuli. *Ann. Ent. Soc. Amer.* **44**, 59–74.

ROTH, L. M. and WILLIS, E. R. (1952. Possible hygroreceptors in *Aëdes aegypti* (L.) and *Blatella germanica* (L.). *J. Morph.* **91**, 1–14.

SAMUEL, E. P. (1935). See in references ch. XXVIII, p. 647.

SHÔZÔ SATO (1950). Compound eyes of *Culex pipiens* var. *pallens* Coquillett. *Sci. Rep. Tôhoku Univ.* (4) (Biol.), **18**, 331–41.

SHÔZÔ SATO (1951*a*).. Development of the compound eye of *Culex pipiens* var. *pallens* Coquillett. *Sci. Rep. Tôhoku Univ.* (4) (Biol.), **19**, 23–8.

SHÔZÔ SATO (1951*b*). Larval eyes of *Culex pipiens* var. *pallens* Coquillett. *Sci. Rep. Tôhoku Univ.* (4) (Biol.), **19**, 29–32.

SHÔZÔ SATO (1953*a*). Structure and development of the compound eye of *Aëdes* (*Finlaya*) *japonicus* Theobald. *Sci. Rep. Tôhoku Univ.* (4) (Biol.), **20**, 33–44.

SHÔZÔ SATO (1953*b*). Structure and development of the compound eye of *Anopheles hyrcanus sinensis* Wiedemann. *Sci. Rep Tôhoku Univ.* (4) (Biol.), **20**, 45–53.

SMITH, K. M. (1919). A comparative study of certain sense organs in the antennae and palpi of Diptera. *Proc. Zool. Soc. Lond.* 1919, 31–69.

SNODGRASS, R. E. (1926). The morphology of insect sense organs and the sensory nervous system. *Smithson. Misc. Coll.* **77**, no. 8.

SNOGDRASS, R. E. (1935). *Principles of Insect Morphology.* New York and London.

THOMPSON, M. T. (1905). The alimentary canal of the mosquito. *Proc. Boston Soc. Nat. Hist.* **32**, 145–202.

TISCHENER, H. (1953). Über den Gehörsinn von Stechmücken. *Acustica.* **3** [Quoted by Risler (1955)].

WALLIS, R. C. (1954). A study of oviposition activity of mosquitoes. *Amer. J. Hyg.* **60**, 135–68.

WEINLAND, E. (1891). Über die Schwinger (Halteren) der Dipteren. *Z. wiss. Zool.* **51**, 55–166.

WIGGLESWORTH, V. B. (1942). *The Principles of Insect Physiology.* Ed. 2 (and subsequent editions). Methuen and Co., London.

XXX

THE IMAGO: THE EXCRETORY AND REPRODUCTIVE SYSTEMS

(*h*) THE EXCRETORY SYSTEM

Excretion may be external or internal. The chief organs of external excretion are the Malpighian tubules, the waste products being passed out of the body with the faeces. The organs of internal excretion are mainly the fat-body cells and the nephrocytes in which waste products are accumulated in the cells and retained in the body.

THE MALPIGHIAN TUBULES (Fig. 83)

The Malpighian tubules in the imaginal mosquito have been described in *Culex pipiens* and in *Theobaldia annulata* by Schindler (1878), who gives a drawing of the gut in the imago with the five tubules. He also notes the unusual number of five, but records a similar number in *Psychoda* and in *Ptychoptera.* It was in the imago also that Lecaillon (1899) observed ciliform processes in *Culex pipiens.* More recent are the observations on the tubules of the imago of *Culex pipiens* by Roubaud (1923); on those of *Anopheles maculipennis* by Nuttall and Shipley (1903) and by Roubaud, *loc. cit.*; and on the tubules of the imago of *Anopheles claviger* (a hibernating species) by Missiroli (1925, 1927).* Especially informative is the paper by Wigglesworth (1931) describing the mechanism and appearances seen in *Rhodnius* (a blood-sucking insect that, like the mosquito, gorges itself at one sitting with a blood meal) and the recent work of Ramsay (1953) on excretion (see under 'Physiology', p. 710).

Unlike the alimentary canal and many other structures which change considerably during passage from larva to imago, the Malpighian tubules remain intact, including their epithelium, throughout metamorphosis and the description already given of the tubules in the larva largely holds good for those of the imago. They have much the same length as in the full-grown larva, about 2·5–3·5 mm., but the more restricted space in the imaginal abdomen causes them to be more crowded together and coiled. Even so their original arrangement as described in the larva is usually to be made out. Thus the two loops of the dorso-lateral and ventro-lateral tubules, for example, show up conspicuously in the gorged insect through the transparent cuticle of the pleura against the black background of the blood-filled mid-gut (Fig. 51 (4)), and the ends of the two ventro-lateral tubules come characteristically into relation, close to the middle line, with the posterior end of the seventh sternite. Their relation to other structures remains also much as in the late larval period, that is they take origin in the fifth abdominal segment from the anterior portion of the pyloric ampulla, one dorsal and two on each side laterally. In the posterior part of the abdomen, especially when the ovaries are enlarged, they come into close relation with the

* See also Veneziani (1904), Bugajew (1928), Saint-Hilaire (1927), Patton (1953).

proximal portion of the rectum, which, swollen with its rectal papillae, holds much the same relation to the tubules as formerly did the colon.

The structure and character of the cells also remains essentially as described for the larva, though some points in this respect require to be considered.

One such point is the decision as to whether in the imago at any stage the striated border of the cells is definitely of the brush type. That it is of this type is probable, since under dark-ground illumination the conspicuous brilliant white line bordering the lumen in the early larval stages is absent or only very imperfectly demonstrable. I have been unable, however, to see in any of the material examined the long ciliform processes one might have expected to see with such a form of striated border. Lecaillon has described such processes in the tubules of *Culex pipiens*, but gives no figures. Also he says that the filaments are of regular length and have the same appearance in the intestine and in the Malpighian tubules. Missiroli states that the cells in *Anopheles claviger* have a border of motionless ciliform appendages, but gives no specific data except in a diagram which shows much the same appearances as seen by me in the imago of *Aëdes aegypti* and which scarcely bear out what the statement signifies. So far I have not been able to find any later author who has definitely described at first hand long ciliform processes in the mosquito.

Another point relates to the accumulation of urate excretion in the tubules. Roubaud noted the accumulation of urate deposit in the tubules of the larva of *Anopheles* under conditions of overcrowding and in the imago in hibernation and considered that this was a cyclical phenomenon. Missiroli, however, showed that such changes were the result of the organs being at rest, as during hibernation, or showing secretory activity as during digestion. He notes that in the first-mentioned condition the cells are swollen and loaded with granules which obscure the nucleus and the lumen. But when actively secreting the volume of the cells is reduced, the granules are cleared from the outer zone exposing the nuclei and the lumen is loaded with excretory matter.

In *Aëdes aegypti* the tubules in the early larval instars are clear and transparent and free from coarse granules. But during the fourth instar the tubules become milky white and opaque and so they remain throughout the pupal period and into the adult stage. In the female they remain so with little change throughout life. Only after a blood meal and at

---

Figure 83. The excretory system.

1 Camera lucida drawing of dissection of Malpighian tubules of imago. Coverglass preparation fixed Schaudinn's fluid. The crosses indicate where a cell as described by Pantel in *Ptychoptera* is present. The two long outlying tubules are the ventro-lateral tubules.

2 Dorsal view of the Malpighian tubules in the larva. Camera lucida drawing. *a*, ventro-lateral tubule; *b*, dorso-lateral tubule; *c*, dorsal tubule.

3 Lateral view of Malpighian tubules in the imago. Reconstruction. Only the tubules of the left side are shown. Lettering as for 2.

4 To show appearance of honey-comb type of striated border as seen under dark ground illumination. Early fourth instar larva.

5 Transverse section of Malpighian tubule of adult showing interstitial cell as described by Pantel in *Ptychoptera*. *a*, normal epithelium cell of tubule; *b*, interstitial cell.

6 As 5.

7 Longitudinal section to show interstitial cell. Lettering as in 5.

8 As 7.

Lettering: *Co*, colon; *Mg*, mid-gut; *Rc*, rectum; V–VII, abdominal segments.

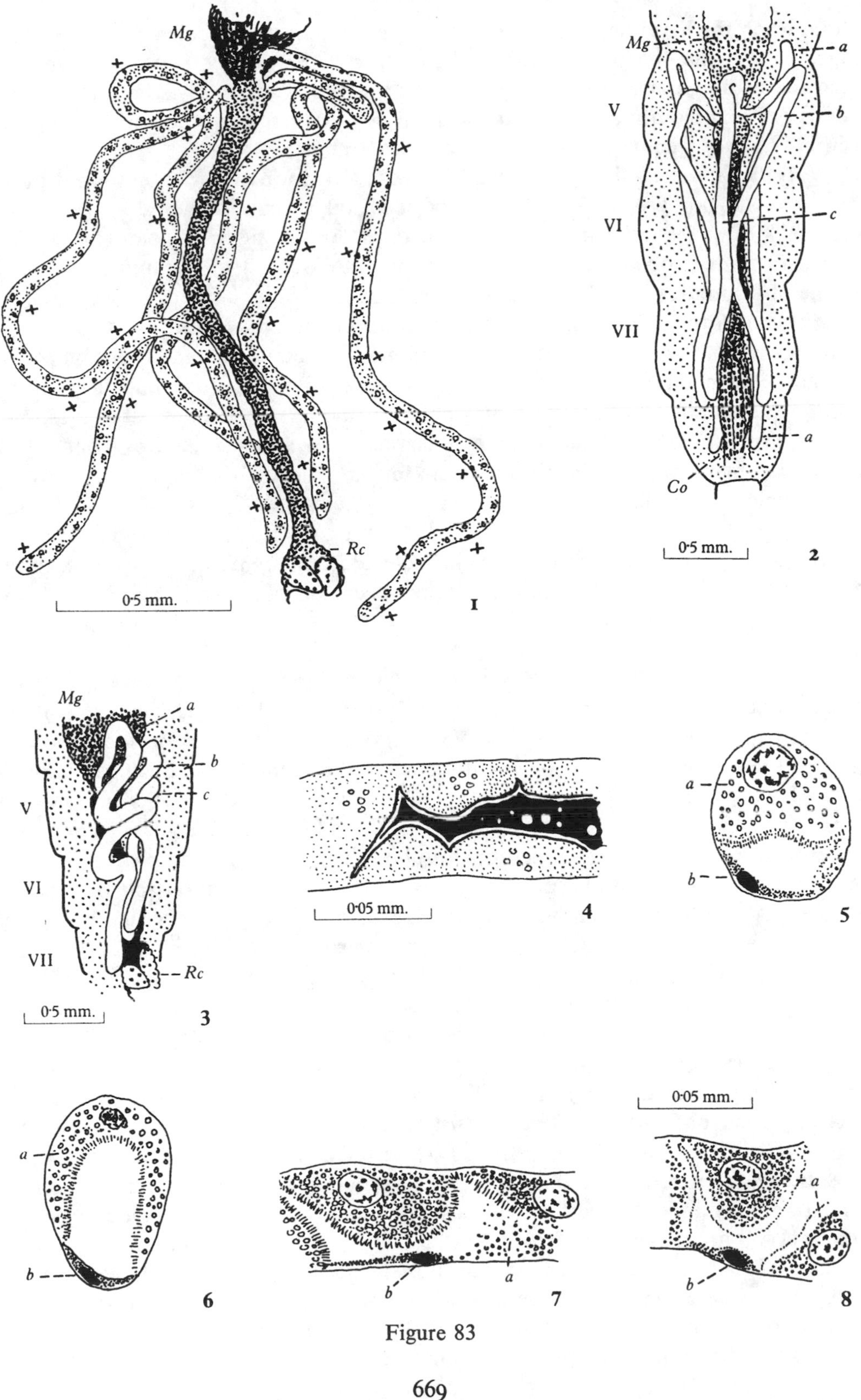

Mg
Rc
0·5 mm.
I
Mg
a
V
b
VI
c
VII
a
Co
0·5 mm.
2
Mg
a
b
c
V
VI
VII
Rc
0·5 mm.
3
0·05 mm.
4
a
b
5
a
b
6
b
a
7
0·05 mm.
a
b
8

Figure 83

the time when the insect is passing copious clear urine do the tubules show any marked change. At this time the lumen is enlarged and the cells, especially their inner portion, are less packed with granules. In some cases after oviposition, especially some days after, the tubules were reduced almost to the transparent condition seen in the early larval stages. But in others they were still heavily charged with granules.

The granules are refractile, and vary in size from a fraction of a micron to four or up to eight microns. They give the appearance of small oil globules. They do not stain with Sudan black or easily with eosin or methylene blue. According to Wigglesworth (1942) in *Rhodnius* they are probably not the same as the urate deposit in the lumen and are not blackened by ammoniacal silver nitrate as are the granules of uric acid in the lumen. In *Aëdes aegypti* in the fresh condition the granules in the cells as noted above have the appearance of fat globules. Secretion in the tubules, on the other hand, appears as amorphous granular material. In sections the granules usually appear as small clear vacuoles, the Malpighian tubules as a result staining only very moderately in contrast to these organs seen in the fresh state or in mounted whole larvae. In sections stained by Holmes' silver impregnation method, however, they stain an intense black.

In the male the tubules are narrower than in the female, 4–7μ as against 7–10μ in the female. They also tend to run a straighter course and are often less packed with granules and more transparent. These differences are present also in the pupal stage and even in the late fourth instar larva, so much so that a rough diagnosis of sex can often be made in the mature larva by the appearance of the tubules.

MECHANISM OF EXCRETION

An important point in connection with the Malpighian tubules in insects is that in many insects, in addition to serving as organs of excretion, they play a role in the conservation of water and in the maintenance of the isotonicity of the haemolymph by abstracting water from the urine before its discharge. Patton (1953) recognises four types of tubule. In two of these, types *B* and *D* (cryptonephridic types), the tubules are organically connected in their terminal portions with the rectum and play a part in abstracting water from urine passed into this organ. In another two types, types *A* and *C* (cryptosolenic types), the tubules are not associated with the rectum in this manner. In type *A* the tubules are solely concerned in excretion, absorption of fluid from the excreta usually being a function of the rectum. In type *C* one portion of the tubule, the most proximal, has an excretory function, whilst a distal portion abstracts water from urine excreted by the proximal portion. The tubules in *A. aegypti* would appear most nearly to approximate to type *A* of Patton.

The mechanism of excretion and the appearances given rise to have been very clearly shown by Wigglesworth (1931) in the blood-sucking bug *Rhodnius*. The tubules in this case clearly come under type *C*. They consist of two quite distinctive portions, a proximal portion characterised by cells with striated border of honey-comb type and a lumen free from solid urate matter, and a distal portion in which the cells have a striated border of brush type and a lumen in which solid urate matter accumulates. After a blood meal, when for the first few hours the insect is secreting clear urine, both distal and proximal portions are distended with clear fluid. Other than this, the proximal portion shows no marked change and by the second day has returned to the normal condition. The distal portion is at first swept free from granules by the initial flow of fluid, but later shows newly formed urate crystals, which increase greatly in number, intermixed with many long ciliform filaments derived from the striated border. Later these filaments contract to the normal resting state, leaving the lumen packed with urate spheres and the cells with a striated border of the normal resting type.

As shown by Wigglesworth the physiological explanation of these appearances is that in the upper proximal portion potassium or sodium urate is excreted in solution. In the lower portion water and alkali are reabsorbed leaving uric acid to crystallise out in characteristic spheres. The absorbed alkali is further available to combine with more uric acid produced in the tissues.

Briefly then in *A. aegypti* it would seem that the tubules as a whole excrete and reabsorb water, or less likely, that the small portion of tubule forming the ascending part of the anterior loop, which differs somewhat in character from the rest, excretes and the rest of the tubule reabsorbs water. Against the latter view is the probability that in the mosquito the rectum with its large rectal papillae is more likely to be the mechanism for water absorption. Whatever the mechanism may be, it can easily be seen how in the mosquito small changes in conditions might give rise to different appearances. An example of this in the larva is given by Wigglesworth (1933). Thus in larvae of *A. aegypti* reared in salt water solid uric acid is seen in the Malpighian tubules making these white and opaque, but in the same larvae a few minutes after transfer to fresh water the tubules have been flushed out by water entering through the anal papillae.

Pantel's cells (Fig. 83 (1), (6), (8)). A feature of some interest that has so far not been noted is the occurrence in the Malpighian tubules of *A. aegypti* of occasional cells, differing from those forming the normal epithelium of the tubule, such as have been described by Pantel (1914) in *Ptychoptera*. In *Ptychoptera* these are thin cells, devoid of granules and without a striated border, that occur scattered among the normal epithelium. Similar cells devoid of granules are described by Missiroli in *Anopheles*. Those in *Aëdes aegypti* occur at more or less regular intervals along the tubule. In sections they appear as thin flat cells with rather deeply-staining homogeneous cytoplasm and are devoid of any striated border. The nucleus is smaller and the chromatin more compact than that of the normal epithelium. In fact they are precisely similar to the cells as depicted in the figure by Pantel (Fig. 83 (6) *b*). In *Aëdes aegypti* they are well shown in the clear Malpighian tubules of the male in Schaudinn-fixed cover-glass preparations of the fresh dissected organs stained by acid fuchsin. They are usually situated opposite to a normal epithelial cell, that is in the gap between the two cells facing this on the opposite side of the tubule (Fig. 83 (6), (7)). They have been supposed to serve some function. It seems possible that they are the nuclei of cells responsible for forming the basement membrane of the tubule, a small portion of cytoplasm left in the neighbourhood of the nucleus giving rise to the appearance of a cell.

Fat-body and nephrocytes. These have already been described in the sections dealing with the circulatory system and associated tissues.

(*i*) THE FEMALE REPRODUCTIVE SYSTEM

GENERAL DESCRIPTION (Fig. 84)

The segmental parts composing the external genitalia of the female have already been described. It remains to describe the internal organs of generation, namely the gonads and efferent ducts with their accessory structures.

A very full account of the histology of the female reproductive system in insects is given by Gross (1903).* On the female organs in Nematocera are: Miall and Hammond (1900, *Chironomus*); Bodenheimer (1923, *Tipula*); Abul-Nasr (1950, post-embryonic development and structure in *Chironomus*, *Anisopus* and *Mycetophilidae*).

Of early papers describing the female organs in the mosquito are: Lecaillon (1900, ovary of *Culex pipiens*); Grandpré and Charmoy (1900); Christophers (1901, 1923); Kulagin (1901); Neveu-Lemaire (1902, spermathecae); Eysell (1924); Herms and Freeborn (1921, mechanism of egg-laying); der Brelje (1924, comparative study of the female organs in different Culicidae). A more recent anatomical and histological study of the female reproductive system and follicular development in *Aëdes aegypti* is that by Parks (1955).

On the changes in the egg follicle are: Christophers (1911, *Anopheles*); Nicholson (1921, a very complete account of follicular structure in *Anopheles*); Vishna Nath (1924, 1929, the egg follicle and Golgi apparatus in *Culex*); Christophers, Sinton and Covell (1939, includes abstract of Christophers, 1911, with original figures); Hosoi (1954, growth and degeneration of ovarian follicles in *Culex pipiens pallens* Coq.); Gillies (1954, 1955, *Anopheles*).

On the oviducts and changes indicative of age are: Mer (1932, 1936); Almazova (1935); Polodova (1941, all on *Anopheles*); Roy and Majundar (1939, *Culex fatigans*). A study

* See also Snodgrass (1933, 1935), Weber (1933), Imms (1938), Wigglesworth (1942).

Figure 84. The female reproductive system.

1 Dorsal view of female organs two days from emergence. The drawing is from a dissection made from a fixed and hardened whole female.

2 Transverse section of ovary. *a*, nucleated peritoneal membrane; *b*, membrane surrounding lumen.

3 Germarium and forming follicle at upper end of an ovariole.

4 Section showing follicles at emergence.

5 Section showing follicles 21 hours after a blood meal. The lower follicle has increased greatly in size, the succeeding follicle remains much as at emergence. The stage in the lower follicle is a little beyond stage 2 as in a fresh preparation the nucleus would be obscured by the yolk granules.

6 Sagittal section at about the median line of the body, showing structures in the region of the genital opening. The spermatheca contains spermatozoa. Ventral to the genital opening the section passes through the sclerite referred to in the text as possibly representing the female gonapophyses.

7 Dissection showing the spermathecae and spermathecal ducts.

8 Coronal section showing arrangement of the spermathecae. The section corresponds to one passing logitudinally in a plane at right-angles to Fig. 84 (6) through the spermatheca and vagina as there shown.

9 Transverse section of spermathecal duct.

10 Posterior portion of first instar larva, showing early ovary in segment VI.

11 Ovarian rudiment in pre-ecdysis third instar larva. *a*, outer layer of cells; *b*, commencing follicle formation.

Lettering: *at*, atrium; *cod*, common oviduct; *cs*, caecus; *cw*, cowl; *f.ep*, follicular epithelium; *fol*, lower follicle; *fol'*, preceding follicle; *gm*, germarium; *lo*, lateral oviduct; *mst*, mesodermal strand; *nsc*, nurse cells; *ooc*, oocyte; *ov*, ovary; *ovl*, ovariole wall; *pdo*, pedicel; *pgp*, postgenital plate; *Rc*, rectum; *slg*, suspensory ligament; *spd*, spermathecal duct; *spm*, spermatheca; *vg*, vagina; *y*, yolk granules; VIII', hollow behind sternite VIII at bottom of which is the genital opening.

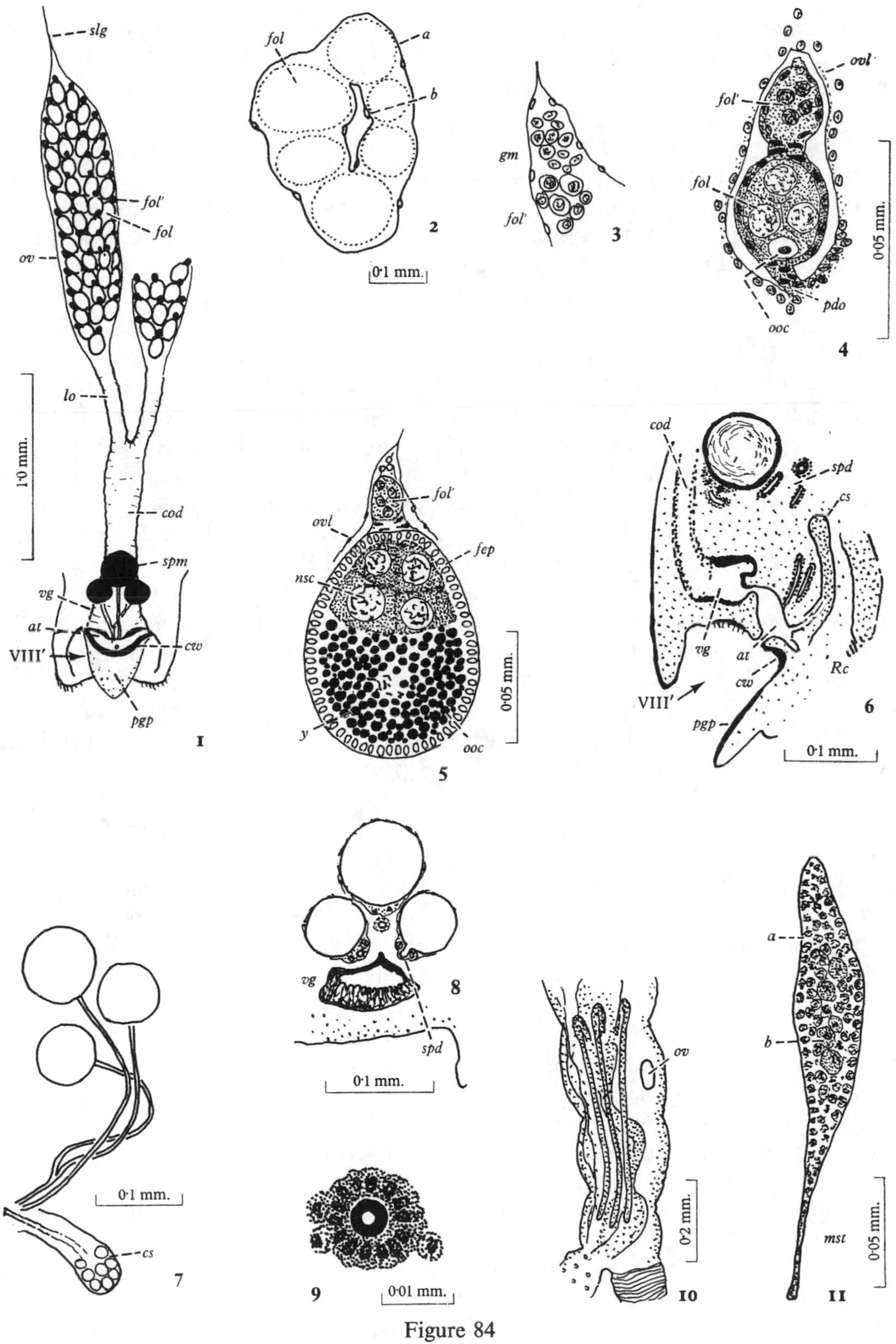
slg
fol
a
b
2
0·1 mm.
gm
fol'
3
ovl
fol'
fol
pdo
ooc
0·05 mm.
4
fol'
fol
ov
lo
1·0 mm.
cod
spm
vg
at
VIII'
cw
pgp
I
fol'
ovl
fep
nsc
0·05 mm.
y
ooc
5
cod
spd
cs
vg
at
cw
Rc
VIII'
pgp
6
0·1 mm.
0·1 mm.
cs
7
vg
8
spd
0·1 mm.
9
0·01 mm.
ov
0·2 mm.
10
a
b
mst
0·05 mm.
II

Figure 84

of insemination and the bursal pouch in *Aëdes aegypti* is given by Hayes (1953). A recent important paper by Möllring (1956) figures the ovary of *Culex* (autogenous and anautogenous) as seen by phase contrast and gives much information regarding ovarian structure and development.

The female generative organs in the mosquito consist of the gonads, *ovaries* (Fig. 84 (1) *ov*) situated one on each side dorso-laterally in the hinder part of the abdomen, the *paired* or *lateral oviducts* (*lo*), one on each side passing from the ovary to the median *common oviduct* (*cod*), which in turn is continued into a short *vagina* (*vg*) and still more distally into the *atrium* (*at*). Into the vagina also open the ducts from the three *spermathecae* (*spm*) and into the atrium the *caecus* (colleterial or accessory gland) (Fig. 84 (6), (7) *cs*).

The *ovary* (Fig. 84 (1–6)). The ovary is a fusiform structure situated when undeveloped in the fifth and sixth abdominal segments, but when fully developed with mature follicles occupying the greater part of the distended abdomen. Anteriorly it is continued as a fine filament, the *suspensory ligament* (*slg*), which in *Aëdes aegypti* can be traced to the posterior end of the fourth segment where it is attached to the body wall in the region of the alary muscle. Posteriorly the ovary joins the lateral oviduct, its lumen being continuous with that in the latter structure. The ovary lies free in the haemocoele and receives large tracheal branches from the spiracle complexes of segments V and VI.

Covering the ovary externally is a thin membranous coat with small scattered nuclei, the *peritoneal membrane* (Fig. 84 (2) *a*). Lining the rather wide central lumen is a similar thin nucleated membrane ((2) *b*). Between the two membranes are some 100–120 *follicular tubes* or *ovarioles* (*fol*). These take origin from the inner aspect of the peritoneal membrane and pass downwards and inwards to the wall of the central lumen. Each consists of an outer tube of thin membrane formed of flattened cells which at the inner end of the ovariole is continuous with the wall of the central lumen (Fig. 84 (4)). Lying within and distinct from the tube is a column of germinative cells, the 'egg-string' of Nicholson. In the apical portion of the tube the cells in the column are still largely undifferentiated, forming the *germarium* (Fig. 84 (3) *gm*). Distal to this part the column consists of a succession of developing follicles, each of which, circumstances permitting, will become an egg, the *vitellarium* (Fig. 84 (4) *fol′*, *fol*). Between each successive follicle is a portion of the column in which the cells are compacted forming a short solid connecting-piece. A similar cord distal to the lowest follicle is referred to in the literature as the *pedicel* (Fig. 84 (4) *pdo*).

Each follicle consists of eight cells surrounded by a single cell layer, the *follicular epithelium* (Fig. 84 (5) *f.ep*). Of the eight cells, seven are the *nurse cells* performing a nutritive function, and one, the most distally situated in the follicle, is the *oocyte* or future ovum (Fig. 84 (4), (5) *ooc*). The lowest follicle is the most advanced in development, very greatly so when final development proceeding to egg formation is in progress. Usually two, or at most three, recognisable follicles are present in *A. aegypti* in each ovariole. The lowest follicle throughout the ovary is always approximately in the same stage of development.

The ovary of the mosquito is peculiar in that it does not consist, as in many insects, of separate individual ovarioles, but of a large number of ovarioles bound into a single organ by a thin containing membrane, which is therefore not correctly termed the *tunica propria*, this term being that applied to the outer covering of an ovariole. Nor do the upper ends of the ovarioles combine to form the suspensory ligament, but take origin from the peritoneal membrane along the length of the ovary. The correctness of the term ovariole has indeed to be considered. It is stated by Imms, making reference to Henneguy, that in *Chironomus*, *Anopheles* and some Braconidae, although there is a differentiation into follicles, ovarioles are wanting. The use of the term ovariole, however, seems justified, even though the tubules lodging the germinative cells in the mosquito are possibly an adventitious tissue, since the column of germinative cells closely resembles that of insects with free ovarioles.

More important is the fact that some confusion seems to exist in the literature regarding the term follicular epithelium in the mosquito. This term is correctly applied to the layer of cells surrounding the follicle. It is, however, sometimes used as though the cells covering the follicle and those lining or forming the walls of the ovariole were identical. The follicular epithelium surrounding the follicle is already part of the future egg of which it will form the chorion and has, when first seen, already lost all connection with the ovariole wall. In *Aëdes aegypti* early development in the pupa clearly shows that the walls of the ovariole are formed around the columns of germinative cells from interstitial cells of the ovary (Fig. 84 (4) *ovl*).

A conspicuous feature of the ovary in the pupa and imago is the abundant tracheal supply and large size of the tracheal vessels. As noted by Nicholson large tracheae form coiled masses at the base of the ovarioles to allow for eventual extension of the tracheae with growth of the follicles. Nicholson also notes the presence of tracheal cells and leashes of tracheal capillaries devoid of taenidia.

DEVELOPMENT OF THE OVARY (Fig. 84 (10), (11))

The ovary of insects develops from thickenings of the mesoderm of the splanchnic wall constituting the *genital ridges*. In the cockroach columns of cells are formed which pass from the genital ridge on the coelomic wall to a ventral strand continuous with the oviduct rudiment. These columns become the ovarioles and eventually appear as opening in series into the proximal portion of the lateral oviduct. A close approach to this primitive condition in Nematocera is seen in *Mycetophila* (see Abul-Nasr, Fig. 26). In the mosquito the lumen of the ovary would appear to represent this proximal portion of the oviduct rudiment. It is continuous with the lumen of the lateral oviduct developed from the mesodermal strand and in an early stage as seen in the pupa the lumen of the ovary tends to lie nearer the internal face of the organ with the young follicles situated for the most part on its external side.

In the first instar larva the ovary is a small body, measuring about 60μ in length, situated in the sixth abdominal segment. It is formed of a small group of cells resembling germ cells with some smaller cells at its periphery. By the second instar the organ has increased to about 70μ with much the same characters. In the third instar it is a small fusiform body about 150μ in length (Fig. 84 (11)). It is continued anteriorly as a fine filament traceable to the intersegmental region IV–V and continued posteriorly as a thicker filament, the *mesodermal strand* (*mst*), which ends blindly towards the posterior end of segment VI. The body of the organ at this stage consists of a mass of cells some of which form a rather regular layer one cell deep at the surface. Among the cells in the interior are some larger cells which show indications of early follicular formation. The organ is solid with no lumen.

Towards the end of the fourth instar the body of the ovary measures about 0·3 mm. and extends into the fifth segment. The mesodermal strand is rather thick and somewhat clubbed. It has a lumen continuous with that now passing through the length of the ovary. The ovary now has a thin outer nucleated membrane that appears to be distinct from the peripheral layer of cells previously noted, which is still recognisable in parts. In the central portion are groups of cells forming early follicles lying in spaces in a matrix of smaller cells. The organ is noticeably solid and compact, not extended as in the pupal stage with enlarged tracheal vessels and the follicles widely separated.

In the pupa the ovary has much the appearance of the organ in the young imago with a central lumen and follicles throughout its whole length. Towards the end of the pupal stage the follicles have already taken on their characteristic appearance. There is a lower follicle about 25μ long and 20μ broad with well-marked nurse cells and oocyte, the latter sometimes distinguishable by its clear nucleus and marked nucleolus. Proximal to this lower follicle is a second follicle and sometimes indications of a third. Between successive follicles there is a connecting column of rather flattened cells. A feature very clearly shown at this stage is the column of cells forming the pedicel passing from the lower follicle to the wall of the central lumen. The exact relation of the pedicel to the wall of the lumen at this stage is, however, difficult to ascertain owing to the aggregation of nuclei about the central lumen. It seems probable that there is no actual opening, and that there is a cellular block at the distal end of the ovariole, the wall of the ovariole and the wall of the lumen being confluent with the substance of the pedicel. Surrounding the follicle, but distinct and separated by a space, is the thin membranous ovariole wall. The follicle is already surrounded by the layer of rather flattened cells that will later develop into the characteristic follicular epithelium.

For some time following emergence the ovary still remains only about 0·6 mm. long with the appearances as described in the pupa. By the fifth day, or later, in the absence of a blood meal, it has increased at most to 1·0 mm. in length and extends to the fourth segment. There are two, or at most three, follicles in each ovariole, the lowest of which has not developed beyond stage 2 as described in the next section.

DEVELOPMENT OF THE FOLLICLE

In the early follicle the nurse cells and oocyte are similar in appearance, the oocyte always being situated distal to the nurse cells. At emergence or shortly after, however, the oocyte becomes conspicuously different by reason of its clear nucleus and prominent nucleolus in contrast with the rather diffusely stained and much enlarged nuclei of the nurse cells apparently lacking a nucleolus. In the fresh condition the oocyte at this stage is still free from granules in the cytoplasm. But shortly afterwards fine granules appear and these are followed by large characteristic yolk granules. As development proceeds the oocyte increases in size, its cytoplasm becoming densely packed with yolk granules, so that it successively occupies half, three-quarters and later practically the whole of the much enlarged follicle. The follicle at first retains its oval shape, but eventually it elongates, taking on more and more the shape of the egg. The nurse cells, which for a time remain very much of their original size and appearance as compared with the rapidly growing oocyte, become eventually inconspicuous bodies at the anterior end of the egg. The follicular epithelium which first appears as rather flattened cells becomes, as the follicle increases in size, more and more epithelium-like eventually forming a definite epithelium of cubical cells. As the follicle elongates to take on the shape of the egg, the follicular epithelium shows changes preparatory to the formation of the chorion and eventually forms the egg-shell.

With due regard to temperature and other modifying conditions the above changes have been much used in determining the age, relation to blood meal and other biological data in *Anopheles* captured under different natural conditions. For this purpose the following stages have been made use of:

Stage 1. Cytoplasm of the oocyte free from granules.

Stage 2. Yolk granules present in the cytoplasm of the oocyte, but not entirely hiding the nucleus.

Stage 3. The nucleus of the oocyte is obscured by yolk granules, but the follicle is still oval in shape.

Stage 4. The follicle is elongate and more or less the shape of the future egg.

Stage 5. Chorionic structure is shown.

Stage 1 is only seen in the newly emerged mosquito and indicates that nearby breeding is in progress. Stage 2 may be seen in a mosquito that has emerged some time previously but has not yet fed on blood, or one that has fed but oviposited and has not since had a blood meal unless a very recent one. Usually if a blood meal has been taken the ovary will have advanced beyond stage 2. The most important fact is that in the absence of a blood meal development of the follicle does not proceed beyond stage 2.

The extent to which development takes place in the absence of a blood meal has been studied by Mer (1936) in *Anopheles elutus*. It has been shown that it is dependent upon the extent to which available reserves are present. Thus in *A. elutus* development up to stage 2 can be achieved either at the expense of reserves carried over from the larval stage or by feeding on raisins. One blood meal will then suffice for oviposition. With poor larval food, as at the onset of adverse conditions prior to hibernation, development to stage 2 is impaired and a single blood meal does not suffice. Hecht (1933) also records that feeding on sugar stimulates egg maturation in *Anopheles maculipennis* previously fed on blood. In *Aëdes aegypti* reared under reasonably good conditions a single blood meal is normally adequate for oviposition.

The nuclear changes in the oocyte and nurse cells have been described by Nicholson in *Anopheles* and by Vishna Nath (1924) in *Culex fatigans*. In the latter species when the mosquito feeds on blood the oocyte quickly becomes differentiated from the nurse cells by a prominent nucleolus in a clearly defined nucleus. Later the nucleus of the oocyte enlarges and after extruding a small mass of chromatin threads (chromatin residue of Nicholson) eventually disorganises in the yolk. The extruded chromatin becomes the female pronucleus situated at the anterior pole of the egg.

The granules in the oocyte are described by Vishna Nath (1929). They are of two kinds only, namely large solid granules of proteid nature *macrospheres*, measuring about 2–4μ and small drop-like granules, *microspheres*, containing free fat. No evidence of mitochondria was found by this author. Under the centrifuge the macrospheres and fatty microspheres (Golgi vesicles) pass to opposite sides. A certain number of microspheres that do not contain fat remain with the cytoplasm in the intermediate zone.

The character of the chromosomes is dealt with in a later section.

THE FEMALE EFFERENT SYSTEM (Fig. 84 (1), (6–9))

The homologies of the different parts of the female efferent system in the mosquito have been much clarified by the developmental studies of Abul-Nasr on different Nematocerous families. It would appear that in the early larval stage in both sexes there are on each side two *mesodermal strands* one passing from the ovary or testis to the seventh segment and one to the ninth segment. In the female the latter disappears before adult life and in the male the former. The *lateral oviducts* may be developed from the female mesodermal strands, or with the backward growth of the ovary and shortening of the mesodermal strands the mesodermal tissue may be replaced or reinforced by ectodermal tissue from the common oviduct invagination. The *common oviduct*, *spermathecae* and a *vaginal apodeme*, when this structure is present, all arise from separate invaginations taking origin from a cavity formed by the fusion of the peripodial cavities of two imaginal buds which appear ventrally on each side of the middle line behind the seventh sternite. The *caecus* is formed from an invagination that takes origin from a cavity formed by fusion of a pair of imaginal buds on the ventral surface of the ninth sternite of the larva. The original peripodial cavities are eventually included in a common depression, the *atrium*.

In *Aëdes aegypti* the mesodermal strand in the female passing to the posterior end of the seventh segment is already present in the third instar larva. Whether there is at any stage a second strand has still to be determined. The lateral oviducts appear to be entirely of mesodermal origin being formed wholly from the mesodermal strands, the swollen ends of which can still be recognised after joining up with the common oviduct. In accord with this view is the fact that, whereas the lumen of the lateral oviduct at emergence is openly continuous with that of the ovary, there is still a block, even after the parts have fused, for some time in the lumen at the point of junction.

In *Anopheles* there are dilatations at the points of junction of the lateral oviducts with the common oviduct that have been termed *ampullae* and the size and shape of these dilatations have been used in determining whether a female has previously oviposited or not. According to Mer (1932) the ampullae in *Anopheles elutus* in the nulliparous female are small and well delimited from the common oviduct, but not from one another. But after ovipositions they are increased in size and clearly divided from each other. They also have the appearance of being more a part of the common oviduct at the expense of the lateral oviducts which have become shorter. For measurement the author uses a figure obtained by multiplying the cross and longitudinal diameters taken at right angles. Almazova and later Polodova have also used this method for *A. maculipennis* var. *atroparvus* and some other varieties of this species, as also for *A. claviger* and *A. superpictus*. The last-mentioned author notes that the parts should be measured in saline isotonic for the mosquito whilst still pulsating and at intervals between the pulsations, or in saline with a narcotic to prevent pulsations. Female *atroparvus* newly emerged gave a figure of 0·01 mm.^{2}. After one oviposition the figure was 0·02 mm.^{2}, after two ovipositions 0·028 mm.^{2} and after three 0·031 mm.^{2}.

In *Aëdes aegypti* dissected in Ringer's fluid at 0·7 per cent salinity the dilatations at the junction of the lateral oviducts with the common oviduct were much less marked than described in *Anopheles*, giving a figure of only 0·004 mm.^{2}. Also such increase in size and appearance as was noted after oviposition did not suggest that the test could be usefully used in the case of this species. Further observations on this point are, however, desirable.*

The *lateral oviducts* (Fig. 84 (1) *lo*) are tubular with rather thin walls. They have an inner lining of small rather flattish cells and a thin coat of muscular fibres, chiefly circular.

* Recently Gillies (1956) has described in *Anopheles gambiae* a new character for recognition of recently fertilised females, a mating-plug of translucent material in the common oviduct derived from the male and absorbed in 36 hours from fertilisation.

The *common oviduct* (*cod*) is very similar in appearance to the lateral oviducts but broader. It has an inner lining of cubical epithelium, external to which is a coat of longitudinal and circular muscle fibres, the former for the most part internal. It lies close behind the eighth sternite passing backwards to its posterior border and then making a sharp bend at right-angles dorsally to join the vagina.

The *vagina* (*vg*), unlike the common oviduct, has a thick transparent cuticular intima. Its proximal portion is dilated forming a well-marked chamber with a thick pad-like roof. Its distal portion is thrown into a sharp transverse fold just beyond which are the openings of the spermathecal ducts and caecus. There is then a short terminal portion ending at the genital opening as described later with a small posterior *cul de sac*. How much of the cavity is to be regarded as atrium is uncertain.

The *spermathecae* (Fig. 84 (7), (8)) are situated in segment VIII. They are three in number a larger one situated medianly and a smaller one on each side of and slightly posterior to this. All three lie close to the sternite, the median spermatheca with its ventral aspect in contact with the common oviduct. Each spermatheca consists of an almost perfectly spherical shell of dark sclerotised cuticular substance. Covering the greater part of this externally is a very thin cellular layer with scattered small flattened nuclei. At or near the posterior pole is a thickened plaque of larger cells through the middle of which the spermathecal duct takes off abruptly. In the unfertilised female the spermathecae are empty of cellular contents, but after fertilisation they are filled with a mass of spermatozoa.

The *spermathecal ducts* (*spd*) take origin abruptly from the spermathecal globes. They loop dorsally and then, bending ventrally, enter the vagina as described above. The duct from the median spermatheca remains distinct and enters the atrium independently of the others. The ducts from the two lateral spermathecae join a little before their termination to form a short length of common duct which enters close to, but separately from, the duct of the median spermatheca. The ducts have a very narrow lumen, less than $2{\cdot}0\,\mu$ in diameter. This is surrounded by a very thick clear sclerotised intima, external to which is a closely packed layer of cells with darkly staining nuclei and, scattered somewhat irregularly on the outer surface of this layer, there are larger cells possibly glandular in nature. The ducts give the appearance of being narrow rigid tubes incapable of expansion and, seeing that their lumen is so narrow, it is remarkable that the spermatozoa apparently so rapidly gain entry to the spermathecae. It is unusual to find a mass of spermatozoa still present in any other part of the efferent system. The ducts end abruptly where they enter the vagina without any special thickening or other noticeable structural feature. This may lead to the more conspicuous entry of the caecal duct being mistaken for that of a common spermathecal duct (see the duct marked in the genital opening in Fig. 84 (1) which is that of the caecus, not the less conspicuous spermathecal duct openings).

The *caecus* (*cs*) in *A. aegypti* is a relatively small structure with the characters of a mucus gland. It lies close to and ventral to the rectum and its duct opens just posterior to the spermathecal ducts. The terminal portion of the duct is sclerotised, forming a short length of rigid tube with a terminal ring-like thickening.

The *genital opening* (Fig. 84 (1), (6)). A word may be said regarding the genital opening, especially as some confusion has occurred in the literature over the structure termed the *insula*. In *A. aegypti* the parts are rather complicated by the extent to which the terminal segments are intussuscepted. This gives rise to a deep pouch between the eighth sternite and the postgenital plate that must be distinguished from the atrium. At the bottom of this pouch, that is at its anterior end, is the characteristic wide transverse genital opening as seen, for example, in *Anopheles* (Christophers, 1923). This has a bow-shaped sclerite corresponding to the *cowl* in *Anopheles* bordering its posterior edge and a sclerotised ridge forming its anterior lip. This ridge has on each side a sclerotised hairy plate and in the mid-line between these is a small, not very conspicuous, hairy area which is clearly the insula as described in *Anopheles* where it is more developed. Where the lips meet at each corner of the opening there is the small characteristic ribbon-like thickening which leaves no doubt as to the identity of the opening, which is in almost every particular similar to that in *Anopheles* and *Culex*. In *Aëdes aegypti*, however, it lies hidden in the outer pouch-like hollow beneath the seventh sternite (Fig. 84 (6)).

In *A. aegypti* the atrium is not large and it is difficult, as already noted, to say how much of the cavity should be considered this structure and how much vagina. On its roof open the spermathecal ducts and the caecus. In the forms studied by Abul-Nasr there crosses the roof between these respective openings a structure homologised by this author as the ninth sternite and termed by him the insula. It should be pointed out, however, that the insula, as so named in *Anopheles*, is a structure on the anterior lip of the genital opening and therefore quite distinct from the insula of Abul-Nasr. The position of the ninth sternite would appear to be correctly as assigned by Abul-Nasr. The insula on the lip of the genital opening cannot be the ninth sternite since the female opening would then be posterior to the ninth sternite, which is very improbable. The sclerotisations on the lip are, however, very suggestive of some segmental structure and it is just possible that the lateral plates in *A. aegypti*, which are hairy, represent in a very obsolescent form the first ovipositor valves (anterior gonapophyses), which in many insects are situated in this position, that is just posterior to the eighth sternite.

(*j*) THE MALE REPRODUCTIVE SYSTEM

GENERAL DESCRIPTION (Fig. 85 (1))

The internal organs of generation in the male consist of the two gonads, *testes* (*te*), a long thin duct (*vd*) passing from each testis to a median rather flask-shaped structure (*ejd*) which is continued distally as a short median duct passing to the intromittent organ (*Ph*) and two sac-like bodies (*ag*) that open by short ducts on each side at the base of this. The nomenclature of the parts requires consideration.

According to Snodgrass (1935) ducts passing from the testicular tubules to a common duct are *vasa efferentia*, the common ducts on each side being the *vasa deferentia*. The lower portions of the vasa deferentia may be swollen, forming the *vesiculae seminales*, and the median duct into which the vasa deferentia open distally is the *ductus ejaculatorius*. In association with the ductus ejaculatorius there may be *accessory glands*. Prashad (1916), who describes and figures the male genital organs in *Anopheles willmori*, *Culex fatigans*, *Aëdes* (*Stegomyia*) *scutellaris*, *Aëdes* (*Ochlerotatus*) *pseudotaeniatus* and *Theobaldia spathipalpis*, finds the organs in all except *Anopheles* to agree in general characters and his descriptions would cover that given above for *Aëdes aegypti*. The ducts leading from the testes he terms *vasa deferentia*, there being he considers no difference in this case between the *vasa efferentia* and the *vasa deferentia*. The median organ into which these open he terms the *sacculus ejaculatorius* and the swollen portions of the vasa deferentia where they enter this organ he names the *receptacula seminis*. The lateral sac-like bodies are the *accessory glands* and the short passage leading to the genital opening is the *ductus ejaculatorius*. Abul-Nasr finds the efferent system in the families of Nematocera studied by him to be formed in part from the mesodermal strands and in part from ectodermal invaginations. In accordance with this he terms the ducts from the testes *vasa efferentia* or *vasa deferentia* depending upon the source from which they are derived. The ectodermal parts commence as a median invagination starting from what will be the gonopore. This forms the *penis cavity* (it may in some forms, for example *Tipula*, be ballooned into a cavity or may take the form of a median duct). From the anterior end of the penis cavity there arise twin invaginations which become the *ejaculatory duct rudiments*. These later become differentiated, their proximal parts becoming glandular and forming the *vesiculae seminales*, whilst their distal parts retain their duct-like character and become joined together in a common muscular coat as the *ejaculatory ducts*.

In selecting appropriate terms for the parts in *Aëdes aegypti* it seems more in keeping with taxonomy to name the ducts leading from the testes the *vasa deferentia*, in spite of the fact that they appear to be derived from the mesodermal strands. The median organ into which they open, which clearly consists of two fused lateral components, appears to represent the *vesiculae seminales*. For reasons that will be obvious later it would seem, however, convenient when referring to the structure as a whole to retain Prashad's term *sacculus ejaculatorius*. The short duct leading from the *sacculus* to the phallosome, at the base of which short ducts enter from the sac-like bodies, would seem to be the *penis cavity* and the sac-like bodies are obviously *accessory glands*. See Addenda, p. 719.

For literature dealing with the male organs as a whole, other than those authors already mentioned, see: Dufour (1851, first to note the testes in the mosquito); Hurst (1890, early description of the testes with note that the hinder end of the testes seem to take the place of the vesiculae seminales); Kulagin (1907, histology of the male organs); Imms (1908, references to testes in the larva); Adie (1912, pigmented testes in differentiation of sex in *Anopheles* larvae); Christophers (1922, development). See also authors mentioned later in section dealing with development of spermatozoa.

THE TESTES (Fig. 85 (2), (3))

The testes in the imago at emergence are fusiform bodies measuring about 0·4 mm. in length and 0·1 mm. in greatest breadth and situated on each side in the sixth abdominal segment. Later they become more swollen and extend some distance into the seventh segment. They lie internal to the peripheral fat-body and, as in previous stages of the life history, are surrounded by a thick coat of special fat-body. Distally they are continued into the vasa deferentia, but proximally they have no definite suspensory ligament as has the ovary.

Externally there is a thin nucleated membrane. Lomen (1914) in *Culex* describes the outer covering as double, a thick connective tissue layer and an inner tunic lining. Within the outer membrane are stratified zones of developing spermatozoa that give to the testis, whether in the fresh condition or in sections, a very characteristic appearance. Apically there is a portion consisting of more or less undifferentiated sexual cells. There follow alternate bands showing various stages in development of the spermatozoa as described later and, occupying the distal third or more of the organ, a mass of mature or nearly mature spermatozoa. At emergence the passage into the vas deferens is blocked by a mass of secretion at the distal end of the testis. Proximal to this spermatozoa tend to be arranged more or less circularly with the heads towards the surface so that in tangential sections the long thin deeply stained heads are seen, while sections passing deeper tend to show the heads in cross-section near the periphery, the more central portions appearing as a mass of granules, many of which are composed of tails cut in cross-section. In older males the greater part of the testis contains mature or nearly mature spermatozoa and the lower portion may be much swollen with accumulation of these forms, so that as observed by Hurst the lower part of the testes appeared to function as the vesiculae seminales.

THE MALE EFFERENT SYSTEM (Fig. 85 (1), (4–8))

The *vasa deferentia* in the newly emerged mosquito are extremely long and thin. Their walls consist of a layer of epithelium (Fig. 85 (5)), but are devoid of obvious muscle fibres, though jerky vermicular movements of their lower ends may be seen when the parts are dissected in Ringer's solution. Later, when packed with spermatozoa, they become thicker and shorter. Each joins the ejaculatory duct of its side at the proximal end of the sacculus. I have not found the lower ends significantly swollen to form vesiculae seminales, even in males several days old.

Figure 85. The male reproductive system.

1 Dorsal view of the male organs shortly after emergence.

2 Section showing testis in fourth instar larva. Most of the organ is in the stage of development corresponding to *b* in no. 3. The organ is surrounded by fat-body.

3 Sagittal section of anterior portion (about one-quarter) of testis half an hour from emergence. *a*, region of spermatogonia; *b*, region of developing spermatocytes; *c*, region of chromosome reduction; *d*, region of developing spermatozoa (spermatids): shows internal portion largely with granules and tails of spermatids. At the periphery are groups of spermatid nuclei for the most part cut transversely as at Fig. 85 (9) A, some cut longitudinally are as at Fig. 85 (9) B. The remaining three-quarters of the testis are very similar to *d*, but with some more mature spermatozoa at the posterior end. As yet no spermatozoa have passed into the vas deferens.

4 Sagittal section showing relation of ejaculatory ducts and accessory gland to genital opening in phallosome.

5 Showing structure of the vas deferens.

6 The same for the ejaculatory duct.

7 Dissection showing conjoined ejaculatory ducts swollen with longitudinally arranged motionless spermatozoa in three days' old male. *a*, active spermatozoa in thickened vas deferens; *b*, swollen ejaculatory ducts; *c*, stream of spermatozoa escaping from rupture.

8 Three spermatocytes.

9 Appearances shown by nuclei of spermatids. A, ring forms of early spermatid nuclei in transverse section; B, the same in longitudinal section; C, cross-sections of more mature forms.

10 Mature spermatozoa from spermatheca of female as seen under dark-ground illumination represented in black, the heads being the more brightly illuminated. Camera lucida drawings after cessation of undulatory movement of the tails.

Lettering: *ag*, accessory gland; *ejd*, conjoined ejaculatory ducts (sacculus ejaculatorius); *pca*, penis cavity; *Ph*, phallosome; *te*, testis; *vd*, vas deferencs.

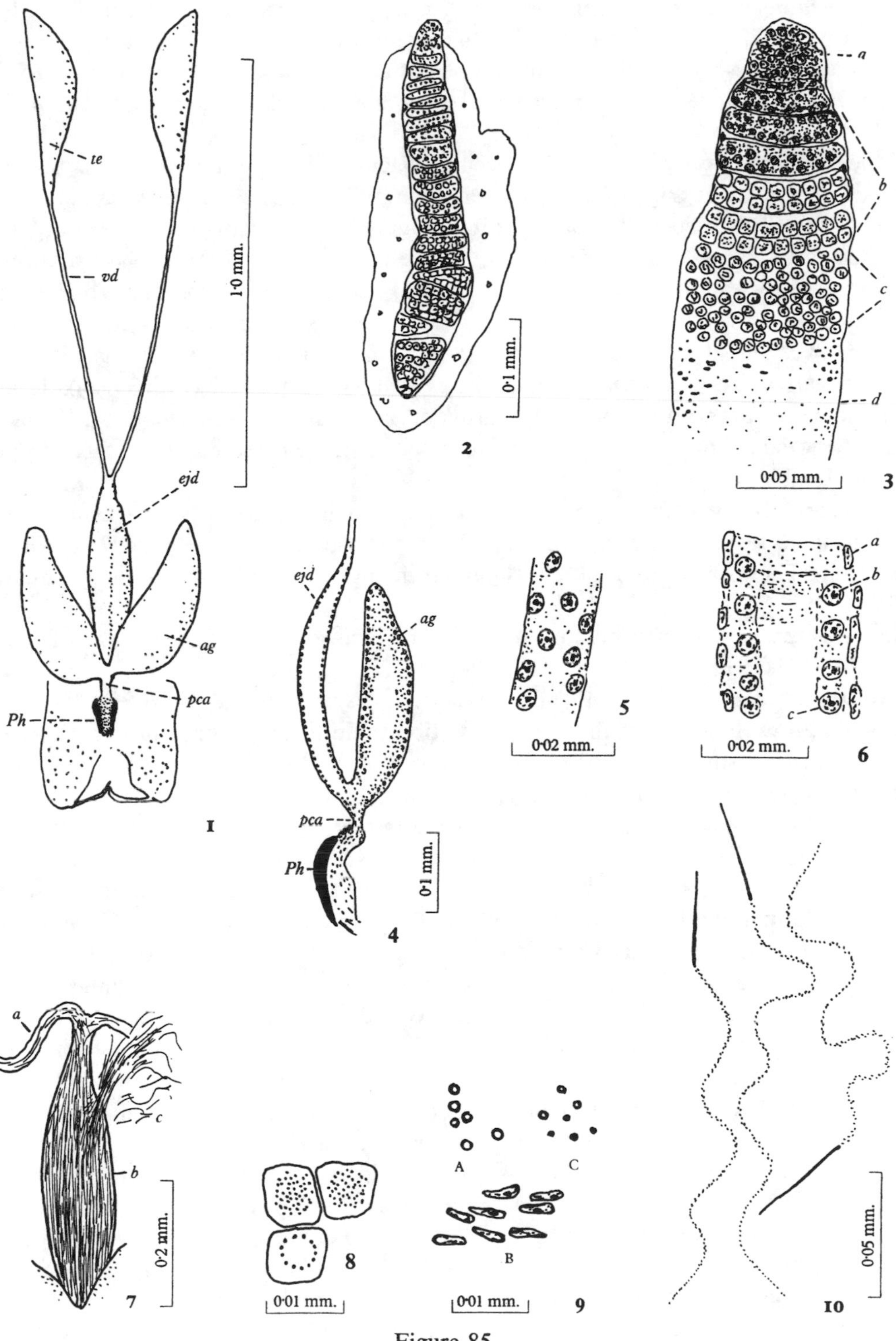
te
vd
1·0 mm.
ejd
ag
pca
Ph
I
0·1 mm.
2
a
b
c
d
0·05 mm.
3
ejd
ag
pca
Ph
0·1 mm.
4
5
0·02 mm.
a
b
c
0·02 mm.
6
a
c
b
0·2 mm.
7
8
0·01 mm.
A
C
B
0·01 mm.
9
0·05 mm.
10

Figure 85

The *ejaculatory ducts* together form the *sacculus ejaculatorius* of Prashad. It has a median longitudinal mark indicating its dual composition. It lies ventral to the bulged bases of the accessory glands appearing in sagittal section as shown in Fig. 85 (5). Its walls have an epithelial lining, external to which is a coat of stout circular muscle fibres (Fig. 85 (6)).

In the newly emerged mosquito the sacculus ejaculatorius is empty as also is the whole of the efferent male passage. But after opening up of the passage to the vasa deferentia it becomes packed with spermatozoa, apparently lying longitudinally so that the organ has a striated appearance. The spermatozoa, before the organ is ruptured, are motionless. On rupture of its wall, however, these issue in a massed stream, the spermatozoa becoming active as they cease to be tightly packed. The mass of spermatozoa so collected in the sacculus is very great, causing this organ to be swollen to twice or more its original size and there can be no doubt that it functions as a vesicula seminalis (Fig. 85 (7)). It seems undesirable, however, on morphological grounds to give it this name, which has been used as already noted for other parts of the efferent system in the mosquito and Prashad's name seems preferable.

The *accessory glands* are large conical sac-like organs. They possess a central lumen and walls lined by columnar epithelial cells the cytoplasm of which is loaded with globules of secretion. They appear to be purely glandular in nature and do not contain spermatozoa at any stage.

The *penis cavity* is a short duct passing into the phallosome and opening at the genital pore within this organ.

The *spermatozoa* (Fig. 85 (7–10)) as present in the spermatheca of the female or in the male parts are well shown in the living condition under dark-ground illumination. They show well also in smears stained by Heidenhain's haematoxylin with counterstain or by Holmes' silver impregnation method. Their development and chromosome structure has been very fully studied and described by Lomen in *Culex pipiens*: see also Stevens (1910, 1911) and Taylor (1914). None of these authors, however, appears to have given a figure or description of the spermatozoon of the mosquito as a whole. Seen as noted above they consist of a long thin refractile and dark-staining *head* measuring about 40–45 μ and a very long less refractile and lighter-staining tail measuring up to 200 μ or more in length. In thickness the head measures little more than a micron. It is somewhat thinner in front, gradually thickening towards its hinder end. No middle piece is to be made out, the tail continuing on from the head. In living material the head exhibits little or no undulatory movement and usually appears straight as shown in the figure. The long tail, on the other hand, exhibits wave-like undulatory movements which under particular conditions appear more or less to synchronise in different individuals. Sometimes the undulations are more rapid and extensive than usual and it has been noted that this may occur when near an air bubble. In cover-glass preparations the undulations eventually cease, the tail remaining in its wavy condition as shown in the figure. Where the spermatozoa are massed, movement may be much like a flowing stream. In spite of the undulating tail, very little forward movement is made by isolated individuals and it was thought that possibly narrow ducts like those of the spermathecae might favour progression by giving purchase to the tail undulations.

Development of the spermatozoon in *Culex* has been described in considerable detail by Lomen, but only a brief account can here be given. The following is the development in broad outline. The cells destined to become spermatozoa, *spermatogonia*, occupy the extreme anterior end of the testis (Fig. 85 (3) *a*). An apical cell, as present in some insects, has not been noted in the present case. The spermatogonia divide repeatedly, forming masses of cells, *spermatocytes*, which become included in a membranous cyst wall formed by somatic cells whose somewhat flattened nuclei may be seen in the earlier cysts. The number of spermatocytes enclosed in a single cyst may in some forms be as many as 256 (the result of eight consectutive divisions) (Wigglesworth, 1942). In *Aëdes aegypti* testis this would appear to be about the number in the larger groups. The cyst wall in the larger groups may be very thin or not apparent, but the groups as they pass down the testis largely retain their individuality. As they do so they tend to become flattened and form stratified lines of spermatocytes that give the testis its characteristic appearance. Eventually the spermatocytes undergo two meiotic divisions with reduction of the chromosomes, the zones so formed being characterised by the appearances shown at *c* (Fig. 85 (3)). The cells so formed, *spermatids*, are the young immature spermatozoa. They have a much reduced and denser nuclear mass, so forming a distinctive zone in the testis (Fig. 85 (3) *d*). They are shown by Lomen as commonly having a nucleus consisting of a U-shaped chromatin thread with a small detached nucleolus. As development advances the U becomes elongated to form a long stretched-out V. Forms such as those depicted in Fig. 85 (9) are common and characteristic of zone *d*. Cross-section of the mature forms are as seen at *C*, earlier forms appearing ring-like as at *A*.

REFERENCES

(*h*) THE EXCRETORY SYSTEM

BUGAJEW, I. I. (1928). Zum Studium des Baues der malpighischen Gefässe bei den Insekten. *Zool. Anz.* **78**, 244–55.

LECAILLON, A. (1899). Sur les prolongements ciliformes de certaines cellules du cousin adulte, *Culex pipiens* L. (Dipt.). *Bull. Soc. Ent. Fr.* 1899, 353–4.

MISSIROLI, A. (1925). I tubuli del Malpighi nell' *Anopheles claviger*. *Annali d'Igiene*, **35**, 113–22.

MISSIROLI, A. (1927). I tubuli del Malpighi nell' *Anopheles claviger*. *Riv. Malariol.* **6**, 1–7.

NUTTALL, G. H. F. and SHIPLEY, A. E. (1903). Studies in relation to malaria. *J. Hyg., Camb.*, **3**, 185–6.

PANTEL, J. (1914). Signification des 'glandes annexes' intestinales des larves des Ptychopteridae et observations sur les tubes de Malpighi de ces Nematocères (larves et adultes). *Cellule*, **29**, 393–429.

PATTON, R. L. (1953). Excretion. In Roeder's *Insect physiology*. John Wiley and Sons, New York; and Chapman and Hall, London.

RAMSAY, J. A. (1953). Exchanges of sodium and potassium in mosquito larvae. *J. Exp. Biol.* **30**, 79–89 (see also other papers listed under excretion in chapter XXXI, Physiology).

ROUBAUD, E. (1923). Les désharmonies de la fonction rénale et leur conséquences biologiques chez les moustiques. *Ann. Inst. Pasteur*, **27**, 627–79.

SAINT-HILAIRE, K. (1927). Vergleichend-histologische Untersuchungen der malpighischen Gefässe bei Insekten. *Zool. Anz.* **73**, 218–29.

SCHINDLER, E. (1878). Beiträge zur Kenntnis der malpighi'schen Gefässe der Insekten. *Z. wiss. Zool.* **30**, 587–660.

VENEZIANI, A. (1904). Valore morphologico e fisiologico dei tubi malpighiani. *Redia*, **2**, 177–230.

WIGGLESWORTH, V. B. (1931). The physiology of excretion in a blood-sucking insect *Rhodnius prolixus* (Hemiptera, Reduviidae). II. Anatomy and histology of the excretory system. *J. Exp. Biol.* **8**, 428–42.

WIGGLESWORTH, V. B. (1933). The adaptation of mosquito larvae to salt water. *J. Exp. Biol.* **10**, 27–37.

WIGGLESWORTH, V. B. (1942). *The Principles of Insect Physiology*. Ed. 2. (and subsequent editions). Methuen and Co., London.

(*i*) THE FEMALE REPRODUCTIVE SYSTEM

ABUL-NASR, S. E. (1950). Structure and development of the reproductive system of some species of Nematocera (order Diptera: suborder Nematocera). *Phil. Trans.* B, **234**, 339–96.

ALMAZOVA, V. V. (1935). Sur la determination de l'âge d'*Anopheles* d'aprēs le grandeur des oviductes (in Russian). *Med. Parasit., Moscow*, **4**, 345–54. (Abstract in *Rev. Appl. Ent.* **24**, 95.)

BODENHEIMER, F. (1923). Beiträge zur Kenntnis der Kohlschnake *Tipula ochracea. Z. wiss. Zool.* **121**, 393–441.

CHRISTOPHERS, S. R. (1901). The anatomy and histology of the adult female mosquito. *Rep. Malar. Comm. R. Soc.* (4), 1–20.

CHRISTOPHERS, S. R. (1911). Development of the egg follicle in Anophelines. *Paludism*, **2**, 73–88.

CHRISTOPHERS, S. R. (1923). The structure and development of the female genital organs and hypopygium of the mosquito. *Ind. J. Med. Res.* **10**, 698–71.

CHRISTOPHERS, S. R., SINTON, J. A. and COVELL, G. (1939). How to do a malaria survey. *Govt. India, Hlth Bull.* no. 14. Ed. 4. Delhi.

DER BRELJE, R. VON (1924). Die Anhangsorgane des weiblichen Geschlechtsganges der Stechmücken (Culicidae). *Zool. Anz.* **61**, 73–80.

EYSELL, A. (1924). Die Stechmücken. In Mense's *Handb. TropenKr.* Ed. 3, **1**, 176–87.

GILLIES, M. T. (1954). The recognition of age groups within populations of *Anopheles gambiae* by the pre-gravid rate and the sporozoite rate. *Ann. Trop. Med. Parasit.* **48**, 58–74.

GILLIES, M. T. (1955). The pre-gravid phase of ovarian development in *Anopheles funestus. Ann. Trop. Med. Parasit.* **49**, 320–5.

GILLIES, M. T. (1956). A new character for the recognition of nulliparous females of *Anopheles gambiae. Bull. World Hlth Org.* **15**, 451–9.

GRANDPRÉ, A. D. DE and CHARMOY, D. D'E. DE (1900). *Les moustiques. Anatomie et biologie.* Port Louis, Mauritius.

GROSS, J. (1903). Untersuchungen über die Histologie des Insekten-ovariums. *Zool. Jb.* **18**, 71–186.

HAYES, R. V. (1953). Studies on the artificial insemination of the mosquito *Aëdes aegypti. Mosquito News*, **13**, 145–52.

HECHT, O. (1933). Die Blutnahrung, die Erzeugung der Eier und die Überwinterung der Stechmücken Weibchen. *Arch. Schiffs. -u. Tropenhyg.* **37**, Beih. 3, 1–87, 125–212.

HERMS, W. B. and FREEBORN, S. B. (1921). The egg-laying habits of Californian Anophelines. *J. Parasit.* **7**, 69–79.

HOSOI, T. (1954). Egg production in *Culex pipiens pallens* Coquillett. III. Growth and degeneration of ovarian follicles. *Jap. J. Med. Sci. Biol.* **7**, 111–27.

IMMS, A. D. (1938). *A General Text Book of Entomology.* Methuen and Co., London.

KULAGIN, N. (1901). Der Bau der weiblichen Geschlechtsorgane bei *Culex. Z. wiss. Zool.* **69**, 578–98.

LECAILLON, A. (1900). Recherches sur la structure et le développement post-embryonnaire de l'ovaire des insectes. I. *Culex pipiens* L. *Bull. Soc. Ent. Fr.* 1900, 96–100.

MER, G. G. (1932). The determination of the age of *Anopheles* by differences in the size of the common oviduct. *Bull. Ent. Res.* **23**, 563–6.

MER, G. G. (1936). Experimental study on the development of the ovary in *Anopheles elutus* Edw. (Dipt. Culic.). *Bull. Ent. Res.* **27**, 351–9.

MIALL, L. C. and HAMMOND, A. P. (1900). *The Structure and Life History of the Harlequin Fly* (Chironomus). Clarendon Press, London.

MÖLLRING, F. K. (1956). Autogene und anautogene Eibildung bei *Culex* L. zugleich ein Beitrag zur Frage der Unterscheidung autogener Weibchen an Hand von Eiröhrenzahl und Flügellänge. *Z. Tropenmed. u. Parasit.* **7**, 15–48.

REFERENCES

NEVEU-LEMAIRE, M. (1902). Sur les réceptacles séminaux de quelque Culicides. *Bull. Soc. Zool. Fr.* **27**, 172–5.

NICHOLSON, A. J. (1921). The development of the ovary and ovarian egg of a mosquito, *Anopheles maculipennis* Meigen. *Quart. J. Micr. Sci.* **65**, 395–448.

PARKS, J. J. (1955). An anatomical and histological study of the female reproductive system and follicular development in *Aëdes aegypti* (L.). Thesis. Univ. Minnesota.

POLODOVA, V. P. (1941). Changes in the oviducts of *Anopheles* with age and the determination of the physiological age of mosquitoes. *Med. Parasit.* **10**, 387. (Abstract in *Rev. Appl. Ent.* **31**, 187.)

ROY, D. N. and MAJUNDAR, S. P. (1939). On mating and egg formation in *Culex fatigans*. *J. Malar. Inst. India*, **2**, 243–51.

SNODGRASS, R. E. (1933). Morphology of the insect abdomen. Part II. The genital ducts and the ovipositor. *Smithson. Misc. Coll.* **89**, no. 8, 148 pp.

SNODGRASS, R. E. (1935). *Principles of insect Morphology*. New York and London.

VISHNA NATH (1924). The egg follicle of *Culex*. *Quart. J. Micr. Sci.* **69**, 151–75.

VISHNA NATH (1929). Studies on the shape of the Golgi apparatus. I. The egg follicle of *Culex*. *Z. Zellforsch.* **8**, 655–70.

WEBER, H. (1933). *Lehrbuch der Entomologie*. Gustav Fischer, Jena.

WIGGLESWORTH, V. B. (1942). See under section (*h*).

(*j*) THE MALE REPRODUCTIVE ORGANS

ABUL-NASR, S. E. (1950). See under section (*i*).

ADIE, H. A. (1912). Note on the sex of mosquito larvae. *Ann. Trop. Med. Parasit.* **6**, 463–6.

CHRISTOPHERS, S. R. (1922). The development and structure of the terminal abdominal segments and hypopygium of the mosquito with observations on the homologies of the terminal segments of the larva. *Ind. J. Med. Res.* **10**, 530–72.

DUFOUR, L. (1851). Recherches anatomiques et physiologiques sur les Diptères. *Mém. Acad. Sci., Paris*, **11**, 171–360.

HURST, C. H. (1890). The post-embryonic development of a gnat (*Culex*). *Proc.* and *Trans. Lpool. Biol. Soc.* **4**.

IMMS, A. D. (1908). On the larval and pupal stages of *Anopheles maculipennis* Meigen. *Parasitology*, **1**, 103–32.

KULAGIN, N. (1907). Zur Naturgesch. d. Mücken. *Zool. Anz.* **31**, 865–81.

LOMEN, F. (1914). Der Hoden von *Culex pipiens*. *Jena Z. Naturwiss.* **52**, 567–628.

PRASHAD, B. (1916). Male generative organs of some Indian mosquitoes. *Ind. J. Med. Res.* **3**, 497–502.

SNODGRASS, R. E. (1935). See under section (*i*).

STEVENS, N. M. (1910). The chromosomes in the germ cells of *Culex*. *J. Exp. Biol.* **8**, 207–17.

STEVENS, N. M. (1911). Further studies in heterochromosomes in mosquitoes. *Biol. Bull.* **20**, 109–20.

TAYLOR, M. (1914). The chromosome complex in *Culex pipiens*. *Quart. J. Micr. Sci.* **60**, 377–98.

WIGGLESWORTH, V. B. (1942). See under section (*h*).

XXXI

THE IMAGO: PHYSIOLOGY

(*a*) GENETICS AND CHROMOSOMAL STRUCTURE

In a paper entitled 'Mosquito genetics and cytogenetics' with a bibliography of 216 references, Kitzmiller (1953) has recently reviewed in detail what has been done in mosquito genetics and chromosomal structure. The author comments on the surprisingly large number of observations that have been made, but laments that these have so largely been merely incidental to other work or lacking in the precise information most desired by the geneticist.

The subject of mosquito genetics is one that has been increasingly engaging attention and has become of considerable practical importance in connection with control measures, since the development of resistance to insecticides is now considered to be the result of the elimination of susceptible forms so that the species comes to be composed of a resistant strain or strains. (See ch. XXV, p. 558.) Another direction in which genetical studies have become of importance is in the systematics of certain groups, such as the *Culex pipiens* complex, where it has been difficult to assign precise specific characterisation. Nor does the question rest here, for if, as may be the case, such groups represent the result of evolution in progress their study offers a wide field for investigation from many points of view, including that of the geneticist. A very helpful review of work in this field is given in the paper entitled 'Species hybrids in mosquitoes' by Mattingly (1956).

Recently Laven (1956) has described hereditarily transmitted changes in the wing venation and antennal characters in *Culex molestus* brought about through the action of X-rays (see ch. II (*c*), p. 33. By using such characters as marker genes a considerable field of investigation has been opened up. Another valuable aid in genetic studies in mosquitoes has been work on the pale-eyed mutant form of *Culex molestus*, the gene for which has been shown by Gilchrist and Haldane (1947) to be partially sex linked.

On *Aëdes aegypti* are the studies by Gillett (1956) on genetic differences affecting egg-laying in two local forms of this species, the Lagos and Newala strains, showing that this character has a genetic basis. In this species too are the results of crossing of hybrid forms discussed in chapter II. In mosquitoes such transmission has certain peculiar characters, fertile offspring being limited to that transmitted by the female parent, at least in many cases, a result ascribed to *cytoplasmic transmission* and independent of genes, though not all authorities agree that this is proven.

Genetical theory deals with the transmission of hereditary characters on the basis that these are (in most cases) dependent on *genes*. It is usual for genes to occur as couples, *alleles*, one of which is *dominant*, that is, if present gives rise to its own characteristic effects, and the other *recessive*, that is, it is only effective in the absence of the dominant gene. Offspring from pairing a 'strain' with the 'wild' form does not therefore usually show the recessive character in the first generation, that is generation F_1, since all the offspring probably possess the dominant gene. But in the next sister-brother generation, F_2, a certain number will show the recessive character. In the case of a single recessive gene, unless there is some interfering effect, the proportion of offspring showing the recessive character in the F_2 generation will be theoretically, as the result of chance distribution of the genes, one to three, the progeny being *BB Bb bB bb*, where *B* is the dominant gene and *b* the recessive, the only individuals showing the recessive effect being *bb*. It is on such numerical results that the geneticist is largely able to form his conclusions.

The simple case as given above is often modified in various ways, for example the recessive gene may be wholly or partly linked with sex, or genic crossing as described under chromosomal structure may modify the result.

In practice, therefore, the original crossing should be followed up by brother-sister pairing at least to the F_2 generation and the numerical results carefully ascertained. Back crossing, that is pairing of selected members of F_2 or later generations with the original, may also give valuable data. Obviously in such cases unless rearing and manipulation of the material is carefully carried out the results will be of little value since mortality or incomplete breeding-out of larvae, or even poor hatching of the original eggs, may give misleading results.

The following are some works, other than those already mentioned, that may be helpful to anyone undertaking experimental work on mosquito genetics and unfamiliar with genetical theory: Goldschmidt (1938); Sinnott and Dunn (1939); Waddington (1939); Sturtevant and Beadle (1940); Snyder (1940); Haldane (1941). Also the account in the Pelican Books by Kalmus and Crump (1948). See also under the next section.

CHROMOSOMAL STRUCTURE (Fig. 85)

The most suitable stages for study of the chromosomes in mosquitoes are for most purposes the larva and pupa. In the imago the testes are too advanced to show division forms well. For the somatic tissues also material is best obtained from the larva or pupa. In *Culex pipiens* and *Aëdes aegypti* Sutton (1942) found the best material to be the fourth instar larva, the pupa or newly emerged adults. For somatic division in different tissues Hance (1917) uses the third instar larva. All authors emphasise the importance of good fixation and also the pre-treatment of material by rearing this under special optimal conditions. Such pre-treatment is especially important for the study of the giant salivary chromosomes. Mattingly (*in lit.*) especially stresses the difficulty of obtaining good preparations of chromosomes in *A. aegypti* and the necessity here of pre-treatment. Sutton found the best material was that kept at low temperatures (10–18° C.) for 2 or 3 days before killing. For the salivary chromosomes of *Anopheles* Frizzi (1947) adds yeast and *vorticella* to spring water used for culture and corrects for salinity and pH. Methods for dissecting the salivary glands in mosquito larvae and pupae are given by Jensen (1955).

Material may be examined in sections, or more usually it is teased out in fixative and examined direct under a cover-glass as a 'squash' preparation or as a smear. Sections should be thick, 15μ. For teasing out tissues many authors use aceto-carmine;* or the tissues are dissected in Ringer's solution and transferred to aceto-carmine, or aceto-carmine is run under the cover-glass (La Cour, 1931; Baker, 1945). Aceto-carmine swells the tissue and

* For aceto-carmine Darlington and La Cour (1942) give the following: saturate boiling 45 per cent glacial acetic acid with excess of carmine, cool and filter. Dr Tate, however, informs me that special methods are required and that the process is difficult to apply. A liquid stain, carmine acetic, is listed by Gurr in his price list C, 1954.

shows the chromosomes larger than in sections (Stevens, 1911). Or tissues may be smeared and fixed (wet) in Flemming's solution. Sutton notes that with the standard treatment (aceto-carmine or aceto-orcein) placing in a drop of stain and flattening by pressure on the cover-glass is ineffective with mosquito chromosomes as these are very soft and tend to be squashed rather than spread out. Tissues in this case are therefore best dissected out and fixed for one minute in acetic alcohol, then stained in acetic orcein (1 per cent orcein in 45 per cent acetic) for about an hour. The tissue is then smeared in the usual way. For sections, fixation by some form of Flemming's solution has usually been used (for formulae see Baker, 1945). For staining, good results are obtained with Heidenhain's haematoxylin, thionin (Stevens, 1911) and safranin (Bolles Lee, 1950).

The chromosomes are only seen as usually figured in cells undergoing division, or in nuclear changes preceding this when actual division of the cell is delayed. Otherwise the ordinary nuclear appearance only is apparent. For the various details in the process see Darlington (1937, who gives a good description); also for changes in the mosquito Whiting (1917, spermatogenesis and somatic division), also Hance (1917, somatic division) and the very full studies in different forms in the earlier literature (Stevens, 1910, 1911; Taylor, 1914, 1917; Carter, 1918; Holt, 1917, who first described multiple complexes during metamorphosis of the ileum as referred to later).

Kitzmiller (1953) notes some thirty species of Culicidae for which the number of chromosomes has been given (*Corethra*, spp. 2; *Anopheles*, spp. 6; *Culex*, spp. 10; *Theobaldia* (*Culiseta*), spp. 2; *Aëdes*, spp. 8; *Armigeres*, sp. 1). Only a few observations relate to *Aëdes aegypti* (Carter, 1918; Sutton, 1942). With some doubtful exceptions the number of chromosomes in all cases has been a diploid number of six, that is three pairs each consisting of two closely approximated chromosomes (chromatids of some authors). This is the number in sexual or tissue cells where reduction (meiosis) has not taken place. For *A. aegypti* Carter (1918) gives the diploid number as four, but Sutton (1942) gives six which is the number now generally accepted. A haploid number (following reduction in meiosis) of three has been described in two species of *Culex*, in *Theobaldia incidens* and in *Anopheles punctipennis*.

In *Culex pipiens* the chromosomes in late prophase appear as three pairs of bent (boomerang, U-, or V-shaped) rod-like structures. These are thicker towards their free ends and narrowest medianly (position of the centromere). Of the pairs, two are of approximately equal length (autosomes), whilst the third pair is composed of shorter rods (Fig. 85 (1), (2)). No indication in *Culex* is given of any separate sex chromosomes. In *Anopheles*, however, Stevens (1911) describes the third shorter pair as consisting of two equal longer portions and two short portions. These short portions in the female are of equal length, but in the male one is longer than the other (hetero-chromosomes) (Fig. 85 (4), (5)).

In *mitosis* the chromosomes in early prophase are seen as fine coiled contorted threads, as a rule evenly distributed throughout the nucleus with a plasmosome also present. The thin coils seen in optical section give the appearance at this stage of granules. The chromosomes in later prophase appear as definite thread-like bodies which shorten and thicken. Later the pairs become arranged equatorially with the centromeres attached to the spindle fibres, which now appear, and with the ends still free (early metaphase). Division and drawing apart of the divided chromosomes to form two polar groups (late metaphase) and final formation of two groups of chromosomes (telophase) complete the process. During prophase the plasmosome, originally present, disappears and in metaphase the nuclear membrane undergoes dissolution until in telophase it is reformed around the two separate chromosome clusters to complete eventually the nuclei of two new cells.

Meiosis has been chiefly studied in spermatogenesis as seen in the testis. In the region of the spermatogonia division is by mitosis. Meiosis with reduction takes place by two successive divisions, each division giving rise to characteristic layers of cells as described in the testis (ch. XXX (*j*)), the final result being the spermatids or developing spermatozoa. Nuclear reduction changes in the ovary, *oogenesis*, is stated by Kitzmiller to have been described only in very few observations.

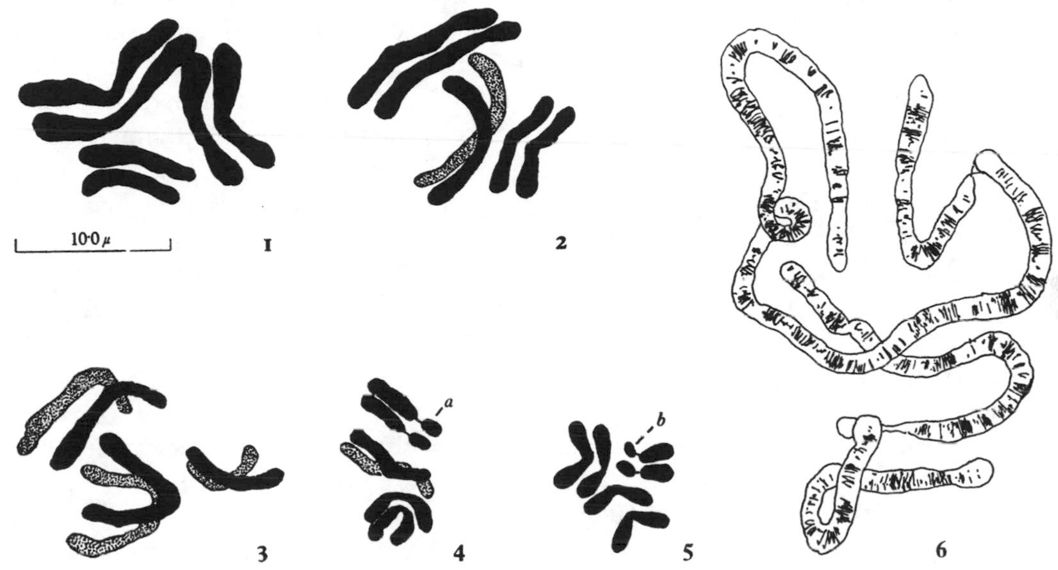

Figure 86. Chromosome structure.

1 Chromosomes of female *Culex pipiens*. Metaphase from oogonia. After Stevens (1911). The magnification given is calculated on that given by the author as ×2000.

2 The same of male *Culex pipiens*. Metaphase from spermatogonia. After Stevens (1911).

3 The same for male *Theobaldia*. After Stevens (1911).

4 The same for female *Anopheles punctipennis*. After Stevens (1911). *a*, equal pair of heterochromosomes.

5 The same for male *Anopheles punctipennis*. After Stevens (1911). *b*, unequal pair of heterochromosomes.

6 Salivary chromosomes of *Anopheles aquasalis*. After Frizzi (1953). Drawn from photograph. The striations are only drawn to give the general effect. Magnification is very approximately that of the other figures.

During meiosis there may be 'crossing over' (chiasmata), the coiled dividing chromosomes becoming adherent at certain points with the formation of characteristic figures. The formation of chiasmata is described in *Culex pipiens* by Moffett (1936) and by Callan and Montalenti (1947). It was not found by the last-mentioned authors in *Theobaldia longiareolata*.

The chromosomes as seen in the testes (spermatogenesis) show mitotic division in the spermatogonal region and meiosis in more distal portions of the organ. In cells of the body tissues division is by mitosis. In the larva and pupa division figures occur in the leg and wing buds, in the hypodermal tissue, in nerve ganglia, in the large nuclei of the Malpighian tubule cells, in the epithelium of the alimentary tract and as the large band forms, described below, in the salivary gland cells.

A peculiar condition is that of the multiple complexes referred to above in the ileum during metamorphosis (Holt, 1917; Berger, 1936, 1938 *a, b*; Grell, 1946 *a, b*; Schuh, 1951). Berger describes multiple complexes of 96, 48, 24 and 12 chromosomes, as also the usual diploid number of 6, present in the epithelial cells of the pupal ileum during metamorphosis. Up to the eighth hour of pupal life the cells of the ileum increase only in size, not number. Multiplication as above then takes place. Schuh notes that in normal development of the larva the epithelium of the ileum becomes highly polyploid, while that of the colon remains diploid. At metamorphosis the colon epithelium degenerates and is replaced by cells from the ileum. These ileac cells divide rapidly to form small cells. Treatment of the larva with colchicine does not prevent degeneration of the colon cells, but does prevent their replacement.

Of special interest are the large (giant) banded chromosomes in the salivary glands of the larva of some Diptera and first described in the larva of *Anopheles maculipennis* by Bogojawlensky (1934). They are described in *Culex pipiens* and *Aëdes aegypti* by Sutton (1942); in *Culex pipiens* by Berger (1936); in different species of *Anopheles* by Frizzi (1947, 1953). These remarkable structures appear as long coiled ribbon-like threads crossed by dark-stained bands and giving the appearance of spectra (Fig. 85 (6)). As noted by Kitzmiller these band-like forms may reach as much as 250μ in length as compared with up to some 6μ of the usual form of chromosome. Maps of the banding in *Culex pipiens* and *Culex fatigans* have been given by Kitzmiller and Clarke (1952), and that in *Anopheles atroparvus* by Frizzi (reproduced in Kitzmiller, 1953).

Sutton notes that large (giant) chromosomes resembling those of the salivary gland cells, but lacking the alternate bands and achromatic regions, are also to be found in the later stages of development in other tissues such as the Malpighian tubule cells and in the gastric caeca and mid-gut of the prepupal stage.

(*b*) HORMONES

Many processes in the development and life activities of insects are initiated or carried through by the action of hormones. These are chemical bodies secreted by special cells or organs which pass into the haemolymph and so reach the tissues on which they act. They are known to be concerned in growth and metamorphosis, in control of ovarian development and in many other vital processes. Research in this field involves delicate manipulations and dissections such as the implantation of insect parts and the linking up during life of parts of one insect with another (parabiosis). For obvious reasons most of the work has been carried out on relatively large insects. But though as yet little has been done in this connection directly upon the mosquito, hormones must here play an important part in the life processes and some brief reference to the results achieved would appear to be desirable. For a very interesting account of recent work in this direction see Wigglesworth (1951 *a*), also the recently published book by this author entitled *The Physiology of Insect Metamorphosis* (1954). A review of the subject is also given by Bodenstein (1953).

GROWTH, MOULTING AND METAMORPHOSIS

Working with the bug *Rhodnius* Wigglesworth (1934) demonstrated in a remarkable series of experiments that two hormones were concerned in moulting, one bringing about growth, moulting and imaginal differentiation, the *growth hormone*, and one that, acting with the growth hormone, maintained nymphal characters, the *juvenile* hormone. There was no hormone special to moulting other than the growth hormone which included moulting as a part of growth. Early experiments seemed to show that the growth hormone was secreted by the brain, since decapitation at a certain critical time following a blood meal in *Rhodnius* prevented moulting, whilst joining on the head of a last-stage nymph caused moulting to take place with imaginal characters. It had been found by Fukuda (1940), however, that in the silkworm pupal moulting was initiated from some source in the thorax which this author identified with the prothoracic gland. This was found also to be so in *Rhodnius*. For the prothoracic gland to act, however, it requires to be activated from the brain. Work on a large number of insects in different orders has amply confirmed the general existence of hormones acting much as in *Rhodnius* and has brought forward a wealth of information regarding the nature and mechanism of growth and metamorphosis which cannot here be dealt with, but which will be found graphically and clearly described in the paper and book by Wigglesworth previously referred to. By implantation and parabiosis giant and minute forms of a number of insect species have been produced through causing extra instar moults or precociously developed imaginal characters. Very briefly it would appear that the body cells, especially those of the cuticle, have latent possibilities of expressing both nymphal and imaginal characters, depending upon the hormones acting upon them. So long as the growth hormone is controlled by the juvenile hormone nymphal or larval characters are maintained by the latter. The growth hormone acting alone causes the latent imaginal characters to appear, even precociously under experimental conditions.

Three sources of hormones are concerned with growth, moulting and metamorphosis, namely (1) the neurosecretory cells of the brain secreting a hormone that activates the prothoracic gland; (2) the prothoracic gland or structures homologous with or representing this which secrete the growth hormone; (3) the *corpora allata* which secrete the juvenile hormone. There is a further structure that should receive mention, namely (4) the *corpus cardiacum* the precise function of which is uncertain (Wigglesworth, 1954).

The neurosecretory cells are large conspicuous cells situated in the *pars cerebralis* bordering upon the median fissure. They were first described in the brain of the honey-bee by Weyer (1928) and later in a similar situation in many insect orders (Hanstrom, 1938; Day, 1940; Thomsen, 1952, and others). The presence of granules gives these cells a characteristic appearance under dark-ground illumination. The granules are fuchsinophilous and stain deep blue with Gomori's stain (chrome-haematoxylin following oxidation with permanganate, Gomori, 1941). Neurosecretory cells giving this characteristic staining with Gomori's stain have recently been described by Clements (1956) in *Culex pipiens*. About ten are present in the *pars cerebralis* and others are present in different parts of the supra-oesophageal and sub-oesophageal ganglia and in the abdominal ganglia of the pupa and adult. Large cells that may be of this character have been figured in the brain of *Aëdes aegypti* in the chapter dealing with this organ.

The prothoracic glands (see review by Williams, 1949) were described by Lyonet in the caterpillar of the goat moth as early as 1762, being termed by him 'granulated vessels'. They were later described in the silkworm by Verson and Bisson (1891) as the hypostigmatic gland and since by other authors in a number of species of Lepidoptera. What appear to be homologous organs have also been recorded in hemimetabolous insects in a number of other orders. In their typical form the glands are situated one on each side in close connection with the tracheae posterior to the prothoracic spiracle and extending along the tracheal trunks as diffuse bead-like strings of large cells. In *Rhodnius* the prothoracic glands are described by Wigglesworth (1951 *b*) as two strands of large cells with enormous lobulated nuclei deeply embedded in the fat-body and extending through the prothorax and mesothorax. They break down and disappear in the adult within 48 hours of moulting. In the higher Diptera what appear to be homologous with the prothoracic glands are the large cells forming the lateral portions of the ring glands of Weismann (peritracheal glands of Possompes, 1953) which secrete the hormone leading to the formation of the puparium.

In the mosquito the large cells surrounding the *corpora allata* have been considered as homologous with the peritracheal cells of Possompes and so probably the source of the growth hormone. As noted by Bodenstein (1945) they undergo dissolution and are not present in the imago.

The *corpora allata* were first recognised as ductless glands by Nabert (1913) who recorded them in a number of insect orders. They usually occur as ganglion-like bodies one on each side in the retrocerebral complex. In the higher Diptera they are present as a single median structure in the dorsal part of the ring gland of Weismann. Their appearance in the mosquito has been described in a previous chapter. They continue to be present with much the same characters as before in the adult when the larger cells forming the peritracheal gland surrounding them have undergone change or dissolution.

The *corpora cardiaca*, also known as the oesophageal or pharyngeal ganglia, are normally small ganglion-like bodies on each side of the hypocerebral ganglion. They receive on each side two nerves from the neurosecretory cells in the brain, and a nerve is continued from them to the *corpus allatum* of their side. They appear to be of mixed nervous and glandular nature and contain cells showing granules that stain selectively with Gomori's stain. In the higher Diptera they appear as a median group of cells in the ventral portion of the ring gland. There has been much controversy as to their presence in the mosquito, and in *Culex* Cazal (1948) considered them to be fused with the hypocerebral ganglion and present as a median group of cells embedded in this ganglion. Recently Clements (1956) has recorded the presence of two groups of previously undescribed cells situated between the *corpora allata* and the tracheal commissure. Each group consists of three or four large cells with large nuclei and extensive cytoplasm containing granular material that stains with Gomori's chrome-haematoxylin. From these cells axons run forward in the cardiacal nerve under the tracheal commissure. The cells are present in the pupa and imago and in less developed form can be detected in the fourth instar larva. The author suggests that the cell groups are the *corpora cardiaca*.

OVARIAN DEVELOPMENT

The *corpora allata*, in addition to their action in metamorphosis, have been shown to be concerned in the control of ovarian development. If the ovaries of the adult female *Rhodnius* are removed, the *corpora allata* undergo hypertrophy as shown also by Thomsen (1952) in *Calliphora* and by Day (1943) in *Lucilia*. Detinova (1945) found that ligation of *Anopheles maculipennis* between the head and the thorax did not prevent development of the ovaries after a blood meal, but if done between the prothorax and mesothorax within 6 hours of a blood meal it did so, and he concluded that the *corpora allata* stimulated ovarian growth by secreting a gonadotrophic hormone after blood feeding. Wigglesworth (1948) in a footnote notes that it is probable that the yolk-forming hormone and the juvenile hormone are identical. Clements, working with the autogenous form of *Culex pipiens* (*C. molestus*), found that no specimens decapitated earlier than 7 hours after emergence developed their ovaries beyond the resting stage and that none ligated at the base of the abdomen earlier than 5 hours did so. But many decapitated and ligated after 7 and 5 hours respectively developed their follicles to maturity or to a lesser degree. In these experiments decapitation removed the neurosecretory cells and cut the cardiacal nerves and the recurrent nerve, whilst ligation isolated the ovaries from all endocrine organs outside the abdomen. There is thus evidence that the development of the ovaries in autogenous mosquitoes is stimulated by a gonadotrophic hormone secreted within a few hours of emergence presumably by the *corpora allata*. On decapitation of anautogenous females (*C. pipiens* form *berbericus*) following a blood meal, a proportion developed their ovaries, as was also found by Detinova in experiments with *Anopheles maculipennis*. A similar result was obtained with *Aëdes aegypti*, as also with *Anopheles labranchiae atroparvus*. *Anopheles stephensi*, however, ligated up to an hour or more following a blood meal did not develop the ovaries. It seemed possible therefore that in the ordinary anautogenous mosquito hormone is secreted before feeding takes place or that some other mechanism is involved.

Gillett (1956), studying development of the follicle in *Aëdes aegypti*, found that development beyond the resting stage (stage 2) depends on the taking of a blood meal. In mosquitoes decapitated before the critical period following this, development of the oocyte begins, but it stops short at or before stage 3, whereas in those decapitated after the critical period development goes on to maturity (stage 5). The critical period in two African strains was at 8–14 hours and in a Malayan strain possibly a little earlier. The author concludes that it is possible that initiation of ovarian development beyond stage 2 is under direct nervous control as a result of a stimulus provided by entry of blood into the stomach (or stretching of the stomach wall), though there was evidence that development in mosquitoes decapitated late in the precritical period goes a little further than in those decapitated early.

(*c*) RESERVES

FORMATION AND PURPOSE OF RESERVES

Except the yolk initially passed from the female in the egg, reserves in mosquitoes are derived from (*a*) food taken in during the larval stage; (*b*) such food as may be taken in by the imago; and (*c*) tissues or organs which in the course of metamorphosis are no longer required and undergo histolysis. In some mosquitoes nectar forms an important source of food of the imago and is capable of allowing for the normal activities of the species without a blood meal (Hocking, 1953). It is improbable that *Aëdes aegypti* females under natural conditions normally take any food other than blood, and in capitivity, if given water, females readily survive for seven days or more without food.

Reserves are used for the purpose of (*a*) maintaining life during periods in the life history when no food is taken; (*b*) providing material for the formation of new tissues and organs; and (*c*) providing energy for such activities as flight.

RESERVES IN METAMORPHOSIS

From the start the egg provides in the form of yolk a mass of reserve substances utilised in the building up of the pre-instar larva. This stage is completed in the first few days from oviposition. But the pre-instar larva may remain viable in the egg for periods which in some circumstances may extend to months. Furthermore at hatching the young larva has to undertake muscular movements and other activities for some hours before it is capable of taking food. As at hatching little or no yolk remains as such, life must be maintained during this period by reserves of some other kind, and, as already noted, the young first-stage larva already has in the fat-body fine granules of an albuminoid nature and some fat globules.

During the larval period when food is taken there is not only growth, but a steady storing up of reserves which reaches a maximum in the fourth instar larval stage and especially is very noticeable just before pupation when the larvae are notably plump and fat. During the pupal stage it is largely on this store of reserves, together with the products of histolysis of no longer required larval structures, that the body of the imago is built up. The large abdominal muscles of the larva are, however, largely retained by the pupa. At emergence of the imago there is still an important period to be passed through before the insect obtains its first blood meal and during which for up to some 2 days it does not feed. During this time important changes are going on. The abdominal muscles are now broken down and absorbed, the salivary glands develop secretion and the ovary and testes undergo changes. Eventually the female takes a blood meal, if available, usually on the third day. If such food is not available, the female imago can still live for a considerable period even actively using its wing muscles and making other expenditure of energy. The male with its more slender resources does not usually live so long. Both sexes, however, at this time have important functions to perform for continuation of the species, the female to develop its ovarian follicles to a suitable stage, the male to provide spermatozoa. It will be convenient to follow these later activities in some detail.

RESERVES AND OVARIAN DEVELOPMENT

Normally in most mosquitoes the ovaries in the absence of a blood meal develop only to a more or less fixed stage, namely with the follicles at stage 2 (Christophers, 1911). In *A. aegypti* a single blood meal is then sufficient to complete ovarian development and allow of oviposition. In certain cases, however, the female mosquito may require more than one blood meal to bring about this result.

Gillies (1955), dealing with the gonotrophic cycle in *Anopheles gambiae* and *A. funestus* as seen in specimens caught in nature, found a proportion of females in which multiple blood meals were required for completion of the first gonotrophic cycle. This proportion he termed the *pre-gravid rate*. Observation showed that this rate was considerable only when the population density of the species was very large or very small. The females showing the condition were found to be mainly newly emerged, the author concluding that a pre-gravid phase of development in which more than one blood meal was required occurred when the ovaries at the time of the first feed had not reached stage 2. In this connection one may recall the observations by Mer (1936) on *Anopheles elutus*, referred to in a previous chapter, in which the author found that development up to stage 2 can be achieved either at the expense of reserves carried over from the larval stage or by feeding on raisins. One blood meal then sufficed for oviposition. With poor larval food, however, development to stage 2 was impaired and a single blood meal did not then suffice. That the pre-gravid rate was high only when the anopheline population density was either very large or very small would seem to support the view that in the first case it was due to overcrowding and in the second to deterioration in the breeding conditions of the species.

A striking instance of the use of material obtained from histolysis of organs to ensure ovarian development and oviposition has been given by Hocking (1952). A species of *Aëdes* in Manitoba normally does not take a blood meal, but utilises the imaginal flight muscles to develop the eggs. The females are found in a flightless condition and the degree of breakdown of the indirect wing muscles is related to that of egg development. The author notes that the nitrogen content of the flight muscles in this case is roughly equivalent to that of about sixty eggs.

RESERVES AND MALE ACTIVITY

The life of the male is usually short, at most 3 or 4 days. But during that time it is not only extremely active in frequent flight, but copulates many times. In an interesting communication Gillett (1955) notes that males of a certain African strain of *A. aegypti* were able to copulate in quick succession with eight females, though only the first three or four were successfully fertilised. Males, however, apparently exhausted of sperms by three or four consecutive matings were again able to fertilise females after a rest of 36 hours. In copulations at lesser periods they failed to fertilise. The formation of new sperms would therefore appear to have been dependent either on reserves or on food taken, should raisins or sugar have been available.

RESERVES AND ENERGY

The relation of reserves to work, fuel consumption and thoracic temperature in flight has been intensively studied in insects by a number of observers (see Wigglesworth, 1942; Hocking, 1953; Sotavalta, 1952, 1954*a*, *b*; Clements, 1955; for other authors see list of references given by Hocking).

The methods of investigation include the use of some form of roundabout, which allows the insect to fly with minimum friction (see description and figure given by Hocking, 1953, p. 276; also Sotavalta, 1954*a*, p. 3; Clements, 1955, p. 547). Flying in the case of mosquitoes is continuous so long as the tarsi are not in contact with any surface.

Reserves in the fat-body consist chiefly of fat, glycogen and protein (Wigglesworth, 1942). In Diptera, however, authors are in agreement that the material used as a source of energy is chiefly, if not entirely, carbohydrates (glycogen and sugar). Thus Clements found no difference in the fat-body fat in flight-exhausted mosquitoes, though glycogen had disappeared. According to Hocking (1953) the exact chemical nature of energy reserves is not required in estimating energy equivalents, as the energy values of the main food materials are essentially similar. The following data are given for four species of *Aëdes* exhausted by flying.

Species	Original weight (mg.)	Exhausted weight (mg.)	Weight lost (mg.)	Equivalent glycogen (mg.)	Equivalent glucose (mg.)	Water loss (mg.)
Aëdes campestris	6·72	5·22	1·50	0·24	0·269	1·27
A. communis	3·11	2·42	0·69	0·11	0·124	0·57
A. nearcticus	3·15	2·45	0·70	0·11	0·124	0·57
A. punctor	5·77	4·49	1·28	0·20	0·225	1·06

The columns given as equivalents of glycogen and of glucose are the weight of tissue reserves expressed as these substances calculated from the duration of flight. The water loss is calculated from the author's figures for water loss per mg. of glycogen, taking the given figure of 5·25 mg. per mg. of glycogen.

Clements, using the flight-mill arm figured by Hocking, gives the following data regarding flight and energy requirements in *Culex pipiens* form *berbericus*. The figures give the mean number of metres flown when flown to exhaustion on successive days.

	Distance flown (in metres)					
Condition in which flown	Day 1	Day 2	Day 3	Day 4	Day 5	Total
Unfed 72–96 hours old, first flight	1292	455	314	242	129	2422
Fed on blood, as above	1564	1809	1055	410	225	5063

The estimated reserves of an unfed female aged 48–72 hours were approximately 0·12 mg. glucose. The excess distance of flight following a blood meal was 2641 metres equivalent to 0·061 mg. glucose as used by an exhausted mosquito fed on glucose. A meal of blood contained only 0·003 mg. glucose so that the mosquitoes were able to utilise products of blood digestion to produce energy reserves for flight.

Hibernating specimens of *Culex pipiens* form *pipiens* taken from a cellar in December and raised to room temperature when flown to exhaustion gave a mean distance flown of 4300 metres (max. 8674; min. 2239) with speeds up to about 1·0 metres per second, equivalent to 0·099 mg. glucose as used by *Culex pipiens* form *berbericus*.

Heat production in insects during flight has been studied by Sotavalta (1954*b*). This author, using a delicate form of thermostat placed in the thorax of certain insects flown on a roundabout, found rises of temperature varying between 3° and 16° C. above that of the outside temperature (in large Diptera 2·5° to 8° C.). It was not found possible to make determinations in small insects such as mosquitoes.

(*d*) THE INTEGUMENT

Physiologically the integument plays a very important and varied role in the life of an insect. Besides being a covering and a skeleton affording attachments for muscles and other structures, it is from the integument that are mainly derived structures controlling respiration, that the organs of tactile and other sensory impressions are largely formed, and it is the integument that mostly determines the very nature of the insect in its different stages of metamorphosis. It is here, however, being considered in the limited sense of a cutaneous covering and from a general point of view. For a description of the insect integument see Wigglesworth (1933, 1947, 1948, 1957) and Richards (1951, 1953). Only such brief account can here be given as will serve to be explanatory of such references as are made to *Aëdes aegypti*.

The insect integument consists of a layer of epithelial cells with basement membrane, the *epidermis* (hypodermis of many authors), and external to this a non-cellular secreted layer, the *cuticle*. The cuticle when normally developed further consists of three layers: (1) the *endocuticle*; (2) the *exocuticle*; and (3) the *epicuticle*. The endocuticle and exocuticle are both chitinous, consisting of a chitin-protein complex, and together constitute the *procuticle*. The epicuticle is non-chitinous.

The endocuticle forms the inner and main thickness of the cuticle. It is usually, as seen in sections, light-coloured and somewhat translucent. It is tough and more or less elastic. It is penetrated by excessively fine canals, *pore canals*, that are shown up by allowing frozen sections to become dry, so that the canals become filled with air, and then mounted. The pore canals pass vertically through its substance and contain fine cytoplasmic extensions from the epidermal cells. The exocuticle is that portion of the procuticle which becomes impregnated with hardening material, 'sclerotin', and forms the sclerites and other hard parts of the integument. The epicuticle is a very thin surface layer. In those insects where it has been studied it consists of a number of layers. These consist of: a thin 'cuticulin' layer, the first to be secreted by the epidermal cells; a 'lipid' or 'wax' layer, also secreted by the cells, which gives water-proofing to the cuticle; and an external 'cement' layer poured out subsequent to moulting as a fluid secretion by the dermal glands and which later hardens (Wigglesworth, 1947).

At ecdysis when the new cuticle is beginning to be formed by the laying down of a sclerotin layer a break occurs between this and the endocuticle of the old cuticle. This break is filled with fluid, *ecdysial fluid*, at first thought to be secreted by the dermal glands, but now known to be secreted by the epidermal cells by way of the pore canals. This fluid contains enzymes which gradually dissolve the old endocuticle, so that ultimately little is left of the old cuticle except the thin epicuticle. It follows that the cast skin, for example at pupation, neither in thickness nor in weight represents that of the whole larval cuticle. Meanwhile the new cuticle is being built up within, but separated from, the old. These changes are more or less common to all insects and apply in general to *A. aegypti*.

Something too should be said regarding hardening of the cuticle following upon moulting. Very commonly the new cuticle when first exposed is pale or white or even transparent as contrasted with the cast old cuticle. Thus in the larva at ecdysis the head of the newly emerged stage is almost transparent, but later becomes dark or almost black. As well as darkening in colour the new cuticle also hardens. This hardening, as first pointed out by Pryor (1940), is now recognised to be a form of tanning. The process as seen in the puparium of *Sarcophaga* is summarised as follows by Richards (1953). Sclerotisation begins at the epicuticle-procuticle interface and spreads inwards giving rise to a mature cuticle composed of exocuticle and endocuticle. The substrate for the tanning reaction, that is tyrosine or products of this substance, diffuses from the blood and becomes oxidised by oxidase located in the epicuticle to form O-quinone, which diffuses inwards transforming the chitin-protein of the outer part of the procuticle into sclerotin and so forming an exocuticle.

Something also should be said about the surface structures borne by the cuticle. The cuticle itself may bear markings, such as polygonal markings, indicating the cell outlines or corrugations. It usually also carries processes in the form of microtrichae, hairs or setae, spines and scales. Microtrichae are simple small spine-like elevations of the epicuticle.

Hairs and cuticular spines are secreted around cytoplasmic extensions from special cells in the epithelial layer, *trichogen cells*, which thrust themselves through a *tormogen cell* from which a socket for the hair is formed. Scales are derived from much enlarged epithelial cells which form club-shaped projections that later flatten. Situated in the epithelial layer are the dermal glands. These consist of large cells and have ducts passing through the cuticle to the surface. Also associated with the integument are the oenocytes (see ch. XIV (*f*), p. 337). Dermal glands are said not to be present in some insects including Diptera and they do not seem to have been described in the mosquito. I have not succeeded in detecting them in the newly formed cuticle of the prepupa or of the imago prior to ecdysis from the pupa, but they may be difficult to detect and the above requires confirmation. Oenocytes, too, in *A. aegypti*, though they may be concerned in formation of the cuticle by providing material for this, are well separated from the integument and are more suitably considered independently.

In *A. aegypti* the cuticle presents different characters in the larva, pupa and imago. It is best studied as developed in the imago, or as being formed prior to ecdysis, especially in the fourth instar pre-ecdysis larva or in the forming imaginal cuticle in the pupa.

In the larva of *A. aegypti* the cuticle is peculiar, even among mosquitoes, in being almost entirely free from sclerotised areas, other than the head capsule, respiratory siphon, anal plate and some areas about, and at the base of, some hairs and spines, notably the pleural hairs. Even the small oval tergites seen in the larvae of *Anopheles* and other mosquitoes are completely absent. Also the cuticle is very thin and the cast skins, even of the fourth instar, are, apart from hairs and the parts mentioned above, almost transparent when mounted. Its thickness is given by Richards and Anderson (1942) as between 0·75μ and 2·0μ. Though very thin, the larval cuticle is nevertheless extremely impermeable to watery fluids as is clearly shown in attempting to mount larvae in say, glycerine, without making some small opening. As shown by Wigglesworth (1933), however, the cuticle of the anal papillae is readily permeable and plays an important part in the osmotic processes connected with the haemolymph.

Though the larval cuticle is so thin and free from marked thickening it still shows areas of cuticle covering the segments as distinct from the thinner and more folded intersegmental areas.

Very different is the cuticle of the pupa where the whole body is encased in hard sclerotised and dark-coloured armour-like cuticle. In this case, namely that of an active obtectate pupa, it is, however, only the exposed portions of the cuticle that are so hardened. Much unexposed cuticle covering the cemented limbs and other parts of the pre-imago are soft and inconspicuous. As described in an earlier chapter such parts of the cuticle are freely exposed at pupal ecdysis through removal of cementing fluid by fixatives, so that the early pupa when fixed no longer retains its shape but shows the wings and other parts much as in the resulting imago. A further point also previously mentioned is the peculiar outer coating of some substance soluble in chloroform which, left intact, gives to the pupal cuticle very resistant properties. Certain colour changes too that occur in the cuticle of the pupa have also been described.

The cuticle of the imago is fully sclerotised, forming the usual exoskeleton of articulated plates and intersegmental membranes. It is formed very early within the pupal skin so that some time before emergence the dark scaling of the imago is already very apparent. Hence on emergence the imago unlike the pupa is already dark-coloured and changes but little in this respect later. The sclerotised areas bear microtrichae, socketed hairs and scales. The microtrichae are small, thin and hair-like, measuring only some 5–10 μ in length. Both hairs and scales have basal sockets (Fig. 50 (7)). The membranous areas carry only microtrichae. These are, however, stouter and larger than are present on the smooth sclerite areas. A very curious feature of the imago is the character of the pleural membrane. As already noted this does not apparently stretch following the great expansion of the abdomen from a blood meal, but is very voluminous and in the unexpanded abdomen is tucked in folds under the edge of the tergites. It is not a simple membrane, but consists of a thin basal membrane on which are closely and regularly arranged small oval or circular plaques each carrying up to five or six long curved thorn-like microtrichae (Fig. 50 (6), A, B, C). The condition does not extend to the intersegmental membrane and is confined to the pleura. The neck membrane is also closely covered with microtrichae which in this case are rather thick and peg-like, but single and arising directly from the membrane (Fig. 50 (6), D). Other membranes carry somewhat similar microtrichae though usually thinner and more sparsely distributed.

(*e*) RESPIRATION

Provision for respiration in the egg of insects is often quite complex with an elaborate arrangement of air channels or spaces. It is possible that the silvery network so conspicuous on the eggs of some anophelines and which have been here described in *A. aegypti* may have such a function. These air spaces are, however, external to the thick endochorion and whether they are concerned with respiration of the egg or not is as yet unknown.

In all other stages arrangements for respiration are through the tracheal system. Respiration is, however, not the only function in which the tracheal system plays a part. In the larva it is largely responsible for maintaining the specific gravity and necessary attitude to enable the larva to attach its siphon to the surface. In the imago it almost certainly helps flight by increasing buoyancy. In the mosquito, however, there is no development of the large air sacs so conspicuous in many insects.

The required oxygen is obtained almost entirely through this system.* Though the larva leads an aquatic life it does not, like some aquatic larvae, possess gills, the anal papillae as shown by Wigglesworth (1938) being concerned almost entirely with osmotic regulation of the haemolymph. It would seem too that, except in the first instar, very little respiration through the integument takes place. With a view to ascertaining to what extent larvae can exist on cutaneous respiration a number of authors have attempted to ascertain the time that larvae can remain alive when submerged, but without very clear results (Da Costa Lima, 1914; Sen, 1915; Macfie, 1917; Koch, 1920). Macfie calculated that a full-grown larva of *A. aegypti* in 8 c.c. of water shaken with air should, if it could utilise the dissolved gas, live for 42 hours, whereas it lived only 12 hours in such water and in boiled water only 4 hours. In the first instar, however, the larva can remain alive for several days without air ever entering its tracheae.

The oxygen required is obtained in the main through the spiracles. Control of air entering the spiracles in the larva is restricted to closure when submerging and establishing air contact when at the surface. Such air contact takes place automatically through the

* For the growth and regeneration of the tracheal system and the method by which tissues receive their tracheal supply see Wigglesworth (1954).

anatomical characters of the spiracular parts and the physical properties of the surface film of water. When these properties are altered, for example in oil, oil enters the tracheae. Much the same occurs with the respiratory trumpets of the pupa, these being fully opened on contact with the surface by surface tension. In the imago there are two large thoracic spiracles and a number of smaller abdominal spiracles in connection with all of which there are muscles capable of controlling the opening. Why this should be necessary is probably more to conserve water than to regulate the supply of oxygen.

In order to reach the tissues it is necessary that oxygen should reach the tracheoles. This it can do only by diffusion. The rate of diffusion is more rapid and effective than might at first be thought. Krogh (1941) estimated the tracheal volume of the *Culex* larva to be 1·5 $mm.^3$ and that enough O_2 diffused through the spiracles to supply the requirements of the insect's tissues. In some insects mechanical ventilation may facilitate diffusion. There may be respiratory movements of the abdomen or rhythmical closing and opening of the spiracles. No such movements appear to take place in either the larva or imago of the mosquito. Babak (1912), however, has described rhythmical pulsations of the tracheal trunks in the larva.

That the necessary amount of oxygen should diffuse to the tracheoles is essential since it is only by this channel that it can be conveyed. The haemolymph contains no haemoglobin or other effective means of doing so. There is, however, in the larva the curious 'tracheal lung' as noted by Imms (1907) and here described, where there would appear to be definite association of a mass of tracheoles with the haemolymph the exact function of which is not very clear. Further, the supply must be adequate to the requirements of the tissues. A very noticeable fact is the much greater volume of the tracheae in the imago, other than the large abdominal trunks, than in the larva, where the muscular system is relatively small in volume and less highly developed than in the imago with its massive wing muscles.

A special study of the physiology of the changes in the tracheoles largely dealing with the mechanism of respiration in the larva of *A. aegypti* has been made by Wigglesworth in a number of papers. Only very brief reference can here be made. Especially important is the permeability of the tracheole wall allowing absorption of fluid to take place. When first hatched the tracheal system of the first instar larva is filled with fluid and remains so until contact is made with air at the surface. When such contact is made, air quickly enters as a result of the fluid being absorbed. The same occurs at ecdysis, the new larger tracheae, which form around the old air-filled tracheae just before ecdysis, being filled with liquid (moulting fluid). In the first instar larva which has been kept from contacting the surface the absorption becomes less and less rapid until by the fourth day the tracheae fail to fill with air. A similar decrease in the absorption of fluid and entry of air replacing this takes place at each ecdysis if larvae are kept from contacting air.

Exchange of O_2 and CO_2 takes place in the tracheoles where these terminate in close connection with the tissue cells. Though the walls of the tracheae are impermeable to fluid, those of the tracheoles are permeable and towards their termination contain fluid permeated from the tissues. The extent to which fluid is present is dependent upon osmotic conditions in the tissues. Muscular activity increases the osmotic power of the tissues and fluid is withdrawn from the tracheoles and air extends further towards their terminations. By observing the extent of fluid in the tracheoles under various conditions, especially in the starved larva, much connected with respiration can be studied. Thus with the aid of an appropriate apparatus Wigglesworth (1930) studied the effect on the larva of various gases and of oil.

An important aspect of respiration is the nature of the respiratory exchange, that is the extent to which oxygen is taken up, *oxygen uptake*, as also the relation of this to CO_2 passed out, *respiratory quotient*, and other features. Determinations of oxygen uptake in the mosquito have been made by Sen. This author found O_2 uptake for the full-grown larva of *Culex microannulatus* (a rather small mosquito) to be 1·1 mm.³ per larva per hour, or, taking the weight of such a larva to be about 3 mg., an uptake of 370 mm.³ per gram per hour. For the pupa the uptake was 536 mm.³ per gram per hour for *C. microannulatus* and 488 for *Aëdes scutellaris*. For the imago of *Culex microannulatus* the rate was 26 mm.³ per mosquito per hour, or about 8000 mm.³ per gram per hour. The amount of CO_2 given off was about 25 mm.³ per mosquito, making the respiratory quotient about 1·0. Judging from the author's description the mosquitoes were able to move about freely.

Table 50. *Oxygen uptake for the larva, pupa and imago of* Aëdes aegypti

Date	Temperature (° C.)	No. used	Weight (mg.)	Time in minutes	O² (mm.³)	O² (mm.³) per gram per hour	Mean uptake (mm.³) per gram per hour
			Larva				
25. vii. 47	25·5	2	10·1	18	3	990	1079
25. vii. 47	25·5	2	—	24	4	—	
25. vii. 47	26·2	4	—	26	8	—	
11. viii. 47	27·9	10	40·0	15	12	1200	
12. viii. 47	27·6	10	45·8	20	16	1048	
			Pupa				
12. viii. 47	28·0	8	31·7	50	24	902	819
2. ix. 47	27·5	3	14·8	94	17	736	
2. ix. 47	27·4	4	—	60	16	—	
3. ix. 47*	27·6	2	—	35	4	—	
3. ix. 47†	27·6	1	—	45	10	—	
			Imago (female)				
5. ix. 47	28·0	2	4·86	105	40	4713	4104
5. ix. 47	28·0	2	6·38	70	26	3495	

* Dark pupa approaching emergence. † Pupa during emergence.

Some observations by the present author for O_2 uptake by *Aëdes aegypti* gave mean values for the full-grown female larva of 1079 mm.³ per gram per hour; for the pupa 819 mm.³ per gram per hour; and for the female imago 4104 mm.³ per gram per hour (Table 50). A pupa in the act of emergence gave an uptake of approximately 3000 mm.³ or about the same as for the adult. The adult mosquitoes were confined in a small space and though able to fly rarely did so and on being liberated were more or less feeble.

Recently Hocking (1953) has given data based on the utilisation of glucose and glycogen. On pp. 287–8 this author gives, along with data for a number of other insects, the metabolic rate at rest and when flying for three species of *Aëdes*, including *A. nearcticus* which was found the most suitable for experiment. The O_2 uptake for resting metabolism in this species was 5·57 cc. per gram per hour, or 5570 mm.³ per gram per hour. For the rate flying at cruising speed the value was 20·6 cc. or 20,600 mm.³ per gram per hour; at bursts of speed it was 118 cc. per gram per hour. The data show a much higher intake by insect wing muscle than by mammalian muscle. Calculated on the weight of the indirect wing muscles oxygen consumption for the bee works out as about forty times that of human muscle. Some useful data are given in table form by the above author on p. 326 including the following taken from Carpenter (1939): 1 gram glucose requires for oxidation 726·2 cc. O_2 at standard temperature and pressure; 1 gram of glycogen 829·3 cc. Other useful data in connection with work on respiration are: 1 mg. O_2 at N.T.P. occupies 699·8 mm.^{3}; 1 mg. CO_2 at N.T.P. occupies 505·9 mm.^{3}; 1 cc. of water at 20° C. and one atmosphere pressure absorbs 0·031 cc. O_2 (Childs, 1934). (See Addenda, p. 719.)

(*f*) DIGESTION

MECHANISM OF DIGESTION IN THE LARVA

In the larva food particles filtered off by the pharyngeal fringes gradually collect at the back of the pharynx to form a bolus which is swallowed at intervals. According to Shipitzima (1930, 1935) the larva of *Anopheles maculipennis* is capable of separating colloidal particles as small as 20 $\mu\mu$ in diameter, but not those of soluble starch which are 5 $\mu\mu$ in diameter or those of haemoglobin which measure 2–4 $\mu\mu$ in diameter. Feeding in this way considerable quantities of water are filtered free from suspended matter as first pointed out by Boyce and Lewis (1910). For *Anopheles maculipennis* Senior White (1928) estimated that as much as 1000 mm.^{3} (1 cc.) could be so filtered in a day. Shipitzima, for the fourth instar of this species at 17–21° C., gives only 100 mm.^{3} The present author found that nine half-grown fourth instar larvae of *Aëdes aegypti* at 26° C. in 24 hours cleared 60 ml. of culture medium as here recommended for culture diluted so that distinct dark type could just not be read at 5 cm. depth. The fluid was still very faintly turbid, but had ceased to yield food as was shown by the behaviour of the larvae which had ceased to feed normally and were rapidly browsing over the bottom. This gave a filtering of some 6·5 ml. per larva in 24 hours.

The food when swallowed passes directly through the oesophagus and proventriculus to the mid-gut, where it becomes compacted to form a column of food material which extends as a long cylindrical mass through the greater part of the abdomen to the pylorus surrounded by a tubular sheath of peritrophic membrane. The food column with its peritrophic sheath lies free in the lumen of the mid-gut with a considerable space between it and the gut wall. In the living larva rhythmical peristaltic contractions of the gut wall directed forwards and driving fluid into the gastric caeca, as described by Wigglesworth (1949), show up very clearly the distinction between the wall of the mid-gut and the motionless tube of peritrophic membrane, as also the considerable space between these structures.

The food material is eventually passed out as the characteristic larval faeces, that is as cylindrical masses resembling portions of the food column, the length of which probably depends in part at least upon the nature of the food. In larvae parasitised by *Penicillium*, as noted in the section on parasites, this character of the faeces is exaggerated, long lengths of cylindrical faeces remaining still attached at the anus. The material forming the food column, as also the faeces, is granular and amorphous with included larger particles of miscellaneous nature and usually with transparent crystalline grains of silica.

The time taken for passage of food through the alimentary canal of the larva as shown by ingested particles of carmine is given by Purdy (1920) for larvae of *Anopheles* collected from ricefields as from 31–45 minutes. Brück (1931), in *Anopheles maculipennis*, gives 40 minutes as the time it took food to pass through, but this depended on the rate that food was taken in. It was not affected by the size of particles or nature of the food. In some observations made with half-grown fourth instar larvae of *Aëdes aegypti* at 26° C. feeding normally in the usual culture medium to which some suspension of carmine had been added the time taken for the carmine to reach the pylorus was about 30 minutes.

Little is given in the literature regarding disposal of the food material and the following observations on the larvae of *A. aegypti* may be given. As well as by dissection and examination of sections, larvae suitably mounted under a cover-glass may be examined in the living state. The most satisfactory stage for this purpose is the third, or early fourth, instar when the parts are relatively transparent. Especially if arranged to give a lateral view of the abdomen movements of the food and concerned parts are well displayed. To aid such observations finely divided carmine may be added to the culture medium. It is also worth noting that, in contradistinction to the method used for drawing out the parts in the adult, the whole alimentary canal in the larva is readily drawn out intact from the thorax, provided only that the rectum is divided. The following observations may be made on the food column.

Anteriorly the column starts abruptly immediately behind and in contact with the invaginated end of the oesophagus into the proventriculus, where it receives steady accessions of swallowed material (Fig. 27 (1), (2)). Since the peritrophic membrane takes origin from the outer fold of the proventriculus it surrounds the food column from its beginning and so cuts off any access of food material to the gastric caeca, which normally remain apparently empty and more or less transparent in contrast to the opaque column of food material. In apparent contradiction to this is the condition seen when suspension of carmine is used. In this case within such a short time as half an hour the gastric caeca appear packed with opaque red material equally with the mid-gut. On crushing them the material is seen to consist of fine carmine granules about $1{\cdot}0\mu$ in diameter. Either in some way carmine particles are able to pass the peritrophic membrane and enter the caeca direct, or more probably carmine is removed in solution from the gut and after being passed forwards is taken up by the epithelium of the gastric caeca and reformed into granules. The remarkable fact is the short time in this latter case that it takes for the caeca to become so massively blocked with carmine. The condition has not been further explored.

At its posterior end the food column also terminates as a rule abruptly, in this case at the pylorus or a little beyond (Fig. 28 (1), (2)). The abruptness here is to some extent exaggerated by the abrupt termination of the thick-walled mid-gut followed by the very thin-walled pyloric ampulla. In the larva, too, there is no sharp constriction at the pylorus as in the adult and the mid-gut wall makes what almost amounts to a narrow ring-like fold. All this serves to give the impression of an abrupt cut-off effect to the food column. Usually the column extends a little beyond the pylorus and in the living larva it may be seen slowly extending more and more into the upper part of the ileum until with a sudden contraction of the parts a length of the food column is detached and rapidly passes along the ileum to be lodged in the colon. Here it remains, usually with two or three similar masses, for a period before, with the emptying of the colon, its contents are passed rapidly down the narrow rectum and discharged as faeces.

The above relates to the food material of the column. It still remains to follow up what happens to the peritrophic sheath. As depicted in figures this is usually shown terminating with the food column at the pylorus. Both dissections and the study of sections, however, show that the membrane is continued at least for some distance into the ileum. When drawn out in dissections it usually shows as a length of empty tube extending beyond the food material, becoming towards its end thinner and ending raggedly. In sections it is more difficult to follow than in the mid-gut, being folded and plicated in the narrow ileum. It appears, however, to be present at least as far as the colon and is sometimes to be made out in the upper part of this structure. In sections the thick-walled colon may be empty with a narrow lumen. At other times it contains food material and may be much distended by this. The rectum in sections is almost always empty and forms a narrow passage from the colon to the anus.

The faeces of the *A. aegypti* larva, as cultured in the laboratory, consist of globular or short cylindrical fragments resembling portions of the food column. Those from the fourth instar larva were about 30μ in diameter and up to 50μ in length. Contrary to what is stated in the literature they showed no trace of peritrophic membrane. Another unexpected feature was the extent to which they were composed of fine silica granules. Similar granules were found to be present in the materials used for culture, but these were clearly much concentrated in the faeces. This and some other facts noted seemed to suggest that digestion in the larva might be more effective than at first sight appeared. Examination of the faeces of larvae of *Anopheles atroparvus* and of *A. stephensi* very kindly sent to me by Mr Shute of the Reference Malaria Laboratory, Horton, showed these to be very similar to those of *Aëdes aegypti* larvae. In neither case was any peritrophic membrane present. The faecal masses in *A. atroparvus* were more or less globular or shortly cylindrical measuring up to about 40μ in diameter and 50μ in greatest length. Those of *A. stephensi* larvae were smaller, being up to 20μ in diameter, possibly from younger larvae.

Briefly, the conclusion was formed that the food column is slowly moved on by pressure of added material past the open pylorus, the extruded portions being detached at intervals by muscular contractions and rapidly passed along the ileum to the colon where such fragments were retained for a period before being discharged as faeces. Regarding the peritrophic sheath there seemed two possibilities, namely that the membrane was continually being formed and digested in the colon, or that it was a relatively static structure which extended to the colon where it ended.

A point that finally may be referred to is the presence of the pyloric spines that have been described in the adult by Trembley (1951), and whether these play any part in the breaking off of the food column at the pylorus. It seems improbable that this could be an important factor since, whilst the spines are large and conspicuous structures covering the inner aspect of the pyloric ampulla in the adult where they measure up to 10μ in length, only occasional transverse lines of exceedingly small and fine spines less than 1·0μ in length can be seen with difficulty in the larva.

DIGESTIVE PROCESSES IN THE LARVA

Very little is known of the essential digestive processes in the larva. In *A. albopictus* Senior White (1926) found the pH of the alimentary canal to be in the alkaline range (7·4–8·8 in the gastric caeca; 9–9·4 in the mid-gut; and 7·6–8·0 in the hind-gut). Richardson and Shepard (1930), in the larva of *Culex*, found the pH inside the peritrophic membrane to be 6·2–8·2, and outside the membrane 8·6–9·8. Other authors dealing with determination of the pH of small quantities of fluid, some of whom give data regarding the mosquito, are

Haas (1919); Felton (1921); Jameson and Atkins (1921); Crozier (1923–4); Pantin (1923); Brown (1923–4).

In the larva of *A. aegypti* Hinman (1933) found alkaline conditions in the gut and ascertained the presence of an amylase, an invertase and a protease. The protease acted in alkaline medium and, as previously found by Senior White, there was no protease acting in acid medium. According to Wigglesworth (1949) digestive secretions probably come from the salivary glands, the gastric caeca and the epithelium of the mid-gut, especially from its posterior portion. Anti-peristaltic movements drive fluid in the gut forward into the gastric caeca and bring about circulation of the digestive juices. Absorption of fat and of glycogen occurs in the gastric caeca as shown by deposits of these substances in this situation. Fat is also taken up by the cells of the anterior portion of the mid-gut, whilst a massive deposit of glycogen occurs in the epithelial cells of the posterior portion.

Little is given in the literature regarding functions of the hind-gut in the larva except in relation to osmotic regulation, more especially by the rectum. The ileum would seem to be chiefly concerned with the passing on of the food material and is very noticeable as a muscular organ and as exhibiting active peristalsis. The colon from its large size and the peculiar character of its cells would appear to have some important function or functions which may include that previously suggested in connection with the peritrophic membrane. The rectum in the larva would appear to play but little part in digestion. As there is some possibility of confusion in the nomenclature of these two last-mentioned portions of the canal it may be well to state that the parts as referred to here are as recognised by Thompson (1905) and most authors dealing with the anatomy of the alimentary canal in the larva.

DIGESTION IN THE ADULT*

The change from the larva to the imago involves profound changes in the structure and functions of the parts concerned with digestion. From dealing in the larva with matter largely bacterial in character digestion in the imago is concerned with a relatively enormous mass of high-class protein matter composing the blood meal. It is not surprising therefore that the mid-gut is one of the relatively few organs the cells of which are completely replaced in metamorphosis, the old epithelium being digested and utilised as reserves. In connection with this replacement a condition of peculiar multiple chromosomal divisions, apparently designed to bring about rapid multiplication of the gut cells, has been described by Holt (1917). In the imaginal gut the ingested blood, unlike food material in the larval gut, is not enclosed in well-marked peritrophic membrane, but has been shown to be surrounded by a very thin membranous sheet which appears to contain chitin (Yaguzhinskaya, 1940; Rajindar Pal, 1943). Changes may be brought about in the ingested blood by the salivary secretion, including agglutination of the red cells and prevention of coagulation (Yorke and Macfie, 1924; Shute, 1948; see also under salivary gland secretion in chapter XX). These changes vary with the species. Even in the same genus some species may show the presence of agglutinins and anti-coagulants, whilst in others these do not occur. In *A. aegypti* the ingested blood undergoes coagulation and there is no agglutination of the red cells. The pH of the gut contents of the adult *Culex pipiens* is given by Popowr and Golzowa (1933) as 6·87–7·05 and in *Anopheles maculipennis* as 6·89.

Little is known of the essential digestive processes. No enzymes appear to have been

* For disposal of food intake to the gut or diverticula see chapter XXI. For stimuli connected with feeding see chapter XXIV.

definitely recorded. Wigglesworth (1949) notes that there is absorption of fluid and that the surface of the blood mass is blackened as the haemoglobin is digested. There appears to be no absorption of unaltered haemoglobin and no breakdown of the haematin nucleus. A description of the appearances is given by Huff (1934). Digestion begins at the stomach wall and proceeds inwardly, the stomach contents showing clear stratification. At 24 hours there is an amorphous serous-like layer just within the stomach wall followed by a layer of pigment and after this a layer of partially digested red cells, internal to which the blood appears normal. At 37 hours cellular blood elements have entirely disappeared. Studies of proteolytic digestion in adult *Aëdes aegypti* have been given by Fisk (1950) and by Fisk and Shambaugh (1952). See also West and Eligh (1952).

(*g*) CIRCULATION AND HAEMOLYMPH

A description of the circulatory system, the haemolymph and haemocytes, together with the associated tissues, has already been given and only a brief review of points connected with the physiology of these structures remains necessary.

Though the larva is an aquatic insect and the imago a terrestrial flying insect, differences in the circulatory system in the two stages are not very great and the same organs are largely present in both. In relation to physiology the system has been studied chiefly from the point of view of the mechanics of circulation and from that of the regulation of osmotic pressure of the haemolymph in the larva.

THE HEART AND CIRCULATION

A very full and detailed account of the heart and circulation in the imago, pupa and larva of *Anopheles quadrimaculatus* has recently been given by Jones (1954) (see also Jones, 1952, 1953). The main organ of propulsion in the mosquito, as in insects generally, is the dorsal vessel. This consists of an abdominal portion with ostia, the *heart*, and a thoracic portion lacking ostia, the *aorta*, which at the level of the prothoracic commissure has a dilatation, the *prothoracic aortic sinus* (Jones, 1952). For further details see chapters XIV and XXVII. In the adult certain other organs and tissues assist in propulsion and direction of movement of the haemolymph, namely (1) the *ventral diaphragm*; (2) the *septa* in the appendages; (3) an *accessory propulsatory organ* in the scutellum. Movements of the alimentary canal and especially of the large ventral diverticulum may play some part in the larva (Jones, 1954). There are also in the imago the two small propulsatory organs at the bases of the antennae.

The ventral diaphragm when exposed in dissections may, as noted by Jones (1954), be seen to contract slowly 3–15 times a minute. It is undeveloped in the larva, but cells that will eventually form it are present in the intersegmental region in the neighbourhood of the transverse nerves. The propulsatory organ in the scutellum according to the same author flutters discontinuously, but with extreme rapidity, at irregular and unpredictable intervals. There was no relation between such movements and the pulsation of the heart.

In the larva pulsation of the heart can be readily observed in the living insect under the binocular. Pulsations are always forward, starting from the posterior end. Haemolymph from the peri-visceral blood sinus is drawn into the dorsal sinus, where it enters the ostial openings, principally those of the terminal chamber, from whence it is pumped into the head and then drawn backwards through the thorax and abdomen for recirculation. In the larva bleeding occurs when the thoracic or abdominal integument is cut, suggesting positive pressure in these regions. No bleeding occurs on decapitation in spite of continuing forward beats of the heart (Jones, 1954).

For observing the heart in the living imago Jones uses one-day-old mosquitoes fed on sugar. These are examined either glued to the slide by the wings and tip of the abdomen, or fixed with the dorsal surface downwards and examined reversed. They may be studied intact, or with the head, legs and wings removed and kept alive in a damp petri dish. The glue used was Resin Adhesive no. 502 manufactured by Southern Adhesives Corporation. In the adult, pulsation in the dorsal vessel is generally about 150 per minute, travelling so rapidly that the heart appears to contract simultaneously. But closer examination shows the movement to be peristalsis. Pulsation in the adult, unlike that in the larva, is characterised by periods of reversal of direction, a period of forward beating being followed by a burst of backward beats. The alary muscles do not contract, but if these are cut the heart twists about from one side of the middle line to the other. In the adult there is possibly a general negative pressure, since bleeding does not occur on decapitation or on amputation of legs, wings or halteres. In the imago Jones (1954) notes that the amount of haemolymph seems to be exceedingly small.

The rate of heart beat. For counting the heart beats Jones notes that it is best to time series of ten beats and take a mean, since counts covering an interval of a minute or more are apt to be interrupted by stoppages. The mean number of beats for *Anopheles quadrimaculatus* was 85·2 per minute. For *A. maculipennis* fourth instar larva Watson (1937) records 50–70 per minute when resting, and 90–100 per minute after much wriggling. Sautet and Audibert (1944) found with larvae of *Culex* and *Theobaldia* that submergence leading to asphyxia caused a rapid decrease in rate of heart beat with irregularity. Jones (1954) found that starvation did not affect the heart beat.

The effect of rise in temperature was to increase the rate. In the larva of *Anopheles quadrimaculatus* the number of beats per minute was: 54·1 at 15° C., 91·97 at 25° C., 90·1 at 28° C. and 139·26 at 35° C. (Jones, 1954). The rate of heart beat at different temperatures has an interest in relation to the *temperature characteristic* (μ) and whether the heart beat rate conforms to the Van t'Hoff-Arrhenius equation. It has been thought that the values found were compatible with this, which might indicate that the heart was not nervously controlled (Jones, 1954). Such a question is, however, beyond the scope of the present work and information if required must be obtained from authors dealing with the subject (see Jahn and Koel (1948) and references given by these authors).

Regulation of the osmotic pressure of the haemolymph. The larva of *Aëdes aegypti* has been much used in research on osmotic pressure of the haemolymph and the means by which this is regulated under different salt conditions in the medium in which the larvae have been reared.

The total quantity of haemolymph in the full-grown larva of *A. aegypti* as drawn off by a capillary pipette coated with paraffin is given by Wigglesworth (1938) as 3–4 $mm.^3$ In a well-fed larva in tap-water the osmotic pressure is given by the same author as 0·75–0·89 per cent NaCl equivalent. Only a relatively small part of this total pressure, as is usual in insect haemolymph, is derived from mineral salts, the chloride equivalent being only about 0·3 per cent NaCl. The difference is made up by organic substances, thought to be largely amino-acids (see Pratt, 1950). Furthermore, whilst the above is the proportion when larvae are grown in tap-water, the larva has the power, where the salt content of the medium is very low, to maintain the osmotic pressure with even smaller amounts of mineral salts. Thus, when grown in distilled water, a total osmotic pressure of 0·65 was found by Wigglesworth to be maintained; but only 7·7 per cent of this was derived from chloride.

Normally the chloride content is kept at the approximate level noted above, that is about 0·3 per cent NaCl equivalent, whilst that of the medium in which the larva lives is much lower. Thus the NaCl per cent equivalent of tap-water is around 0·006.

The larva also has the power, within limits, to maintain a normal haemolymph pressure in a medium in which the salt content is higher than that of the haemolymph. Thus Wigglesworth found that larvae of *A. aegypti* maintained a normal salt equivalent of the haemolymph even when that of the medium was gradually raised to 0·75 per cent when control began to break, and at 1·6 per cent the larvae were unable to live. The larvae of both *A. aegypti* and of *Culex pipiens* were thus able: (1) to retain chloride in a medium with low salt content, and (2) keep down chloride in a medium with a high salt content. In both these respects the larvae of *Aëdes aegypti* were found more efficient than those of *Culex pipiens*.

A particularly clear account of the nature of the mechanism at work in this respect is given by Ramsay (1953*a*) (see also other papers listed under this author whose work has been largely responsible for what is known on this subject). The following is a brief résumé:

The general body surface of the larva is for practical purposes impermeable to water and salts. But the anal papillae are permeable and act as organs of absorption for water and salt (Wigglesworth, 1933; Koch, 1938). Fluids and salts enter the body through the papillae, pass into the haemolymph and are excreted by the Malpighian tubules. In the larva grown in tap-water the fluid so secreted by the tubules is slightly hypotonic to the haemolymph. Some, after passing into the gut, seems to be carried forward through the mid-gut and absorbed in the gastric caeca (Ramsay, 1953*a*). Some passes down the intestine to the rectum where salt is absorbed from it before it is passed out of the body. The net result, therefore, is retention of chloride. When, however, the salt content of the medium is gradually increased, the salt content of the fluid secreted by the tubules is also increased. But it is never greater, so long as control can be maintained, than that of the haemolymph and thus, since less chloride is correspondingly being taken up in the rectum, the total effect is elimination of salt.

That the anal papillae play a part in such control also receives support from the fact that when larvae are grown in distilled water the anal papillae are increased in size, whereas as the salt content of the medium is progressively greater the size of the papillae is decreased. This is especially marked in larvae of *Culex pipiens* and to a less extent in the larvae of *Aëdes aegypti* which are more efficient in obtaining salt from a medium with low salt content. The same effect occurs in nature, the anal papillae of some species normally living in brackish water, for example *A. detritus*, being reduced in many cases to mere stumps.

The work of Ramsay and his collaborators has also shown some interesting facts in regard to the regulation of salts of potassium. When larvae are reared in tap-water (assuming that this contains some potassium) both sodium and potassium salts are taken up through the anal papillae and passed into the haemolymph. Both Na and K are also present in the fluid excreted by the Malpighian tubules. Both metal salts are also absorbed from such fluid in the rectum. When Na is in excess the concentration in the tubules is increased and absorption from the rectum held in abeyance as described above. The concentration of K in the urine, however, is always greater than in the haemolymph and consequently K is taken up from the rectum, so that there is a steady circulation, potassium being secreted by the tubules, being reabsorbed from the rectum and mid-gut and again secreted (Ramsay, 1953*a*, *b*).

In addition to sodium chloride the concentration of some other salts in the living medium may have importance and may affect different species differently (Bates, 1939; Woodhill, 1942). Of interest also are the analysis by Bradford and Ramsay (1949) of mosquito tissue for salts and the formula given by these authors for an insect Ringer's solution based on the tissues of *Aëdes aegypti*. The following is the composition of the solution in grams per litre: NaCl, 7·5; KCl, 0·35; $CaCl_2$, 0·21; M/150 K phosphate buffers to pH 6·8. For information on amino-acids of insects' blood see Pratt (1950).

REFERENCES

(*a*) GENETICS AND CHROMOSOMAL STRUCTURE

BAKER, J. R. (1945). *Cytological Technique*. Ed. 2. Methuen and Co., London.

BERGER, C. A. (1936). Observations on the relation between salivary gland chromosomes and multiple chromosome complexes. *Proc. Nat. Acad. Sci. Wash.* **22**, 186–7.

BERGER, C. A. (1938*a*). Cytology of metamorphosis in the Culicinae. *Nature, Lond.*, **141**, 834–5.

BERGER, C. A. (1938*b*). Multiplication and reduction of somatic chromosome groups as a regular development process in the mosquito, *Culex pipiens*. *Carnegie Inst. Wash. Publ.* **496**, 209–32.

BOGOJAWLENSKY, K. S. (1934). Studien über Zellengrössen und Zellenwachstum. *Z. Zellforsch.* **22**, 47–53.

BOLLES LEE (1950). *The Microtomist's Vade Mecum*. Ed. 11. J. and A. Churchill, London.

CALLAN, H. G. and MONTALENTI, G. (1947). Chiasma interference in mosquitoes. *J. Genet.* **48**, 119–34.

CARTER, L. A. (1918). The somatic mitosis of *Stegomyia fasciata*. *Quart. J. Micr. Sci.* **63**, 375–86.

DARLINGTON, C. D. (1937). *Recent Advances in Cytology*. Ed. 2. J. and A. Churchill, London.

DARLINGTON, C. D. and LA COUR, L. F. (1942). *The Handling of Chromosomes*. George Allen and Unwin Ltd., London.

FRIZZI, G. (1947). Salivary gland chromosomes of *Anopheles*. *Nature, Lond.*, **160**, 226–7.

FRIZZI, G. (1953). Extension of the salivary chromosome method to *Anopheles claviger*, *quadrimaculatus* and *aquasalis*. *Nature, Lond.*, **171**, 1072.

GILCHRIST, B. M. and HALDANE, J. B. S. (1947). Sex linkage and sex determination in a mosquito, *Culex molestus*. *Hereditas, Lund*, **33**, 175–90.

GILLETT, J. D. (1956). Genetic differences affecting egg-laying in the mosquito *Aëdes* (*Stegomyia*) *aegypti* (Linnaeus). *Ann. Trop. Med. Parasit.* **50**, 362–74.

GOLDSCHMIDT, R. (1938). *Physiological Genetics*. New York.

GRELL, S. M. (1946*a*). Cytological studies in *Culex*. I. Somatic reduction divisions. *Genetics*, **31**, 60–76.

GRELL, S. M. (1946*b*). Cytological studies in *Culex*. II. Diploid and meiotic divisions. *Genetics*, **31**, 77–94.

HALDANE, J. B. S. (1941). *New Paths in Genetics*. London.

HANCE, R. T. (1917). The somatic mitoses in the mosquito *Culex pipiens*. *J. Morph.* **28**, 579–88.

HOLT, C. M. (1917). Multiple complexes in the alimentary tract of *Culex pipiens*. *J. Morph.* **29**, 607–17.

JENSEN, D. V. (1955). Methods for dissecting salivary glands in mosquito larvae and pupae. *Mosquito News*, **15**, 215–16.

REFERENCES

KALMUS, H. and CRUMP, L. M. (1948). *Genetics*. Pelican Book.

KITZMILLER, J. B. (1953). Mosquito genetics and cytogenetics. *Rev. Bras. Malariol.* **5**, 285–359.

KITZMILLER, J. B. and CLARKE, C. C. (1952). Salivary gland chromosomes in *Culex* mosquitoes. *Genetics*, **37**, 596.

LA COUR, L. (1931). (Quoted by Bolles Lee, Ed. 10, 1937, p. 681.) *J. R. Micr. Soc.* **51**, 119.

LAVEN, H. (1956). X-Ray induced mutations in mosquitoes. *Proc. R. Ent. Soc. Lond.* (A), **31**, 17.

MATTINGLY, P. F. (1956). Species hybrids in mosquitoes. *Trans. R. Ent. Soc. Lond.* **108**, 21–36.

MOFFETT, A. A. (1936). The origin and behaviour of chiasmata. XIII. Diploid and tetraploid *Culex pipiens*. *Cytologia*, **7**, 184–97.

SCHUH, J. E. (1951). Some effects of colchicine on the metamorphosis of *Culex pipiens* Linn. *Chromosoma*, **4**, 456–69.

SINNOTT, E. W. and DUNN, L. C. (1939). *Principles of Genetics*. New York.

SNYDER, L. H. (1940). *The Principles of Heredity*. Ed. 2. New York.

STEVENS, N. M. (1910). The chromosomes in the germ cells of *Culex*. *J. Exp. Zool.* **8**, 207–17.

STEVENS, N. M. (1911). Further studies on heterochromosomes in mosquitoes. *Biol. Bull.* **20**, 109–20.

STURTEVANT, A. H. and BEADLE, G. W. (1940). *Introduction to Genetics.*

SUTTON, E. (1942). Salivary gland type chromosomes in mosquitoes. *Proc. Nat. Acad. Sci. Wash.* **28**, 268–72.

TAYLOR, M. (1914). The chromosome complex of *Culex pipiens*. *Quart. J. Micr. Sci.* **60**, 377–98.

TAYLOR, M. (1917). The chromosome complex of *Culex pipiens*. II. Fertilisation. *Quart. J. Micr. Sci.* **62**, 287–301.

WADDINGTON, G. H. (1939). *Introduction to Modern Genetics*. London, George Allen & Unwin Ltd.

WHITING, P. W. (1917). The chromosomes of the common house mosquito *Culex pipiens* L. *J. Morph.* **28**, 523–63.

(*b*) HORMONES

BODENSTEIN, D. (1945). The *corpora allata* of mosquitoes. *44th Rep. Connect. State Entomologist. Connect. Agric. Exp. Sta.* Bull. 488, pp. 396–405.

BODENSTEIN, D. (1953). The role of hormones in moulting and metamorphosis. In Roeder's *Insect Physiology*. See under Roeder, K. D.

CAZAL, P. (1948). Les glandes endocrines rétrocérébrales des insectes (Étude morphologique). *Bull. Biol. Fr. Belg.* Suppl. **32**, 227 pp.

CLEMENTS, A. N. (1956). Hormonal control of ovary development in mosquitoes. *J. Exp. Biol.* **33**, 211–23.

DAY, M. F. (1940). Neurosecretory cells in the ganglia of Lepidoptera. *Nature, Lond.*, **145**, 264.

DAY, M. F. (1943). The functions of the corpus allatum in muscoid Diptera. *Biol. Bull.* **84**, 127–40.

DETINOVA, T. S. (1945). On the influence of glands of internal secretion upon the ripening of the gonads and the imaginal diapause of *Anopheles maculipennis* (in Russian). *Zool. Zh.* **34**, 291–8. Quoted by Clements (1955).

FUKUDA, S. (1940). Induction of pupation in silkworm by transplanting the prothoracic gland. *Proc. Imp. Acad. Tokyo*, **16**, 414–16.

FUKUDA, S. (1944). The hormonal mechanism of larval moulting and metamorphosis in the silkworm. *J. Fac. Sci. Tokyo Imp. Univ.* Sec. IV, **6**, 477–532.

GILLETT, J. D. (1956). Initiation and promotion of ovarian development in the mosquito, *Aëdes* (*Stegomyia*) *aegypti* (Linnaeus). *Ann. Trop. Med. Parasit.* **50**, 375–80.

GOMORI, G. (1941). Observations with differential stains on human islets of Langerhaus. *Amer. J. Path.* **17**, 395–406.

HANSTROM, B. (1938). Zwei Probleme betreffs der hormonal Lokalisation im Insektenkopf. *Acta Univ. Lund.* avd. 2, **39**, 1–17.

LYONET, P. (1762). *Traité anatomique de la chenille qui ronge le bois de Saule.* Amsterdam.

NABERT, A. (1913). Die corpora allata der Insekten. *Z. wiss. Zool.* **104**, 181–358.

REFERENCES

POSSOMPES, B. (1953). Recherches expérimentales sur la détermination de la métamorphose de *Calliphora erythrocephala* Meig. *Arch. Zool. Exp. Gén.* **89**, 203–64.

THOMSEN, E. (1952). Functional significance of the neurosecretory brain cells and the corpus cardiacum in the female blowfly, *Calliphora erythrocephala* Meig. *J. Exp. Biol.* **29**, 137–72.

VERSON, E. and BISSON, E. (1891). Cellule glandulaire ipostigmatiche nel *Bombyx mori*. *Bull. Soc. Ent. Ital.* **23**, 3–20.

WEYER, F. (1928). Untersuchungen über die Keimdrüsen bei Hymenopterenarbeiterinnen. *Z. wiss. Zool.* **131**, 345–501.

WIGGLESWORTH, V. B. (1934). The physiology of ecdysis in *Rhodnius prolixus* (Hemiptera). II. Factors controlling moulting and metamorphosis. *Quart. J. Micr. Sci.* **77**, 191–222.

WIGGLESWORTH, V. B. (1948). The functions of the corpus allatum in *Rhodnius prolixus* (Hemiptera). *J. Exp. Biol.* **25**, 1–14.

WIGGLESWORTH, V. B. (1951*a*). Source of moulting hormones in *Rhodnius*. *Nature, Lond.*, **168**, 558.

WIGGLESWORTH, V. B. (1951*b*). Metamorphosis in insects. *Proc. R. Ent. Soc. London* (C), **15**, 78.

WIGGLESWORTH, V. B. (1954). *The Physiology of Insect Metamorphosis*. Cambridge Univ. Press.

WILLIAMS, C. M. (1949). The prothoracic glands of insects in retrospect and in prospect. *Biol. Bull.* **97**, 111–14.

(*c*) RESERVES

CHRISTOPHERS, S. R. (1911). Development of the egg follicle in Anophelines. *Paludism*, **2**, 73–88.

CLEMENTS, A. N. (1955). The source of energy for flight in mosquitoes. *J. Exp. Biol.* **32**, 547–54.

GILLETT, J. D. (1955). Behaviour difference in two strains of *Aëdes aegypti*. *Nature, Lond.*, **176**, 124–6.

GILLIES, M. T. (1955). The pre-gravid phase of ovarian development in *Anopheles funestus*. *Ann. Trop. Med. Parasit.* **49**, 320–5.

HOCKING, B. (1952). Autolysis of flight muscles in a mosquito. *Nature, Lond.*, **169**, 1101.

HOCKING, B. (1953). The intrinsic range and speed of flight of insects. *Trans. R. Ent. Soc. London*, **104**, 223–345.

MER, G. G. (1936). Experimental study on the development of the ovary in *Anopheles elutus* Edw. (Dipt. Culic.). *Bull. Ent. Res.* **27**, 351–9.

SOTAVALTA, O. (1952). Flight-tone and wing-stroke frequency of insects and the dynamics of insect flight. *Nature, Lond.*, **170**, 1057.

SOTAVALTA, O. (1954*a*). On the fuel consumption of the honeybee (*Apis mellifica* L.) in flight experiments. *Ann. Zool. Soc. Vanamo*, **16**, no. 5, 27 pp.

SOTAVALTA, O. (1954*b*). On the thoracic temperature of insects in flight. *Ann. Zool. Soc. Vanamo*, **16**, no. 8, 22 pp.

WIGGLESWORTH, V. B. (1942). The storage of protein, fat, glycogen and uric acid in the fatbody and other tissues of mosquito larvae. *J. Exp. Biol.* **19**, 56–77.

(*d*) THE INTEGUMENT

PRYOR, M. G. M. (1940). On the hardening of the cuticle of insects. *Proc. Roy. Soc.* B, **128**, 393–407.

RICHARDS, A. G. (1951). *The Integument of Arthropods*. Univ. of Minnesota Press, Minneapolis, and Oxford Univ. Press, London.

RICHARDS, A. G. (1953). The integument. In Roeder's *Insect Physiology*. See Roeder, K. D., under section (*b*).

RICHARDS, A. G. and ANDERSON, T. F. (1942). Electron microscope studies of insect cuticle with a discussion of electron optics to this problem. *J. Morph.* **71**, 135–71.

WIGGLESWORTH, V. B. (1933). The function of the anal gills of the mosquito larva. *J. Exp. Biol.* **10**, 16–26.

WIGGLESWORTH, V. B. (1947). The epicuticle of an insect, *Rhodnius prolixus* (Hemiptera). *Proc. Roy. Soc.* B, **134**, 163–81.

REFERENCES

WIGGLESWORTH, V. B. (1948). The insect epicuticle. *Proc. 8th Internat. Congr. Entom. 1948*, pp. 307–9.

WIGGLESWORTH, V. B. (1957). The physiology of insect cuticle. *Ann. Rev. Ent.* **2**, 37–54.

(*e*) RESPIRATION

BABAK, E. (1912). Zur Physiologie der Atmung bei *Culex*. *Int. Rev. Hydrobiol.* **5**, 81–90.

CARPENTER, T. M. (1939). Tables, factors and formulae for computing respiratory exchange and biological transformations of energy. *Publ. Carnegie Inst.* **303**, B, 142 pp.

CHILDS, W. H. J. (1934). *Physical Constants*. Methuen and Co., London.

DA COSTA LIMA, S. (1914). Contributions to the biology of the Culicidae. Observations on the respiratory process of the larva. *Mem. Inst. Osw. Cruz*, **6**, 18–34.

HOCKING, B. (1953). See under section (*c*).

IMMS, A. D. (1907). On the larval and pupal stages of *Anopheles maculipennis* Meigen. *J. Hyg., Camb.*, **7**, 291–316.

KOCH, A. (1920). Messende Untersuchungen über den Einfluss von Sauerstoff und Kohlensäure auf *Culex*-larven der Submersion. *Zool. Jb., Physiol.*, **37**, 361–492.

KROGH, A. (1941). *The Comparative Physiology of Respiratory Mechanisms*. Philadelphia.

MACFIE, J. W. S. (1917). The limitations of kerosene as a larvicide with some observations on the cutaneous respiration of mosquito larvae. *Bull. Ent. Res.* **7**, 277–95.

SEN, S. K. (1915). Observations on respiration of Culicidae. *Ind. J. Med. Res.* **2**, 681–97.

WIGGLESWORTH, V. B. (1930). A theory of tracheal respiration in insects. *Proc. Roy. Soc.* B, **106**, 229–50.

WIGGLESWORTH, V. B. (1938). The regulation of osmotic pressure and chloride concentration in the haemolymph of mosquito larvae. *J. Exp. Biol.* **15**, 235–47.

WIGGLESWORTH, V. B. (1954). Growth and regeneration in the tracheal system of an insect *Rhodnius prolixus* (Hemiptera). *Quart. J. Micr. Sci.* **95**, 115–37.

(*f*) DIGESTION

BOYCE, R. and LEWIS, F. C. (1910). The effect of mosquito larvae upon drinking water. *Ann. Trop. Med. Parasit.* **3**, 591–4.

BROWN, J. H. (1923–4). The colorimetric determination of the hydrogen ion concentration of small amounts of fluid. *J. Lab. Clin. Med.* **9**, 239.

BRÜCK, R. (1931). Zur Frage über die Ernährung der Larven von *Anopheles maculipennis*. *Trav. Soc. Nat. Lening.* **60**, 15. Abstract in *Rev. Appl. Ent.* **19**, 144.

CROZIER, W. J. (1923–4). Hydrogen ion concentration within the alimentary tract of insects. *J. Gen. Physiol.* **6**, 289–93.

FELTON, J. D. (1921). A colorimetric method of determining the hydrogen ion concentration in small quantities of solution. *J. Biol. Chem.* **46**, 299.

FISK, F. W. (1950). Studies on proteolytic digestion in adult *Aëdes aegypti* mosquitoes. *Ann. Ent. Soc. Amer.* **43**, 555–72.

FISK, F. W. and SHAMBAUGH, G. F. (1952). Protease activity in adult *Aëdes aegypti* mosquitoes as related to feeding. *Ohio J. Sci.* **52**, 80–8.

HAAS, A. R. C. (1919). Colorimetric determination of hydrogen ion concentration in small quantities of solution. *J. Biol. Chem.* **38**, 49.

HINMAN, E. H. (1933). The presence of enzymes in the digestive tract of mosquito larvae. *Ann. Ent. Soc. Amer.* **26**, 45–52.

HOLT, C. M. (1917). See under section (*a*).

HUFF, C. G. (1934). Comparative studies on susceptible and insusceptible *Culex pipiens* in relation to infections with *Plasmodium cathemerium* and *P. relictum*. *Amer. J. Hyg.* **19**, 123–47.

JAMESON, A. P. and ATKINS, W. R. (1921). On the physiology of the silkworm. *Biochem. J.* **15**, 209–12.

PANTIN, C. F. A. (1923). The determination of pH of microscopical bodies. *Nature, Lond.*, **111**, 81.

REFERENCES

POPOWR, P. P. and GOLZOWA, R. D. (1933). Zur Kenntnis der Wasserstoffionenkonzentration im Darmkanale einiger blutsaugender Arthropoden. *Arch. Schiffs.- u. Tropenhyg.* **37**, 465–6.

PURDY, M. S. (1920). Biological investigation of Californian ricefields relative to mosquito breeding. *Publ. Hlth Rep. Wash.* **35**, 2556–70.

RAJINDAR PAL (1943). On the histological structure of the midgut of mosquitoes. *J. Malar. Inst. India*, **5**, 247–50.

RICHARDSON, C. H. and SHEPARD, H. H. (1930). The effect of hydrogen-ion concentration on the toxicity of nicotine, pyridine and methyl pyrrolidine to mosquito larvae. *J. Afric. Res.* **41**, 337–48.

SENIOR WHITE, R. (1926). Physical factors in mosquito oecology. *Bull. Ent. Res.* **16**, 187–248.

SENIOR WHITE, R. (1928). Algae and the food of Anopheline larvae. *Ind. J. Med. Res.* **15**, 969–88.

SHIPITZIMA, N. K. (1930). On the rôle of the organic colloids of water in the feeding of the larva of *Anopheles maculipennis* (in Russian). *Bull. Inst. Tech. Biol. Univ. Perm.* **7**, 171. Abstract in *Rev. Appl. Ent.* **19**, 25.

SHIPITZIMA, N. K. (1935). Grandeur maximum et minimum des particles pouvant être avalées par les larves d'*Anopheles maculipennis. Med. Parasit., Moscow*, **4**, 381–9. Abstract in *Rev. Appl. Ent.* **24**, 70.

SHUTE, P. G. (1948). The comparative distribution of oocysts of human malaria parasites on the stomach wall of *Anopheles. Trans. R. Soc. Trop. Med. Hyg.* **42**, 324.

THOMPSON, M. T. (1905). The alimentary canal of the mosquito. *Proc. Boston Soc. Nat. Hist.* **32**, 145–202.

TREMBLEY, H. L. (1951). Pyloric spines in mosquitoes. *J. Nat. Malar. Soc.* **10**, 213–15.

WEST, A. S. and ELIGH, G. S. (1952). The rate of digestion of blood in mosquitoes: precipitin test studies. *Canad. J. Zool.* **30**, 267–72.

WIGGLESWORTH, V. B. (1949). *Physiology of Mosquitoes.* In Boyd's *Malariology*, **1**. W. B. Saunders, Philadelphia and London.

YAGUZHINSKAYA, L. W. (1940). See references in ch. XXVI (*a*), p. 599.

YORKE, W. and MACFIE, J. W. S. (1924). The action of the salivary secretion of the mosquito and *Glossina tachinoides* on human blood. *Ann. Trop. Med. Parasit.* **18**, 103–8.

(*g*) CIRCULATION AND HAEMOLYMPH

BATES, M. (1939). Use of salt solutions for the demonstration of physiological differences between the larvae of certain European *Anopheles* mosquitoes. *Amer. J. Trop. Med.* **19**, 357–83.

BRADFORD, S. and RAMSAY, R. W. (1949). Analysis of mosquito tissues for sodium and potassium for a physiological salt solution. *Fed. Proc.* **8**, 1.

JAHN, T. J. and KOEL, B. S. (1948). The effect of temperature in the frequency of beat of the grasshopper heart. *Ann. Ent. Soc. Amer.* **41**, 258–66.

JONES, J. C. (1952). Prothoracic aortic sinuses in *Anopheles, Culex* and *Aëdes. Proc. Ent. Soc. Wash.* **54**, 244–6.

JONES, J. C. (1953*a*). On the heart in relation to circulation of haemocytes in insects. *Ann. Ent. Soc. Amer.* **46**, 366–72.

JONES, J. C. (1953*b*). A chamber for observations on living Anopheline mosquitoes. *Science*, **117**, 42.

JONES, J. C. (1954). The heart and associated tissues of *Anopheles quadrimaculatus* Say (Diptera: Culicidae). *J. Morph.* **94**, 71–124.

KOCH, H. J. (1938). The absorption of chloride ions by the anal papillae of Diptera larvae. *J. Exp. Biol.* **15**, 152–60.

PRATT, J. J. (1950). A qualitative analysis of free amino-acids in insect blood. *Ann. Ent. Soc. Amer.* **43**, 573–80.

RAMSAY, J. A. (1949). A new method of freezing-point determination for small quantities. *J. Exp. Biol.* **26**, 57–64.

RAMSAY, J. A. (1950*a*). Osmotic regulation in mosquito larvae. *J. Exp. Biol.* **27**, 145–57.

RAMSAY, J. A. (1950*b*). The determination of sodium in small volumes of fluid by flame photometry. *J. Exp. Biol.* **27**, 417–19.

REFERENCES

RAMSAY, J. A. (1951). Osmotic regulation in mosquito larvae: the role of the maliphighian tubules. *J. Exp. Biol.* **28**, 62–73.

RAMSAY, J. A. (1952). The excretion of sodium and potassium by the malpighian tubules of *Rhodnius*. *J. Exp. Biol.* **29**, 110–26.

RAMSAY, J. A. (1953*a*). Exchanges of sodium and potassium in mosquito larvae. *J. Exp. Biol.* **30**, 79–89.

RAMSAY, J. A. (1953*b*). The active transport of potassium by the malpighian tubules of insects. *J. Exp. Biol.* **30**, 358–69.

SAUTET, J. and AUDIBERT, Y. (1944). Le rythme cardiaque des larves de moustiques en asphyxie. *C.R. Soc. Biol., Paris*, **138**, 679–80.

WATSON, G. I. (1937). Some observations on mosquito larvae dying in anti-malarial oils and other substances. *Ann. Trop. Med. Parasit.* **31**, 417–26.

WIGGLESWORTH, V. B. (1933). The adaptation of mosquito larvae to salt water. *J. Exp. Biol.* **10**, 27–37.

WIGGLESWORTH, V. B. (1938). See under section (*e*).

WOODHILL, A. R. (1942). A comparison of factors affecting the development of three species of mosquitoes (*Aëdes (Pseudoskusea) concolor*; *Aëdes* (*Stegomyia*) *aegypti* Linnaeus and *Culex* (*Culex*) *fatigans* Wied.). *Proc. Linn. Soc. N.S.W.* **67**, 95–7.

ADDENDA

CHAPTER I (*a*), p. 1. *The name mosquito*

Under the name 'mosquito' something might have been said of the generic name *Aëdes*, which appears in the title of this book. *Aedes* was the name for a third genus additional to *Culex* and *Anopheles* designated by Meigen (1818), solely for the European species *Culex cinereus*. In the brief note establishing the genus the heading appears as 'III. Schnackenmukke. AEDES. Hoffmgg.' and after description of the generic characters follows a description of the species under the heading 'I. Aed. cinereus Hoffmgg.'. In this, after a few lines of description, Meigen says: 'Diese ist alles, was mir Hr. Justizrath Wiedemann von dieser Art bemerkt hat, die ich weiter nicht kenne. Den Gattungsnamen hat der Hr. Graf v. Hoffmansegg, in dessen Sammlung sie sich befindet, aus dem Griechischen Aedes beschwerlich, gebildet.' The meaning of the Greek word *Aedes* without the diaeresis is 'a building'; with the diaeresis it means, as stated by Meigen, 'troublesome' (beschwerlich), which was clearly the meaning intended. But Meigen gave the name without the diaeresis, and as a result there was later some controversy as to whether in strict taxonomy the name of the genus should be given without the diaeresis and some even so used it.

CHAPTER II (*d*), p. 39. *Temperature limit to the species*

A recent contribution to the limiting temperature for distribution of *Aëdes aegypti* is that by Smith and Love (1958), who, studying survival of *Aëdes aegypti* in south-west Georgia, U.S.A., found that an average weekly temperature below 60° F. (15·6° C.) killed most larvae, but in sheltered positions some survived when the weekly temperature was as low as 48° F. (9·1° C.). This is very close to the limit assigned in the text: Smith, W. W. and Love, G. J. (1958). Winter and spring survival of *Aëdes aegypti* in South Western Georgia. *Amer. J. Trop. Med.* **7**, 309–11.

CHAPTER IV (*a*), p. 79. *Sylvan yellow fever*

Among recent accounts of mosquitoes concerned in sylvan yellow fever are: Trapido, H. and Galindo, P. (1957). Mosquitoes associated with sylvan yellow fever near Almirante, Panama. *Amer. J. Trop. Med.* **6**, 114–44; Galindo, P. and Trapido, H. (1957). Forest mosquitoes associated with sylvan yellow fever in Nicaragua. *Amer. J. Trop. Med.* **6**, 145–52.

CHAPTER IV (*a*), p. 82. *Virus diseases*

A special trap for collecting mosquitoes in connection with virus research is described by Lumsden, W. H. R. (1958). A trap for insects biting small vertebrates. *Nature, Lond.* **18**, 809–20.

CHAPTER V, p. 104. *Technique for rearing*

Dr Ann Bishop, F.R.S., informs me that some modifications have been made by her in the rearing cage here described as used at the Molteno Institute. As originally employed it was quite adequate for rearing and manipulating *Culex pipiens* and *Culex molestus*, but there were apt to be escapes with *Aëdes aegypti*. This is entirely avoided by using a broad band of elastic material fitting tightly around the bottom of the cylinder and holding cylinder and dish closely together. This also makes the apparatus more secure and easily handled. The band is formed from a strip of fine meshed nylon net (corset elastic) about 5 in. wide stitched into a band with strong thread and double seam. About a yard is required to make the band. Other improvements are: lining the metal cylinder with white paper held in place with surgical plaster; lining the hole through the muslin top with surgical plaster stuck both underneath and above the muslin; fixing the muslin netting with two or three strands of round hat elastic.

ADDENDA

CHAPTER V, p. 124. *Preparing celloidin double-embedded sections for mounting*

In the procedure given for treating double-embedded sections for mounting it is stated that carbol-xylol is largely responsible for removing the celloidin. It is to be noted, however, that normally celloidin is not soluble either in carbol-xylol or in a mixture of carbol-xylol and alcohol. If, therefore, this is found necessary, a rinse of equal parts alcohol and ether should be given after the xylol. Possibly the use of a warm slide or some other detail as used by the author may be the reason this has not usually been found necessary.

CHAPTER V, p. 126. *Special techniques*

Among papers giving information on the use of radio-isotopes in research on and control of mosquitoes may be noted: Bruce-Chwatt, L. J. (1956). Radio-isotopes for research on and control of mosquitoes. *Bull. Org. Mond. Santé*, **15**, 491–511; Ascher, K. R. S. (1957). Investigations on a fluorocarbon as 'O.I.T.C.-agent' (Oviposition-inhibitory-tarsal-contact-agent) in mosquitoes. *Riv. Malariol.* **26**, 209–15.

CHAPTER VIII, p. 177. *Embryology*

Professor Colvard Jones has since drawn my attention to a paper by Ivanova-Kazas, O.M. (1949). Embryological development of *Anopheles maculipennis* Mg. (in Russian) *Akad. Nauk. S.S.S.R. Inv. Ser. Biol.* 1949, 160–70. Some embryological changes in the egg are also described by Gander (1951), see refs. p. 286.

CHAPTER X (*a*), p. 227 (footnote). *Larval instar mouthparts*

The mouthparts of the larval instars of *Anopheles quadrimaculatus* are described by Shalaby, A. M. (1956). On the mouthparts of the larval instars of *Anopheles quadrimaculatus* (Say). *Soc. Ent. d'Egypte Bull.* **40**, 137–74.

CHAPTER XI (*b*), p. 255. *Surface tension*

The effects of surface tension on larval development of six species of mosquito including *Aëdes aegypti* have been studied by: Singh, K. R. P. and Micks, Don W. (1957). The effects of surface tension on mosquito development. *Mosq. News*, **17**, 70–3. None of the forms experimented with emerged at 41 or less or 78 or more dynes per cm., though with *Aëdes aegypti* at 41 dynes some emergences were incomplete.

CHAPTER XI (*g*), p. 261. *Food requirements of the larva*

Besides the authors given in the footnote, the following deal with the part played by amino acids in nutrition of the larva. Singh, K. R. P. and Micks, Don W. (1957). Synthesis of amino-acids in *Aëdes aegypti* L. *Mosq. News*, **17**, 248–9; Singh, K. R. P. and Brown, A. W. A. (1957). Nutritional requirements of *Aëdes aegypti* L. *J. Insect. Physiol.* **1**, 199–220; Terzion, L. A., Irreverre, F. and Stahler, N. (1957). A study of nitrogen and uric acid patterns in the excreta and body tissues of adult *Aedes aegypti*. *J. Insect. Physiol.* **1**, 221–8.

CHAPTER XV (*b*), p. 355. *Pupal chaetotaxy of* Aëdes aegypti

Senevet and Andarelli have dealt with the systematics of pupae of the *Aëdes* of North Africa and describe the pupal chaetotaxy of *Aëdes aegypti* there occurring: Senevet, G. and Andarelli, L. (1958). Le genus *Aëdes* en Afrique du Nord. II. Les nymphes. *Ann. Inst. Pasteur d'Algérie*, **36**, 266–93. A further short note on *Aëdes aegypti* in Algeria is also given by: Senevet, G., Andarelli, L. and Monpere, H. (1958). *Aëdes aegypti* à Alger. *Ann. Inst. Pasteur d'Algérie*, **36**, 506–7. The authors note that though the species appears to be rarer than formerly in towns in Algeria it is present always in crowded Algerian areas and probably in most coastal towns of North Africa.

ADDENDA

CHAPTER XXI (*i*), p. 496. *Treatment of mosquito bites*

An account of anti-histamine compounds including a reference to 'anthisan' is given by Burn, J. H. (1958). The anti-histamine compounds. *Brit. Med. J.* 4 October 1958, 845–56.

CHAPTER XXV (*e*), p. 561. *Insecticide resistance*

Among a number of recent papers on insecticide resistance may be noted: Busvine, J. R. (1956). A survey of measurements of the susceptibility of different mosquitoes to insecticides. *Bull. World Hlth Org.* **15**, 787–91; Busvine, J. R. (1957). A critical review of techniques for testing of insecticides. *Commonw. Inst. Entom. London*; Brown, A. W. A. (1958). The insecticide-resistance problem. A review of developments in 1956 and 1957. *Bull. World Hlth Org.* **18**, 309–21. The last mentioned gives species resistant to particular insecticides and description of resistance in Trinidad of *Aëdes aegypti*.

CHAPTER XXVI (*a*), p. 568. *The labral canal*

In the paper by Bhatia and Wadia (1957), referred to in the footnote on this page, the authors refer to the reason for the binding of the stylets in the proboscis and consider that surface tension alone is unlikely to be so effective as observation shows the binding to be. They note that in some insects the parts are effectively bound by their interlocking and they think it possible that the septa now described by them may serve the function of keeping the hypopharynx, mandibles and maxillae firmly applied against the ventral gutter-like opening of the labrum-epipharynx. The rings have been observed in both sexes of *Aëdes aegypti* and in a number of species of *Anopheles* and *Culex fatigans*. They are best seen in fresh preparations examined in glycerine or water.

CHAPTER XXX (*i*), p. 678. *The mating plug and ovariole funicle*

Gillies (1956) found that the mating plug described by him and referred to in the footnote on this page can be observed in females recently fertilised and was absorbed during the 36 hr. following. It therefore gives a very precise indication of time of fertilization.

Recently a still further means towards ascertaining the age and condition of captured females has been described by Lewis (1958). This relates to the condition of the ovariole funicle or the structure linking the ovariole to the oviduct. After oviposition this is much enlarged, giving evidence that the insect has already passed through at least one oviposition. Lewis, D. J. (1958). The recognition of nulliparous and parous *Anopheles gambiae* by examining the ovarioles. *Trans. R. Soc. Trop. Med. Hyg.* **52**, 456–61.

CHAPTER XXX (*j*), p. 681. *Male reproductive system*

The nomenclature given in the text is mainly based on the work of Abdul Nasr on the organs in Nematocera. Prof. Colvard Jones, however, informs me that he and Prof. Snodgrass feel that the duct labelled *penis cavity* in the text on p. 681 and in Fig. 85 (1) and (7), would be more appropriately termed the true *ejaculatory duct*, and the organ labelled *ejd* the true *seminal vesicles*, this being more in keeping with what is known for other insects.

CHAPTER XXXI (*e*) p. 704. *Oxygen uptake*

An important paper on respiration of mosquito larvae dealing with oxygen uptake is: Mercado, T.I., Trembley, H. L. and Von Brand, T. (1956). Observations on the oxygen consumption of some adult mosquitoes. *Physiol. Comp. et Oecolog.* **4**, 200–8.

The present author's observations were made under conditions as near as possible to those made by him in previous communications, see Christophers and Fulton (1938). *Ann. Trop. Med. Parasit.* **32**, 43.

INDEX

The figures in bold type refer to the principal references

INDEX

INDEX

INDEX

INDEX

INDEX

INDEX